AQA GCSE

MATHEMATICS
for Higher sets

Series editor: **Glyn Payne**

Authors:
Gwenllian Burns
Greg Byrd
Lynn Byrd
Crawford Craig
Janet Crawshaw
Fiona Mapp
Avnee Morjaria
Catherine Murphy
Katherine Pate
Glyn Payne
Ian Robinson
Harry Smith

www.pearsonschools.co.uk
✓ Free online support
✓ Useful weblinks
✓ 24 hour online ordering
0845 630 33 33

Longman is an imprint of Pearson Education Limited, a company incorporated in England and Wales, having its registered office at Edinburgh Gate, Harlow, Essex, CM20 2JE. Registered company number: 872828

www.pearsonschoolsandfecolleges.co.uk

Longman is a registered trademark of Pearson Education Limited

First published 2010
14 13 12 11 10
10 9 8 7 6 5 4 3 2 1

British Library Cataloguing in Publication Data
A catalogue record for this book is available from the British Library.
ISBN 978 1 408 23278 1

Edited by Gwenllian Burns, Nicola Morgan, Jim Newall, Maggie Rumble, Alex Sharpe, Laurice Suess and Christine Vaughan
Designed by Pearson Education Limited
Typeset by Tech-Set Ltd, Gateshead
Original illustrations © Pearson Education Ltd 2010
Illustrated by Tech-Set Ltd
Cover design by Wooden Ark
Picture research by Chrissie Martin
Cover photo © Getty Images / Aurora
Index created by Indexing Specialists (UK) Ltd
Printed in Italy by Rotolito

Acknowledgements

The author and publisher would like to thank the following individuals and organisations for permission to reproduce photographs.

Alamy / Alex Segre p262; Alamy / Bruce McGowan p348; Alamy / Jon Arnold Images Ltd p475; Alamy / Kevin Wheal p257; Alamy / Mark Sunderland p538; Alamy / Martin Pick p372; Alamy / Robert Convery p107; Alamy / Sinibomb p450; Alamy / Steve Allen p557; Alamy / Yiap Views p192; Corbis / Arcaid / Clive Nichols p408; Corbis / Crush p297; Corbis / Darrell Gulin p190; Corbis / Frank Lukasseck p272; Corbis / Herbert Pfarrhofer p91; Digital Vision p33; Getty Images / AFP / Martin Bureau p336; Getty Images / Iconica p156; Getty Images / PhotoDisc pp10, 157, 168, 203, 381, 485, 506; Getty Images / Photographers Choice pp79, 316; Getty Images / Riser p410; Getty Images / Stone p1; ImageState / John Foxx Imagery p442; Masterfile / Allan Davey p128; Masterfile / Jeremy Maude p496; Masterfile / R. Ian Lloyd p601; NASA / Goddard Space Flight Center p242; Pearson Education Ltd p274; Pearson Education Ltd / Gareth Boden p183; Pearson Education Ltd / Jules Selmes p524; Photolibrary / Digital Light Source p424; Photolibrary / Imagestate Media p215; Photolibrary / Pacific Stock p588; Rex Features p302; Science Photo Library / Antonia Reeve p68; Science Photo Library / Chris Priest p173; Science Photo Library / D. van Ravenswaay p75; Science Photo Library / Eye of Science p225; Science Photo Library / Friedrich Saurer p134; Science Photo Library / Skyscan p601; Science Photo Library / Susumu Nishinaga p462; Science Photo Library / TEK-Image p565; Science Photo Library / TRL Ltd p52; Shutterstock / 0833379753 p390; Shutterstock / Afaizal p165; Shutterstock / Anyka p595; Shutterstock / BlueOrange Studio p126; Shutterstock / CBPix p540; Shutterstock / Charles Taylor p379; Shutterstock / Elena Elisseva p35; Shutterstock / Erhan Daly p274; Shutterstock / Freebird p153; Shutterstock / Galyna Andrushko p332; Shutterstock / IcemanJ p364; Shutterstock / Ioana Drutu p246; Shutterstock / J & S Photography p121; Shutterstock / Jan van der Hoeven p56; Shutterstock / Jenny T p289; Shutterstock / Johann Helgason p353; Shutterstock / Kapu p389; Shutterstock / Karin Lau p149; Shutterstock / Konstantin Chagin p208; Shutterstock / Liew Weng Keong p218; Shutterstock / Mary Katherine Donovan p71; Shutterstock / Monika 23 p195; Shutterstock / Monkey Business Images pp26, 432; Shutterstock / New Photo Service p447; Shutterstock / Olga Lyubkina p105; Shutterstock / Olivier le Queinec p401; Shutterstock / Pavel K p20; Shutterstock / Prism68 p497; Shutterstock / Rachelle Burnside p356; Shutterstock / Russ Witherington p413; Shutterstock / Sas Partout p70; Shutterstock / Tompet p177; Shutterstock / Vasilly Koval p342; Shutterstock / Vladimir Melnik p411; Shutterstock / Wendy Kaveney Photography p422; Shutterstock / Teresa Kasprzycka p19; Superstock / Greer & Associates p523.

Every effort has been made to contact copyright holders of material reproduced in this book. Any omissions will be rectified in subsequent printings if notice is given to the publishers.

Quick contents guide

Contents

All set to make the grade!

AQA GCSE Mathematics for Higher sets is specially written to help you get your best grade in the exams.

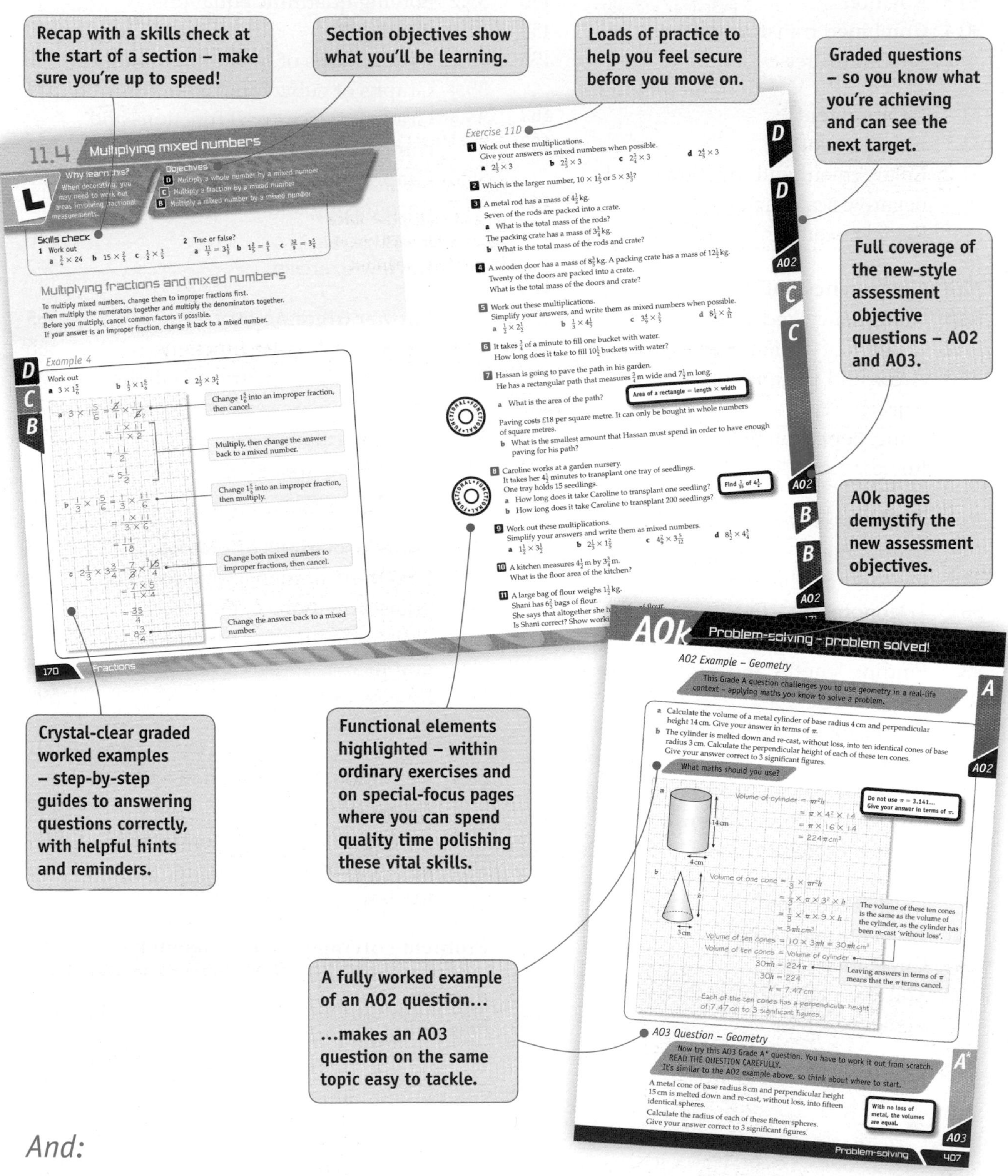

Recap with a skills check at the start of a section – make sure you're up to speed!
Section objectives show what you'll be learning.
Loads of practice to help you feel secure before you move on.
Graded questions – so you know what you're achieving and can see the next target.
Full coverage of the new-style assessment objective questions – AO2 and AO3.
AOk pages demystify the new assessment objectives.
Crystal-clear graded worked examples – step-by-step guides to answering questions correctly, with helpful hints and reminders.
Functional elements highlighted – within ordinary exercises and on special-focus pages where you can spend quality time polishing these vital skills.
A fully worked example of an AO2 question...
...makes an AO3 question on the same topic easy to tackle.

And:

- A pre-check at the start of each chapter helps you recall what you know!
- End-of-chapter graded review exercises consolidate your learning and include exam-style questions.

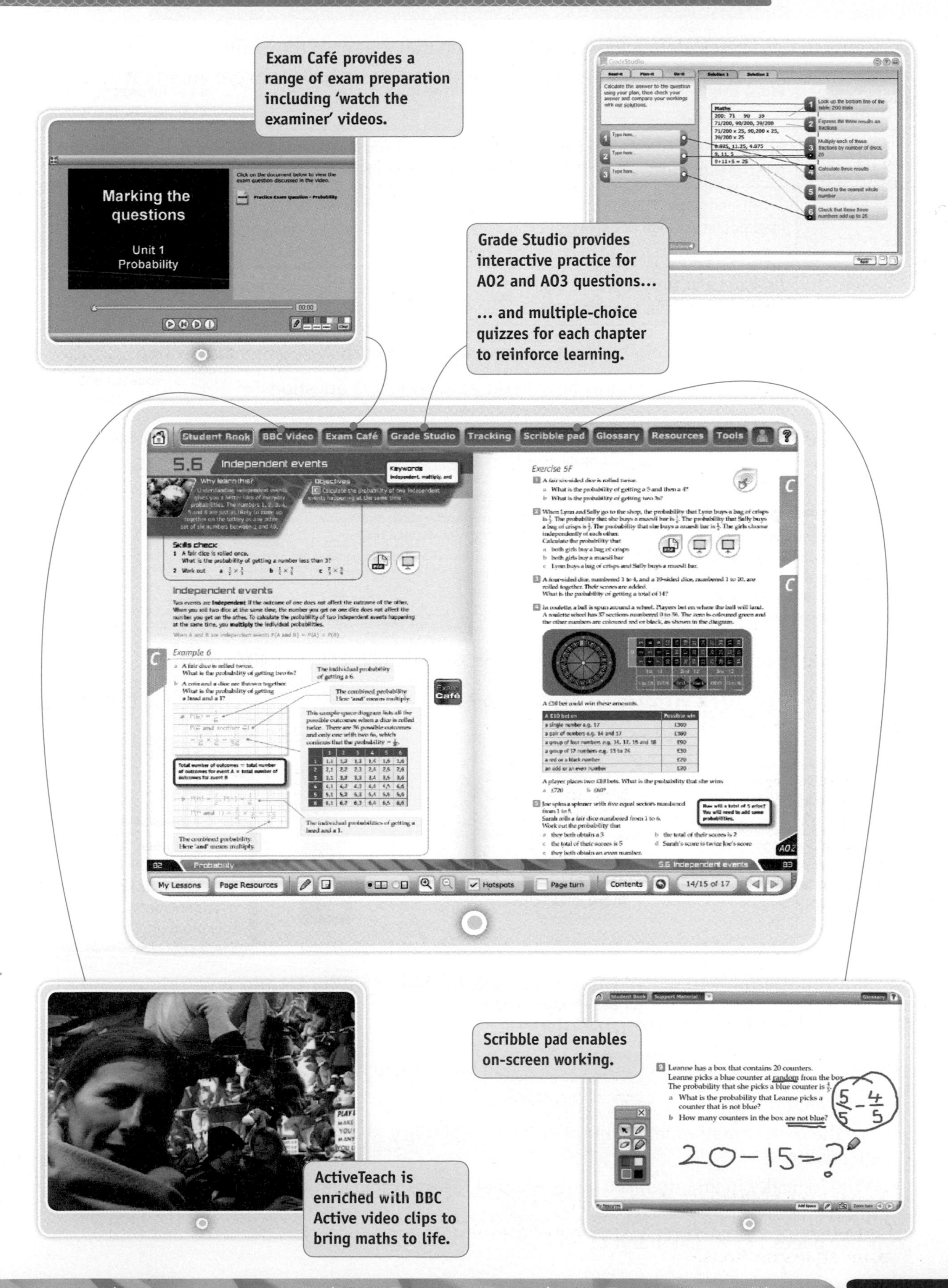
Exam Café provides a range of exam preparation including 'watch the examiner' videos.
Marking the questions
Unit 1
Probability
Grade Studio provides interactive practice for AO2 and AO3 questions...
... and multiple-choice quizzes for each chapter to reinforce learning.
Student Book
BBC Video
Exam Café
Grade Studio
Tracking
Scribble pad
Glossary
Resources
Tools
Scribble pad enables on-screen working.
ActiveTeach is enriched with BBC Active video clips to bring maths to life.

'Assessment Objectives' define the types of question that are set in the exam:

Assessment Objective	What it is	What this means	Approx % of marks in the exam
AO1	Recall and use knowledge of the prescribed content.	Ordinary questions testing your knowledge of each topic.	50
AO2	Select and apply mathematical methods in a range of contexts.	Problem-solving: find the ordinary maths you need to get to the correct answer.	30
AO3	Interpret and analyse problems and generate strategies to solve them.	A step up from AO2. There could be more than one way to tackle these.	20

The proportion of marks available in the exam varies with each Assessment Objective.

So it's worth making sure you know how to do AO2 and AO3 questions!

What does an AO2 question look like?

This just needs you to (a) read and understand the question and (b) recall and apply the correct formulae. Simple!

3 A cone has a base radius of 6.5 cm and a slant height of 14 cm.

a Calculate the curved surface area of the cone. Give your answer to one decimal place.

b Calculate the volume of the cone, giving your answer to four significant figures.

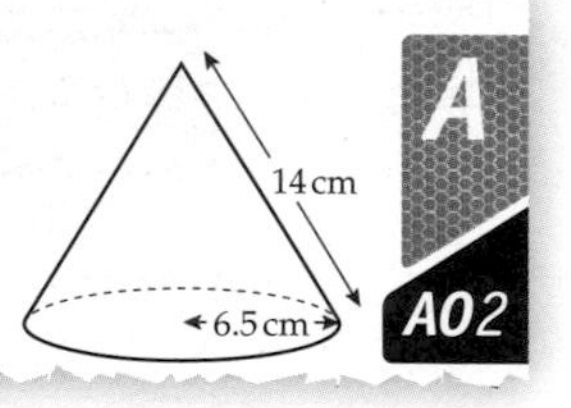

We give you special help with AO2s on pages 66, 67, 145, 147, 271, 330, 407 and 613.

What does an AO3 question look like?

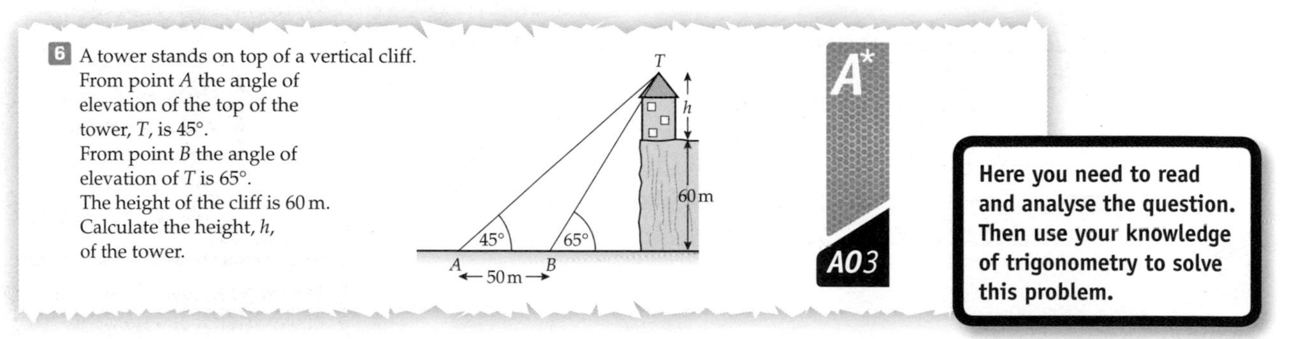
6 A tower stands on top of a vertical cliff. From point A the angle of elevation of the top of the tower, T, is 45°. From point B the angle of elevation of T is 65°. The height of the cliff is 60 m. Calculate the height, h, of the tower.

Here you need to read and analyse the question. Then use your knowledge of trigonometry to solve this problem.

We give you special help with AO3s on pages 66, 67, 146, 148, 271, 331, 407 and 614.

Quality of written communication

There are a few extra marks in the exam if you take care to write your working 'properly'.

- Write legibly.
- Use the correct mathematical notation and vocabulary, showing that you can communicate effectively.
- Write logically in an organised sequence that the examiner can follow.

In the exam paper, such questions will be marked with a star (☆) – see the problem-solving practice pages 145–8, 330–1 and 613–14.

About functional elements

Functional maths means using maths effectively in a wide range of real-life contexts.

There are three 'key processes' for maths:

Key process	What it is	What this means
Representing	Understanding real-life problems and selecting the mathematics to solve them.	• Understanding the information in the question. • Working out what maths you need to use. • Planning the best order in which to do your working.
Analysing	Applying a range of mathematics within realistic contexts.	• Being organised and following your plan. • Using appropriate maths to work out the answer. • Checking your calculations. • Explaining your plan.
Interpreting	Communicating and justifying solutions and linking solutions back to the original context of the problem.	• Explaining what you've worked out. • Explaining how it relates to the question.

The proportion of functional maths marks in the GCSE exam depends on which tier you are taking:

GCSE tier	Approx % of marks in the exam
Foundation	30 to 40
Higher	20 to 30

So it's worth making sure you know how to do functional maths questions!

What does a question with functional maths look like?

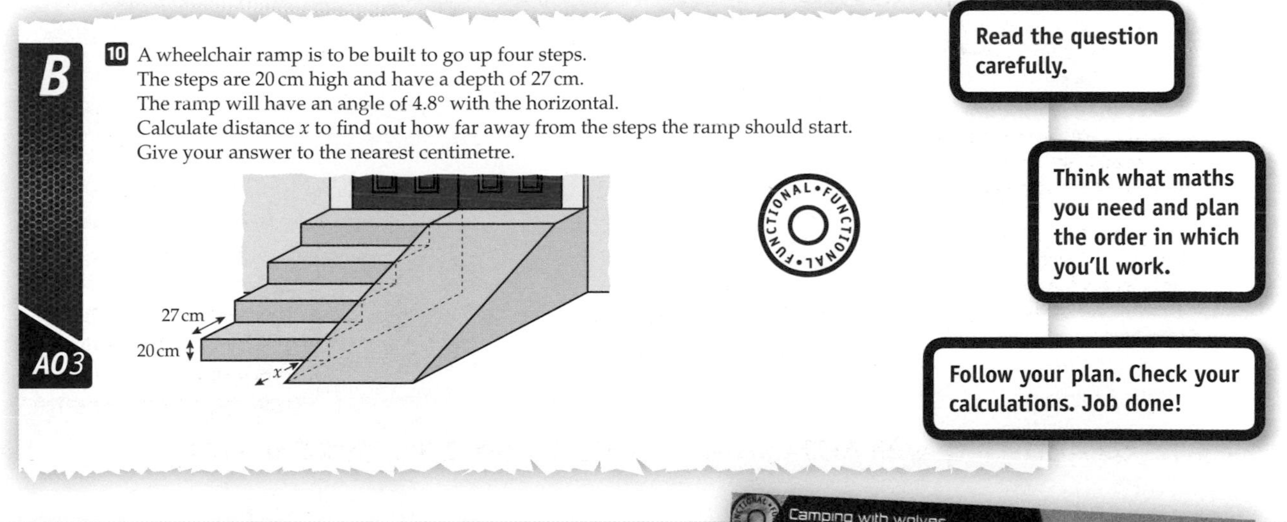

B
10 A wheelchair ramp is to be built to go up four steps.
The steps are 20 cm high and have a depth of 27 cm.
The ramp will have an angle of 4.8° with the horizontal.
Calculate distance x to find out how far away from the steps the ramp should start.
Give your answer to the nearest centimetre.
27 cm
20 cm
x
A03
FUNCTIONAL
Read the question carefully.
Think what maths you need and plan the order in which you'll work.
Follow your plan. Check your calculations. Job done!

Don't miss the fun on our special functional maths pages: 68, 126, 190, 272, 408 and 422! These are not like the functional questions you'll get in GCSE but they'll give you more practice with the key processes.

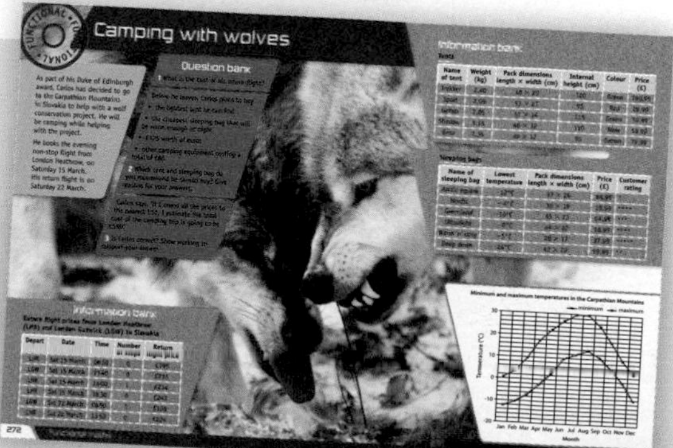

1

Data collection

BBC Video

This chapter is about collecting information.

Do more people attend rock concerts now than 5 years ago? Does it vary by age group?

Objectives

This chapter will show you how to

- understand the data handling cycle D
- state a hypothesis D
- collect information D
- display information D
- design a questionnaire C
- look at sampling techniques C A A*

Before you start this chapter

1 How will you know if rock concerts are better attended now than 5 years ago?

2 What kind of questions do you need to ask?

3 Where can you find information about the age of those who attend?

4 Will you sample a section of the population to get the views of those who attend rock concerts?

5 What are the best techniques to use when taking a sample?

1.1 The data handling cycle

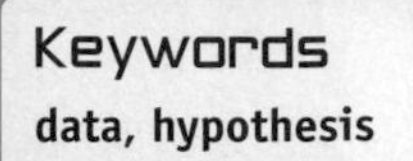

Keywords
data, hypothesis

Why learn this?

You need to ask the right questions to get meaningful results.

Objectives

D Learn about the data handling cycle

D Know how to write a hypothesis

Skills check

How can you find out the following information?

a The average amount of savings for your classmates.

b What times trains leave your local station to go to Manchester.

c The number of votes cast for each party in the last local council elections.

The data handling cycle

When you carry out a statistical investigation, you deal with lots of factual information, called **data**.

The investigation follows the data handling cycle:

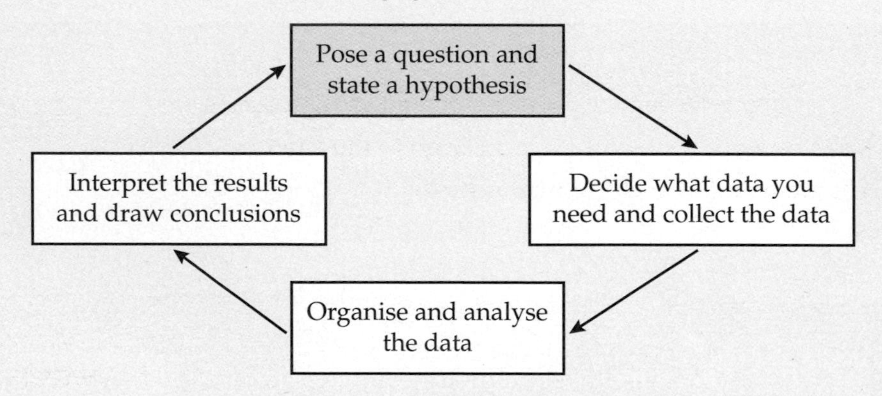

Stating a hypothesis

A **hypothesis** is a statement that helps to answer a question.

A hypothesis must be written so that the answer is 'true' or 'false'.

When you write a hypothesis, use words with clear meaning.

Always make statements about things that can be measured.

D

Example 1

You are playing a game that involves rolling a dice.
You think that a score of 6 occurs fewer times than it ought to.

State a hypothesis to investigate this.

'A score of 6 happens fewer times than any other score.'

You can easily test whether this hypothesis is true or false by carrying out a large number of trials (>100) and recording the number of times each score occurs.

Exercise 1A

1 Write a hypothesis to investigate each question. D

a Who can run faster, boys in Year 10 or girls in Year 10?

b Do people do most of their shopping at the supermarket or at their local shop?

Example 2

D

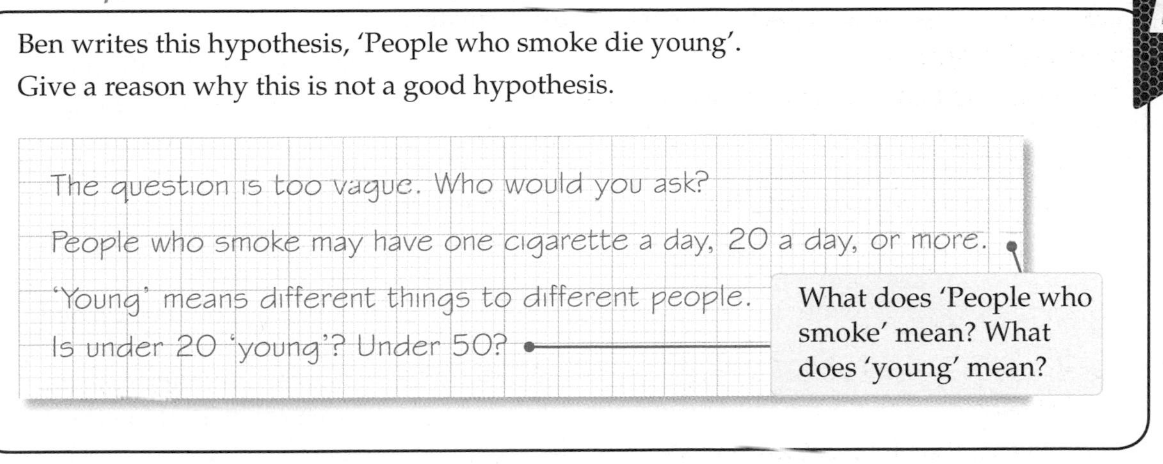

Ben writes this hypothesis, 'People who smoke die young'.

Give a reason why this is not a good hypothesis.

The question is too vague. Who would you ask?

People who smoke may have one cigarette a day, 20 a day, or more.

'Young' means different things to different people.

Is under 20 'young'? Under 50?

What does 'People who smoke' mean? What does 'young' mean?

Exercise 1B

1 Give one reason why each of these is not a good hypothesis. D

a 'More young people than old people go to the cinema.'

b 'Girls are better at spelling than boys.'

1.2 Gathering information

Keywords
primary, secondary

Why learn this?

You need to find out some facts if you are going to test your hypothesis.

Objectives

D Know where to look for information

Skills check

1. Where do you go to find out the times of trains to Manchester?
2. Where and how do you find out the voting patterns in the last local council elections?

Data sources

After writing a hypothesis you need to think about how you are going to test it.

- What information do you need?
- Does the information exist already?
- How easy is it to get the information if it doesn't exist already?
- Where do you try to find the information if it does exist?

There are two types of data source: **primary** and **secondary**.

Primary data is data you collect yourself. You can ask people questions or carry out an experiment.

Secondary data is data that has already been collected. You can look at newspapers, magazines, the internet and many other sources.

D

Example 3

For each hypothesis state
- whether you need primary data or secondary data
- how you would find or collect the data
- how you would use the data.

a 'The yearly total rainfall in Plymouth is greater than in Norwich.'

b 'Girls in Year 10 at my school prefer English to maths.'

a Secondary data

Find records of rainfall for Plymouth and Norwich on the internet.

It will then be easy to see which of them has more rainfall.

You can't record the data yourself so you need data collected by someone else.

b Primary data

Carry out a survey of the girls in Year 10.

Count the votes for each subject to find out which one is preferred.

This information needs to be collected.
A survey could be carried out during registration time by putting a note in each register asking for the two totals from each tutor group.

Exercise 1C

D

1 For each hypothesis state
- whether you need primary data or secondary data
- how you would find or collect the data
- how you would use the data.

a 'In 2008, in England, more cars with petrol engines were bought than cars with diesel engines.'

b 'Tenby has more hours of sunshine in June than Southend.'

c 'People living in my street prefer Chinese takeaway to Indian.'

d 'There are more students in Year 11 who would prefer to go to London on an end-of-term trip than to a theme park or to Blackpool.'

e 'Attendance at the local cinema has fallen steadily over the last 12 months.'

1.3 Types of data

Why learn this?

You need to know the correct terms to describe the data you collect.

Objectives

D Be able to identify different types of data

Keywords

qualitative, quantitative, discrete, continuous

Skills check

How will the times of trains to Manchester be displayed?

Types of data

There are two types of data: **qualitative** and **quantitative.**

Qualitative data can only be described in words. It is usually organised into categories such as colour (red, green, ...) or breed of dog (Labrador, greyhound, ...).

Quantitative data can be given numerical values, such as shoe size or temperature.

Example 4

D

Is this data qualitative or quantitative? The

a number of factory employees **b** taste of a tangerine **c** weight of a car.

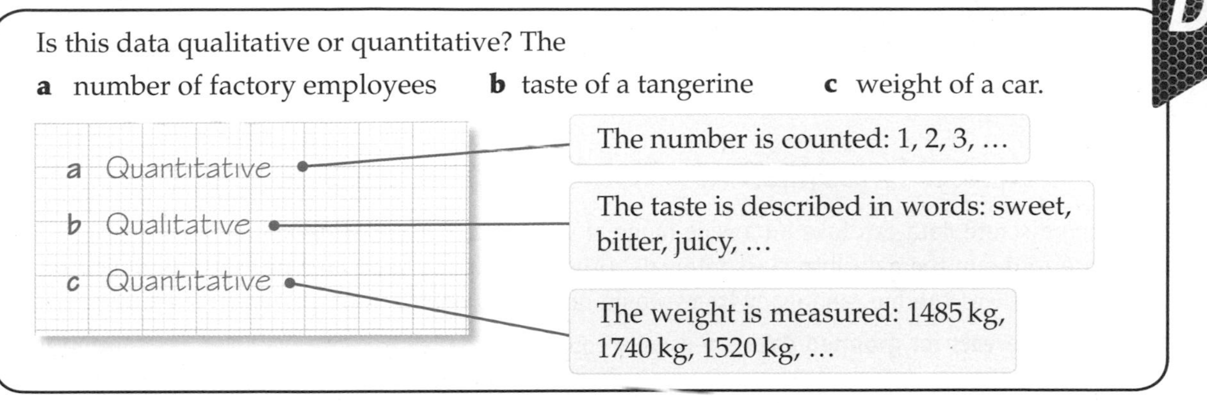

a Quantitative
b Qualitative
c Quantitative
The number is counted: 1, 2, 3, ...
The taste is described in words: sweet, bitter, juicy, ...
The weight is measured: 1485 kg, 1740 kg, 1520 kg, ...

All quantitative data is either **discrete** or **continuous.**

Discrete data can have only certain values, usually whole numbers (the number of goals scored in a match), but may include fractions (shoe sizes).

Continuous data can take any value in a range and can be measured (tree heights).

Example 5

D

For each example, write whether the data is qualitative or quantitative.
If it is quantitative say whether it is discrete or continuous.

a The names of Formula 1 racing car teams.

b The sizes of women's clothes in a boutique.

c The time taken to run 800 metres.

a Qualitative — Formula 1 teams will have names such as Ferrari, McLaren, Brawn GP, ...

b Quantitative; discrete — Women's clothes come in whole number sizes: 12, 14, 16, ...

c Quantitative; continuous — Time can take any value: 1 minute 52.4 seconds, ...

Exercise 1D

D

1 For each question, write whether the data is qualitative or quantitative.
If it is quantitative say whether it is discrete or continuous.
For each one give an example of a typical item of data.

a The score when you throw three darts at a dartboard.

b The masses of eggs in a carton.

c The brands of cereal on sale in a supermarket.

d The times recorded at the Olympic Games for the men's 200 metres.

Keywords
grouped data, class intervals

Why learn this?
Grouping data can make it easier to collect and analyse.

Objectives
D Work out methods for recording data that can take a wide range of values

HELP Section 1.3

Skills check
What is the difference between discrete and continuous data?

Grouped frequency tables for discrete data

When discrete data can take on a wide range of values, such as examination marks, it makes sense to group the data into class intervals. Otherwise, listing each item of data would give a table that was too big, and many items would occur only once.

Class intervals for **grouped data** are usually equal.

D

Example 6

In a quiz, there are 40 questions, each worth one mark.

These are the scores of 30 people who entered the quiz.

23 14 17 36 25 31 20 38 33 28 29 25 19 22 36
30 34 35 36 28 19 21 26 30 32 35 28 31 27 25

Design a frequency table to show this information.

Score	Tally	Frequency
11–15	|	1
16–20	||||	4
21–25	~~||||~~ |	6
26–30	~~||||~~ |||	8
31–35	~~||||~~ ||	7
36–40	||||	4
	Total	30

The groups must not overlap. 11–15, 16–20 etc. are called class intervals.

The scores are discrete (see section 1.3) and can be grouped. In this example, it's sensible to put them into groups of five marks.

After the data has been put into the frequency table, you can't identify individual scores unless you look back at the original data.

Grouped frequency tables for continuous data

Continuous data can take values anywhere in a range. The range can be wide.

Height, weight and time are examples of continuous data.

To group continuous data, the class intervals must use inequality symbols $\leqslant$ and $\geqslant$.

$160 \leqslant h < 170$ means a height from 160 cm up to but not including a height of 170 cm.
A person of height 170 cm would belong to the next class interval, $170 \leqslant h < 180$.

Example 7

Here are the heights, to the nearest centimetre, of 25 basketball players.

215 220 211 212 198 190 210 208 206 212 208 218 210
199 204 206 207 188 209 207 210 200 203 205 222

Put these heights into a grouped frequency table.

Use the class intervals $180 \leqslant h < 190$, $190 \leqslant h < 200$ etc.

Height, h cm	Tally	Frequency
$180 \leqslant h < 190$	|	1
$190 \leqslant h < 200$	|||	3
$200 \leqslant h < 210$	卌 卌 |	11
$210 \leqslant h < 220$	卌 |||	8
$220 \leqslant h < 230$	||	2
	Total	25

210 goes in this class interval.

220 goes in this one.

Check that the total of the frequencies is 25.

Be careful where you put heights that are on the boundary of a class interval.

D

Exercise 1E

1 Mrs Fisher gave her class a mental arithmetic test.
There were 28 questions.
Here are the students' marks.

7 13 18 9 5 12 14 11 16 8 19 11 6 16
17 15 18 10 11 15 4 10 7 15 6 14 12 9

a Put these marks into a grouped frequency table using these class intervals: 1–5, 6–10, 11–15, 16–20.

b Three students were absent for the test.
How many students does Mrs Fisher have in her class?

c In which class interval do most marks occur?

d Did more students score 0–50% or 50–100%?

D

2 Here is a list of the pocket money received in one particular week by 27 students.
All amounts are in £s.

18.40	9.50	12.00	12.25	14.40	5.50	7.80
11.45	16.30	12.45	9.60	13.60	10.00	16.75
13.20	12.80	9.80	10.20	7.75	15.50	8.70
13.50	12.00	8.80	6.65	11.25	5.60	

Design a grouped frequency table to illustrate this data.
Choose suitable class intervals.

Look for the smallest and largest amounts and choose equal class intervals. Four or five equal class intervals ought to be enough.

D

3 This grouped frequency table shows the heights of some plants.

Height, h cm	Frequency
$10 \leqslant h < 16$	9
$16 \leqslant h < 22$	11
$22 \leqslant h < 28$	15
$28 \leqslant h < 34$	28
$34 \leqslant h < 40$	17

a How many plants are in this survey?

b How many plants are less than 22 cm high?

c How many plants are at least 28 cm high?

d Imagine all the plants lined up in a row from the smallest to the tallest.
In which class interval would the plant in the middle of the row lie?

AO2

D

4 The weights of 30 people attending a fitness class were recorded to the nearest kilogram. The results were

46 57 49 66 82 64 55 61 69 68 75 53 94 60 89
64 80 55 83 85 74 53 65 59 54 75 70 78 90 72

a Put the results into a grouped frequency table using these class intervals: $45 \leqslant w < 55$, $55 \leqslant w < 65$ etc.

b How many people weighed less than 75 kilograms?

c In which class interval are there the most people?

5 Here are the times, to the nearest minute, of some runners in a 10 kilometre race.

44 52 58 34 41 55 42 50 48 37 39 46 45 33
40 46 38 50 45 44 49 39 42 40 57 38 55 43

a Show the times in a grouped frequency table using these class intervals: $30 \leqslant t < 35$, $35 \leqslant t < 40$ etc.

b Which class interval contains the most runners?

c How many runners took at least 45 minutes?

d How many runners are shown in this table?

e What percentage of runners took less than 40 minutes?

1.5 Two-way tables

Keywords
two-way tables

Why learn this?
You need to be able to record two sets of related data in a clear way.

Objectives
D Work out methods for recording related data

Skills check
Each row has the same total and each column has the same total.
Work out the numbers represented by A, B, C and D.

4	9	8	3	16
2	9	A	11	B
18	C	4	D	2

Recording data in a two-way table

Data such as the eye colour of boys and girls in a particular class or year group needs to be presented in a way that makes it easy to answer simple questions.

A **two-way table** helps you to do this.

It shows how eye colour and gender are related.

	Blue eyes	Brown eyes	Other eye colour
Boys			
Girls			

Boys with blue eyes.

Girls with neither blue nor brown eyes.

The two-way table can be extended to show the totals for each row and column.

	Blue eyes	Brown eyes	Other eye colour	Total
Boys				
Girls				
Total				

The total number of boys.

The total number of brown-eyed students.

The total number of people.

D

Example 8

This two-way table shows the eye colour and gender of students in Mr Jamir's tutor group.

	Blue eyes	Brown eyes	Other eye colour	Total
Boys	3	8	2	13
Girls	7	6	5	18
Total	10	14	7	31

Including the row and column totals is a good idea.

a How many girls have blue eyes?

b How many boys have neither blue nor brown eyes?

c How many students have brown eyes?

d How many students are in the class?

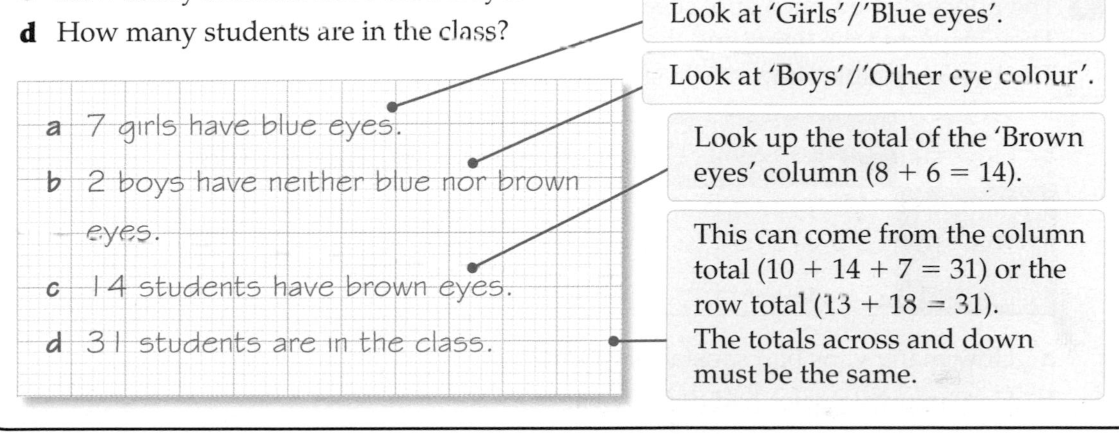

Look at 'Girls'/'Blue eyes'.
Look at 'Boys'/'Other eye colour'.
a 7 girls have blue eyes.
b 2 boys have neither blue nor brown eyes.
Look up the total of the 'Brown eyes' column (8 + 6 = 14).
c 14 students have brown eyes.
This can come from the column total (10 + 14 + 7 = 31) or the row total (13 + 18 = 31). The totals across and down must be the same.
d 31 students are in the class.

Exercise 1F

D

1 The two-way table shows the results of the games played by a football team.

	Won	Drawn	Lost	Total
Home games	9		1	
Away games		5	3	14
Total	15	9		28

a Copy the two-way table and fill in the missing numbers.

b How many home games were played?

c How many games were lost?

d If three points are awarded for a win, one point for a draw and zero points for a loss, how many points did the team gain after playing these games?

2 The two-way table shows the number of adults and the number of cars in 45 houses in a street.

	0 car	1 car	2 cars	3 cars
1 adult	1	3	0	0
2 adults	2	14	6	0
3 adults	0	2	8	3
4 adults	0	1	3	2

a How many houses have exactly three adults and three cars?

b How many houses have two cars?

c How many houses have two adults living in the house?

d What percentage of the houses with one car have three adults living in the house?

D

3 In a survey of 50 teachers about how they travel to work, 22 men came by car, one man came on the bus and five men walked to school.
There were 16 women altogether.
Ten women came by car, one cycled and three came by bus.

- **a** Design a two-way table to show this information.
- **b** Complete the table showing the totals for men and women and for each method of travel.
- **c** How many teachers cycled to school?
- **d** What percentage of teachers walked to school?

AO2

D

4 The police carried out a spot check on vehicles passing through a town.
They inspected the lights and the tyres on each vehicle.
The two-way table shows the results of these checks.

	Satisfactory lights	Defective lights	Satisfactory tyres	Defective tyres
Motorbikes	13	7	15	5
Cars	35	11	42	4
Vans	19	3	16	6

- **a** How many vans had satisfactory tyres?
- **b** How many vehicles had defective lights?
- **c** How many vehicles had satisfactory tyres?
- **d** How many vehicles were stopped and checked?

1.6 Questionnaires

Why learn this?

The best surveys start with a good questionnaire.

Objectives

C Learn how to write good questions to find out information

Keywords

survey, questionnaire, response, leading question

Skills check

Describe a good method for recording data on a data collection sheet.

Questionnaires

When you want to find out information you might want to do a **survey**.

A survey collects primary data. One way to collect this data is to use a **questionnaire**.

A questionnaire is a form that people fill in. On the form are a number of questions.

Asking the right questions is important if you are to find out what you want to know.

Key points

- Use simple language. Ask short questions that can be answered easily.

  Ask questions where the **response** is 'Yes' or 'No'.

Do you have breakfast every day?

Yes ☐ No ☐

- Always give a choice of answers with tick boxes.

The answer options provided must cover all possibilities.

How many sisters do you have?

0 ☐ 1 ☐ 2 ☐ 3 ☐ 4 ☐ More than 4 ☐

- Make sure the responses do not overlap and do not give too many choices.

  These choices are unsuitable for the previous question because they overlap:

0 ☐ 1–2 ☐ 2–3 ☐ 3–4 ☐ 4 or more ☐

Which box would someone tick who has 2 or 3 or 4 sisters?

- Ask a specific question. Make it clear and easy to answer.

How often do you use the internet?

Sometimes ☐ Occasionally ☐ Often ☐

'Sometimes', 'occasionally' and 'often' mean different things to different people, so don't use them.

This response choice is much better.

Never ☐ 1–2 times a week ☐ 3–6 times a week ☐ Every day ☐

- Never ask a personal question. Never ask people to put their names on the questionnaire.

How old are you? ☐ years

Many people will refuse to answer this question. Some may give a false age or a false name!

This is a much better question.

How old are you?

Under 18 years ☐ 18–30 years ☐ Over 30 years ☐

- Never ask a **leading question**.

Watching too much TV is bad for you.

Don't you agree? Yes ☐ No ☐

A leading question encourages people to give a particular answer.

- Don't ask too many questions.
  If your questionnaire is too long people won't want to answer it.

You can also use a two-way table to gather information. It is useful for recording responses to two related questions.

C

Example 9

Four athletics clubs, A, B, C and D enter runners into a half-marathon.

Design a data collection sheet to show the finishing times of runners from all four clubs.

A data collection sheet is sometimes called an observation sheet.

Time (min)	A	B	C	D
$t < 70$		I		
$70 \leqslant t < 80$				I
$80 \leqslant t < 90$	I	II	I	I
$90 \leqslant t$	III	II	~~IIII~~	III

In an exam question you may be asked to make up data, say 20 responses, to put into your two-way table.

Exercise 1G

C

1 Ann wants to find out what people think of their local health centre. She includes these three questions in her questionnaire.

1 What is your date of birth?

2 Don't you agree that it takes too long to get an appointment to see the doctor?

3 How many times did you visit the doctor last year?

Fewer than 3 times ☐ 3–7 times ☐ 7–10 times ☐ More than 10 times ☐

a Say why each question is unsuitable.

b Rewrite each question to make it suitable for a questionnaire.

2 These are questions about diet.

a Eating plenty of vegetables each day is good for you. Don't you agree?

Strongly agree ☐ Agree ☐ Don't know ☐

Give two reasons why this question is unsuitable.

b Do you eat vegetables? Yes ☐ No ☐

If yes, how many times in a week, on average, do you eat vegetables?

Once or less ☐ 2 or 3 times ☐ 4–7 times ☐ More than 7 times ☐

Give two reasons why this is a good question.

C

3 Many of the workers at a factory use a car to travel to work.
The manager of the factory wants to find out if they are willing to car-share.
Write a suitable question with a response section to find out which days, from Monday to Friday, the workers are willing to car-share.

4 Julie is carrying out a survey about how much exercise students in her school do.
One of her questions is 'How many days a week do you exercise for 30 minutes or more?'

a Design a response section for Julie's question.

b Write a question that she can use to find out what activities the students take part in.

5 **a** Design an observation sheet to show how far students travel to school.
It must show data for year groups 7, 8, 9, 10 and 11.

b Make up data for 20 students. Show their responses on your observation sheet.

AO2

1.7 Sampling

Keywords
census, population, sample, random, representative, bias

Why learn this?
You want a sample to give a fair and balanced range of views.

Objectives
C Know the techniques to use to get a reliable sample

Skills check
It takes 30 seconds to get a response to a question from one person.
How long will it take to ask the question to every student in your school?

Sampling techniques

A **census** gathers information from all possible sources.

When you want to carry out an investigation it is too time consuming to ask everyone.

The total number of people you *could* be investigating is called the **population**. It might be 952 students in your school or 1478 people who live in your local area.

Instead of asking everyone, you ask some of them. This smaller group of people you *do* ask is called a **sample**.

You need to make sure that you obtain a **representative** sample. A representative sample is a sample that will give you a fair and balanced range of people's opinions.

Random sampling allows every member of the population an equal chance of being selected. You could choose people by picking names out of a hat or by using a random number generator.

The national census that takes place every 10 years gathers information from every household in the country.

A sample that is not representative of the population will be biased.

Picking every 10th person (for example) does not give a random sample because those who are not 10th, 20th, 30th etc. have no chance of being chosen.

C

Example 10

AO2

Some students at the local school hear that the local council want to make the centre of the town a pedestrian-only area. They decide to find out how much support there is for this plan.

a Why will the students have to use a sample?

b The students decide to ask people their views. They design a questionnaire to do this. They carry out their survey in the town centre between 10 am and midday. Give reasons why this sample will not be representative.

a It is impossible or too time consuming to ask all the people who live in the town.

b Many people are at work during the day so they will not go to the town centre at this time. These people will not be able to give their opinion.

Also, the people who are in the town centre are probably doing their shopping and will be likely to support a traffic-free area because it will make their shopping experience more pleasant.

The opinions of the people who are in the town centre in the late morning might not be the same as the people who are there at other times.
The sample is unlikely to include a representative number of motorists.
The sample is therefore likely to be biased in favour of the pedestrian-only area.

Exercise 1H

C

1 To find out what people think of the new refuse collection arrangements, a telephone survey was carried out between 9 am and 5 pm.
Give reasons why the sample might not be representative.

Who might not be contactable by phone at these times?

2 A survey about the need to provide more parking spaces for people in the town centre was carried out. This was done by asking the opinions of people using the existing car parks in the town.
Why might this sample be unrepresentative?

3 A survey about car ownership was carried out at an out-of-town shopping centre.
Is this likely to give a representative sample? Give reasons for your answer.

4 A survey into how much exercise young people take was carried out at the local sports centre.
Why is this sample likely to be unrepresentative?

5 Ying wants to know how people travel to work.
He interviews people waiting at the bus station.
Is this likely to give a representative sample? Give reasons for your answer.

AO2

6 Gwen wants to know how much people spend on entertainment each week.
She carried out a survey by interviewing people in the town centre in the evening.
Do you think this will be a representative sample? Give reasons for your answer.

1.8 Stratified sampling

Keywords
stratified, sampling fraction

Why learn this?
You will be able to take a better sample and get better results.

Objectives
A A* Know how to take a stratified sample

Skills check
There are 1786 houses in a town.
Work out 5% of this. (Round down the answer.)

Stratified sampling

A simple random sample might not be representative. If a population contains 120 men and 240 women, a random sample might not contain twice as many women as men. This won't matter if gender is irrelevant to the investigation but could be important otherwise.

A **stratified** sample first divides the population into groups. Then it takes a simple random sample from each group. The number chosen from each group is the same fraction of the group as the sample size is of the population. This is called the **sampling fraction**.

Example 11

A

The number of students in each year group in a school is given below.

Year group	7	8	9	10	11
Number of students	254	262	279	213	192

A sample of 100 students is required.
Calculate the number of students that should be chosen from each year group.

Any sample should reflect the numbers in each year group.

Calculate the total number: 254 + 262 + 279 + 213 + 192 = 1200
There are 1200 students in this school.

The sampling fraction for 100 students in this school is $\frac{100}{1200} = \frac{1}{12}$.

Take $\frac{1}{12}$ of each year group for your sample. Rounding up or down is often necessary.

Number of students from Year 7
= 254 ÷ 12 = 21.166...
= 21 (round down)

Number of students from Year 8
= 262 ÷ 12 = 21.833...
= 22 (round up)

Number of students from Year 9
= 279 ÷ 12 = 23.25
= 23 (round down)

Number of students from Year 10
= 213 ÷ 12 = 17.75
= 18 (round up)

Number of students from Year 11
= 192 ÷ 12 = 16 (exactly)

Sample size:
21 + 22 + 23 + 18 + 16 = 100

Check the sample size. Students are then selected at random from each group.

Exercise 1I

A

1 The table shows the ages of members of a gym.

Age (years)	Under 18	18–30	31–50	51–65	Over 65
Number of members	78	112	146	86	58

A sample of 60 people is required.
Calculate the number of people that should be chosen from each age group.

2 A survey of 1834 science students was carried out at a university.
The number of students studying each science is shown in the table below.

Subject	Biology	Chemistry	Geology	Physics
Number of students	679	403	271	481

A 10% stratified sample is required.
Write down the number of students from each category that should be chosen.

3 A secondary school of 1245 students has the following numbers of students in its year groups.

Year group	7	8	9	10	11
Number of students	221	266	224	273	261

How many students from each year group should be in a 5% stratified sample?

4 The table shows the number of students enrolled in some of the departments of a college.

Department	Catering	Engineering	Business	Sport	Humanities	Science
Number of students	80	45	101	94	61	69

A stratified sample of 50 students is to be chosen.
How many students from each department should be chosen?

A*

5 The table shows the numbers of boys and girls in each year group of a school.

Year group	Boys	Girls	Total
7	82	74	156
8	75	69	144
9	50	64	114
10	63	73	136
11	70	80	150
Total	340	360	700

The head teacher wants to take a stratified sample of 100 students for a survey about school uniform.

a Calculate the number of boys and girls in each year group that should be chosen for the sample.

b Explain how to choose a random sample of students.

Review exercise

D

1 Write a hypothesis to investigate whether more people go to France or Spain for their holiday. **[1 mark]**

2 'More men read *Classic Cars* magazine than read *Top Gear* magazine.'
For this hypothesis state
- whether you need primary data or secondary data
- how you would find or collect the data
- how you would use the data. **[2 marks]**

3 For each question, write whether the data is qualitative or quantitative.
If it is quantitative say whether it is discrete or continuous.
For each one give an example of a typical item of data.

a The countries in South America. **[2 marks]**
b The colours of flowers in a garden. **[2 marks]**
c The times recorded at the Olympic Games for the men's 100 metres. **[2 marks]**
d The number of people who attend Glastonbury Festival. **[2 marks]**
e The rainfall in Windermere on each day in April. **[2 marks]**

4 The sizes of shoes sold in a shop during one busy Saturday are given below.

3	5	7	5	9	12	11	5	7	4	8	8	13	6	14
6	8	9	5	8	10	6	7	4	12	11	3	7	9	10
6	10	7	12	10	9	8	7	4	10	7	5	9	10	5
11	8	9	6	10	4	10	7	11	14	7	6	8	10	9

a Design a grouped frequency table to illustrate this data.
Use the class intervals 3–5, 6–8, 9–11 and 12–14. **[3 marks]**
b Which class interval has the most shoes? **[1 mark]**
c How can you tell which size of shoe was the most common? **[1 mark]**

5 These are the heights (h) of some plants at a garden centre, measured to the nearest centimetre.

12	15	21	22	14	31	34	17	25	30	24	13	26	17	20	10	15	38
29	35	19	20	33	23	30	14	33	26	31	17	39	22	16	32	11	14

a Put this information into a grouped frequency table.
Use the class intervals $10 \leqslant h < 15$, $15 \leqslant h < 20$ etc. **[3 marks]**
b Which class interval contained the most plants? **[1 mark]**

6 People who had been on holiday during the summer months were asked what accommodation they had stayed in. The information was put into this two-way table.

	Caravan	B&B	Apartment	Hotel	Total
June		4	5	14	
July	8		12		37
August	11	10	15		46
September	6	7	13	13	39
Total	28	36	45	48	

a Copy the table and complete it by filling in all the missing numbers. **[5 marks]**
b How many people stayed in B&B? **[1 mark]**
c How many people stayed in a hotel in August? **[1 mark]**
d How many people stayed in an apartment in September? **[1 mark]**
e How many people took part in this survey? **[1 mark]**

C

7 Write a question that Ian could use to find out how much time students spend doing homework at the weekend. Include a response section. **[3 marks]**

AO2

8 Linda wants to find out which sport is most popular amongst 14- to 16-year-olds in her school.
She asks a group of Year 10 girls.
Give two reasons why this sample is not representative. **[2 marks]**

A

9 A golf club has 300 members. The members are classified by age as follows.

Age (years)	Under 16	16–25	26–40	41–60	Over 60
Number of members	32	42	70	92	64

A stratified sample of 50 people is planned.
Calculate the number of people that should be sampled from each age group. **[4 marks]**

Chapter summary

In this chapter you have learned how to

- understand the data handling cycle D
- write a hypothesis D
- decide where to look for information D
- identify different types of data D
- work out methods for recording data that can take a wide range of values D
- work out methods for recording related data D
- write good questions to find out information C
- get a reliable sample C
- take a stratified sample A A*

2 Fractions, decimals and percentages

This chapter is about working with fractions, decimals and percentages.

If you cook a batch of 40 pancakes, you need these skills to help share them equally with 5 friends.

Objectives

This chapter will show you how to

- find a fraction of an amount using a calculator **D**
- write one quantity as a fraction of another **D**
- calculate a percentage increase or decrease **D**
- write one quantity as a percentage of another **D** **C**
- understand and use the retail prices index (RPI) **D** **C**
- calculate with fractions using a calculator **D** **C**

Before you start this chapter

1 Write down the common factors of 18 and 27.

2 Write down the highest common factor of 48 and 36.

3 Simplify

a $\frac{4}{8}$ b $\frac{8}{20}$ c $\frac{28}{32}$

4 Find the missing number in each of these conversions.

a 7.5 litres = ☐ m*l*   b 32 mm = ☐ cm

c 0.6 km = ☐ m = ☐ cm   d $\frac{1}{2}$ hour = ☐ minutes

5 Work out 15% of £50.

6 Anna needs £250 to buy a new bicycle.
She saves £18 each week.
For how many weeks must she save so that she has enough money?

7 Use your calculator to work out $\frac{125}{800}$ as a decimal.

8 a Convert 24% to a decimal.

b Convert 124% to a decimal.

9 Calculate $40 \times \frac{60}{100}$.

2.1 Fraction of an amount

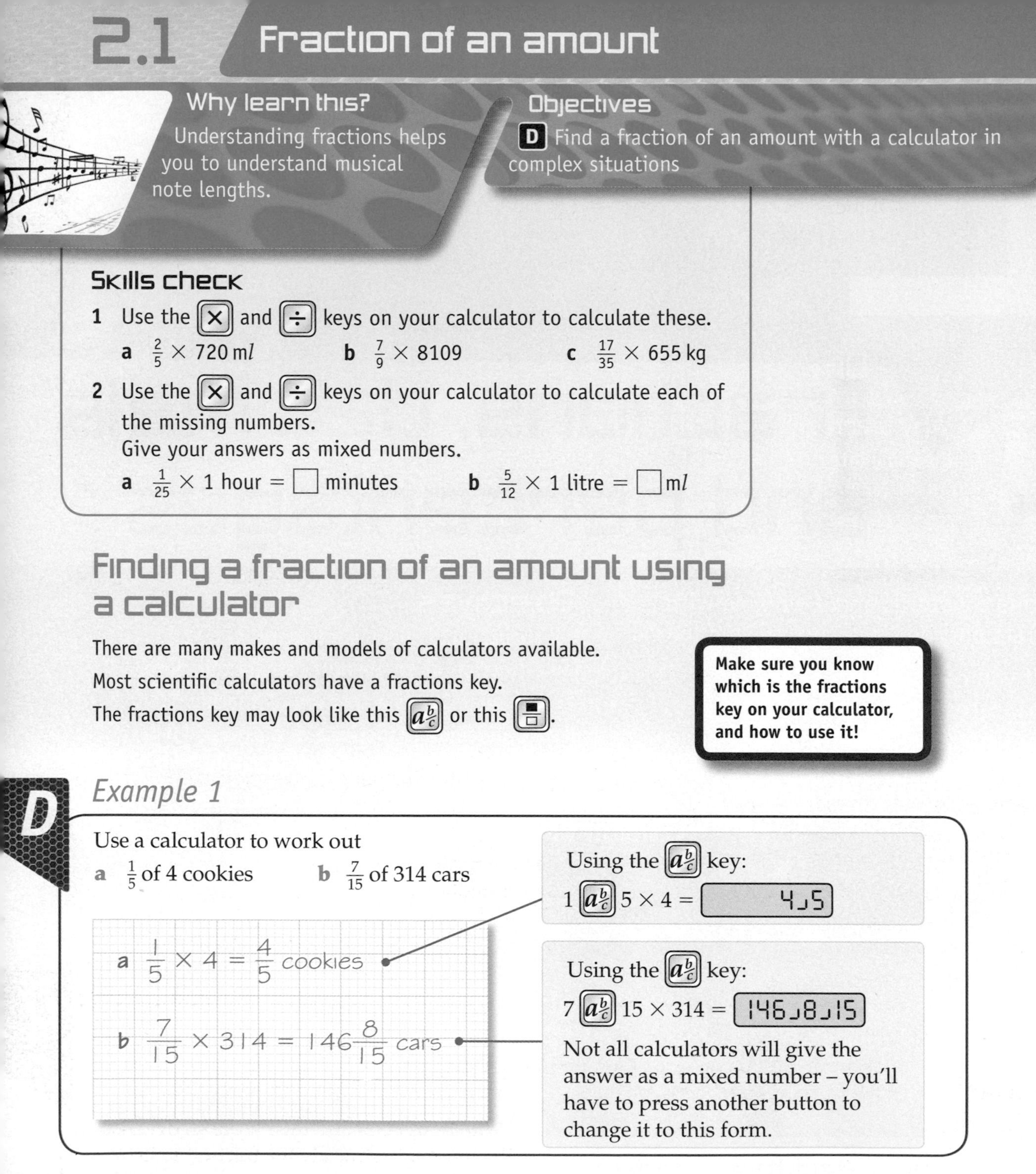

Why learn this?

Understanding fractions helps you to understand musical note lengths.

Objectives

D Find a fraction of an amount with a calculator in complex situations

Skills check

1 Use the [×] and [÷] keys on your calculator to calculate these.

a $\frac{2}{5} \times 720$ ml b $\frac{7}{9} \times 8109$ c $\frac{17}{35} \times 655$ kg

2 Use the [×] and [÷] keys on your calculator to calculate each of the missing numbers.
Give your answers as mixed numbers.

a $\frac{1}{25} \times 1$ hour = ☐ minutes b $\frac{5}{12} \times 1$ litre = ☐ ml

Finding a fraction of an amount using a calculator

There are many makes and models of calculators available.

Most scientific calculators have a fractions key.

The fractions key may look like this [$a\frac{b}{c}$] or this [▭/▭].

Make sure you know which is the fractions key on your calculator, and how to use it!

Example 1

Use a calculator to work out

a $\frac{1}{5}$ of 4 cookies b $\frac{7}{15}$ of 314 cars

a $\frac{1}{5} \times 4 = \frac{4}{5}$ cookies

b $\frac{7}{15} \times 314 = 146\frac{8}{15}$ cars

Using the [$a\frac{b}{c}$] key:
1 [$a\frac{b}{c}$] 5 × 4 = [4⌟5]

Using the [$a\frac{b}{c}$] key:
7 [$a\frac{b}{c}$] 15 × 314 = [146⌟8⌟15]

Not all calculators will give the answer as a mixed number – you'll have to press another button to change it to this form.

Exercise 2A

1 Ryan runs a dog rehoming centre.
On average he feeds each dog $\frac{3}{4}$ of a tin of food per day.
This week there are 23 dogs at the centre.
How many tins of dog food does Ryan need for this week?

D AO2

2 Heather buys 80 duvets for £595.
She sells $\frac{2}{5}$ of them for £12 each and $\frac{7}{16}$ of them for £10 each.

a How much profit does Heather make?

b How many duvets does she have left over?

D AO3

3 In a school election Jack got $\frac{2}{9}$ of the votes and Rebecca got $\frac{4}{9}$ of the votes.
Rebecca had 42 more votes than Jack.
How many students didn't vote for Jack or Rebecca?

4 There are 48 members in a scout group.
Of these members, $\frac{5}{8}$ are boys. Of the boys, $\frac{1}{3}$ are 12 years old or over.
Half of all the scouts are 12 years old or over.
How many of the scouts are girls under the age of 12?

D

AO2

2.2 One quantity as a fraction of another

Why learn this?
This topic often comes up in an exam.

Objectives
D Write one quantity as a fraction of another

Skills check

1 Find the missing number in each of these conversions.
  a 1 kg = □ g b 1 km = □ m c 1 m = □ cm

2 Simplify each fraction to its lowest terms.
  a $\frac{2}{6}$ b $\frac{8}{12}$ c $\frac{15}{20}$

Writing one quantity as a fraction of another

To write one quantity as a fraction of another
- first write both quantities in the same units
- then write the first quantity over the second quantity
- finally, simplify the fraction.

Example 2

D

a Write 40p as a fraction of £6.
b In Diane's class there are 13 boys and 14 girls.
What fraction of the class are boys?

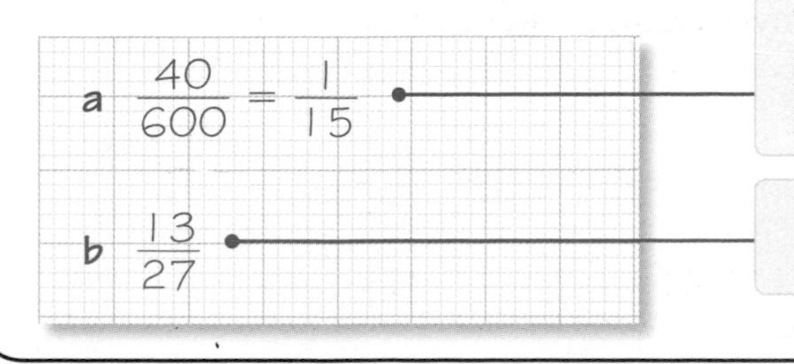

Write £6 as 600p and then write 40 over 600. Remember to simplify the fraction to its lowest terms.

13 + 14 = 27 students altogether. 13 out of the 27 are boys.

Exercise 2B

1 In Polly's dog training class, three of the dogs pass their elementary certificate.
The other four do not pass.
What fraction of the dogs pass?

D

2 During April it was sunny for 14 days.
What fraction of the days in April were not sunny?

AO2

3 Donna said, 'I've got 5 red sweets and 9 blue sweets so the fraction of my sweets that are red is $\frac{5}{9}$.'
Is Donna correct? Explain your answer.

D

4 In each case, write the first quantity as a fraction of the second.

a £2, £7 **b** 1 hour, 24 hours **c** 6 weeks, 8 weeks

D

5 In each case, write the first quantity as a fraction of the second.

a 20p, £1 **b** 5 minutes, 1 hour
c 5 days, 2 weeks **d** 30 cm, 1 m
e 750 g, 1 kg **f** 12 m, 1 km

6 Lauren gets £5 pocket money per week.
Each week she buys a magazine for £1.75 and she saves £1.50.
The rest is left over for her to spend on other things.

What fraction of her pocket money does Lauren

a spend on a magazine
b save
c have left over?

AO2

2.3 Calculating with fractions

Why learn this?

By learning how to enter fractions into a calculator you can save a lot of time.

Objectives

D C Use the fractions key on a calculator with mixed numbers

Skills check

1 Use the fractions key on a calculator to simplify these fractions.

a $\frac{24}{36}$ **b** $\frac{55}{75}$ **c** $\frac{52}{65}$

2 Use the fractions key on a calculator to convert these mixed numbers to improper fractions.

a $3\frac{8}{9}$ **b** $4\frac{12}{17}$ **c** $16\frac{18}{25}$

3 Use the fractions key on a calculator to convert these improper fractions to mixed numbers.

a $\frac{130}{7}$ **b** $\frac{196}{15}$ **c** $\frac{223}{32}$

Calculating with fractions using a calculator

To carry out any calculations involving fractions, enter the fractions on your calculator using the fractions key. Make sure you practise using your calculator so you become skilled.

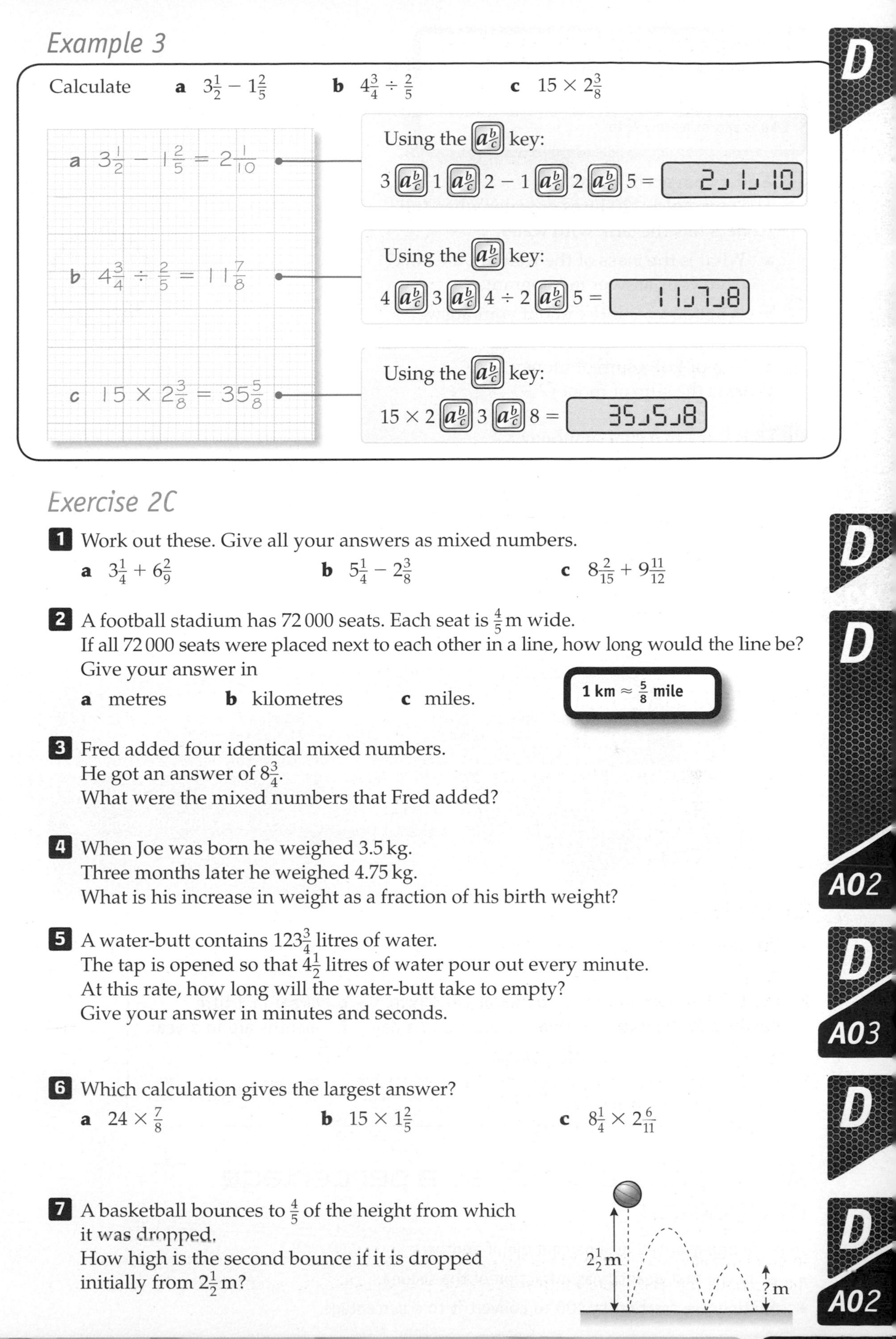

Example 3

Calculate **a** $3\frac{1}{2} - 1\frac{2}{5}$ **b** $4\frac{3}{4} \div \frac{2}{5}$ **c** $15 \times 2\frac{3}{8}$

D

a $3\frac{1}{2} - 1\frac{2}{5} = 2\frac{1}{10}$

Using the $a\frac{b}{c}$ key:

3 $a\frac{b}{c}$ 1 $a\frac{b}{c}$ 2 − 1 $a\frac{b}{c}$ 2 $a\frac{b}{c}$ 5 = 2⌟1⌟10

b $4\frac{3}{4} \div \frac{2}{5} = 11\frac{7}{8}$

Using the $a\frac{b}{c}$ key:

4 $a\frac{b}{c}$ 3 $a\frac{b}{c}$ 4 ÷ 2 $a\frac{b}{c}$ 5 = 11⌟7⌟8

c $15 \times 2\frac{3}{8} = 35\frac{5}{8}$

Using the $a\frac{b}{c}$ key:

15 × 2 $a\frac{b}{c}$ 3 $a\frac{b}{c}$ 8 = 35⌟5⌟8

Exercise 2C

D

1 Work out these. Give all your answers as mixed numbers.

a $3\frac{1}{4} + 6\frac{2}{9}$ **b** $5\frac{1}{4} - 2\frac{3}{8}$ **c** $8\frac{2}{15} + 9\frac{11}{12}$

D

2 A football stadium has 72 000 seats. Each seat is $\frac{4}{5}$ m wide.
If all 72 000 seats were placed next to each other in a line, how long would the line be?
Give your answer in

a metres **b** kilometres **c** miles.

1 km ≈ $\frac{5}{8}$ mile

3 Fred added four identical mixed numbers.
He got an answer of $8\frac{3}{4}$.
What were the mixed numbers that Fred added?

AO2

4 When Joe was born he weighed 3.5 kg.
Three months later he weighed 4.75 kg.
What is his increase in weight as a fraction of his birth weight?

D

5 A water-butt contains $123\frac{3}{4}$ litres of water.
The tap is opened so that $4\frac{1}{2}$ litres of water pour out every minute.
At this rate, how long will the water-butt take to empty?
Give your answer in minutes and seconds.

AO3

D

6 Which calculation gives the largest answer?

a $24 \times \frac{7}{8}$ **b** $15 \times 1\frac{2}{5}$ **c** $8\frac{1}{4} \times 2\frac{6}{11}$

D

7 A basketball bounces to $\frac{4}{5}$ of the height from which it was dropped.
How high is the second bounce if it is dropped initially from $2\frac{1}{2}$ m?

$2\frac{1}{2}$ m

? m

AO2

Metric–Imperial conversions

1 litre is approximately $1\frac{3}{4}$ pints

1 pint of water weighs approximately $1\frac{1}{4}$ pounds (lb)

1 kg is approximately $2\frac{1}{5}$ lb

D

AO2

8 Anders buys a fish tank.
The tank holds 120 litres of water when full.
Anders fills the tank with water.

a What is the mass of the water in the tank?
Give your answer in kilograms.

b What do you notice about your answer?

C

9 £10 is $\frac{1}{3}$ of $\frac{1}{2}$ of a sum of money.
What is the sum of money?

10 £5 is $\frac{3}{4}$ of $\frac{2}{3}$ of a sum of money.
What is the sum of money?

11 £5 is half of four-sevenths of a sum of money.
What is the sum of money?

AO3

12 £6 is one-sixth of three-fifths of a sum of money.
What is the sum of money?

2.4 One quantity as a percentage of another

Why learn this?

So you can work out your test results as a percentage.

Objectives

D Write one quantity as a percentage of another

C Write one quantity as a percentage of another in more complex situations

Skills check

1 Write down how many

a cm are in 1 m **b** m*l* are in 1 litre **c** c*l* are in 1 litre

d g are in 1 kg **e** hours are in a day **f** months are in a year.

2 Copy and complete these divisions.

a $\frac{3}{5} = 3 \div 5 = \square$ **b** $\frac{7}{28} = 7 \div \square = \square$ **c** $\frac{25}{40} = \square \div \square = \square$

Writing one quantity as a percentage of another

To write one quantity as a percentage of another

- write the first quantity as a fraction of the second.
- multiply the fraction by 100 to convert it to a percentage.

Example 4

a Express £5 as a percentage of £25.

b Express 8 mm as a percentage of 2 cm.

c Daren bought a car for £8000 and sold it for £6500.
What percentage of the price that he paid has he lost?

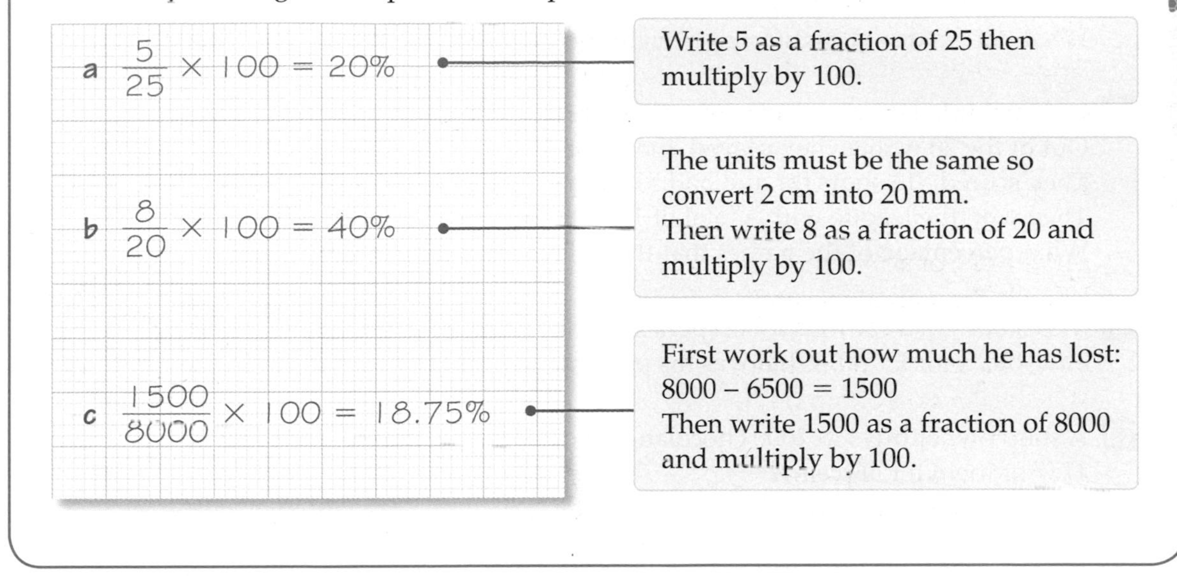

a $\frac{5}{25} \times 100 = 20\%$
Write 5 as a fraction of 25 then multiply by 100.
b $\frac{8}{20} \times 100 = 40\%$
The units must be the same so convert 2 cm into 20 mm.
Then write 8 as a fraction of 20 and multiply by 100.
c $\frac{1500}{8000} \times 100 = 18.75\%$
First work out how much he has lost:
8000 – 6500 = 1500
Then write 1500 as a fraction of 8000 and multiply by 100.

Exercise 2D

1 In each case, express the first quantity as a percentage of the second.

a £5, £50 **b** £5, £80 **c** £25, £75

d 4 hours, 1 day **e** 36 minutes, 1 hour **f** 125 g, 1 kg

g 13 weeks, 1 year **h** 14.5 cm, 1 m **i** 275 *ml*, 1 litre

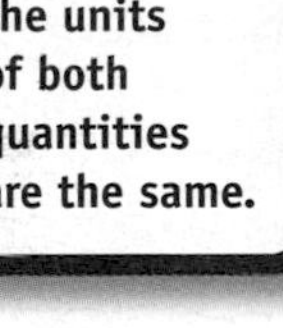

Make sure the units of both quantities are the same.

2 Fathe goes to the shops with £12. He buys a book for £4.50.
What percentage of his money has he got left?

3 Hari scores 26 out of 30 in an English test.
What percentage of the test does Hari get wrong?

4 A badminton club has 63 members.
The table shows the membership numbers.

Men	Women	Girls	Boys
17	21	12	13

What percentage of the members are

a men

b female?

5 Billy weighed 102 kg at the start of his diet.
He now weighs 82 kg.
What percentage of his starting weight has he lost?
Give your answer to one decimal place.

C

6 Last year Llanreath Divers took £1800 in membership fees.
This year they took £2100 in membership fees.
What is the percentage increase in the amount taken in membership fees?

C

7 Jon Brower Minnoch was the heaviest man recorded in history.
In 16 months he lost 419 kg in weight.
His final weight was 216 kg.
What percentage of his starting weight did he lose?

8 In 2007 Little Haven won the South Pembrokeshire short mat bowls league.
Out of the 20 games they played, they won 11 and drew 2.
They scored 813 shots for and had 515 shots against.
They won the league with a total of 102 points.
What percentage of the games that they played did they lose?

FUNCTIONAL

9 Mosel bought a car for £2500, spent £400 on improvements and sold it for £3750. What was his percentage profit?

AO2

10 A shop owner buys in 2000 chocolate bars at 37p each and sells 1777 of them for 45p each.
The rest were not sold and were discarded.
Work out the shop owner's percentage profit.

2.5 Percentage increase and decrease

Keywords
percentage increase, percentage decrease, original amount, reduce

Why learn this?
You can use this to work out how much you could save in a sale.

Objectives
- D Calculate a percentage increase or decrease
- D Perform calculations using VAT

Skills check
1 Convert 73% to a decimal.
2 Convert 119% to a decimal.
3 Work out **a** 5% of £200 **b** 13% of £420 **c** 17% of £333

Percentage increase and decrease

Method A

1 Work out the value of the increase (or decrease).
2 Add it to (or subtract it from) the **original amount**.

This method is most commonly used when working without a calculator.

Method B

1 Add the **percentage increase** to 100% (or subtract the **percentage decrease** from 100%).
2 Convert this percentage to a decimal.
3 Multiply it by the original amount.

This method is especially useful when using a calculator.

Example 5

D

John used to earn £320 a week. He has had an 8% pay rise.
What does he earn now?

Method A

1% of £320 = £320 ÷ 100
= £3.20

So an 8% rise = £3.20 × 8
= £25.60

So John gets £320 + £25.60
= £345.60

Method B

Rise = 8%

New salary = 8% + 100%
= 108%
= 1.08

So John gets £320 × 1.08
= £345.60

Divide by 100 to convert a percentage to a decimal.

Example 6

D

The price of a bike is reduced by 20% in a sale. The original price was £150.
What is the sale price?

Method A

Decrease = £150 ÷ 5
= £30

Sale price is £150 − £30
= £120

20% is $\frac{1}{5}$, so divide by 5.

Method B

Decrease = 20%

New price = 100% − 20%
= 80%
= 0.8

Sale price is £150 × 0.8
= £120

Exercise 2E

D

1 **a** Increase 20 by 5%
b Increase 56 by 12%
c Decrease 3400 m*l* by 17%
d Decrease 480 by 32%
e Increase £135 by $7\frac{1}{2}$%
f Decrease 890 m*l* by 8.4%

2 Firefighters are given a 4% pay rise. If John earned £420 per week, how much will he earn after the pay rise?

3 There has been a 6% decrease in the number of reported thefts in Walton this year. There were 350 reported thefts last year. How many were there this year?

4 A new car costs £8450. After two years the value of the car will have decreased by 43%. How much will the car be worth?

D

5 Sales of a magazine, costing £1.95, decreased by 11% this week. They sold 4300 copies last week. How much money have they lost on their sales this week?

6 Jane starts a new job earning £12 000 per year, increasing by 3% after 3 months. Milly starts on £11 500 per year, increasing by 5% after 3 months. Who has the greater salary after 3 months?

D

7 Ali weighed 84 kg. He lost 4% of his body weight when he started running but then put 2% of his new weight back on.

a How much was Ali's lowest weight?

b How much does Ali weigh now?

AO2

8 In 2007 the number of pairs of breeding sparrows was estimated as 200 000. This decreased by 38% in 2008. In 2009 there was a slight recovery with an increase of 4%. How many breeding pairs were estimated at the end of 2009?

VAT

VAT stands for value added tax. It is a tax that is added to the price of most items in shops and to many other services.

VAT is calculated as a percentage. Generally it is 17.5% in the UK.

VAT at 17.5% can be worked out by finding 10% + 5% + 2.5%.

Example 7

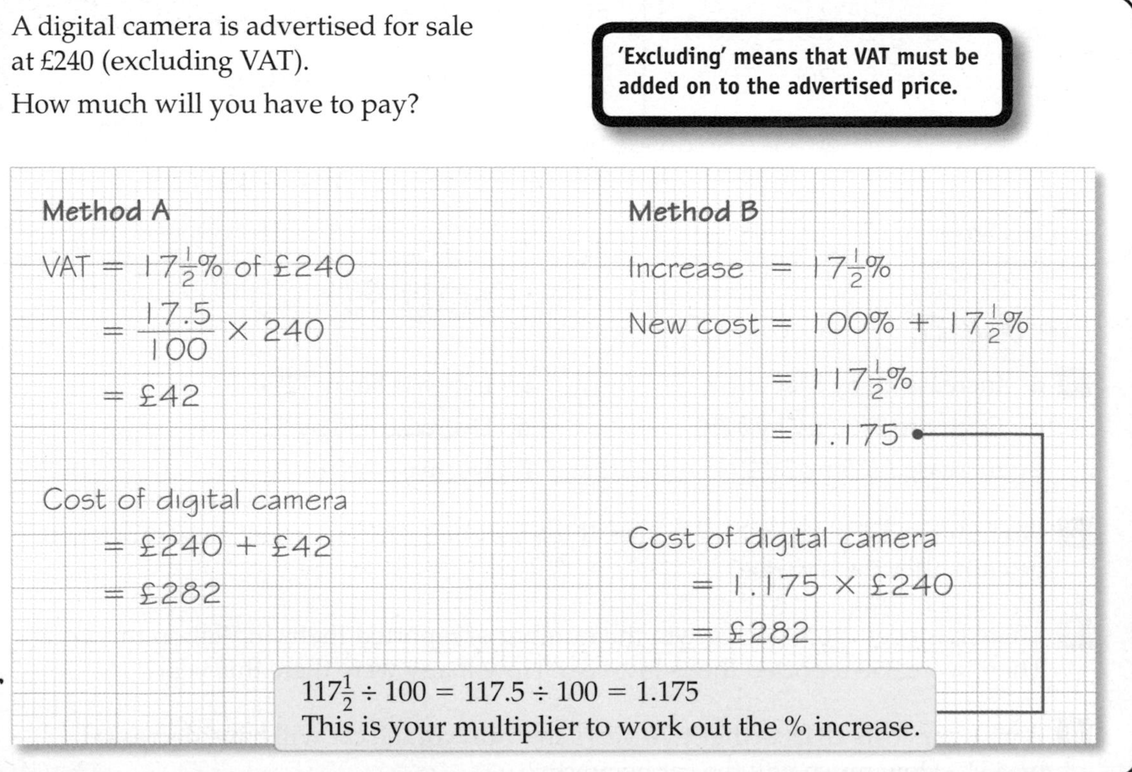

A digital camera is advertised for sale at £240 (excluding VAT).

How much will you have to pay?

'Excluding' means that VAT must be added on to the advertised price.

Method A

VAT $= 17\frac{1}{2}\%$ of £240

$= \frac{17.5}{100} \times 240$

$= £42$

Cost of digital camera

$= £240 + £42$

$= £282$

Method B

Increase $= 17\frac{1}{2}\%$

New cost $= 100\% + 17\frac{1}{2}\%$

$= 117\frac{1}{2}\%$

$= 1.175$

Cost of digital camera

$= 1.175 \times £240$

$= £282$

$117\frac{1}{2} \div 100 = 117.5 \div 100 = 1.175$

This is your multiplier to work out the % increase.

Exercise 2F

For questions 1 to 5, assume the rate of VAT is 17.5%.

1 A DVD player costs £130 (excluding VAT).
A phone costs £40 (excluding VAT).
Work out **a** the VAT and **b** the total cost of each item.

2 What is the total cost of a TV sold for £126 + VAT?

3 One car is sold at £4000 (including VAT) and another is sold at £3800 (excluding VAT). Which car is the more expensive?

4 A meal for four costs £73.60 plus VAT. If the four people decide to share the bill equally, how much will each pay?

D

5 The cost of a scooter is £376 + VAT. Jas decides to pay by credit. He pays an initial deposit of 15% and then 12 monthly payments of £35. How much more does Jas pay by buying on credit?

D
AO2

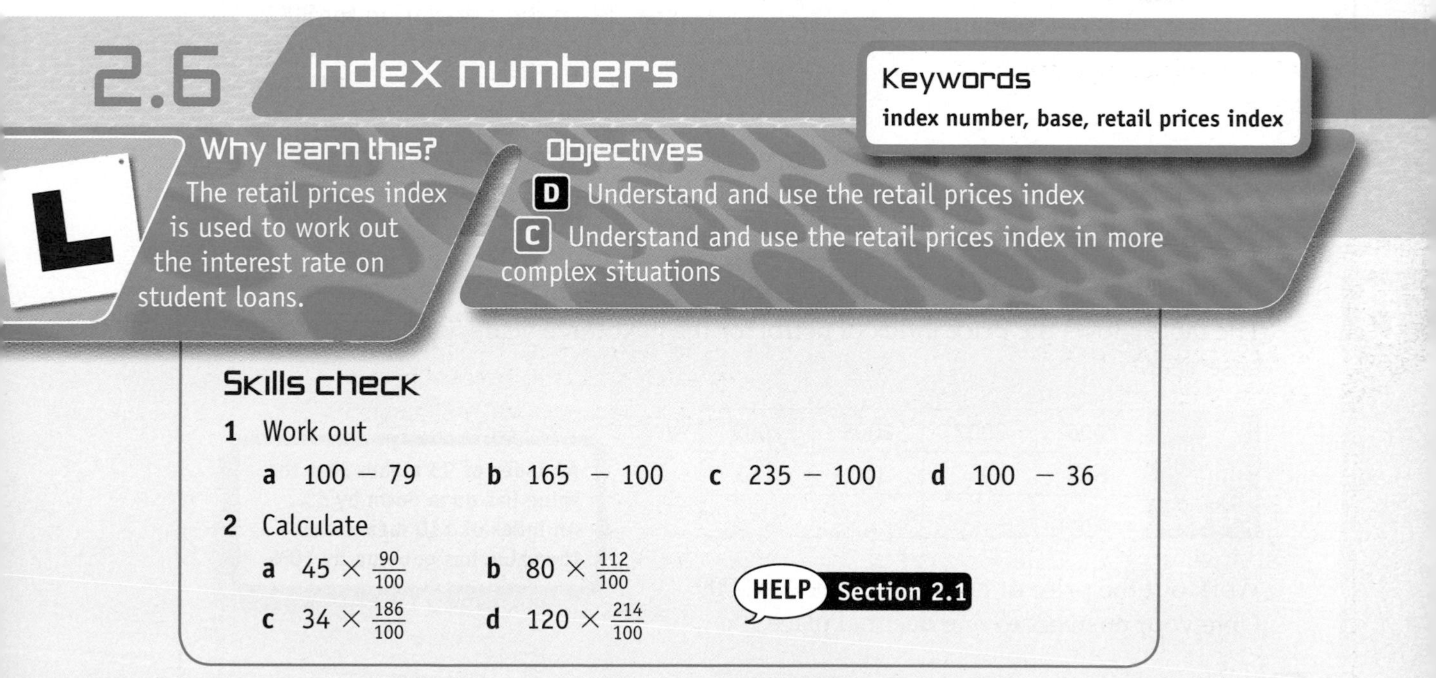

2.6 Index numbers

Keywords
index number, base, retail prices index

Why learn this?
The retail prices index is used to work out the interest rate on student loans.

Objectives
D Understand and use the retail prices index
C Understand and use the retail prices index in more complex situations

Skills check

1 Work out

a 100 − 79 **b** 165 − 100 **c** 235 − 100 **d** 100 − 36

2 Calculate

a $45 \times \frac{90}{100}$ **b** $80 \times \frac{112}{100}$

c $34 \times \frac{186}{100}$ **d** $120 \times \frac{214}{100}$

HELP Section 2.1

Retail prices index

An **index number** compares one quantity, usually a price, with another.

The figure that the quantities are compared with is called the **base.**

The index number is a percentage of the base, but the percentage sign is left out.

The base usually starts at 100.

The UK **retail prices index** started at base 100 in 1987.

In May 2009 the UK retail prices index was 211.3.
This means that average retail prices increased by 111.3% between 1987 and 2009.

D

Example 8

In 2000 the price of a litre of petrol was 69p.
Using the year 2000 as the base year, the price indices of petrol for 1999, 2001, 2002 and 2003 are given in the table.

Year	1999	2000	2001	2002	2003
Index	92	100	103	107	110
Price		69p			

Work out the price of petrol in 1999 and in 2001 to 2003.
Give your answers to one decimal place.

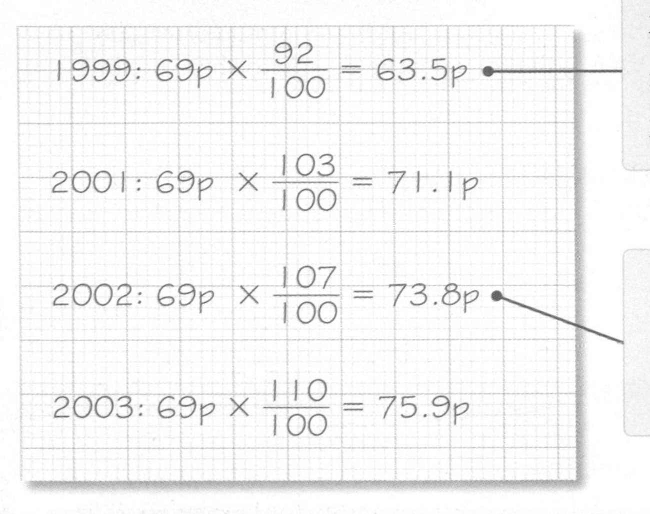

1999: $69p \times \frac{92}{100} = 63.5p$
2001: $69p \times \frac{103}{100} = 71.1p$
2002: $69p \times \frac{107}{100} = 73.8p$
2003: $69p \times \frac{110}{100} = 75.9p$

In 1999 the index was less than the base, so the price of a litre of petrol must be less. In the years 2001 to 2003 the index was more than the base, so the price must be more.

Notice that for each year's calculation, you always use 69p as this is the price of petrol in the base year. So in the calculation for 2002, you don't use the 71.1p from 2001.

Exercise 2G

D

1 In 2006 the price of a litre of petrol was 95p.
The table shows the price index of petrol for the next three years, using 2006 as the base year.

Year	2006	2007	2008	2009
Index	100	95	100	110
Price	95p			

Work out the price of petrol from 2007 to 2009.
Give your answers to one decimal place.

An index of 95 means that the value has gone down by 5%. An index of 110 means that the value has gone up by 10%.

2 The index for the price of laptop computers, compared with 2005 as base, is 74.
- **a** Has the price of laptop computers gone up or down?
- **b** By what percentage has the price of laptop computers changed?

3 This year the price of a 4 GB memory stick is 40% lower than last year.
What is the index for the price of a 4 GB memory stick this year, using last year as base?

4 In 1990 an average box of tissues cost 30p.
Taking 1990 as the base year of 100, work out how much an average box of tissues costs today, when the price index is 240.

5 The retail prices index was introduced in January 1987.
It was given a base number of 100.
In May 2009 the index number was 211.3.
In January 1987 the 'standard weekly shopping basket' cost £38.50.
How much did the 'standard weekly shopping basket' cost in May 2009?

6 In 1990 the price of 1 kg of bananas was £1.14.
Using 1990 as the base year, the price index of 1 kg of bananas in 2008 was 75.
Peter says, 'The price of bananas in 2008 is $\frac{1}{4}$ of the price they were in 1990.'
Is Peter correct? Explain your answer.

7 The graph shows the exchange rates for the euro (€) and the pound (£) in 2008.

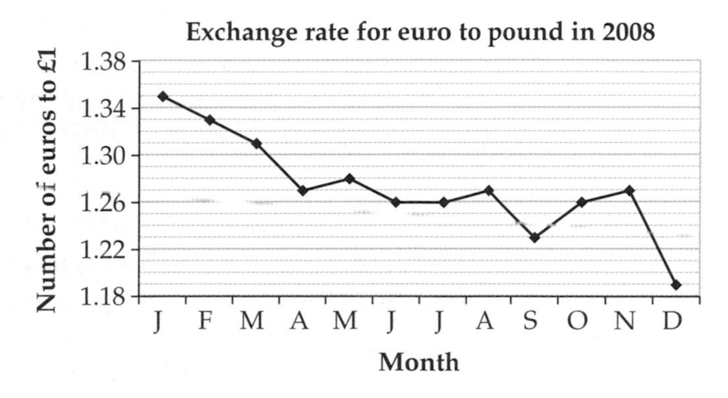

a What was the exchange rate in January?

b Using January 2008 as the base of 100, work out the index for December 2008.

8 A toy factory produced 455 000 toys in November 2008.
This represented an index of 78, using November 2002 as the base year with an index of 100.
How many toys did this factory produce in November 2002?

Review exercise

1 Rosa says '2 mm as a fraction of 50 cm can be written like this: $\frac{2}{50} = \frac{1}{25}$.'
Is Rosa correct?
Explain your answer. **[2 marks]**

2 Nigel is given £40 for his birthday.
He spends £16.40 on clothes.
What percentage of his money has he got left? **[3 marks]**

3 This year the price of a DVD is 8% lower than last year.
What is the index for the price of a DVD this year, using last year as base? **[2 marks]**

4 In 1978 1 kg of mushrooms cost £1.53.
Taking 1978 as the base year of 100, work out how much 1 kg of mushrooms cost in 2008, when the price index was 172. **[2 marks]**

D

5 All train fares are to increase by $7\frac{1}{2}\%$ next year.
At the moment, Jane pays £28 for her ticket.
How much will she have to pay next year? **[3 marks]**

D

6 Gaynor plants 60 daffodil bulbs.
48 of the bulbs grow.
What percentage of the bulbs do not grow? **[2 marks]**

AO2

7 Heather sleeps for 8 hours every day.
What percentage of a week is Heather awake? **[3 marks]**

D

8 Tom invested £250 last year. His money has decreased by $2\frac{1}{2}\%$.
How much is Tom's investment worth now? **[2 marks]**

D

AO2

9 Two posters advertise the same TV.
Which method is cheaper?
Find how much can be saved using the cheaper method. **[6 marks]**

TVs R US
£240
+ VAT ($17\frac{1}{2}\%$)
Buy now!

A.A. Electricals
deposit *22%* of £280
plus
12 monthly payments of
£19.75

C

10 Last year a summer fête raised £1420 for charity.
This year it raised £1650.
What is the percentage increase in the amount raised for charity?
Give your answer to the nearest whole number. **[3 marks]**

11 Last year Amy bought a laptop computer.
This year the price of the same laptop computer fell by £80 to £370.
By what percentage of last year's price has the price fallen?
Give your answer to one decimal place. **[3 marks]**

Chapter summary

In this chapter you have learned how to

- find a fraction of an amount with a calculator in complex situations **D**
- write one quantity as a fraction of another **D**
- calculate a percentage increase or decrease **D**
- perform calculations using VAT **D**
- write one quantity as a percentage of another **D** **C**
- understand and use the retail prices index **D** **C**
- use the fractions key on a calculator with mixed numbers **D** **C**

3 Interpreting and representing data

This chapter is about interpreting and representing data.

Is there a relationship between global temperature and size of penguin colonies?

Objectives

This chapter will show you how to

- **draw stem-and-leaf diagrams** D
- **draw frequency diagrams** D
- **draw and interpret scatter graphs** D C
- **draw frequency polygons for grouped data** C
- **draw histograms for grouped continuous data** A
- **intepret histograms and draw conclusions** A A*

Before you start this chapter

1 Work out these calculations.

a 15×26 b $360 \div 24$
c 12×28 d $360 \div 72$
e 8×24 f $360 \div 80$
g 18×6 h $360 \div 45$

2 a Work out the mid-point between 17 and 25.
b Work out the mid-point between 20 and 32.

3 Write down one possible value of x for each of these.

a $x > 10$ b $2 \geqslant x$
c $1 < x \leqslant 9$ d $5 > x \geqslant 4$
e $25 < x \leqslant 35$ f $3.5 < x \leqslant 6.5$

3.1 Stem-and-leaf diagrams

Keywords
stem-and-leaf diagram, key

Why learn this?
A stem-and-leaf diagram is a good way to organise jumbled up data.

Objectives
D Draw a stem-and-leaf diagram

Skills check

Write each set of numbers in order of size.

a 27, 36, 42, 28 **b** 7.3, 7.1, 7.5, 7.6 **c** 142, 147, 140, 143 **d** 9.6, 9.2, 9.8, 9.4

Stem-and-leaf diagrams

Stem-and-leaf diagrams can be used to organise discrete data, so that analysis is easier. The data is grouped and ordered according to size, from smallest to largest. A **key** is needed to explain the numbers in the diagram.

Example 1

A maths test was marked out of 50. Here are the scores.

27 28 36 42 50 18 25 31 39 25 49 31
33 27 37 25 47 40 7 31 26 36 9 42

a Draw an ordered stem-and-leaf diagram to represent this data.

b How many students had a maths score of less than 20?

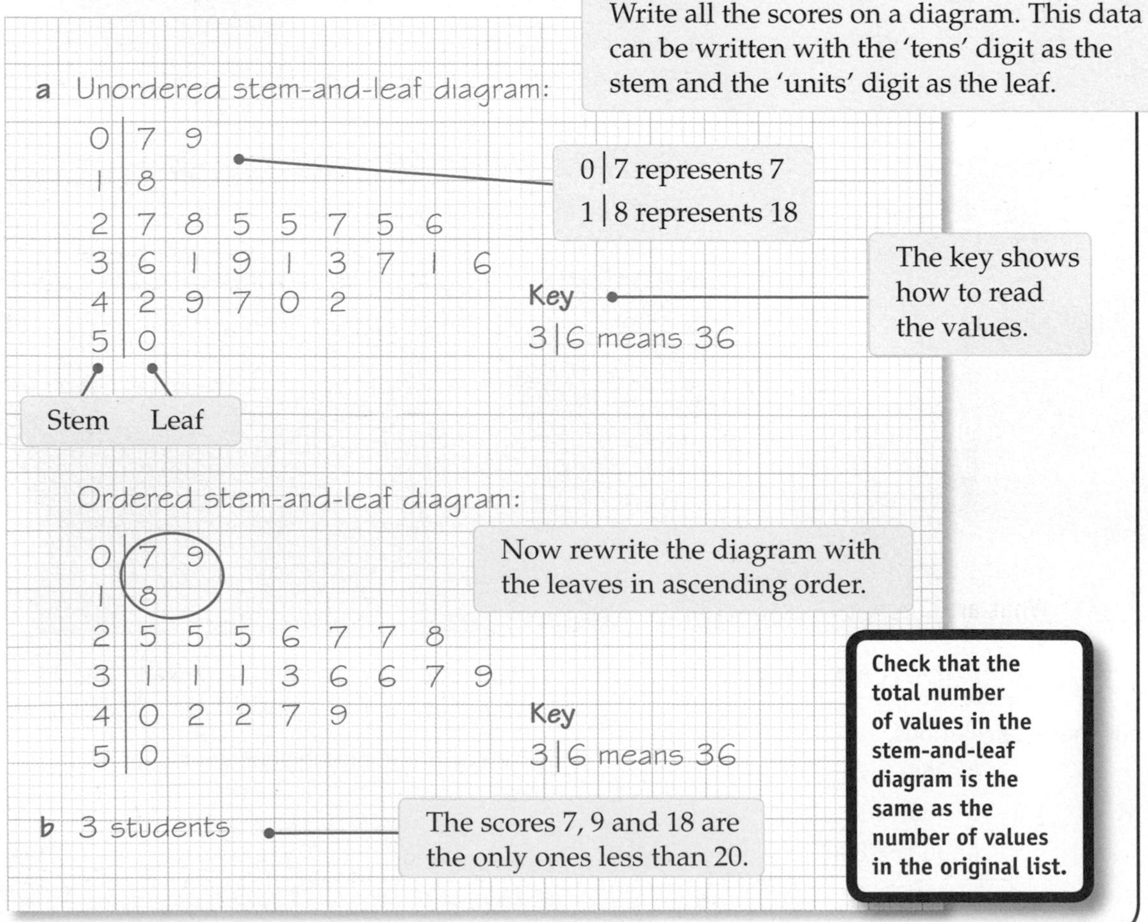

a Unordered stem-and-leaf diagram:
0 | 7 9
1 | 8
2 | 7 8 5 5 7 5 6
3 | 6 1 9 1 3 7 1 6
4 | 2 9 7 0 2
5 | 0
Key
3|6 means 36
Stem Leaf
Write all the scores on a diagram. This data can be written with the 'tens' digit as the stem and the 'units' digit as the leaf.
0|7 represents 7
1|8 represents 18
The key shows how to read the values.
Ordered stem-and-leaf diagram:
0 | 7 9
1 | 8
2 | 5 5 5 6 7 7 8
3 | 1 1 1 3 6 6 7 9
4 | 0 2 2 7 9
5 | 0
Key
3|6 means 36
Now rewrite the diagram with the leaves in ascending order.
b 3 students
The scores 7, 9 and 18 are the only ones less than 20.

Check that the total number of values in the stem-and-leaf diagram is the same as the number of values in the original list.

Exercise 3A

D

1 The temperatures in °C are recorded in 20 towns on one day.

15	13	7	15	21	16	13	18	20	9
17	9	12	19	20	12	13	21	19	8

Draw a stem-and-leaf diagram for this data.

2 Here are the recording times (in minutes) of some CDs.

56	27	39	51	46	62	59	47	49
58	36	45	47	53	60	51	36	58

a How many CDs were recorded?

b Draw a stem-and-leaf diagram for this data.

3 Twenty girls were timed over a 10 metre sprint. Here are their times to the nearest tenth of a second.

2.8	4.6	3.7	3.1	4.7	2.9	3.2	4.0
4.1	2.9	3.6	4.3	3.9	2.8	3.3	3.9

a Draw a stem-and-leaf diagram for this data.

b How many girls had a time greater than 3.3 seconds?

4 The customers in a coffee shop spent the following amounts.

£4.63	£4.91	£3.62	£5.25	£2.61	£4.86
£5.27	£3.75	£4.70	£2.93	£3.81	£5.23

a Draw a stem-and-leaf diagram for this data.

b How many people spent less than £3?

3.2 Scatter diagrams

Keywords
scatter diagram, correlation, line of best fit, linear correlation, strong, weak

Why learn this?

You can use a scatter diagram to find out if there is a correlation between temperature and ice cream sales.

Objectives

D Draw and interpret points on a scatter diagram

D Draw a line of best fit

C Use a line of best fit

C Describe types of correlation

What are the values of points A to H?

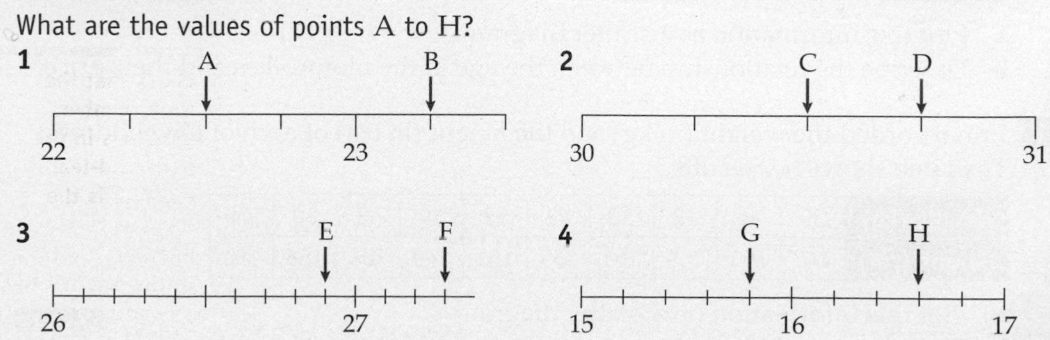

Using and plotting scatter diagrams

Scatter diagrams are used to compare two sets of data. They show if there is a connection or relationship, called a **correlation**, between the two quantities plotted.

D

Example 2

The table below shows the exam results in maths and science for 15 students.

Maths mark	40	50	56	62	67	43	74	57	75	48	83	50	64	80	70
Science mark	29	34	40	43	48	33	50	44	55	37	62	39	46	57	52

a Draw a scatter diagram for this data. Plot the maths mark along the horizontal (x) axis. Plot the science mark on the vertical (y) axis.

b Describe the relationship between the maths and science marks.

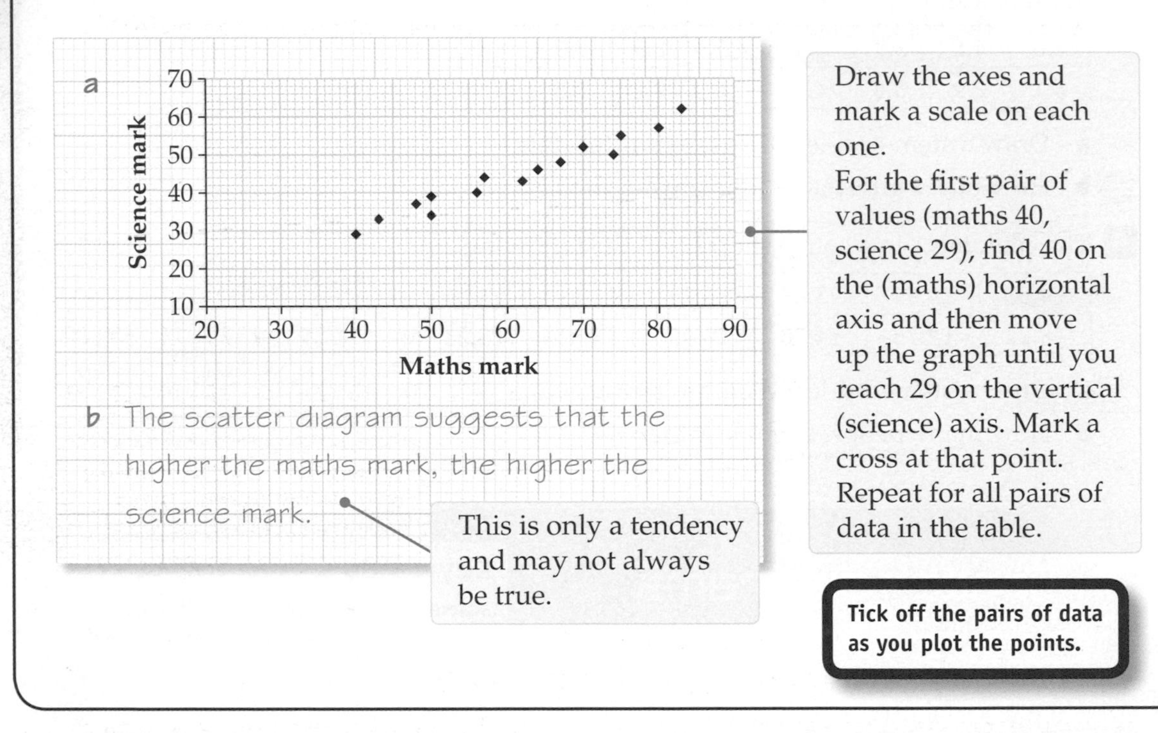

b The scatter diagram suggests that the higher the maths mark, the higher the science mark.

This is only a tendency and may not always be true.

Draw the axes and mark a scale on each one.
For the first pair of values (maths 40, science 29), find 40 on the (maths) horizontal axis and then move up the graph until you reach 29 on the vertical (science) axis. Mark a cross at that point.
Repeat for all pairs of data in the table.

Tick off the pairs of data as you plot the points.

Exercise 3B

D

1 This table shows information about the age and price of some motorbikes.

Age (years)	2	6	2	3	4	5	4	7	9	8
Price (£)	1300	1000	1800	1600	1200	1000	1400	600	200	400

a Plot this information as a scatter diagram.
b Describe the relationship between the age of the motorbikes and their price.

2 Erin recorded the weight (in kg) and the height (in cm) of each of ten children. The table shows her results.

Weight (kg)	39	46	42	50	49	52	39	53	50	44
Height (cm)	145	153	149	161	159	161	149	164	156	154

a Plot this information on a scatter diagram.
b Describe the relationship between the weight and height of the ten children.

3 Mr Gray is investigating this claim.
'The greater the percentage attendance at his maths lessons, the higher the mark in the maths test.'
Mr Gray collects some data and draws a scatter diagram to show his results.

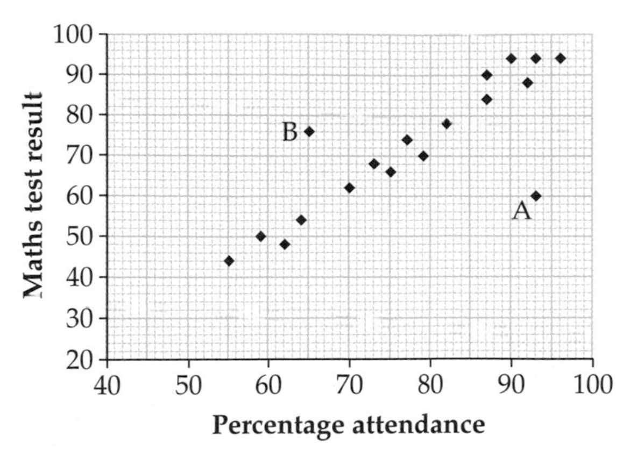

a Decide whether the claim, 'The greater the percentage attendance at his maths lessons, the higher the mark in the maths test', is correct.

b Students A and B do not fit the general trend. What can you say about
i student A **ii** student B?

D

AO2

4 Kushal is investigating this hypothesis.
'The greater your hand span, the higher your maths test result.'
He collects the following data.

Maths test result (%)	30	34	35	38	42	45	50	50	57	57	58	65	65	70	75	75	80	80
Hand span (cm)	23	21	13	18	21	19	14	18	17	22	14	16	21	13	17	19	13	23

Decide whether Kushal's hypothesis is correct.
Give reasons for your answer.

A key part of the data handling cycle is drawing conclusions from the data.

D

AO3

Lines of best fit and correlation

On a scatter diagram a **line of best fit** is a straight line that passes through the data, with an approximately equal number of points on either side of the line.

If a line of best fit can be drawn then there is some form of **linear correlation** between the two sets of data.

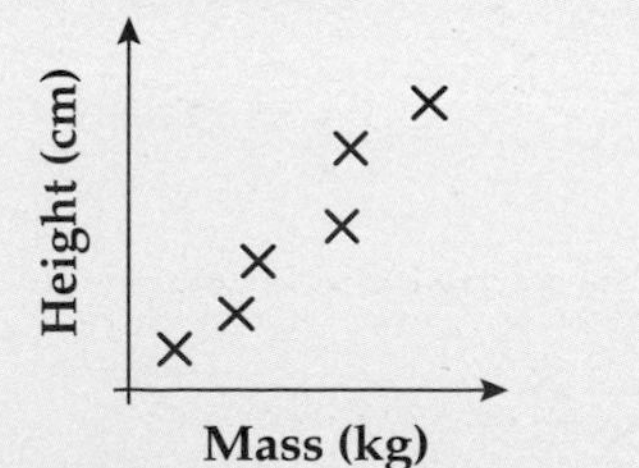

Positive correlation

As one quantity increases, the other one tends to increase.

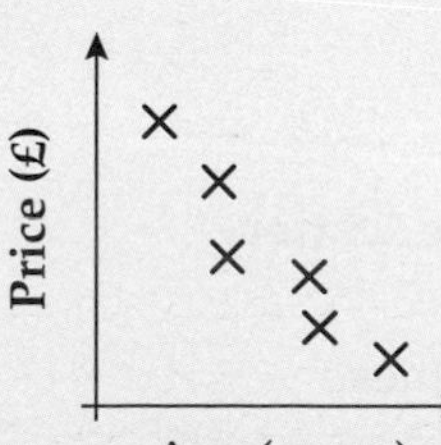

Negative correlation

As one quantity increases, the other one tends to decrease.

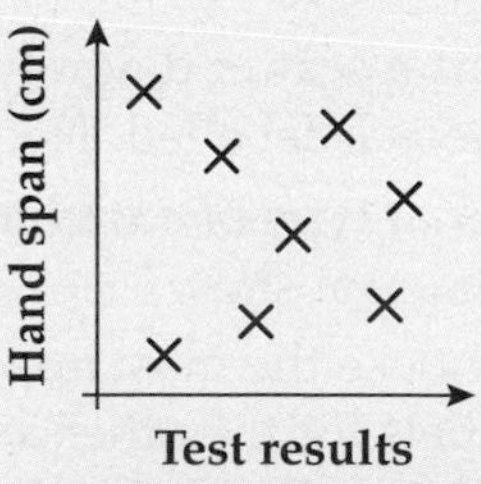

No (linear) correlation

These points are scattered randomly across the diagram. This is sometimes known as zero correlation.

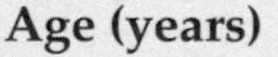

When the points are close to the line of best fit, there is a **strong** linear correlation.

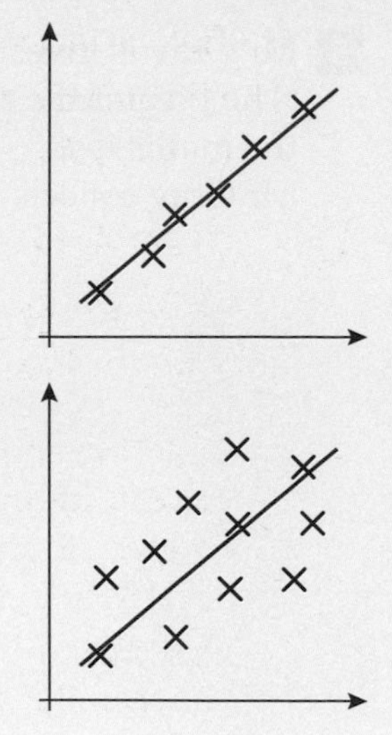

When the points are not all close to the line of best fit, there is a **weak** linear correlation.

A line of best fit can be used to estimate the value of one quantity if the other value is known and there is a correlation.

Example 3

Robert scores 65 marks in his maths test.
Use a line of best fit to estimate his science mark.

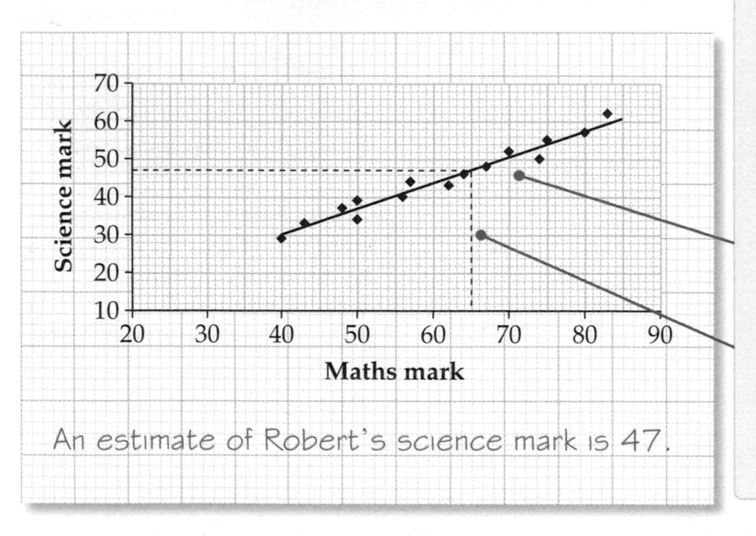

An estimate of Robert's science mark is 47.

The scatter diagram shows that there is a positive correlation between the maths and science marks. Therefore a line of best fit can be used to estimate a mark when the other mark is known.

A line of best fit has been drawn on the scatter diagram used in Example 2.

Draw vertically up from 65 to the line of best fit. Draw a horizontal line across and read off the science mark, 47.

Exercise 3C

1 Here is a scatter diagram.
One axis is labelled 'Height'.

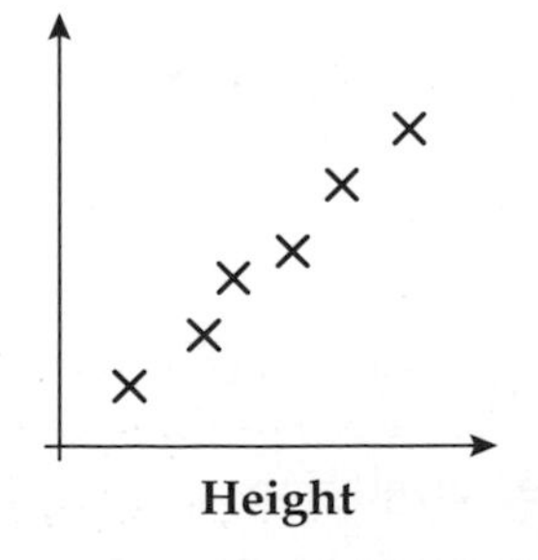

a What type of correlation does the scatter diagram show?

b Choose the most appropriate of these labels for the other axis.

A maths mark　　**B** arm span
C number of brothers　　**D** length of hair

2 Would you expect there to be a positive correlation, a negative correlation or no correlation between each of these pairs of quantities?

a Heights of men and their IQ.

b The hat sizes of children and their height.

c The outside temperature and the number of cups of tea sold in a café.

d A car's fuel consumption and its speed.

3 The table shows the marks scored by students in their two maths exams.

Paper 1	10	68	80	46	24	84	60	16	90	32	26	94	56	76	80
Paper 2	8	66	78	44	22	74	56	18	86	24	30	92	48	72	78

a Copy the axes on to graph paper and draw a scatter diagram to show the information in the table.

b Describe the correlation between the marks scored in the two exams.

c Draw a line of best fit on your scatter diagram.

d Use your line of best fit to estimate

i the Paper 2 score of a student whose score on Paper 1 is 64

ii the Paper 1 score of a student whose score on Paper 2 is 84.

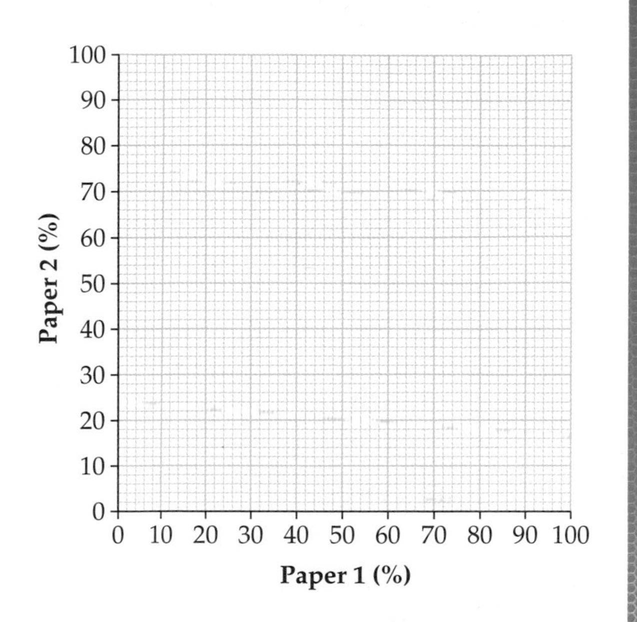

4 The table shows the average price of a two-bedroom flat at certain distances from the mainline train station.

Distance from train station (km)	0	6	2	7	1	8	5	2	10	8	3	9	11	12
Average price (£000s)	213	198	207	197	211	192	203	210	186	194	207	190	184	182

a Copy the axes on to graph paper and draw a scatter diagram to show the information in the table.

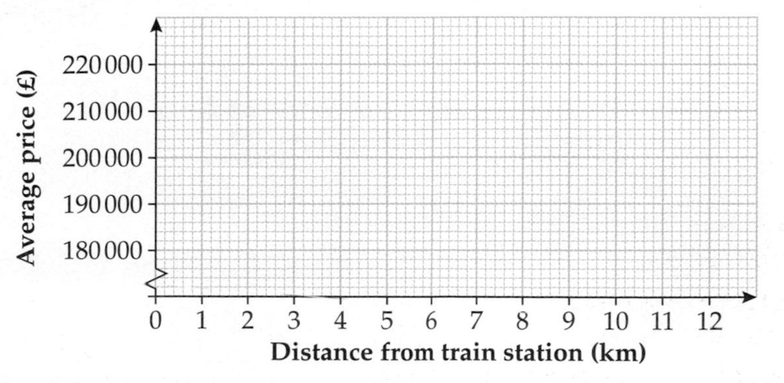

b Use your scatter diagram to estimate the price of a two-bedroom flat that is 4 km from the mainline train station.

c Describe the correlation between the distance from the mainline train station and the average price of a two-bedroom flat.

d Lucy needs to catch the train to work.
She has £195 000 to spend on a two-bedroom flat.
Approximately how far will her flat be from the station?

C

AO2

C

5 The scatter diagram shows the age and price of some used cars.
A line of best fit has been drawn.

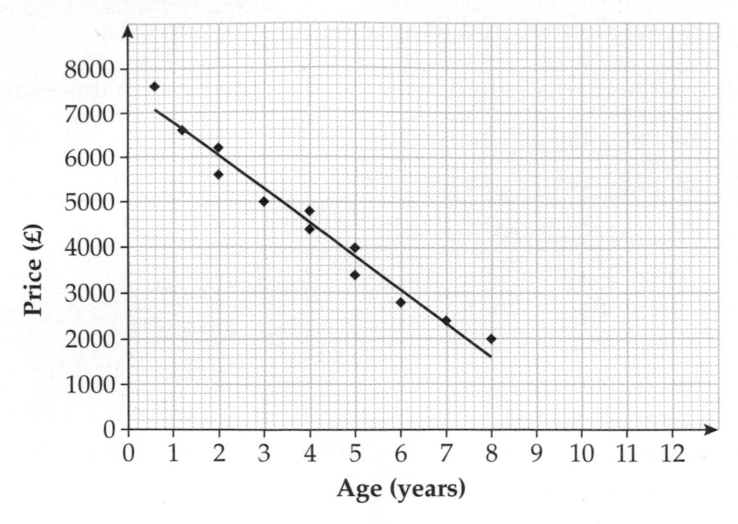

a Describe the correlation between the age of used cars and their price.

b Use the line of best fit to find

i the price of a 3-year-old used car

ii the age of a used car that costs £2200.

AO2

c Why would it not be useful to use the line of best fit to estimate the price of a 12-year-old car?

C

6 The table shows the number of female competitors in each Olympic Games from 1948 to 1984.

Year	1948	1952	1956	1960	1964	1968	1972	1976	1980	1984
Female competitors	685	518	384	610	683	781	1070	1251	1088	1620

a Draw a scatter graph for this data.

b Draw a line of best fit.

c State what type of correlation you see.

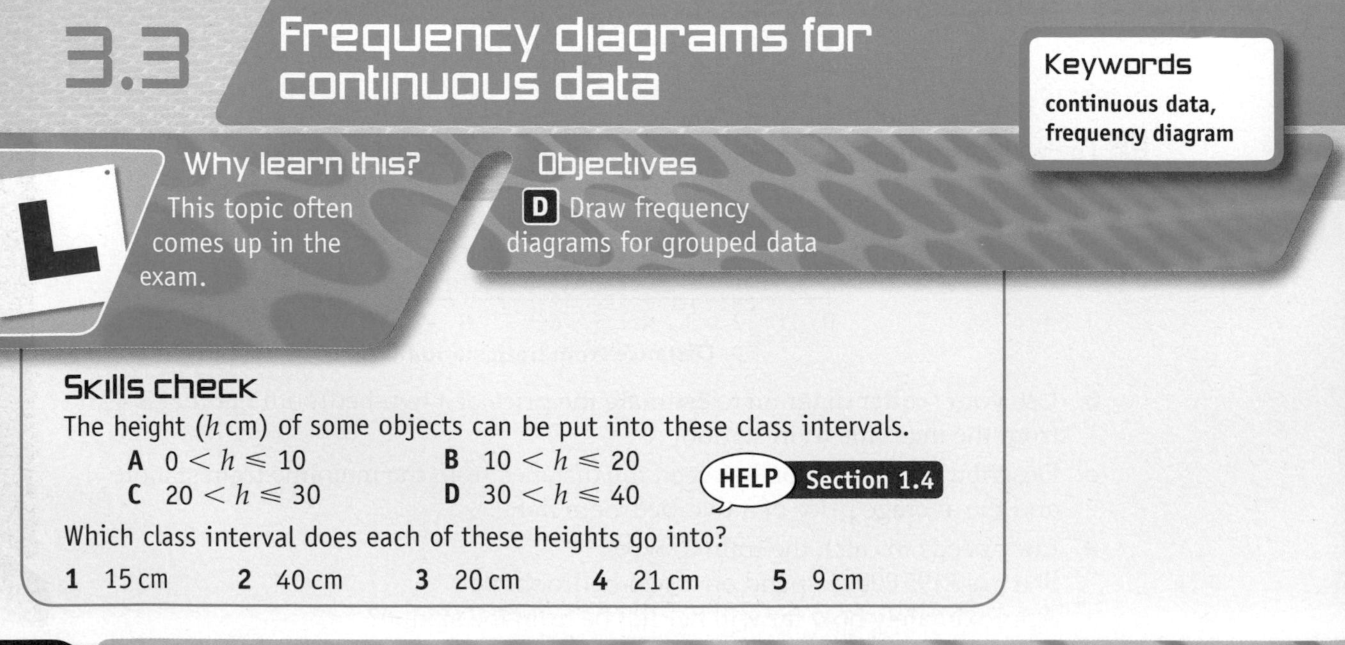

3.3 Frequency diagrams for continuous data

Keywords
continuous data, frequency diagram

Why learn this?
This topic often comes up in the exam.

Objectives
D Draw frequency diagrams for grouped data

Skills check

The height (h cm) of some objects can be put into these class intervals.

A $0 < h \leqslant 10$ **B** $10 < h \leqslant 20$
C $20 < h \leqslant 30$ **D** $30 < h \leqslant 40$

HELP Section 1.4

Which class interval does each of these heights go into?

1 15 cm **2** 40 cm **3** 20 cm **4** 21 cm **5** 9 cm

Frequency diagrams for continuous data

Continuous data can be represented by a **frequency diagram.** A frequency diagram is similar to a bar chart except that it has no gaps between the bars.

Example 4

D

The heights of some swimmers were measured and are recorded in the table.

Draw a frequency diagram to show this data.

Height, h (cm)	Frequency
$150 \leqslant h < 155$	5
$155 \leqslant h < 160$	8
$160 \leqslant h < 165$	11
$165 \leqslant h < 170$	7
$170 \leqslant h < 175$	3

Frequency always goes on the vertical axis.

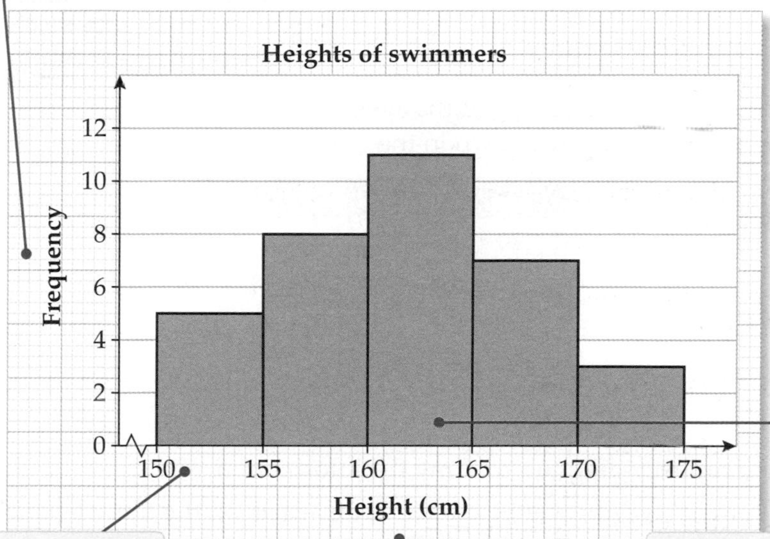

Heights of swimmers

Frequency

12
10
8
6
4
2
0

150 155 160 165 170 175

Height (cm)

The width of each bar is the same as the class interval.

The scale on the horizontal axis must be a continuous scale.

For continuous data there are no gaps between the bars.

Exercise 3D

D

1 The weights, w, of some potatoes (to the nearest gram) were recorded.

Weight, w (g)	Frequency
$0 \leqslant w < 20$	7
$20 \leqslant w < 40$	25
$40 \leqslant w < 60$	30
$60 \leqslant w < 80$	36
$80 \leqslant w < 100$	21
$100 \leqslant w < 120$	15
$120 \leqslant w < 140$	2

Draw a frequency diagram to show this information.

D

2 The table gives information about the ages (y, in years) of members at a leisure centre.

Age, y (years)	Frequency
$0 \leqslant y < 10$	23
$10 \leqslant y < 20$	45
$20 \leqslant y < 30$	56
$30 \leqslant y < 40$	36
$40 \leqslant y < 50$	49
$50 \leqslant y < 60$	32
$60 \leqslant y < 70$	16

a Draw a frequency diagram to show the spread of ages.

b The members are offered a special discount on nightclub tickets.
Is this offer likely to be very popular?
Give reasons for your answer.

It is important that you can interpret the data and draw conclusions from it as part of the data-handling cycle.

D

3 A magazine carried out a survey of the ages (x, in years) of its readers.
The results of the survey are shown in the table.

Age, x (years)	Frequency
$20 \leqslant x < 30$	5
$30 \leqslant x < 40$	38
$40 \leqslant x < 50$	30
$50 \leqslant x < 60$	15
$60 \leqslant x < 70$	12

AO2

a Draw a frequency diagram to show the information.

b The magazine's editor wants to include an article on the film *High School Musical*.
Do you think this is a good idea? Give reasons for your answer.

3.4 Frequency polygons

Keywords
frequency polygon, continuous data, mid-point

Why learn this?

A frequency polygon shows patterns or trends in the data.

Objectives

C Draw frequency polygons for grouped data

Skills check

Work out the number halfway between

1 7 and 11
2 10 and 15
3 20 and 23
4 52 and 59.

Frequency polygons for grouped data

A **frequency polygon** shows patterns or trends in the data.

When drawing a frequency polygon for grouped or **continuous data**, the **mid-point** of each class interval is plotted against the frequency.

Example 5

The frequency table shows the times taken by a sample of students to solve a maths problem. Draw a frequency polygon for this data.

Time, t (minutes)	Frequency	Mid-point
$0 \leqslant t < 5$	1	2.5
$5 \leqslant t < 10$	4	7.5
$10 \leqslant t < 15$	10	12.5
$15 \leqslant t < 20$	7	17.5
$20 \leqslant t < 25$	2	22.5

Before a frequency polygon can be drawn, the mid-point of each class interval needs to be calculated. Add a column to the table for these values.

2.5 is halfway between 0 and 5.

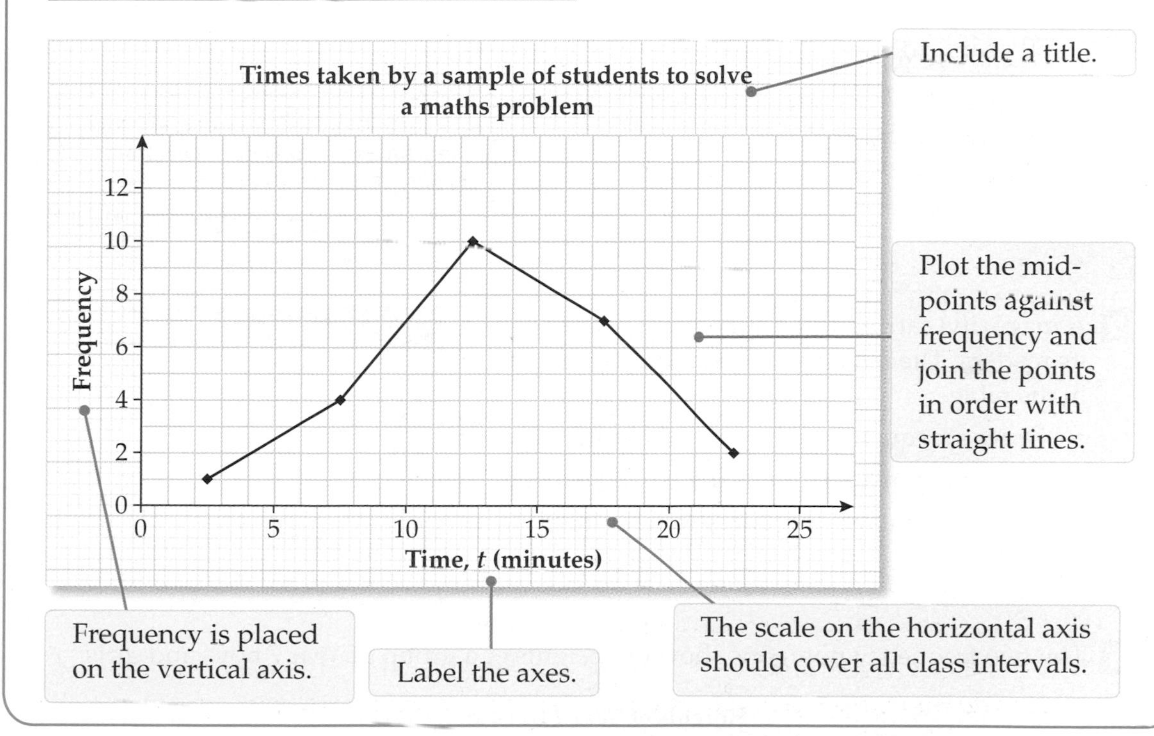

Frequency is placed on the vertical axis.

Label the axes.

The scale on the horizontal axis should cover all class intervals.

Exercise 3E

1 The heights of some seedlings were recorded in the table.

a Copy and complete the table.

b Draw a frequency polygon for this data.

Height, h (cm)	Number of seedlings	Mid-point
$5 \leqslant h < 10$	6	
$10 \leqslant h < 15$	10	
$15 \leqslant h < 20$	12	
$20 \leqslant h < 25$	9	
$25 \leqslant h < 30$	3	

2 The frequency table shows the weights of some Year 9 students.

Weight, w (kg)	Frequency
$30 \leqslant w < 40$	5
$40 \leqslant w < 50$	12
$50 \leqslant w < 60$	20
$60 \leqslant w < 70$	14
$70 \leqslant w < 80$	6

Draw a frequency polygon for this data.

C

3 The table shows the times, in minutes, that some patients waited in a doctors' surgery. Draw a frequency polygon for this data.

Time, t (minutes)	Frequency
$0 \leqslant t < 10$	10
$10 \leqslant t < 20$	5
$20 \leqslant t < 30$	4
$30 \leqslant t < 40$	1

4 The table gives the age range of the members of a local leisure centre.

a Draw a frequency polygon for this data.

b How many members were less than 50 years old?

c How many members were are least 30 but less than 60 years old?

Age, x (years)	Frequency
$0 \leqslant x < 10$	18
$10 \leqslant x < 20$	27
$20 \leqslant x < 30$	36
$30 \leqslant x < 40$	38
$40 \leqslant x < 50$	42
$50 \leqslant x < 60$	40
$60 \leqslant x < 70$	23
$70 \leqslant x < 80$	17

5 A magazine carried out a survey of the ages of its readers. The results of the survey are shown in the table.

a Draw a frequency polygon to show this data.

b How many readers were at least 35 years old?

c What percentage of readers were in their thirties?

Age, x (years)	Frequency
$25 \leqslant x < 30$	25
$30 \leqslant x < 35$	38
$35 \leqslant x < 40$	17
$40 \leqslant x < 45$	12
$45 \leqslant x < 50$	12
$50 \leqslant x < 55$	6

C

6 The two frequency polygons show the heights of a group of Year 7 boys and girls.

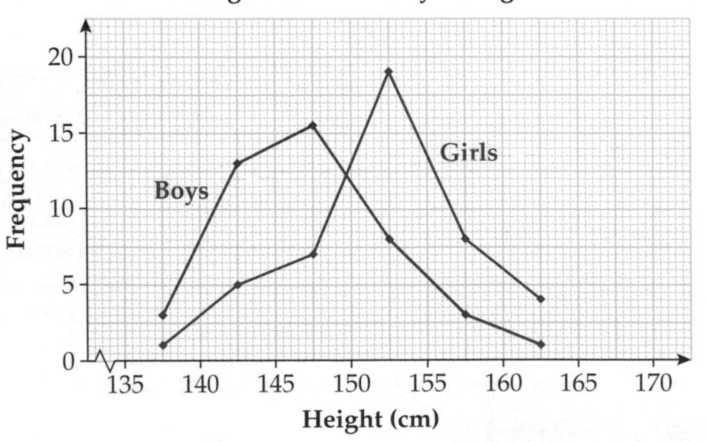

AO2

Compare the heights of the two groups. Give a reason for your answer.

Keywords
histogram, class interval, continuous data, class width, frequency density

Why learn this?
Histograms are the best way to display continuous data with varying class intervals.

Objectives
A Draw a histogram for grouped continuous data
A A* Interpret histograms and make conclusions

Skills check
What is the width of each class interval?
a $0 \leqslant h < 4$ **b** $4 \leqslant h < 20$
c $20 \leqslant h < 25$ **d** $25 \leqslant h < 50$

HELP Section 1.4

Drawing and interpreting histograms

A **histogram** is a diagram that represents **continuous data.** It is similar to a bar chart except

- the data is continuous so there are no gaps between the bars
- the bars can be different widths, to represent different **class widths** or **class intervals**
- the areas of the bars are proportional to the frequencies they represent.

As in a frequency diagram, the scale on the horizontal axis should be continuous.
On a histogram, the vertical axis shows the **frequency density** and always starts from zero.
The frequency density is given by

$$\text{frequency density} = \frac{\text{frequency}}{\text{class width}}$$

so $\text{frequency} = \text{frequency density} \times \text{class width}$

Example 6

A

The table gives the heights of some students. Draw a histogram to illustrate the data.

Height, h (cm)	Frequency	Class width	Frequency density
$145 \leqslant h < 150$	10	5	$10 \div 5 = 2$
$150 \leqslant h < 160$	22	10	$22 \div 10 = 2.2$
$160 \leqslant h < 165$	15	5	$15 \div 5 = 3$
$165 \leqslant h < 170$	12	5	$12 \div 5 = 2.4$
$170 \leqslant h < 180$	8	10	$8 \div 10 = 0.8$

The class widths are not equal. Extend the table to include the class width.

Add an extra column to calculate the frequency density.
$$\text{Frequency density} = \frac{\text{frequency}}{\text{class width}}$$

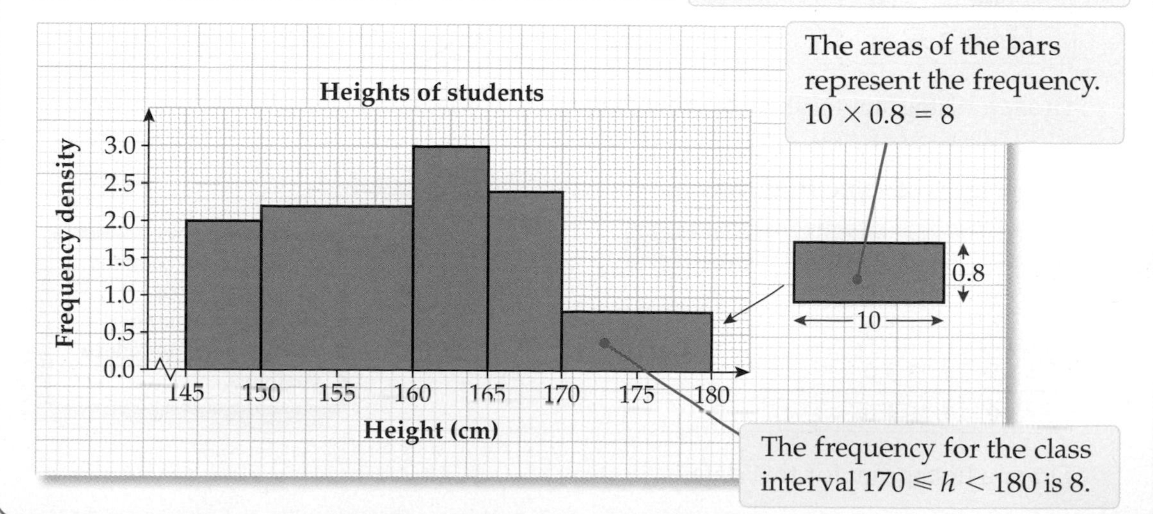

Exercise 3F

A

1 The frequency table shows the ages of some people who attend a charity walk.

Age, n (years)	Frequency	Class width	Frequency density
$0 \leq n < 20$	10	20	$10 \div 20 = 0.5$
$20 \leq n < 25$	25		
$25 \leq n < 30$	15		
$30 \leq n < 40$	29		
$40 \leq n < 60$	48		
$60 \leq n < 70$	12		
$70 \leq n < 100$	3		

a Copy and complete the table.
b Draw a histogram to illustrate the data.

Mark multiples of 10 on the horizontal scale.

2 The table shows the times taken to solve a crossword puzzle.

Time, t (minutes)	Frequency	Class width	Frequency density
$0 \leq t < 15$	9		
$15 \leq t < 25$	21		
$25 \leq t < 30$	25		
$30 \leq t < 50$	20		

a Copy and complete the table.
b Draw a histogram to illustrate the data.

Look at the start and end of the class intervals to help choose scale markings.

3 The table shows some information about the areas of some carpets.

Carpet area, A (m^2)	Frequency
$10 < A \leq 12$	16
$12 < A \leq 15$	12
$15 < A \leq 20$	9
$20 < A \leq 30$	6

Draw a histogram to illustrate this data.

4 The heights of some fruit trees are shown in the table below.

Height, h (metres)	Frequency
$0 < h \leq 1.0$	3
$1.0 < h \leq 1.5$	2
$1.5 < h \leq 3.0$	6
$3.0 < h \leq 4.0$	2

Drawing extra columns on your table to calculate frequency density helps to reduce mistakes.

Draw a histogram for this data.

Example 7

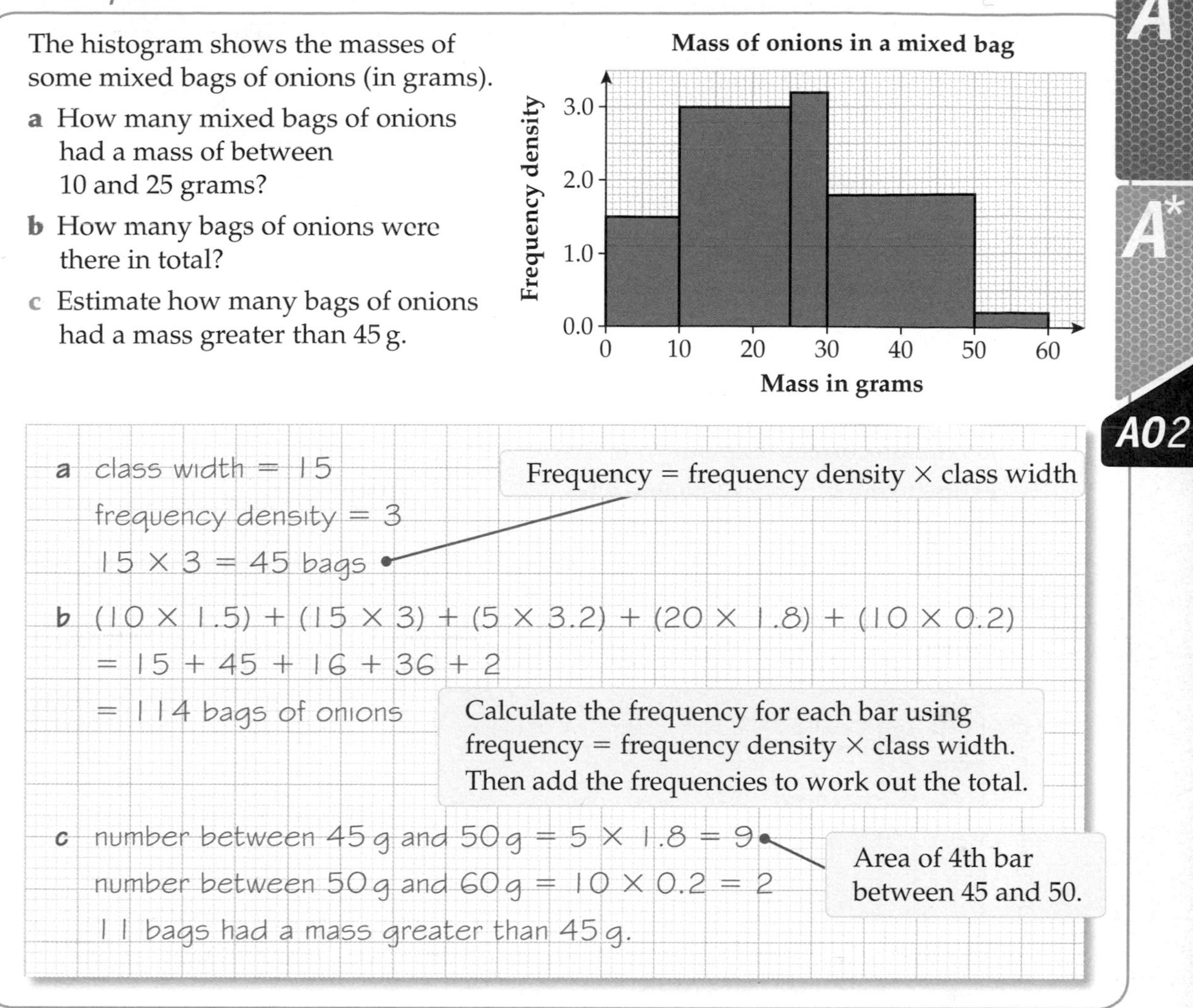

The histogram shows the masses of some mixed bags of onions (in grams).

a How many mixed bags of onions had a mass of between 10 and 25 grams?

b How many bags of onions were there in total?

c Estimate how many bags of onions had a mass greater than 45 g.

a class width = 15

frequency density = 3

$15 \times 3 = 45$ bags

Frequency = frequency density × class width

b $(10 \times 1.5) + (15 \times 3) + (5 \times 3.2) + (20 \times 1.8) + (10 \times 0.2)$

$= 15 + 45 + 16 + 36 + 2$

$= 114$ bags of onions

Calculate the frequency for each bar using frequency = frequency density × class width. Then add the frequencies to work out the total.

c number between 45 g and 50 g $= 5 \times 1.8 = 9$

number between 50 g and 60 g $= 10 \times 0.2 = 2$

11 bags had a mass greater than 45 g.

Area of 4th bar between 45 and 50.

Exercise 3G

1 The histogram shows the lifetimes of some batteries to the nearest hour.

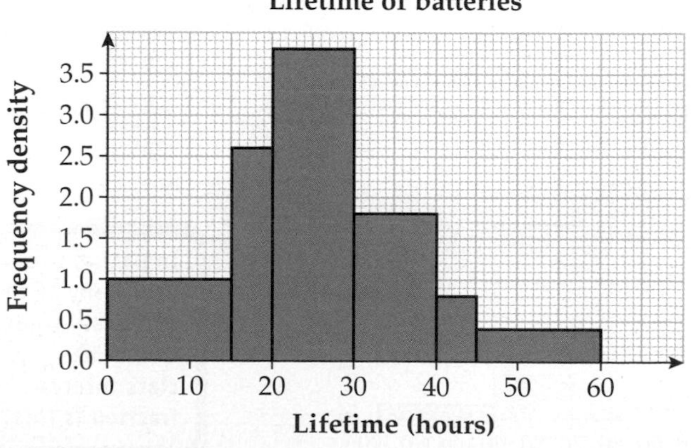

a Calculate the total number of batteries.

b How many batteries had a lifetime of less than 20 hours?

A

2 The histogram shows information about the ages of some swimmers in a swimming gala.

a Calculate the total number of swimmers in the swimming gala.

b Calculate the number of swimmers up to 25 years old.

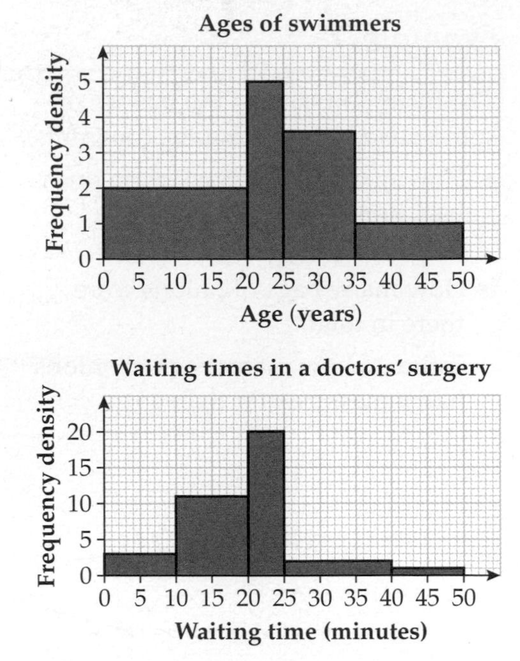

A*

3 The histogram summarises the waiting times in a doctors' surgery, during a week.

a William says, 'More than half the people waited less than 20 minutes.' Is William correct? Explain your answer.

b Estimate the number of people who waited for more than 32 minutes.

4 The histogram and the frequency table below are both incomplete. They represent the same information about the ages of people living in a small village.

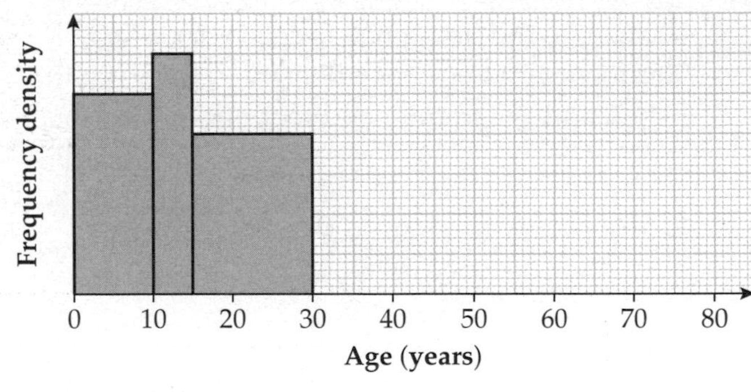

Age, x (years)	Frequency
$0 \leqslant x < 10$	50
$10 \leqslant x < 15$	
$15 \leqslant x < 30$	
$30 \leqslant x < 50$	60
$50 \leqslant x < 75$	25
$75 \leqslant x < 80$	20

a Use the information in the histogram to complete the table.

b Copy and complete the histogram.

c Estimate how many people are older than 45.

5 The table shows the weights of 100 red apples.

a Draw a histogram to represent this data.

b Twenty-five of these apples are classified as small. Calculate an estimate of the maximum weight of a small red apple.

c This histogram represents the weights of 100 green apples.

Weight of apple, w (grams)	Frequency
$30 \leqslant w < 50$	18
$50 \leqslant w < 60$	28
$60 \leqslant w < 70$	36
$70 \leqslant w < 90$	15
$90 \leqslant w < 120$	3

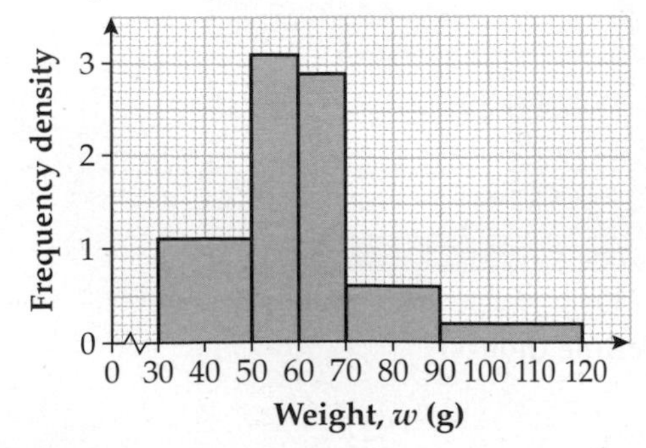

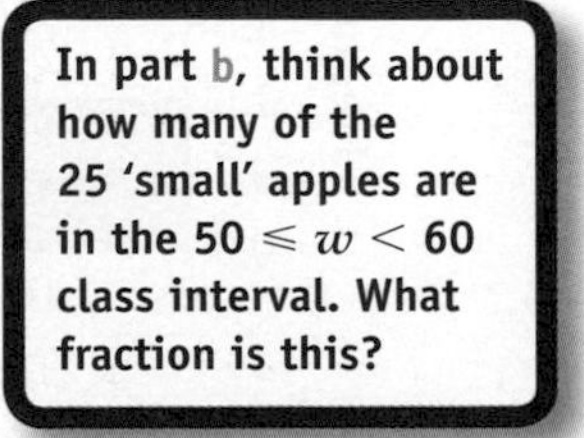

In part b, think about how many of the 25 'small' apples are in the $50 \leqslant w < 60$ class interval. What fraction is this?

AO2

How many more green apples than red apples weighed less than 60 g?

Review exercise

1 Here are the times, in minutes, taken to walk around a park.

36	27	35	47	45	47	29
42	21	43	49	50	41	37
51	37	56	54	56	40	35

Draw a stem-and-leaf diagram to show these times. **[3 marks]**

2 The table shows the times taken by a group of students to complete a challenge, called 'Challenge 1'.

Time, t (minutes)	Number of students
$10 \leqslant t < 20$	6
$20 \leqslant t < 30$	10
$30 \leqslant t < 40$	11
$40 \leqslant t < 50$	17
$50 \leqslant t < 60$	8
$60 \leqslant t < 70$	2

a Draw a frequency polygon for this data. **[2 marks]**

b The table shows the times taken for the same group of students to complete a different challenge, called 'Challenge 2'.

Time, t (minutes)	Number of students
$0 \leqslant t < 10$	4
$10 \leqslant t < 20$	14
$20 \leqslant t < 30$	11
$30 \leqslant t < 40$	17
$40 \leqslant t < 50$	8

Draw a frequency polygon for this data on the same grid as part **a**. **[2 marks]**

c Compare the times taken for the students to complete the two challenges. Give reasons for your answer. **[2 marks]**

3 **a** Here is a scatter diagram. One axis is labelled 'Weight'.

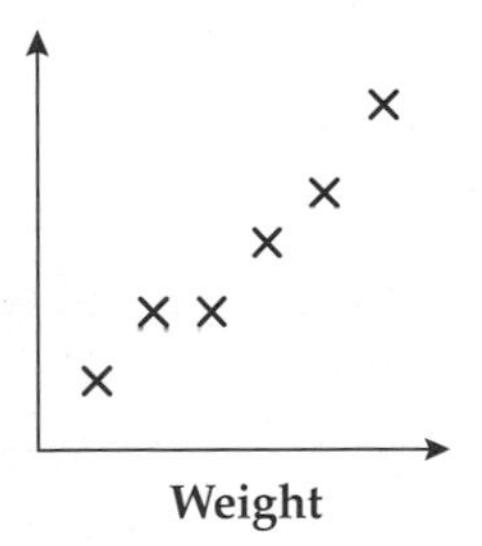

i For this graph, state the type of correlation.

ii From this list choose an appropriate label for the other axis.

A maths score **B** length of hair **C** waist measurement **D** hat size **[2 marks]**

b Here is another scatter diagram. One axis is labelled 'Height'.

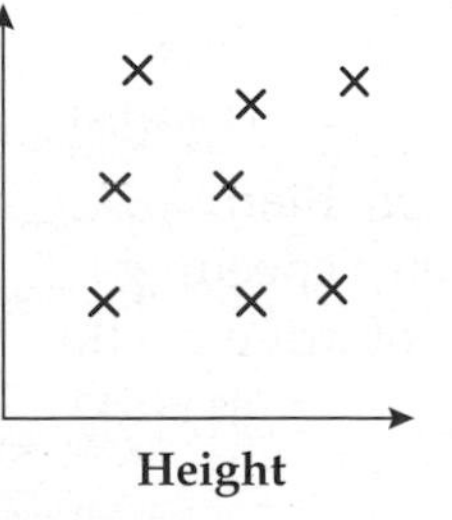

i For this graph, state the type of correlation.

ii From this list choose an appropriate label for the other axis.

A weight **B** length of arm **C** hat size **D** maths score **[2 marks]**

C

AO2

C

4 The table shows the midday temperatures of some cities and the number of hours of sunshine one day in July.

City	Hours of sunshine	Midday temperature (°C)
London	6	18
Manchester	7	19
Sydney	3	12
Paris	9	22
Berlin	5	16
New York	11	24
Montreal	6	17
Abu Dhabi	12	26
Rio de Janeiro	8	20
Wellington	2	11

a Plot the data on a scatter diagram on a grid similar to this.

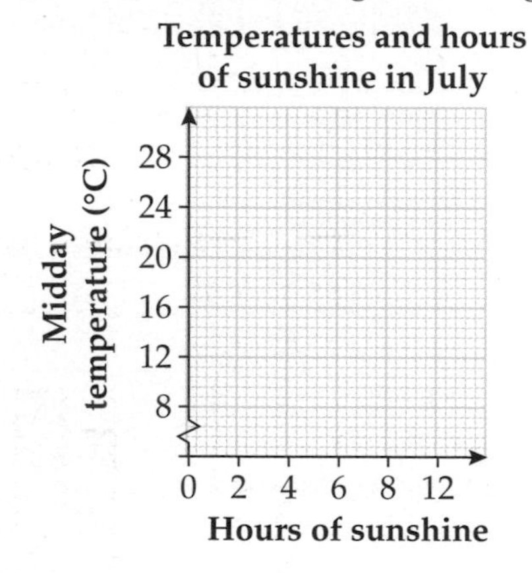

[2 marks]

b Describe the strength and type of correlation. **[1 mark]**

c Draw a line of best fit on your scatter diagram. **[1 mark]**

d Use your line of best fit to estimate the number of hours of sunshine at a place with an average temperature of 21°C. **[1 mark]**

AO2

e It is not sensible to use your line of best fit to estimate the number of hours of sunshine in the Sahara Desert, which had a midday temperature of 36°C. Explain why. **[1 mark]**

A

5 A speed camera recorded the speed of some vehicles on a motorway. The table shows the results.

Speed, s (mph)	Frequency
$0 \leqslant s < 30$	18
$30 \leqslant s < 50$	38
$50 \leqslant s < 60$	118
$60 \leqslant s < 70$	122
$70 \leqslant s < 80$	84
$80 \leqslant s < 120$	20
Total	400

a Draw a histogram to illustrate the data. **[3 marks]**

b Drivers of vehicles doing more than 77 miles per hour were given a speeding ticket. Estimate the number of drivers who received a ticket. **[1 mark]**

6 The table shows the heights of some seedlings from Batch A of an experiment.

Height of seedling, h (cm)	Frequency
$0 \leqslant h < 10$	16
$10 \leqslant h < 15$	24
$15 \leqslant h < 20$	28
$20 \leqslant h < 35$	15
$35 \leqslant h < 40$	3

a Draw a histogram to represent this data, using a grid similar to this. **[3 marks]**

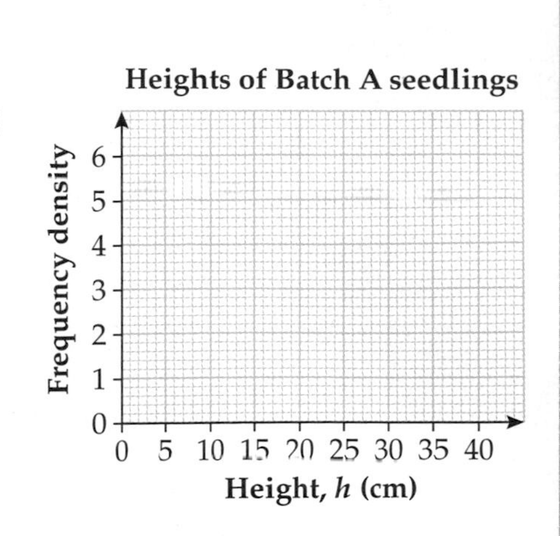

b This histogram represents 86 seedlings from Batch B of an experiment.

How many seedlings had height $20 \leqslant h < 35$? **[1 mark]**

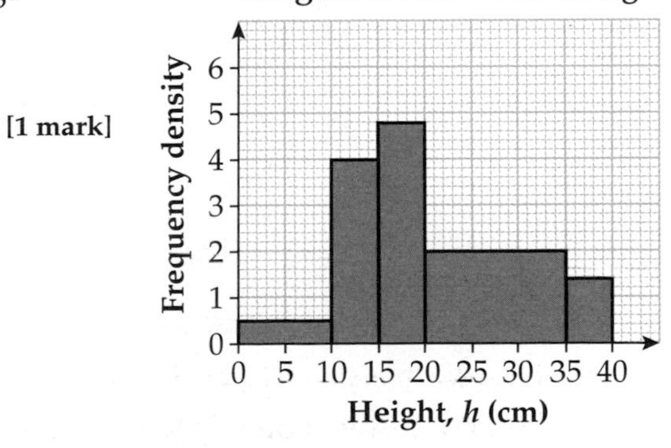

c How many more seedlings from Batch B than Batch A had a height of 20 cm or more? **[3 marks]**

d Which batch of seedlings were treated with *SupaGro* to enhance their rate of growth? Give reasons for your answer. **[2 marks]**

A* A02

Chapter summary

In this chapter you have learned how to

- draw a stem-and-leaf diagram **D**
- draw and interpret points on a scatter diagram **D**
- draw a line of best fit **D**
- draw frequency diagrams for grouped data **D**
- use a line of best fit **C**
- describe types of correlation **C**
- draw frequency polygons for grouped data **C**
- draw a histogram for grouped continuous data **A**
- interpret histograms and make conclusions **A** **A***

4

Range and averages

This chapter is about finding averages.

Car manufacturers use average-sized dummies to test the safety features of their new designs.

Objectives

This chapter will show you how to

- identify the modal class in a grouped frequency table D
- calculate the mean, median, mode and range from a frequency table D C
- estimate the range and mean from a grouped frequency table C
- work out which class interval contains the median C
- choose the most appropriate average and give reasons for your choice A

Before you start this chapter

1 Imran compared the prices of flights from London to Mumbai.

£550 £490 £470 £600 £580 £720 £520 £720

a Work out the range and mode of this data.

b Calculate the mean price per flight.

c Work out the median price per flight.

2 This stem-and-leaf diagram shows the results from the men's 400 m final at the Sydney Olympics.

Stem	Leaf
43	8
44	4 7
45	0 1 3 4

Key
43 | 8 means 43.8 seconds

a Work out the range for this data.

b Work out the median time.

c Calculate the mean time. Round your answer to one decimal place.

4.1 Calculating the mean, median, mode and range from a frequency table

Why learn this?

When you collect data from a survey or experiment your results will often be in a frequency table.

Objectives

D C Calculate the mean, median, mode and range from a frequency table

Keywords

average, mean, median, mode, range, modal value

Skills check

Kieran wrote down the test scores of 11 members of his class.

17 20 11 19 15 12 20 8 12 20 18

1 Work out the range, median and mode of this data.

2 Calculate the mean mark, correct to two significant figures.

Mean, median and mode from a frequency table

The **mean** is the most commonly used **average**.

$$\text{Mean} = \frac{\text{sum of all the data values}}{\text{number of data values}}$$

The **median** is the middle value when the data is written in order. You can work out the median using this formula.

$$\text{Median} = \left(\frac{n+1}{2}\right)\text{th value}$$

The **mode**, or **modal value**, of a set of data is the number or item that occurs most often.

The **range** of a set of data is the difference between the largest value and the smallest value. The range tells you how spread out the data is.

Range = largest value − smallest value

Example 1

D

This frequency table shows the numbers of goals scored per match by a football team for one season.

a Work out the range and mode of this data.

b Work out the median number of goals scored per match.

c Calculate the mean number of goals scored per match.

Number of goals scored	Frequency
0	8
1	15
2	12
3	6
4	3

a The range is 4 goals.
The mode is 1 goal.

Range = largest value − smallest value = 4 − 0 = 4 goals

The mode has the highest frequency.

b There are 44 data values.
(44 + 1) ÷ 2 = 22.5
The median is halfway between the 22nd and 23rd values.
The median is 1 goal.

$$\text{Median} = \left(\frac{n+1}{2}\right)\text{th value}$$

The first eight values are 0. The next 15 values are 1. So the 22nd and 23rd values are both 1.

Add an extra column to your frequency table to show the total number of goals scored in each row.

c

Number of goals scored	Frequency	Number of goals × frequency
0	8	$0 \times 8 = 0$
1	15	$1 \times 15 = 15$
2	12	$2 \times 12 = 24$
3	6	$3 \times 6 = 18$
4	3	$4 \times 3 = 12$
Total	44	69

The team scored 3 goals in a match 6 times. The total number of goals they scored in these matches is $3 \times 6 = 18$ goals.

The team scored 69 goals in the season.

Mean $= \frac{69}{44} = 1.57$ goals per match (2 d.p.).

$$\text{Mean} = \frac{\text{total number of goals scored}}{\text{number of matches played}}$$

Exercise 4A

D

1 Darren counted the people in each checkout queue at a supermarket.

Number of people in queue	Frequency	Number of people × frequency
0	4	
1	6	
2	13	
3	2	$3 \times 2 = 6$
4	0	
Total		

a Work out the range and mode of this data.

b Work out the median number of people in a queue.

c Copy and complete the table.
Use it to calculate the mean number of people in a queue.

2 Alison counted the buses passing a road junction each minute for an hour.

Number of buses	Frequency
0	8
1	17
2	12
3	14
4	5
5	4

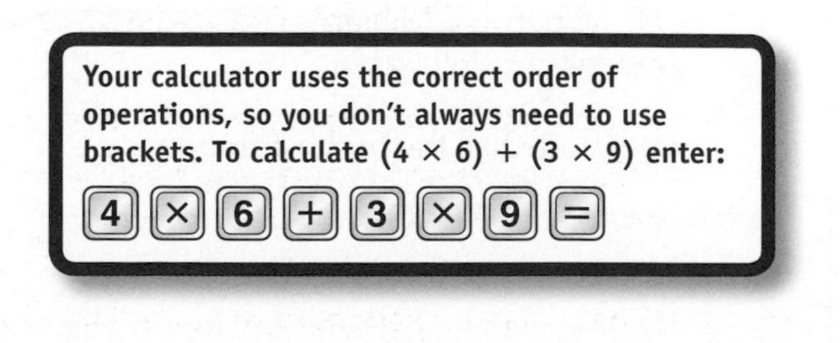

Your calculator uses the correct order of operations, so you don't always need to use brackets. To calculate (4 × 6) + (3 × 9) enter:
4 × 6 + 3 × 9 =

a Work out the median number of buses per minute.

b Calculate the mean number of buses per minute.

 Amit asked a group of people to predict the results of five coin flips, and recorded the number of correct predictions in a frequency table.

Number of correct predictions	Frequency
0	1
1	2
2	10
3	7
4	5
5	0

a Work out the range of this data.

b Write down the modal number of correct predictions.

c Work out the median of Amit's data.

d Calculate the mean number of correct predictions.

e On the next trial, Amit's friend Chloe correctly predicts four out of five coin flips. Amit adds her result to his data. Explain how this result will affect
i the range **ii** the mode **iii** the median **iv** the mean.

4 This table shows the number of hours of TV watched each night by the members of a class.

Number of hours of TV watched	0	1	2	3	4	More than 4
Frequency	3	8	11	5	4	2

a Is it possible to calculate the mean of this data? Give a reason for your answer.

b Jacob said, 'It is impossible to calculate the median of this data.'
Give reasons to show that he is wrong.

5 Eric and Megan each asked a group of their friends what size shoes they wear.
Eric recorded his results in a frequency table.
Megan recorded hers in a stem-and-leaf diagram.

Stem	Leaf
5	5 5 5
6	0 0 0 5 5
7	0 0 0 0 5 5
8	0 0 0 0 0
9	0 0

Key
5 | 5 means 5.5

Shoe size	Frequency
6	5
6.5	3
7	8
7.5	3
8	6
8.5	1

a How many of Megan's friends had a shoe size of 8 or larger?

b What was the median shoe size for Megan's friends?

c What was the mean shoe size for Eric's friends? Give your answer to one decimal place.

d Megan and Eric combined their results.
What was the median of their combined results?

e What was the mean of their combined results? Give your answer to one decimal place.

D

AO2

C

AO2

C

6 In a school, students are awarded gold stars for good work. This table shows the number of gold stars awarded in three different year groups in one month.

Number of gold stars	Year 7 frequency	Year 8 frequency	Year 9 frequency
0	12	2	22
1	15	20	36
2	40	28	18
3	22	31	11
4	6	15	9
5	2	14	5
6	13	6	8

a How many pupils are in Year 8?

b How many gold stars were awarded in total to students in Year 9?

c How many more students were awarded 6 gold stars in Year 7 than in Year 8?

d Calculate the mean number of gold stars awarded per student for the students in Year 7. Give your answer correct to two significant figures.

AO2

e The year group with the highest mean number of gold stars awarded per student is awarded a trophy. Which year group won the trophy?

4.2 Calculating the range, median and mode from a grouped frequency table

Why learn this?

When you are collecting continuous data, such as times taken to run 100 m, you need to have class intervals in your frequency table.

Objectives

D Identify the modal class from a grouped frequency table

D Estimate the range from a grouped frequency table

C Work out the class interval which contains the median from data given in a grouped frequency table

Keywords

class intervals, grouped frequency table, modal class, estimate

Skills check

1 Write down four different numbers with a mean of 6, a range of 8 and a mode of 5.

2 Say whether each type of data is discrete or continuous.

a Heights of some trees. **b** Number of items in a pencil case.

c Number of visits to the cinema. **d** Weights of the eggs in a box.

HELP Section 1.3

3 Write $<$, $>$ or $=$ between each pair of values.

a 4.6 cm ☐ 4.39 cm **b** 300 m ☐ 0.3 km

c 15 cm ☐ 125 mm **d** 500 g ☐ 5 kg

Range, median and mode from a grouped frequency table

Sometimes the data in a frequency table is grouped into different **class intervals.**

When the data is arranged in a **grouped frequency table** you don't know the exact data values. This means you can't calculate the mode, median and range exactly.

The class interval with the highest frequency is called the **modal class.**

You can **estimate** the range using the formula:

Estimated range = highest value of largest class interval − lowest value of smallest class interval

With n data values, the median is the $\left(\frac{n+1}{2}\right)$th data value.

You can use this formula to work out which class interval contains the median.

Example 2

This frequency table shows the heights of some plant seedlings.

a Write down the modal class.

b Estimate the range of this data.

c Which class interval contains the median?

Height, h (cm)	Frequency
$5 \leq h < 10$	14
$10 \leq h < 15$	11
$15 \leq h < 20$	8
$20 \leq h < 25$	2

a The modal class is $5 \leq h < 10$.

The class interval $5 \leq h < 10$ has the highest frequency.

b $25 - 5 = 20$

An estimate for the range is 20 cm.

Range = highest value − lowest value

c There are 35 data values, so the median is the $\left(\frac{35+1}{2}\right) = 18$th data value. The median is in the class interval $10 \leq h < 15$.

Add up the frequencies to find the number of data values.
The first 14 values are in the class interval $5 \leq h < 10$. The next 11 values are in the class interval $10 \leq h < 15$.

Exercise 4B

D

1 Roselle weighed some eggs, and recorded her results in a frequency table.

a Write down the modal class.

b Estimate the range of this data.

c How many eggs did Roselle weigh in total?

d Which class interval contains the median?

Weight, w (g)	Frequency
$45 \leq w < 50$	3
$50 \leq w < 55$	8
$55 \leq w < 60$	11
$60 \leq w < 65$	7

C

2 A café recorded the number of portions of spaghetti bolognese it sold each day for a month.

a Write down the modal class.

b Estimate the range of this data.

c Which class interval contains the median?

Number of portions	Frequency
0–9	2
10–19	7
20–29	14
30–39	7
40–49	1

3 Andrew used a computer to record the reaction times of a group of his friends for an experiment.

Reaction time, t (milliseconds)	$100 \leqslant t < 200$	$200 \leqslant t < 300$	$300 \leqslant t < 400$	$400 \leqslant t < 500$
Frequency	2	8	5	4

a Write down the modal class.
b Estimate the range of this data.
c Which class interval contains the median?

4 This frequency table shows the number of hits on a website each day for eight weeks.

Number of hits	0–999	1000–1999	2000–2999	3000–3999	4000–4999	5000–5999
Frequency	0	6	18	21	9	2

a Write down the modal class.
b Estimate the range of this data.
c Which class interval contains the median?
d In the next week, the website server was broken, and the website received no hits each day. What effect will this new data have on
i the modal class **ii** the estimated range
iii the class interval containing the median?
e Aaron says that it is impossible to calculate the mean number of hits per day exactly. Is he correct? Give a reason for your answer.

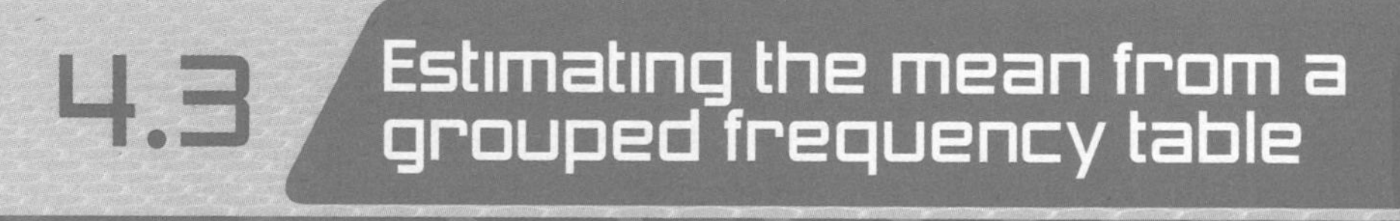

4.3 Estimating the mean from a grouped frequency table

Keywords
estimate, mid-point

Why learn this?

Grouping data and then calculating an estimate of the mean saves time when there is a lot of data.

Objectives

C Estimate the mean from a grouped frequency table

Skills check

1 Work out these calculations, giving your answers to two significant figures.
a $106 \div 6$ **b** $2260 \div 28$ **c** $18.65 \div 12$ **d** $416.2 \div 30$

2 What value is halfway between
a 4 and 5 **b** 10 and 19 **c** 115 and 120 **d** 4.5 and 4.9?

Estimating the mean from a grouped frequency table

When data is arranged in a grouped frequency table, you don't know the exact data values. This means you can't calculate the mean exactly.

You can **estimate** the mean by assuming that every data value lies exactly in the middle of a class interval. You need to work out the **mid-point** of each class interval.

$$\text{Mid-point} = \frac{\text{minimum class interval value} + \text{maximum class interval value}}{2}$$

$$\text{Estimate of mean} = \frac{\text{total of 'mid-point} \times \text{frequency' column}}{\text{total frequency}}$$

Example 3

This frequency table shows the heights of the trees in a park.

Height, h (m)	Frequency
$0 \leqslant h < 5$	10
$5 \leqslant h < 10$	18
$10 \leqslant h < 15$	6
$15 \leqslant h < 20$	2

a Work out the mid-point of the class interval $10 \leqslant h < 15$.

b Estimate the total height of all the trees in the park.

c Calculate an estimate for the mean height of the trees in the park.

Mid-point = $\frac{\text{minimum class interval value + maximum class interval value}}{2}$

a $\frac{10 + 15}{2} = 12.5$

The mid-point of the class interval $10 \leqslant h < 15$ is 12.5 m.

Add an extra column to your frequency table to show the mid-point of each class interval.

b

Height, h (m)	Frequency	Mid-point	Mid-point × frequency
$0 \leqslant h < 5$	10	2.5	$2.5 \times 10 = 25$
$5 \leqslant h < 10$	18	7.5	$18 \times 7.5 = 135$
$10 \leqslant h < 15$	6	12.5	$6 \times 12.5 = 75$
$15 \leqslant h < 20$	2	17.5	$2 \times 17.5 = 35$
Total	36		270

Calculate 'mid-point × frequency' for each row and write it in a fourth column.

An estimate for the total height of all the trees in the park is 270 m.

This total is an estimate for the total height of all the trees in the park.

c Estimate of mean = $\frac{270}{36} = 7.5$ m

Estimate of mean = $\frac{\text{total of 'mid-point × frequency' column}}{\text{total frequency}}$

C

Exercise 4C

1 This frequency table shows the times taken by members of a class to solve a puzzle.

Time taken, t (minutes)	Frequency	Mid-point	Mid-point × frequency
$0 \leqslant t < 5$	3		
$5 \leqslant t < 10$	15		
$10 \leqslant t < 15$	8		
$15 \leqslant t < 20$	2		
$20 \leqslant t < 25$	5		
Total			

a Copy and complete the table to work out an estimate for the total time taken by the whole class.

b Calculate an estimate for the mean time taken, correct to one decimal place.

C

C

2 Jaden used a frequency table to record the number of times each student in her year group logged in to the school's intranet in a week.

Number of log-ins	Frequency
0–4	22
5–9	31
10–14	17
15–19	20
20–24	6
Total	

a Use mid-points to calculate an estimate for the mean number of log-ins per student. Give your answer correct to one decimal place.

> **The mid-point of the class interval 5–9 is $\frac{5+9}{2} = 7$.**

b Explain why your answer to part **a** is an estimate.

C

3 Archie carried out an experiment to find out how far the members of his tennis club could throw a tennis ball. He used a tally chart to record his results.

Distance thrown, d (m)	Boys' tally	Girls' tally
$0 \leqslant d < 8$	|||	||
$8 \leqslant d < 16$	~~||||~~ |	~~||||~~ ~~||||~~ ||
$16 \leqslant d < 24$	~~||||~~ ~~||||~~ ||	~~||||~~
$24 \leqslant d < 32$	||||	|||
$32 \leqslant d < 40$	|	~~||||~~ ||
$40 \leqslant d < 48$	|	

a Will Archie be able to use his tally chart to calculate the exact mean distance thrown for the boys? Give a reason for your answer.

b Draw a frequency table for the girls' results. Use it to calculate an estimate of the mean distance thrown by the girls. Give your answer to one decimal place.

c Calculate an estimate for the mean distance thrown by the whole tennis club. Give your answer to one decimal place.

AO2

d In his conclusion Archie wrote, 'The girls' results were below average. This means that the boys could throw the tennis ball further.'
Do you agree with this statement? Give evidence for your answer.

4.4 Which average?

Keywords
extreme value

Why learn this?

You need to understand the differences between mean, median and mode so that you can decide which is the most appropriate for your data.

Objectives

A Choose the most appropriate average and give reasons for your choice

Skills check

1 Say whether each type of data is qualitative or quantitative.

HELP Section 1.3

a Heights of students **b** Favourite colour
c Most popular A-level subject **d** Number of red cars passing school gates

2 Work out the mean, median and mode of these numbers:

4 5 3 9 11 6 22 9.

Appropriate averages

The mean, median and mode are all different types of average. You need to be able to choose the most appropriate average to use in different situations.

The mode is the only average you can use for qualitative data. The mode tells you which data value is most likely to occur.

The median tells you the middle value. Half the data values are greater than the median and half are less than the median.

The mean is the only average that takes every value into account. The mean can be affected by very high or low values. These values are called **extreme values**.

Example 4

A

Nick wants to buy a digital camera. He gathers some data about the prices of different cameras.

£95 £120 £80 £170 £80 £150 £130 £155 £790 £180

a Calculate the mean price of these digital cameras.

b What is the mode of Nick's data?

c Work out the median price.

d Nick wants to know the price of an average camera. Which average should he choose? Give reasons for your answer.

Mean = $\frac{\text{sum of all the data values}}{\text{number of data values}}$

a $\frac{95 + 120 + 80 + 170 + 80 + 150 + 130 + 155 + 790 + 180}{10} = 195$

The mean price is £195.

b The modal price is £80.

£80 appears twice. The other prices only appear once each.

c The median is $\frac{130 + 150}{2}$ = £140.

Median = $\left(\frac{n+1}{2}\right)$th value. There are 10 pieces of data so the median is half way between the 5th and the 6th values when the data is ordered.

d Nine of the cameras are cheaper than the mean. The mean is not a good representation of the price of an average camera.

£790 is an extreme value. This value has affected the mean, but not the median.

The modal value is the smallest value.
This is not a good average for Nick to use.

The median tells you that half of the cameras cost more than £140 and half of the cameras cost less than £140. This is the best average for Nick to use.

Exercise 4D

A

1 Alix asked 10 people how many CDs they own.

42 10 15 30 225 28 15 25 51 46

In her conclusion, Alix wrote, 'The average person owns about 50 CDs (to one significant figure).'

a Calculate the mean, median and mode of this data.

b Which average has Alix used for her conclusion?

c Do you agree with Alix's conclusion? Give reasons for your answer.

d Choose an appropriate average and write your own conclusion for Alix's data.

2 Jonathan was investigating the annual salaries of workers at a design agency. He surveyed a manager, four designers, an administrator and two trainees.

Manager:	£31 000			
Designers:	£22 900	£28 000	£24 500	£21 000
Administrator:	£22 500			
Trainees:	£6500	£4200		

a Calculate the mean annual salary. Give your answer to a suitable degree of accuracy.

b Is the mean a good representation of the average annual salary? Give a reason for your answer.

c The company takes on three new trainees. What effect will this have on the mean salary?

d Calculate the median before the new trainees arrive.

e Estimate what the median is likely to be after the new trainees arrive. Write a conclusion about how the new trainees have affected the average salary.

3 For each investigation, choose the most suitable average. Give reasons for your answers.

a Identifying the most popular games console among members of a year group.

b Finding out how much an average supermarket employee earns.

c Comparing average growth for two sets of plant seedlings.

d Predicting the next roll on two six-sided dice.

A

4 This bar chart shows the distribution of weekly incomes in the UK in 2006/07.

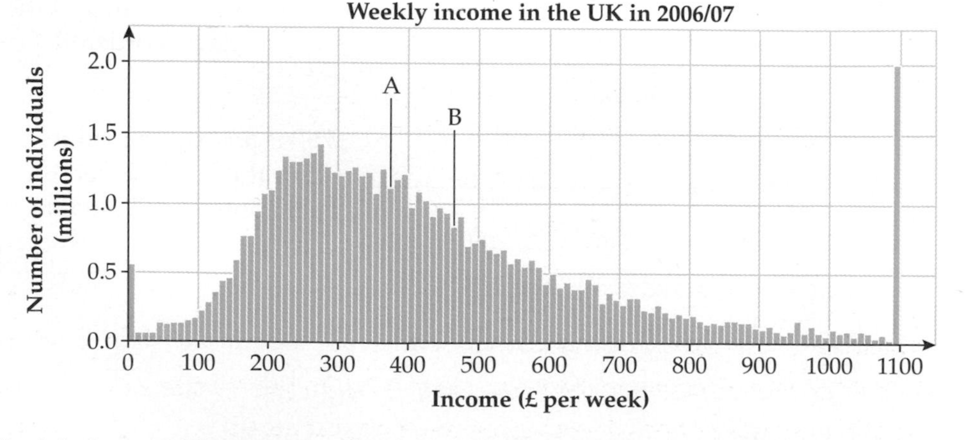

The labels A and B represent the mean and the median incomes.

a Say which label represents which average. Give justifications for your answers.

b Which average would you use to represent the salary of an average worker?

A02

Review exercise

1 This frequency table shows the number of people in each car entering a supermarket car park.

Number of people	Frequency
1	44
2	31
3	19
4	6
5	2

a Work out the range and the mode of this data. **[2 marks]**

b Work out the median number of people per car. **[1 mark]**

c Calculate the mean number of people in each car.
Give your answer to one decimal place. **[3 marks]**

D

2 Beth is gathering data on song downloads. She chooses a sample of songs and times how long they each take to download. This frequency table shows the download times for her sample.

Download time, t (seconds)	Frequency
$10 \leqslant t < 20$	8
$20 \leqslant t < 30$	18
$30 \leqslant t < 40$	22
$40 \leqslant t < 50$	2

a How many downloads did Beth complete for her sample? **[1 mark]**

b What is the modal class interval for Beth's data? **[1 mark]**

c Which class interval will contain the median? **[1 mark]**

d Use mid-points to estimate the mean download time for Beth's sample. **[3 marks]**

D

C

3 The table shows the amounts spent on food by 25 students on a school trip to a theme park.

Amount spent, x (£)	Frequency
$0 \leqslant x < 5$	12
$5 \leqslant x < 10$	8
$10 \leqslant x < 15$	3
$15 \leqslant x < 20$	2

a Which class interval contains the median? **[1 mark]**

b Use mid-points to estimate the mean amount spent per student. **[3 marks]**

c Explain why it is not possible to calculate the exact mean. **[1 mark]**

C

A

4 This frequency table shows the numbers of goals scored per match by a hockey team in one year.

Number of goals scored	Frequency
0	2
1	7
2	16
3	14
4	8
5	6
6	1

a Calculate the mean number of goals scored per match. Give your answer to two decimal places. **[3 marks]**

b Work out the median number of goals scored per match. **[1 mark]**

c Katrina wants to predict how many goals the team will score in their next match. Which average should she use for her prediction? Give a reason for your answer. **[2 marks]**

5 The carbon dioxide emissions of six different cars are given below.

110 g/km 140 g/km 155 g/km
319 g/km 102 g/km 129 g/km

Carbon dioxide emissions for cars are measured in grams per kilometre (g/km).

a Does this data contain any extreme values? Give reasons for your answer. **[2 marks]**

b Holly wants to know about the emissions of an average car. Which average should she use? Give a reason for your answer. **[2 marks]**

c Lydia wants to calculate an average that takes into account any extreme values. Which average should she use? Give a reason for your answer. **[2 marks]**

AO2

6 Chris rolled two dice 11 times and recorded the total score each time.

5 3 6 11 7 3 7 6 7 6 7

Chloe wants to predict what the next roll will be. Which average should she use? Give a reason for your answer. **[2 marks]**

A

7 A newspaper reported the results of a survey.

AVERAGE FAMILY
HAS 2.4 CHILDREN

a Which average did the newspaper use? **[1 mark]**

b Explain why this statistic would not be suitable for describing the number of children in an average family. **[1 mark]**

8 In a shareholders' report, a company claimed that 70% of its branches had above average profits.

a Is the company's claim possible? **[1 mark]**

b If the company's claim is true, which averages could they have been using? Give reasons for your answer. **[2 marks]**

9 James calculated these statistics for students in his class.

Modal shoe size: 8
Median shoe size: 8.75
Mean shoe size: 8.62 (3 s.f.)

Write 'true' or 'false' for each statement. Give reasons for your answers.

Shoes only come in half sizes: 5, 5.5, 6, 6.5, 7 etc.

A If I pick a student at random, there is a 50% chance they will have size 8 shoes. **[2 marks]**

B Half the students wear size 8.5 shoes or smaller. **[2 marks]**

C The most common shoe size in the class is 8. **[2 marks]**

D There were an odd number of students in the class. **[2 marks]**

A

Chapter summary

In this chapter you have learned how to

- identify the modal class from a grouped frequency table **D**
- estimate the range from a grouped frequency table **D**
- calculate the mean, median, mode and range from a frequency table **D** **C**
- work out the class interval which contains the median from data given in a grouped frequency table **C**
- estimate the mean from a grouped frequency table **C**
- choose the most appropriate average and give reasons for your choice **A**

A

A*

AO2

AO2 Example – Statistics

This Grade A/A* question challenges you to use statistics in a real-life context – working with the data to solve a problem.

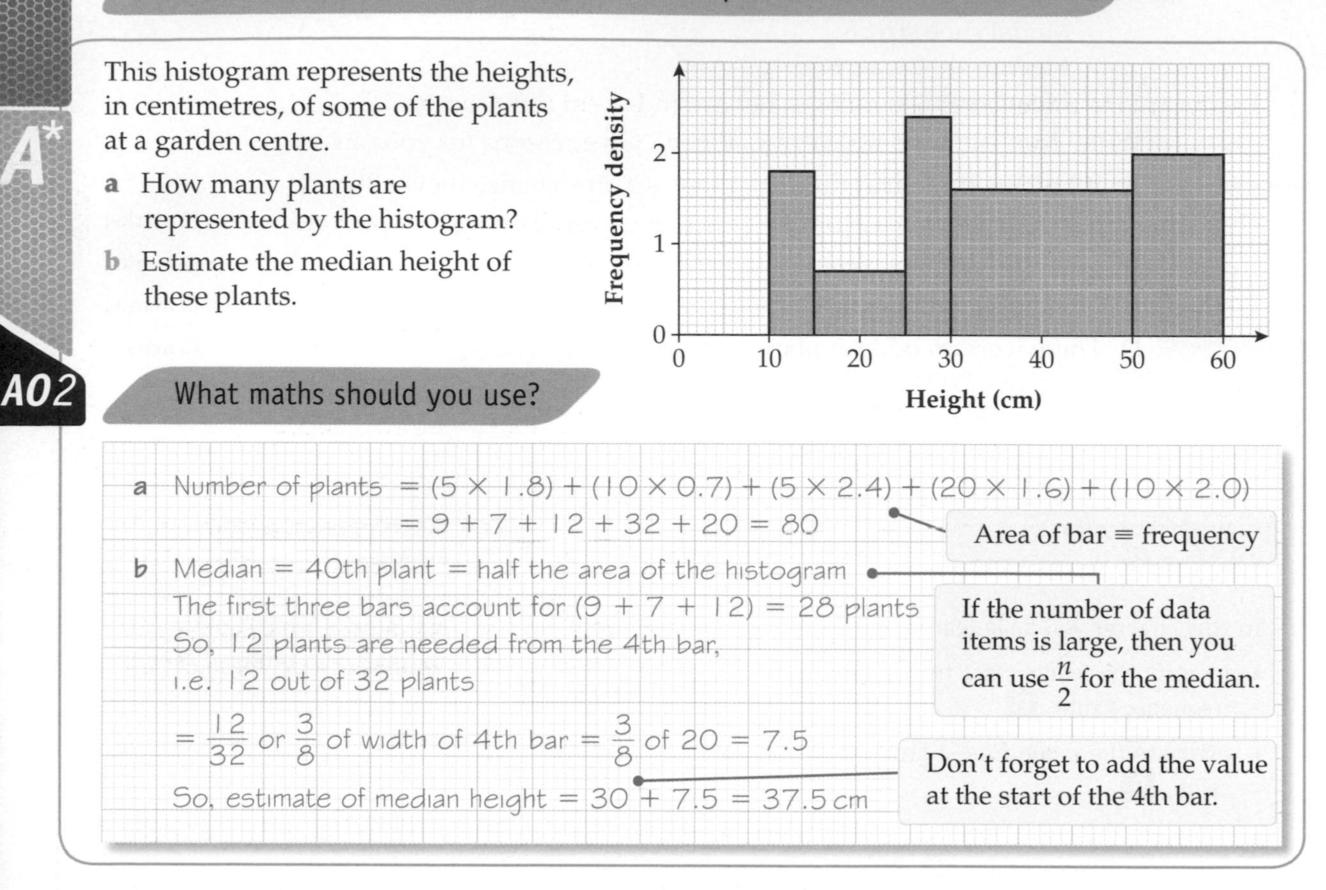

This histogram represents the heights, in centimetres, of some of the plants at a garden centre.

a How many plants are represented by the histogram?

b Estimate the median height of these plants.

What maths should you use?

a Number of plants $= (5 \times 1.8) + (10 \times 0.7) + (5 \times 2.4) + (20 \times 1.6) + (10 \times 2.0)$
$= 9 + 7 + 12 + 32 + 20 = 80$

Area of bar ≡ frequency

b Median = 40th plant = half the area of the histogram

The first three bars account for $(9 + 7 + 12) = 28$ plants

So, 12 plants are needed from the 4th bar,

i.e. 12 out of 32 plants

$= \frac{12}{32}$ or $\frac{3}{8}$ of width of 4th bar $= \frac{3}{8}$ of $20 = 7.5$

So, estimate of median height $= 30 + 7.5 = 37.5$ cm

If the number of data items is large, then you can use $\frac{n}{2}$ for the median.

Don't forget to add the value at the start of the 4th bar.

A*

AO3

AO3 Question – Statistics

Now try this AO3 Grade A* question. You have to work it out from scratch.
READ THE QUESTION CAREFULLY.
It's similar to the AO2 example above, so think about where to start.

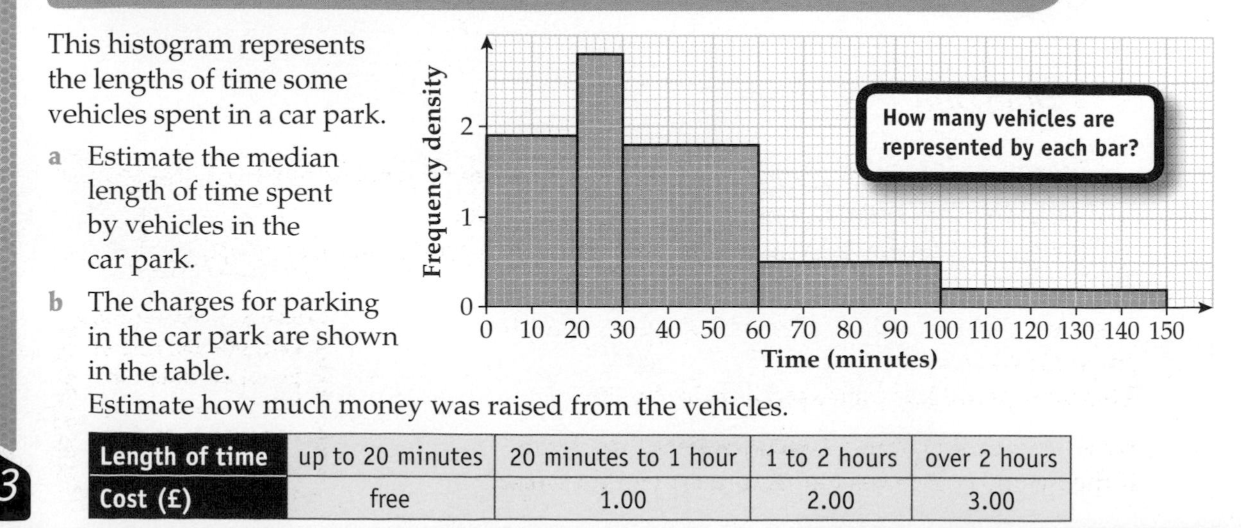

This histogram represents the lengths of time some vehicles spent in a car park.

a Estimate the median length of time spent by vehicles in the car park.

b The charges for parking in the car park are shown in the table.
Estimate how much money was raised from the vehicles.

Length of time	up to 20 minutes	20 minutes to 1 hour	1 to 2 hours	over 2 hours
Cost (£)	free	1.00	2.00	3.00

A02 Example – Number

This Grade C question challenges you to use number in a real-life context – applying maths you know to solve a problem.

C

A02

Last year Jane travelled 15 000 miles in her car and spent £1800 on petrol.
This year petrol has increased in price by 20%.
Jane has cut back her mileage by 20%.

Calculate the percentage increase or decrease in the overall cost of her petrol.

What maths should you use?

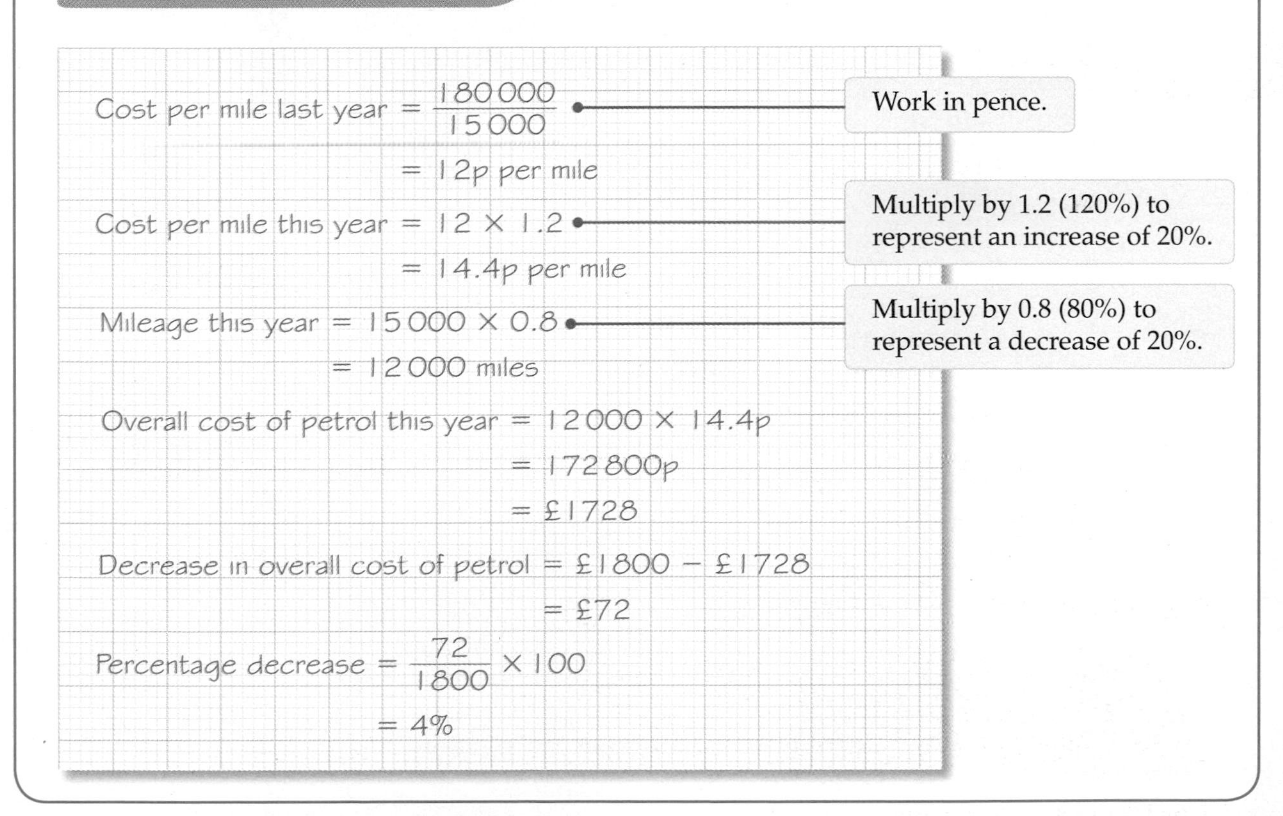

Cost per mile last year = 180 000 / 15 000
= 12p per mile
Cost per mile this year = 12 × 1.2
= 14.4p per mile
Mileage this year = 15 000 × 0.8
= 12 000 miles
Overall cost of petrol this year = 12 000 × 14.4p
= 172 800p
= £1728
Decrease in overall cost of petrol = £1800 − £1728
= £72
Percentage decrease = 72 / 1800 × 100
= 4%
Work in pence.
Multiply by 1.2 (120%) to represent an increase of 20%.
Multiply by 0.8 (80%) to represent a decrease of 20%.

A03 Question – Number

C

Now try this A03 Grade C question. You have to work it out from scratch.
READ THE QUESTION CAREFULLY.
It's similar to the A02 example above, so think about where to start.

Last year Julio travelled 16 000 miles in his car and spent £2240 on petrol.
This year petrol has increased in price by 25%.

By what percentage must Julio's mileage decrease if the overall cost of his petrol is to remain the same?

Calculate the cost of petrol per mile for this year and set up an equation to find this year's mileage.

A03

Missed appointments

At a dental surgery, patients make an appointment to see either a dentist or a hygienist or both. If the patient misses their appointment they are fined according to the length of the appointment they missed. They are fined at the rate of £60 per hour.

Question bank

1 How long is a dentist appointment?

2 In 2009, in which month were the most dentist appointments missed?

3 How much is a patient fined for missing a hygienist appointment?

At the end of 2009, the practice manager writes a report on the number of appointments that patients missed. The practice manager writes, 'Altogether the total money due from fines comes to £28 620.'

4 Is the practice manager's total correct? Show working to justify your answer.

5 Calculate the mean, median and modal monthly number of dentist appointments missed in 2009.

In his report, the practice manager also writes, 'On average, more dentist appointments were missed per month than hygienist appointments.'

6 Is the practice manager correct? Explain clearly the reasons for making your decision. You could use some of your previous working to help you.

Information bank

Surgery facts

The surgery opens from 9 am to 6 pm Monday to Friday.

The surgery closes for lunch from 1 pm to 2 pm.

There are six dentists at the surgery.

A dentist appointment lasts 15 minutes.

There are three hygienists at the surgery.

A hygienist appointment lasts 10 minutes.

The total number of hygienist appointments missed in 2009 is 1008.

The median monthly number of hygienist appointments missed in 2009 is 81.

The modal monthly number of hygienist appointments missed in 2009 is 96.

There are two receptionists at the surgery.

Salary facts

A dentist at the surgery earns £86 000 per year.

A hygienist at the surgery earns £26 000 per year.

A receptionist at the surgery earns £15 000 per year.

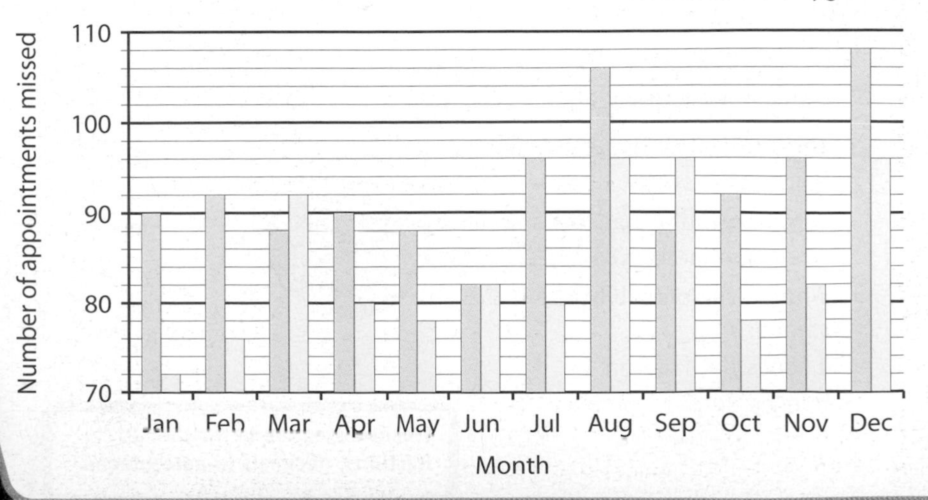

5 Probability

This chapter is about predicting the chance of things happening.

Jelly beans come in 60 different flavours! If there are 65 beans in a bag, what is the chance of picking your favourite?

Objectives

This chapter will show you how to

- **identify mutually exclusive events** D
- **predict the number of times an event is likely to happen** D
- **calculate relative frequencies and estimate probabilities** C
- **calculate the probability of two independent events happening at the same time** C
- **draw and use tree diagrams for independent events** B A
- **draw and use tree diagrams for more complex problems** A*

BBC Video

Before you start this chapter

1 A fair six-sided dice is rolled.
Work out the probability of

a rolling a 1 b rolling a number less than 7

c rolling an even number d rolling a 12.

2 One letter is chosen at random from the word

P R O B A B I L I T Y

Work out the probability that the letter is

a the letter B b a vowel

c made up entirely of straight lines.

3 The probability that this spinner lands on 1 is 0.7.

The probability that this spinner lands on blue is 0.85.

What is the probability that the spinner

a does not land on 1

b does not land on blue?

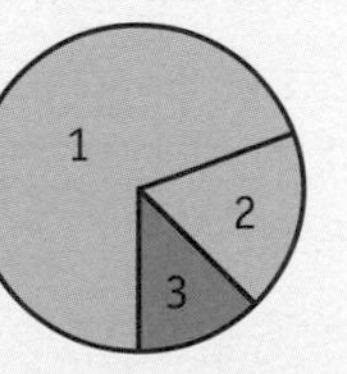

Probabilities can be written as fractions, decimals or percentages.

4 A fair three-sided spinner has sections labelled 1, 2 and 3. The spinner is spun and a fair six-sided dice is rolled at the same time. The two scores are added to give a total score.

a Copy and complete the sample space diagram to show all the possible total scores.

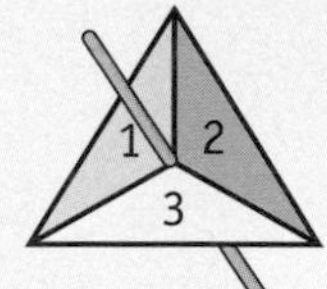

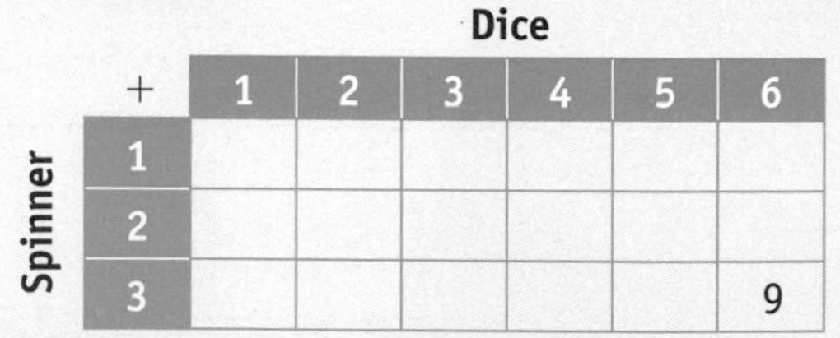

Dice

Spinner +	1	2	3	4	5	6
1						
2						
3						9

b What is the probability that the total score is 6?

c What is the probability that the total score is more than 6?

5 Copy and complete this table.

Fraction	Decimal	Percentage
$\frac{1}{10}$		
		50%
$\frac{3}{4}$		
	0.8	

HELP Chapter 2

5.1 Mutually exclusive events

Keywords
mutually exclusive, or, certain, add

Why learn this?
It could help you win if you remember what cards have already been played.

Objectives

D Understand and use the fact that the sum of the probabilities of all mutually exclusive outcomes is 1

Skills check

1 Work out **a** $\frac{1}{5} + \frac{1}{5}$ **b** $1 - \frac{1}{4}$ **c** $1 - \frac{2}{3}$

2 Work out **a** $0.4 + 0.3$ **b** $1 - 0.82$ **c** $0.4 \div 2$

Mutually exclusive events

Mutually exclusive events cannot happen at the same time. When you roll a dice you cannot get a 1 and a 2 at the same time.

For a fair dice, the probability of rolling a 2 is $\frac{1}{6}$. You can write this as $P(2) = \frac{1}{6}$

Also, $P(1) = \frac{1}{6}$

To calculate the probability of 1 **or** 2 you **add** the probabilities.

$P(1 \text{ or } 2) = \frac{1}{6} + \frac{1}{6} = \frac{2}{6}$ (or $\frac{1}{3}$)

For any two events, A and B, which are mutually exclusive

$P(A \text{ or } B) = P(A) + P(B)$

P(A) means the probability of event A occurring.

Example 1

This spinner has four sections numbered 5 to 8.

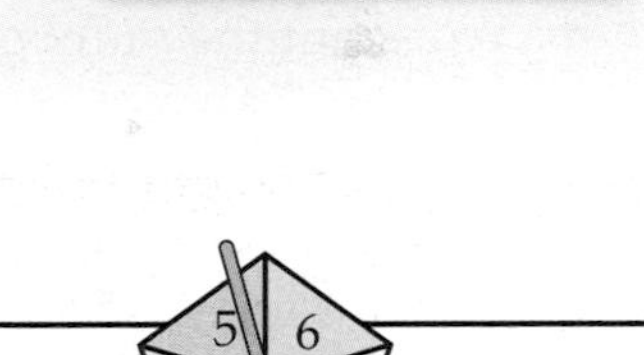

The table shows the probability of the spinner landing on each number.

What is the probability that the spinner lands on 8?

Number	5	6	7	8
Probability	0.2	0.2	0.2	?

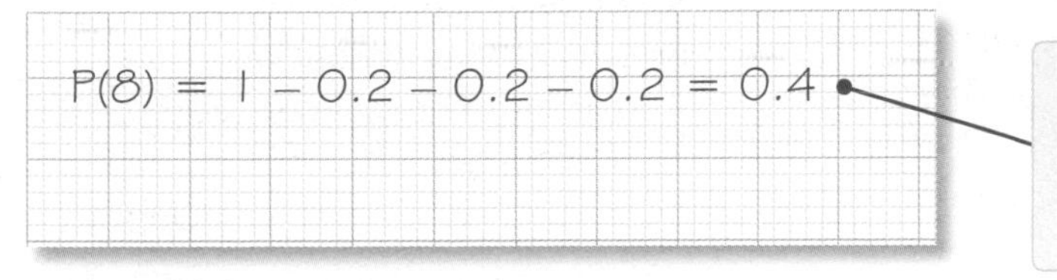

The events 5, 6, 7 and 8 are mutually exclusive, so $P(5) + P(6) + P(7) + P(8) = 1$, as you are **certain** to get one of them.

D

Exercise 5A

1 A box of chocolates contains 15 identical looking chocolates.

Six of the chocolates have toffee centres, four are solid chocolate, three have soft centres and two have nut centres.

One chocolate is taken from the box at random.

What is the probability that the chocolate

a has a toffee or a solid centre

b has a toffee or a soft centre

c has a toffee or a nut centre

d doesn't have a toffee or a nut centre

e doesn't have a soft or a nut or a toffee centre?

D

D

2 A tin contains biscuits.

One biscuit is taken from the tin at random.

The table shows the probabilities of taking each type of biscuit.

a What is the probability that the biscuit is a digestive or a cookie?

b What is the probability that the biscuit is a wafer?

Biscuit	Probability
digestive	0.4
wafer	
cookie	0.15
ginger	0.25

D

3 A bag contains cosmetics.

One cosmetic is taken from the bag at random.

The table shows the probabilities of taking each type of cosmetic. There are three times as many eyeshadows as blushers.

What is the probability that the cosmetic is an eyeshadow?

Cosmetic	Probability
eyeliner	0.3
lipgloss	0.3
eyeshadow	
blusher	

AO2

4 A bag contains 36 marbles of three different colours, red (R), blue (B) and yellow (Y).

$P(R) = \frac{5}{12}$ $\quad P(Y) = \frac{1}{4}$

a Work out the probability of picking a blue marble.

b Work out the number of marbles of each colour in the bag.

D

AO3

5 David puts 15 CDs into a bag.
Elliot puts 9 computer games into the same bag.
Fern puts some DVDs into the bag.
The probability of taking a DVD from the bag at random is $\frac{1}{3}$.
How many DVDs did Fern put in the bag?

5.2 Expectation

Keywords
likely, estimate, trial

Why learn this?

Knowing the expected number of 6s in a number of rolls could help you work out if a dice is fair.

Objectives

D Predict the likely number of successful events given the probability of any outcome and the number of trials or experiments

Skills check

1 Alice rolls a fair six-sided dice.
What is the probability that she rolls

a 2 **b** an odd number **c** a number less than 3?

2 Work out

a $\frac{1}{2} \times 40$ **b** $\frac{1}{3} \times 15$ **c** $\frac{2}{5} \times 30$

HELP Section 2.1

The number of times an event is likely to happen

Sometimes you will want to know the number of times an event is **likely** to happen.

You can work out an **estimate** of the frequency using this formula:

expected frequency = probability of the event happening × number of **trials**

Example 2

In a game a fair six-sided dice is rolled 30 times.

a How many 6s would you expect to get?

b How many even numbers would you expect to get?

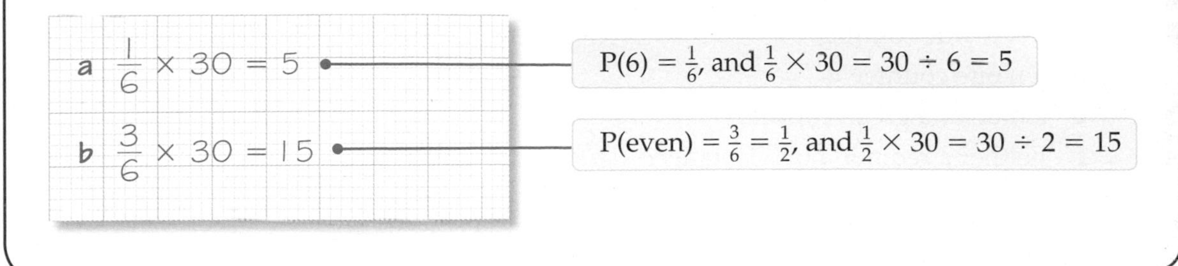

P(6) $= \frac{1}{6}$, and $\frac{1}{6} \times 30 = 30 \div 6 = 5$

P(even) $= \frac{3}{6} = \frac{1}{2}$, and $\frac{1}{2} \times 30 = 30 \div 2 = 15$

D

Exercise 5B

1 A fair dice is rolled 60 times.
How many times would you expect it to land on 2?

2 In an experiment, a card is drawn at random from a normal pack of playing cards.
This is done 520 times.
How many times would you expect to get

a a red card

b a heart

c a king

d the king of hearts?

3 Asif buys 60 scratch-cards. Each scratch-card costs £1.
The probability of winning the £20 prize with each scratch-card is $\frac{1}{30}$.

a How many times is Asif likely to win?

b How much money is Asif likely to win?

c Overall, how much money is Asif likely to lose?

4 In a bag there are 15 red, 5 blue, 5 green and 5 orange counters.
Moira takes a counter at random from the bag, notes the colour, then puts the counter back in the bag. She does this 150 times.
How many times would you expect her to take a blue counter from the bag?

5 The probability that a slot machine pays out its £10 jackpot is $\frac{1}{80}$.
The rest of the time it pays out nothing.
Jimmy plays the slot machine 400 times.

a How many times is Jimmy likely to win?

b How much money is Jimmy likely to win?

Each game costs 20p to play.

c How much does it cost Jimmy to play the 400 games?

d Is Jimmy likely to make a profit? Give a reason for your answer.

D

D

AO2

D

6 In a bag there are 20 counters. Ten of the counters are red, five are blue, four are green and one is gold. A counter is taken at random from the bag then replaced. This is done 300 times.

How many times would you expect to get

a a gold counter

b a red counter

c not a blue counter

d a white counter

AO2

e a blue or a green counter

f neither a red nor a blue counter?

D

7 At a summer fête, Alun runs a charity 'Wheel of fortune' game.

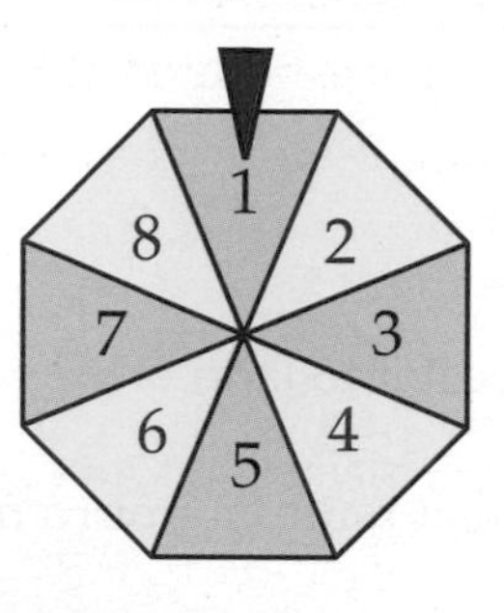

He charges £1 to spin the wheel.
If the arrow lands on a square number he gives a prize of £2.
Altogether 200 people play the game.
How much money would you expect Alun to make for charity?

AO3

D

8 This table shows the probability of selecting coloured balls from a bag.

Colour	Probability
black	0.2
white	0.3
yellow	?
pink	?

a The probability of choosing yellow is four times the probability of choosing pink. Work out the missing values.

b 400 balls are taken at random from the bag, one at a time, and replaced each time. Estimate how many of these are white.

9 John bought 20 tickets for a raffle. The probability of him winning is 0.04. How many raffle tickets were sold?

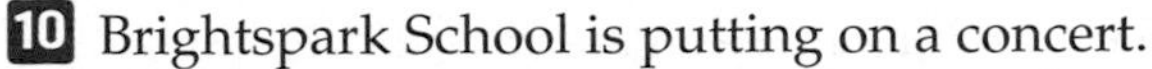

10 Brightspark School is putting on a concert.

The weather on the day of the concert is bad – frequent snow showers and very cold.

Statistics show that in bad weather some parents will not come, even though they have bought tickets.

The table shows the number of tickets sold and the probabilities that parents will come to the concert.

Distance of parents from school	Number of tickets sold	Probability of coming
2 miles or less	870	0.9
more than 2 miles	240	0.6

The school kitchen staff are providing refreshments for the concert.

The cost per head of providing food is estimated at 85p.

Work out the likely cost of providing food for the concert.

D

AO3

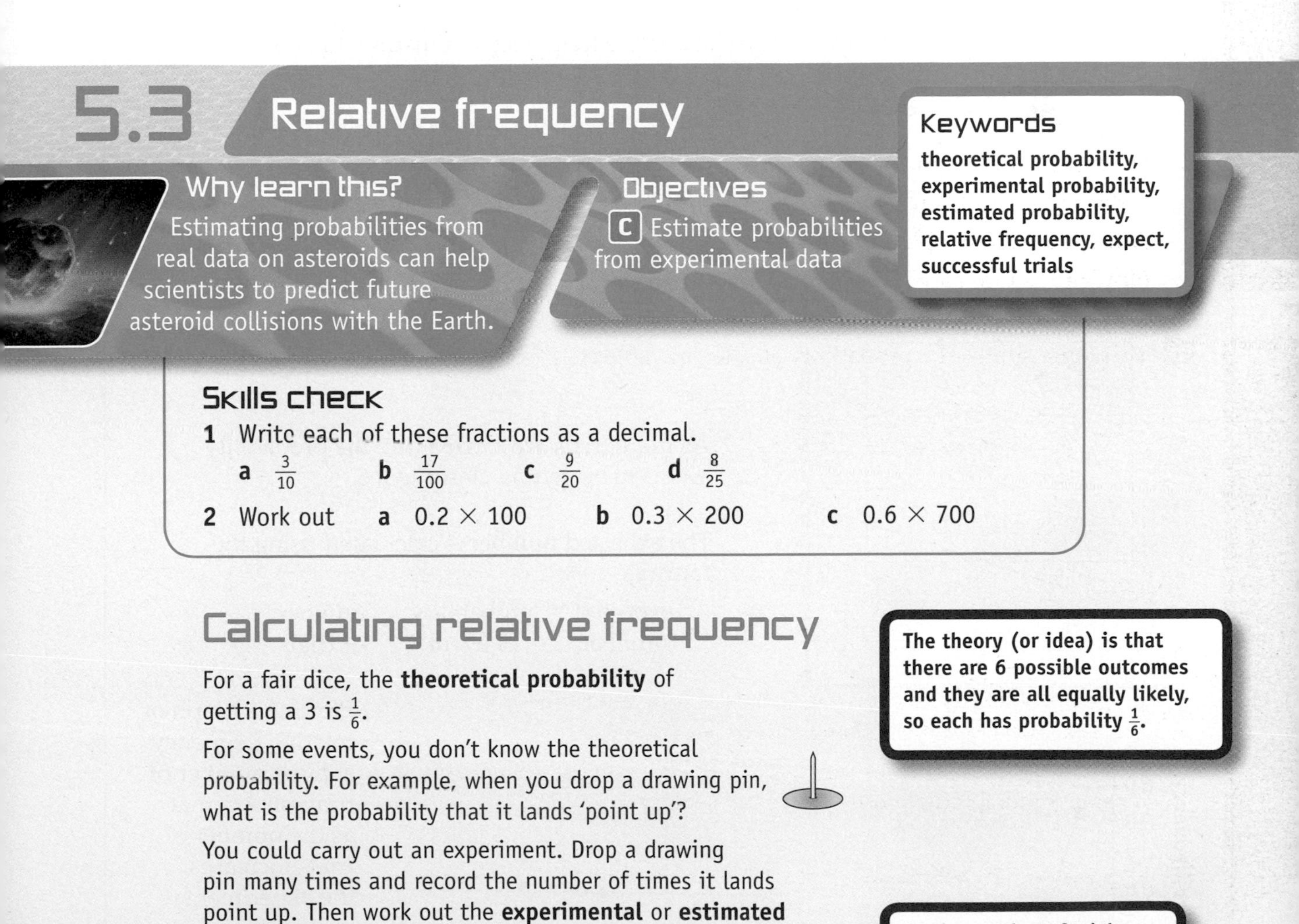

5.3 Relative frequency

Keywords
theoretical probability, experimental probability, estimated probability, relative frequency, expect, successful trials

Why learn this?

Estimating probabilities from real data on asteroids can help scientists to predict future asteroid collisions with the Earth.

Objectives

C Estimate probabilities from experimental data

Skills check

1 Write each of these fractions as a decimal.

a $\frac{3}{10}$ b $\frac{17}{100}$ c $\frac{9}{20}$ d $\frac{8}{25}$

2 Work out a 0.2×100 b 0.3×200 c 0.6×700

Calculating relative frequency

For a fair dice, the **theoretical probability** of getting a 3 is $\frac{1}{6}$.

> The theory (or idea) is that there are 6 possible outcomes and they are all equally likely, so each has probability $\frac{1}{6}$.

For some events, you don't know the theoretical probability. For example, when you drop a drawing pin, what is the probability that it lands 'point up'?

You could carry out an experiment. Drop a drawing pin many times and record the number of times it lands point up. Then work out the **experimental** or **estimated probability.** This estimated probability is called the **relative frequency**.

$$\text{Relative frequency} = \frac{\text{number of \textbf{successful trials}}}{\text{total number of trials}}$$

> As the number of trials increases, the relative frequency approaches the theoretical probability.

C

Example 3

Anil carries out an experiment to work out the probability that when he drops a drawing pin it will land 'point up'.
The table shows his results at different stages of his 2000 trials.

Number of times pin is dropped	Number of times pin lands 'point up'	Relative frequency
100	82	
200	101	
500	326	
1000	586	
1500	882	
2000	1194	

a Calculate the relative frequency at each stage of the testing.

b What do you think is the probability of this drawing pin landing 'point up'?

c If 15 000 of these drawing pins were dropped, how many would you **expect** to land 'point up'?

d Draw a graph of number of trials against relative frequency to illustrate the results.

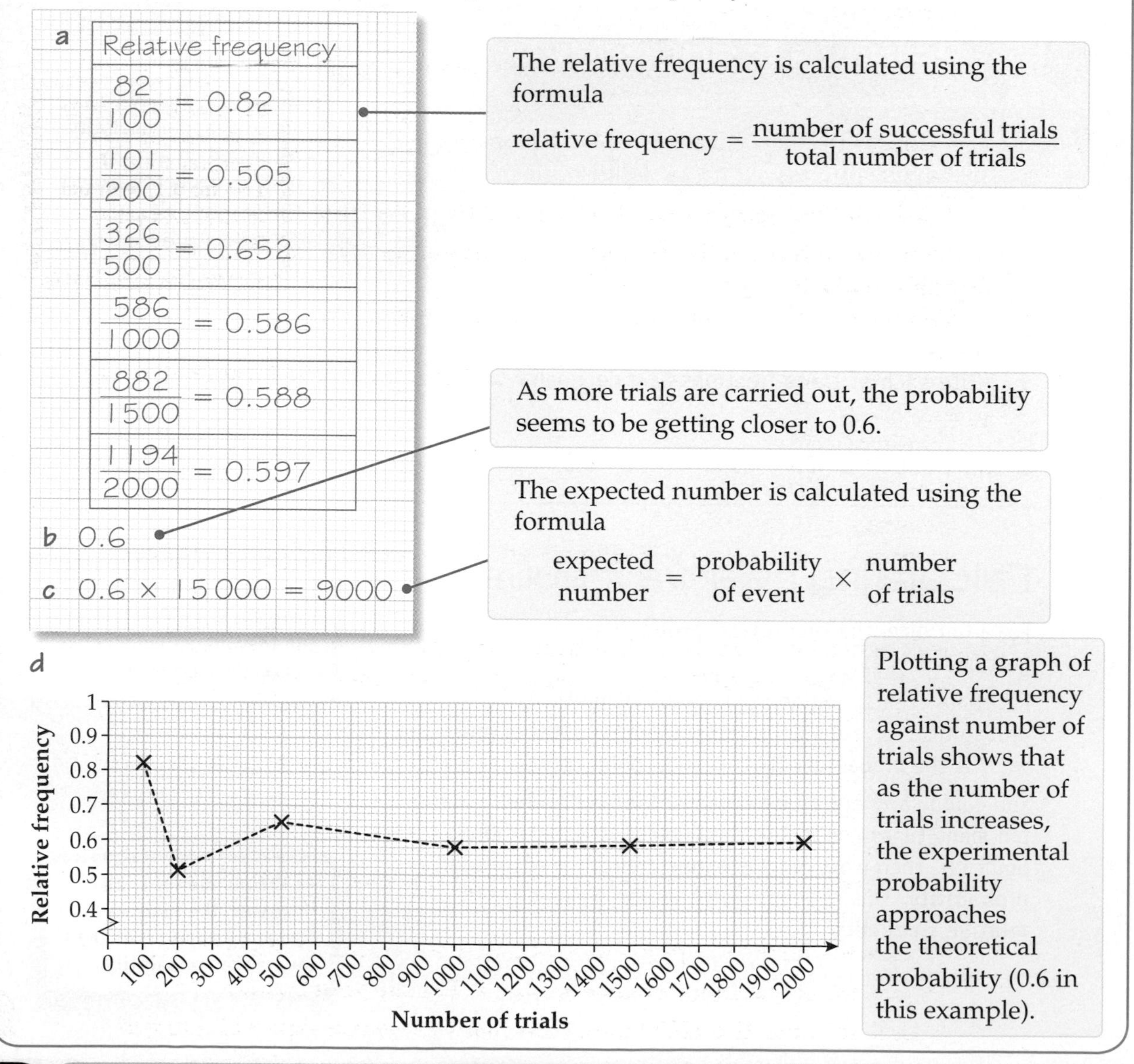

a
Relative frequency
82/100 = 0.82
101/200 = 0.505
326/500 = 0.652
586/1000 = 0.586
882/1500 = 0.588
1194/2000 = 0.597
b 0.6
c 0.6 × 15 000 = 9000
The relative frequency is calculated using the formula
relative frequency = number of successful trials / total number of trials
As more trials are carried out, the probability seems to be getting closer to 0.6.
The expected number is calculated using the formula
expected number = probability of event × number of trials
d
Relative frequency
1
0.9
0.8
0.7
0.6
0.5
0.4
0
100
200
300
400
500
600
700
800
900
1000
1100
1200
1300
1400
1500
1600
1700
1800
1900
2000
Number of trials
Plotting a graph of relative frequency against number of trials shows that as the number of trials increases, the experimental probability approaches the theoretical probability (0.6 in this example).

Exercise 5C

C

1 A bag contains 50 coloured discs. The discs are either blue or red. Jake conducts an experiment to see if he can work out how many of each colour there are.
He takes out a disc, records its colour then replaces it in the bag.
He keeps a tally of how many of each colour there are after different numbers of trials.
The table shows his results.

Number of trials	Number of blue discs	Relative frequency of blue	Number of red discs	Relative frequency of red
20	15		5	
50	29		21	
100	56		44	
200	124		76	
500	341		159	
750	480		270	
1000	631		369	
1500	996		504	
2000	1316		684	

a Calculate the relative frequency for each colour at each stage of the experiment.

b Estimate the theoretical probability of obtaining
i a blue disc **ii** a red disc.

c Work out how many of each colour there are in the bag.

d Draw a graph of relative frequency against number of trials to illustrate your results.
Plot the graphs for blue and red discs on the same axes.

Use a horizontal scale as in Example 3 and a vertical scale of 1 cm for 0.1, with the vertical axis going from 0 to 1.

2 200 drivers in Swansea were asked if they had ever parked their car on double yellow lines.
47 answered 'yes'.

a What is the relative frequency of 'yes' answers?

b There are 130 000 drivers in Swansea.
How many of these do you estimate will have parked their car on double yellow lines?

C

3 Maleek thinks his dice is biased, as he never gets a 6 when he wants to.
To test this theory, he rolls the dice and records the number of 6s he gets.
The table shows his results.

Number of rolls	20	50	100	150	200	500
Number of 6s	1	11	14	24	32	84
Relative frequency						

a Calculate the relative frequency of scoring a 6 at each stage of Maleek's experiment.

b What is the theoretical probability of rolling a 6 with a fair dice?

c Do you think that Maleek's dice is biased? Explain your answer.

d Maleek rolls the dice 1200 times. How many 6s do you expect him to get?

A02

C

4 Hollie thinks her dice is biased. To test this theory she rolls the dice 200 times and records the scores. Her results are shown in the table below.

Score on the dice	1	2	3	4	5	6
Frequency	35	22	25	27	51	40
Relative frequency						

a Calculate the relative frequency of rolling each number.

AO2

b What is the theoretical probability of rolling each number on a fair dice?

c Do you think Hollie's dice is fair? Explain your answer.

C

5 George and Zoe each carry out an experiment with the same four-sided spinner. The tables show their results.

George's results

Number on spinner	1	2	3	4
Frequency	3	14	10	13

Zoe's results

Number on spinner	1	2	3	4
Frequency	45	53	48	54

George thinks the spinner is biased. Zoe thinks the spinner is fair.
Who is correct? Explain your answer.

6 Peter wants to test if a spinner is biased.
The spinner has five equal sections labelled 1, 2, 3, 4, 5.
Peter spins the spinner 20 times. Here are his results.

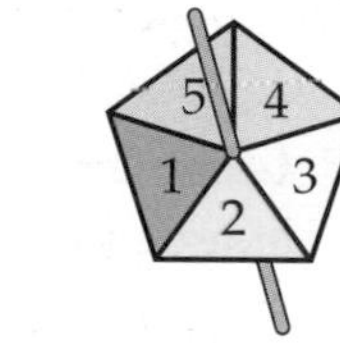

2 1 3 1 5 5 1 4 3 5
4 2 1 5 1 4 3 1 2 1

a Copy and complete the relative frequency table.

Number	1	2	3	4	5
Relative frequency					

b Peter thinks that the spinner is biased.
Write down the number you think the spinner is biased towards.
Explain your answer.

c What could Peter do to make sure his results are more reliable?

7 A fair six-sided dice is repeatedly rolled 10 times.
The number of 6s is counted for each set of 10 rolls. Here are the results.

Set of 10 rolls	Number of 6s	Total number of 6s	Total rolls	Relative frequency
1	2	2	10	0.2
2	2	4	20	0.2
3	5	9	30	0.3
4	5	14	40	0.35
5	2	16	50	0.32
6	4	20	60	0.33
7	1	21	70	0.3
8	3			
9	3			
10	2			

a Complete the table.

AO3

b Do these results suggest that the dice is biased towards the number 6?
Explain your answer.

5.4 Independent events

Keywords
independent, multiply, and

Why learn this?

Understanding independent events gives you a better idea of everyday probabilities. The numbers 1, 2, 3, 4, 5 and 6 are just as likely to come up together on the lottery as any other set of six numbers between 1 and 49.

Objectives

C Calculate the probability of two independent events happening at the same time

Skills check

1 A fair dice is rolled once.
What is the probability of getting a number less than 3?

2 Work out

a $\frac{1}{2} \times \frac{1}{3}$ **b** $\frac{1}{4} \times \frac{3}{5}$ **c** $\frac{2}{3} \times \frac{5}{9}$

Independent events

Two events are **independent** if the outcome of one does not affect the outcome of the other. When you roll two dice at the same time, the number you get on one dice does not affect the number you get on the other.

To calculate the probability of two independent events happening at the same time, you **multiply** the individual probabilities.

When A and B are independent events

P(A **and** B) = P(A) × P(B)

Example 4

a A fair dice is rolled twice. What is the probability of getting two 6s?

b A coin and a dice are thrown together. What is the probability of getting a head and a 1?

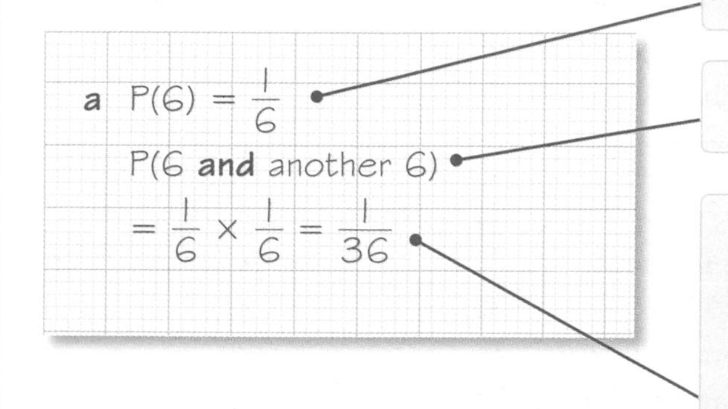

The individual probability of getting a 6.

The events are independent, so P(A **and** B) = P(A) × P(B)

This sample space diagram lists all the possible outcomes when a dice is rolled twice. There are 36 possible outcomes and only one with two 6s, which confirms that the probability is $\frac{1}{36}$.

	1	2	3	4	5	6
1	1,1	1,2	1,3	1,4	1,5	1,6
2	2,1	2,2	2,3	2,4	2,5	2,6
3	3,1	3,2	3,3	3,4	3,5	3,6
4	4,1	4,2	4,3	4,4	4,5	4,6
5	5,1	5,2	5,3	5,4	5,5	5,6
6	6,1	6,2	6,3	6,4	6,5	6,6

Total number of outcomes = total number of outcomes for event A × total number of outcomes for event B

b $P(H) = \frac{1}{2}$, $P(1) = \frac{1}{6}$

$P(H \text{ and } 1) = \frac{1}{2} \times \frac{1}{6} = \frac{1}{12}$

The individual probabilities of getting a head and a 1.

P(A **and** B) = P(A) × P(B)

Exercise 5D

C

1 An ordinary coin is flipped twice. What is the probability of getting two heads?

2 Box A contains 10 identical looking chocolates. Four of them are truffles.
Box B contains 20 identical looking chocolates. Ten of them are truffles.
Carol takes one chocolate from each box.
What is the probability that she gets two truffles?

3 When Lynn and Sally go to the shop, the probability of Lynn buying a bag of crisps is $\frac{1}{3}$, and a muesli bar is $\frac{1}{4}$. The probability of Sally buying a bag of crisps is $\frac{1}{2}$, and a muesli bar is $\frac{1}{4}$. The girls choose independently of each other.
Calculate the probability that

a both girls buy a bag of crisps

b both girls buy a muesli bar

c Lynn buys a bag of crisps and Sally buys a muesli bar.

C

A02

4 Sam plays the National Lottery 'Thunderball' every week.
The probability of winning a £5 prize is 0.03.
The probability of winning a £10 prize is 0.009.
Work out the probability that

a Sam wins £5 one week and £10 the next week

b Sam wins £5 two weeks in a row.

C

5 Zoe takes two exams: history and French.
The probability that she passes history is $\frac{3}{4}$.
The probability that she passes French is $\frac{1}{3}$.

a What is the probability that she passes both exams?

b What is the probability that she passes one exam?

6 A box contains 5 yellow and 7 green tennis balls.
A ball is chosen, its colour noted, then it is replaced.
A second ball is chosen.

a What is the probability that both balls are yellow?

b What is the probability that both balls are the same colour?

7 Ethan and Jerry go to the canteen for lunch.
Main meals come with either chips or salad.
The probability that Ethan chooses chips is 0.7.
The probability that Jerry chooses chips is 0.4.

a What is the probability that they both choose chips?

b What is the probability that they both choose salad?

c What is the probability that one chooses chips and the other chooses salad?

d What do you notice about your answers to parts **a**, **b** and **c**?

8 Joe spins a spinner with five equal sectors numbered from 1 to 5.
Sarah rolls a fair dice numbered from 1 to 6.
Work out the probability that

a they both obtain a 3

b the total of their scores is 2

c the total of their scores is 5

d Sarah's score is twice Joe's score

e they both obtain an even number.

How will a total of 5 arise? You will need to add some probabilities.

C

9 Bernie has an ordinary pack of 52 cards.
She shuffles the pack then selects a card at random.
She replaces the card, shuffles the pack again and selects another card.
What is the probability that

a both cards are red

b neither card is a spade

c both cards are queens

d exactly one card is a queen?

In d, which card will be a queen, the first or the second?

AO2

5.5 Tree diagrams

Keywords
tree diagram, combined events

Why learn this?

If you know the probability of your team winning, losing or drawing matches, a tree diagram is an easy way to see the possible outcomes of future matches.

Objectives

B Use and understand tree diagrams in simple contexts

A A* Use the 'and' and 'or' rules in tree diagrams

Skills check

1 Write 'true' or 'false' for each of these.

a $\frac{1}{3} \times \frac{1}{3} = \frac{1}{6}$ **b** $\frac{1}{4} \times \frac{1}{5} = \frac{1}{20}$ **c** $\frac{3}{4} \times \frac{1}{2} = \frac{3}{8}$ **d** $\frac{4}{5} \times \frac{2}{7} = \frac{6}{12}$

2 Work out **a** 0.2×0.2 **b** 0.4×0.8 **c** $0.3 \times 0.3 \times 0.5$

Drawing tree diagrams

A **tree diagram** can show all the possible outcomes of two or more **combined events**, and their probabilities.

B

Example 5

Imagine you have a bag of discs. Each disc is either red or blue.
You pick out a disc, record its colour then replace it in the bag.
Then you pick out a second disc.

You can show the possible results by drawing a tree diagram.

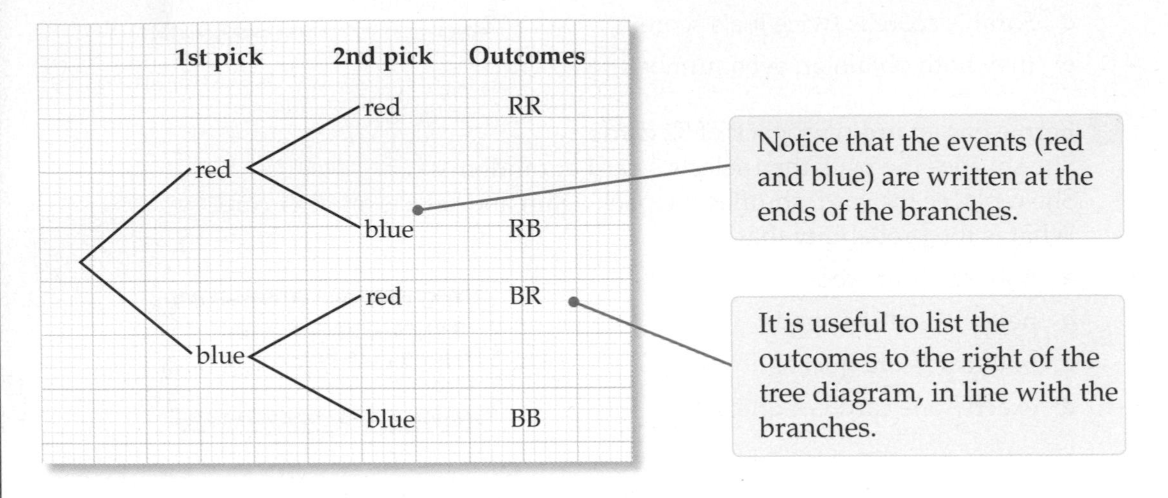

There is only one way of obtaining the outcomes 'red, red' or 'blue, blue' (usually written 'both red' or 'both blue').

There are two ways of obtaining the outcome 'one of each colour' because both 'red, blue' and 'blue, red' fit this description.

B

Example 6

Suppose there are 3 red discs and 2 blue discs.
The probability of picking a red on the first pick is $\frac{3}{5}$ and picking a blue is $\frac{2}{5}$.
Since the discs are replaced, these probabilities always remain the same.

There are always 5 discs in the bag, 3 red and 2 blue.

You can write the probabilities on the appropriate branches, so the completed tree diagram looks like this.

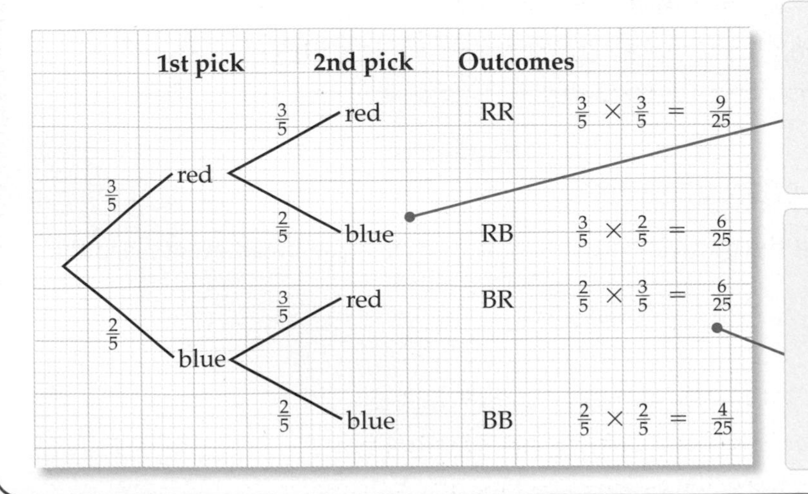

The events are independent so you can multiply the probabilities on the branches to calculate the probability of each final outcome.

The sum of the probabilities of the final outcomes is
$\frac{9}{25} + \frac{6}{25} + \frac{6}{25} + \frac{4}{25} = \frac{25}{25} = 1$
This is always the case since the final outcomes represent everything that can possibly happen.

Example 7

Use the tree diagram in Example 6 to answer these questions.

a What is the probability that the two discs are the same colour?

b What is the probability that the two discs are different colours?

c What is the probability of picking at least one red disc?

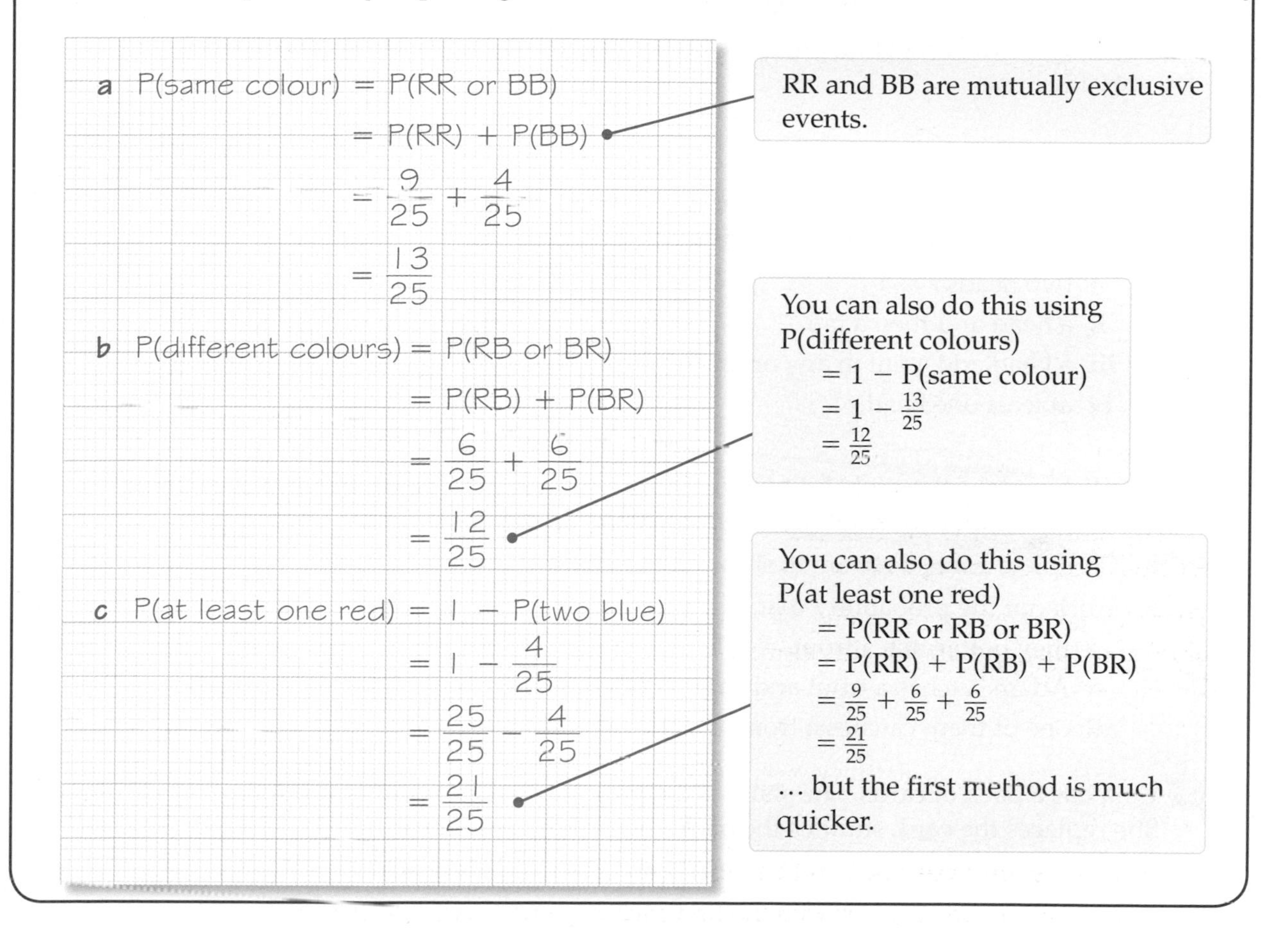

a P(same colour) = P(RR or BB)
= P(RR) + P(BB)
$= \frac{9}{25} + \frac{4}{25}$
$= \frac{13}{25}$
RR and BB are mutually exclusive events.
b P(different colours) = P(RB or BR)
= P(RB) + P(BR)
$= \frac{6}{25} + \frac{6}{25}$
$= \frac{12}{25}$
You can also do this using P(different colours)
= 1 − P(same colour)
$= 1 - \frac{13}{25}$
$= \frac{12}{25}$
c P(at least one red) = 1 − P(two blue)
$= 1 - \frac{4}{25}$
$= \frac{25}{25} - \frac{4}{25}$
$= \frac{21}{25}$
You can also do this using P(at least one red)
= P(RR or RB or BR)
= P(RR) + P(RB) + P(BR)
$= \frac{9}{25} + \frac{6}{25} + \frac{6}{25}$
$= \frac{21}{25}$
… but the first method is much quicker.

B

Exercise 5E

1 Siobhan has two maths tests next week.
She estimates the probability of her passing the geometry test is 0.8, but the probability of her passing the statistics test is only 0.2.

a Copy and complete the tree diagram to show all the possible outcomes.

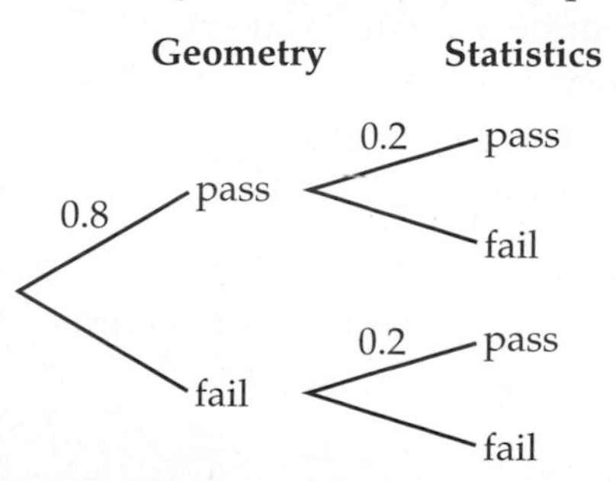

b Work out the probability that Siobhan passes both tests.

c Work out the probability that Siobhan passes neither test.

d Work out the probability that Siobhan passes only one test.

B

B

2 On Megan's way to school there are two sets of traffic lights. The bus driver knows that the probability of the first set being green is $\frac{1}{4}$ and the probability of the second set being green is $\frac{1}{3}$.

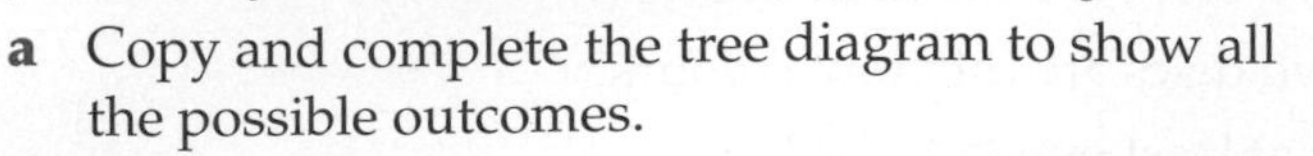

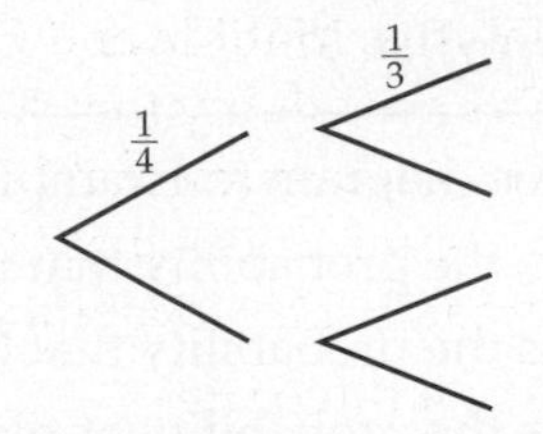

a Copy and complete the tree diagram to show all the possible outcomes.

b Work out the probability that the bus gets held up at both sets of traffic lights.

A02

c Work out the probability that the bus only gets held up at the second set of traffic lights.

A

3 **a** David flips a coin twice.
Complete a tree diagram to show all the possible outcomes.

b Work out the probability that David flips

i two heads

ii a head and then a tail

iii a head and a tail in any order

iv at least one head.

4 At a fishing lake the probability of catching a trout is 0.2. The probability of catching a carp is 0.8. Adam and Tabitha both catch one fish each.

a Complete a tree diagram to show all the possible outcomes.

b Work out the probability that

i they both catch a trout

ii Adam catches a trout and Tabitha catches a carp

iii one of them catches a trout and the other catches a carp.

A

5 Zina has a pack of cards. She shuffles the pack then takes a card at random. She replaces the card, shuffles the pack, then takes another card.

a What is the probability that the first card Zina takes is a jack?

b What is the probability that the first card Zina takes isn't a jack?

c Work out the probability that

i both cards are a jack

ii neither card is a jack

iii at least one card is a jack.

6 Luke, Matthew and Sophie all work for the same company.
The probability that Luke is late for work is 0.3.
The probability that Matthew is late for work is 0.4.
The probability that Sophie is late for work is 0.2.
On any day, what is the probability that

a all three are late for work

b all three are on time

c Sophie is late but Luke and Matthew are on time

A02

d at least one of them is late?

7 Tabitha, Matilda and Oscar are taking an English exam.
The probability that Tabitha passes is 0.9.
The probability that Matilda passes is 0.7.
The probability that Oscar passes is 0.6.
What is the probability that

- **a** all three pass
- **b** Tabitha and Oscar pass but Matilda fails
- **c** all three fail
- **d** at least one of them passes
- **e** any two of them pass?

A

8 A bag contains 10 coloured beads.
4 are red, 5 are white and 1 is blue.
Simon takes a bead from the bag at random, records its colour then replaces it in the bag.
He then takes out another bead and records its colour.
What is the probability that

- **a** the first bead is white and the second is red
- **b** both beads are blue
- **c** both beads are the same colour
- **d** neither bead is red
- **e** at least one is red?

AO2

9 Amy, Beth and Clare take their cars for an MOT.
The probabilities of their cars passing on lights, brakes and tyres are shown in the table.

	Lights	Brakes	Tyres
Amy's car	0.8	0.4	0.9
Beth's car	0.7	0.8	0.6
Clare's car	0.9	0.7	0.3

What is the probability that

- **a** all three cars pass on lights
- **b** all three cars fail on tyres
- **c** Beth's car passes all three tests
- **d** two of the cars pass on brakes but one car fails on brakes
- **e** Amy's car passes only one of the three tests?

A*

AO2

10 At a fairground you can play 'Spin and Win'.
The game involves spinning a spinner with five equal sections numbered from 1 to 5.
This poster on the fairground stall shows how you can win.

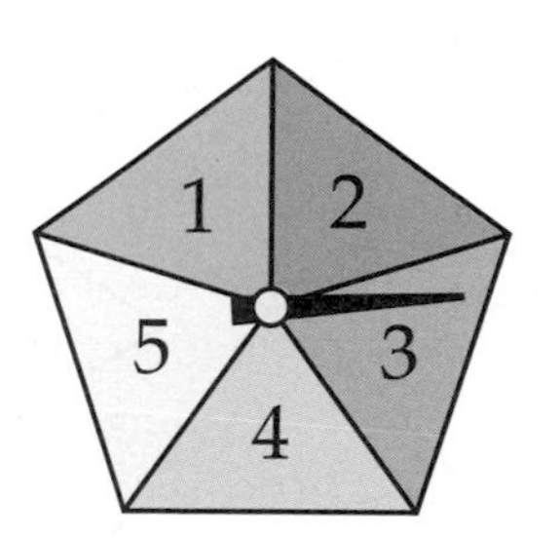

Spin and Win
£2 a go
Spin the spinner three times
Three 5s = £15
Two 5s = £10

Throughout the day, 250 people play this game.

- **a** Estimate how much profit the stallholder will make if he charges £2 a go.
- **b** Estimate the minimum that the stallholder should charge if he is not to make a loss.

A*

AO2

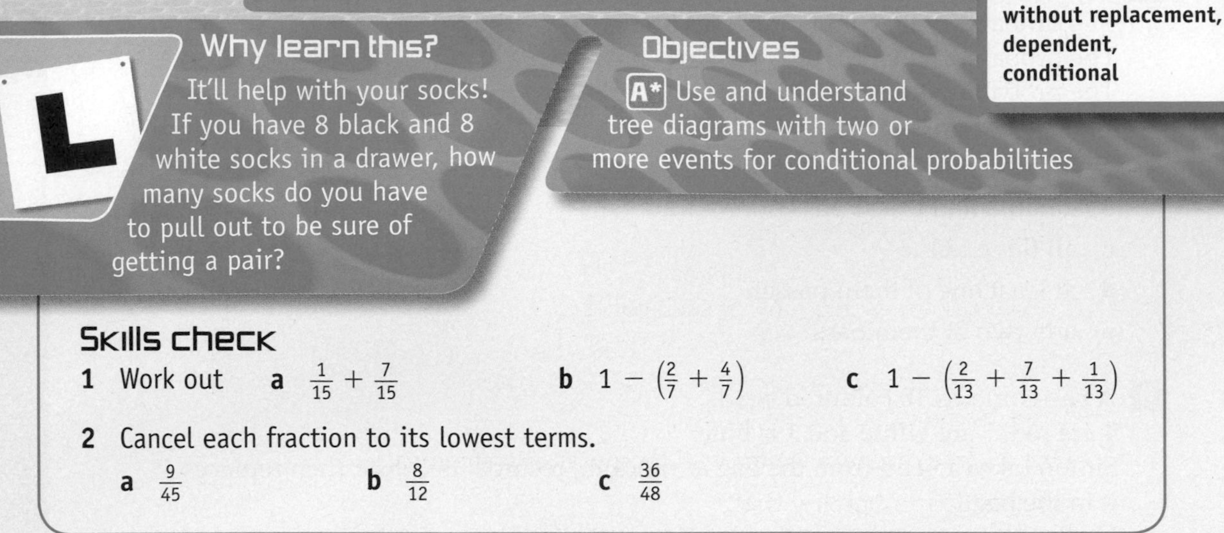

5.6 Conditional probability

Keywords
without replacement, dependent, conditional

Why learn this?

It'll help with your socks! If you have 8 black and 8 white socks in a drawer, how many socks do you have to pull out to be sure of getting a pair?

Objectives

A* Use and understand tree diagrams with two or more events for conditional probabilities

Skills check

1 Work out **a** $\frac{1}{15} + \frac{7}{15}$ **b** $1 - \left(\frac{2}{7} + \frac{4}{7}\right)$ **c** $1 - \left(\frac{2}{13} + \frac{7}{13} + \frac{1}{13}\right)$

2 Cancel each fraction to its lowest terms.
a $\frac{9}{45}$ **b** $\frac{8}{12}$ **c** $\frac{36}{48}$

Tree diagrams when one outcome affects the next

In real life, the outcome of one event often affects the outcome of the next.

When you pick a card from a pack, P(black) $= \frac{26}{52}$.
If you don't replace it, the probability of picking a black card next time depends on what colour you picked the first time.

The probability of the second event is **conditional** on (**dependent** on) the outcome of the first event.

Not replacing the card changes the probability of picking a black. Mathematicians call this a '**without replacement**' problem.

Example 8

A*

Oscar picks a card at random from a normal pack.
He does not replace it.
He then picks another card at random. Work out the probability that he picks

a two black cards
b one red and one black card.

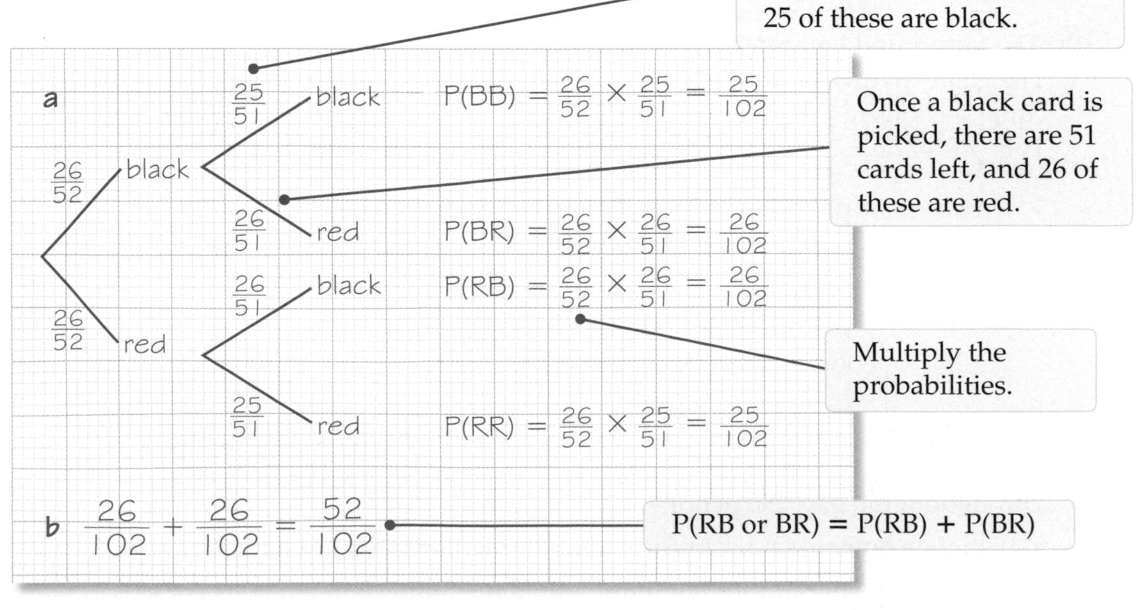
a
26/52 black
25/51 black
P(BB) = 26/52 × 25/51 = 25/102
26/51 red
P(BR) = 26/52 × 26/51 = 26/102
26/52 red
26/51 black
P(RB) = 26/52 × 26/51 = 26/102
25/51 red
P(RR) = 26/52 × 25/51 = 25/102
b 26/102 + 26/102 = 52/102
Once a black card is picked, there are 51 cards left, and 25 of these are black.
Once a black card is picked, there are 51 cards left, and 26 of these are red.
Multiply the probabilities.
P(RB or BR) = P(RB) + P(BR)

Exercise 5F

A*

1 Sandra has three blue mugs and four green mugs in her cupboard.
She takes out two mugs at random.

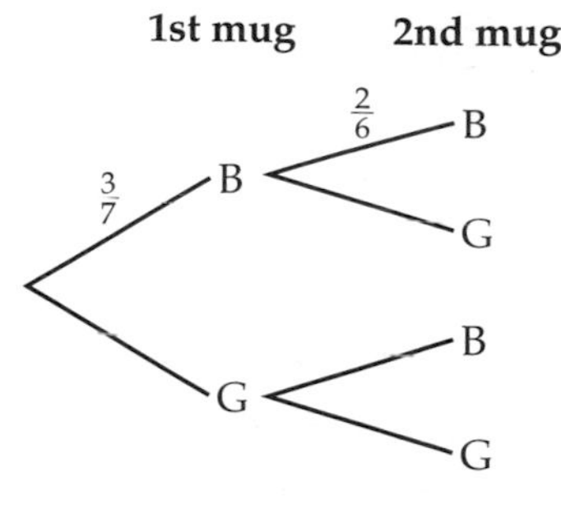

a Copy and complete the tree diagram to show all the possible outcomes.

b What is the probability that

  i Sandra takes two blue mugs

  ii Sandra takes one mug of each colour?

2 John has four red and two green socks in a drawer.
He takes two socks out at random.

a Draw a tree diagram to show all the possible outcomes.

b What is the probability that

  i John takes a matching pair of socks

  ii John doesn't take a matching pair of socks?

3 A tin contains 12 shortbread and 10 coconut biscuits.
Brendan takes two biscuits at random and eats them.
Work out the probability that

a both biscuits are coconut

b the first biscuit is shortbread and the second biscuit is coconut.

4 Tina has these cards.

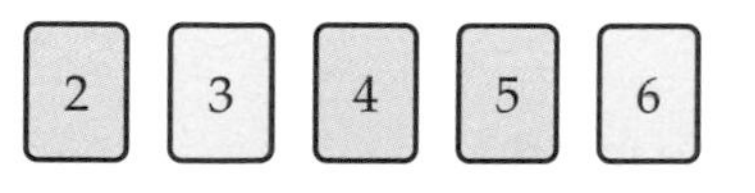

This is just like taking the first one, then the other.

She takes two of the cards at random.

Without drawing a tree diagram, work out the probability that

a both cards are yellow

b both cards have an even number

c the first card is blue and the second card is yellow.

5 Jane has a bag containing 15 identical cubes.
Eight are red, four are white and three are blue.
Jane takes two cubes at random from the bag.

a Draw a tree diagram to show all the possible outcomes.

b What is the probability that

  i both cubes are the same colour

  ii the cubes are different colours

  iii at least one cube is white?

6 Debbie has 10 tins of food in her cupboard, but all the labels are missing.
She knows that six of them contain soup and four contain baked beans.
She opens three tins at random.
What is the probability that she opens one tin of soup and two tins of baked beans?

A* AO2

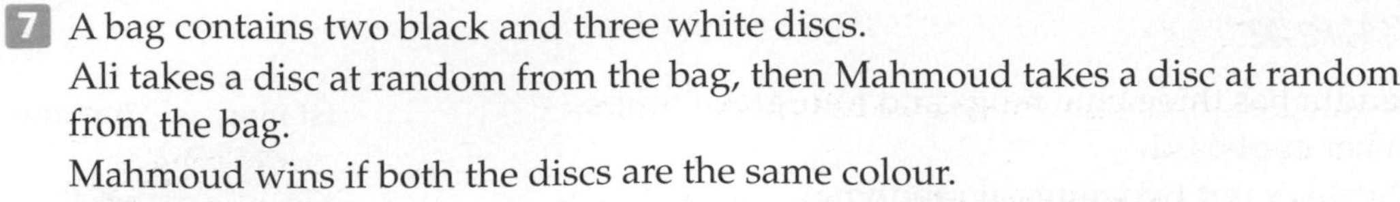

7 A bag contains two black and three white discs.
Ali takes a disc at random from the bag, then Mahmoud takes a disc at random from the bag.
Mahmoud wins if both the discs are the same colour.
Who has the better chance of winning? Give a reason for your answer.

A

8 Ian and Steve play three sets of tennis.
The probability that Ian wins the first set is 0.6.
When Ian wins a set, the probability that he wins the next set is 0.7.
When Steve wins a set, the probability that he wins the next set is 0.5.
The first person to win two sets wins the match.

a Copy and complete the tree diagram to show all the probabilities.

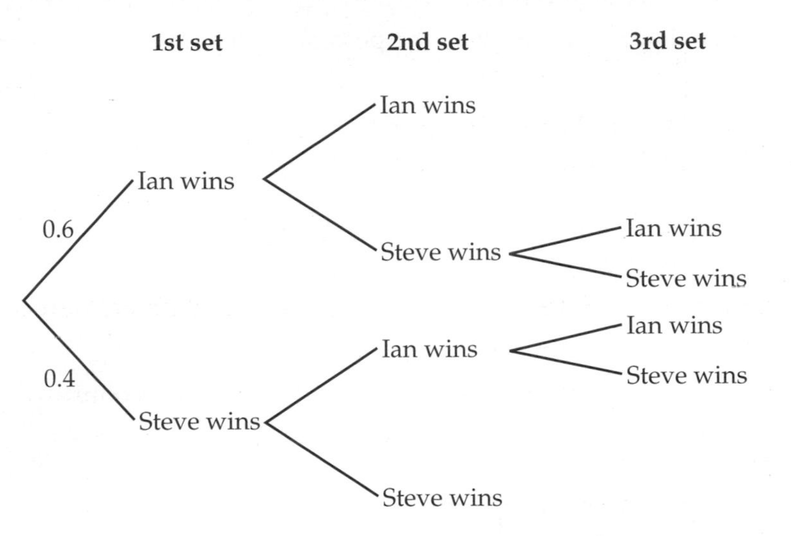

A* AO2

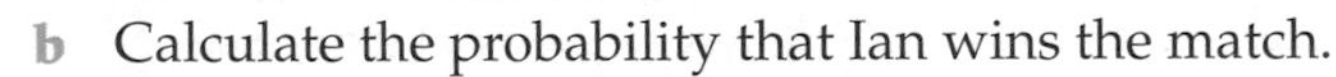

b Calculate the probability that Ian wins the match.

A*

9 At the end of a training course students must pass a test to gain a diploma.
The probability of passing first time is 0.8.
Those who fail first time are allowed just one more attempt.
The probability of passing the re-sit is 0.65.
Two friends, Sam and Tim, follow the training course and take the test.
What is the probability that they both gain a diploma?

10 Bag A contains 5 red and 3 blue counters. Bag B contains 2 red and 6 blue counters.

Bag A

Bag B

Step 1 A counter is taken, at random, from Bag A and placed in Bag B.
Step 2 A counter is taken, at random, from Bag B and placed in Bag A.
Calculate the probability that Bag A has more red counters than blue counters after these two steps.

AO2

Review exercise

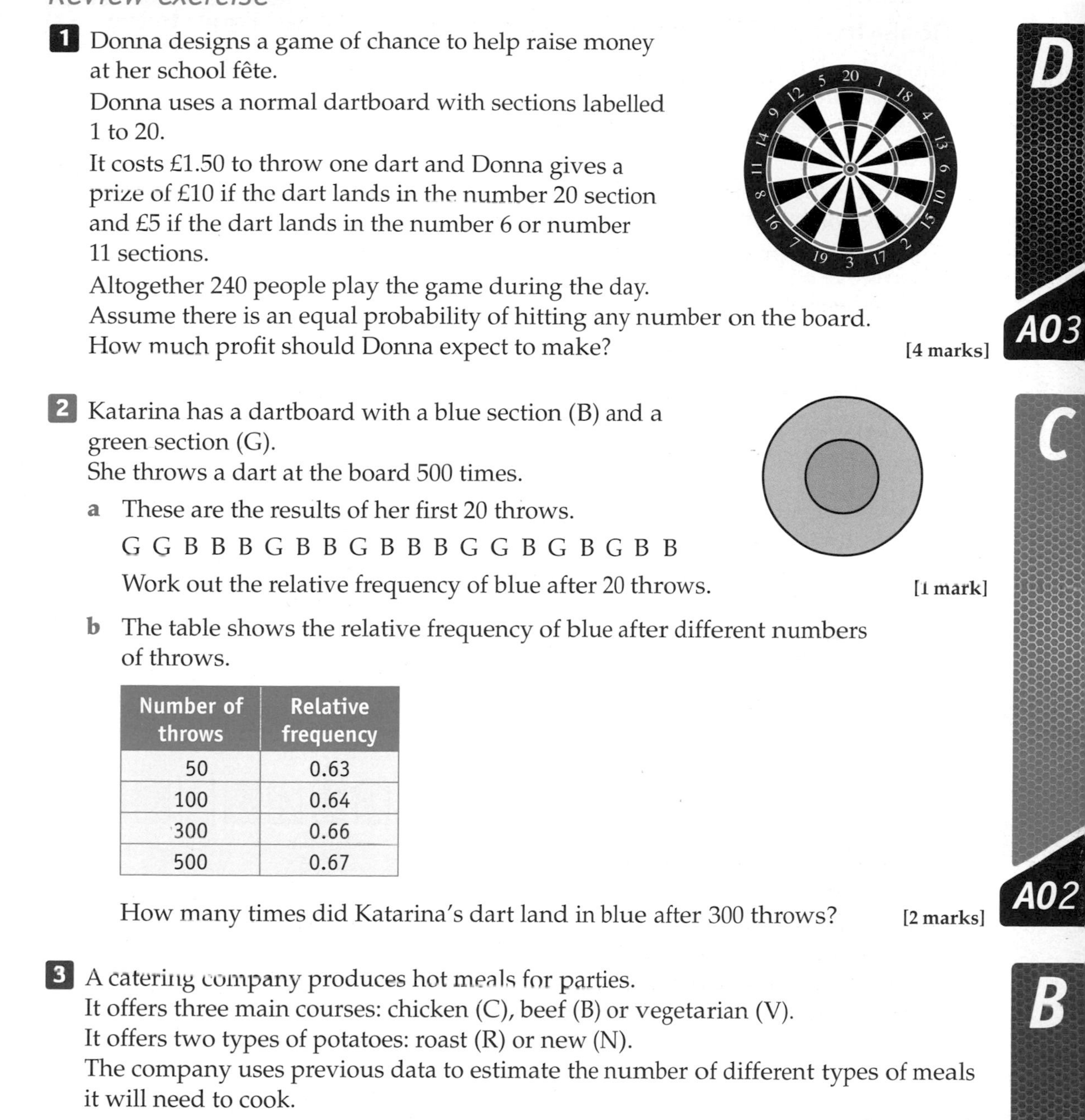

1 Donna designs a game of chance to help raise money at her school fête.
Donna uses a normal dartboard with sections labelled 1 to 20.
It costs £1.50 to throw one dart and Donna gives a prize of £10 if the dart lands in the number 20 section and £5 if the dart lands in the number 6 or number 11 sections.
Altogether 240 people play the game during the day.
Assume there is an equal probability of hitting any number on the board.
How much profit should Donna expect to make? **[4 marks]**

D AO3

2 Katarina has a dartboard with a blue section (B) and a green section (G).
She throws a dart at the board 500 times.

a These are the results of her first 20 throws.

G G B B B G B B G B B B G G B G B G B B

Work out the relative frequency of blue after 20 throws. **[1 mark]**

b The table shows the relative frequency of blue after different numbers of throws.

Number of throws	Relative frequency
50	0.63
100	0.64
300	0.66
500	0.67

How many times did Katarina's dart land in blue after 300 throws? **[2 marks]**

C AO2

3 A catering company produces hot meals for parties.
It offers three main courses: chicken (C), beef (B) or vegetarian (V).
It offers two types of potatoes: roast (R) or new (N).
The company uses previous data to estimate the number of different types of meals it will need to cook.
Previous data shows that the probability of a person choosing chicken is $\frac{1}{2}$, beef is $\frac{1}{3}$ and vegetarian is $\frac{1}{6}$.
The probability of a person choosing roast potatoes is $\frac{1}{4}$ and new potatoes is $\frac{3}{4}$.

FUNCTIONAL • FUNCTIONAL •

a Copy and complete the tree diagram to show all the possible outcomes. **[2 marks]**

$\frac{1}{2}$ C — $\frac{1}{4}$ R, N
B — R, N
V — R, N

b Work out the probability that a person chooses chicken and new potatoes. **[2 marks]**

c Work out the probability that a person chooses meat and roast potatoes. **[3 marks]**

d At the next party, 120 guests are expected.
Estimate the number of 'vegetarian and new potato' meals the company needs to cook. **[2 marks]**

B AO2

A

AO2

4 Nick always has cereal or toast for breakfast.
He also has tea, coffee or fruit juice.
The probabilities of what he chooses for breakfast are given in the table.

Item	Probability
toast	$\frac{1}{4}$
cereal	$\frac{3}{4}$
tea	$\frac{1}{4}$
coffee	$\frac{5}{8}$
fruit juice	$\frac{1}{8}$

a What is the probability that Nick has toast and coffee? **[2 marks]**

b What is the most likely combination Nick will choose? **[1 mark]**

c What is the probability of the combination in part **b**? **[2 marks]**

A*

5 For breakfast, Magda always has cereal or toast followed by orange juice or apple juice.
The probability that Magda has cereal is $\frac{1}{3}$.
When Magda has cereal, the probability that she then has apple juice is $\frac{5}{6}$.
When Magda has toast, the probability that she then has orange juice is $\frac{7}{8}$.
Work out the probability that Magda has toast and apple juice for breakfast. **[3 marks]**

AO2

6 Alice takes maths and further maths at sixth-form college.
The probability that she will pass maths is 0.7.
If she passes maths, the probability that she will pass further maths is 0.6.
If she fails maths, the probability that she will pass further maths is 0.2.
Work out the probability that

a she passes both **[2 marks]**

b she only passes one **[2 marks]**

c she passes at least one. **[2 marks]**

A*

7 Steven and Rob have these cards.

L I V E R P O O L

Steven shuffles the cards then selects one at random. He does not replace it.
Rob then selects a card at random.
Steven wins if both the cards are vowels or both the cards are consonants.
Who has the better chance of winning? Explain your answer. **[3 marks]**

AO3

8 Rachel has a credit card but sometimes has trouble remembering the PIN.
The probability that she gets it right first time is 0.8.
If she gets it wrong the first time, the probability that she gets it right the second time is 0.65.
If she gets it wrong twice, the probability that she gets it right the third time is 0.15.
A cash machine will keep her card after three wrong attempts.
What is the probability that the machine doesn't keep Rachel's card? **[3 marks]**

Chapter summary

In this chapter you have learned how to

- understand and use the fact that the sum of the probabilities of all mutually exclusive outcomes is 1 **D**
- predict the likely number of successful events given the probability of any outcome and the number of trials or experiments **D**
- estimate probabilities from experimental data **C**
- calculate the probability of two independent events happening at the same time **C**
- use and understand tree diagrams in simple contexts **B**
- use the 'and' and 'or' rules in tree diagrams **A** **A***
- use and understand tree diagrams with two or more events for conditional probabilities **A***

6 Cumulative frequency

This chapter is about analysing information.

These runners are crossing the Reichbruecke bridge in the Vienna City Marathon. Analysing data of the runners' times will show what percentage took between 2 hours 30 minutes and 3 hours 30 minutes.

Objectives

This chapter will show you how to

- compile a cumulative frequency table for continuous (grouped) data **B**
- draw cumulative frequency diagrams to find the median and quartiles of grouped data **B**
- use cumulative frequency diagrams to analyse other features of grouped data **B**
- draw box plots from a cumulative frequency diagram **B**
- use cumulative frequency diagrams and box plots to compare data and draw conclusions **B**

Before you start this chapter

1 a What must you do before you can find the median of these numbers?

8 31 4 20 49 12 6 34 5 16 22 9 43

b What is the median?

c The number 49 is removed. What happens to the median?

2 Write down six different numbers with a median of 7.5 and a range of 7.

3 This frequency table shows the scores of a group of students who took a mental maths test.

Test score	2	3	4	5	6	7	8	9	10
Frequency	4	2	3	2	5	8	6	5	1

a What is the range of these marks? b Work out the median test score.

6.1 Cumulative frequency

Keywords
lower quartile, upper quartile, inter-quartile range, median, cumulative frequency

Why learn this?

You can use a cumulative frequency diagram to find extra features of your data.

Objectives

B Compile a cumulative frequency table and draw cumulative frequency diagrams

B Use cumulative frequency diagrams to analyse data

Skills check

1 Look at this list of numbers.

3 4 4 6 9 11 11 13 15 18 20 23 24 24 27

Which number is

a half way along the list

b one quarter of the way along the list

c three quarters of the way along the list?

2 Look at this set of axes.
What does one small square represent on

a the x-axis

b the y-axis?

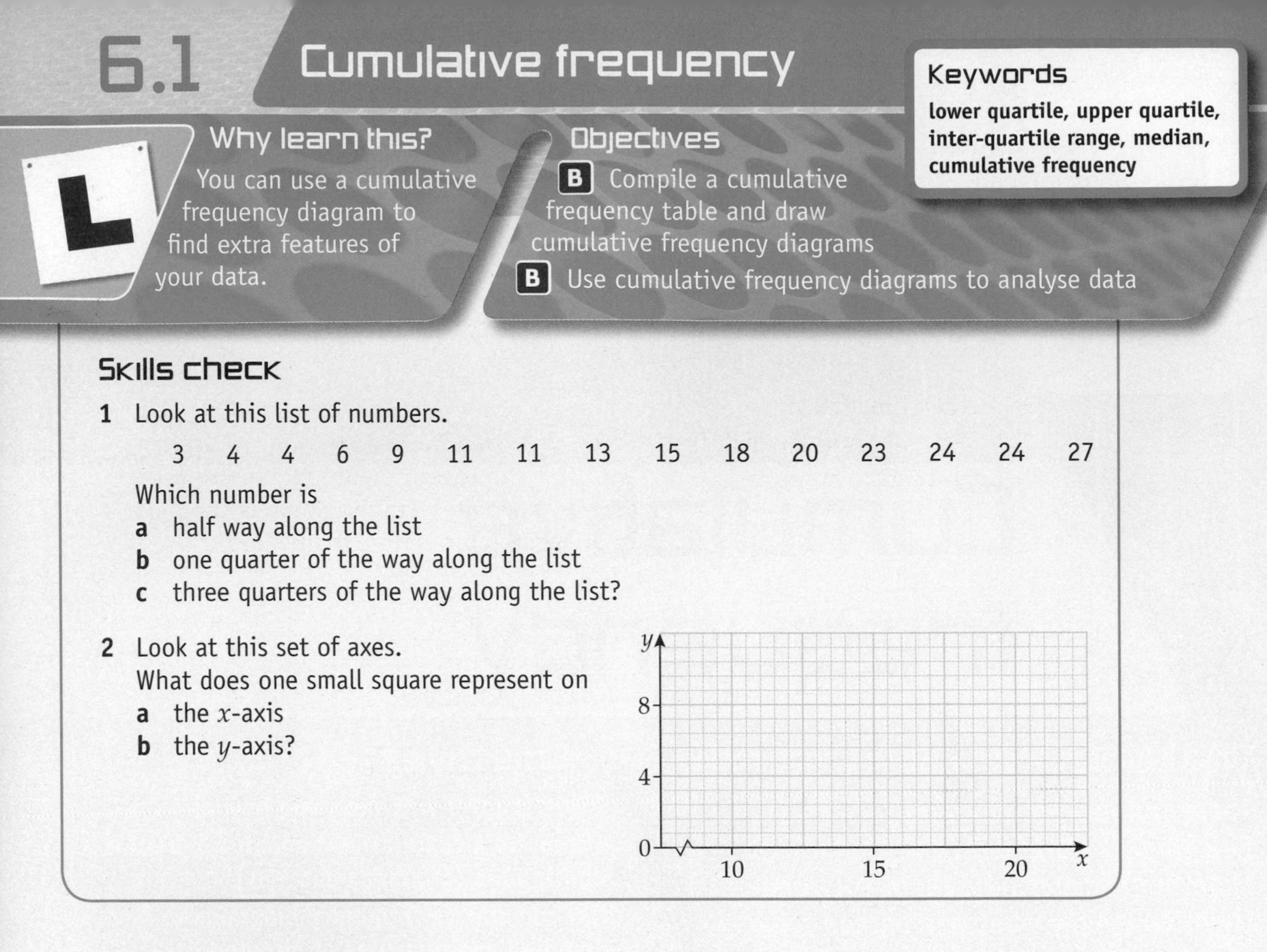

Drawing and using a cumulative frequency diagram for grouped data

The **lower quartile** is the data item one quarter of the way along the list when the data is in ascending order.

The **upper quartile** is the data item three quarters of the way along the list when the data is in ascending order.

The **inter-quartile range** is the difference between the values of the upper and lower quartiles of the data.

Inter-quartile range = upper quartile − lower quartile

It is a single value and is often used instead of the range to indicate the spread of the data.

When data is grouped into class intervals you can work out which class interval contains the **median** or quartiles, but you cannot give the exact values (see Chapter 1). Continuous data is always grouped into class intervals in a frequency table.

Cumulative frequency is a running total of the frequencies.

You can use a cumulative frequency diagram to estimate the median, the lower quartile and the upper quartile of the data. They can only be estimates because when data is grouped some of the detail of the original data is lost.

In cumulative frequency questions there is usually a large amount of data.

If the number of data items, n, is large ($n \geqslant 25$) you can use these formulae for locating the positions of the median and quartiles.

Median $= \frac{n}{2}$th value Lower quartile $= \frac{n}{4}$th value Upper quartile $= \frac{3n}{4}$th value

Example 1

This frequency table shows the times taken by some students to work out the answer to a mathematical puzzle.

Time taken, t (minutes)	$0 < t \leqslant 1$	$1 < t \leqslant 2$	$2 < t \leqslant 3$	$3 < t \leqslant 4$	$4 < t \leqslant 5$	$5 < t \leqslant 6$	$6 < t \leqslant 7$
Frequency	3	4	7	11	9	4	2

a Draw a cumulative frequency diagram to illustrate this data.

b Use the cumulative frequency diagram to estimate

  i the median time taken

  ii the lower and upper quartiles

  iii the inter-quartile range.

First create a cumulative frequency table. Add an extra row or column to the frequency table to calculate a running total of the frequencies (cumulative frequency).

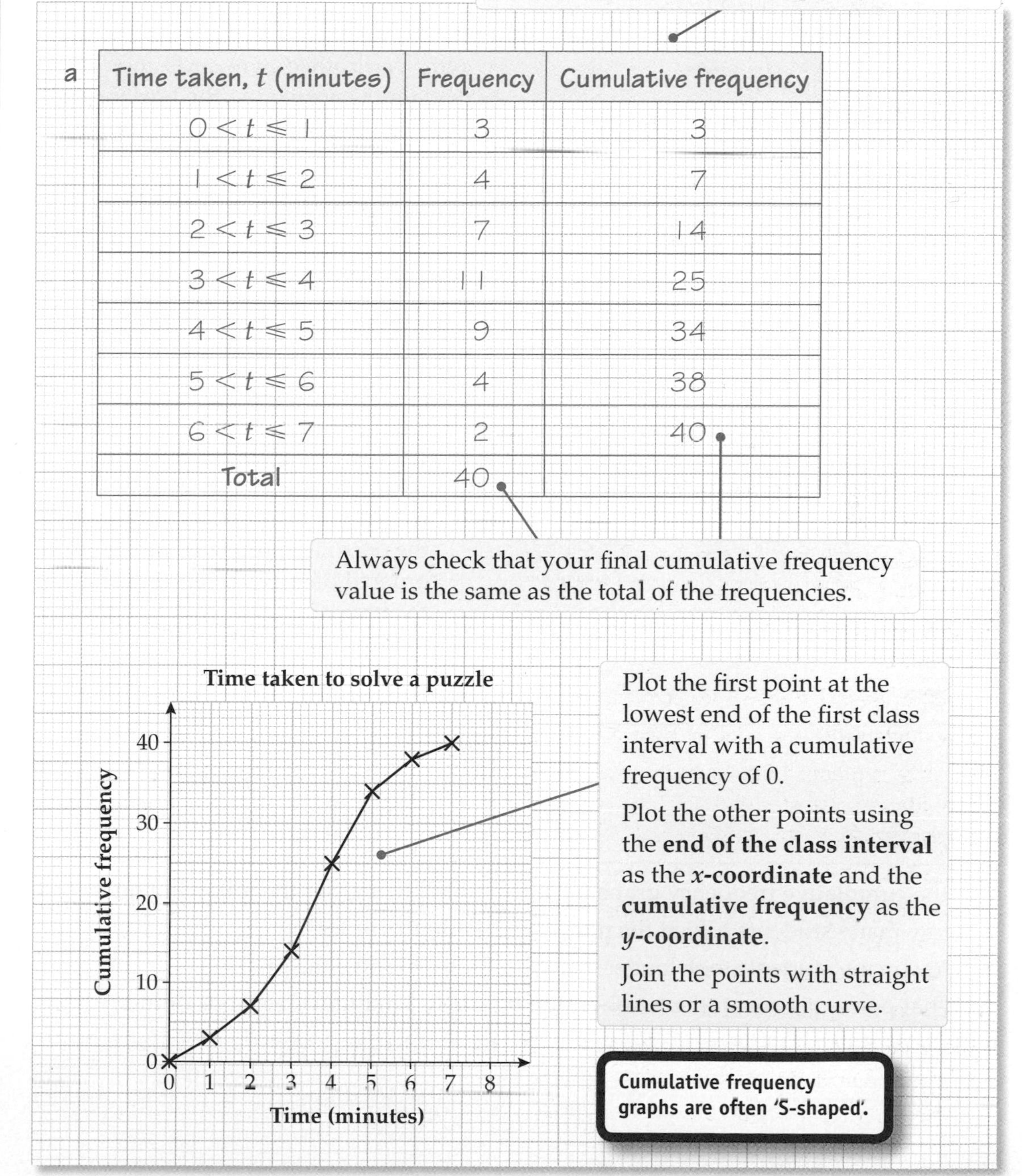

a

Time taken, t (minutes)	Frequency	Cumulative frequency
$0 < t \leqslant 1$	3	3
$1 < t \leqslant 2$	4	7
$2 < t \leqslant 3$	7	14
$3 < t \leqslant 4$	11	25
$4 < t \leqslant 5$	9	34
$5 < t \leqslant 6$	4	38
$6 < t \leqslant 7$	2	40
Total	40	

Always check that your final cumulative frequency value is the same as the total of the frequencies.
Time taken to solve a puzzle
Cumulative frequency
0
10
20
30
40
0 1 2 3 4 5 6 7 8
Time (minutes)
Plot the first point at the lowest end of the first class interval with a cumulative frequency of 0.
Plot the other points using the **end of the class interval** as the ***x*-coordinate** and the **cumulative frequency** as the ***y*-coordinate**.
Join the points with straight lines or a smooth curve.
Cumulative frequency graphs are often 'S-shaped'.

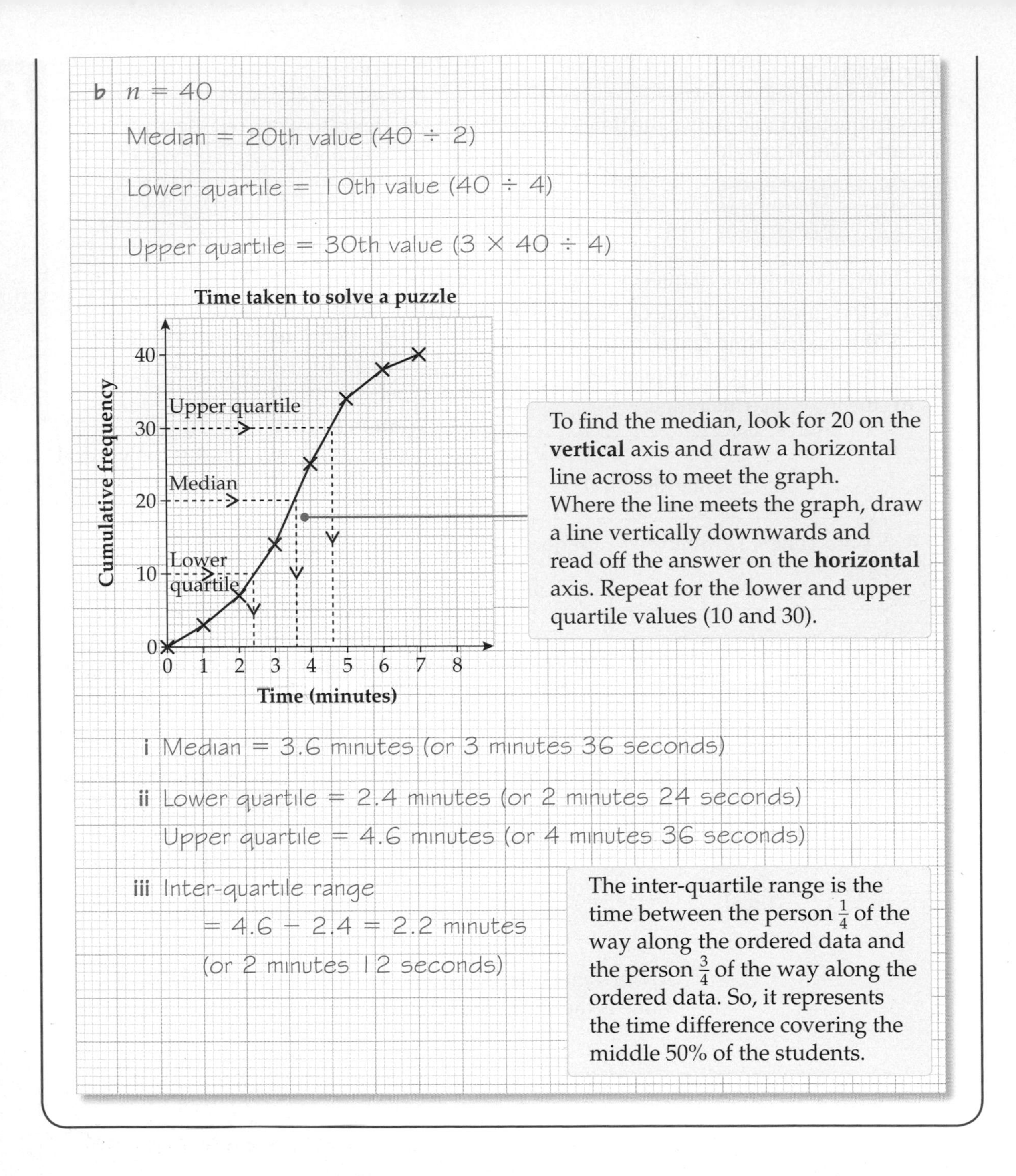

b $n = 40$

Median = 20th value (40 ÷ 2)

Lower quartile = 10th value (40 ÷ 4)

Upper quartile = 30th value (3 × 40 ÷ 4)

To find the median, look for 20 on the **vertical** axis and draw a horizontal line across to meet the graph. Where the line meets the graph, draw a line vertically downwards and read off the answer on the **horizontal** axis. Repeat for the lower and upper quartile values (10 and 30).

i Median = 3.6 minutes (or 3 minutes 36 seconds)

ii Lower quartile = 2.4 minutes (or 2 minutes 24 seconds)
Upper quartile = 4.6 minutes (or 4 minutes 36 seconds)

iii Inter-quartile range
= 4.6 − 2.4 = 2.2 minutes
(or 2 minutes 12 seconds)

The inter-quartile range is the time between the person $\frac{1}{4}$ of the way along the ordered data and the person $\frac{3}{4}$ of the way along the ordered data. So, it represents the time difference covering the middle 50% of the students.

Example 2

Use the cumulative frequency graph from Example 1 to answer the following questions.

a How many students solved the puzzle in 2.4 minutes or less?

b How many students took longer than $5\frac{1}{2}$ minutes to solve the puzzle?

c What percentage of students solved the puzzle in 1 minute 30 seconds or less?

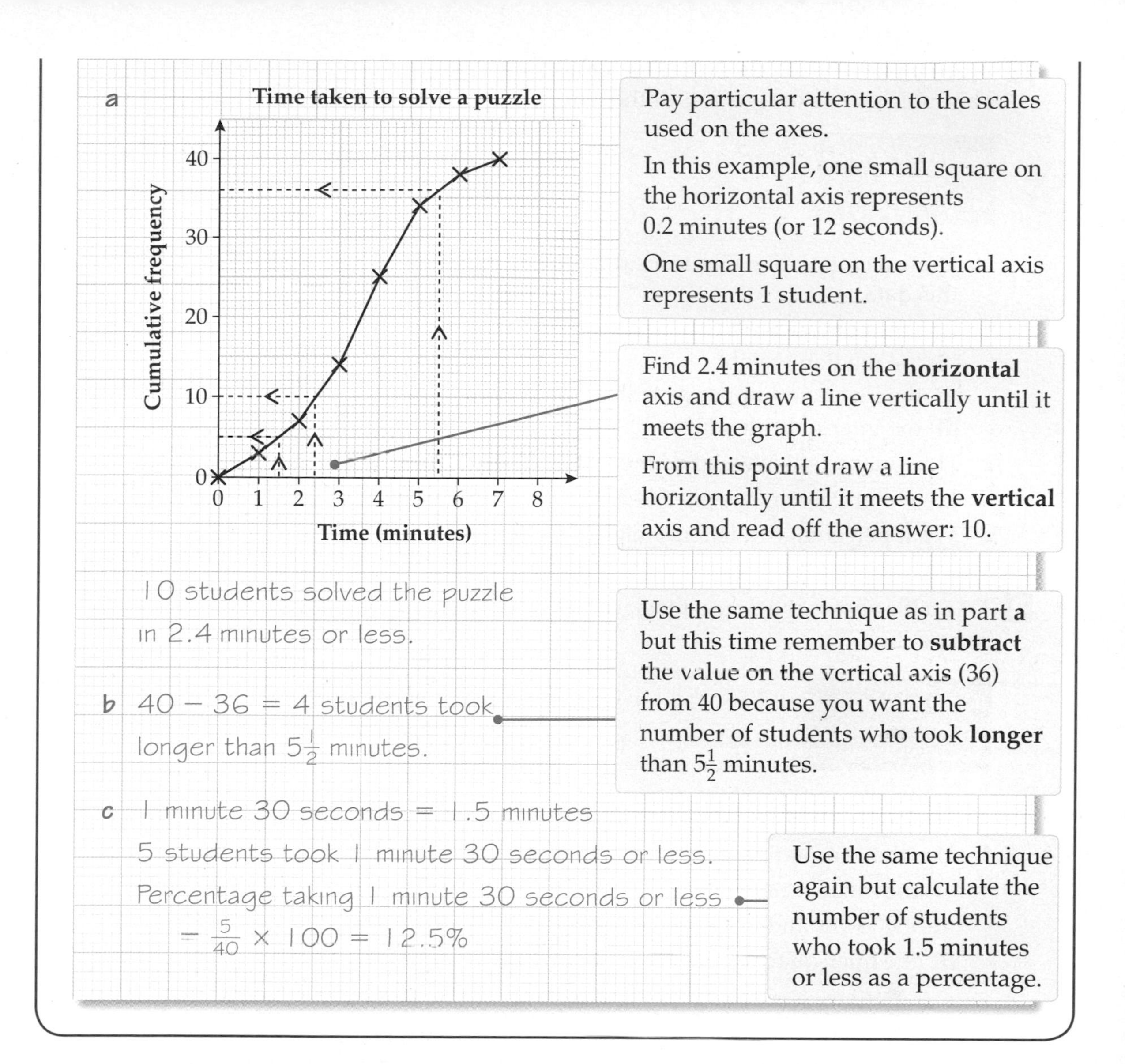

a
Time taken to solve a puzzle
Cumulative frequency
Time (minutes)
0
1
2
3
4
5
6
7
8
10
20
30
40
Pay particular attention to the scales used on the axes.
In this example, one small square on the horizontal axis represents 0.2 minutes (or 12 seconds).
One small square on the vertical axis represents 1 student.
Find 2.4 minutes on the horizontal axis and draw a line vertically until it meets the graph.
From this point draw a line horizontally until it meets the vertical axis and read off the answer: 10.
10 students solved the puzzle in 2.4 minutes or less.
Use the same technique as in part a but this time remember to subtract the value on the vertical axis (36) from 40 because you want the number of students who took longer than 5½ minutes.
b 40 − 36 = 4 students took longer than 5½ minutes.
c 1 minute 30 seconds = 1.5 minutes
5 students took 1 minute 30 seconds or less.
Percentage taking 1 minute 30 seconds or less
= 5/40 × 100 = 12.5%
Use the same technique again but calculate the number of students who took 1.5 minutes or less as a percentage.

Exercise 6A

1 The times spent by a group of students using their mobile phones in one day are shown in the table.

Time, t (minutes)	$0 < t \leqslant 10$	$10 < t \leqslant 20$	$20 < t \leqslant 30$	$30 < t \leqslant 40$	$40 < t \leqslant 50$	$50 < t \leqslant 60$
Frequency	6	10	21	46	11	6

a Draw a cumulative frequency diagram to illustrate this data.

b Use the cumulative frequency diagram to estimate

i the median time

ii the lower and upper quartiles

iii the inter-quartile range.

c How many students used their mobile for 16 minutes or less?

d How many students used their mobile for more than 48 minutes?

e How many students used their mobile for between 27 and 36 minutes?

B

B

2 Grace took some letters to the Post Office.
This table shows the weights of the letters.

Weight, g (grams)	$20 < g \leqslant 30$	$30 < g \leqslant 40$	$40 < g \leqslant 50$	$50 < g \leqslant 60$	$60 < g \leqslant 70$	$70 < g \leqslant 80$	$80 < g \leqslant 90$
Frequency	8	14	17	8	6	4	3

You can start the horizontal axis at 20 but the vertical axis must start at zero.

a Draw a cumulative frequency diagram to illustrate this data.

b Use the cumulative frequency diagram to estimate
  i the median weight
  ii the lower and upper quartiles
  iii the inter-quartile range.

c How many letters weighed more than 75 grams?

d How many letters weighed between 62 and 84 grams?

e What percentage of letters weighed 41 grams or less?

3 This table shows the times it took competitors to complete a crossword in a competition.

Time, t (minutes)	$10 < t \leqslant 15$	$15 < t \leqslant 20$	$20 < t \leqslant 25$	$25 < t \leqslant 30$	$30 < t \leqslant 35$	$35 < t \leqslant 40$
Frequency	12	15	27	50	13	3

a Draw a cumulative frequency diagram to illustrate this data.

b Use the cumulative frequency diagram to estimate
  i the median time taken
  ii the lower and upper quartiles
  iii the inter-quartile range.

c Competitors were graded A, B, C or D according to the time they took to complete the crossword.

Grade	A	B	C	D
Time, t (minutes)	$t \leqslant 13$	$13 < t \leqslant 22$	$22 < t \leqslant 33$	$33 < t$

How many competitors achieved each of the grades?

4 In France, speed limits are given in kilometres per hour.
This table shows the speeds of 140 motorists on a French toll motorway.

Speed, v (km/h)	$65 < v \leqslant 75$	$75 < v \leqslant 85$	$85 < v \leqslant 95$	$95 < v \leqslant 105$
Frequency	14	18	10	16

Speed, v (km/h)	$105 < v \leqslant 115$	$115 < v \leqslant 125$	$125 < v \leqslant 135$	$135 < v \leqslant 145$
Frequency	38	26	12	6

a Draw a cumulative frequency diagram to illustrate this data.

b Use the cumulative frequency diagram to estimate

i the median speed

ii the lower and upper quartiles

iii the inter-quartile range.

c The speed limit on French toll motorways is 130 km/h in dry weather and 110 km/h in wet weather.

i Assume the weather is dry. Use your cumulative frequency diagram to find out how many drivers are breaking the speed limit.

ii How many more drivers would be breaking the speed limit if this diagram represented data recorded in wet weather?

d To convert miles per hour to kilometres per hour you multiply by 1.6.
In Britain, the speed limit on motorways is 70 mph.
How many of the French drivers were travelling at a speed greater than the speed limit on British motorways?

B

AO2

6.2 Box plots

Keywords
box plot, whiskers

Why learn this?
A box plot is a quick method of looking at the important features of a frequency distribution.

Objectives
B Learn how to draw a box plot from a cumulative frequency diagram

Skills check

1 Look at this list of numbers.

3 4 4 6 9 11 11 13
15 18 20 23 24 24 27

Write down the median, the lower and upper quartiles, the inter-quartile range and the range.

Drawing a box plot from a cumulative frequency diagram

A **box plot** (sometimes called a box-and-whisker diagram) shows the range, the lower and upper quartiles (and hence the inter-quartile range) and the median of a set of data.

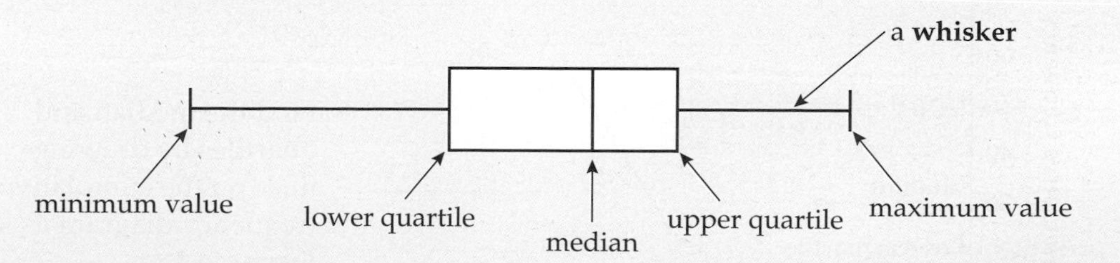

Box plots are usually drawn on a scale. Use the same scale as the cumulative frequency diagram from which the data comes.

A box plot has advantages and disadvantages. It is easy to read and provides information at a glance, but it gives no details about the number of data items in the distribution.

B Example 3

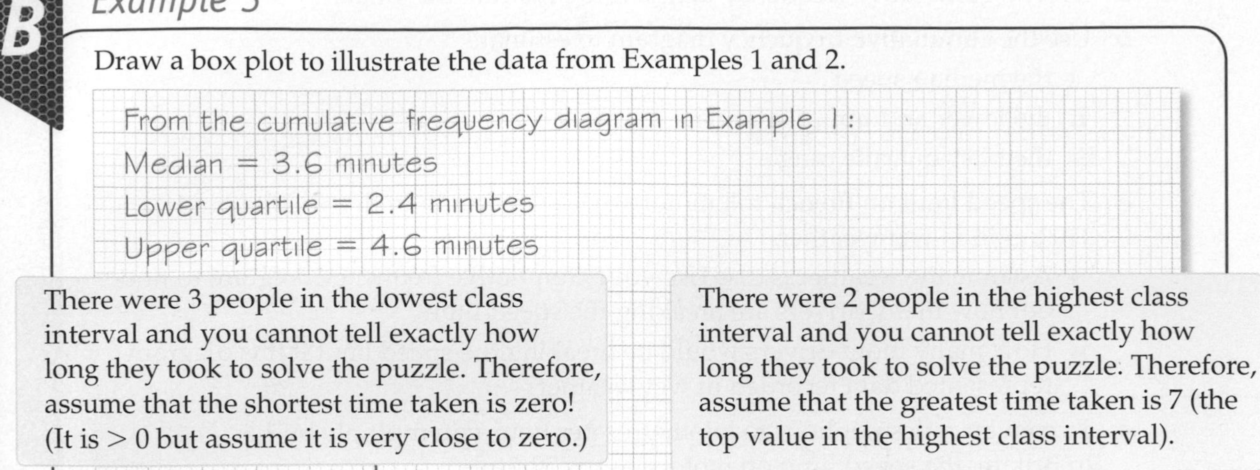

Draw a box plot to illustrate the data from Examples 1 and 2.

From the cumulative frequency diagram in Example 1:

Median = 3.6 minutes

Lower quartile = 2.4 minutes

Upper quartile = 4.6 minutes

There were 3 people in the lowest class interval and you cannot tell exactly how long they took to solve the puzzle. Therefore, assume that the shortest time taken is zero! (It is > 0 but assume it is very close to zero.)

There were 2 people in the highest class interval and you cannot tell exactly how long they took to solve the puzzle. Therefore, assume that the greatest time taken is 7 (the top value in the highest class interval).

This is the box plot:

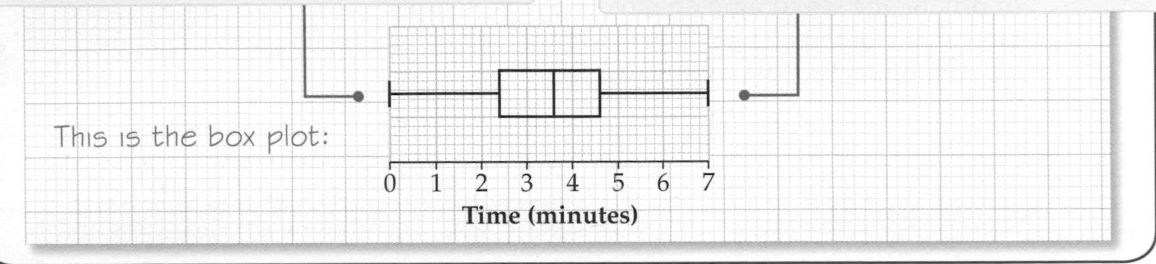

B Example 4

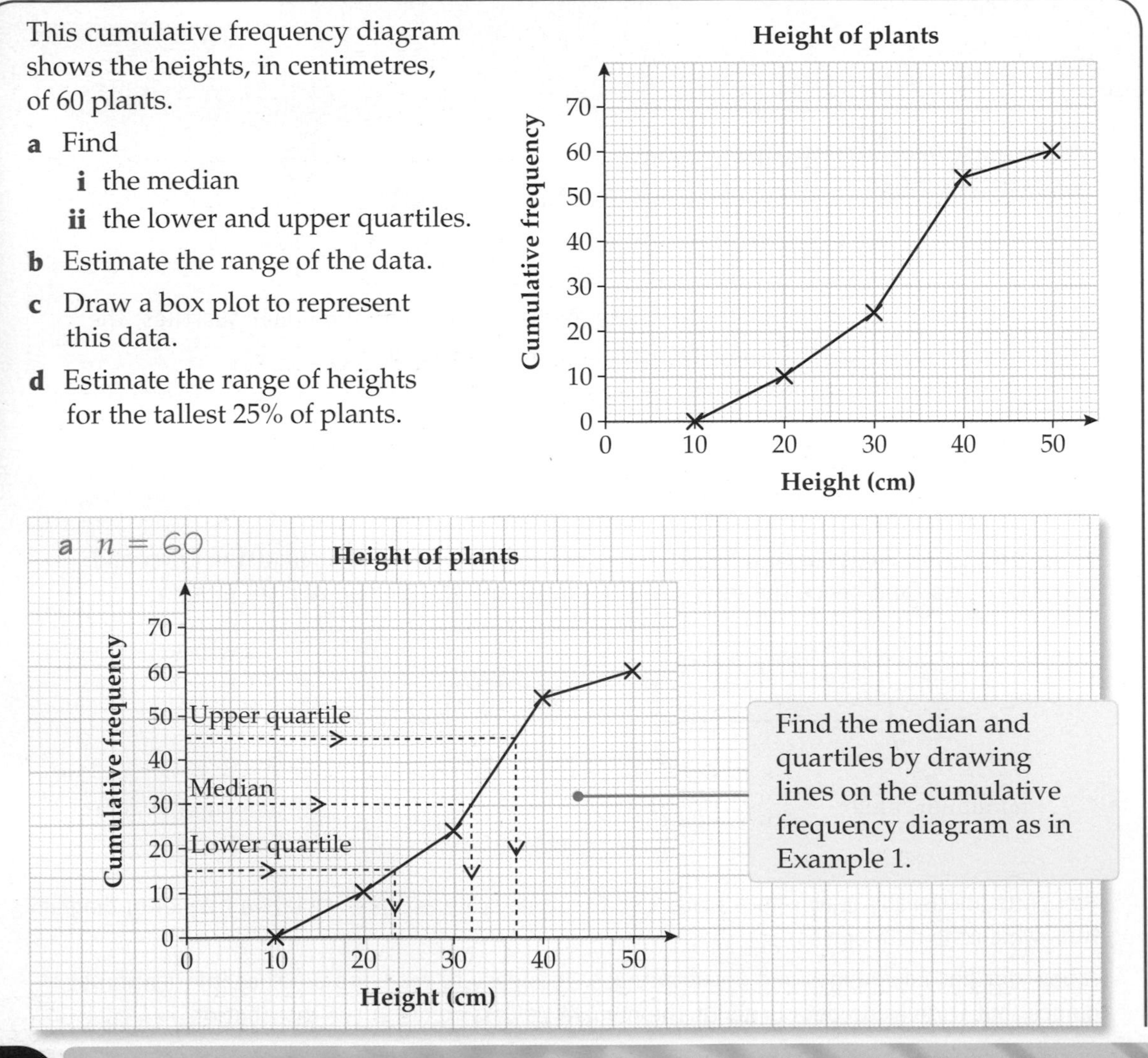

This cumulative frequency diagram shows the heights, in centimetres, of 60 plants.

a Find

  i the median

  ii the lower and upper quartiles.

b Estimate the range of the data.

c Draw a box plot to represent this data.

d Estimate the range of heights for the tallest 25% of plants.

a $n = 60$

Find the median and quartiles by drawing lines on the cumulative frequency diagram as in Example 1.

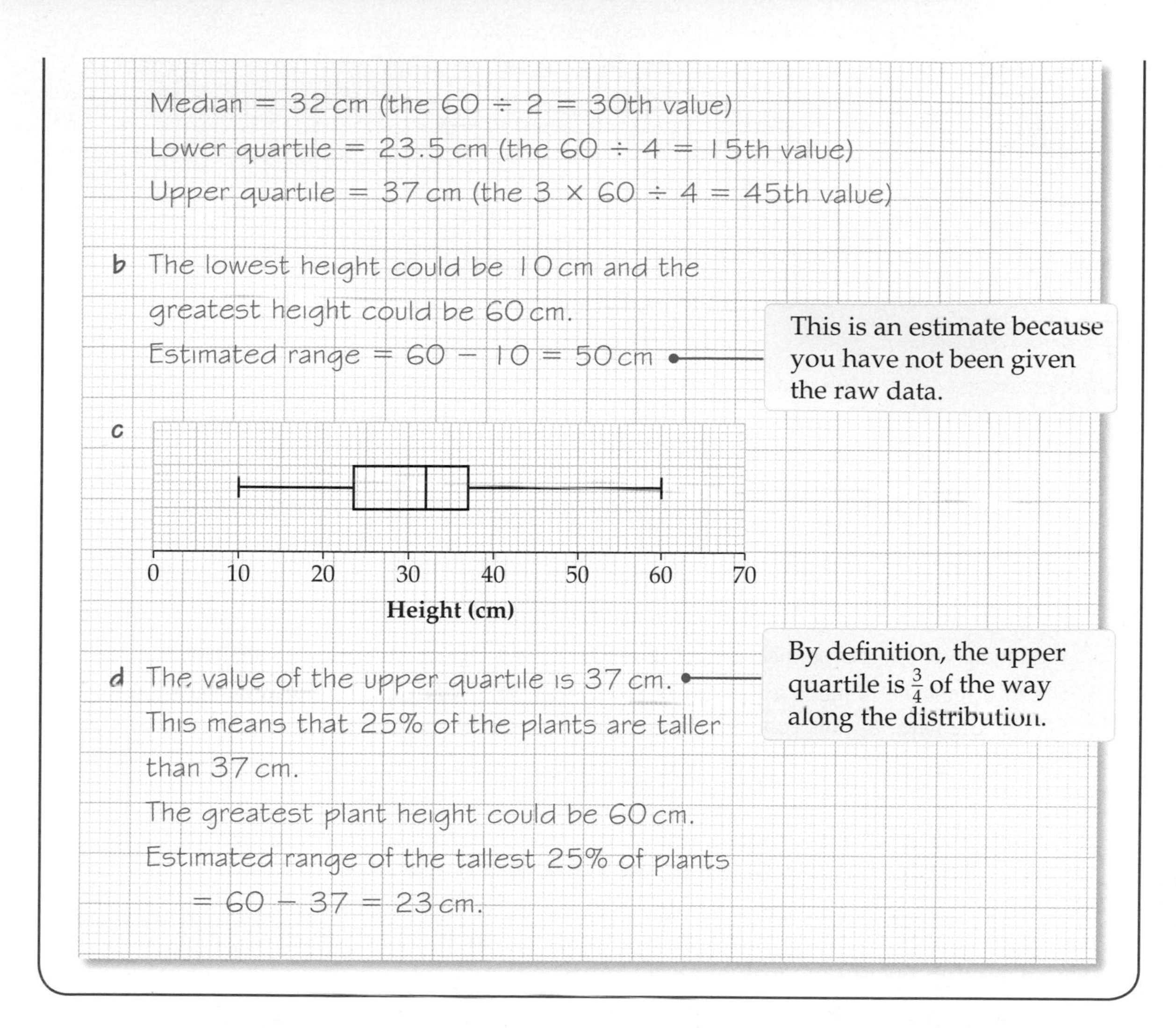
Median = 32 cm (the 60 ÷ 2 = 30th value)
Lower quartile = 23.5 cm (the 60 ÷ 4 = 15th value)
Upper quartile = 37 cm (the 3 × 60 ÷ 4 = 45th value)

b The lowest height could be 10 cm and the greatest height could be 60 cm.
Estimated range = 60 − 10 = 50 cm

This is an estimate because you have not been given the raw data.

c

d The value of the upper quartile is 37 cm.
This means that 25% of the plants are taller than 37 cm.
The greatest plant height could be 60 cm.
Estimated range of the tallest 25% of plants
= 60 − 37 = 23 cm.

By definition, the upper quartile is $\frac{3}{4}$ of the way along the distribution.

Exercise 6B

1 **a** to **d** Draw a box plot for each of the four questions in Exercise 6A. Remember to clearly label the scale on the x-axis.

2 For each of these box plots, work out

i the median

ii the inter-quartile range

iii the range.

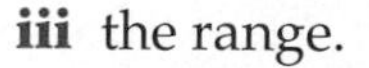

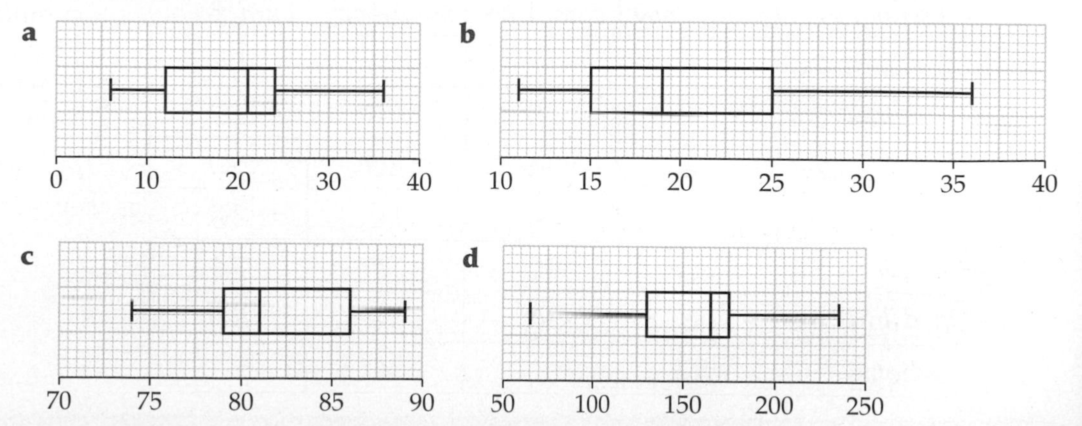

B

6.3 Comparing data sets and drawing conclusions

Why learn this?

This topic often comes up in the exam.

Objectives

B Use cumulative frequency diagrams and box plots to compare data and draw conclusions

Skills check

1 Look at the two lists of numbers.

A	3	7	9	13	15	20	21
B	7	8	11	12	13	17	20

a Which list has the higher median?

b Which list has the larger inter-quartile range?

Comparing data sets and drawing conclusions

When comparing frequency distributions remember that

- the median is one of the measures of 'average'. Always look to compare the medians.
- the inter-quartile range is a measure of consistency and tells you the spread of the middle 50% of the data. Always comment on the size of the inter-quartile range.

B

Example 5

Eighty of each of two types of battery, Superamp and Powerplus, are tested.

The minimum life of each type was 16 hours.
The maximum life of Superamp was 36 hours.
The maximum life of Powerplus was 40 hours.

This frequency table shows the test results for each type of battery.

Battery life, h (hours)	Superamp	Powerplus
$16 < h \leq 20$	7	14
$20 < h \leq 24$	11	16
$24 < h \leq 28$	16	20
$28 < h \leq 32$	40	16
$32 < h \leq 36$	6	8
$36 < h \leq 40$	0	6
Total	80	80

a Draw a cumulative frequency diagram for each type of battery.

b Estimate the median and the inter-quartile range for each type of battery.

c Draw a box plot for each type of battery.

AO3 **d** Which type of battery is the better buy? Explain your answer.

a

Battery life, h (hours)	Superamp frequency	Superamp cum. freq.	Powerplus frequency	Powerplus cum. freq.
$16 < h \leq 20$	7	7	14	14
$20 < h \leq 24$	11	18	16	30
$24 < h \leq 28$	16	34	20	50
$28 < h \leq 32$	40	74	16	66
$32 < h \leq 36$	6	80	8	74
$36 < h \leq 40$	0	80	6	80
Total	80		80	

Both sets of cumulative frequencies can be written in the same table.

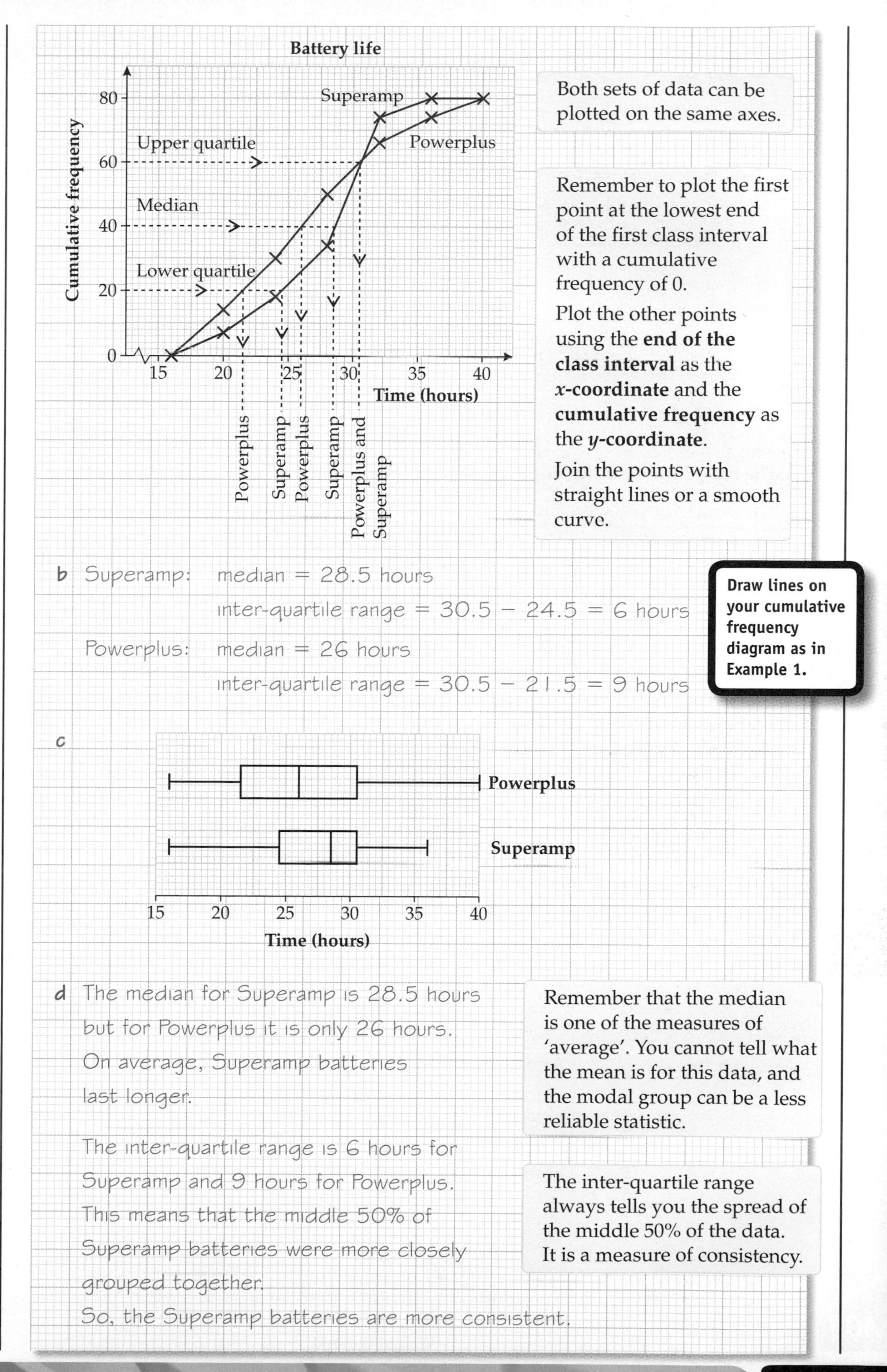

Both sets of data can be plotted on the same axes.

Remember to plot the first point at the lowest end of the first class interval with a cumulative frequency of 0.

Plot the other points using the **end of the class interval** as the ***x*-coordinate** and the **cumulative frequency** as the ***y*-coordinate.**

Join the points with straight lines or a smooth curve.

b Superamp: median = 28.5 hours

inter-quartile range = 30.5 − 24.5 = 6 hours

Powerplus: median = 26 hours

inter-quartile range = 30.5 − 21.5 = 9 hours

Draw lines on your cumulative frequency diagram as in Example 1.

c

d The median for Superamp is 28.5 hours but for Powerplus it is only 26 hours. On average, Superamp batteries last longer.

The inter-quartile range is 6 hours for Superamp and 9 hours for Powerplus. This means that the middle 50% of Superamp batteries were more closely grouped together.

So, the Superamp batteries are more consistent.

Remember that the median is one of the measures of 'average'. You cannot tell what the mean is for this data, and the modal group can be a less reliable statistic.

The inter-quartile range always tells you the spread of the middle 50% of the data. It is a measure of consistency.

Comparing the lower quartiles,
Superamp = 24.5 hours and
Powerplus = 21.5 hours.
This means that 60 out of 80 (75%) of the Superamp batteries lasted over 24.5 hours.
Using the cumulative frequency diagram you can see that the number of Powerplus batteries lasting over 24.5 hours is only about 48.

Although some of the Powerplus batteries lasted more than 36 hours, overall they are less reliable.
The Superamp batteries are the better buy.

The lower quartile is the point 25% into the data, so 75% of the data lies above it.

Exercise 6C

B

1 The box plots show the lengths jumped by 50 boys and 40 girls in the triple jump.

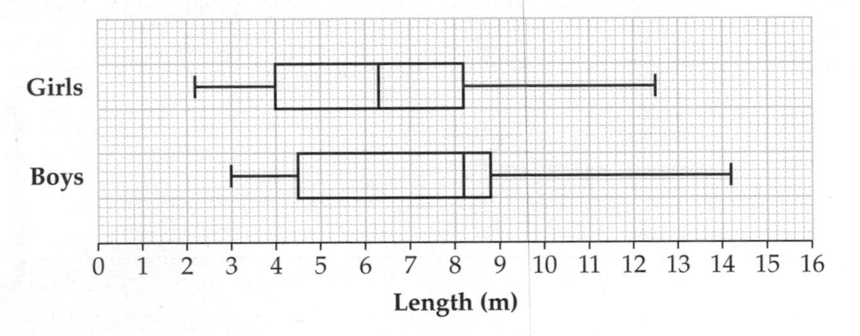

a What is the median length jumped by the girls?
b What was the longest jump made by a boy?
c Give **two** differences between the lengths jumped by the boys and the lengths jumped by the girls.
d How many students jumped more than 8.2 metres?

AO2

B

2 Potatoes of variety X and Y are weighed.
The frequency distributions of the two varieties are shown on the diagram.

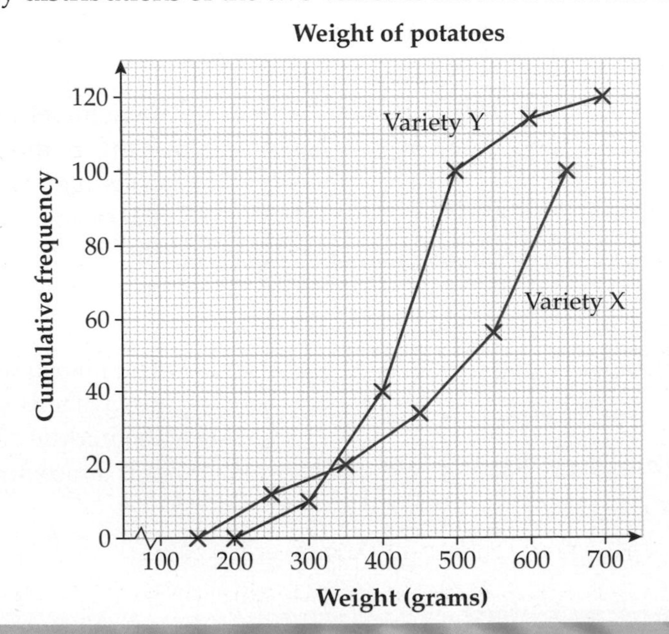

AO3

a How many potatoes of variety X were weighed?

b Estimate the minimum and maximum weight of potatoes from variety X.

c Work out the median and the inter-quartile range for variety X.

d How many potatoes of variety Y were weighed?

e Estimate the minimum and maximum weight of potatoes from variety Y.

f Work out the median and the inter-quartile range for variety Y.

g Draw box plots for both varieties, using a scale of 100 to 800 on the horizontal axis.
Draw one box plot directly beneath the other.

h Compare the two varieties, giving reasons for any statements you make.

B

AO3

Review exercise

B

1 A group of physics university students and a group of history university students were given 20 long multiplication calculations to work out.
The times taken (in minutes) by the physics students are shown in the box plot.

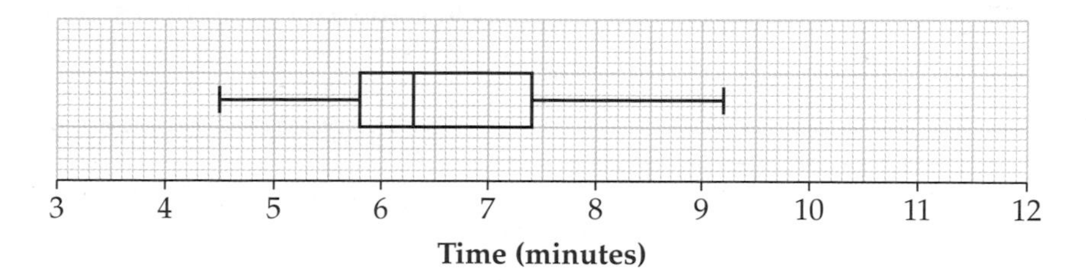

The history students had a shortest time of 4.2 minutes, a longest time of 10.5 minutes, a median of 7.2 minutes, a lower quartile of 6 minutes and an upper quartile of 8.5 minutes.

a Copy the box plot for the physics students and on the same scale draw the box plot for the history students. **[2 marks]**

b Comment on the differences between the two frequency distributions. **[2 marks]**

2 Joe is planning his summer holidays.
He looks at data about two holiday destinations, A and B.
The box plots show the temperatures from May to September.

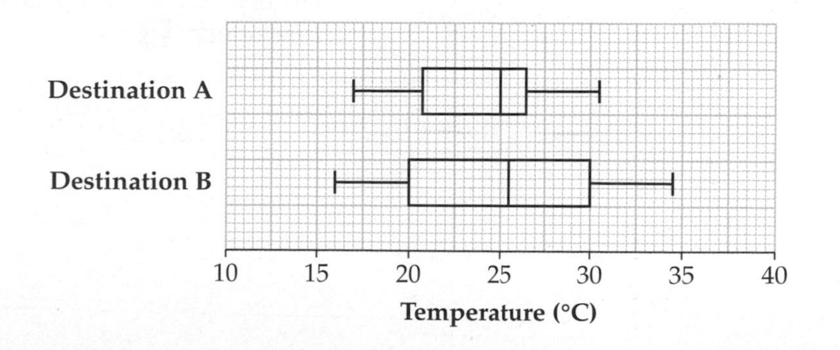

a Give **two** reasons why Joe should go to destination A. **[2 marks]**

b Give **two** reasons why Joe should go to destination B. **[2 marks]**

AO2

B

3 Two products to improve the growth of tomato plants are being tested, *Lottagrow* and *Supertom*.

The heights of plants are measured after treatment by each product.
These frequency tables show the results.

Lottagrow

Height, h (cm)	Frequency
$50 < h \leqslant 55$	4
$55 < h \leqslant 60$	20
$60 < h \leqslant 65$	30
$65 < h \leqslant 70$	13
$70 < h \leqslant 75$	15
$75 < h \leqslant 80$	10

Supertom

Height, h (cm)	Frequency
$50 < h \leqslant 55$	2
$55 < h \leqslant 60$	13
$60 < h \leqslant 65$	25
$65 < h \leqslant 70$	28
$70 < h \leqslant 75$	18
$75 < h \leqslant 80$	2

The *Lottagrow* tomatoes had a minimum height of 50 cm and a maximum height of 80 cm.

The *Supertom* tomatoes had a minimum height of 53 cm and a maximum height of 77 cm.

a Draw a cumulative frequency diagram for each product. **(6 marks)**

> **Use a scale of 2 cm for 5 units on the x-axis, with values from 50 to 80.**
> **Use a scale of 2 cm for 20 units on the y-axis with values from 0 to 100.**

b Draw a box plot for each product. **[4 marks]**

c Estimate the median and inter-quartile range for each product. **[6 marks]**

AO3

d Which product would you recommend?
Give at least **two** reasons for your choice. **[2 marks]**

Chapter summary

In this chapter you have learned how to

- compile a cumulative frequency table and draw cumulative frequency diagrams **B**
- use cumulative frequency diagrams to analyse data **B**
- draw a box plot from a cumulative frequency diagram **B**
- use cumulative frequency diagrams and box plots to compare data and draw conclusions **B**

7

Ratio and proportion

This chapter shows you how to use ratio and proportion to solve problems.

Whatever fruits, ice and yogurt you use, they need to be in the correct ratio to make a delicious smoothie.

Objectives

This chapter will show you how to

- use ratio notation, including reduction to its simplest form and its various links to fraction notation **D**
- solve word problems involving ratio and proportion, including using informal strategies and the unitary method of solution **D** **C**
- divide a quantity in a given ratio **D** **C**
- solve problems involving quantities that vary in direct or inverse proportions **D** **C** **B**
- set up and use equations to solve word and other problems involving direct proportion or inverse proportion **A**

Before you start this chapter

1 What is the highest common factor of
 a 4 and 12 b 9 and 15
 c 36 and 48?

2 Convert
 a 4 km to metres b 20 mm to cm
 c 3.5 kg to grams d 2750 m*l* to litres
 e 240 minutes to hours.

3 Work out
 a $225 \div 5$ b £17.60 ÷ 5 c £1.75 ÷ 7

4 Work out
 a $\frac{1}{9} + \frac{4}{9}$ b $1 + \frac{3}{4}$ c $1 - \frac{3}{4}$

7.1 Using ratios

Keywords
ratio

Why learn this?
You need ratios to adapt recipes for different numbers of people.

Objectives
D Use a ratio in practical situations

Skills check

1 Simplify these ratios.

a 4 : 6 b 60 : 12 c 40 m*l* : 15 m*l*
d 25p : £2 e 50 minutes : 1 hour f $1\frac{1}{2}:3$
g 2.5 m : 3.5 m h 1.6 kg : 0.4 kg

2 A map has a scale of 1 : 50 000.

a A road on the map is 12 cm long. How long is the road in real life? Give your answer in kilometres.

b The real-life distance between two towns is 16 km. What is the distance between the two towns on the map?

Using ratios

A **ratio** compares two or more quantities.

In a cranberry and raspberry smoothie, you use 50 g of cranberries and 100 g of raspberries. You can write this using the ratio symbol.

cranberries : raspberries = 50 : 100

Ratios are often used in real-life situations, for example on maps and scale drawings, when cooking and entertaining, and when describing the steepness of a hill or wheelchair ramp.

D

Example 1

To make 2 berry smoothies you need 100 g of raspberries and 50 g of blackberries.
How much of each ingredient do you need to make

a 6 smoothies b 5 smoothies?

a For 6 smoothies: — The recipe makes 2 smoothies. 6 smoothies is 3 lots of 2. Multiply all the quantities by 3.

3 × 100 g = 300 g raspberries

3 × 50 g = 150 g blackberries

b For 1 smoothie: — First work out the quantities for 1 smoothie.

100 g ÷ 2 = 50 g raspberries

50 g ÷ 2 = 25 g blackberries

For 5 smoothies: — Then multiply the quantities for 1 smoothie by 5.

50 g × 5 = 250 g raspberries

25 g × 5 = 125 g blackberries

Exercise 7A

D

1 This recipe makes 10 pancakes.

500 g flour
2 eggs
300 m*l* milk

Work out the quantities for **a** 30 pancakes **b** 5 pancakes **c** 15 pancakes.

2 It takes a photocopier 20 seconds to produce 15 copies.

a How long will it take the photocopier to produce 21 copies?

b How many copies could it produce in 44 seconds?

D

AO2

3 Here is a recipe for potato layer bake.

I have 1.75 kg of potatoes and I want to use them all.

a How much of each of the other ingredients should I use?

b How many people will my potato layer bake serve?

Potato layer bake (serves 4)
1 kg potatoes
100 g onion
50 g butter
180 g cheddar cheese
300 m*l* milk

D

AO3

4 Tony is doing some DIY. He already has a 5 kg sack of cement.
He is going to use half the cement for some general building work (above ground).
The other half will be used to lay some paving in the garden.
Tony finds this table on the different mixes of concrete on the internet.

Concrete mixes	Ballast	Sand	Cement
General building (above ground)	—	5 kg	1 kg
General building (below ground)	—	3 kg	1 kg
Internal walls	—	8 kg	1 kg
Driveways	5 kg	—	1 kg
Footpaths	3.25 kg	—	1 kg
Paving	—	3 kg	1 kg

How much sand does Tony need to order if he is to use up all of his cement?

7.2 Ratios and fractions

Why learn this?

If you know the ratio of men : women expected at a festival, you can work out the fraction of the toilets that need to be male and female.

Objectives

D Write a ratio as a fraction

D **C** Use a ratio to find one quantity when the other is known

Skills check

1 Work out

a $\frac{1}{6} + \frac{1}{6}$ **b** $\frac{3}{8} + \frac{1}{8}$ **c** $\frac{9}{11} - \frac{7}{11}$

2 True or false?

a $\frac{3}{5} + \frac{2}{5} = 1$ **b** $1 - \frac{3}{13} = \frac{9}{13}$ **c** $\frac{1}{9} + \frac{4}{9} + \frac{5}{9} = 1$

Writing a ratio as a fraction

You can write a ratio as a fraction by simply changing the whole numbers in the ratio into fractions with the same denominator.

In a class of students, if the ratio of boys : girls is 2 : 3, then $\frac{2}{5}$ of the class are boys and $\frac{3}{5}$ of the class are girls.

The denominator is found by adding the whole numbers in the ratio. In this case, $2 + 3 = 5$.

Thinking of ratios as fractions can help when solving ratio problems.

D C

Example 2

a Trish and Del share a pie in the ratio 1 : 3.
What fraction of the pie does Trish eat?
What fraction of the pie does Del eat?

b A supermarket sells cheese pizzas and meat pizzas in the ratio 4 : 3.
One day 120 cheese pizzas are sold.
How many meat pizzas are sold?

a Trish : Del
1 : 3

Trish has 1 part and Del has 3 parts.

Total number of parts = 1 + 3 = 4

Work out the total number of parts.

Trish: 1 out of 4 = $\frac{1}{4}$

Del: 3 out of 4 = $\frac{3}{4}$

b $\frac{4}{7}$ are cheese and $\frac{3}{7}$ are meat.

$4 + 3 = 7$, so the denominator of the fractions is 7.

120 pizzas are cheese $\left(= \frac{4}{7}\right)$

$120 \div 4 = 30 \quad \left(= \frac{1}{7}\right)$

120 represents $\frac{4}{7}$, so divide 120 by 4 to find the value of $\frac{1}{7}$.

$3 \times 30 = 90 \quad \left(= \frac{3}{7}\right)$

Then multiply by 3 to find the value of $\frac{3}{7}$.

So 90 meat pizzas are sold.

There are other methods you could use for part b.

Exercise 7B

D

1 Jo and Sam share a cash prize in the ratio 2 : 5.

a What fraction of the prize does Jo have?

b What fraction of the prize does Sam have?

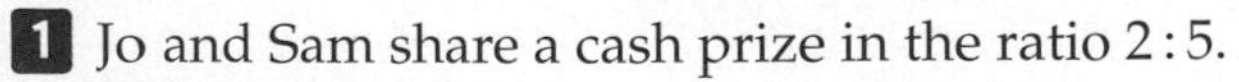

2 The ratio of pedigree dogs to mongrel dogs at a training class is 5 : 6.
Meryl says, '$\frac{5}{6}$ of the class are pedigree dogs.'
Is Meryl correct? Explain your answer.

3 Peter and David share an inheritance from their grandmother in the ratio 2 : 3.
Peter receives £600.
How much does David receive?

C

4 In a bag of apples the ratio of bruised to non-bruised apples is 1 : 6.
There are 8 bruised apples in the bag.
How many apples are there in the bag altogether?

C AO2

7.3 Ratios in the form 1 : n or n : 1

Why learn this?
Ratios can be used to compare the copper content of different colours of 18 carat gold.

Objectives
C Write a ratio in the form 1 : n or n : 1

Skills check

1 Copy and complete these conversions.

a 7.2 km = ☐ m b 6.2 m = ☐ cm c 3600 m = ☐ km
d 22 m = ☐ cm e 0.62 km = ☐ m f 50 m = ☐ km

2 Work out

a 21 ÷ 4 b 21 ÷ 5 c 21 ÷ 6

Writing ratios in the form 1 : n or n : 1

To write a ratio in the form 1 : n or n : 1

- first decide which number in the ratio you want to be 1
- then divide all the numbers in the ratio by that number.

Example 3

C

a Write the ratio 15 : 10 in the form n : 1.
b Write the ratio 5 : 3 in the form 1 : n.

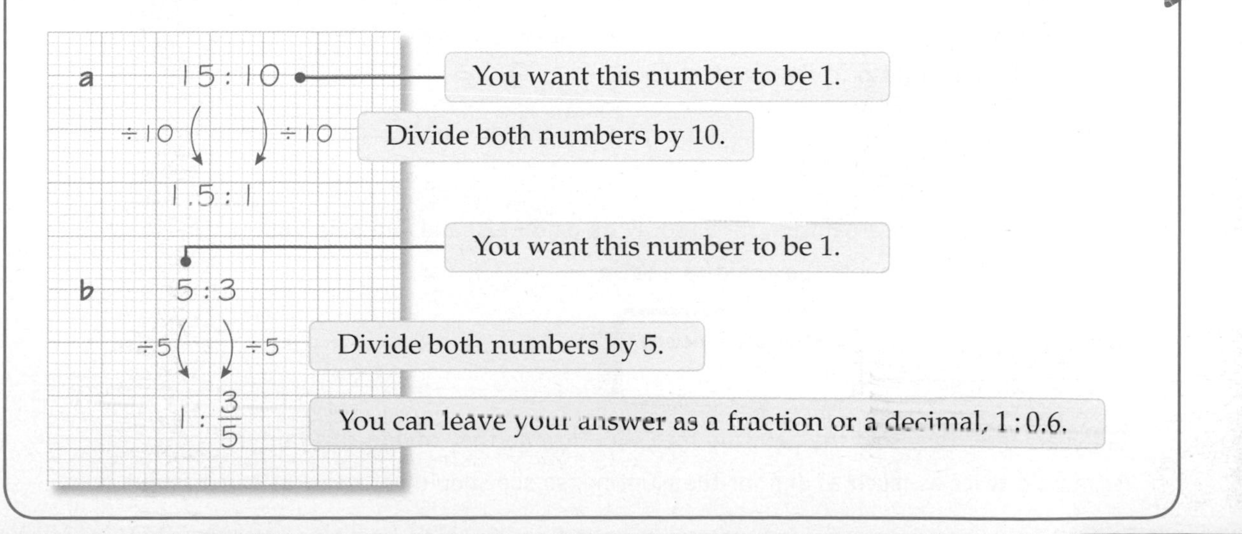

a 15 : 10
÷10 ÷10
1.5 : 1
b 5 : 3
÷5 ÷5
1 : 3/5
You want this number to be 1.
Divide both numbers by 10.
You want this number to be 1.
Divide both numbers by 5.
You can leave your answer as a fraction or a decimal, 1 : 0.6.

Exercise 7C

1 Write these ratios in the form $n:1$.

a 6 : 2   b 12 : 6   c 21 : 3   d 5 : 4

2 Write these ratios in the form $1:n$.

a 3 m : 12 m   b 400 cm : 15 m   c 700 g : 1.2 kg

d 5 litres : 7500 ml e 2 years : 8 months   f 2 days : 15 hours

Write both numbers in the same units first.

C

3 The ratio of gold : silver in necklace A is 229 : 21.

a Write this as a ratio in the form $n:1$.

b The ratio of gold : silver in necklace B is 117 : 83.
Write this as a ratio in the form $n:1$.

c Compare your answers to parts **a** and **b**.
Which necklace has the higher proportion of gold?
Explain your answer.

AO2

4 The ratio of gold : silver : copper in one type of gold bar is 75 : 3 : 22.
The ratio of gold : silver : copper in a different type of gold bar is 15 : 1 : 4.
Which gold bar, the first or the second, has the higher proportion of copper?
Show workings to support your answer.

AO3

7.4 Working with ratios

Keywords
ratio, divide

Why learn this?

The government decides safe adult to child ratios for childcare. For children under two, the ratio of adults must be 1 : 3.

Objectives

D C Share a quantity in a given ratio

C Solve word problems involving ratio

Skills check

1 Helen and Bas share a chocolate bar in the ratio 5 : 3.
  a What fraction of the bar does Helen have?
  b What fraction of the bar does Bas have?

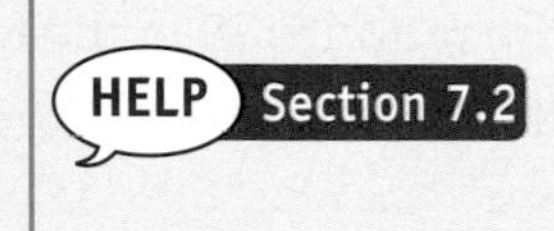

2 A farmer has 36 black lambs and 60 white lambs.
Write this as a ratio in its simplest form.

Fair shares

Petra and Jim bought a painting for £180. Petra paid £120 and Jim paid £60.

You can write the amounts they paid as a **ratio**.

Petra Jim
£120 : £60
2 : 1

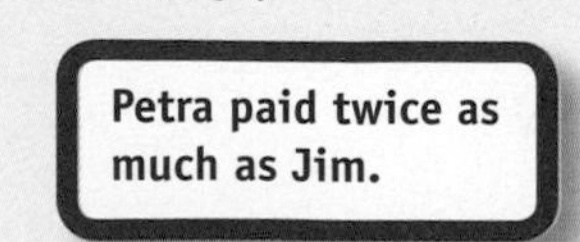

Ten years later they sold the painting for £540. How do they **divide** this fairly?

Petra paid twice as much as Jim for the painting, so she should get twice as much as Jim does.

D

Example 4

Share £540 between Petra and Jim in the ratio 2 : 1.

Petra has 2 parts. Jim has 1 part.

Total = 1 + 2 = 3 parts — Work out the total number of parts.

£540 ÷ 3 = £180 — Work out the value of 1 part.

Jim has £180 — 1 part for Jim.

Petra has £360 — 2 parts for Petra = 2 × £180 = £360

To share in a given ratio

- work out the total number of parts to share into
- work out the value of 1 part
- work out the value of each share.

Exercise 7D

D

1 Share these amounts in the ratios given.

a £36 in the ratio 2 : 1

b £12 in the ratio 3 : 1

c 32 litres in the ratio 5 : 3

d 66 kg in the ratio 4 : 7

e £500 in the ratio 2 : 2 : 1

f 250 cm in the ratio 10 : 6 : 9

Work out the total number of parts first.

C

2 Dan and Tris buy a lottery ticket. Dan pays 30p and Tris pays 70p.

a Write the amounts they pay as a ratio.

Dan and Tris win £100 000.

b Work out how much prize money each should have.

C

3 Alix and Liberty buy a racehorse. Alix pays £4000 and Liberty pays £5000.
The racehorse wins a £27 000 prize.
Work out how much money Alix and Liberty should each receive.

4 Pip, Wilf and Anna share 300 g of fudge in the ratio 1 : 2 : 3.
How much fudge does each have?

5 Fred, Sid and Ruby inherit £50 000.
They share it in the ratio of their ages.
Ruby is 10, Sid is 14 and Fred is 19.
How much does each receive?

Work out the amounts to the nearest pound.

Word problems

For 3–7-year-olds on a playscheme, the law says that the ratio of adults to children must be 1 : 8.

This means that for every 8 children there must be at least 1 adult.

You can use ratios to work out the numbers of adults for different numbers of children.

C

Example 5

On a playscheme, the minimum ratio of adults to children is 1 : 8.

a There are 7 adults. How many children can they look after?

b How many adults do you need for 40 children?

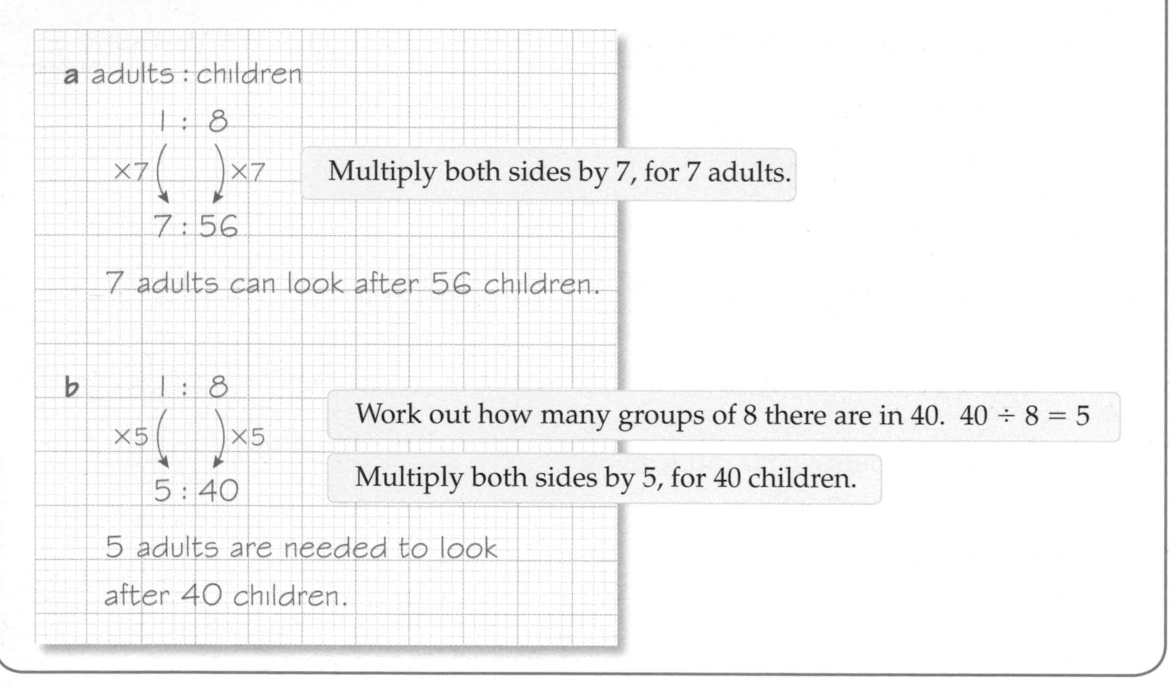

a adults : children
1 : 8
×7 ×7
7 : 56
Multiply both sides by 7, for 7 adults.
7 adults can look after 56 children.
b 1 : 8
×5 ×5
5 : 40
Work out how many groups of 8 there are in 40. 40 ÷ 8 = 5
Multiply both sides by 5, for 40 children.
5 adults are needed to look after 40 children.

Exercise 7E

C

1 For children under two years old, the ratio of adults to children must be at least 1 : 3. How many under-two-year-olds can these numbers of adults care for?

a 4 adults **b** 6 adults **c** 9 adults

2 Julie makes celebration cakes.
Her recipe uses 3 eggs for every 150 g of flour.

a Write the amounts of eggs and flour as a ratio.

b Simplify your ratio.

c How much flour does she need for 12 eggs?

d How many eggs does she need for 900 g of flour?

C

AO2

3 Bronze is made from 88% copper and 12% tin.

a Write the amounts of copper and tin as a ratio.

b Write your ratio in its simplest form.

c Tom has 55 kg of copper to make bronze. How much tin does he need?

C

4 To make strawberry jam, you need 3 kg sugar for every 3.5 kg of strawberries.
Penny has 10.5 kg of strawberries.
How much sugar does she need?

5 On a playscheme, the ratio of adults to children must be at least 1 : 8.
Work out how many adults you need for 36 children.

> **Your answer must be a whole number of people.**

6 A fruit cocktail recipe uses 5 peaches for 200 g of cherries.
How many peaches do you need for 1 kg of cherries?

AO2

C

7 The table shows the adult : child ratios for different ages of children in daycare.

Age of children	Adult : child ratio
0–2 years	1 : 3
2 years	1 : 4
3–7 years	1 : 8

These are the ages of children registered at a daycare centre in the school holidays.

Babies less than one year old	4
Toddlers over 1 year, but less than 2	7
Two-year-olds	8
Three-year-olds	6
Four-year-olds	12
Five-year-olds	9
Six-year-olds	5

Work out the number of adults needed to care for these children.

AO3

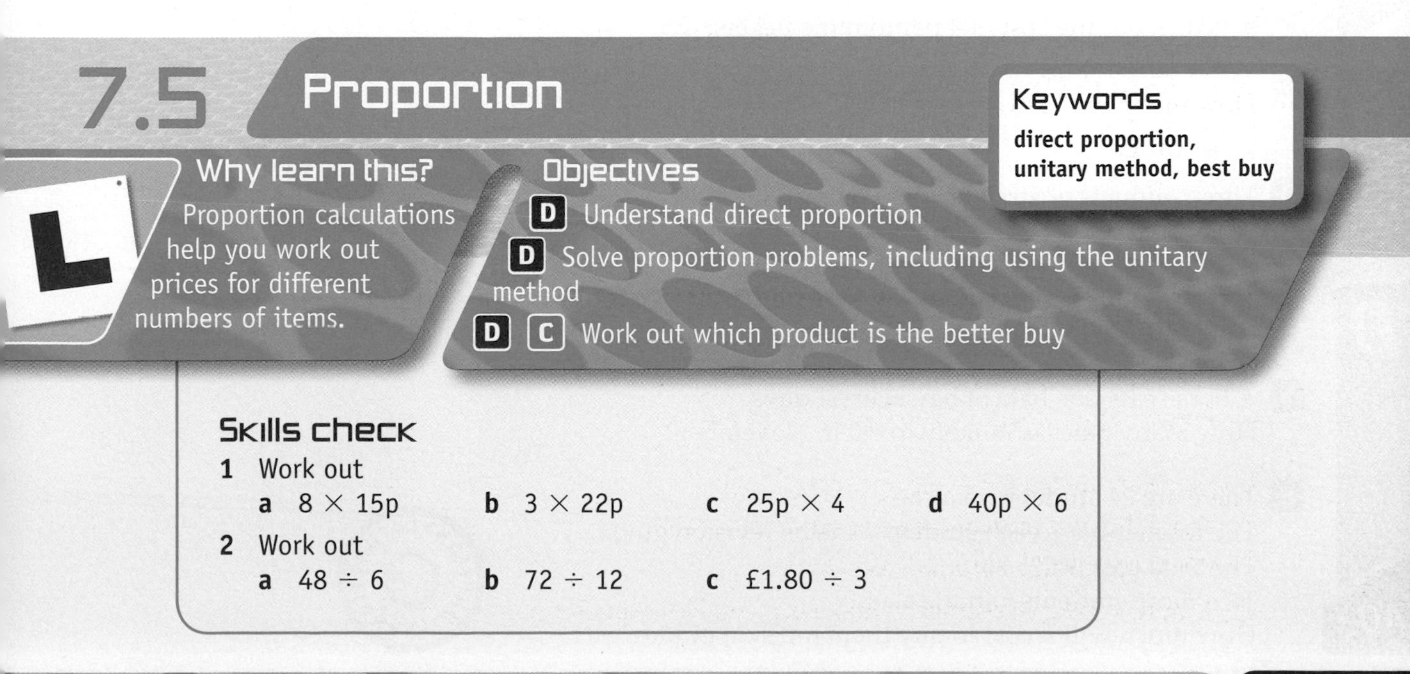

7.5 Proportion

Keywords
direct proportion, unitary method, best buy

Why learn this?

Proportion calculations help you work out prices for different numbers of items.

Objectives

- **D** Understand direct proportion
- **D** Solve proportion problems, including using the unitary method
- **D** **C** Work out which product is the better buy

Skills check

1 Work out
 a 8 × 15p b 3 × 22p c 25p × 4 d 40p × 6

2 Work out
 a 48 ÷ 6 b 72 ÷ 12 c £1.80 ÷ 3

Direct proportion

Oranges cost 30p each.
The number of oranges and the cost are in **direct proportion**.

When two values are in direct proportion:

- if one value is zero, so is the other
- if one value doubles, so does the other.

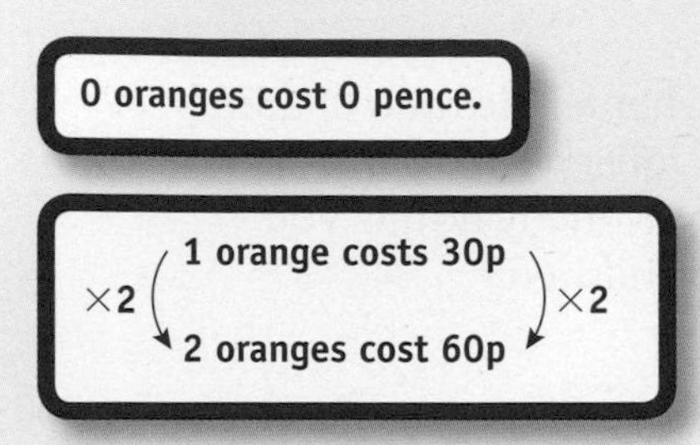

When two quantities are in direct proportion, their ratio stays the same as they increase or decrease.

In the **unitary method**, you solve proportion problems by working out the value of 1 unit first.

D

Example 6

Jim works 12 hours a week for £78.
He increases his hours to 15 per week.
How much will he now be paid?

£78 ÷ 12 = £6.50 — Work out his pay for 1 hour.

£6.50 × 15 = £97.50 — Multiply his pay for 1 hour by 15.

Exercise 7F

D

1 Five pantomime tickets cost £75.

  a Work out the cost of 1 pantomime ticket.

  b Work out the cost of 4 pantomime tickets.

D

2 Tanya is paid £40.95 for 7 hours' work.
How much will she be paid for 10 hours' work?

AO2

3 Three punnets of strawberries weigh 1.35 kg.
Work out the cost of 7 punnets of strawberries.

D

4 A catering firm charges £637.50 for a buffet for 75 people.
How much will it charge for 225 people?

5 A horse eats one bale of hay every 4 days.
How many bales is it likely to eat in November?

6 There are 24 students in a class.
The teacher buys each student a maths revision guide.
The total cost is £95.76.
Two more students join the class.
How much will it cost to buy them revision guides?

AO3

Use this table of exchange rates for Q7 to Q9.

Exchange rates one day in 2009	
£1	€1.2 (euros)
£1	US$1.6
£1	AUS$1.96 (Australian dollars)
£1	4.6 Polish zloty
£1	124 Pakistani rupees

7 Convert

a £25 to euros
b £40 to US dollars
c £30 to Australian dollars
d €35 to pounds
e US$30 to pounds
f 400 Polish zloty to pounds
g 5000 Pakistani rupees to pounds
h AUS$500 to pounds

8 Which is worth more, 7000 Pakistani rupees or US$70?

AO3

9 Sadie goes to New York. She sees a games console in a shop for US$399.99.
Sadie calls her friend Jamila in London.
Jamila says the same console costs £279.99 in London.
Where should Sadie buy the console?
Explain why.

Getting value for money

The **best buy** means the product that gives you the best value for money.
To compare two prices and sizes, work out the price for one unit for each size.

D

Example 7

Shampoo comes in two sizes.
Which is the better buy?

75 m*l* for £2.25

120 m*l* for £3.25

Small: £2.25 = 225p — Convert the price to pence.

225 ÷ 75 = 3p for 1 m*l* — Divide price by quantity to get the price for 1 m*l*.

Large: £3.25 = 325p

325 ÷ 120 = 2.7p for 1 m*l* — Compare the prices for 1 m*l* and decide which is cheaper.

The larger bottle is the better buy.

Exercise 7G

1 A large pack of cereal costs £2.34 for 500 g.
A small pack of cereal costs 98p for 200 g.
Which pack is the better buy?
Explain your answer.

D

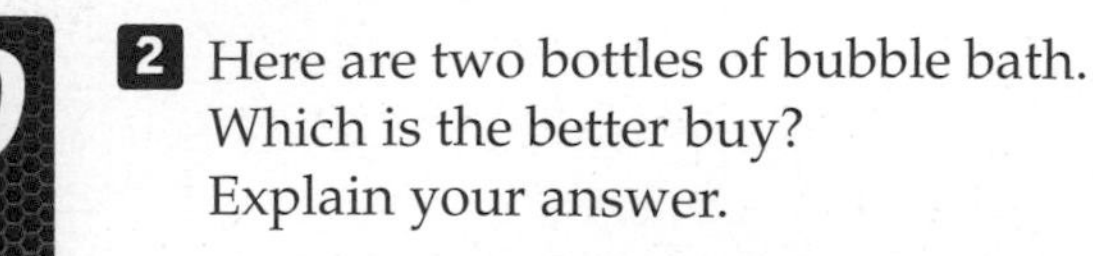

2 Here are two bottles of bubble bath.
Which is the better buy?
Explain your answer.

100 m*l* for £1.75

250 m*l* for £4.40

3 Which bottle of water is the best value for money?

A

75 cl
65p

B

2 litres
£1.80

C

5 litres
£4.20

AO3

7.6 Inverse proportion

Keywords
inverse proportion

Why learn this?
Builders use inverse proportion to work out how many people they need for a job.

Objectives
C B Understand inverse proportion

Skills check

1 Work out
   a $33 \div 4$ b $48 \div 5$ c $76 \div 8$

2 Convert
   a $1\frac{1}{2}$ hours to minutes
   b 225 minutes to hours and minutes.

Inverse proportion

It takes 2 people 1 day to put up a fence.

Working at the same rate

- it would take 1 person 2 days to put up the same fence
- it would take 4 people $\frac{1}{2}$ a day to put up the same fence.

As the number of people goes down, the time taken goes up.

As the number of people goes up, the time taken goes down.

When two values are in **inverse proportion**, one increases at the same rate as the other decreases.

Example 8

It takes 4 people 8 hours to paint a wall along one side of a large garden.
There is an identical wall along the other side of the garden.

a How long will it take 3 people to paint this wall?

b How long will it take 5 people to paint this wall?

Give your answers to the nearest hour.

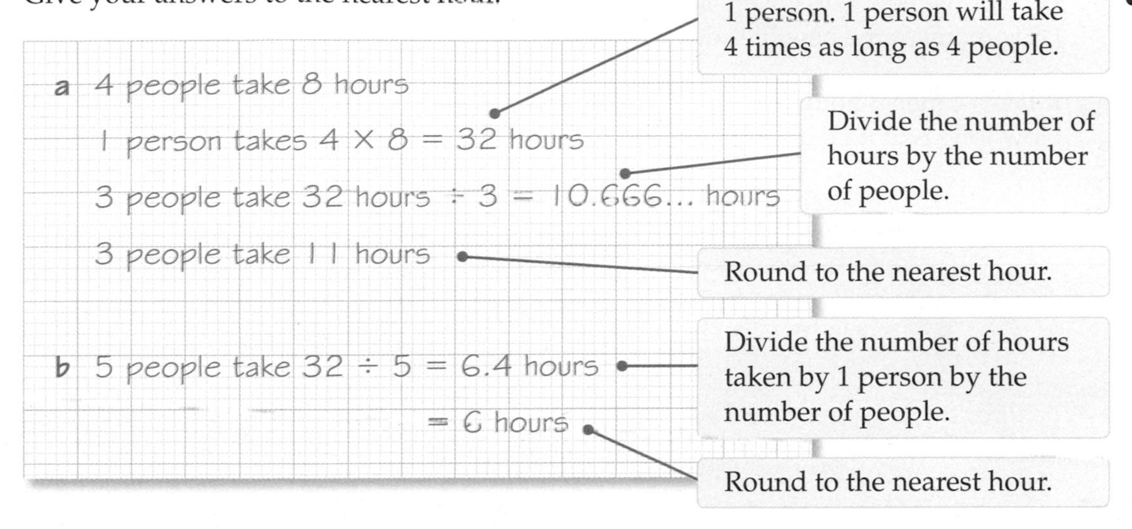

a 4 people take 8 hours
1 person takes 4 × 8 = 32 hours
3 people take 32 hours ÷ 3 = 10.666... hours
3 people take 11 hours
b 5 people take 32 ÷ 5 = 6.4 hours
= 6 hours
Work out the time for 1 person. 1 person will take 4 times as long as 4 people.
Divide the number of hours by the number of people.
Round to the nearest hour.
Divide the number of hours taken by 1 person by the number of people.
Round to the nearest hour.

C AO2

Exercise 7H

1 It takes 5 children 45 minutes to build a sandcastle.
How long will it take 3 children to build an identical sandcastle?

2 It takes 2 electricians 3 days to rewire a house.
How long will it take 5 electricians to rewire an identical house?

3 In one day, 10 woodcutters chop 18 trees into logs.

a How many trees could 5 woodcutters chop into logs in one day?

b How many days will it take 5 woodcutters to chop 36 trees into logs?

4 Three women dig a vegetable plot in 4 hours.

a How long will it take 1 woman to dig the same size plot?

b How long will it take 5 women?
Give your answer in hours and minutes.

1 hour = 60 minutes

C AO2

5 A farmer wants to lay 320 hawthorn plants to make a hedge.
He works out that 1 person can lay 6 plants in an hour.
He wants the whole hedge planted in 8 hours or less.
How many people does the farmer need for the job?

C AO3

6 It takes 60 workers 1 hour 20 minutes to erect a stage for a concert.
How long will it take 45 workers to erect the stage?
Give your answer in hours and minutes.

B AO2

B

7 In a hotel, it takes one cleaner 90 minutes to clean 5 rooms.
The hotel has 48 rooms.
All the rooms have to be cleaned between 10 am and 2 pm.
How many cleaners does the hotel need?

8 Pen is retiling her bathroom.
There were 200 old tiles on the walls. Each tile covered 225 cm^2.
Each of the new tiles covers 300 cm^2.
How many of the new tiles does she need?

AO3

9 It takes 4 men 3 hours to turf a 20 m by 30 m lawn.
How long will it take 5 men to turf a 25 m by 40 m lawn?

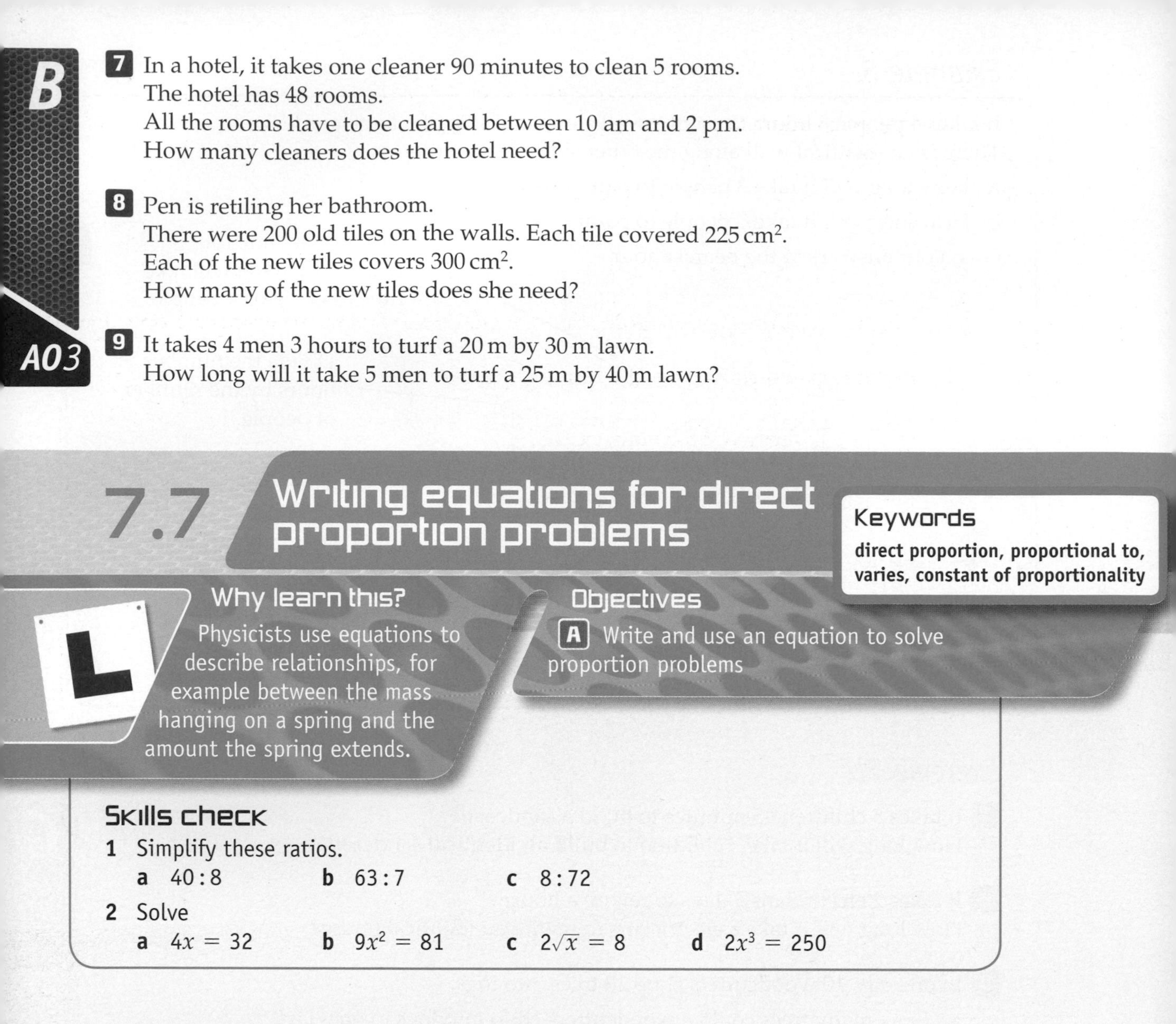

7.7 Writing equations for direct proportion problems

Keywords
direct proportion, proportional to, varies, constant of proportionality

Why learn this?
Physicists use equations to describe relationships, for example between the mass hanging on a spring and the amount the spring extends.

Objectives
A Write and use an equation to solve proportion problems

Skills check

1 Simplify these ratios.
 a 40 : 8 b 63 : 7 c 8 : 72

2 Solve
 a $4x = 32$ b $9x^2 = 81$ c $2\sqrt{x} = 8$ d $2x^3 = 250$

Direct proportion equations

Two values are in **direct proportion** if

- when one is zero, so is the other
- their ratio stays the same as they increase or decrease.

For example, the number of oranges and their cost are in direct proportion.

 1 orange costs 30p ratio 1 : 30
 2 oranges cost 60p ratio 2 : 60 = 1 : 30

You could write this as an equation

$$y = 30x$$

where y = total cost and x = number of oranges.

Here are four ways of saying that two variables, x and y, are in direct proportion:

A variable is a letter that can take different values.

- y is directly **proportional to** x
- y **varies** directly as x
- $y \propto x$
- $y = kx$ for some value of k.
 k is called the **constant of proportionality**.

$\propto$ means 'is proportional to'.

Example 9

y is directly proportional to x.
When $y = 12$, $x = 4$.
Work out **a** y when $x = 7$ **b** x when $y = 15$.

$y \propto x$ — Write the statement of proportionality.

$y = kx$ — Write the equation linking x and y, with k as the constant of proportionality.

$12 = k \times 4$ — Substitute the values of x and y.

$k = 3$ — Work out the value of k.

$y = 3x$ — Write the equation linking x and y.

a When $x = 7$
$y = 3 \times 7 = 21$ — Substitute $x = 7$ into the equation.

b When $y = 15$
$15 = 3x$ — Substitute $y = 15$ into the equation and solve to find x.
$x = 5$

Exercise 7I

A

1 y is directly proportional to x.
When $y = 45$, $x = 9$.
Work out
a y when $x = 14$ **b** x when $y = 75$.

2 x varies as t.
When $x = 4.5$, $t = 3$.
Work out
a x when $t = 7$ **b** t when $x = 15.3$.

A
AO2

3 The extension, l, of a spring is directly proportional to the mass, m, hanging on the spring.
A mass of 20 g gives an extension of 3 cm.
a Write an equation connecting l and m.
b Work out the value of l when $m = 45$ g.
c Work out the value of m when $l = 10$ cm.

4 The table shows values of two variables, x and y.
x is directly proportional to y.
Work out the missing values.

x	21	14.25	
y		4.75	3.2

A

5 The table shows values of two variables, a and b.

a	5	8	11
b	18	28.8	39.6

Show that a is directly proportional to b.

Show that the ratio $a:b$ is always the same.

6 The table shows values of two variables, d and e.

d	25.2	42	60
e	6	10	12

Is d directly proportional to e?
Explain your reasoning.

Other kinds of variation

Sometimes one quantity increases as another increases, but not at the same rate.

You can write an equation using k as the constant of proportionality.

y varies as x^2	$y \propto x^2$	equation $y = kx^2$
y varies as $\sqrt{x}$	$y \propto \sqrt{x}$	equation $y = k\sqrt{x}$
y varies as x^3	$y \propto x^3$	equation $y = kx^3$

A

Example 10

y varies as x^2.

When $x = 4$, $y = 56$.

Work out **a** y when $x = 3$ **b** x when $y = 70$.

$y \propto x^2$ — Write the statement of proportionality.

$y = kx^2$ — Write the equation linking x and y, with k as the constant.

$56 = k \times 4^2$ — Substitute the values of x and y.

$56 = 16k$

$k = 3.5$ — Work out the value of k.

$y = 3.5x^2$ — Write the equation linking x and y.

a When $x = 3$

$y = 3.5 \times 3^2 = 3.5 \times 9 = 31.5$ — Substitute $x = 3$.

b When $y = 70$

$70 = 3.5x^2$ — Substitute $y = 70$.

$x^2 = 20$ — Rearrange and solve to find x.

$x = 4.47$ (2 d.p.)

Exercise 7J

1 y varies as x^2.
When $x = 3$, $y = 22.5$.
Work out **a** y when $x = 5$ **b** x when $y = 40$.

2 y is proportional to x^3.
When $x = 5$, $y = 700$.
Work out **a** y when $x = 4$ **b** x when $y = 520$.

3 y varies as $\sqrt{x}$.
When $x = 4$, $y = 17$.
Work out **a** y when $x = 9$ **b** x when $y = 40$.

4 The area of a circle, A, is proportional to the square of its radius, r.
When the radius is 10 cm, the area is 314.2 cm^2 (to 1 d.p.).

a Write an equation linking A and r.

b What is another name for the constant of proportionality in your equation?

5 The volume of a sphere is proportional to the cube of its radius.
A sphere of radius 5 cm has a volume of 523.6 cm^3.
Work out

a the volume of a sphere of radius 8 cm

b the radius of a sphere of volume 4188.8 cm^3.

6 In an experiment, F is found to be proportional to r^2.
Work out the missing values for the table.

F	22.5		62.5
r	3	4	

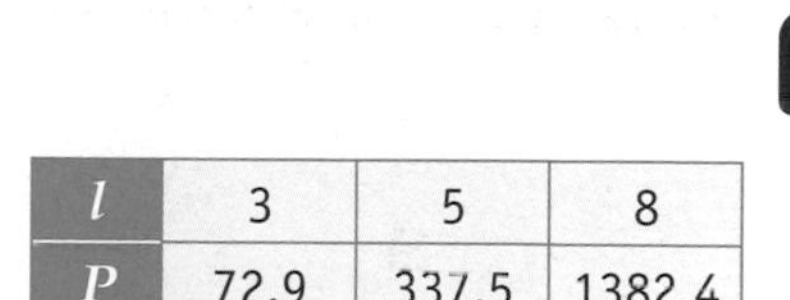

7 A scientist measured values of l and P.
The values are shown in this table.
Which of these rules fits the results?

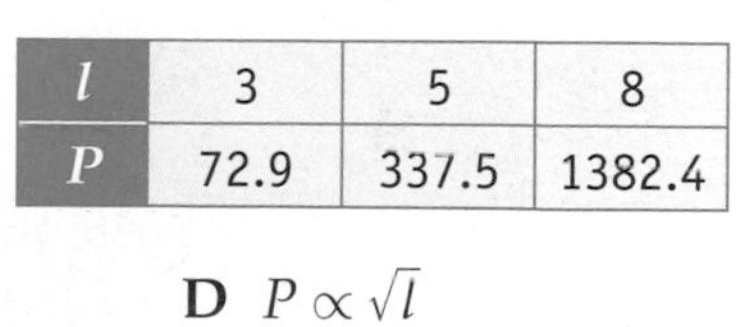

l	3	5	8
P	72.9	337.5	1382.4

A $P \propto l$ **B** $P \propto l^2$ **C** $P \propto l^3$ **D** $P \propto \sqrt{l}$

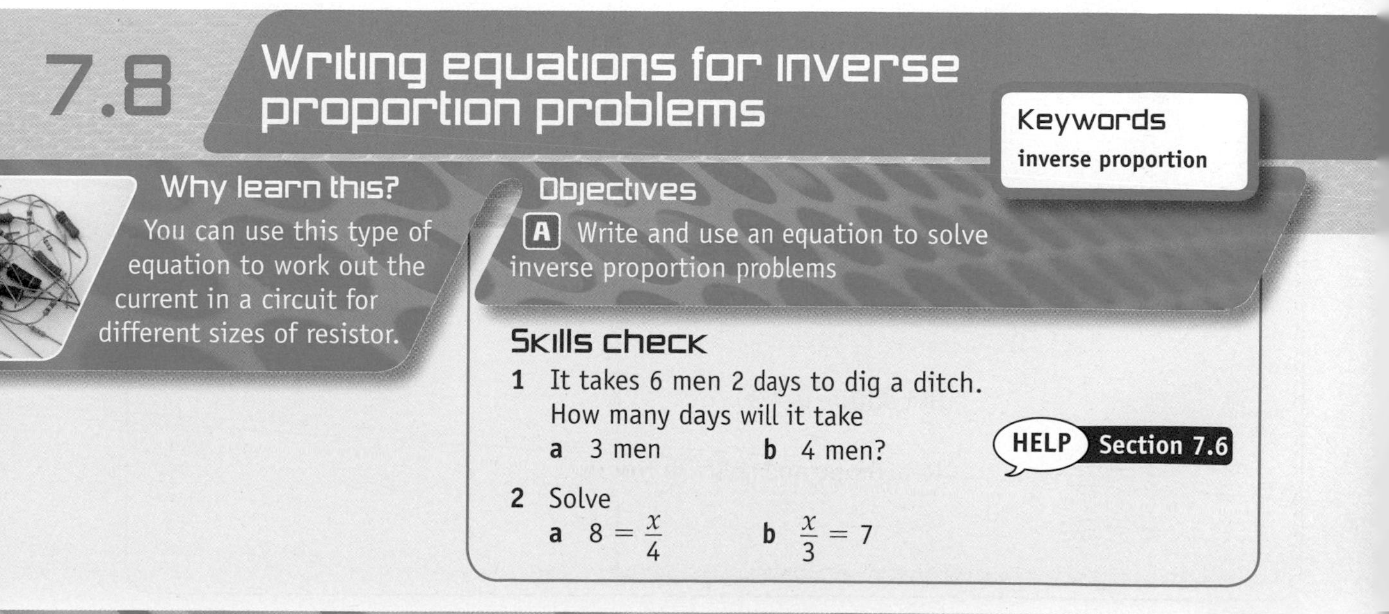

7.8 Writing equations for inverse proportion problems

Keywords
inverse proportion

Why learn this?
You can use this type of equation to work out the current in a circuit for different sizes of resistor.

Objectives
A Write and use an equation to solve inverse proportion problems

Skills check

1 It takes 6 men 2 days to dig a ditch.
How many days will it take
a 3 men **b** 4 men?

HELP Section 7.6

2 Solve
a $8 = \frac{x}{4}$ **b** $\frac{x}{3} = 7$

Inverse proportion equations

When two values are in **inverse proportion**, one increases at the same rate as the other decreases.

Here are four ways of saying that two variables, x and y, are in inverse proportion:

- y is inversely proportional to x
- y varies as the inverse of x
- $y \propto \frac{1}{x}$

$\propto$ means 'is proportional to'.

- $y = \frac{k}{x}$ for some value of k.

  k is called the constant of proportionality.

Sometimes one quantity increases as another decreases, but not at the same rate.

You can write an equation using k as the constant of proportionality.

y varies as the inverse of x^2	$y \propto \frac{1}{x^2}$	equation $y = \frac{k}{x^2}$
y varies as the inverse of $\sqrt{x}$	$y \propto \frac{1}{\sqrt{x}}$	equation $y = \frac{k}{\sqrt{x}}$
y varies as the inverse of x^3	$y \propto \frac{1}{x^3}$	equation $y = \frac{k}{x^3}$

A

Example 11

y is inversely proportional to x.

When $x = 4$, $y = 3$.

Work out **a** y when $x = 2$ **b** x when $y = 30$.

$y \propto \frac{1}{x}$ — Write the statement of proportionality.

$y = \frac{k}{x}$ — Write the equation linking x and y, with k as the constant.

$3 = \frac{k}{4}$ — Substitute the values of x and y.

$k = 12$ — Work out the value of k.

$y = \frac{12}{x}$ — Write the equation linking x and y.

a When $x = 2$

$y = \frac{12}{2} = 6$ — Substitute $x = 2$.

b When $y = 30$

$30 = \frac{12}{x}$ — Substitute $y = 30$.

$x = \frac{12}{30}$ — Rearrange and solve to find x.

$x = 0.4$

Exercise 7K

A

1 y is inversely proportional to x.
When $x = 7, y = 8$.
Work out

a y when $x = 4$ **b** x when $y = 40$.

2 y varies as the inverse of x.
When $x = 4.2, y = 17$.
Work out

a y when $x = 2.7$ **b** x when $y = 32$.

3 y is proportional to $\frac{1}{\sqrt{x}}$.
When $x = 9, y = 10$.
Work out

a y when $x = 16$ **b** x when $y = 7$.

A

4 The time taken to build a wall is inversely proportional to the number of people building it.
Four people take 18 hours to build the wall.

a Write an equation relating the number of people, P, to the time taken, t.

b How many hours will it take 3 people to build the wall?

c How many people would you need to build the wall in 10 hours?

5 The pressure of a gas is inversely proportional to its volume.
The table shows values for the pressure and volume of a gas, from a physics experiment.

Pressure (bar)	1.6	2.5	
Volume (cm³)	180		250

Work out the missing values for the table.

6 The pressure of water from a hose is inversely proportional to the square of the hose radius.
For a hose of radius 1.5 cm, the water pressure is 40 Pa.
What hose radius do you need for a pressure of 70 Pa?

Pa is short for pascal. It is the unit used to measure pressure.

7 m varies inversely as the cube of t
Copy and complete this table of values for m and t.

m		50	70
t	0.5	2	

8 The table shows values of p and Q.
Is Q inversely proportional to p?
Give reasons for your answer.

p	0.6	2.4	9	30
Q	18	4.5	1.2	0.36

9 In a science experiment, p is found to vary inversely as t^2.
Copy and complete the table of values for p and t.

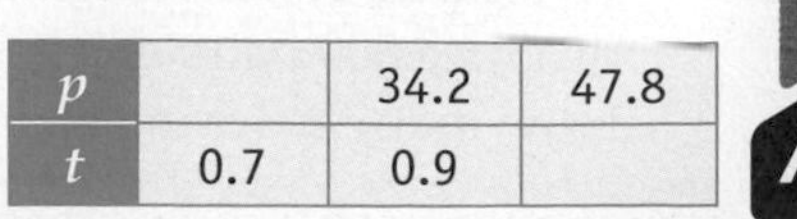

p		34.2	47.8
t	0.7	0.9	

AO2

Review exercise

1 $\frac{9}{20}$ of the people at a pub quiz are men.
What is the ratio of men : women at the pub quiz? **[2 marks]**

C

2 In a class of 27 pupils, the ratio of boys to girls is 4 : 5.
How many of the pupils are

a girls **[1 mark]**

b boys? **[1 mark]**

3 Two sisters share £45 in the ratio 2 : 3.
How much is the larger share? **[2 marks]**

4 Ana and Berwyn share a lottery win in the ratio 3 : 4.
Berwyn receives £2000.
How much does Ana receive? **[2 marks]**

5 Mr Jones leaves £10 000 to his grandchildren, Al, Deb and Cat, to be shared in the ratio 5 : 3 : 2.
Work out how much each receives. **[3 marks]**

6 A chemical reaction uses chemical A and chemical B in the ratio 3 : 7.
A chemist has 450 mg of chemical A.
How much of chemical B does she need? **[2 marks]**

C

7 1.5 litres of paint covers an area of 36 m^2.

a How many litres of paint do you need to cover 94 m^2? **[2 marks]**

b Paint comes in 1.5 litre tins.
How many 1.5 litre tins do you need to buy? **[1 mark]**

You may need to round up to get a sensible answer.

AO2

8 It takes 3 plumbers 7 hours to install a central heating system.
How long will it take 2 plumbers to do the same job? **[2 marks]**

B

9 Builders remove 4000 tiles from the school hall floor.
Each tile measures 50 cm by 50 cm.
The builders lay a new laminate floor in the hall.
Laminate comes in strips measuring 30 cm by 2 metres.
How many laminate strips will they need? **[4 marks]**

10 A factory makes jeans.
Four machinists produce 21 pairs of jeans in one day.
To fill a large order, the factory needs to produce 800 pairs of jeans in two days.
How many machinists do they need? **[4 marks]**

The number of machinists must be a whole number.

AO3

11 m varies as n.
When $m = 63$, $n = 7$.
Work out

a the value of m when $n = 10$

b the value of n when $m = 18$. **[4 marks]**

A

12 y is directly proportional to the square of x.
When $x = 7$, $y = 98$.
Work out an equation connecting x and y. **[3 marks]**

A

13 p is inversely proportional to q.
Copy and complete this table of values for p and q.

p	0.2	0.5	
q		125	200

[4 marks]

14 Match each statement to a table.

a y is inversely proportional to x^2.

b y is inversely proportional to x.

c y is directly proportional to x^2.

Table A

x	1	2	3
y	0.5	2	4.5

Table B

x	1	2	3
y	40	20	13.33 (2 d.p.)

Table C

x	1	2	3
y	7	1.75	$0.\dot{7}$

AO2

[3 marks]

Chapter summary

In this chapter you have learned how to

- use a ratio in practical situations **D**
- write a ratio as a fraction **D**
- understand direct proportion **D**
- solve proportion problems, including using the unitary method **D**
- use a ratio to find one quantity when the other is known **D** **C**
- share a quantity in a given ratio **D** **C**
- work out which product is the better buy **D** **C**
- write a ratio in the form $1:n$ or $n:1$ **C**
- solve word problems involving ratio **C**
- understand inverse proportion **C** **B**
- write and use an equation to solve proportion problems **A**
- write and use an equation to solve inverse proportion problems **A**

Come dine with Brian

It's Brian's birthday, and he wants to have some fun. He decides to cook a meal for himself and seven friends.

Here is the food he plans to serve.

menu
stilton soup
leek and macaroni bake
rhubarb and orange crumble

Question bank

1 How much butter does Brian need for the Stilton soup?

2 What is the total amount of butter that Brian needs for the meal?

3 Write a list showing the total amounts of all the ingredients that Brian needs for the rhubarb and orange crumble.

Brian already has some of the ingredients he needs in his kitchen cupboard.

Brian says, 'I only need to buy seven ingredients as I already have the rest.'

4 Is Brian correct? Explain your answer.

Brian also says, 'It should only take about 3 hours to prepare and cook the whole meal.'

5 Is Brian correct? Show working to support your answer.

6 What other things do you think Brian needs to buy to go with the meal?

Information bank

Chef's top tips

- Soup can be prepared in advance and re-heated before serving.
- One vegetable stock cube will make 200 m*l* of vegetable stock.
- Leek and macaroni bake must be served straight from the oven.
- Organic leeks have a better flavour than normal leeks.
- Make your own breadcrumbs from slices of white bread.
- Tinned rhubarb can be used instead of fresh rhubarb.
- For an extra-sweet finish, sprinkle icing sugar on top of the crumble.

stilton soup (serves 4)

125 g Stilton cheese
50 g butter
40 g plain flour
900 ml vegetable stock
300 ml skimmed milk
60 ml double cream
45 ml white wine
1 onion

Cooking time: 45 minutes
Preparation time: 25 minutes

leek and macaroni bake (serves 6)

450 g leeks
300 g Cheddar cheese
240 g macaroni
75 g butter
60 g breadcrumbs
60 g plain flour
900 ml full-fat milk
30 g chives

Cooking time: 45 minutes
Preparation time: 20 minutes
Oven temperature: 190°C

rhubarb and orange crumble (serves 10)

1.4 kg rhubarb
225 g butter
150 g caster sugar
150 g self-raising flour
150 g plain flour
125 g granulated sugar
2 eggs
1 orange

Cooking time: 1 hour
Preparation time: 30 minutes
Oven temperature: 200°C

In his kitchen, Brian already has:

250 g butter	400 g Cheddar cheese
1 kg plain flour	½ dozen eggs
1 kg self-raising flour	8 vegetable stock cubes
1 *l* skimmed milk	500 g granulated sugar
1 *l* full-fat milk	½ bottle of red wine
250 m*l* single cream	½ loaf of white bread
250 m*l* double cream	1 onion
500 g spaghetti	3 oranges

8 Complex calculations and accuracy

BBC Video

This chapter is about using calculators to solve problems.

The computers that got Apollo 11 to the Moon were *less* powerful than today's average scientific calculator!

Objectives

This chapter will show you how to

- **solve problems involving repeated percentage change** C
- **solve reverse percentage problems** B
- **handle very large and very small numbers using standard form** B
- **understand limits of accuracy** C B A
- **calculate absolute error and percentage error** A A*

Before you start this chapter

1 Work out these without using a calculator.

a Increase £250 by 8%. b What is $266 \div 0.7$? c Divide 45 000 000 by 0.005.

2 Write these numbers correct to one decimal place.

a 5.681 b 11.06 c 21.974

3 Write these numbers correct to two significant figures.

a 6842 b 153 945 c 0.000 382

8.1 Repeated percentage change

Keywords
compound interest, invest, multiplier

Why learn this?
Savings accounts are all about repeated percentage change.

Objectives
C Perform calculations involving repeated percentage change

Skills check

1 Pete puts £1700 into a savings account. The interest rate is 3% per annum. Work out the amount gained through simple interest after one year.

HELP Section 2.5

2 Morgan puts £2700 into a savings account. The interest rate is 4% per annum. Work out the amount gained through simple interest after four years.

Compound interest

Examples of repeated percentage change include **compound interest** and population changes.

When you **invest** money, the interest you earn is usually calculated using compound interest.

In compound interest

- the rate of interest is fixed
- the amount of interest you receive each year is not the same
- each year's interest is calculated on the sum of money you invested in the first place *plus* any interest you have already received.

When you put money into a savings account, the interest you earn is a percentage of the amount in your account.

The amount you pay back on a loan is also calculated using compound interest.

Example 1

C

Venetia puts £800 into a savings account earning 6% per annum interest.
How much does she have after two years?

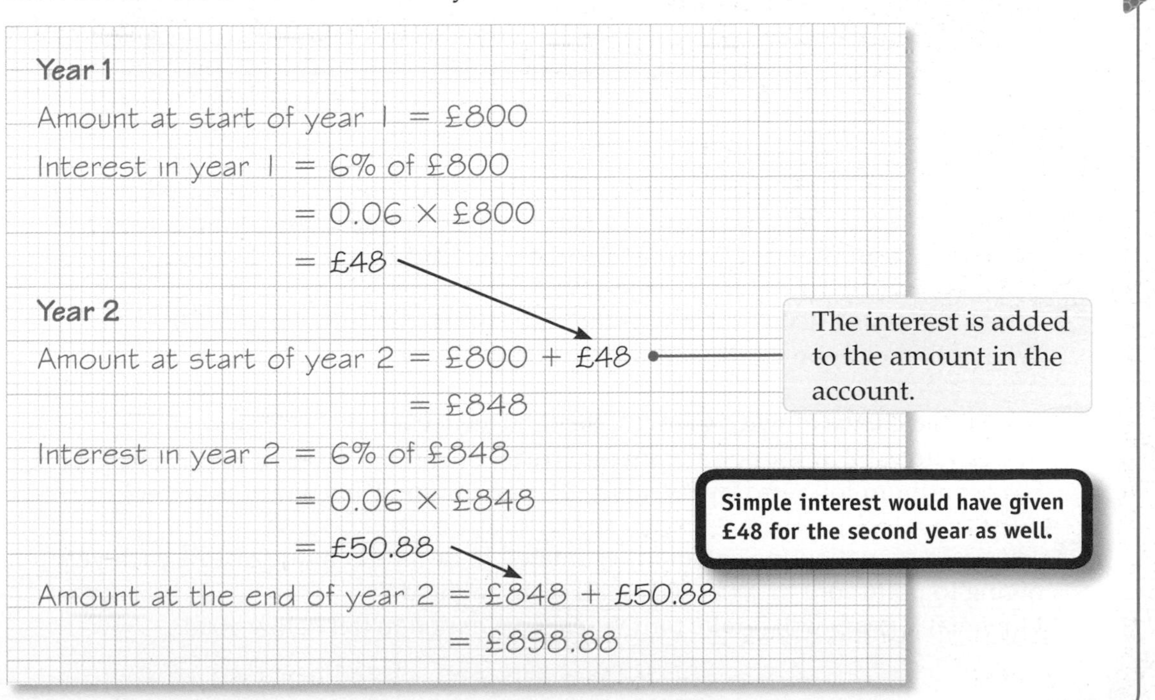

Year 1
Amount at start of year 1 = £800
Interest in year 1 = 6% of £800
= 0.06 × £800
= £48
Year 2
Amount at start of year 2 = £800 + £48
= £848
Interest in year 2 = 6% of £848
= 0.06 × £848
= £50.88
Amount at the end of year 2 = £848 + £50.88
= £898.88
The interest is added to the amount in the account.
Simple interest would have given £48 for the second year as well.

Repeated percentage change

When the same percentage change is repeated, as in the example above, you can simplify the calculations.

1 Add the rate of interest to 100%.

2 Convert this percentage to a decimal to get the **multiplier**.

3 Multiply the original amount by the multiplier as many times as the number of years for which the money is invested.

Using this method for Example 1 above...

1 6% + 100% = 106%

2 106% = 1.06

3 £800 × 1.06 × 1.06 = £898.88

1.06 is called the multiplier.

Multiply by 1.06 *twice* because the money is invested for *two* years.

C

Example 2

There are approximately 5000 whales of a certain species, but scientists think that their numbers are reducing by 8% each year.

Estimate how many of these whales there will be in three years.

The multiplier is 100% − 8% = 92%
= 0.92

This time there is a loss each year so you must subtract the percentage from 100%.

Number in three years' time = 5000 × 0.92 × 0.92 × 0.92
= 3893.44

The estimated population in three years' time is 3893 whales.

Exercise 8A

C

1 Amy invests £135 for two years at a rate of 4% per annum compound interest. How much will she have at the end of the two years?

2 A new car costs £12 000. Each year the value of the car depreciates by 8%. What will the car be worth at the end of three years?

3 Sami has a choice of ways to invest the £250 legacy left by her grandfather. She can invest it at 4% per annum for 5 years or she can invest it for 2 years at 5% per annum followed by 3 years at 3% per annum. Which way is better?

C AO2

4 The seal population in Scotland is estimated to decline at a rate of 16% each year owing to pollution. In 2007, 3000 seals were counted. How many years will it take for the seal population to fall below 1000?

5 A company leases a car at £260 per month. In the leasing agreement, the price is reduced by 2% each month after the first 6 months.

a Work out the difference between the amounts paid in the 6th and 7th months.

b How much will the car cost in total in the first 9 months?

6 Mamet earns £14 250 per year. His contract says this will increase each year in line with the annual rate of inflation. An estimate for the annual rate of inflation is 3%.

a Find his salary at the end of three years.

b Find his monthly pay at the end of five years.

c If the rate of inflation is 5% instead of 3%, how much more is his monthly pay at the end of five years?

Compare the monthly pay for the two rates of inflation.

7 How long will it take each of the following investments to reach £1 000 000?

a £200 000 invested at a rate of 15%.

b £150 000 invested at a rate of 19%.

C

AO2

8 If £7400 grows to £10 873 in five years, what is the rate of compound interest?

8.2 Reverse percentages

Keywords

original quantity, final quantity, reverse percentage

Why learn this?

Knowing the rate at which a population is decreasing, scientists can work backwards and estimate how big a population was in the past.

Objectives

B Perform reverse percentage calculations

Skills check

1 Increase £32 by 10%.

2 Decrease £130 by 6%.

3 Increase 27 mm by 5%.

4 Increase 78 km by 4%.

5 Decrease 1592 m*l* by 2%.

6 Decrease £47 by 23%.

HELP Section 2.5

Finding the original quantity 1

In some problems you are told the quantity after a percentage increase or decrease and you have to work out the **original quantity**.

Remember that the original quantity in any calculation is always taken to be 100% and the **final quantity** will be a given percentage above or below 100%, depending on whether there has been an increase or a decrease.

Consider this question.

A bike is reduced by 20% in a sale. The sale price is £135.
What was the original price?

You have not been told the original amount, the 100% figure, and you need to work it out.

These questions are called **reverse percentage** problems.

There are two methods for answering these questions.

Method A

1. Work out what percentage the final quantity represents.
2. Divide by this percentage to find 1%.
3. Multiply by 100 to get the 100% figure.

Method B

1. Work out what percentage the final quantity represents.
2. Divide by 100 to get the multiplier.
3. Divide the final quantity by the multiplier.

Example 3

A magazine's circulation is 30% down on last month's figure.
This month they sold 490 copies of the magazine.
How many copies did they sell last month?

You do not have last month's figure, which is the original amount.

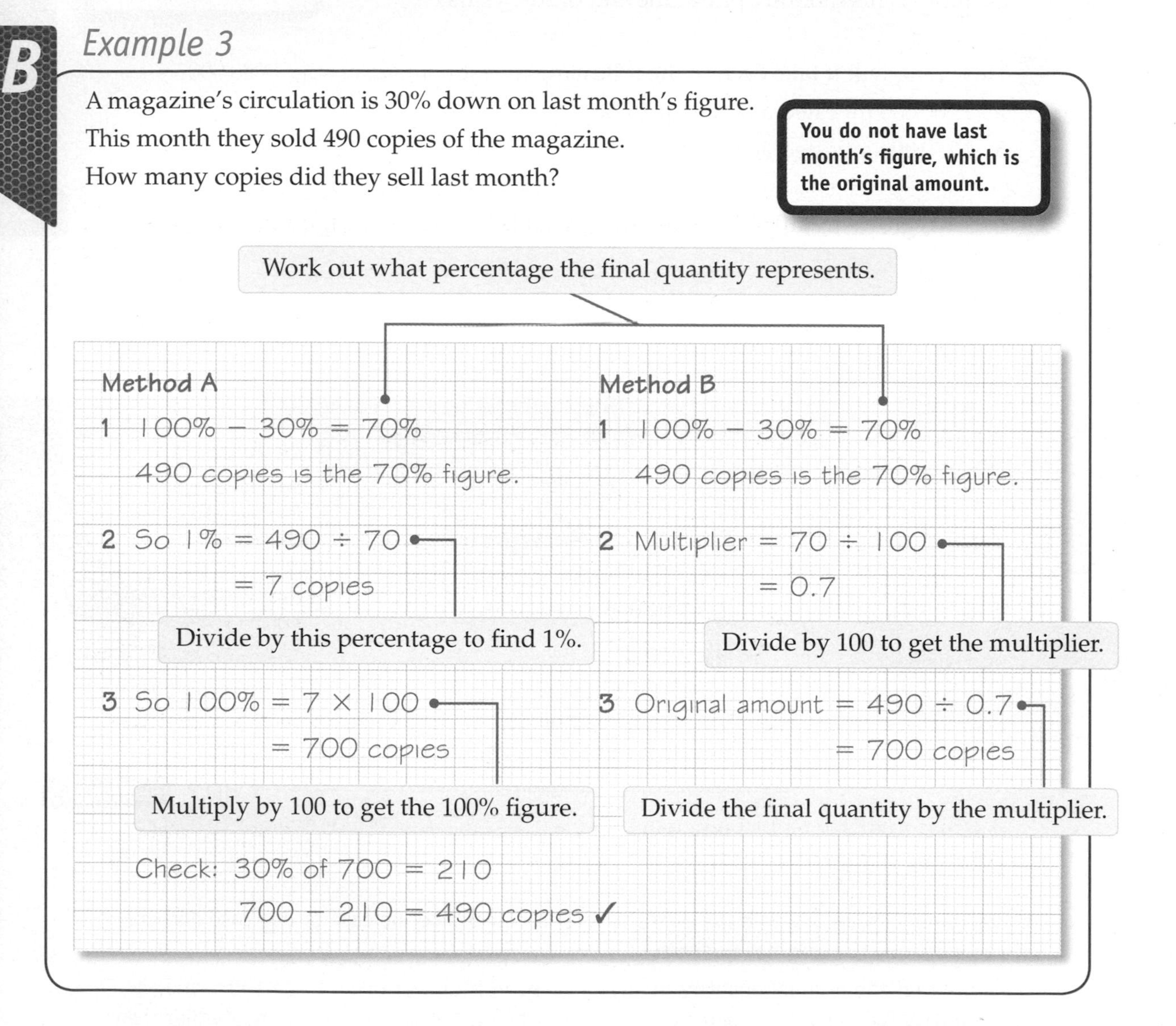

Work out what percentage the final quantity represents.
Method A
1 100% − 30% = 70%
490 copies is the 70% figure.
2 So 1% = 490 ÷ 70
= 7 copies
Divide by this percentage to find 1%.
3 So 100% = 7 × 100
= 700 copies
Multiply by 100 to get the 100% figure.
Check: 30% of 700 = 210
700 − 210 = 490 copies ✓
Method B
1 100% − 30% = 70%
490 copies is the 70% figure.
2 Multiplier = 70 ÷ 100
= 0.7
Divide by 100 to get the multiplier.
3 Original amount = 490 ÷ 0.7
= 700 copies
Divide the final quantity by the multiplier.

Finding the original quantity 2

You can use the same methods for a percentage increase.

Example 4

The number of students applying to Intelligentsia University has increased by 15% compared to last year.

This year the university had 9890 applications.

How many applications did it have last year?

Again you do not have the original amount, the number of applications before the rise.

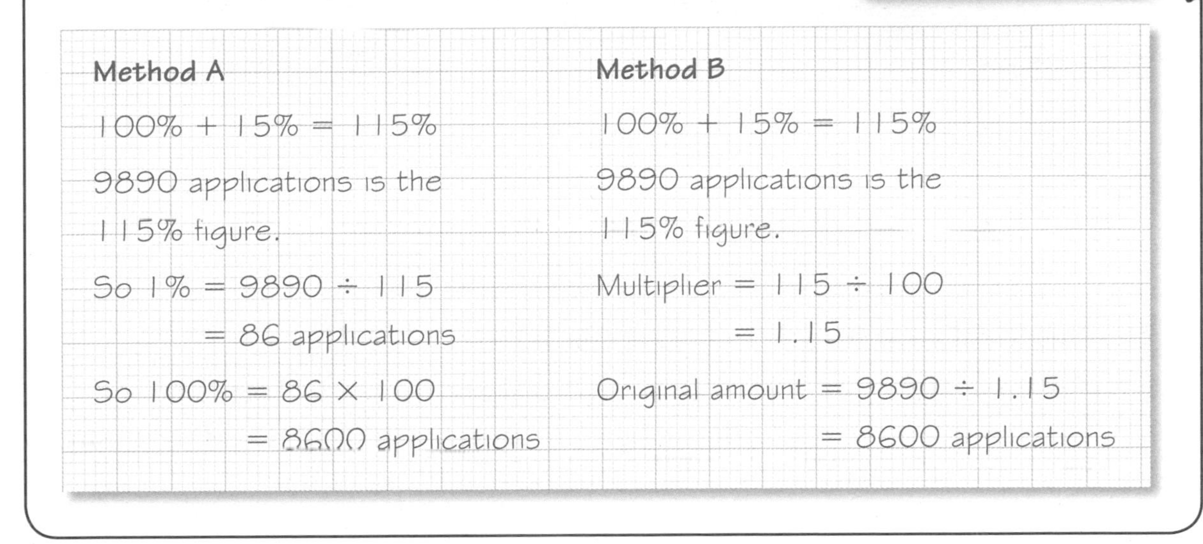

Method A
100% + 15% = 115%
9890 applications is the 115% figure.
So 1% = 9890 ÷ 115
= 86 applications
So 100% = 86 × 100
= 8600 applications
Method B
100% + 15% = 115%
9890 applications is the 115% figure.
Multiplier = 115 ÷ 100
= 1.15
Original amount = 9890 ÷ 1.15
= 8600 applications

B

Exercise 8B

1 A laptop is reduced by 12% in a sale. It now costs £334.40.
What was the original price of the laptop?

2 After a pay rise of 4%, Kate earns £14 560. What did she earn before the pay rise?

3 A mobile phone was priced at £135 including VAT.
What would be the price without VAT?

Assume VAT = 17.5%.

4 A spring is stretched by 32% to a length of 22 cm. What length was it before it was stretched? Give your answer to the nearest millimetre.

5 A seal pup increases its body weight by $4\frac{1}{2}$% in one day. It now weighs 6.27 kg.
What did it weigh yesterday?

6 A shop advertises, 'We will pay your VAT'.
How much would I pay for a bike priced at £87.50 inclusive of VAT?

B

7 After paying a 12% deposit on a house, Ben needs a mortgage for £171 160.
The estate agent will charge him $2\frac{1}{4}$% of the full cost of the house for a survey of the property. How much will he have to pay for the survey?

8 During heating, a metal rod expands by $1\frac{1}{4}$%.

a If the rod measures 40.5 cm when hot, what was its length originally?

b After 10 minutes, the metal rod is still 0.5% longer than its original length.
What is the length of the rod at this time?

B

AO2

8.3 Standard form

Keywords
standard form

Why learn this?

You can use standard form to write very large and very small numbers. The distance from the Earth to the Sun is approximately 1.5×10^{11} metres.

Objectives

B Interpret and use standard form

Skills check

1 Work out

a $325 \div 100\,000$ b $0.008\,57 \times 1000$ c $3.665 \times 10\,000$

2 Work out

a 1.5×2 b $3.6 \div 20$ c $3.2 \div 0.08$

Using and interpreting standard form

It is often convenient to write very large numbers or very small numbers in **standard form.**

All numbers can be expressed as $x \times 10^n$, where x is a number between 1 and 10 and n is an integer.

Numbers greater than 1

B

Example 5

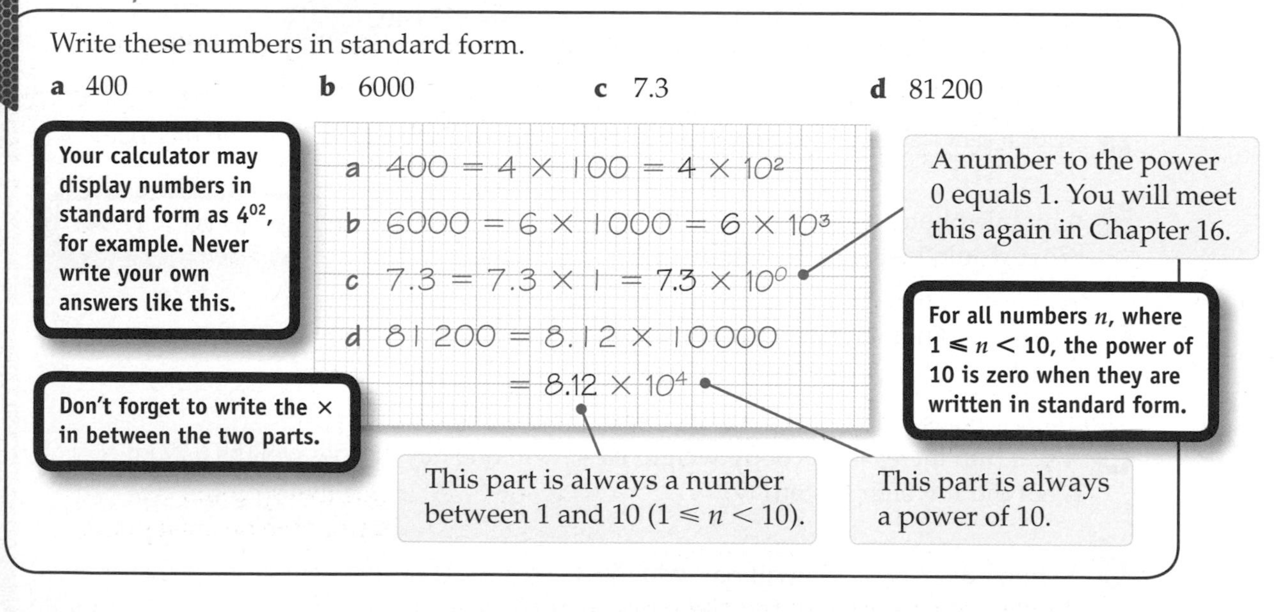

Write these numbers in standard form.
a 400
b 6000
c 7.3
d 81 200
Your calculator may display numbers in standard form as 4⁰², for example. Never write your own answers like this.
a 400 = 4 × 100 = 4 × 10²
b 6000 = 6 × 1000 = 6 × 10³
c 7.3 = 7.3 × 1 = 7.3 × 10⁰
d 81 200 = 8.12 × 10 000
= 8.12 × 10⁴
A number to the power 0 equals 1. You will meet this again in Chapter 16.
For all numbers n, where 1 ≤ n < 10, the power of 10 is zero when they are written in standard form.
Don't forget to write the × in between the two parts.
This part is always a number between 1 and 10 (1 ≤ n < 10).
This part is always a power of 10.

Exercise 8C

B

1 Write these numbers in standard form.

a 3 000 000 b 7400 c 32 000 d 603 500 e 108
f 68 g 650.5 h 99.9 i 5 j 2.04

2 Write down the power of 10 for each of these numbers when they are written in standard form.

a 45.7 b 3.002 c 8293
d 94 000 e five million f one hundred thousand

Numbers less than 1

Example 6

Write these numbers in standard form.

a 0.6 **b** 0.03 **c** 0.0078 **d** 0.000 419

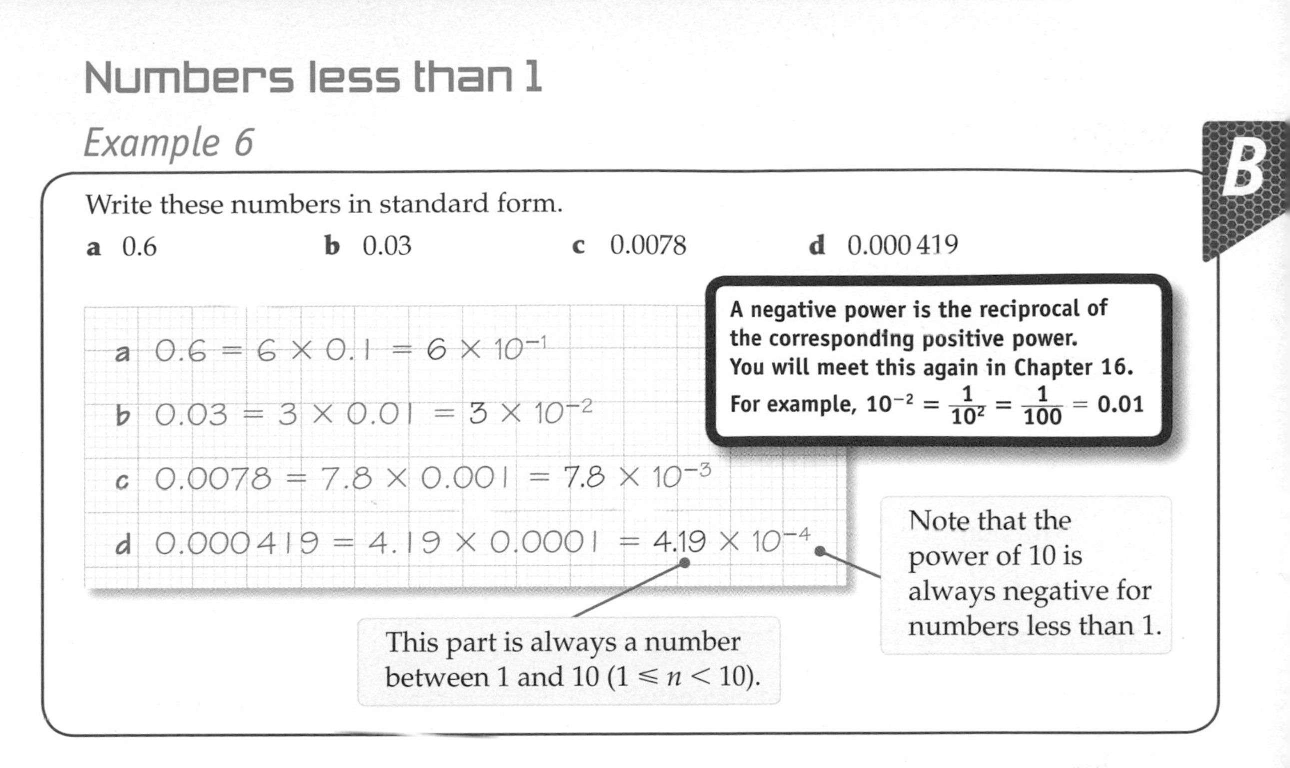

a $0.6 = 6 \times 0.1 = 6 \times 10^{-1}$
b $0.03 = 3 \times 0.01 = 3 \times 10^{-2}$
c $0.0078 = 7.8 \times 0.001 = 7.8 \times 10^{-3}$
d $0.000\,419 = 4.19 \times 0.0001 = 4.19 \times 10^{-4}$
A negative power is the reciprocal of the corresponding positive power. You will meet this again in Chapter 16.
For example, $10^{-2} = \frac{1}{10^2} = \frac{1}{100} = 0.01$
Note that the power of 10 is always negative for numbers less than 1.
This part is always a number between 1 and 10 ($1 \leq n < 10$).

Exercise 8D

1 Write these numbers in standard form.

a 0.0005 **b** 0.006 **c** 0.4 **d** 0.000 12
e 0.0717 **f** 0.000 197 5 **g** 0.9009 **h** 0.001 000 3

2 Write down the power of 10 for each of these numbers when they are written in standard form.

a 0.0745 **b** 0.003 090 4 **c** 0.000 054 6
d 0.000 000 027 1 **e** one thousandth **f** three-quarters

Changing between standard form and a decimal number

Example 7

Write these standard form numbers as decimal numbers.

a 3.709×10^5 **b** 4.19×10^{-4}

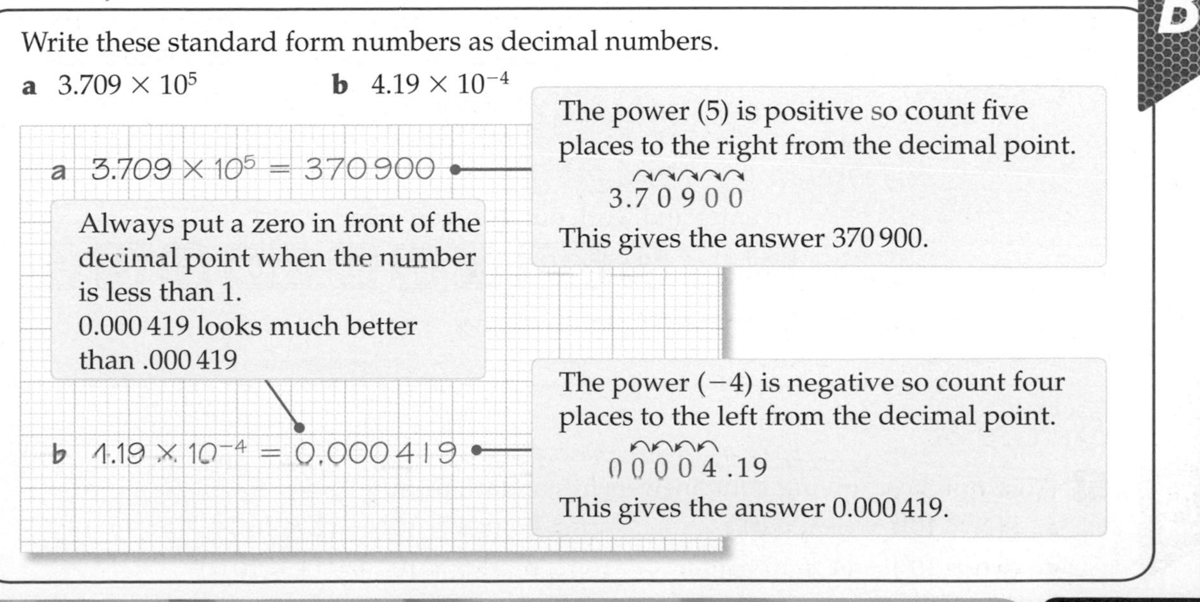

a $3.709 \times 10^5 = 370\,900$
The power (5) is positive so count five places to the right from the decimal point.
3.7 0 9 0 0
This gives the answer 370 900.
Always put a zero in front of the decimal point when the number is less than 1.
0.000 419 looks much better than .000 419
b $4.19 \times 10^{-4} = 0.000\,419$
The power (−4) is negative so count four places to the left from the decimal point.
0 0 0 0 4 .19
This gives the answer 0.000 419.

Exercise 8E

B

1 Write these as decimal numbers.

a 5×10^4 **b** 3.8×10^3 **c** 6×10^{-3} **d** 7.26×10^9
e 8.492×10^{-2} **f** 4.37×10^6 **g** 1.006×10^{-4} **h** 6.2387×10^3

2 Write these numbers given in standard form as decimal numbers.

a The distance from the Earth to the Sun is approximately 1.488×10^8 kilometres.
b The average width of an iris of an eye is 1×10^{-2} metres.
c The mass of a billion molecules of water is 3×10^{-14} grams.

3 These numbers are *not* in standard form. Rewrite them in standard form.

a 123×10^2 **b** 0.8×10^7 **c** 17×10^{-2} **d** 0.25×10^{-4}
e 18 million **f** $\frac{1}{8}$ **g** $36 \times 10^4 \times 0.006$ **h** $\sqrt{40 \times 10}$

Calculators and standard form

To enter a number given in standard form on your calculator you can use the button marked

[×10ˣ] or [EXP]

So to enter 1.86×10^5 you press

[1] [·] [8] [6] [×10ˣ] [5]

This is much quicker than pressing

[1] [·] [8] [6] [×] [1] [0] [$x^■$] [5]

Never press [×] or [1] [0] when using [×10ˣ] or [EXP].

Example 8

B

Work out

a $(2.2 \times 10^3) \times (4 \times 10^5)$ **b** $(3.3 \times 10^6) \div (3 \times 10^{-4})$

a $(2.2 \times 10^3) \times (4 \times 10^5)$

To enter and work out this, you press

[2] [·] [2] [×10ˣ] [3] [×] [4] [×10ˣ] [5] [=]

The answer is 880 000 000 or 8.8×10^8.

b $(3.3 \times 10^6) \div (3 \times 10^{-4})$

To enter and work out this, you press

[3] [·] [3] [×10ˣ] [6] [÷] [3] [×10ˣ] [–] [4] [=]

The answer is 11 000 000 000 or 1.1×10^{10}.

Exercise 8F

B

1 Work out these, giving your answers in standard form.

a $(4.2 \times 10^4) \times (5.3 \times 10^3)$ **b** $(8.4 \times 10^5) \div (2 \times 10^4)$
c $(5.6 \times 10^3) \div (4 \times 10^{-2})$ **d** $(1.6 \times 10^{-2}) \div (4 \times 10^{-4})$

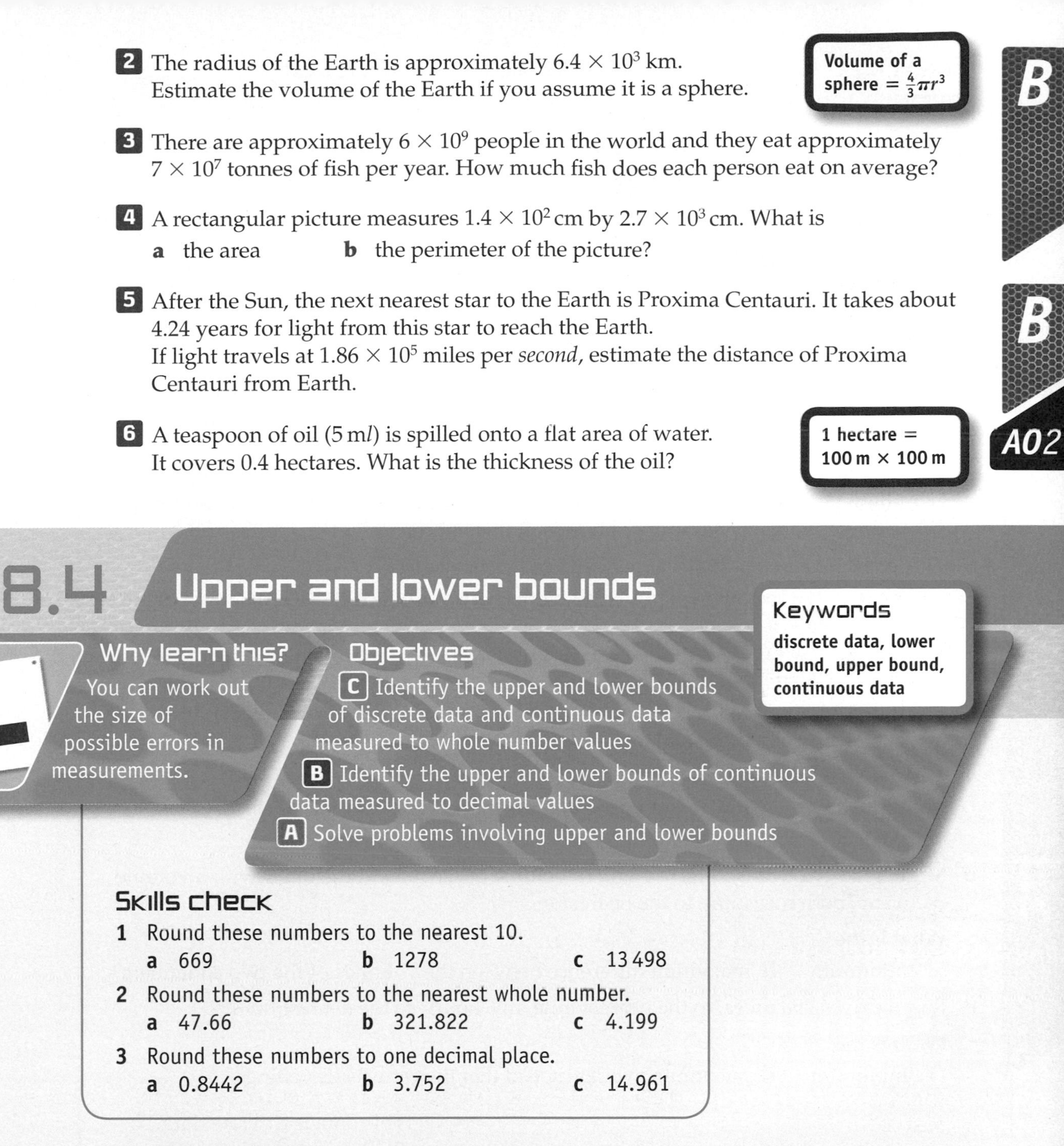

2 The radius of the Earth is approximately 6.4×10^3 km.
Estimate the volume of the Earth if you assume it is a sphere.

> **Volume of a sphere $= \frac{4}{3}\pi r^3$**

3 There are approximately 6×10^9 people in the world and they eat approximately 7×10^7 tonnes of fish per year. How much fish does each person eat on average?

4 A rectangular picture measures 1.4×10^2 cm by 2.7×10^3 cm. What is

a the area **b** the perimeter of the picture?

B

5 After the Sun, the next nearest star to the Earth is Proxima Centauri. It takes about 4.24 years for light from this star to reach the Earth.
If light travels at 1.86×10^5 miles per *second*, estimate the distance of Proxima Centauri from Earth.

B

6 A teaspoon of oil (5 m*l*) is spilled onto a flat area of water.
It covers 0.4 hectares. What is the thickness of the oil?

> **1 hectare = 100 m × 100 m**

AO2

8.4 Upper and lower bounds

Keywords
discrete data, lower bound, upper bound, continuous data

Why learn this?
You can work out the size of possible errors in measurements.

Objectives

- **C** Identify the upper and lower bounds of discrete data and continuous data measured to whole number values
- **B** Identify the upper and lower bounds of continuous data measured to decimal values
- **A** Solve problems involving upper and lower bounds

Skills check

1 Round these numbers to the nearest 10.

a 669 **b** 1278 **c** 13 498

2 Round these numbers to the nearest whole number.

a 47.66 **b** 321.822 **c** 4.199

3 Round these numbers to one decimal place.

a 0.8442 **b** 3.752 **c** 14.961

Finding the upper and lower bounds of discrete data

Discrete data can have only certain values, usually whole numbers (the number of goals scored in a match), but may include fractions (shoe sizes).

If there are 1400 students in a school, counted to the nearest 100, then the number of students in the school could be as few as 1350 (as this rounds up to 1400) or as many as 1449 (as this rounds down to 1400).

You say the **lower bound** is 1350 and the **upper bound** is 1449.
We write it like this:

$1350 \leqslant \text{number of students} \leqslant 1449$

This shows that the number of students is greater than or equal to 1350 but less than or equal to 1449.

Finding the upper and lower bounds of continuous data

Continuous data is data that can take any value, such as weight or length.

A length of 25 cm measured to the nearest cm could be as short as 24.5 cm (as this rounds up to 25 cm) or as long as 25.499 999 999... cm (as this rounds down to 25 cm).

But 25.499 999 999 cm is just 0.000 000 001 cm away from 25.5 cm – less than the width of one atom! 25.499 999 999... is so close to 25.5 that we use 25.5 for the upper bound. We write it like this:

24.5 cm ≤ distance < 25.5 cm

This shows that the distance is greater than or equal to 24.5 cm but less than 25.5 cm.

C

Example 9

a 78 million people, to the nearest million, watched the 2008 Brazilian F1 Grand Prix. Write down the lower and upper bounds of this figure.

b The length of a pencil is measured as 18 cm to the nearest cm.

i Write down the range of possible lengths between which the actual length of the pencil lies.

B

ii Lindsey has two pencils, one with a length of 18 cm and the other with a length of 14 cm, both measured to the nearest cm.
If they were placed end to end, what would be the range of possible lengths between which the actual total length of the two pencils lies?

c A square has sides of length 6.4 cm, measured correct to one decimal place.

i Write down the lower and upper bounds for the length of the sides.

ii Write down the lower and upper bounds for the area of the square.

A

d Miguel has two containers, one with a volume of 125 m*l* and the other with a volume of 165 m*l*, both measured to the nearest m*l*.

What is the

i minimum **ii** maximum difference between the volumes of the two containers?

e Rahul drives 250 miles, to the nearest mile, in 6 hours, to the nearest hour.

What is the

i minimum **ii** maximum average speed that the car was travelling?

a lower bound = 77 500 000
upper bound = 78 499 999

This data is discrete data, so the upper bound can take the exact value of 78 499 999.

b **i** 17.5 cm ≤ length < 18.5 cm

This data is continuous so you need to use '< 18.5 cm'.

ii lower bound = 17.5 + 13.5 = 31 cm
upper bound = 18.5 + 14.5 = 33 cm
31 cm ≤ total length < 33 cm

The lower bound for the length of the two pencils is the sum of the lower bounds for each pencil. The upper bound for the length of the two pencils is the sum of the upper bounds for each pencil.

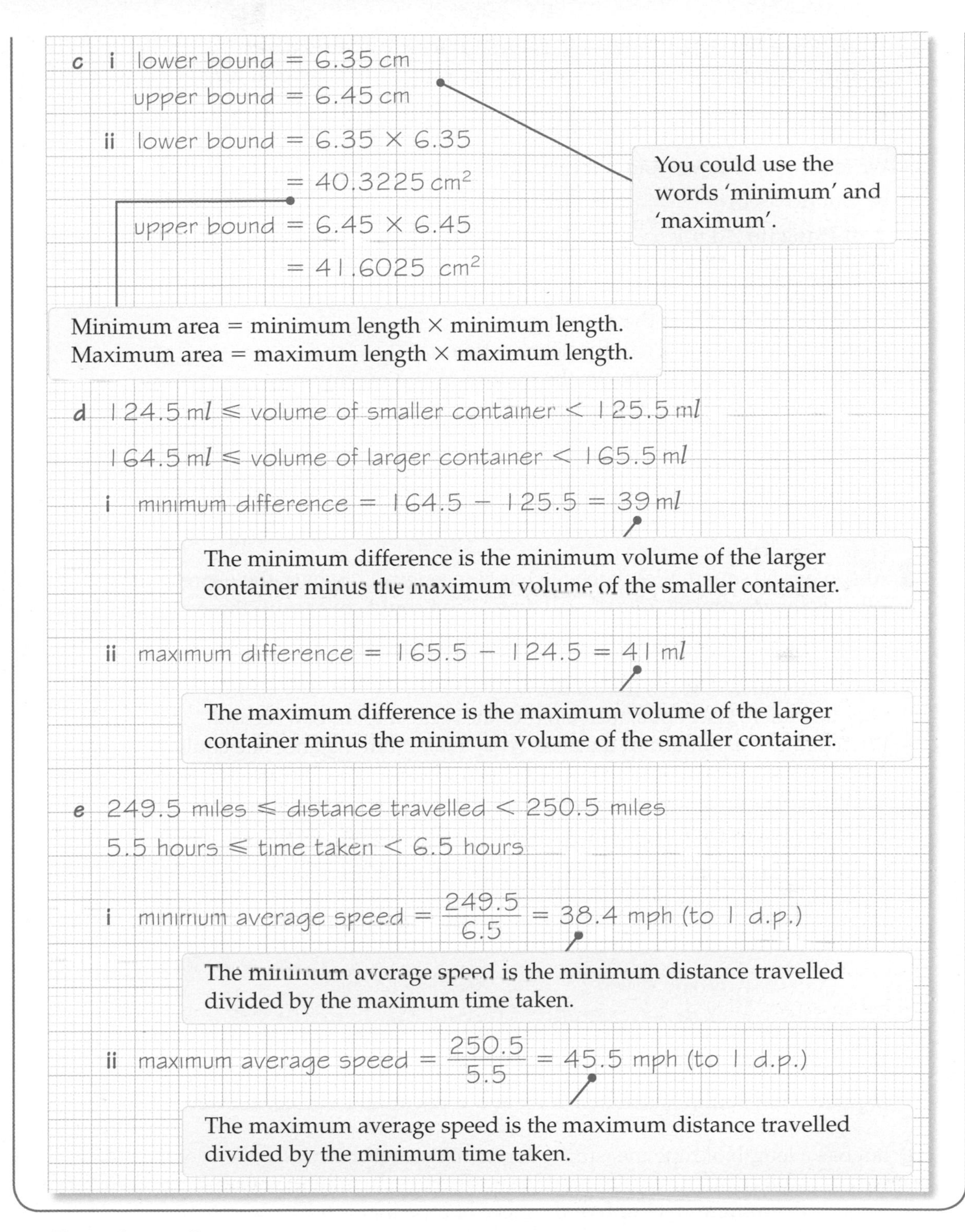

c i lower bound = 6.35 cm
upper bound = 6.45 cm
ii lower bound = 6.35 × 6.35
= 40.3225 cm²
upper bound = 6.45 × 6.45
= 41.6025 cm²
You could use the words 'minimum' and 'maximum'.
Minimum area = minimum length × minimum length.
Maximum area = maximum length × maximum length.
d 124.5 ml ≤ volume of smaller container < 125.5 ml
164.5 ml ≤ volume of larger container < 165.5 ml
i minimum difference = 164.5 − 125.5 = 39 ml
The minimum difference is the minimum volume of the larger container minus the maximum volume of the smaller container.
ii maximum difference = 165.5 − 124.5 = 41 ml
The maximum difference is the maximum volume of the larger container minus the minimum volume of the smaller container.
e 249.5 miles ≤ distance travelled < 250.5 miles
5.5 hours ≤ time taken < 6.5 hours
i minimum average speed = 249.5/6.5 = 38.4 mph (to 1 d.p.)
The minimum average speed is the minimum distance travelled divided by the maximum time taken.
ii maximum average speed = 250.5/5.5 = 45.5 mph (to 1 d.p.)
The maximum average speed is the maximum distance travelled divided by the minimum time taken.

Exercise 8G

1 Write down the upper and lower bounds for each of these quantities.

- **a** 16 000 people, to the nearest 1000
- **b** 250 dogs, to the nearest 10
- **c** 3400 pens, to the nearest 100
- **d** 850 children, to the nearest 50

2 Write down the range of possible values for these measurements.

- **a** 60 mph, measured to the nearest 1 mph
- **b** 13 400 km, measured to the nearest 100 km
- **c** 130 m*l*, measured to the nearest 10 m*l*
- **d** 21.5 cm, measured to the nearest tenth of a cm

C

B

3 Write down the range of possible values for these measurements.

a 8.9 kg (to 1 d.p.) **b** 100.2 m*l* (to 1 d.p.) **c** 8.39 m (to 2 d.p.)

d 1110 g (to 3 s.f.) **e** 1100 g (to 2 s.f.) **f** 1000 g (to 1 s.f.)

4 Write down the lower and upper bounds of each of these measurements.

a 7 m (to 1 s.f.) **b** 9.3 days (to 2 s.f.) **c** 4.8 kg (to 1 d.p.)

d 0.75 mg (to 2 d.p.) **e** 3.10 c*l* (to 3 s.f.) **f** 19.999 cm (to 3 d.p.)

B

5 Alfie has a bag containing 25 identical marbles.
The bag weighs 15 g to the nearest gram.
Each marble weighs 72 g to the nearest gram.
What is the difference between the minimum and maximum possible weights of the bag of marbles?

6 This cube has a side length of 5 cm to the nearest cm.
What is the smallest possible volume of the cube?

5 cm
5 cm
5 cm

7 Alison has 50 square tiles. Each tile has side length 12.5 cm to the nearest mm.
What is the maximum area of wall that Alison could cover with these tiles?
Assume the tiles touch edge to edge.

FUNCTIONAL • FUNCTIONAL •

A

8 Abby can run at 7.3 m/s (to 1 d.p.).
She runs for 13.7 seconds, measured to the nearest tenth of a second.
What is the **a** minimum distance **b** maximum distance she could have run?

9 $x = 11.2$ (to 1 d.p.) $y = 18$ (to 2 s.f.) $z = 0.55$ (to 2 d.p.)
What is the **a** minimum **b** maximum value of $2x^2(y + z)$?

A

10 This is part of Paulin's homework.

Q1: Find the minimum side length of a cube with volume 450 cm³ (to 2 s.f.).

Length = $\sqrt[3]{450}$ = 7.663 09... = 7.7 (2 s.f.)

Minimum = 7.65 cm

AO2

Explain what Paulin has done wrong and work out the correct answer.

A

11 Ian has a length of rope measuring 1 m to the nearest cm.
He cuts a piece off the end of the rope measuring 12 cm to the nearest cm.
What is the minimum length of the piece of rope that is left?

12 Jade ran 100 m, measured to the nearest m, in 14.1 seconds, measured to the nearest tenth of a second.
What is the maximum average speed she could have ran?
Give your answer in m/s.

A

13 Charlie is going to pave his garden path. The path is 30 m long to the nearest metre.
He plans to use paving slabs that are 40 cm long to the nearest 5 cm.
Charlie says, '3000 ÷ 40 = 75, so if I buy 80 slabs I'm sure to have enough'.
Is Charlie correct? Explain your answer.

FUNCTIONAL • FUNCTIONAL •

AO2

8.5 Error

Keywords
nominal value, absolute error, percentage error

Why learn this?
Machine parts need to be made accurately – the percentage error must be very small.

Objectives
A Calculate absolute error and percentage error
A* Calculate absolute error and percentage error in more complex situations

Skills check

1 Write down the range of possible values for these measurements.
 a 12 cm, measured to the nearest 1 cm
 b 50 litres, measured to the nearest 10 litres

HELP Section 8.4

2 Write down the lower and upper bounds of each of these measurements.
 a 20 m (to 1 s.f.) b 120 m (to 2 s.f.)
 c 5.2 kg (to 1 d.p.) d 3.45 mg (to 2 d.p.)
 e 15.0 m*l* (to 1 d.p.) f 600 g (to 1 s.f.)

Calculating absolute and percentage error

The **nominal value** is the value that a quantity is supposed to be if there were no errors.

The **absolute error** is the difference between the measured value and the nominal value.

The **percentage error** is found by finding the absolute error as a percentage of the nominal value.

$$\text{Percentage error} = \frac{\text{absolute error}}{\text{nominal value}} \times 100\%$$

Example 10

A A* AO2

a A bag of flour should weigh 1.5 kg, but actually weighs 1.45 kg.
Calculate the absolute error and the percentage error.

b A rectangle measures 4.5 cm by 6.7 cm.
Both measurements are correct to one decimal place.
Calculate the maximum percentage error in the area of the rectangle.

a Absolute error $= 1.5 - 1.45 = 0.05$ kg

Absolute error = nominal value − measured value

Percentage error $= \frac{0.05}{1.5} \times 100\% = 3\frac{1}{3}\%$

Percentage error $= \frac{\text{absolute error}}{\text{nominal value}} \times 100\%$

b Nominal area $= 4.5\text{ cm} \times 6.7\text{ cm}$
$= 30.15\text{ cm}^2$

First, work out the nominal area.

For the shorter side:
$4.45\text{ cm} \leqslant \text{length} < 4.55\text{ cm}$

For the longer side:
$6.65\text{ cm} \leqslant \text{length} < 6.75\text{ cm}$

Least area $= 4.45\text{ cm} \times 6.65\text{ cm}$
$= 29.5925\text{ cm}^2$

Greatest area $= 4.55\text{ cm} \times 6.75\text{ cm}$
$= 30.7125\text{ cm}^2$

Then work out the upper and lower bounds of the side lengths of the rectangle, then the least and greatest areas of the rectangle.

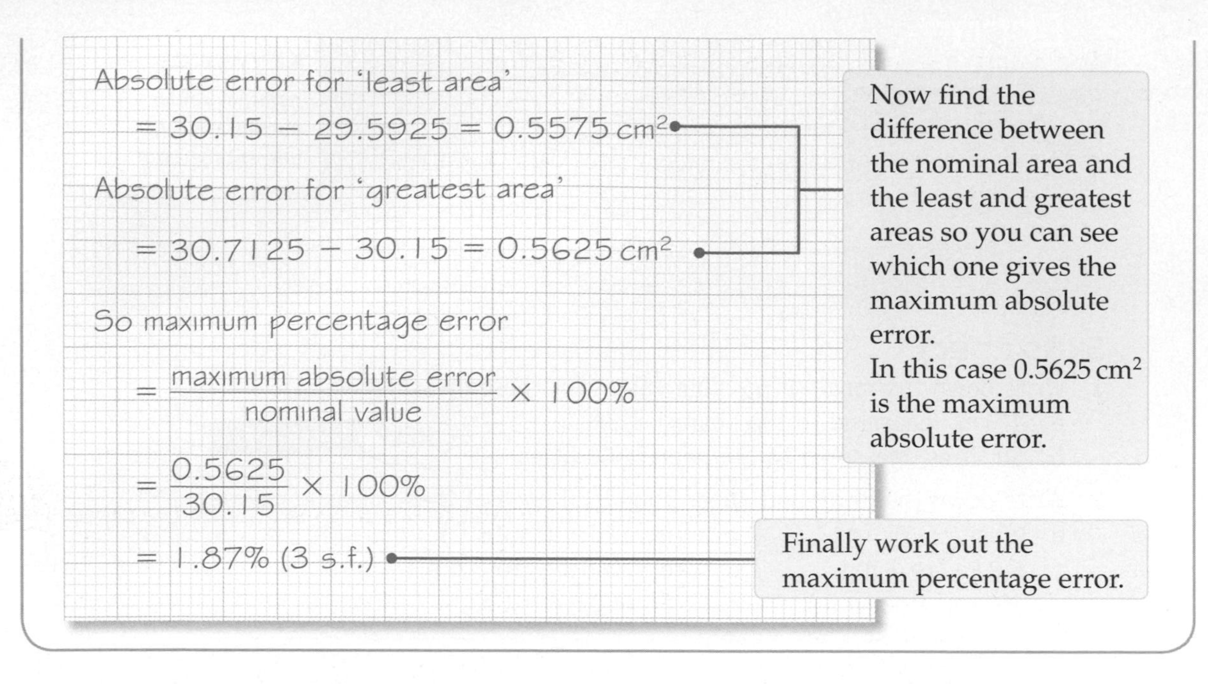

Absolute error for 'least area'
= 30.15 − 29.5925 = 0.5575 cm²
Absolute error for 'greatest area'
= 30.7125 − 30.15 = 0.5625 cm²
So maximum percentage error
= maximum absolute error / nominal value × 100%
= 0.5625 / 30.15 × 100%
= 1.87% (3 s.f.)
Now find the difference between the nominal area and the least and greatest areas so you can see which one gives the maximum absolute error.
In this case 0.5625 cm² is the maximum absolute error.
Finally work out the maximum percentage error.

Exercise 8H

A

1 A sculpture is supposed to be 6 m tall but when checked measures 6.05 m.
Calculate the absolute error and the percentage error.

2 A packet of biscuits weighs 194 g. It should weigh 200 g.
Work out the percentage error.

A AO2

3 Natalie is told that the capacity of a 750 m*l* bottle of squash may have up to a 1.5% error. Natalie says, 'The bottle of squash could contain any amount between 738.75 m*l* and 761.25 m*l*.'
Is Natalie correct? Show working to support your answer.

A*

4 A triangle has a base length of 8.2 cm and a height of 4.6 cm.
Both measurements are correct to one decimal place.
Calculate

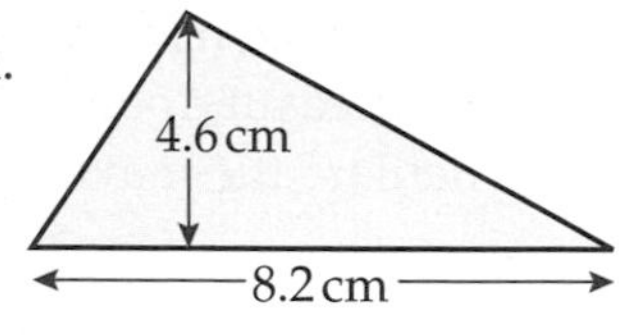

- **a** the nominal area of the triangle
- **b** the upper and lower bounds of the base length and height
- **c** the least and greatest possible areas of the triangle
- **d** the maximum absolute error
- **e** the maximum percentage error.

5 A circle has a radius of 14 cm, correct to 2 s.f.
Calculate the maximum percentage error in the area of the circle.

Area of circle $= \pi r^2$

AO2

6 Jill has two pieces of material that she is joining together to make some curtains. One piece is 1.4 m wide and the other is 70 cm wide, both measurements correct to the nearest 10 centimetres. When she joins the pieces of material she will lose exactly 3 cm in width altogether.
What is the maximum percentage error in the total width of the curtains?

7 The diagram shows the cross-section of a pipe.
The external diameter of the pipe is 6.5 cm.
The internal diameter of the pipe is 5.0 cm.
Both diameters are measured correct to one decimal place.

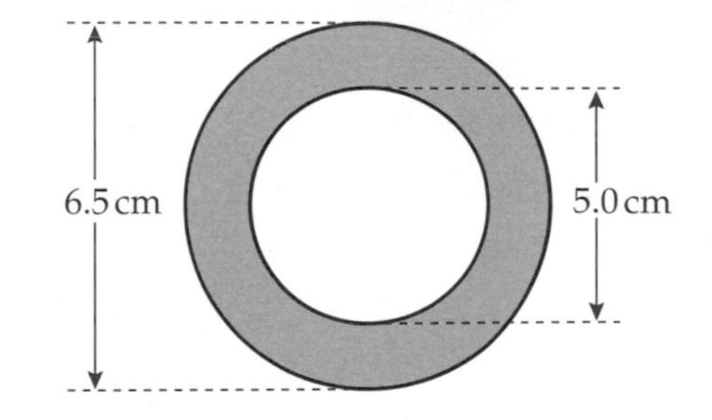

Calculate the maximum percentage error in the area of the cross-section of the pipe.

AO3

Review exercise

1 Amy invests £1025 for four years at a rate of 4.1% per annum compound interest.
George invests £997 for four years at a rate of 4.9% per annum compound interest.
Who will have more money after four years?
Show your working. **[3 marks]**

C

2 The value of a car depreciates by 17% in its first year.

Assume VAT = 17.5%.

a If it is worth £12 035 after one year, what was its original price? **[2 marks]**

b The original price included VAT.
How much was the car excluding VAT? **[2 marks]**

B

3 The number of krill (a type of plankton) is estimated to be 6×10^{15} and their mass is about 6.5×10^{8} tonnes.
What is the mass of one krill in grams? **[2 marks]**

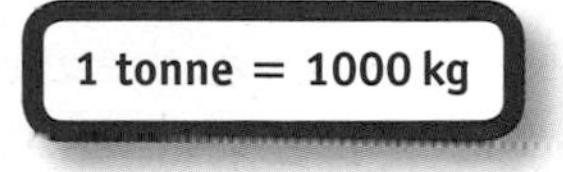

4 A cube has a side length of 14 cm measured to two significant figures.
What is the smallest possible volume of the cube? **[3 marks]**

AO2

5 Mark can run at 7.9 m/s, correct to two significant figures.
He runs for 25.2 seconds, correct to the nearest tenth of a second.
Calculate the

a minimum distance **[2 marks]**

b maximum distance he can run. **[2 marks]**

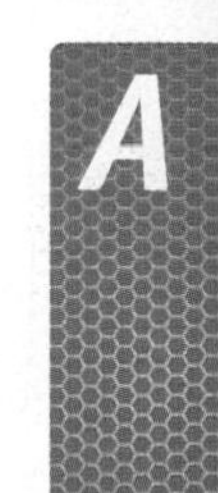

6 A formula used in applied mathematics is $v = u + at$.
Calculate the minimum value of t when $v = 6.4$, $u = 1.8$ and $a = 0.8$.
All values are given correct to one decimal place. **[4 marks]**

A

7 Anton runs 200 m in 24.8 seconds.
The distance is measured to the nearest metre and the time is measured to three significant figures.
Anton says, 'My maximum average speed could have been 8.06 metres per second'.
Is Anton correct? Show working to support your answer. **[3 marks]**

8 $x = 11.2$ measured to one decimal place.

$y = 18$ measured to two significant figures.

$z = 0.55$ measured to two decimal places.

What is the maximum value of $\frac{2x^2}{y + z}$? **[3 marks]**

A*

9 A cuboid is supposed to have dimensions of length = 8.5 cm, width = 4.5 cm and height = 4.0 cm. When it is checked, these measurements are found to be correct to the nearest millimetre.
Calculate the maximum percentage error in the volume of the cuboid.
Give your answer correct to three significant figures. **[5 marks]**

10 The diagram shows the length of a plastic pipe.
The external diameter of the pipe is 11.0 cm.
The internal diameter of the pipe is 8.5 cm.
The length of the pipe is 2.3 m.
All measurements are given correct to one decimal place.

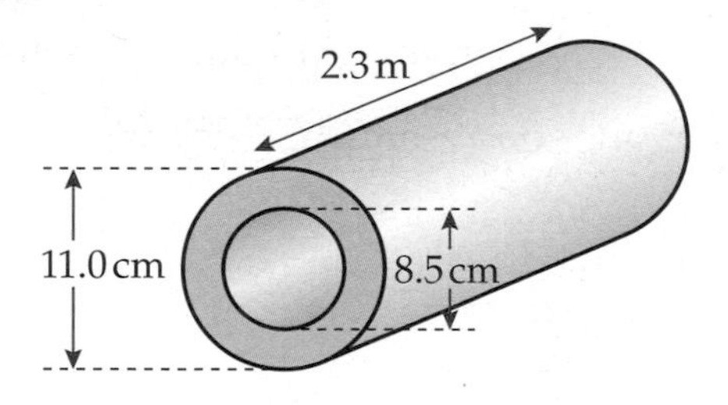

AO3

Calculate the maximum percentage error in the volume of plastic used to make the pipe. **[5 marks]**

Chapter summary

In this chapter you have learned how to

- perform calculations involving repeated percentage change **C**
- identify the upper and lower bounds of discrete data and continuous data measured to whole-number values **C**
- perform reverse percentage calculations **B**
- interpret and use standard form **B**
- identify the upper and lower bounds of continuous data measured to decimal values **B**
- solve problems involving upper and lower bounds **A**
- calculate absolute error and percentage error **A** **A***

Quality of written communication: Some questions on this page are marked with a star ☆. In the exam, this sort of question may earn you some extra marks if you
- use correct and accurate maths notation and vocabulary
- organise your work clearly, showing that you can communicate effectively.

C

☆ **1** Anselm wants to survey ten members of his class to find out how often they watch the news. He chooses the first ten names in the register.

a Is Anselm's sample likely to be representative. Give a reason for your answer. [2]

b Is this a random sample? Give a reason for your answer. [2]

c Describe one way of choosing a random sample of ten students from a class. [1]

AO2

B

2 The probability that it rains in the morning is 0.6. The probability that it rains in the afternoon is 0.8.

a Draw a tree diagram to show all the possible outcomes. [3]

b Work out the probability that it rains at some point during the day. [2]

c Calculate the probability that it does not rain all day. [2]

AO2

A

☆ **3** Eisha measured the heights of the boys and girls in her year group. She recorded her results in a grouped frequency table.

Height, h (cm)	Boys' frequency	Girls' frequency
$140 \leqslant h < 150$	9	23
$150 \leqslant h < 160$	14	26
$160 \leqslant h < 170$	32	17
$170 \leqslant h < 180$	20	6
$180 \leqslant h < 190$	5	0

a Draw a histogram to represent the boys' heights. [3]

b Draw a histogram to represent the girls' heights. [3]

c Write a statement comparing the boys' and girls' heights. Use evidence from your histograms to support your statement. [2]

AO2

A*

4 Beccy has two bags of numbered balls.

She chooses a ball at random from bag A and places it in bag B. Then she chooses a ball at random from bag B and places it in bag A.

Calculate the probability that the sum of the balls in bag A will be 18 or greater. [4]

AO2

Quality of written communication: Some questions on this page are marked with a star ☆. In the exam, this sort of question may earn you some extra marks if you
- use correct and accurate maths notation and vocabulary
- organise your work clearly, showing that you can communicate effectively.

D

☆ **1** Kirsten is investigating this hypothesis:

'There is no relationship between the amount of time you spend in full-time education and your salary.'

She asked 15 people about their starting salary in their first job.

Number of years of post-16 education	2	2	0	5	5	4	2	3	0	6	4	0	5	2	1
Starting salary in first job (£1000s)	17	21	12	24.5	23	20.5	18.5	18	10.5	25	19.5	13	20	14.5	15

a Draw a scatter graph to display this data. [3]

AO3 **b** Do you agree with Kirsten's hypothesis? Give reasons for your answer. [2]

C

☆ **2** Pippa and Nisha are playing a game. They each roll a fair 6-sided dice.
If the difference between the two numbers is odd then Pippa wins a point.
If the difference between the two numbers is even, Nisha wins a point.
If the two numbers are the same nobody wins a point.

a Is this game fair? Give a reason for your answer. [3]

AO3 **b** Calculate Pippa and Nisha's expected scores after 150 rolls. [2]

A

☆ **3** This table shows the numbers of students at five different primary schools in Manchester.

Southfield	Harcourt	Mulberry	Ridgeway	Seven Sisters
243	397	679	186	492

The council wants to carry out a survey about road safety.
It needs to select a stratified sample of 100 students.

a Calculate the number of students from each school that should be included in the sample. [5]

AO3 **b** Explain why it would be easier for the council to use a stratified sample for this survey than to select a random sample from the entire population. [2]

A*

4 Aaron has invented a game for the school fête. He shuffles these number cards.

Each player pays £1 to play and picks two cards.
If the total on the two cards is 15 or more the player wins £10.

AO3 Will Aaron make a profit on his game? Show all of your working. [4]

Quality of written communication: Some questions on this page are marked with a star ☆. In the exam, this sort of question may earn you some extra marks if you

- use correct and accurate maths notation and vocabulary
- organise your work clearly, showing that you can communicate effectively.

☆ **1** Ian and Maya have eaten a pizza. Ian ate 5 slices and Maya ate 7 slices. Ian says that he ate $\frac{5}{7}$ of the pizza. Is he correct? Give a reason for your answer. [2]

2 A mining company calculates price indexes for iron ore using 2004 as the base year.

Year	2004	2005	2006	2007	2008
Price per tonne (pence)	60	72	90	140	126

a Calculate the price index for iron ore in 2005. [2]

b Calculate the price index for iron ore in 2008. [1]

c In 2009 the price index was 182. Calculate the price per tonne of iron ore in 2009. [2]

3 At a sixth form college, the ratio of male to female students is 7 : 6. There are 144 female students at the college. How many students are there in total? [2]

4 David invests £2600 in a bank account which pays compound interest. After four years he has £3221 in his account. Calculate the annual rate of compound interest. Give your answer as a percentage to 1 decimal place. [3]

5 A tennis ball has a mass of 57 g to the nearest gram. During the Wimbledon tennis tournament 32 000 balls are used. The total mass of all the tennis balls used during the tournament is m kg. Calculate

a the upper bound for m [3]

b the lower bound for m. [2]

6 A ream of paper contains 500 sheets. It is 4.2 cm thick. Calculate the thickness of one sheet of paper in metres. Give your answer in standard form. [2]

7 Supraj is going on holiday. He has one piece of baggage that weighs 11 kg and two pieces of baggage that each weigh 4 kg, all correct to the nearest kg. The airline has a weight limit of 20 kg. Supraj says that because 11 + 4 + 4 is 19 he will definitely be within the weight limit. Show working to explain why Supraj is wrong. [3]

8 An electrical shop sells soldering wire. The cost of the wire is directly proportional to the length of wire sold. Nadia buys 2.8 m of wire for £8.96.

a Using c to represent the cost of the wire, and l to represent the length sold, write a formula for c in terms of l. [2]

b Work out the cost of 65 cm of wire. [2]

c Nadia buys another piece of wire for £5.12. How long was this piece of wire? [2]

Quality of written communication: Some questions on this page are marked with a star ☆. In the exam, this sort of question may earn you some extra marks if you
- use correct and accurate maths notation and vocabulary
- organise your work clearly, showing that you can communicate effectively.

D AO3

1 A bank offers this exchange rate from pounds to euros. Jamal changes £250 into euros. He receives €272.25. How much has the bank charged him to make the exchange? Give your answer in pounds. [3]

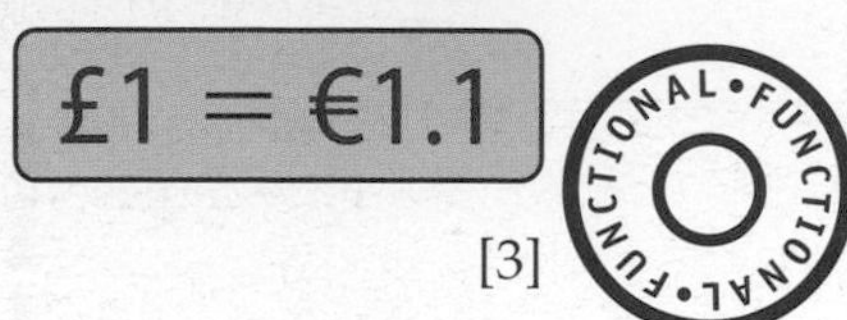

C AO3

2 Dennis, Poppy and Max get pocket money each week. Max receives half his age in pounds. Poppy's pocket money is $\frac{4}{5}$ of Max's. Dennis receives $\frac{2}{3}$ as much as Poppy. Dennis receives £3.20 a week. How old is Max? [4]

B AO3

3 One fisherman can mend 2 m^2 of net in 45 minutes. The captain of a boat wants to mend this rectangular fishing net in less than three hours.
How many fishermen will he need? [3]

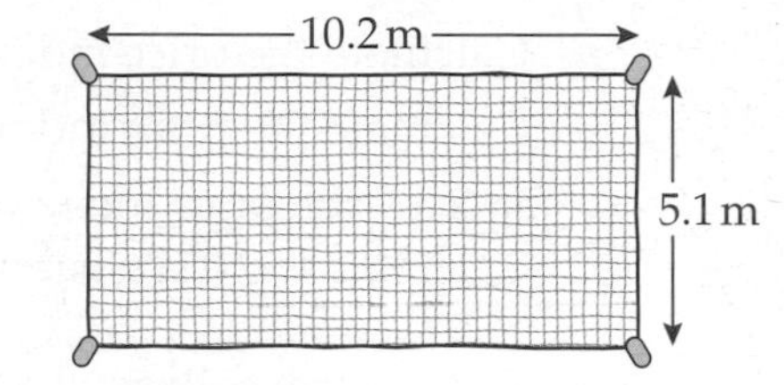

A AO3

4 Marta is laying tiles in her bathroom. She measures the width of one tile as 19 cm, correct to the nearest cm. She needs to lay a row of 27 tiles. She says her row will be 5.1 m long.

a Calculate the maximum possible percentage error for the length of the row of tiles. [3]

b Do you think Marta has given her answer to a suitable degree of accuracy? [2]

5 Finian is investigating the relationship between the period of a pendulum and the length of the string.
He uses T to represent the period and l to represent the length of the string.
He records these results.

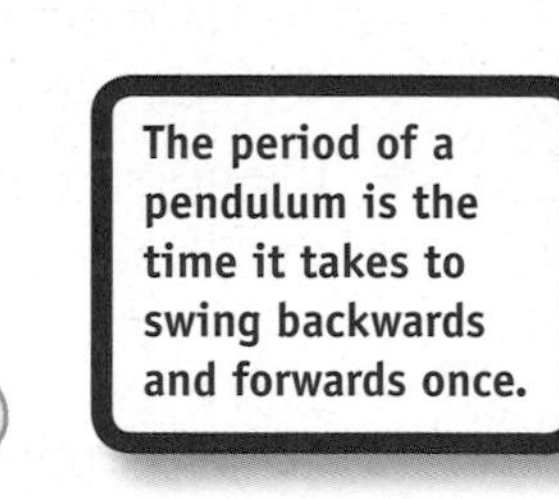

The period of a pendulum is the time it takes to swing backwards and forwards once.

l (m)	0.8	1.2	2.7
T (s)	1.8	2.2	3.3

a Which of these relationships best describes Finian's results? [3]

A $l \propto T$ **B** $l \propto T^2$ **C** $T \propto l^2$ **D** $T \propto l^3$

b Estimate the period of a pendulum with length 50 cm. [1]

A* AO3

☆ **6** Archie is making a box to store Christmas decorations.
He wants to pack a glass sphere with a diameter of 6.6 cm (correct to 1 decimal place) into a box shaped like an equilateral triangle with a side length of 12 cm (correct to the nearest cm).
Will the sphere definitely fit?
Explain your answer. [6]

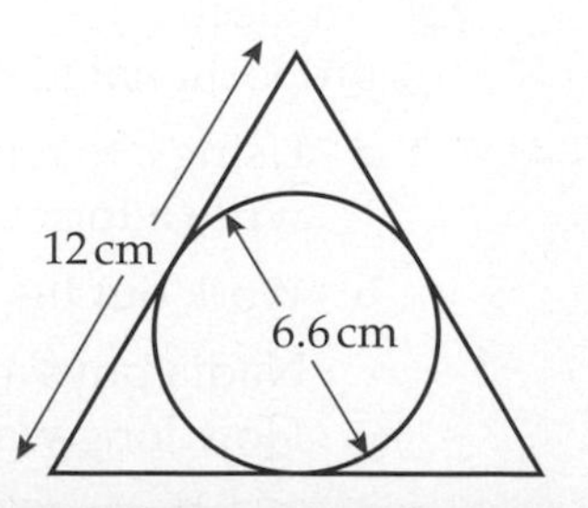

9 Estimation and currency conversion

This chapter is about maths strategies and real-life applications of maths.

Currencies are bought and sold on the foreign exchange market. In 2001, an estimated \$1.2 trillion was traded on the foreign exchange market each day!

Objectives

This chapter will show you how to

- **check and estimate answers to problems** D C
- **estimate answers to problems involving decimals** D C
- **recall integer squares from 2 × 2 to 15 × 15 to make mental estimates of the answers to calculations** D C
- **select and use suitable strategies and techniques to solve problems and word problems** D C

Before you start this chapter

Put your calculator away!

1 Work out

a $24 \div (3 \times 4) + 5$ b $4 \times 6 - 3^2$ c $12 - (2^3 + 3)$

2 Work out

a 12^2 b 20^2 c 400×1.6 d $\frac{260}{13}$

P

Number skills: multiplying and dividing

1. Work out

   a 36×24 b 108×72 c $486 \div 27$ d $1008 \div 36$

2. Prize money of £29 805 is shared equally between 15 people. How much does each person receive?

AO3

3. Two consecutive numbers have a product of 182 and a sum of 27. What are the two numbers?

4. You are told that $87 \times 132 = 11\,484$.

   Work out $\frac{11\,484}{870}$.

5. What is the remainder when 258 is divided by 19?

AO2

6. 475 football supporters are travelling to a match by coach. Each coach seats 45 people.

   a How many coaches will be required?

   b How many spare seats will there be?

FUNCTIONAL • FUNCTIONAL •

Number skills: negative numbers

1. Arrange each set of numbers in size order, with the smallest first.

   a 2, 0, −4, 3, −6 b −11, −19, −3, −7, −9

2. The temperature in Istanbul is 3°C. The temperature in Moscow is −11°C. What is the temperature difference between these two cities?

3. Copy and complete these calculations.

   a $4 + \square = -7$ b $-8 + \square = 2$ c $6 + \square = -2$ d $\square - -4 = 7$

4. Copy and complete these calculations.

   a $6 \times \square = -18$ b $24 \div \square = -3$ c $\square \times -7 = 21$

   d $-40 \div \square = -10$ e $\square \div -5 = -9$ f $-2 \times -3 \times 2 = \square$

   g $\square \times 4 \times -3 = -36$ h $-2 \times \square \times -3 = -30$

5. What do you need to multiply −9 by to get 72?

6. Find two different pairs of numbers that have a difference of 7 and a product of −12.

9.1 Estimation

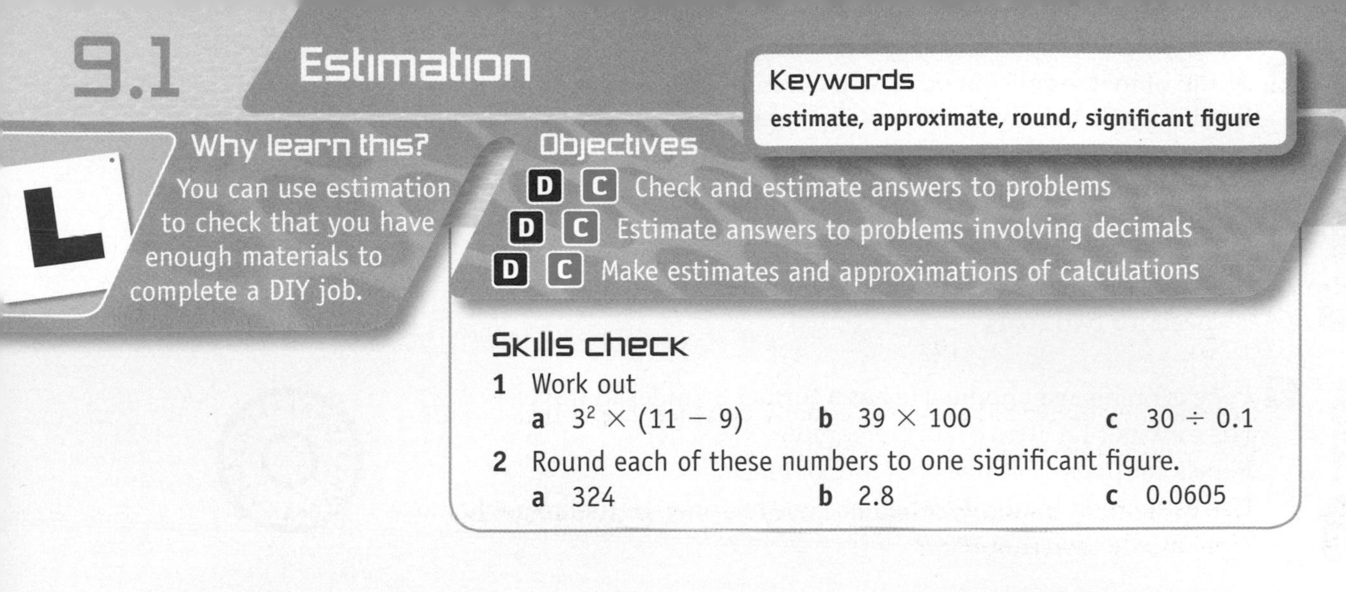

Keywords
estimate, approximate, round, significant figure

Why learn this?
You can use estimation to check that you have enough materials to complete a DIY job.

Objectives

D **C** Check and estimate answers to problems
D **C** Estimate answers to problems involving decimals
D **C** Make estimates and approximations of calculations

Skills check

1 Work out
 a $3^2 \times (11 - 9)$ b 39×100 c $30 \div 0.1$
2 Round each of these numbers to one significant figure.
 a 324 b 2.8 c 0.0605

Estimating

Estimating the answer to a calculation gives you an **approximate** answer.

You can use estimation to check that an answer is about right.

To estimate

- **round** all the numbers to one **significant figure**
- do the calculation using these approximations.

Example 1

D C

Use approximation to estimate the answer to each of these calculations.

a $\frac{119 \times 5.4}{46}$ **b** $\frac{5.3 \times 19.8}{6.2 - 1.7}$ **c** $\frac{560 \times 5.45}{0.534}$

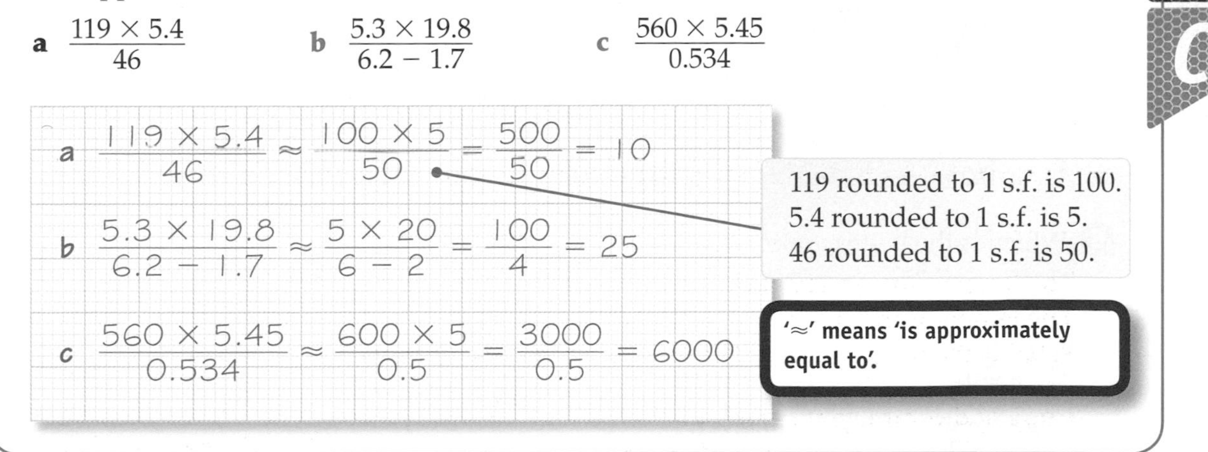

a $\frac{119 \times 5.4}{46} \approx \frac{100 \times 5}{50} = \frac{500}{50} = 10$

b $\frac{5.3 \times 19.8}{6.2 - 1.7} \approx \frac{5 \times 20}{6 - 2} = \frac{100}{4} = 25$

c $\frac{560 \times 5.45}{0.534} \approx \frac{600 \times 5}{0.5} = \frac{3000}{0.5} = 6000$

119 rounded to 1 s.f. is 100.
5.4 rounded to 1 s.f. is 5.
46 rounded to 1 s.f. is 50.

'≈' means 'is approximately equal to'.

Exercise 9A

D

1 Estimate the answer to each of these calculations.

a $\frac{436 + 394}{109}$ **b** $\frac{27 \times 105}{55}$ **c** $\frac{40.26 \times 8.49}{16.4}$ **d** $\frac{324 \times 7.63}{75.9}$

2 Estimate the answer to each of these calculations.

a $\frac{5.4 \times 19.8}{4.3 - 2.2}$ **b** $\frac{17.32 + 14.29}{4.08 - 1.79}$ **c** $\frac{584 + 829}{749 - 485}$ **d** $\frac{294 + 149}{842 - 385}$

C

3 A litre of matt emulsion paint will cover an area of about 15.5 m^2.
Rashid needs to paint a room with a total wall area of 645 m^2.

a Which is the best approximate calculation to use for this problem?

A $600 \div 20$ **B** $600 \div 15$ **C** $600 \div 10$

Give a full reason for your answer.

AO2

b Use estimation to decide how many litres of paint Rashid will need for two coats.

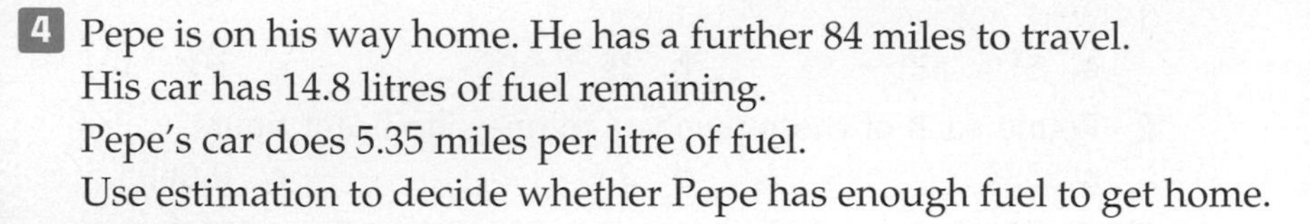

C

4 Pepe is on his way home. He has a further 84 miles to travel.
His car has 14.8 litres of fuel remaining.
Pepe's car does 5.35 miles per litre of fuel.
Use estimation to decide whether Pepe has enough fuel to get home.

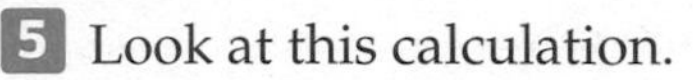

C

5 Look at this calculation.

$$\frac{1.8 \times 584.2}{7.43 + 84.89 + 7.68}$$

Steven and Dirk use calculators to work it out.
Steven gets an answer of 10.52. Dirk gets 7.04.
Use approximation to decide who is more likely to be correct.

6 Estimate the value of each of these calculations.

a $\dfrac{394.25 \times 22}{83.6 - 37.25}$

b $\dfrac{3 \times 584.7}{(39.25 + 18.5) \times 10.4}$

c $\dfrac{75.5 \times 32.6}{(527.5 + 149.25) \times 7.6}$

7 Use approximation to estimate the value of each of these calculations.

a $\dfrac{325 \times 4.34}{0.237}$ **b** $\dfrac{26.79}{3.51 \times 0.48}$ **c** $\dfrac{5.62 \times 478}{64.5 \times 0.527}$

C

8 Which is the better approximation for this calculation?

$$\frac{7.25^2}{1.86 \times 12.42}$$

A $\dfrac{7^2}{2 \times 12}$ **B** $\dfrac{7^2}{2 \times 10}$

AO2

Give a reason for your answer.

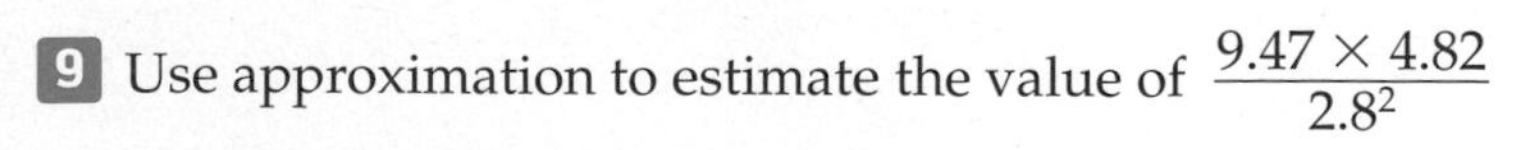

C

9 Use approximation to estimate the value of $\dfrac{9.47 \times 4.82}{2.8^2}$

10 Estimate the answer to each of these calculations.

a $\dfrac{19.5 \times 8.3}{0.53}$ **b** $\dfrac{84.9 \times 3.3}{0.13}$

c $\dfrac{98.6 + 9.6}{0.17}$ **d** $\dfrac{98.6 \times 9.6}{0.17}$

9.2 Conversions

Keywords
currency, exchange rate

Why learn this?
The exchange rate lets you work out how many euros you can get for your pounds.

Objectives

D **C** Convert between different currencies

Skills check

1 Work out
   a 480 ÷ 10   b 326 × 20   c 30 × 1.5
2 Work out
   a 23 × 19   b 40 × 1.6   c 384 ÷ 16

Currency conversion

Most countries in the world have their own **currency.** An exception is the eurozone, where the euro is the single currency. The eurozone covers most of Europe apart from the UK.

To convert one currency to another, you need to know the **exchange rate.**

The exchange rate tells you what one unit of currency is equal to in another country's currency.

Every currency has an exchange rate with all other currencies.

You either multiply or divide by the exchange rate depending on whether you are converting to or from the given currency.

Example 2

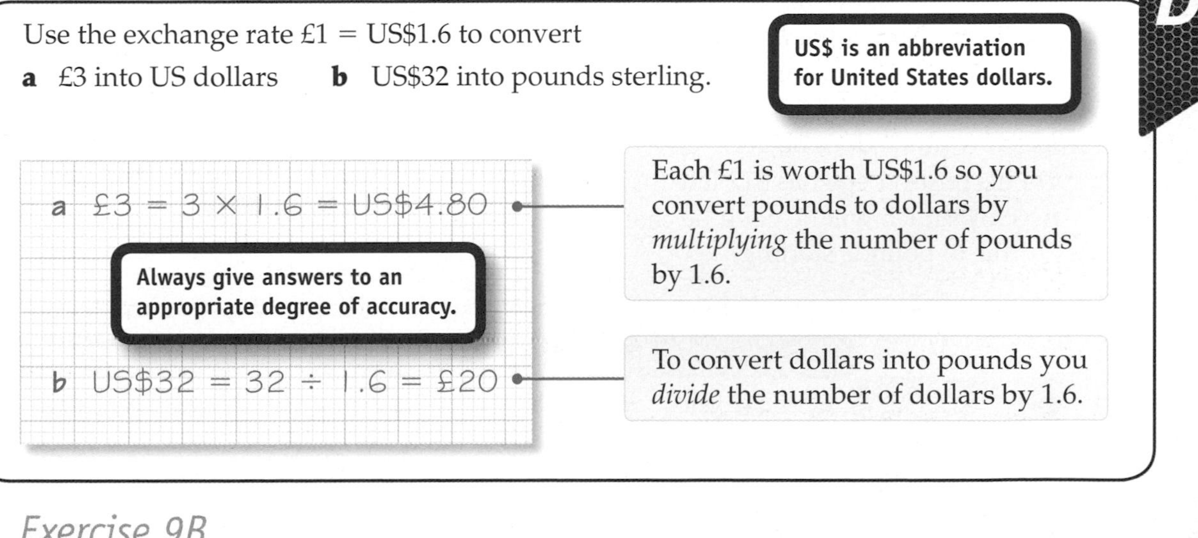

Use the exchange rate £1 = US$1.6 to convert
a £3 into US dollars b US$32 into pounds sterling.
US$ is an abbreviation for United States dollars.
a £3 = 3 × 1.6 = US$4.80
Each £1 is worth US$1.6 so you convert pounds to dollars by *multiplying* the number of pounds by 1.6.
Always give answers to an appropriate degree of accuracy.
b US$32 = 32 ÷ 1.6 = £20
To convert dollars into pounds you *divide* the number of dollars by 1.6.

D

Exercise 9B

1 The exchange rate from pounds to CAN$ is £1 = CAN$2.
Convert each of these sums of money into CAN$.
   a £300   **b** £930   **c** £8250   **d** £15.75

2 The currency exchange rate between Swedish kronor and pounds is £1 = 11 kronor.
Convert each of these sums of money into pounds.
   a 198 kronor   **b** 253 kronor   **c** 385 kronor   **d** 814 kronor

3 Harry is going to Mexico.
He wants to change £275 into pesos.
The exchange rate is 19 pesos for £1.
How many pesos will Harry get?

D

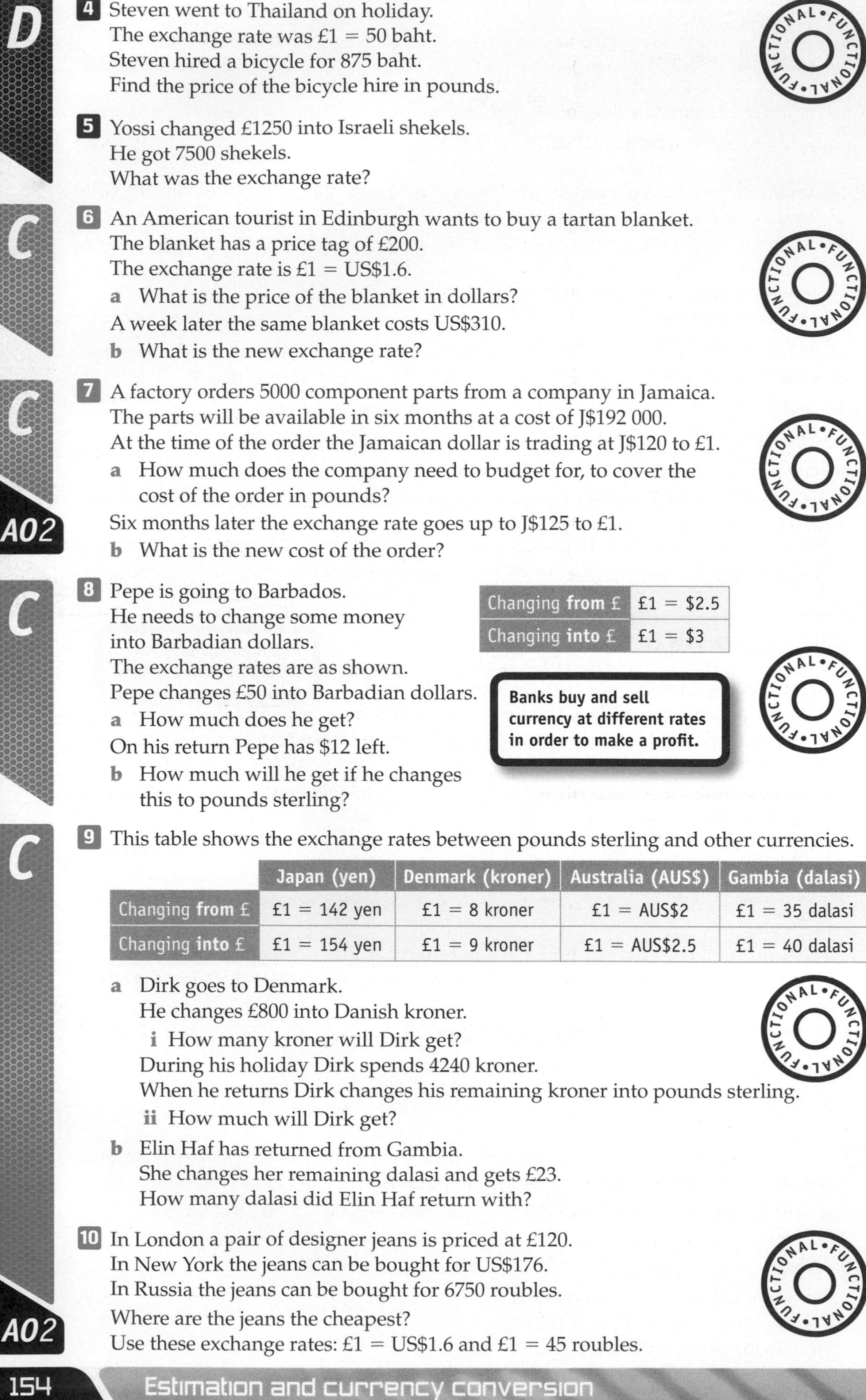

4 Steven went to Thailand on holiday.
The exchange rate was £1 = 50 baht.
Steven hired a bicycle for 875 baht.
Find the price of the bicycle hire in pounds.

5 Yossi changed £1250 into Israeli shekels.
He got 7500 shekels.
What was the exchange rate?

6 An American tourist in Edinburgh wants to buy a tartan blanket.
The blanket has a price tag of £200.
The exchange rate is £1 = US$1.6.

a What is the price of the blanket in dollars?

A week later the same blanket costs US$310.

b What is the new exchange rate?

7 A factory orders 5000 component parts from a company in Jamaica.
The parts will be available in six months at a cost of J$192 000.
At the time of the order the Jamaican dollar is trading at J$120 to £1.

a How much does the company need to budget for, to cover the cost of the order in pounds?

Six months later the exchange rate goes up to J$125 to £1.

b What is the new cost of the order?

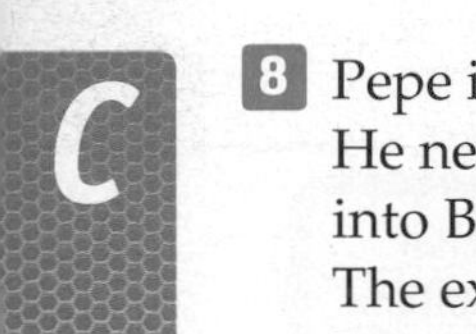

8 Pepe is going to Barbados.
He needs to change some money into Barbadian dollars.
The exchange rates are as shown.

Changing **from** £	£1 = $2.5
Changing **into** £	£1 = $3

Banks buy and sell currency at different rates in order to make a profit.

Pepe changes £50 into Barbadian dollars.

a How much does he get?

On his return Pepe has $12 left.

b How much will he get if he changes this to pounds sterling?

9 This table shows the exchange rates between pounds sterling and other currencies.

	Japan (yen)	Denmark (kroner)	Australia (AUS$)	Gambia (dalasi)
Changing **from** £	£1 = 142 yen	£1 = 8 kroner	£1 = AUS$2	£1 = 35 dalasi
Changing **into** £	£1 = 154 yen	£1 = 9 kroner	£1 = AUS$2.5	£1 = 40 dalasi

a Dirk goes to Denmark.
He changes £800 into Danish kroner.

i How many kroner will Dirk get?

During his holiday Dirk spends 4240 kroner.
When he returns Dirk changes his remaining kroner into pounds sterling.

ii How much will Dirk get?

b Elin Haf has returned from Gambia.
She changes her remaining dalasi and gets £23.
How many dalasi did Elin Haf return with?

10 In London a pair of designer jeans is priced at £120.
In New York the jeans can be bought for US$176.
In Russia the jeans can be bought for 6750 roubles.
Where are the jeans the cheapest?
Use these exchange rates: £1 = US$1.6 and £1 = 45 roubles.

Review exercise

1 Estimate the value of $\frac{385 \times 54}{1010}$ **[2 marks]**

2 Find an approximate value for $\frac{39 \times 196}{83}$

Show all your working. **[2 marks]**

3 The currency exchange rate between South African rands and pounds is £1 = 12 rand.
Convert each of these sums of money into pounds.

a 156 rand **b** 336 rand **c** 936 rand **d** 1506 rand **[4 marks, 1 mark each]**

D

FUNCTIONAL

4 Samuel is going on holiday to the Philippines.
The currency of the Philippines is the peso.
The exchange rate is £1 = 65 pesos.

a Samuel changes £50 into pesos.
How many pesos does Samuel receive? **[2 marks]**

b After the holiday Samuel has 585 pesos remaining.
Convert 585 pesos into pounds. **[2 marks]**

FUNCTIONAL

5 A jacket in Cardiff costs £80.
The same jacket in Mexico costs 999 pesos.
The exchange rate is £1 = 18 pesos.
In which country is the jacket cheaper and by how much? **[4 marks]**

D AO2

C

6 Look at this calculation. $\frac{34.96}{3.61 \times 0.54}$

Pepe, Daniel and Xabi use calculators to work it out.
Pepe gets an answer of 179.3, Daniel gets 17.93 and Xabi gets 5.23.
Use approximation to decide who is most likely to be correct. **[3 marks]**

7 Use approximation to estimate the value of $\frac{178 \times 8.49}{0.425}$ **[3 marks]**

8 Use approximation to estimate the value of $\frac{5.26 \times 27.6}{4.7^2}$ **[3 marks]**

FUNCTIONAL

9 In Bahrain a bottle of water costs 2.5 dinar when the exchange rate is 0.5 dinar to the pound.
The same bottle of water costs £2.75 in Birmingham.
In Mexico the bottle of water costs 27 pesos when the exchange rate is 18 pesos to the pound.
In which country is the bottle of water cheapest? **[3 marks]**

C AO2

Chapter summary

In this chapter you have learned how to

- check and estimate answers to problems D C
- estimate answers to problems involving decimals D C
- make estimates and approximations of calculations D C
- convert between different currencies D C

10 Factors, powers and roots

This chapter is about multiples, factors, powers and roots.

In the Chinese calendar two separate cycles interact. There are 10 heavenly stems and 12 zodiac animals. You can use lowest common multiples to work out when a certain year will come round again.

Objectives

This chapter will show you how to

- **use the terms positive and negative square root** D
- **calculate common factors, highest common factors and lowest common multiples** D C
- **use powers and roots to solve problems** D C
- **write a number as a product of prime factors** C

Before you start this chapter

Put your calculator away!

1 Which numbers from the cloud are

a multiples of 10

b multiples of 7?

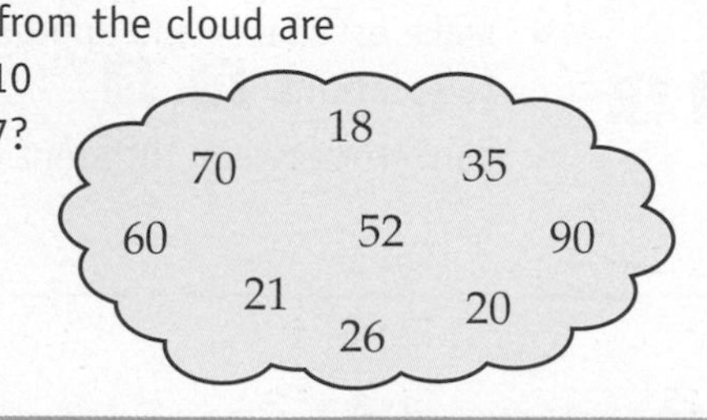

2 Work out

a 6^2 b $(-1)^2$ c $(-12)^2$ d 13^2

3 List all the factors of

a 24 b 20 c 15 d 50

4 List all the prime numbers between 20 and 40.

10.1 Lowest common multiples

Keywords
lowest common multiple (LCM), common multiple

Why learn this?
Astronomers use lowest common multiples to calculate when the Sun and Moon will be aligned in an eclipse.

Objectives
D **C** Find lowest common multiples

Skills check

1 Write down the next two terms in each sequence.
 a 4, 8, 12, 16, 20, ...
 b 27, 36, 45, 54, 63, ...

2 Write down all the multiples of 7 between 40 and 60.

Lowest common multiples

The **lowest common multiple (LCM)** of two numbers is the smallest number that is a multiple of both numbers.

The multiples of 3 are 3, 6, 9, **12**, 15, 18, 21, **24**, ...

The multiples of 4 are 4, 8, **12**, 16, 20, **24**, ...

The **common multiples** of 3 and 4 are 12, 24, ...

So the lowest common multiple of 3 and 4 is 12.

> **The lowest common multiple is also known as the least common multiple.**

> **Another way to find LCMs is covered in Section 10.4.**

Example 1

What is the lowest common multiple of 6 and 8?

Write down the multiples of both numbers and circle the common multiples.

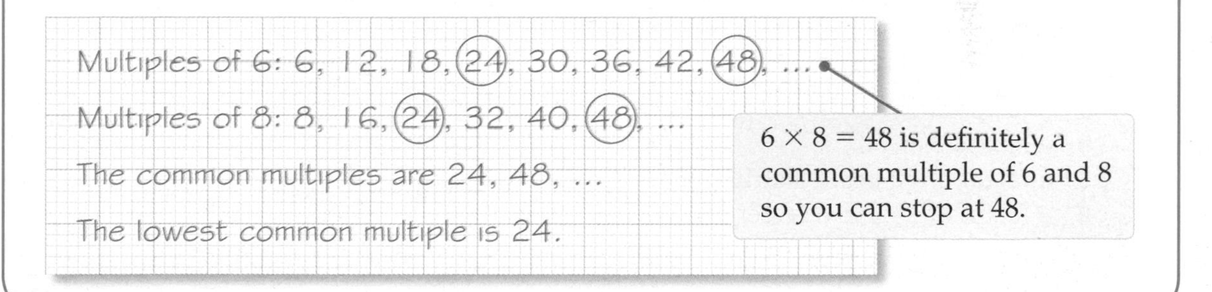

Multiples of 6: 6, 12, 18, (24), 30, 36, 42, (48), ...
Multiples of 8: 8, 16, (24), 32, 40, (48), ...
The common multiples are 24, 48, ...
The lowest common multiple is 24.
6 × 8 = 48 is definitely a common multiple of 6 and 8 so you can stop at 48.

Exercise 10A

1 a Write down the first ten multiples of 6.
 b Write down the first ten multiples of 9.
 c What is the LCM of 6 and 9?

> **LCM stands for lowest common multiple.**

2 Write down two common multiples of
 a 2 and 3 b 7 and 5 c 2 and 10 d 9 and 3.

3 Work out the lowest common multiple of
 a 5 and 6 b 8 and 10 c 2 and 5
 d 10 and 15 e 12 and 15 f 20 and 30.

C

4 Shazia says that you can find the LCM of two numbers by multiplying them together. Give an example to show that Shazia is wrong.

C

5 Carla and Guy have each got the same number of CDs.
Carla has arranged her CDs into 8 equal piles.
Guy has arranged his CDs into 12 equal piles.
What is the smallest number of CDs they could each have?

6 Fred is building a wall. He uses red bricks which are 12 cm long and yellow bricks which are 14 cm long. The bricks must line up at the start and end of the wall.

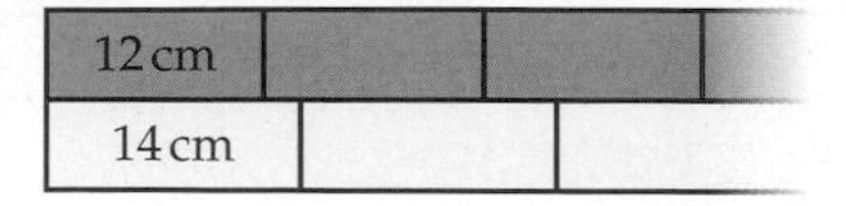

AO2

Work out the shortest length of wall that Fred could build.

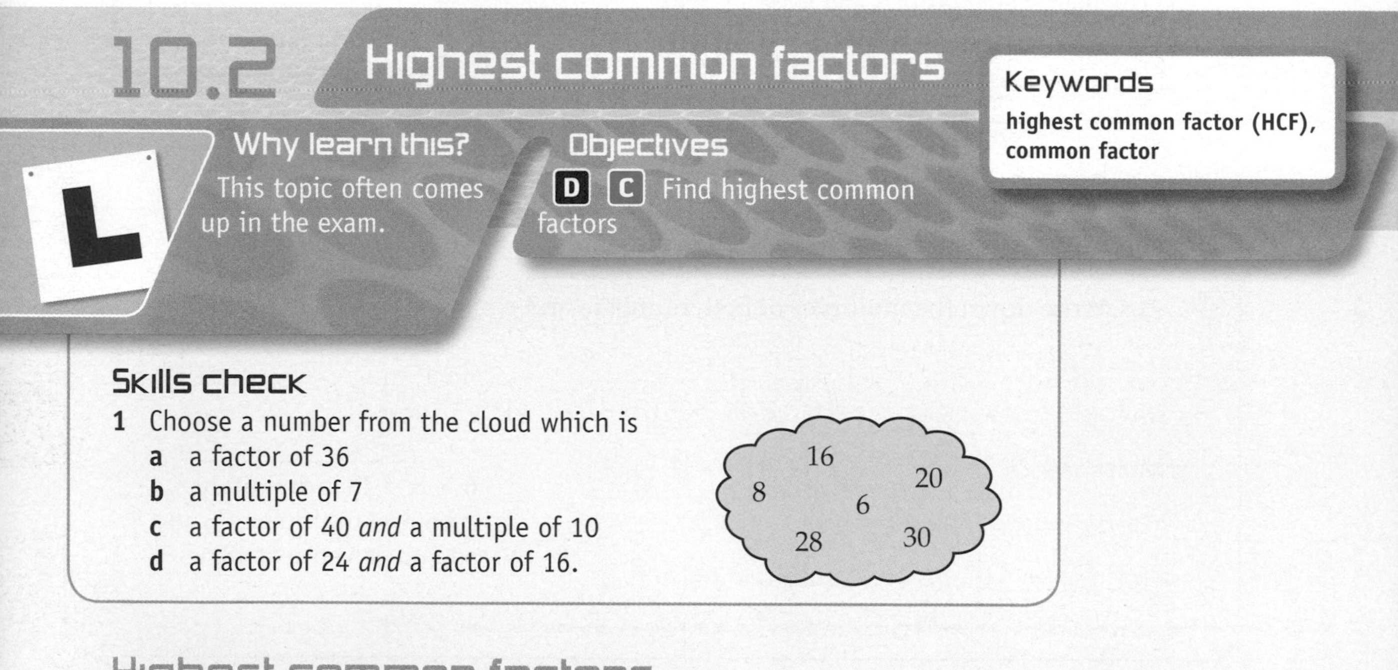

10.2 Highest common factors

Keywords
highest common factor (HCF), common factor

Why learn this?
This topic often comes up in the exam.

Objectives
D C Find highest common factors

Skills check

1 Choose a number from the cloud which is
- **a** a factor of 36
- **b** a multiple of 7
- **c** a factor of 40 *and* a multiple of 10
- **d** a factor of 24 *and* a factor of 16.

Highest common factors

The **highest common factor (HCF)** of two numbers is the largest number that is a factor of both numbers.

The factors of 12 are **1**, **2**, 3, **4**, 6 and 12.

The factors of 16 are **1**, **2**, **4**, 8 and 16.

The **common factors** of 12 and 16 are 1, 2 and 4.

So the highest common factor of 12 and 16 is 4.

Another way to find HCFs is covered in Section 10.4.

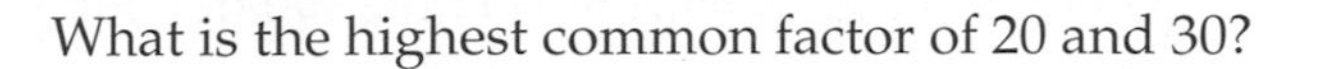

Example 2

What is the highest common factor of 20 and 30?

Write down the factors of both numbers and circle the common factors.

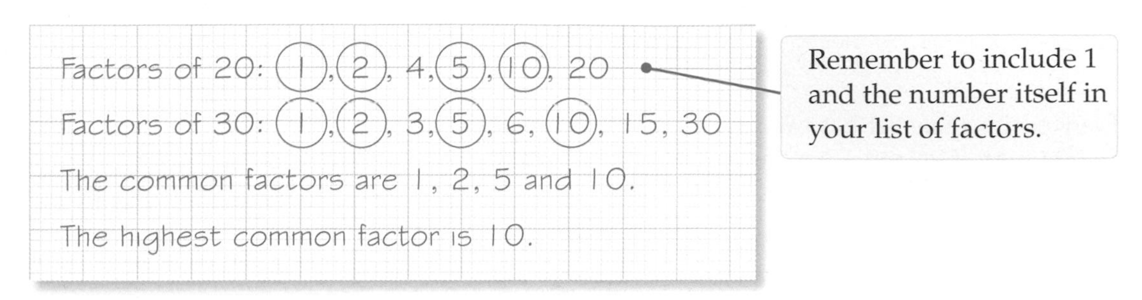

Factors of 20: 1, 2, 4, 5, 10, 20

Factors of 30: 1, 2, 3, 5, 6, 10, 15, 30

The common factors are 1, 2, 5 and 10.

The highest common factor is 10.

Remember to include 1 and the number itself in your list of factors.

Exercise 10B

1 **a** Write down the factors of 12.

b Write down the factors of 8.

c What is the HCF of 12 and 8?

HCF stands for highest common factor.

2 Work out the highest common factor of

a 15 and 25 **b** 14 and 12 **c** 21 and 15

d 24 and 20 **e** 8 and 10 **f** 8 and 16.

3 Write down two numbers larger than 10 with an HCF of 8.

4 Lydia is making Christmas decorations. She needs to cut identical squares out of a rectangular sheet of paper.

a What is the largest square size Lydia can use without wasting any paper?

b How many of these squares will Lydia be able to cut from this sheet of paper?

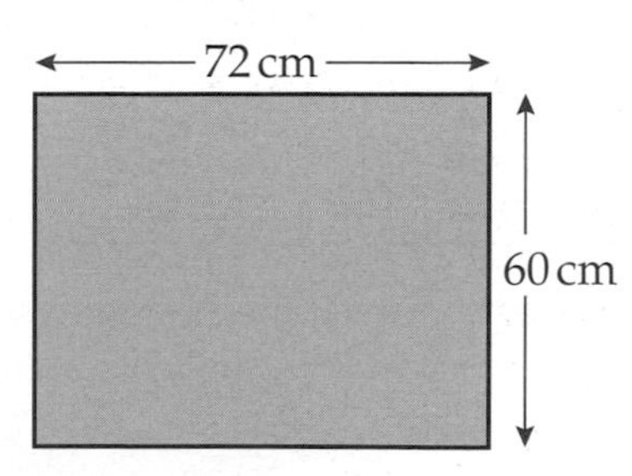

FUNCTIONAL

D C C AO2

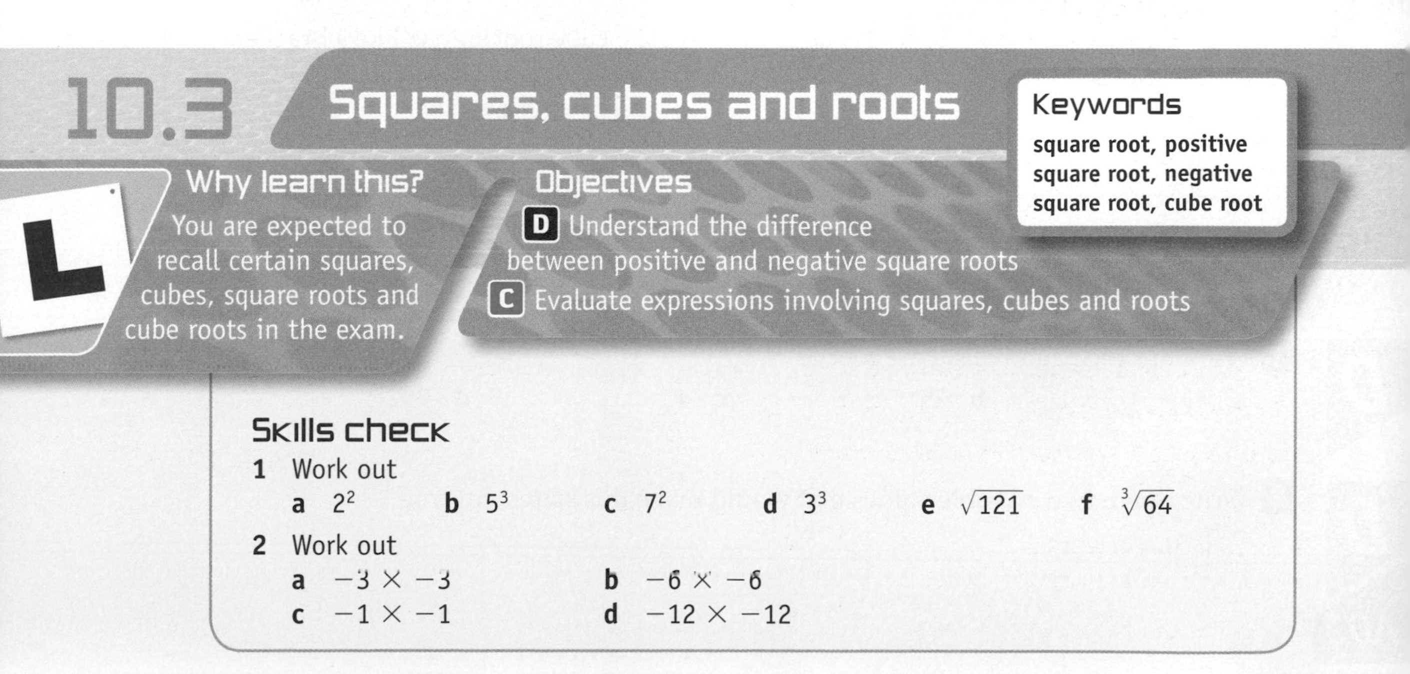

10.3 Squares, cubes and roots

Keywords
square root, positive square root, negative square root, cube root

Why learn this?
You are expected to recall certain squares, cubes, square roots and cube roots in the exam.

Objectives

D Understand the difference between positive and negative square roots

C Evaluate expressions involving squares, cubes and roots

Skills check

1 Work out

a 2^2 **b** 5^3 **c** 7^2 **d** 3^3 **e** $\sqrt{121}$ **f** $\sqrt[3]{64}$

2 Work out

a -3×-3 **b** -6×-6

c -1×-1 **d** -12×-12

Squares and square roots

To square a number you multiply it by itself. The inverse of squaring is finding the **square root**.

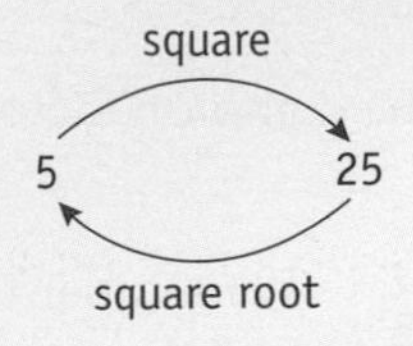

Every positive number has two square roots. 5 is the **positive square root** of 25 and −5 is the **negative square root** of 25.

$5^2 = 5 \times 5 = 25$
$(-5)^2 = -5 \times -5 = 25$

The symbol $\sqrt{\ }$ is used to represent the positive square root of a number. You write $\sqrt{25} = 5$.

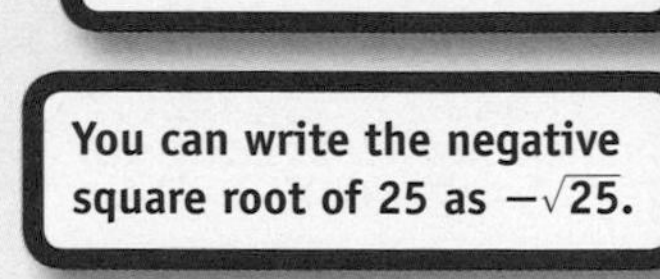

You can write the negative square root of 25 as $-\sqrt{25}$.

You need to know the squares of integers up to 15 and their corresponding square roots.

Cubes and cube roots

The inverse operation of cubing is finding the **cube root**. Every number has exactly one cube root. The symbol $\sqrt[3]{\ }$ is used to represent the cube root of a number.

$4^3 = 64$
$\sqrt[3]{64} = 4$

You need to know the cubes of 1, 2, 3, 4, 5 and 10 and their corresponding cube roots.

D
C

Example 3

a Write down the negative square root of 81. **b** Work out $\sqrt[3]{5^2 + 2}$.

a $\sqrt{81} = 9$ so the negative square root of 81 is −9.

$(-9)^2 = 81$

b $\sqrt[3]{5^2 + 2} = \sqrt[3]{25 + 2}$
$= \sqrt[3]{27}$
$= 3$

The cube root sign is like a bracket. You have to work out the value underneath the cube root first.

Exercise 10C

D

1 Work out the negative square root of

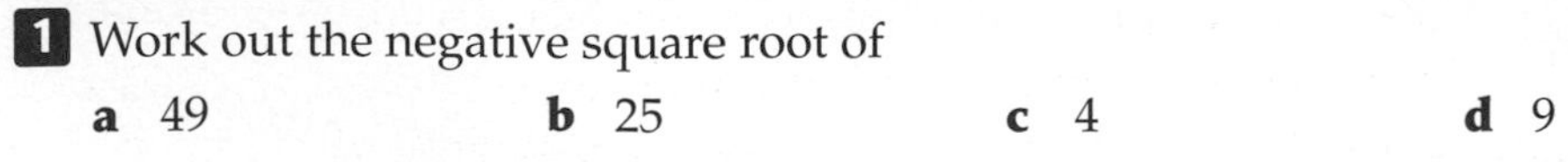

a 49 **b** 25 **c** 4 **d** 9

D
AO2

2 Write down two possible values that would make this statement true.

$\square^2 + 9 = 45$

3 Work out

a $\sqrt{2^2 + 5}$ b $\sqrt{6^2 + 8^2}$ c $\sqrt{5^2 - 4^2}$ d $\sqrt{13^2 - 12^2}$

4 Work out

a $\sqrt{5^3 - 5^2}$ b $\sqrt{2^3 + 2^3}$ c $\sqrt[3]{11^2 + 2^2}$ d $\sqrt[3]{3^2 - 1}$

5 Estimate the answers to these calculations by rounding each value to the nearest whole number.

a $6.7^2 + 3.1^2$ b $5.04^3 - 3.28^3$ c $\sqrt{19.8 - 4.1}$ d $\sqrt[3]{2.1^2 + 1.7^2}$

C

6 Amber and Simon have each made a 10 cm square pattern using 1 cm square tiles. Amber gives some tiles to Simon. They are both able to arrange their tiles exactly into square patterns.

How many tiles did Amber give to Simon?

C A03

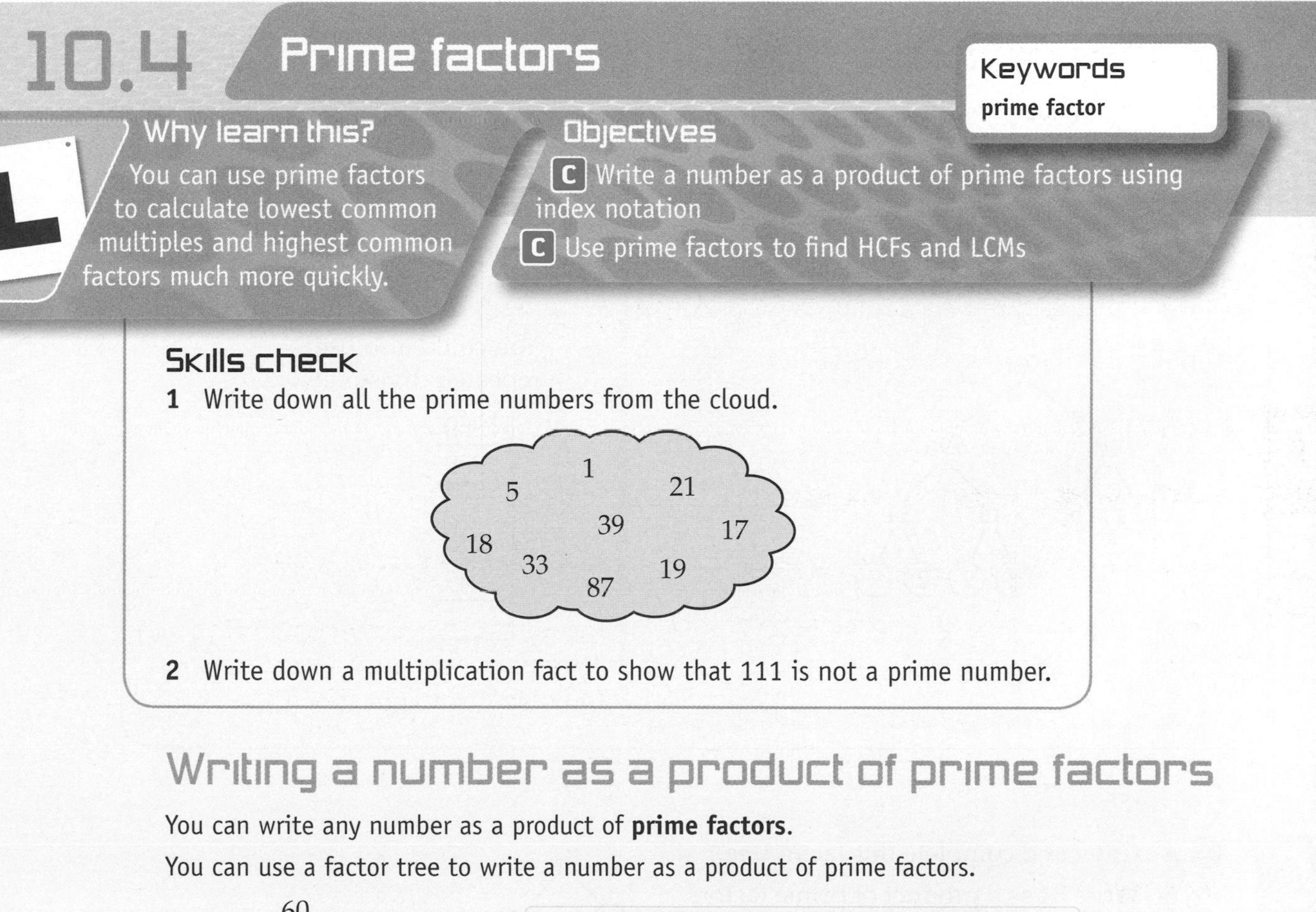

10.4 Prime factors

Keywords
prime factor

Why learn this?
You can use prime factors to calculate lowest common multiples and highest common factors much more quickly.

Objectives
C Write a number as a product of prime factors using index notation
C Use prime factors to find HCFs and LCMs

Skills check

1 Write down all the prime numbers from the cloud.

2 Write down a multiplication fact to show that 111 is not a prime number.

Writing a number as a product of prime factors

You can write any number as a product of **prime factors.**

You can use a factor tree to write a number as a product of prime factors.

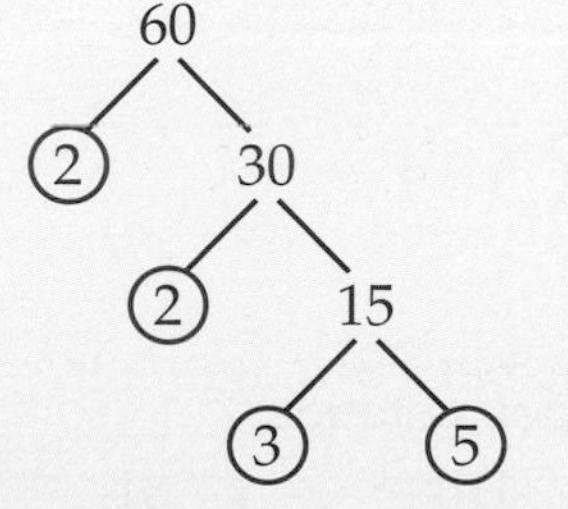

- Split each number up into factor pairs.
- When you reach a prime number, draw a circle around it. These are the ends of the branches.
- The answer is the product of the prime numbers on the branches.

$60 = 2 \times 2 \times 3 \times 5$

You can write this using index notation as $60 = 2^2 \times 3 \times 5$.

You can also use repeated division to write a number as a product of prime factors.

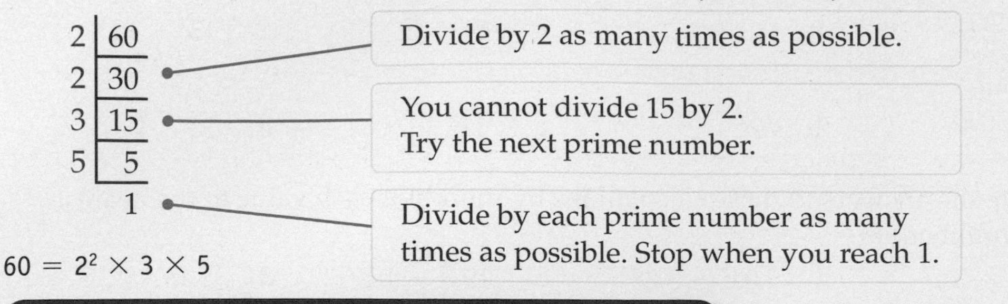

2 | 60
2 | 30
3 | 15
5 | 5
1
Divide by 2 as many times as possible.
You cannot divide 15 by 2.
Try the next prime number.
Divide by each prime number as many times as possible. Stop when you reach 1.

$60 = 2^2 \times 3 \times 5$

When writing a number as a product of prime factors, write the prime factors in order from smallest to largest.

C

Example 4

Write each number as a product of prime factors using index notation.

a 50 **b** 1960

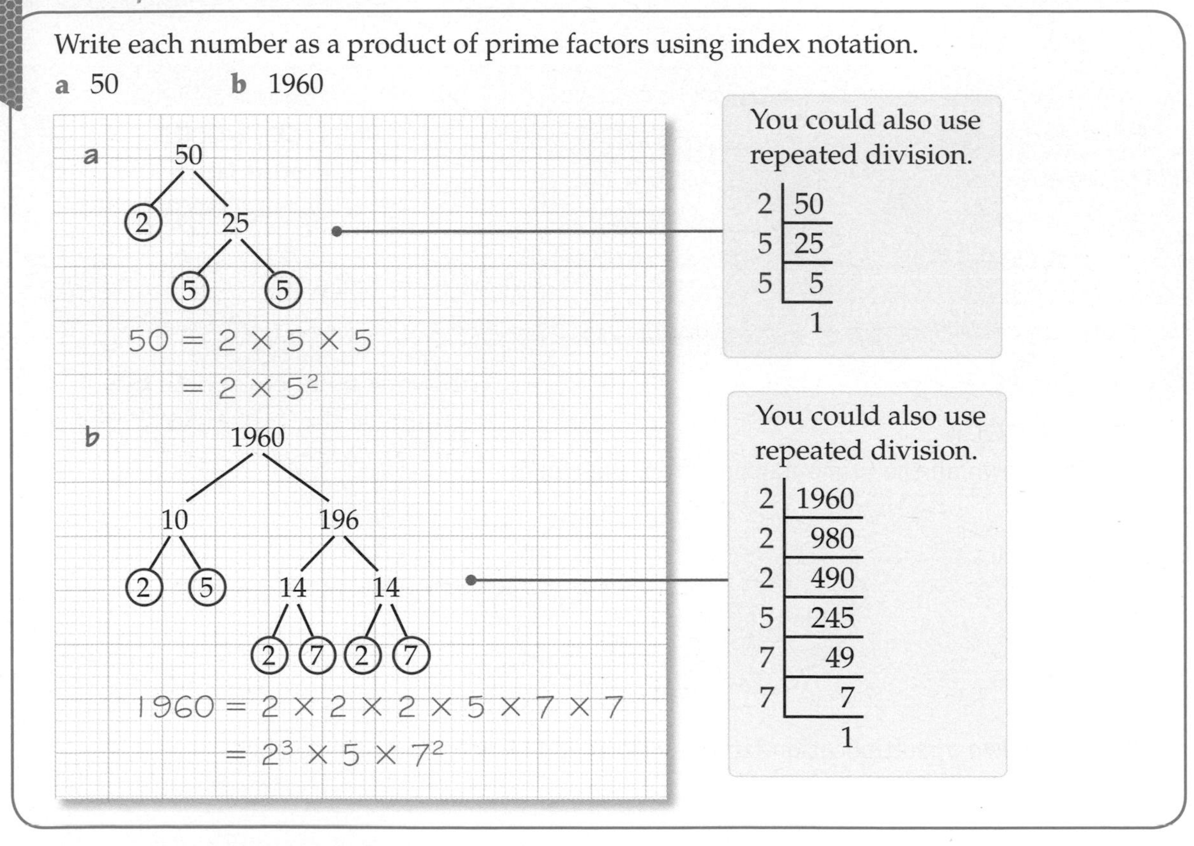

Exercise 10D

C

1 **a** Copy and complete this factor tree.

b Write 84 as a product of prime factors using index notation.

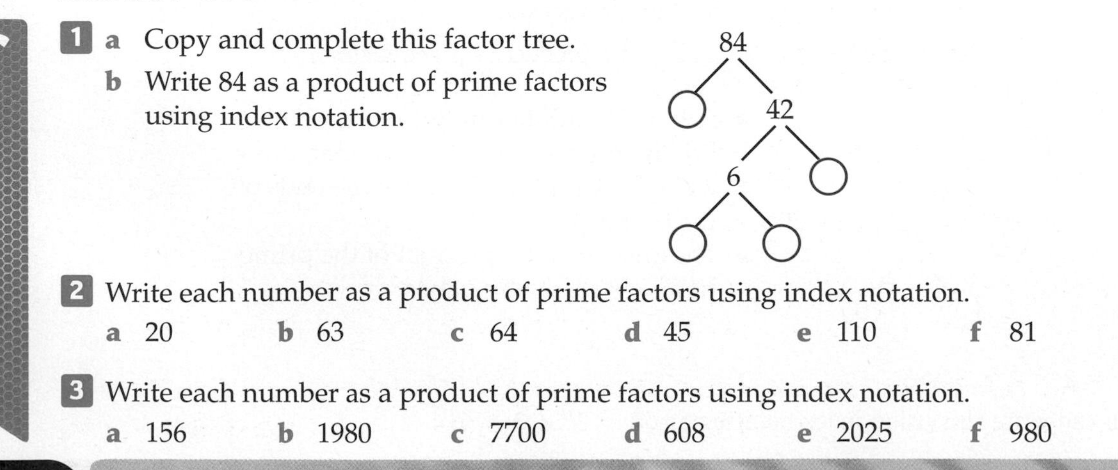

2 Write each number as a product of prime factors using index notation.

a 20 **b** 63 **c** 64 **d** 45 **e** 110 **f** 81

3 Write each number as a product of prime factors using index notation.

a 156 **b** 1980 **c** 7700 **d** 608 **e** 2025 **f** 980

4 a Copy and complete these two factor trees for 100.

100 → 10, 10
100 → (2), 50

b Jamie says that the prime factors will be different depending on which factor pair you choose first. Do you agree with him? Demonstrate your answer using another number.

C

AO2

Using prime factors to find the HCF

- Write each number as the product of prime factors.
- If a prime number is in *both* lists, circle the *lowest* power.
- Multiply these to find the HCF.

If a prime number is only in one of the lists you can't include it in your HCF.

Using prime factors to find the LCM

- Write each number as the product of prime factors.
- Circle the *highest* power of each prime number.
- Multiply these to find the LCM.

Example 5

C

Work out **a** the HCF of 180 and 168 **b** the LCM of 24 and 60.

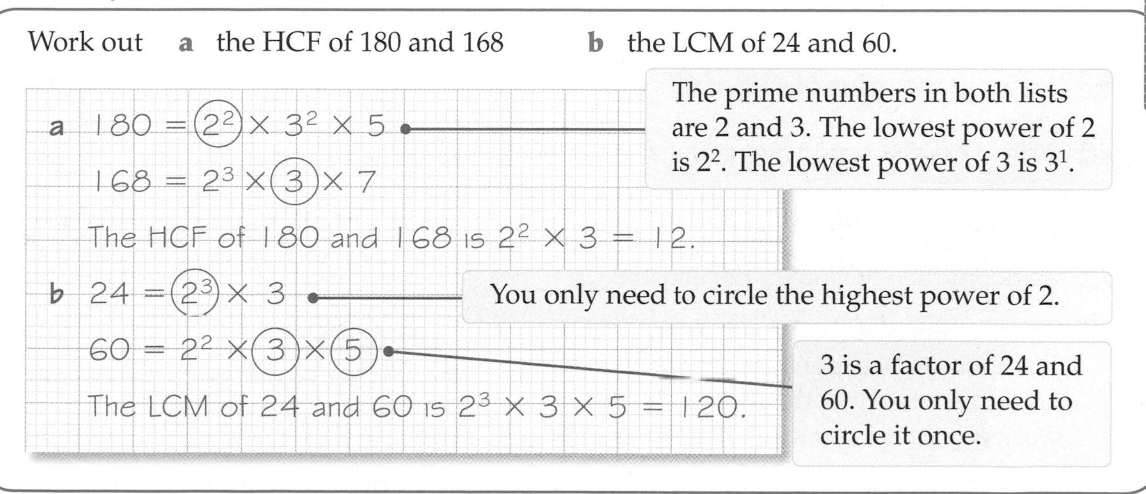

a $180 = 2^2 \times 3^2 \times 5$

$168 = 2^3 \times 3 \times 7$

The HCF of 180 and 168 is $2^2 \times 3 = 12$.

The prime numbers in both lists are 2 and 3. The lowest power of 2 is 2^2. The lowest power of 3 is 3^1.

b $24 = 2^3 \times 3$

You only need to circle the highest power of 2.

$60 = 2^2 \times 3 \times 5$

The LCM of 24 and 60 is $2^3 \times 3 \times 5 = 120$.

3 is a factor of 24 and 60. You only need to circle it once.

Exercise 10E

C

1 a Write 90 as a product of prime factors.
b Write 165 as a product of prime factors.
c Find the HCF of 90 and 165.

2 a Write 42 as a product of prime factors.
b Write 30 as a product of prime factors.
c Find the LCM of 42 and 30.

3 Work out the highest common factor of each pair of numbers.
a 32 and 56 b 80 and 72 c 27 and 45 d 100 and 75

4 Work out the lowest common multiple of each pair of numbers.
a 18 and 20 b 6 and 32 c 27 and 15 d 9 and 75

C

5 Work out the highest common factor of 2016 and 1512.

6 Amy is investigating the relationship between the LCM and the HCF.

a Work out the HCF of 18 and 30.

b Amy says that she can find the LCM of 18 and 30 using the rule LCM $= \frac{18 \times 30}{\text{HCF}}$.
Show working to check that Amy's rule works.

c Show that Amy's rule will also work for 16 and 40.

AO2

C

7 David has a pack of playing cards with some missing.
He arranges his playing cards into 15 rows of equal length.
He then rearranges his cards into 9 rows of equal length.
How many cards are missing from David's pack?

A normal pack of playing cards contains 52 cards.

AO3

Review exercise

D

1 Write down the negative square root of

a 144 b 81 c 225 d 16 **[1 mark each]**

D

2 Write down two different solutions to each equation.

a $x^2 = 64$ b $x^2 + 1 = 50$ c $x^2 - 10 = 90$ d $2x^2 = 8$ **[2 marks each]**

AO2

C

3 Work out the value of $\sqrt{10^2 - 6^2}$. **[2 marks]**

4 Write 3300 as a product of prime factors using index notation. **[3 marks]**

5 Work out a the HCF of 80 and 96 b the LCM of 45 and 54. **[3 marks each]**

6 Work out the HCF of 96, 120 and 150. **[3 marks]**

C

7 Daisy is tiling her bathroom floor.
She wants to use identical square tiles
to completely cover the floor with no overlap.
Work out the largest size of square tile Daisy can use. **[3 marks]**

←260 cm→

180 cm

FUNCTIONAL • FUNCTIONAL •

8 A clock tower has three different bells. The largest bell rings once every 12 seconds.
The middle bell rings once every 10 seconds. The smallest bell rings once every
8 seconds. The bells all ring at the same time.

a How long will it take before the bells all ring at the same time again?
Give your answer in minutes. **[3 marks]**

b How many times has each bell rung? **[3 marks]**

AO2

Chapter summary

In this chapter you have learned how to

- understand the difference between positive and negative square roots D
- find lowest common multiples D C
- find highest common factors D C
- evaluate expressions involving squares, cubes and roots C
- write a number as a product of prime factors using index notation C
- use prime factors to find HCFs and LCMs C

11 Fractions

This chapter is about calculating with fractions.

In Formula 1, the difference between first and second place is incredibly small. As Jenson Button said, 'You're qualifying on pole by half a tenth of a second.'

Objectives

This chapter will show you how to

- compare fractions **D**
- add and subtract fractions **D**
- add and subtract mixed numbers **C**
- work out reciprocals **C**
- multiply and divide fractions and mixed numbers **D** **C** **B**

Before you start this chapter

Put your calculator away!

1 Nikki has these fraction cards.

Nikki says, 'All the fractions on the cards are equivalent fractions.'
Explain why Nikki is wrong.

2 Greg starts with one fraction, subtracts another fraction, and gets an answer of $\frac{4}{11}$.
Suggest two fractions that Greg may be using.

3 a Work out $\frac{2}{3} \times \frac{1}{4}$ and $\frac{4}{5} \times \frac{1}{6}$.
Write your answers in their simplest form.

b Write your answers to part a as equivalent fractions with denominators of 30.

c Which is the smaller fraction, $\frac{2}{3} \times \frac{1}{4}$ or $\frac{4}{5} \times \frac{1}{6}$?

Comparing fractions

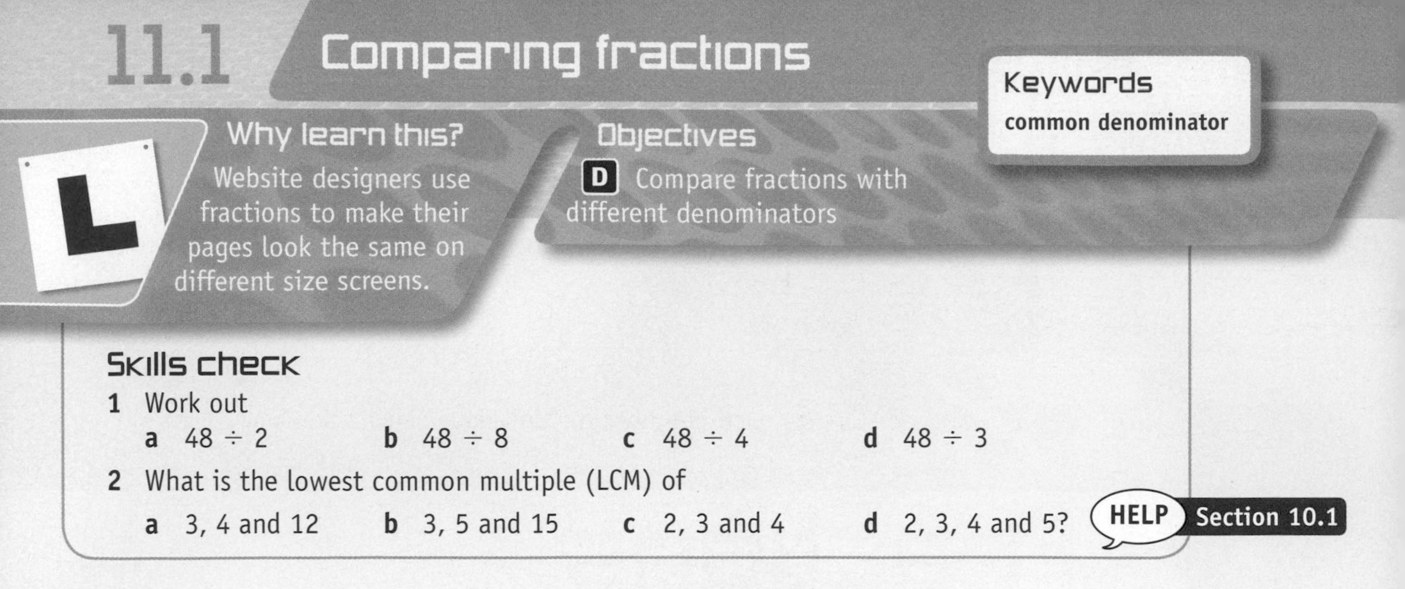

Keywords
common denominator

Why learn this?
Website designers use fractions to make their pages look the same on different size screens.

Objectives
D Compare fractions with different denominators

Skills check

1 Work out

a $48 \div 2$ b $48 \div 8$ c $48 \div 4$ d $48 \div 3$

2 What is the lowest common multiple (LCM) of

a 3, 4 and 12 b 3, 5 and 15 c 2, 3 and 4 d 2, 3, 4 and 5?

HELP Section 10.1

Comparing two or more fractions

To compare fractions with different denominators, change them into equivalent fractions with the same denominator. The 'same denominator' is often called the **common denominator**.

Example 1

D

Put these fractions in order of size, smallest first.

$\frac{2}{3}, \frac{3}{5}, \frac{11}{15}$

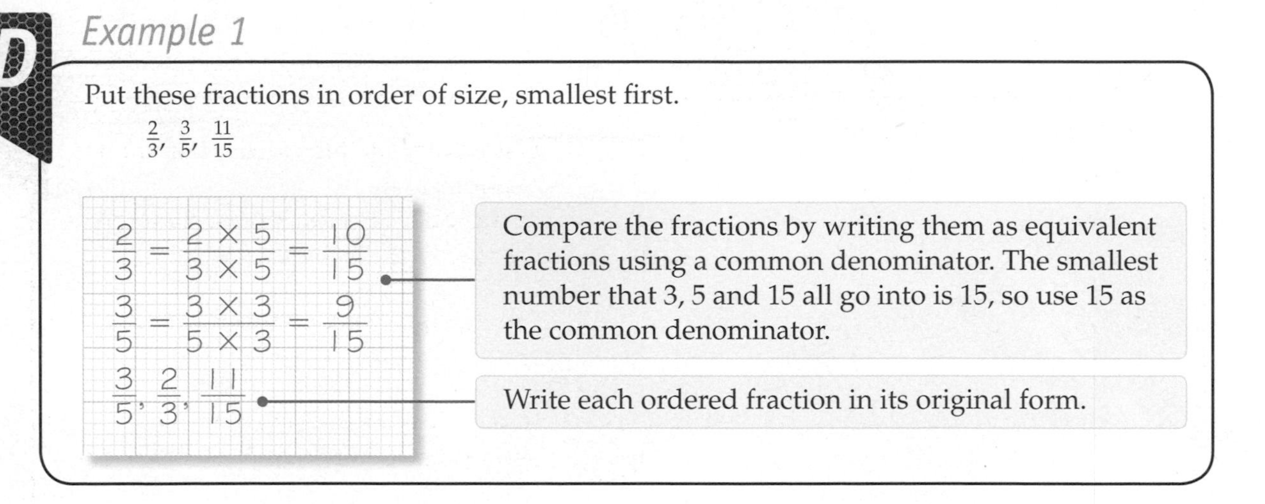

$\frac{2}{3} = \frac{2 \times 5}{3 \times 5} = \frac{10}{15}$

$\frac{3}{5} = \frac{3 \times 3}{5 \times 3} = \frac{9}{15}$

Compare the fractions by writing them as equivalent fractions using a common denominator. The smallest number that 3, 5 and 15 all go into is 15, so use 15 as the common denominator.

$\frac{3}{5}, \frac{2}{3}, \frac{11}{15}$

Write each ordered fraction in its original form.

Exercise 11A

D AO2

1 Put each of these sets of fractions in order of size, smallest first.

a $\frac{3}{4}, \frac{10}{12}, \frac{2}{3}$ b $\frac{2}{3}, \frac{17}{24}, \frac{3}{4}$ c $\frac{1}{4}, \frac{3}{8}, \frac{1}{3}$ d $\frac{2}{3}, \frac{4}{5}, \frac{11}{15}, \frac{5}{6}$

D

2 Which of these fractions is nearest to $\frac{1}{4}$?

A $\frac{1}{5}$ B $\frac{3}{10}$ C $\frac{7}{30}$ D $\frac{11}{40}$

You must show your working.

D AO3

3 Derry says:

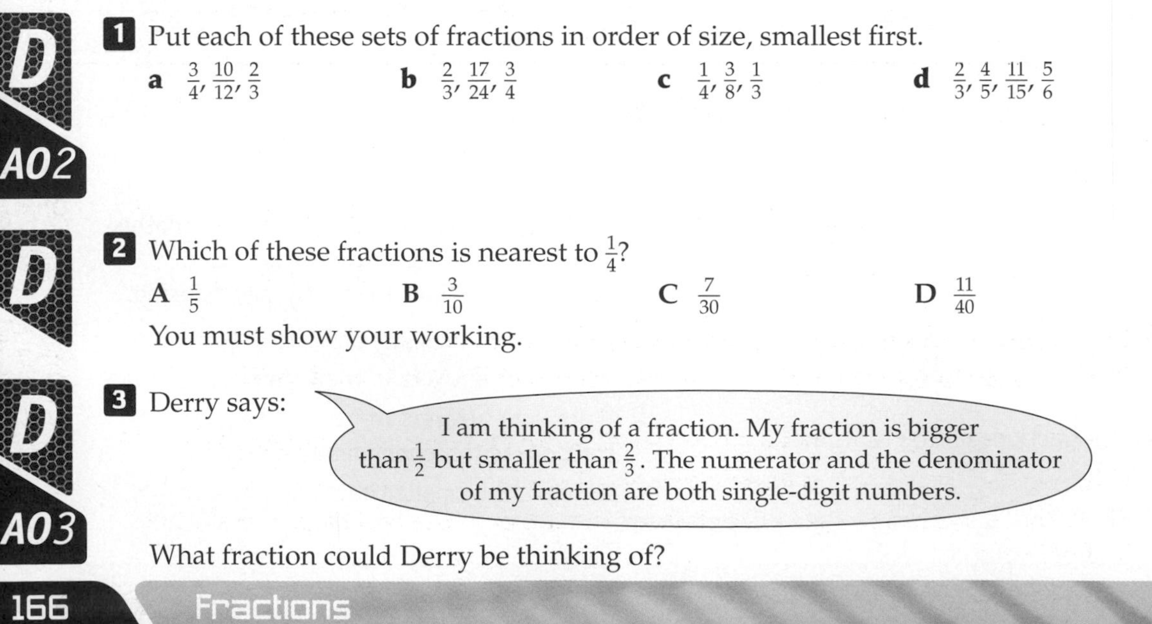

I am thinking of a fraction. My fraction is bigger than ½ but smaller than ⅔. The numerator and the denominator of my fraction are both single-digit numbers.

What fraction could Derry be thinking of?

Adding and subtracting fractions

Why learn this?

Carpenters often have to add or subtract lengths measured with fractions.

Objectives

D Add and subtract fractions when both denominators have to be changed

Keywords
lowest common denominator

Skills check

1 Work out

a $\frac{1}{3} + \frac{1}{3}$ **b** $\frac{5}{7} - \frac{2}{7}$ **c** $\frac{2}{3} + \frac{1}{6}$

2 Change each improper fraction into a mixed number.

a $\frac{3}{2}$ **b** $\frac{9}{5}$ **c** $\frac{12}{7}$

Adding and subtracting fractions

To add or subtract fractions with different denominators, write them as equivalent fractions with the same denominator, then add or subtract their numerators.

If you use the **lowest common denominator**, you will avoid having to simplify your answer at the end.

Example 2

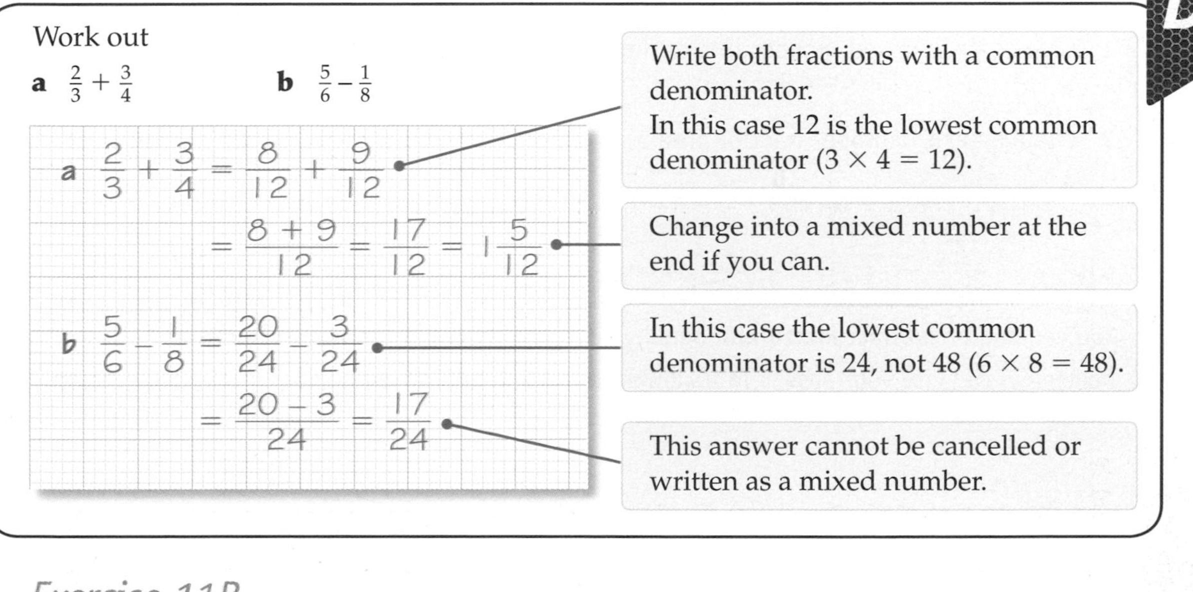

Work out

a $\frac{2}{3} + \frac{3}{4}$ **b** $\frac{5}{6} - \frac{1}{8}$

a $\frac{2}{3} + \frac{3}{4} = \frac{8}{12} + \frac{9}{12}$

$= \frac{8+9}{12} = \frac{17}{12} = 1\frac{5}{12}$

b $\frac{5}{6} - \frac{1}{8} = \frac{20}{24} - \frac{3}{24}$

$= \frac{20-3}{24} = \frac{17}{24}$

Write both fractions with a common denominator. In this case 12 is the lowest common denominator ($3 \times 4 = 12$).

Change into a mixed number at the end if you can.

In this case the lowest common denominator is 24, not 48 ($6 \times 8 = 48$).

This answer cannot be cancelled or written as a mixed number.

D

Exercise 11B

1 Work out

a $\frac{1}{3} - \frac{1}{4}$ **b** $\frac{2}{5} + \frac{1}{3}$ **c** $\frac{3}{4} - \frac{1}{5}$ **d** $\frac{1}{3} - \frac{1}{7}$

e $\frac{3}{5} - \frac{2}{7}$ **f** $\frac{3}{4} + \frac{3}{5}$ **g** $\frac{5}{6} + \frac{2}{7}$ **h** $\frac{1}{4} + \frac{6}{7}$

D

2 In a driving reaction test, one driver reacted in $\frac{1}{8}$ of a second. Another driver took $\frac{2}{3}$ of a second to react. What is the difference in their reaction times?

D

3 Clive has two bags of flour. One bag contains $\frac{1}{8}$ kg and the other contains $\frac{1}{3}$ kg. Clive needs $\frac{1}{2}$ kg of flour to make one loaf of bread.
Show that Clive has not got enough flour altogether in the two bags to make one loaf of bread.

AO2

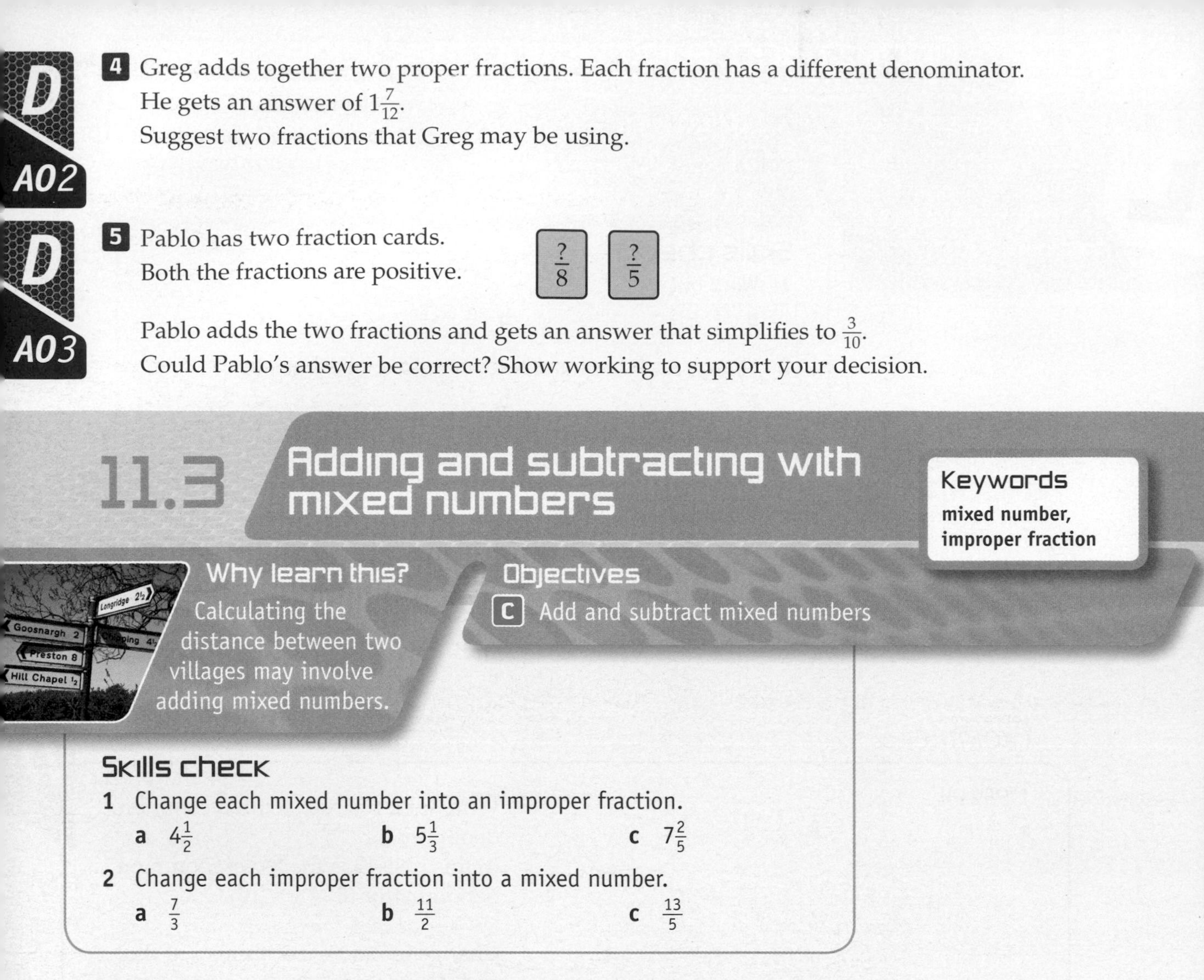

D AO2

4 Greg adds together two proper fractions. Each fraction has a different denominator.
He gets an answer of $1\frac{7}{12}$.
Suggest two fractions that Greg may be using.

D AO3

5 Pablo has two fraction cards.
Both the fractions are positive.

$\frac{?}{8}$ $\frac{?}{5}$

Pablo adds the two fractions and gets an answer that simplifies to $\frac{3}{10}$.
Could Pablo's answer be correct? Show working to support your decision.

11.3 Adding and subtracting with mixed numbers

Keywords
mixed number, improper fraction

Why learn this?
Calculating the distance between two villages may involve adding mixed numbers.

Objectives
C Add and subtract mixed numbers

Skills check

1 Change each mixed number into an improper fraction.
   a $4\frac{1}{2}$ b $5\frac{1}{3}$ c $7\frac{2}{5}$
2 Change each improper fraction into a mixed number.
   a $\frac{7}{3}$ b $\frac{11}{2}$ c $\frac{13}{5}$

Adding and subtracting mixed numbers

You can add or subtract **mixed numbers** by changing them into **improper fractions**.
Remember to write your answer in its lowest terms and as a mixed number if possible.

C

Example 3

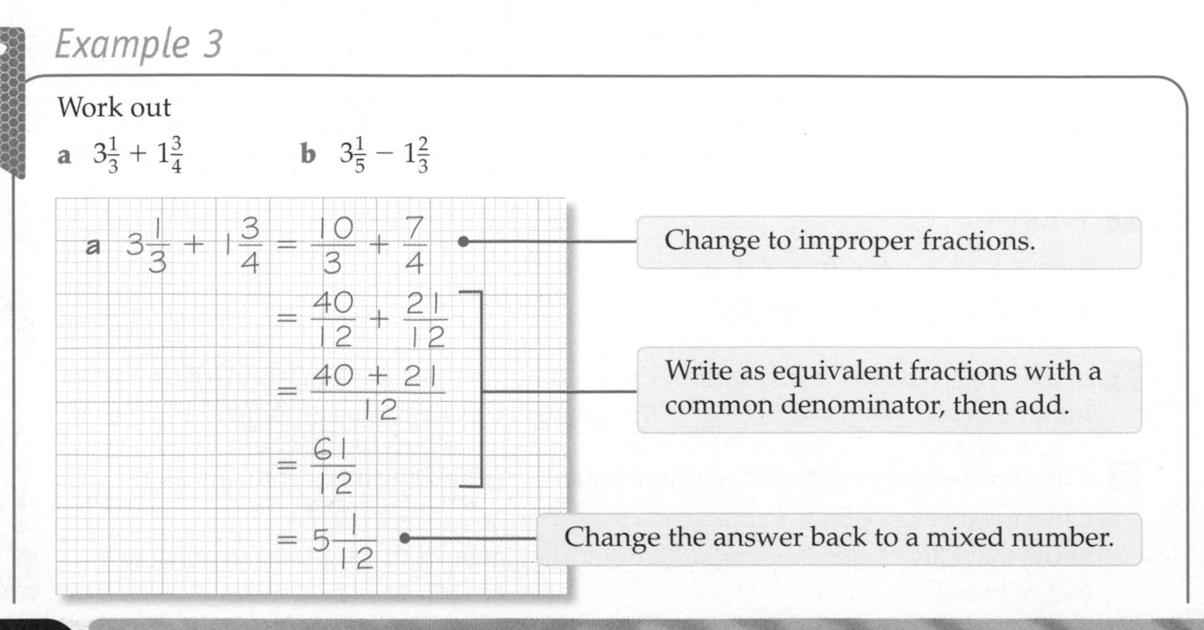

Work out

a $3\frac{1}{3} + 1\frac{3}{4}$ b $3\frac{1}{5} - 1\frac{2}{3}$

a $3\frac{1}{3} + 1\frac{3}{4} = \frac{10}{3} + \frac{7}{4}$ — Change to improper fractions.

$= \frac{40}{12} + \frac{21}{12}$

$= \frac{40 + 21}{12}$

$= \frac{61}{12}$ — Write as equivalent fractions with a common denominator, then add.

$= 5\frac{1}{12}$ — Change the answer back to a mixed number.

b $3\frac{1}{5} - 1\frac{2}{3} = \frac{16}{5} - \frac{5}{3}$ — Change to improper fractions.

$= \frac{48}{15} - \frac{25}{15}$

$= \frac{48 - 25}{15}$ — Write as equivalent fractions with a common denominator, then subtract.

$= \frac{23}{15}$

$= 1\frac{8}{15}$ — Change the answer back to a mixed number.

Exercise 11C

1 Work out

a $3\frac{2}{3} + 1\frac{1}{4}$

b $3\frac{7}{10} - 1\frac{3}{40}$

c $2\frac{1}{2} - 1\frac{3}{4} + \frac{5}{8}$

d $1\frac{1}{9} + 1\frac{1}{3} + 2\frac{1}{18} - 3\frac{1}{6}$

C

2 Sandra cycles from her home in Witney to Bampton.
The distance is $6\frac{1}{2}$ miles.
On her way home she cycles $3\frac{1}{5}$ miles and then her bike has a puncture.
She pushes her bike the rest of the way home.
How far does she push her bike?

C

3 Erin adds together two mixed numbers. The fractions have different denominators.
Erin gets an answer of $3\frac{17}{30}$.
Suggest two different sets of fractions that Erin may have added.

AO2

4 Patrick has two mixed-number cards. Both the mixed numbers are positive.

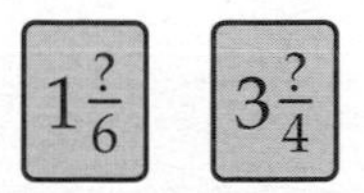

Patrick adds the two numbers and gets an answer that simplifies to $4\frac{1}{4}$.
Could Patrick's answer be correct? Show working to support your decision.

C

AO3

5 A small lorry is allowed to carry a maximum weight of 7 tonnes.
The lorry is loaded with four pallets of machine parts.
The pallets weigh $1\frac{5}{6}$ tonnes, $1\frac{7}{8}$ tonnes, $1\frac{11}{12}$ tonnes and $1\frac{1}{2}$ tonnes.
Show that the lorry is overloaded.

C

AO2

11.4 Multiplying mixed numbers

Why learn this?

When decorating, you may need to work out areas involving fractional measurements.

Objectives

D Multiply a whole number by a mixed number

C Multiply a fraction by a mixed number

B Multiply a mixed number by a mixed number

Skills check

1 Work out

a $\frac{3}{4} \times 24$ **b** $15 \times \frac{2}{5}$ **c** $\frac{1}{2} \times \frac{3}{5}$

2 True or false?

a $\frac{11}{3} = 3\frac{1}{3}$ **b** $1\frac{2}{5} = \frac{6}{5}$ **c** $\frac{32}{9} = 3\frac{5}{9}$

Multiplying fractions and mixed numbers

To multiply mixed numbers, change them to improper fractions first.
Then multiply the numerators together and multiply the denominators together.
Before you multiply, cancel common factors if possible.
If your answer is an improper fraction, change it back to a mixed number.

Example 4

Work out

a $3 \times 1\frac{5}{6}$ **b** $\frac{1}{3} \times 1\frac{5}{6}$ **c** $2\frac{1}{3} \times 3\frac{3}{4}$

a $3 \times 1\frac{5}{6} = \frac{\cancel{3}^{1}}{1} \times \frac{11}{\cancel{6}_{2}}$ — Change $1\frac{5}{6}$ into an improper fraction, then cancel.

$= \frac{1 \times 11}{1 \times 2}$

$= \frac{11}{2}$

$= 5\frac{1}{2}$ — Multiply, then change the answer back to a mixed number.

b $\frac{1}{3} \times 1\frac{5}{6} = \frac{1}{3} \times \frac{11}{6}$ — Change $1\frac{5}{6}$ into an improper fraction, then multiply.

$= \frac{1 \times 11}{3 \times 6}$

$= \frac{11}{18}$

c $2\frac{1}{3} \times 3\frac{3}{4} = \frac{7}{\cancel{3}_{1}} \times \frac{\cancel{15}^{5}}{4}$ — Change both mixed numbers to improper fractions, then cancel.

$= \frac{7 \times 5}{1 \times 4}$

$= \frac{35}{4}$

$= 8\frac{3}{4}$ — Change the answer back to a mixed number.

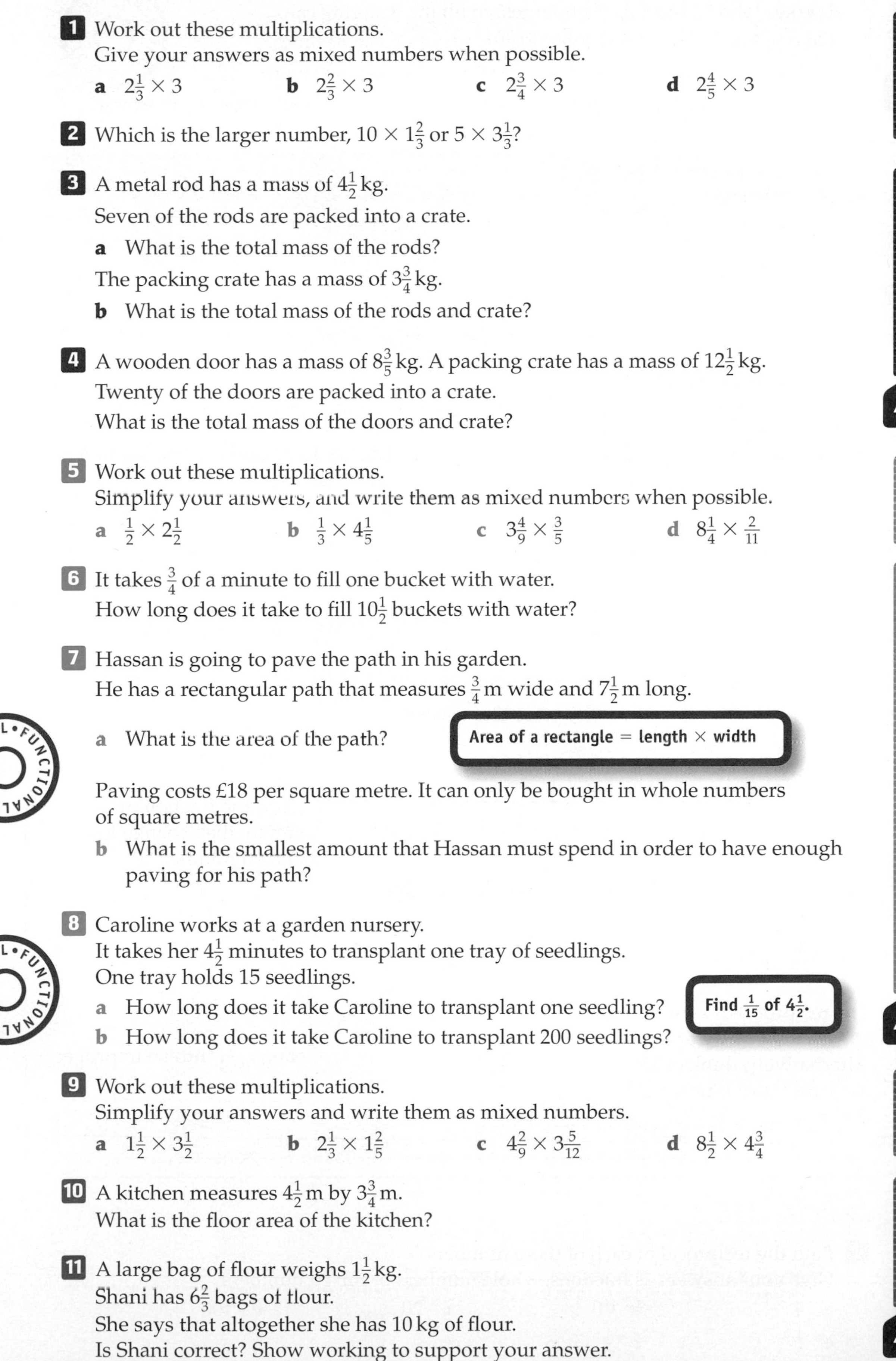

Exercise 11D

1 Work out these multiplications.
Give your answers as mixed numbers when possible.

a $2\frac{1}{3} \times 3$ **b** $2\frac{2}{3} \times 3$ **c** $2\frac{3}{4} \times 3$ **d** $2\frac{4}{5} \times 3$

2 Which is the larger number, $10 \times 1\frac{2}{3}$ or $5 \times 3\frac{1}{3}$?

3 A metal rod has a mass of $4\frac{1}{2}$ kg.
Seven of the rods are packed into a crate.

a What is the total mass of the rods?

The packing crate has a mass of $3\frac{3}{4}$ kg.

b What is the total mass of the rods and crate?

4 A wooden door has a mass of $8\frac{3}{5}$ kg. A packing crate has a mass of $12\frac{1}{2}$ kg.
Twenty of the doors are packed into a crate.
What is the total mass of the doors and crate?

5 Work out these multiplications.
Simplify your answers, and write them as mixed numbers when possible.

a $\frac{1}{2} \times 2\frac{1}{2}$ **b** $\frac{1}{3} \times 4\frac{1}{5}$ **c** $3\frac{4}{9} \times \frac{3}{5}$ **d** $8\frac{1}{4} \times \frac{2}{11}$

6 It takes $\frac{3}{4}$ of a minute to fill one bucket with water.
How long does it take to fill $10\frac{1}{2}$ buckets with water?

7 Hassan is going to pave the path in his garden.
He has a rectangular path that measures $\frac{3}{4}$ m wide and $7\frac{1}{2}$ m long.

a What is the area of the path?

> Area of a rectangle = length × width

Paving costs £18 per square metre. It can only be bought in whole numbers of square metres.

b What is the smallest amount that Hassan must spend in order to have enough paving for his path?

8 Caroline works at a garden nursery.
It takes her $4\frac{1}{2}$ minutes to transplant one tray of seedlings.
One tray holds 15 seedlings.

a How long does it take Caroline to transplant one seedling?

b How long does it take Caroline to transplant 200 seedlings?

> Find $\frac{1}{15}$ of $4\frac{1}{2}$.

9 Work out these multiplications.
Simplify your answers and write them as mixed numbers.

a $1\frac{1}{2} \times 3\frac{1}{2}$ **b** $2\frac{1}{3} \times 1\frac{2}{5}$ **c** $4\frac{2}{9} \times 3\frac{5}{12}$ **d** $8\frac{1}{2} \times 4\frac{3}{4}$

10 A kitchen measures $4\frac{1}{2}$ m by $3\frac{3}{4}$ m.
What is the floor area of the kitchen?

11 A large bag of flour weighs $1\frac{1}{2}$ kg.
Shani has $6\frac{2}{3}$ bags of flour.
She says that altogether she has 10 kg of flour.
Is Shani correct? Show working to support your answer.

B

AO2

12 John has a watering can that holds $6\frac{4}{5}$ litres of water.
It takes John 70 seconds ($1\frac{1}{6}$ minutes) to fill the watering can.
On one Monday morning John fills the watering can $8\frac{1}{2}$ times.

a How much water does John use on this Monday morning?

b How long does John spend filling the watering can on this Monday morning?

FUNCTIONAL • FUNCTIONAL •

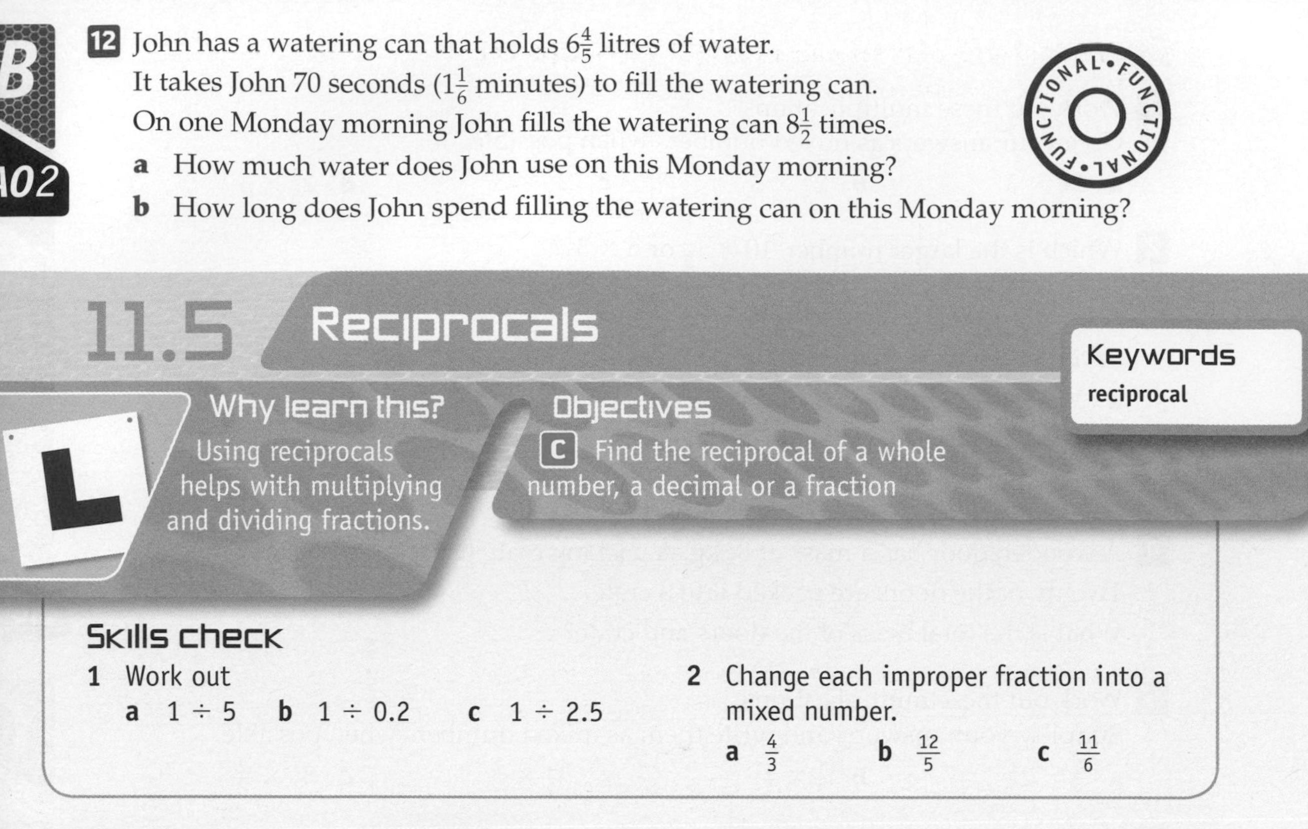

11.5 Reciprocals

Keywords
reciprocal

Why learn this?
Using reciprocals helps with multiplying and dividing fractions.

Objectives
C Find the reciprocal of a whole number, a decimal or a fraction

Skills check

1 Work out
a $1 \div 5$ **b** $1 \div 0.2$ **c** $1 \div 2.5$

2 Change each improper fraction into a mixed number.
a $\frac{4}{3}$ **b** $\frac{12}{5}$ **c** $\frac{11}{6}$

Finding reciprocals

When two numbers can be multiplied together to give the answer 1, then each number is called the **reciprocal** of the other.

The reciprocal of a fraction is found by turning the fraction upside down.

The reciprocal of a number is 1 divided by that number.

C

Example 5

Find the reciprocal of

a 25 **b** 0.4 **c** $\frac{6}{7}$

Give your answers as mixed numbers when possible.

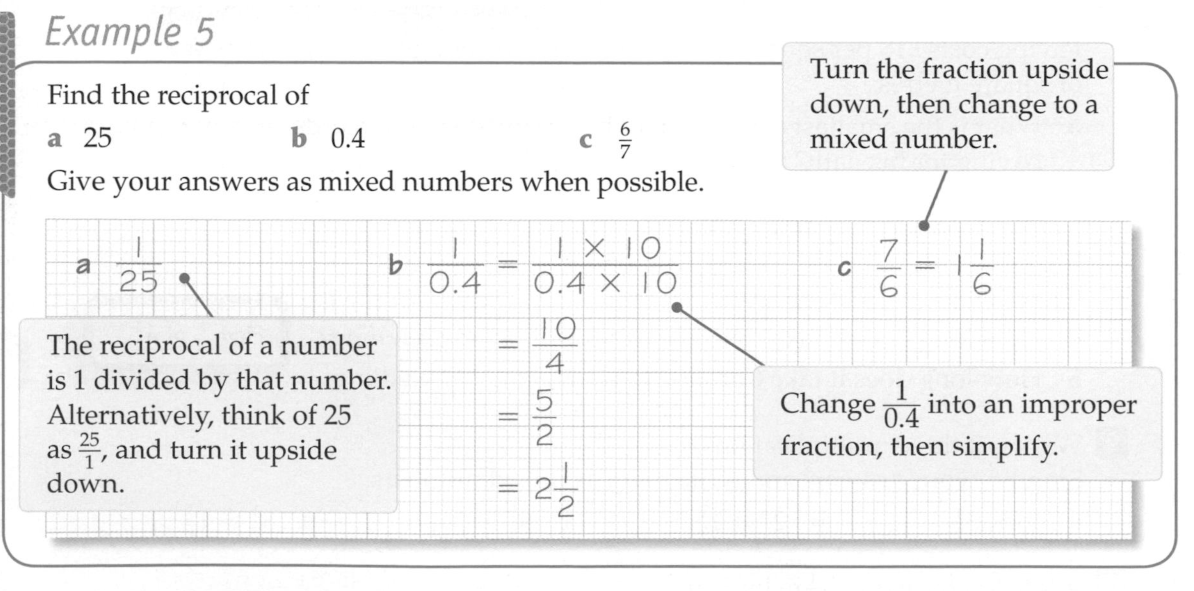

a $\frac{1}{25}$

The reciprocal of a number is 1 divided by that number. Alternatively, think of 25 as $\frac{25}{1}$, and turn it upside down.

b $\frac{1}{0.4} = \frac{1 \times 10}{0.4 \times 10} = \frac{10}{4} = \frac{5}{2} = 2\frac{1}{2}$

Change $\frac{1}{0.4}$ into an improper fraction, then simplify.

c $\frac{7}{6} = 1\frac{1}{6}$

Turn the fraction upside down, then change to a mixed number.

Exercise 11E

C

1 Find the reciprocal of each of these numbers.
Give your answers as fractions, whole numbers or mixed numbers.

a 4 **b** 10 **c** 20 **d** 100

e $\frac{1}{2}$ **f** $\frac{1}{5}$ **g** $\frac{2}{3}$ **h** $\frac{3}{10}$

2 Multiply each of the numbers in Q1 by its reciprocal.
What do you notice?

C
AO2

3 Kedar says, 'The reciprocal of $\frac{5}{12}$ is $\frac{7}{12}$ because $\frac{5}{12} + \frac{7}{12} = \frac{5+7}{12} = 1$.'
Is Kedar correct? Explain your answer.

C
AO3

11.6 Dividing fractions

Why learn this?

A nurse uses fractions to work out how many $\frac{1}{2}$ m*l* doses she can give from a 5 m*l* vial.

Objectives

D Divide a whole number or a fraction by a fraction

C Divide mixed numbers by whole numbers or fractions

B Divide mixed numbers by mixed numbers

Skills check

1 Cancel each fraction to its lowest terms.
  a $\frac{4}{16}$ **b** $\frac{12}{15}$ **c** $\frac{32}{40}$

2 Work out **a** $\frac{2}{5} \times \frac{3}{7}$ **b** $\frac{3}{8} \times \frac{2}{9}$ **c** $1\frac{1}{2} \times \frac{3}{10}$

Dividing by a fraction

To divide by a fraction, turn the fraction upside down and multiply.
If the division involves mixed numbers, change them to improper fractions first.

Example 6

D

Work out **a** $18 \div \frac{2}{3}$ **b** $\frac{3}{20} \div \frac{9}{40}$

a $18 \div \frac{2}{3} = \frac{18}{1} \div \frac{2}{3}$ — Write 18 as $\frac{18}{1}$.

$= \frac{\overset{9}{\cancel{18}}}{1} \times \frac{3}{\cancel{2}_1}$ — Turn $\frac{2}{3}$ upside down, change ÷ to ×, and cancel the fractions.

$= \frac{9 \times 3}{1 \times 1}$ — Multiply the numerators and multiply the denominators.

$= \frac{27}{1} = 27$

b $\frac{3}{20} \div \frac{9}{40} = \frac{\overset{1}{\cancel{3}}}{\cancel{20}_1} \times \frac{\overset{2}{\cancel{40}}}{\cancel{9}_3}$ — Turn $\frac{9}{40}$ upside down, change ÷ to ×, then cancel the fractions.

$= \frac{1 \times 2}{1 \times 3}$ — Multiply the numerators and multiply the denominators.

$= \frac{2}{3}$

Example 7

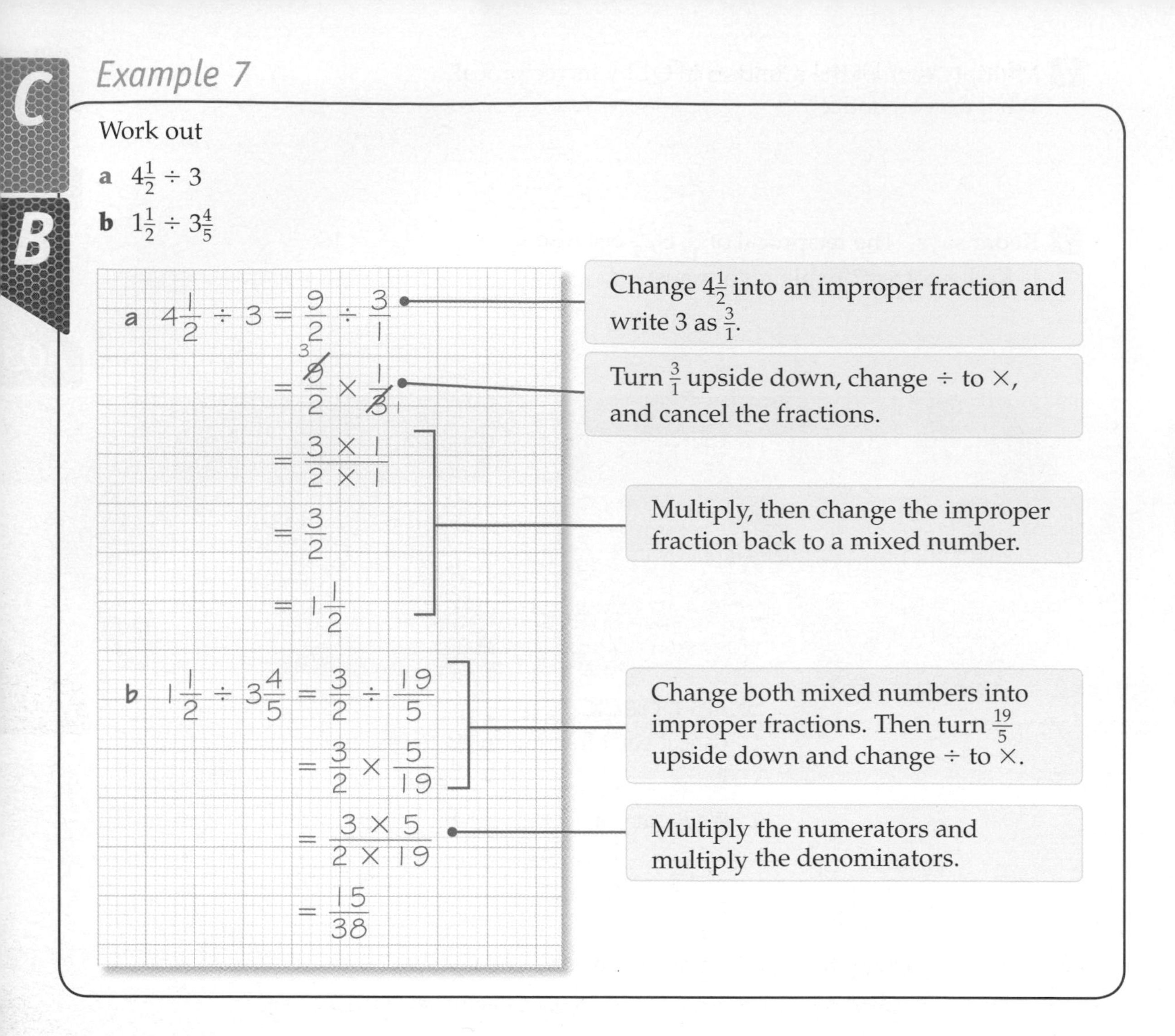

C
B
Work out
a $4\frac{1}{2} \div 3$
b $1\frac{1}{2} \div 3\frac{4}{5}$
a $4\frac{1}{2} \div 3 = \frac{9}{2} \div \frac{3}{1}$
$= \frac{\cancel{9}^{3}}{2} \times \frac{1}{\cancel{3}_{1}}$
$= \frac{3 \times 1}{2 \times 1}$
$= \frac{3}{2}$
$= 1\frac{1}{2}$
Change $4\frac{1}{2}$ into an improper fraction and write 3 as $\frac{3}{1}$.
Turn $\frac{3}{1}$ upside down, change ÷ to ×, and cancel the fractions.
Multiply, then change the improper fraction back to a mixed number.
b $1\frac{1}{2} \div 3\frac{4}{5} = \frac{3}{2} \div \frac{19}{5}$
$= \frac{3}{2} \times \frac{5}{19}$
$= \frac{3 \times 5}{2 \times 19}$
$= \frac{15}{38}$
Change both mixed numbers into improper fractions. Then turn $\frac{19}{5}$ upside down and change ÷ to ×.
Multiply the numerators and multiply the denominators.

Exercise 11F

D

1 Work out

a $15 \div \frac{1}{2}$ **b** $8 \div \frac{2}{5}$ **c** $3 \div \frac{2}{9}$

D C

2 Work out

a $\frac{2}{3} \div \frac{4}{5}$ **b** $\frac{6}{7} \div \frac{12}{13}$ **c** $\frac{8}{11} \div \frac{24}{33}$

d $3\frac{3}{4} \div 10$ **e** $1\frac{2}{7} \div 12$ **f** $7\frac{1}{5} \div 6$

C

3 Work out

a $2\frac{1}{4} \div \frac{7}{12}$ **b** $1\frac{2}{7} \div \frac{3}{4}$ **c** $3\frac{2}{3} \div \frac{20}{21}$

C

4 Adam shares $3\frac{1}{2}$ pizzas between four people.
How much do they each receive?

A02

5 Matthew pours $4\frac{2}{3}$ litres of lemonade into $\frac{1}{4}$ litre glasses.
How many full glasses does he pour?

6 This is part of David's homework.

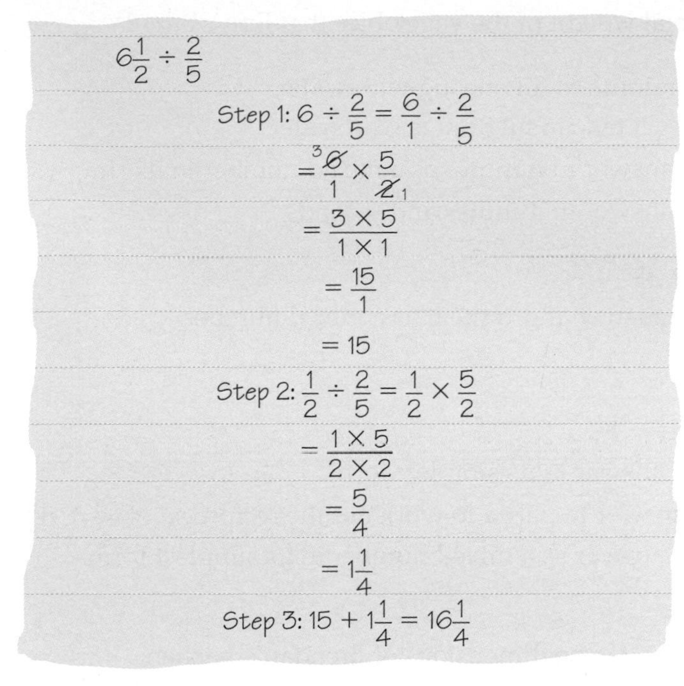

$6\frac{1}{2} \div \frac{2}{5}$

Step 1: $6 \div \frac{2}{5} = \frac{6}{1} \div \frac{2}{5}$

$= \frac{\not{6}^{3}}{1} \times \frac{5}{\not{2}_{1}}$

$= \frac{3 \times 5}{1 \times 1}$

$= \frac{15}{1}$

$= 15$

Step 2: $\frac{1}{2} \div \frac{2}{5} = \frac{1}{2} \times \frac{5}{2}$

$= \frac{1 \times 5}{2 \times 2}$

$= \frac{5}{4}$

$= 1\frac{1}{4}$

Step 3: $15 + 1\frac{1}{4} = 16\frac{1}{4}$

Are his workings correct? Explain your answer.

7 Work out

a $1\frac{1}{3} \div 4\frac{1}{2}$ **b** $2\frac{3}{5} \div 3\frac{1}{3}$ **c** $5\frac{4}{7} \div 6\frac{3}{8}$

d $4\frac{1}{4} \div 1\frac{1}{5}$ **e** $6\frac{3}{5} \div 1\frac{5}{7}$ **f** $4\frac{1}{2} \div 3\frac{2}{15}$

8 **a** Work out

i $3\frac{1}{2} \div 4\frac{2}{3}$ **ii** $5\frac{1}{3} \div 7\frac{1}{9}$

Write your answers in their simplest form.

b Use your answers to part **a** to work out $(3\frac{1}{2} \div 4\frac{2}{3}) - (5\frac{1}{3} \div 7\frac{1}{9})$.

9 Ryan says, '$6\frac{2}{3} \div 3\frac{3}{4}$ is more than half of $3\frac{1}{3}$.'
Is Ryan correct? Show workings to support your answer.

Review exercise

1 Work out $5 \div \frac{3}{4}$. **[2 marks]**

2 Brian spends $\frac{2}{3}$ of his wages on bills.
He spends $\frac{1}{5}$ of his wages on clothes.
The rest of his wages he saves.
What fraction of his wages does Brian save? **[3 marks]**

3 Find the reciprocal of 0.2.
Give your answer as a whole number. **[2 marks]**

C

4 Emyr buys $1\frac{1}{4}$ kg of carrots, $\frac{2}{3}$ kg of parsnips and $2\frac{3}{5}$ kg of potatoes.
What is the total weight of the vegetables that Emyr buys? **[3 marks]**

5 It takes $\frac{2}{3}$ of a minute to fill one jug with water.
How long does it take to fill $8\frac{1}{2}$ jugs with water?

AO2

a Give your answer in minutes as a mixed number in its simplest form. **[3 marks]**

b Give your answer in minutes and seconds. **[1 mark]**

B

6 Work out $4\frac{1}{5} \times 3\frac{6}{7}$.
Simplify your answer and write it as a mixed number. **[3 marks]**

B

7 **a** Work out $4\frac{2}{5} \div 8\frac{1}{4}$.
Write your answer in its simplest form. **[3 marks]**

b Use your answer to part **a** to work out the reciprocal of $4\frac{2}{5} \div 8\frac{1}{4}$.
Write your answer as a mixed number in its simplest form. **[2 marks]**

8 The diagram shows the dimensions of Brendan's kitchen.

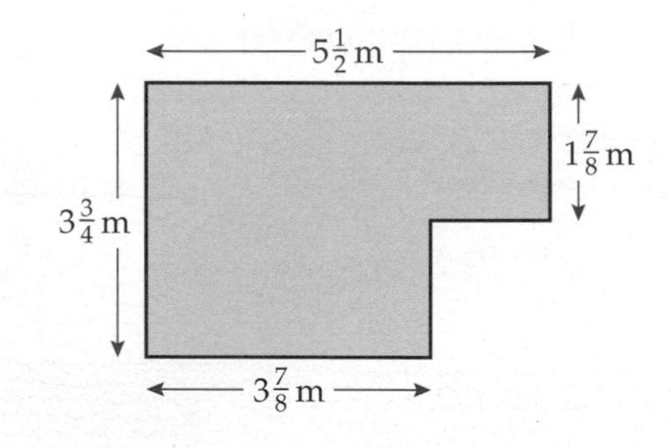

AO2

Brendan buys flooring for the kitchen that costs £16.50 per square metre.
The flooring can only be bought in whole numbers of square metres.
How much does the flooring cost Brendan? **[5 marks]**

Chapter summary

In this chapter you have learned how to

- compare fractions with different denominators **D**
- add and subtract fractions when both denominators have to be changed **D**
- multiply a whole number by a mixed number **D**
- divide a whole number or a fraction by a fraction **D**
- add and subtract mixed numbers **C**
- multiply a fraction by a mixed number **C**
- find the reciprocal of a whole number, a decimal or a fraction **C**
- divide mixed numbers by whole numbers or fractions **C**
- multiply a mixed number by a mixed number **B**
- divide mixed numbers by mixed numbers **B**

12

Basic rules of algebra

This chapter is about how to manipulate algebraic expressions.

Text messaging uses abbreviations to keep messages short. Mathematicians use algebra to communicate their ideas in a short form.

Objectives

This chapter will show you how to

- **solve linear equations involving brackets** D
- **multiply a single term over a bracket** D C
- **simplify expressions involving brackets** D C
- **understand the mathematical meaning of 'expression' and 'identity'** C
- **take out common factors** D C B
- **expand the product of two linear expressions** C B

Before you start this chapter

Put your calculator away!

1 Work out

a -3×2 b 4×-3 c -6×-5 d $3 \times x$

e $x \times x$ f $5x \times 2x$ g $a \times a^2$ h $4m \times 5n$

2 Simplify

a $3y + 2y$ b $3x + 4y + 2x + 5y$ c $6x + 2y - 4x - 3y$

3 Solve

a $x + 9 = 15$ b $2y + 5 = 13$

12.1 Expanding brackets

Keywords
expand, brackets, expression

Why learn this?

For every expression with brackets, there is an equivalent expression without brackets.

Objectives

D **C** Multiply terms in a bracket by a single term outside the bracket

C Understand the mathematical meaning of 'expression' and 'identity'

Skills check

1 Simplify

a $5 \times t$ b $7 \times 4x$ c $6 \times y^2$

2 Work out

a 2×-4 b 3×-6 c -2×-2

Multiplying out brackets

An **expression** is a collection of algebraic terms without an equals sign. Some algebraic expressions have **brackets**.

$6(x + 4)$ means $6 \times (x + 4)$.

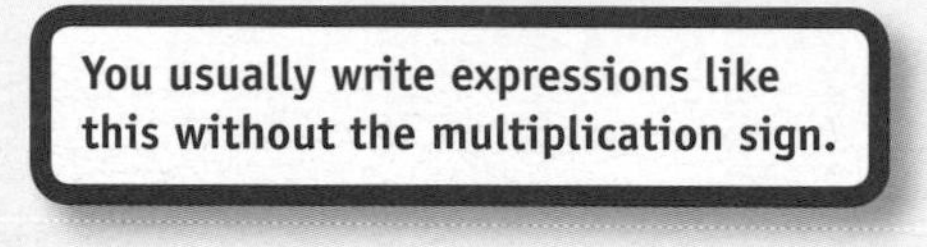

You usually write expressions like this without the multiplication sign.

x 4

6

You multiply each term inside the bracket by 6.

$$6(x + 4) = 6 \times x + 6 \times 4$$
$$= 6x + 24$$

To **expand** a bracket, you multiply each term inside the bracket by the term outside the bracket. It is also sometimes called 'multiplying out the brackets'.

D

Example 1

Expand the brackets.

a $5(a + 6)$

b $2(x - 8)$

c $3(2c - d)$

a $5(a + 6) = 5 \times a + 5 \times 6$
$= 5a + 30$

Multiply each term inside the bracket by the term outside the bracket.

b $2(x - 8) = 2 \times x - 2 \times 8$
$= 2x - 16$

$2 \times -8 = -2 \times 8 = -16$
A common mistake is to forget to multiply the second term.

c $3(2c - d) = 3 \times 2c - 3 \times d$
$= 6c - 3d$

$3 \times -d = -3d$

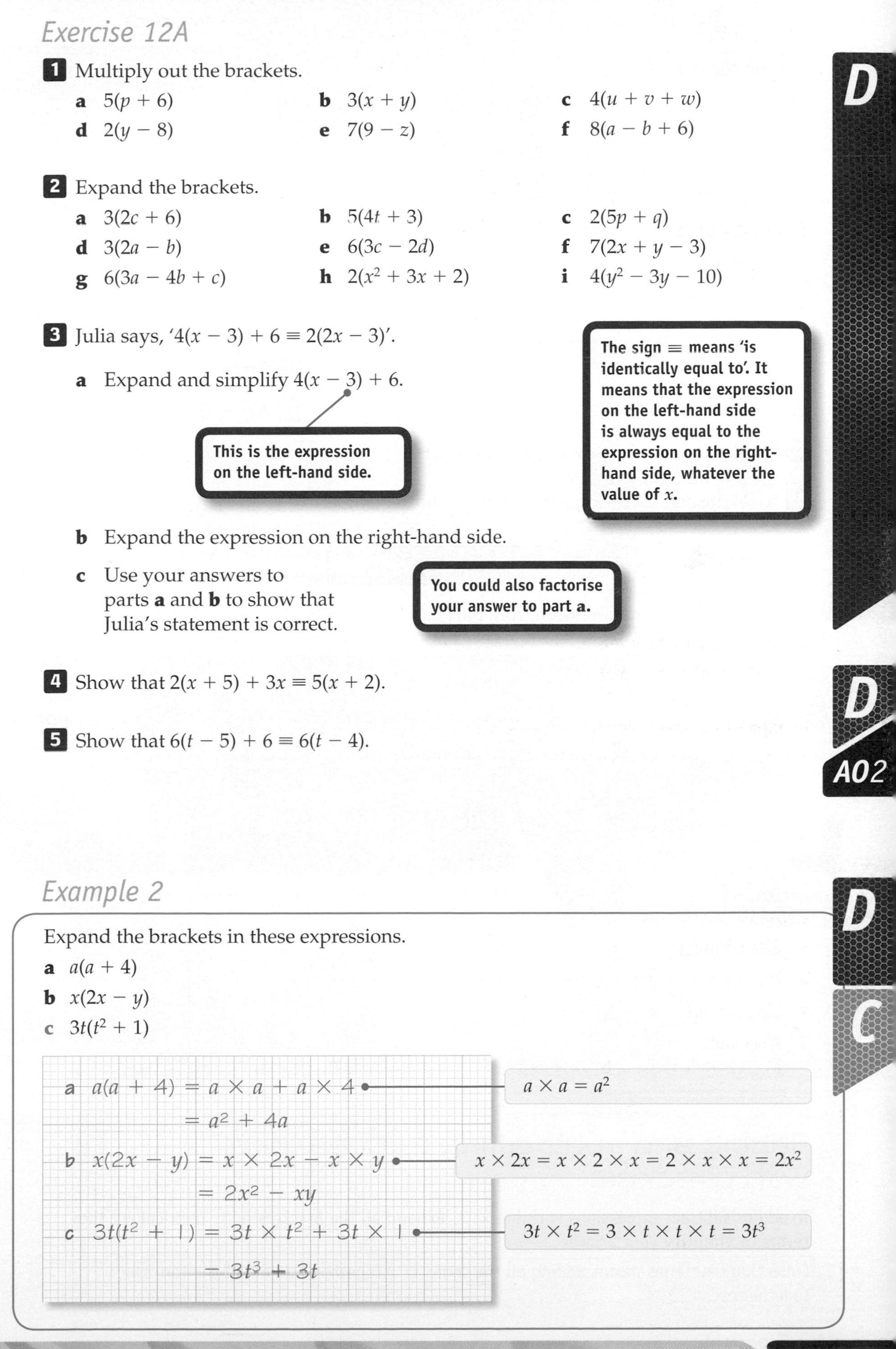

Exercise 12A

D

1 Multiply out the brackets.

a $5(p + 6)$ **b** $3(x + y)$ **c** $4(u + v + w)$

d $2(y - 8)$ **e** $7(9 - z)$ **f** $8(a - b + 6)$

2 Expand the brackets.

a $3(2c + 6)$ **b** $5(4t + 3)$ **c** $2(5p + q)$

d $3(2a - b)$ **e** $6(3c - 2d)$ **f** $7(2x + y - 3)$

g $6(3a - 4b + c)$ **h** $2(x^2 + 3x + 2)$ **i** $4(y^2 - 3y - 10)$

3 Julia says, '$4(x - 3) + 6 \equiv 2(2x - 3)$'.

a Expand and simplify $4(x - 3) + 6$.

This is the expression on the left-hand side.

The sign $\equiv$ means 'is identically equal to'. It means that the expression on the left-hand side is always equal to the expression on the right-hand side, whatever the value of x.

b Expand the expression on the right-hand side.

c Use your answers to parts **a** and **b** to show that Julia's statement is correct.

You could also factorise your answer to part **a**.

D AO2

4 Show that $2(x + 5) + 3x \equiv 5(x + 2)$.

5 Show that $6(t - 5) + 6 \equiv 6(t - 4)$.

Example 2

D C

Expand the brackets in these expressions.

a $a(a + 4)$

b $x(2x - y)$

c $3t(t^2 + 1)$

a $a(a + 4) = a \times a + a \times 4$
$= a^2 + 4a$

$a \times a = a^2$

b $x(2x - y) = x \times 2x - x \times y$
$= 2x^2 - xy$

$x \times 2x = x \times 2 \times x = 2 \times x \times x = 2x^2$

c $3t(t^2 + 1) = 3t \times t^2 + 3t \times 1$
$= 3t^3 + 3t$

$3t \times t^2 = 3 \times t \times t \times t = 3t^3$

Exercise 12B

Expand the brackets.

D

1 $b(b + 4)$
2 $a(5 + a)$
3 $k(k - 6)$
4 $m(9 - m)$
5 $a(2a + 3)$
6 $g(4g + 1)$
7 $p(2p + q)$
8 $t(t + 5w)$
9 $m(m + 3n)$
10 $x(2x - y)$
11 $r(4r - t)$
12 $a(a - 4b)$
13 $2t(t + 5)$
14 $3x(x - 8)$
15 $5k(k + l)$
16 $3a(2a + 4)$

$3a \times 2a = 3 \times 2 \times a \times a = 6a^2$

17 $2g(4g + h)$
18 $5p(3p - 2q)$
19 $3x(2y + 5z)$

C

20 $r(r^2 + 1)$
21 $a(a^2 + 3)$
22 $t(t^2 - 7)$
23 $2p(p^2 + 3q)$
24 $4x(x^2 + x)$

C AO2

25 **a** Write an expression for the area of this rectangle.

$3x + 2$ (length), x (width)

b Expand your expression from part **a**.

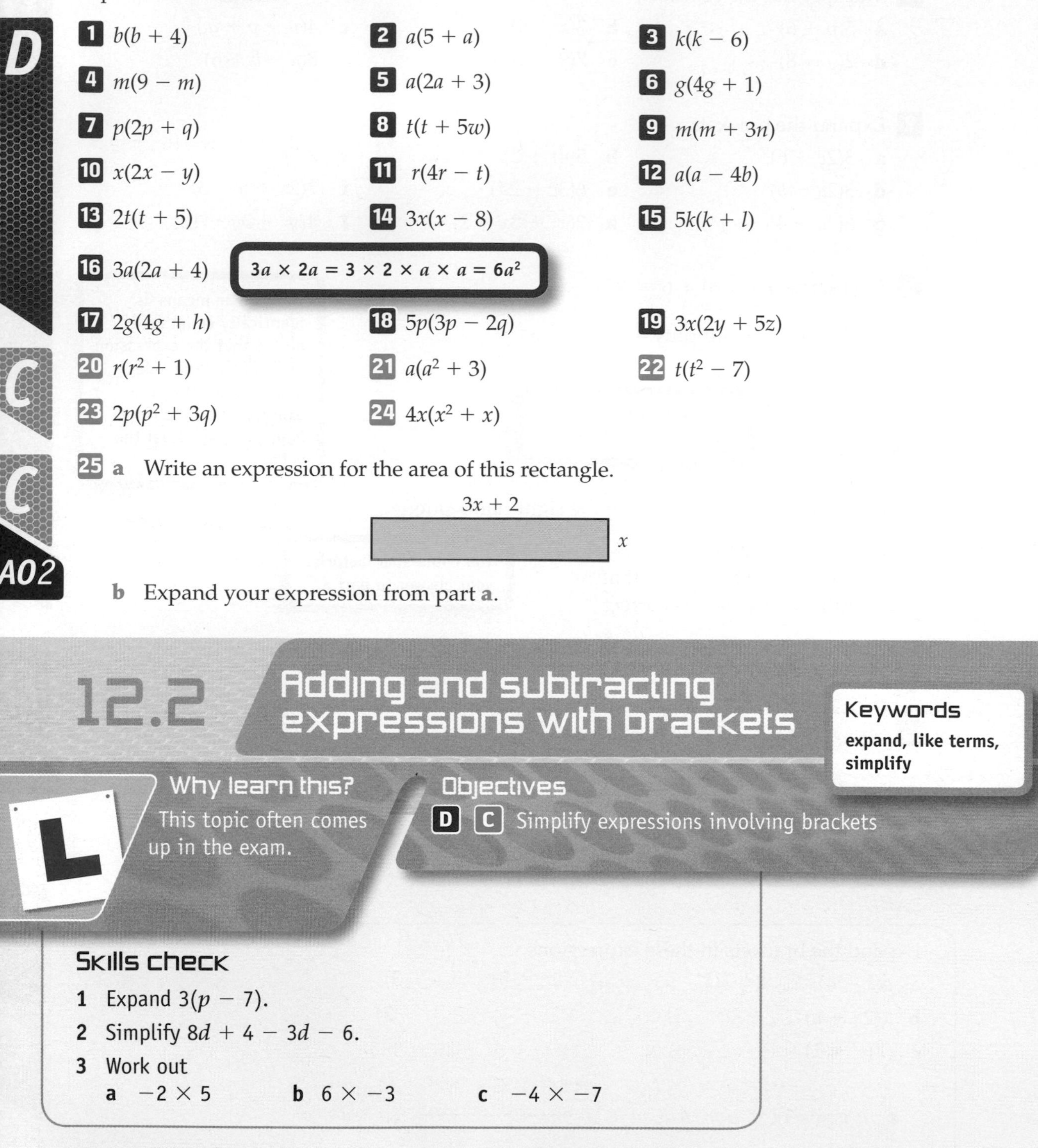

12.2 Adding and subtracting expressions with brackets

Keywords
expand, like terms, simplify

Why learn this?
This topic often comes up in the exam.

Objectives
D C Simplify expressions involving brackets

Skills check

1 Expand $3(p - 7)$.
2 Simplify $8d + 4 - 3d - 6$.
3 Work out
a -2×5 **b** 6×-3 **c** -4×-7

Adding and subtracting expressions with brackets

To add or subtract expressions with brackets, **expand** all the brackets first. Then collect **like terms** to **simplify** your answer.

Collecting like terms means adding all the terms in x together, adding all the terms in y together etc.

Example 3

Expand and simplify.

a $3(a + 4) + 2(a + 5)$

b $3(2t + 1) - 2(2t + 4)$

c $8(x + 1) - 3(2x - 5)$

a $3(a + 4) + 2(a + 5) = 3a + 12 + 2a + 10$ — Expand the brackets first.

$= 3a + 2a + 12 + 10$ — Collect like terms.

$= 5a + 22$

b $3(2t + 1) - 2(2t + 4) = 6t + 3 - 4t - 8$ — Expand both sets of brackets, multiplying *both* terms in the second bracket by -2.

$= 6t - 4t + 3 - 8$

$= 2t - 5$

c $8(x + 1) - 3(2x - 5) = 8x + 8 - 6x + 15$ — Expand the brackets. Remember that $-3 \times -5 = +15$.

$= 8x - 6x + 8 + 15$ — Collect like terms.

$= 2x + 23$

Exercise 12C

D

1 Expand the brackets.

a $-2(2k + 4)$ **b** $-3(2x + 6)$ **c** $-5(3n - 1)$

d $-4(3t + 5)$ **e** $-3(4p - 1)$ **f** $-2(3x - 7)$

D C

2 Expand and simplify.

a $3(y + 4) + 2y + 10$ **b** $2(k + 6) + 3k + 9$

c $4(a + 3) - 2a + 6$ **d** $3(t - 2) + 4t - 10$

e $3(2y + 3) - 2(y + 5)$ **f** $4(x + 7) + 3(x + 4)$

g $3(2x + 5) + 2(x - 4)$ **h** $2(4n + 5) - 5(n - 3)$

i $3(x - 5) + 2(x - 3)$ **j** $4(2x - 1) + 2(3x - 2)$

k $3(2b + 1) - 2(2b + 4)$ **l** $4(2m + 3) - 2(2m + 5)$

m $2(5t + 3) - 2(3t + 1)$ **n** $5(2k + 2) - 4(2k + 6)$

o $8(a + 1) - 3(2a - 5)$ **p** $2(4p + 1) - 4(p - 3)$

q $5(2g - 4) - 2(4g - 6)$ **r** $2(w - 4) - 3(2w - 1)$

s $x(x + 3) + 4(x + 2)$ **t** $x(2x + 1) - 3(x - 4)$

C AO2

3 Show that $9(x + 1) + 3(x + 2) \equiv 3(4x + 5)$.

12.3 Solving equations involving brackets

Keywords
solve, equation

Why learn this?
Brackets in algebra are like punctuation: they help you interpret algebra correctly.

Objectives
D Solve linear equations with brackets on one side

Skills check
Solve these equations.

1 $2x + 5 = 11$ 2 $3x + 10 = 1$

Finding the unknown

When **solving equations** involving brackets, you usually expand the brackets first.

Example 4

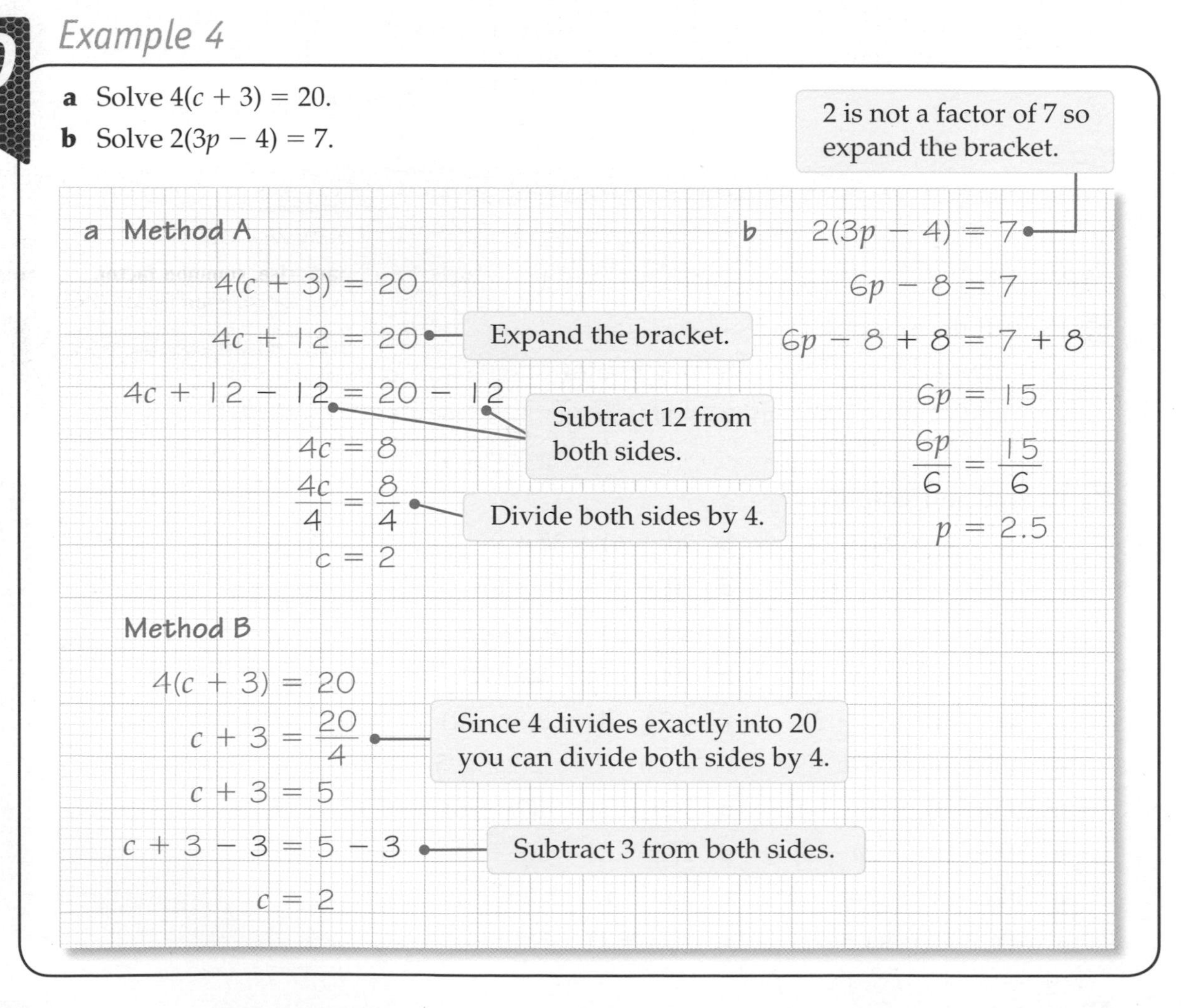

D
a Solve $4(c + 3) = 20$.
b Solve $2(3p - 4) = 7$.
2 is not a factor of 7 so expand the bracket.
a Method A
$4(c + 3) = 20$
$4c + 12 = 20$
Expand the bracket.
$4c + 12 - 12 = 20 - 12$
Subtract 12 from both sides.
$4c = 8$
$\frac{4c}{4} = \frac{8}{4}$
Divide both sides by 4.
$c = 2$
Method B
$4(c + 3) = 20$
$c + 3 = \frac{20}{4}$
Since 4 divides exactly into 20 you can divide both sides by 4.
$c + 3 = 5$
$c + 3 - 3 = 5 - 3$
Subtract 3 from both sides.
$c = 2$
b $2(3p - 4) = 7$
$6p - 8 = 7$
$6p - 8 + 8 = 7 + 8$
$6p = 15$
$\frac{6p}{6} = \frac{15}{6}$
$p = 2.5$

Exercise 12D

D

1 Solve these equations.

a $4(g + 6) = 32$ b $7(k + 1) = 21$ c $5(s + 10) = 65$ d $2(n - 4) = 6$
e $3(f - 2) = 24$ f $6(v - 3) = 42$ g $4(m + 7) = 14$ h $2(w + 11) = 19$

2 Solve these equations.

a $4(5t + 2) = 48$ **b** $3(2r + 4) = 30$

c $2(2b + 15) = 22$ **d** $2(3w - 6) = 27$

e $3(4x - 2) = 24$ **f** $5(2y + 11) = 40$

g $6(2k - 1) = 36$ **h** $3(2a + 13) = 15$

D

3 A regular hexagon has sides of length $7x + 4$ centimetres.

D

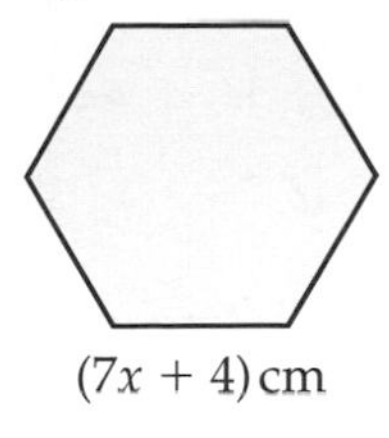

a Complete this expression for its perimeter: $\square(7x + \square)$.

The hexagon has a perimeter of 45 cm.

b Complete this equation in x: $\square(7x + \square) = \square$.

c Solve the equation to find the value of x.

d What is the length of one side of the hexagon?

AO2

12.4 Factorising algebraic expressions

Keywords
factorise, common factor, highest common factor (HCF)

Why learn this?
It's useful to know how to write expressions in equivalent ways. Buying two dustpans and two brushes, $2d + 2b$, is the equivalent of buying two dustpan and brush sets, $2(d + b)$.

Objectives
D C B Simplify algebraic expressions by taking out common factors

Skills check

1 Write down the HCF (highest common factor) of

a 6 and 10 **b** $5x$ and 15 **c** x^2 and $6x$

d $6x^2$ and $10x$ **e** $5pq$ and $15p^2$

Finding the factors

Factorising an algebraic expression is the opposite of expanding brackets.

expanding

$2(3x + 5) = 6x + 10$

factorising

Start by writing a **common factor** of both terms outside a bracket.
Then work out the terms inside the bracket.

D

Example 5

Factorise

a $5a + 20$ **b** $4x - 12$

a $5a + 20 = 5 \times a + 5 \times 4$
$= 5(a + 4)$

5 is a factor of $5a$: $5a = 5 \times a$
5 is a factor of 20: $20 = 5 \times 4$
So 5 is a common factor of $5a$ and 20, and is the term outside the bracket.

Check: $5(a + 4) = 5 \times a + 5 \times 4$
$= 5a + 20$ ✓

Check your answer by expanding the bracket.

b $4x - 12 = 4 \times x - 4 \times 3$
$= 4(x - 3)$

2 is a factor of $4x$ and 12. 4 is also a common factor of $4x$ and 12. Use 4 because it is the **HCF** of $4x$ and 12.

Exercise 12E

D

1 Copy and complete these.

a $3x + 15 = 3(\square + 5)$ **b** $5a + 10 = 5(\square + 2)$
c $2x - 12 = 2(x - \square)$ **d** $4m - 16 = 4(m - \square)$
e $4t + 12 = \square(t + 3)$ **f** $3n + 18 = \square(n + 6)$
g $2b - 14 = \square(b - 7)$ **h** $4t - 20 = \square(t - 5)$

Always look for the HCF.

Remember to check all your answers by expanding the brackets.

2 Factorise these expressions.

a $5p + 20$ **b** $2a + 12$ **c** $3y + 15$ **d** $7b + 21$
e $4q + 12p$ **f** $6k + 24j$ **g** $4t - 12$ **h** $3x - 9$
i $5n - 20$ **j** $2b - 8$ **k** $6a - 18b$ **l** $7k - 7$

3 Factorise these expressions.

a $y^2 + 7y$ **b** $x^2 + 5x$ **c** $n^2 + n$
d $x^2 - 7x$ **e** $p^2 - 8p$ **f** $a^2 - ab$

Remember, $n = n \times 1$.

4 Factorise these expressions.

a $6p + 4$ **b** $4a + 10$ **c** $6 - 4t$
d $12m - 8n$ **e** $25x + 15y$ **f** $12y - 9z$

Remember, $6p = 2 \times 3p$.

C B

Example 6

Factorise

a $2x^2 + 14x$ **b** $6p^2q^2 - 9pq^3$

a $2x^2 + 14x = 2x \times x + 2x \times 7$
$= 2x(x + 7)$

$2x$ is the HCF of $2x^2$ and $14x$.

b $6p^2q^2 - 9pq^3 = 3pq^2 \times 2p - 3pq^2 \times 3q$
$= 3pq^2(2p - 3q)$

The HCF is $3pq^2$.

Exercise 12F

> When an expression is factorised completely, the terms inside the brackets have no common factor other than 1.

1 Factorise these completely.

a $3x^2 - 6x$ **b** $8x^2 - xy$

c $8a + 4ab$ **d** $p^3 - 5p^2$

e $3t^3 + 6t^2$ **f** $10yz - 15y^2$

g $18a^2 + 12ab$ **h** $16p^2 - 12pq$

B

2 Factorise these expressions.

a $4ab^2 + 6ab^3$ **b** $10xy - 5x^2$

c $3p^2q - 6p^3q^2$ **d** $8mn^3 + 4n^2 - 6m^2n$

e $6h^2k - 12hk^3 - 18h^2k^2$ **f** $6ab^2 + 9a^2b - 3ab$

g $8xy^2 + 6xy - 4x^2y$ **h** $12p^2r + 18p^2r^2 + 6p^3r$

12.5 Expanding two brackets

Why learn this?

Knowing how to write expressions in different ways means you can choose the easiest one to work with.

Objectives

C B Multiply two algebraic expressions with brackets

C B Square a linear expression

Skills check

1 Simplify

a $5x - x$ **b** $4x - 7x$ **c** $-6x + 4x$ **d** $-x - 4x$

2 Simplify

a $t \times t$ **b** $t \times -3$ **c** $2t \times 5t$ **d** $-t \times -4t$

Multiplying two bracketed expressions

You can use a grid method to multiply two bracketed expressions. You multiply each term in one bracket by each term in the other bracket.

Example 7

C

Expand and simplify $(x + 2)(x + 5)$.

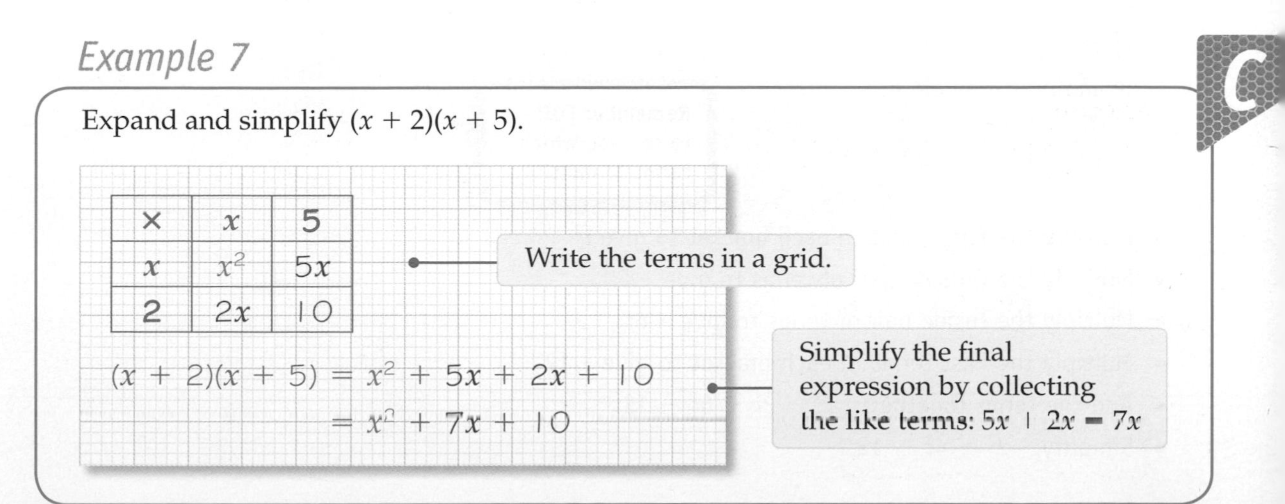

× x 5
x x² 5x
2 2x 10
Write the terms in a grid.
(x + 2)(x + 5) = x² + 5x + 2x + 10
= x² + 7x + 10
Simplify the final expression by collecting the like terms: 5x + 2x = 7x

It is like working out the area of a rectangle of length $x + 5$ and width $x + 2$.

Total area $= (x + 2)(x + 5)$
$= x^2 + 5x + 2x + 10$
$= x^2 + 7x + 10$

	x	5
x	area $= x \times x = x^2$	area $= x \times 5 = 5x$
2	area $= 2 \times x = 2x$	area $= 2 \times 5 = 10$

C

Example 8

Expand and simplify $(t + 6)(t - 2)$.

×	t	-2
t	t^2	$-2t$
6	$6t$	-12

Remember that you are multiplying by −2.
positive × negative = negative

$(t + 6)(t - 2) = t \times t + t \times (-2) + 6 \times t + 6 \times (-2)$
$= t^2 - 2t + 6t - 12$
$= t^2 + 4t - 12$

Exercise 12G

Use the grid method to expand and simplify these expressions.

C

1 $(a + 2)(a + 7)$

2 $(x + 5)(x + 5)$

3 $(t + 5)(t - 2)$

4 $(h - 3)(h - 8)$

5 $(y - 3)(y - 3)$

6 $(x - 12)(x - 7)$

7 $(p - 4)(p + 4)$

8 $(4 + a)(a + 7)$

Look again at Example 8.

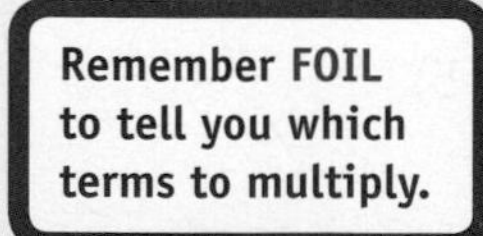

Remember FOIL to tell you which terms to multiply.

- Multiply the **F**irst terms in each bracket to give t^2
- Multiply the **O**utside pair of terms to give $-2t$
- Multiply the **I**nside pair of terms to give $+6t$
- Multiply the **L**ast terms in each bracket to give -12
- Add the terms together: $t^2 - 2t + 6t - 12$
- Simplify: $t^2 + 4t - 12$

Exercise 12H

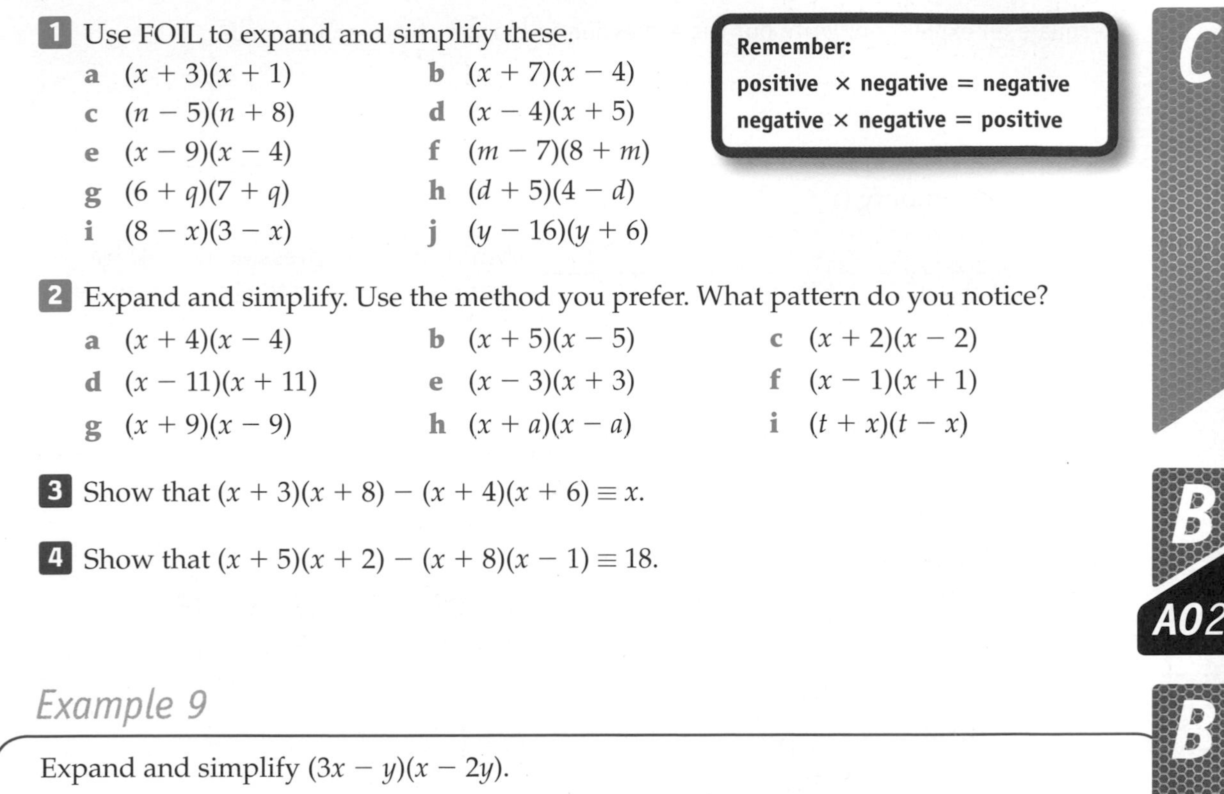

C

1 Use FOIL to expand and simplify these.

> **Remember:**
> positive × negative = negative
> negative × negative = positive

a $(x + 3)(x + 1)$ b $(x + 7)(x - 4)$
c $(n - 5)(n + 8)$ d $(x - 4)(x + 5)$
e $(x - 9)(x - 4)$ f $(m - 7)(8 + m)$
g $(6 + q)(7 + q)$ h $(d + 5)(4 - d)$
i $(8 - x)(3 - x)$ j $(y - 16)(y + 6)$

2 Expand and simplify. Use the method you prefer. What pattern do you notice?

a $(x + 4)(x - 4)$ b $(x + 5)(x - 5)$ c $(x + 2)(x - 2)$
d $(x - 11)(x + 11)$ e $(x - 3)(x + 3)$ f $(x - 1)(x + 1)$
g $(x + 9)(x - 9)$ h $(x + a)(x - a)$ i $(t + x)(t - x)$

B AO2

3 Show that $(x + 3)(x + 8) - (x + 4)(x + 6) \equiv x$.

4 Show that $(x + 5)(x + 2) - (x + 8)(x - 1) \equiv 18$.

Example 9

B

Expand and simplify $(3x - y)(x - 2y)$.

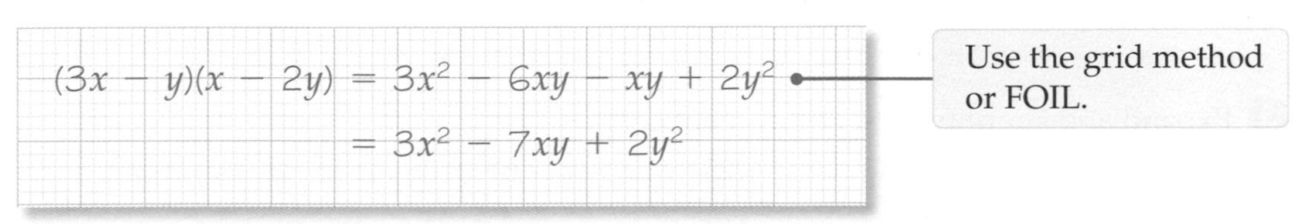

$(3x - y)(x - 2y) = 3x^2 - 6xy - xy + 2y^2$
$= 3x^2 - 7xy + 2y^2$

Use the grid method or FOIL.

Exercise 12I

B

Expand and simplify these. What is the connection in Q16 to Q21?

1 $(3a + 2)(a + 4)$ **2** $(5x + 3)(x + 2)$ **3** $(2t + 3)(3t + 5)$

4 $(4y + 1)(2y + 7)$ **5** $(6x + 5)(2x + 3)$ **6** $(4x + 3)(x - 1)$

7 $(2z + 5)(3z - 2)$ **8** $(y + 1)(7y - 8)$ **9** $(3n - 5)(n + 8)$

10 $(3b - 5)(2b + 1)$ **11** $(p - 4)(7p + 3)$ **12** $(2z - 3)(3z - 4)$

13 $(5x - 9)(2x - 1)$ **14** $(2y - 3)(2y - 3)$ **15** $(2 + 3a)(4a + 5)$

16 $(2x + 1)(2x - 1)$ **17** $(3y + 2)(3y - 2)$ **18** $(5n + 4)(5n - 4)$

19 $(3x + 5)(3x - 5)$ **20** $(1 + 2x)(1 - 2x)$ **21** $(3t + 2x)(3t - 2x)$

B AO2

22 Show that $(2x + 3)(2x - 5) \equiv 4(x^2 - x - 4) + 1$.

23 a Write an expression for the area of this rectangle.
b Expand and simplify your expression from part **a**.

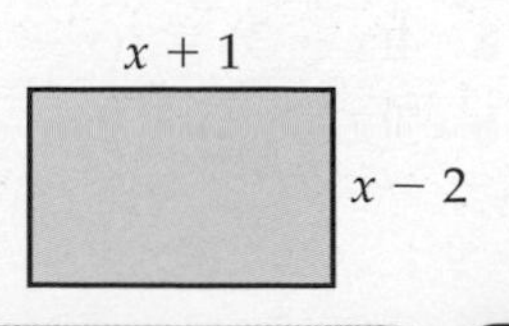

Squaring an expression

To square an expression, write out the expression in brackets twice and expand.

C

Example 10

Expand and simplify $(x + 4)^2$.

$$(x + 4)^2 = (x + 4)(x + 4)$$
$$= x^2 + 4x + 4x + 16$$
$$= x^2 + 8x + 16$$

Multiply the expression $(x + 4)$ by itself. Write down the bracket twice and expand.

Exercise 12J

C

1 Expand and simplify.

a $(x + 5)^2$ **b** $(x + 6)^2$ **c** $(x - 3)^2$ **d** $(x - 8)^2$

e $(3 + x)^2$ **f** $(2 + x)^2$ **g** $(5 - x)^2$ **h** $(x + a)^2$

C AO2

2 Copy and complete these.

a $(x + \square)^2 = x^2 + \square x + 36$ **b** $(x - \square)^2 = x^2 - \square x + 49$

c $(x + \square)^2 = x^2 + 18x + \square$ **d** $(x - \square)^2 = x^2 - 20x + \square$

C

3 Expand and simplify. What do you notice?

a $(x + 3)^2 - 9$ **b** $(x + 5)^2 - 25$ **c** $(x - 6)^2 - 36$ **d** $(x + a)^2 - a^2$

B

4 Expand and simplify.

a $(2x + 1)^2$ **b** $(2y - 7)^2$ **c** $(2 + 5z)^2$ **d** $(7 - 4m)^2$

e $(x + 2y)^2$ **f** $(2a - b)^2$ **g** $(3x + 2y)^2$ **h** $(4z - 5t)^2$

B AO3

5 **a** Show that $(10x + 5)^2 \equiv 100x(x + 1) + 25$.

b Use $x = 5$ in part **a** to work out 55^2.

c Calculate 75^2 using this method.

6 **a** Show that $(a + b)^2 - (a - b)^2 \equiv 4ab$.

b Use this result to calculate $57^2 - 43^2$.

c Use this method to work out $48^2 - 32^2$.

Review exercise

D

1 Expand

a $6(a + 2)$ **b** $p(p - 8)$ **[2 marks]**

D AO2

2 Show that

a $4(x - 3) \equiv 3(x - 4) + x$ **[2 marks]**

b $3(2x + 5) + 2(x - 3) \equiv 8(x + 1) + 1$ **[3 marks]**

3 Solve

a $6(4y - 1) = 42$ b $5(3m + 2) = 40$ **[6 marks]**

D

4 Expand and simplify

a $3(2t - 5) + 4(t + 3)$ b $4(3x + 2) - 5(2x - 1)$ **[4 marks]**

C

5 Expand and simplify

a $(x + 3)(x + 1)$ b $(y - 4)(y - 2)$ c $(x + 4)^2$ **[6 marks]**

C

6 Here is part of a number grid.

The shaded shape is called T_{10} because it has the number 10 on the left.

The sum of the numbers in T_{10} is 43.

1	2	3	4	5	6	7
8	9	10	11	12	13	14
15	16	17	18	19	20	21
22	23	24	25	26	27	28

a This is T_n.

Copy T_n and write expressions in the empty boxes. **[3 marks]**

n	

b Find the sum of all the numbers in T_n in terms of n.
Give your answer in its simplest form. **[2 marks]**

C AO3

7 Show that

a $(x + 3)(x + 2) - 5x \equiv x^2 + 6$ **[3 marks]**

b $(x - 2)(x + 8) \equiv (x + 3)^2 - 25$ **[3 marks]**

C AO2

8 Factorise

a $5y + 10$ **[1 mark]**

D

b $m^2 - 6m$ **[1 mark]**

c $3t^3 - 9t^2$ **[2 marks]**

C

d $4x^2 + 8xy$ **[2 marks]**

e $p^2q - 8pq + 6pq^2$ **[2 marks]**

B

9 Expand and simplify

a $(4x - 3)(x + 5)$ b $(2t - 5r)(2t + 5r)$ c $(4n - 5)^2$ **[6 marks]**

B

10 Show that $(2x - 1)^2 - 16 \equiv (2x + 3)(2x - 5)$. **[3 marks]**

B AO2

Chapter summary

In this chapter you have learned how to

- solve linear equations with brackets on one side D
- multiply terms in a bracket by a single term outside the bracket D C
- simplify expressions involving brackets D C
- understand the mathematical meaning of 'expression' and 'identity' C
- simplify algebraic expressions by taking out common factors D C B
- multiply two algebraic expressions with brackets C B
- square a linear expression C B

FUNCTIONAL • FUNCTIONAL

eBay business

Shani lives in Liverpool. She runs a business that imports wooden carvings from Africa. She then sells these to customers via eBay.

Shani sells carvings of elephants and rhinos in three sizes, small (S), medium (M) and large (L). One day she receives three large orders.

Name of customer	Number of elephants			Number of rhinos			Type of delivery
	S	M	L	S	M	L	
Mrs Read	6	3	3	4	2	2	next day
Mr Owen	4	4	0	2	2	0	standard
Mrs Patel	4	3	1	3	2	1	standard

Question bank

1 What is the total cost of the elephants that Mrs Read orders, excluding any delivery charge?

Shani uses two delivery companies. Both companies offer a 'standard' and 'next day' delivery service.

2 Which type of delivery service does Mrs Patel order?

Shani uses two types of boxes to pack the carvings. She always chooses the cheapest delivery company.

Both companies work out the delivery charge by measuring the weight of the package Shani sends.

3 How much does it cost to send a 9 kg package by 'next day' delivery with Liverpool Xpress?

Shani completes a bill for each of her customers. The bill shows the total cost of the carvings and the cost of delivery. Shani says, 'Mrs Read's bill is more than Mr Owen's and Mrs Patel's bills added together.'

4 Is Shani correct? Show working to support your answer.

One day Shani asks her brother to post a package for her. She tells him it contains three identical carvings, but doesn't tell him which animal. Her brother weighs the package and it comes to 950 g. He knows that the packaging weighs 50 g.

5 Solve this equation to work out which animal is in the package:

$$3x + 50 = 950$$

where x is the weight of the mystery animal in grams.

Information bank

Price list

Animal	Weight of carving	Price
Elephant (S)	400 g	£8.50
Rhino (S)	300 g	£6.95
Elephant (M)	1 kg	£12.50
Rhino (M)	900 g	£9.95
Elephant (L)	1.5 kg	£17.50
Rhino (L)	1.3 kg	£13.95

Delivery charges

Liverpool Xpress prices		
Weight (up to)	Standard delivery	Next day delivery
5 kg	£8.70	£12.50
10 kg	£14.40	£24.65
12 kg	£16.80	£25.70
14 kg	£18.55	£28.95
16 kg	£22.95	£30.78

UK Connect prices		
Weight (up to)	Standard delivery	Next day delivery
2 kg	£4.20	£8.25
4 kg	£6.85	£13.60
6 kg	£9.30	£22.00
8 kg	£11.40	£24.30
10 kg	£12.24	£26.50
15 kg	£14.26	£32.20
20 kg	£21.95	£40.54

Types of boxes

A box that can hold up to 9 kg weighs 350 g.
This box measures 30 cm long by 21 cm wide by 25 cm high.
A box that can hold up to 16 kg weighs 450 g.
This box measures 40 cm long by 30 cm wide by 25 cm high.
All boxes containing glass must have a 'Fragile' label.
All the boxes are made from 80% recycled cardboard.

13 Decimals

This chapter is about calculating with decimals.

Your money may be safe if you keep it in a money box, but you don't earn any interest. When you save your money with a bank, the bank will calculate your interest using decimals.

Objectives

This chapter will show you how to

- multiply and divide decimal numbers D C
- convert decimals to fractions and fractions to decimals D C
- understand terminating and recurring decimals and their relationships to fractions C B
- convert a recurring decimal to a fraction A

Before you start this chapter

Put your calculator away!

1 Work out
  a 82×100 b $82 \div 100$
  c 2.65×1000 d $2.65 \div 1000$

2 Work out
  a 25×174 b $2736 \div 4$

3 Work out
  a $3.62 + 0.071$ b $1.25 + 4.6 + 8.92$

4 Work out
  a $88.88 - 4.56$ b $90 - 3.67$

Multiplying and dividing decimals

Why learn this?
Most engineers calculate with decimals all the time in their work.

Objectives
D C Multiply and divide decimal numbers

Skills check
1 Work out
a 245 × 31 b 1016 ÷ 8

Multiplying decimals

You can multiply decimals in the same way as whole numbers.

Example 1

D

Work out 1.25 × 7.3.

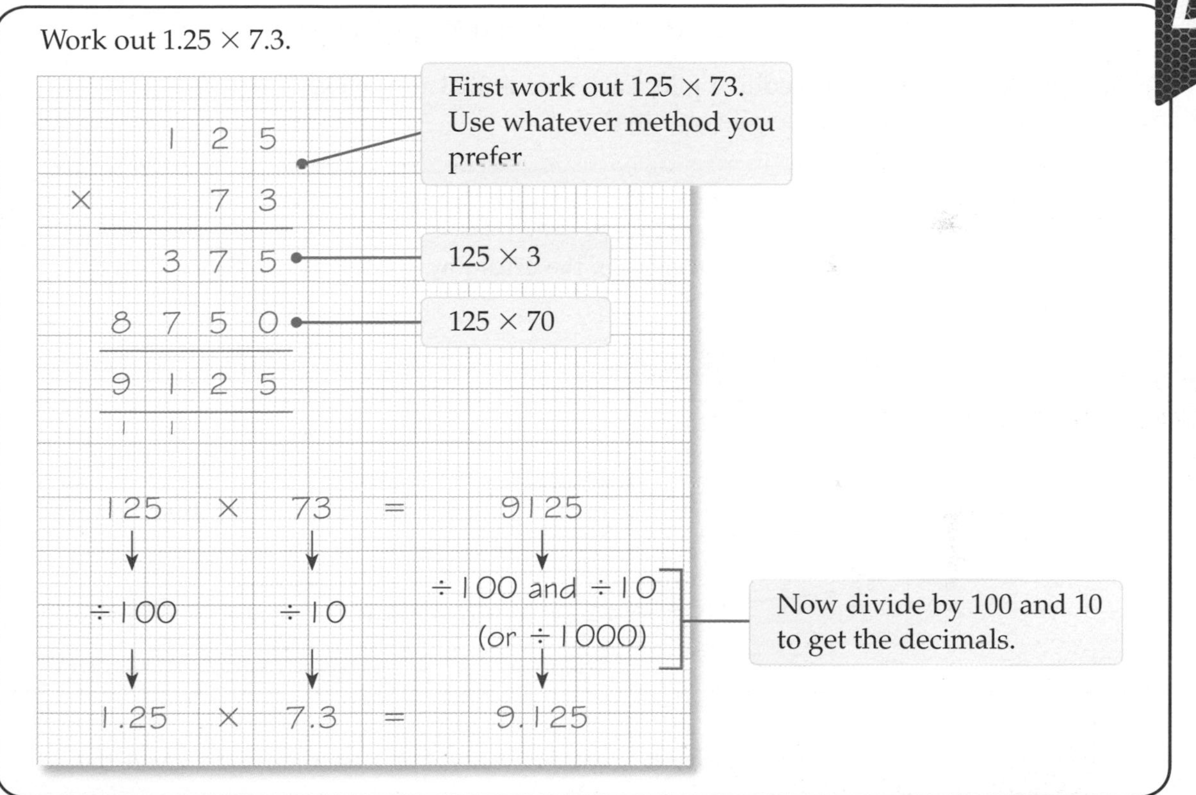

1 2 5
× 7 3
3 7 5
8 7 5 0
9 1 2 5
1 1
First work out 125 × 73. Use whatever method you prefer.
125 × 3
125 × 70
125 × 73 = 9125
÷100
÷10
÷100 and ÷10
(or ÷1000)
Now divide by 100 and 10 to get the decimals.
1.25 × 7.3 = 9.125

Here is a useful rule for multiplying decimals.

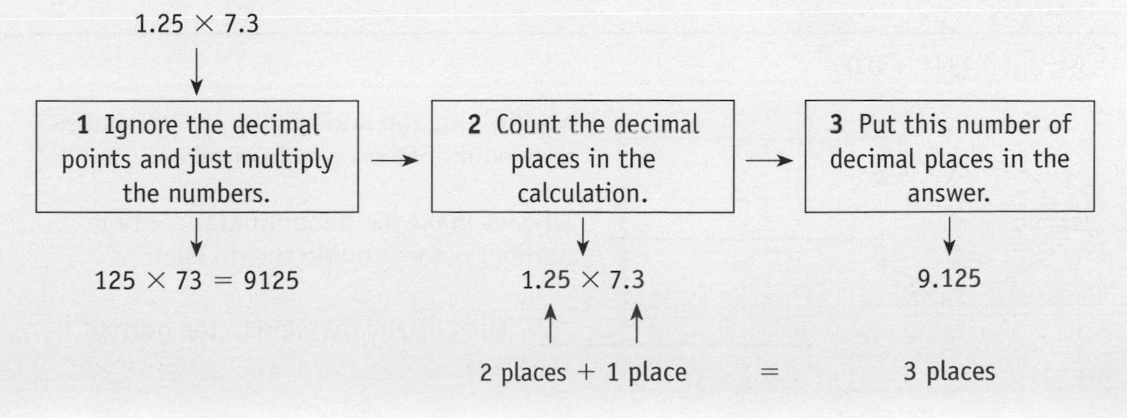

1.25 × 7.3
1 Ignore the decimal points and just multiply the numbers.
2 Count the decimal places in the calculation.
3 Put this number of decimal places in the answer.
125 × 73 = 9125
1.25 × 7.3
9.125
2 places + 1 place = 3 places

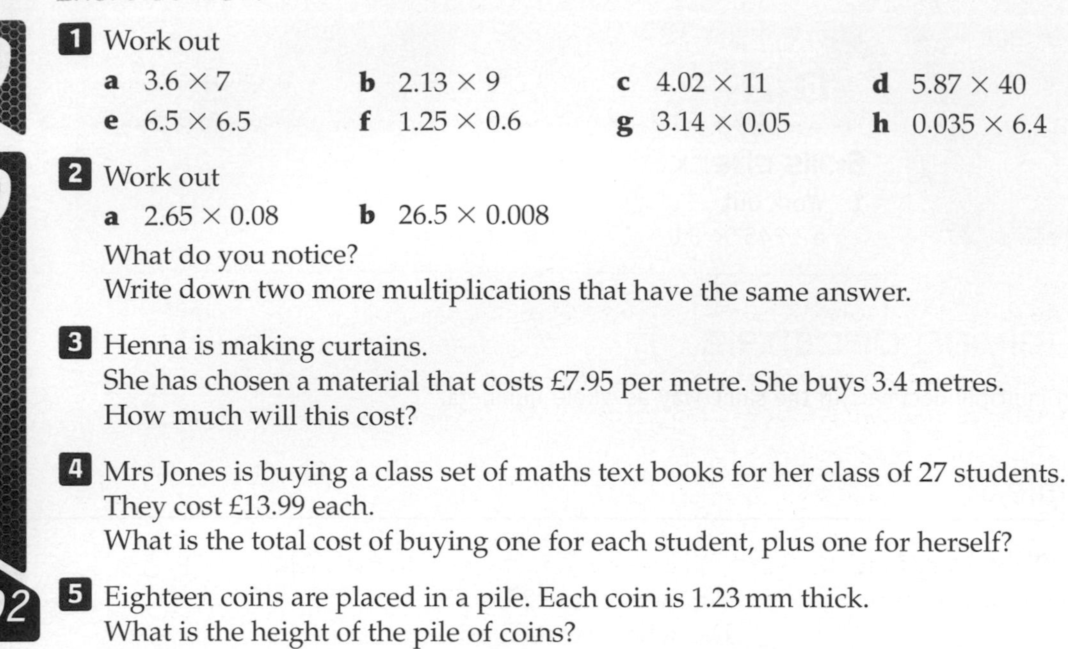

Exercise 13A

D

1 Work out

a 3.6×7 b 2.13×9 c 4.02×11 d 5.87×40

e 6.5×6.5 f 1.25×0.6 g 3.14×0.05 h 0.035×6.4

D

2 Work out

a 2.65×0.08 b 26.5×0.008

What do you notice?

Write down two more multiplications that have the same answer.

3 Henna is making curtains.

She has chosen a material that costs £7.95 per metre. She buys 3.4 metres.

How much will this cost?

4 Mrs Jones is buying a class set of maths text books for her class of 27 students.

They cost £13.99 each.

What is the total cost of buying one for each student, plus one for herself?

AO2

5 Eighteen coins are placed in a pile. Each coin is 1.23 mm thick.

What is the height of the pile of coins?

Dividing decimals

You can divide a decimal by a whole number in the usual way.

D

Example 2

Work out $43.41 \div 6$.

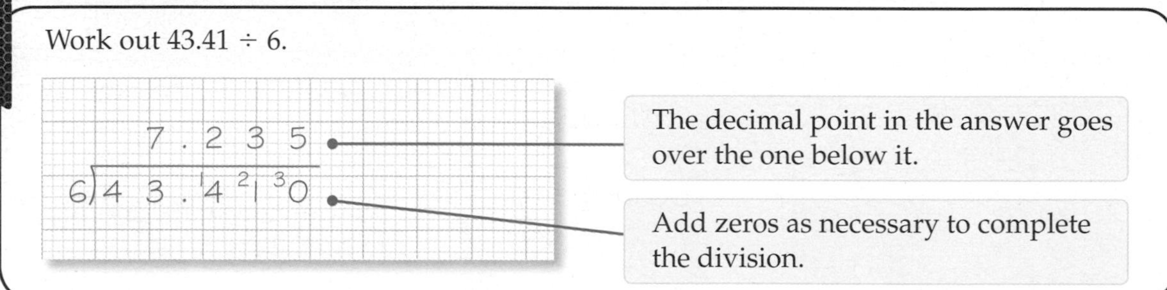

The decimal point in the answer goes over the one below it.

Add zeros as necessary to complete the division.

If you are dividing a decimal by another decimal, first write it as a fraction, and convert the denominator to a whole number.

C

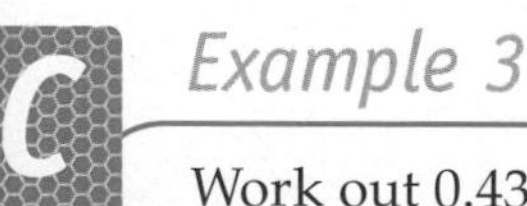

Example 3

Work out $0.4341 \div 0.06$.

$$\frac{0.4341}{0.06} = \frac{43.41}{6} \quad (\times 100 \text{ top and bottom})$$

$$= 7.235$$

Multiplying top and bottom by 100 makes the denominator a whole number.

Always make the denominator a whole number before you do the division.

Then do the division in the normal way.

Exercise 13B

1 Work out

a $8 \div 0.4$ b $72 \div 0.06$ c $12.78 \div 0.3$ d $0.125 \div 0.005$

e $3.6 \div 0.12$ f $4.5 \div 0.03$ g $26.72 \div 0.008$ h $0.56 \div 0.004$

C

2 Work out

a $6.95 \div 0.05$ b $69.5 \div 0.5$

What do you notice?

Write down two more divisions that have the same answer.

C

FUNCTIONAL

3 George is putting a new fence down the side of his garden.
The fence panels are 0.9 metres long. The total length of fence is 7.2 metres.

a How many panels does he need to buy?

b How many fence posts does he need?

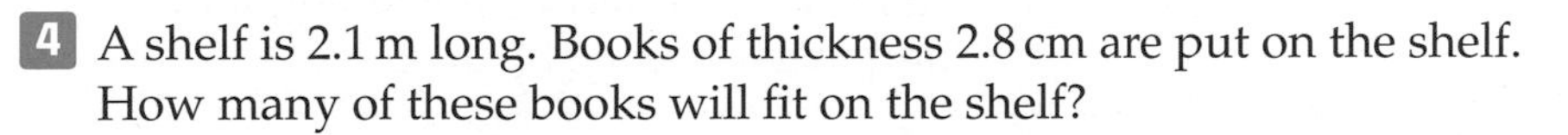

4 A shelf is 2.1 m long. Books of thickness 2.8 cm are put on the shelf.
How many of these books will fit on the shelf?

5 A small business signs up for 'BISTEXT', a service that charges 4.8p per SMS text message to customers in the UK. Their first bill is for £103.20.
How many text messages did they send?

AO2

13.2 Converting decimals and fractions

Why learn this?

This micrometer measures thickness in hundredths of a millimetre. Such measurements are usually witten as decimals.

Objectives

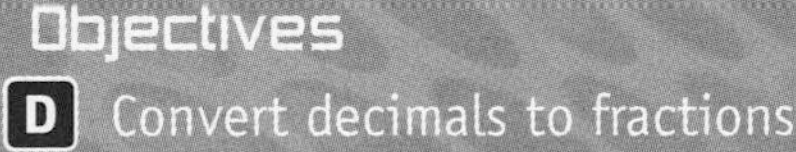

D Convert decimals to fractions

D **C** Convert fractions to decimals

Keywords
terminating, recurring

Skills check

1 Round each of these numbers to one decimal place.

a 24.27 b 9.772 c 0.4499

2 Cancel each of these fractions to its lowest terms.

a $\frac{8}{10}$ b $\frac{15}{25}$ c $\frac{28}{42}$

Converting decimals to fractions

You can use a place value diagram to convert a decimal to a fraction.

To convert 0.475 to a fraction, look at the place value of the last decimal place.

The value of the '5' is $\frac{5}{1000}$.

So $0.475 = \frac{475}{1000} = \frac{95}{200} = \frac{19}{40}$

Cancel by dividing top and bottom by 5, and then dividing by 5 again.

Remember to give your answer in its simplest form.

D

Example 4

a Convert each of these numbers to a fraction.
i 0.7 **ii** 0.48

b Convert 2.125 to a mixed number.

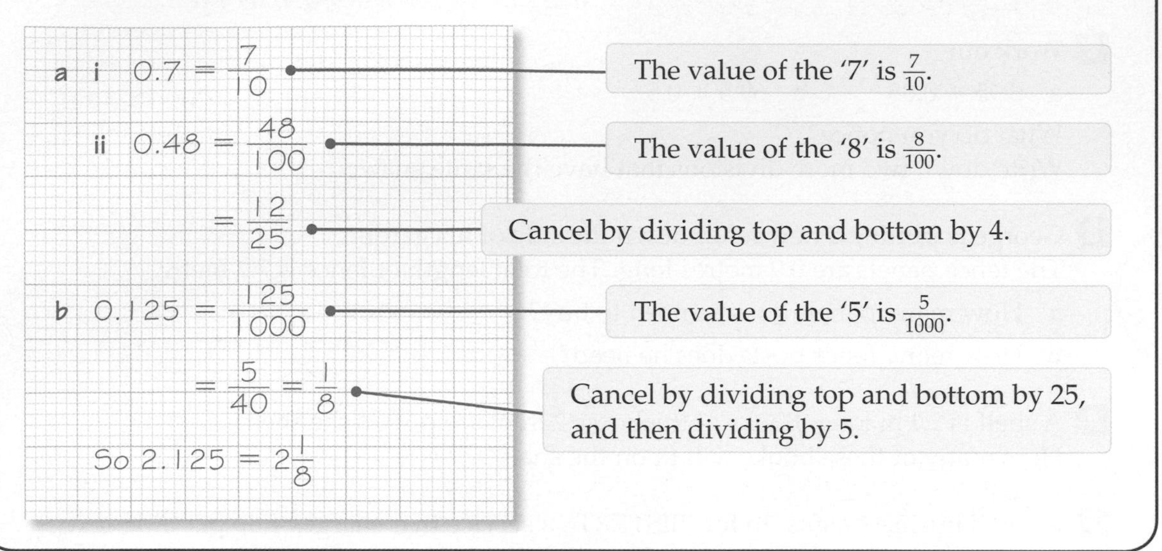

a i $0.7 = \frac{7}{10}$
ii $0.48 = \frac{48}{100}$
$= \frac{12}{25}$
b $0.125 = \frac{125}{1000}$
$= \frac{5}{40} = \frac{1}{8}$
So $2.125 = 2\frac{1}{8}$
The value of the '7' is $\frac{7}{10}$.
The value of the '8' is $\frac{8}{100}$.
Cancel by dividing top and bottom by 4.
The value of the '5' is $\frac{5}{1000}$.
Cancel by dividing top and bottom by 25, and then dividing by 5.

Exercise 13C

D

1 Convert each of these decimals to a fraction in its lowest terms.

a 0.1	**b** 0.5	**c** 0.07	**d** 0.05
e 0.75	**f** 0.28	**g** 0.04	**h** 0.65
i 0.52	**j** 0.79	**k** 0.24	**l** 0.35

2 Convert each of these decimals to a fraction in its lowest terms.

a 0.0001	**b** 0.002	**c** 0.084	**d** 0.009
e 0.035	**f** 0.025	**g** 0.375	**h** 0.425

3 Convert each of these to a mixed number.

a 3.5	**b** 14.8	**c** 5.64	**d** 4.85

4 A decimal fraction of 0.72 of the cats observed in a survey liked Moggymeat best. What fraction of the cats liked Moggymeat best?

5 Michael does a survey and finds that a decimal fraction of 0.44 of the cars in the school car park are diesel. What fraction of the cars are diesel?

6 Jasmine did a survey at her school and found that a decimal fraction of 0.36 of the students took a packed lunch. What fraction of the students took a packed lunch?

Converting fractions to decimals

To convert a fraction to a decimal, divide the numerator by the denominator.

Some fractions stop after a number of decimal places.
These are known as **terminating** decimals.

When you convert some fractions to decimals you get answers that never end.
These are known as **recurring** decimals.

'Recurring dots' are used to make the pattern clear. A dot over a digit shows that it recurs.

$\frac{1}{3} = 0.333\,333\,333... = 0.\dot{3}$

$\frac{7}{12} = 0.583\,333\,33... = 0.58\dot{3}$

In these two examples, a single digit (3) recurs, so there is a dot over the 3.

$\frac{5}{11} = 0.454\,545\,45... = 0.\dot{4}\dot{5}$

$\frac{4}{7} = 0.571\,428\,571\,428\,57... = 0.\dot{5}7142\dot{8}$

In these two examples, two or more digits recur. There is a dot over the first and last digits in the sequence that repeats.

Example 5

D

Convert each of these fractions to a decimal.

a $\frac{5}{8}$ **b** $\frac{5}{12}$ **c** $\frac{6}{7}$

Remember the rule 'Top divided by bottom'.

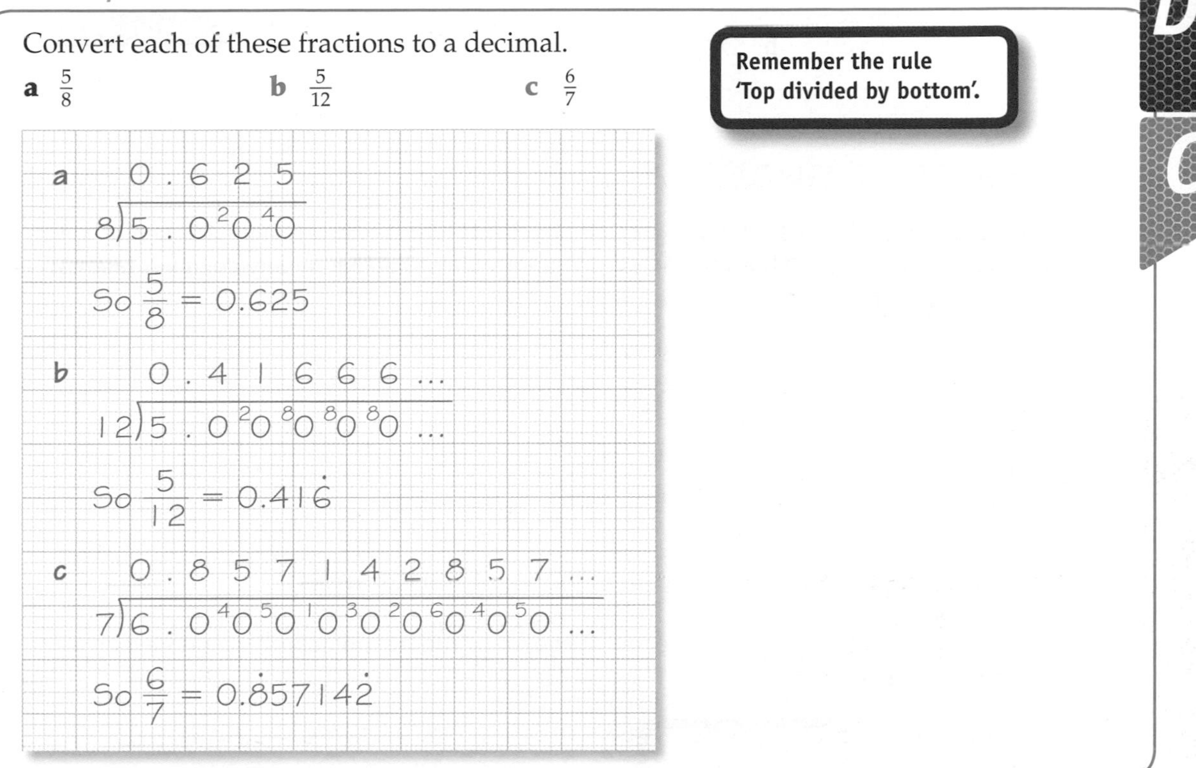

C

Exercise 13D

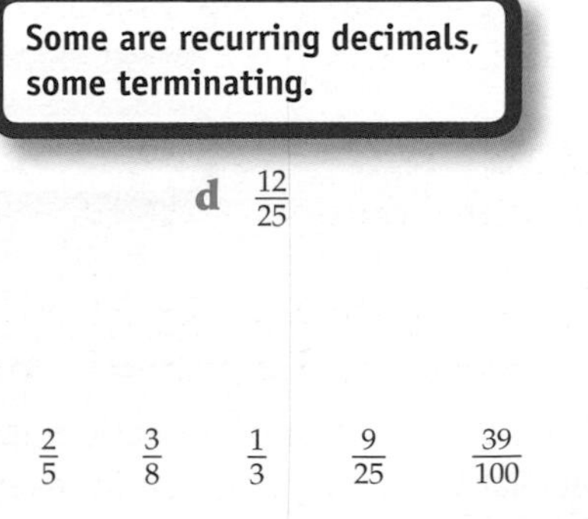
Some are recurring decimals, some terminating.

C

1 Convert each of these fractions to a decimal.

a $\frac{1}{8}$ **b** $\frac{1}{6}$ **c** $\frac{2}{9}$ **d** $\frac{12}{25}$

2 Convert each of these fractions to a decimal.

a $\frac{7}{30}$ **b** $\frac{7}{300}$ **c** $\frac{7}{3000}$

3 Convert these fractions to decimals and put them in order of size, starting with the smallest.

$\frac{2}{5}$ $\frac{3}{8}$ $\frac{1}{3}$ $\frac{9}{25}$ $\frac{39}{100}$

C

4 Convert each of these fractions to a decimal.

$\frac{1}{7}$, $\frac{2}{7}$, $\frac{3}{7}$, $\frac{4}{7}$, $\frac{5}{7}$, $\frac{6}{7}$

What do you notice?

AO2

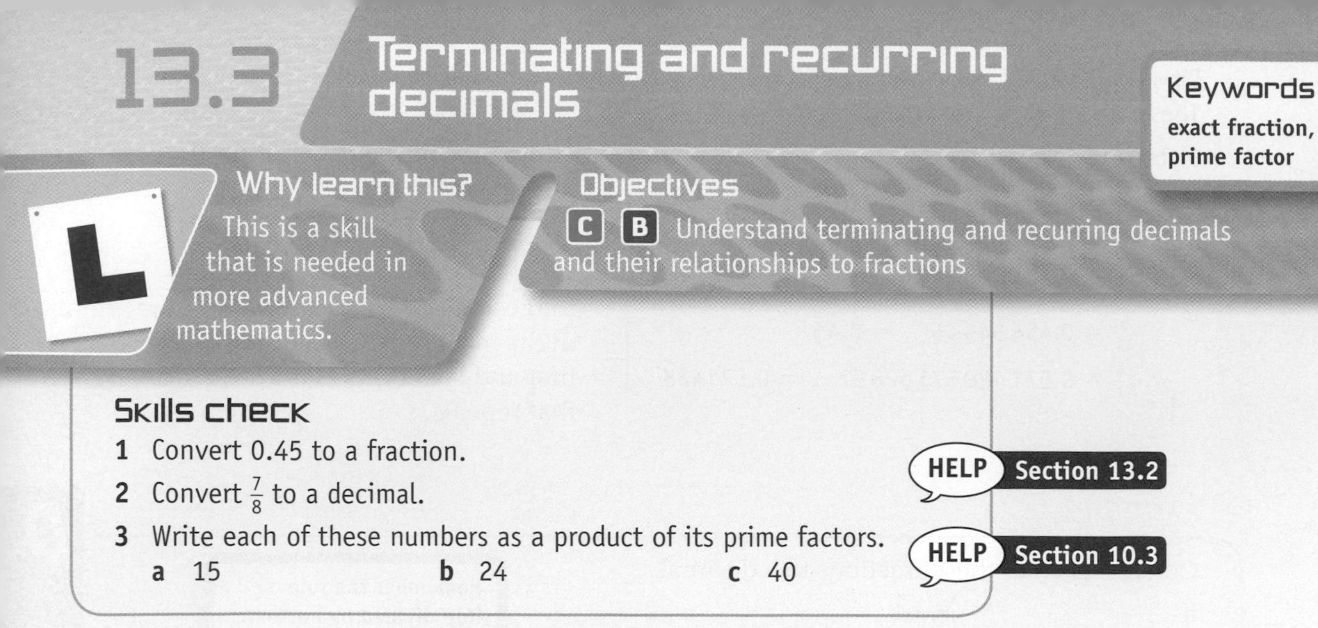

13.3 Terminating and recurring decimals

Keywords
exact fraction, prime factor

Why learn this?
This is a skill that is needed in more advanced mathematics.

Objectives
C B Understand terminating and recurring decimals and their relationships to fractions

Skills check

1 Convert 0.45 to a fraction.

2 Convert $\frac{7}{8}$ to a decimal.

3 Write each of these numbers as a product of its prime factors.
a 15 **b** 24 **c** 40

HELP Section 13.2
HELP Section 10.3

Terminating decimals

You can express any terminating decimal as an **exact fraction** whose denominator is a power of 10. You can then simplify the fraction by dividing top and bottom by any common factors: the **prime factors** of powers of 10 are 2 and 5.

For example: $0.5 = \frac{5}{10} = \frac{1}{2}$ (cancelling by 5)

$0.06 = \frac{6}{100} = \frac{3}{50}$ (cancelling by 2)

$0.375 = \frac{375}{1000} = \frac{3}{8}$ (cancelling by 125, $125 = 5^3$)

$0.00256 = \frac{256}{100\,000} = \frac{8}{3125}$ (cancelling by 32, $32 = 2^5$)

This means that you can tell whether you can convert a fraction to a terminating decimal by looking at the denominator.

If the denominator only has prime factors of 2 and/or 5, the fraction can be expressed as a terminating decimal.

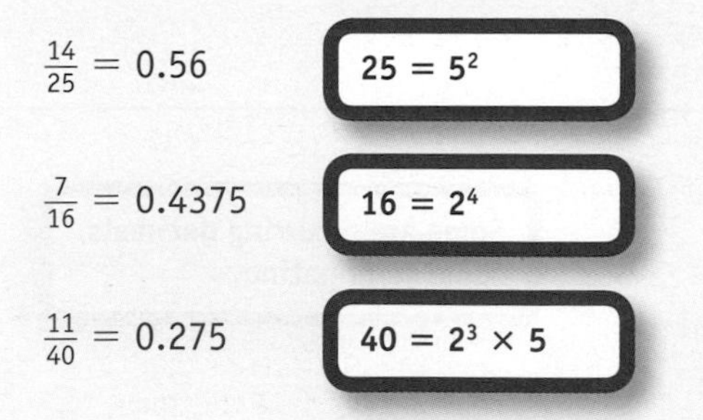

$\frac{14}{25} = 0.56$ **$25 = 5^2$**

$\frac{7}{16} = 0.4375$ **$16 = 2^4$**

$\frac{11}{40} = 0.275$ **$40 = 2^3 \times 5$**

Recurring decimals

If the denominator of a fraction has any prime factors other than 2 and 5, it will give a decimal whose digits go on for ever (i.e. they recur).

For example, $\frac{5}{11} = 0.454\,545\,454\,5\ldots = 0.\dot{4}\dot{5}$

You need to put the fraction in its simplest form first.

Example 6

a Convert each of these fractions to a decimal.

i $\frac{17}{40}$

ii $\frac{1}{12}$

b Does $\frac{127}{600}$ convert to a terminating or a recurring decimal?

a i

$$\begin{array}{r} 0.425 \\ 40\overline{)17.0^{10}0^{20}0} \end{array}$$

So $\frac{17}{40} = 0.425$

ii

$$\begin{array}{r} 0.08333\ldots \\ 12\overline{)1.0^{10}0^{4}0^{4}0^{4}0\ldots} \end{array}$$

So $\frac{1}{12} = 0.08\dot{3}$

b $600 = 2^3 \times 3 \times 5^2$, so the denominator has a factor other than 2 and 5, namely 3.

Therefore $\frac{127}{600}$ will give a recurring decimal.

C

B

Exercise 13E

1 Convert each of these fractions to a decimal.

a $\frac{2}{3}$ **b** $\frac{3}{5}$ **c** $\frac{6}{11}$ **d** $\frac{12}{25}$

e $\frac{5}{8}$ **f** $\frac{19}{40}$ **g** $\frac{5}{9}$ **h** $\frac{11}{18}$

C

2 Put the fractions in Q1 in order of size, starting with the smallest.

3 By looking at the denominators, predict which of these fractions will give terminating decimals and which will give recurring decimals.

$\frac{29}{64}$ $\frac{3}{17}$ $\frac{13}{125}$ $\frac{29}{72}$ $\frac{3}{16}$ $\frac{169}{500}$ $\frac{49}{320}$ $\frac{1}{36}$ $\frac{66}{125}$

B

4 Look at the patterns in the decimals for $\frac{1}{13}, \frac{2}{13}, \frac{3}{13}, \ldots$
What do you notice?

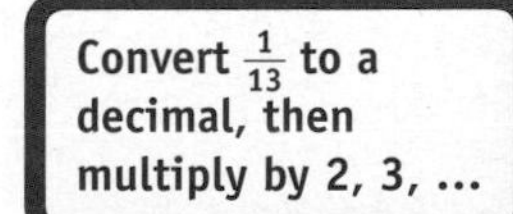
Convert $\frac{1}{13}$ to a decimal, then multiply by 2, 3, ...

B

5 Investigate what happens

a when you multiply two fractions together that each give a terminating decimal. Will your answer always be a terminating decimal?

b when you multiply two fractions together that each give a recurring decimal. Will your answer always be a recurring decimal?

c when you multiply one of each type of fraction together. What can you conclude in this case?

AO2

13.4 Converting recurring decimals to fractions

Keywords
recurring

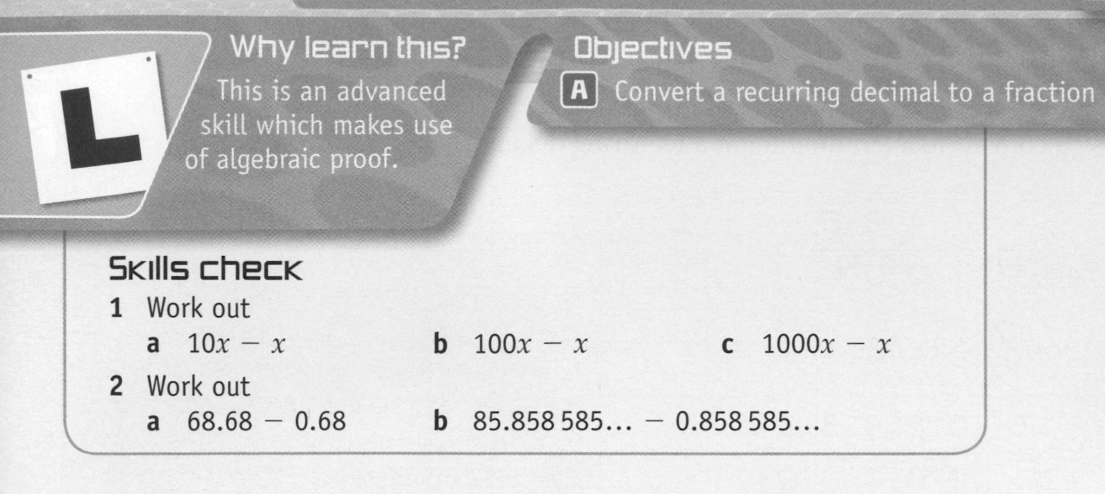

Why learn this?
This is an advanced skill which makes use of algebraic proof.

Objectives
A Convert a recurring decimal to a fraction

Skills check

1 Work out
  a $10x - x$ b $100x - x$ c $1000x - x$

2 Work out
  a $68.68 - 0.68$ b $85.858\,585... - 0.858\,585...$

Converting recurring decimals to fractions

Converting **recurring** decimals to fractions requires a special technique.

Example 7

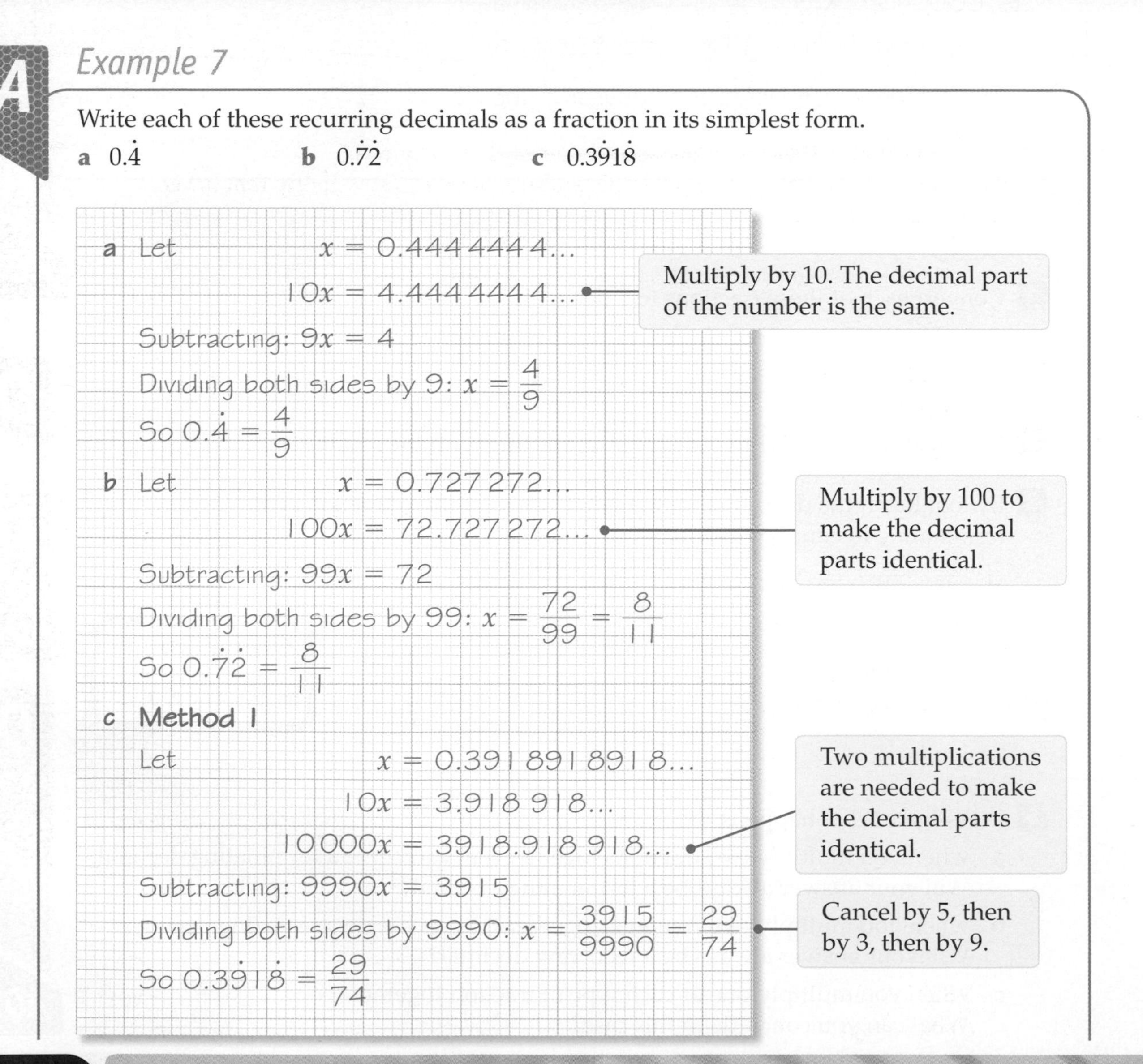

Write each of these recurring decimals as a fraction in its simplest form.

a $0.\dot{4}$ b $0.\dot{7}\dot{2}$ c $0.3\dot{9}1\dot{8}$

a Let $x = 0.444\,444\,4...$
$10x = 4.444\,444\,4...$ — Multiply by 10. The decimal part of the number is the same.
Subtracting: $9x = 4$
Dividing both sides by 9: $x = \frac{4}{9}$
So $0.\dot{4} = \frac{4}{9}$

b Let $x = 0.727\,272...$
$100x = 72.727\,272...$ — Multiply by 100 to make the decimal parts identical.
Subtracting: $99x = 72$
Dividing both sides by 99: $x = \frac{72}{99} = \frac{8}{11}$
So $0.\dot{7}\dot{2} = \frac{8}{11}$

c **Method 1**
Let $x = 0.391\,891\,891\,8...$
$10x = 3.918\,918...$
$10\,000x = 3918.918\,918...$ — Two multiplications are needed to make the decimal parts identical.
Subtracting: $9990x = 3915$
Dividing both sides by 9990: $x = \frac{3915}{9990} = \frac{29}{74}$ — Cancel by 5, then by 3, then by 9.
So $0.3\dot{9}1\dot{8} = \frac{29}{74}$

Method 2

Let $x = 0.391\,891\,891\,8\ldots$

$1000x = 391.891\,891\,8\ldots$ — Be careful, the decimal parts are not identical.

Subtracting: $999x = 391.5$

Dividing both sides by 999: $x = \frac{391.5}{999}$

$= \frac{3915}{9990}$ — Multiply by 10 to get whole numbers.

$= \frac{29}{74}$ — Always simplify fully.

So $0.3\dot{9}1\dot{8} = \frac{29}{74}$

For **one** recurring digit, multiply by **10**.
For **two** recurring digits, multiply by **100**.
For **three** recurring digits, multiply by **1000**.

Exercise 13F

1 Convert each of these recurring decimals to a fraction in its simplest form.

a $0.\dot{6}$ **b** $0.\dot{2}$ **c** $0.0\dot{4}$ **d** $0.\dot{5}\dot{4}$ **e** $0.\dot{0}\dot{3}$
f $0.\dot{2}1\dot{6}$ **g** $0.6\dot{1}$ **h** $0.58\dot{3}$ **i** $0.2\dot{4}0\dot{5}$ **j** $0.6\dot{3}8\dot{1}$

A AO2

2 Amy says that $0.\dot{9}$ is equivalent to 1. Jenny says she is wrong.
How can Amy use recurring decimals and fractions to show Jenny that $0.\dot{9}$ is equivalent to 1?

Review exercise

1 Work out

a 3.4×3.4 [2 marks]

b 0.024×62 [2 marks]

D AO2

2 Zak went shopping with £50. He bought two T-shirts at £7.99 each, a three-pack of socks for £8.49 and a pair of sunglasses for £22.35.

a How much did he spend altogether? [2 marks]

b How much money did he have left? [2 marks]

3 Convert each of these decimals to a fraction in its lowest terms.

a 0.078 [1 mark]

b 0.95 [2 marks]

c 0.645 [2 marks]

d 0.3125 [2 marks]

4 Work out

a $37.04 \div 0.8$ [2 marks]

b $3.822 \div 0.3$ [2 marks]

c $0.075 \div 0.06$ [2 marks]

C

AO2

5 Ahmed has a lot of empty oil cans that hold 1.2 litres each.
He has an oil drum with 42 litres of oil in it.
How many cans can he fill from the drum? **[2 marks]**

C

6 Convert each of these fractions to a decimal.

> **Some are recurring decimals, some terminating.**

a $\frac{3}{5}$ **[1 mark]**

b $\frac{63}{1000}$ **[1 mark]**

c $\frac{2}{11}$ **[1 mark]**

d $\frac{11}{12}$ **[1 mark]**

7 Convert these fractions to decimals and put them in order of size, starting with the smallest.

$\frac{2}{3}$ $\frac{3}{5}$ $\frac{4}{7}$ $\frac{5}{8}$ $\frac{7}{10}$ **[3 marks]**

B

8 By looking at the denominators, predict which of these fractions will give terminating decimals and which will give recurring decimals.

$\frac{1}{11}$ $\frac{15}{25}$ $\frac{13}{25}$ $\frac{23}{60}$ $\frac{3}{32}$ $\frac{3}{18}$ $\frac{9}{21}$ $\frac{17}{125}$ $\frac{13}{16}$ **[4 marks]**

A

9 Convert each of these recurring decimals to a fraction in its simplest form.

a $0.\dot{3}$ **[2 marks]**

b $0.\dot{6}\dot{3}$ **[2 marks]**

c $0.\dot{2}1\dot{5}$ **[2 marks]**

d $0.0\dot{7}$ **[2 marks]**

e $0.9\dot{8}\dot{4}$ **[3 marks]**

f $0.5\dot{4}2\dot{3}$ **[3 marks]**

Chapter summary

In this chapter you have learned how to

- convert decimals to fractions **D**
- multiply and divide decimal numbers **D** **C**
- convert fractions to decimals **D** **C**
- understand terminating and recurring decimals and their relationships to fractions **C** **B**
- convert a recurring decimal to a fraction **A**

14

Equations and inequalities

This chapter is about solving equations and inequalities.

For a hot air balloon to rise off the ground, the upthrust (acting upwards) must be greater than gravity (acting downwards).

Objectives

This chapter will show you how to

- solve equations involving brackets D C
- solve equations where the unknown appears on both sides of the equation D C
- solve equations which have negative, decimal or fractional solutions C
- solve equations involving fractions C
- solve simple linear inequalities and represent the solution on a number line D C B
- solve simultaneous linear equations B

Before you start this chapter

Put your calculator away!

HELP Chapter 12

1 Expand and simplify

a $2(x + 2)$ b $-3(2x + 3)$ c $2(3 - x) + 4x$ d $7x - 2(3 - 2x) + 2$

2 Factorise

a $7t + 35$ b $64 - 4x$ c $y^2 + 4y$ d $5n^2 - n$

14.1 Equations with brackets

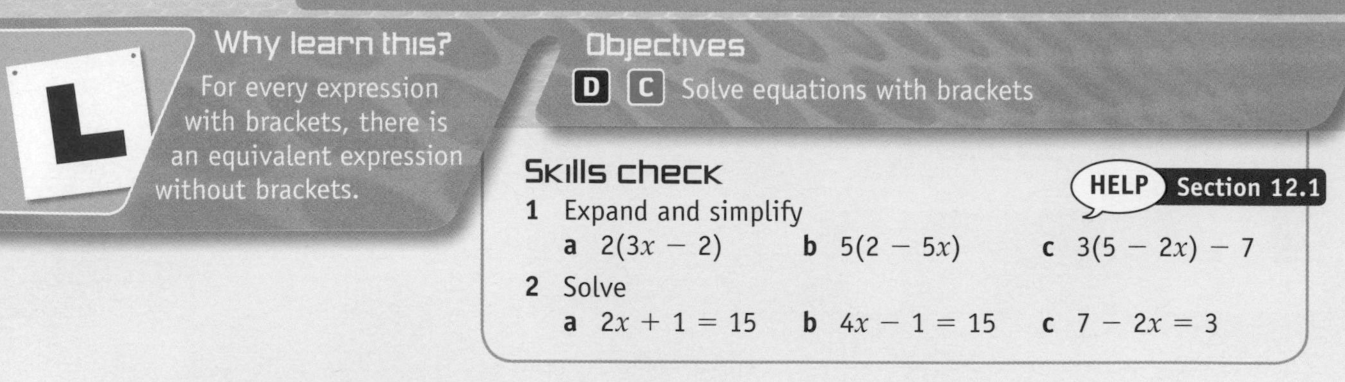

Why learn this?
For every expression with brackets, there is an equivalent expression without brackets.

Objectives

D **C** Solve equations with brackets

Skills check

HELP Section 12.1

1 Expand and simplify
  a $2(3x - 2)$ b $5(2 - 5x)$ c $3(5 - 2x) - 7$

2 Solve
  a $2x + 1 = 15$ b $4x - 1 = 15$ c $7 - 2x = 3$

Equations with brackets

When solving equations involving brackets, you usually expand the brackets first.

D

Example 1

Solve $4(2a - 1) = 12$

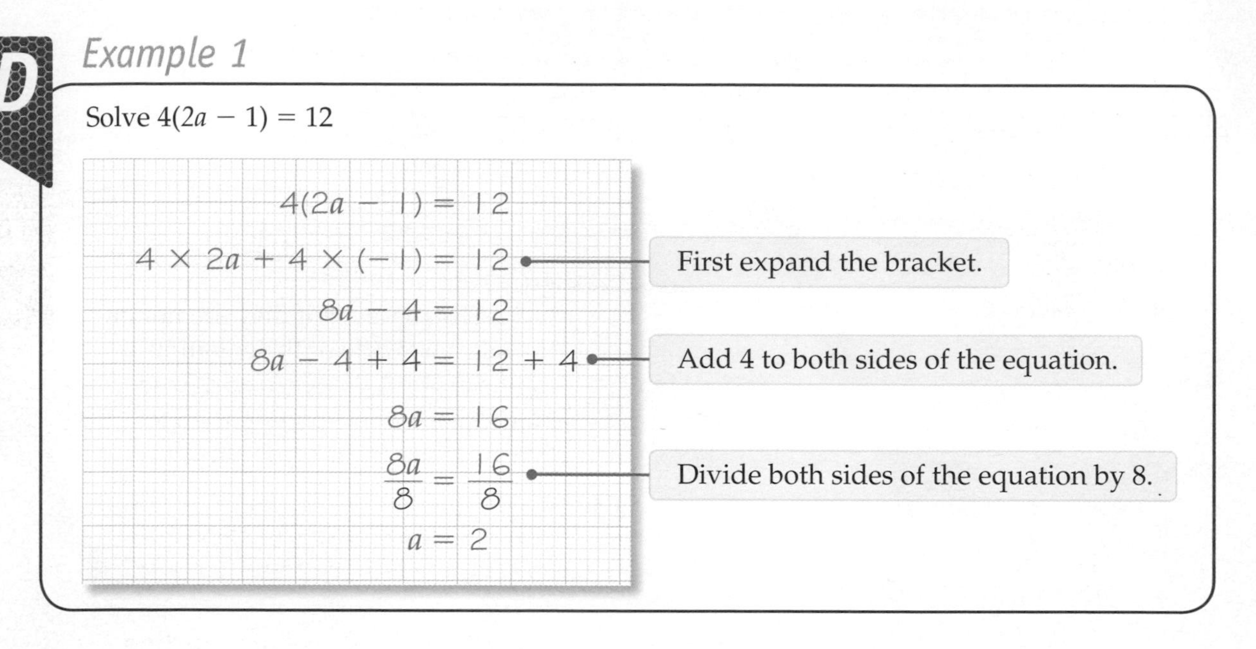

$4(2a - 1) = 12$
$4 \times 2a + 4 \times (-1) = 12$
First expand the bracket.
$8a - 4 = 12$
$8a - 4 + 4 = 12 + 4$
Add 4 to both sides of the equation.
$8a = 16$
$\frac{8a}{8} = \frac{16}{8}$
Divide both sides of the equation by 8.
$a = 2$

Exercise 14A

Solve these equations.

D

1 $4(a + 2) = 20$
2 $2(b + 1) = 12$
3 $3(c + 4) = 21$
4 $7(d - 2) = 56$
5 $9(k - 3) = 18$
6 $7(2h + 3) = 70$
7 $2(5i + 3) = 12$
8 $50 = 5(3j + 1)$
9 $2(k - 3) = -10$
10 $5(l + 2) = -40$
11 $10 = 2(2f + 1)$
12 $2(x + 7) = 42$
13 $4(4x - 3) = 68$
14 $54 = 3(2f - 2)$

C

15 $2(m + 9) + 3 = 17$
16 $8(t + 3) + 3 + 8t = 11$
17 $2(2q + 1) + 3q = 9$
18 $7(n + 11) + n + 9 = 46$
19 $6(2p + 20) + 6p = 30$
20 $3(3r + 1) - 2r = 17$

14.2 Equations with an unknown on both sides

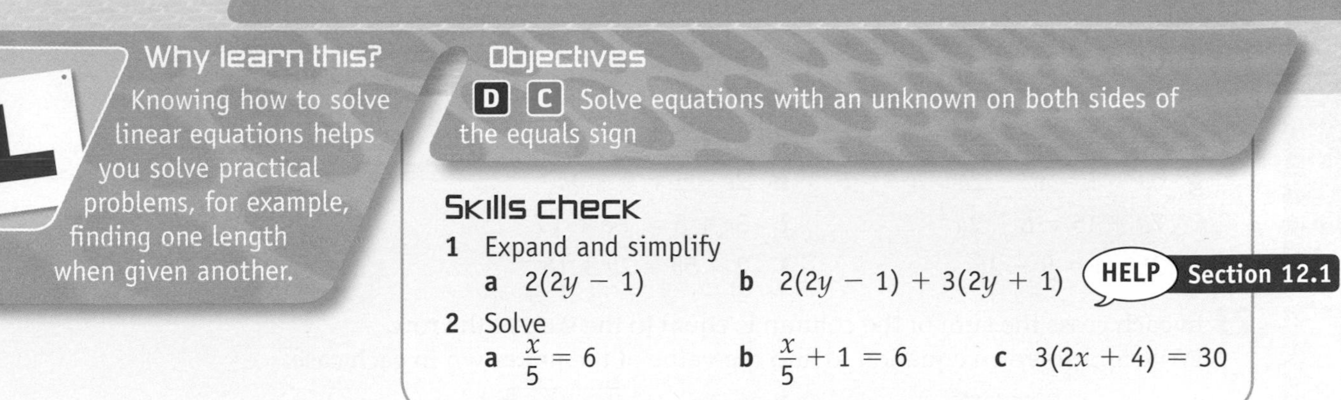

Why learn this?
Knowing how to solve linear equations helps you solve practical problems, for example, finding one length when given another.

Objectives
D **C** Solve equations with an unknown on both sides of the equals sign

Skills check
1 Expand and simplify
   a $2(2y - 1)$ b $2(2y - 1) + 3(2y + 1)$ HELP Section 12.1
2 Solve
   a $\frac{x}{5} = 6$ b $\frac{x}{5} + 1 = 6$ c $3(2x + 4) = 30$

Equations with an unknown on both sides

Sometimes equations have an unknown letter on both sides of the equals sign.

To solve an equation like this, add or subtract unknowns from both sides of the equation so that the unknown appears on only one side of the equation.

Example 2

Solve

a $4 - 2x = 3x - 6$ **b** $3(x + 1) = 4x - 2$

a

$4 - 2x = 3x - 6$

$4 - 2x + 2x = 3x - 6 + 2x$ — Try to end up with a positive number of unknowns, even if they are not on the left-hand side of the equation. In this case it is better to add $2x$ to both sides of the equation than to take $3x$ away from both sides.

$4 = 5x - 6$

$5x - 6 = 4$ — You can swap the sides of the equation, keeping everything else the same.

$5x - 6 + 6 = 4 + 6$ — Add 6 to both sides of the equation.

$5x = 10$

$\frac{5x}{5} = \frac{10}{5}$ — Divide both sides by 5.

$x = 2$

b

$3(x + 1) = 4x - 2$

$3x + 3 = 4x - 2$ — Expand the bracket.

$3x + 3 - 3x = 4x - 2 - 3x$ — Subtract $3x$ from both sides.

$3 = x - 2$

$3 + 2 = x - 2 + 2$ — Add 2 to both sides.

$5 = x$ or $x = 5$

D C

Exercise 14B

D

1 Solve

a $2a + 3 = a + 8$

b $4h + 6 = 2h + 2$

c $2b + 1 = b - 1$

d $7i + 11 = 4i + 2$

e $3c + 3 = 2c + 9$

f $4e + 2 = 2e + 7$

g $5a + 3 = 17 - 2a$

h $2c - 1 = 9 - 3c$

i $7d + 15 = 6 - 2d$

j $3e + 5 = -e + 1$

k $3f - 1 = 9 - 2f$

l $2 - 5b = 2b + 16$

Some solutions are negative numbers, decimals or fractions.

C AO2

2 In each cross the sum of the column is equal to the sum of the row.
Write and solve an equation to find the value of the unknown in each case.

a

	$2e + 3$	
$6e$	$2e + 3$	$2e + 2$
	$2e + 3$	

b

	3	
$20x - 3$	$-10x + 1$	$20x - 4$
	$-10x$	

C

3 Parts of the solution to each equation have been covered in ink.
Rewrite the solution in full.

a
$7 - j = 3(5 - j)$
$7 - j = ■ - 3j$
$7 + 2j = ■$
$2j = ■$
$j = ■$

b
$10(d - 2) = 6(d + 4)$
$10d ■ = 6d + 24$
$■ - 20 = 24$
$■ = 44$
$d = ■$

4 Solve

a $5a + 16 = 3(a + 6)$

b $4(b + 3) = 3b + 13$

c $2(2c + 3) = 3c + 12$

d $3(2e + 3) = 5(2e + 1)$

e $3(f - 5) = 2(f - 3)$

f $-3(2x + 2) = 2(x + 5)$

g $4(1 - 5h) = 6(5h - 1)$

h $3(2i - 6) = 18 - 2i$

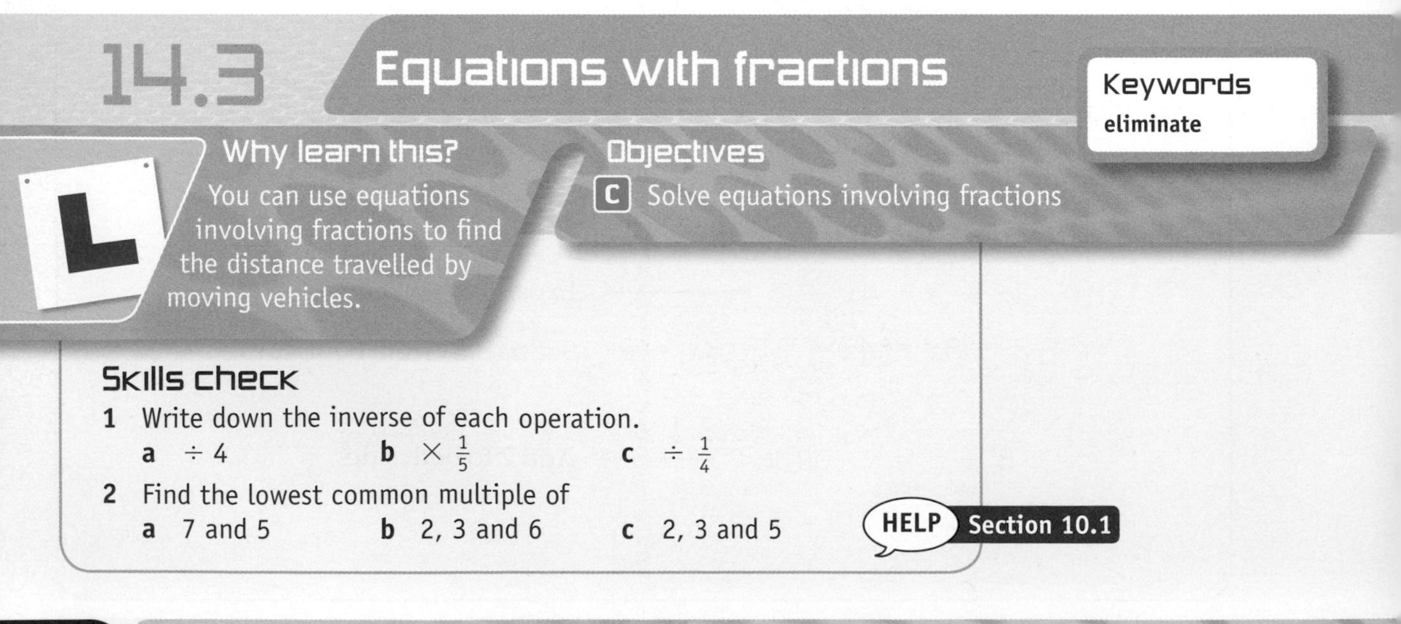

14.3 Equations with fractions

Keywords
eliminate

Why learn this?

You can use equations involving fractions to find the distance travelled by moving vehicles.

Objectives

C Solve equations involving fractions

Skills check

1 Write down the inverse of each operation.

a $\div 4$ b $\times \frac{1}{5}$ c $\div \frac{1}{4}$

2 Find the lowest common multiple of

a 7 and 5 b 2, 3 and 6 c 2, 3 and 5

HELP Section 10.1

Equations with fractions

Some equations involve fractions. When solving equations with fractions, find a way to **eliminate** the denominator.

Example 3

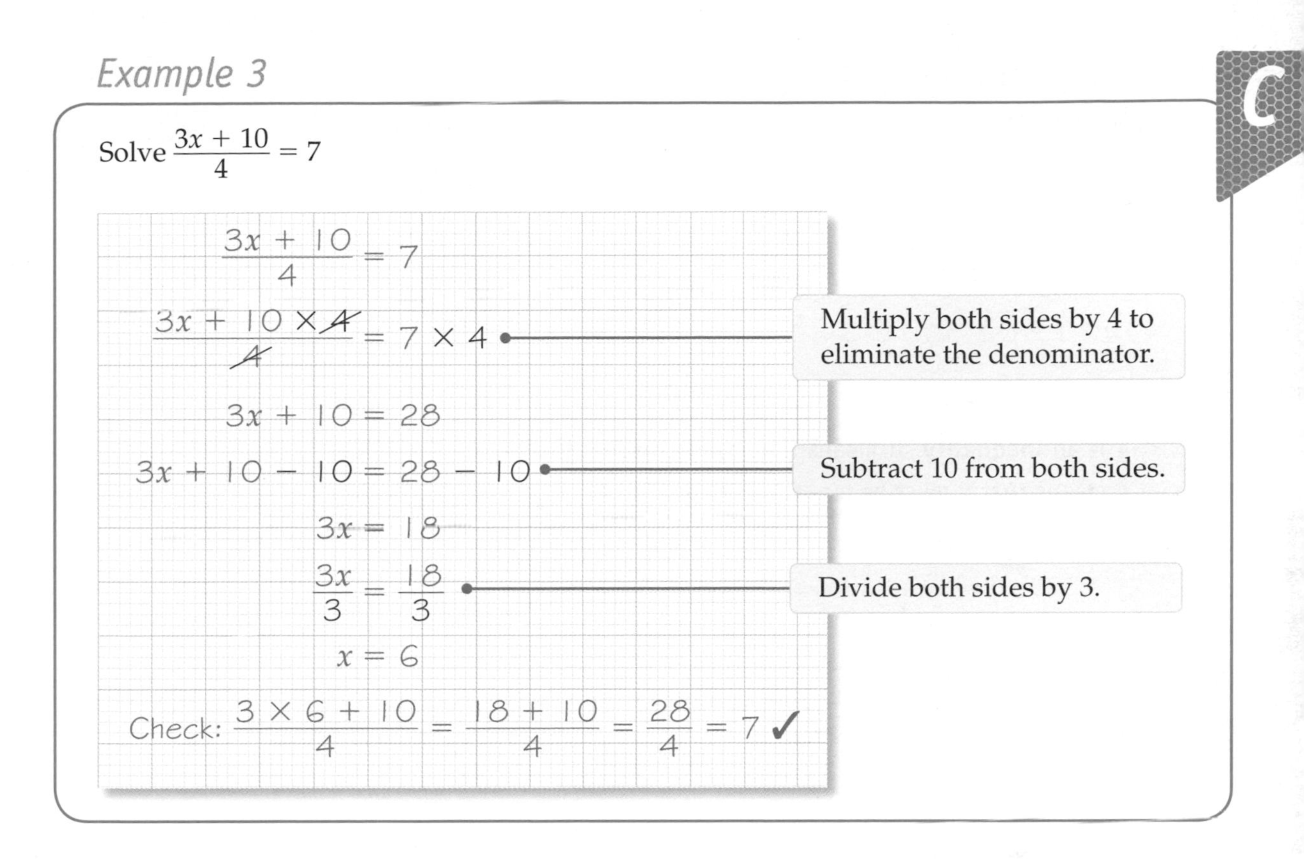

Solve $\frac{3x + 10}{4} = 7$
$\frac{3x + 10}{4} = 7$
$\frac{3x + 10 \times 4}{4} = 7 \times 4$
Multiply both sides by 4 to eliminate the denominator.
$3x + 10 = 28$
$3x + 10 - 10 = 28 - 10$
Subtract 10 from both sides.
$3x = 18$
$\frac{3x}{3} = \frac{18}{3}$
Divide both sides by 3.
$x = 6$
Check: $\frac{3 \times 6 + 10}{4} = \frac{18 + 10}{4} = \frac{28}{4} = 7$ ✓

C

Exercise 14C

Solve these equations.

1 $\frac{3x + 2}{4} = 2$

2 $\frac{2x + 4}{7} = 2$

3 $\frac{5x + 3}{4} = -3$

4 $1 = \frac{6 - x}{3}$

5 $\frac{3x - 1}{5} = 4$

6 $\frac{12y - 1}{5} = 7$

7 $\frac{4 - 3y}{2} = -1$

8 $\frac{2w + 11}{3} = -4$

9 $\frac{y + 5}{3} = \frac{2y - 1}{5}$

Multiply both sides by the LCM of both denominators.

10 $\frac{3x - 1}{4} = \frac{x - 2}{3}$

11 $\frac{w + 6}{2} = \frac{4w - 5}{7}$

12 $\frac{5t + 3}{4} = \frac{2t - 3}{3}$

C

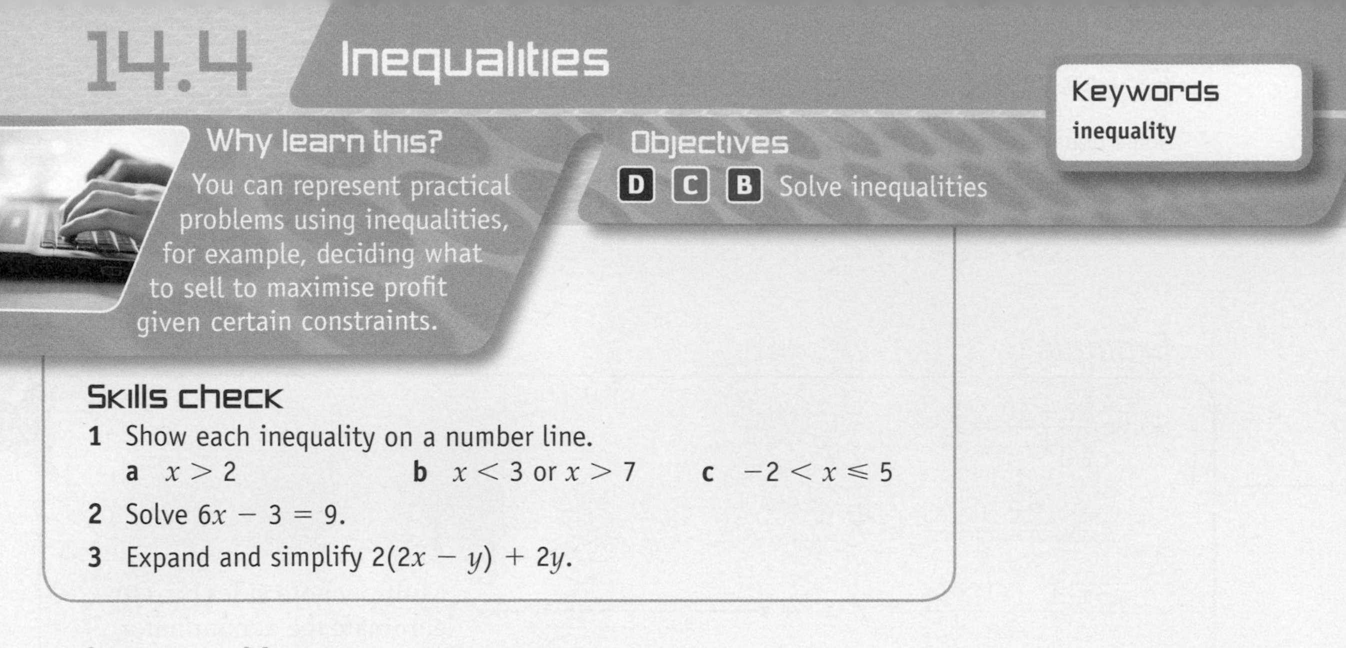

14.4 Inequalities

Keywords
inequality

Why learn this?

You can represent practical problems using inequalities, for example, deciding what to sell to maximise profit given certain constraints.

Objectives

D **C** **B** Solve inequalities

Skills check

1. Show each inequality on a number line.
   a $x > 2$ **b** $x < 3$ or $x > 7$ **c** $-2 < x \leqslant 5$
2. Solve $6x - 3 = 9$.
3. Expand and simplify $2(2x - y) + 2y$.

Inequalities

$x < 4$ is an **inequality**. It means that x must be less than 4.

$x \leqslant 4$ means that x must be less than or equal to 4.

Example 4

Write down all the whole-number values for the x in the inequality $-3 < x \leqslant 2$ and show the inequality on a number line.

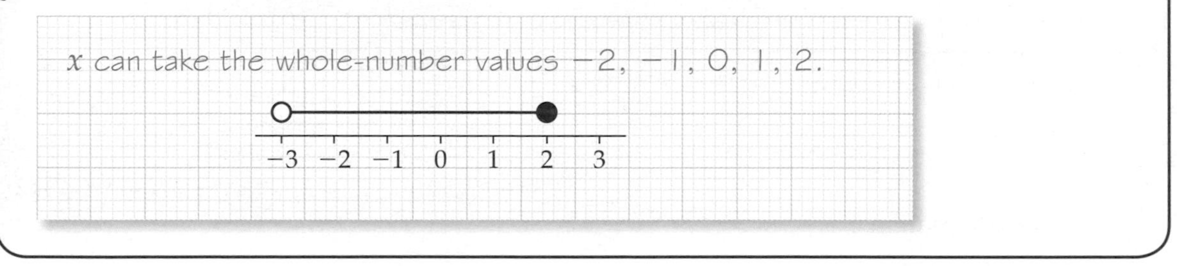

x can take the whole-number values −2, −1, 0, 1, 2.
−3 −2 −1 0 1 2 3

Solving inequalities

You can use the balance method to solve inequalities.

When you multiply or divide both sides of an inequality by a positive number, the inequality sign remains unchanged.

When you multiply or divide both sides of an inequality by a negative number, you need to change the direction of the sign.

For example, if both sides of the inequality $2 < 3$ are multiplied by -1, the inequality becomes $-2 > -3$. The sign has been reversed to make the inequality true.

Example 5

Solve each inequality and show the solution on a number line.

a $4x > 3x - 3$ **b** $6 \leqslant 3x - 3 < 8$ **c** $\frac{x+7}{2} < 5$

a $4x > 3x - 3$

$4x - 3x > 3x - 3 - 3x$ — Subtract $3x$ from both sides to keep the balance.

$x > -3$

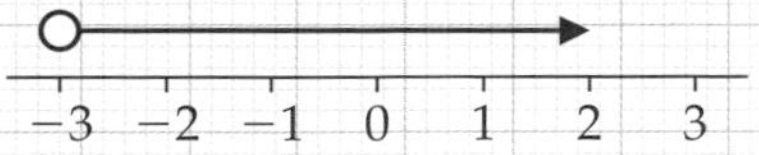

b $6 \leqslant 3x - 3 < 8$

$6 \leqslant 3x - 3$ and $3x - 3 < 8$ — Split into two inequalities and solve them separately.

$9 \leqslant 3x$ $3x < 11$

$3 \leqslant x$ $x < 3\frac{2}{3}$

$3 \leqslant x < 3\frac{2}{3}$ — Rewrite the two solved inequalities as a single inequality.

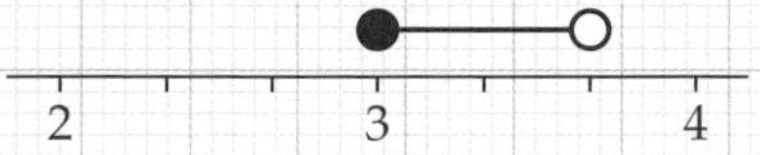

c $\frac{x+7}{2} < 5$ — Multiply both sides by 2 to eliminate the denominator.

$x + 7 < 10$ — Subtract 7 from both sides.

$x < 3$

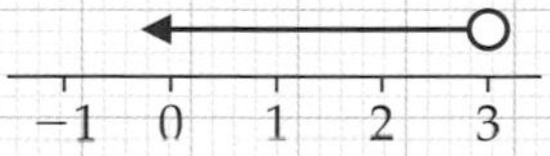

The direction of the sign does not need changing in Example 5 because we did not need to multiply or divide by a negative number.

Exercise 14D

1 Solve each inequality and show the solution on a number line.

a $4x < 20$ **b** $x + 7 < 10$ **c** $\frac{x}{2} < 10$

d $x - 3 \leqslant 5$ **e** $5x > 30$ **f** $x - 3 > -3$

2 Solve each inequality.

a $5x - 3 < 7$ **b** $2x + 1 > 11$

c $\frac{x}{3} + 1 \geqslant 3$ **d** $2(x - 70) \leqslant 8$

e $3(2y + 1) \leqslant -15$ **f** $3x < 2x + 7$

g $5x > 3x + 10$ **h** $2x \leqslant 6x + 10$

3 Solve each inequality and show the solution on a number line.

a $6 < 2x \leqslant 18$ **b** $9 < 3x < 21$ **c** $2x < 4x + 6$

B

4 Solve each inequality.

a $\frac{x+2}{3} > 7$ b $\frac{2x+2}{3} < 10$

c $\frac{x}{4} + 1 \leqslant 11$ d $\frac{x+1}{-2} > 4$

e $\frac{3x-1}{2} < 4$ f $7x + 3 < 5x - 3$

g $8 - 2x \geqslant 1 - 4x$ h $4 \leqslant 2n + 1 < 9$

i $2(p - 3) > 4 + 3p$ j $2(n + 5) \geqslant 3(2n - 2)$

Remember to change the direction of the sign when multiplying or dividing by a negative number.

5 List all integer solutions for each inequality.

a $-3 < 3(x - 4) < 6$

b $-6 \leqslant 5(y + 1) < 11$

c $-17 < 2(2x - 3) \leqslant 10$

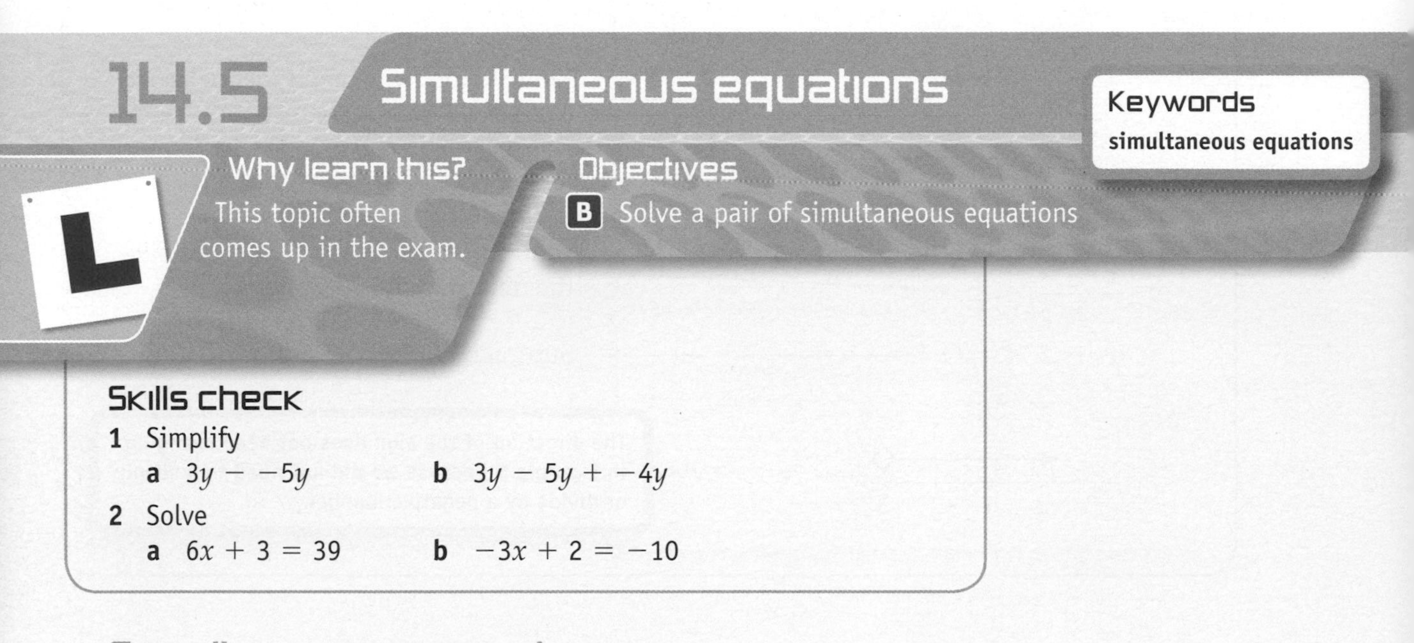

14.5 Simultaneous equations

Keywords
simultaneous equations

Why learn this?
This topic often comes up in the exam.

Objectives
B Solve a pair of simultaneous equations

Skills check

1 Simplify
a $3y - -5y$ b $3y - 5y + -4y$

2 Solve
a $6x + 3 = 39$ b $-3x + 2 = -10$

Simultaneous equations

Simultaneous equations are pairs of equations which have two unknowns and two solutions. Both equations are true at the same time, so they have the same solutions.

To solve simultaneous equations you eliminate one of the unknowns using algebraic steps. You should then have a linear equation with one unknown that you can solve.

Example 6

Solve each pair of simultaneous equations.

a $5x + y = 19$
$2x + y = 10$

b $3x - y = 9$
$2x + y = 1$

a

$5x + y = 19 \quad (1)$

$2x + y = 10 \quad (2)$

$3x = 9$ — You can eliminate one of the unknowns (y) by subtracting equation 2 from equation 1.

$x = 3$

$5 \times 3 + y = 19 \quad (1)$ — Now you need to find the value of y. Substitute the value of x in one of the equations.

$15 + y = 19$

$y = 4$

Check: $2 \times 3 + 4 = 10 \quad (2)$ ✓ — Check your answer by substituting for both unknowns in the other equation.

b

$3x - y = 9 \quad (1)$

$2x + y = 1 \quad (2)$

$5x = 10$ — In part **a**, equation 2 was subtracted from equation 1. If you did this here, you would end up with $x - 2y = 8$. This does not eliminate an unknown. Instead *add* the two equations together to eliminate y.

$x = 2$

$6 - y = 9 \quad (1)$ — Find the value of y by substituting the value of x.

$y = -3$

Check: $2 \times 2 + -3 = 1 \quad (2)$ ✓ — Check your answer by substituting for both unknowns in the other equation.

Exercise 14E

1 Solve each pair of simultaneous equations.

a $2x + y = 10$
$-2x + 3y = 6$

b $2x - 3y = -5$
$5x + 3y = 19$

c $4x + y = 13$
$2y - 4x = 14$

d $3x + 4y = 25$
$5x - 4y = -33$

e $3x + y = 30$
$-2x + y = -5$

f $2x - y = -4$
$x + y = -5$

B

2 The sum of Reena's parents' ages is 83. The difference is 3.

a Write equations for the sum of their ages and the difference of their ages.

b Solve them simultaneously. How old are Reena's parents?

B

AO2

Manipulating simultaneous equations 1

Sometimes you cannot add or subtract the equations to eliminate one of the unknowns because they do not have the same number of either of them.

In these cases multiply one or both equations so that one of the unknowns appears the same number of times in both equations.

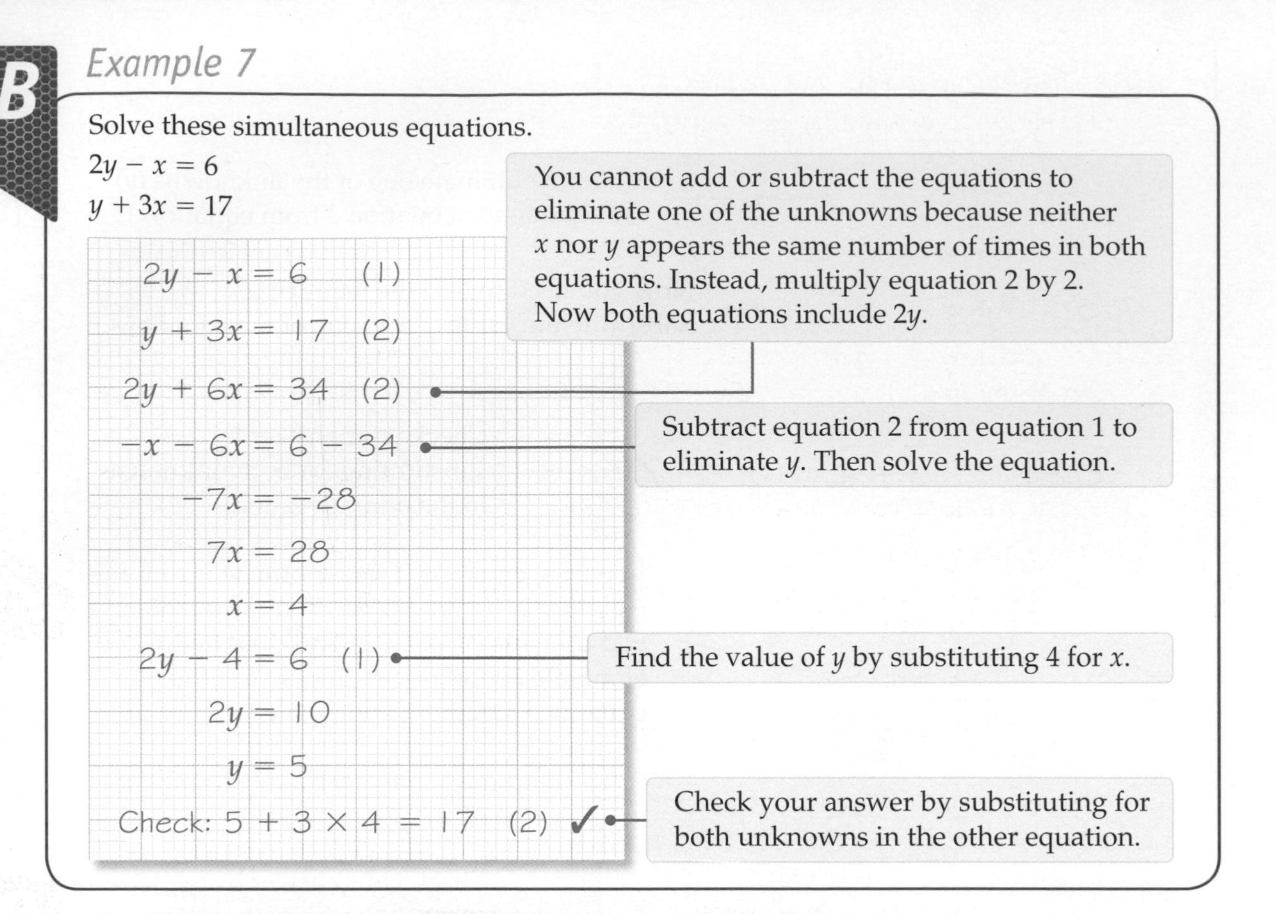

B
Example 7
Solve these simultaneous equations.
2y − x = 6
y + 3x = 17
2y − x = 6 (1)
y + 3x = 17 (2)
2y + 6x = 34 (2)
−x − 6x = 6 − 34
−7x = −28
7x = 28
x = 4
2y − 4 = 6 (1)
2y = 10
y = 5
Check: 5 + 3 × 4 = 17 (2) ✓
You cannot add or subtract the equations to eliminate one of the unknowns because neither x nor y appears the same number of times in both equations. Instead, multiply equation 2 by 2. Now both equations include 2y.
Subtract equation 2 from equation 1 to eliminate y. Then solve the equation.
Find the value of y by substituting 4 for x.
Check your answer by substituting for both unknowns in the other equation.

Manipulating simultaneous equations 2

Sometimes you need to multiply *both* equations to get the same number of unknowns.

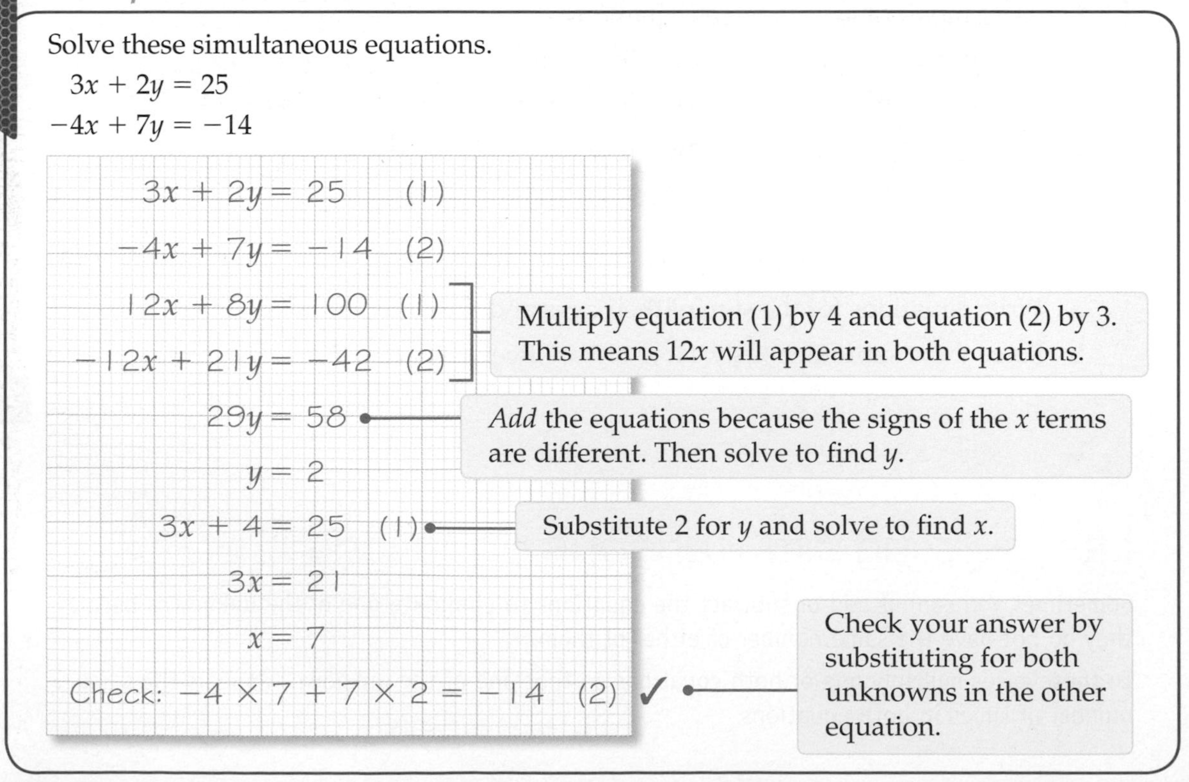

B
Example 8
Solve these simultaneous equations.
3x + 2y = 25
−4x + 7y = −14
3x + 2y = 25 (1)
−4x + 7y = −14 (2)
12x + 8y = 100 (1)
−12x + 21y = −42 (2)
29y = 58
y = 2
3x + 4 = 25 (1)
3x = 21
x = 7
Check: −4 × 7 + 7 × 2 = −14 (2) ✓
Multiply equation (1) by 4 and equation (2) by 3. This means 12x will appear in both equations.
Add the equations because the signs of the x terms are different. Then solve to find y.
Substitute 2 for y and solve to find x.
Check your answer by substituting for both unknowns in the other equation.

Exercise 14F

1 Solve each pair of simultaneous equations.

a $3x + 2y = 12$
$2x + y = 7$

b $x + 2y = 8$
$3x + 5y = 19$

c $2x - 3y = -11$
$9x + y = 23$

d $2x + 2y = 6$
$3x - 4y = -26$

e $-2x - y = 2$
$3x - 4y = -25$

f $2x + 3y = 27$
$5x + 9y = 114$

g $2x - 4y = 2$
$9x + 16y = 77$

h $2x - 3y = -19$
$5x + 2y = -19$

i $2x + 3y = 12$
$5x + 2y = 19$

B

2 A music download site makes a total of 1000 sales in a day. The revenue for the day is £6300. Albums cost £7.50 and individual songs cost £1.50.

FUNCTIONAL

a How many albums were sold?

b An album has twelve songs on it on average.
Estimate the total number of tracks sold.

Set up a pair of simultaneous equations.

Review exercise

1 Solve

a $2(m + 9) = 14$ **[3 marks]**

b $6j - 1 = 2j + 14$ **[3 marks]**

c $2 - 5b = 2b + 16$ **[3 marks]**

D

2 Write an equation and solve it to find the size of the smallest angle in the triangle.

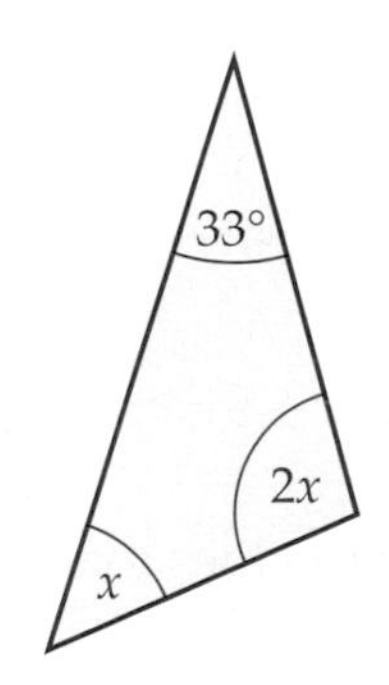

[3 marks]

3 **a** x is an integer.
$0 < x \leqslant 4$
Write down all the possible values of x. **[2 marks]**

b x and y are integers.
$0 < x \leqslant 4$ $\quad y < x \quad$ $x + y < 6$
Write down **two** pairs of values of x and y which satisfy all three inequalities. **[2 marks]**

4 Solve

a $2x + 1 = 3(x - 2)$ **[3 marks]**

b $\dfrac{2x - 3}{5} = 3$ **[3 marks]**

C

C

AO3

5 A cube has faces with an area of $(9x + 1)\ \text{cm}^2$.
The cuboid shown is made of four of these cubes.
The surface area of the cuboid is $736\ \text{cm}^2$.
What is the value of x? **[4 marks]**

C

6 a Solve the inequality $3(x - 1) \leqslant 6$. **[3 marks]**

b The inequality $x \leqslant 3$ is shown on the number line below.

−4 −3 −2 −1 0 1 2 3 4 5 6 x

Copy the number line and draw another inequality on it so that only the following integers satisfy both inequalities.
−2, −1, 0, 1, 2, 3 **[1 mark]**

7 Solve each inequality and represent the solution on a number line.

a $14 < 7x \leqslant 28$ **[3 marks]**

b $3x - 1 > 14$ **[2 marks]**

c $4x < 3x + 8$ **[2 marks]**

B

8 Solve each pair of equations.

a $2x + 3y = -6$
$4x - 3y = 24$ **[3 marks]**

b $2x + 3y = 28$
$3x - y = 9$ **[4 marks]**

c $3x + 4y = 14$
$2x - 3y = -19$ **[4 marks]**

9 Solve each inequality.

a $2x + 1 \leqslant 3x - 4$ **[2 marks]**

b $3x + 9 \geqslant 2(x + 9)$ **[3 marks]**

10 List the integer values that satisfy $-3 < 2(x - 3) \leqslant 7$. **[3 marks]**

Chapter summary

In this chapter you have learned how to

- solve equations with brackets D C
- solve equations with an unknown on both sides of the equals sign D C
- solve equations involving fractions C
- solve inequalities D C B
- solve a pair of simultaneous equations B

15 Formulae

BBC Video

This chapter is about formulae.

The trebuchet is a medieval weapon designed to throw things at or over castle walls. If you know the angle and speed of launch, you can write a formula to work out the range of your weapon.

Objectives

This chapter will show you how to

- use algebra to write formulae in different situations **D**
- substitute numbers into algebraic expressions and formulae **D** **C** **B**
- rearrange a formula to make a different variable the subject of the formula **C** **B**

Before you start this chapter

Put your calculator away!

1 Use algebra to write an expression for each of these.

a 6 more than p b 2 taken away from y c p added to q

d 6 lots of d e x divided by 2

HELP Chapter 12

2 Expand the brackets.

a $3(x + 4)$ b $2(d - 3)$ c $5(x + y - 6)$

3 Solve these equations.

a $3x + 3 = 9$ b $5x - 3 = 2x + 7$ c $2x = \frac{3(x - 5)}{4}$

Writing your own formulae

Keywords
formula, variable

Why learn this?
Organisations often create formulae to use for allocating funding fairly. The money given to run your school may have been decided by a funding formula.

Objectives
D Use algebra to write formulae in different situations

Skills check
Write an algebraic expression for
1 d multiplied by 16
2 the total cost of 6 applies at c pence each.

Writing your own formula

A **formula** is a general rule that shows how quantities are related to each other.

The quantities in a formula can vary in size. These quantities are called **variables.**

The formula $v = u + at$ shows the relationship between an object's final velocity, v, its initial velocity, u, its acceleration, a, and the time it has been moving, t.

A formula must contain an equals sign. For example, the formula for an object's final velocity is $v = u + at$, not just the expression $u + at$.

D

Example 1

Alex buys x packets of sweets.
Each packet of sweets costs 45 pence.
Alex pays with a £5 note.
Write a formula for the change in pence, C, that Alex should receive.

$C = 500 - 45x$

Remember: £5 = 500p. The sweets cost 45p per packet so the cost, in pence, for x packets is $45x$.

Exercise 15A

D

1 Nilesh buys y packets of sweets.
Each packet of sweets costs 48 pence.
Nilesh pays with a £5 note.
Write a formula for the change in pence, C, that Nilesh should receive.

2 Apples cost r pence each and bananas cost s pence each.
Sam buys 7 apples and 5 bananas.
Write a formula for the total cost in pence, t, of these fruit.

3 To roast a chicken you allow 45 minutes per kg and then a further 20 minutes.
Write a formula for the time in minutes, t, to roast a chicken that weighs w kg.

4 A rectangle has length $3x + 1$ and width $x + 2$.
Write a formula for the perimeter, p, of this rectangle.

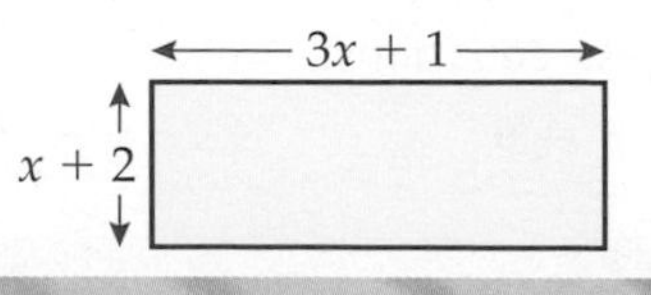

Substituting into expressions

Why learn this?
You will need to substitute into an expression when using the quadratic formula in Chapter 20.

Objectives
D C Substitute numbers into algebraic expressions

Keywords
substitute, evaluate

Skills check
Work out

1 $5 \times 3^2 - 7$

2 $3(2 \times 5 + 4 \times 3)$

3 $\frac{5 \times (-4)^2}{8}$

4 $\frac{\sqrt{3 \times 2^2 + 2 \times 12}}{2 - (-1)}$

Substituting into expressions

You can **substitute** numbers for the variables in an expression, including expressions involving brackets and powers.

This is called **evaluating** the expression.

Example 2

D

Evaluate these expressions when $a = 5$, $b = 4$ and $c = 3$.

a $3b^2 - 1$ **b** $\frac{6(c^2 + 2c - a)}{b}$

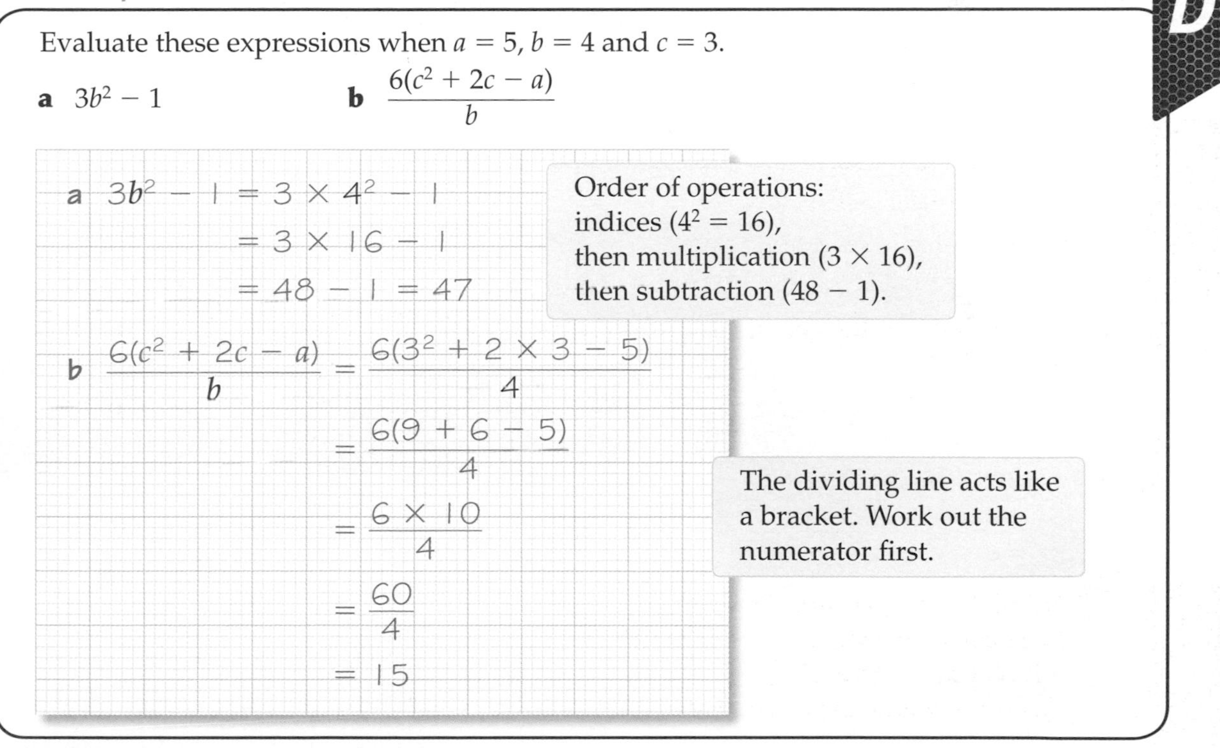

a $3b^2 - 1 = 3 \times 4^2 - 1$
$= 3 \times 16 - 1$
$= 48 - 1 = 47$

Order of operations: indices ($4^2 = 16$), then multiplication (3×16), then subtraction ($48 - 1$).

b $\frac{6(c^2 + 2c - a)}{b} = \frac{6(3^2 + 2 \times 3 - 5)}{4}$
$= \frac{6(9 + 6 - 5)}{4}$
$= \frac{6 \times 10}{4}$
$= \frac{60}{4}$
$= 15$

The dividing line acts like a bracket. Work out the numerator first.

Exercise 15B

D

1 Evaluate each of these expressions when $r = 5$, $s = 4$ and $t = 3$.

a $3r^2 + 1$

$3r^2 = 3 \times r^2$
$= 3 \times r \times r$

b $4t^2 - 6$

c $2s^2 + r$

d $4(5s + 1)$

Order of operations: brackets first.

e $t^2(r + s)$ **f** $5(2s - 3t^2)$ **g** $\frac{5t + 1}{s^2}$ **h** $\frac{4r^2 - 2}{t}$ **i** $\frac{3s + t^2}{r}$

D

2 Copy and complete this table.

x	2	3	4	5	6
$x^3 - x$			60		

$4^3 - 4 = 64 - 4$
$= 60$

3 If $A = 6$, $B = -4$, $C = 3$ and $D = 30$, work out the value of each of these expressions.

a $D(B + 7)$ **b** $A(B + 1)$ **c** $\frac{2A + 3}{C}$

d $\frac{4B + D}{2}$ **e** $A^2 + 2B + C$ **f** $\frac{A^2 + 3B}{C}$

C

4 Evaluate each of these expressions when $d = 5$, $e = -2$ and $f = 3$.

a $f^2(2d + 3e)$ **b** $e(d^2 - 2f^2)$ **c** $\sqrt{d^2 - f^2}$

d $\sqrt{5d + 3f - e}$ **e** $\frac{d + f^2}{d - e}$ **f** $\frac{\sqrt{2df - 3e}}{e}$

5 Calculate the value of each of these expressions when $a = 1$, $b = 2$ and $c = -3$.

a $\frac{-b + \sqrt{b^2 - 4ac}}{2a}$ **b** $\frac{-b - \sqrt{b^2 - 4ac}}{2a}$

6 Calculate the value of each of the expressions in Q5 when $a = 2$, $b = -7$ and $c = -3$.

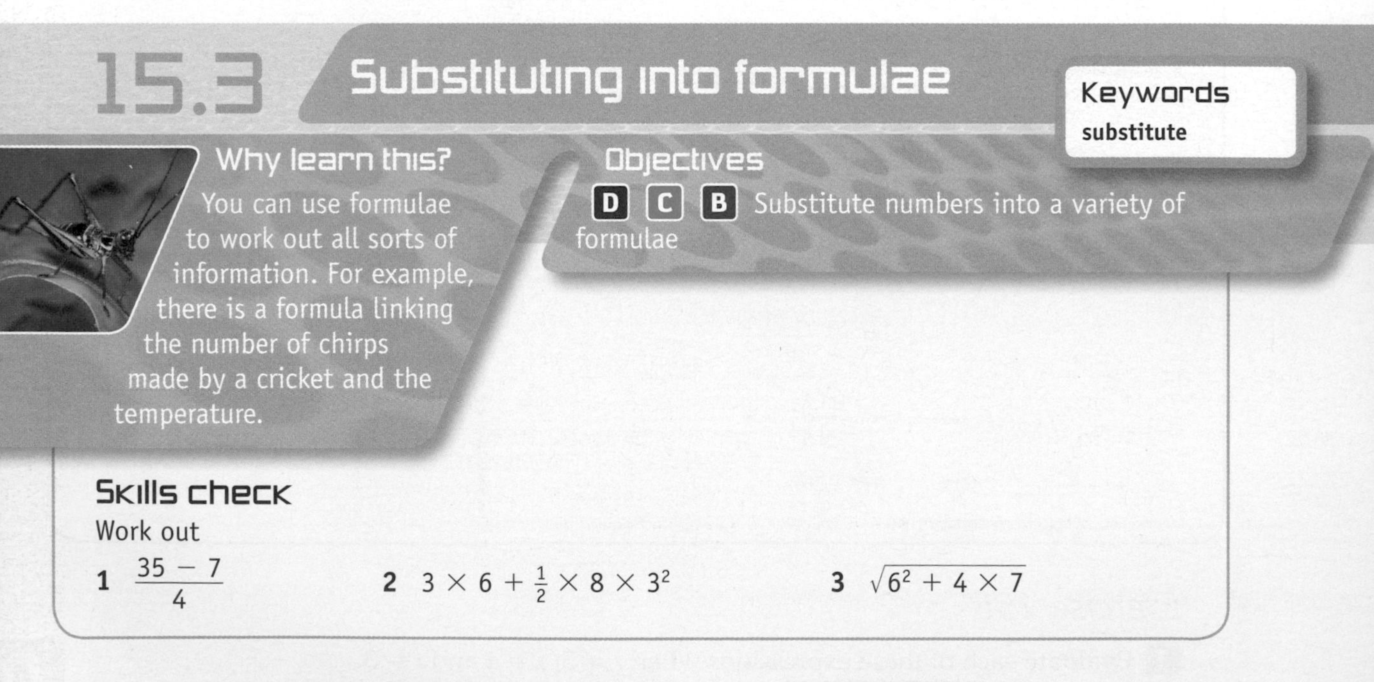

15.3 Substituting into formulae

Keywords
substitute

Why learn this?

You can use formulae to work out all sorts of information. For example, there is a formula linking the number of chirps made by a cricket and the temperature.

Objectives

D **C** **B** Substitute numbers into a variety of formulae

Skills check

Work out

1 $\frac{35 - 7}{4}$ **2** $3 \times 6 + \frac{1}{2} \times 8 \times 3^2$ **3** $\sqrt{6^2 + 4 \times 7}$

Substituting into formulae

You can **substitute** values into a formula to work out the value of a variable.

Example 3

D

A formula for working out distance travelled is

$$s = ut + \tfrac{1}{2}at^2$$

where u is the initial velocity, a is the acceleration and t is the time taken.

Work out the value of s when $u = 3$, $a = 8$ and $t = 5$.

$$s = ut + \tfrac{1}{2}at^2$$
$$= 3 \times 5 + \tfrac{1}{2} \times 8 \times 5^2$$
$$= 15 + 4 \times 25$$
$$= 15 + 100$$
$$= 115$$

Order of operations:
indices ($5^2 = 25$),
then multiplication ($3 \times 5 = 15$, $\frac{1}{2} \times 8 = 4$, 4×25),
then addition ($15 + 100 = 115$).

Exercise 15C

D

1 Use the formula $s = ut + \frac{1}{2}at^2$ to work out the value of s when

a $u = 3$, $a = 10$ and $t = 2$

b $u = -4$, $a = 8$ and $t = 3$

2 A formula for working out the velocity of a car is

$$v = \sqrt{u^2 + 2as}$$

where u is the initial velocity, a is the acceleration and s is the distance travelled.
Work out the value of v when

a $u = 3$, $a = 4$ and $s = 5$

b $u = 7$, $a = 4$ and $s = 15$

3 An approximate formula for the surface area of a cone, including the base, is

$$A = 3r(r + l)$$

where r is the radius and l is the slant height.

Work out the approximate surface area of a cone with dimensions

a $r = 2$ cm and $l = 13$ cm

b $r = 4$ cm and $l = 10$ cm

Don't forget to include the correct units in your answers.

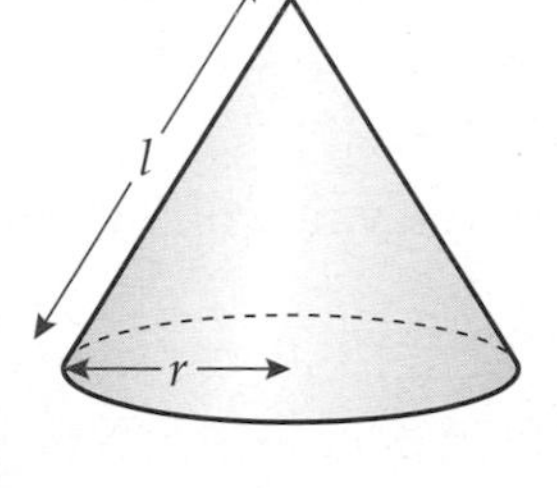

4 A formula for the approximate total surface area of a cylinder is

$$A = 6r(r + h)$$

where r is the radius and h is the height.

Work out the approximate surface area of a cylinder with dimensions

a $r = 4$ cm and $h = 11$ cm

b $r = 5$ cm and $h = 4$ cm

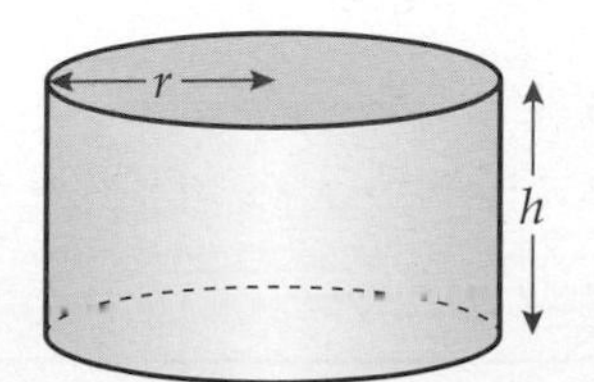

5 Use the formula $r = \sqrt{\frac{A}{3}}$ to work out the approximate radius, r, of a circle with area $A = 27$.

C

6 Use the formula $T = 6\sqrt{\frac{l}{g}}$ to calculate the approximate time, T, of a pendulum swing for a pendulum of length $l = 40$ when $g = 10$.

B

7 Heron's formula (also known as Hero's formula) for calculating the area of a triangle with sides of length a, b and c is

$$A = \sqrt{s(s-a)(s-b)(s-c)}$$

where $s = \frac{a+b+c}{2}$

> s is known as the semi-perimeter.

Use Heron's formula to calculate the area of a triangle with sides 3 cm, 4 cm and 5 cm.

8 Another formula, discovered by mathematicians in China, for the area of a triangle with sides a, b and c is

$$A = \frac{1}{2}\sqrt{a^2c^2 - \left(\frac{a^2 + c^2 - b^2}{2}\right)^2}$$

Show that this gives the same result as Heron's formula for a triangle with sides 3 cm, 4 cm and 5 cm.

Sometimes the value you want is not on the left of the equals sign. Substitute the values you know into the formula. Then solve the equation to find the value of the unknown.

D

Example 4

The perimeter of a rectangle is given by $P = 2l + 2w$, where l is the length and w is the width.

Work out the value of l when $P = 24$ and $w = 5$.

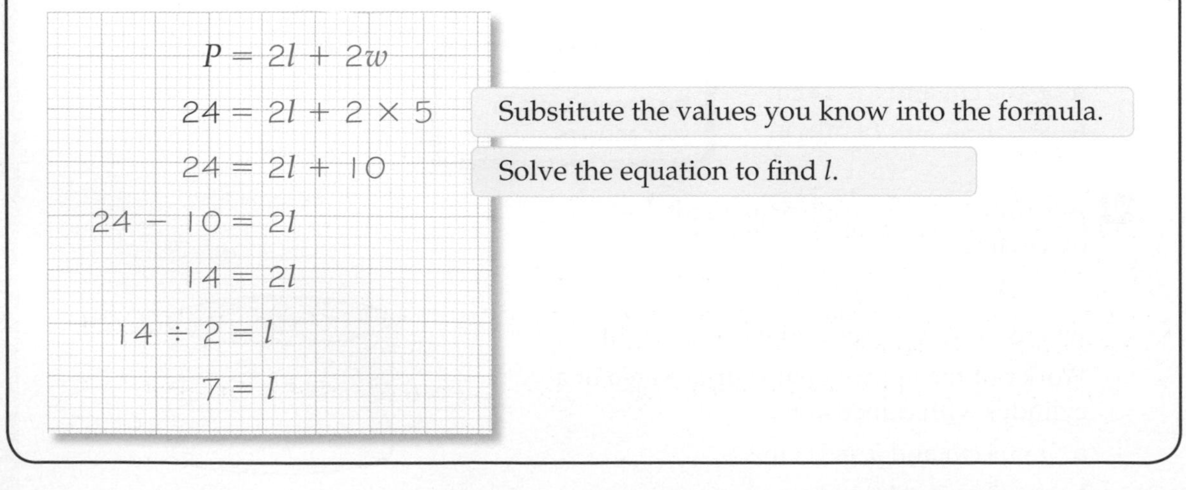

P = 2l + 2w
24 = 2l + 2 × 5
Substitute the values you know into the formula.
24 = 2l + 10
Solve the equation to find l.
24 − 10 = 2l
14 = 2l
14 ÷ 2 = l
7 = l

Exercise 15D

D

1 The perimeter of a rectangle is given by $P = 2l + 2w$, where l is the length and w is the width.

Use the formula to find

a l when $P = 18$ and $w = 4$

b l when $P = 32$ and $w = 7$

c w when $P = 60$ and $l = 17$

d w when $P = 50$ and $l = 13.5$

e l when $P = 40.5$ and $w = 0.25$

2 Use the formula $v = u + at$ to find

a u when $v = 30$, $a = 8$ and $t = 3$

b u when $v = 47$, $a = 4$ and $t = 9$

c a when $v = 54$, $u = 19$ and $t = 7$

d t when $v = 60$, $u = 15$ and $a = 5$

e u when $v = 20$, $a = 7$ and $t = 4$

15.4 Changing the subject of a formula

Keywords
subject, rearrange

Why learn this?
In physics, you have to be able to use lots of different formulae confidently.

Objectives

C B Rearrange a formula to make a different variable the subject of the formula

Skills check

Solve

1 $15 = 9 + x$

2 $42 = 6t$

3 $23 = 2d + 9$

4 $11 = \frac{1}{2}m - 5$

Rearranging formulae

In the formula $v = u + at$ the variable v is called the **subject** of the formula.

The subject of a formula is always the letter on its own on one side of the equation. This letter only appears once during the formula.

For example, P is the subject of the formula $P = 2l + 2w$.

You can **rearrange** a formula to make a different variable the subject. This is called 'changing the subject' of the formula.

You can use the same techniques to rearrange a formula as when you solve an equation.

Example 5

a Make x the subject of the formula $y = 5x - 2$.

b Rearrange $P = 4g + 2h$ to make g the subject.

c Rearrange $V = \sqrt{w + y}$ to make w the subject.

d Rearrange $p = \frac{q^2}{r} - s$ to make q the subject.

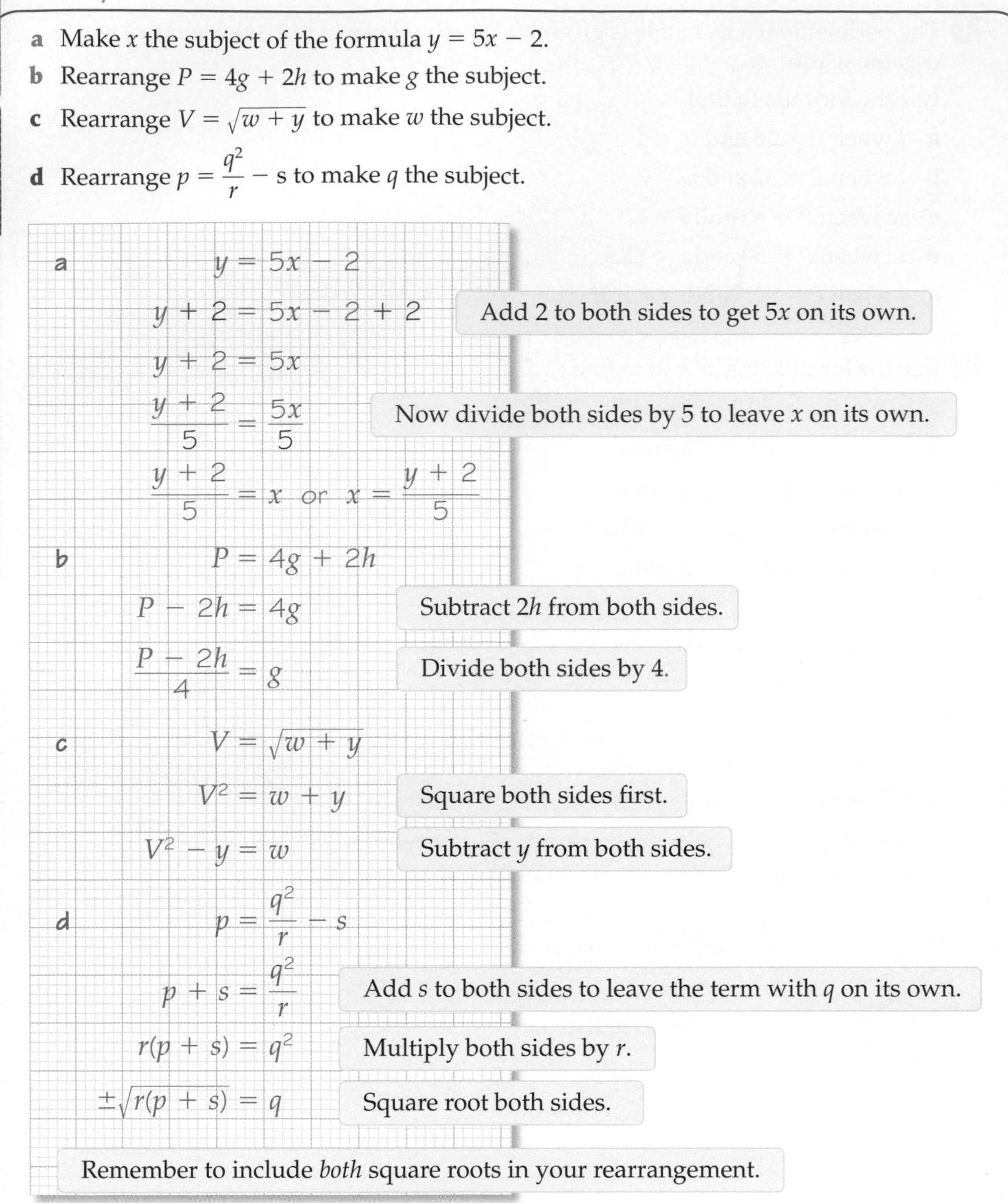

a

$$y = 5x - 2$$

$$y + 2 = 5x - 2 + 2$$

Add 2 to both sides to get $5x$ on its own.

$$y + 2 = 5x$$

$$\frac{y + 2}{5} = \frac{5x}{5}$$

Now divide both sides by 5 to leave x on its own.

$$\frac{y + 2}{5} = x \text{ or } x = \frac{y + 2}{5}$$

b

$$P = 4g + 2h$$

$$P - 2h = 4g$$

Subtract $2h$ from both sides.

$$\frac{P - 2h}{4} = g$$

Divide both sides by 4.

c

$$V = \sqrt{w + y}$$

$$V^2 = w + y$$

Square both sides first.

$$V^2 - y = w$$

Subtract y from both sides.

d

$$p = \frac{q^2}{r} - s$$

$$p + s = \frac{q^2}{r}$$

Add s to both sides to leave the term with q on its own.

$$r(p + s) = q^2$$

Multiply both sides by r.

$$\pm\sqrt{r(p + s)} = q$$

Square root both sides.

Remember to include *both* square roots in your rearrangement.

Exercise 15E

1 Rearrange each of these formulae to make a the subject.

a $c = a + 5$ **b** $t = a + 12$

c $w = 3a$ **d** $P = wa$

2 Make x the subject of each of these formulae.

a $y = 5x - 6$ **b** $y = 2x + 1$

c $p = 4x + 2t$ **d** $y = 6x - 5p$

3 A formula used to calculate velocity is

$$v = u - at$$

where u is the initial velocity, a is the acceleration and t is the time taken.

a Rearrange the formula to make a the subject.

b Rearrange the formula to make t the subject.

4 Rearrange these formulae to make a the subject.

Multiply both sides by 2 to get a on its own.

a $b = \frac{1}{2}a + 6$

b $b = \frac{1}{3}a - 1$

c $b = \frac{1}{4}a - 3$

d $b = 2(a + 1)$

e $b = 3(a - 5)$

5 Make y the subject of these formulae.

a $3x + 2y = 10$

b $6x + 3y = 8$

c $2x - 3y = 2$

d $5x - 4y = 10$

6 Rearrange these formulae to make x the subject.

a $3(x + y) = 5y$

b $2(x - y) = y + 5$

c $z = \frac{x}{y} - 5$

d $3p = \frac{2x}{q} - s$

7 Rearrange these formulae to make w the subject.

a $K = \sqrt{w + t}$

b $A = \sqrt{w - a}$

c $h = 2\sqrt{w} + l$

d $T = \sqrt{wr} + 5$

8 Rearrange these formulae to make r the subject:

a $t = \frac{r^2}{g} - m$

b $h = \frac{r^2}{4} + 3a$

c $V = \frac{1}{3}\pi r^2 h$

d $A = \pi(r^2 - s^2)$

9 A formula for the total surface area of a cylinder is

$$A = 2\pi r(r + h)$$

where r is the radius and h is the height.
Rearrange the formula to make h the subject.

10 A formula for the period of a pendulum is

$$T = 2\pi\sqrt{\frac{l}{g}}$$

where T is the time for one complete swing, l is the length and g is a constant.
Rearrange the formula to make l the subject.

C

B

Review exercise

D

1 If $a = -4$ and $b = 8$, find the value of

a $2a^2 + b$ **[2 marks]**

b $\frac{b + 12}{a}$ **[2 marks]**

2 Copy and complete this table.

x	2	3	4	5	6
$x^3 + 2x$			72		

[3 marks]

3 The total number of triangles in a pattern is given by the formula

$$T = \frac{p(p + 3)}{2}$$

where T is the total number of triangles and p is the pattern number.
Work out the value of T when $p = 40$. **[2 marks]**

D C

4 **a** Use the formula $C = 5T + 2R$ to find R when $C = 58$ and $T = 6$. **[3 marks]**

b Make T the subject of the formula $C = 5T + 2R$. **[2 marks]**

C

5 Given that $r = 9$, $s = -4$, $t = 1$ and $u = -5$, work out the value of

a $\frac{r + s}{tu}$ **[3 marks]**

b $\sqrt{r^2 + 4s - 1}$ **[3 marks]**

c $\frac{u^2 + r}{r - 2s}$ **[3 marks]**

6 **a** Make x the subject of the formula $w = \frac{1}{2}x - y$. **[2 marks]**

b Make x the subject of the formula $w = x^2 + y$. **[2 marks]**

7 Given that $a = 1$, $b = 3$ and $c = -10$, calculate the value of

$$\frac{-b + \sqrt{b^2 - 4ac}}{2a}$$ **[3 marks]**

B

8 Rearrange $7(n - p) = 3p + 4$ to make n the subject. **[3 marks]**

9 Rearrange $f = \sqrt{\frac{e}{g + 7}}$ to make g the subject. **[4 marks]**

Chapter summary

In this chapter you have learned how to

- use algebra to write formulae in different situations D
- substitute numbers into increasingly complex algebraic expressions D C
- substitute numbers into a variety of formulae D C B
- rearrange a formula to make a different variable the subject of the formula C B

16

Indices and standard form

This chapter is about indices, standard form and surds.

Standard form is useful for writing very small and very large numbers. Scientists estimate that there are between 1×10^{18} and 1×10^{19} insects alive in the world at any one time.

Objectives

This chapter will show you how to

- **use laws of indices for multiplication and division of integer powers** C B
- **calculate using negative integer powers** B
- **write numbers in standard index form** B
- **calculate using standard index form** B
- **use laws of indices to calculate powers and roots of numbers written using index notation** A
- **use laws of indices to work out positive and negative fractional powers of whole numbers** A A*
- **use surds to give exact answers to calculations** A A*
- **simplify expressions involving surds** A A*

Before you start this chapter

Put your calculator away!

1 Work out

a 5^3 b 2^4 c 7^2 d $(-1)^3$

2 Work out

a $\sqrt{3^3 - 2}$ b $\sqrt{2^3 + 1}$

c $\sqrt[3]{8^2}$ d $\sqrt{5^3 - 2^2}$

3 Write these numbers using standard form.

a 4200

b 8 700 000

c 0.000 69

d 0.008 04

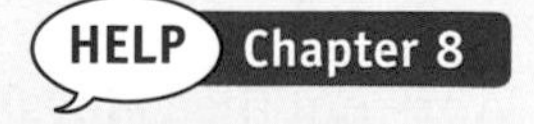

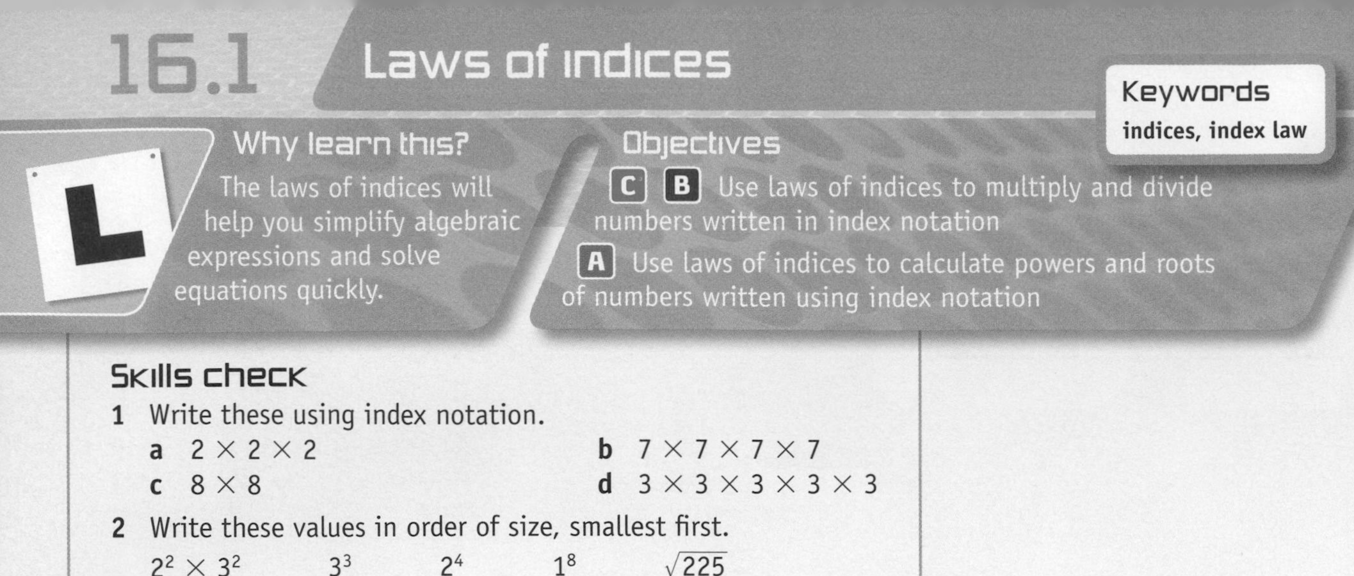

16.1 Laws of indices

Keywords
indices, index law

Why learn this?

The laws of indices will help you simplify algebraic expressions and solve equations quickly.

Objectives

C **B** Use laws of indices to multiply and divide numbers written in index notation

A Use laws of indices to calculate powers and roots of numbers written using index notation

Skills check

1 Write these using index notation.

a $2 \times 2 \times 2$ b $7 \times 7 \times 7 \times 7$

c 8×8 d $3 \times 3 \times 3 \times 3 \times 3$

2 Write these values in order of size, smallest first.

$2^2 \times 3^2$ 3^3 2^4 1^8 $\sqrt{225}$

Laws of indices

- To *multiply* powers of the same number you *add* the **indices**.

$$4^2 \times 4^5 = (4 \times 4) \times (4 \times 4 \times 4 \times 4 \times 4) = 4^7$$

Using the **laws of indices:**

$$4^2 \times 4^5 = 4^{2+5} = 4^7$$

You can multiply in any order so the brackets don't matter.

- To *divide* powers of the same number you *subtract* the indices.

$$6^5 \div 6^3 = \frac{6 \times 6 \times \cancel{6} \times \cancel{6} \times \cancel{6}}{\cancel{6} \times \cancel{6} \times \cancel{6}} = \frac{6 \times 6}{1} = 6^2$$

You can cancel 6 three times from the top and bottom of the fraction.

Using the laws of indices:

$$6^5 \div 6^3 = 6^{5-3} = 6^2$$

Powers of different numbers

You can only use the laws of indices to multiply and divide powers of the same number.

To simplify powers of different numbers you need to look at each base separately.

$$(2^4 \times 3^2) \times (2 \times 3^7) = (2^4 \times 2) \times (3^2 \times 3^7) = 2^5 \times 3^9$$

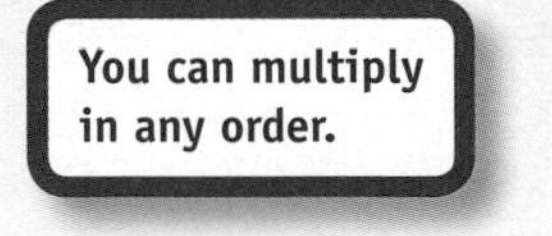

You can multiply in any order.

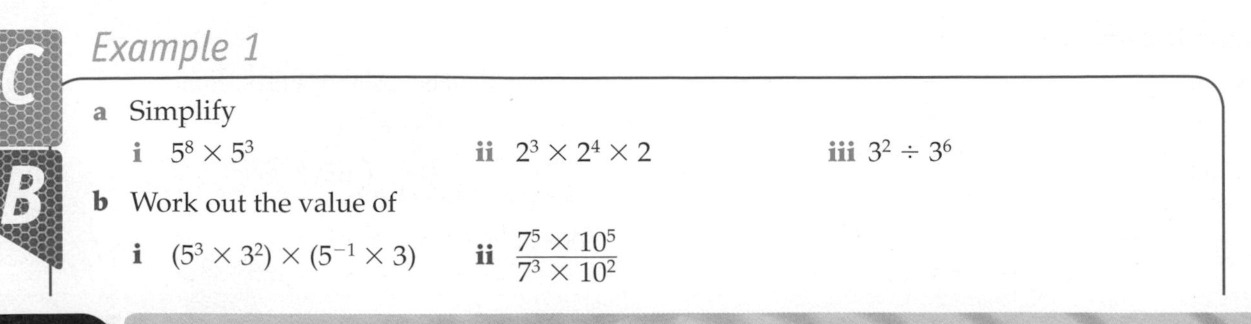

Example 1

C **B**

a Simplify

i $5^8 \times 5^3$ ii $2^3 \times 2^4 \times 2$ iii $3^2 \div 3^6$

b Work out the value of

i $(5^3 \times 3^2) \times (5^{-1} \times 3)$ ii $\dfrac{7^5 \times 10^5}{7^3 \times 10^2}$

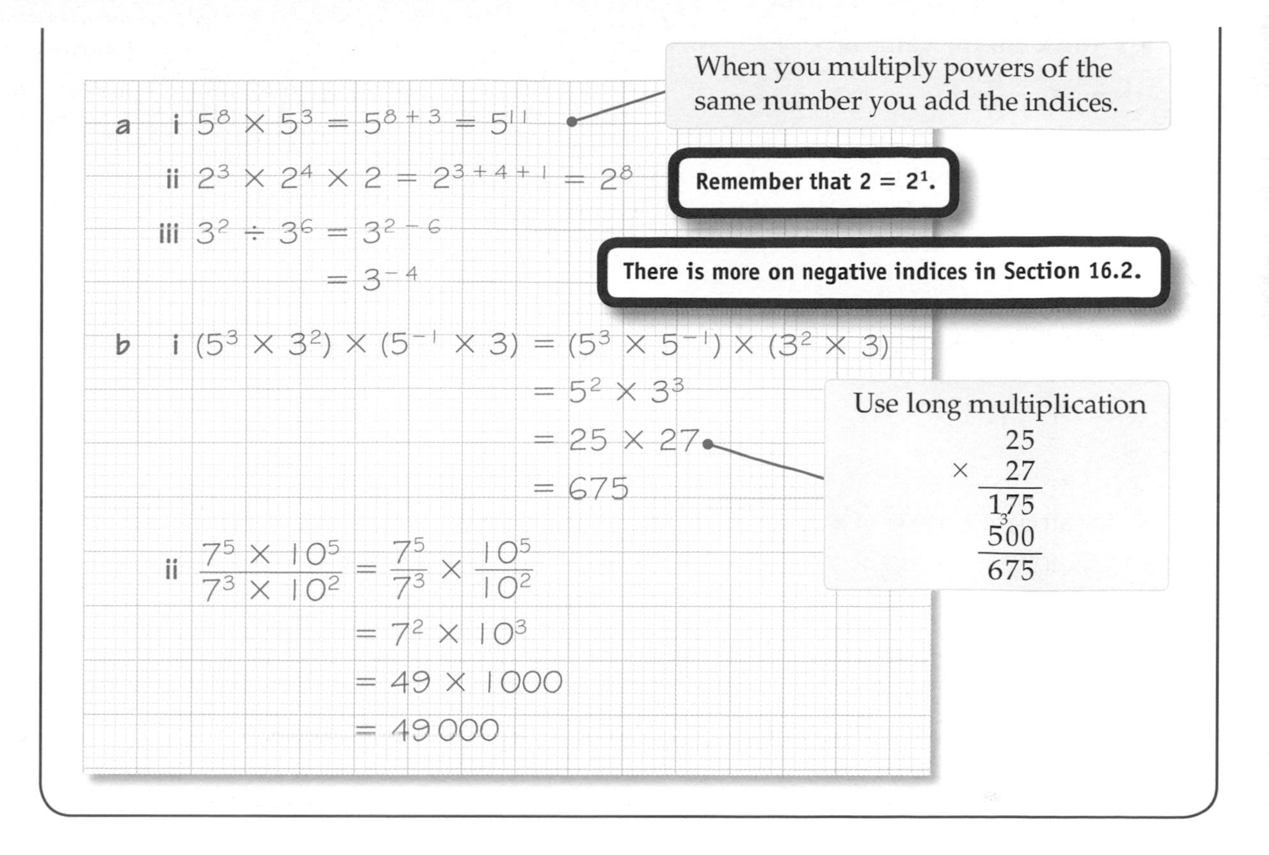

a i $5^8 \times 5^3 = 5^{8+3} = 5^{11}$
When you multiply powers of the same number you add the indices.
ii $2^3 \times 2^4 \times 2 = 2^{3+4+1} = 2^8$
Remember that $2 = 2^1$.
iii $3^2 \div 3^6 = 3^{2-6}$
$= 3^{-4}$
There is more on negative indices in Section 16.2.
b i $(5^3 \times 3^2) \times (5^{-1} \times 3) = (5^3 \times 5^{-1}) \times (3^2 \times 3)$
$= 5^2 \times 3^3$
$= 25 \times 27$
$= 675$
Use long multiplication
25
× 27
175
500
675
ii $\frac{7^5 \times 10^5}{7^3 \times 10^2} = \frac{7^5}{7^3} \times \frac{10^5}{10^2}$
$= 7^2 \times 10^3$
$= 49 \times 1000$
$= 49\,000$

Exercise 16A

C

1 Write each expression as a single power.

a $9^7 \times 9^3$ **b** $4^7 \times 4^3$ **c** 5×5^3 **d** $10^2 \times 10^6$
e $8^6 \div 8^3$ **f** $6^{10} \div 6^5$ **g** $4^{17} \div 4^{12}$ **h** $5^3 \div 5$

2 Write each expression as a single power.

a $9^4 \times 9 \times 9^6$ **b** $2^2 \times 2^7 \times 2^4$ **c** $8^3 \times 8^5 \times 8$ **d** $4 \times 4^3 \times 4$

3 Write down the value of

a $5^8 \div 5^5$ **b** $4^{13} \div 4^{11}$ **c** $2^{10} \div 2^6$ **d** $4^6 \div 4^3$

Give your answers as whole numbers.

4 Alison writes that $7^{10} \div 7^2 = 7^5$. Is Alison correct? Give a reason for your answer.

B

5 Write each expression as a single power.

a $5^2 \div 5^9$ **b** $\frac{3^2}{3^3}$ **c** $7 \div 7^5$ **d** $\frac{9^2}{9^6}$
e $4^3 \div 4^5$ **f** $2 \div 2^4$ **g** $\frac{3^2 \times 3}{3^5}$ **h** $\frac{16}{4^4}$

6 Write down the value of

a $6^{-2} \times 6^4$ **b** $3^{-3} \times 3^6$ **c** $\frac{5^{-1} \times 5^6}{25}$ **d** $\frac{6^6 \times 6^{-1}}{6^4}$

7 Work out the value of

a $(5^2 \times 2^2) \times (2^{-1} \times 5)$ **b** $(4^5 \times 10^2) \times (10 \times 4^{-2})$
c $(2^5 \times 6^3) \times (2^{-3} \times 6^{-1})$ **d** $(5 \times 10^2) \times (5^2 \times 10^{-1})$
e $(2 \times 3^2 \times 5^2) \times (2^2 \times 3)$ **f** $(5^3 \times 4^2 \times 10) \times (5^{-2} \times 4^{-1} \times 10^2)$

8 Work out the value of

a $\dfrac{3^5 \times 6^6}{3^3 \times 6^4}$ b $\dfrac{2^6 \times 3^5}{2^3 \times 3^3}$ c $\dfrac{10^4 \times 5^9}{10 \times 5^6}$ d $\dfrac{2^4 \times 10^7}{2 \times 10^4}$

9 Salma says that you can write $2^{11} \times 8^2$ as a single power of 2.
Show working to explain why Salma is correct.

Powers of powers

- To calculate a power of a power you multiply the indices.

$(7^3)^2 = (7 \times 7 \times 7) \times (7 \times 7 \times 7)$
$= 7^6$

$(7^3)^2 = 7^3 \times 7^3$

Using the laws of indices:

$(7^3)^2 = 7^{3 \times 2} = 7^6$

You can use this index law to find roots of numbers written in index notation.

$(8^5)^3 = 8^{15}$ so $\sqrt[3]{8^{15}} = 8^5$

$(3^7)^5 = 3^{35}$ so $\sqrt[5]{3^{35}} = 3^7$

$\sqrt[5]{\ }$ means 'fifth root'. $3^5 = 243$ so $\sqrt[5]{243} = 3$.

Example 2

Write each expression as a single power of 4.

a $(4^6)^2$ b $\sqrt[4]{4^8}$

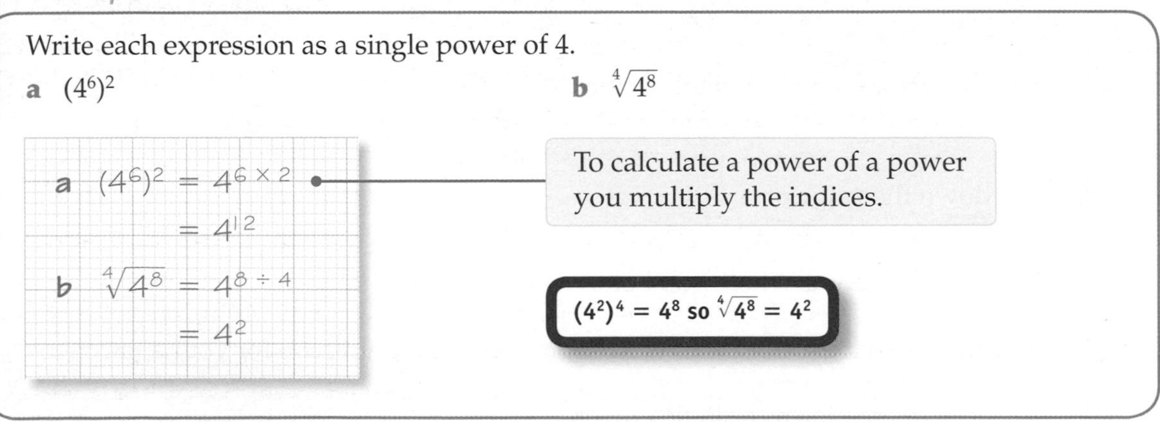

a $(4^6)^2 = 4^{6 \times 2}$
$= 4^{12}$
b $\sqrt[4]{4^8} = 4^{8 \div 4}$
$= 4^2$
To calculate a power of a power you multiply the indices.
$(4^2)^4 = 4^8$ so $\sqrt[4]{4^8} = 4^2$

Exercise 16B

1 Use the laws of indices to write each expression as a single power.

a $(2^3)^6$ b $(7^2)^5$ c $(6^3)^3$ d $(5^7)^2$
e $(9^2)^3$ f $(7^{-1})^2$ g $(15^4)^{-3}$ h $(8^4)^{-1}$

2 a Write $(5^3)^2$ as a single power of 5.
b Use your answer to part **a** to work out the value of $\sqrt{5^6}$.

3 Write each expression as a single power.

a $\sqrt{2^8}$ b $\sqrt{6^4}$ c $\sqrt{10^{10}}$ d $\sqrt{3^{12}}$

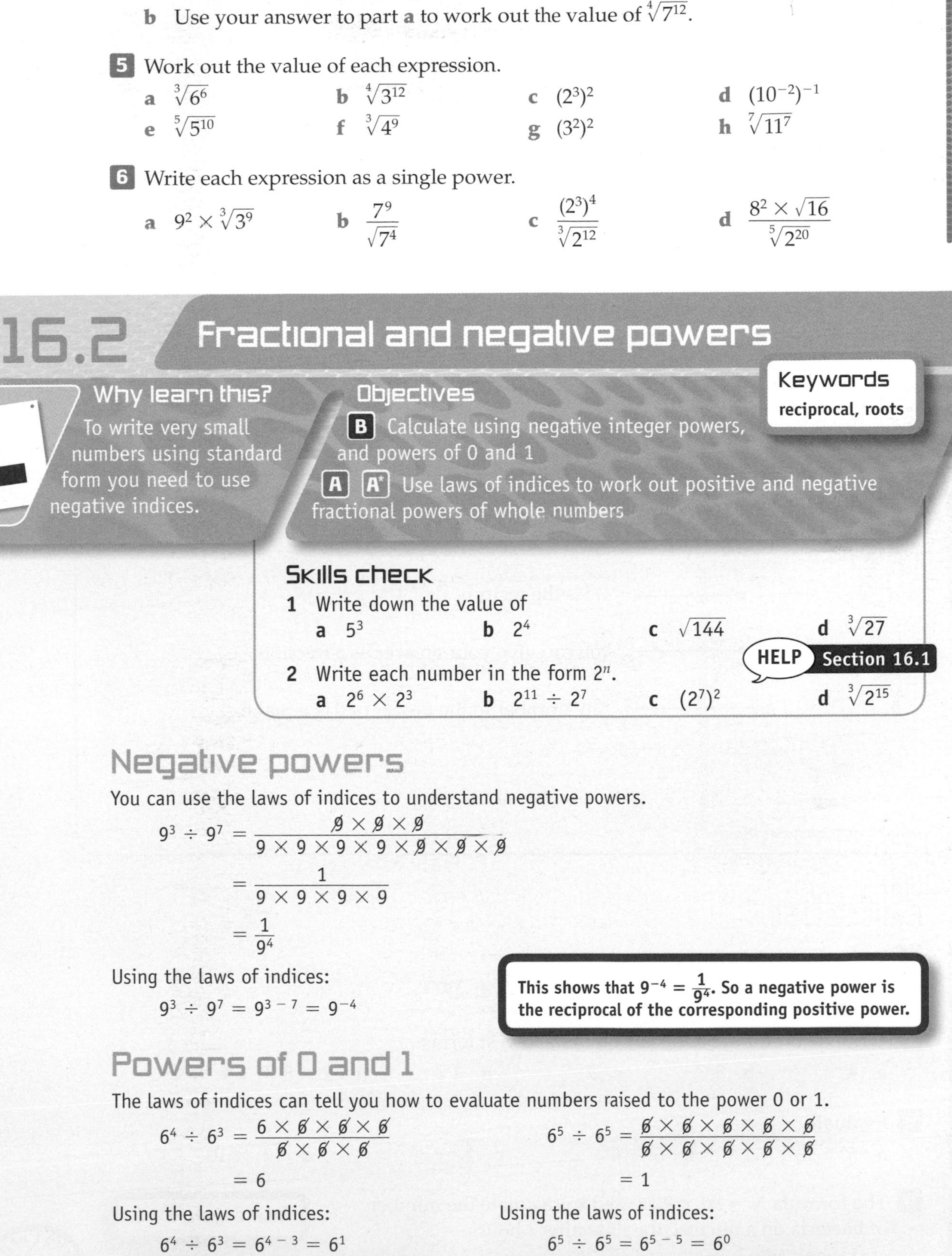

4 **a** Write $(7^3)^4$ as a single power of 7.

b Use your answer to part **a** to work out the value of $\sqrt[4]{7^{12}}$.

5 Work out the value of each expression.

a $\sqrt[3]{6^6}$ **b** $\sqrt[4]{3^{12}}$ **c** $(2^3)^2$ **d** $(10^{-2})^{-1}$

e $\sqrt[5]{5^{10}}$ **f** $\sqrt[3]{4^9}$ **g** $(3^2)^2$ **h** $\sqrt[7]{11^7}$

6 Write each expression as a single power.

a $9^2 \times \sqrt[3]{3^9}$ **b** $\dfrac{7^9}{\sqrt{7^4}}$ **c** $\dfrac{(2^3)^4}{\sqrt[3]{2^{12}}}$ **d** $\dfrac{8^2 \times \sqrt{16}}{\sqrt[5]{2^{20}}}$

16.2 Fractional and negative powers

Why learn this?

To write very small numbers using standard form you need to use negative indices.

Objectives

B Calculate using negative integer powers, and powers of 0 and 1

A **A*** Use laws of indices to work out positive and negative fractional powers of whole numbers

Keywords

reciprocal, roots

Skills check

1 Write down the value of

a 5^3 b 2^4 c $\sqrt{144}$ d $\sqrt[3]{27}$

2 Write each number in the form 2^n.

a $2^6 \times 2^3$ b $2^{11} \div 2^7$ c $(2^7)^2$ d $\sqrt[3]{2^{15}}$

HELP Section 16.1

Negative powers

You can use the laws of indices to understand negative powers.

$$9^3 \div 9^7 = \frac{\cancel{9} \times \cancel{9} \times \cancel{9}}{9 \times 9 \times 9 \times 9 \times \cancel{9} \times \cancel{9} \times \cancel{9}}$$

$$= \frac{1}{9 \times 9 \times 9 \times 9}$$

$$= \frac{1}{9^4}$$

Using the laws of indices:

$$9^3 \div 9^7 = 9^{3-7} = 9^{-4}$$

This shows that $9^{-4} = \frac{1}{9^4}$. So a negative power is the reciprocal of the corresponding positive power.

Powers of 0 and 1

The laws of indices can tell you how to evaluate numbers raised to the power 0 or 1.

$$6^4 \div 6^3 = \frac{6 \times \cancel{6} \times \cancel{6} \times \cancel{6}}{\cancel{6} \times \cancel{6} \times \cancel{6}}$$

$$= 6$$

Using the laws of indices:

$$6^4 \div 6^3 = 6^{4-3} = 6^1$$

This shows that $6^1 = 6$. Any number to the power 1 is the number itself.

$$6^5 \div 6^5 = \frac{\cancel{6} \times \cancel{6} \times \cancel{6} \times \cancel{6} \times \cancel{6}}{\cancel{6} \times \cancel{6} \times \cancel{6} \times \cancel{6} \times \cancel{6}}$$

$$= 1$$

Using the laws of indices:

$$6^5 \div 6^5 = 6^{5-5} = 6^0$$

This shows that $6^0 = 1$. Any number to the power 0 is equal to 1.

Fractional powers and roots

If the index is a fraction then you can evaluate a number using **roots.**

$$9^{\frac{1}{2}} \times 9^{\frac{1}{2}} = 9^{\frac{1}{2} + \frac{1}{2}}$$
$$= 9^1$$
$$= 9$$

But the number that equals 9 when multiplied by itself is called the square root of 9.

$$9^{\frac{1}{2}} = \sqrt{9} = 3$$

You can use this method for other roots.

$$8^{\frac{1}{3}} \times 8^{\frac{1}{3}} \times 8^{\frac{1}{3}} = 8^{\frac{1}{3} + \frac{1}{3} + \frac{1}{3}}$$
$$= 8^1$$
$$= 8$$

So $8^{\frac{1}{3}} = \sqrt[3]{8} = 2$

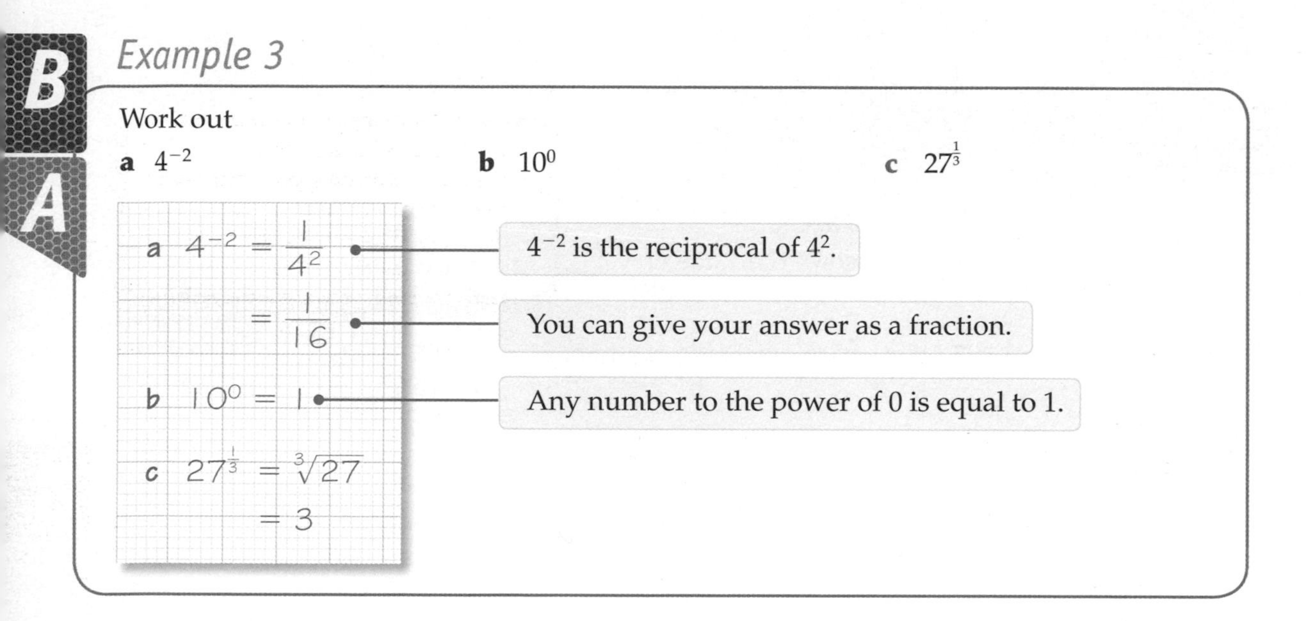

Example 3

Work out

a 4^{-2} **b** 10^0 **c** $27^{\frac{1}{3}}$

a $4^{-2} = \frac{1}{4^2}$ — 4^{-2} is the reciprocal of 4^2.

$= \frac{1}{16}$ — You can give your answer as a fraction.

b $10^0 = 1$ — Any number to the power of 0 is equal to 1.

c $27^{\frac{1}{3}} = \sqrt[3]{27}$

$= 3$

Exercise 16C

1 Write down the value of

a 8^0 **b** 6^1 **c** 100^0 **d** 17^1 **e** 9^1 **f** 83^0

2 Write each of these as a fraction in its lowest terms.

a 4^{-1} **b** 3^{-2} **c** 7^{-1} **d** 2^{-2} **e** 2×8^{-1} **f** 3×6^{-2}

3 Evaluate

a $5^4 \times 5^{-5}$ **b** $6^2 \div 6^4$ **c** $\frac{2^6}{2^9}$ **d** $10^5 \times 10^{-5}$

4 The formula $N = 80 \times 2^t$ is used to calculate the number of bacteria on a microscope slide after t hours.

After 2 hours, $t = 2$.

a Use this formula to calculate the number of bacteria on the microscope slide after

i 2 hours **ii** 6 hours.

b How many bacteria do you think were placed on the microscope slide at the beginning of the experiment? Give a reason for your answer.

AO2

FUNCTIONAL

5 Write each of these as a fraction in its lowest terms.

a $(5^2)^{-1}$　　b $3 \times (3^{-1})^3$　　c $\dfrac{\sqrt{2^6}}{(2^3)^2}$　　d $\sqrt[4]{5^8} \times (5^2)^{-2}$

6 Work out the value of

a $4^{\frac{1}{2}}$　　b $49^{\frac{1}{2}}$　　c $64^{\frac{1}{3}}$　　d $1^{\frac{1}{2}}$

e $196^{\frac{1}{2}}$　　f $100^{0.5}$　　g $16^{\frac{1}{4}}$　　h $81^{0.25}$

7 Write $\sqrt{11} \times \sqrt[3]{11}$ as a single power of 11.

A

A

AO2

More complicated fractional indices

You can combine the laws of indices to evaluate more complicated powers.

$$8^{\frac{2}{3}} = \left(8^{\frac{1}{3}}\right)^2 = \left(\sqrt[3]{8}\right)^2 = 2^2 = 4$$

or

$$8^{\frac{2}{3}} = (8^2)^{\frac{1}{3}} = \sqrt[3]{8^2} = \sqrt[3]{64} = 4$$

When the index is a fraction:
- **the numerator tells you what power to raise the number to**
- **the denominator tells you what root to calculate.**

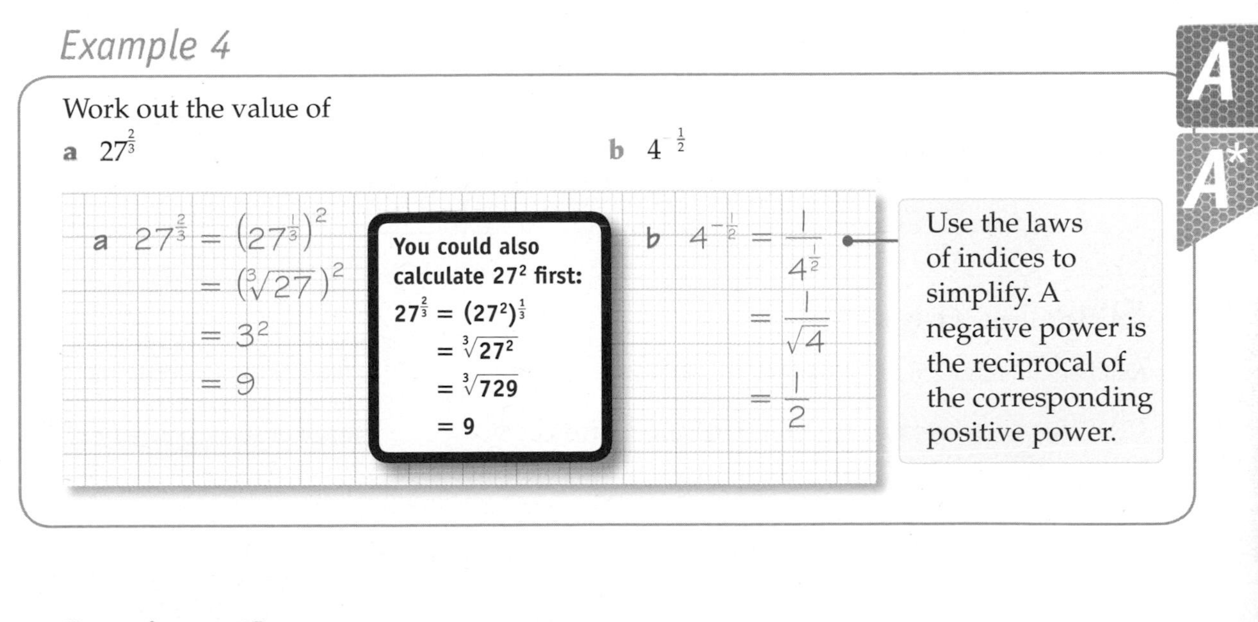

Example 4

Work out the value of

a $27^{\frac{2}{3}}$　　b $4^{-\frac{1}{2}}$

a $27^{\frac{2}{3}} = \left(27^{\frac{1}{3}}\right)^2$
$= \left(\sqrt[3]{27}\right)^2$
$= 3^2$
$= 9$

You could also calculate 27^2 first:
$27^{\frac{2}{3}} = (27^2)^{\frac{1}{3}}$
$= \sqrt[3]{27^2}$
$= \sqrt[3]{729}$
$= 9$

b $4^{-\frac{1}{2}} = \dfrac{1}{4^{\frac{1}{2}}}$
$= \dfrac{1}{\sqrt{4}}$
$= \dfrac{1}{2}$

Use the laws of indices to simplify. A negative power is the reciprocal of the corresponding positive power.

A

A*

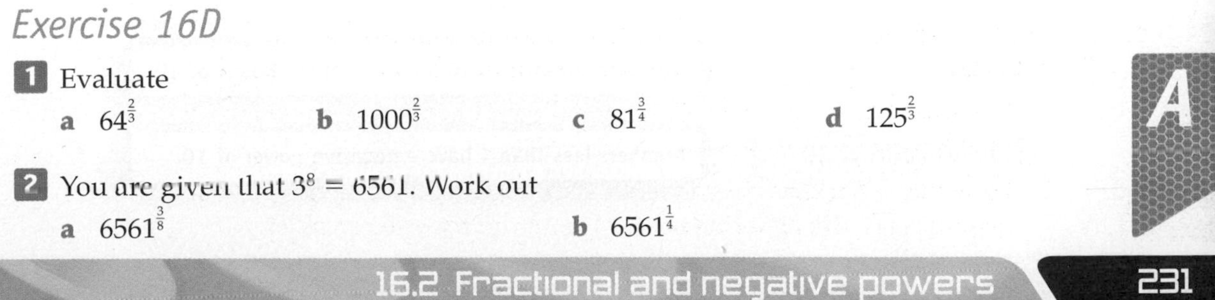

Exercise 16D

1 Evaluate

a $64^{\frac{2}{3}}$　　b $1000^{\frac{2}{3}}$　　c $81^{\frac{3}{4}}$　　d $125^{\frac{2}{3}}$

2 You are given that $3^8 = 6561$. Work out

a $6561^{\frac{3}{8}}$　　b $6561^{\frac{1}{4}}$

A

A

3 Work out the value of $16^{\frac{3}{2}}$.

4 Write each of the following using index notation.

a $(\sqrt[3]{11})^2$ **b** $\sqrt[4]{2^3}$ **c** $\sqrt[5]{7^3}$ **d** $(\sqrt[4]{13})^2$

A*

5 Work out the value of

a $25^{-\frac{1}{2}}$ **b** $27^{-\frac{1}{3}}$ **c** $64^{-\frac{1}{3}}$

d $125^{-\frac{2}{3}}$ **e** $16^{-\frac{3}{4}}$ **f** $49^{-\frac{1}{2}}$

A* **AO3**

6 Catelyn says that if a is bigger than 1 and n is a fraction between 0 and 1 then $1 < a^n < a$.
Do you agree with Catelyn's statement?
Give a reason for your answer.

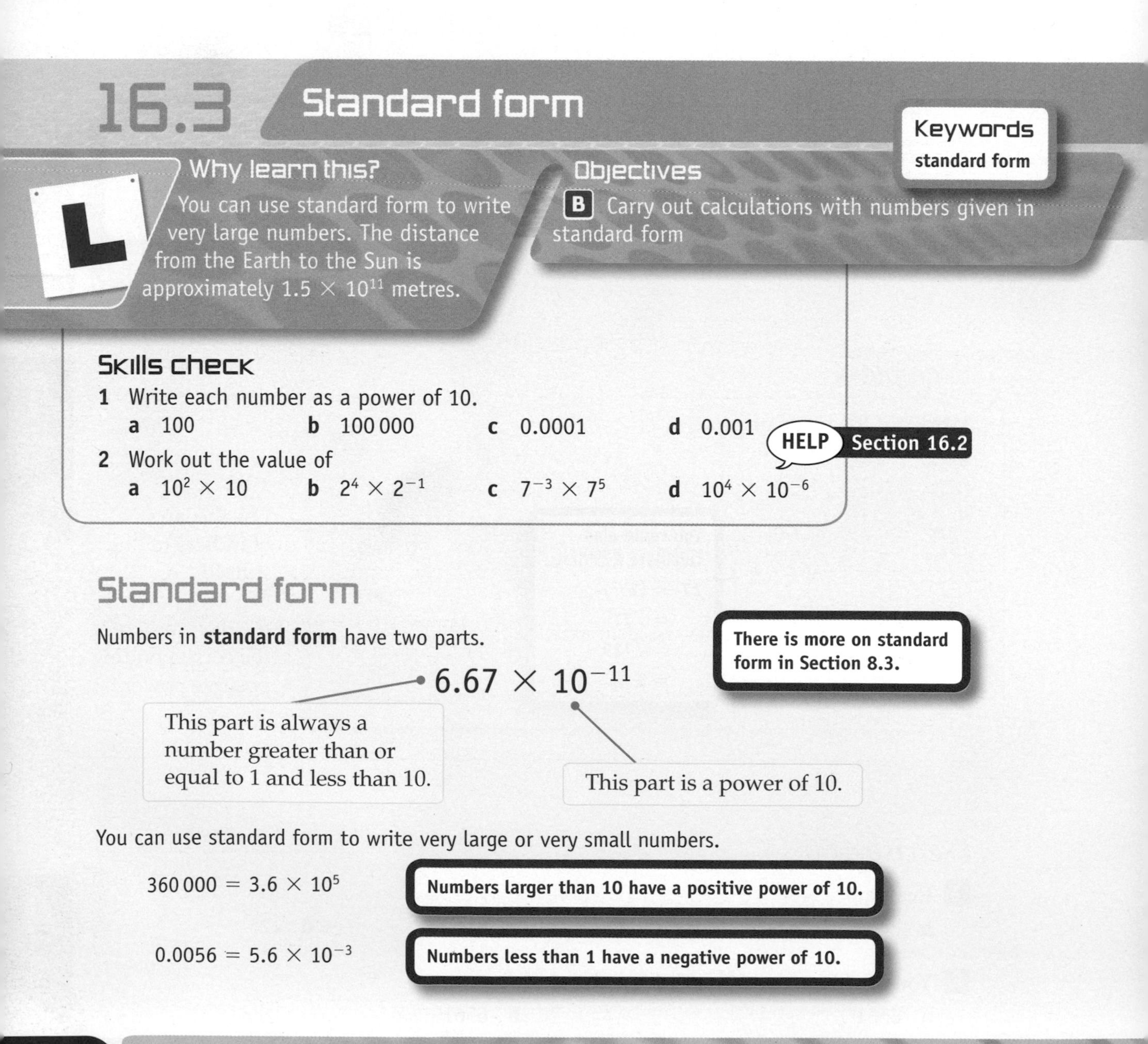

16.3 Standard form

Keywords
standard form

Why learn this?

You can use standard form to write very large numbers. The distance from the Earth to the Sun is approximately 1.5×10^{11} metres.

Objectives

B Carry out calculations with numbers given in standard form

Skills check

1 Write each number as a power of 10.

a 100 **b** 100 000 **c** 0.0001 **d** 0.001

HELP Section 16.2

2 Work out the value of

a $10^2 \times 10$ **b** $2^4 \times 2^{-1}$ **c** $7^{-3} \times 7^5$ **d** $10^4 \times 10^{-6}$

Standard form

Numbers in **standard form** have two parts.

There is more on standard form in Section 8.3.

$$6.67 \times 10^{-11}$$

This part is always a number greater than or equal to 1 and less than 10.

This part is a power of 10.

You can use standard form to write very large or very small numbers.

$360\,000 = 3.6 \times 10^5$

Numbers larger than 10 have a positive power of 10.

$0.0056 = 5.6 \times 10^{-3}$

Numbers less than 1 have a negative power of 10.

Example 5

Write these numbers in order of size, smallest first.

6.4×10^3 1×10^5 7.4×10^{-6} 5.2×10^3 4.3×10^5

7.4×10^{-6} — Look at the indices first. 7.4×10^{-6} has the lowest index, so it is the smallest number.

5.2×10^3

6.4×10^3 — 5.2×10^3 and 6.4×10^3 have the same index. 5.2 is smaller than 6.4, so 5.2×10^3 is smaller.

1×10^5

4.3×10^5

B

Exercise 16E

1 Write these numbers in standard form.

a 34 000 **b** 2600 **c** 740

d 200 000 **e** 6 030 000 **f** 22.6

2 Write these numbers in standard form.

a 0.0006 **b** 0.0032 **c** 0.003 09

d 0.445 **e** 0.01 **f** 0.000 009 8

3 Write each of these as a decimal number.

a 2×10^4 **b** 6.2×10^3 **c** 4.54×10^{-3}

d 2.07×10^{-1} **e** 7.9×10^7 **f** 4.551×10^{-6}

4 Write these numbers in order of size, smallest first.

8.04×10^3 4.8×10^4 9.31×10^{-1} 3.2×10^{-2} 8.66×10^{-1} 1×10^7

5 In the 2001 census the population of the UK was recorded as 58 789 194. Round this number to 3 significant figures and write your answer in standard form.

6 This table shows the masses of atoms of different elements.
Write the elements in order of mass, smallest first.

Substance	Mass of molecule (kg)
carbon	2.0×10^{-26}
plutonium	4.1×10^{-25}
gold	3.3×10^{-25}
iron	9.1×10^{-26}
hydrogen	1.7×10^{-27}
tungsten	3.1×10^{-25}

B

7 A human hair has a diameter of 0.002 54 cm.
Write this in metres using standard form.

B

AO2

Calculating with numbers in standard form

To multiply or divide numbers in standard form you multiply or divide each part separately. You can use the laws of indices to convert your answers back into standard form if necessary.

To add or subtract numbers in standard form you write them as decimal numbers first.

B

Example 6

Work these out. Give your answers in standard form.

a $(5 \times 10^3) \times (7 \times 10^5)$

b $\dfrac{4.5 \times 10^3}{5 \times 10^7}$

c $(3.1 \times 10^2) + (6.8 \times 10^3)$

d $(9.2 \times 10^4) - (1.6 \times 10^3)$

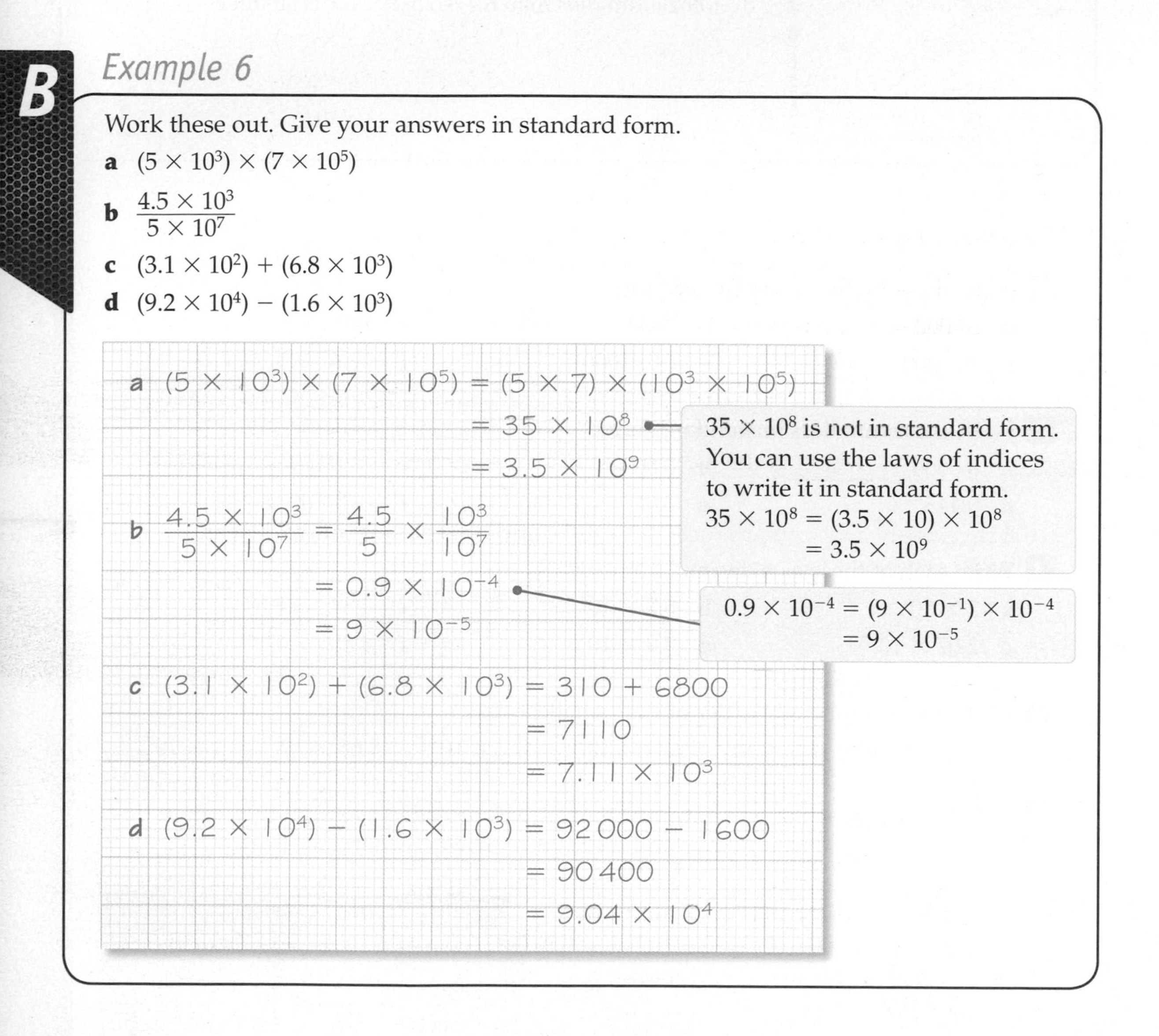

Exercise 16F

B

1 Work these out. Give your answers in standard form.

a $(2 \times 10^6) \times (4 \times 10^2)$

b $(6 \times 10^4) \times (8 \times 10^5)$

c $(2.5 \times 10^2) \times (8 \times 10^7)$

d $(9 \times 10^{-6}) \times (1 \times 10^{11})$

e $(2.5 \times 10^4) \times (2 \times 10^{-9})$

f $(3 \times 10^{-1}) \times (6.2 \times 10^{-6})$

g $(8 \times 10^9) \times (4.4 \times 10^3)$

h $(1.6 \times 10^{-5}) \times (7 \times 10^{-3})$

B

2 Work these out. Give your answers in standard form.

a $\dfrac{8 \times 10^{12}}{4 \times 10^{7}}$

b $\dfrac{8.4 \times 10^{5}}{2 \times 10^{2}}$

c $\dfrac{3 \times 10^{8}}{6 \times 10^{3}}$

d $\dfrac{2 \times 10^{-6}}{5 \times 10^{4}}$

e $(4.2 \times 10^{4}) \div (2 \times 10^{9})$

f $(1 \times 10^{3}) \div (4 \times 10^{11})$

g $(3.5 \times 10^{4}) \div (5 \times 10^{8})$

h $(2 \times 10^{7}) \div (8 \times 10^{-1})$

3 Work these out. Give your answers in standard form.

a $(7 \times 10^{3}) + (6 \times 10^{4})$

b $(1.7 \times 10^{5}) + (3.5 \times 10^{4})$

c $(2.2 \times 10^{4}) - (9 \times 10^{3})$

d $(8.45 \times 10^{5}) - (2 \times 10^{4})$

e $(8 \times 10^{3}) + (6.9 \times 10^{5})$

f $(9.4 \times 10^{4}) - (3 \times 10^{2})$

g $(1.2 \times 10^{5}) + (7 \times 10^{3})$

h $(4.09 \times 10^{4}) - (2.6 \times 10^{2})$

B

4 A water molecule has a mass of 3×10^{-26} kg.
A glass of water contains 1.2×10^{25} molecules of water.
Work out the mass of the water in the glass.
Give your answer in grams.

5 An adult brain has a mass of 1.4×10^{3} grams. It contains 7×10^{10} brain neurons.
Work out the average mass of a neuron.
Give your answer in grams in standard form.

B

AO3

6 A machine at a factory produces 4.5×10^{4} computer components each year.
Each component has a mass of 8.1×10^{-3} kg.
2% of all the components produced by the machine are faulty.
Work out the total mass of all the faulty components produced in one year.

16.4 Surds

Why learn this?

You can use surds to give exact answers to problems involving Pythagoras' theorem.

Objectives

A Multiply and divide expressions involving surds

A A* Use surds to give exact answers to calculations

A A* Simplify expressions involving surds

Keywords

rational number, irrational number, surd, rationalising the denominator

Skills check

1 Work out

a $\frac{1}{2} + \frac{2}{3}$ b $\frac{3}{5} + \frac{1}{3}$ c $\frac{4}{5} - \frac{3}{8}$ d $\frac{3}{4} + \frac{2}{3} + \frac{2}{5}$

2 Multiply out the brackets.

a $x(x - 5)$ b $(a + 6)(a + 1)$

c $(x - 3)(x + 2)$ d $(a + b)(2a + b)$

Rational and irrational numbers

Numbers that can be written in the form $\frac{a}{b}$

where a and b are both integers are called **rational numbers.**

3, −6, $\frac{2}{3}$ and $-\frac{8}{5}$ are all rational numbers.

Numbers that can't be written in this way are called **irrational numbers.**

$\sqrt{2}$, π and $\sqrt[3]{10}$ are all irrational numbers.

If you try to write an irrational number as a decimal it never terminates and never repeats itself. You can write some irrational numbers exactly using roots. Irrational numbers written this way are called **surds.**

Rational numbers can be written as decimals. They always have a finite number of decimal places or recurring digits.

You can give the exact value for an answer by using surds.

Simplifying surds

To simplify expressions involving surds, use these rules.

$$\sqrt{a \times b} = \sqrt{a} \times \sqrt{b}$$

$$\sqrt{\frac{a}{b}} = \frac{\sqrt{a}}{\sqrt{b}}$$

Example 7

Simplify each of these expressions as much as possible.
Leave your answers in surd form where necessary.

a $2\sqrt{2} \times 5\sqrt{18}$

b $\frac{4\sqrt{21}}{2\sqrt{7}}$

a $2\sqrt{2} \times 5\sqrt{18} = 2 \times 5 \times \sqrt{2} \times \sqrt{18}$
$= 10 \times \sqrt{2 \times 18}$
$= 10 \times \sqrt{36}$
$= 10 \times 6$
$= 60$

You can multiply in any order.

Use the rules of surds to simplify your answer.
$\sqrt{a} \times \sqrt{b} = \sqrt{ab}$
so $\sqrt{2} \times \sqrt{18} = \sqrt{36}$

b $\frac{4\sqrt{21}}{2\sqrt{7}} = \frac{4}{2} \times \frac{\sqrt{21}}{\sqrt{7}}$
$= 2 \times \sqrt{\frac{21}{7}}$
$= 2\sqrt{3}$

$\frac{\sqrt{a}}{\sqrt{b}} = \sqrt{\frac{a}{b}}$ so $\frac{\sqrt{21}}{\sqrt{7}} = \sqrt{3}$

Exercise 16G

1 Write these as a single square root.

a $\sqrt{3} \times \sqrt{6}$ b $\sqrt{2} \times \sqrt{5}$ c $\sqrt{5} \times \sqrt{6}$

d $\sqrt{11} \times \sqrt{10}$ e $\sqrt{12} \times \sqrt{2} \times \sqrt{3}$ f $\sqrt{3} \times \sqrt{6} \times \sqrt{5}$

2 Work out the value of

a $\sqrt{8} \times \sqrt{2}$ b $\sqrt{20} \times \sqrt{5}$ c $\sqrt{2} \times \sqrt{32}$

d $\sqrt{6} \times \sqrt{24}$ e $\sqrt{5} \times \sqrt{2} \times \sqrt{10}$ f $\sqrt{7} \times \sqrt{2} \times \sqrt{14}$

3 Work out the value of

a $4\sqrt{18} \times 2\sqrt{2}$ b $5\sqrt{2} \times 7\sqrt{8}$ c $10\sqrt{24} \times 5\sqrt{6}$

d $2\sqrt{10} \times \sqrt{40}$ e $6\sqrt{32} \times 2\sqrt{2}$ f $5\sqrt{50} \times 2\sqrt{2}$

4 Work out the value of

a $\sqrt{125} \div \sqrt{5}$ b $\sqrt{8} \div \sqrt{2}$ c $\sqrt{500} \div \sqrt{5}$

d $\frac{\sqrt{20}}{\sqrt{5}}$ e $\frac{\sqrt{45}}{\sqrt{5}}$ f $\frac{\sqrt{450}}{\sqrt{2}}$

5 Simplify these expressions. Write your answers in surd form.

a $5\sqrt{2} \times 2\sqrt{3}$ b $7\sqrt{8} \times 3\sqrt{3}$ c $\sqrt{5} \times 2\sqrt{7}$

d $\frac{3\sqrt{15}}{\sqrt{5}}$ e $\frac{20\sqrt{6}}{2\sqrt{2}}$ f $\frac{36\sqrt{70}}{4\sqrt{5}}$

6 Evaluate

a $\frac{12}{\sqrt{3}} \times \frac{\sqrt{12}}{6}$ b $\frac{\sqrt{3} \times \sqrt{21}}{\sqrt{7}}$

c $\frac{12\sqrt{150}}{\sqrt{3}} \times \frac{5}{4\sqrt{2}}$ d $\frac{8\sqrt{6}}{\sqrt{3}} \times \frac{\sqrt{16}}{2\sqrt{2}}$

Simplified surd form

You can simplify surds by looking for factors which are square numbers.

$$\sqrt{8} = \sqrt{4 \times 2} = \sqrt{4} \times \sqrt{2} = 2\sqrt{2}$$

You can simplify surds in fractions by **rationalising the denominator.**

$$\frac{3}{\sqrt{7}} = \frac{3}{\sqrt{7}} \times \frac{\sqrt{7}}{\sqrt{7}} = \frac{3\sqrt{7}}{\sqrt{7} \times \sqrt{7}} = \frac{3\sqrt{7}}{7}$$

In simplified surd form, all the square roots have no factors which are square numbers, and all the fractions have rational denominators.

A

Example 8

a Write $\sqrt{63}$ in the form $a\sqrt{b}$ where a and b are prime numbers.

b Simplify $\frac{6\sqrt{3}}{\sqrt{2}}$ by rationalising the denominator.

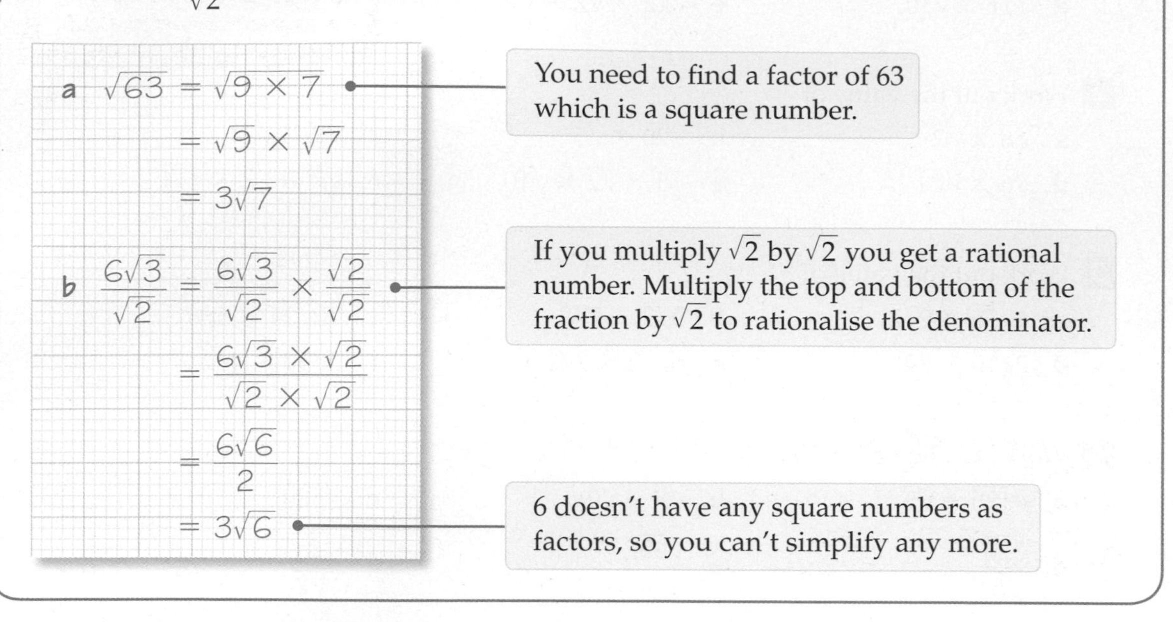

a √63 = √9 × 7
= √9 × √7
= 3√7
You need to find a factor of 63 which is a square number.
b 6√3/√2 = 6√3/√2 × √2/√2
= (6√3 × √2)/(√2 × √2)
= 6√6/2
= 3√6
If you multiply √2 by √2 you get a rational number. Multiply the top and bottom of the fraction by √2 to rationalise the denominator.
6 doesn't have any square numbers as factors, so you can't simplify any more.

Exercise 16H

A

1 Write $\sqrt{45}$ in the form $a\sqrt{b}$ where a and b are prime numbers.

2 Write each of these expressions in the form $a\sqrt{2}$.

a $\sqrt{18}$ **b** $\sqrt{50}$ **c** $\sqrt{200}$ **d** $\sqrt{32}$

3 Simplify each expression as much as possible.

a $\sqrt{12} + \sqrt{75}$ **b** $\sqrt{45} + \sqrt{20}$ **c** $2\sqrt{8} + \sqrt{18}$ **d** $\sqrt{175} - 2\sqrt{28}$

4 Simplify each expression by rationalising the denominator.

a $\frac{10}{\sqrt{2}}$ **b** $\frac{6}{\sqrt{3}}$ **c** $\frac{1}{\sqrt{5}}$ **d** $\frac{\sqrt{5}}{\sqrt{2}}$

e $\frac{15\sqrt{3}}{\sqrt{5}}$ **f** $\frac{8\sqrt{6}}{2\sqrt{2}}$ **g** $\frac{5\sqrt{3}}{2\sqrt{75}}$ **h** $\frac{12\sqrt{6}}{\sqrt{32}}$

A AO2

5 Aisha has written this working in her exercise book.

a What mistake has Aisha made?

b Simplify $\frac{12}{\sqrt{3}}$ by rationalising the denominator.

$$\frac{12}{\sqrt{3}} = \sqrt{\frac{12}{3}}$$
$$= \sqrt{4}$$
$$= 2$$

A AO3

6 Write $\frac{1}{2\sqrt{20}} + \frac{1}{3\sqrt{5}}$ as a single fraction with a rational denominator.

Multiplying more complicated surds

You can work with more complicated surds by multiplying out brackets.

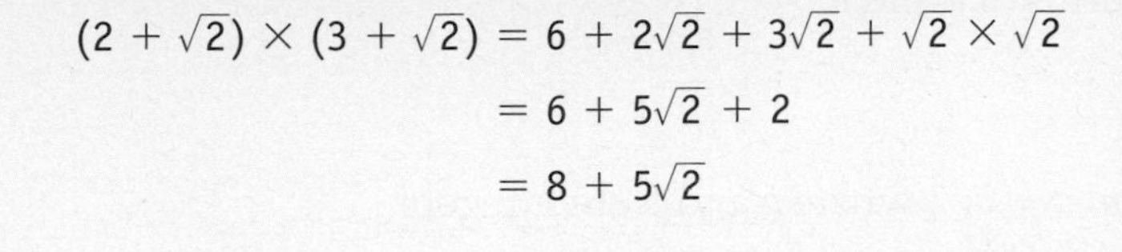

$$(2 + \sqrt{2}) \times (3 + \sqrt{2}) = 6 + 2\sqrt{2} + 3\sqrt{2} + \sqrt{2} \times \sqrt{2}$$
$$= 6 + 5\sqrt{2} + 2$$
$$= 8 + 5\sqrt{2}$$

There is more on multiplying out brackets in Section 12.5.

Example 9

Write these expressions in the form $a - b\sqrt{c}$ where a, b and c are integers.

a $(4 - \sqrt{12})(6 + \sqrt{3})$ **b** $(\sqrt{5} - 7)^2$

a $(4 - \sqrt{12})(6 + \sqrt{3}) = 24 + 4\sqrt{3} - 6\sqrt{12} - \sqrt{12} \times \sqrt{3}$
$= 24 + 4\sqrt{3} - 12\sqrt{3} - \sqrt{36}$
$= 24 - 8\sqrt{3} - 6$
$= 18 - 8\sqrt{3}$

$\sqrt{12} = \sqrt{4 \times 3} = \sqrt{4} \times \sqrt{3} = 2\sqrt{3}$

Remember FOIL for multiplying out brackets. First, Outside, Inside, Last.

b $(\sqrt{5} - 7)^2 = (\sqrt{5} - 7)(\sqrt{5} - 7)$
$= \sqrt{5} \times \sqrt{5} - 7\sqrt{5} - 7\sqrt{5} + 49$
$= 5 - 14\sqrt{5} + 49$
$= 54 - 14\sqrt{5}$

Remember that $\sqrt{5} \times \sqrt{5} = 5$.

A*

Exercise 16I

1 Write each expression in the form $a + b\sqrt{2}$ where a and b are integers.

a $(1 + \sqrt{2})(4 + \sqrt{2})$ **b** $(3 + \sqrt{2})(10 + \sqrt{2})$ **c** $(\sqrt{2} + 5)(\sqrt{2} + 3)$
d $(1 - \sqrt{2})(3 - \sqrt{2})$ **e** $(\sqrt{2} - 6)(\sqrt{2} + 1)$ **f** $(2 + \sqrt{2})^2$
g $(\sqrt{8} - 6)^2$ **h** $(6 - \sqrt{2})(2 + \sqrt{32})$

2 Simplify

a $(2 + \sqrt{3})(3 + \sqrt{3})$ **b** $(1 + 2\sqrt{5})(1 - \sqrt{5})$ **c** $(\sqrt{7} - 5)(1 + \sqrt{28})$
d $(4 + 2\sqrt{3})(2 + 5\sqrt{12})$ **e** $(1 - 2\sqrt{7})^2$ **f** $(4 + 2\sqrt{3})^2$

3 Write each expression as an integer.

a $(1 + \sqrt{2})(1 - \sqrt{2})$ **b** $(7 - \sqrt{2})(14 + \sqrt{8})$ **c** $(\sqrt{20} - 6)(\sqrt{5} + 3)$
d $(12 + \sqrt{7})(12 - \sqrt{7})$ **e** $(\sqrt{32} + 12)(\sqrt{2} - 3)$ **f** $(\sqrt{12} + 6)(\sqrt{27} - 9)$

4 Show that $(\sqrt{13} + \sqrt{11})(\sqrt{13} - \sqrt{11}) = 2$.

A*

A02

5 Ganesh is investigating the rule $(a + b\sqrt{5})^2 = p + q\sqrt{5}$. Work out

a an expression for p in terms of a and b

b an expression for q in terms of a and b.

A* AO2

6 Write $(3 + 2\sqrt{3})^3$ in the form $a + b\sqrt{3}$ where a and b are integers.

A* AO3

Review exercise

C

1 Write each expression as a single power of 2.

a $2^5 \times 2^3$ **[1 mark]**

b $2^7 \div 2^4$ **[1 mark]**

B

2 Write each of these as a fraction in its lowest terms.

a $\frac{2 \times 4^2}{4^4}$ **[1 mark]**

b $\frac{5^5 \times 10^5}{5^3 \times 10^7}$ **[2 marks]**

3 Write each of these as a decimal number.

a 17^0 **[1 mark]**

b 2^{-1} **[1 mark]**

4 Work out the value of $(4^3 \times 10^{-1}) \times (4^{-4} \times 10^3)$. **[2 marks]**

5 Work these out. Give your answers in standard form.

a $(6.2 \times 10^6) \times (3 \times 10^3)$ **[2 marks]**

b $(2.4 \times 10^3) \div (2 \times 10^{11})$ **[2 marks]**

B AO2

6 An ant colony contains 5×10^5 ants.
The total mass of all the ants in the colony is 1.5 kg.
Calculate the average mass of an ant.
Give your answer in kg in standard form. **[3 marks]**

7 Anselm wants to take these files to a friend's house.

Filename	File size (kb)
project_ideas.doc	7.1×10^2
science1.mpg	8.34×10^5
blues_in_C.wav	9.42×10^4
me_bowling.jpg	2.41×10^3

He has a memory stick which can hold 1 000 000 kb of data.
Will all of his files fit on the memory stick? Show all of your working. **[4 marks]**

8 a Write $(5^3)^5$ as a single power of 5. **[1 mark]**

b Use your answer to part **a** to work out the value of $\sqrt[5]{5^{15}}$. **[1 mark]**

9 Write down the value of

a $9^{\frac{1}{2}}$ **[1 mark]**

b $225^{\frac{1}{2}}$ **[1 mark]**

c $125^{\frac{1}{3}}$ **[1 mark]**

d $216^{\frac{2}{3}}$ **[2 marks]**

10 Write $\sqrt{147}$ in the form $p\sqrt{q}$ where p and q are prime numbers. **[2 marks]**

11 Work out the value of

a $\dfrac{\sqrt{3} \times \sqrt{33}}{\sqrt{11}}$ **[2 marks]**

b $\dfrac{10\sqrt{2}}{\sqrt{5}} \times \dfrac{\sqrt{125}}{4\sqrt{8}}$ **[3 marks]**

12 Simplify $\dfrac{40\sqrt{2}}{\sqrt{10}}$ by rationalising the denominator. **[2 marks]**

13 Write each expression in the form $a + b\sqrt{7}$ where a and b are integers.

a $(1 + \sqrt{7})(4 + \sqrt{7})$ **[3 marks]**

b $(\sqrt{7} + 3)^2$ **[3 marks]**

c $(\sqrt{28} - 2)(\sqrt{63} + 1)$ **[4 marks]**

A

A*

Chapter summary

In this chapter you have learned how to

- use laws of indices to multiply and divide numbers written in index notation C B
- calculate using negative integer powers, and powers of 0 and 1 B
- carry out calculations with numbers given in standard form B
- use laws of indices to calculate powers and roots of numbers written using index notation A
- multiply and divide expressions involving surds A
- use laws of indices to work out positive and negative fractional powers of whole numbers A A*
- use surds to give exact answers to calculations A A*
- simplify expressions involving surds A A*

17 Sequences and proof

This chapter is about finding the next terms in a sequence, predicting patterns and justifying them.

Meteorologists look at trends and patterns, along with other data, to predict how fast weather fronts move.

Objectives

This chapter will show you how to

- find any term of a sequence given a formula for the nth term **D**
- find the nth term of a linear or simple quadratic sequence **C**
- use counter-examples to disprove a statement **C**
- show step-by-step deduction when proving results **D** **C** **B** **A** **A***

Before you start this chapter

Put your calculator away!

1 Work out the value of each expression when $n = 4$.

a $2n + 8$ b $5n - 12$ c $7n^2$ d $8n^2 - 2n$

HELP Chapter 15

2 Here is a list of numbers.

8 9 20 100 7

Using each of these numbers only once, identify

a an even number b an odd number c a square number

d a power of 2 e a power of 10.

Finding the nth term

Keywords
sequence, term, consecutive, linear sequence, nth term

Why learn this?

Environmental scientists spot trends and form algebraic expressions to predict what will happen in the future.

Objectives

C Find the nth term of a linear sequence

Skills check

1 By looking at differences, find the next two terms in each of these sequences.
   a $-8, -6, -4, \ldots$ b 7, 12, 17, 22, …   c 2.5, 3, 3.5, 4, …

2 The nth term of a sequence is $3n - 5$. Find
   a the 1st term   b the 10th term   c the 50th term.

Sequences

A **sequence** is a list of numbers in a given order.

The numbers in a sequence are called **terms**.

Terms next to one another are called **consecutive** terms.

A **linear sequence** goes up (or down) in equal sized steps.

Multiples

The sequence of multiples of 3 has nth term $3n$.

1st term $= 3 \times 1 = 3$

2nd term $= 3 \times 2 = 6$

3rd term $= 3 \times 3 = 9$

4th term $= 3 \times 4 = 12$

…

The sequence of multiples of 4 has nth term $4n$.

The sequence of multiples of 5 has nth term $5n$.

And so on.

The nth term

The sequence of multiples of 3 goes up in 3s.

3, 6, 9, 12, …
+3 +3 +3

To find the **nth term** of a linear sequence, look at the differences between the terms.

If the difference is 2, compare the sequence with the sequence $2n$.

If the difference is 3, compare the sequence with the sequence $3n$.

C

Example 1

Find the nth term of the sequence 5, 9, 13, 17, …

	1st term	2nd term	3rd term	4th term
Sequence	5	9	13	17
$4n$ (multiples of 4)	4	8	12	16

Work out the difference between consecutive terms (= 4).

Compare the sequence with the sequence for $4n$.

The nth term is $4n + 1$.

Each term in the sequence is 1 more than the term in the sequence $4n$.

C

Example 2

Find the nth term of the sequence 13, 11, 9, 7, …

	1st term	2nd term	3rd term	4th term
Sequence	13	11	9	7
$-2n$ (multiples of -2)	-2	-4	-6	-8

The difference between consecutive terms is -2. Compare the sequence with the sequence for $-2n$.

The nth term is $-2n + 15$ or $15 - 2n$.

Each term in the sequence is 15 more than the term in the sequence $-2n$.

Exercise 17A

C

1 The first four terms of a linear sequence are 3, 5, 7, 9.

a Write down the next three terms.

b Copy and complete the table.

	1st term	2nd term	3rd term	4th term
Sequence	3	5	7	9
Multiples of …				

c Use your table to find the nth term of the sequence.

2 Find the nth term of each sequence.

a 5, 9, 13, 17, …

b 8, 10, 12, 14, …

c 4, 9, 14, 19, …

d 9, 20, 31, 42, …

e 75, 175, 275, 375, …

f $-5, -10, -15, -20, \dots$

g 10, 7, 4, 1, …

h 19, 15, 11, 7, …

i 77, 67, 57, 47, …

j 50, 44, 38, 32, …

Using the nth term

You can use the nth term to calculate any term in a sequence or to check if a particular number is in a sequence.

Example 3

C

Here is a sequence.

5, 9, 13, 17, 21, …

a Find the 50th term of the sequence.

b Is 98 a term in the sequence?

a 5, 9, 13, 17, 21, …

+4 +4 +4 +4

Term	5	9	13	17	21
$4n$	4	8	12	16	20

The sequence goes up in 4s, so compare with the sequence for $4n$.

Each term in the sequence is 1 more than $4n$.

The nth term is $4n + 1$.

50th term $= 4 \times 50 + 1 = 201$

Substitute $n = 50$ in the nth term.

b If 98 is in the sequence, then $4n + 1 = 98$.

$4n = 97$

Solve the equation.

$n = 97 \div 4 = 24.25$

So 98 is not in the sequence.

n must be an integer for 98 to be in the sequence.

Exercise 17B

In each of the questions in this exercise, you will have to work out a formula for the nth term.

C

1 The first four terms of a sequence are 8, 13, 18, 23.

  a Find the nth term of the sequence.

  b Use the nth term to find the 100th term.

2 A botanist records the diameter of a plant's leaf on five consecutive days.

Day	1	2	3	4	5
Diameter (mm)	21	23	25	27	29

  a Find the nth term of the sequence of diameters.

  b What will the diameter of the leaf be on day 10?

C

3 **a** Find the 50th term of the sequence whose first four terms are 1, 3, 5, 7.

  b Find the 15th term of the sequence whose first four terms are 4, 10, 16, 22.

  c Find the 20th term of the sequence whose first four terms are 8, 15, 22, 29.

4 Find the 59th term of the sequence which begins −8, −6, −4, −2, …

AO2

5 Find the 25th term of the sequence which begins −8, −10, −12, −14, …

C

6 The first five terms of a linear sequence are shown below.

3.5, ☐, ☐, 5, 5.5

Find the 220th term.

AO3

7 For each sequence, find the term closest to 150.

a 4, 11, 18, 25, 32, …

b 5, 8, 11, 14, 17, …

c 23, 32, 41, 50, 59, …

d −7, −2, 3, 8, 13, …

C

8 A sequence begins 7, 11, 15, 19, …
One of the terms in the sequence is 47.
Which number term is this?

9 Here is a sequence.

8, 17, 26, 35, 44, …

a Is 134 a term in this sequence?

b Is 198 a term in this sequence?

10 Here is a sequence.

19, 23, 27, 31, 35, …

a 75 is a term in this sequence. What number term is it?

b Explain why 52 cannot be a term in this sequence.

C

11 For each sequence, find the first negative term.

a 112, 101, 90, 79, 68, …

b 47, 41, 35, 29, 23, …

c 220, 208, 196, 184, 172, …

d 58, 51, 44, 37, 30, …

AO3

17.2 Sequences of patterns

Why learn this?

Tiles can be arranged in patterns. If you can find the nth term for the number of tiles needed you can work out how many will be needed for any area.

Objectives

C Find the nth term for a sequence of diagrams

Skills check

1 Find the nth term for each of these sequences.

a 8, 12, 16, 20, …

b 32, 35, 38, 41, …

c −12, −10, −8, −6, …

HELP Section 17.1

Patterns

Sequences of patterns can lead to number sequences.

Example 4

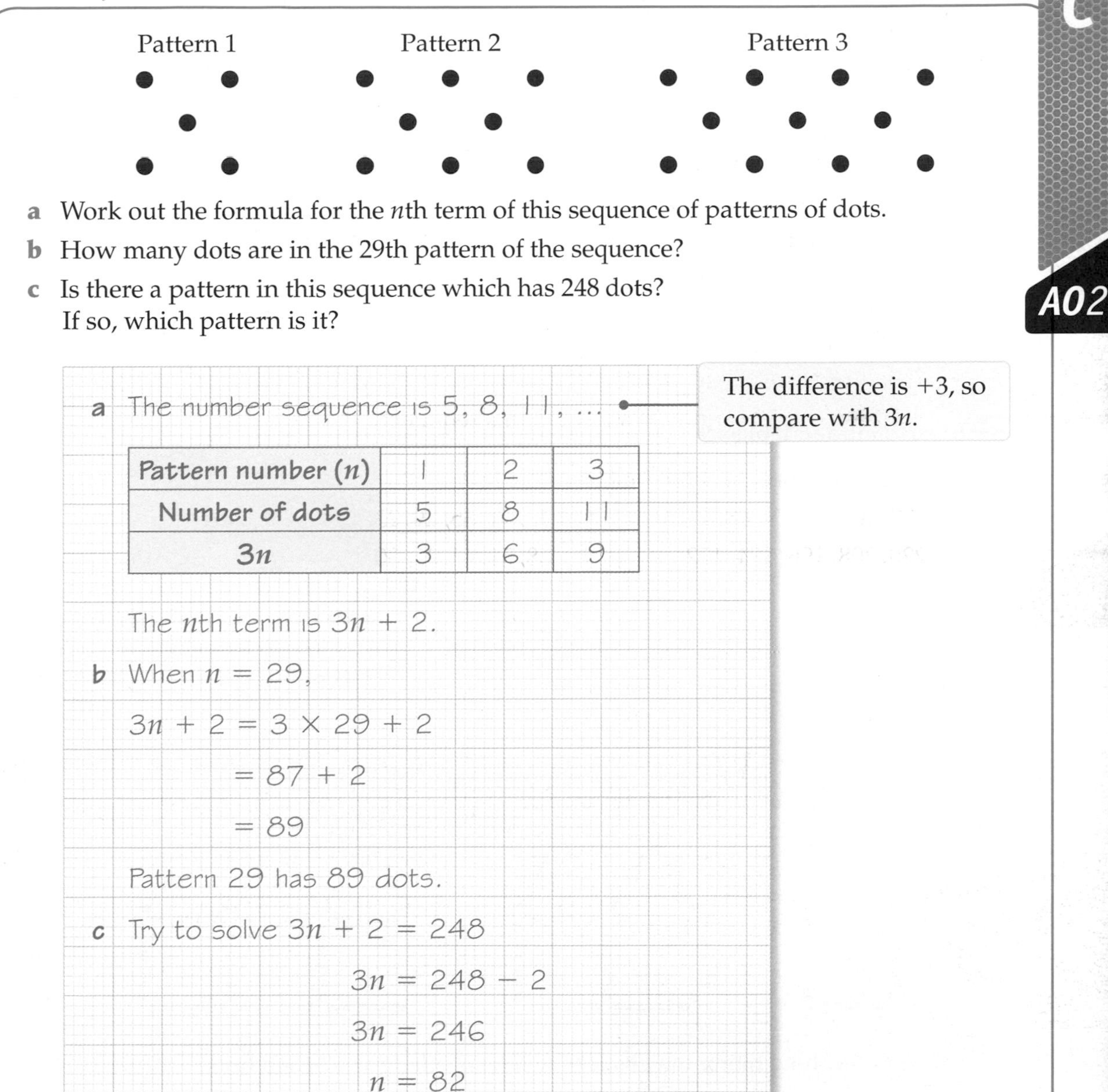

a Work out the formula for the nth term of this sequence of patterns of dots.

b How many dots are in the 29th pattern of the sequence?

c Is there a pattern in this sequence which has 248 dots?
If so, which pattern is it?

a The number sequence is 5, 8, 11, …

The difference is +3, so compare with $3n$.

Pattern number (n)	1	2	3
Number of dots	5	8	11
$3n$	3	6	9

The nth term is $3n + 2$.

b When $n = 29$,

$$3n + 2 = 3 \times 29 + 2$$
$$= 87 + 2$$
$$= 89$$

Pattern 29 has 89 dots.

c Try to solve $3n + 2 = 248$

$$3n = 248 - 2$$
$$3n = 246$$
$$n = 82$$

The pattern with 248 dots is the 82nd pattern.

C

AO2

Exercise 17C

C

1 Matchsticks are used to make the patterns below.

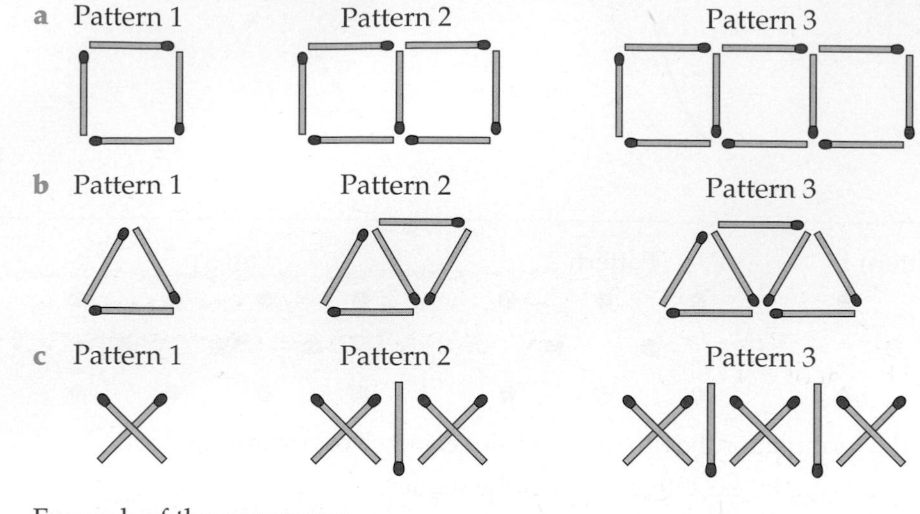

For each of the sequences

- **i** work out how many matchsticks will be needed for the 9th term
- **ii** find a formula for the nth term.

C

2 In a restaurant, tables can be pushed together for larger parties.

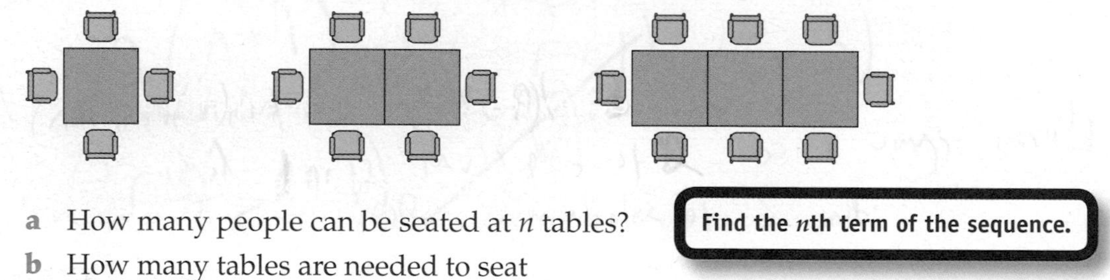

a How many people can be seated at n tables?

b How many tables are needed to seat 24 people?

Find the nth term of the sequence.

3 Beach huts are made using matchsticks.

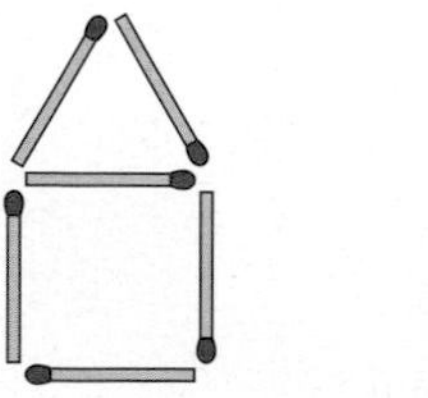 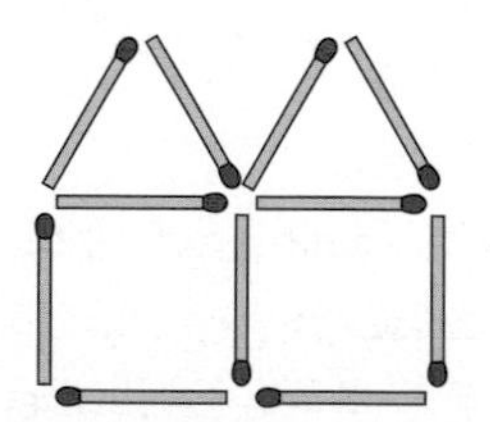 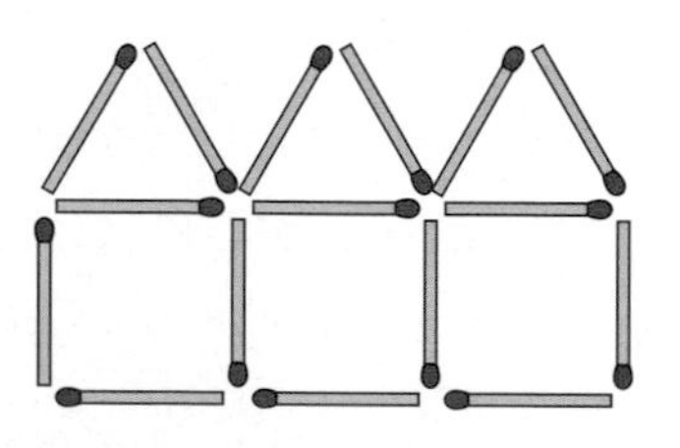

a Work out how many matchsticks will be needed for n beach huts.

b How many beach huts can I make with 90 matches?

4 Here is a pattern of dots.

AO2

One of the patterns contains 132 dots. Which pattern is it?

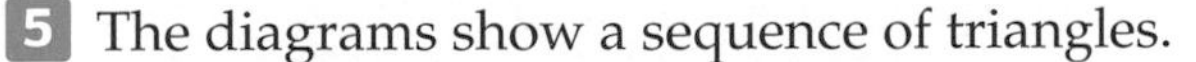

5 The diagrams show a sequence of triangles.

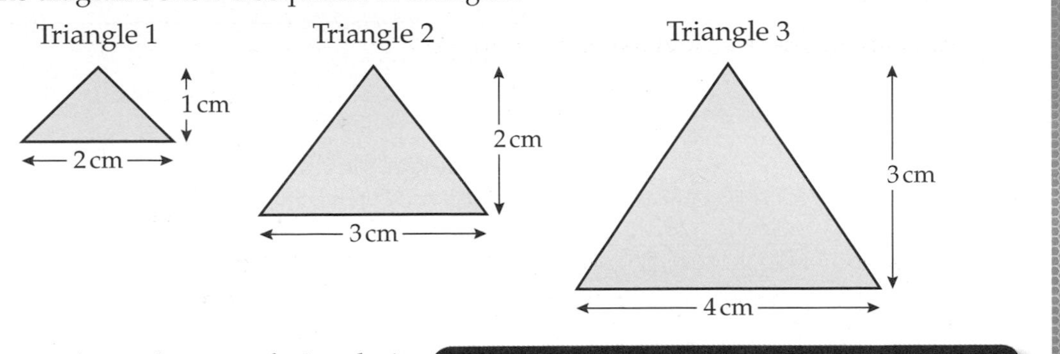

a Work out the area of triangle 4.

b Write down an expression for

i the height of triangle n

ii the base length of triangle n

iii the area of triangle n.

c A triangle in the sequence has area $78\,\text{cm}^2$. What is the height of this triangle?

> **Area of triangle $= \frac{1}{2} \times$ base $\times$ perpendicular height**

C

AO2

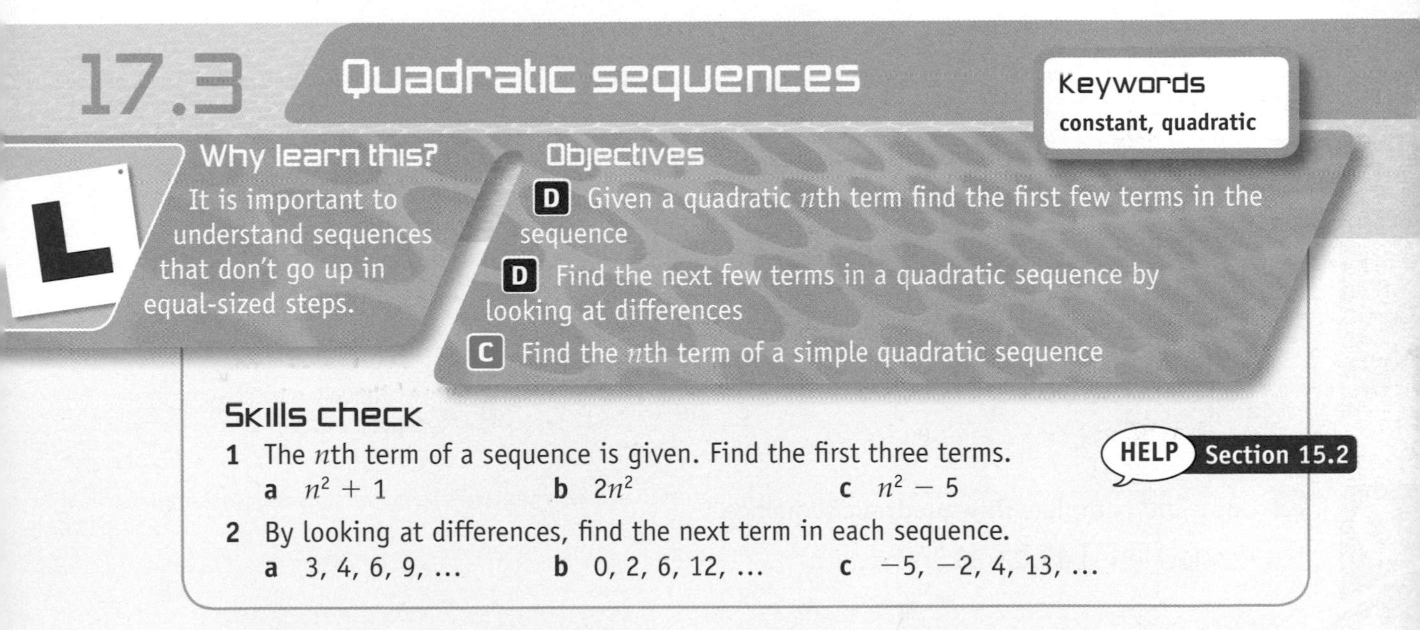

17.3 Quadratic sequences

Keywords
constant, quadratic

Why learn this?
It is important to understand sequences that don't go up in equal-sized steps.

Objectives

D Given a quadratic nth term find the first few terms in the sequence

D Find the next few terms in a quadratic sequence by looking at differences

C Find the nth term of a simple quadratic sequence

Skills check

1 The nth term of a sequence is given. Find the first three terms.

a $n^2 + 1$ **b** $2n^2$ **c** $n^2 - 5$

HELP Section 15.2

2 By looking at differences, find the next term in each sequence.

a 3, 4, 6, 9, … **b** 0, 2, 6, 12, … **c** −5, −2, 4, 13, …

Quadratic sequences

In a linear sequence the terms go up or down in equal size steps.

The difference between consecutive terms is **constant** (the same).

3, 7, 11, 15, … is a linear sequence because the difference between consecutive terms is 4.

The nth term of a linear sequence includes an 'n'.

$3n + 2$, $4n - 5$, $7 - 2n$ are nth terms for linear sequences.

The nth term of a **quadratic** sequence includes an 'n^2'.

Here are the first five terms of the quadratic sequence n^2.

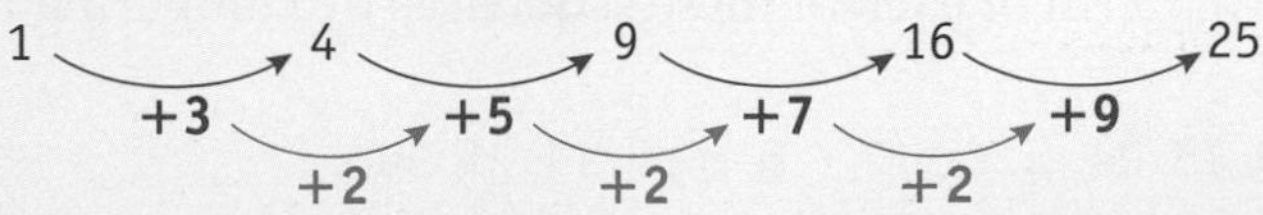

In a quadratic sequence, the difference between consecutive terms is *not* constant.

But the second difference *is* constant.

If the second difference is 2, the nth term includes n^2.

Example 5

Find the nth term of the sequence 5, 8, 13, 20, 29, …

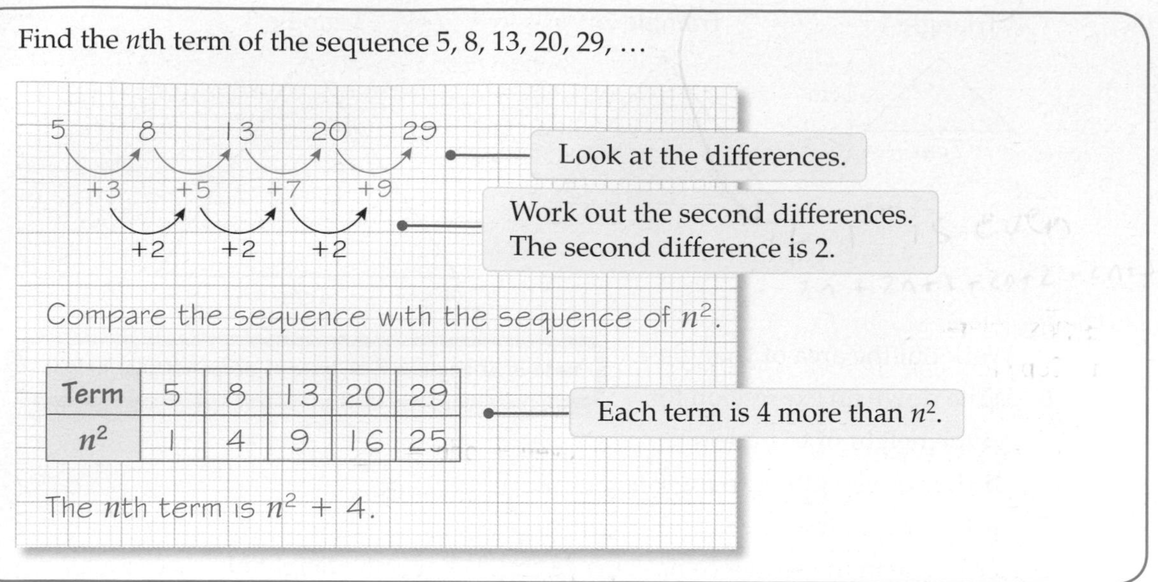

5 8 13 20 29
+3 +5 +7 +9
+2 +2 +2
Look at the differences.
Work out the second differences.
The second difference is 2.
Compare the sequence with the sequence of n^2.
Term 5 8 13 20 29
n^2 1 4 9 16 25
Each term is 4 more than n^2.
The nth term is $n^2 + 4$.

Exercise 17D

1 Find the next three terms in each of these sequences.

a 1, 4, 9 16, … **b** 8, 9, 12, 17, …

c 0, 5, 12, 21, … **d** 102, 104, 108, 114, …

2 Find the first three terms and the 10th term of these sequences given the nth terms.

a n^2 **b** $n^2 - 1$ **c** $2n^2$ **d** $n^2 + 12$

3 Copy and complete this quadratic sequence.

3, ☐, ☐, 15, 23, 33

4 Here is a sequence.

2, 5, 10, 17, 26, …

a Write down the next three terms in the sequence.

b Find the nth term of the sequence.

Work out the differences. What do you notice?

5 Find the nth term of each of these sequences by comparing them with the square numbers.

a 0, 3, 8, 15, 24, … **b** 6, 9, 14, 21, 30, …

c −4, −1, 4, 11, 20, … **d** 2, 8, 18, 32, 50, …

e $\frac{1}{2}$, 2, $4\frac{1}{2}$, 8, $12\frac{1}{2}$, … **f** 10, 40, 90, 160, 250, …

17.4 Proof or verification?

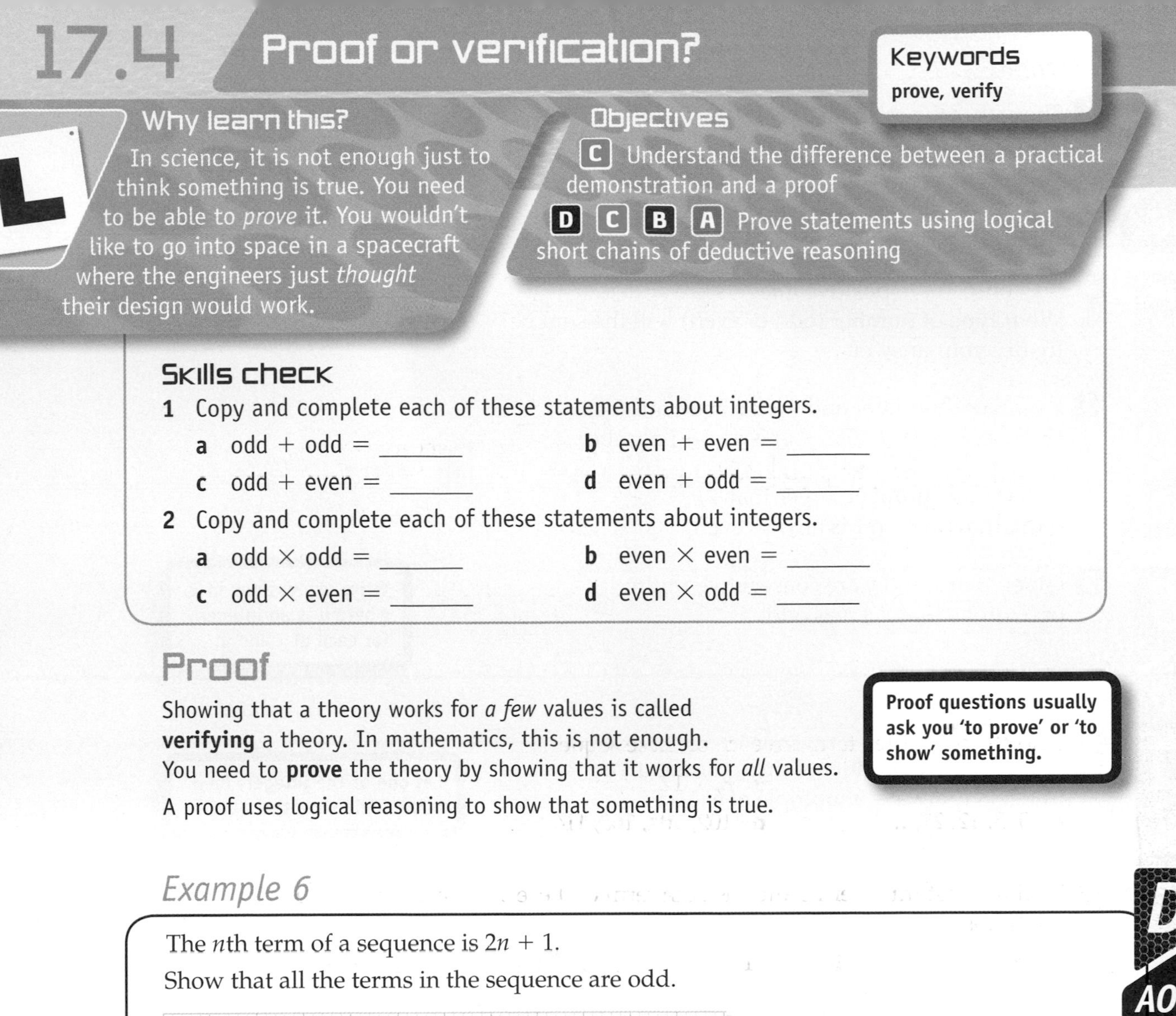

Keywords
prove, verify

Why learn this?

In science, it is not enough just to think something is true. You need to be able to *prove* it. You wouldn't like to go into space in a spacecraft where the engineers just *thought* their design would work.

Objectives

C Understand the difference between a practical demonstration and a proof

D C B A Prove statements using logical short chains of deductive reasoning

Skills check

1 Copy and complete each of these statements about integers.

a odd + odd = ______ **b** even + even = ______

c odd + even = ______ **d** even + odd = ______

2 Copy and complete each of these statements about integers.

a odd × odd = ______ **b** even × even = ______

c odd × even = ______ **d** even × odd = ______

Proof

Showing that a theory works for *a few* values is called **verifying** a theory. In mathematics, this is not enough. You need to **prove** the theory by showing that it works for *all* values.

A proof uses logical reasoning to show that something is true.

Proof questions usually ask you 'to prove' or 'to show' something.

Example 6

D AO2

The nth term of a sequence is $2n + 1$.

Show that all the terms in the sequence are odd.

1st term = $2 \times 1 + 1 = 2 + 1 = 3$
2nd term = $2 \times 2 + 1 = 4 + 1 = 5$
3rd term = $2 \times 3 + 1 = 6 + 1 = 7$

Work out the first few terms of the sequence.

The first three terms are odd. They go up in 2s. 'Odd number' + 2 = 'odd'.

Explain why all the terms will be odd. Use facts you know about odd and even numbers.

Example 7

C AO2

t represents an odd number greater than 1.

Prove that $t^3 - 1$ is even.

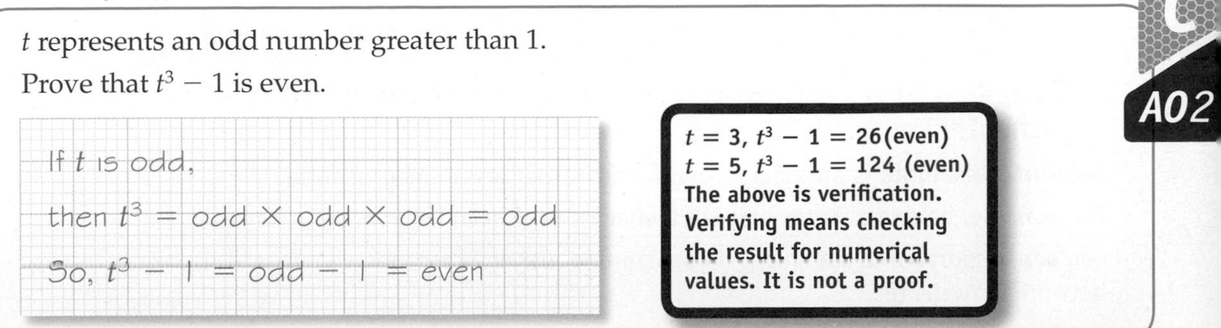

If t is odd,
then t^3 = odd × odd × odd = odd
So, $t^3 - 1$ = odd − 1 = even

$t = 3$, $t^3 - 1 = 26$ (even)
$t = 5$, $t^3 - 1 = 124$ (even)
The above is verification. Verifying means checking the result for numerical values. It is not a proof.

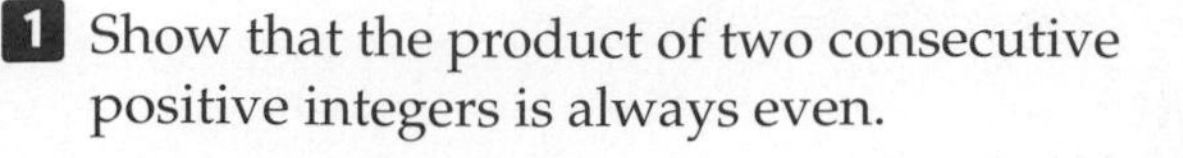

Exercise 17E

D
AO2

1 Show that the product of two consecutive positive integers is always even.

> **Consider whether the numbers involved are odd or even and what will happen when they are added or multiplied.**

C

2 Six consecutive positive integers are added.
What type of number (odd or even) will the sum be?
Justify your answer.

3 x is an odd number and y is an even number.
Explain why $(x + y)(x - y)$ is an odd number.

AO2

4 n is a prime number larger than 2.
Explain why $n^2 + 1$ is always even.

B

5 Given that x and y are consecutive multiples of 3, prove that $x + y$ is odd.

> **Write expressions in n, where n is an integer, for each of x and y.**

AO3

A

6 Show that the product of three consecutive integers is always a multiple of 6.

> **Let one of the integers be n. What could the others be?**

AO3

17.5 Using counter-examples

Keywords
counter-example

'All teenagers use social networking websites'

Why learn this?

Generalisations are often made in the media. Finding a counter-example is a good way of disproving statements like these.

Objectives

C Show that something is false by using a counter-example

Skills check

HELP Section 17.3

1. The general term of a sequence is $n^2 - 5$.
   Which of the following numbers are *not* terms in the sequence?
   4, −2, 6.5, 20, 31, 7
2. t represents a number. Write expressions in t for the next four consecutive numbers.

Counter-examples

A **counter-example** is an example which shows that a statement is false.

For example, here is a statement: 'All teenagers use social networking websites.'

You can disprove this statement if you can find *just one* teenager who doesn't use social networking websites.

Example 8

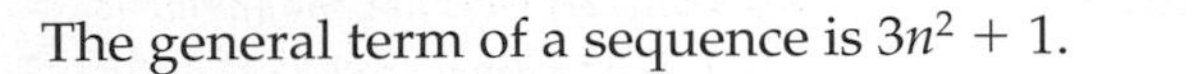

The general term of a sequence is $3n^2 + 1$.

Jeremy says that all the terms in the sequence are even.

Prove that he is wrong.

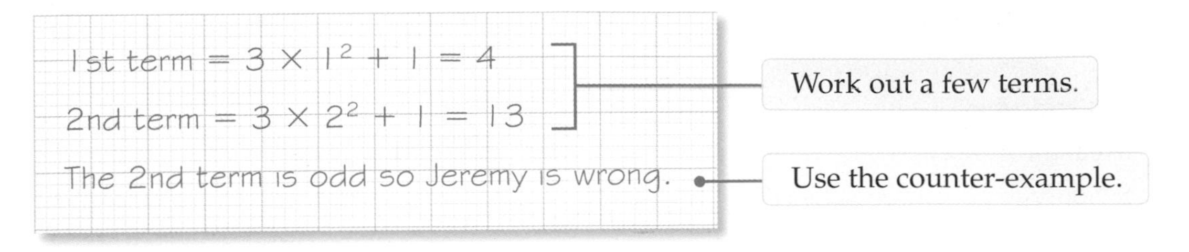

1st term = 3 × 1² + 1 = 4
2nd term = 3 × 2² + 1 = 13
The 2nd term is odd so Jeremy is wrong.
Work out a few terms.
Use the counter-example.

Exercise 17F

1 Give a counter-example to show that each of these statements is false.

- **a** Adding a positive number to a negative number always gives a positive result.
- **b** The product of an integer and a decimal is always a decimal.
- **c** The sum of two decimals is always a decimal.

2 An answer, rounded to one decimal place is 7.6.
Iram claims that the number *must* have been larger than 7.55.
Give a counter-example to show that she is wrong.

3 Decide whether each of these statements is true or false.
If they are false give a counter-example.

- **a** All the terms in the sequence with nth term $n^2 + 1$ are odd.
- **b** If p is odd, $(p + 1)(p - 1)$ is always even.
- **c** The cube of any number is greater than 0.
- **d** All odd numbers are prime.
- **e** If n is an integer, then $n(n + 1)(n + 2)$ is always even.

4 A website claims, 'If x and y are prime, then $x^2 + y^2$ is an even number'.
Explain why the claim is false.

5 'The square of any number is smaller than the cube.'
Give a counter-example to show that this is false.

6 Show that the statement $\sqrt{x} < x^2$ is not true for all values of x.

7 The nth term of a sequence is $n^2 + n + 1$.
Eric claims that all the terms in the sequence are prime.
Prove that he is wrong.

Proof using algebra

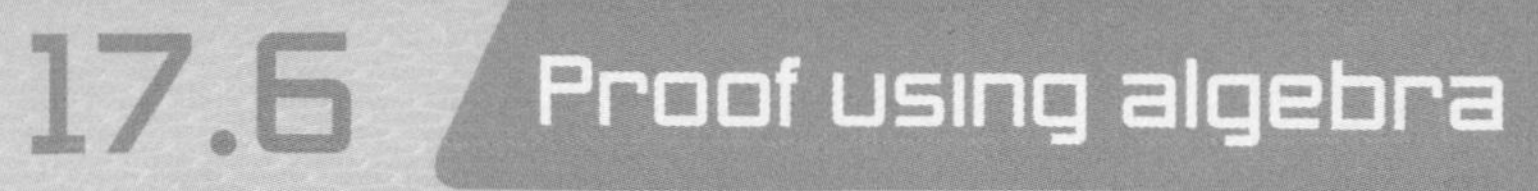

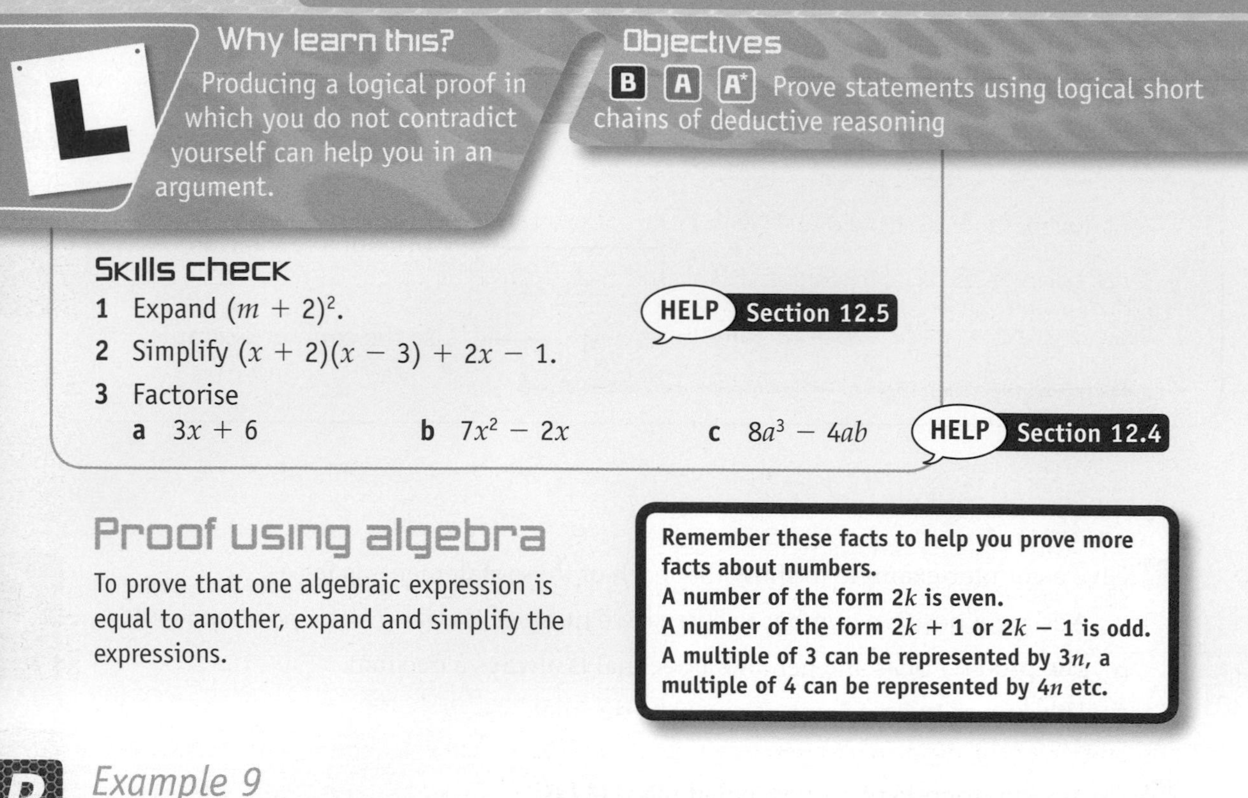

Why learn this?

Producing a logical proof in which you do not contradict yourself can help you in an argument.

Objectives

B A A* Prove statements using logical short chains of deductive reasoning

Skills check

1 Expand $(m + 2)^2$.

2 Simplify $(x + 2)(x - 3) + 2x - 1$.

3 Factorise

a $3x + 6$ b $7x^2 - 2x$ c $8a^3 - 4ab$

HELP Section 12.5

HELP Section 12.4

Proof using algebra

To prove that one algebraic expression is equal to another, expand and simplify the expressions.

Remember these facts to help you prove more facts about numbers.
A number of the form $2k$ is even.
A number of the form $2k + 1$ or $2k - 1$ is odd.
A multiple of 3 can be represented by $3n$, a multiple of 4 can be represented by $4n$ etc.

B AO3

Example 9

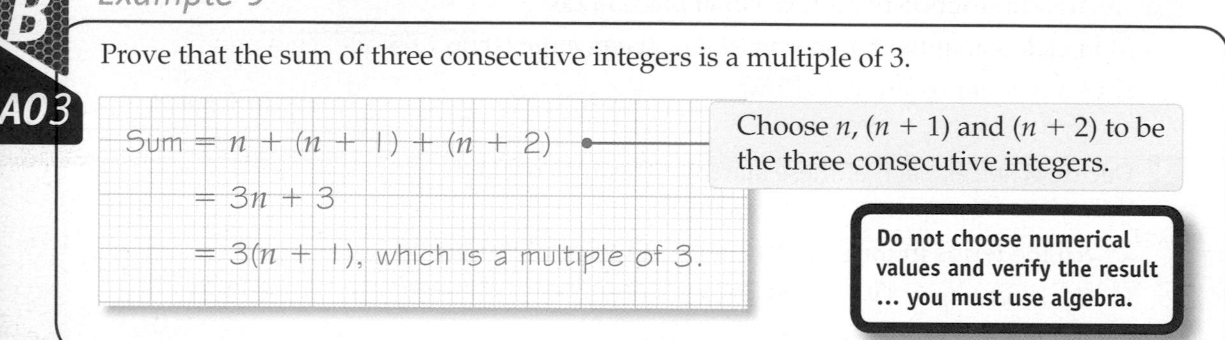

Prove that the sum of three consecutive integers is a multiple of 3.

$$\text{Sum} = n + (n + 1) + (n + 2)$$
$$= 3n + 3$$
$$= 3(n + 1), \text{ which is a multiple of 3.}$$

Choose n, $(n + 1)$ and $(n + 2)$ to be the three consecutive integers.

Do not choose numerical values and verify the result … you must use algebra.

B AO2

Example 10

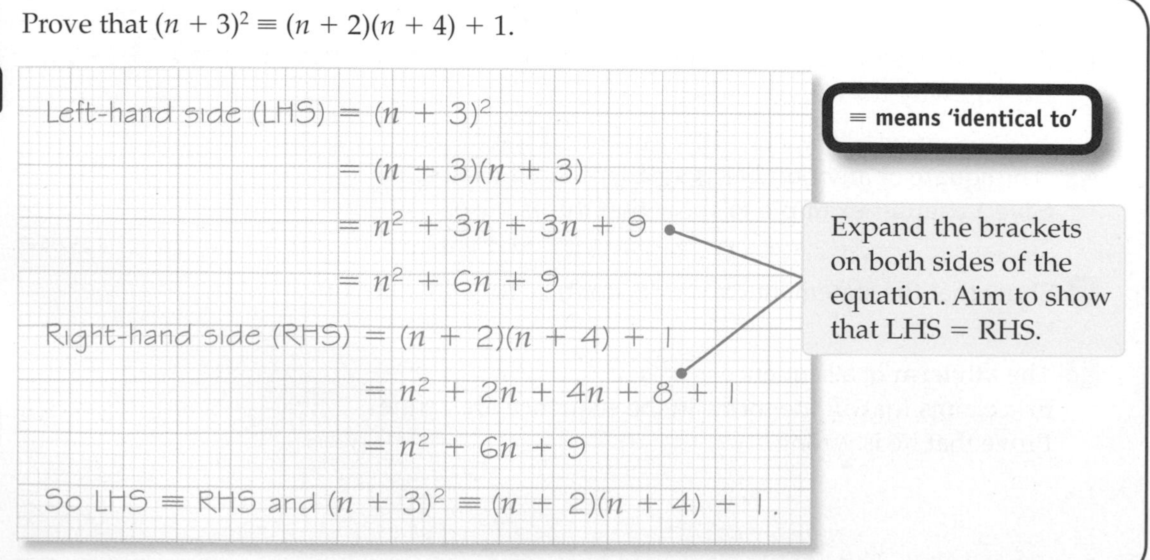

Prove that $(n + 3)^2 \equiv (n + 2)(n + 4) + 1$.

$$\text{Left-hand side (LHS)} = (n + 3)^2$$
$$= (n + 3)(n + 3)$$
$$= n^2 + 3n + 3n + 9$$
$$= n^2 + 6n + 9$$
$$\text{Right-hand side (RHS)} = (n + 2)(n + 4) + 1$$
$$= n^2 + 2n + 4n + 8 + 1$$
$$= n^2 + 6n + 9$$

So LHS $\equiv$ RHS and $(n + 3)^2 \equiv (n + 2)(n + 4) + 1$.

$\equiv$ means 'identical to'

Expand the brackets on both sides of the equation. Aim to show that LHS = RHS.

Exercise 17G

1 Prove that

a the sum of two odd numbers is even

b the sum of two consecutive multiples of 3 is odd

c the difference of two consecutive square numbers is odd

d the product of an odd and an even number is always even.

B

AO2

2 Prove that $(2n + 1)^2 \equiv (n + 3)(n + 1) + 3n^2 - 2$.

3 Prove that the square of any odd number is odd.

A

4 The sum of ten consecutive positive integers, where the smallest integer is a, is $10a + 45$.

a Show that the statement is true for $a = 1$.

b Given that the first term of the sequence is a, write down algebraic expressions for the other nine terms.

c Hence prove the statement above.

AO2

5 The first four terms of sequence A are $-1, 3, 7, 11, \ldots$

The first four terms of sequence B are $8, 11, 14, 17, \ldots$

Sequence C is found by adding the corresponding terms in sequence A and sequence B.

a Verify that the terms in sequence C are all multiples of 7.

b Prove this result.

A

AO3

6 The nth term for the sequence of triangular numbers is $\frac{1}{2}n(n + 1)$.

Prove that the sum of two consecutive triangular numbers is a square number.

Review exercise

1 Aimee is playing a game with her brother Jon.
He thinks of an integer, she doubles it and subtracts 1.
Aimee wins if the outcome is odd.
Jon says the game is unfair as he cannot win. Explain why this is true. **[2 marks]**

D

2 Here is a sequence.

$-8, 2, 12, 22, \ldots$

What is the nth term of the sequence? **[2 marks]**

C

3 A pattern is made using tiles.

Pattern 1 Pattern 2 Pattern 3

a Copy and complete the table.

Pattern number	1	2	3	4	5	10	n
Number of tiles	$1 \times 3 = 3$	$2 \times \square = \square$					

[4 marks]

b Which number pattern contains 80 tiles? [2 marks]

AO2

4 n and q are prime numbers. Adrian says that nq will always be odd.
Find a counter-example to show that he is wrong. [2 marks]

C

5 Here is a sequence.

$-1, 2, 7, 14, \ldots$

a Write down the next three terms in the sequence. [1 mark]

b Find the nth term of the sequence. [1 mark]

C

6 Rachel says that $m^3 + 1$ cannot be a multiple of 5.
Explain why she is wrong. [2 marks]

AO2

B

7 Prove that $(2n - 3)^2 \equiv (n + 2)^2 + (3n - 1)(n - 5)$ [4 marks]

AO2

A

8 Prove that an odd number added to its square will always be even. [4 marks]

AO3

A*

9 Prove that the product of two consecutive even numbers is a multiple of 8. [4 marks]

10 n is an integer. Prove that

$$(n + 3)(2n + 1) + (n - 2)(2n + 1)$$

is always 1 more than a multiple of 8. [5 marks]

AO3

Chapter summary

In this chapter you have learned how to

- find the first few terms in the sequence given a quadratic nth term **D**
- find the next few terms in a quadratic sequence by looking at differences **D**
- find the nth term of a linear sequence **C**
- find the nth term for a sequence of diagrams **C**
- find the nth term of a simple quadratic sequence **C**
- understand the difference between a practical demonstration and a proof **C**
- show that something is false by using a counter-example **C**
- prove statements using logical short chains of deductive reasoning **D** **C** **B** **A** **A***

18 Percentages

This chapter is about percentages.

If you don't pay off your credit card bill each month, you will be charged a high percentage interest on the amount you owe.

This chapter will show you how to

- calculate a percentage increase or decrease **D**
- perform calculations involving VAT **D**
- calculate a percentage profit or loss **C**
- perform calculations involving a repeated percentage change **C**
- perform calculations involving finding the original quantity **B**

Before you start this chapter

Put your calculator away!

1 Work out

a 0.27×100

b $42.5 \div 100$

2 Work out $5 \div 8$.

3 Cancel $\frac{15}{75}$ to its lowest terms.

4 Convert $\frac{2}{5}$ to a decimal.

5 Convert 0.36 to a fraction.

6 Work out $\frac{17}{200} \times 100$.

7 Hazel buys a trombone with a cash price of £1500. The credit terms are a deposit of 20% and 36 monthly payments of £45.

a What is the total credit price?

b How much extra is paid for credit compared with the cash price?

P Number skills: percentage calculations

1 a Write 57% as a decimal.
b Write 0.008 as a percentage.
c Write 65% as a fraction.
d Write $\frac{5}{200}$ as a percentage.

2 Write 0.3, $\frac{1}{3}$, and 32% in order of size, starting with the smallest.

3 Write 0.38, 37%, $\frac{2}{5}$, 0.36, $\frac{3}{8}$ in order of size, starting with the smallest.

4 Ava scored 15 out of 20 in a test.
Ben scored 80% in the same test.
Who got more questions correct?

5 Copy and complete this table of equivalent fractions, decimals and percentages.

Percentage	Fraction	Decimal
35%		
	$\frac{1}{20}$	
		0.125

6 Work out

a 4% of 56 kg
b 75% of £800
c 6% of 30 litres
d 2% of £6000

Remember 1% = $\frac{1}{100}$, 10% = $\frac{1}{10}$, 25% = $\frac{1}{4}$, 50% = $\frac{1}{2}$, 75% = $\frac{3}{4}$

7 Ali earns £27 600 per year and pays 21% of this in income tax.
How much income tax does she pay?

8 In Market Street there are 160 houses. 85% of them have broadband.
How many houses in the street do not have broadband?

9 Work out

a 6 as a percentage of 10
b 3 as a percentage of 20
c 48 as a percentage of 60
d 54 as a percentage of 300
e 6 mm as a percentage of 2 cm
f 380 g as a percentage of 0.4 kg.

Always cancel fractions when you can.

Make sure the units of both quantities are the same.

10 A school has 300 students in Years 10 and 11.

Year group	Boys	Girls
Year 10	72	69
Year 11	75	84

Use the information in the table to work out what percentage of the students are

a Year 10 boys
b Year 11 girls
c Year 10
d boys.

18.1 Percentage increase and decrease

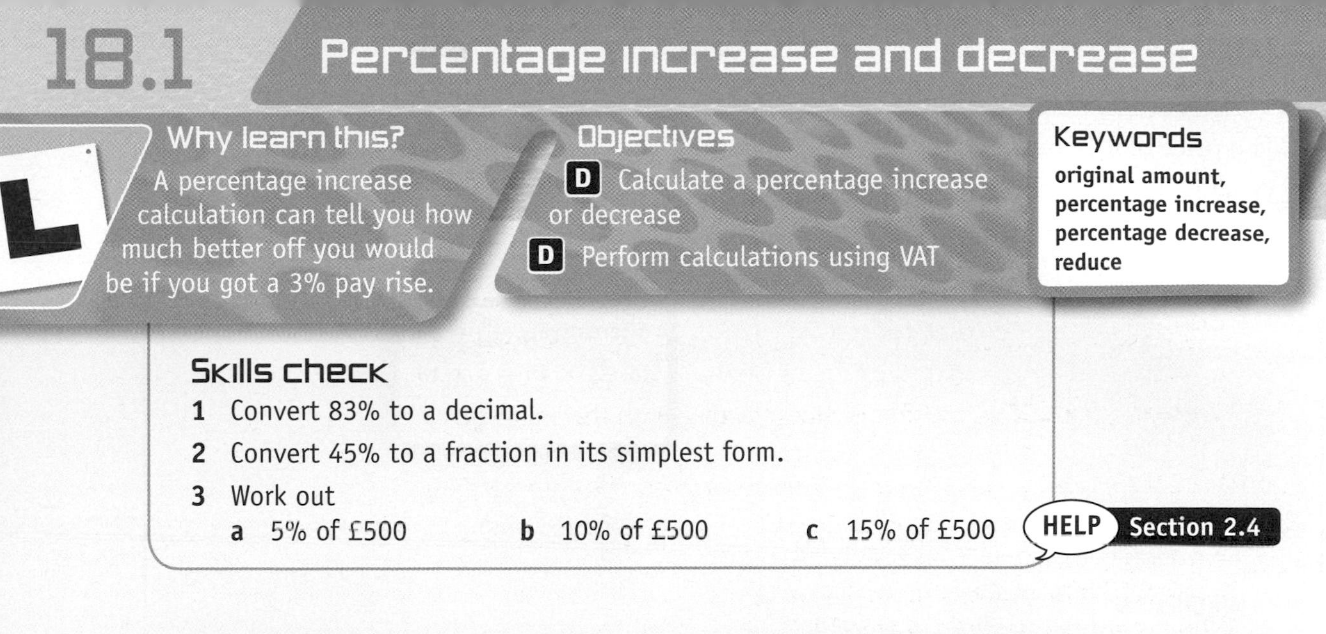

Why learn this?
A percentage increase calculation can tell you how much better off you would be if you got a 3% pay rise.

Objectives

D Calculate a percentage increase or decrease

D Perform calculations using VAT

Keywords
original amount, percentage increase, percentage decrease, reduce

Skills check

1 Convert 83% to a decimal.

2 Convert 45% to a fraction in its simplest form.

3 Work out
 a 5% of £500 b 10% of £500 c 15% of £500

HELP Section 2.4

Percentage increase and decrease

Method A

1 Work out the value of the increase (or decrease).

2 Add it to (or subtract it from) the **original amount**.

This method is most commonly used when working without a calculator.

Method B

1 Add the **percentage increase** to 100% (or subtract the **percentage decrease** from 100%).

2 Convert this percentage to a decimal.

3 Multiply it by the original amount.

This method is especially useful when using a calculator.

Example 1

D

Tim used to earn £460 a week. He has had a 4% pay rise.
What does he earn now?

Method A

Pay rise = 4% of £460

$= \frac{4}{100} \times 460$

$= \frac{1840}{100}$

= £18.40

New pay = £460 + £18.40

= £478.40

Method B

Increase = 4%

New pay = 100% + 4% of old pay

= 104% of old pay

= 1.04 × old pay

= 1.04 × 460

= £478.40

Divide by 100 to convert a percentage to a decimal.

$$\begin{array}{r} 460 \\ \times\ 104 \\ \hline 1840 \\ 46000 \\ \hline 47840 \end{array}$$

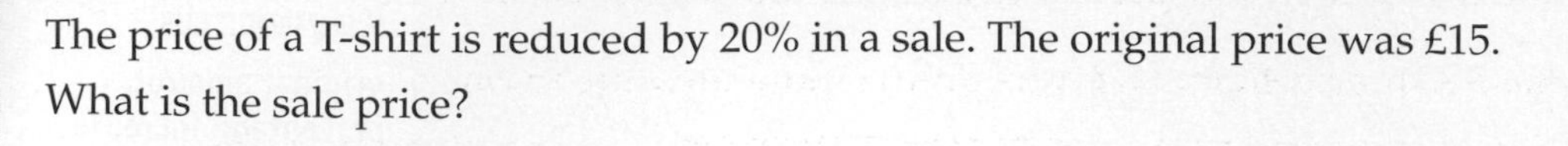

Example 2

The price of a T-shirt is reduced by 20% in a sale. The original price was £15.
What is the sale price?

Method A

Decrease $= 20\%$ of £15

$= \frac{20}{100} \times 15$

$= \frac{300}{100}$

$=$ £3

Sale price $=$ £15 $-$ £3 $=$ £12

> **Alternatively,** $\frac{20}{100} = \frac{1}{5}$
> **so** $\frac{20}{100} \times 15 = \frac{1}{5} \times 15$
> $= \frac{15}{5} = 3$

Method B

Decrease $= 20\%$

Sale price $= 100\% - 20\%$

$= 80\%$

$= 0.8$

Sale price $= 0.8 \times 15 =$ £12

Divide by 100 to convert a percentage to a decimal.

> **Alternatively,** $0.8 = \frac{8}{10} = \frac{4}{5}$
> **so** $0.8 \times 15 = \frac{4}{5} \times 15 = 12$

Exercise 18A

D

1 **a** Increase £95 by 10%. **b** Increase 42 mm by 5%.
c Decrease £160 by 10%. **d** Increase £300 by 7%.
e Decrease 75 m*l* by 8%. **f** Decrease 50 miles by 12%.

2 Sanjeev wants to buy a CD priced at £12.
He gets a 10% discount with his student card.
How much does he have to pay?

3 Nigel is fitting new skirting board round his living room.
He measures the total length he needs as 17 m.
He decides to buy 10% more than this to allow for cutting and wastage.
How much skirting board should he buy?

4 Nita earns £285.00 per week. She gets a 4% pay rise. How much will she earn now?

5 A new car costs £9600. After two years it has lost 40% of its value.
What is it worth now?

6 A pair of jeans, priced at £35, is reduced by 15% in a sale.
What is the sale price?

7 In April 2008, 175 000 new cars were sold in the UK.
In April 2009, the figure was 25% down on the previous year.
How many cars were sold in April 2009?

8 A sum of £200 is increased by 10%, and then the new amount is decreased by 10%. Will the final amount be greater or less than the original £200?

D

9 Simon sees three different adverts for the same pressure washer.

Dumbo's DIY	Rock Bottom	*Suit you, Sir*
Pressure washer	**Pressure washer**	**Pressure washer**
Normally £99	**£79**	Normally £120
20% off	*Unbeatable value*	35% off

Which is the best buy?

AO2

VAT

VAT stands for Value Added Tax. It is a tax that is added to the price of most items in shops and to many other services.

VAT is calculated as a percentage. Generally it is 17.5% in the UK.

> **VAT at 17.5% can be worked out by finding 10% + 5% + 2.5%.**

Example 3

D

A plasma screen television is advertised for sale at £800 + 17.5% VAT.
How much will you have to pay?

10% of £800 = £80 — To work out 17.5% of £800, start by finding 10%.

5% of £800 = £40 — Halve it to find 5%.

2.5% of £800 = £20 — Halve it again to find 2.5%.

So 17.5% of £800 = £80 + £40 + £20
= £140

So the total price = £800 + £140
= £940

Exercise 18B

D

1 A builder is calculating the total bill for a new conservatory that she has recently installed. It comes to £15 500 + 17.5% VAT.
What is the total bill?

2 A child car seat is advertised at £110 + 5% VAT. What is the total price?

3 A car service comes to £280 + 17.5% VAT.
What is the total bill?

4 A meal for four people comes to £84.80 + 17.5% VAT.

a What is the total bill?

b The four people share the bill equally. How much will each pay?

18.2 Percentage profit or loss

Why learn this?
Shop managers use this to compare the profit made on different items.

Objectives
C Calculate a percentage profit or loss

Keywords
cost price, selling price, profit, loss, percentage profit, percentage loss, depreciation

Skills check

HELP Section 2.4

1 Work out

a 9 as a percentage of 20 **b** 72 as a percentage of 80

c 294 as a percentage of 300 **d** 1050 as a percentage of 1750.

Profit and loss

A shopkeeper buys items from a wholesaler at **cost price**.

The shopkeeper sells the items at the **selling price**.

When you make money on the sale of an item you make a **profit**.

When you lose money on the sale of an item you make a **loss**.

You can use **percentage profit** (or **loss**) to compare the profitability of items.

$$\text{Percentage profit (or loss)} = \frac{\text{actual profit (or loss)}}{\text{cost price}} \times 100\%$$

where actual profit = selling price − cost price

and actual loss = cost price − selling price

The cost price is the original price of the item. It goes on the bottom of the fraction.

C

Example 4

A DIY store buys 5 litre cans of white emulsion paint for £14.00 each and sells them for £17.50.

What is the store's percentage profit?

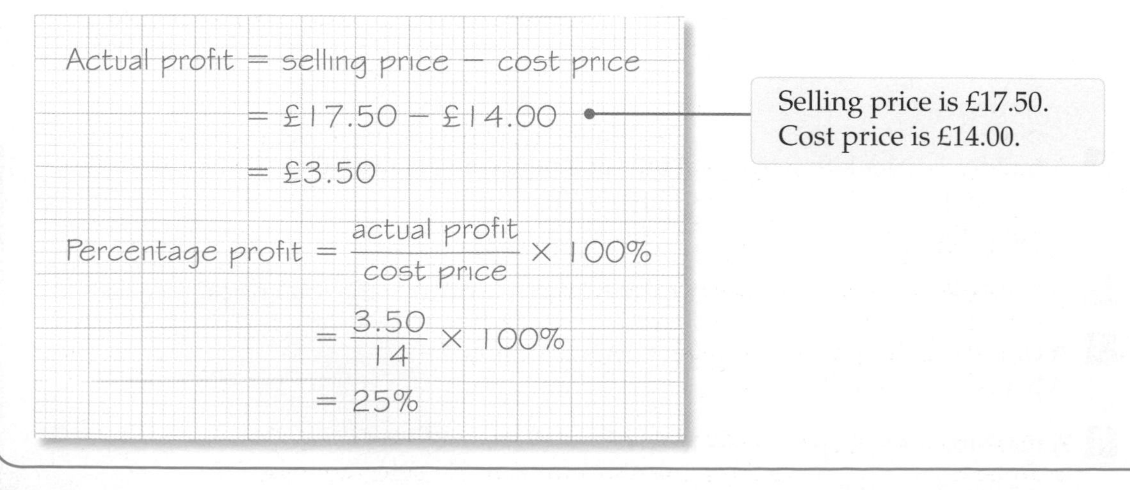

Actual profit = selling price − cost price

= £17.50 − £14.00

= £3.50

$$\text{Percentage profit} = \frac{\text{actual profit}}{\text{cost price}} \times 100\%$$

$$= \frac{3.50}{14} \times 100\%$$

= 25%

Selling price is £17.50. Cost price is £14.00.

Example 5

Omar bought a car for £7500 and sold it two years later for £4500.

What was the percentage depreciation in the value of the car?

> When objects lose value over time, the loss is called **depreciation**.

Actual loss = cost price − selling price

= £7500 − £4500

= £3000

Cost price is £7500.
Selling price is £4500.

$$\text{Percentage depreciation} = \frac{\text{actual loss}}{\text{cost price}} \times 100\%$$

$$= \frac{3000}{7500} \times 100\%$$

$$= \frac{2}{5} \times 100\%$$

$$= 40\%$$

So the percentage depreciation is 40%.

C

Exercise 18C

1 Find the percentage profit or loss for each item.

	Pair of trousers	House	Barbecue set	Car	Television
Cost price	£30	£125 000	£40	£8000	£350
Selling price	£42	£118 750	£60	£6800	£455

2 A collector bought an antique table for £360 and sold it for £420.
What was her percentage profit?

3 Phoebe bought a games console for £160 and later sold it to her friend for £140.
What was her percentage loss?

4 Colin restores lawnmowers. He bought a petrol mower for £5 and sold it for £45.
What was his percentage profit?

C

5 Two friends are restoring old cars.
Sarah buys one for £400, restores it and sells it for £750.
George buys one for £1200, restores it and sells it for £2000.
Which of them makes the bigger percentage profit?

C

6 Gardens-R-Us buys Wellington boots for £15 and sells them for £21.
Yuppies Shoe Shop buys the same boots for £16 and sells them for £22.
Which shop makes the larger percentage profit?

AO2

18.3 Repeated percentage change

Keywords
compound interest

Why learn this?
Demographers use repeated percentage change to predict population growth.

Objectives
C Perform calculations involving repeated percentage changes

Skills check

1. Jon puts £3000 into a savings account. The interest rate is 4% per annum. Work out the amount gained through simple interest after one year.
2. Lucy puts £2500 into a savings account. The interest rate is 5% per annum. Work out the amount gained through simple interest after four years.

HELP Section 18.1

Compound interest

Generally when you invest money, the interest is calculated using **compound interest.**

In Section 8.1 you learned how to work out repeated percentage change using a calculator. Here you will practise these calculations without a calculator.

Compound interest is interest paid on the amount *plus* on the interest already earned.

C

Example 6

Helen puts £1000 into a savings account earning 5% per annum compound interest. How much does she have after two years?

Year 1

Amount at start of year 1 = £1000

Interest in year 1 = 5% of £1000
= 0.05 × £1000
= £50

Or 5% of £1000 = $\frac{5}{100}$ × £1000

Year 2

The interest is added to the amount in the account.

Amount at start of year 2 = £1000 + £50
= £1050

Interest in year 2 = 5% of £1050
= 0.05 × £1050
= £52.50

Or 5% of £1050 = $\frac{5}{100}$ × £1050

So after two years the amount in her account
= £1050 + £52.50
= £1102.50

Alternatively, you could use 1.05 as a multiplier.
£1000 × 1.05 × 1.05 = £1102.50
See Section 8.1.

Example 7

It is estimated that a mountain bike loses 10% of its value each year.

A new mountain bike costs £650.

Calculate its value after two years.

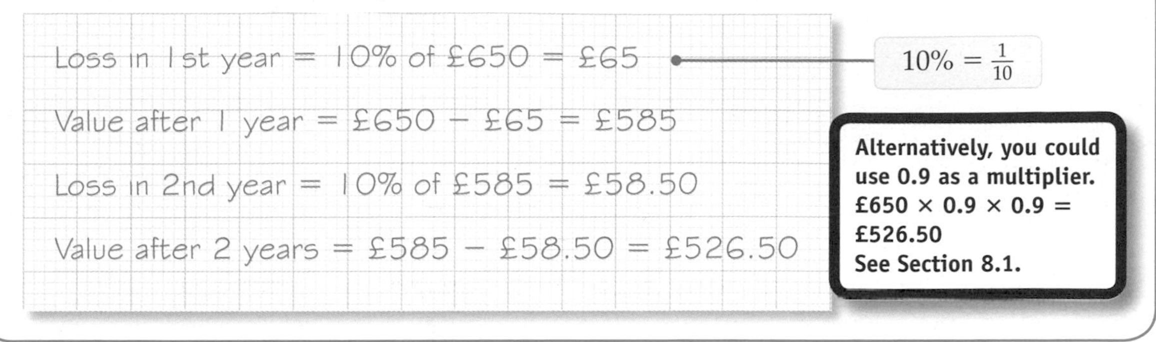

Loss in 1st year = 10% of £650 = £65
Value after 1 year = £650 − £65 = £585
Loss in 2nd year = 10% of £585 = £58.50
Value after 2 years = £585 − £58.50 = £526.50
10% = 1/10
Alternatively, you could use 0.9 as a multiplier.
£650 × 0.9 × 0.9 = £526.50
See Section 8.1.

Exercise 18D

1 Jenny invests £500 at a compound interest rate of 4% per annum.
How much will she have at the end of two years?

2 Paul invests £300 at a compound interest rate of 10% per annum.
How much will he have after three years?

3 Work out the compound interest on

a £120 invested for two years at a rate of 10% per annum

b £2000 invested for three years at a rate of 5% per annum.

4 The number of rabbits in a field increases at the rate of 20% each month.
If there were 50 rabbits two months ago, how many are there now?

5 The value of a car depreciates by 20% each year.
It was worth £10 000 when it was new three years ago.
How much is it worth now?

'Depreciates' means it loses value.

6 There are 5000 whales of a certain species.
Scientists believe that their numbers are reducing by about 10% each year.
Estimate how many whales there will be in two years' time.

7 A colony of bacteria is increasing in a Petri dish at the rate of 50% each hour.
A biologist estimates that there are one million bacteria when she starts her clock.
How many does she estimate there will be three hours later?

8 Hiroshi invests £400 at a rate of 8% per annum compound interest.
Amy invests £380 at a rate of 10% per annum compound interest.
Who has more money after two years?

9 A bank offers an account where interest at a rate of 10% is added every six months.
What would be the equivalent percentage rate if the interest was added once a year?

This is known as the Annual Percentage Rate (APR).

C AO3

10 Sarah wants to invest £100. She compares two investment accounts.
Account A: Compound interest rate is 3%, added every six months.
Account B: Compound interest rate is 6% per annum.
Which account should she choose?

18.4 Reverse percentages

Keywords
reverse percentage, original quantity

Why learn this?

If you know the sale price of a pair of jeans, you can work backwards and find the original price.

Objectives

B Perform calculations involving finding the original quantity

Skills check

1 a Increase £78 by 10%.
b Decrease £80 by 6%.
c Increase 25 mm by 5%.
d Increase 75 km by 4%.
e Decrease 1500 m*l* by 2%.
f Decrease £48 by 20%.

HELP Section 18.1

Finding the original quantity

In Section 8.2 you learned how to work out **reverse percentage** problems using a calculator. Here you will practise these calculations without a calculator.

In reverse percentage problems, you are told the quantity after a percentage increase or decrease and you have to work out the **original quantity**.

Consider this question, for example.

A dress is reduced by 20% in a sale. The sale price is £36.
What was the original price?

There are two methods for answering these questions.

Method A

1 Work out what percentage the final quantity represents.
2 Divide by this percentage to find 1%.
3 Multiply by 100 to get the 100% figure.

This method is most commonly used when working without a calculator.

Method B

1 Work out what percentage the final quantity represents.
2 Divide by 100 to get the multiplier.
3 Divide the final quantity by the multiplier.

This method is especially useful when using a calculator.

Example 8

B

Attendance at a school prom was 30% down on last year's figure.

This year there were 140 students.

How many attended last year?

You do not have last year's figure, which is the original amount.

Method A

1 $100\% - 30\% = 70\%$

140 students is the 70% figure.

Work out what percentage the figure you are given represents. '30% down' means this figure represents 70%.

2 So $1\% = 140 \div 70$

$= 2$ students

Divide by this percentage to find 1%.

3 So $100\% = 2 \times 100$

$= 200$ students

Multiply by 100 to get the 100% figure.

Check: 30% of $200 = 60$

$200 - 60 = 140$ students

Method B

1 $100\% - 30\% = 70\%$

140 students is the 70% figure.

Work out what percentage the figure you are given represents.

2 Multiplier $= 70 \div 100$

$= 0.7$

Divide by 100 to find the multiplier.

3 Original amount $= 140 \div 0.7$

$= 200$ students

Divide the original amount by the multiplier.

B

Example 9

George has had a 10% pay rise.
His new salary is £352 per week.
What was his salary before his pay rise?

Again you do not have the original amount, the salary before the rise.

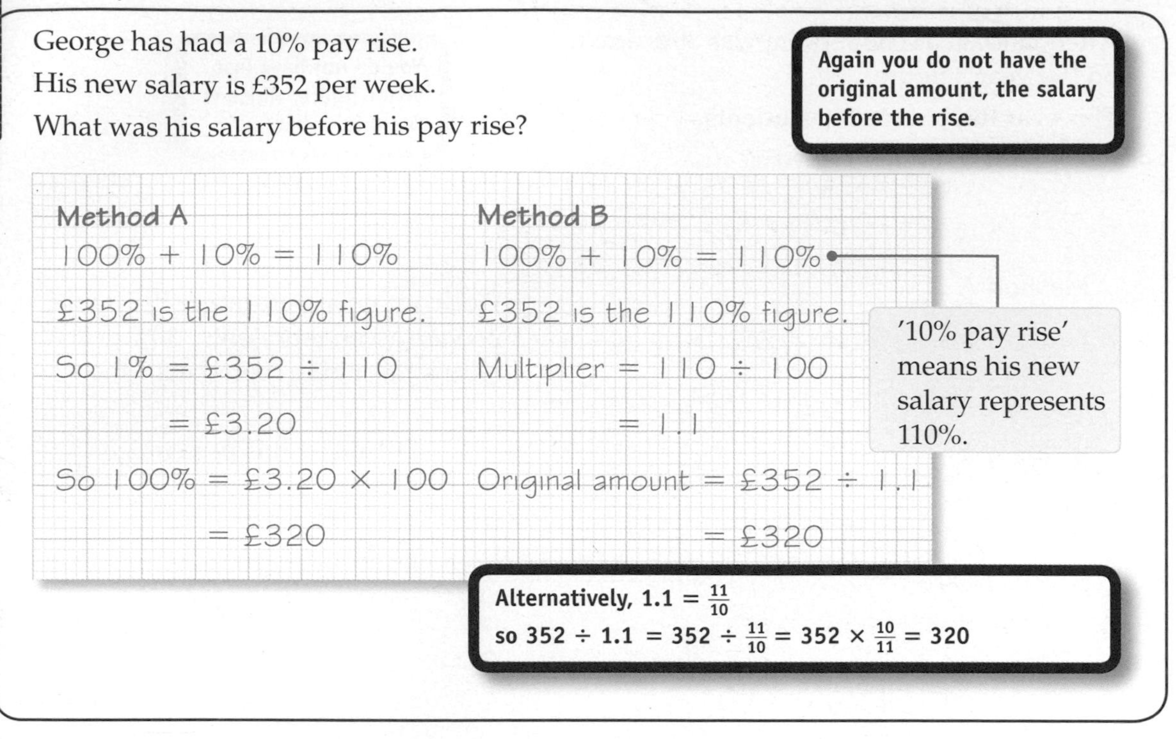

Method A
100% + 10% = 110%
£352 is the 110% figure.
So 1% = £352 ÷ 110
= £3.20
So 100% = £3.20 × 100
= £320
Method B
100% + 10% = 110%
£352 is the 110% figure.
Multiplier = 110 ÷ 100
= 1.1
Original amount = £352 ÷ 1.1
= £320
'10% pay rise' means his new salary represents 110%.

Alternatively, $1.1 = \frac{11}{10}$
so $352 \div 1.1 = 352 \div \frac{11}{10} = 352 \times \frac{10}{11} = 320$

Exercise 18E

B

1 A shop has reduced all its items by 20% in a sale.
Work out the full price of each of these items.

- **a** shirt, sale price £15.60
- **b** jeans, sale price £31.20
- **c** scarf, sale price £16.80
- **d** three pairs of socks, sale price £7.92

2 Helen's car insurance is £350.
This includes a 60% 'no claims' discount.
How much would the insurance cost without any discount?

3 The value of a car depreciates by 20% each year.
A one-year-old car is worth £7680.
How much was it worth when new?

'Depreciates' means that it loses value.

4 A train company has put up all its fares by 10%.
A ticket for a particular journey now costs £28.60.
What was the ticket price before the rise?

5 Peter has booked an eight-day package holiday for £1020.
The price includes a 20% surcharge for peak season.
What is the price without the surcharge?

A 'surcharge' is an additional amount added to the price.

B
AO2

6 John bought a trumpet in a sale for £405.
He knew it had been reduced by 10%.
John said, 'I saved £40.50 by buying the trumpet in the sale.'
Explain why John is wrong.

7 On the island of Mathia, a new 20% tax is placed on petrol.
A litre of petrol now costs 132 Mathia dollars.
What was the price before the new tax?

8 Scientists believe that the population of wildebeest in a particular game reserve is declining at 10% each year.
They estimate that the present population is about 55 000 animals.

a Estimate what the population will be in one year's time.

b Estimate what the population was one year ago.

9 Two years ago, Andy opened a savings account with a sum of money.
The account earned 10% per annum compound interest.
He now has £6050 in his account.
What was the original sum he opened the account with?

B

B

AO2

Review exercise

D

1 Sarah books a train ticket that would normally cost her £45.
She has a rail card that gives her a 20% discount.
How much will she have to pay? **[2 marks]**

2 A football club had 120 members.
After a publicity campaign, the number of members increased by 15%.
How many members did the club have after the campaign? **[2 marks]**

3 A bill for plumbing repairs comes to £360 + 17.5% VAT.
What is the total bill? **[2 marks]**

4 In 2008, the population of a town was 84 000.
By 2009 the population had decreased by 3%.
Work out the population of the town in 2009. **[3 marks]**

C

5 Lars bought a house for £120 000. Three years later, he sold it for £156 000.
What was the percentage profit? **[3 marks]**

6 Will bought a car for £6400 and sold it three years later for £3840.
What was the percentage depreciation in the value of the car? **[3 marks]**

7 The mayor is worried.
The population of rats in his town is increasing at the rate of about 20% each year.
There are about 100 000 rats now.
Estimate how many there will be in three years' time. **[3 marks]**

8 The value of a car depreciates by 20% each year.
It was worth £8000 when it was new three years ago.
How much is it worth now? **[3 marks]**

B

9 The attendance at a school dance is 20% up on last year.
This year 150 students attended.
How many attended last year? **[2 marks]**

B

10 The numbers of fish in a lake are increasing at the rate of 20% each year.
There are estimated to be 600 fish in the lake this year.

a Estimate how many there will be in one year's time. **[2 marks]**

b Estimate how many there were one year ago. **[2 marks]**

c Estimate how many there were two years ago. **[2 marks]**

11 An LCD TV costs £616.
It has decreased in price by 23% from its price last year.
What did it cost last year? **[2 marks]**

B
AO2

12 Nita has opened a savings account with her bank with some money her granny has given her. The compound interest rate is 10% per annum.
She has calculated that in two years' time she will have £2420.
How much has she put in her account? **[3 marks]**

Chapter summary

In this chapter you have learned how to

- calculate a percentage increase or decrease **D**
- perform calculations using VAT **D**
- calculate a percentage profit or loss **C**
- perform calculations involving repeated percentage changes **C**
- perform calculations involving finding the original quantity **B**

A02 Example – Algebra

A

This Grade A question challenges you to use your algebra – setting up simple equations to represent a problem algebraically, then finding its solution.

The numbers in this sequence increase by equal amounts.

☐ a ☐ ☐ 30 ☐ ☐ ☐

a Show that the first term in the sequence is $\frac{4a}{3} - 10$.

b Write an expression, in terms of a, for the 32nd term in the sequence.

A02

What maths should you use?

a There are three **gaps** between the **terms** a and 30.

The gap between each term $= \frac{30 - a}{3}$

$= 10 - \frac{a}{3}$ — Identify the structure of the sequence.

1st term $= a - \left(10 - \frac{a}{3}\right)$

$= a - 10 + \frac{a}{3}$

$= \frac{4a}{3} - 10$ — Subtract the gap from the 2nd term to get the 1st.

b 32nd term = 1st term + 31 gaps — Be careful … work out the number of gaps that need to be added on.

= 2nd term + 30 gaps

$= a + 30\left(10 - \frac{a}{3}\right)$ — The algebra is easier if you add 30 gaps to the 2nd term rather than 31 gaps to the 1st term.

$= a + 300 - 10a$

$= 300 - 9a$

Take care with the algebra.

A03 Question – Algebra

A

Now try this A03 Grade A question. You have to work it out from scratch. READ THE QUESTION CAREFULLY.
It's similar to the A02 example above, so think about where to start.

The numbers in this sequence decrease by equal amounts.

200 ☐ ☐ ☐ ☐ t ☐ ☐ ☐

Write an expression, in terms of t, for the 46th term in the sequence.

Be careful with the signs when dealing with terms inside brackets.

A03

FUNCTIONAL • FUNCTIONAL • FUNCTIONAL •

Camping with wolves

As part of his Duke of Edinburgh award, Carlos has decided to go to the Carpathian Mountains in Slovakia to help with a wolf conservation project. He will be camping while helping with the project.

He books the evening non-stop flight from London Heathrow, on Saturday 15 March. His return flight is on Saturday 22 March.

Question bank

1 What is the cost of his return flight?

Before he leaves, Carlos plans to buy

- the lightest tent he can find
- the cheapest sleeping bag that will be warm enough at night
- £125 worth of euros
- other camping equipment costing a total of £86.

2 Which tent and sleeping bag do you recommend he should buy? Give reasons for your answers.

Carlos says, 'If I round all the prices to the nearest £10, I estimate the total cost of the camping trip is going to be £580.'

3 Is Carlos correct? Show working to support your answer.

Information bank

Return flight prices from London Heathrow (LHR) and London Gatwick (LGW) to Slovakia

Depart	Date	Time	Number of stops	Return flight price
LHR	Sat 15 March	06 50	0	£195
LGW	Sat 15 March	13 40	0	£215
LHR	Sat 15 March	14 00	1	£234
LHR	Sat 15 March	18 30	0	£242
LGW	Sat 22 March	08 50	1	£199
LHR	Sat 22 March	13 50	0	£224

Information bank

Tents

Name of tent	Weight (kg)	Pack dimensions length × width (cm)	Internal height (cm)	Colour	Price (£)
Trekker	2.40	48 × 20	120	Green	199.99
Sport	2.05	53 × 27	95	Red	29.99
Getup	2.05	51 × 14	115	Green	39.99
Shadow	3.15	46 × 32	110	Blue	59.99
Beta	2.35	49 × 12	95	Brown	79.99

Sleeping bags

Name of sleeping bag	Lowest temperature	Pack dimensions length × width (cm)	Price (£)	Customer rating
Arctic square	−12°C	37 × 24	84.99	*
Nordic	−4°C	36 × 19	19.99	****
Greenland	−10°C	45 × 23	54.99	***
Ultralight	0°C	44 × 22	16.99	****
Warm 'n' cosy	−5°C	28 × 17	32.99	*****
Deep down	−24°C	42 × 20	99.99	**

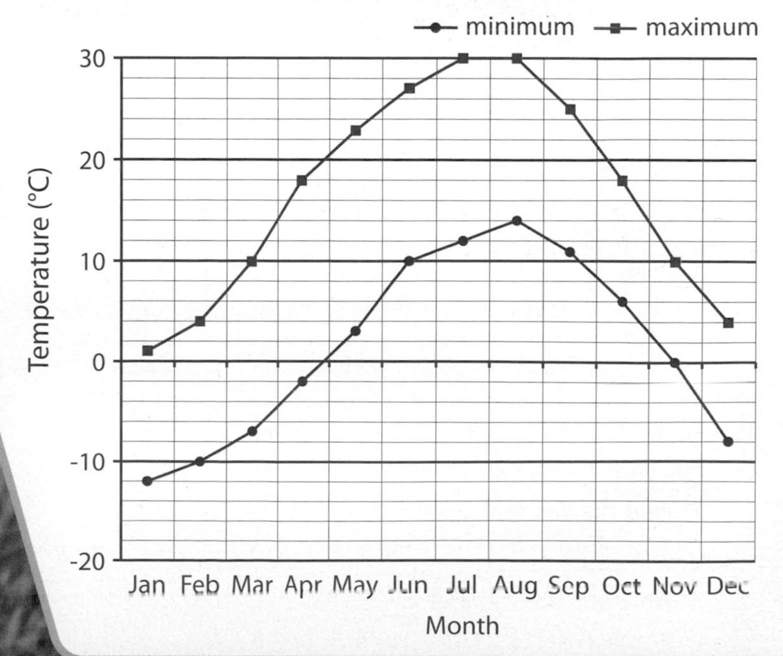

19

Linear graphs

BBC Video

This chapter explores coordinates and graphs and their real-life applications.

Video game designers use coordinates and the equations of straight lines to define characters' movements within a game world.

Objectives

This chapter will show you how to

- find the coordinates of the mid-point of a line segment D C
- recognise (when values are given for m and c) that equations of the form $y = mx + c$ correspond to straight-line graphs in the coordinate plane D C
- plot graphs of functions in which y is given explicitly in terms of x, or implicitly (no table or axes given) D C
- discuss and interpret graphs modelling real situations D C B
- interpret simultaneous equations as lines and their common solution as the point of intersection B
- solve linear inequalities in two variables and find the solution set B
- construct linear functions and plot the corresponding graphs arising from real-life problems D C A
- explore the gradients of parallel lines and lines perpendicular to each other D C B A
- understand that the form $y = mx + c$ represents a straight line and that m is the gradient of the line and c is the value of the y-intercept C B A

Before you start this chapter

Put your calculator away!

1 Find the value of y when $x = -2$.

a $y = 2x$ b $y = x + 3$ c $y = x - 1$ d $y = 2x - 4$

2 Work out the value of c.

a $c + 4 = 13$ b $5 + c = 3$ c $7 = -3 + c$ d $3 = 3(2) + c$

HELP Chapter

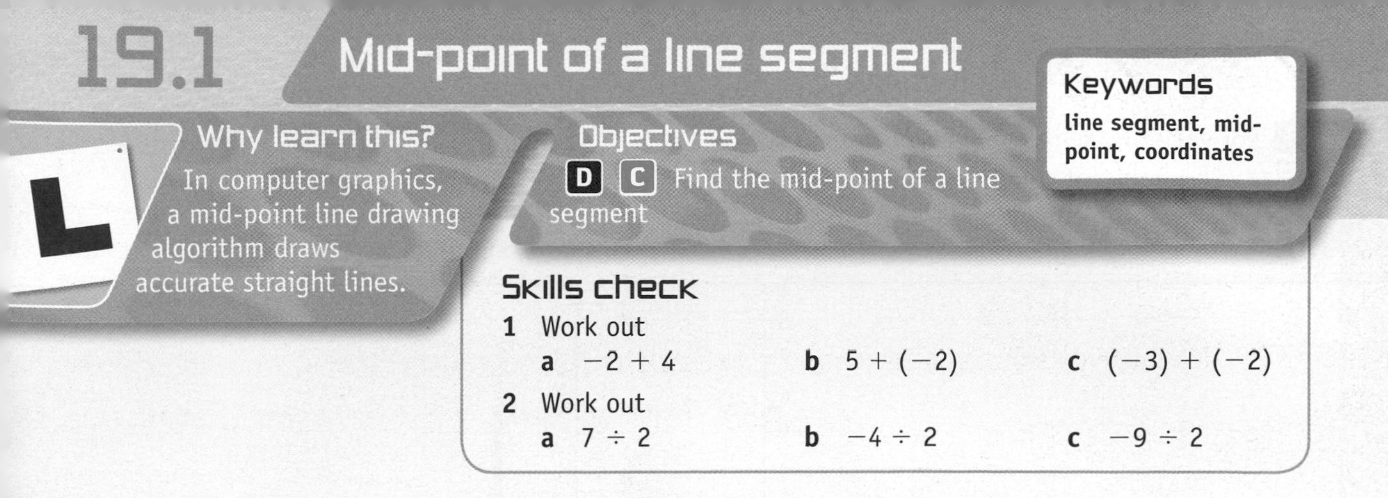

19.1 Mid-point of a line segment

Keywords
line segment, mid-point, coordinates

Why learn this?
In computer graphics, a mid-point line drawing algorithm draws accurate straight lines.

Objectives
D C Find the mid-point of a line segment

Skills check

1 Work out
 a $-2 + 4$ b $5 + (-2)$ c $(-3) + (-2)$

2 Work out
 a $7 \div 2$ b $-4 \div 2$ c $-9 \div 2$

Mid-point of a line segment

A **line segment** is the line between two points.

The **mid-point** of a line segment is exactly half way along the line.

You can find the coordinates of the mid-point using the **coordinates** of the end-points of the line segment.

$$\text{Mid-point } (x, y) = \left(\frac{x_1 + x_2}{2}, \frac{y_1 + y_2}{2}\right)$$

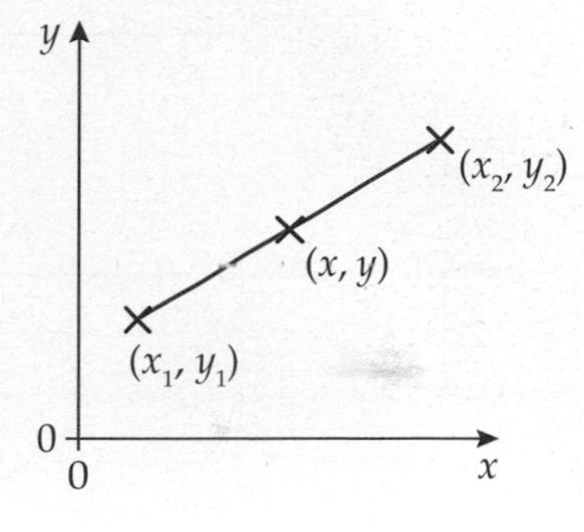

Example 1

Work out the coordinates of the mid-point, M, of the line AB.

Coordinates of A: (1, 7)

Coordinates of B: (5, 6)

$$\text{Mid-point } (x, y) = \left(\frac{1+5}{2}, \frac{7+6}{2}\right)$$

$$= \left(\frac{6}{2}, \frac{13}{2}\right)$$

$$= \left(3, 6\frac{1}{2}\right)$$

The coordinates of a point can be fractions.

Exercise 19A

D

1 For each line segment shown on the grid

a write the coordinates of the end-points

b work out the coordinates of the mid-point.

You can use the same formula with coordinates that have negative values.

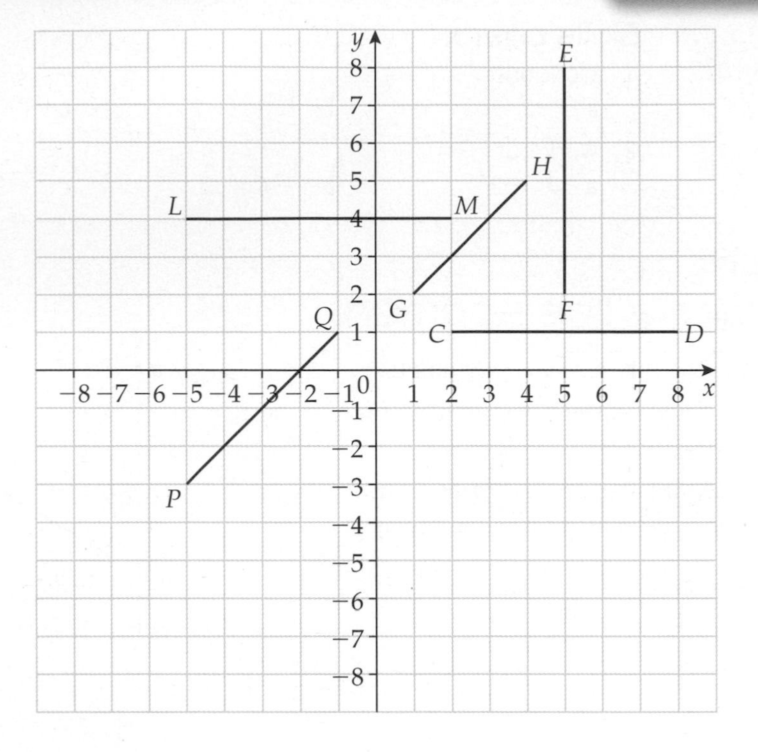

C

2 Work out the coordinates of the mid-point of the line segment joining each pair of points.

a $G(1, 1)$ and $H(8, 1)$

b $L(2, 3)$ and $M(5, 9)$

c $S(-2, 2)$ and $T(3, -3)$

d $U(-3, -5)$ and $V(7, -5)$

You can draw a sketch to check that your answer is correct.

C AO2

3 $ABCD$ is a quadrilateral with coordinates $A(-1, 2)$, $B(4, 2)$, $C(4, -3)$, $D(-1, -3)$.

a Work out the coordinates of the mid-point of the diagonal AC.

b Work out the coordinates of the mid-point of the diagonal BD.

c What can you say about quadrilateral $ABCD$?

C AO3

4 The coordinates of the mid-point of a line segment PQ are $(1, 2)$. Work out the coordinates of Q when P is at

a $(1, 4)$ **b** $(-1, 2)$

c $(6, 2)$ **d** $(1, -3)$

Plotting straight-line graphs

Keywords
linear function, coordinate pair, straight-line graph

Why learn this?
Straight-line graphs can help us model and analyse real-life situations such as mobile phone tariffs.

Objectives

D C Plot graphs of linear functions

D Work out coordinates of points of intersection when two graphs cross

Skills check

HELP Section 15.3

1 Work out the value of each expression when $x = -2$.

a $4x$ **b** $3x + 2$ **c** $2x - 4$

Graphs of linear functions

$y = 2x + 1$ is a **linear function.**

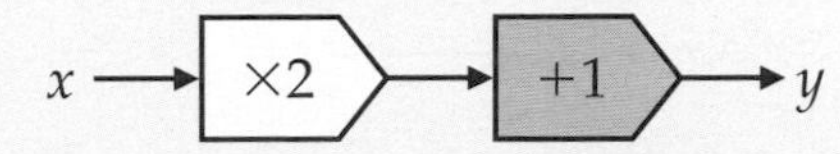

Put in a value of x and you get a value for y.

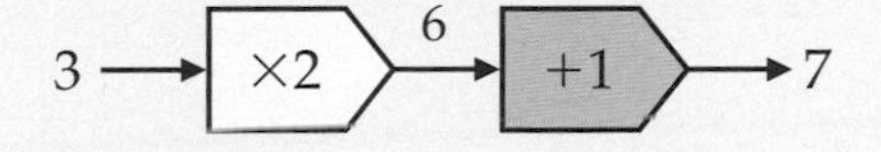

You can draw a graph of the function $y = 2x + 1$.

1. Substitute values for x.
2. Write the values of x and the corresponding values of y in a table.
3. Plot the (x, y) **coordinate pairs** on a grid.
4. Join the points with a **straight line.**

Example 2

D

Draw the graph of $y = 2x + 2$ for values of x from -3 to $+2$.

Step 1: Draw a table of values with values of x from -3 to $+2$.

x	−3	−2	−1	0	1	2
y						

Step 2: Substitute the values of x into the equation and work out the corresponding values of y. Write the values in the table.

x	−3	−2	−1	0	1	2
y	−4	−2	0	2	4	6

Substituting $x = -3$ into $y = 2x + 2$ gives $-6 + 2 = -4$.

Substituting $x = 2$ into $y = 2x + 2$ gives $4 + 2 = 6$.

This is the coordinate pair (0, 2).

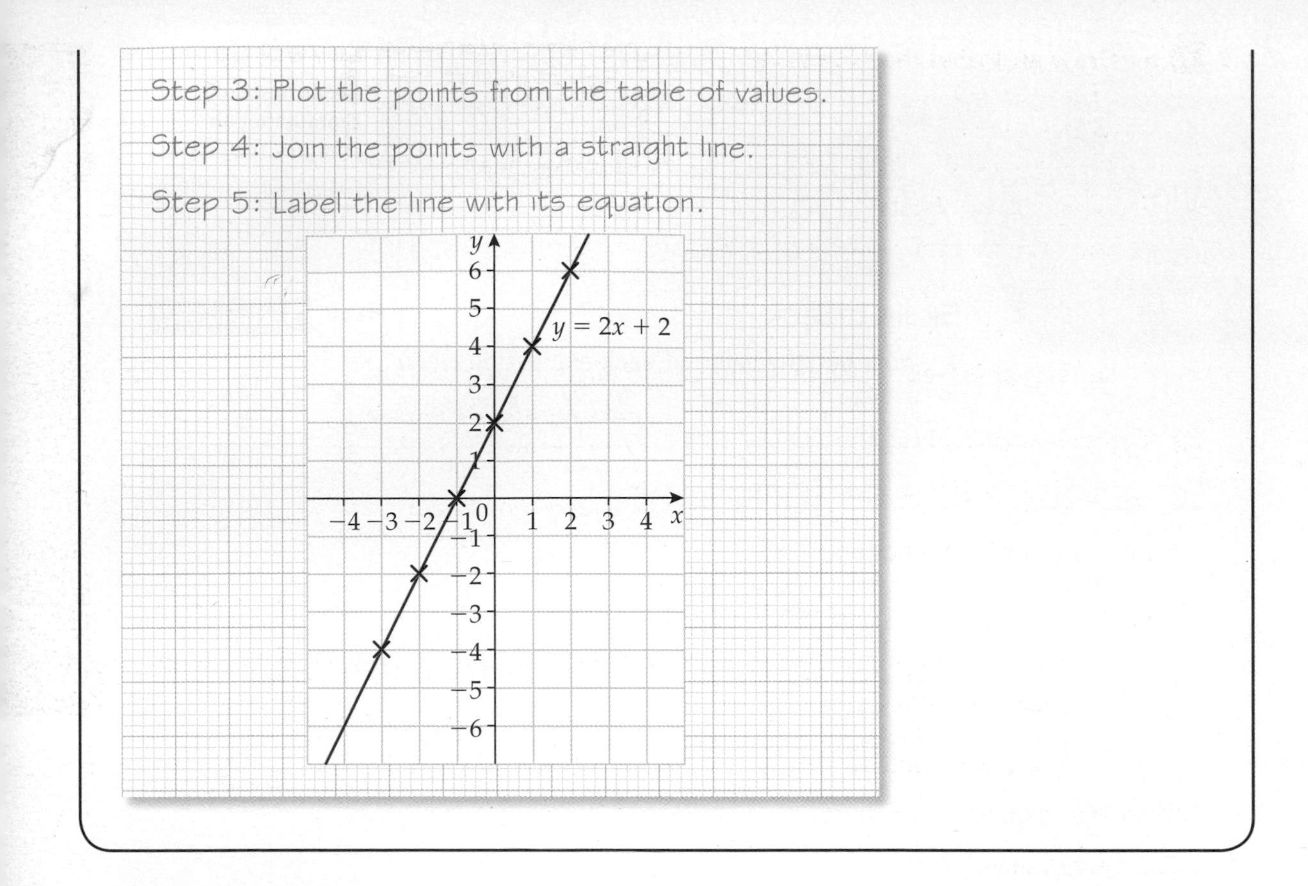

Step 3: Plot the points from the table of values.
Step 4: Join the points with a straight line.
Step 5: Label the line with its equation.
y
x
y = 2x + 2

Exercise 19B

D

1 Draw the graph of $y = 2x - 1$ for $-2 \leqslant x \leqslant 2$.

> **$-2 \leqslant x \leqslant 2$ means x-values from -2 to $+2$. Draw a coordinate grid with both x- and y-axes from -5 to $+5$.**

2 Draw the graph of $y = \frac{1}{2}x - 1$ for $-4 \leqslant x \leqslant 4$.

> **Draw x- and y-axes from -4 to $+4$.**

D

3 **a** Copy and complete this table for $y = 3 - x$.

x	−2	−1	0	1	2
y	5	4			

b Draw the graph of $y = 3 - x$.

c Draw the line $x = 2$ on your graph.

d A is the point where the two lines cross.
Mark the point A and write its coordinates.

A02

4 **a** Draw the line $y = x$.

b Draw the line $y = -x$.

c What do you notice about the two lines?

D

5 The line $y = x + 2$ crosses the line $x = 4$ at the point A.
Work out the coordinates of point A.

A03

6 Work out the point where the lines $y = 2x$ and $x = 3$ cross.

7 **a** Draw and label the graphs of
   i $y = 2x + 1$
   ii $y = 3x + 1$
   iii $y = 4x + 1$

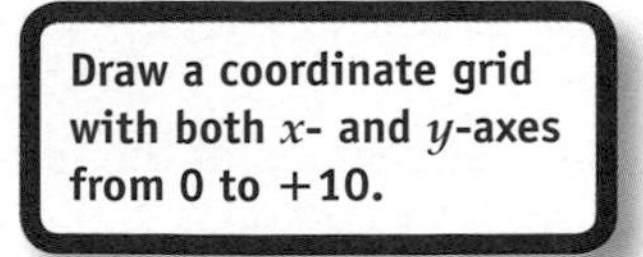
Draw a coordinate grid with both x- and y-axes from 0 to +10.

b Which line is the steepest?

c How can you tell which line is steepest from the equations?

d Draw the graph of $y = 2x + 2$ on your coordinate grid.

e Which lines are parallel to each other?

f How can you tell which lines are parallel from the equations?

D

8 **a** Draw and label the graphs of
   i $y = 2x$
   ii $y = x - 4$
   iii $y = 2x + 3$
   iv $y = x - 2$

Draw a coordinate grid with both x- and y-axes from −5 to +5.

b Where does each line intercept the y-axis?

c By looking at the equations of the lines, how can you tell where each line will intercept the y-axis?

AO2

9 **a** Copy and complete this table of values for $2x + y = 6$.

x	0	
y		0

b Draw the graph of $2x + y = 6$.

10 Draw and label the graphs of

a $3x + y = 9$ **b** $4x - y = 8$

c $3y + 2x = 12$ **d** $3y - 5x = 15$

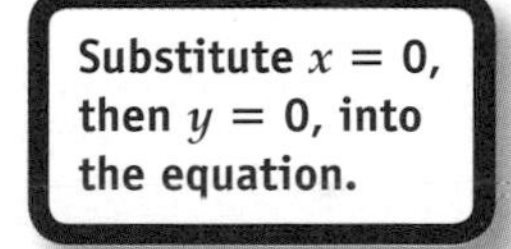
Substitute $x = 0$, then $y = 0$, into the equation.

AO2

19.3 Equations of straight-line graphs

Why learn this?

Computer graphics programmers use coordinates and equations of straight lines to make solid objects move on screen.

Objectives

C Understand the meaning of m and c in the equation $y = mx + c$

B **A** Find the equation of a line

Keywords

equation, gradient, slope, y-intercept, parallel, perpendicular

Skills check

1 Work out the value of c in these equations.
   a $c + 8 = 12$
   b $3 = -2 + c$
   c $5 = 2(3) + c$

HELP Section 15.4

2 What is the reciprocal of each of these numbers?
   a 5 **b** $\frac{1}{3}$ **c** $\frac{3}{4}$

HELP Section 11.5

Using the general equation $y = mx + c$

Straight line graphs have **equations** of the form $y = mx + c$, where

- the number (m) in front of x is the **gradient** (**slope**)
- the value of c tells you where the line intercepts the y-axis (c is the **y-intercept**).

You may have to rearrange the equation to '$y = \ldots$' first.

Example 3

Work out the gradient and y-intercept of each of these straight lines.

a $2y = x + 4$

b $2x - y = 3$

c $2x - 4y = 6$

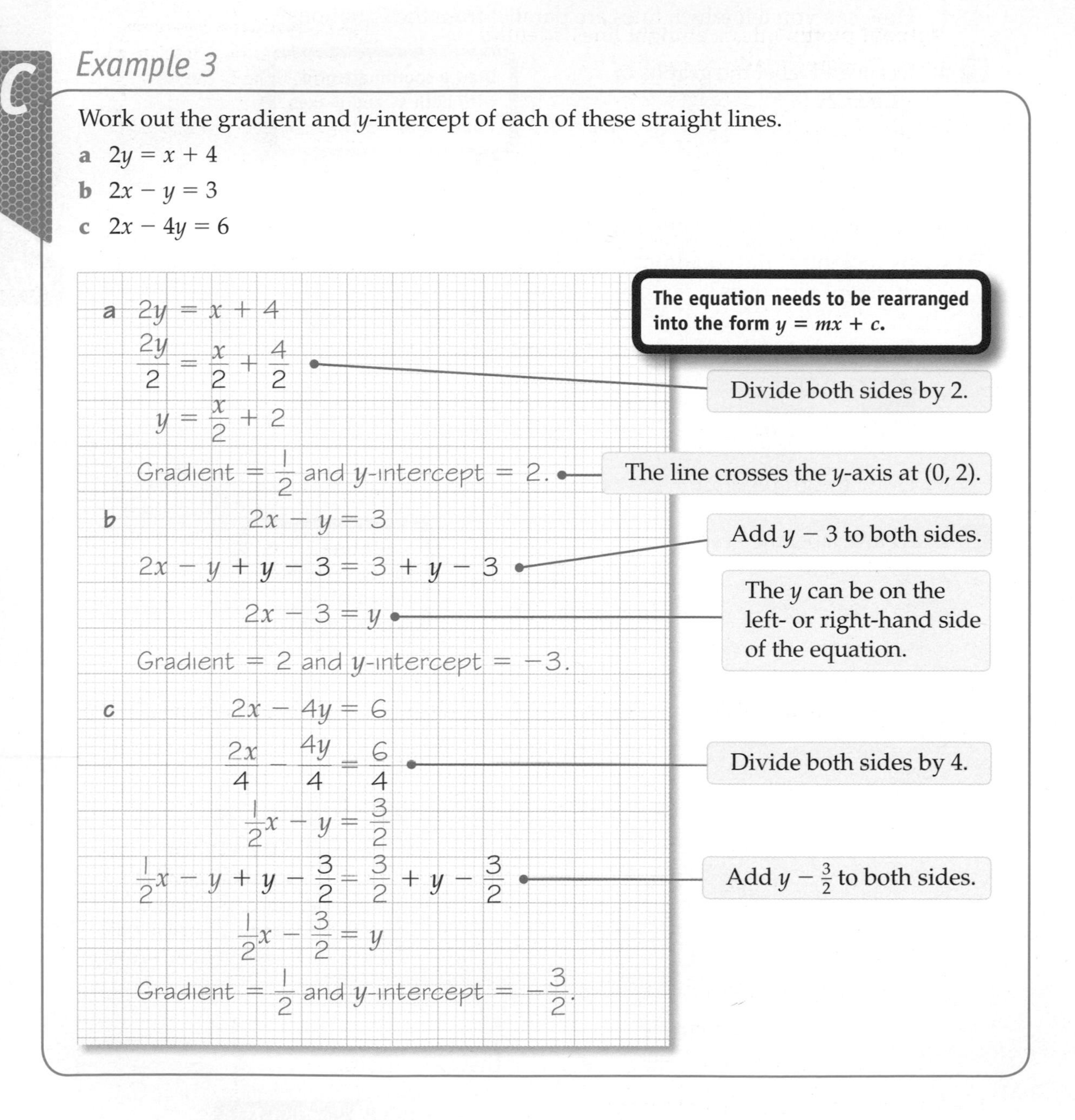

a $2y = x + 4$
$\frac{2y}{2} = \frac{x}{2} + \frac{4}{2}$
$y = \frac{x}{2} + 2$
Gradient $= \frac{1}{2}$ and y-intercept $= 2$.
b $2x - y = 3$
$2x - y + y - 3 = 3 + y - 3$
$2x - 3 = y$
Gradient $= 2$ and y-intercept $= -3$.
c $2x - 4y = 6$
$\frac{2x}{4} - \frac{4y}{4} = \frac{6}{4}$
$\frac{1}{2}x - y = \frac{3}{2}$
$\frac{1}{2}x - y + y - \frac{3}{2} = \frac{3}{2} + y - \frac{3}{2}$
$\frac{1}{2}x - \frac{3}{2} = y$
Gradient $= \frac{1}{2}$ and y-intercept $= -\frac{3}{2}$.
The equation needs to be rearranged into the form $y = mx + c$.
Divide both sides by 2.
The line crosses the y-axis at (0, 2).
Add $y - 3$ to both sides.
The y can be on the left- or right-hand side of the equation.
Divide both sides by 4.
Add $y - \frac{3}{2}$ to both sides.

Exercise 19C

1 Write down the gradient of each line.

a $y = 3x + 9$ **b** $y = 2x + 1$ **c** $y = 4 + x$

d $y = -\frac{1}{2}x - 6$ **e** $y = 5 - 2x$ **f** $2x - 4 = y$

2 Which of these lines are parallel to each other?

A $y = 2x - 3$ **B** $y = 3 - 5x$ **C** $y = 3x - 1$

D $y = 4 + 3x$ **E** $y = -2x + 3$ **F** $2x + 3 = y$

Straight lines that are parallel have the same gradient.

C

3 Write down the y-intercept of each line.

a $y = 2x + 4$ **b** $y = -2x + 1$ **c** $y = x - 4$

d $y = 2x$ **e** $y = 3 + x$ **f** $2 + 3x = y$

4 Without plotting these straight lines, identify those that are parallel to the line $y = -3x - 2$.

A $y = 3x - 2$ **B** $y = -3x + 2$

C $y = 2 + 3x$ **D** $2y = -2 + 3x$

E $x = -3y - 2$ **F** $3x = -y - 2$

Remember the equation needs to be in the form $y = mx + c$.

C

AO2

5 Work out the gradient and y-intercept of each line.

a $2y = 2x + 6$ **b** $2y + 4 = x$

c $y - 2 = -2x$ **d** $4y = -4x + 8$

e $2y - 7 = -2x$ **f** $-3x = 4 + y$

g $2y + 3x = 2$ **h** $4y + 2 + 3x = 0$

C

Equations of straight lines

In an equation of a straight line in the form $y = mx + c$, the x- and y-coefficients of *any* point on the line satisfy the equation of the line.

The gradient, m, of a straight line is the change in the value of y divided by the change in the value of x.

$$\text{Gradient, } m = \frac{\text{change in } y}{\text{change in } x} = \frac{y_2 - y_1}{x_2 - x_1}$$

The gradient can be positive or negative.

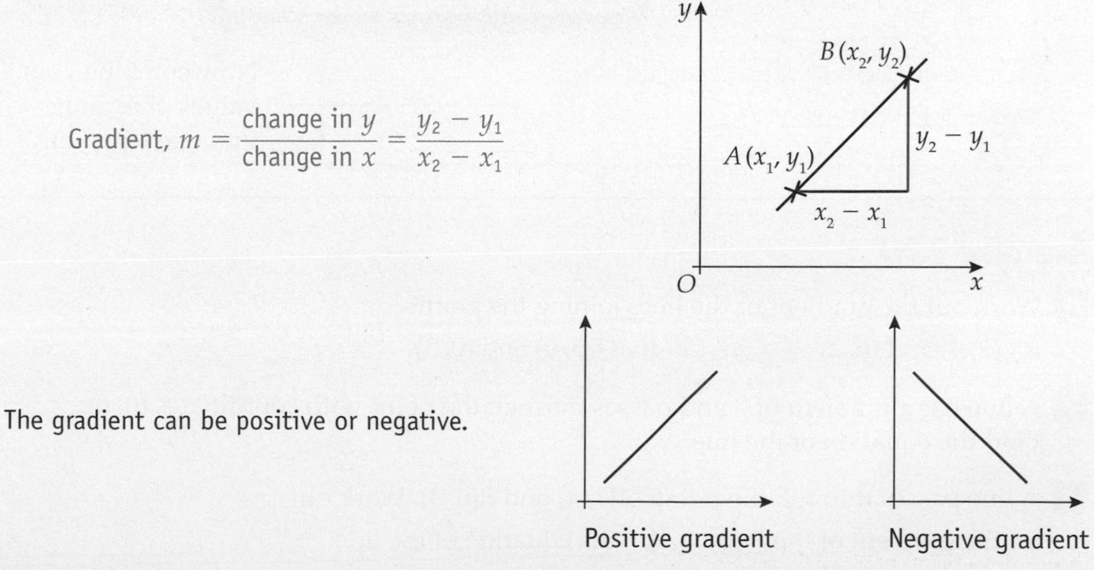

Straight lines that are **parallel** have the same gradient.

AO2

Example 4

A line passes through the points $A(4, 0)$ and $B(5, 2)$.

Work out

a the gradient of the line **b** the equation of the line.

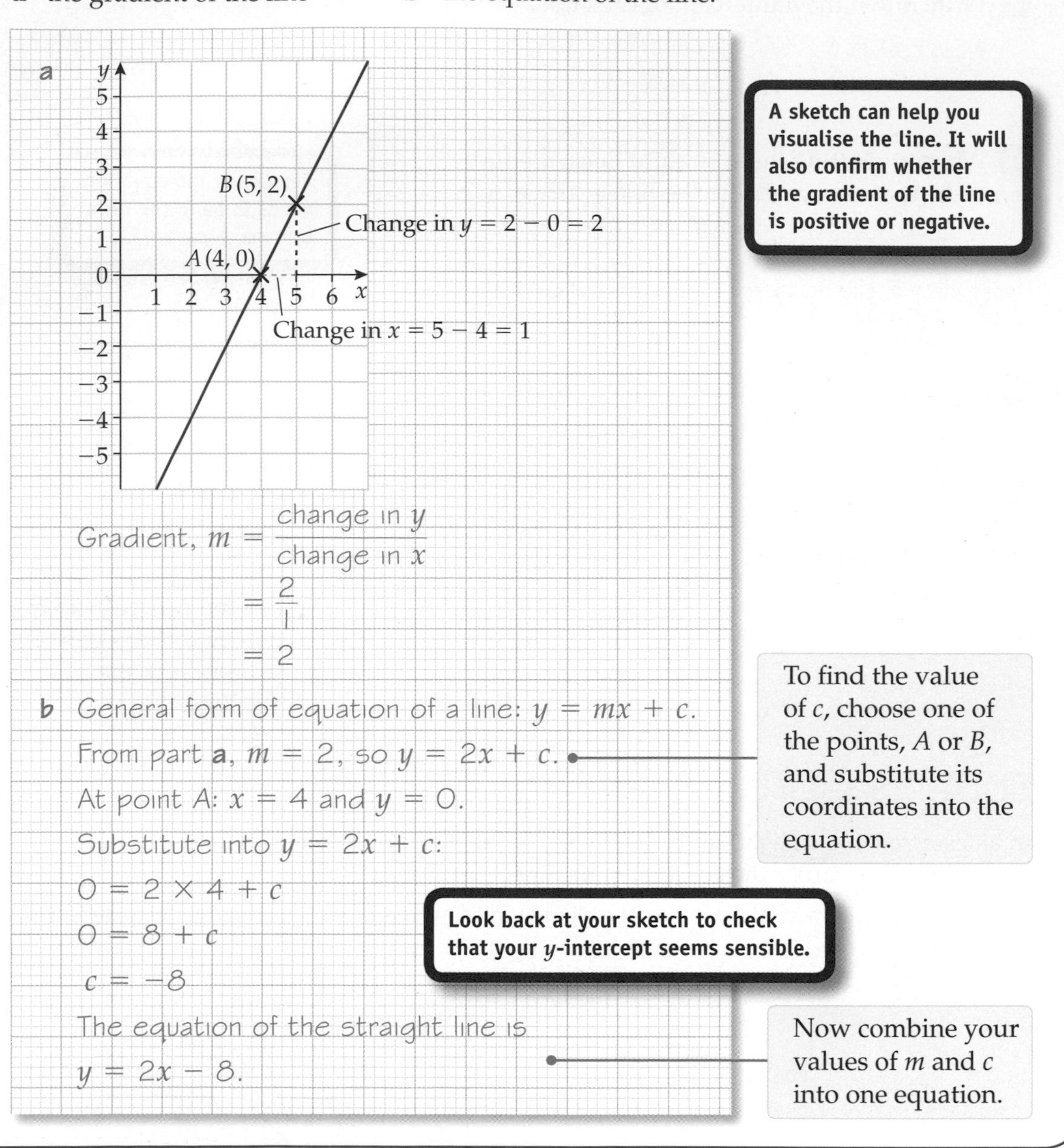

a
y
5
4
3
2
1
0
−1
−2
−3
−4
−5
1 2 3 4 5 6 x
B (5, 2)
A (4, 0)
Change in y = 2 − 0 = 2
Change in x = 5 − 4 = 1
A sketch can help you visualise the line. It will also confirm whether the gradient of the line is positive or negative.
Gradient, m = change in y / change in x
= 2/1
= 2
b General form of equation of a line: y = mx + c.
From part a, m = 2, so y = 2x + c.
At point A: x = 4 and y = 0.
Substitute into y = 2x + c:
0 = 2 × 4 + c
0 = 8 + c
c = −8
The equation of the straight line is
y = 2x − 8.
To find the value of c, choose one of the points, A or B, and substitute its coordinates into the equation.
Look back at your sketch to check that your y-intercept seems sensible.
Now combine your values of m and c into one equation.

Exercise 19D

B

1 Work out the gradient of the lines joining the points

a $(2, 4)$ and $(6, 2)$ **b** $(-3, 0)$ and $(0, 3)$.

2 A line has a gradient of 3 and passes through the point with coordinates $(0, 6)$. Find the equation of the line.

B

AO2

3 A line passes through the points $A(1, 4)$ and $B(6, 3)$. Work out

a the gradient of the line **b** the equation of the line.

4 A straight line parallel to $y = 3x + 2$ passes through the point $(0, -2)$. What is the equation of this line?

5 The line segment AB has end-points $A(2, -1)$ and $B(4, 5)$.
Find the equation of the line parallel to AB that passes through the point $(2, 8)$.

6 A video game character is at the point $(3, 5)$ on a computer screen.
The game player wants the character to go to the point $(-1, -3)$.
Find the equation of the straight line that will take the character to this point.

Perpendicular lines

Perpendicular lines are straight lines that cross each other at right angles.

A straight line has the equation $y = mx + c$. Any line perpendicular to it has gradient $-\frac{1}{m}$.

The product of the gradients of two perpendicular lines is always -1.

$$m \times -\frac{1}{m} = -1$$

Example 5

A

Write the gradient of a line that is perpendicular to

a a line with gradient 6 **b** a line with gradient -4.

a Gradient of line, $m = 6$

Gradient of perpendicular line is $-\frac{1}{m} = -\frac{1}{6}$

For the perpendicular line, find the negative reciprocal of the gradient.

b Gradient of line, $m = -4$.

Gradient of perpendicular line is $-\frac{1}{m} = -\frac{1}{-4} = \frac{1}{4}$

Example 6

A

Find the equation of the line perpendicular to $y = -\frac{1}{3}x - 2$ that passes through the point $(2, 11)$.

Gradient of line, $m = -\frac{1}{3}$

Gradient of perpendicular line is $-\frac{1}{m} = 3$

3 is the negative reciprocal.

Equation of perpendicular line is $y = mx + c$ with $m = 3$, so $y = 3x + c$

Perpendicular line passes through the point $(2, 11)$, so substitute $x = 2$ and $y = 11$.

$11 = 3 \times 2 + c$

$11 = 6 + c$

$c = 5$

The equation of the perpendicular line is $y = 3x + 5$.

Now combine the values of m and c into one equation.

Exercise 19E

A

1 Write the gradient of a line that is perpendicular to a line with gradient

a 2 **b** -3 **c** $\frac{1}{3}$ **d** $-\frac{1}{4}$ **e** $1\frac{1}{2}$

2 A straight line perpendicular to $y = 4x - 3$ passes through the point (0, 7).
Work out the equation of the line.

3 Work out the equation of the line perpendicular to $y = 2x$ that passes through the point (0, 3).

A

4 A is the point $(2, -1)$ and B is the point (4, 5).
Find the equation of the line perpendicular to AB which passes through its mid-point.

5 Find the equation of the perpendicular bisector of the line segment joining the points $P(-2, 5)$ and $Q(4, 5)$.

AO3

6 A character in a game world is moving along the line $y = 3x + 2$.
When it gets to the point $(3, -2)$, the player wants the character to turn 90 degrees to the right and continue along this new path.
Find the equation for the line of the new path.

19.4 Using graphs to solve simultaneous equations

Keywords
simultaneous equations, intersect

Why learn this?
Oil refineries use simultaneous equations to optimise production from crude oil.

Objectives
B Use a graphical method to solve simultaneous equations

Skills check

HELP Section 19.2

1 Draw the graph of

a $x + 3y = 6$ **b** $4x + y = 8$ **c** $3x - y = 9$

Using graphs to solve simultaneous equations

You can solve a pair of **simultaneous equations** by drawing their graphs on the same set of axes.

The point on the graph where the two lines **intersect** (cross) is their solution.

Solving simultaneous equations algebraically was covered in Section 14.5.

Graphical solutions to simultaneous equations are only approximate, as they depend on the accuracy of the drawing.

B

Example 7

Solve these simultaneous equations graphically.

$x + 2y = 8$
$3x + y = 9$

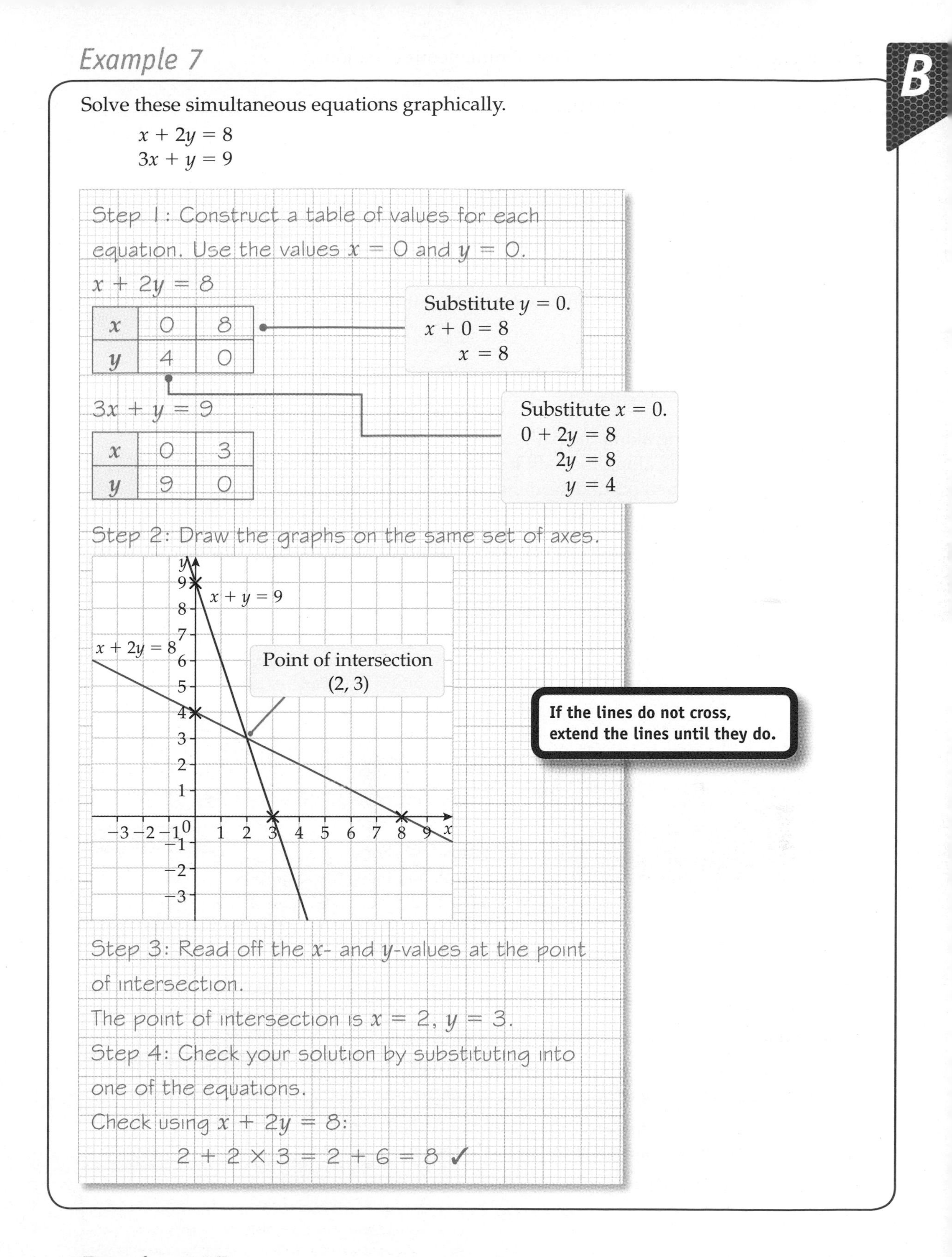

Step 1: Construct a table of values for each equation. Use the values $x = 0$ and $y = 0$.
$x + 2y = 8$
| x | 0 | 8 |
| y | 4 | 0 |
Substitute $y = 0$.
$x + 0 = 8$
$x = 8$
$3x + y = 9$
| x | 0 | 3 |
| y | 9 | 0 |
Substitute $x = 0$.
$0 + 2y = 8$
$2y = 8$
$y = 4$
Step 2: Draw the graphs on the same set of axes.
$x + y = 9$
$x + 2y = 8$
Point of intersection (2, 3)
Step 3: Read off the x- and y-values at the point of intersection.
The point of intersection is $x = 2$, $y = 3$.
Step 4: Check your solution by substituting into one of the equations.
Check using $x + 2y = 8$:
$2 + 2 \times 3 = 2 + 6 = 8$ ✓

If the lines do not cross, extend the lines until they do.

Exercise 19F

1 Solve each pair of simultaneous equations graphically.

a $x + y = 4$
$x + 2y = 3$

b $3x - y = -3$
$x + y = -1$

2 Find a graphical solution to these simultaneous equations.

$2x + y = 6$
$x - 2y = 8$

3 **a** Solve this pair of simultaneous equations graphically.

$x + 2y = 4$
$2x - y = 3$

b What can you say about these two lines?

4 Use a graphical method to show why these simultaneous equations cannot be solved.

$2x + 2y = 5$
$3y = 4 - 3x$

5 Boat A is travelling along the path $-x + 2y = 10$.
Boat B is travelling along the path $y = 2x + 2$.
Use a graphical method to determine whether their paths will cross.
If appropriate, state the coordinates of where they cross.

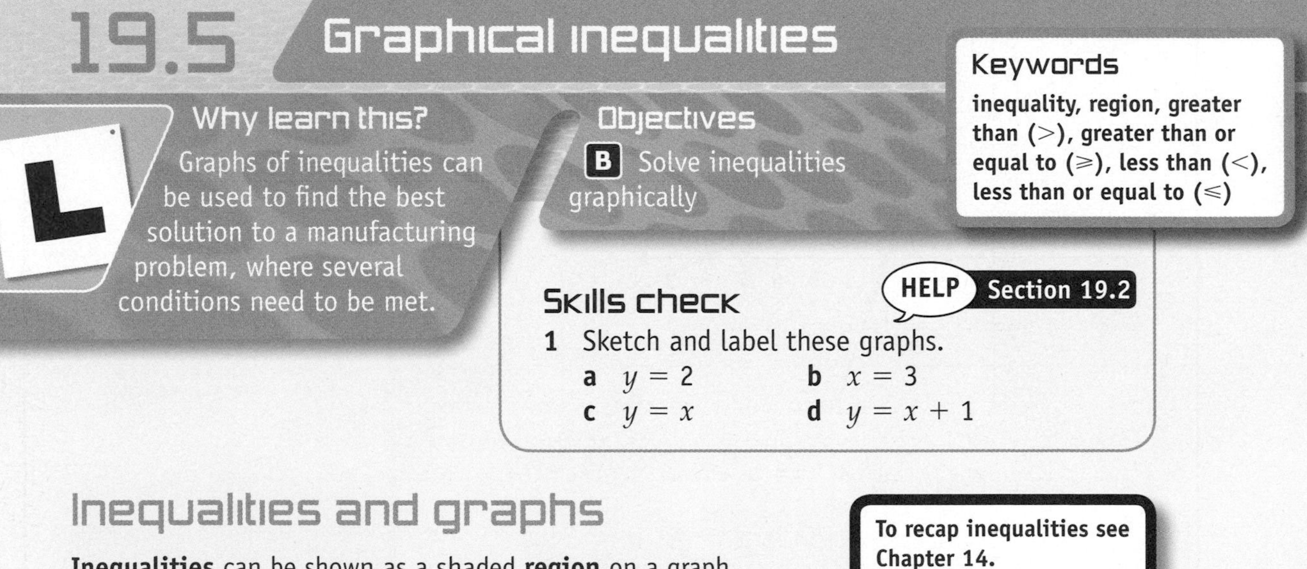

19.5 Graphical inequalities

Why learn this?
Graphs of inequalities can be used to find the best solution to a manufacturing problem, where several conditions need to be met.

Objectives
B Solve inequalities graphically

Keywords
inequality, region, greater than (>), greater than or equal to (≥), less than (<), less than or equal to (≤)

Skills check

HELP Section 19.2

1 Sketch and label these graphs.

a $y = 2$ **b** $x = 3$
c $y = x$ **d** $y = x + 1$

Inequalities and graphs

To recap inequalities see Chapter 14.

Inequalities can be shown as a shaded **region** on a graph.

- For the inequalities $\leqslant$ and $\geqslant$ the boundary is a solid line. The points on the line *are* included in the region.
- For the inequalities $<$ and $>$ the boundary is a dashed line. The points on the line are *not* included in the region.

Example 8

Show the regions defined by these inequalities.

a $x \geqslant 3$ **b** $y < 1$ **c** $x + y > 2$

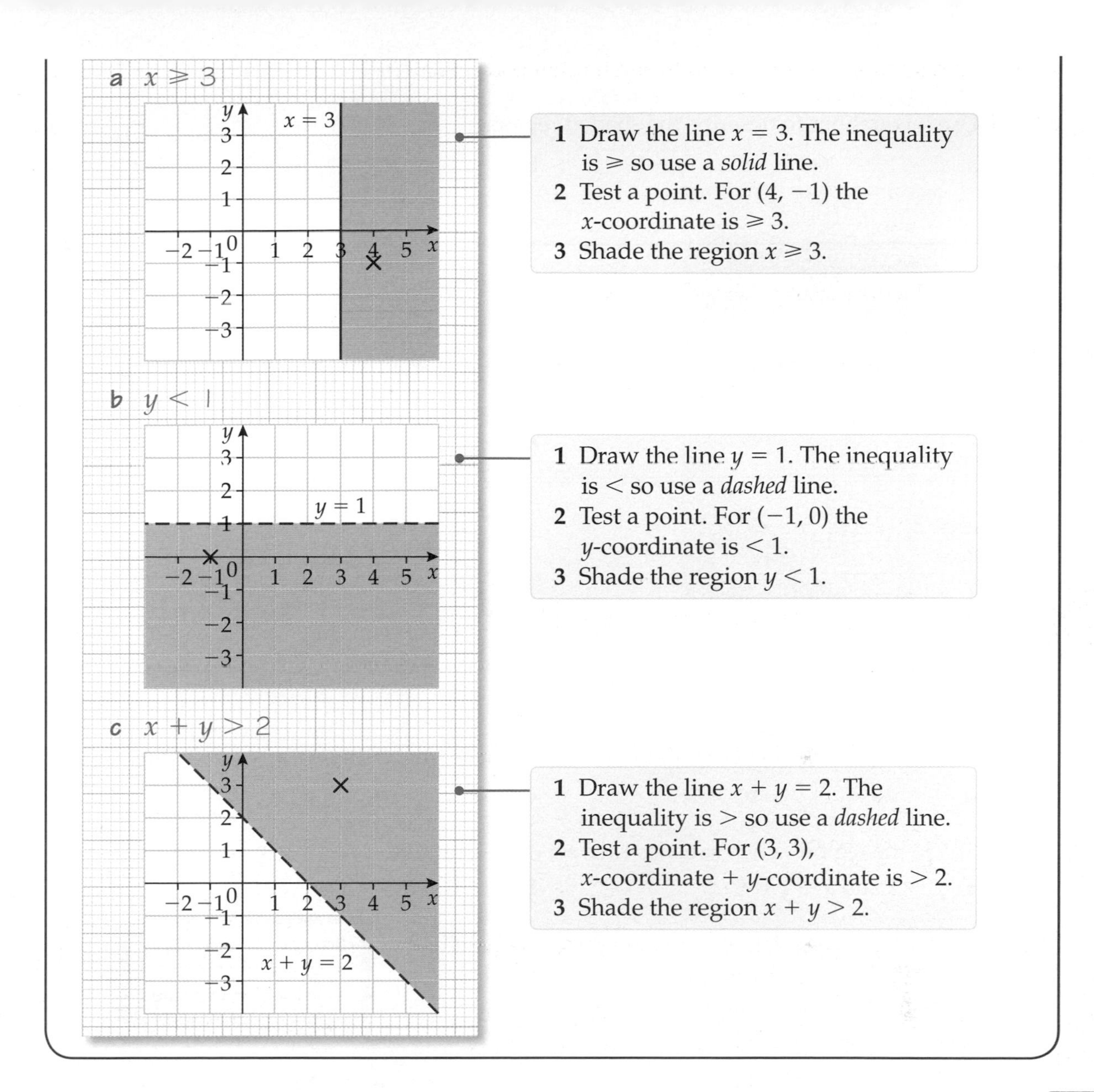

a
1. Draw the line $x = 3$. The inequality is $\geqslant$ so use a *solid* line.
2. Test a point. For $(4, -1)$ the x-coordinate is $\geqslant 3$.
3. Shade the region $x \geqslant 3$.

b
1. Draw the line $y = 1$. The inequality is $<$ so use a *dashed* line.
2. Test a point. For $(-1, 0)$ the y-coordinate is < 1.
3. Shade the region $y < 1$.

c
1. Draw the line $x + y = 2$. The inequality is $>$ so use a *dashed* line.
2. Test a point. For $(3, 3)$, x-coordinate + y-coordinate is > 2.
3. Shade the region $x + y > 2$.

Example 9

On a grid show the region that is defined by these three inequalities.

$x \geqslant 1 \qquad y \leqslant 3 \qquad x + y \leqslant 5$

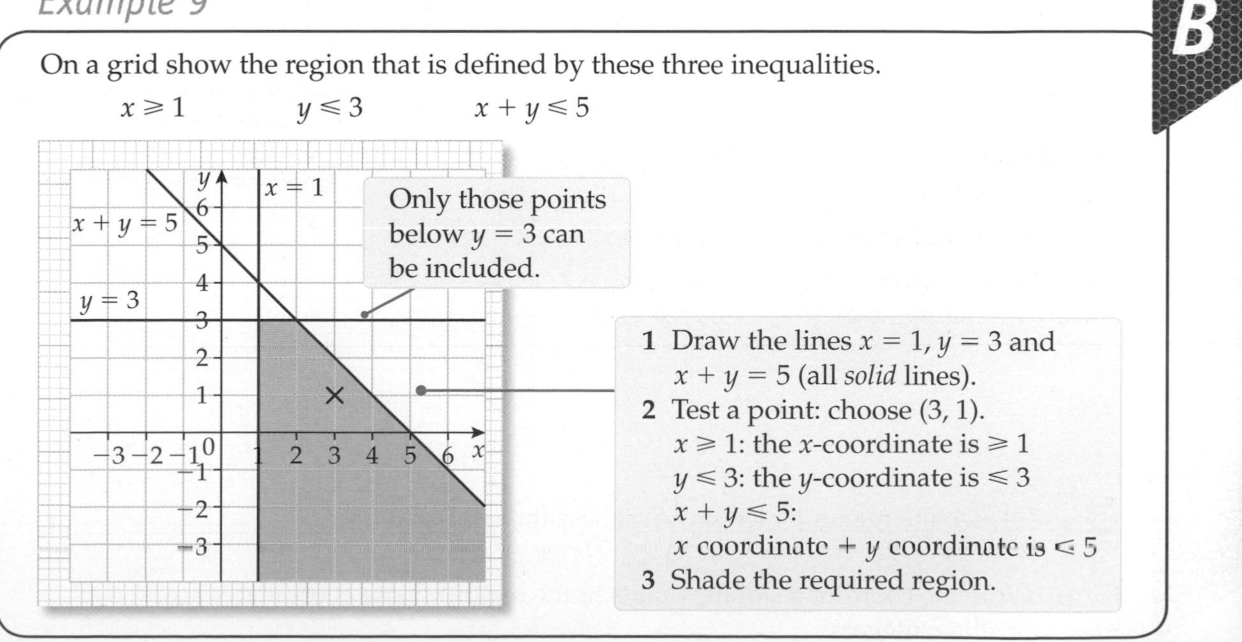

1. Draw the lines $x = 1$, $y = 3$ and $x + y = 5$ (all *solid* lines).
2. Test a point: choose $(3, 1)$.
   $x \geqslant 1$: the x-coordinate is $\geqslant 1$
   $y \leqslant 3$: the y-coordinate is $\leqslant 3$
   $x + y \leqslant 5$:
   x coordinate + y coordinate is $\leqslant 5$
3. Shade the required region.

Exercise 19G

B

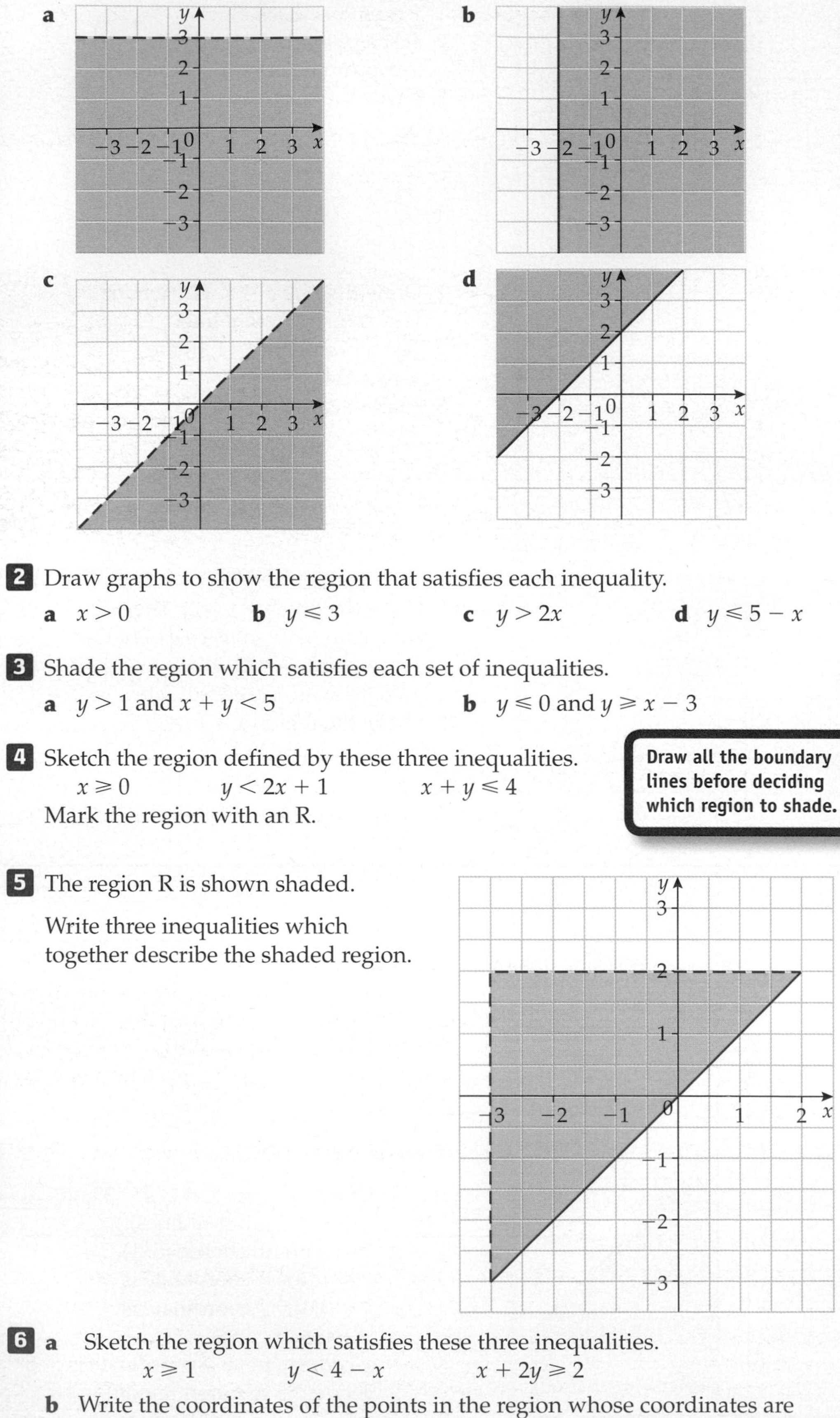

1 Use inequalities to describe the shaded regions.

a

b

c

d

2 Draw graphs to show the region that satisfies each inequality.

a $x > 0$ **b** $y \leqslant 3$ **c** $y > 2x$ **d** $y \leqslant 5 - x$

3 Shade the region which satisfies each set of inequalities.

a $y > 1$ and $x + y < 5$ **b** $y \leqslant 0$ and $y \geqslant x - 3$

4 Sketch the region defined by these three inequalities.

$x \geqslant 0$ $y < 2x + 1$ $x + y \leqslant 4$

Mark the region with an R.

Draw all the boundary lines before deciding which region to shade.

5 The region R is shown shaded.

Write three inequalities which together describe the shaded region.

6 **a** Sketch the region which satisfies these three inequalities.

$x \geqslant 1$ $y < 4 - x$ $x + 2y \geqslant 2$

b Write the coordinates of the points in the region whose coordinates are positive integers.

Real-life graphs

Why learn this?

Companies can use distance–time graphs to work out journey times and help plan deliveries.

Objectives

D **C** **A** Draw, read and interpret distance–time and velocity–time graphs

C Sketch and interpret real-life graphs

Keywords

distance–time graph, time, distance, average speed, velocity, velocity–time graph, acceleration, deceleration

Skills check

1. A bus travels 6 miles in $\frac{1}{4}$ hour. What is its speed in mph?
2. What is the formula for
   a speed **b** area of a triangle?

Distance–time graphs

A **distance–time graph** represents a journey.

The x-axis (horizontal) represents the **time** taken. The y-axis (vertical) represents the **distance** from the starting point.

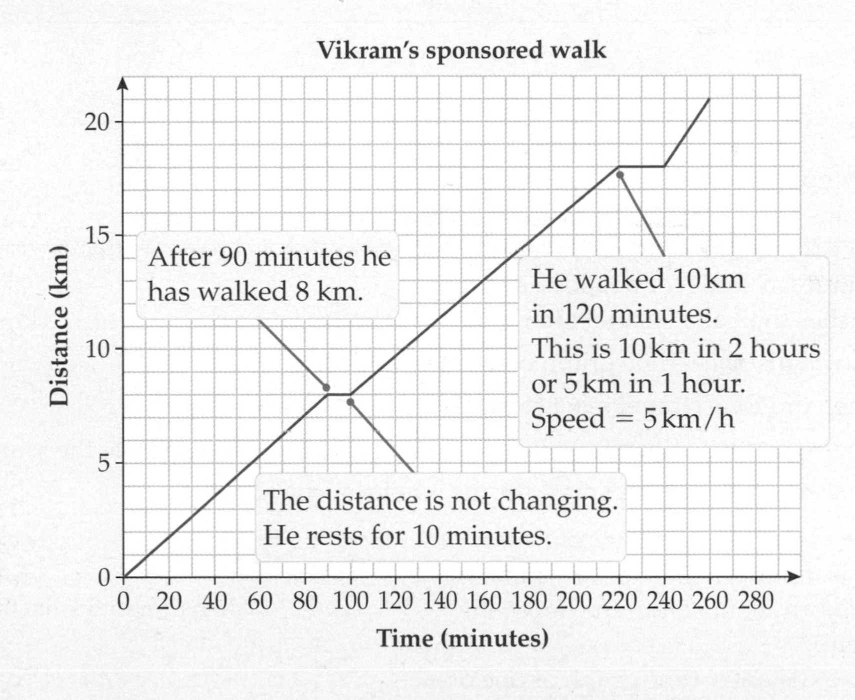
Vikram's sponsored walk
Distance (km)
Time (minutes)
0
5
10
15
20
0 20 40 60 80 100 120 140 160 180 200 220 240 260 280
After 90 minutes he has walked 8 km.
The distance is not changing. He rests for 10 minutes.
He walked 10 km in 120 minutes.
This is 10 km in 2 hours or 5 km in 1 hour.
Speed = 5 km/h

$$\text{Speed} = \frac{\text{distance}}{\text{time}}$$

The units of speed can be metres per second (m/s), kilometres per hour (km/h) or miles per hour (mph).

You can work out the **average speed** of a complete journey using this formula.

$$\text{Average speed} = \frac{\text{total distance}}{\text{total time}}$$

The gradient of a distance–time graph represents the speed of the journey.

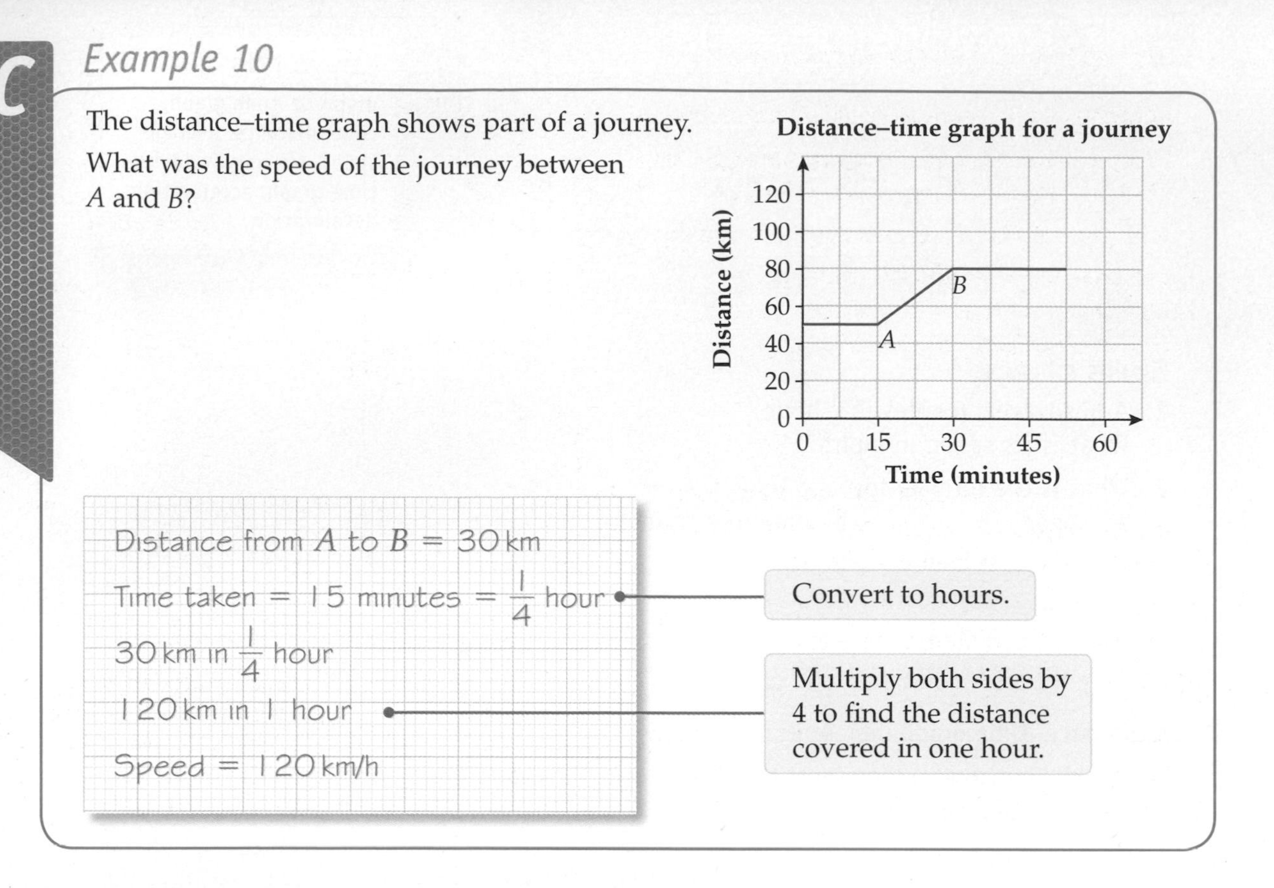

C
Example 10
The distance–time graph shows part of a journey.
What was the speed of the journey between A and B?
Distance–time graph for a journey
Distance (km)
Time (minutes)
0
20
40
60
80
100
120
15
30
45
60
A
B
Distance from A to B = 30 km
Time taken = 15 minutes = 1/4 hour
30 km in 1/4 hour
120 km in 1 hour
Speed = 120 km/h
Convert to hours.
Multiply both sides by 4 to find the distance covered in one hour.

Exercise 19H

1 Ryan is travelling to the Lake District for his holiday.
He begins his journey at 1 pm and travels 50 km in the first hour.
Between 2 pm and 3 pm he travels only 25 km owing to heavy traffic.
At 3 pm Ryan stops for a half-hour break.
He reaches the Lake District after a further 2 hours and a distance of 150 km.

a Draw a distance–time graph of Ryan's journey.

b When did Ryan travel most slowly?

c What speed was Ryan travelling at during the slowest section of the journey?

d What was the average speed for the whole journey?

2 Llinos is a courier. She delivers parcels across North Wales from the depot in Llangefni.
Here is a distance–time graph of one of her journeys. She left the depot at 8 am.

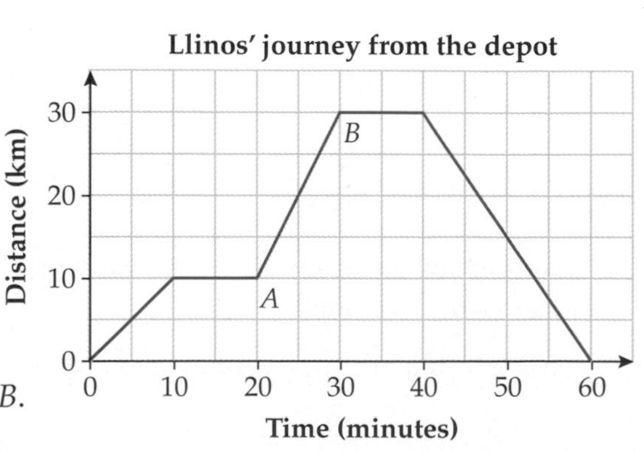

a How far did Llinos travel in the first 10 minutes?

b At what time did Llinos arrive at her first delivery point?

c Calculate Llinos' speed during section *AB*.

d At what time did Llinos begin the return journey to the depot?

e Calculate Llinos' average speed for the whole journey.

Velocity–time graphs

Velocity is a measure of speed and direction.

On a velocity–time graph, the x-axis represents time and the y-axis represents velocity.

A horizontal line represents constant velocity.

The gradient of a **velocity–time graph** represents the rate of change of velocity (the **acceleration**) of the object.

$$\text{Acceleration} = \frac{\text{change in velocity}}{\text{time}}$$

The units of acceleration and **deceleration** can be metres per second per second (m/s^2) or kilometres per hour per hour (km/h^2).

The **area** under a velocity–time graph represents the **distance** travelled.

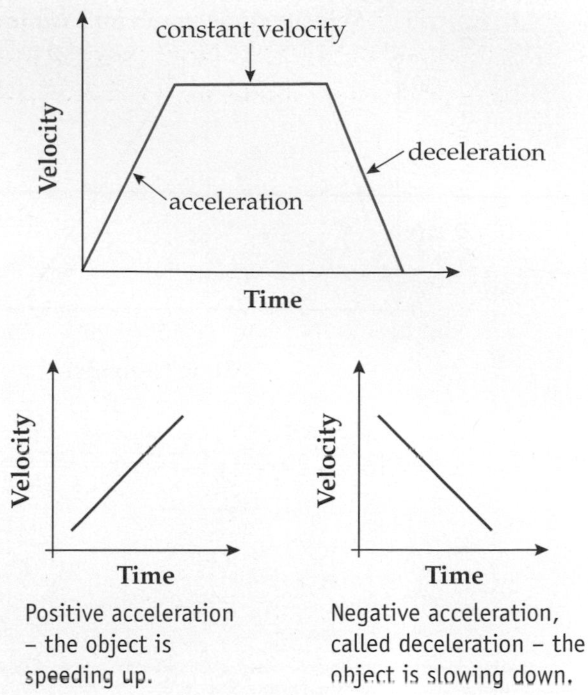

Positive acceleration – the object is speeding up.

Negative acceleration, called deceleration – the object is slowing down.

Example 11

The diagram shows the velocity–time graph for a car journey.

Work out

a the acceleration in the first 5 seconds

b the deceleration

c the total distance travelled.

Velocity–time graph for a car journey

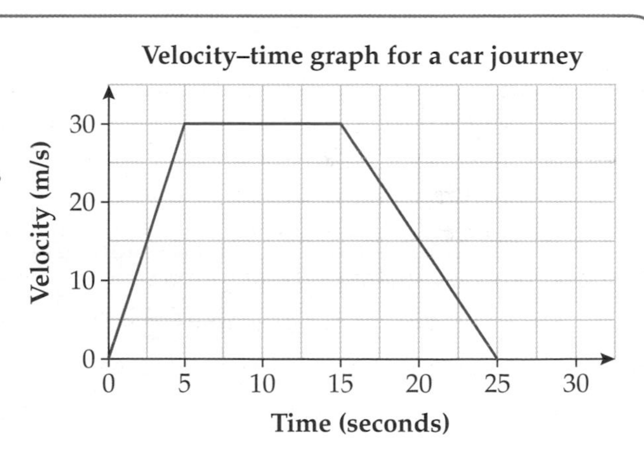

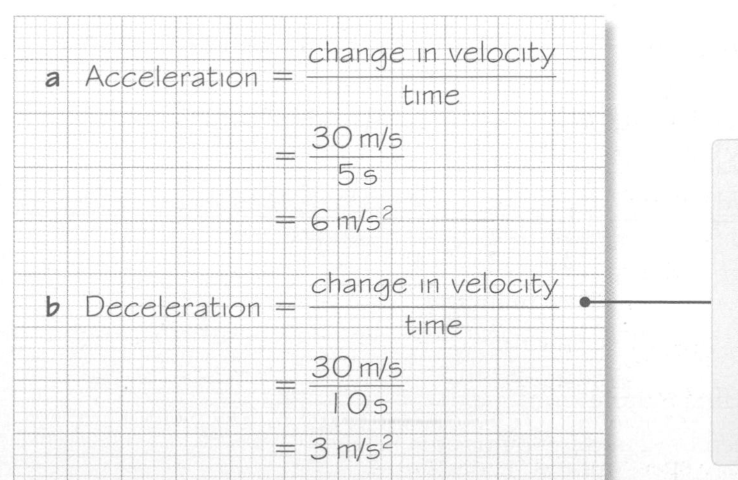

a $\text{Acceleration} = \frac{\text{change in velocity}}{\text{time}}$

$= \frac{30\text{ m/s}}{5\text{ s}}$

$= 6\text{ m/s}^2$

b $\text{Deceleration} = \frac{\text{change in velocity}}{\text{time}}$

$= \frac{30\text{ m/s}}{10\text{ s}}$

$= 3\text{ m/s}^2$

A downward sloping line indicates deceleration. Deceleration is taking place from 15 to 25 seconds. A deceleration can also be written as an acceleration with a negative sign. In this case, the acceleration is -3 m/s^2.

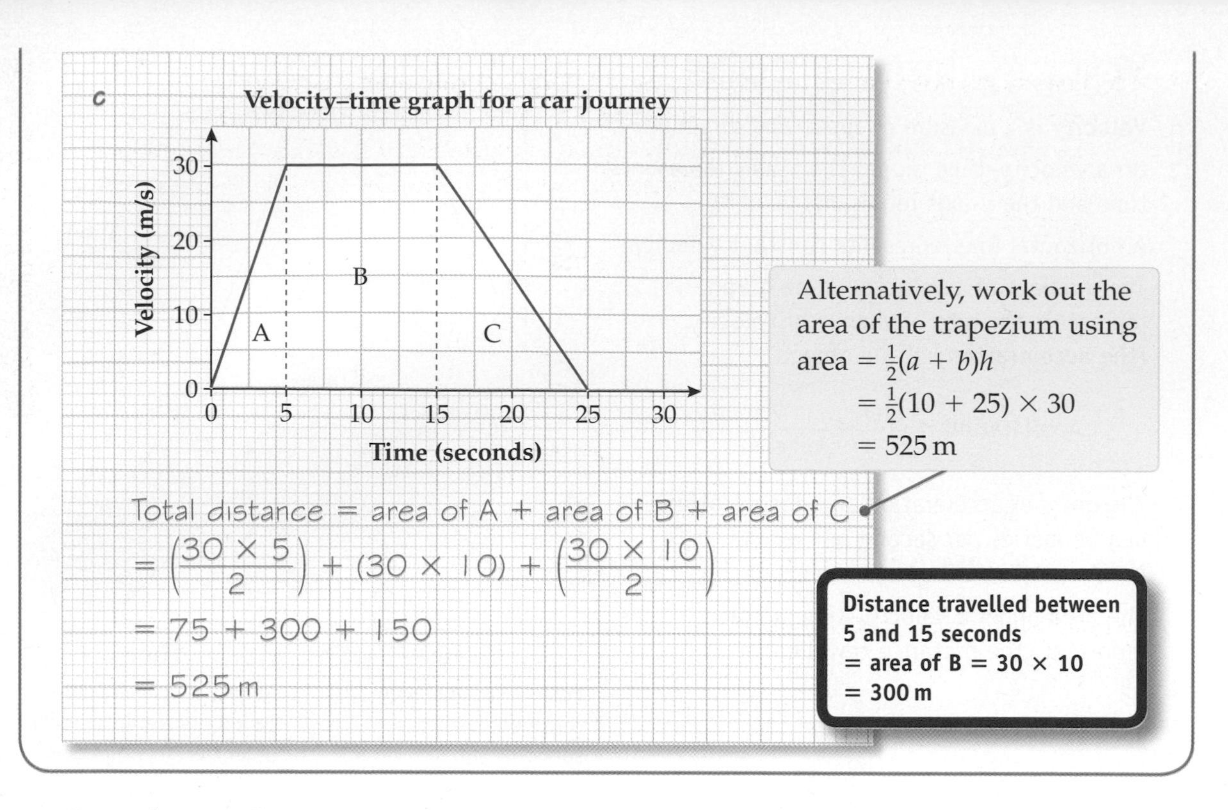

c
Velocity–time graph for a car journey
Velocity (m/s)
30
20
10
0
0
5
10
15
20
25
30
Time (seconds)
A
B
C
Alternatively, work out the area of the trapezium using
area = $\frac{1}{2}(a + b)h$
= $\frac{1}{2}(10 + 25) \times 30$
= 525 m
Total distance = area of A + area of B + area of C
= $\left(\frac{30 \times 5}{2}\right) + (30 \times 10) + \left(\frac{30 \times 10}{2}\right)$
= 75 + 300 + 150
= 525 m
Distance travelled between 5 and 15 seconds
= area of B = 30 × 10
= 300 m

Exercise 19I

A

1 The diagram shows the velocity–time graph for a car journey.

Use the diagram to calculate

a the acceleration in the first 40 seconds

b the total distance travelled.

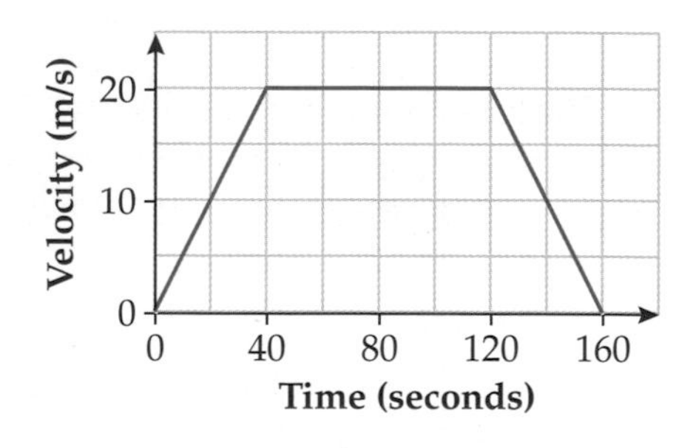

2 The diagram shows the velocity–time graph for a train journey.

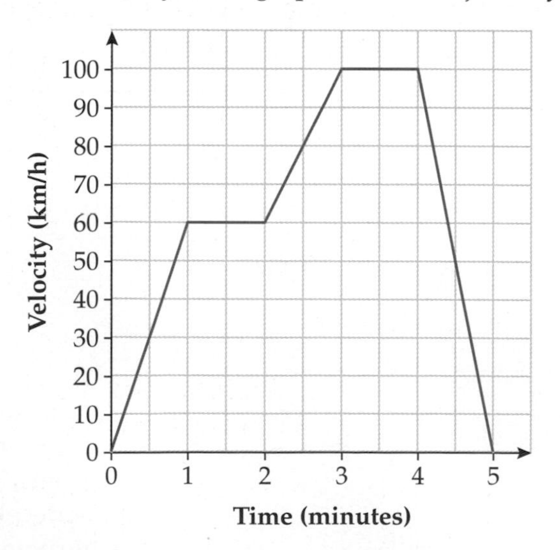

Work out

a the acceleration in the first minute

b the acceleration between $t = 2$ and $t = 3$ minutes

c the distance travelled in the first two minutes.

3 A car starts out at rest and accelerates at a constant rate over 5 seconds to reach 30 m/s. The car continues at a constant speed of 30 m/s for 5 seconds. The car accelerates over the next 10 seconds to reach 50 m/s, which it maintains for a further 10 seconds. The car then slows down, taking 10 seconds to come to a complete stop.

a Draw the velocity–time graph for this journey.

b Use your graph to work out

i the acceleration in the first 5 seconds

ii the distance travelled in the first 20 seconds

iii the deceleration.

A

Interpreting real-life graphs

A straight line shows that the rate of change is steady.

A curved line shows that the rate of change varies.

The steeper the line, the faster the rate of change.

Example 12

C

Water is poured at a constant rate into each of these containers.

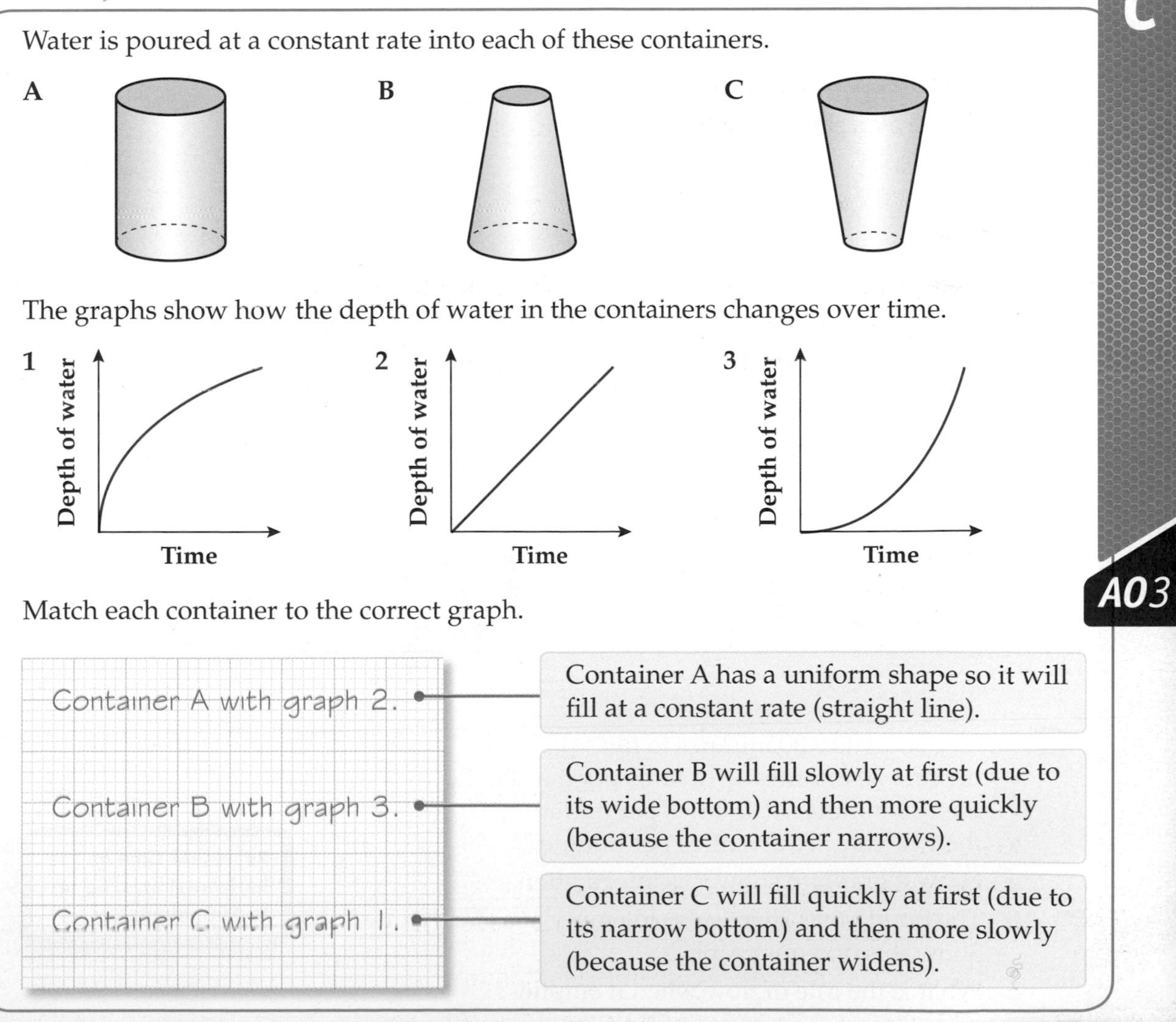

A
B
C
The graphs show how the depth of water in the containers changes over time.
1
2
3
Depth of water
Time
Match each container to the correct graph.
Container A with graph 2.
Container B with graph 3.
Container C with graph 1.
Container A has a uniform shape so it will fill at a constant rate (straight line).
Container B will fill slowly at first (due to its wide bottom) and then more quickly (because the container narrows).
Container C will fill quickly at first (due to its narrow bottom) and then more slowly (because the container widens).

AO3

Exercise 19J

C

1 Water is poured at a steady rate into these jars.

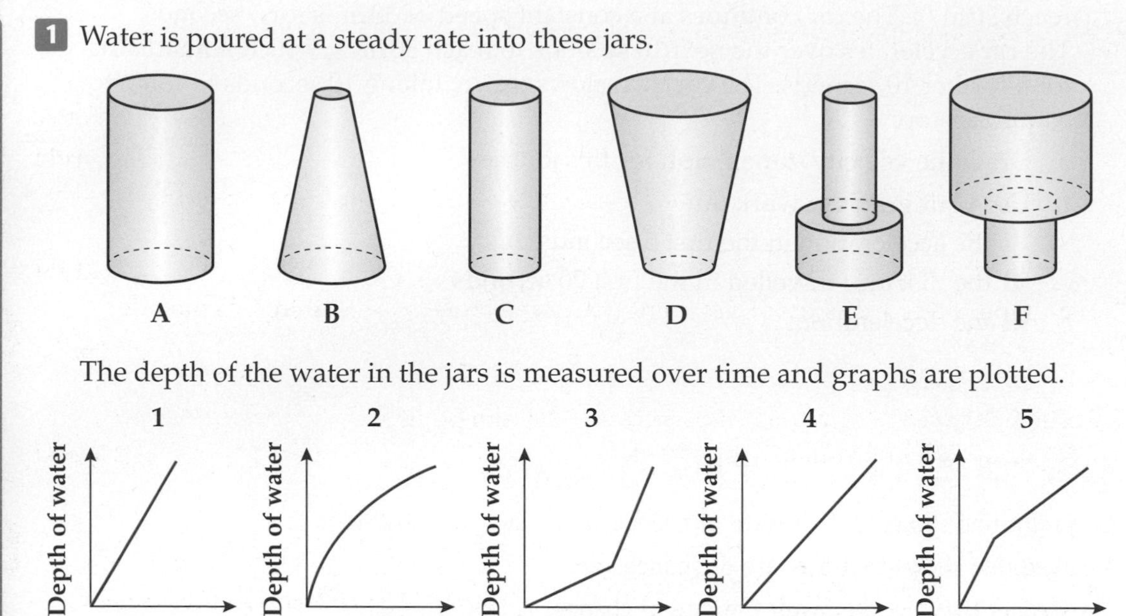

The depth of the water in the jars is measured over time and graphs are plotted.

- **a** Match the jars to the graphs.
- **b** One jar has not been matched. Which one is it?
- **c** Sketch a graph for this jar.

2 Look at this vase.

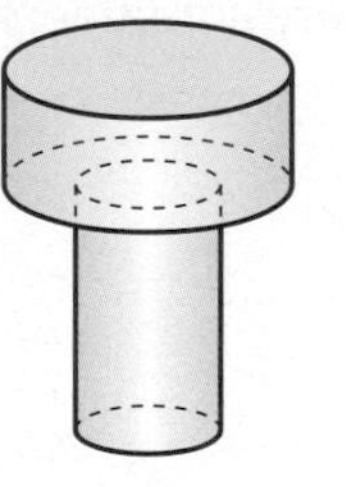

Malik has sketched a graph to show how the depth of water changes over time, as water drips steadily into the vase.

Is Malik's graph correct?
Give a full reason for your answer.

Depth of water
Time

3 The graph shows a 1500 m race between Nathan and John.
Describe what happens in the race.

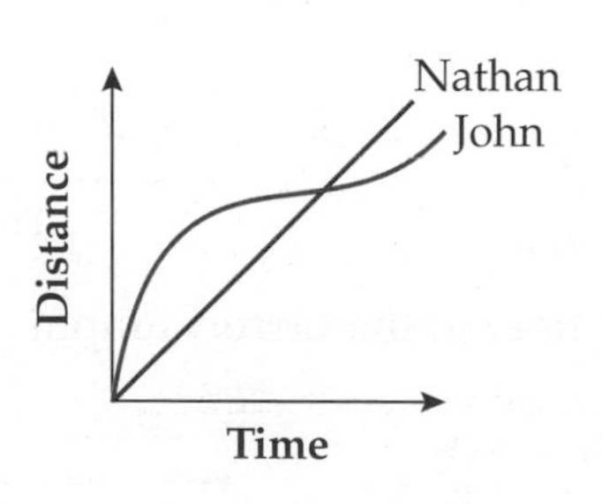

4 A fish pond takes 80 minutes to fill:

- the first 30 minutes at 10 litres per minute
- then 40 minutes at 15 litres per minute
- then 10 minutes at 6 litres per minute.

- **a** Draw a graph to show this information.
- **b** The pond takes 20 minutes to empty at a steady rate.
  Show this on your graph.
- **c** What is the rate of flow when it empties?

> Represent time on the horizontal axis. Go up to 100 minutes.

AO3

Review exercise

1 **a** Complete the table of values for $y = 2x + 1$.

x	-3	-2	-1	0	1	2	3
y		-3	-1				7

[1 mark]

b Draw a coordinate grid with the x-axis from -4 to $+4$ and the y-axis from -8 to $+8$.
Draw the graph of $y = 2x + 1$. **[1 mark]**

c Find the coordinates of the point where the line $y = 2x + 1$ crosses the line $y = -4$. **[1 mark]**

D
AO2

2 The line $y = 3x - 1$ crosses the line $x = -2$ at the point A.
Work out the coordinates of point A. **[3 marks]**

D
AO3

3 The distance–time graph shows the journey of a train between two stations.
The stations are 6 km apart.

a During the journey the train had to stop at a red signal.
How long was the train stopped? **[1 mark]**

b What was the average speed of the train for the whole journey?
Give your answer in kilometres per hour. **[2 marks]**

C

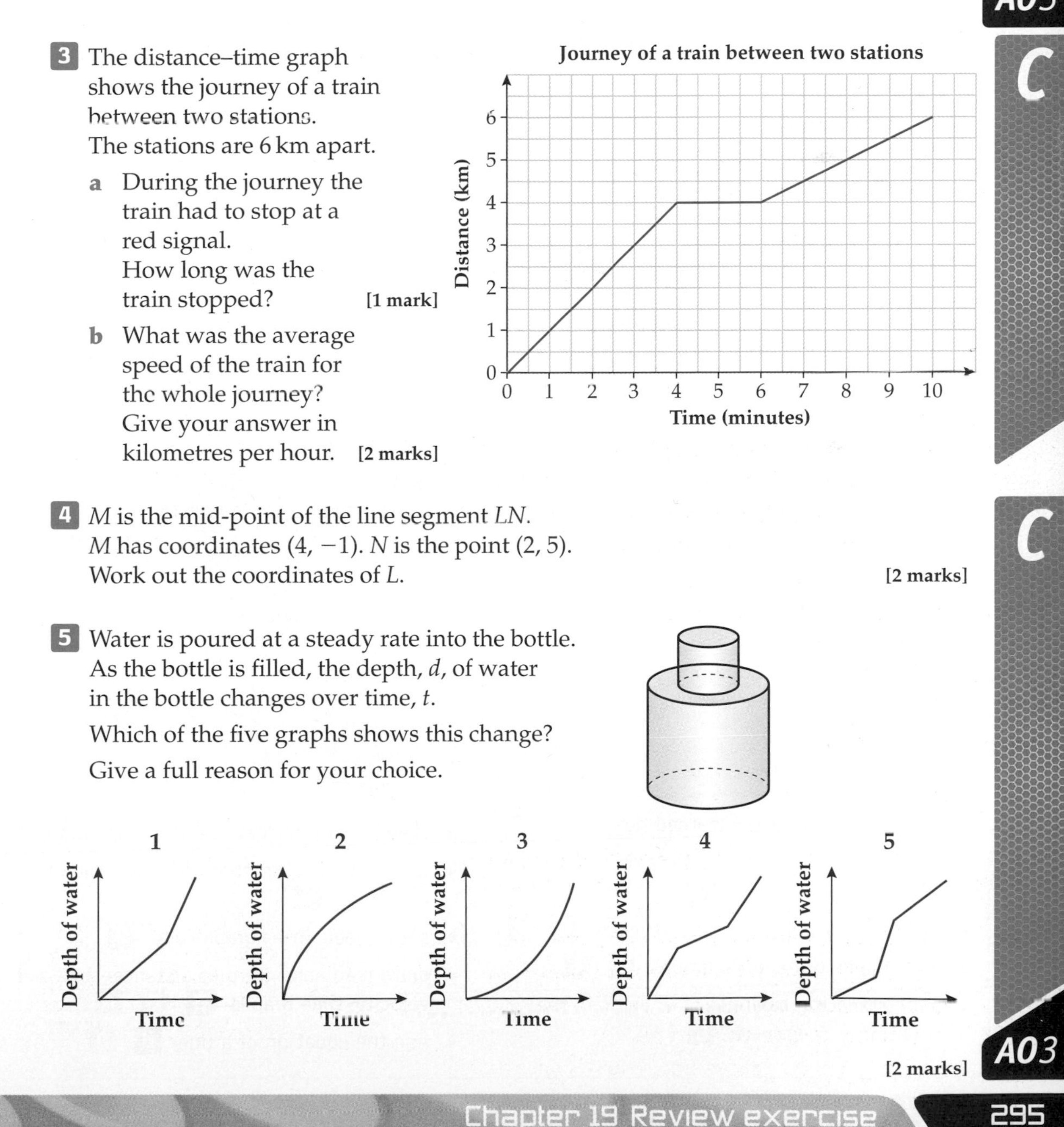

4 M is the mid-point of the line segment LN.
M has coordinates $(4, -1)$. N is the point $(2, 5)$.
Work out the coordinates of L. **[2 marks]**

C

5 Water is poured at a steady rate into the bottle.
As the bottle is filled, the depth, d, of water in the bottle changes over time, t.

Which of the five graphs shows this change?

Give a full reason for your choice.

[2 marks]

AO3

AO3

6 Work out the equation of the line parallel to the line $y = 3x + 5$ that passes through the point (2, 9). **[2 marks]**

B

7 Solve this pair of simultaneous equations graphically.

$2x - y = 1$

$3x + 2y = 12$ **[4 marks]**

8 Work out the equation of the line which passes through the points (−5, 6) and (−1, −2). **[2 marks]**

9 Draw x- and y-axes from −6 to +6.
Indicate the region defined by these three inequalities.

$x \geqslant -3$ $y < 4$ $y \geqslant x + 2$

Mark the region with an R. **[3 marks]**

A

10 **a** Work out the equation of the line PQ. **[2 marks]**

b A point on the line has x-coordinate 8. What is the y-coordinate of this point? **[2 marks]**

c State the gradient of a line perpendicular to line PQ. **[2 marks]**

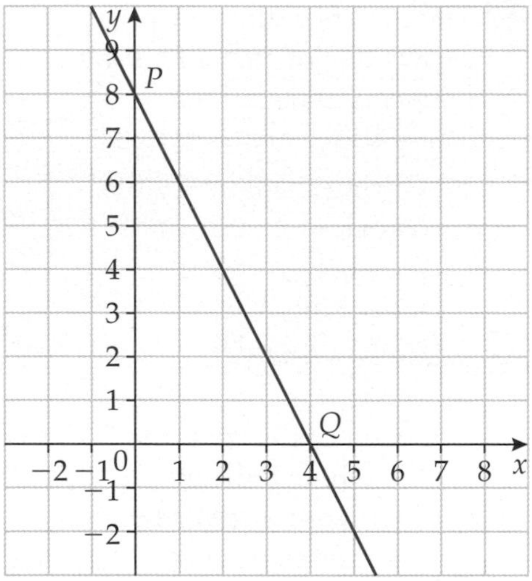

11 Work out the equation of the line that is perpendicular to $y = \frac{1}{3}x + 4$ and which passes through (2, −8). **[3 marks]**

12 Write the equation of the straight line perpendicular to $3x - 6y + 8 = 0$ and which has y-intercept −2. **[3 marks]**

Chapter summary

In this chapter you have learned how to

- work out coordinates of points of intersection when two graphs cross **D**
- find the mid-point of a line segment **D** **C**
- plot graphs of linear functions **D** **C**
- understand the meaning of m and c in the equation $y = mx + c$ **C**
- sketch and interpret real-life graphs **C**
- use a graphical method to solve simultaneous equations **B**
- solve inequalities graphically **B**
- draw, read and interpret distance–time and velocity–time graphs **D** **C** **A**
- find the equation of a line **B** **A**

20

Quadratic equations

This chapter is about quadratics.

In nature, the growth of a population of rabbits can be modelled by a quadratic equation.

Objectives

This chapter will show you how to

- **factorise quadratic expressions, including the difference of two squares** B A
- **solve quadratic equations by rearranging** B
- **factorise quadratics and solve quadratic equations of the form $ax^2 + bx + c = 0$** B A A*
- **use the quadratic equations formula** A A*
- **complete the square** A*

Before you start this chapter

Put your calculator away!

1 Factorise

a $3x + 6y$ b $8x + x^2$ c $3m^2 + mn$

d $5r^2 + 15rt$ e $12xyz + 6xy + 18y^2$

HELP Chapter 12

2 Work out

a -3×4 b $8 \times -6 \times -1$ c $-7 \times 5 + 12$

d $(2 + -4) \times -3$ e $(-6)^2 \times 3$ f $\dfrac{7 \times -4 + 4}{-12}$

3 Work out the value of these expressions when $a = -3$, $b = -4$ and $c = 2$.

a abc b $\sqrt{4b}$

c $ab^2 - c$ d $b^2 - 4ac$

HELP Chapter 15

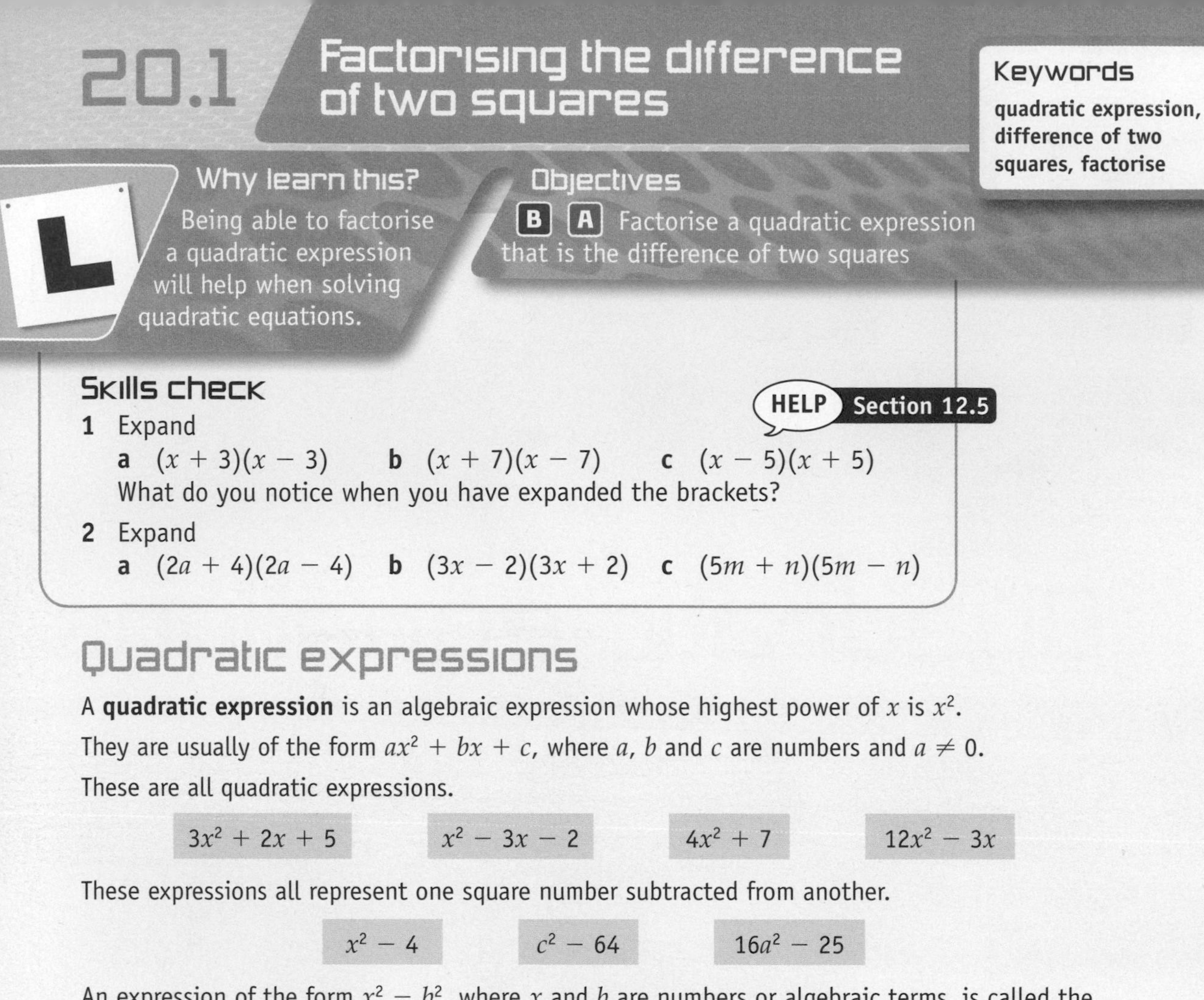

20.1 Factorising the difference of two squares

Keywords
quadratic expression, difference of two squares, factorise

Why learn this?
Being able to factorise a quadratic expression will help when solving quadratic equations.

Objectives
B A Factorise a quadratic expression that is the difference of two squares

Skills check

HELP Section 12.5

1 Expand
a $(x + 3)(x - 3)$ b $(x + 7)(x - 7)$ c $(x - 5)(x + 5)$
What do you notice when you have expanded the brackets?

2 Expand
a $(2a + 4)(2a - 4)$ b $(3x - 2)(3x + 2)$ c $(5m + n)(5m - n)$

Quadratic expressions

A **quadratic expression** is an algebraic expression whose highest power of x is x^2.

They are usually of the form $ax^2 + bx + c$, where a, b and c are numbers and $a \neq 0$.

These are all quadratic expressions.

$3x^2 + 2x + 5$ $x^2 - 3x - 2$ $4x^2 + 7$ $12x^2 - 3x$

These expressions all represent one square number subtracted from another.

$x^2 - 4$ $c^2 - 64$ $16a^2 - 25$

An expression of the form $x^2 - b^2$, where x and b are numbers or algebraic terms, is called the **difference of two squares**.

In general, $x^2 - b^2 = (x - b)(x + b)$.

Check this by multiplying out $(x - b)(x + b)$.

B

Example 1

Factorise $x^2 - 9$.

Remember that **factorising** is the inverse of expanding brackets.

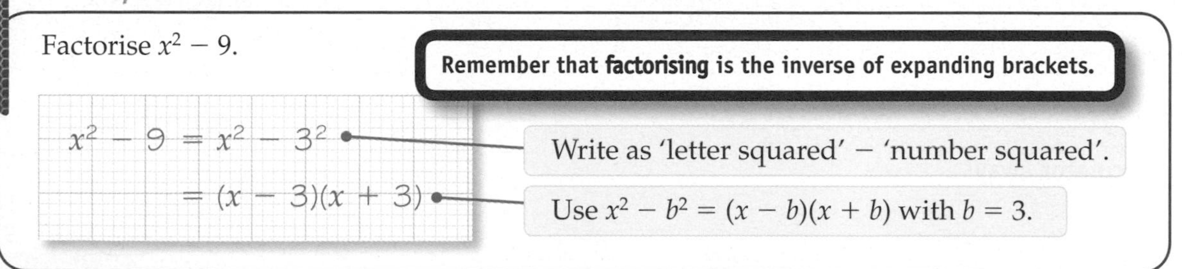

$x^2 - 9 = x^2 - 3^2$ — Write as 'letter squared' − 'number squared'.

$= (x - 3)(x + 3)$ — Use $x^2 - b^2 = (x - b)(x + b)$ with $b = 3$.

Exercise 20A

B

1 Factorise

a $x^2 - 16$ b $x^2 - 25$ c $x^2 - 100$ d $x^2 - 144$
e $a^2 - 1$ f $m^2 - 64$ g $n^2 - 36$ h $t^2 - 121$

B A02

2 Joe thinks of a number, squares it and subtracts 16.

a Write down an algebraic expression to illustrate this.

b Factorise your answer to part **a**.

Example 2

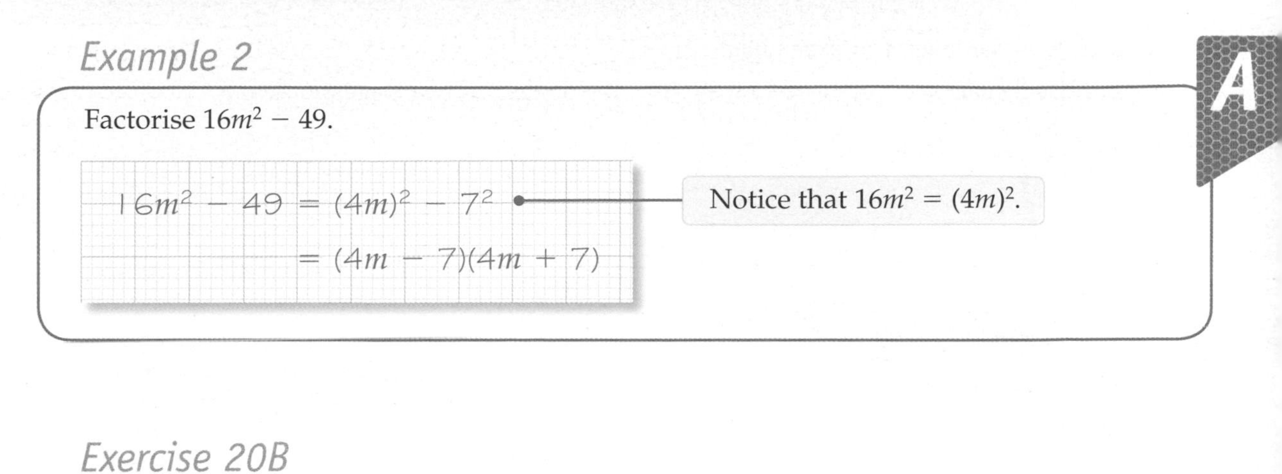

Factorise $16m^2 - 49$.

$$16m^2 - 49 = (4m)^2 - 7^2$$
$$= (4m - 7)(4m + 7)$$

Notice that $16m^2 = (4m)^2$.

Exercise 20B

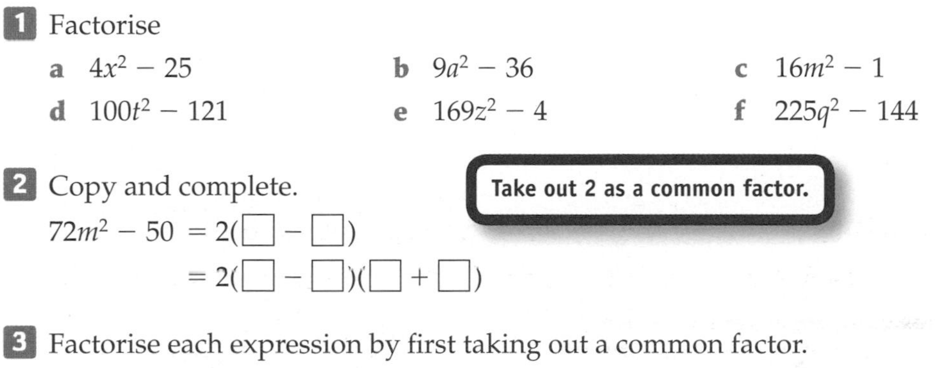

1 Factorise

a $4x^2 - 25$ **b** $9a^2 - 36$ **c** $16m^2 - 1$

d $100t^2 - 121$ **e** $169z^2 - 4$ **f** $225q^2 - 144$

2 Copy and complete.

Take out 2 as a common factor.

$72m^2 - 50 = 2(\square - \square)$

$= 2(\square - \square)(\square + \square)$

3 Factorise each expression by first taking out a common factor.

a $50x^2 - 200$ **b** $27m^2 - 3$ **c** $80t^2 - 45$

d $2a^2 - 18b^2$ **e** $3h^2 - 75k^2$ **f** $600x^2 - 6y^2$

20.2 Factorising quadratics of the form $x^2 + bx + c$

Keywords
product, sum, coefficient

Why learn this?

Understanding how an algebraic expression is constructed can tell you much more about the expression.

Objectives

B Factorise a quadratic expression of the form $x^2 + bx + c$

Skills check

1 Find two positive numbers whose

a product is 12 and sum is 7

b product is 20 and sum is 12

c product is 12 and sum is 13

d product is −12 and sum is 1

e product is −12 and sum is −1

f product is −12 and sum is −4.

Factorising quadratics of the form $x^2 + bx + c$

Expanding a **product** of two expressions, like $(x + 2)$ and $(x + 3)$, gives a quadratic expression.

$$(x + 2)(x + 3) = x^2 + 5x + 6$$

Factorising is the inverse of expanding.

To factorise a quadratic, you need to write it as the product of two expressions.

5 is the sum of 2 and 3.

$$x^2 + 5x + 6 = x^2 + 2x + 3x + 6$$
$$= (x + 2)(x + 3)$$

6 is the product of 2 and 3.

In general, to factorise the equation $x^2 + bx + c$, find two numbers whose **sum** is b (the **coefficient** of x) and whose product is c.

The coefficient of x is the number multiplying the x.

Example 3

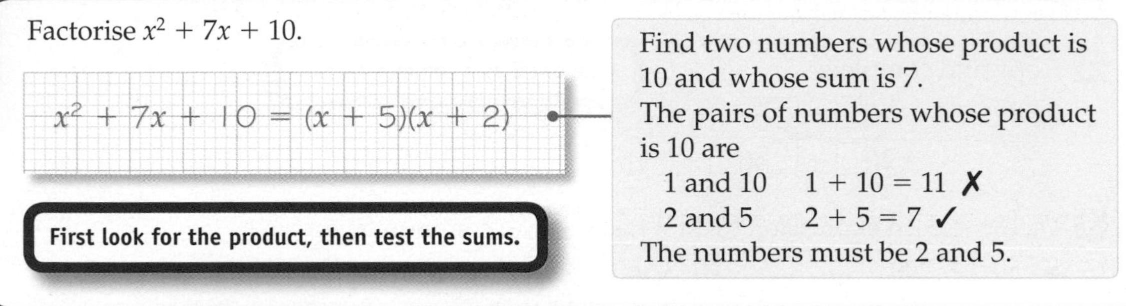

Factorise $x^2 + 7x + 10$.

$x^2 + 7x + 10 = (x + 5)(x + 2)$

Find two numbers whose product is 10 and whose sum is 7.
The pairs of numbers whose product is 10 are
1 and 10 $\quad 1 + 10 = 11$ ✗
2 and 5 $\quad 2 + 5 = 7$ ✓
The numbers must be 2 and 5.

First look for the product, then test the sums.

Exercise 20C

1 Factorise each quadratic expression.

a $x^2 + 5x + 6$ **b** $x^2 + 6x + 8$ **c** $z^2 + 6z + 5$
d $a^2 + 11a + 10$ **e** $n^2 + 8n + 15$ **f** $f^2 + 12f + 36$
g $m^2 + 8m + 12$ **h** $x^2 + 14x + 24$ **i** $b^2 + 11b + 30$

Example 4

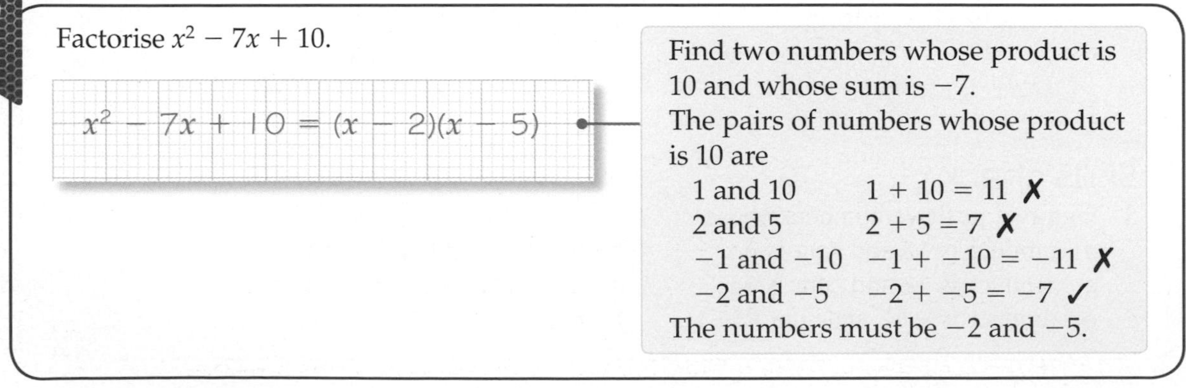

Factorise $x^2 - 7x + 10$.

$x^2 - 7x + 10 = (x - 2)(x - 5)$

Find two numbers whose product is 10 and whose sum is -7.
The pairs of numbers whose product is 10 are
1 and 10 $\quad 1 + 10 = 11$ ✗
2 and 5 $\quad 2 + 5 = 7$ ✗
-1 and -10 $\quad -1 + -10 = -11$ ✗
-2 and -5 $\quad -2 + -5 = -7$ ✓
The numbers must be -2 and -5.

Exercise 20D

1 Factorise each quadratic expression.

a $x^2 - 5x + 6$ **b** $x^2 - 9x + 8$ **c** $z^2 - 7z + 12$
d $a^2 - 9a + 18$ **e** $n^2 - 10n + 25$ **f** $f^2 - 8f + 16$
g $x^2 - 13x + 30$ **h** $b^2 - 11b + 28$ **i** $p^2 - 10p + 24$

Example 5

Factorise **a** $x^2 - 6x - 7$ **b** $x^2 + x - 12$

a $x^2 - 6x - 7 = (x - 7)(x + 1)$ **b** $x^2 + x - 12 = (x - 3)(x + 4)$

Find two numbers whose product is -7 and whose sum is -6.
The pairs of numbers whose product is -7 are

-1 and 7 $\quad -1 + 7 = 6$ ✗
1 and -7 $\quad 1 + -7 = -6$ ✓

The numbers must be 1 and -7.

With practice, you will become better at spotting the correct combination.

Find two numbers whose product is -12 and whose sum is 1.
The pairs of numbers whose product is -12 are

-1 and 12 $\quad -1 + 12 = 11$ ✗
1 and -12 $\quad 1 + -12 = -11$ ✗
-2 and 6 $\quad -2 + 6 = 4$ ✗
2 and -6 $\quad 2 + -6 = -4$ ✗
-3 and 4 $\quad -3 + 4 = 1$ ✓
3 and -4 $\quad 3 + -4 = -1$ ✗

The numbers must be -3 and 4.

B

General rules for factorising quadratics

In general

- if c is positive and b is positive, both numbers in the brackets will be positive
- if c is positive and b is negative, both numbers in the brackets will be negative
- if c is negative, one number will be negative, one will be positive.

Exercise 20E

B

1 Factorise

a $x^2 + 4x - 12$ **b** $x^2 - x - 20$ **c** $z^2 - 2z - 15$
d $a^2 + 6a - 7$ **e** $n^2 + 6n - 16$ **f** $f^2 - f - 30$
g $m^2 + m - 30$ **h** $t^2 - 6t - 72$ **i** $y^2 + 19y - 120$

2 Copy and complete these statements.

a $t^2 + 7r - \square = (t + 10)(t - \square)$
b $m^2 - \square + 15 = (t\square\square)(t - 5)$
c $q^2 - 12q\square\square = (q\square\square)(q - 2)$

3 **a** Factorise each expression. Simplify your answers as much as possible.

i $x^2 + 6x + 9$ **ii** $x^2 - 8x + 16$ **iii** $x^2 + 4x + 4$
iv $x^2 - 14x + 49$ **v** $x^2 - 10x + 25$ **vi** $x^2 + 16x + 64$

b What do you notice about all the answers to part **a**?

c Copy and complete these statements, where m and n are numbers.

i $(x + m)^2 = x^2 + \square x + \square$
ii $(x - n)^2 = x^2 - \square x + \square$

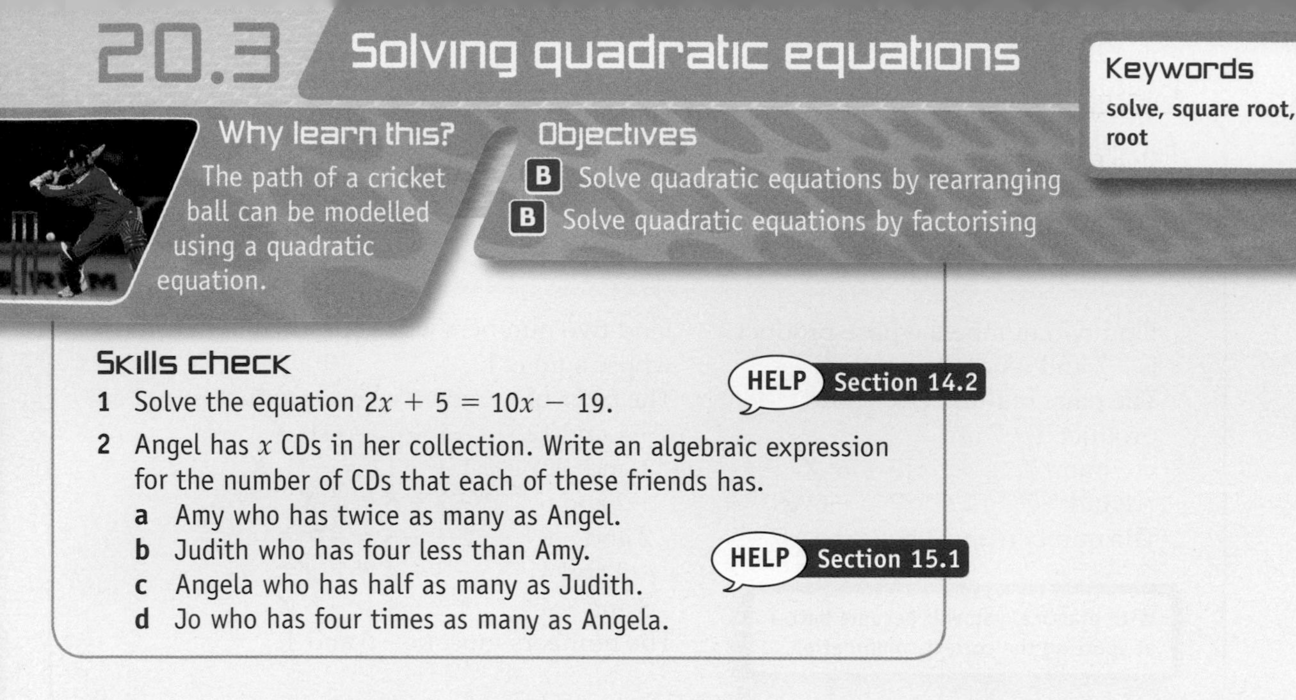

20.3 Solving quadratic equations

Keywords
solve, square root, root

Why learn this?
The path of a cricket ball can be modelled using a quadratic equation.

Objectives

B Solve quadratic equations by rearranging

B Solve quadratic equations by factorising

Skills check

1 Solve the equation $2x + 5 = 10x - 19$.

HELP Section 14.2

2 Angel has x CDs in her collection. Write an algebraic expression for the number of CDs that each of these friends has.

a Amy who has twice as many as Angel.

b Judith who has four less than Amy.

c Angela who has half as many as Judith.

d Jo who has four times as many as Angela.

HELP Section 15.1

Solving quadratic equations by rearranging

You can **solve** some quadratic equations by rearranging them to make x the subject.

B

Example 6

Solve the quadratic equation $3x^2 - 27 = 0$.

$3x^2 - 27 = 0$

$3x^2 = 27$ — Add 27 to both sides of the equation.

$x^2 = 9$ — Divide both sides by 3.

$x = \pm 3$ — Find the **square root** of both sides. Remember, when you find the root there are *two* solutions: positive and negative.

Exercise 20F

B

1 Solve these equations.

a $r^2 = 169$

b $x^2 + 5 = 14$

c $18 = 2t^2$

d $y^2 - 20 = -19$

e $\frac{m^2}{4} = 6.25$

f $\frac{p^2}{5} = 20$

2 Find the roots of

a $2x^2 + 7 = 39$

b $5y^2 - 100 = -80$

c $3r^2 = 5r^2 - 98$

d $3t^2 = t^2 + 18$

e $100 - 5y^2 = 95$

f $2x^2 + 2 = 130$

Root is another name for a solution.

Example 7

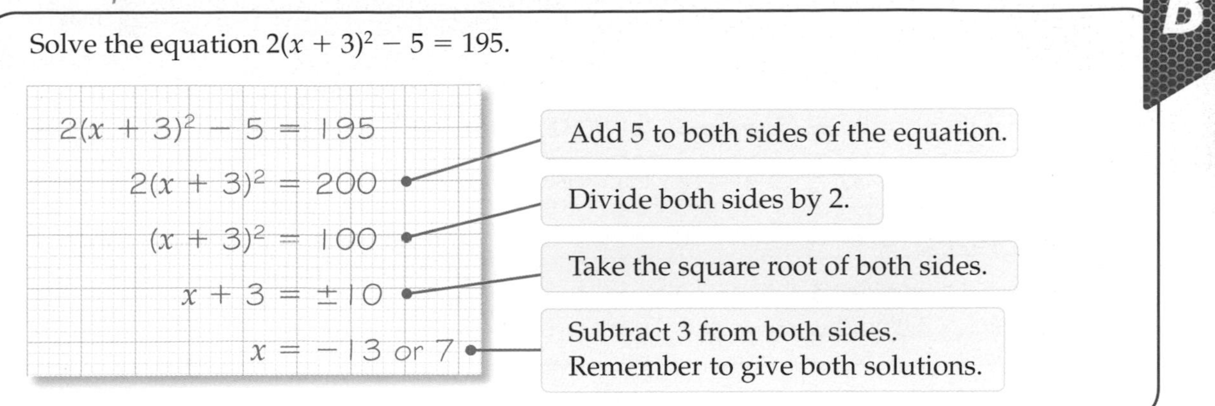

Solve the equation $2(x + 3)^2 - 5 = 195$.
$2(x + 3)^2 - 5 = 195$
$2(x + 3)^2 = 200$
$(x + 3)^2 = 100$
$x + 3 = \pm 10$
$x = -13$ or 7
Add 5 to both sides of the equation.
Divide both sides by 2.
Take the square root of both sides.
Subtract 3 from both sides.
Remember to give both solutions.

Exercise 20G

1 Find the roots of

a $2(x + 1)^2 = 8$

b $4.5 = \dfrac{(r - 7)^2}{2}$

2 Solve these equations.

a $(x + 1)^2 - 16 = 20$

b $100 = 4t^2 + 36$

c $150 - 3t^2 = 42$

d $6 + 3t^2 = 2t^2 + 15$

e $4(x + 3)^2 = 100$

f $7(y - 2)^2 = 700$

3 A field is three times as long as it is wide.

a Using x for the width of the field, write an expression for its length.

b Write an expression for the area of the field, in terms of x.

Use your answer to part a.

c The field has an area of 1200 m^2.
Write an equation for the area of the field.

Use your answer to part b.

d Solve your equation to find x.

e What are the length and the width of the field?

4 Explain why you cannot find a solution to $x^2 + 20 = 5$.

5 A rectangle has length five times its width.
The area of the rectangle is 845 mm^2.
What is the width of the rectangle?

Solving quadratic equations by factorising

Solving quadratic equations by factorising relies on the fact that when $a \times b = 0$, a is 0, b is 0 or both are 0.

So if $(x + 2)(x - 4) = 0$, either $x + 2 = 0$, which means $x = -2$, or $x - 4 = 0$, which means $x = 4$.

If the product of two things is zero, one of them *must* be zero.

To solve a quadratic equation

Step 1: Rearrange the equation so that one side is zero.

Step 2: Factorise the quadratic expression.

Step 3: Find the solutions.

Usually there are two solutions. However, when the expression factorises to $(x + m)^2 = 0$, there is only one solution.

B

Example 8

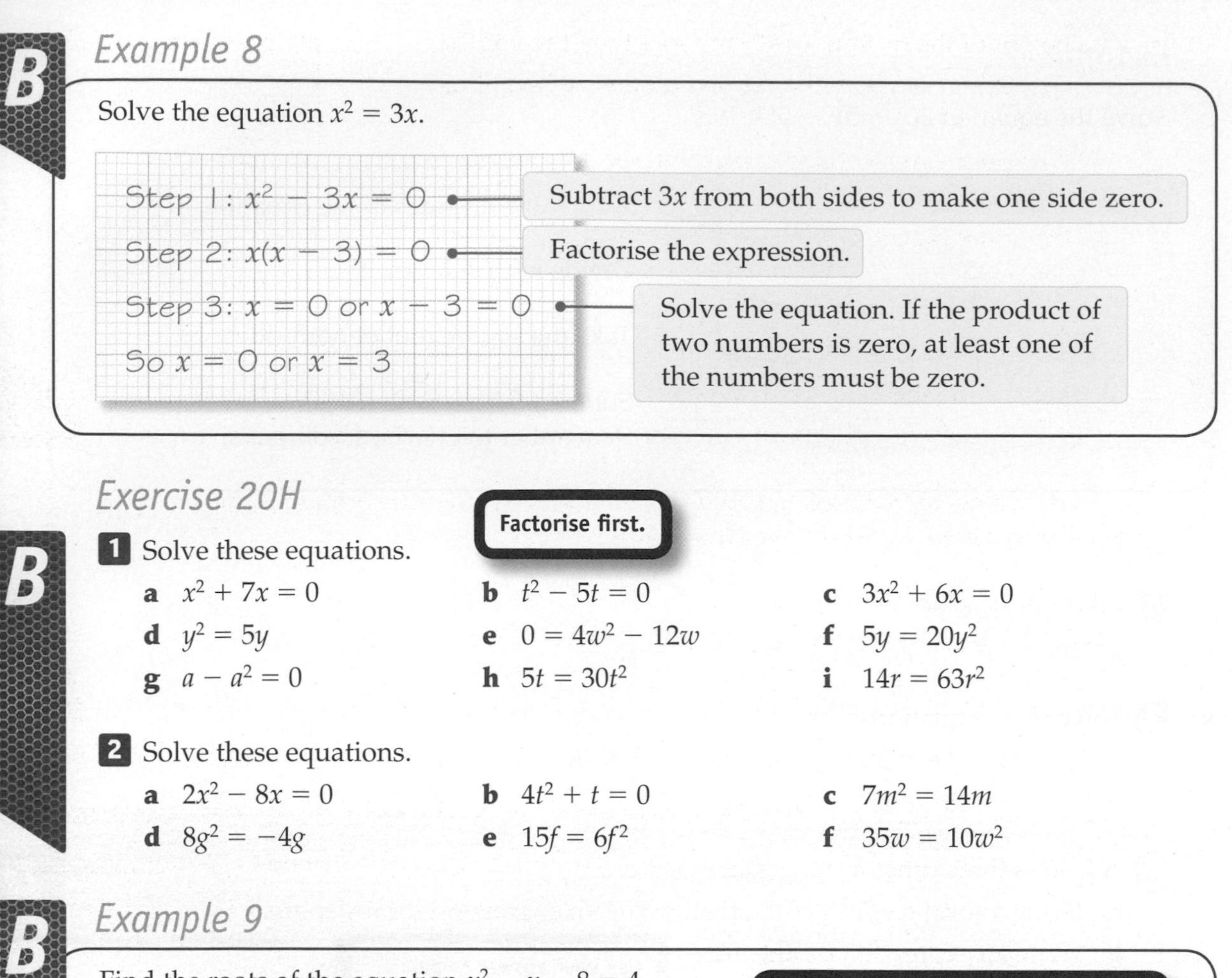

Solve the equation $x^2 = 3x$.

Step 1: $x^2 - 3x = 0$ — Subtract $3x$ from both sides to make one side zero.

Step 2: $x(x - 3) = 0$ — Factorise the expression.

Step 3: $x = 0$ or $x - 3 = 0$ — Solve the equation. If the product of two numbers is zero, at least one of the numbers must be zero.

So $x = 0$ or $x = 3$

Exercise 20H

B

Factorise first.

1 Solve these equations.

a $x^2 + 7x = 0$ **b** $t^2 - 5t = 0$ **c** $3x^2 + 6x = 0$

d $y^2 = 5y$ **e** $0 = 4w^2 - 12w$ **f** $5y = 20y^2$

g $a - a^2 = 0$ **h** $5t = 30t^2$ **i** $14r = 63r^2$

2 Solve these equations.

a $2x^2 - 8x = 0$ **b** $4t^2 + t = 0$ **c** $7m^2 = 14m$

d $8g^2 = -4g$ **e** $15f = 6f^2$ **f** $35w = 10w^2$

B

Example 9

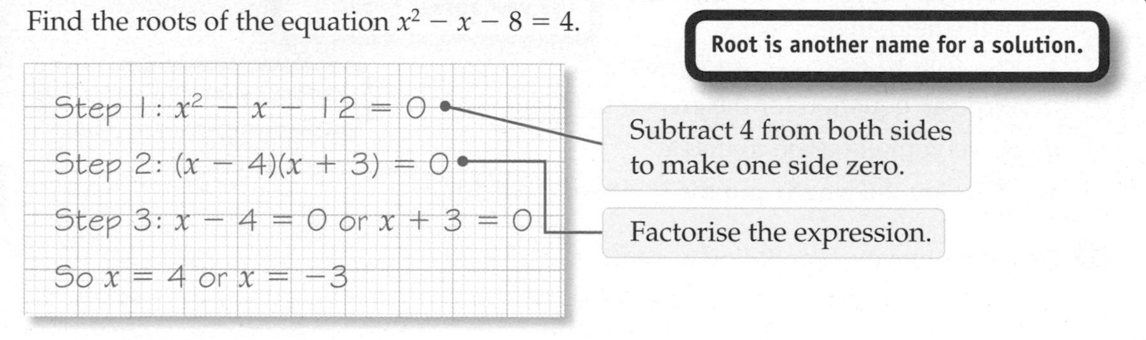

Find the roots of the equation $x^2 - x - 8 = 4$.

Root is another name for a solution.

Step 1: $x^2 - x - 12 = 0$ — Subtract 4 from both sides to make one side zero.

Step 2: $(x - 4)(x + 3) = 0$ — Factorise the expression.

Step 3: $x - 4 = 0$ or $x + 3 = 0$

So $x = 4$ or $x = -3$

Exercise 20I

B

1 Find the roots of these equations.

a $x^2 + 4x + 3 = 0$ **b** $x^2 - x - 6 = 0$ **c** $x^2 - 6x + 8 = 0$

d $x^2 + x = 12$ **e** $x^2 = x + 20$ **f** $x^2 + 2x = -1$

g $z^2 = 3z + 4$ **h** $2q + q^2 = 15$ **i** $w^2 = 4w - 4$

j $6t + 7 = t^2$ **k** $6p + 9 = -p^2$ **l** $10x - 25 = x^2$

B

2 Jane is three years younger than her older sister. The product of their ages is 54. Use x to represent Jane's age.

Remember that Jane is younger.

a Write down an algebraic expression for her sister's age.

b Write down and simplify an algebraic expression for the product of their ages.

c Form and solve an algebraic equation to find the value of x.

d Explain why only one of the solutions makes sense.

3 The height of the rectangle is 3 cm more than the width.

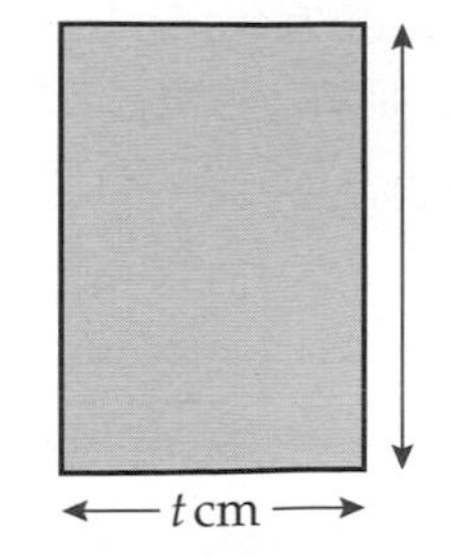

a Write down an algebraic expression for the height of the rectangle.

b Write down an algebraic expression for the area of the rectangle.

c Given that the area of the rectangle is 40 cm², form and solve a quadratic equation to work out the value of t.

B

4 A rectangular garden is 4 m longer than it is wide. Its area is 165 m².

a Sketch and label a diagram to show the area.

b Form and solve a quadratic equation to work out the dimensions of the garden.

AO2

5 I think of a negative number, square it and add five times the original number. My answer is 24. What number did I think of?

B

6 I think of a positive number.
I square it, then subtract six times the number.
The answer is 27.
What was my original number?

AO3

20.4 Factorising quadratics of the form $ax^2 + bx + c$

Why learn this?

By breaking down an algebraic expression you can discover some of the properties of the expression.

Objectives

A Solve quadratic equations by factorising

A* Factorise quadratic expressions of the form $ax^2 + bx + c$

Skills check

1 Write down all the pairs of numbers whose product is

a 10 **b** 12 **c** -36

Don't forget negative numbers.

2 Factorise

a $x^2 + 3x - 4$ **b** $x^2 + 5x + 6$ **c** $x^2 - 8x + 7$

Factorising quadratics of the form $ax^2 + bx + c$

In the expression $ax^2 + bx + c$, the a is the coefficient of x^2 and b is the coefficient of x.

In the quadratic expression $3x^2 + 13x + 4$, the coefficient of x^2 is 3. The first terms in the brackets must multiply to give $3x^2$.

The first terms in the brackets must be $3x$ and $1x$.

$$3x^2 + 13x + 4 = (3x \square\square)(x \square\square)$$

3 is a prime number – the only factors are 3 and 1.

The product of the last two terms must be +4.

Possible pairs of numbers are 1 and 4, −1 and −4, 2 and 2, or −2 and −2.

The coefficient of x is positive (+13), so the two numbers must be positive.

Possible factorisations are

$(3x + 4)(x + 1)$ $(3x + 1)(x + 4)$ $(3x + 2)(x + 2)$

Try expanding each one.

$(3x + 4)(x + 1) = 3x^2 + 3x + 4x + 4$ ✗

$(3x + 1)(x + 4) = 3x^2 + 12x + x + 4$ ✓

$(3x + 2)(x + 2) = 3x^2 + 6x + 2x + 4$ ✗

Always check using FOIL to expand the brackets.

So $3x^2 + 13x + 4 = (3x + 1)(x + 4)$

A

Example 10

Factorise $2x^2 - 7x - 4$.

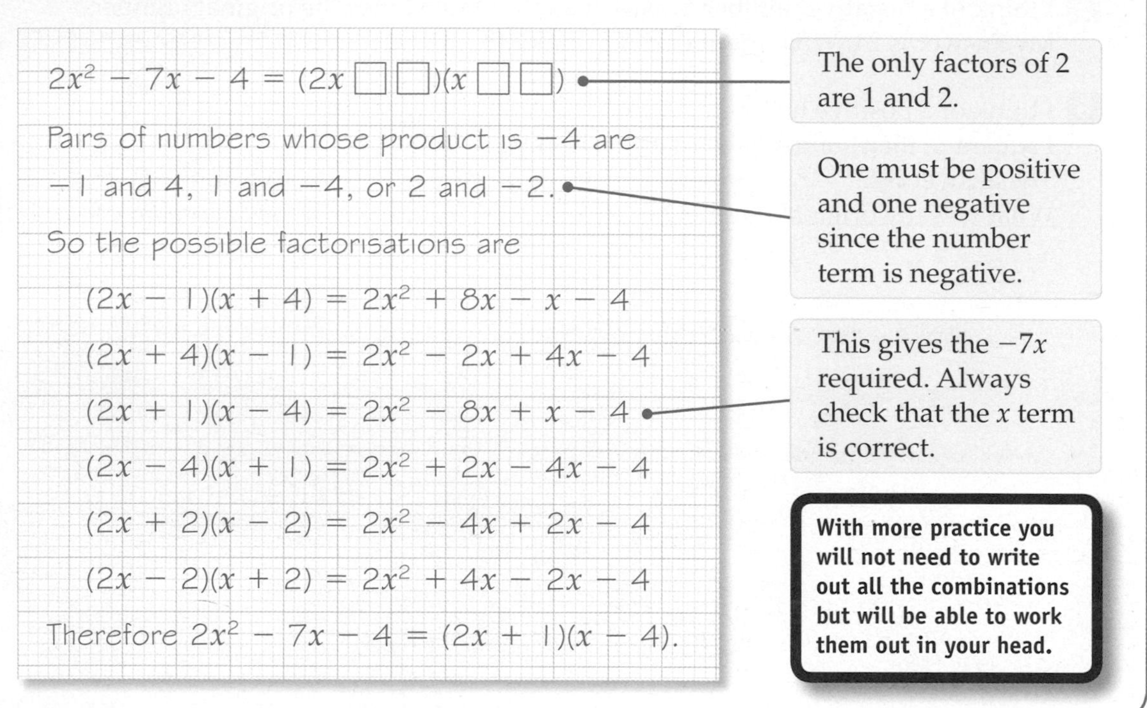

2x² − 7x − 4 = (2x □□)(x □□)
Pairs of numbers whose product is −4 are
−1 and 4, 1 and −4, or 2 and −2.
So the possible factorisations are
(2x − 1)(x + 4) = 2x² + 8x − x − 4
(2x + 4)(x − 1) = 2x² − 2x + 4x − 4
(2x + 1)(x − 4) = 2x² − 8x + x − 4
(2x − 4)(x + 1) = 2x² + 2x − 4x − 4
(2x + 2)(x − 2) = 2x² − 4x + 2x − 4
(2x − 2)(x + 2) = 2x² + 4x − 2x − 4
Therefore 2x² − 7x − 4 = (2x + 1)(x − 4).
The only factors of 2 are 1 and 2.
One must be positive and one negative since the number term is negative.
This gives the −7x required. Always check that the x term is correct.
With more practice you will not need to write out all the combinations but will be able to work them out in your head.

Exercise 20J

A

Factorise each quadratic expression.

1 $2x^2 + 5x + 3$

2 $3x^2 + 14x + 8$

3 $5x^2 + 12x + 4$

4 $7x^2 + 26x + 15$

5 $5x^2 + 19x - 4$

6 $3x^2 - 4x - 4$

7 $11x^2 - 13x + 2$

8 $2x^2 - 5x + 2$

9 $3x^2 - 19x + 20$

10 $5x^2 - 39x - 8$

11 $2x^2 - 14x + 24$

12 $7x^2 - 8x - 12$

Factorising quadratics of the form $ax^2 + bx + c$ when the coefficient of x^2 is not prime

When the coefficient of x^2 is not prime, there are more possible cases to consider.

Example 11

Factorise $6x^2 + 11x + 4$.

$(3x \square\square)(2x \square\square)$ or $(6x \square\square)(x \square\square)$ — Factors of 6 are 1 and 6 or 2 and 3.

Pairs of numbers whose product is +4 are 2 and 2 or 1 and 4. — All terms are positive, so only consider positive numbers.

So the possible factorisations are

$(3x + 2)(2x + 2) = \ldots + 6x + 4x + \ldots = \ldots + 10x + \ldots$

$(6x + 2)(x + 2) = \ldots + 12x + 2x + \ldots = \ldots + 14x + \ldots$

$(3x + 1)(2x + 4) = \ldots + 12x + 2x + \ldots = \ldots + 14x + \ldots$

$(3x + 4)(2x + 1) = \ldots + 3x + 8x + \ldots = \ldots + 11x + \ldots$ — This gives $+11x$ required.

$(6x + 1)(x + 4) =$

$(6x + 4)(x + 1) =$ — You can stop trying once you have found the correct pair.

Therefore $6x^2 + 11x + 4 = (3x + 4)(2x + 1)$.

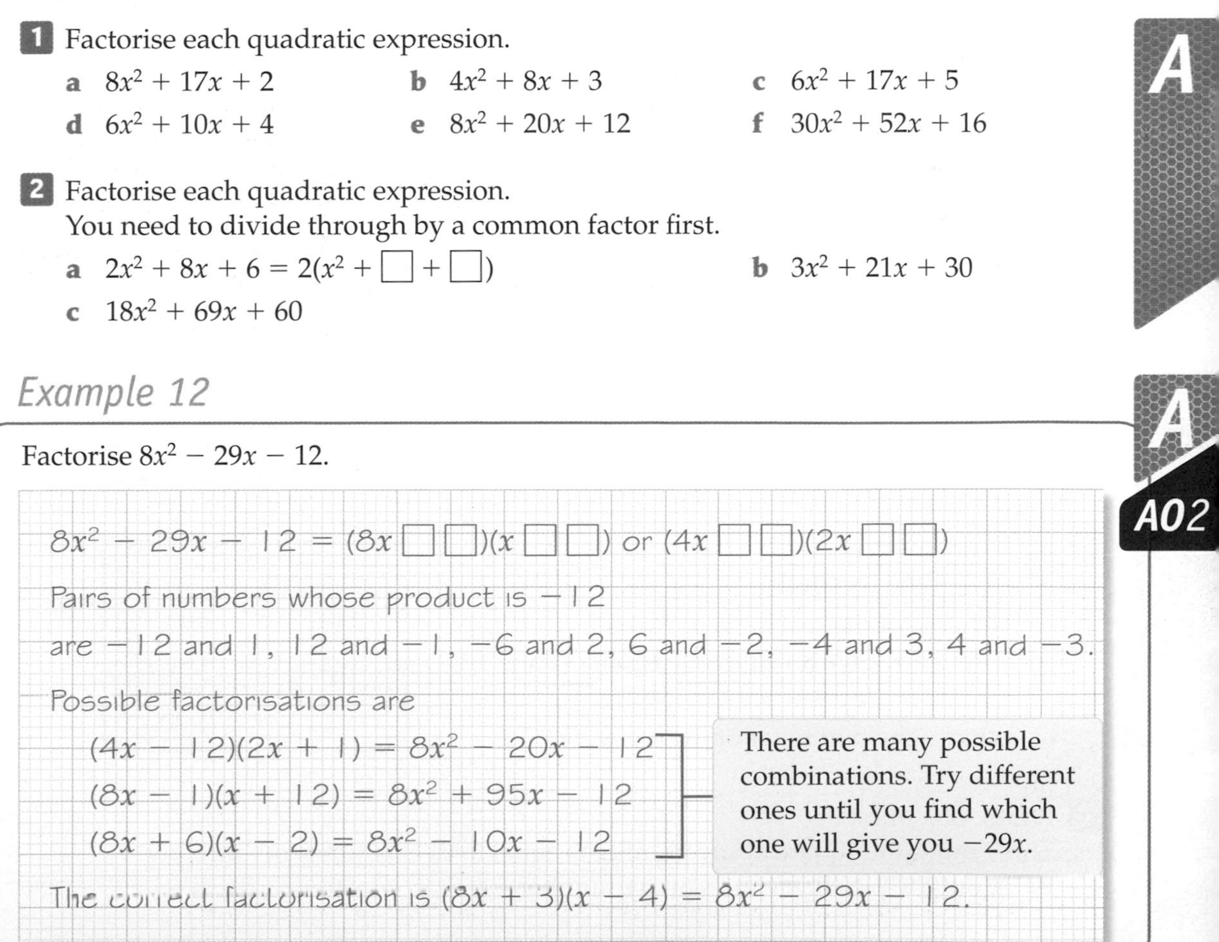

Exercise 20K

1 Factorise each quadratic expression.

a $8x^2 + 17x + 2$ **b** $4x^2 + 8x + 3$ **c** $6x^2 + 17x + 5$

d $6x^2 + 10x + 4$ **e** $8x^2 + 20x + 12$ **f** $30x^2 + 52x + 16$

2 Factorise each quadratic expression.
You need to divide through by a common factor first.

a $2x^2 + 8x + 6 = 2(x^2 + \square + \square)$ **b** $3x^2 + 21x + 30$

c $18x^2 + 69x + 60$

Example 12

A02

Factorise $8x^2 - 29x - 12$.

$8x^2 - 29x - 12 = (8x \square\square)(x \square\square)$ or $(4x \square\square)(2x \square\square)$

Pairs of numbers whose product is -12 are -12 and 1, 12 and -1, -6 and 2, 6 and -2, -4 and 3, 4 and -3.

Possible factorisations are

$(4x - 12)(2x + 1) = 8x^2 - 20x - 12$

$(8x - 1)(x + 12) = 8x^2 + 95x - 12$

$(8x + 6)(x - 2) = 8x^2 - 10x - 12$

There are many possible combinations. Try different ones until you find which one will give you $-29x$.

The correct factorisation is $(8x + 3)(x - 4) = 8x^2 - 29x - 12$.

Exercise 20L

A

1 Factorise each quadratic expression.

a $10x^2 + x - 3$
b $12x^2 - x - 6$
c $4x^2 + 2x - 6$
d $20x^2 + 19x - 28$
e $30x^2 - 52x + 16$

2 Factorise each quadratic expression.

You need to divide through by a common factor first.

a $5x^2 + 5x - 10$
b $14x^2 + 35x - 84$
c $15x^2 - 72x - 15$
d $28x^2 - 88x + 12$
e $24x^2 + 4x - 4$
f $50x^2 - 70x - 60$
g $36x^2 - 42x + 12$

A

Example 13

Find the values of x which satisfy the equation $8x^2 = 14x + 4$.

Step 1: $8x^2 - 14x - 4 = 0$ — Rearrange the equation to make one side zero.

Step 2: $(2x - 4)(4x + 1) = 0$ — Factorise the equation.

Step 3: $2x - 4 = 0$ or $4x + 1 = 0$ — Solve the two linear equations.

$2x = 4$ or $4x = -1$

So $x = 2$ or $x = -\frac{1}{4}$

Exercise 20M

A

1 Find the roots of these quadratic equations.
Leave your answers as fractions where necessary.

a $2a^2 + 5a - 3 = 0$
b $3b^2 + 5b + 2 = 0$
c $4c^2 - c - 5 = 0$
d $0 = 5d^2 - 8d - 4$
e $6e^2 - 16e + 8 = 0$
f $4f^2 - 6f - 4 = 0$
g $6g^2 + 19g + 10 = 0$
h $0 = 4h^2 + 8h + 4$
i $7i^2 - 3i - 4 = 0$

2 Find the values of x which satisfy these equations.

a $2x^2 = 4x + 6$
b $9x^2 + 10 = 21x$
c $10x^2 + 13x = 9$
d $4x + 16 - 6x^2 = 0$
e $15x^2 = -30x - 15$
f $(x + 2)(x - 2) = 3x$

A
AO2

3 a Write down an algebraic expression for the area of the rectangle.

b The area of the rectangle is 108 cm^2.
Form and solve an algebraic equation to find the value of x.

c What is the perimeter of the rectangle?

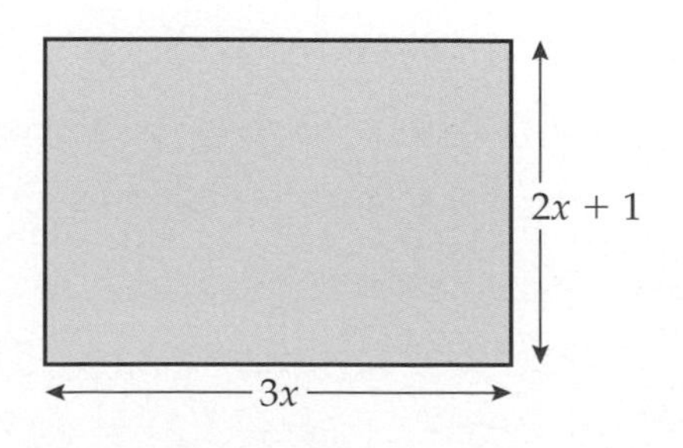

A
AO3

4 I think of a number.
Three times the square of my number is equal to twelve times my number.
Work out the possible values of my number.

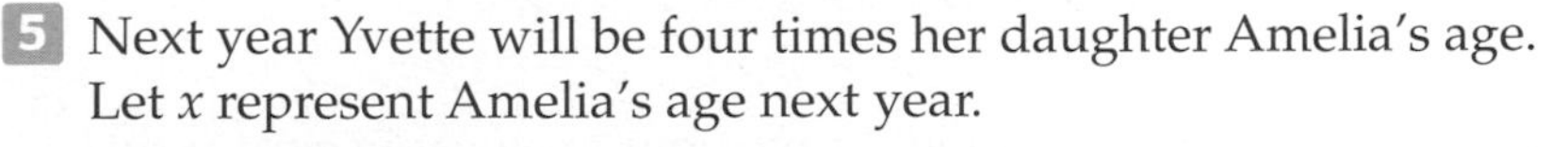

5 Next year Yvette will be four times her daughter Amelia's age.
Let x represent Amelia's age next year.

a Write down an algebraic expression for
i Yvette's age next year **ii** Amelia's age this year **iii** Yvette's age this year.

b The product of their ages is 351.
Form and solve an algebraic equation to work out Amelia's age.

c How old is Yvette this year?

A* AO2

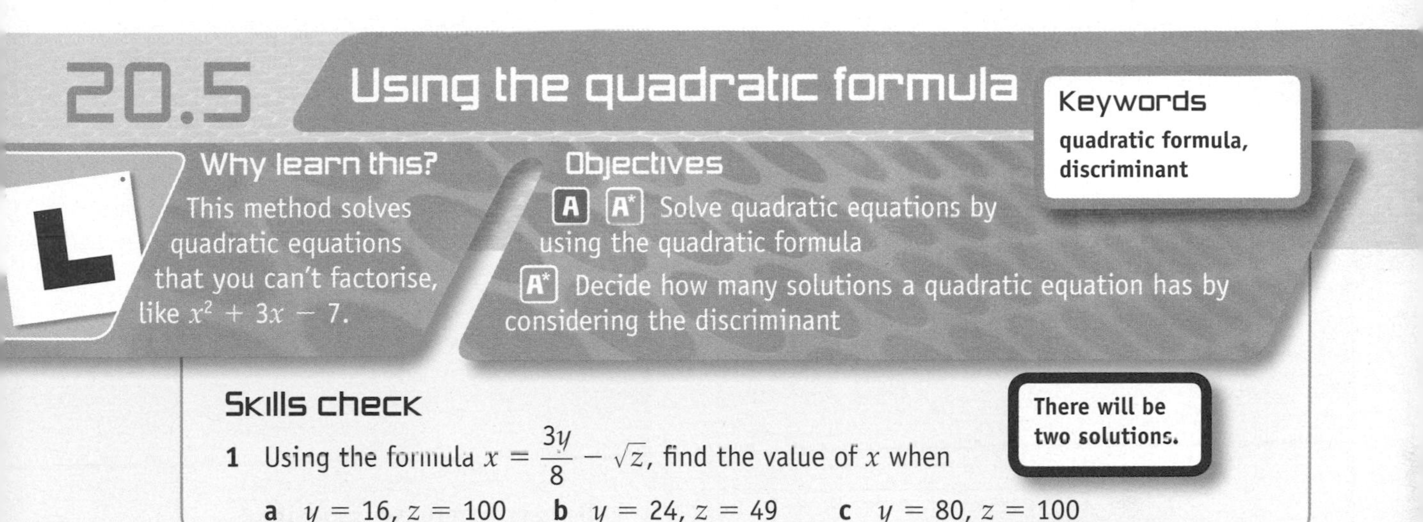

20.5 Using the quadratic formula

Keywords
quadratic formula, discriminant

Why learn this?
This method solves quadratic equations that you can't factorise, like $x^2 + 3x - 7$.

Objectives
A A* Solve quadratic equations by using the quadratic formula
A* Decide how many solutions a quadratic equation has by considering the discriminant

Skills check

1 Using the formula $x = \frac{3y}{8} - \sqrt{z}$, find the value of x when

a $y = 16, z = 100$ **b** $y = 24, z = 49$ **c** $y = 80, z = 100$

There will be two solutions.

The quadratic formula

Sometimes a quadratic expression cannot be factorised.

You can use the **quadratic formula** to solve a quadratic equation of the form $ax^2 + bx + c = 0$, where $a \neq 0$.

$$x = \frac{-b \pm \sqrt{b^2 - 4ac}}{2a}$$

You do not need to learn the formula – it will be on the exam formula sheet.

Be careful! You cannot use the quadratic formula until you have made one side of the equation zero.

Example 14

A

Use the quadratic formula to solve the equation $x^2 + 3x - 7 = 0$.

$a = 1, b = 3, c = -7$ — Write down the values of a, b and c.

$$x = \frac{-b \pm \sqrt{b^2 - 4ac}}{2a}$$

$$= \frac{-3 \pm \sqrt{3^2 - 4 \times 1 \times -7}}{2 \times 1}$$

Substitute these values into the quadratic formula. Be careful with the negative value.

$$= \frac{-3 \pm \sqrt{9 + 28}}{2}$$

Simplify the calculation. Follow the order of operations.

$$= \frac{-3 \pm \sqrt{37}}{2}$$

$$x = \frac{-3 + \sqrt{37}}{2} \text{ or } x = \frac{-3 - \sqrt{37}}{2}$$

Leave your answer in surd form.

Exercise 20N

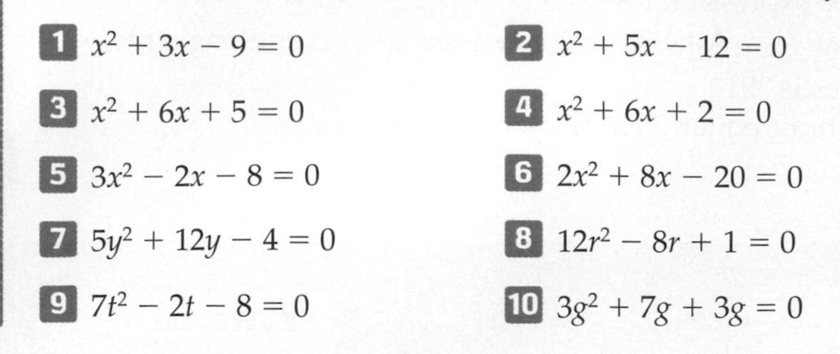

Use the quadratic formula to solve each equation. Leave your answers in surd form.

1 $x^2 + 3x - 9 = 0$

2 $x^2 + 5x - 12 = 0$

3 $x^2 + 6x + 5 = 0$

4 $x^2 + 6x + 2 = 0$

5 $3x^2 - 2x - 8 = 0$

6 $2x^2 + 8x - 20 = 0$

7 $5y^2 + 12y - 4 = 0$

8 $12r^2 - 8r + 1 = 0$

9 $7t^2 - 2t - 8 = 0$

10 $3g^2 + 7g + 3g = 0$

Example 15

Solve the quadratic equation $2x^2 = 6x + 12$.

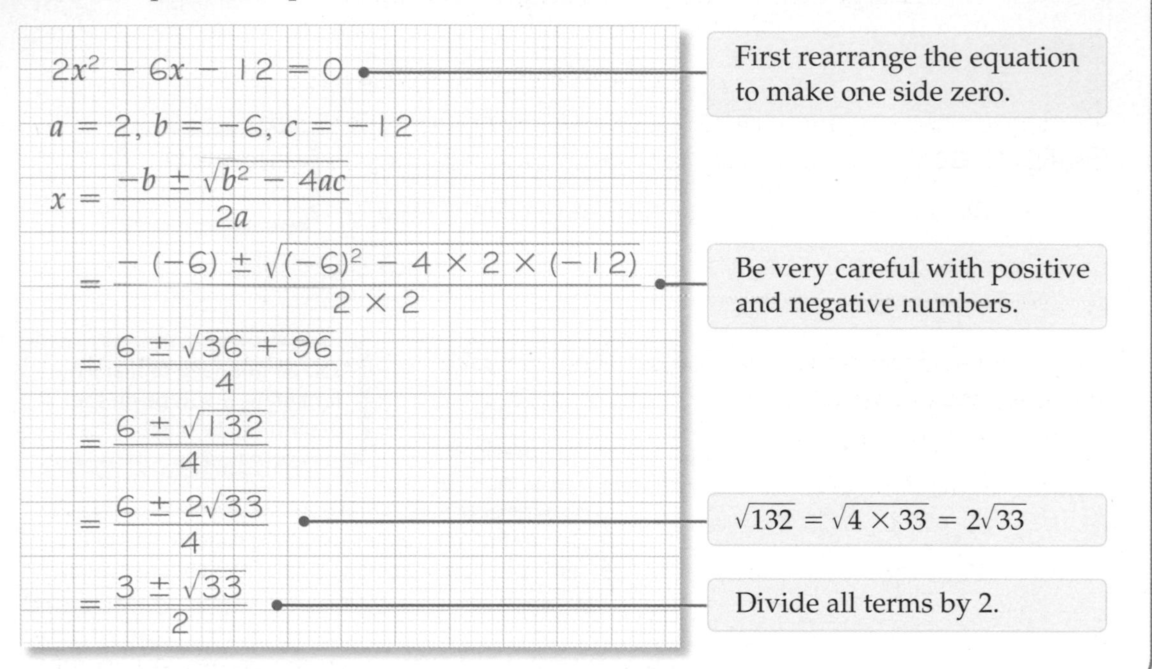

$2x^2 - 6x - 12 = 0$ — First rearrange the equation to make one side zero.

$a = 2, b = -6, c = -12$

$x = \frac{-b \pm \sqrt{b^2 - 4ac}}{2a}$

$= \frac{-(-6) \pm \sqrt{(-6)^2 - 4 \times 2 \times (-12)}}{2 \times 2}$ — Be very careful with positive and negative numbers.

$= \frac{6 \pm \sqrt{36 + 96}}{4}$

$= \frac{6 \pm \sqrt{132}}{4}$

$= \frac{6 \pm 2\sqrt{33}}{4}$ — $\sqrt{132} = \sqrt{4 \times 33} = 2\sqrt{33}$

$= \frac{3 \pm \sqrt{33}}{2}$ — Divide all terms by 2.

Exercise 20O

Make sure you rearrange the equations first.

1 Use the quadratic formula to solve these equations. Leave your answers in surd form.

a $x^2 = 4x + 1$ **b** $x^2 + 16 = 12x$ **c** $x^2 - 8x = 6$

d $x^2 = 1 - 6x$ **e** $4 + 2x = x^2$ **f** $x^2 + 8x + 2 = 0$

2 Try to solve this quadratic equation using the quadratic formula.
$x^2 + x + 1 = 0$
Explain why you cannot find a solution.

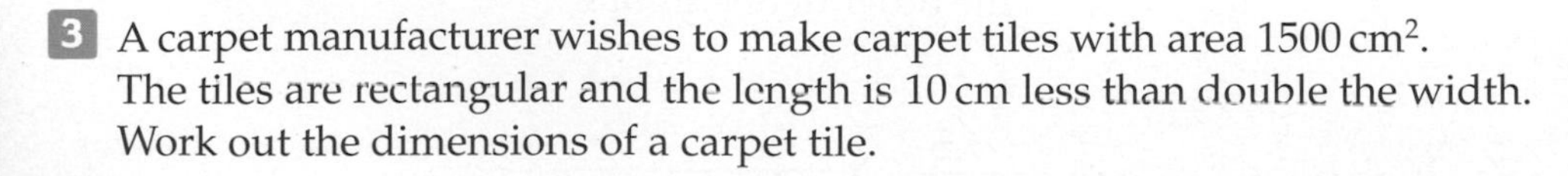

3 A carpet manufacturer wishes to make carpet tiles with area 1500 cm². The tiles are rectangular and the length is 10 cm less than double the width. Work out the dimensions of a carpet tile.

4 Look at your answer to **Q3**. Did you need to use the quadratic formula? Explain your answer.

The discriminant

Question 2 in Exercise 200 asked you to try to solve the equation $x^2 + x + 1 = 0$.

Using the quadratic formula

$$a = 1, b = 1, c = 1$$

$$x = \frac{-b \pm \sqrt{b^2 - 4ac}}{2a}$$

$$= \frac{-1 \pm \sqrt{1^2 - 4 \times 1 \times 1}}{2 \times 1}$$

$$= \frac{-1 \pm \sqrt{-3}}{2}$$

The calculations result in trying to find the square root of a negative number. This has no real solutions – you will learn more about this if you do A-level maths.

$b^2 - 4ac$ in the quadratic formula is known as the **discriminant.**

In general,

- when $b^2 - 4ac > 0$, there are two distinct solutions to the quadratic equation
- when $b^2 - 4ac < 0$, there are no real solutions to the quadratic equation
- when $b^2 - 4ac = 0$, there is one solution (sometimes called a repeated root).

Example 16

A*

By considering the discriminant, decide whether each of these quadratic equations has zero, one or two solutions.

a $3x^2 + 2x - 5 = 0$

b $7x^2 = 10x - 8$

c $9x^2 + 16 = 24x$

a $3x^2 + 2x - 5 = 0$

$a = 3, b = 2, c = -5$ — Write down the values of a, b and c.

$b^2 - 4ac = 2^2 - 4 \times 3 \times -5$ — Work out $b^2 - 4ac$

$= 4 + 60$

$= 64$

Since $64 > 0$ there are two solutions.

b $7x^2 = 10x - 8$

$7x^2 - 10x + 8 = 0$ — Rearrange to the form $ax^2 + bx + c = 0$

$a = 7, b = -10, c = 8$

$b^2 - 4ac = (-10)^2 - 4 \times 7 \times 8$

$= 100 - 224$

$= -124$

Since $-124 < 0$ there are no solutions.

c $9x^2 + 16 = 24x$

$9x^2 - 24x + 16 = 0$ — Rearrange to the form $ax^2 + bx + c = 0$

$a = 9, b = -24, c = 16$

$b^2 - 4ac = (-24)^2 - 4 \times 9 \times 16$

$= 576 - 576$

$= 0$

There is one (repeated) solution.

Exercise 20P

A*

1 For each quadratic equation, decide if there are zero, one or two solutions.

a $3x^2 + 2x - 4 = 0$ **b** $5m^2 + 9m + 6 = 0$ **c** $3t^2 + 6t + 3 = 0$

d $4d^2 - 5d + 6 = 0$ **e** $0 = 2z^2 + 5z + 1$ **f** $4x^2 = 3x - 1$

g $9t = 5t^2 - 12$ **h** $2q^2 = -8q - 8$

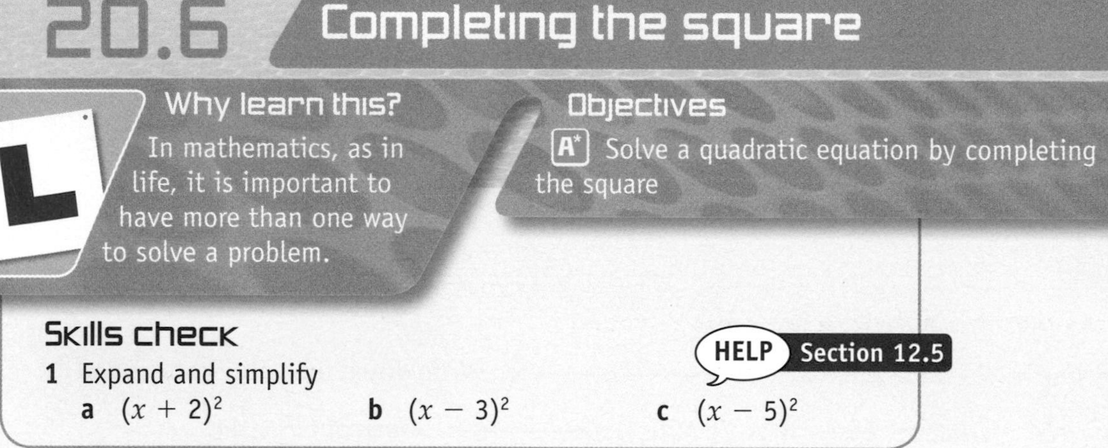

20.6 Completing the square

Why learn this?

In mathematics, as in life, it is important to have more than one way to solve a problem.

Objectives

A* Solve a quadratic equation by completing the square

Skills check

1 Expand and simplify

a $(x + 2)^2$ **b** $(x - 3)^2$ **c** $(x - 5)^2$

HELP Section 12.5

Completing the square

Completing the square is another way to solve a quadratic equation which cannot be factorised.

Expanding an expression of the form $(x + a)^2$ gives

$$(x + a)(x + a) = x^2 + 2ax + a^2$$

Working backwards, this can be used to 'complete the square'.

Consider the equation $x^2 + 4x + 10 = 0$.

For the coefficient of x to be 4 the squared bracket must be $(x + 2)^2$. — This is half the coefficient of x.

But $(x + 2)^2 = x^2 + 4x + 4$.

To get from $(x + 2)^2$ to $x^2 + 4x + 10$ you need to subtract 4 and then add 10.

$$x^2 + 4x + 10 = 0$$

$$(x + 2)^2 - 4 + 10 = 0$$

$$(x + 2)^2 + 6 = 0$$

Example 17

Write the expression $x^2 + 10x + 9$ in completed square form.

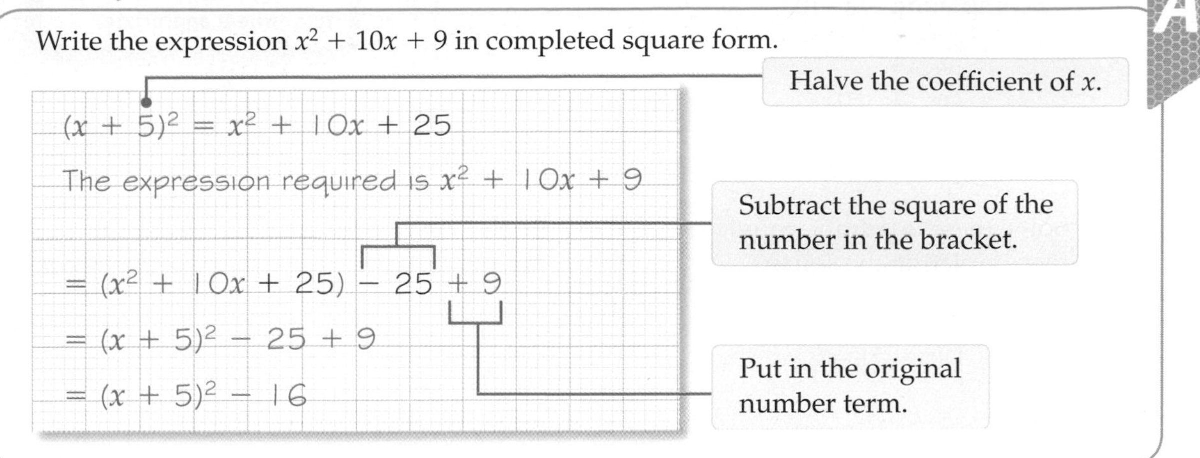

(x + 5)² = x² + 10x + 25
The expression required is x² + 10x + 9
= (x² + 10x + 25) − 25 + 9
= (x + 5)² − 25 + 9
= (x + 5)² − 16
Halve the coefficient of x.
Subtract the square of the number in the bracket.
Put in the original number term.

Exercise 20Q

1 Write each expression in completed square form.

a $x^2 + 6x + 3$ **b** $x^2 + 2x + 7$ **c** $x^2 - 8x + 5$

d $x^2 - 12x + 12$ **e** $x^2 - 4x - 7$ **f** $x^2 - 10x - 1$

2 Write each algebraic expressions in the form $(x + p)^2 + q$, giving the values of p and q.

Be careful with the values of p and q. Are they positive or negative?

a $x^2 + 10x + 32$ **b** $x^2 + 2x + 2$ **c** $x^2 - 4x + 20$

d $x^2 - 14x + 10$ **e** $x^2 - 6x - 3$ **f** $x^2 - 4x - 2$

Example 18

By completing the square, solve the equation $x^2 - 8x + 5 = 0$.

Leave your answer in surd form.

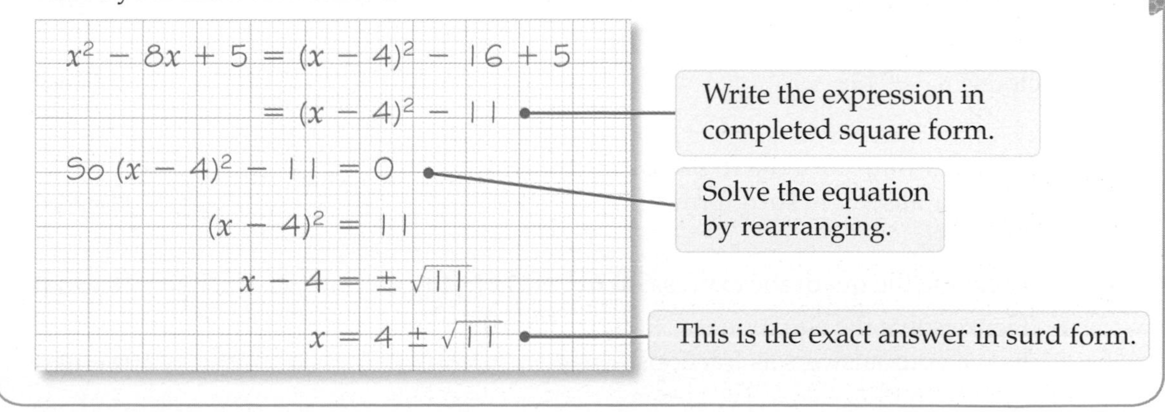

x² − 8x + 5 = (x − 4)² − 16 + 5
= (x − 4)² − 11
So (x − 4)² − 11 = 0
(x − 4)² = 11
x − 4 = ±√11
x = 4 ± √11
Write the expression in completed square form.
Solve the equation by rearranging.
This is the exact answer in surd form.

Exercise 20R

1 Solve the quadratic equations by completing the square.

a $x^2 + 10x + 9 = 0$ **b** $x^2 + 2x - 8 = 0$

c $x^2 - 8x + 10 = 0$ **d** $x^2 - 12x + 16 = 0$

e $x^2 - 4x - 4 = 0$ **f** $x^2 + 6x - 7 = 0$

2 Give the exact solution to these quadratic equations by completing the square.

> Leave your answers in surd form where apprpriate.

a $x^2 + 8x - 9 = 0$

b $x^2 + 4x - 8 = 0$

c $x^2 - 2x - 1 = 0$

d $x^2 - 8x + 10 = 0$

e $x^2 - 20x + 50 = 0$

f $x^2 - 14x + 41 = 0$

3 Solve these quadratic equations by completing the square.

> Don't forget to rearrange the equations first.

a $x^2 = 6x - 4$

b $3x(x + 6)= 6$

c $(x + 1)(x - 5) = 7$

d $(x - 2)(x + 8) = 7$

e $\dfrac{2}{x + 6} = x$

f $\dfrac{3}{(r - 1)(r + 2)} = 1$

Review exercise

1 Factorise the expression $x^2 + 6x + 5$. **[2 marks]**

2 I think of a number, square it, then subtract three times the number. The result is 108. Form and solve an algebraic equation to work out the possible values of the number I thought of. **[4 marks]**

3 I think of a number, square it, then add it to 5 times the number. The answer is 24. Form and solve an algebraic equation to work out the possible values of the number I thought of. **[4 marks]**

4 a Factorise $2x^2 - 15x - 8$. **[2 marks]**

b Hence solve the equation $2x^2 - 15x - 8 = 0$. **[2 marks]**

5 Factorise $6y^2 + 13y - 5$. **[2 marks]**

6 Use the quadratic formula to solve $2x^2 - 6x + 1 = 0$. Leave your answer in surd form. **[3 marks]**

7 a Factorise the quadratic expression $6x^2 - 11x - 10$. **[2 marks]**

b Hence solve the equation $6x^2 - 11x - 10 = 0$. Leave your answers as fractions. **[2 marks]**

8 A rectangular piece of land has length 3 m more than double the width. The area of the rectangle is 170 m². Work out the dimensions of the rectangle. **[5 marks]**

9 A rectangular rug is 6 m longer than its width. The area of the rug is 16 m². Calculate the dimensions of the rug. **[5 marks]**

10 a Find the values of m and n such that $x^2 + 4x - 6 = (x + m)^2 - n$. **[2 marks]**

b Hence solve the equation $x^2 + 4x - 6 = 0$ by rearranging, leaving your answer in the form $a \pm \sqrt{b}$. **[3 marks]**

A*

11 How many roots does each of these quadratic equations have?

a $5x^2 - 2x - 7 = 0$ b $3x^2 - 11x + 12 = 0$ c $4x^2 - 12x + 9 = 0$

[6 marks]

12 a Write the following algebraic expression in completed square form.

$x^2 - 4x + 2$ **[2 marks]**

b Hence find the exact solution to the equation $x^2 - 4x + 2 = 0$. **[2 marks]**

Chapter summary

In this chapter you have learned how to

- factorise a quadratic expression of the form $x^2 + bx + c$ **B**
- solve quadratic equations by rearranging **B**
- factorise a quadratic expression that is the difference of two squares **B** **A**
- solve quadratic equations by factorising **B** **A**
- factorise quadratic expressions of the form $ax^2 + bx + c$ **A** **A***
- solve quadratic equations by using the quadratic formula **A** **A***
- decide how many solutions a quadratic equation has by considering the discriminant **A***
- solve a quadratic equation by completing the square **A***

21

Further algebra

This chapter extends your knowledge of algebra.

There are several different types of volcanic eruption. The photo shows a Strombolian eruption in which blobs of lava are ejected from the volcano and fly in a parabolic path. You could model this path using a quadratic equation.

Objectives

This chapter will show you how to

- **simplify, add, subtract, multiply and divide algebraic fractions and use them to solve equations** B A A*
- **deal with harder 'change of subject' questions where the 'subject' appears on both sides of the formula** A A*
- **solve simultaneous equations when one is linear and one is quadratic** A*

Before you start this chapter

Put your calculator away!

1 Work out

a $\frac{3}{4} + \frac{2}{9}$ b $\frac{5}{6} - \frac{3}{5}$

2 Make x the subject of

a $y = 2x - 1$ b $2y = 1 - 3x$

3 Expand and simplify

a $(x - 3)^2$ b $(x + 6)^2$

c $(2 - x)^2$ d $(2x - 1)(3x + 2)$

HELP Chapter 20

4 Factorise

a $x^2 - 49$ b $2x^2 - 32$

c $x^2 + 3x - 18$ d $3x^2 - 2x - 8$

5 Solve

a $x^2 - 7x + 10 = 0$ b $2x^2 + 7x - 30 = 0$

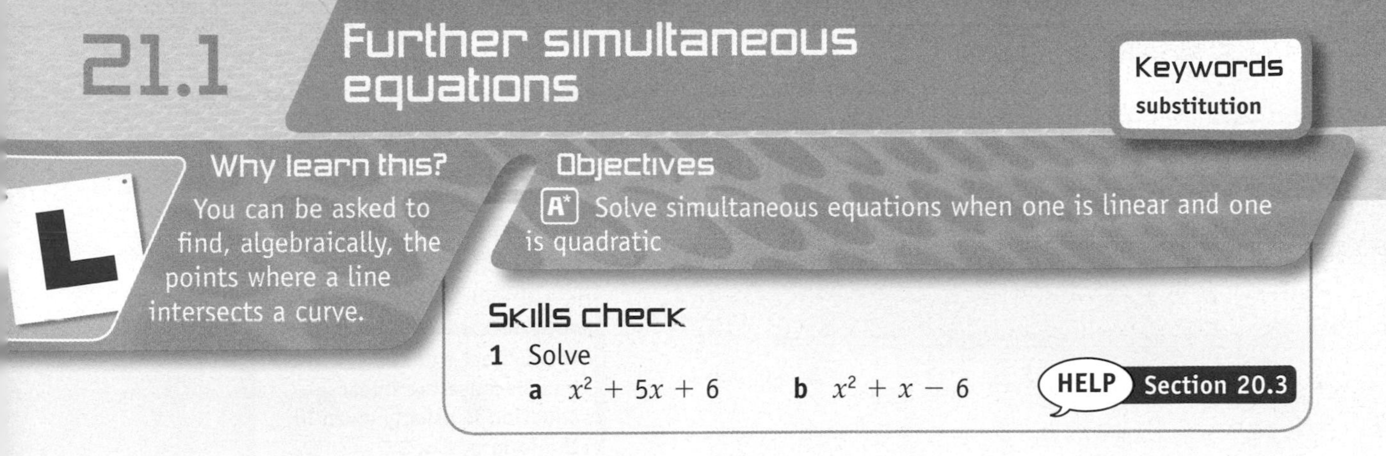

21.1 Further simultaneous equations

Keywords
substitution

Why learn this?
You can be asked to find, algebraically, the points where a line intersects a curve.

Objectives
A* Solve simultaneous equations when one is linear and one is quadratic

Skills check

1 Solve
a $x^2 + 5x + 6$ **b** $x^2 + x - 6$

HELP Section 20.3

Linear and quadratic graphs

When two straight lines intersect they do so in only one point.

When a quadratic graph and a linear graph are plotted on the same axes there are three possible situations:

- the linear graph *intersects* the quadratic graph at *two points*
- the linear graph *touches* the quadratic graph at *one point*
- the linear graph and the quadratic graph *do not intersect*.

This diagram shows the quadratic graph of $y = x^2 - 2x - 7$ and the graph of the straight line $y = x + 3$.

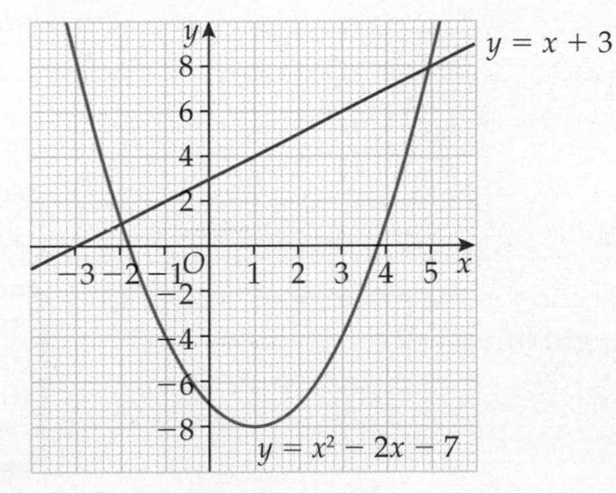

There are two points of intersection, $(-2, 1)$ and $(5, 8)$.

As well as plotting the graphs to find a solution, you can also solve these simultaneous equations algebraically. To find the solutions by an algebraic method means that you have to find values of x and y that satisfy the quadratic equation and the linear equation simultaneously.

To solve one linear and one quadratic simultaneous equation you need to use the method of **substitution**.

You cannot use the method of elimination (which you used to solve pairs of linear equations in Section 14.5) because you cannot add or subtract the equations to eliminate a variable.

A* A02

Example 1

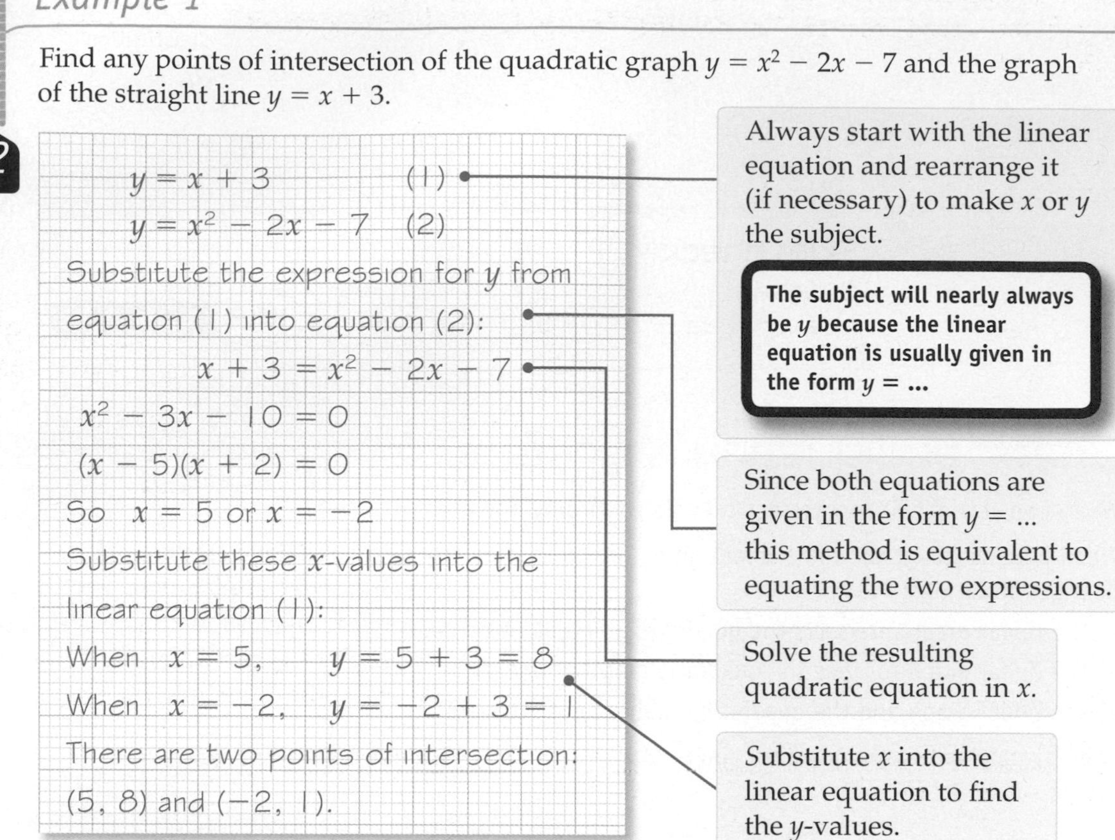

Find any points of intersection of the quadratic graph $y = x^2 - 2x - 7$ and the graph of the straight line $y = x + 3$.

$y = x + 3 \quad (1)$

$y = x^2 - 2x - 7 \quad (2)$

Substitute the expression for y from equation (1) into equation (2):

$x + 3 = x^2 - 2x - 7$

$x^2 - 3x - 10 = 0$

$(x - 5)(x + 2) = 0$

So $x = 5$ or $x = -2$

Substitute these x-values into the linear equation (1):

When $x = 5$, $y = 5 + 3 = 8$

When $x = -2$, $y = -2 + 3 = 1$

There are two points of intersection: (5, 8) and (−2, 1).

Always start with the linear equation and rearrange it (if necessary) to make x or y the subject.

The subject will nearly always be y because the linear equation is usually given in the form $y = ...$

Since both equations are given in the form $y = ...$ this method is equivalent to equating the two expressions.

Solve the resulting quadratic equation in x.

Substitute x into the linear equation to find the y-values.

A* A02

Example 2

Find any points of intersection of the graph of $y^2 = 4x + 13$ and the graph of the straight line $y = 2x - 1$.

Notice that this quadratic is of a different type from the ones you have met before.

$y = 2x - 1 \quad (1)$

$y^2 = 4x + 13 \quad (2)$

From (1) $\quad y + 1 = 2x$

so $\quad 2y + 2 = 4x$

Substitute this expression for $4x$ in (2):

$y^2 = 2y + 2 + 13$

$y^2 - 2y - 15 = 0$

$(y + 3)(y - 5) = 0$

So $y = -3$ or $y = 5$

Substituting these y-values into the linear equation (1):

When $y = -3$, $-3 = 2x - 1$, giving $x = -1$

When $y = 5$, $5 = 2x - 1$, giving $x = 3$

There are two points of intersection: (−1, −3) and (3, 5).

One way to solve this is to spot that one equation contains $2x$ and the other one contains $4x$.

Since these are the only x-terms in each equation, you can easily eliminate x and solve the resulting quadratic equation in y.

Here is another solution to Example 2. It involves squaring the first equation and substituting for y^2 in the second equation. It is a method frequently used in solving simultaneous equations of this type.

Example 2 (revisited)

A* AO2

$$y = 2x - 1 \quad (1)$$
$$y^2 = 4x + 13 \quad (2)$$

Square equation (1) and substitute for y^2 in equation (2):

Remember that squaring a linear expression will give a quadratic expression with three terms.

From (1) $y^2 = (2x - 1)^2$
$= 4x^2 - 2x - 2x + 1$
$= 4x^2 - 4x + 1$

Always put a bracket round the linear term before you square it. This will help you to remember the correct method for expanding the bracket.

Substitute in (2):

$4x^2 - 4x + 1 = 4x + 13$
$4x^2 - 8x - 12 = 0$

Divide each term by 4.

$x^2 - 2x - 3 = 0$
$(x + 1)(x - 3) = 0$

So $x = -1$ or $x = 3$

From equation (1) when $x = -1$ $y = -3$
and when $x = 3$ $y = 5$

The two solutions are $(-1, -3)$ and $(3, 5)$.

Exercise 21A

Use the method of substitution to solve each of these pairs of simultaneous equations. State clearly the points of intersection of the straight line graph and the quadratic graph in each case.

A*

1 $y = x + 1$
$y = x^2 - 5$

2 $y = x + 4$
$y = x^2 - 2x$

3 $y = 4x$
$y = x^2 - 2x + 5$

4 $y = 1 - x$
$y = 2x^2 - 5$

5 $y = 2x - 1$
$y = x^2 + 4x$

6 $y = 2x - 3$
$y = 2x^2 - x - 23$

7 $y = 5x + 2$
$y = x^2 + x - 10$

8 $3y = x + 6$
$y^2 = 2x + 7$

A*

9 A straight line has the equation $y = 2x - 3$.
A curve has the equation $y^2 = 8x - 16$.

a Solve these simultaneous equations to find any points of intersection of the line and the curve.
Do **not** use trial and improvement. You must show all your working.

b Here are three sketches showing the curve $y^2 = 8x - 16$ and three possible positions of the line $y = 2x - 3$.
Which is the correct sketch? You must explain your answer.

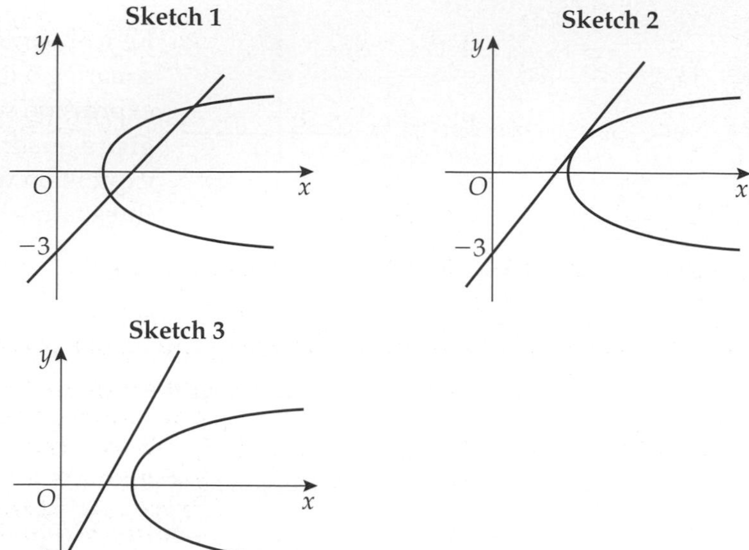

10 A straight line has the equation $y = -x - 3$.
A curve has the equation $y^2 + 2x + 6 = 0$.

a Solve these simultaneous equations to find any points of intersection of the line and the curve.
Do **not** use trial and improvement. You must show all your working.

b Here are three sketches showing the curve $y^2 + 2x + 6 = 0$ and three possible positions of the line $y = -x - 3$.
Which is the correct sketch? You must explain your answer.

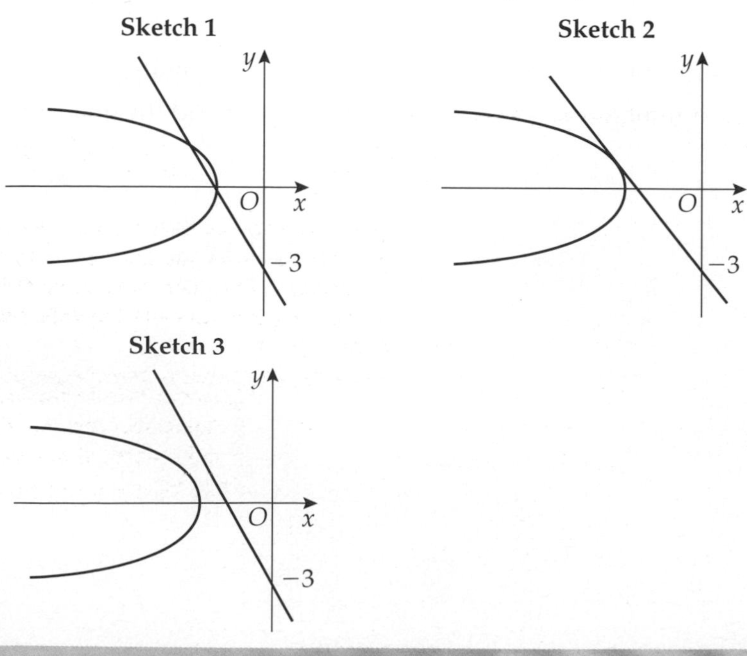

AO2

Why learn this?
These are key skills if you wish to take a course in A-level mathematics.

Objectives
B A A* Simplify, add, subtract, multiply and divide algebraic fractions and use them to solve equations

Keywords
common factor, quadratic, difference of two squares, HCF, LCM

Skills check

1 Expand and simplify
 a $-3(x + 1)$ b $(x - 5)^2$ HELP Section 12.1

2 Factorise
 a $6x + 18$ b $y = x^2 - 25$ HELP Section 20.1

Algebraic fractions

Algebraic fractions have letters, or combinations of letters and numbers, in their numerator and their denominator.

Algebraic fractions can sometimes be simplified. Look for a **common factor** of the numerator and the denominator.

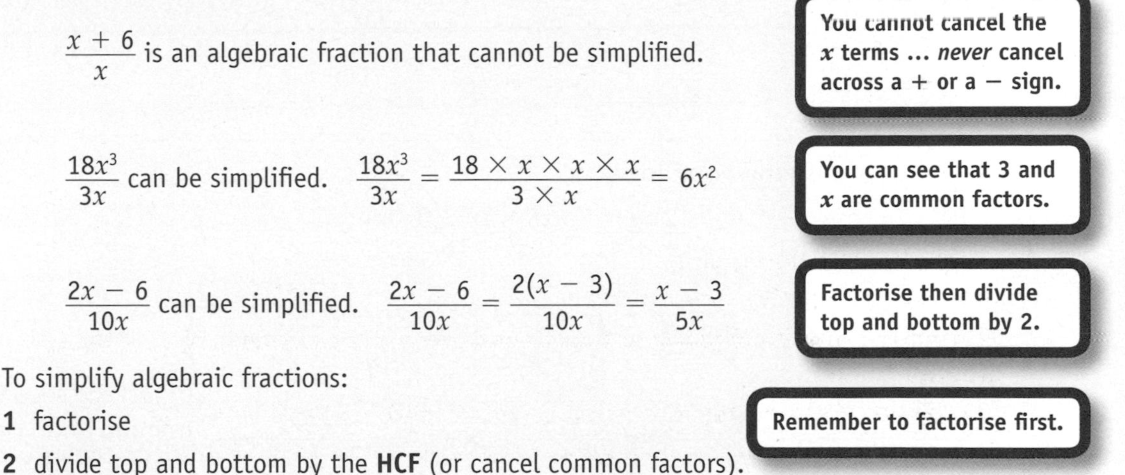

$\frac{x + 6}{x}$ is an algebraic fraction that cannot be simplified.

You cannot cancel the x terms … *never* cancel across a + or a − sign.

$\frac{18x^3}{3x}$ can be simplified. $\frac{18x^3}{3x} = \frac{18 \times x \times x \times x}{3 \times x} = 6x^2$

You can see that 3 and x are common factors.

$\frac{2x - 6}{10x}$ can be simplified. $\frac{2x - 6}{10x} = \frac{2(x - 3)}{10x} = \frac{x - 3}{5x}$

Factorise then divide top and bottom by 2.

To simplify algebraic fractions:

1 factorise

2 divide top and bottom by the **HCF** (or cancel common factors).

Remember to factorise first.

Example 3

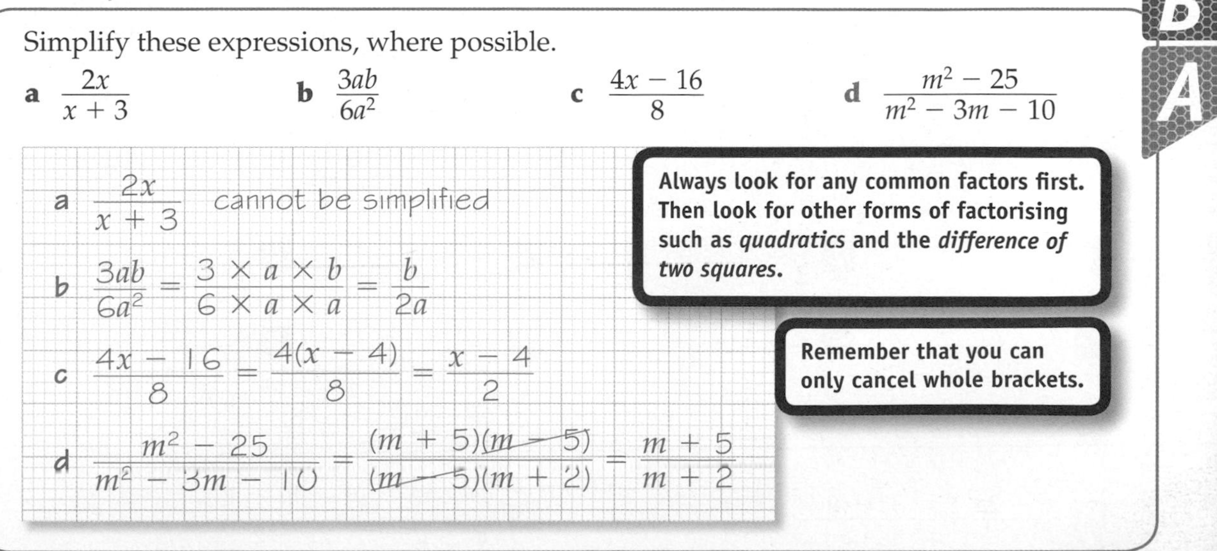

Simplify these expressions, where possible.

a $\frac{2x}{x + 3}$ **b** $\frac{3ab}{6a^2}$ **c** $\frac{4x - 16}{8}$ **d** $\frac{m^2 - 25}{m^2 - 3m - 10}$

a $\frac{2x}{x + 3}$ cannot be simplified

b $\frac{3ab}{6a^2} = \frac{3 \times a \times b}{6 \times a \times a} = \frac{b}{2a}$

c $\frac{4x - 16}{8} = \frac{4(x - 4)}{8} = \frac{x - 4}{2}$

d $\frac{m^2 - 25}{m^2 - 3m - 10} = \frac{(m + 5)(m - 5)}{(m - 5)(m + 2)} = \frac{m + 5}{m + 2}$

Always look for any common factors first. Then look for other forms of factorising such as *quadratics* and the *difference of two squares*.

Remember that you can only cancel whole brackets.

B A

Exercise 21B

Simplify these expressions, where possible.

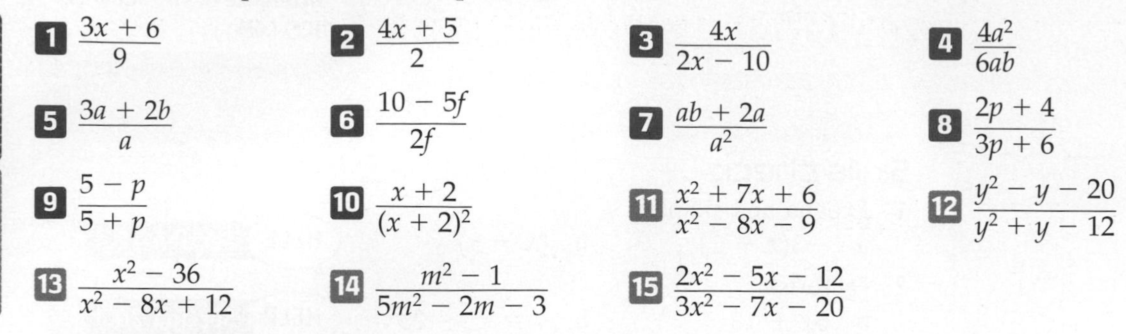

1 $\dfrac{3x+6}{9}$ **2** $\dfrac{4x+5}{2}$ **3** $\dfrac{4x}{2x-10}$ **4** $\dfrac{4a^2}{6ab}$

5 $\dfrac{3a+2b}{a}$ **6** $\dfrac{10-5f}{2f}$ **7** $\dfrac{ab+2a}{a^2}$ **8** $\dfrac{2p+4}{3p+6}$

9 $\dfrac{5-p}{5+p}$ **10** $\dfrac{x+2}{(x+2)^2}$ **11** $\dfrac{x^2+7x+6}{x^2-8x-9}$ **12** $\dfrac{y^2-y-20}{y^2+y-12}$

13 $\dfrac{x^2-36}{x^2-8x+12}$ **14** $\dfrac{m^2-1}{5m^2-2m-3}$ **15** $\dfrac{2x^2-5x-12}{3x^2-7x-20}$

Multiplying and dividing algebraic fractions

To multiply numerical fractions you multiply the numerators and multiply the denominators, simplifying if possible. To divide numerical fractions you invert the second fraction and multiply, simplifying if possible. The rules are exactly the same for algebraic fractions, but always remember to look for any common factors before you simplify.

Example 4

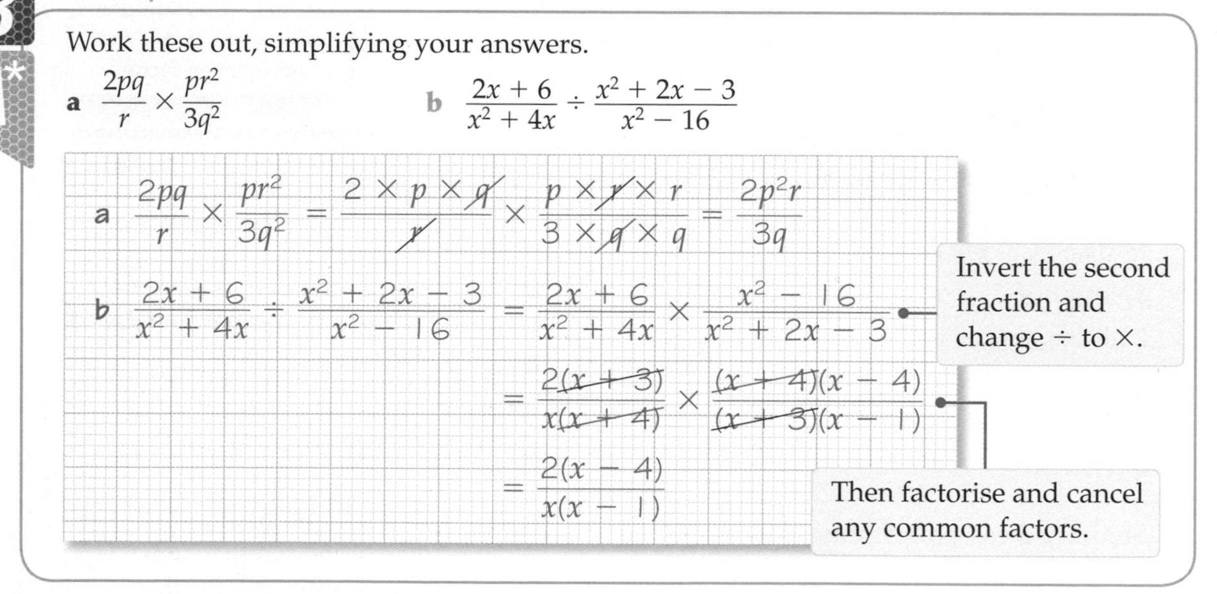

Work these out, simplifying your answers.

a $\dfrac{2pq}{r} \times \dfrac{pr^2}{3q^2}$ **b** $\dfrac{2x+6}{x^2+4x} \div \dfrac{x^2+2x-3}{x^2-16}$

a $\dfrac{2pq}{r} \times \dfrac{pr^2}{3q^2} = \dfrac{2 \times p \times q}{r} \times \dfrac{p \times r \times r}{3 \times q \times q} = \dfrac{2p^2r}{3q}$

b $\dfrac{2x+6}{x^2+4x} \div \dfrac{x^2+2x-3}{x^2-16} = \dfrac{2x+6}{x^2+4x} \times \dfrac{x^2-16}{x^2+2x-3}$

$= \dfrac{2(x+3)}{x(x+4)} \times \dfrac{(x+4)(x-4)}{(x+3)(x-1)}$

$= \dfrac{2(x-4)}{x(x-1)}$

Invert the second fraction and change ÷ to ×.

Then factorise and cancel any common factors.

Exercise 21C

Work these out, simplifying your answers.

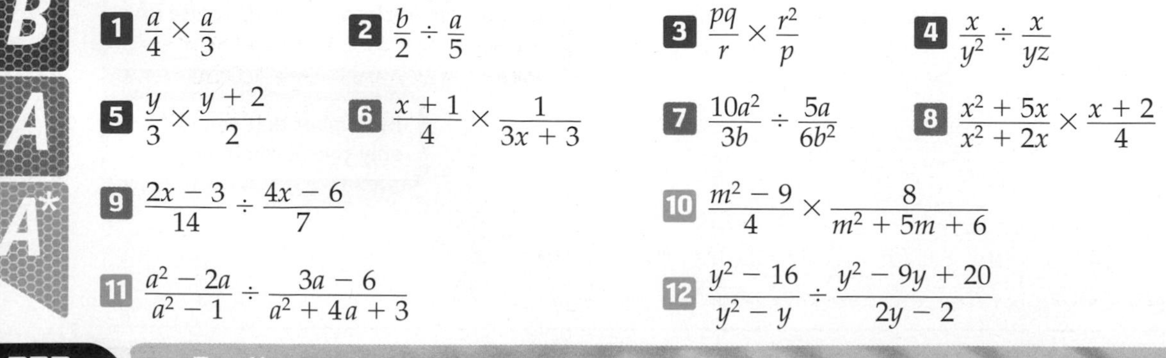

1 $\dfrac{a}{4} \times \dfrac{a}{3}$ **2** $\dfrac{b}{2} \div \dfrac{a}{5}$ **3** $\dfrac{pq}{r} \times \dfrac{r^2}{p}$ **4** $\dfrac{x}{y^2} \div \dfrac{x}{yz}$

5 $\dfrac{y}{3} \times \dfrac{y+2}{2}$ **6** $\dfrac{x+1}{4} \times \dfrac{1}{3x+3}$ **7** $\dfrac{10a^2}{3b} \div \dfrac{5a}{6b^2}$ **8** $\dfrac{x^2+5x}{x^2+2x} \times \dfrac{x+2}{4}$

9 $\dfrac{2x-3}{14} \div \dfrac{4x-6}{7}$ **10** $\dfrac{m^2-9}{4} \times \dfrac{8}{m^2+5m+6}$

11 $\dfrac{a^2-2a}{a^2-1} \div \dfrac{3a-6}{a^2+4a+3}$ **12** $\dfrac{y^2-16}{y^2-y} \div \dfrac{y^2-9y+20}{2y-2}$

Adding and subtracting algebraic fractions

When adding or subtracting algebraic fractions, the rules are the same as for numerical fractions. Convert each fraction to an equivalent fraction with the **LCM** as denominator.

Example 5

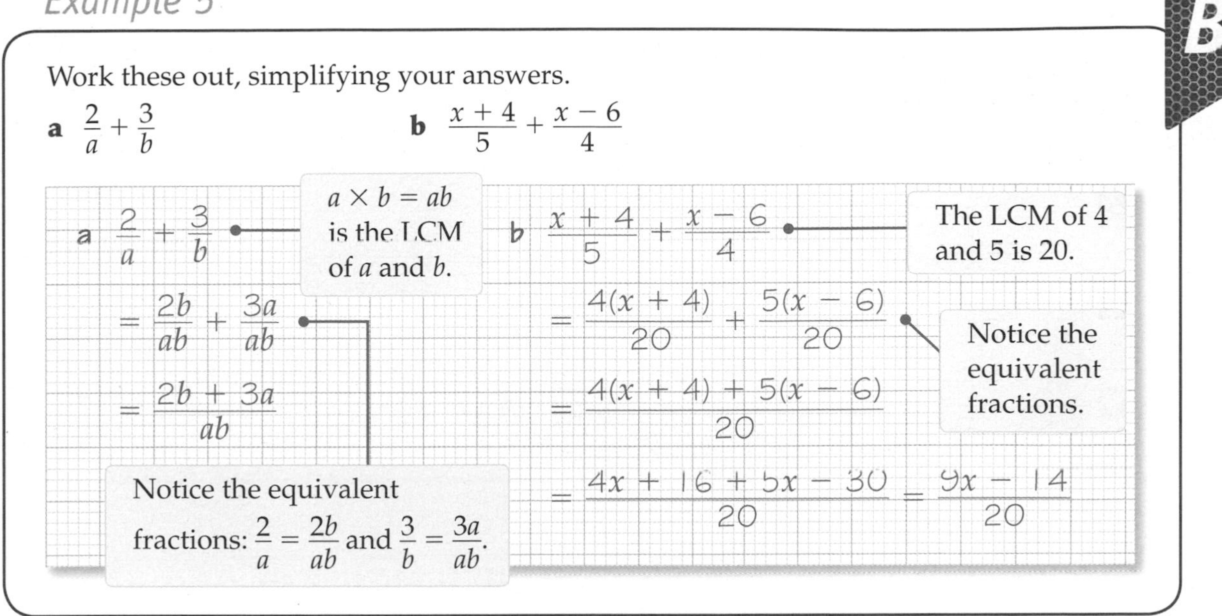

B
Work these out, simplifying your answers.
a $\frac{2}{a} + \frac{3}{b}$
b $\frac{x+4}{5} + \frac{x-6}{4}$
a $\frac{2}{a} + \frac{3}{b}$
$a \times b = ab$ is the LCM of a and b.
$= \frac{2b}{ab} + \frac{3a}{ab}$
$= \frac{2b + 3a}{ab}$
Notice the equivalent fractions: $\frac{2}{a} = \frac{2b}{ab}$ and $\frac{3}{b} = \frac{3a}{ab}$.
b $\frac{x+4}{5} + \frac{x-6}{4}$
The LCM of 4 and 5 is 20.
$= \frac{4(x+4)}{20} + \frac{5(x-6)}{20}$
Notice the equivalent fractions.
$= \frac{4(x+4) + 5(x-6)}{20}$
$= \frac{4x + 16 + 5x - 30}{20} = \frac{9x - 14}{20}$

Example 6

A A*

Work these out, simplifying your answers.

a $\frac{5}{2x} + \frac{3}{8y}$ **b** $\frac{4}{x+1} - \frac{3}{x-5}$

Notice that in a the LCM of $2x$ and $8y$ is $8xy$. You do not need to use $2x \times 8y = 16xy$ although it is not wrong to do so. But you *will* need to simplify later if you do.

a $\frac{5}{2x} + \frac{3}{8y}$

$= \frac{20y}{8xy} + \frac{3x}{8xy}$

$= \frac{20y + 3x}{8xy}$

b $\frac{4}{(x+1)} - \frac{3}{(x-5)}$

When a denominator has more than one term, put brackets round the expression. It helps you to spot the LCM and get the correct equivalent fractions.

$= \frac{4(x-5)}{(x+1)(x-5)} - \frac{3(x+1)}{(x+1)(x-5)}$

$= \frac{4(x-5) - 3(x+1)}{(x+1)(x-5)}$

Take care with the signs when you expand the brackets in the numerator.

$= \frac{4x - 20 - 3x - 3}{(x+1)(x-5)}$

$= \frac{x - 23}{(x+1)(x-5)}$

Exercise 21D

Work these out, simplifying your answers.

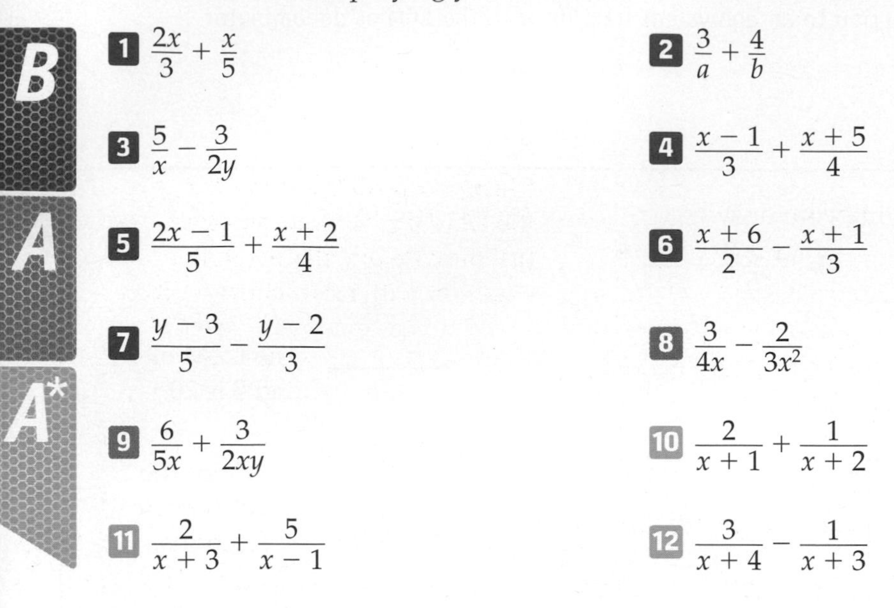

1 $\frac{2x}{3} + \frac{x}{5}$

2 $\frac{3}{a} + \frac{4}{b}$

3 $\frac{5}{x} - \frac{3}{2y}$

4 $\frac{x-1}{3} + \frac{x+5}{4}$

5 $\frac{2x-1}{5} + \frac{x+2}{4}$

6 $\frac{x+6}{2} - \frac{x+1}{3}$

7 $\frac{y-3}{5} - \frac{y-2}{3}$

8 $\frac{3}{4x} - \frac{2}{3x^2}$

9 $\frac{6}{5x} + \frac{3}{2xy}$

10 $\frac{2}{x+1} + \frac{1}{x+2}$

11 $\frac{2}{x+3} + \frac{5}{x-1}$

12 $\frac{3}{x+4} - \frac{1}{x+3}$

Equations involving algebraic fractions

You can use algebraic fractions to solve quite complex equations.

B

Example 7

Solve the equation $\frac{3x+1}{2} - \frac{x+4}{3} = 5$

Method A

$$\frac{3x+1}{2} - \frac{x+4}{3} = 5$$

$$\frac{3(3x+1) - 2(x+4)}{6} = 5$$

$$\frac{9x+3-2x-8}{6} = 5$$

$$\frac{7x-5}{6} = 5$$

$$7x - 5 = 30$$

$$7x = 35$$

$$x = 5$$

Method B

$$\frac{3x+1}{2} - \frac{x+4}{3} = 5$$

The LCM of the terms in the denominator is 6.

So multiply all the terms by 6.

$$\frac{6(3x+1)}{2} - \frac{6(x+4)}{3} = 5 \times 6$$

$$3(3x+1) - 2(x+4) = 30$$

$$9x + 3 - 2x - 8 = 30$$

$$7x - 5 = 30$$

$$7x = 35$$

$$x = 5$$

Multiplying all the terms by the LCM of the denominators (in this case 6) will give you an equation with no fractions in it.

Example 8

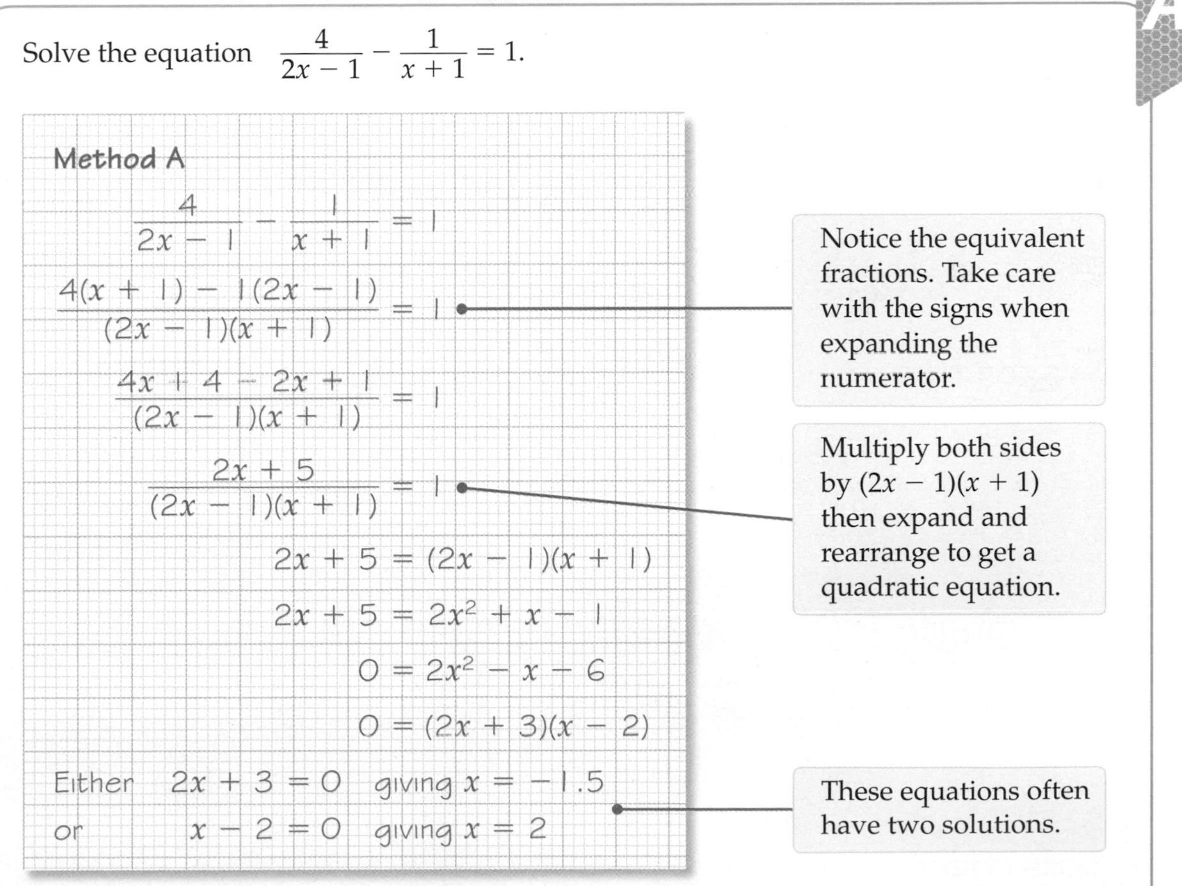

Solve the equation $\frac{4}{2x-1} - \frac{1}{x+1} = 1$.

Method A

$$\frac{4}{2x-1} - \frac{1}{x+1} = 1$$

$$\frac{4(x+1) - 1(2x-1)}{(2x-1)(x+1)} = 1$$

$$\frac{4x+4-2x+1}{(2x-1)(x+1)} = 1$$

$$\frac{2x+5}{(2x-1)(x+1)} = 1$$

$$2x + 5 = (2x-1)(x+1)$$

$$2x + 5 = 2x^2 + x - 1$$

$$0 = 2x^2 - x - 6$$

$$0 = (2x+3)(x-2)$$

Either $2x + 3 = 0$ giving $x = -1.5$

or $x - 2 = 0$ giving $x = 2$

Notice the equivalent fractions. Take care with the signs when expanding the numerator.

Multiply both sides by $(2x - 1)(x + 1)$ then expand and rearrange to get a quadratic equation.

These equations often have two solutions.

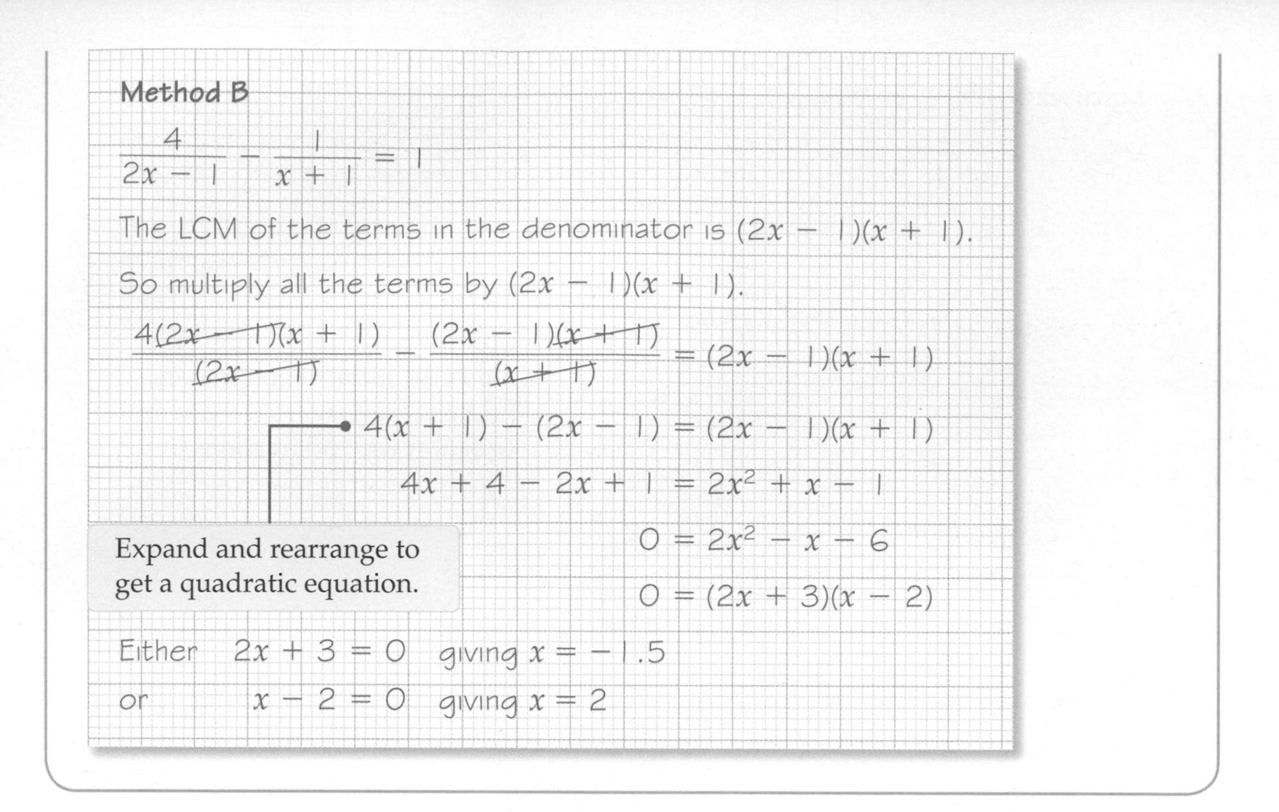

Method B

$$\frac{4}{2x-1} - \frac{1}{x+1} = 1$$

The LCM of the terms in the denominator is $(2x-1)(x+1)$.

So multiply all the terms by $(2x-1)(x+1)$.

$$\frac{4(2x-1)(x+1)}{(2x-1)} - \frac{(2x-1)(x+1)}{(x+1)} = (2x-1)(x+1)$$

$$4(x+1) - (2x-1) = (2x-1)(x+1)$$

$$4x + 4 - 2x + 1 = 2x^2 + x - 1$$

$$0 = 2x^2 - x - 6$$

$$0 = (2x+3)(x-2)$$

Expand and rearrange to get a quadratic equation.

Either $2x + 3 = 0$ giving $x = -1.5$

or $x - 2 = 0$ giving $x = 2$

Exercise 21E

Solve each of these equations.

B

1 $\frac{x+1}{2} + \frac{x-4}{3} = 5$

2 $\frac{x+5}{4} + \frac{x-1}{2} = 3$

3 $\frac{x-2}{3} - \frac{x+1}{4} = 1$

4 $\frac{2x-3}{3} + \frac{3x+1}{4} = 12$

5 $\frac{5x+1}{4} - \frac{x-4}{6} = 2$

6 $\frac{2x-5}{2} - \frac{4x-1}{3} = 0.5$

A*

7 $\frac{3}{x+1} + \frac{2}{2x-3} = 1$

8 $\frac{9}{x-2} + \frac{5}{x+2} = 2$

9 $\frac{7}{x-3} - \frac{6}{x-1} = 2$

10 $\frac{9}{2x-7} + \frac{6}{x-1} = 3$

11 $\frac{11}{4x-5} - \frac{8}{x+1} = 1$

12 $\frac{7}{2x-1} + \frac{11}{x+1} = 6$

21.3 Changing the subject of more complicated formulae

Why learn this?

This is a skill that is essential if you wish to take A-level mathematics.

Objectives

A A* Change the subject of a formula when the subject appears more than once and/or on both sides of the formula

Skills check

1 Make x the subject of

a $y = 3x + 5$ b $y = 7 - 2x$

HELP Section 15.4

Changing the subject of more complicated formulae

In this section you will see how to change the subject of a formula when the subject appears more than once and on both sides of the formula.

You use the same methods as for solving equations. You will see that a pattern emerges for the order in which you do the operations.

You learned how to change the subject of a formula in Section 15.4. But in all of the questions in this section, the subject only appeared *once*.

A

Example 9

a Solve the equation $2(2m + 1) = 7(m - 1)$.

b Make m the subject of the formula: $a(bm + c) = d(m - e)$.

a

$2(2m + 1) = 7(m - 1)$

$4m + 2 = 7m - 7$

$2 + 7 = 7m - 4m$

$9 = 3m$

$\frac{9}{3} = m$

$3 = m$

b

$a(bm + c) = d(m - e)$

$abm + ac = dm - de$ — Expand.

$ac + de = dm - abm$ — Rearrange.

$ac + de = m(d - ab)$ — Factorise.

$\frac{ac + de}{d - ab} = m$ — Divide through by $(d - ab)$.

A*

Example 10

a Solve the equation $\sqrt{\frac{2a}{a-1}} = 3$.

b Make a the subject of the formula $\sqrt{\frac{2a}{a-x}} = 3x$.

a

$\sqrt{\frac{2a}{a-1}} = 3$

$\frac{2a}{a-1} = 9$ — Square both sides.

$2a = 9(a - 1)$ — Multiply through by $(a - 1)$.

$2a = 9a - 9$ — Expand.

$9 = 9a - 2a$ — Rearrange to separate the a terms on to one side.

$9 = 7a$

$\frac{9}{7} = a$ — Divide through by 7.

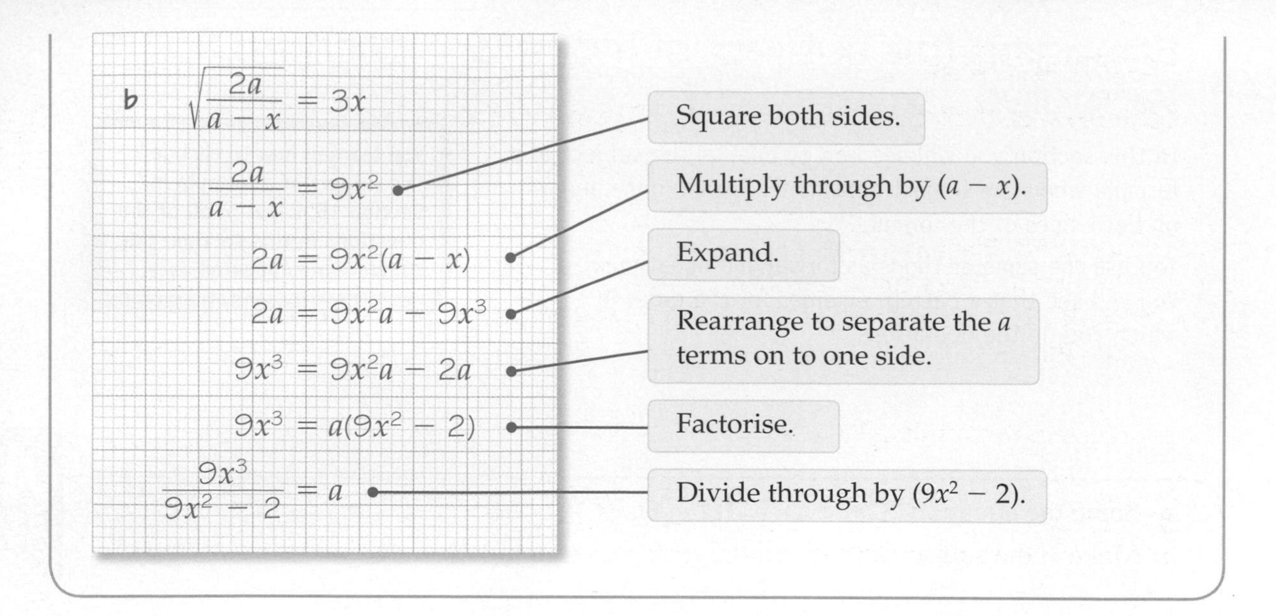

b $\sqrt{\frac{2a}{a-x}} = 3x$
$\frac{2a}{a-x} = 9x^2$
$2a = 9x^2(a-x)$
$2a = 9x^2a - 9x^3$
$9x^3 = 9x^2a - 2a$
$9x^3 = a(9x^2 - 2)$
$\frac{9x^3}{9x^2-2} = a$
Square both sides.
Multiply through by $(a - x)$.
Expand.
Rearrange to separate the a terms on to one side.
Factorise.
Divide through by $(9x^2 - 2)$.

To change the subject of a formula, follow this order of operations.

1. Square both sides.
2. Multiply through by ...
3. Expand.
4. Rearrange.
5. Factorise.
6. Divide through by ...
7. Square root both sides.

You may not need to use all these operations.

Exercise 21F

Make y the subject of each of these formulae.

A

1 $my + b = d - my$

2 $py + 3 = 7 + qy$

3 $ay - x = w + by$

4 $w(y + a) = k(y + b)$

5 $\frac{ay - c}{m} = \frac{y + d}{h}$

6 $\frac{h - y}{h + y} = k$

A*

7 $\frac{y + 3}{y - 2} = \frac{d}{e}$

8 $\sqrt{\frac{2y + a}{y}} = m$

9 $\sqrt{\frac{3y}{m - y}} = 2w$

10 $T = 2k\sqrt{\frac{y}{g}}$

11 $ky^2 + 2e = f - hy^2$

12 $m(2y^2 + a) = h(w - 3y^2)$

Review exercise

1 Rearrange the formula $3(a - b) = 2b + 7$ to make b the subject. **[3 marks]**

2 A straight line has the equation $y = 2x - 3$.
A curve has the equation $y^2 = 8x - 16$.
Work out the coordinates of the points of intersection of the line and the curve.
You must show your working. **[5 marks]**

3 Solve these simultaneous equations.

$$y = 2x - 5$$
$$x^2 + y^2 = 25$$

You must show your working. **[6 marks]**

4 Simplify fully $\dfrac{x^2 - 16}{3x^2 + 10x - 8}$ **[4 marks]**

5 Simplify fully $\dfrac{2x^2 - 9x - 18}{x^2 - 36}$ **[4 marks]**

6 Solve the equation $\dfrac{6}{x - 1} - \dfrac{4}{x + 3} = 1$ **[5 marks]**

7 Solve the equation $\dfrac{x}{x + 1} - \dfrac{2}{x - 1} = 1$ **[5 marks]**

8 Make x the subject of the formula $\dfrac{t}{m} = \dfrac{x}{k(a - x)}$ **[5 marks]**

9 Make x the subject of the formula $\sqrt{\dfrac{u}{x + b}} = c$ **[4 marks]**

Chapter summary

In this chapter you have learned how to

- simplify, add, subtract, multiply and divide algebraic fractions and use them to solve equations **B** **A** **A***
- change the subject of a formula when the subject appears more than once and/or on both sides of the formula **A** **A***
- solve simultaneous equations when one is linear and one is quadratic **A***

Quality of written communication: Some questions on this page are marked with a star ☆. In the exam, this sort of question may earn you some extra marks if you

- use correct and accurate maths notation and vocabulary
- organise your work clearly, showing that you can communicate effectively.

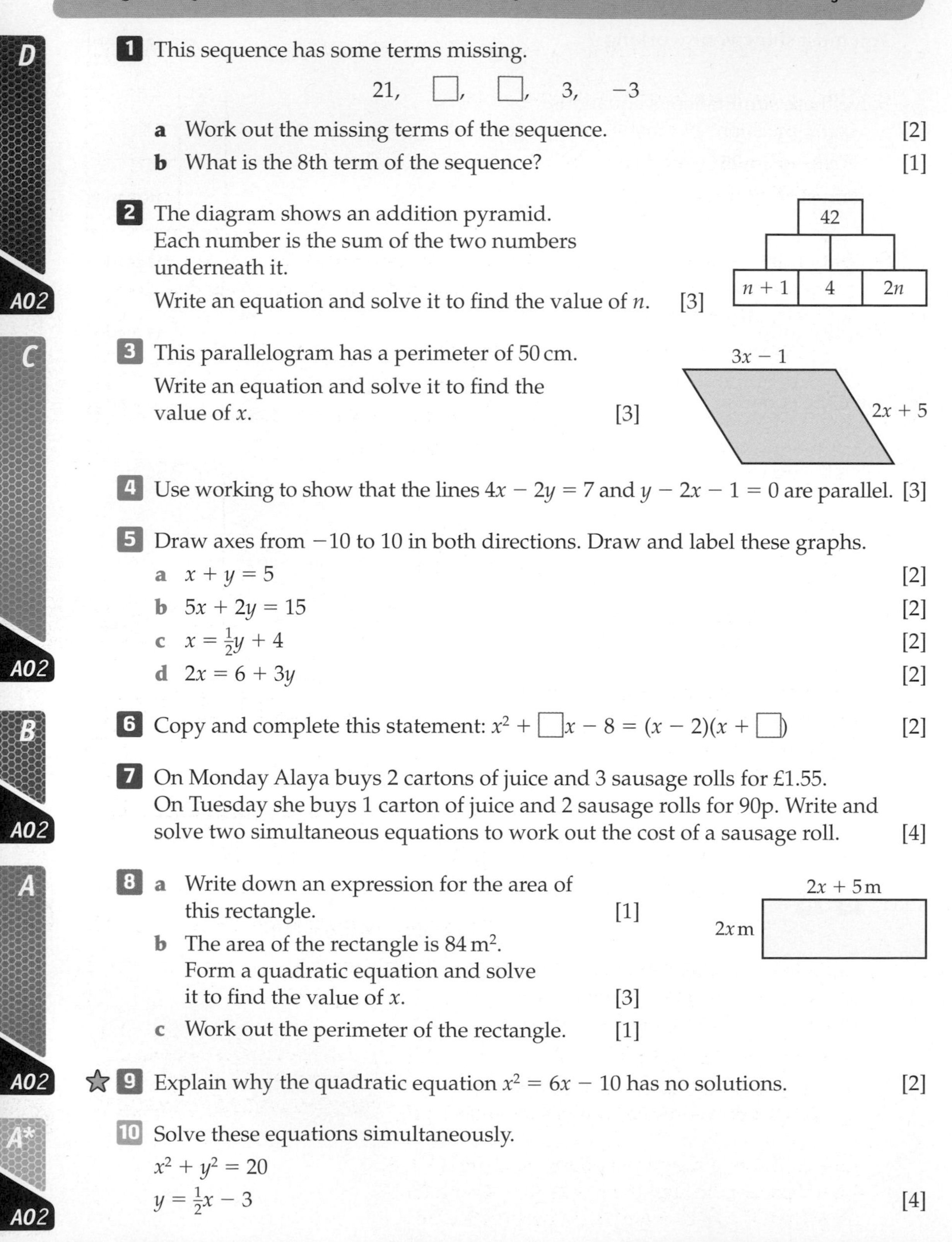

D

1 This sequence has some terms missing.

21, □, □, 3, −3

a. Work out the missing terms of the sequence. [2]

b. What is the 8th term of the sequence? [1]

2 The diagram shows an addition pyramid. Each number is the sum of the two numbers underneath it.

Write an equation and solve it to find the value of n. [3]

AO2

C

3 This parallelogram has a perimeter of 50 cm.

Write an equation and solve it to find the value of x. [3]

4 Use working to show that the lines $4x - 2y = 7$ and $y - 2x - 1 = 0$ are parallel. [3]

5 Draw axes from −10 to 10 in both directions. Draw and label these graphs.

a. $x + y = 5$ [2]

b. $5x + 2y = 15$ [2]

c. $x = \frac{1}{2}y + 4$ [2]

d. $2x = 6 + 3y$ [2]

AO2

B

6 Copy and complete this statement: $x^2 + \square x - 8 = (x - 2)(x + \square)$ [2]

7 On Monday Alaya buys 2 cartons of juice and 3 sausage rolls for £1.55. On Tuesday she buys 1 carton of juice and 2 sausage rolls for 90p. Write and solve two simultaneous equations to work out the cost of a sausage roll. [4]

AO2

A

8 a. Write down an expression for the area of this rectangle. [1]

b. The area of the rectangle is 84 m². Form a quadratic equation and solve it to find the value of x. [3]

c. Work out the perimeter of the rectangle. [1]

AO2

☆ **9** Explain why the quadratic equation $x^2 = 6x - 10$ has no solutions. [2]

A*

10 Solve these equations simultaneously.

$x^2 + y^2 = 20$

$y = \frac{1}{2}x - 3$ [4]

AO2

Quality of written communication: Some questions on this page are marked with a star ☆. In the exam, this sort of question may earn you some extra marks if you
- use correct and accurate maths notation and vocabulary
- organise your work clearly, showing that you can communicate effectively.

C

☆ **1** Explain why the sum of four consecutive integers is always an even number. [2]

2 This square is made from rectangular tiles.
The dimensions of one tile are shown.
Write an equation and solve it to find the value of n. [4]

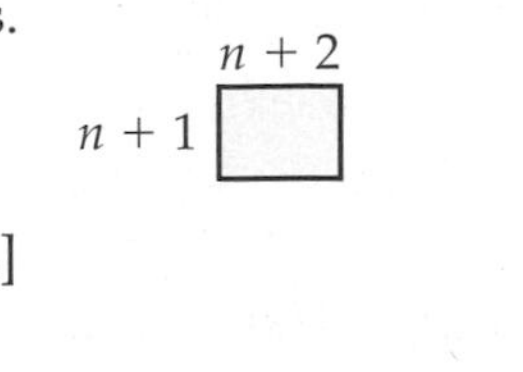

3 Ellie is comparing three different batteries. The description of each battery is given below. For each battery, choose the graph that most accurately describes how the battery charges.

a Super-cell: Charges at a constant rate with no maximum. [1]

b Electronizer: Charges slowly at first then more quickly. [1]

c Powerpack: Charges at a constant rate until a maximum charge is reached. [1]

d Stay-brite: Charges quickly at first then more slowly. [1]

FUNCTIONAL • FUNCTIONAL •

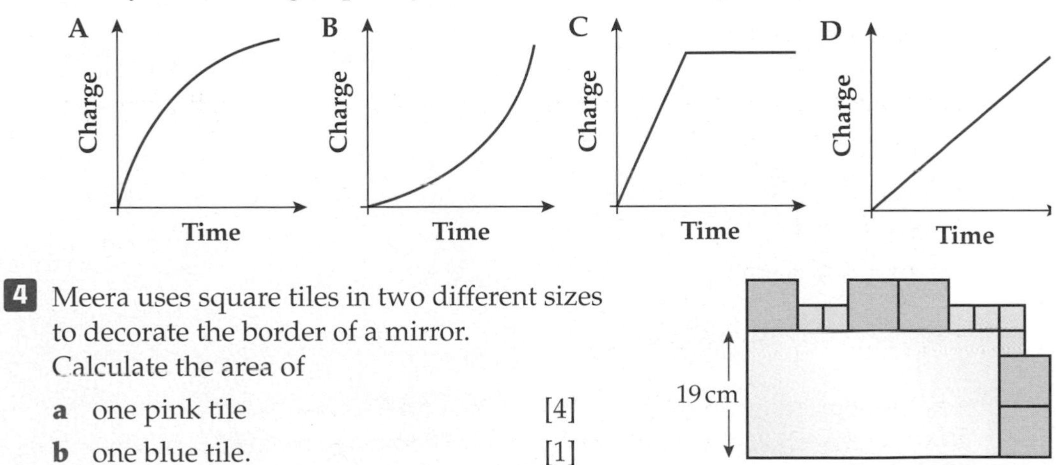

AO3

B

4 Meera uses square tiles in two different sizes to decorate the border of a mirror.
Calculate the area of

a one pink tile [4]

b one blue tile. [1]

19 cm
41 cm

5 Find the equation of the straight line which

a passes through the points $(3, -2)$ and $(9, 1)$ [3]

b passes through the point $(0, 10)$ and is parallel to the line $y = 1 - x$. [2]

AO3

A

6 Carla uses T_n to represent the nth term of a sequence. She forms a sequence using these rules, where a and b are integers.

a Prove that $T_7 = 5a + 8b + 12$. [3]

b Prove that the sum of the first six terms of the sequence is an even number. [3]

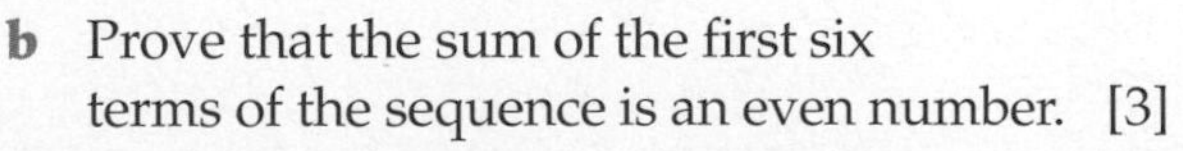

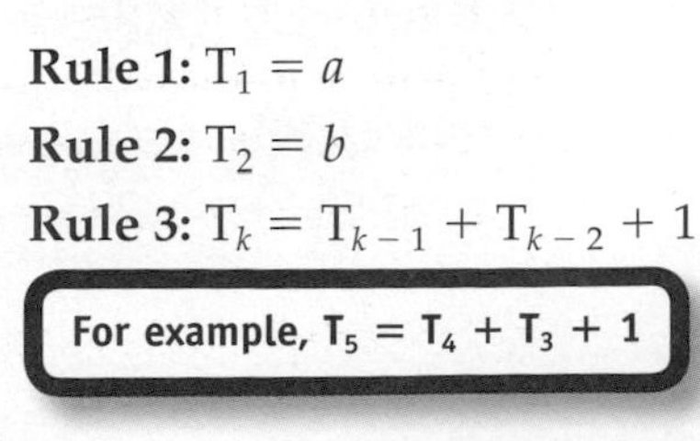

Rule 1: $T_1 = a$

Rule 2: $T_2 = b$

Rule 3: $T_k = T_{k-1} + T_{k-2} + 1$

For example, $T_5 = T_4 + T_3 + 1$

AO3

A*

7 Line A passes through the points $(5, 2)$ and $(-1, 6)$.
Line B passes through the points $(0, -4)$ and $(8, 8)$.
Show that the lines A and B are perpendicular. [4]

AO3

22 Number skills

BBC Video

This chapter is about revising essential number skills.

When you go on holiday with friends, you need to be able to work out if you have enough money for accommodation and food, and how to share the bills fairly.

Objectives

This chapter will remind you how to

- understand equivalent fractions
- use ratio notation, including reduction to its simplest form
- recognise that each terminating decimal is a fraction
- convert simple fractions to percentages and vice versa
- use brackets and the hierarchy of operations
- use calculators effectively and efficiently, using an extended range of function keys
- select and justify appropriate degrees of accuracy for answers to problems
- round to a given number of significant figures
- simplify a fraction by cancelling all common factors
- understand that 'percentage' means 'number of parts per 100' and use this to compare proportions
- understand 'reciprocal' as multiplicative inverse

P Number skills: revision exercise 1

1 Write each of these as an equivalent fraction with a denominator of 24.

a $\frac{1}{2}$ b $\frac{3}{4}$ c $\frac{2}{3}$ d $\frac{5}{6}$ e $\frac{9}{12}$ f $\frac{7}{8}$

2 Write each fraction in its simplest form.

a $\frac{10}{15}$ b $\frac{4}{20}$ c $\frac{12}{27}$ d $\frac{18}{24}$ e $\frac{30}{36}$

3 Write each of these percentages as a fraction in its lowest terms.

a 15% b 32% c 76% d $10\frac{1}{2}$%

4 a Write $\frac{9}{16}$ as a decimal. b Write 32% as a decimal.

5 Which is larger, $\frac{3}{7}$ or 37%? Give a reason for your answer.

6 Nine out of 15 of the text messages that Steven sends are to friends.
Twelve out of 20 of the text messages that Fernando sends are to friends.
Who sends the greater proportion of text messages to friends?

7 Find the reciprocal of each of these.

a 2 b 5 c $\frac{1}{3}$ d $\frac{2}{3}$ e $\frac{7}{10}$ f $2\frac{2}{5}$

8 a Work out $\frac{1}{2} \div \frac{3}{2}$. b Work out $\frac{1}{2} \times \frac{2}{3}$.

c Look at your answers to parts **a** and **b**. What can you say about these calculations?

9 In a youth group there are 36 girls and 24 boys.
Write the ratio of girls to boys in its simplest form.

HELP Chapter 7

10 Simplify the ratio 5 m : 40 cm.

11 Work out

a $3 \times 4 - 2$ b $3 + 2 \times 4$ c $12 \div (3 + 3)$ d $(12 - 6) \times 3^2$ e $18 + 4^2 \div 2$

12 Calculate $\frac{3.7 \times 2.6}{2.3 - 1.1}$

a Write your full calculator display.

b Write your answer to part **a** to one decimal place.

c Write your answer to part **a** to two decimal places.

13 Round each of these numbers to the degree of accuracy indicated.

a 285 (1 s.f.) b 3470 (2 s.f.) c 8.24 (1 s.f.)

d 0.4508 (2 s.f.) e 19.0956 (3 s.f.) f 0.009 072 (2 s.f.)

14 Round the number in each of these statements to an appropriate degree of accuracy.
Give a reason for your decision.

a Nathan is 1.7587 m tall. b The temperature is 28.345°C.

15 A room measures 14.27 m by 8.9 m. What is its area?
Give your answer to an appropriate degree of accuracy.

16 Calculate these, giving your answers correct to two significant figures.

a 2.88^2 b $\sqrt{14}$ c 1.4^3 d $\sqrt[3]{32}$

Number skills: revision exercise 2

1 Write each of these as an equivalent fraction with a denominator of 12.

a $\frac{1}{2}$ b $\frac{3}{4}$ c $\frac{5}{6}$ d $\frac{2}{3}$

2 Which of these fractions is not equivalent?

A $\frac{3}{8}$ B $\frac{6}{16}$ C $\frac{12}{32}$ D $\frac{21}{58}$ E $\frac{27}{72}$

3 Write each fraction in its simplest form.

a $\frac{12}{15}$ b $\frac{15}{18}$ c $\frac{30}{36}$ d $\frac{18}{27}$ e $\frac{42}{54}$

4 a Write $\frac{2}{5}$ as a decimal. b Write 47% as a decimal.

5 Write each of these percentages as a fraction in its lowest terms.

a 28% b 49% c $40\frac{1}{2}\%$ d 23.2%

6 Express each of these as a percentage.

a $\frac{1}{8}$ b $\frac{8}{40}$ c 0.32 d 0.05

7 Jenson has won a competition. He can choose one prize from:

70% of £3500 $\frac{18}{20}$ of £3500

Which prize should he choose?

8 In a group of teenagers, 40% own a laptop computer.
$\frac{4}{5}$ of the group own a mobile phone.
Does the group own a greater proportion of laptop computers or mobile phones?

9 Find the reciprocal of each of these.

a 7 b 13 c $\frac{2}{5}$ d $\frac{6}{8}$ e $4\frac{4}{9}$

10 What is the reciprocal of 1.125?

11 In an animal rescue centre there are 18 cats and 45 dogs.
Write the ratio of dogs to cats in its simplest form.

12 A garden lawn is 1.75 m long and 85 cm wide.
Write the ratio of length to width in its simplest form.

13 Work out

a $5 \times 3 + 2^2$ b $15 + 8 \div 4$ c $3 + 2 \times 3^2 - 3$ d $21 \div 3 \times (2 + 4)$

14 Look at this calculation.

$24 \times 36.8 = 883.2$

Write down the inverse calculation you would use to check that the answer is correct.

15 Calculate

$$\frac{5.95 \times 4.8}{3.6 - 0.8}$$

16 What is the product of 18.2 and 3.05?

17 $6\frac{1}{2}$ litre bottles are to be filled with water. There are 1100 litres of water available.
How many full bottles of water will there be?

18 Llinos has been told to eat $\frac{1}{4}$ kg of fruit each day.
She buys 3 kg of fruit.

a How many days will it take Llinos to eat all this fruit?

Llinos increases her fruit and vegetable portions to $\frac{2}{5}$ kg per day.

b What is the least number of whole kilograms of fruit and vegetables that Llinos would need for 7 days?

19 What is 3.957 m rounded to one decimal place?

20 Calculate

a $18.6 \times (10.9 - 3.2)$ **b** the square of 14 **c** the square root of 256

d 2.17^2

i Write your full calculator display.

ii Write your answer to two decimal places.

e $\sqrt[3]{56}$

i Write your answer to three decimal places.

ii Write your answer to two significant figures.

21 Car insurance for 12 months costs £680. Payment is to be made by 12 equal instalments.
Calculate the cost of each payment to the nearest penny.

22 Mount Kilimanjaro is 5900 m high to two significant figures.
What is the smallest height that Mount Kilimanjaro could be?

23 A picture frame measures 2.35 m by 147 cm. What is its area?
Give your answer to an appropriate degree of accuracy.

24 Calculate

a $(-1.8)^2$ **b** 2.4^3 **c** $(-3.5)^3$ **d** $\sqrt[3]{9.261}$

25 Copy and complete

a $\square \div 42 = 26.35$ **b** $65.8 \times \square = 1250.2$

c $\square^2 = 196$ **d** $\sqrt[3]{\square} = 3.7$

23 Angles

This chapter is about angles, which measure the amount of turn.

Olympic divers take care to turn the right amount in the final somersault before straightening out to enter the water.

Objectives

This chapter will show you how to

- **recall and use properties of angles at a point, angles on a straight line (including right angles), perpendicular lines, and opposite angles at a vertex D**
- **recognise alternate angles, corresponding angles and interior angles D**
- **understand and use bearings D C**

Before you start this chapter

1 Describe these turns by giving the fraction of a whole turn, the angle in degrees and the direction.

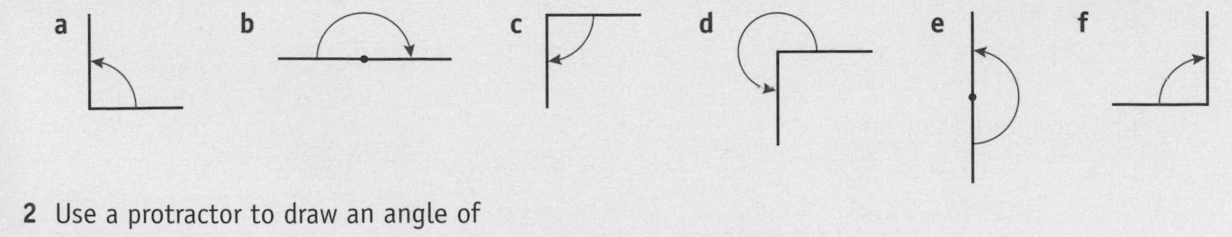

2 Use a protractor to draw an angle of

a 35° **b** 100° **c** 190° **d** 280°

Naming angles

Acute angle		Less than 90°
Right angle		Exactly 90° This symbol means 'right angle'.
Obtuse angle		More than 90° but less than 180°
Straight line		Exactly 180°
Reflex angle		More than 180° but less than 360°

Labelling angles

You can use letters on a diagram to label angles.

AB and *BC* are line segments. They meet at *B*.

The arrow is pointing at angle *ABC*.

You can write this angle as $\angle ABC$ or $A\widehat{B}C$ or angle *ABC*.

The point of the angle is at the middle letter.

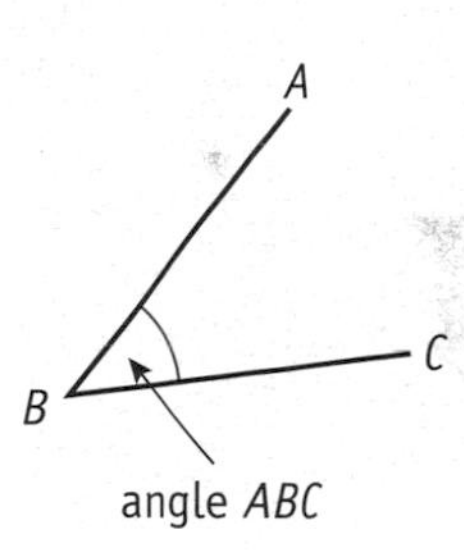

Angle facts

Angles on a straight line add up to 180°.

Angles around a point add up to 360°.

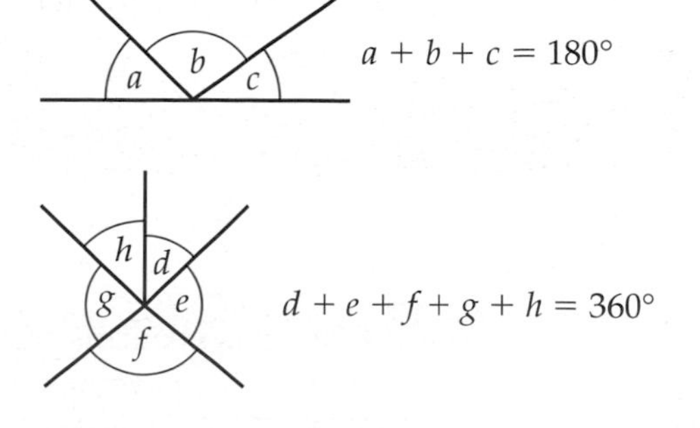

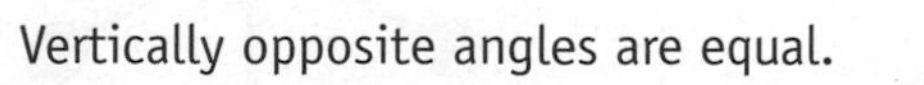

Vertically opposite angles are equal.

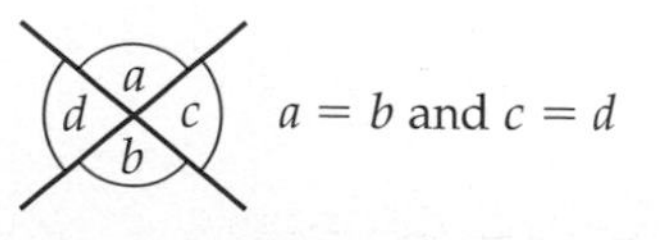

P Geometry skills: revision exercise

1. Work out the sizes of the angles marked with letters.

a

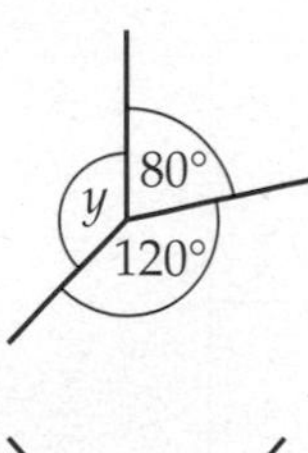

b

c

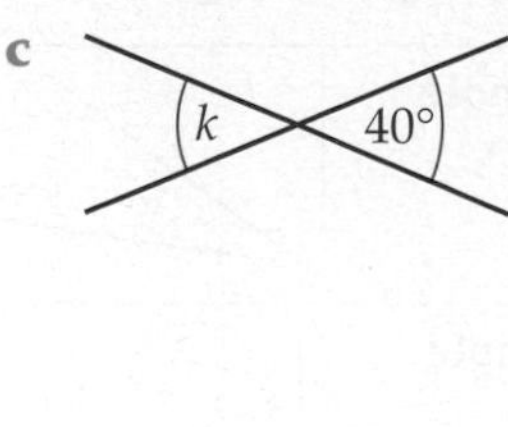

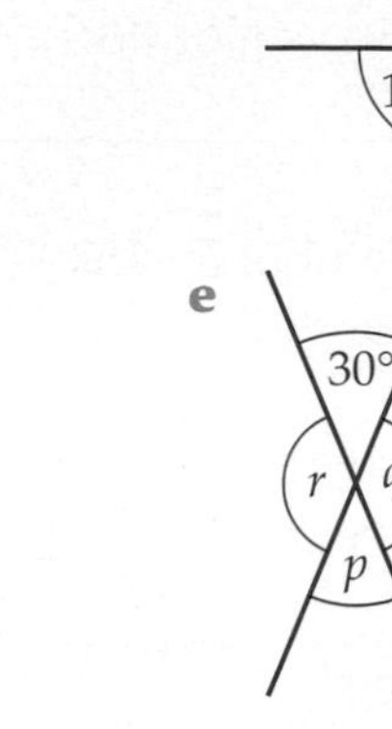

d

e

f

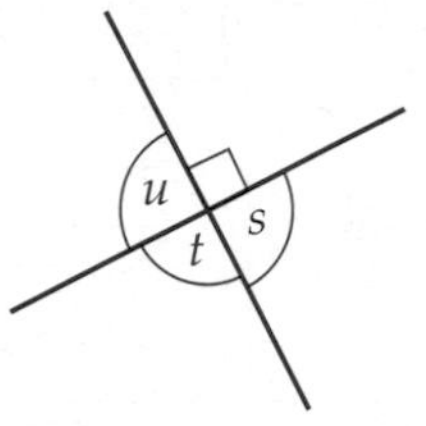

2. Work out the sizes of the angles marked with letters.

a

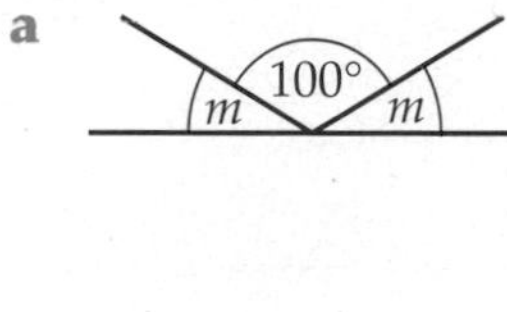

$m + m + 100° = 180°$
$2m + 100° = 180°$
$2m = \square$
$m = \square$

b

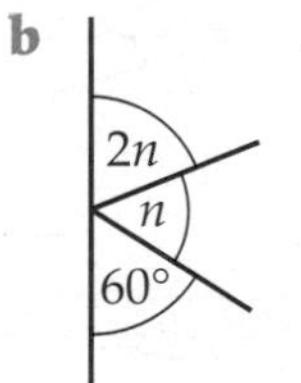

c

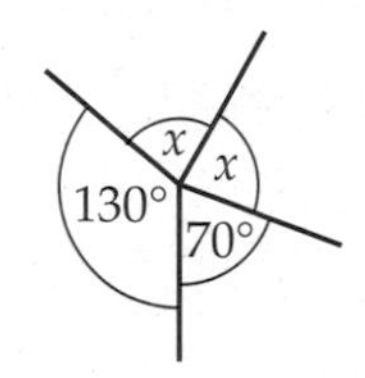

d

3. The diagram shows the design of a roof truss.
Work out the size of angle x and angle y.

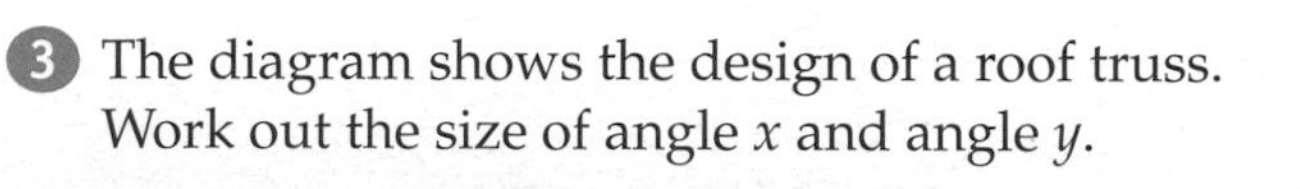

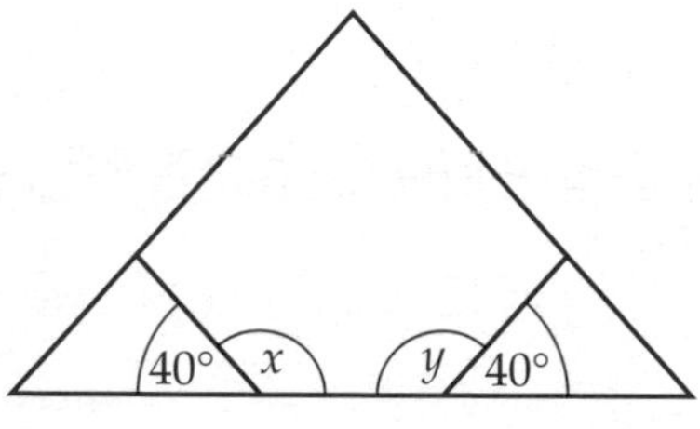

4. A children's roundabout has six arms from a central pole.
The angle between each pair of arms is the same.
Work out the angle between each pair of arms.

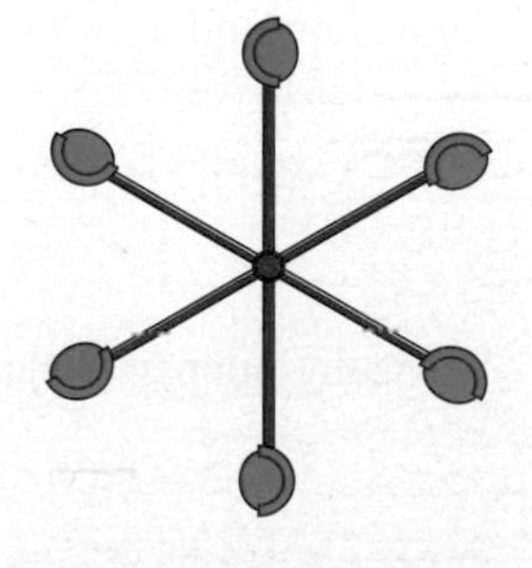

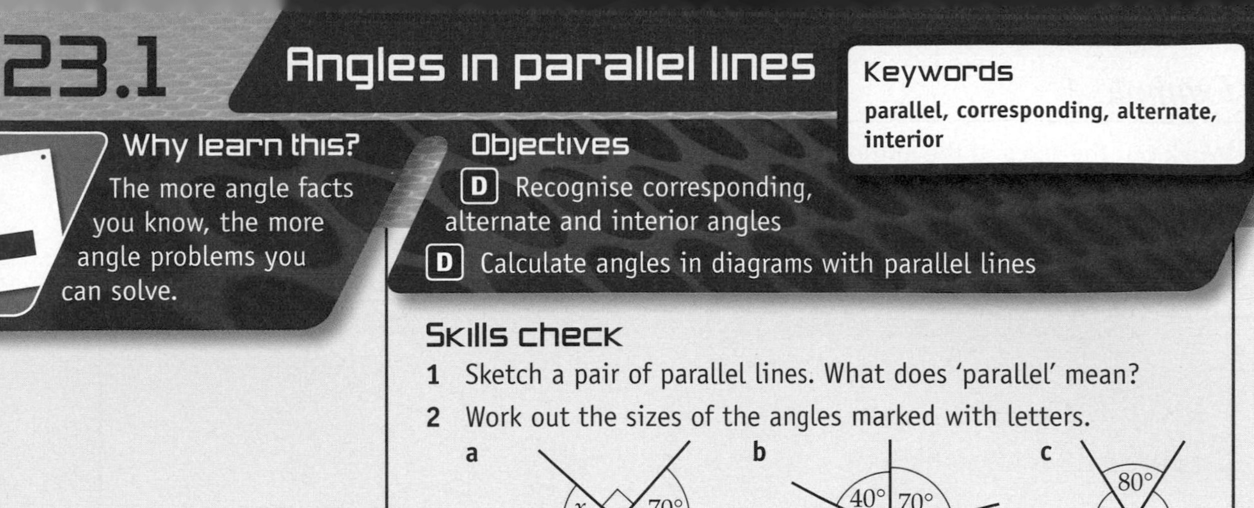

23.1 Angles in parallel lines

Keywords
parallel, corresponding, alternate, interior

Why learn this?

The more angle facts you know, the more angle problems you can solve.

Objectives

D Recognise corresponding, alternate and interior angles

D Calculate angles in diagrams with parallel lines

Skills check

1 Sketch a pair of parallel lines. What does 'parallel' mean?

2 Work out the sizes of the angles marked with letters.

Angle facts

Arrowheads are used to show that two lines are **parallel**.

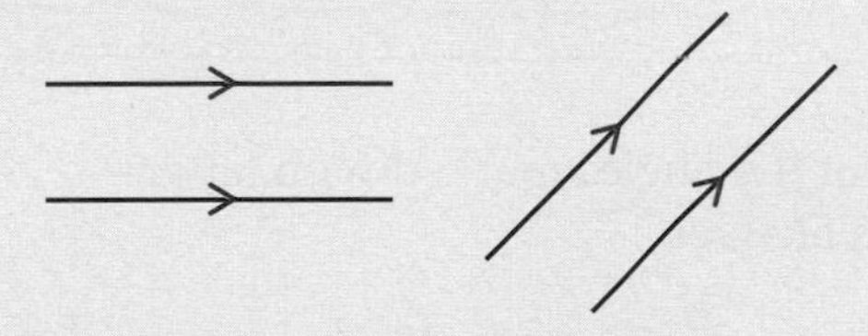

A line crossing two parallel lines creates pairs of equal angles.

a and b are **corresponding** angles.
c and d are also corresponding angles.
The lines make an F shape.

Corresponding angles are equal.

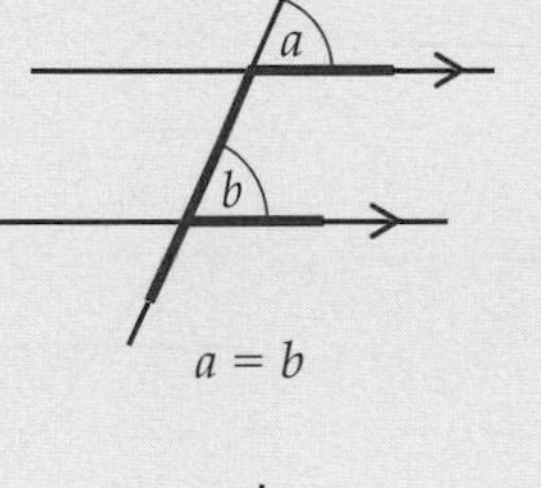

$a = b$

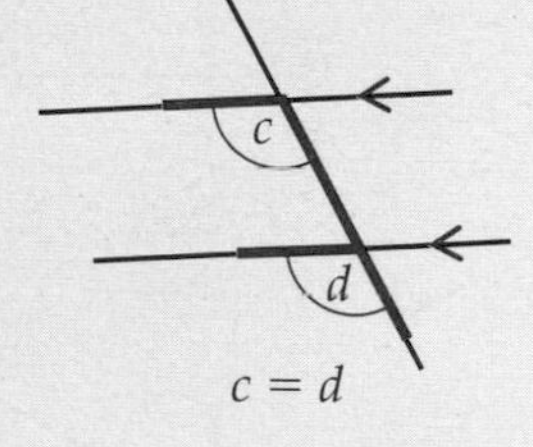

$c = d$

m and n are **alternate** angles.
p and q are also alternate angles.
The lines make a Z shape.

Alternate angles are equal.

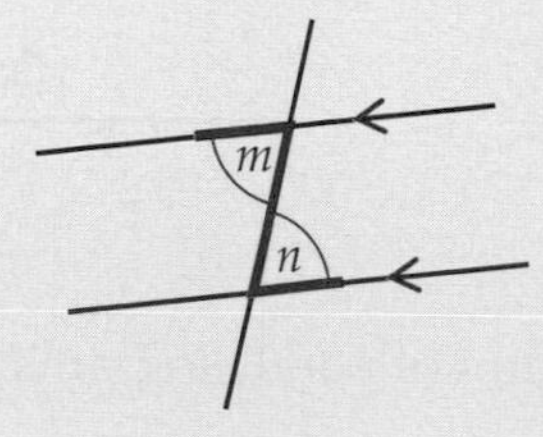

$m = n$

$p = q$

r and s are **interior** angles.

Interior angles add up to 180°.

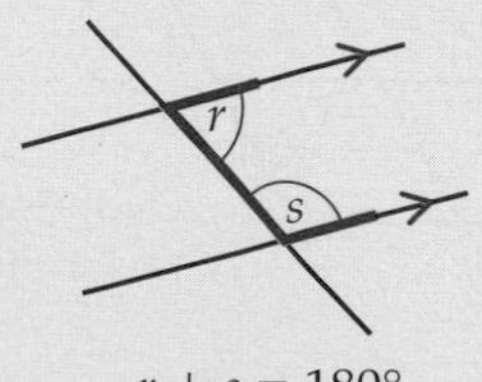

$r + s = 180°$

To solve angle problems, you may need to use more than one angle fact.
Always write down the facts you use, to show your reasoning.

D

Example 1

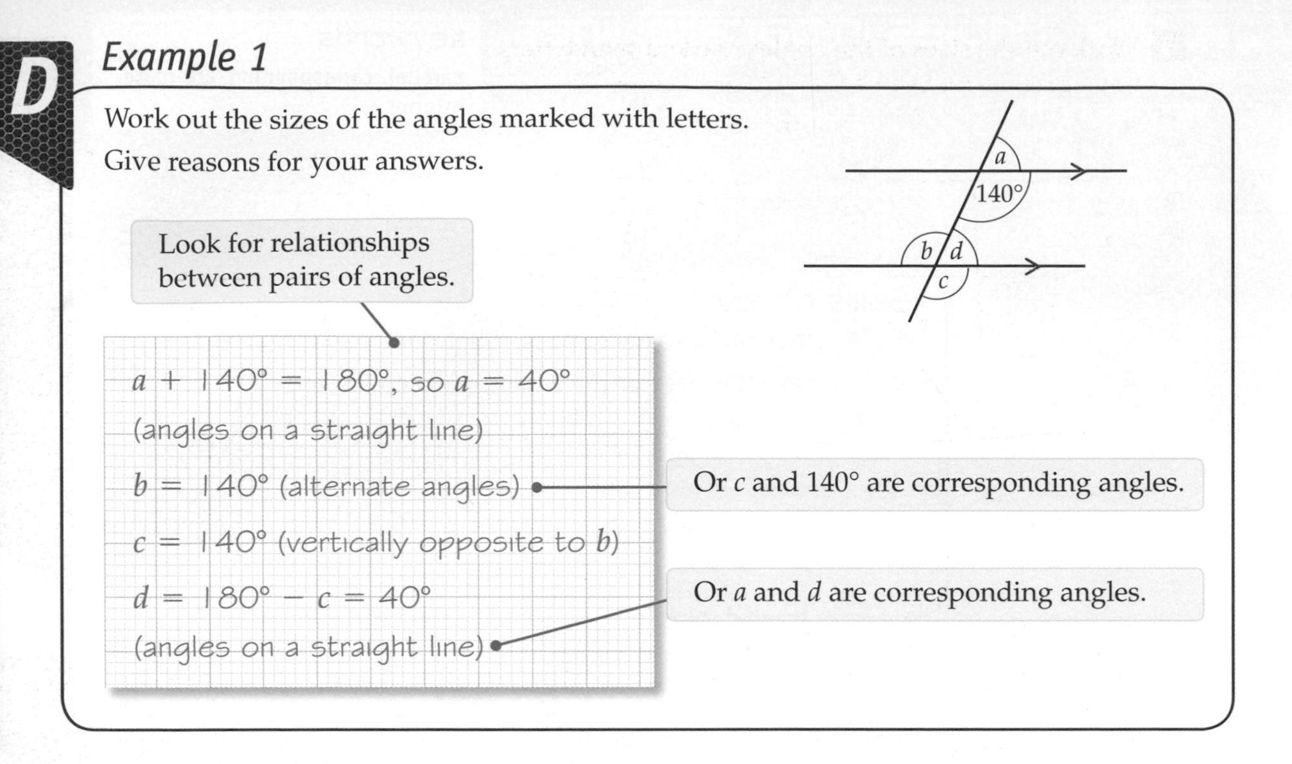

Work out the sizes of the angles marked with letters.
Give reasons for your answers.

Look for relationships between pairs of angles.

$a + 140° = 180°$, so $a = 40°$
(angles on a straight line)

$b = 140°$ (alternate angles)

$c = 140°$ (vertically opposite to b)

$d = 180° - c = 40°$
(angles on a straight line)

Or c and 140° are corresponding angles.

Or a and d are corresponding angles.

Exercise 23A

D

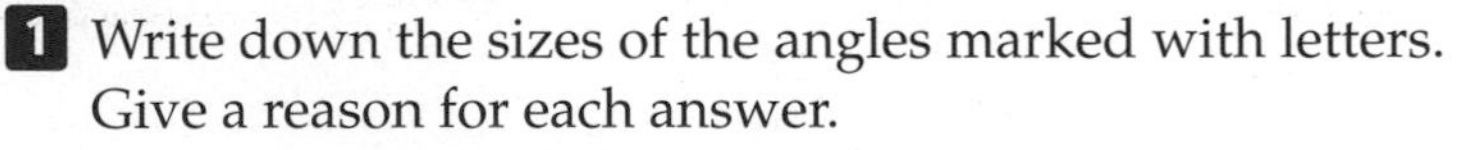

1 Write down the sizes of the angles marked with letters.
Give a reason for each answer.

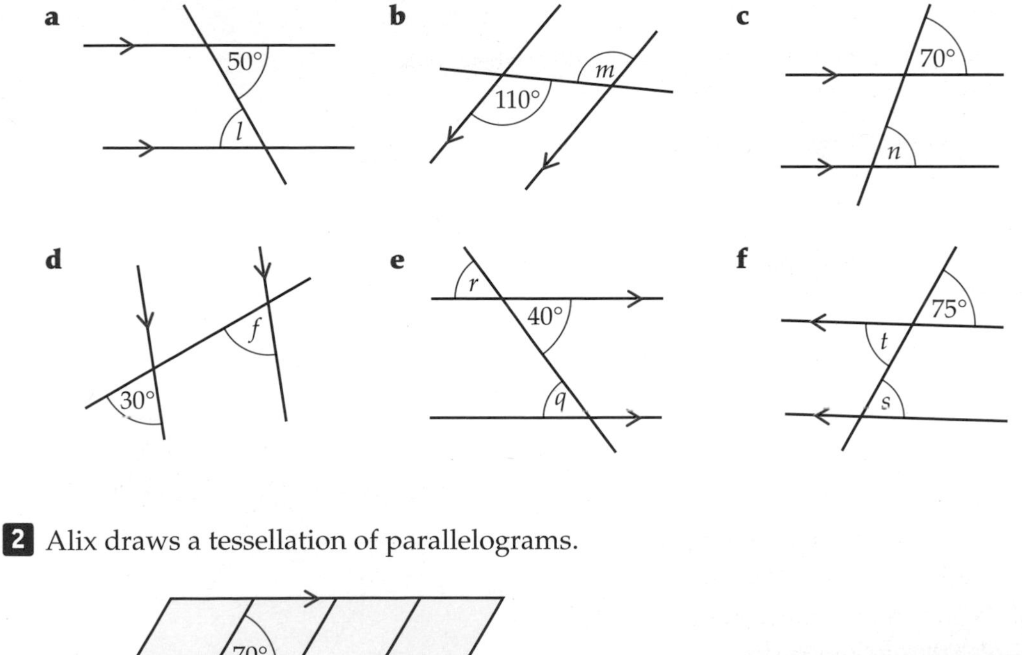

D

2 Alix draws a tessellation of parallelograms.

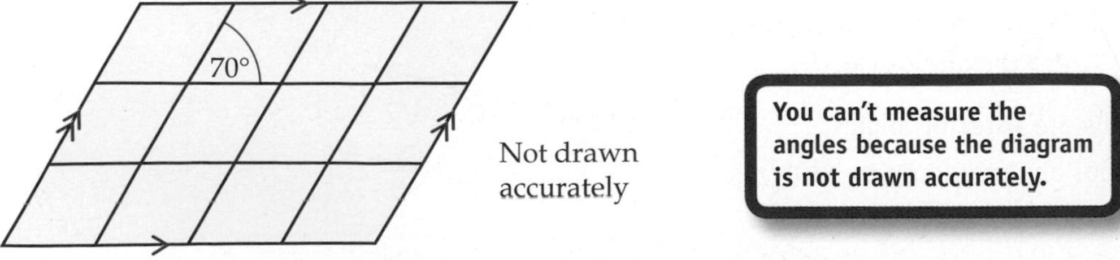

You can't measure the angles because the diagram is not drawn accurately.

a Copy a two-by-two block of four parallelograms from her tessellation.

b Use the angle facts for parallel lines to label all the angles that are 70°.

c How could you use the angle facts you know to work out the other angles?

AO2

3 Work out the sizes of the angles marked with letters.
Give reasons for your answers.

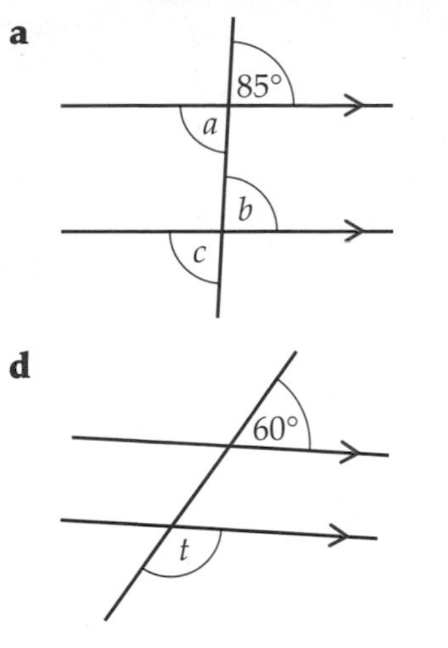

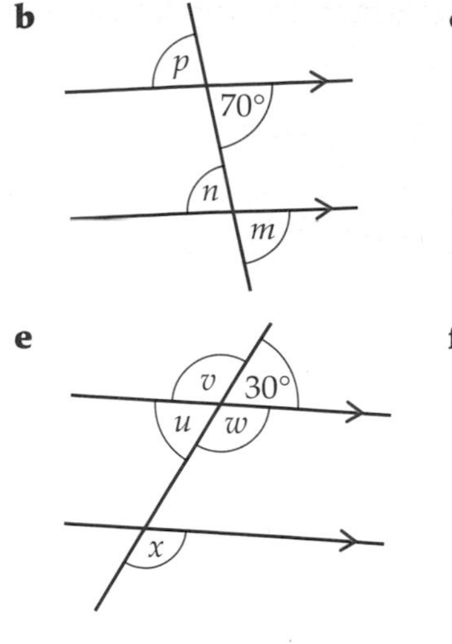

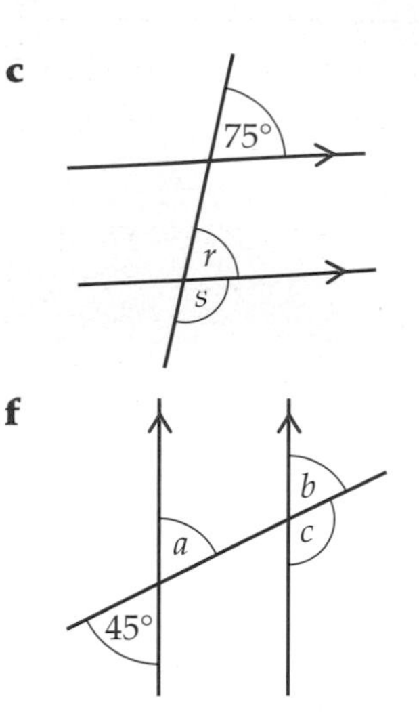

4 Show why angle $x = 60°$.

'Show' here means 'give a reason'. Use angle facts.

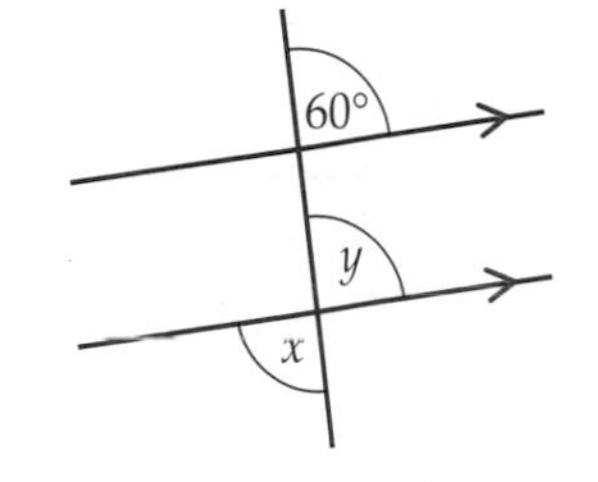

5 Show why angle $m = 120°$.

Copy the diagram and label any other angles you need.

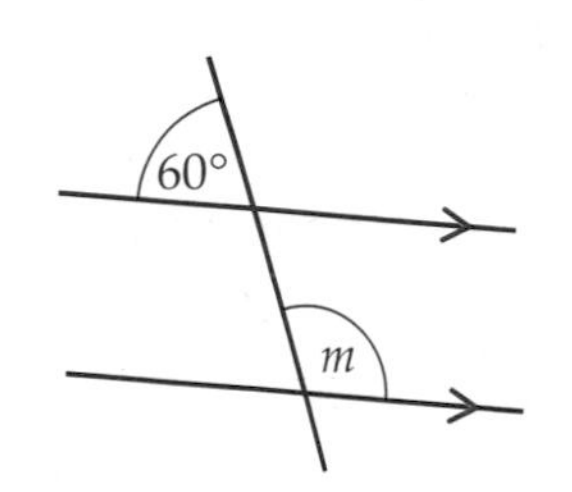

6 *PQRS* is a parallelogram.
Angle $PQR = 120°$.

a Copy the parallelogram and extend each side as shown.

b Use angle facts to work out the other angles in the parallelogram.

c What do you notice about the angles in the parallelogram?

d Draw more parallelograms. Does part **c** work for all of them?

7 *ABCD* is a trapezium.
Angle $BCD = 80°$ and angle $CDA = 70°$.
Work out the size of each of the remaining angles.

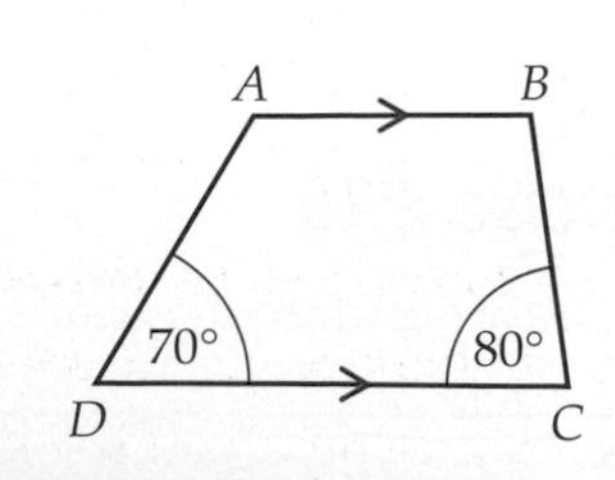

23.2 Bearings

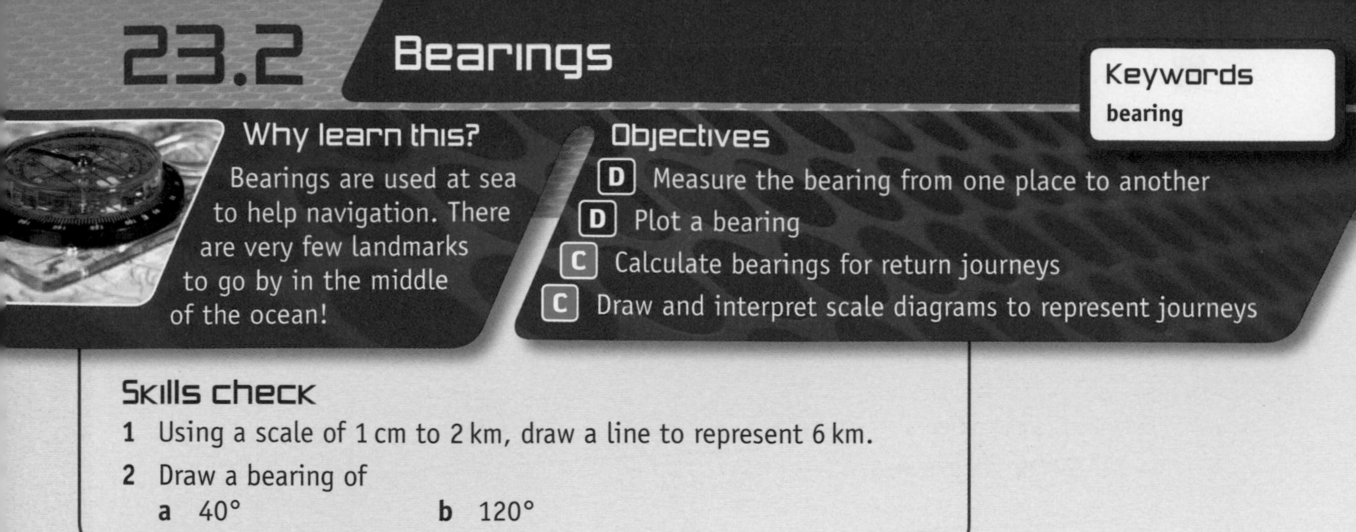

Why learn this?
Bearings are used at sea to help navigation. There are very few landmarks to go by in the middle of the ocean!
Objectives
D Measure the bearing from one place to another
D Plot a bearing
C Calculate bearings for return journeys
C Draw and interpret scale diagrams to represent journeys

Keywords
bearing

Skills check

1 Using a scale of 1 cm to 2 km, draw a line to represent 6 km.

2 Draw a bearing of

a 40° b 120°

Understanding bearings

A **bearing** tells you the direction to travel. It is an angle measured clockwise from north.

A bearing can have any value from 0° to 360°. It is always written with *three* figures. For a two-digit number, like 72, add a zero in front: 072°.

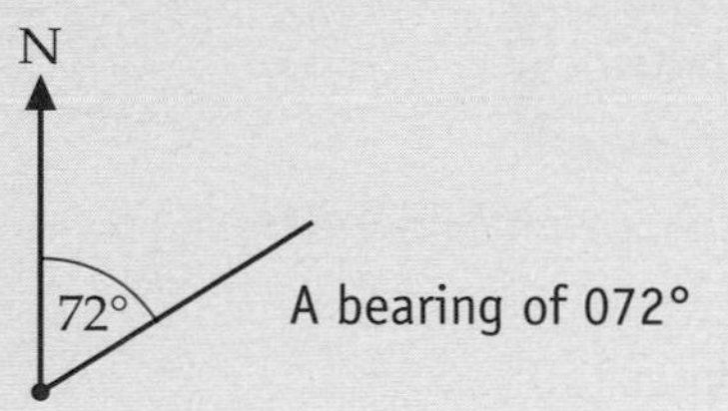

To draw or measure a bearing from a point, draw the north line at that point first, straight up the page.

D

Example 2

The diagram shows the positions of two towns: Heretown and Thereville.

Measure the bearing of Thereville from Heretown.

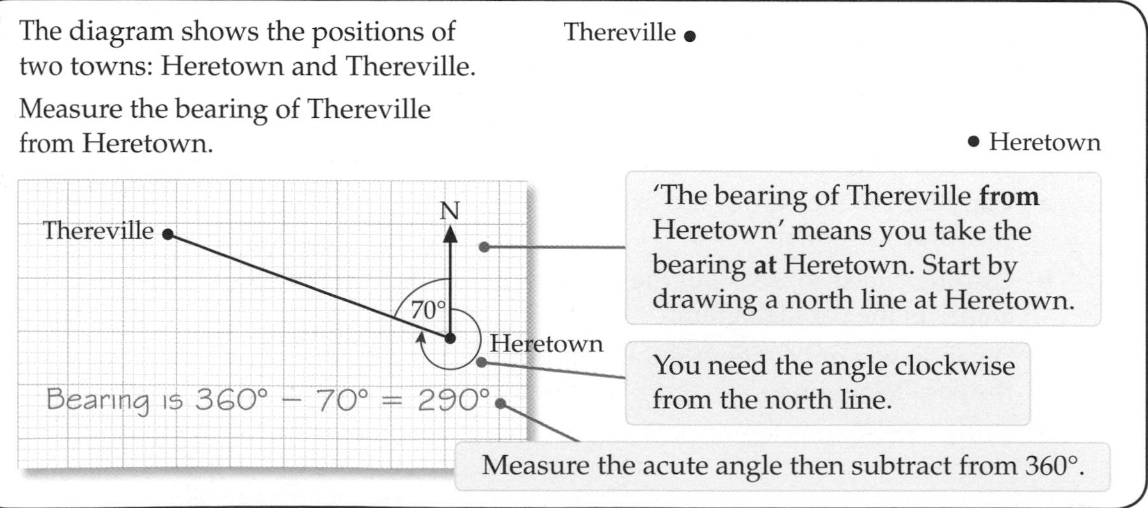

Thereville
Heretown
Thereville
N
70°
Heretown
Bearing is 360° − 70° = 290°
'The bearing of Thereville **from** Heretown' means you take the bearing **at** Heretown. Start by drawing a north line at Heretown.
You need the angle clockwise from the north line.
Measure the acute angle then subtract from 360°.

Exercise 23B

D

1 Draw accurate diagrams to show these bearings.

a 200° b 235° c 330°

2 A compass is shown.
It shows eight directions.
Write each direction as a bearing.

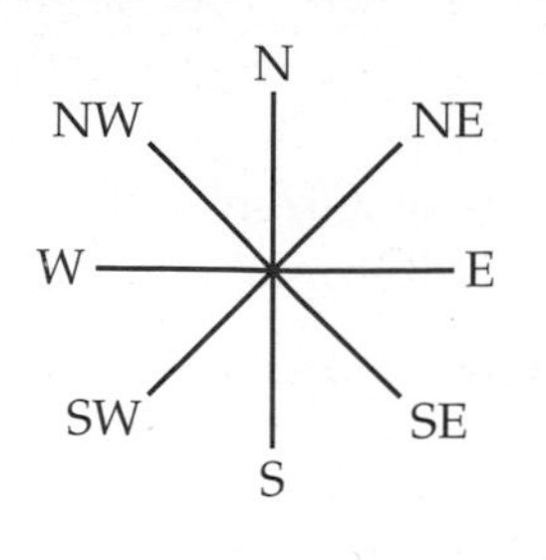

3 The diagram shows the positions of two hikers: Tim and Sue.
Tim walks on a bearing of 050°.

Tim

Sue

a Copy the diagram. Plot and draw Tim's route.

b Sue walks on a bearing of 340°.
Plot and draw Sue's route.

c Mark the point where Sue and Tim meet with an X.

4 The diagram shows a ship and a speed boat.

The ship sails on a bearing of 140°.

The speed boat travels on a bearing of 320°.

Could the ship and the speed boat collide?
Explain your answer.

Ship

Speed boat

5 A ship leaves a harbour and sails 6 km east.

a Draw a scale diagram to represent this.
Use a scale of 1 cm to 1 km.

b The ship then changes course and sails 4 km on a bearing of 200°.
Add this part of the journey to your scale drawing.

Use a protractor.

c The ship sails directly back to the harbour.
i What bearing must it take?
ii How far must it sail?

Measure it on your diagram.

6 A dinghy sails 3 km west.
It then changes course and sails 7 km on a bearing of 050°.

a How far is the dinghy from its starting point?

b What bearing should it take to return to the starting point?

Parallel lines and bearings

Two north lines are always parallel to each other.

You can use the angle facts for parallel lines to help you work out bearings.

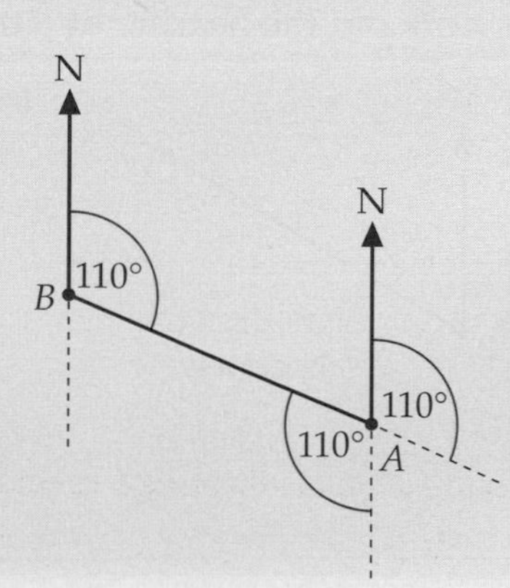

Example 3

a Write down the bearing of A from B.
b Work out the bearing of B from A.

You can't measure the angles because the diagram is not drawn accurately.

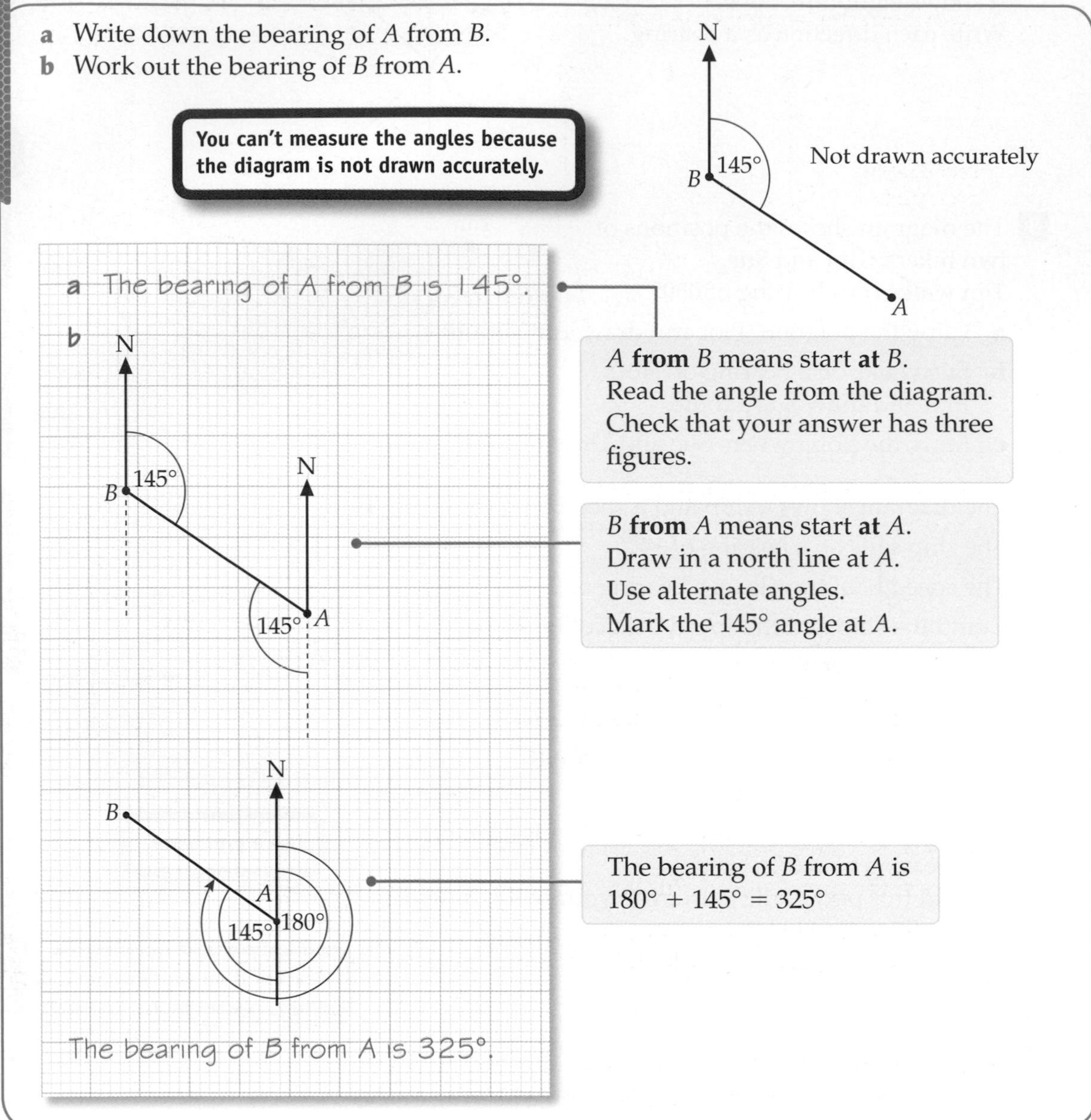

a The bearing of A from B is 145°.

b

A **from** B means start **at** B.
Read the angle from the diagram.
Check that your answer has three figures.

B **from** A means start **at** A.
Draw in a north line at A.
Use alternate angles.
Mark the 145° angle at A.

The bearing of B from A is $180° + 145° = 325°$

The bearing of B from A is 325°.

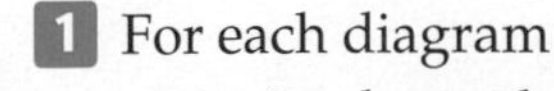

Exercise 23C

1 For each diagram
  i write down the bearing of A from B
  ii work out the bearing of B from A.

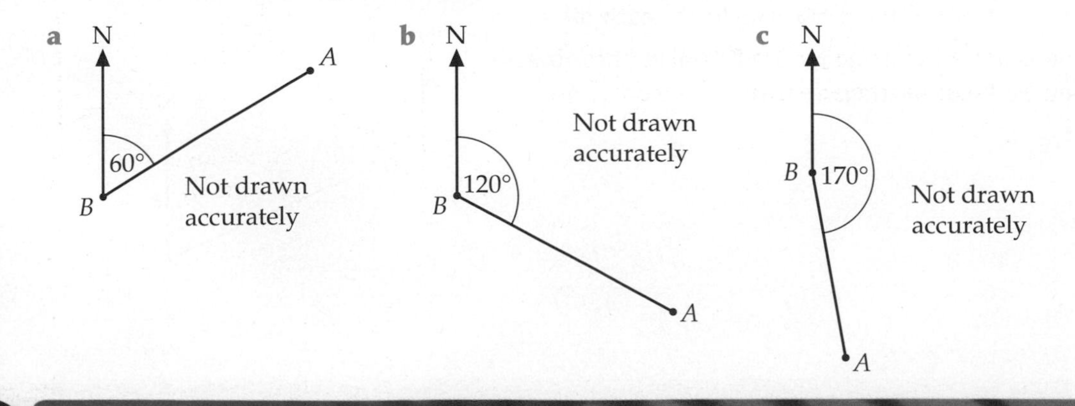

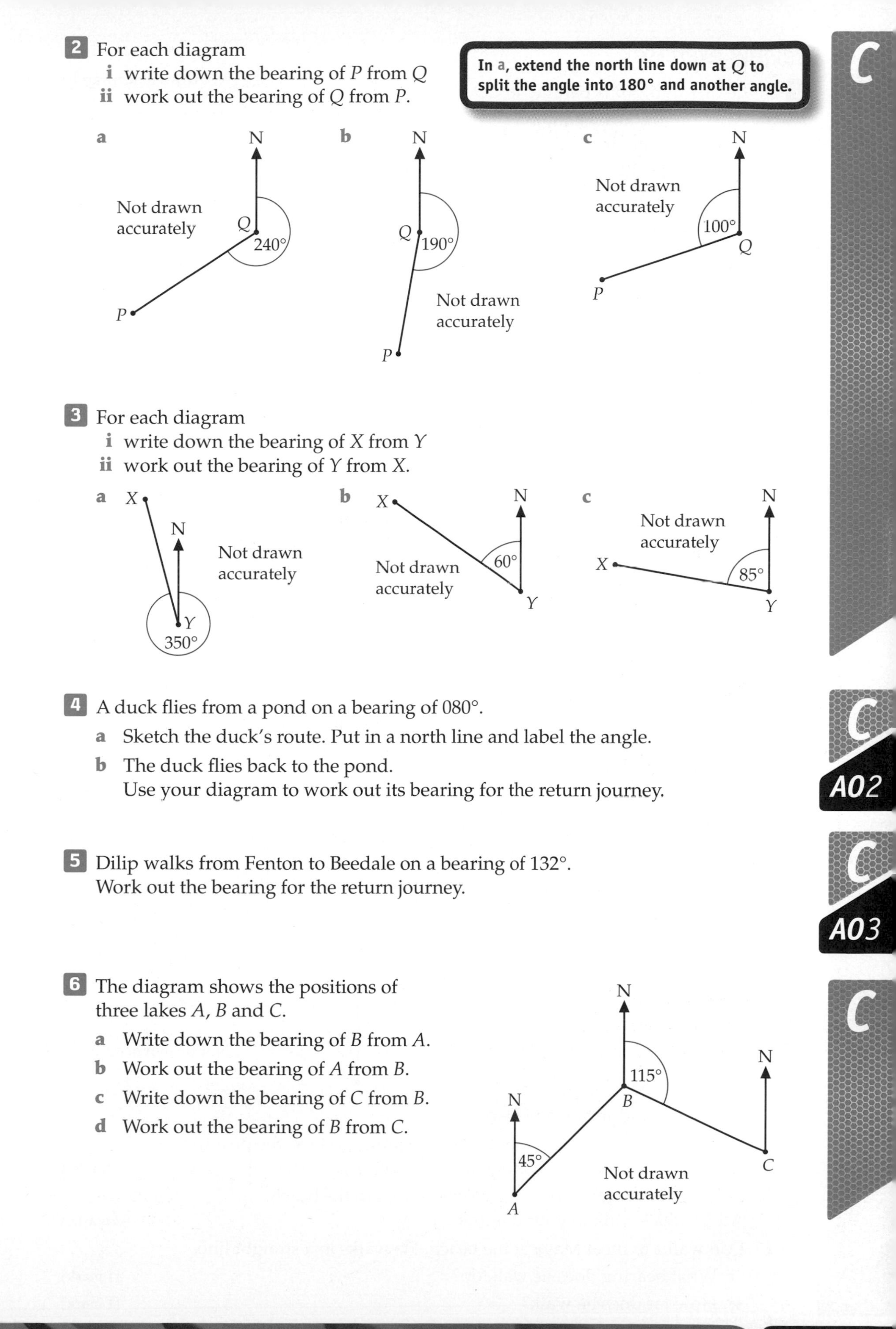

2 For each diagram

i write down the bearing of P from Q

ii work out the bearing of Q from P.

> In **a**, extend the north line down at Q to split the angle into 180° and another angle.

a **b** **c**

3 For each diagram

i write down the bearing of X from Y

ii work out the bearing of Y from X.

a **b** **c**

4 A duck flies from a pond on a bearing of 080°.

a Sketch the duck's route. Put in a north line and label the angle.

b The duck flies back to the pond.
Use your diagram to work out its bearing for the return journey.

5 Dilip walks from Fenton to Beedale on a bearing of 132°.
Work out the bearing for the return journey.

6 The diagram shows the positions of three lakes A, B and C.

a Write down the bearing of B from A.

b Work out the bearing of A from B.

c Write down the bearing of C from B.

d Work out the bearing of B from C.

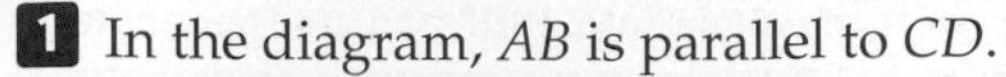

Review exercise

1 In the diagram, AB is parallel to CD.

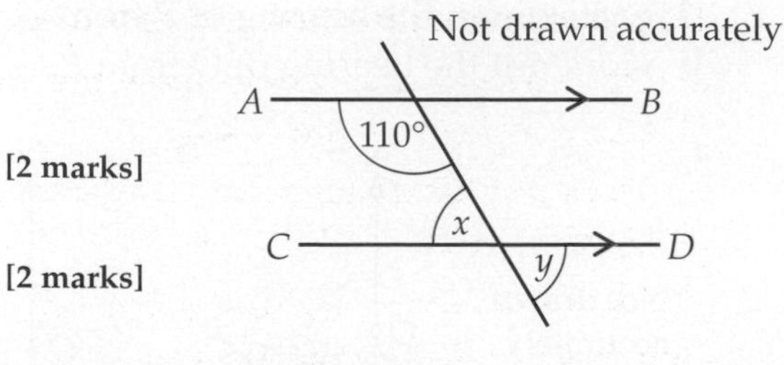

a Work out the size of angle x.
Give a reason for your answer. **[2 marks]**

b Write down the size of angle y.
Give a reason for your answer. **[2 marks]**

2 $PQRS$ is a parallelogram.
PQ is parallel to SR.
PS is parallel to QR.
Angle $PQR = 75°$.
Work out the size of each of the other angles in the parallelogram.

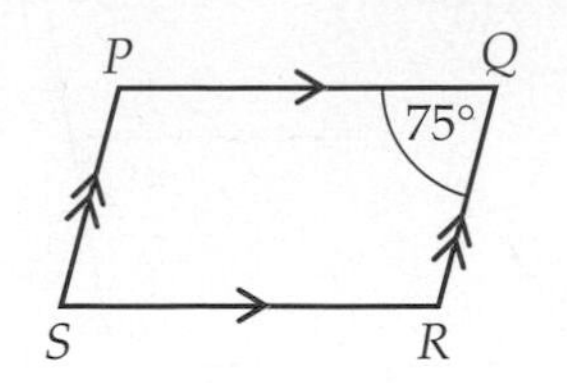

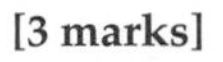

[3 marks]

C

3 A is due south of B.
The bearing of C from A is 080°.
The bearing of C from B is 120°.
Trace the diagram and mark the position of C.

N
B
A

[3 marks]

AO2

C

4 The diagram shows three points A, B and C.
The bearing of B from A is 125°.

a Write down the bearing of A from C. **[1 mark]**

b Work out the bearing of A from B. **[2 marks]**

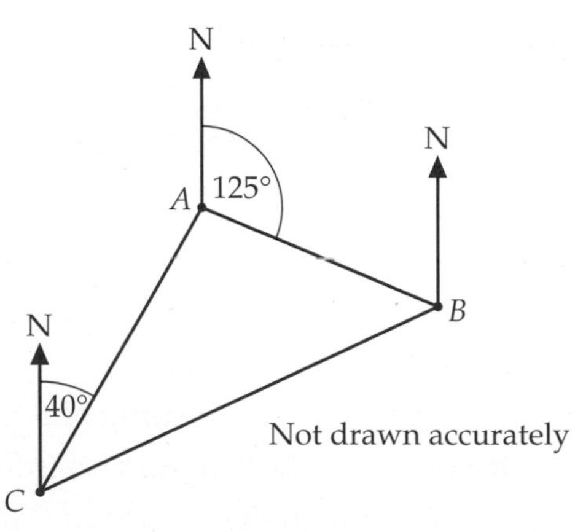

C

5 Maya lives 5 km due north of Dan.

a Using a scale of 1 cm to 1 km, draw a diagram to show the position of Maya's house and the position of Dan's house. **[2 marks]**

b Maya leaves home and walks 5 km due east to the beach.
Add Maya's walk to your diagram. **[2 marks]**

c Dan walks to meet Maya at the beach. He walks in a straight line.

i What bearing does he walk on? **[1 mark]**

ii How far does he walk? **[1 mark]**

AO2

C

6 An aircraft flies 8 km on a bearing of 60°.
Then it changes direction and flies 6 km on a bearing of 290°.

a Draw a scale diagram to show the aircraft's journey.
Use a scale of 1 cm to 1 km. **[4 marks]**

b Work out the distance and the bearing of the return journey to the starting point. **[2 marks]**

7 Here is a parallelogram *ABCD*.

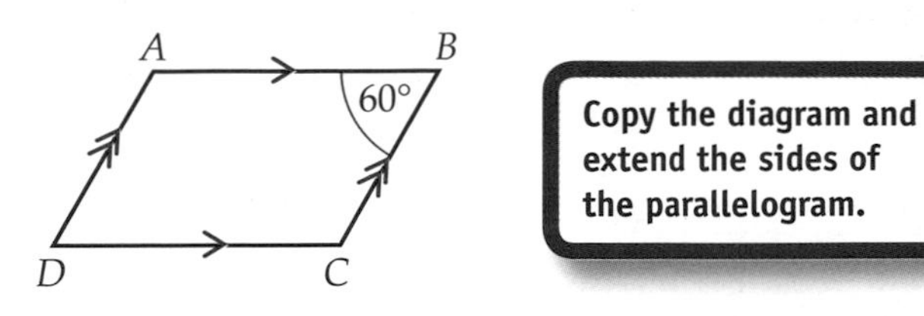

Copy the diagram and extend the sides of the parallelogram.

Angle $ABC = 60°$.
Show that angle $BCD = 120°$. **[2 marks]**

AO2

8 Show that for any parallelogram *ABCD*,
angle ABC + angle $BCD = 180°$.

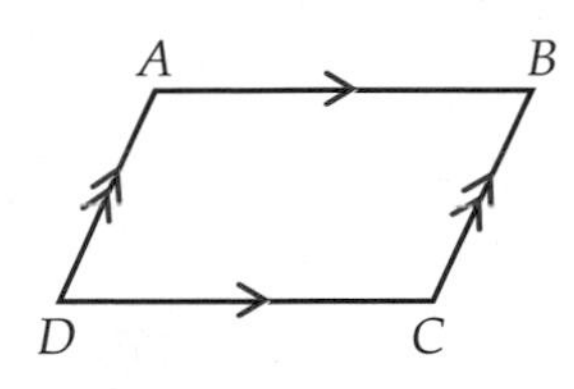

[2 marks]

Chapter summary

In this chapter you have learned how to

- recognise corresponding, alternate and interior angles **D**
- calculate angles in diagrams with parallel lines **D**
- measure the bearing from one place to another **D**
- plot a bearing **D**
- calculate bearings for return journeys **C**
- draw and interpret scale diagrams to represent journeys **C**

24 Triangles, polygons and constructions

BBC Video

This chapter is about working with triangles and polygons.

Certain triangles and polygons are often used in architecture since they are very strong shapes that are difficult to deform.

Objectives

This chapter will show you how to

- use angle properties of equilateral, isosceles and right-angled triangles D
- draw triangles using a ruler and protractor given information about their side lengths and angles D
- recall the essential properties and definitions of special types of quadrilateral, including square, rectangle, parallelogram, trapezium and rhombus D
- understand that inscribed regular polygons can be constructed by equal division of a circle D
- calculate and use the sums of the interior and exterior angles of quadrilaterals, pentagons and hexagons D C
- calculate and use the angles of regular polygons D C
- show step-by-step deduction in solving a geometrical problem D C

Before you start this chapter

1 a Copy the shapes below.

b Draw in one or more straight lines to divide each shape into the number of triangles required.

i Four right-angled triangles **ii** Two scalene triangles

2 Use a protractor to draw an angle of **a** 75° **b** 135°

3 Use a pair of compasses to construct triangle ABC, where $AB = 6$ cm, $BC = 5$ cm and $AC = 4$ cm.

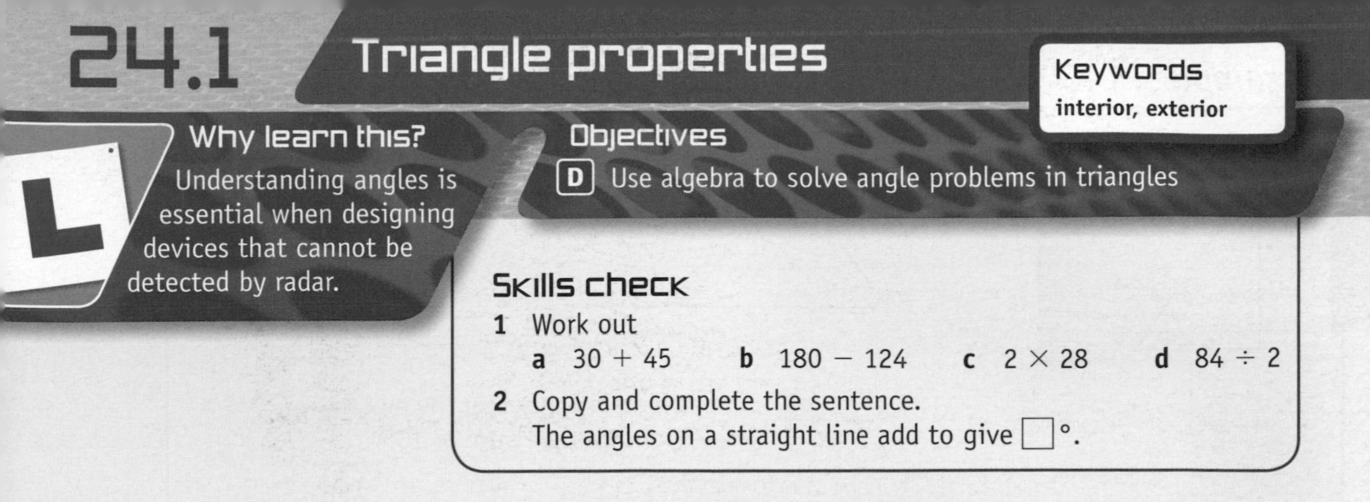

24.1 Triangle properties

Keywords
interior, exterior

Why learn this?
Understanding angles is essential when designing devices that cannot be detected by radar.

Objectives

D Use algebra to solve angle problems in triangles

Skills check

1 Work out
   a $30 + 45$ **b** $180 - 124$ **c** 2×28 **d** $84 \div 2$

2 Copy and complete the sentence.
   The angles on a straight line add to give □°.

Interior and exterior angles of a triangle

Look at this triangle.

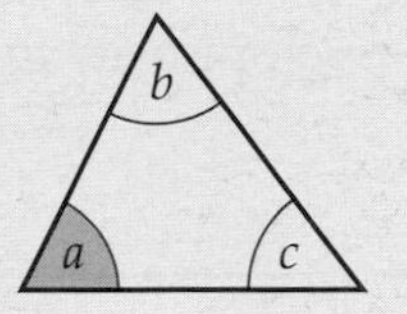

Angles a, b and c are called **interior** angles because they are *inside* the triangle.

If you tear off the three interior angles and place them alongside each other, they fit exactly onto a straight line.

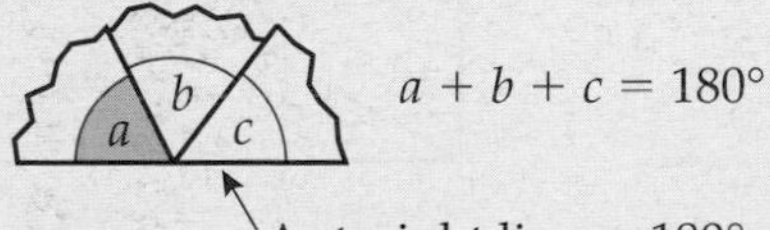

The sum of the angles of a triangle is 180°.

Look at this triangle.

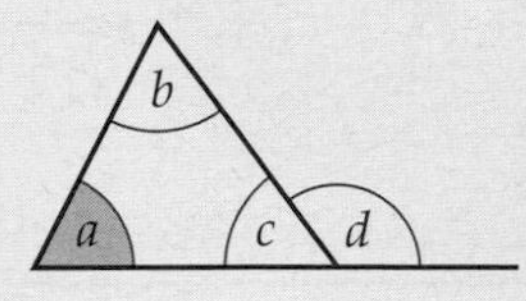

Angle d is called an **exterior** angle because it lies *outside* the triangle, on a straight line formed by extending one of the sides of the triangle.

If you tear off the two interior angles a and b, then place them on top of the exterior angle d, they fit exactly.

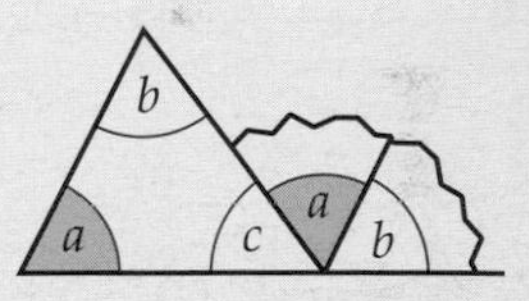

The exterior angle of a triangle is equal to the sum of the two opposite interior angles.

You can use this diagram to prove both of these properties as follows.

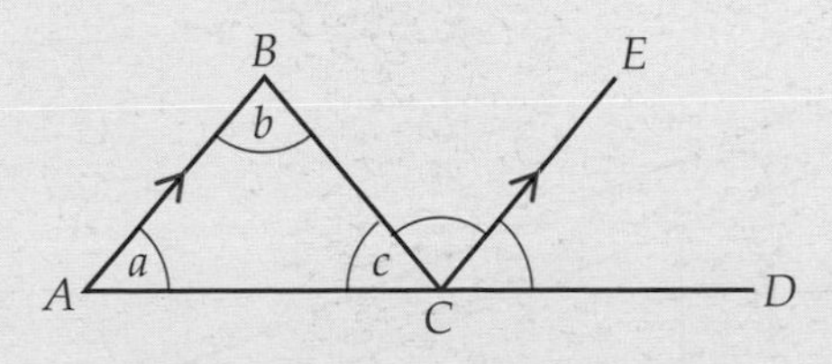

AB is parallel to CE, so $\angle DCE$ = angle a (corresponding angles) and $\angle BCE$ = angle b (alternate angles).

$\angle DCE + \angle BCE + \angle ACB = 180°$ (angles on a straight line),

so $a + b + c = 180°$.

Also $\angle BCD = \angle ECD + \angle BCE$,

so $\angle BCD = a + b$

D

Example 1

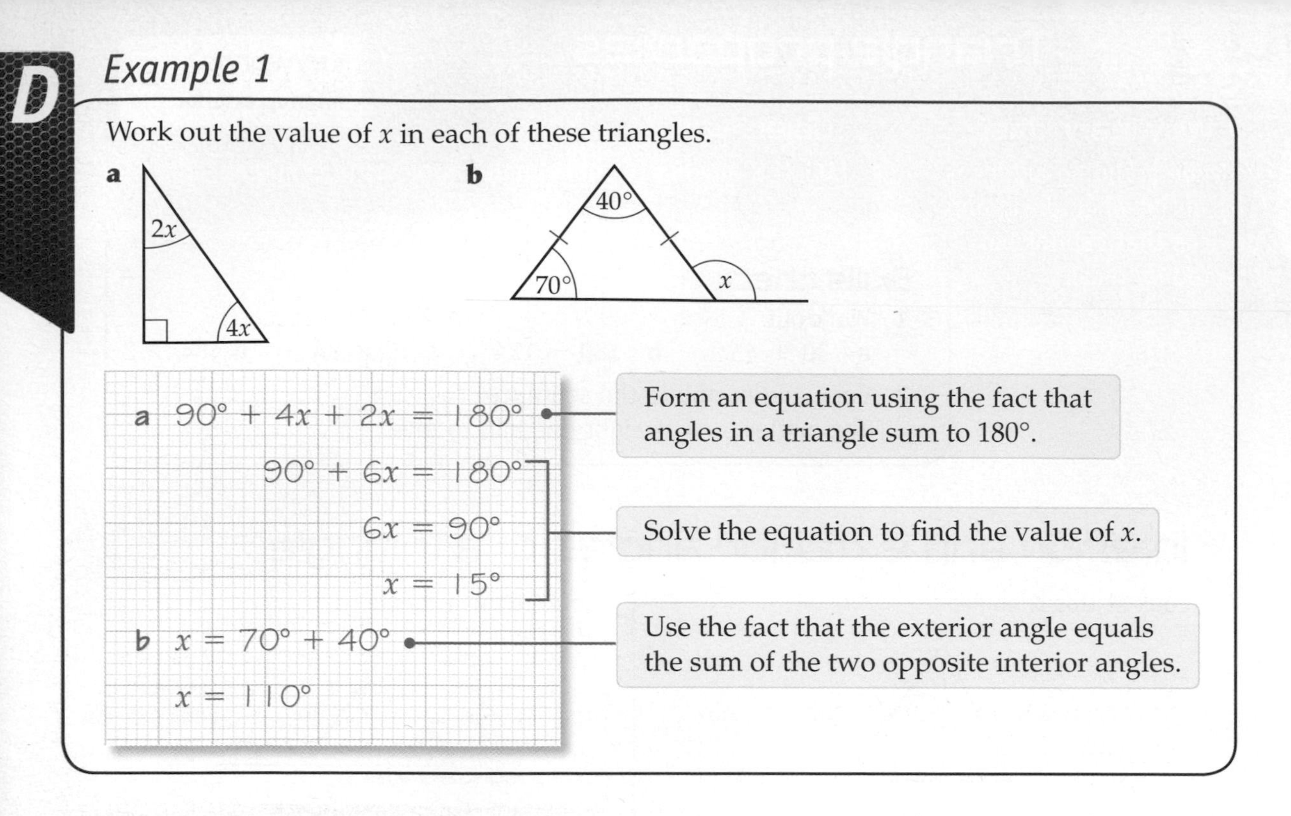

Work out the value of x in each of these triangles.

a $90° + 4x + 2x = 180°$

$90° + 6x = 180°$

$6x = 90°$

$x = 15°$

b $x = 70° + 40°$

$x = 110°$

Form an equation using the fact that angles in a triangle sum to 180°.

Solve the equation to find the value of x.

Use the fact that the exterior angle equals the sum of the two opposite interior angles.

Exercise 24A

D

1 Work out the value of x in each diagram.

a

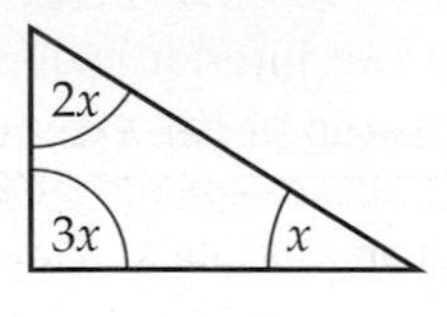

b

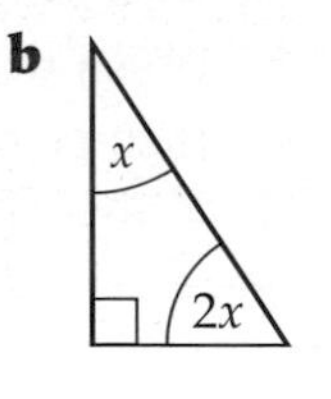

c

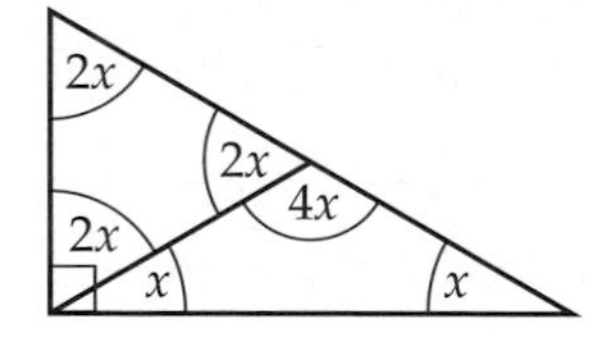

d

65°

x

e

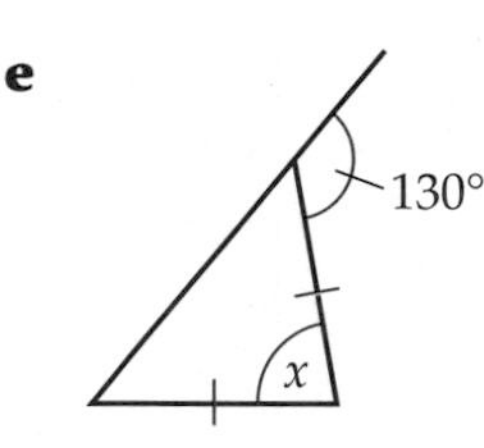

2 The diagram shows a section of a roof truss.

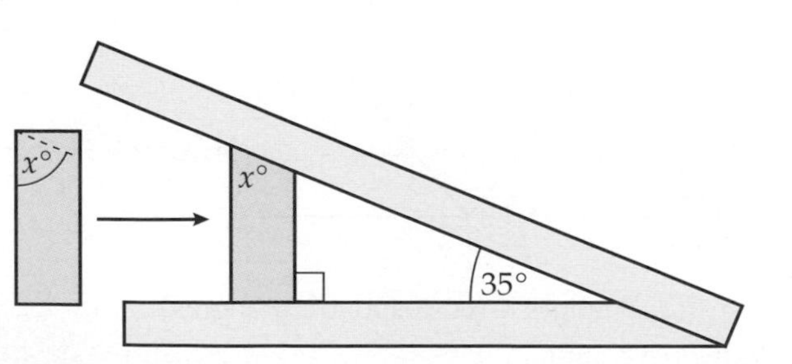

The vertical uprights for the roof truss must be cut accurately so that they support the mass of the tiles on the roof.
The angle x must be calculated before cutting the uprights.
The roof is at an angle of 35° to the horizontal.
Work out the size of angle x.

AO2

Constructing triangles

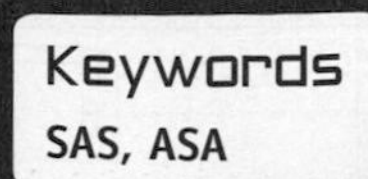

Keywords
SAS, ASA

Why learn this?

Architects' plans need to be very accurate since a small error on the plan could mean a big mistake in real life.

Objectives

D Draw triangles accurately when at least one angle is given

Skills check

1 Use a ruler to draw a line of length 8.2 cm accurately.
2 At one end use a protractor to draw an angle of 45°.
3 At the other end use a protractor to draw an angle of 68°.

Constructing triangles given SAS or ASA

There are two types of triangle constructions that you need to be able to do.

The first is known as **SAS.** This stands for Side Angle Side.

The second is known as **ASA.** This stands for Angle Side Angle.

Remember that to construct a triangle given all three sides:
1 **Draw the longest side of the triangle.**
2 **Using a pair of compasses, put the point on each end of the line in turn and draw arcs to intersect. The radii of the arcs are the lengths of the other two sides of the triangle.**
3 **Complete the triangle.**

Example 2

D

Use a ruler and a protractor to make an accurate drawing of each of these triangles.

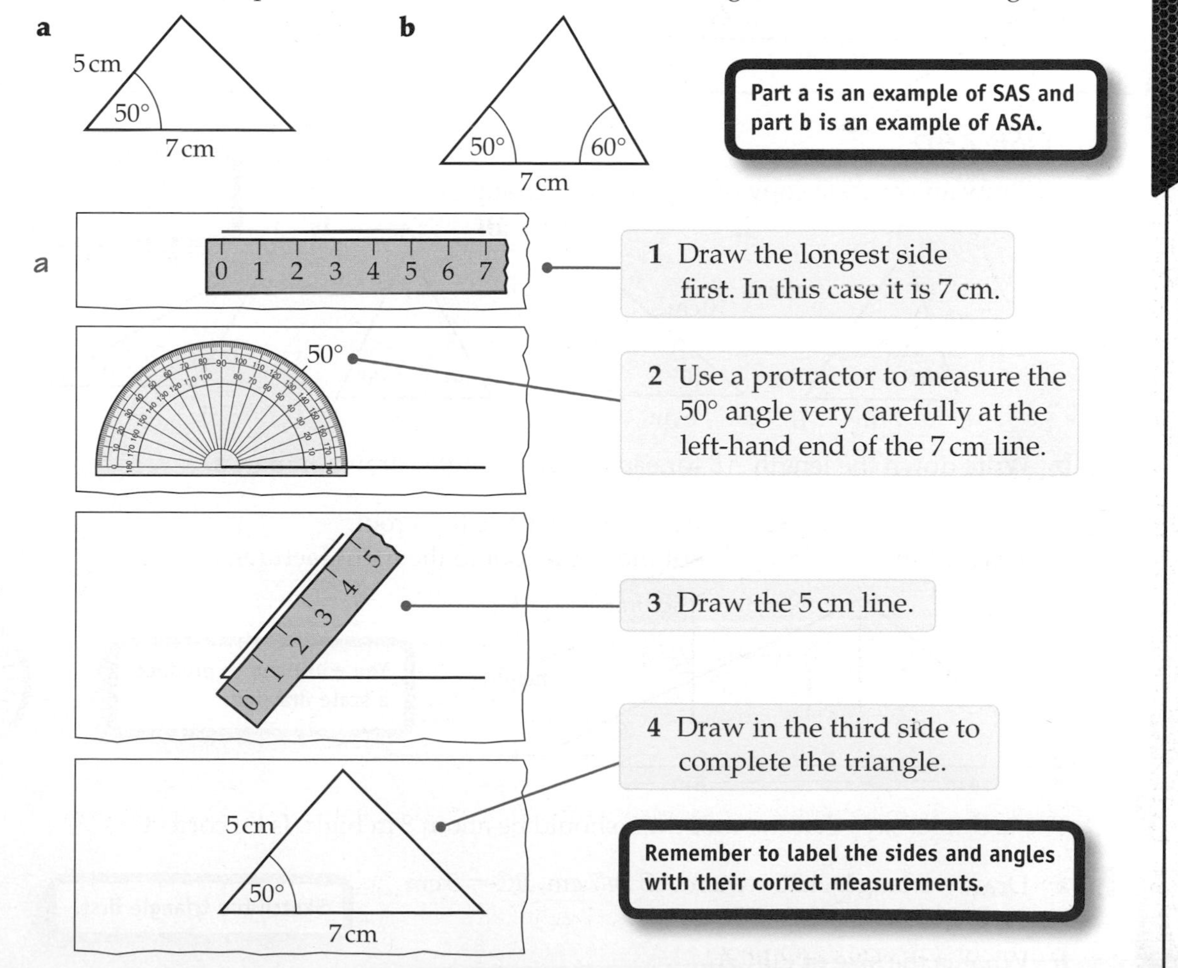

Part a is an example of SAS and part b is an example of ASA.

1 Draw the longest side first. In this case it is 7 cm.

2 Use a protractor to measure the 50° angle very carefully at the left-hand end of the 7 cm line.

3 Draw the 5 cm line.

4 Draw in the third side to complete the triangle.

Remember to label the sides and angles with their correct measurements.

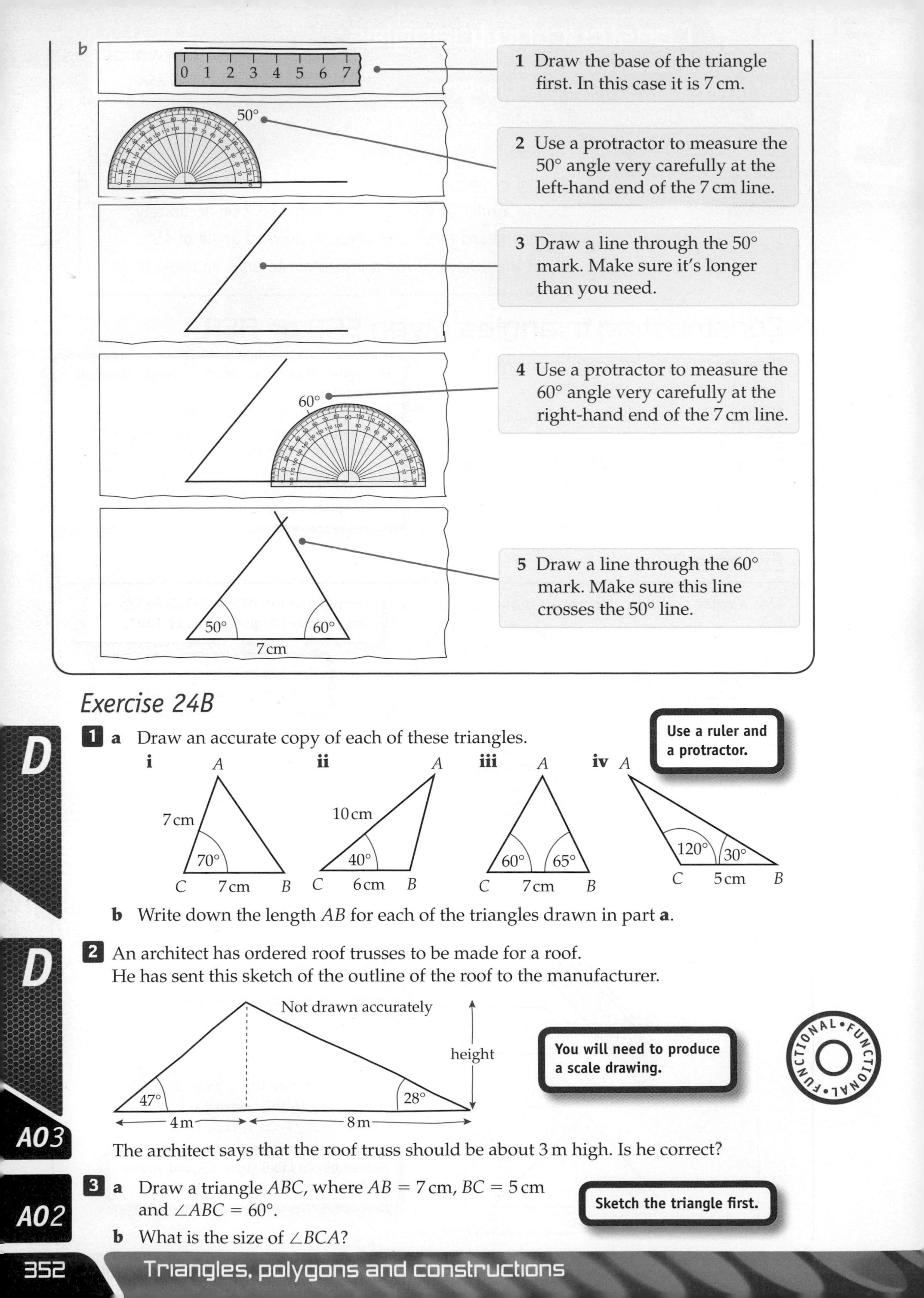

b

1 Draw the base of the triangle first. In this case it is 7 cm.

2 Use a protractor to measure the 50° angle very carefully at the left-hand end of the 7 cm line.

3 Draw a line through the 50° mark. Make sure it's longer than you need.

4 Use a protractor to measure the 60° angle very carefully at the right-hand end of the 7 cm line.

5 Draw a line through the 60° mark. Make sure this line crosses the 50° line.

Exercise 24B

D

1 **a** Draw an accurate copy of each of these triangles.

Use a ruler and a protractor.

b Write down the length AB for each of the triangles drawn in part **a**.

D

A03

2 An architect has ordered roof trusses to be made for a roof.
He has sent this sketch of the outline of the roof to the manufacturer.

You will need to produce a scale drawing.

FUNCTIONAL • FUNCTIONAL •

The architect says that the roof truss should be about 3 m high. Is he correct?

A02

3 **a** Draw a triangle ABC, where $AB = 7$ cm, $BC = 5$ cm and $\angle ABC = 60°$.

b What is the size of $\angle BCA$?

Sketch the triangle first.

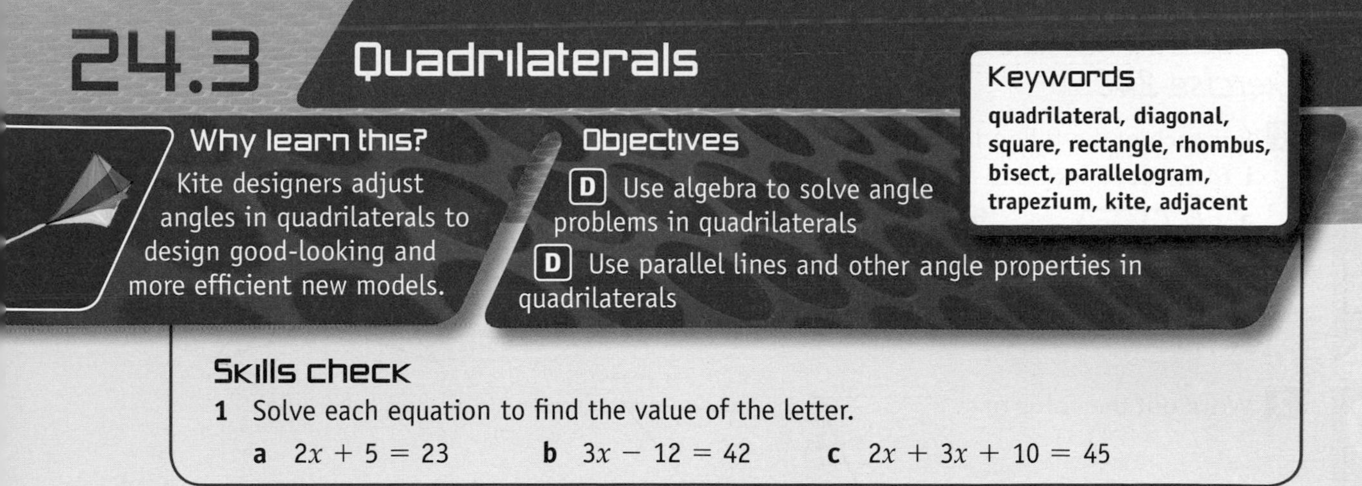

24.3 Quadrilaterals

Keywords
quadrilateral, diagonal, square, rectangle, rhombus, bisect, parallelogram, trapezium, kite, adjacent

Why learn this?

Kite designers adjust angles in quadrilaterals to design good-looking and more efficient new models.

Objectives

D Use algebra to solve angle problems in quadrilaterals

D Use parallel lines and other angle properties in quadrilaterals

Skills check

1 Solve each equation to find the value of the letter.

a $2x + 5 = 23$ **b** $3x - 12 = 42$ **c** $2x + 3x + 10 = 45$

Properties of quadrilaterals

A **quadrilateral** is a 2-D shape bounded by four straight lines.

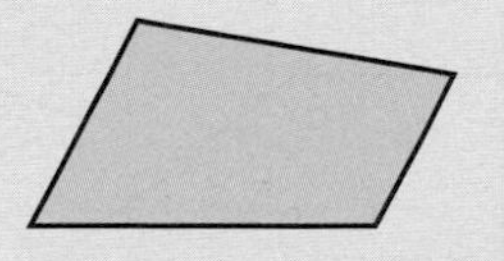

A **diagonal** is a line joining two opposite corners (vertices).
The diagonal divides the quadrilateral into two triangles.

diagonal

The three yellow angles add up to 180°.
The three red angles add up to 180°.

The six angles from the two triangles add up to 360°.
The sum of the interior angles of a quadrilateral is 360°.

Example 3

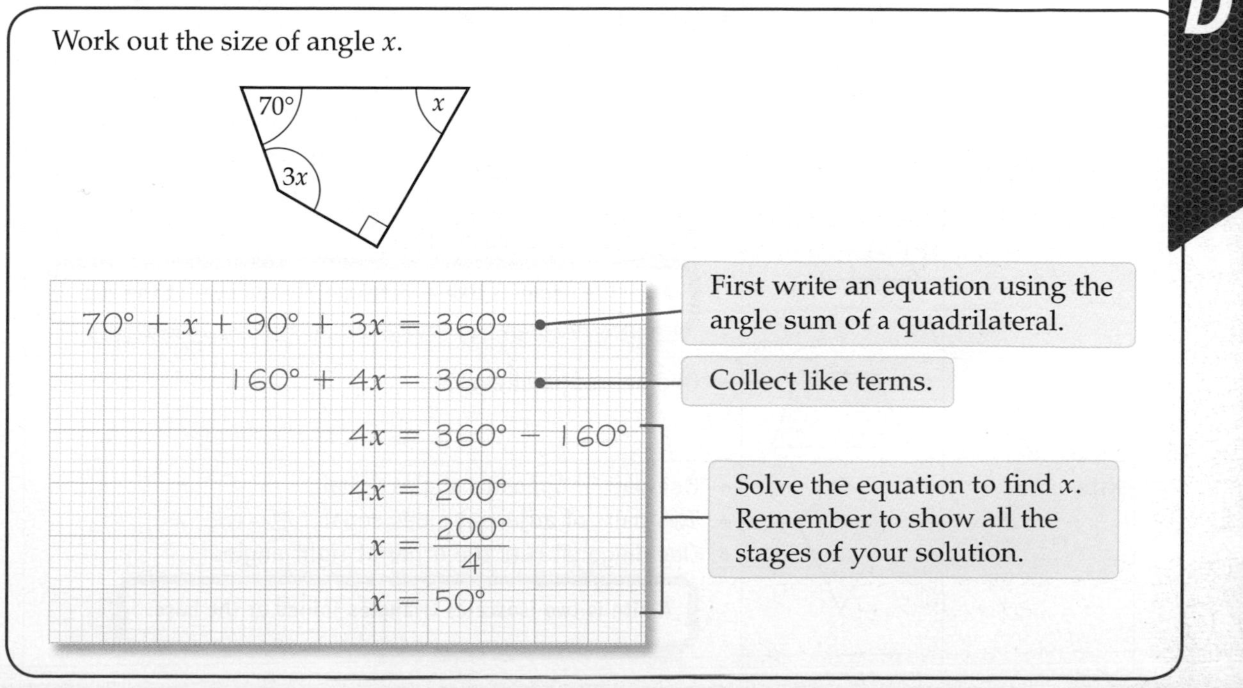

Work out the size of angle x.

$$70° + x + 90° + 3x = 360°$$
$$160° + 4x = 360°$$
$$4x = 360° - 160°$$
$$4x = 200°$$
$$x = \frac{200°}{4}$$
$$x = 50°$$

First write an equation using the angle sum of a quadrilateral.

Collect like terms.

Solve the equation to find x. Remember to show all the stages of your solution.

D

Exercise 24C

1 For each of the following quadrilaterals

i form an equation in x **ii** solve the equation to find the value of x.

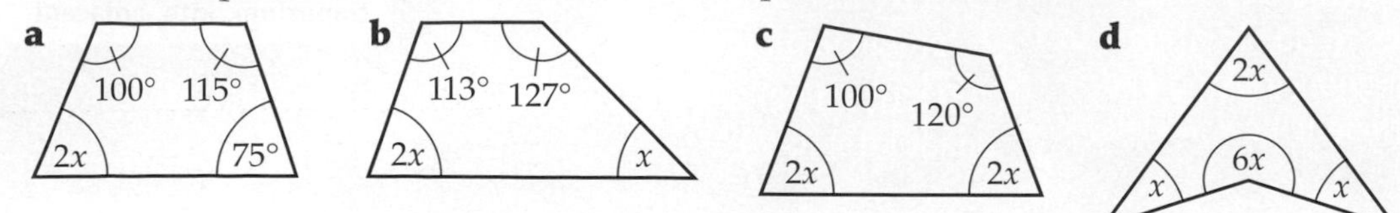

2 Work out the value of x.

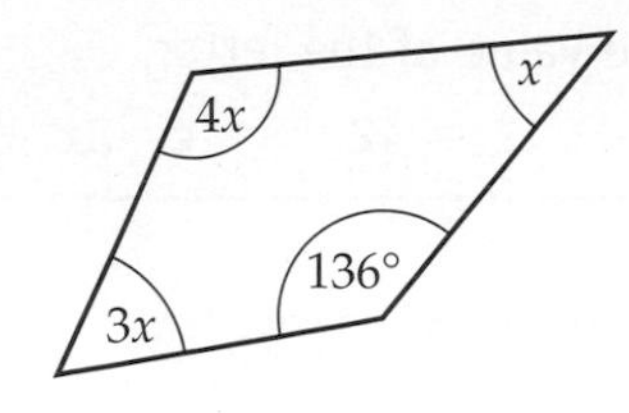

3 Work out the value of $2x + 3y$.

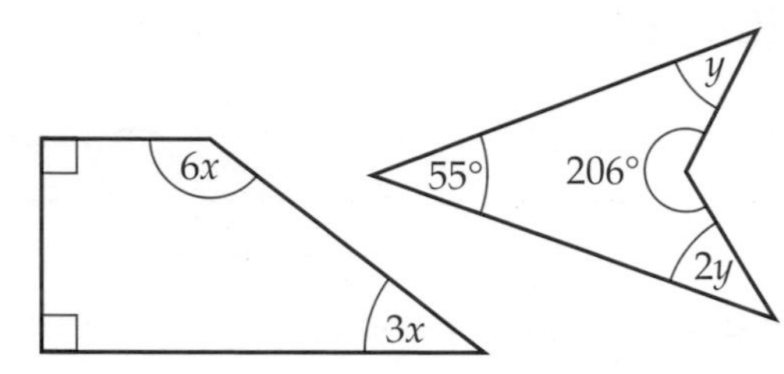

The names of special quadrilaterals

Some quadrilaterals have special names.

Square		• All four sides the same length • Four right angles
Rectangle		• Opposite sides equal • Four right angles
Rhombus		• Opposite angles equal • Opposite sides parallel • All four sides the same length • Diagonals **bisect** each other at right angles
Parallelogram		• Opposite angles equal • Opposite sides equal and parallel • Diagonals bisect each other **A parallelogram is like a rectangle pushed over at the top.**
Trapezium		• One pair of parallel sides
Kite		• One pair of opposite angles equal • Two pairs of **adjacent** sides equal • One diagonal cuts the other at right angles **A kite is two isosceles triangles joined at the base.**

Example 4

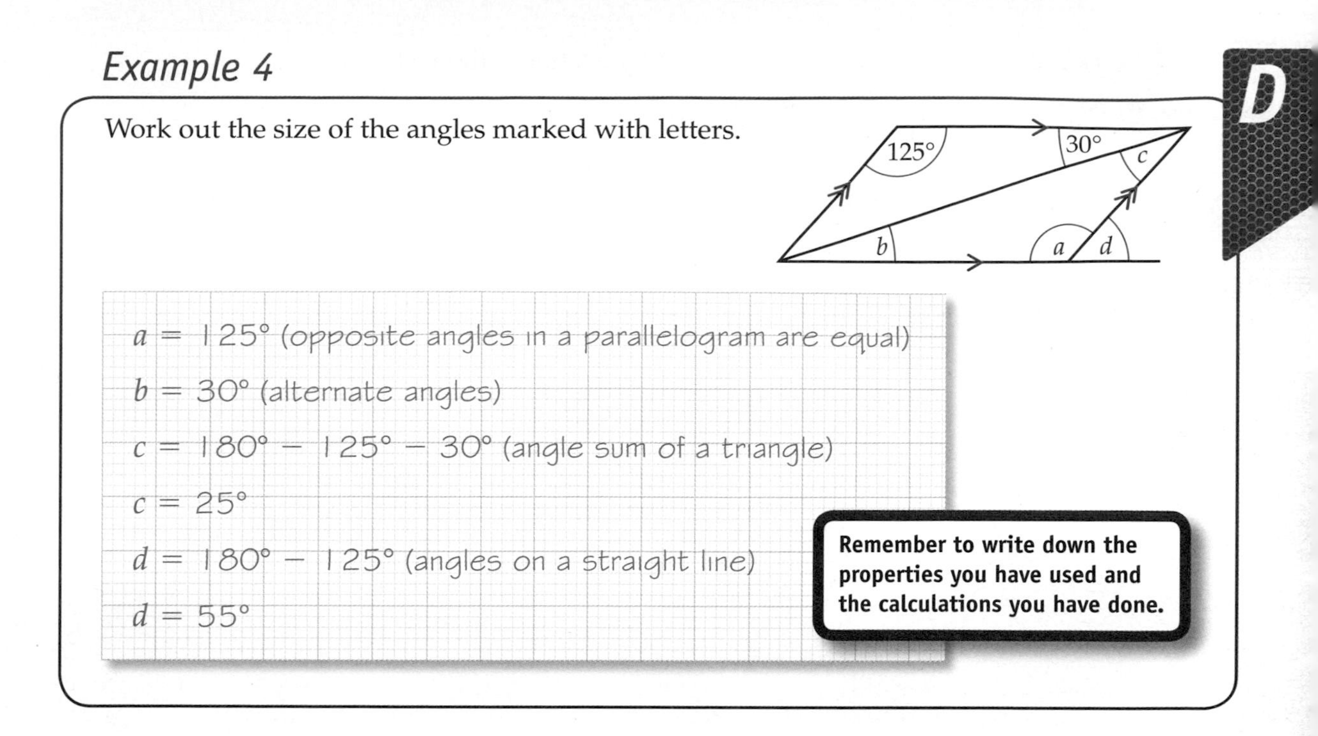

Work out the size of the angles marked with letters.
125°
30°
c
b
a
d
a = 125° (opposite angles in a parallelogram are equal)
b = 30° (alternate angles)
c = 180° − 125° − 30° (angle sum of a triangle)
c = 25°
d = 180° − 125° (angles on a straight line)
d = 55°
Remember to write down the properties you have used and the calculations you have done.
D

Exercise 24D

1 Work out the size of the angles marked with letters.

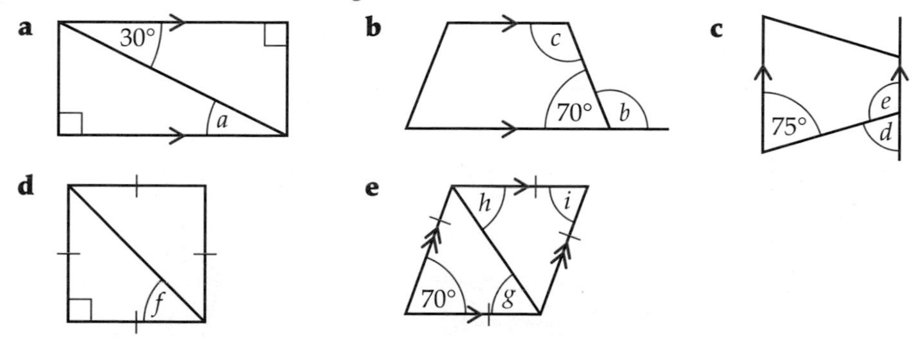

D

2 Work out the size of the angles marked with letters.

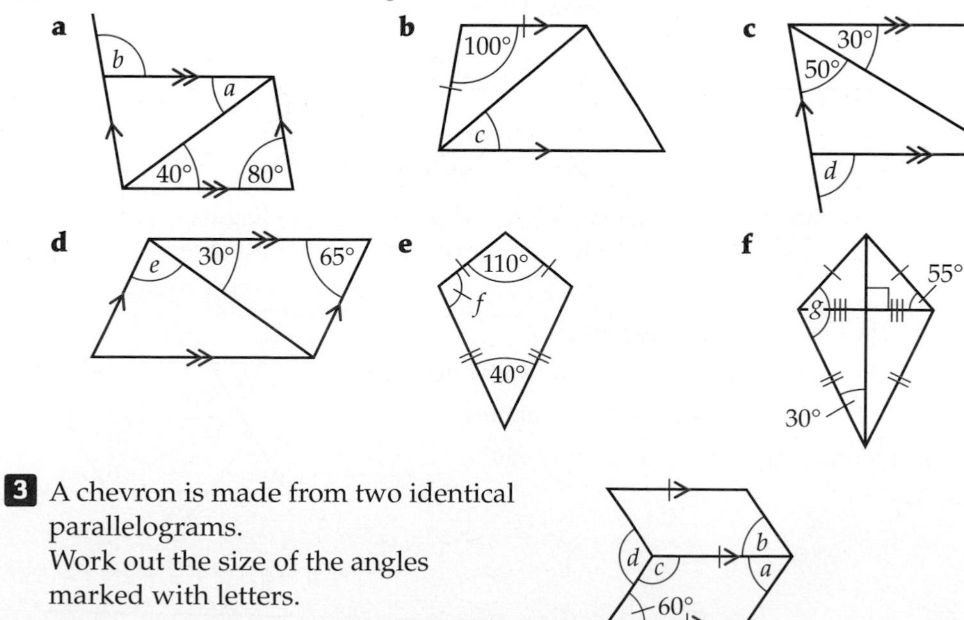

D

3 A chevron is made from two identical parallelograms.
Work out the size of the angles marked with letters.

AO2

4 Work out the size of each angle marked by a letter in these chevrons.

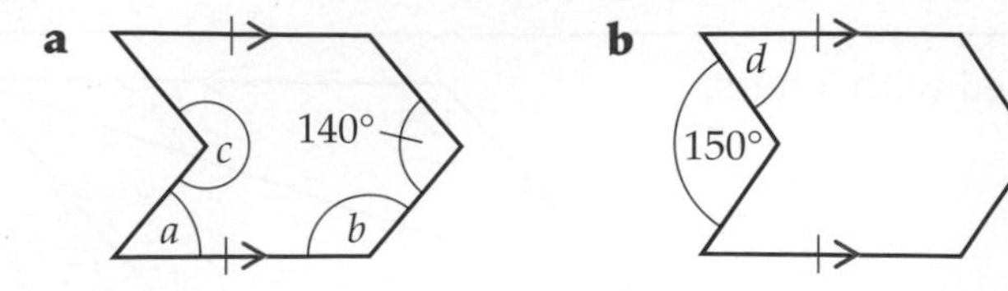

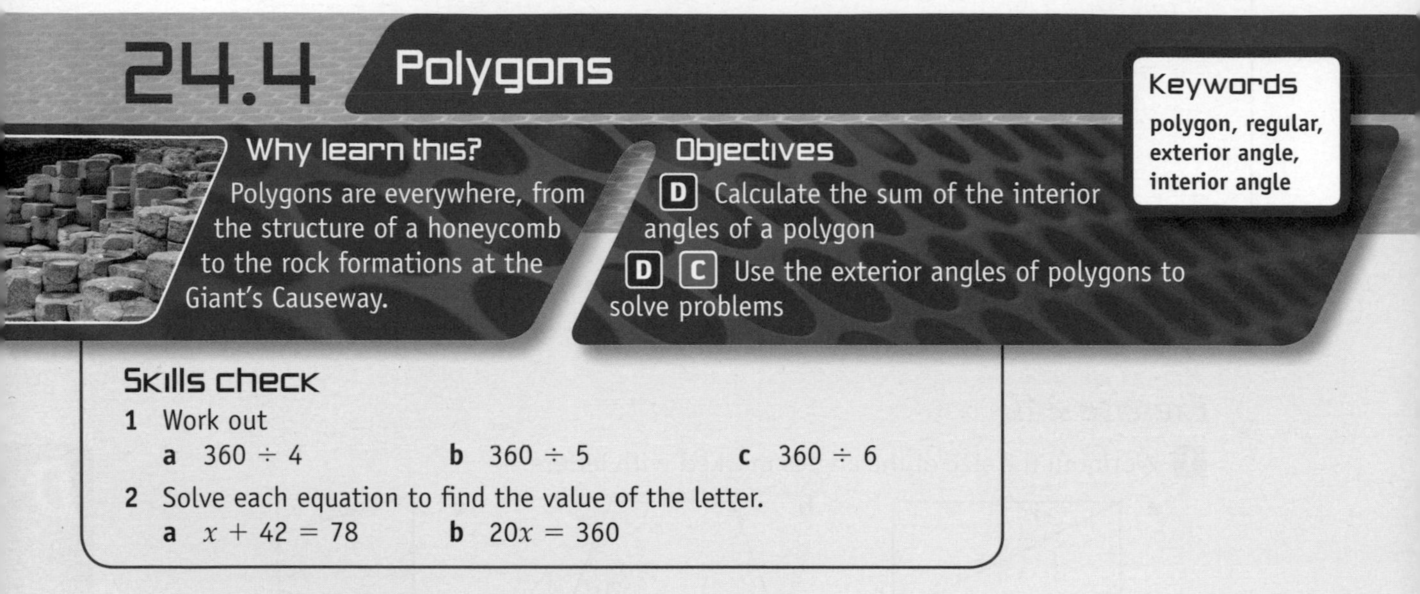

24.4 Polygons

Keywords
polygon, regular, exterior angle, interior angle

Why learn this?

Polygons are everywhere, from the structure of a honeycomb to the rock formations at the Giant's Causeway.

Objectives

D Calculate the sum of the interior angles of a polygon

D **C** Use the exterior angles of polygons to solve problems

Skills check

1 Work out

a $360 \div 4$ b $360 \div 5$ c $360 \div 6$

2 Solve each equation to find the value of the letter.

a $x + 42 = 78$ b $20x = 360$

Exterior and interior angles of a polygon

A **polygon** is a 2-D shape bounded by straight lines.

All the sides in a **regular** polygon are the same length.

All the angles in a regular polygon are the same size.

Regular pentagon

Here are some other polygons you need to know.

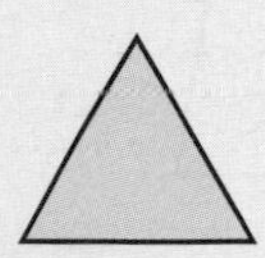

Equilateral triangle

Square

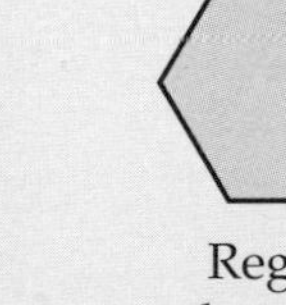

Regular hexagon

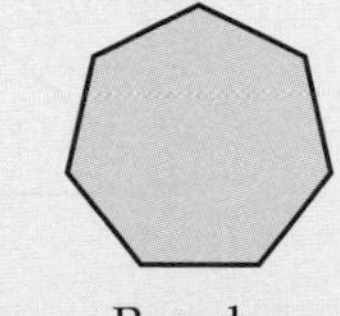

Regular heptagon

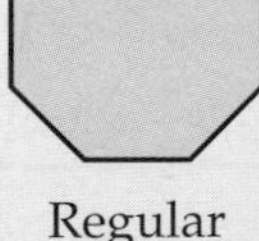

Regular octagon

This polygon has six sides, so it is a hexagon.

The sides are not the same length so it is *not* a regular hexagon.

The angles a, b, c, d, e and f are called **exterior angles.**

The sum of the exterior angles of any polygon is 360°.

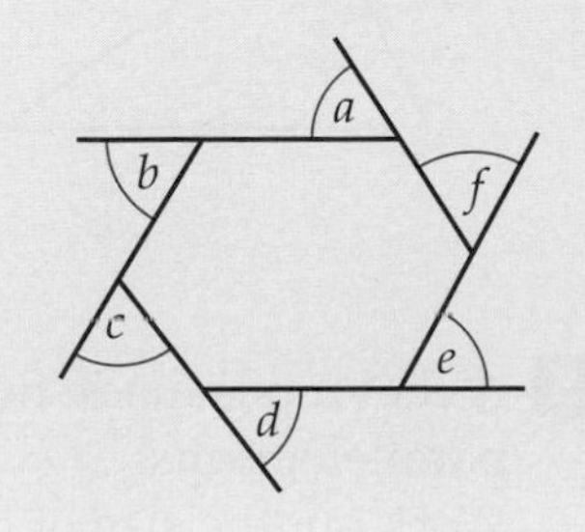

$a + b + c + d + e + f = 360°$

In a *regular* polygon, all the exterior angles are the same size.

You can find the exterior angle of a regular polygon using the formula

$$\text{Exterior angle of a regular polygon} = \frac{360°}{\text{number of sides}}$$

This polygon has four sides so it is a quadrilateral.

It is one of the special quadrilaterals, a trapezium.

The angles v, w, x and y are called **interior angles**.

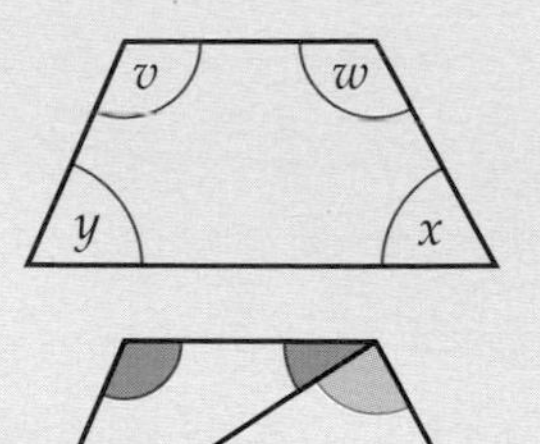

The trapezium can be divided into two triangles as shown.

The sum of the interior angles = 2 × 180° = 360°.

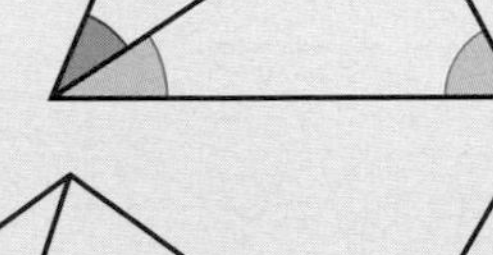

By dividing any polygon into triangles, you can find the sum of the interior angles.

The sum of the interior angles of any polygon is

(number of sides − 2) × 180°

(5 − 2) × 180°
= 3 × 180°
= 540°

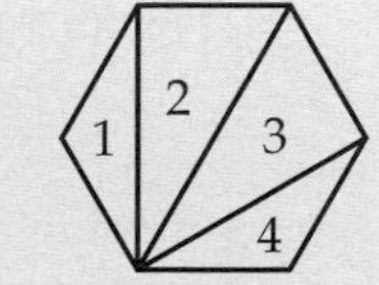

(6 − 2) × 180°
= 4 × 180°
= 720°

You can also use the fact that the interior and exterior angles of a polygon lie along a straight line to give the formula

Interior angle = 180° − exterior angle

Example 5

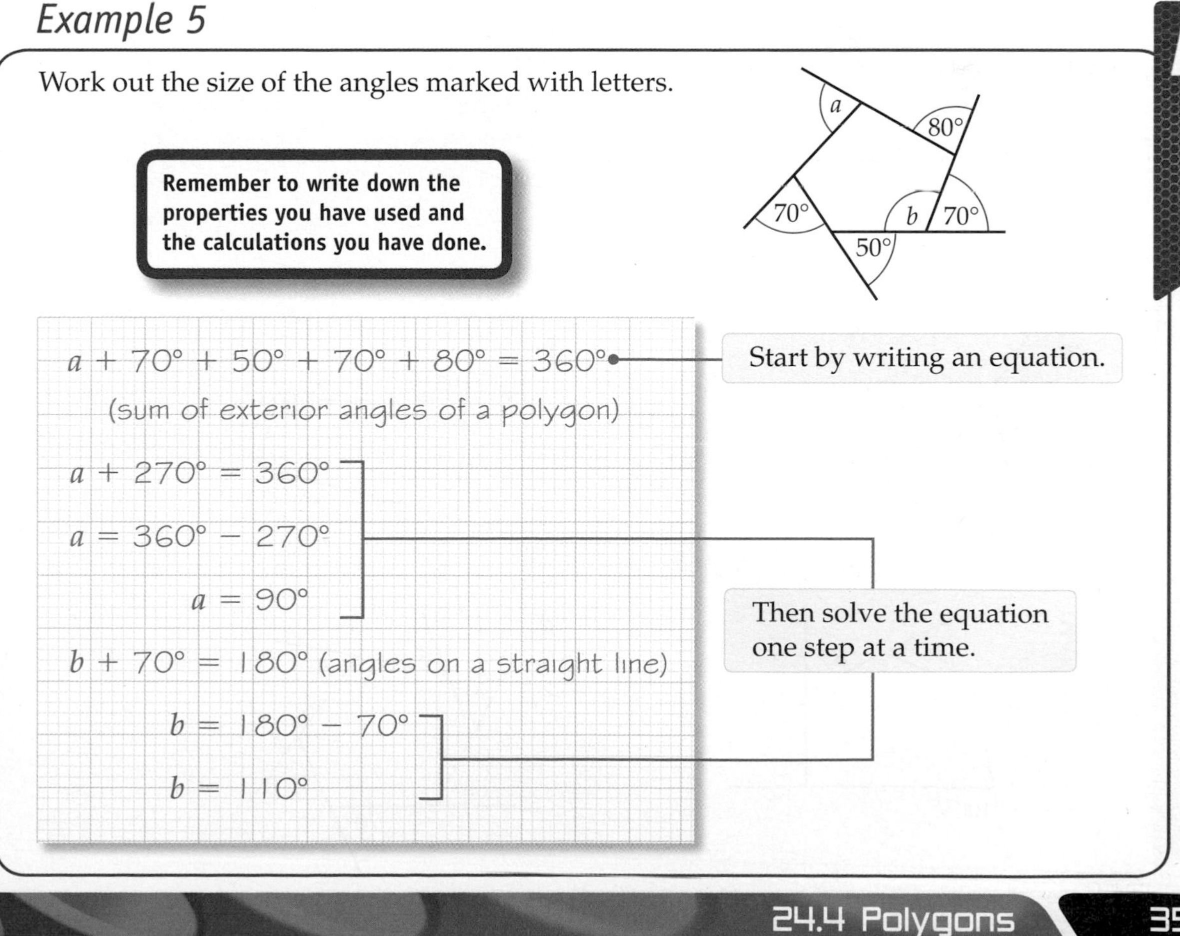

Work out the size of the angles marked with letters.
a
80°
70°
b
70°
50°
Remember to write down the properties you have used and the calculations you have done.
$a + 70° + 50° + 70° + 80° = 360°$
(sum of exterior angles of a polygon)
$a + 270° = 360°$
$a = 360° - 270°$
$a = 90°$
$b + 70° = 180°$ (angles on a straight line)
$b = 180° - 70°$
$b = 110°$
Start by writing an equation.
Then solve the equation one step at a time.
D

D

Example 6

Work out the size of angle x.

Sum of interior angles of a hexagon
$= (6 - 2) \times 180° = 720°$

$x + 110° + 115° + 120° + 125° + 130° = 720°$

$x + 600° = 720°$

$x = 720° - 600°$

$x = 120°$

Start by working out the sum of the interior angles.

Then set up an equation and solve the equation one step at a time.

C

Example 7

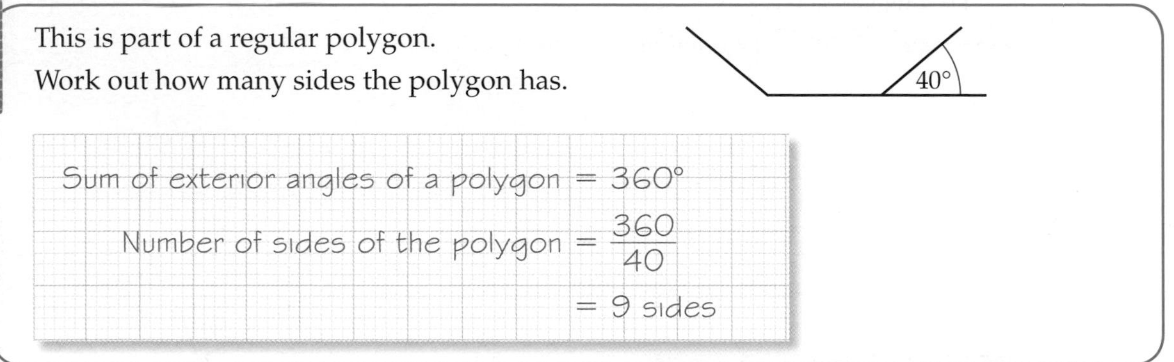

This is part of a regular polygon.
Work out how many sides the polygon has.

Sum of exterior angles of a polygon $= 360°$

Number of sides of the polygon $= \frac{360}{40}$

$= 9$ sides

Exercise 24E

D

1 Work out the size of the angles marked with letters.

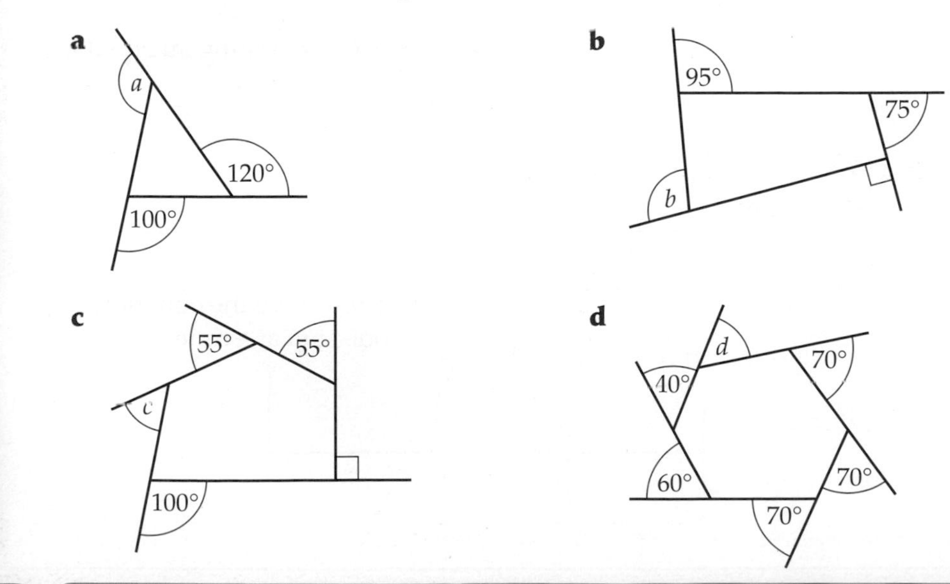

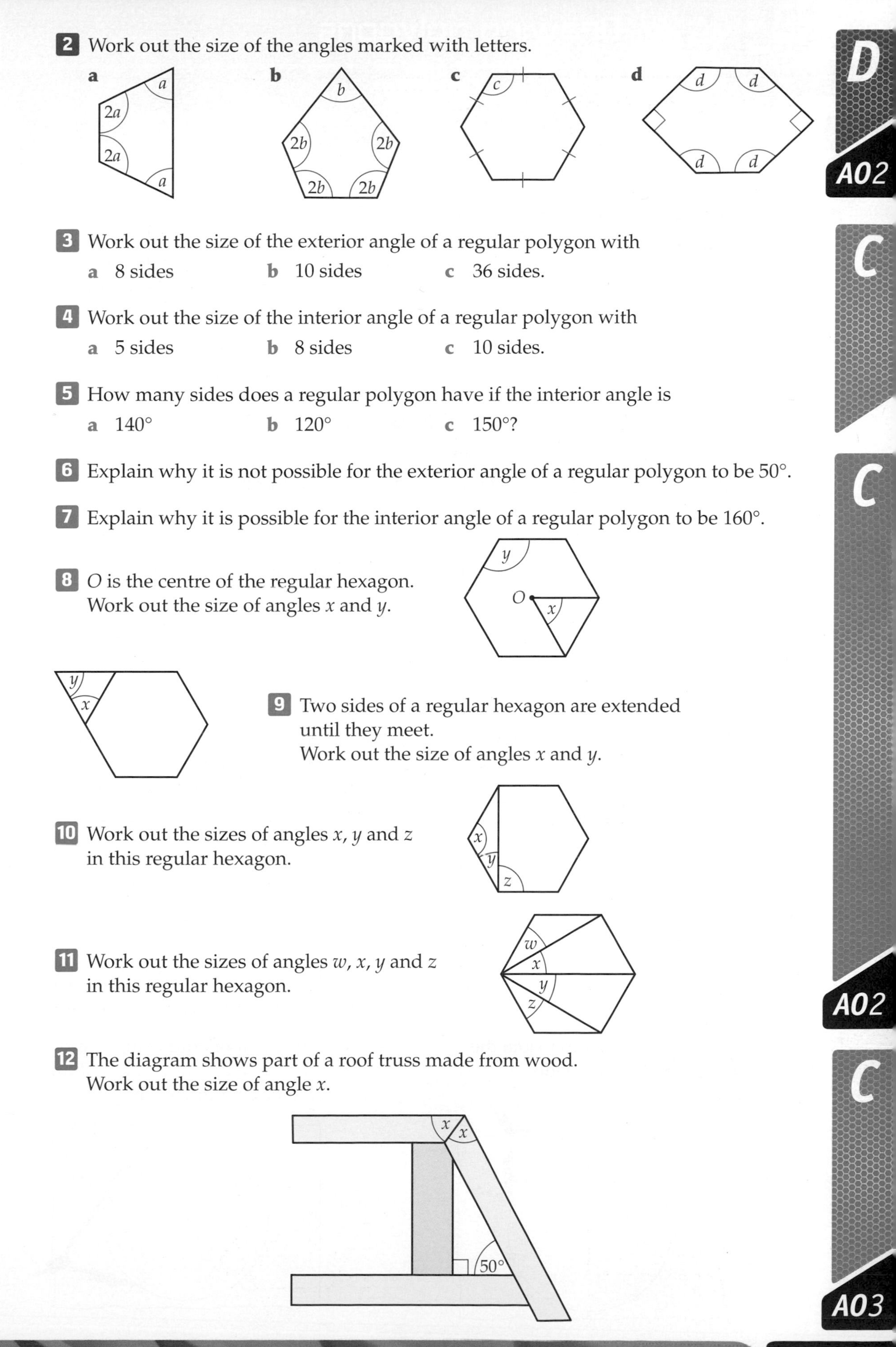

2 Work out the size of the angles marked with letters.

3 Work out the size of the exterior angle of a regular polygon with

a 8 sides **b** 10 sides **c** 36 sides.

4 Work out the size of the interior angle of a regular polygon with

a 5 sides **b** 8 sides **c** 10 sides.

5 How many sides does a regular polygon have if the interior angle is

a 140° **b** 120° **c** 150°?

6 Explain why it is not possible for the exterior angle of a regular polygon to be 50°.

7 Explain why it is possible for the interior angle of a regular polygon to be 160°.

8 O is the centre of the regular hexagon. Work out the size of angles x and y.

9 Two sides of a regular hexagon are extended until they meet. Work out the size of angles x and y.

10 Work out the sizes of angles x, y and z in this regular hexagon.

11 Work out the sizes of angles w, x, y and z in this regular hexagon.

12 The diagram shows part of a roof truss made from wood. Work out the size of angle x.

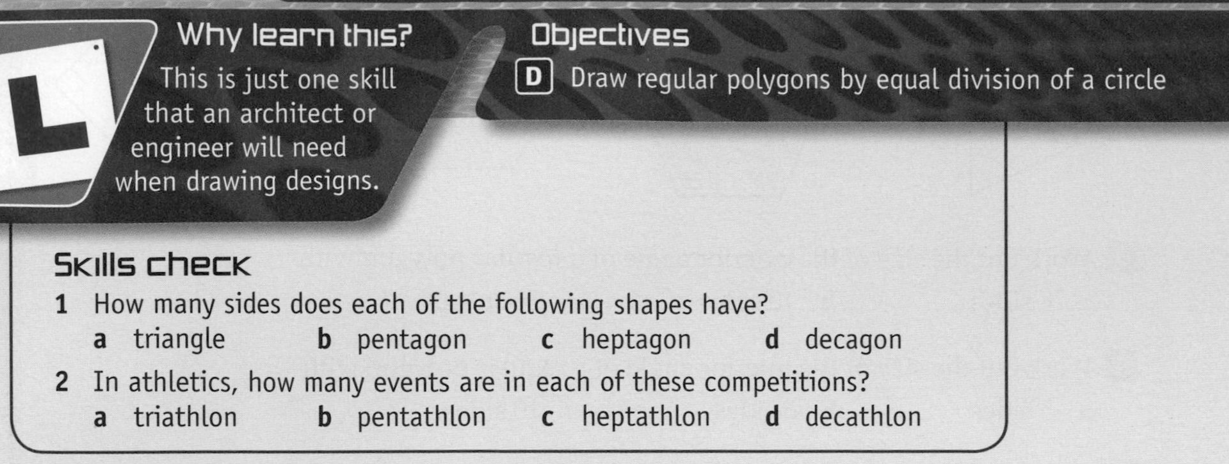

Why learn this?

This is just one skill that an architect or engineer will need when drawing designs.

Objectives

D Draw regular polygons by equal division of a circle

Skills check

1 How many sides does each of the following shapes have?
  a triangle **b** pentagon **c** heptagon **d** decagon

2 In athletics, how many events are in each of these competitions?
  a triathlon **b** pentathlon **c** heptathlon **d** decathlon

Drawing regular polygons

You can use a circle to draw a regular polygon.

One method of doing this is by division of the circumference of the circle.

Example 8

Draw a regular pentagon by division of a circle.

When you join each vertex of a regular pentagon to its centre, you get five identical triangles.

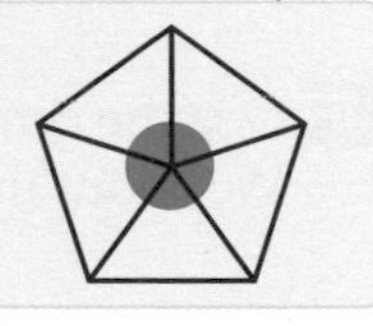

Angles at the centre of pentagon $= \frac{360°}{5} = 72°$

1 Draw a circle. Mark the centre and a point on the circumference.

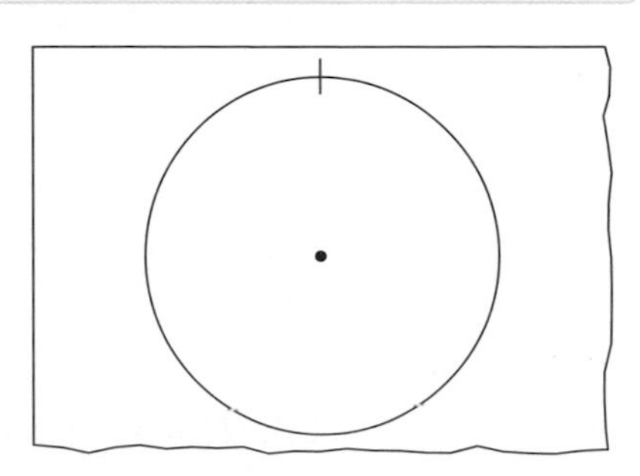

2 From the point on the circumference, use a protractor to measure an angle of 72°. Put another mark on the circumference.

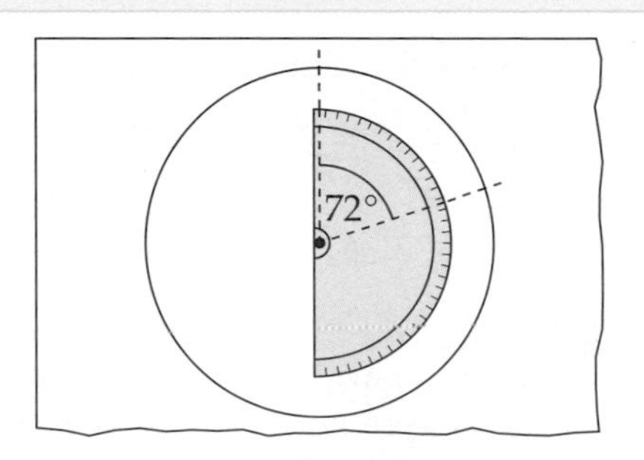

3 Open your compasses to join the two marks already made.

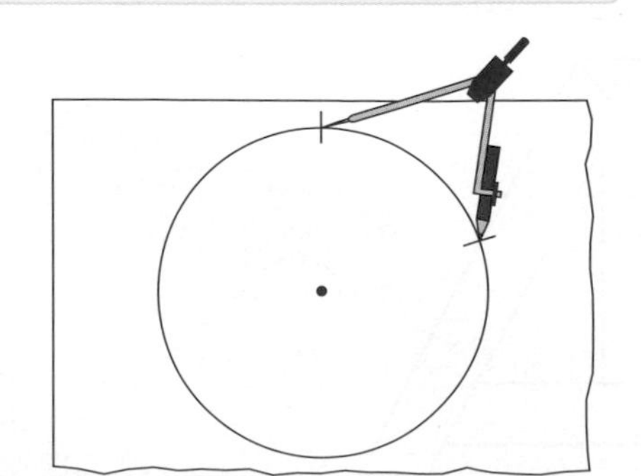

4 Move the point of your compasses to the second mark and make a third mark on the circumference.

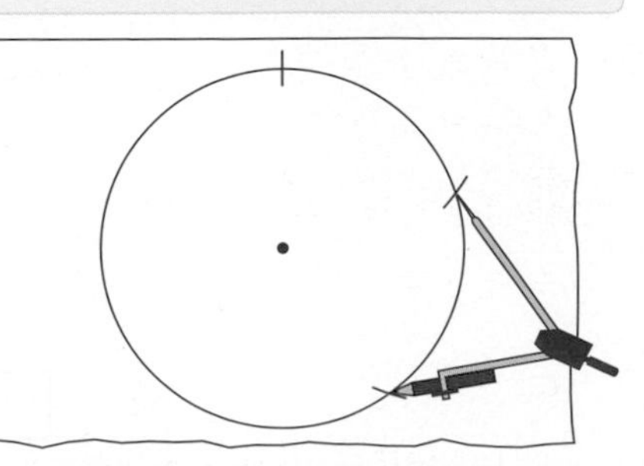

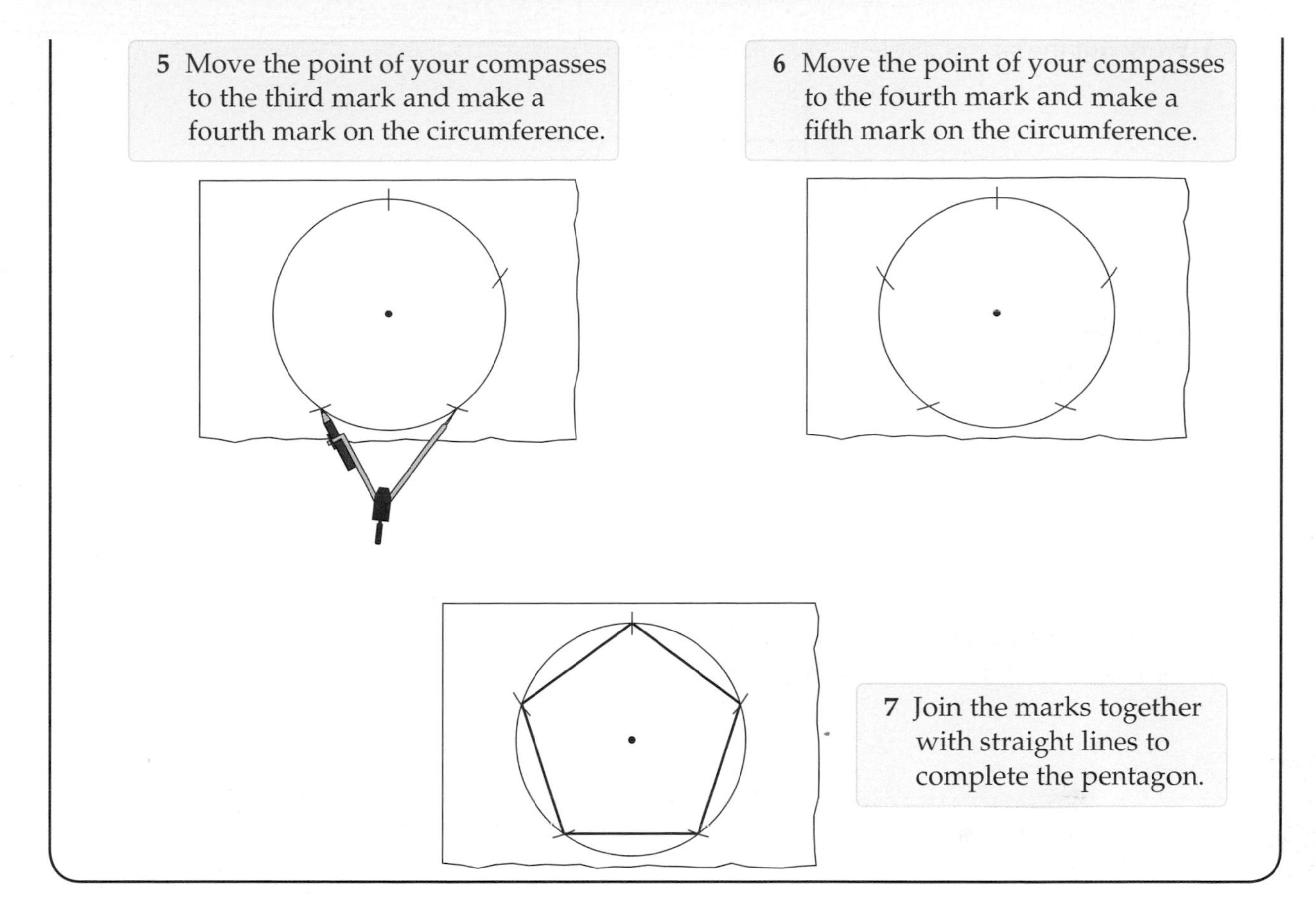

5 Move the point of your compasses to the third mark and make a fourth mark on the circumference.
6 Move the point of your compasses to the fourth mark and make a fifth mark on the circumference.
7 Join the marks together with straight lines to complete the pentagon.

Exercise 24F

D

1 For parts **a** to **d**

i draw a circle of radius 6 cm

ii by equal division of a circle, draw the polygon required.

a an equilateral triangle

b a square

c a regular hexagon

d a regular octagon

D AO2

2 Tomas draws a regular pentagon inside a circle of radius 6 cm.

Tanya draws a regular heptagon inside a circle of radius 10 cm.

Tomas says, 'The side length of my pentagon is longer than the side length of your heptagon because my shape has fewer sides.'

Tanya says, 'The side length of my heptagon is longer than the side length of your pentagon because I have drawn a bigger circle.'

Who is correct? Show working to support your answer.

Review exercise

D

1 **a** Make an accurate drawing of this triangle.

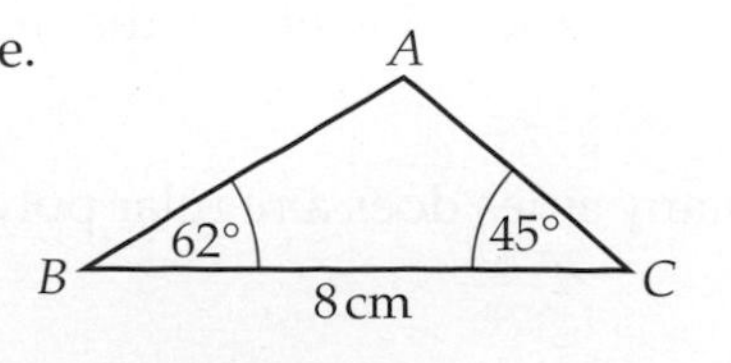

[3 marks]

b Measure the length of *AC*. **[1 mark]**

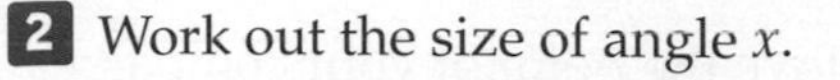

2 Work out the size of angle x.

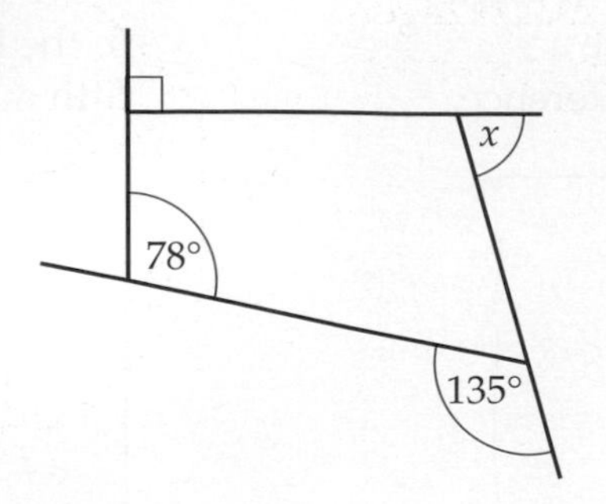

D AO2

[3 marks]

3 Accurately draw a regular pentagon. **[2 marks]**

4 Work out the value of x.

2x
81°
5x
2x

[3 marks]

D

5 **a** Jamie said that a quadrilateral with both diagonals the same length must be a rectangle.
Is Jamie correct?
Give a reason for your answer. **[1 mark]**

b Colleen said that if a parallelogram has one angle equal to 90°, all the angles must be 90°.
Is Colleen correct?
Give a reason for your answer. **[1 mark]**

6 A chevron is made from two identical parallelograms placed edge to edge.

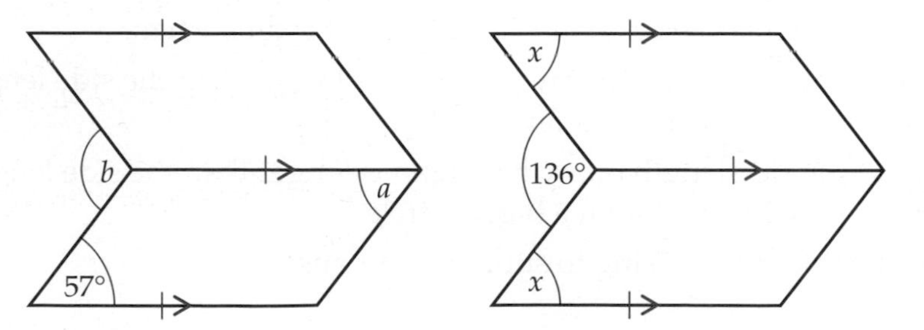

AO2

a Work out the values of a and b in the first chevron. **[2 marks]**

b Work out the value of x in the second chevron. **[1 mark]**

C

7 How many sides does a regular polygon have if the interior angle is 170°? **[2 marks]**

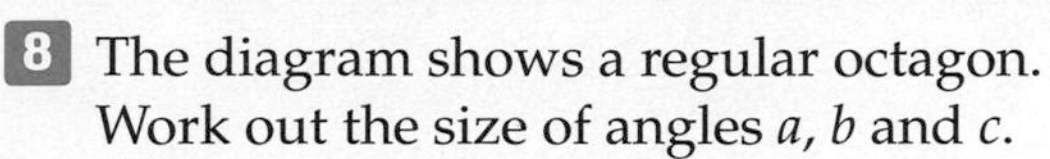

8 The diagram shows a regular octagon.
Work out the size of angles a, b and c.

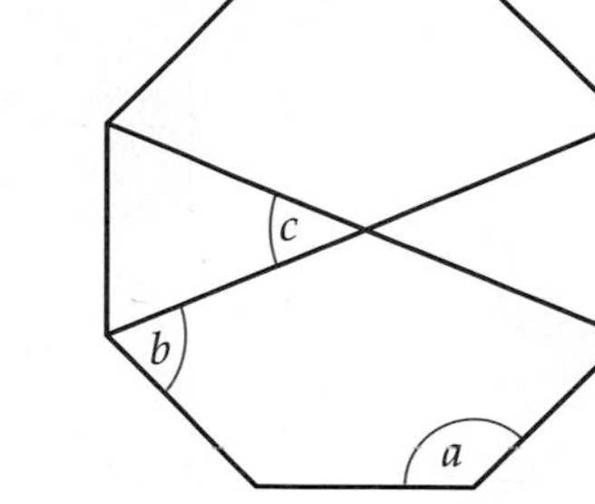
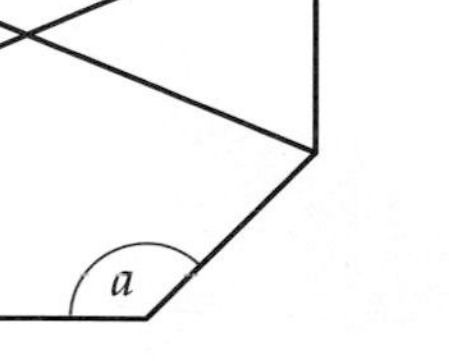

[3 marks]

9 Explain why it is not possible for the exterior angle of a regular polygon to be 70°.

[1 mark]

10 *ABCDEFGH* is a regular octagon.

Work out the size of the angles marked

a x **[2 marks]**

b y **[2 marks]**

c z. **[2 marks]**

C

AO2

Chapter summary

In this chapter you have learned how to

- use algebra to solve angle problems in triangles **D**
- draw triangles accurately when at least one angle is given **D**
- use algebra to solve angle problems in quadrilaterals **D**
- use parallel lines and other angle properties in quadrilaterals **D**
- calculate the sum of the interior angles of a polygon **D**
- draw regular polygons by equal division of a circle **D**
- use the exterior angles of polygons to solve problems **C**

25 More equations and formulae

This chapter is about using formulae and solving equations.

There is even an equation to determine how long to immerse your biscuit when dunking it in your drink!

Objectives

This chapter will show you how to

- derive formulae and equations **D**
- substitute into a formula to solve problems **D** **C**
- use trial and improvement to find solutions to equations **C**
- change the subject of a formula **C** **B**
- set up and solve equations **C** **B** **A**

Before you start this chapter

1 Write down whether each of these is an expression, an equation or a formula.

a $6x + 1$ **b** $v = u + at$ **c** $\frac{a + b}{2}$

d $E = \frac{1}{2}mv^2$ **e** $2n + 1 = 11$ **f** $2p^2 + pq$

2 Simplify by collecting like terms

a $2x + x^2 - 5x + 3x^2$ **b** $mn + mn^2 - m + 6mn + 4m$

c $2a + ab - 3a + 4ab + a^2$ **d** $4b^2 + 6b - 2b^2 + 3b^3 - 2b$

3 Simplify

a $2a \times 4b$ **b** $6x \times 2y \times x$ **c** $3p^2 \times q^3 \times 2p \times q$ **d** $4 \times 2m^2 \times n \times 3m \times n^4$

4 Factorise

a $16 - 4x$ **b** $21 - 60y$ **c** $y + 11y^2$ **d** $2s^2 + 4s$

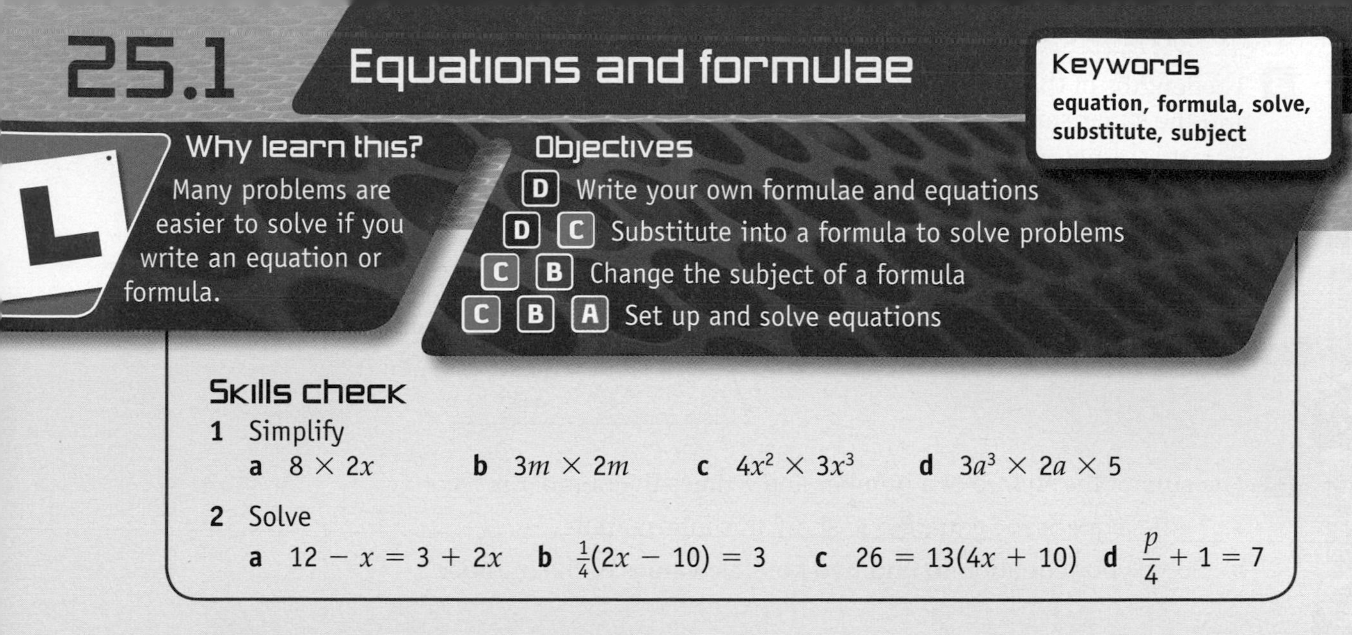

25.1 Equations and formulae

Keywords
equation, formula, solve, substitute, subject

Why learn this?

Many problems are easier to solve if you write an equation or formula.

Objectives

D Write your own formulae and equations
D C Substitute into a formula to solve problems
C B Change the subject of a formula
C B A Set up and solve equations

Skills check

1 Simplify

a $8 \times 2x$ **b** $3m \times 2m$ **c** $4x^2 \times 3x^3$ **d** $3a^3 \times 2a \times 5$

2 Solve

a $12 - x = 3 + 2x$ **b** $\frac{1}{4}(2x - 10) = 3$ **c** $26 = 13(4x + 10)$ **d** $\frac{p}{4} + 1 = 7$

Writing an equation

You can solve problems by writing and solving **equations**.

Example 1

Write an equation and solve it to find the value of a in this triangle.

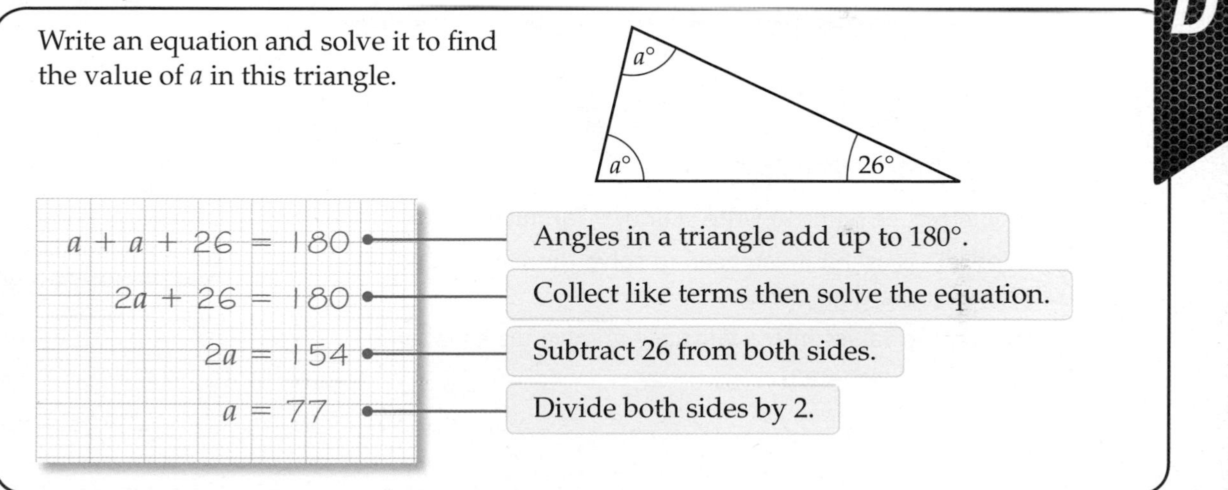

$a + a + 26 = 180$ — Angles in a triangle add up to 180°.

$2a + 26 = 180$ — Collect like terms then solve the equation.

$2a = 154$ — Subtract 26 from both sides.

$a = 77$ — Divide both sides by 2.

Exercise 25A

1 Find the value of the letter in each of these triangles.

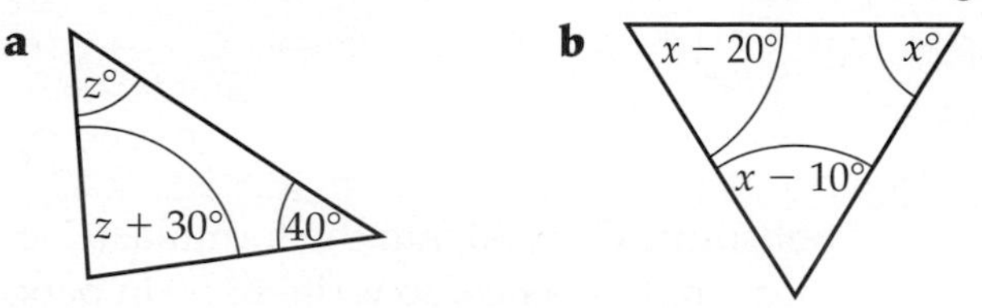

2 Find the value of the letter in each of these diagrams.

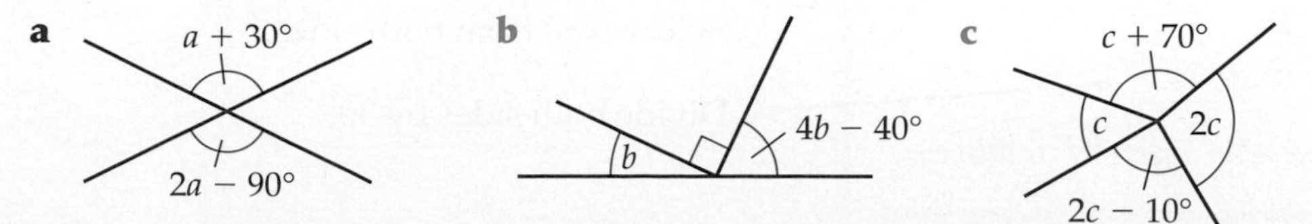

3 The lengths of the perimeters of the regular hexagon and the square are the same.
The area of the square is $144\,\text{cm}^2$.
Find the length of one side of the hexagon.

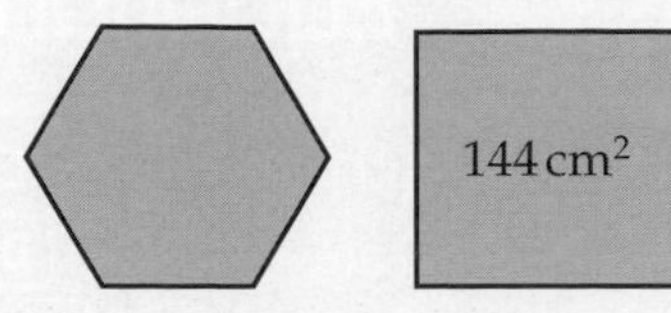

C
AO2

4 Work out the value of x.

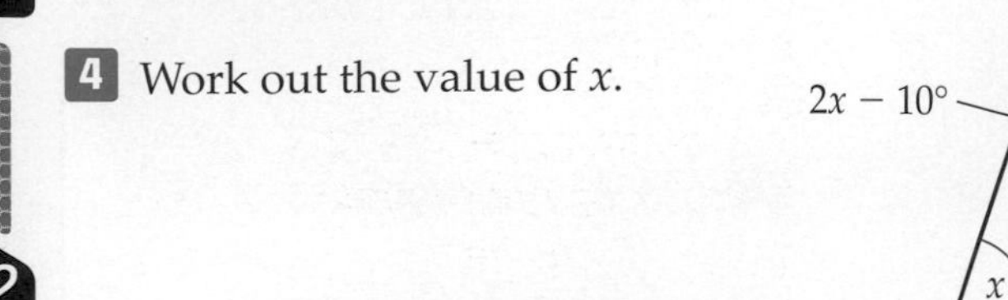

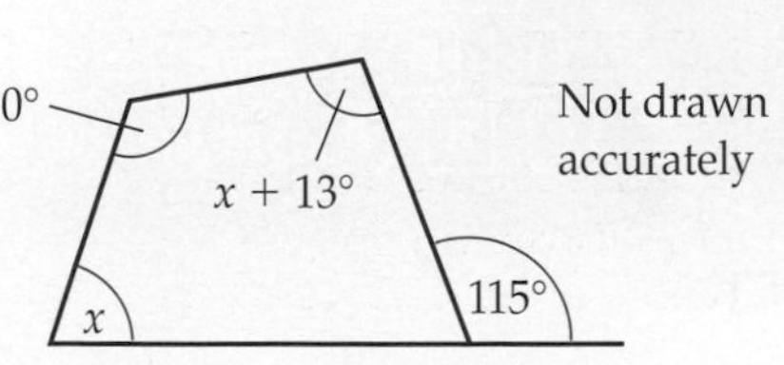

B
AO2

5 The sum of the square of a number and 7 times the number is 78.

a Write a quadratic equation to show this information.

b Solve your equation to find two possible values of the number.

A
AO2

6 The area of this trapezium is $50\,\text{cm}^2$.

a Write a quadratic equation using this information.

b Solve your equation to find the value of x.

$2x$ cm
$x + 1$ cm
$2x + 4$ cm

Area of trapezium = $\frac{1}{2}$ × (sum of parallel sides) × perpendicular height

Writing a formula

You can use letters or words to write your own **formulae.**

Writing a formula is useful if you need to **solve** the same type of problem more than once.

You can **substitute** values into a formula and solve the equation to find the value of an unknown.

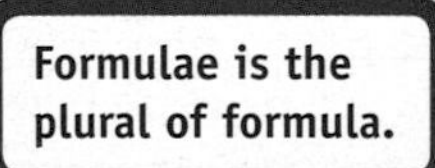

Formulae is the plural of formula.

D

Example 2

A telephone news service charges 30p a minute plus a connection fee of 80p.
The price of a call lasting x minutes is C pence.

a Write a formula for C in terms of x.

b Find the length of a call costing £3.50.

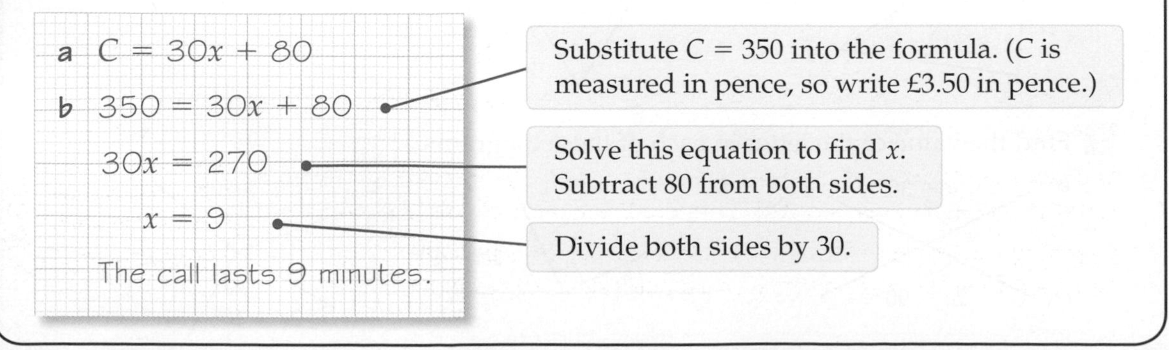

a C = 30x + 80
b 350 = 30x + 80
30x = 270
x = 9
The call lasts 9 minutes.
Substitute C = 350 into the formula. (C is measured in pence, so write £3.50 in pence.)
Solve this equation to find x. Subtract 80 from both sides.
Divide both sides by 30.

Exercise 25B

D

1 Misha is comparing mobile phone plans. She uses m minutes each month and sends t text messages. The Easy-talk plan will cost her £A per month and the Text-tastic plan will cost her £B per month.

Easy-talk
Calls: 4p per minute
Texts: 5p each

Text-tastic
Monthly fee: £8
Calls: 6p per minute
Texts: free

a Write a formula for A in terms of m and t.

b Write a formula for B in terms of m and t.

c If $m = 90$ and $t = 210$ which plan is cheaper?

AO2

D

2 A rectangle has length l and width w.

a Write a formula for the perimeter of the rectangle, P, in terms of l and w.

b Find the value of l when $P = 22$ and $w = 8$.

c Find the value of w when $P = 18$ and $l = 4$.

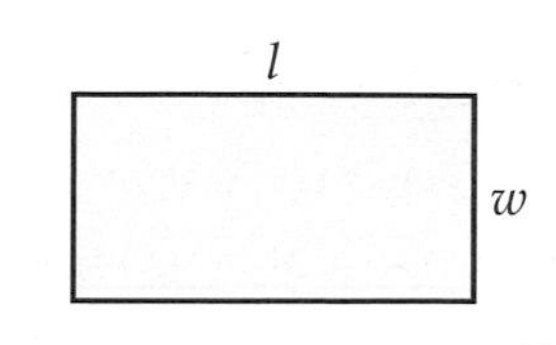

3 The length of a rectangle is 3 more than twice its width, w.

a Write a formula for the perimeter of the rectangle, P.

b Find the value of w when $P = 39$.

C

4 Small bags of nails contain x nails each.
Large bags of nails contain three times as many.
Polly buys two small bags and one large bag of nails.
She uses 12 nails to fix her window boxes and has n nails left.

a Write a formula for n in terms of x.

b Calculate x when $n = 63$.

5 Beth has n sweets. She eats three then shares the rest equally among x friends. Each friend gets s sweets.

a Write a formula for s in terms of n and x.

b Calculate n when $s = 5$ and $x = 8$.

AO2

Rearranging formulae to change the subject

The **subject** of a formula is the letter on its own. For example, P is the subject of the formula $P = 2l + 2w$.

You can use the rules of algebra to rearrange a formula to make a different letter the subject. This is called changing the subject of the formula.

Changing the subject of a formula is like solving an equation.
You need to get a letter on its own on one side of the formula.

C

Example 3

The formula for the perimeter of a rectangle is $P = 2l + 2w$.

a Rearrange this formula to make w the subject.

b Find the value w when $P = 42$ and $l = 12$.

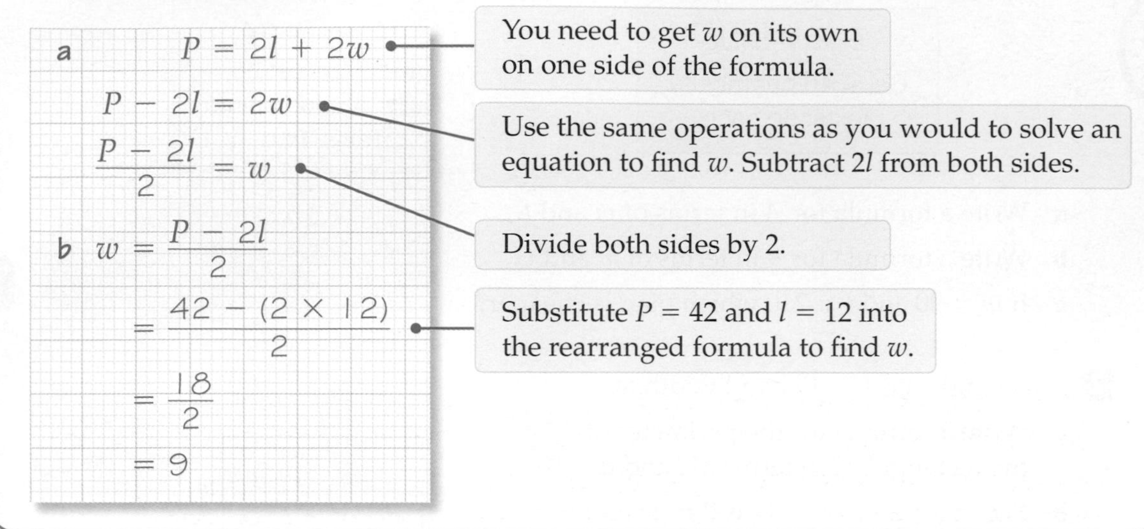

a $P = 2l + 2w$
$P - 2l = 2w$
$\frac{P - 2l}{2} = w$
b $w = \frac{P - 2l}{2}$
$= \frac{42 - (2 \times 12)}{2}$
$= \frac{18}{2}$
$= 9$
You need to get w on its own on one side of the formula.
Use the same operations as you would to solve an equation to find w. Subtract $2l$ from both sides.
Divide both sides by 2.
Substitute $P = 42$ and $l = 12$ into the rearranged formula to find w.

Exercise 25C

C

1 Write down the letter that is the subject of each formula.

a $M = 2(a + b)$ **b** $ha = R$ **c** $v = u + at$ **d** $K = \frac{2B}{a}$ **e** $W = 7a - 15$

2 Make a the subject of each of the formulae in Q1.

C

3 The formula for the area of a triangle is: area $= \frac{1}{2} \times$ base $\times$ height

a Rearrange this formula to make 'base' the subject.

b Use your rearranged formula to find the base length (x) of each triangle.

i area = $9\,\text{cm}^2$

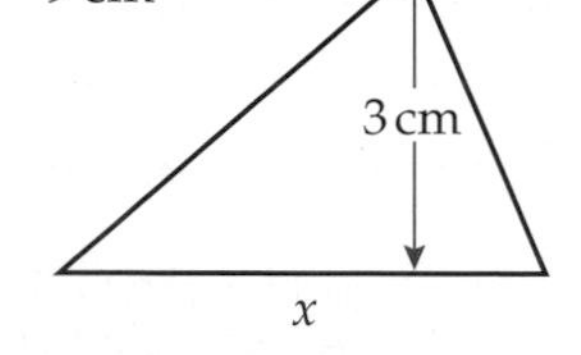

ii area = $6\,\text{cm}^2$

4 cm

x

iii area = $7.5\,\text{m}^2$

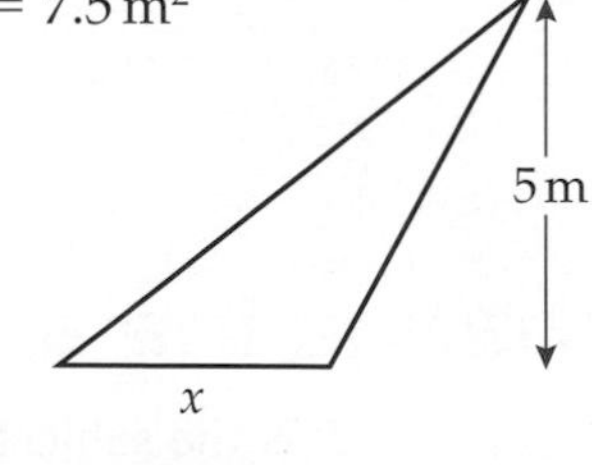

iv area = $6.3\,\text{cm}^2$

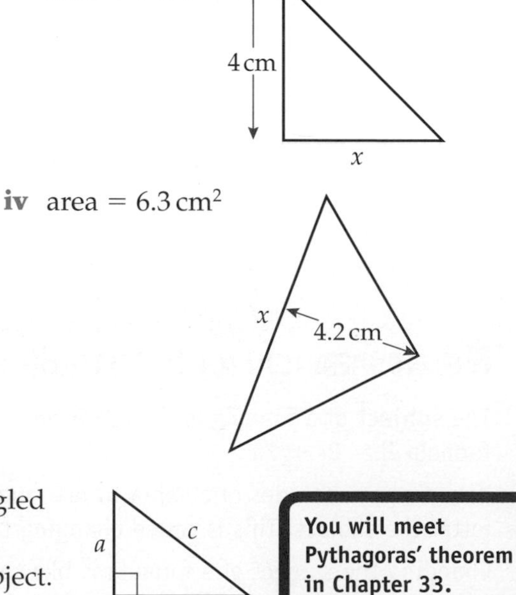

AO2

B

4 Pythagoras' theorem says that, in a right-angled triangle, $a^2 + b^2 = c^2$.

a Rearrange the equation to make b the subject.

b Find the value of b when $a = 7$ and $c = 9$.
Give your answer to three significant figures.

a c b

You will meet Pythagoras' theorem in Chapter 33.

AO2

Trial and improvement

Keywords
trial and improvement

Why learn this?
Learning by trial and error is part of life. Increasing accuracy by trial and improvement is part of maths.

Objectives
C Use trial and improvement to find solutions to equations

Skills check

1 Use a calculator to work out
 a $6.8^3 + 6.8^2$ b $(1.52 - 4)^3$
 c $3.25^3 - 3.25$ d $223.4^3 + 16.1^2$

2 Write the number that is exactly half way between these pairs.
 a 8 and 9 b 19.2 and 19.3
 c 1.87 and 1.88 d 8.34 and 8.42

Trial and improvement

Some equations cannot be solved using algebra. Instead, you can solve them by **trial and improvement**.

Solve an equation by substituting values of x in the equation to see if they give the correct solution. The more values you try, the closer you can get to the solution.

Example 4

C

Use trial and improvement to find a solution to the equation $2x^3 - x = 70$.
Give your answer to one decimal place.

Draw a table to record your trials.

x	$2x^3 - x$	Comment
3	51	Too low
4	124	Too high
3.5	82.25	Too high
3.3	68.574	Too low
3.4	75.208	Too high
3.35	71.840...	Too high

Choose a starting value for x and use your calculator to work out $2x^3 - x$.

Compare your result with 70.

3 is too low and 4 is too high, so try 3.5.

The solution is between 3.3 and 3.4. To find the answer to one decimal place you need to know which is closer, so try 3.35.

The solution is $x = 3.3$ to one decimal place.

3.35 is too high. This means the solution is closer to 3.3 than 3.4.

Exercise 25D

C

1 Nisha is using trial and improvement to solve the equation $x^3 - 2x = 50$.
Her first two trials are shown in this table.

x	$x^3 - 2x$	Comment
3	21	Too low
5	115	Too high

Copy the table and add as many rows to it as you need to find the solution.

Copy and complete the table to find a solution to the equation.
Give your answer to one decimal place.

C

2 Use trial and improvement to find a solution to the equation $\frac{x^3}{2} + x = 300$.
Give your answer correct to one decimal place.

x	$\frac{x^3}{2} + x$	Comment
5	67.5	Too low
10	510	Too high

3 Use trial and improvement to find a solution to the equation $x^3 + \frac{20}{x} = 50$.
Give your answer correct to one decimal place.

x	$x^3 + \frac{20}{x} = 50$	Comment
2	18	Too low
5	129	Too high

4 Use trial and improvement to find a solution to the equation $x^3 + 3x = 100$.
Give your answer correct to one decimal place.

x	$x^3 + 3x$	Comment
5	140	Too high

5 The equation $x^3 + \frac{200}{x^2} = 90$ has a solution between 3 and 6.

a Find this solution using trial and improvement.
Give your answer correct to one decimal place.

b This equation has another solution between −5 and 0. Find this solution using trial and improvement. Give your answer correct to one decimal place.

6 Michelle is using trial and improvement to solve the equation $x^3 - x = 600$.
She says the solution is $x = 8.4$ to one decimal place.

a Show working to explain why Michelle is wrong.

b What is the correct solution?

7 Use trial and improvement to solve the equation $2x^3 + 8x = -90$.
Give your answer correct to one decimal place.

C

8 The volume of water in a barrel is given by the formula $V = \frac{t^3 - t}{10}$.
Use trial and improvement to find the value of t when $V = 30$.
Give your answer correct to one decimal place.

A02

9 The area of this rectangle is $800\,\text{cm}^2$.

a Write an equation showing this information.

b Use trial and improvement to find the value of x correct to one decimal place.

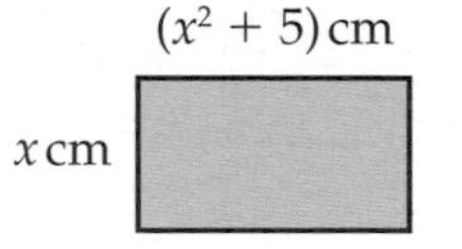

Review exercise

D

1 Work out the value of the letter in each of these triangles.

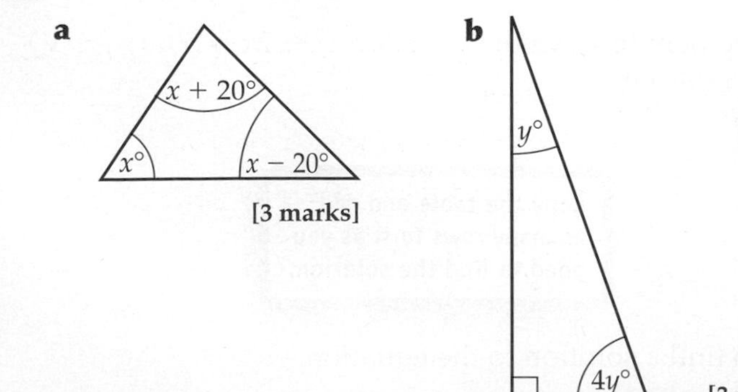

a [3 marks]

b [3 marks]

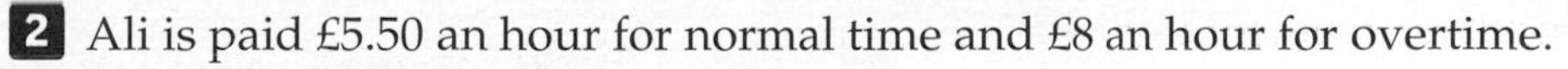

2 Ali is paid £5.50 an hour for normal time and £8 an hour for overtime.

a How much does Ali receive when he works for 7 hours of normal time and 2 hours of overtime? **[1 mark]**

b Construct a formula for working out Ali's pay when he works for p hours of normal time and q hours of overtime. **[2 marks]**

c Use your formula to find Ali's pay when he works for 5 hours of normal time and 6 hours of overtime. **[1 mark]**

C

3 Use trial and improvement to find a solution to the equation $2x^3 - 20x = 80$.
Give your answer correct to one decimal place.
You can use a copy of this table to help you.

x	$2x^3 - 20x$	Comment
4	48	Too low

[4 marks]

4 Use trial and improvement to solve the equation $x^2 + \sqrt{x + 5} = 50$.
Give your answer correct to one decimal place. **[4 marks]**

5 Alex is using trial and improvement to solve the equation $\frac{x^2}{\sqrt{x}} = 50$.

Her first two trials are shown in the table.
Copy and complete the table to find a solution to the equation.
Give your answer to one decimal place.

x	$\frac{x^2}{\sqrt{x}}$	Comment
10	31.62...	Too low
20	89.44...	Too high

[4 marks]

B

6 Rearrange the formulae $p = \frac{q + 1}{r} + 5$ to make q the subject. **[2 marks]**

7 The area of this trapezium is given by the formula $A = \frac{ah + bh}{2}$.
Rearrange this formula to make b the subject.

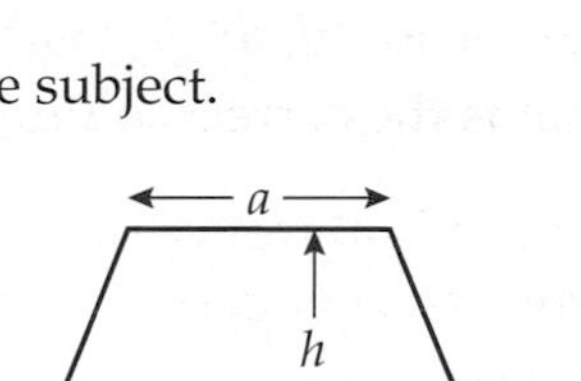

[3 marks]

8 Leanne and Barry have written down the same positive number.
Leanne squares the number then multiplies by 3 and then subtracts 12.
Barry adds 4 to the number then multiplies by 7.
Both finish with the same answer.
What was the mystery number? **[4 marks]**

Chapter summary

In this chapter you have learned how to

- write your own formulae and equations **D**
- substitute into a formula to solve problems **D** **C**
- use trial and improvement to find solutions to equations **C**
- change the subject of a formula **C** **B**
- set up and solve equations **C** **B** **A**

26

Compound shapes and 3-D objects

This chapter is about compound shapes and 3-D objects.

The Sage, Gateshead, is a live music venue. To work out the number of stainless steel panels for the curved roof, the architects had to estimate and calculate the areas of complex curved shapes.

Objectives

This chapter will show you how to

- **calculate perimeters and areas of shapes made from triangles and rectangles** D
- **use 2-D representations of 3-D shapes and analyse 3-D shapes through 2-D projections and cross-sections, including plan and elevation** D
- **solve problems involving surface areas and volumes of prisms and pyramids** D C A

Before you start this chapter

Calculate the perimeter and area of each these shapes.
All lengths are in centimetres.

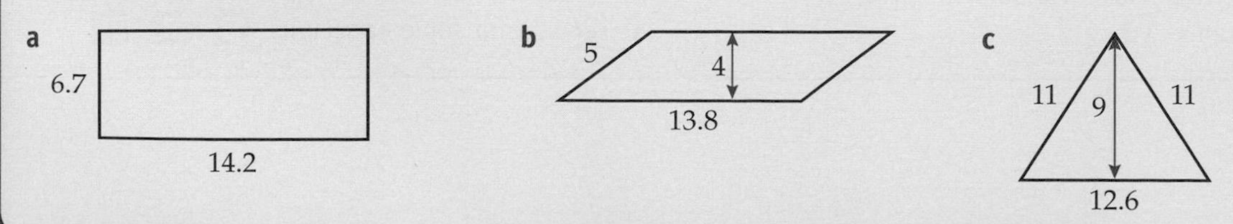

Geometry skills: area of simple shapes

Rectangle

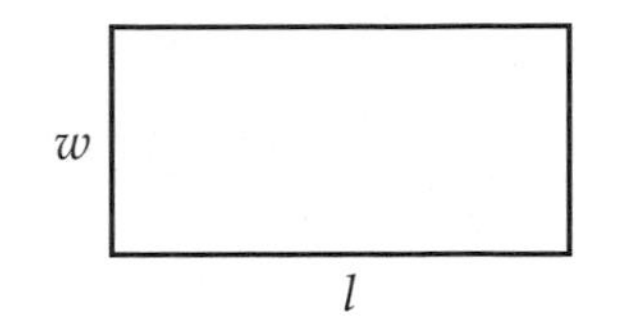

Area of a rectangle = length × width
$= l \times w$

The perimeter of a rectangle
$= l + w + l + w$
$= 2l + 2w$

Parallelogram

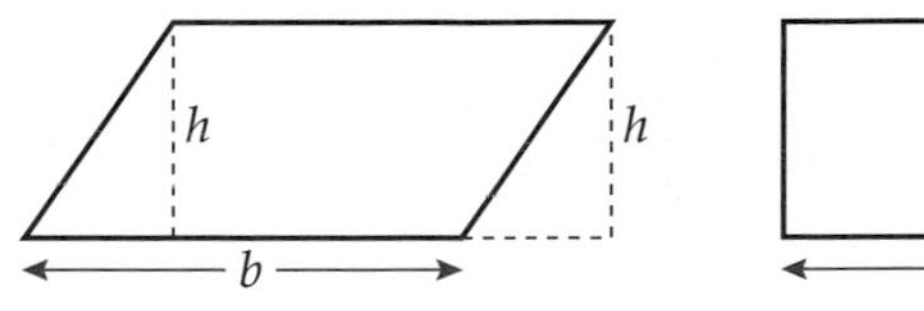

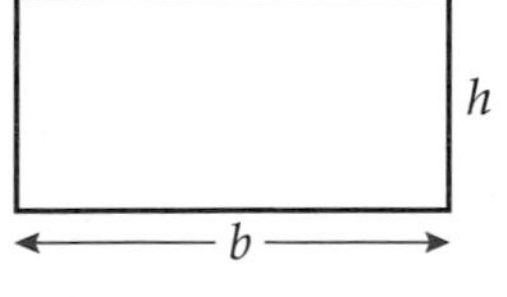

Area of a parallelogram = base × perpendicular height
$= b \times h$

Remember to use the perpendicular height, h, not the slant height.

Triangle

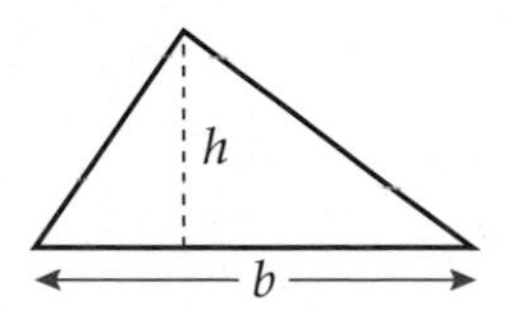

Area of a triangle $= \frac{1}{2}$ × base × perpendicular height
$= \frac{1}{2} \times b \times h$

Trapezium

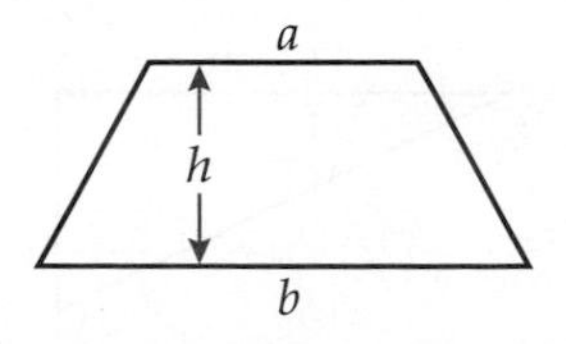

Area of a trapezium $= \frac{1}{2}$ × (sum of parallel sides) × perpendicular height
$= \frac{1}{2} \times (a + b) \times h$

1 Calculate the area of these shapes.
All lengths are in centimetres.

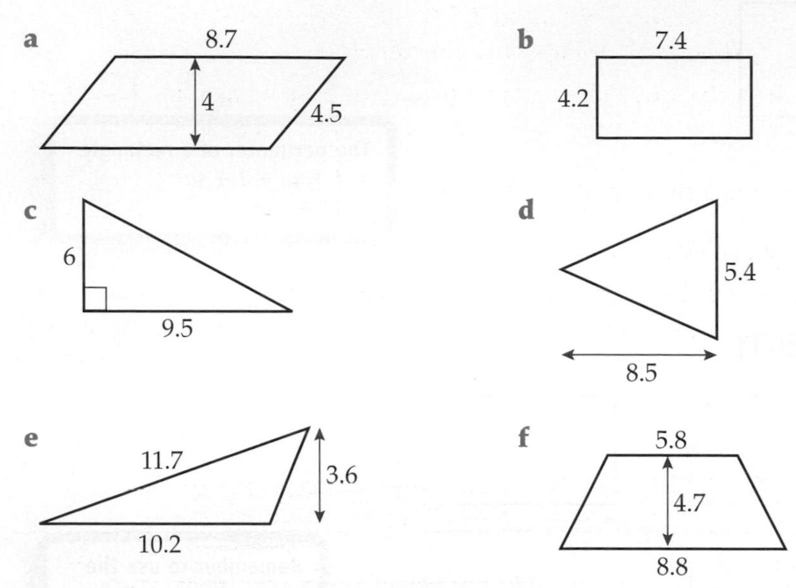

2 Tom has a rectangular area of garden that he wants to re-turf.

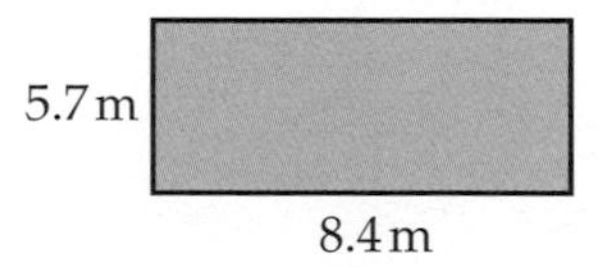

The price of turf is £4.60 per square metre.
It can only be bought in whole numbers of square metres.
The price for standard delivery is £15.95.
How much will it cost Tom for the turf and delivery?

3 Each of these shapes has an area of 30 cm^2.
Calculate the lengths marked by letters.

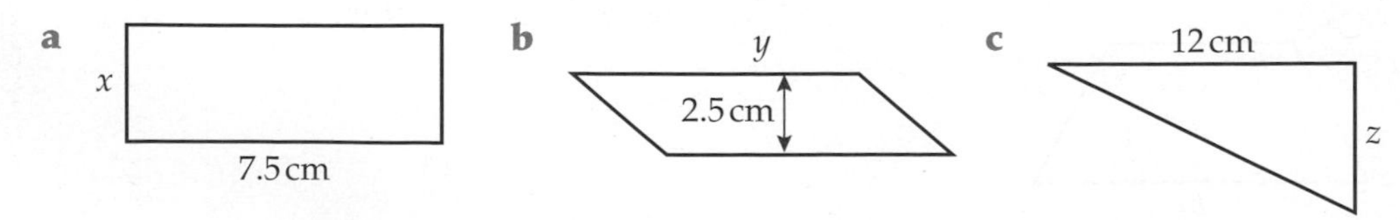

26.1 Perimeter and area of compound shapes

Keywords
compound shape

Why learn this?
Real-life area problems often involve compound shapes, for example, carpeting an L-shaped room.

Objectives

D Find the perimeter and area of compound shapes

Skills check

1. A rectangle of length 8.5 cm has a perimeter of 25.4 cm. What is the width of the rectangle?
2. Calculate the area of a triangle of base 45 mm and height 6 cm.

Compound shapes

A **compound shape** is a shape made up of simple shapes.

To find the area of a compound shape, you split it into simple shapes.

Then use the formulae for areas of simple shapes.

Example 1

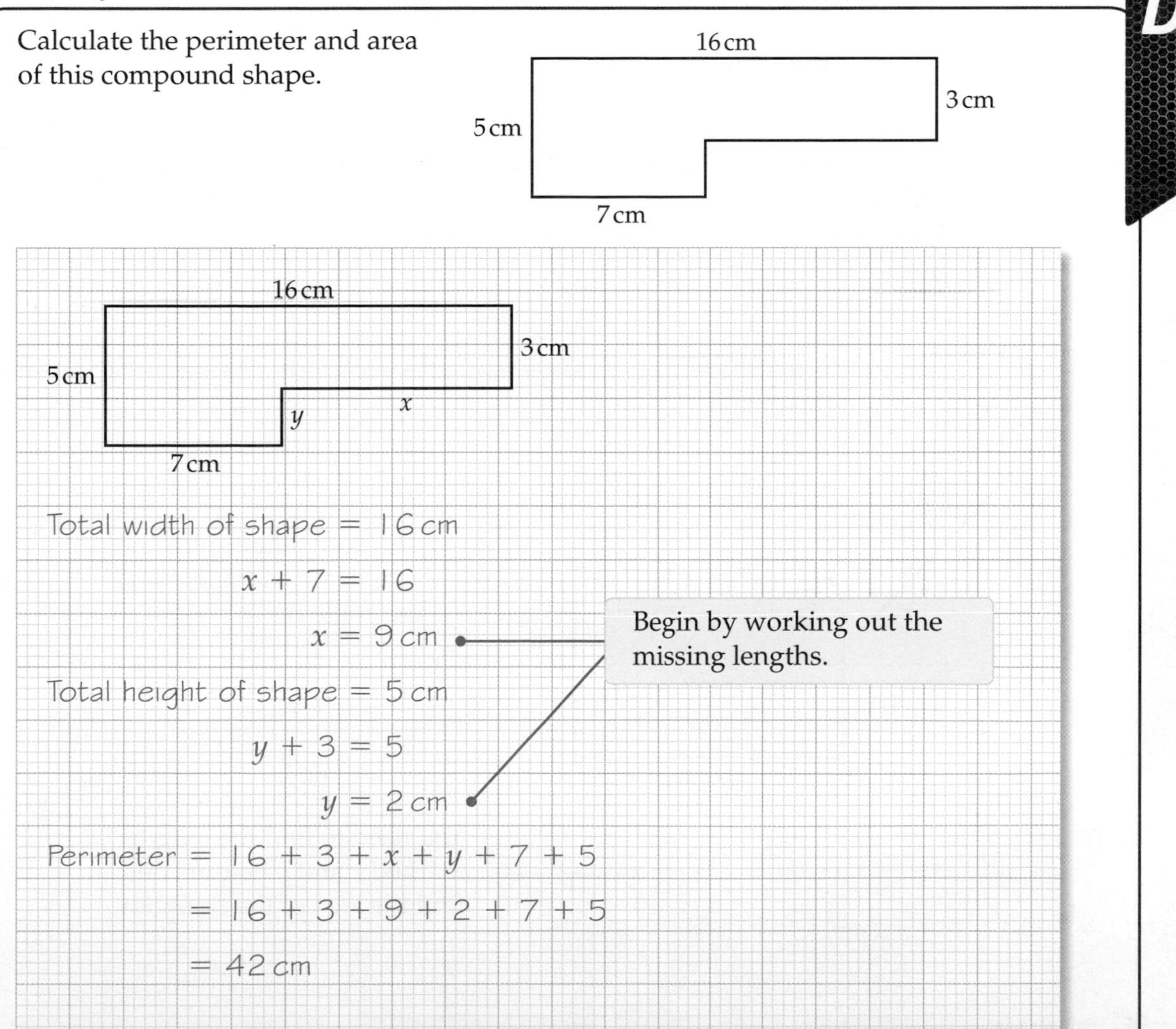

Calculate the perimeter and area of this compound shape.

Total width of shape = 16 cm

$x + 7 = 16$

$x = 9$ cm

Total height of shape = 5 cm

$y + 3 = 5$

$y = 2$ cm

Begin by working out the missing lengths.

Perimeter $= 16 + 3 + x + y + 7 + 5$

$= 16 + 3 + 9 + 2 + 7 + 5$

$= 42$ cm

D

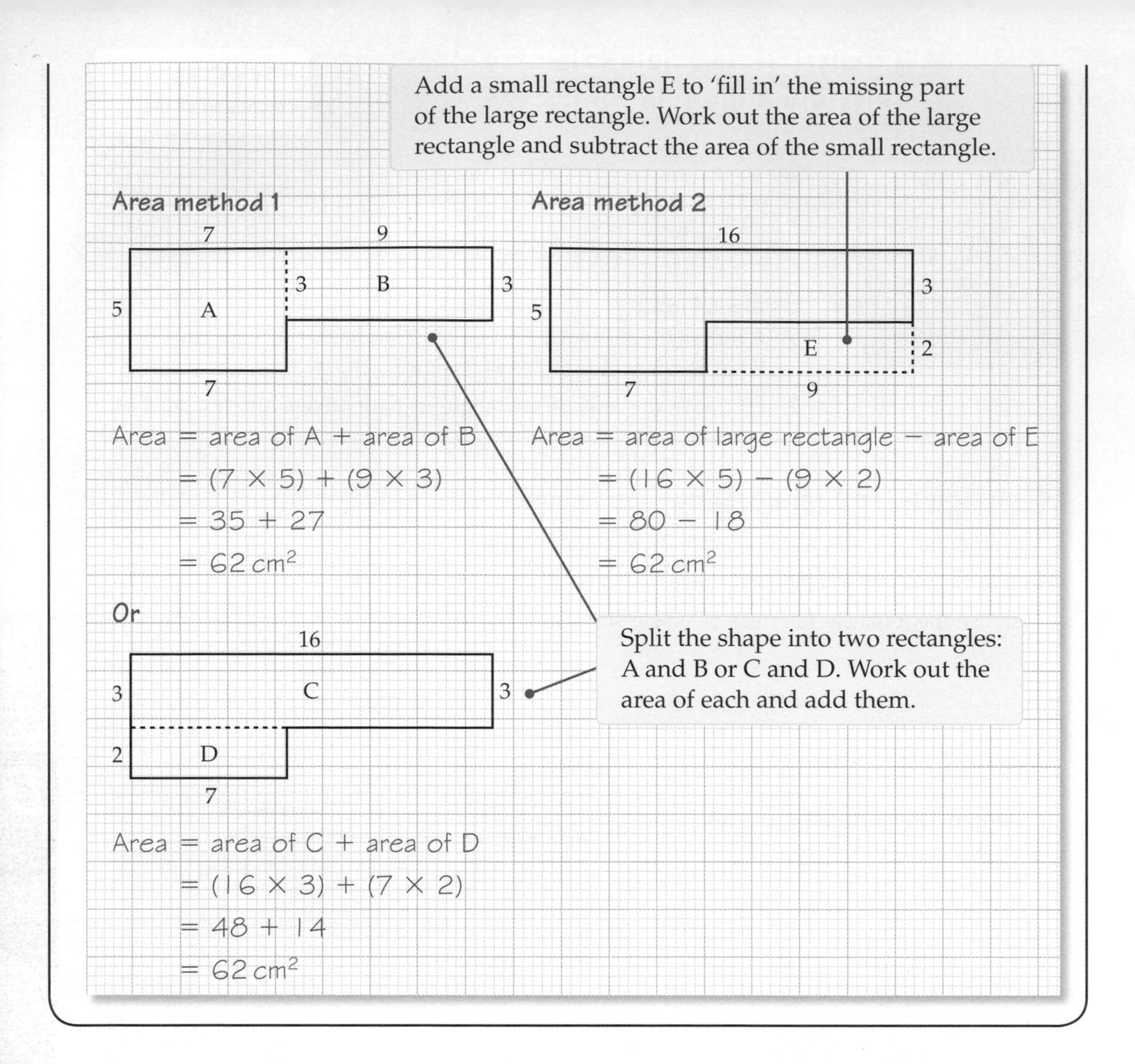
Add a small rectangle E to 'fill in' the missing part of the large rectangle. Work out the area of the large rectangle and subtract the area of the small rectangle.
Area method 1
7
9
3
B
3
5
A
7
Area method 2
16
3
5
E
2
7
9
Or
16
3
C
3
2
D
7
Split the shape into two rectangles: A and B or C and D. Work out the area of each and add them.

Area method 1

Area = area of A + area of B
$= (7 \times 5) + (9 \times 3)$
$= 35 + 27$
$= 62\text{ cm}^2$

Area method 2

Area = area of large rectangle − area of E
$= (16 \times 5) - (9 \times 2)$
$= 80 - 18$
$= 62\text{ cm}^2$

Or

Area = area of C + area of D
$= (16 \times 3) + (7 \times 2)$
$= 48 + 14$
$= 62\text{ cm}^2$

D

Example 2

Calculate the area of this shape.

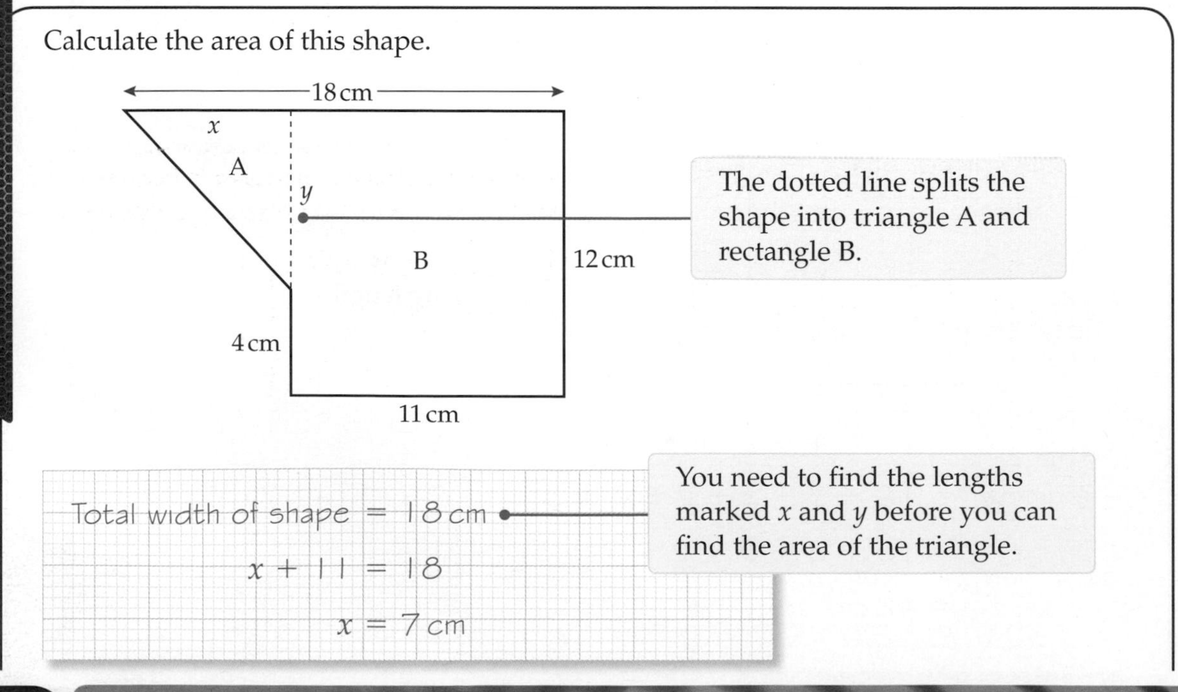
18 cm
x
A
y
B
12 cm
4 cm
11 cm
The dotted line splits the shape into triangle A and rectangle B.
You need to find the lengths marked x and y before you can find the area of the triangle.

Total width of shape = 18 cm
$x + 11 = 18$
$x = 7\text{ cm}$

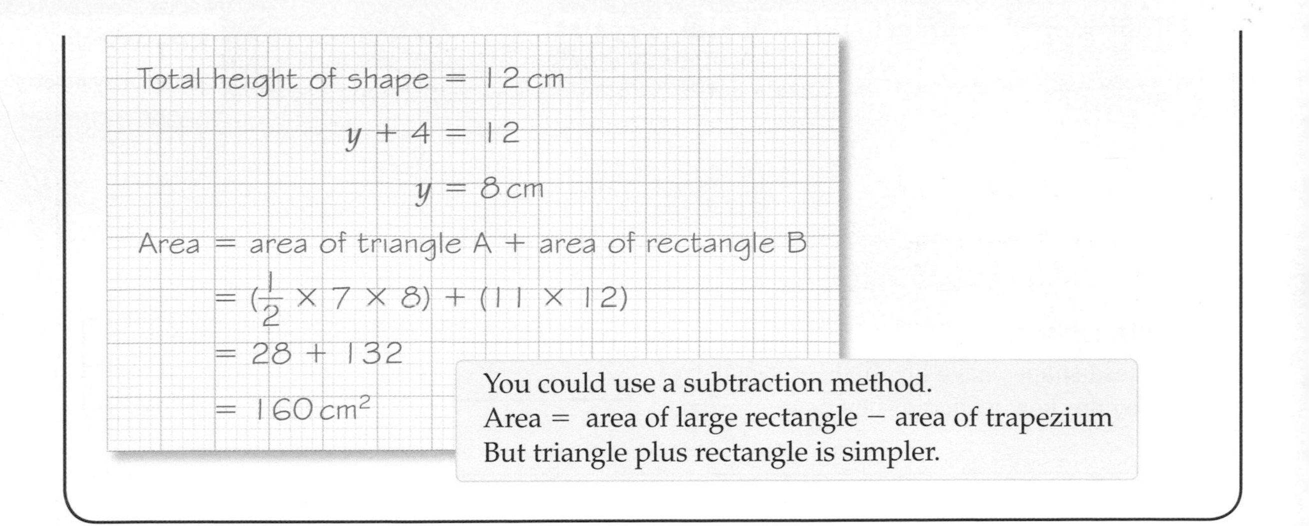

Total height of shape = 12 cm
$y + 4 = 12$
$y = 8$ cm
Area = area of triangle A + area of rectangle B
$= (\frac{1}{2} \times 7 \times 8) + (11 \times 12)$
$= 28 + 132$
$= 160\text{ cm}^2$
You could use a subtraction method.
Area = area of large rectangle − area of trapezium
But triangle plus rectangle is simpler.

Exercise 26A

1 Calculate the perimeter and area of each of these compound shapes.
All lengths are in centimetres.

Choose one of the methods used in Example 1.

a

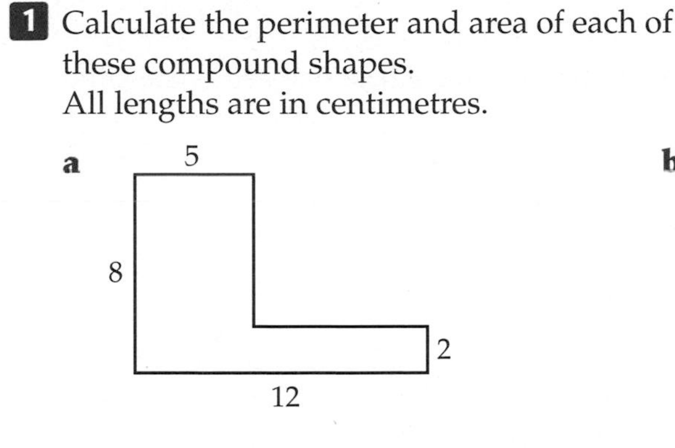

b

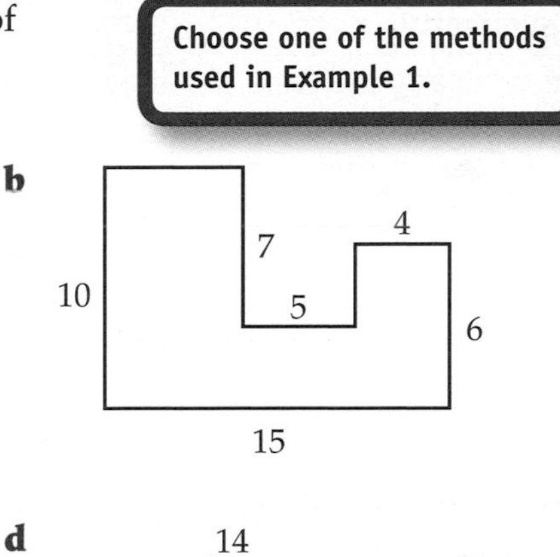

c

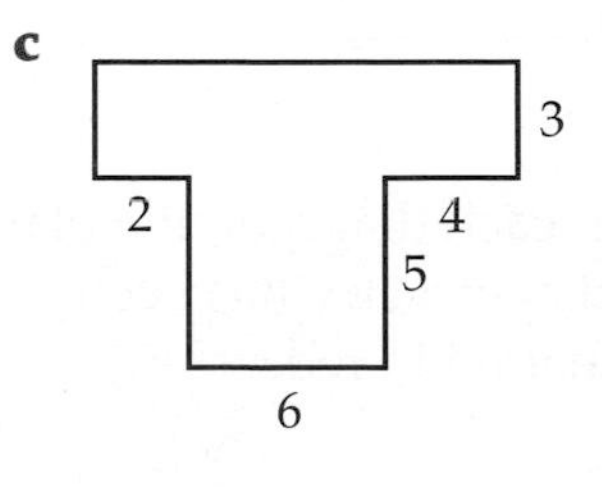

d

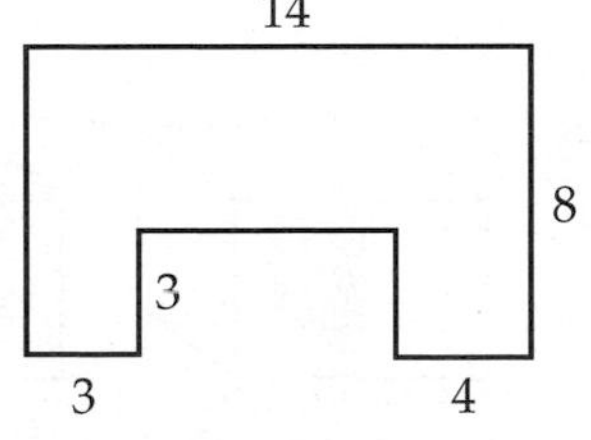

2 Calculate the area of these shapes.
All lengths are in centimetres.

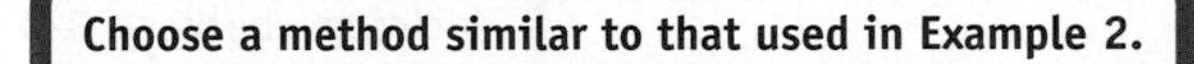

Choose a method similar to that used in Example 2.

a

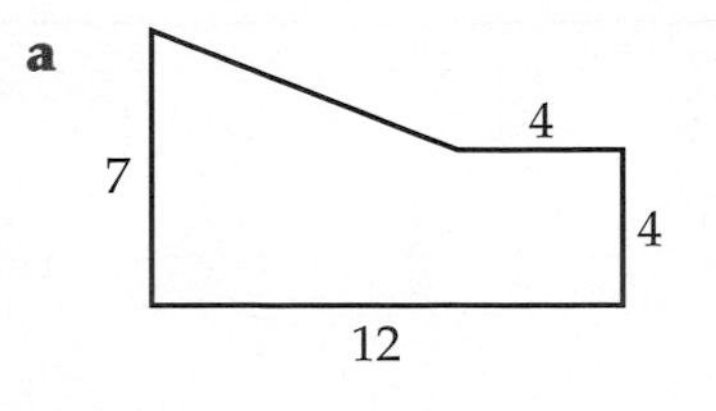

b

d

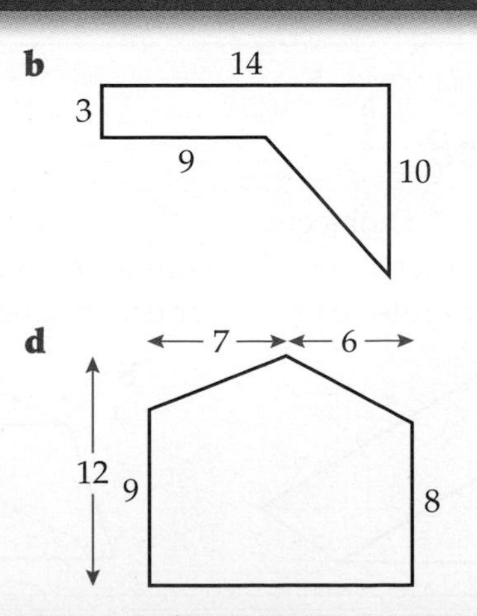

c

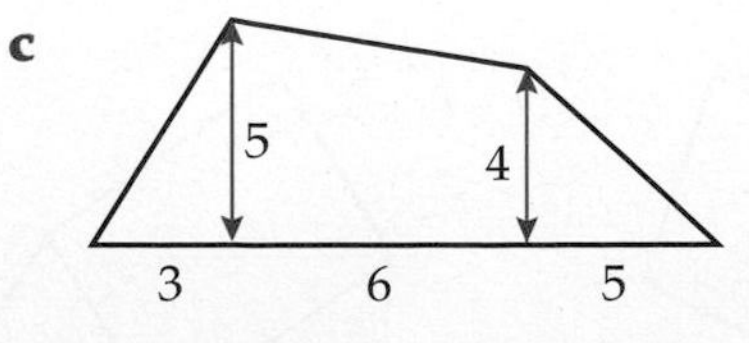

D

26.2 Planes of symmetry

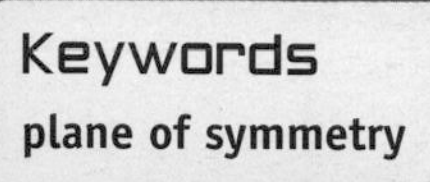

Why learn this?
You need to recognise symmetry in 3-D objects.

Objectives
D Identify the planes of symmetry of 3-D objects

Skills check

Do these shapes have any lines of symmetry?
If they do, how many do they have?

Symmetry in 3-D

Symmetry can exist in 3-D objects as well as in 2-D shapes.

If symmetry exists in 3-D objects, you don't call it a line of symmetry, it is a **plane of symmetry**.

A plane of symmetry divides a 3-D object into two equal halves, where one half is the mirror image of the other.

Example 3

Copy this cuboid.
Show all its planes of symmetry.
Draw a separate diagram for each plane of symmetry.

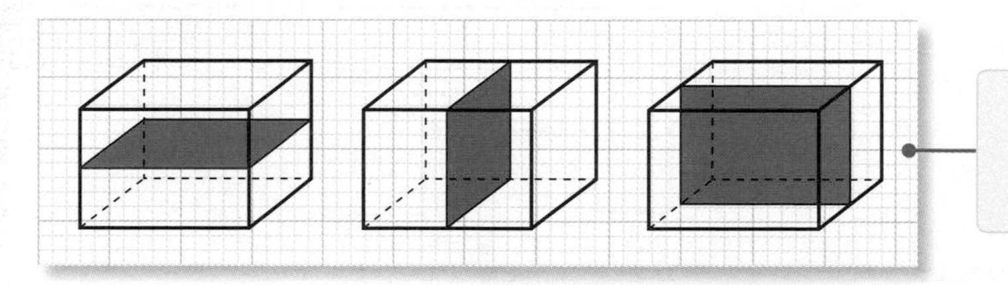

In each diagram, the plane of symmetry has been shaded in red.

When the cuboid is cut along the planes marked in red, it will be cut into two identical halves. This is the test you should apply when deciding whether a 3-D object has any planes of symmetry.

Exercise 26B

D

1 Copy these 3-D objects.
For each object show all its planes of symmetry.
Draw a separate diagram for each plane of symmetry.

a

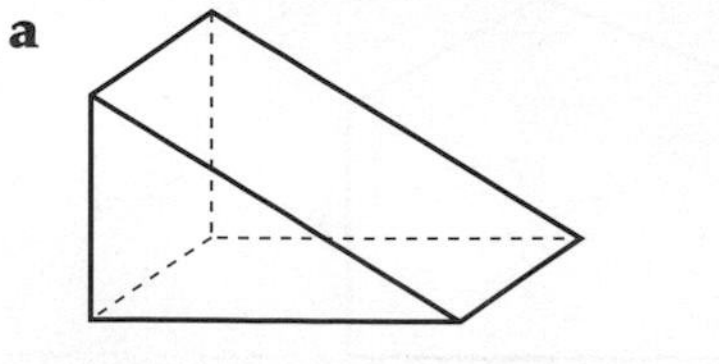

b

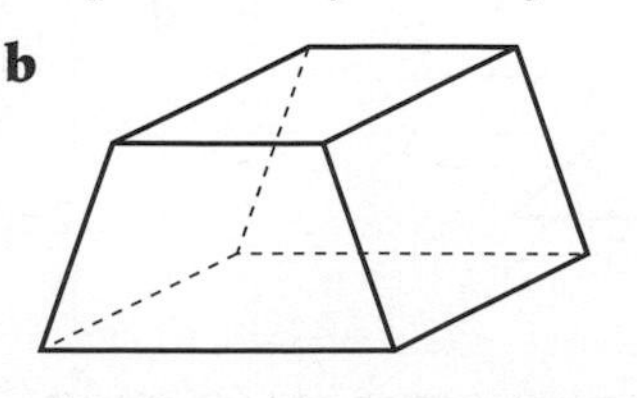

c

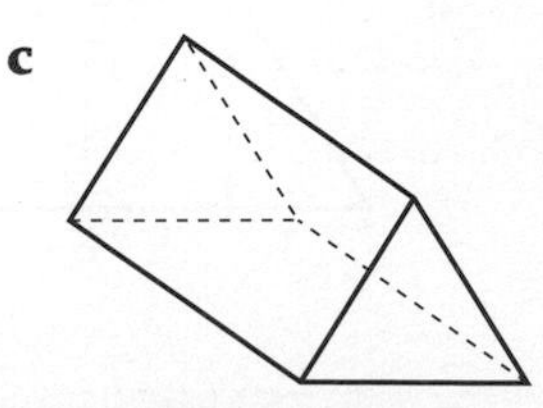

26.3 Plans and elevations

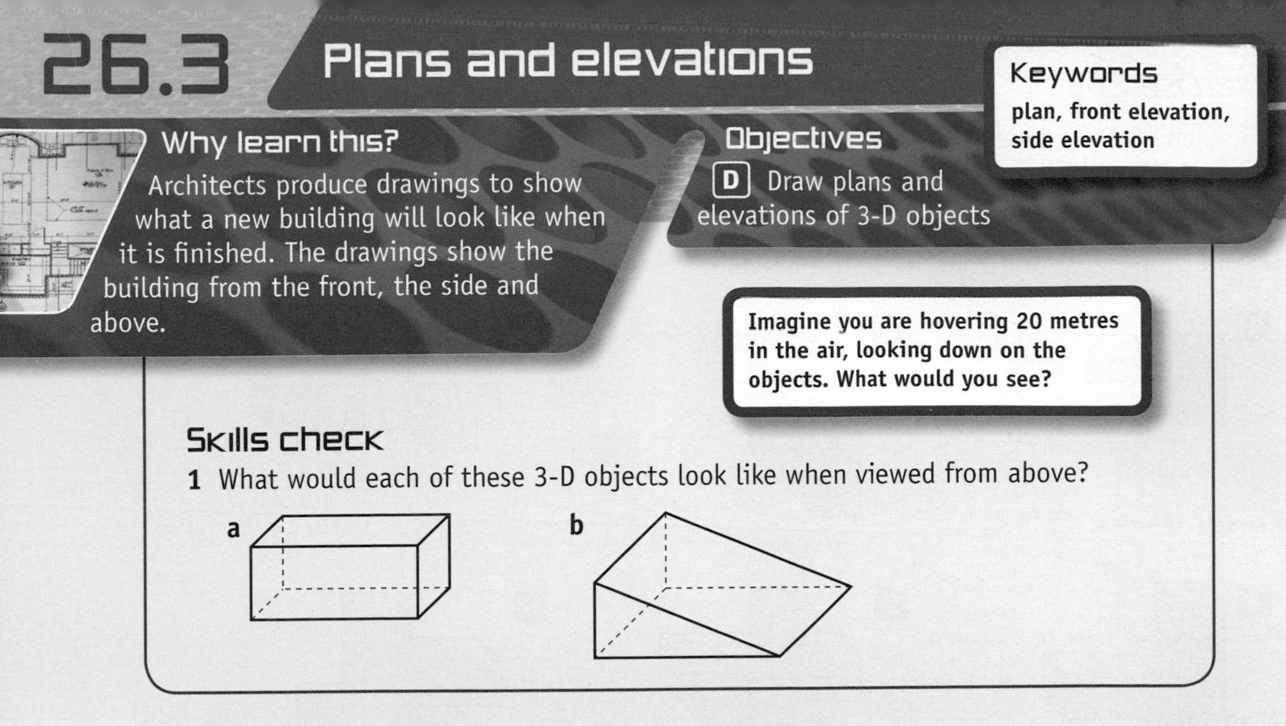

Keywords
plan, front elevation, side elevation

Why learn this?

Architects produce drawings to show what a new building will look like when it is finished. The drawings show the building from the front, the side and above.

Objectives

D Draw plans and elevations of 3-D objects

Imagine you are hovering 20 metres in the air, looking down on the objects. What would you see?

Skills check

1 What would each of these 3-D objects look like when viewed from above?

a

b

Drawing plans and elevations of a 3-D object

Look at a 3-D object from above, from the front and from the side and think about exactly what you can see and what you can't see.

The **plan** is the view from above the object.

The **front elevation** is the view from the front of the object.

The **side elevation** is the view from the side of the object.

Example 4

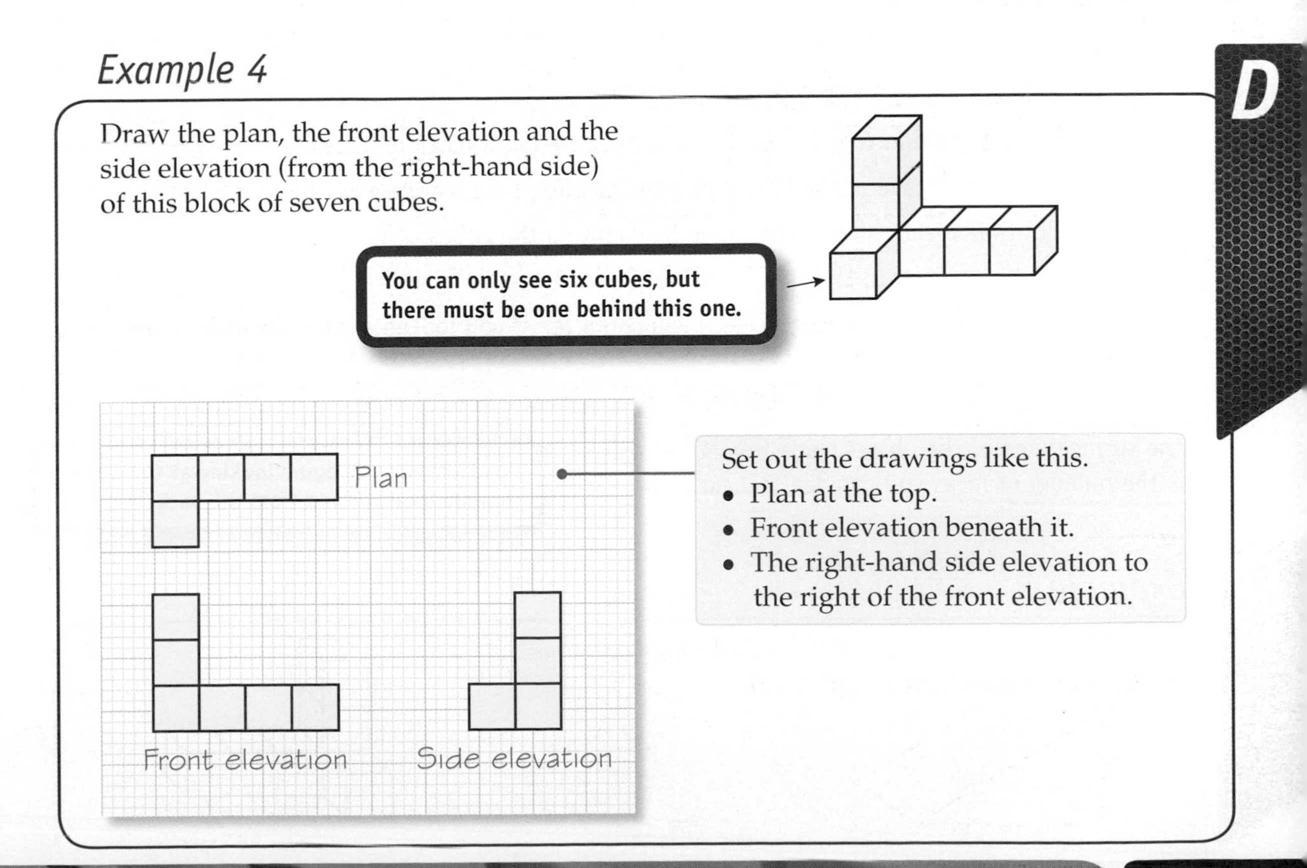

Draw the plan, the front elevation and the side elevation (from the right-hand side) of this block of seven cubes.

You can only see six cubes, but there must be one behind this one.

Set out the drawings like this.

- Plan at the top.
- Front elevation beneath it.
- The right-hand side elevation to the right of the front elevation.

Exercise 26C

D

For each block of cubes, draw

a the plan

b the front elevation

c the side elevation (from the right-hand side).

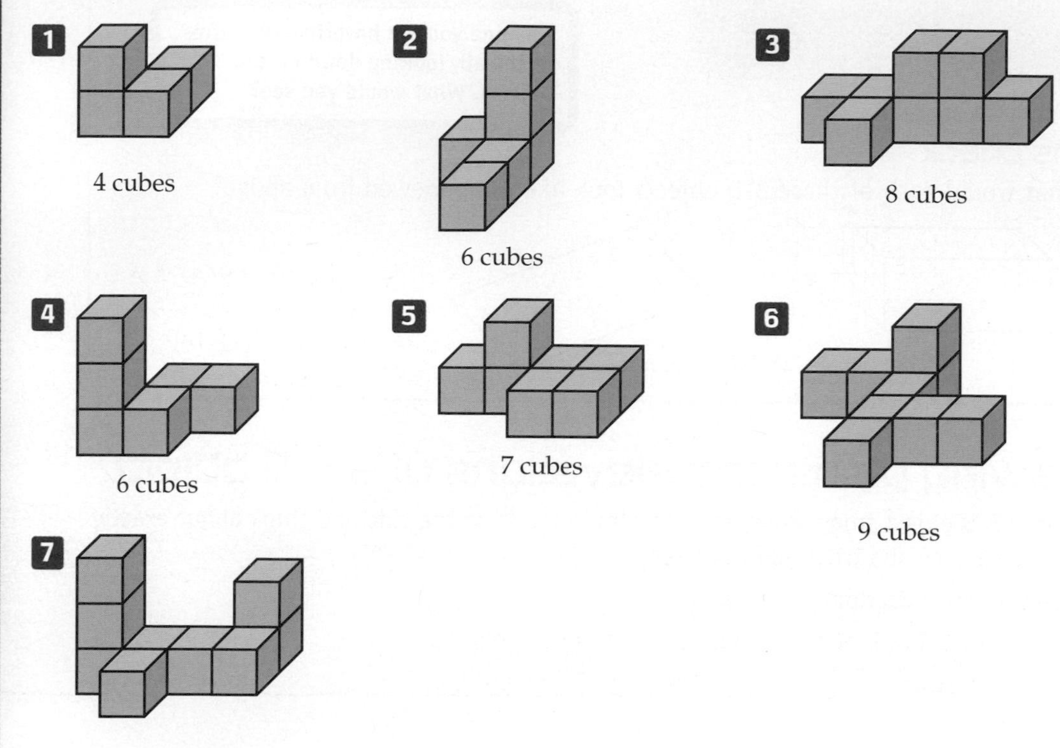

Using plans and elevations

You could be asked to find the volume or the surface area of a block of cubes.

Usually the cubes will have a side length of 1 cm, so each cube will have a volume of 1 cm^3.

The volume of the object can then be found by counting the cubes.

For the seven 1 cm^3 cubes in Example 4, the volume is $7 \times 1 = 7\text{ cm}^3$.

The surface area is the sum of all the faces you could see if you looked at the object from all sides.

The area of each face of a cube of side length 1 cm is 1 cm^2.

The surface area of the object made from 1 cm^2 cubes is the number of faces you can see $\times$ 1 cm^2.

Don't forget to imagine looking at the object from underneath and from the back.

D

Example 5

Calculate the surface area of this block of seven cubes.
All the cubes have side length 1 cm.

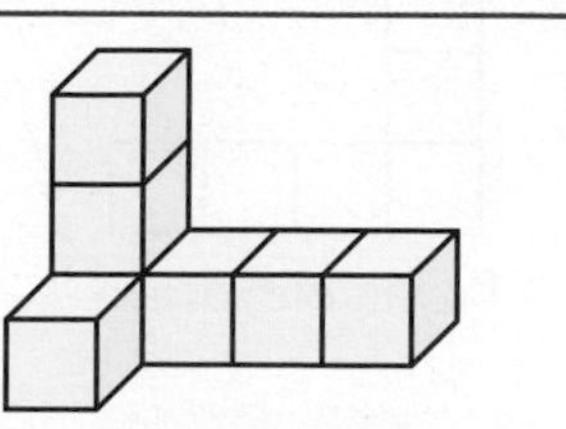

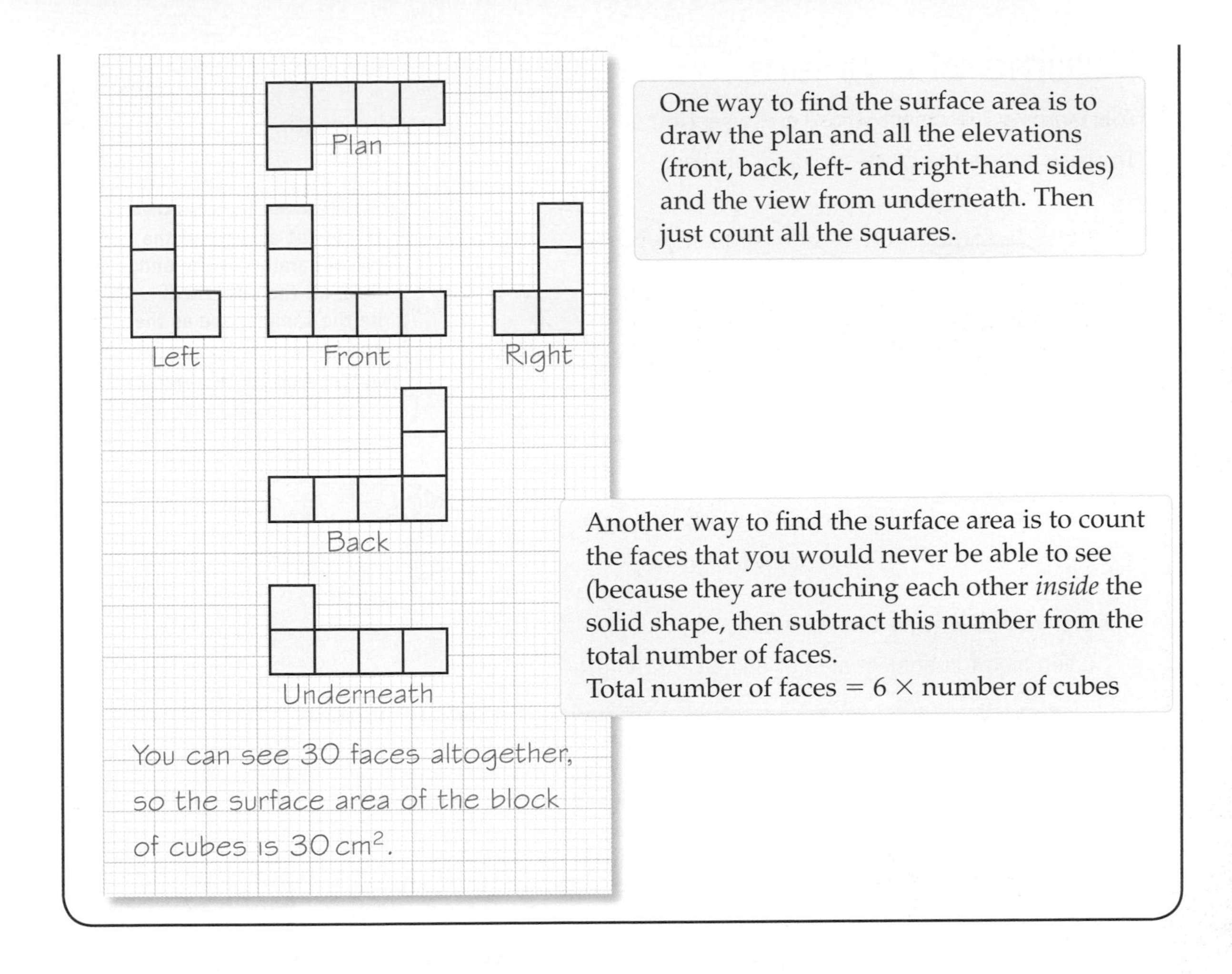

One way to find the surface area is to draw the plan and all the elevations (front, back, left- and right-hand sides) and the view from underneath. Then just count all the squares.

Another way to find the surface area is to count the faces that you would never be able to see (because they are touching each other *inside* the solid shape, then subtract this number from the total number of faces.
Total number of faces $= 6 \times$ number of cubes

Exercise 26D

Work out the volume and surface area of each block of cubes in Exercise 26C.
All the cubes have side length 1 cm.

26.4 Volume and surface area of prisms

Keywords
prism, cross-section, cuboid, net

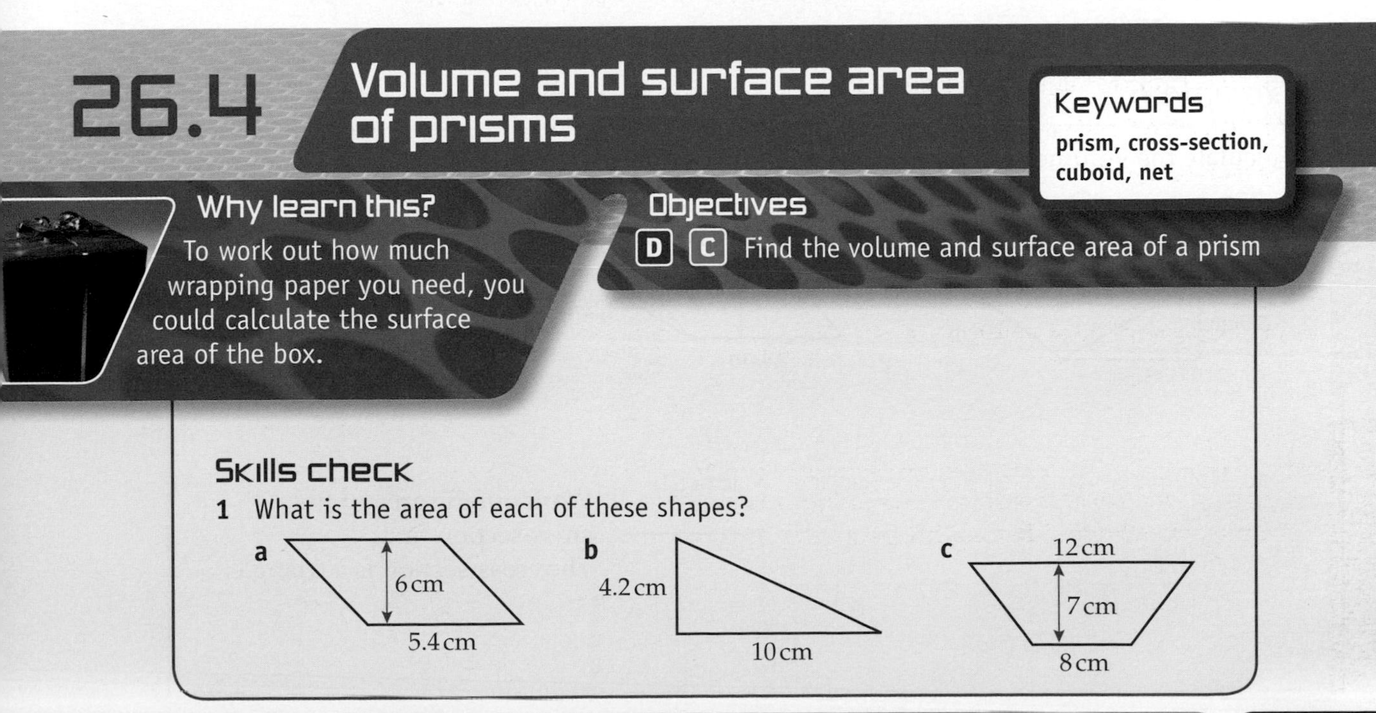

Why learn this?

To work out how much wrapping paper you need, you could calculate the surface area of the box.

Objectives

D C Find the volume and surface area of a prism

Skills check

1 What is the area of each of these shapes?

Volume of a prism

A **prism** is a 3-D object whose **cross-section** is the same all through its length.

In these prisms the cross-section is shaded.

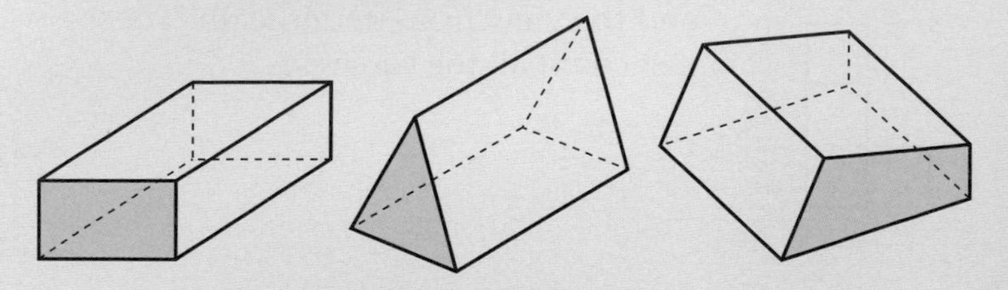

If you cut a 'slice' of the prism parallel to the end face, all the slices will be the same shape as the end face.

A **cuboid** is a 3-D object. Its cross-section is a rectangle.

To calculate the volume you use the formula

Volume of cuboid = length × width × height

$= l \times w \times h$

h l w

Another way of writing this formula is

Volume of cuboid = area of end face × length

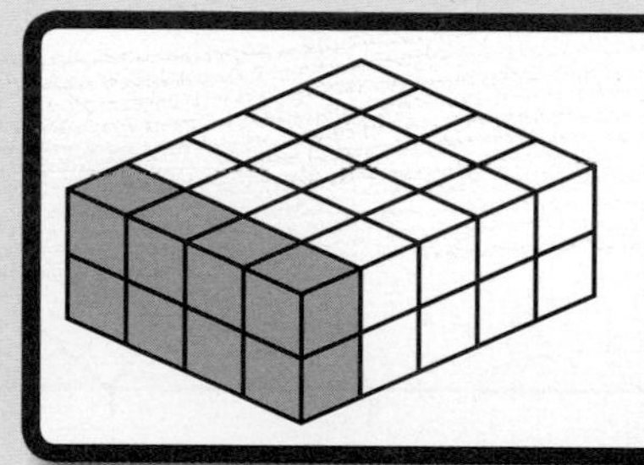

Imagine a cuboid of length 5 cm, width 4 cm and height 2 cm made from 1 cm cubes.
The end face has 4 × 2 cubes.
There are 5 'slices' of 4 × 2 cubes, so there are 5 × 4 × 2 = 40 cubes in total.

You can use a similar formula to calculate the volume of any prism.

Volume of prism = area of cross-section × length

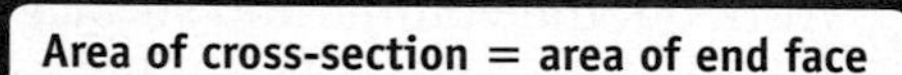

Area of cross-section = area of end face

D

C

Example 6

Calculate the volume of these prisms.

a

3.5 cm
20 cm
5 cm

b

5 cm
12 cm
9.4 cm

a Volume = area of cross-section × length

$= (\frac{1}{2} \times 5 \times 3.5) \times 20$

$= 8.75 \times 20$

$= 175\ \text{cm}^3$

Work out the area of the cross-section first.
The cross-section is a triangle.

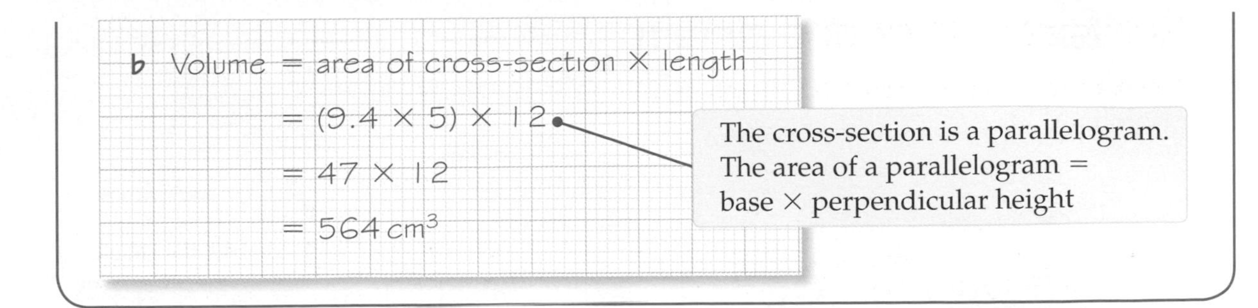

b Volume = area of cross-section × length
= (9.4 × 5) × 12
= 47 × 12
= 564 cm³
The cross-section is a parallelogram.
The area of a parallelogram = base × perpendicular height

Exercise 26E

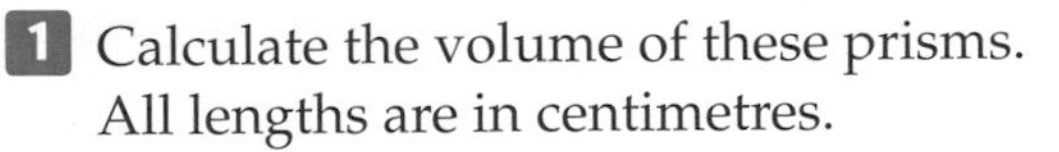

1 Calculate the volume of these prisms.
All lengths are in centimetres.

a

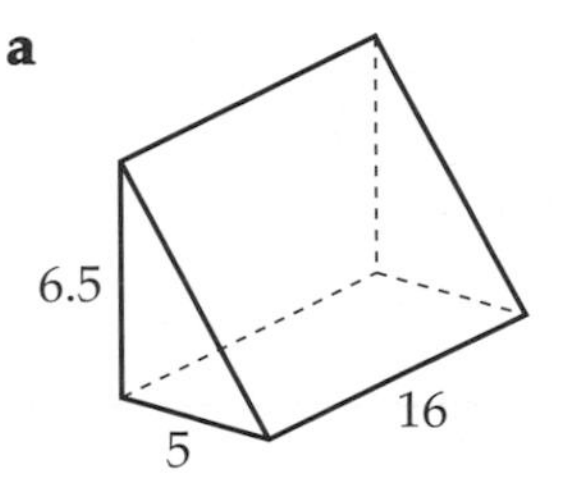

b

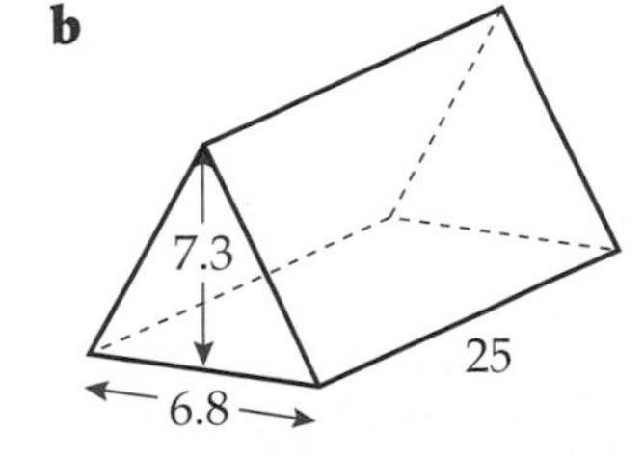

c

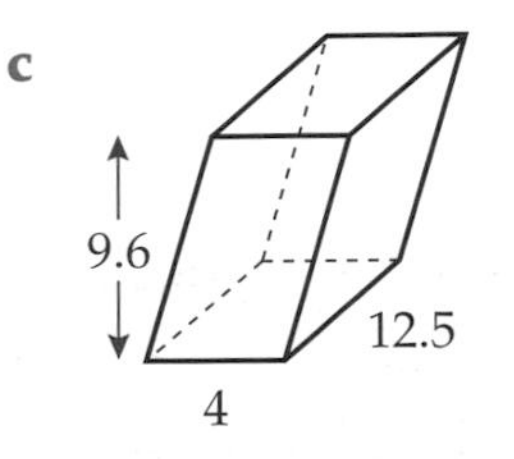

d

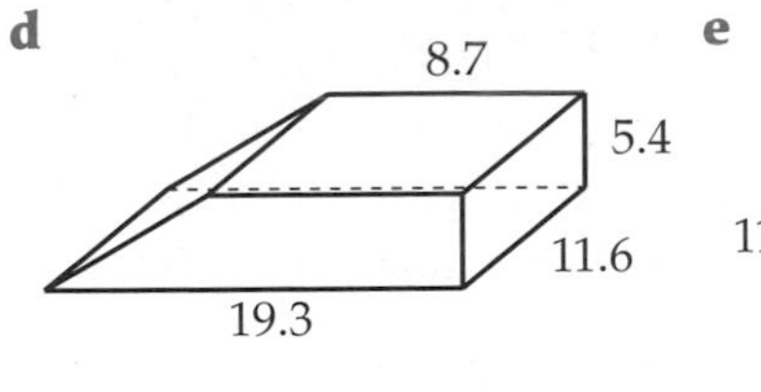

e

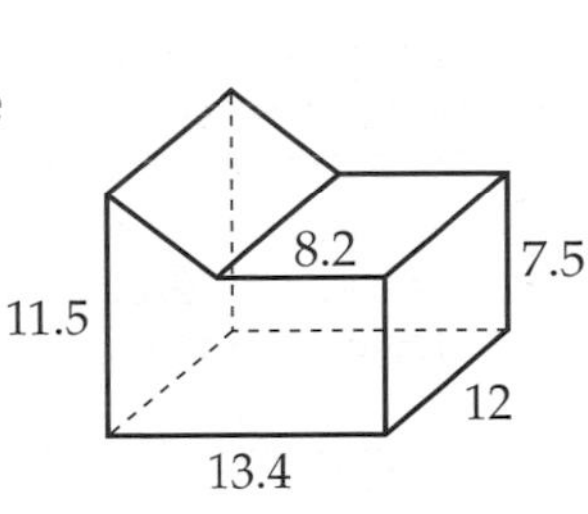

f

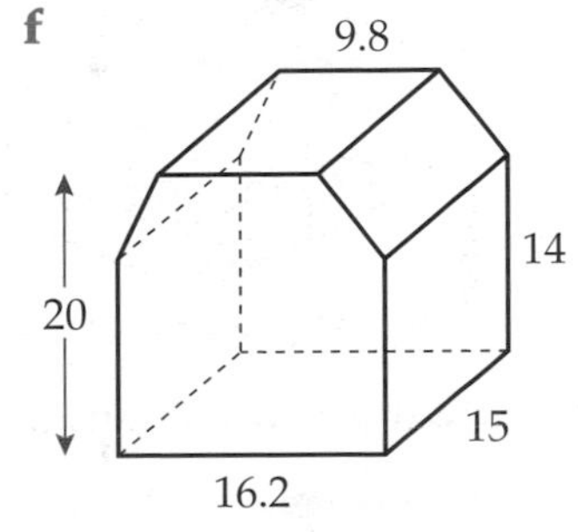

g

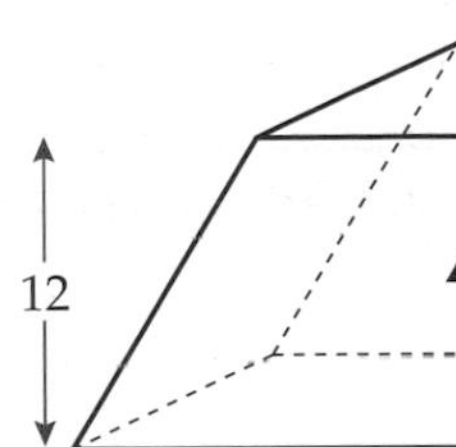

D

C

2 Ben has a pond in the shape of a cuboid.
The cuboid is 5.5 m long, 1.8 m wide and 1 m deep.

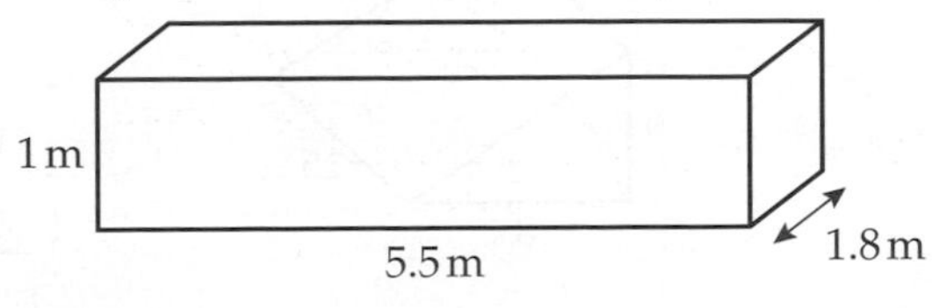

Ben wants to empty the pond and fill it in with hardcore and pebbles.
Hardcore and pebbles are sold in 1 m³ bags.
The price of a 1 m³ bag of hardcore is £31.65.
The price of a 1 m³ bag of pebbles is £52.20.
Ben estimates that he will need two 1 m³ bags of pebbles for the top surface.
He will fill the rest of the pond with hardcore.
Calculate how much it will cost him to do the work.

C

AO3

Surface area of a prism

This is the **net** of a cuboid.

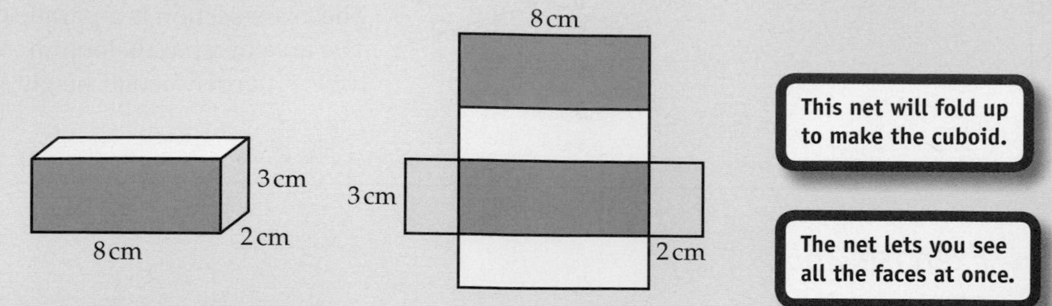

This net will fold up to make the cuboid.

The net lets you see all the faces at once.

The surface area of a 3-D object is the sum of the area of all its surfaces.

D

Example 7

Calculate the surface area of the cuboid shown above.

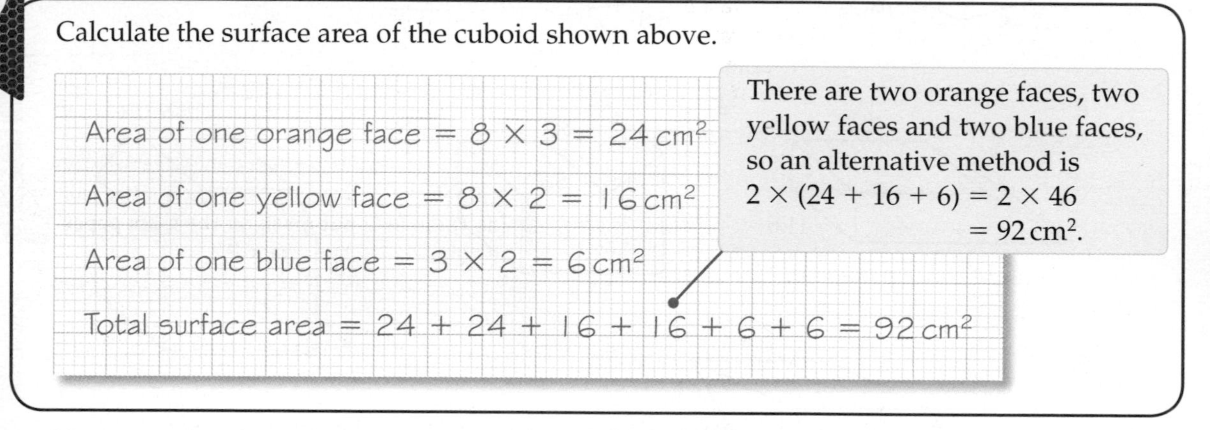

Area of one orange face = 8 × 3 = 24 cm²
Area of one yellow face = 8 × 2 = 16 cm²
Area of one blue face = 3 × 2 = 6 cm²
Total surface area = 24 + 24 + 16 + 16 + 6 + 6 = 92 cm²

There are two orange faces, two yellow faces and two blue faces, so an alternative method is $2 \times (24 + 16 + 6) = 2 \times 46$ $= 92\,\text{cm}^2$.

Exercise 26F

D

Calculate the surface area of these prisms.
All lengths are in centimetres.

a

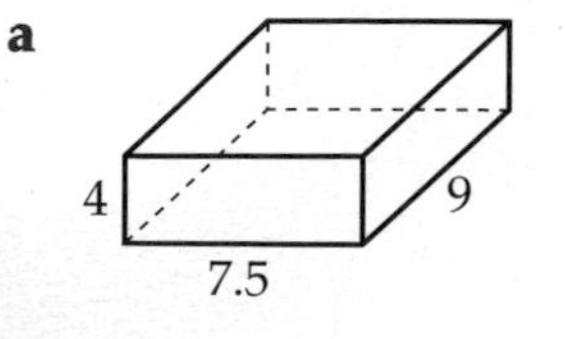

b

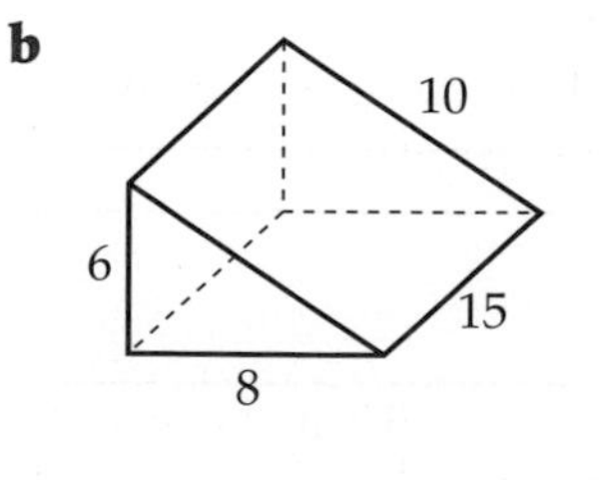

c

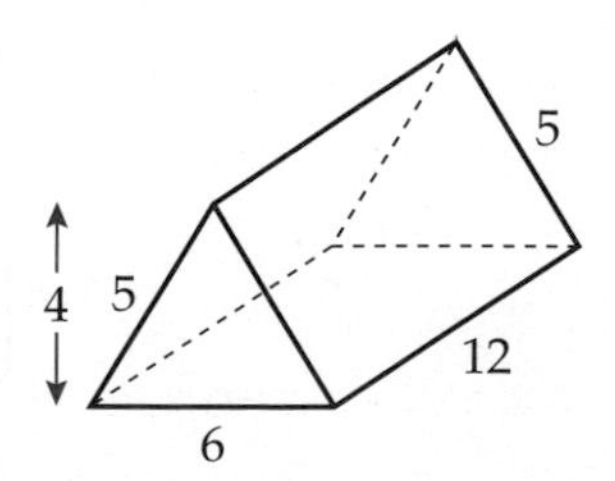

C

d

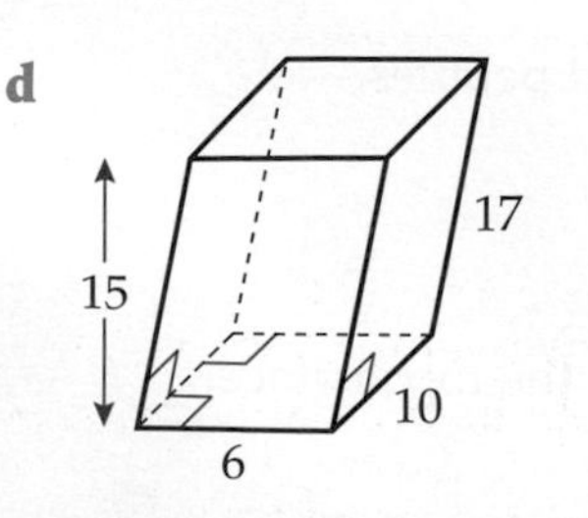

e

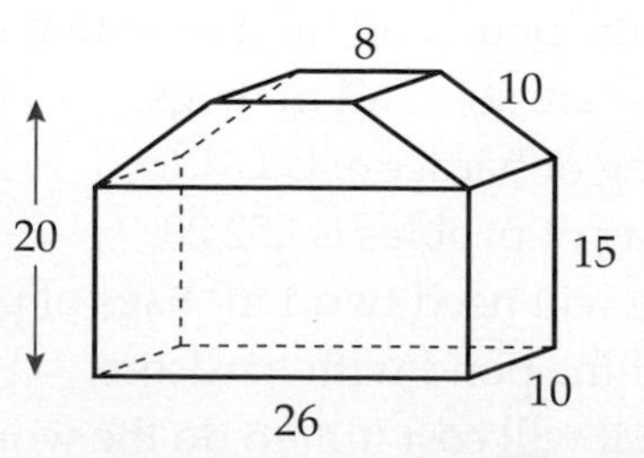

Keywords
pyramid, polyhedron

Why learn this?
Pyramids have been used right through from early civilisations to contemporary designs such as the Louvre in Paris.

Objectives

A Find the volume and surface area of a pyramid

Skills check

A triangle of height 12 cm has an area of 48 cm^2.
What is the base length of the triangle?

Volume and surface area of a pyramid

A **pyramid** is a **polyhedron** (a solid with flat faces).

Its base is usually a square, a rectangle or a triangle but it could be any polygon.

Its sides are triangles that meet at the top (the apex).

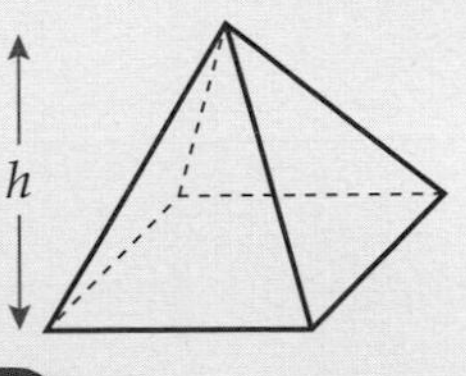

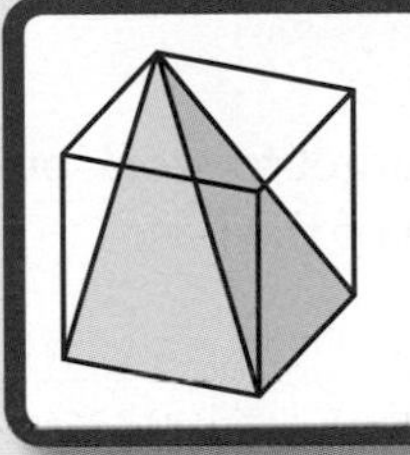

This diagram shows a cube with a square-based pyramid inside it.
The base of the pyramid is one face of the cube.
The apex of the pyramid is at another of the vertices of the cube.
You can fit three of these pyramids in the cube.

The volume of a pyramid is given by the formula

Volume $= \frac{1}{3} \times$ area of base $\times$ perpendicular height

A proof of this formula is very complicated. The formula is not given on the exam paper so you need to remember it.

The surface area of a pyramid is the sum of the areas of all its faces.

If the base is square and the apex is directly above its centre, the four triangular faces will be identical.

If the base is rectangular and the apex is directly above its centre, there will be two equal pairs of triangular faces.

Triangular bases are usually right-angled, equilateral or isosceles triangles.

Example 8

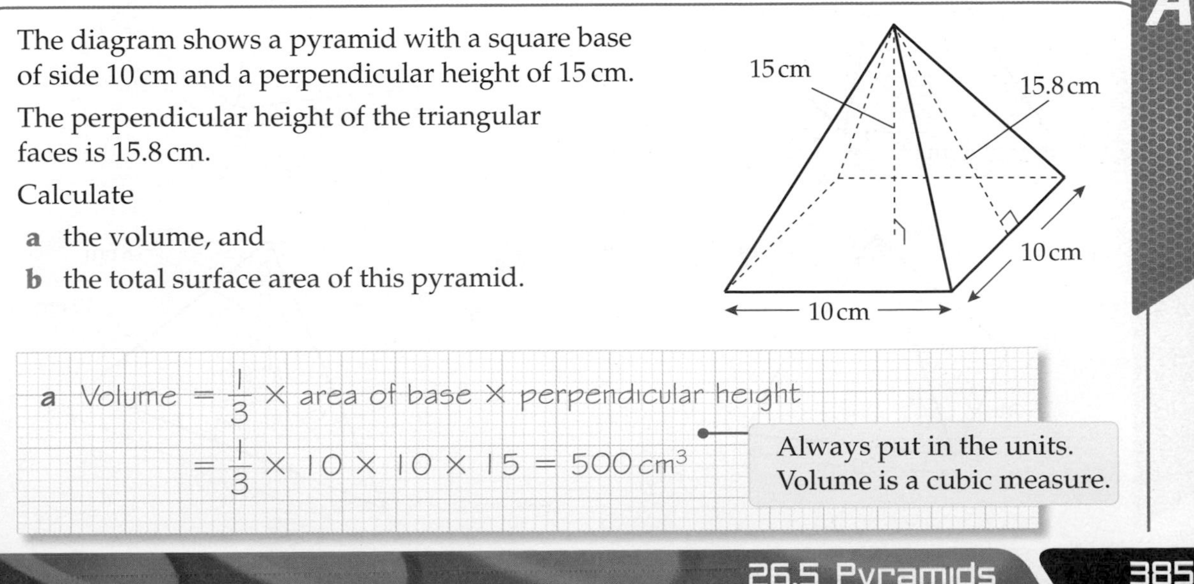

The diagram shows a pyramid with a square base of side 10 cm and a perpendicular height of 15 cm.

The perpendicular height of the triangular faces is 15.8 cm.

Calculate

a the volume, and

b the total surface area of this pyramid.

a Volume $= \frac{1}{3} \times$ area of base $\times$ perpendicular height

$= \frac{1}{3} \times 10 \times 10 \times 15 = 500\ cm^3$

Always put in the units. Volume is a cubic measure.

A

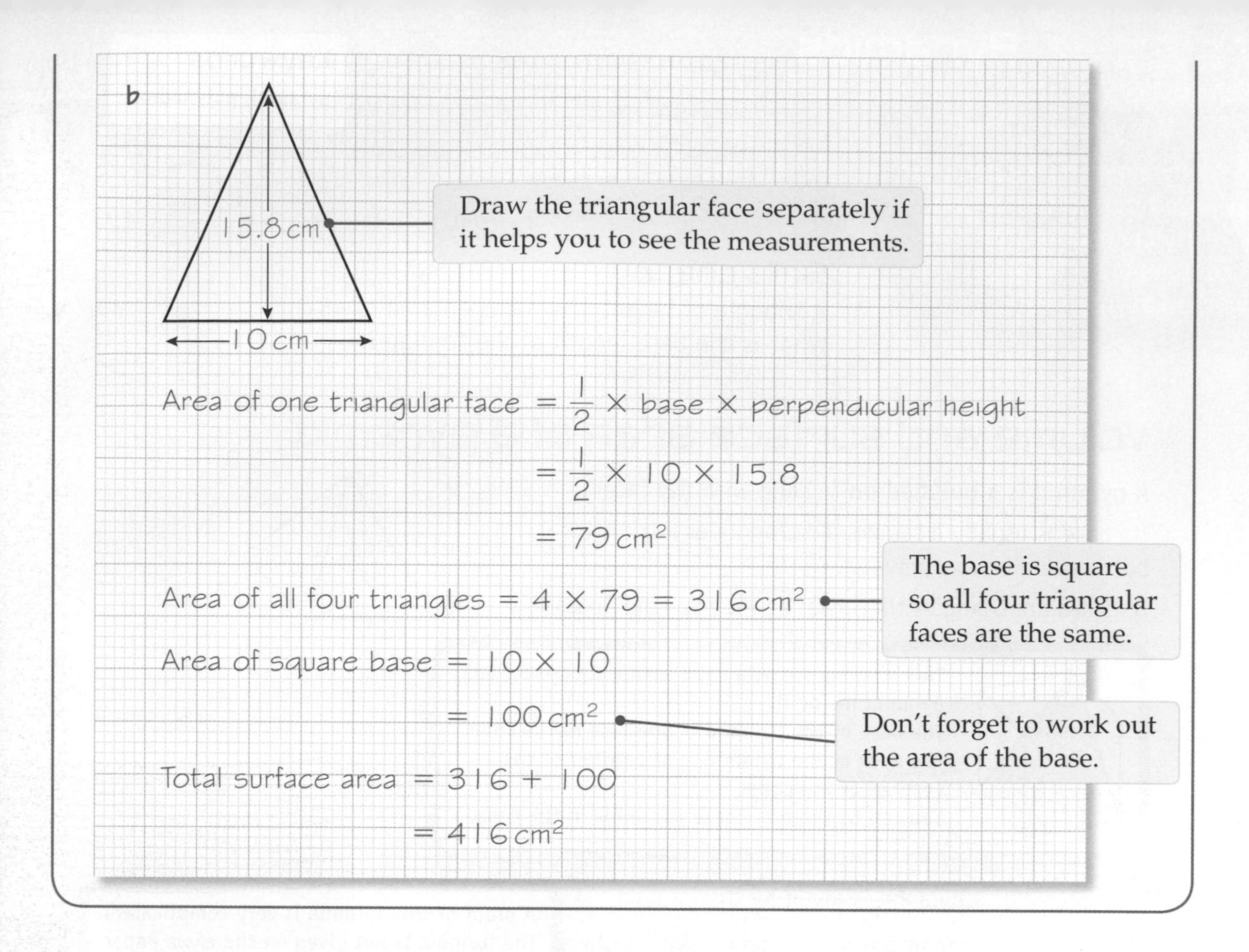

b
15.8 cm
Draw the triangular face separately if it helps you to see the measurements.
10 cm
Area of one triangular face = $\frac{1}{2}$ × base × perpendicular height
= $\frac{1}{2}$ × 10 × 15.8
= 79 cm²
Area of all four triangles = 4 × 79 = 316 cm²
The base is square so all four triangular faces are the same.
Area of square base = 10 × 10
= 100 cm²
Don't forget to work out the area of the base.
Total surface area = 316 + 100
= 416 cm²

Exercise 26G

Calculate

a the volume, and

b the total surface area of each of these pyramids.

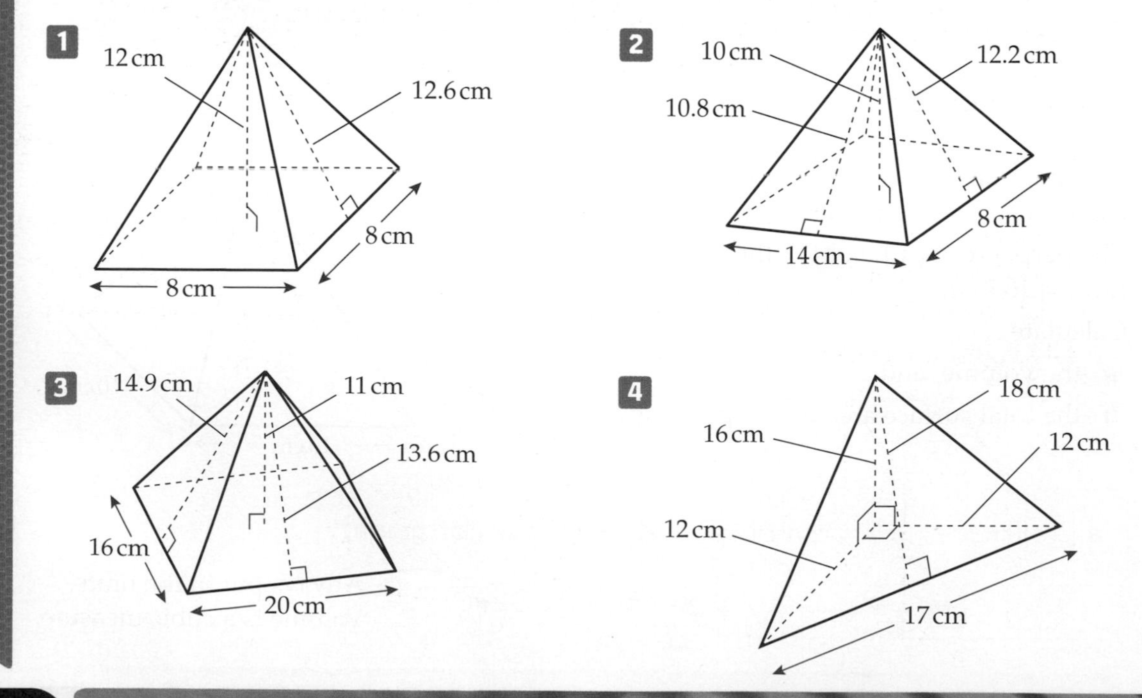

Review exercise

1 Calculate

i the area, and **[3 marks each]**

ii the perimeter of these compound shapes. **[2 marks each]**

All lengths are in centimetres.

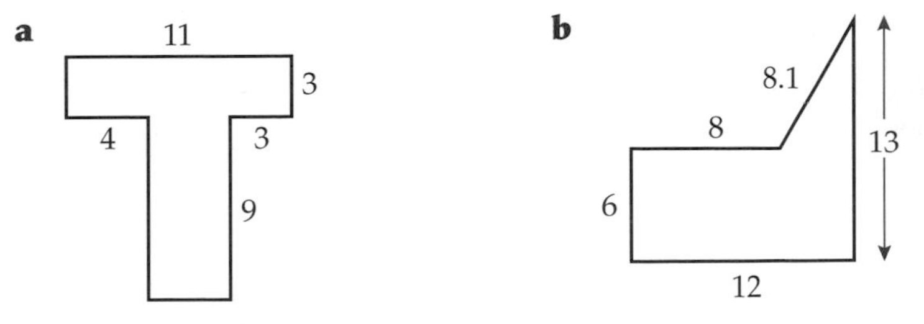

2 Copy this 3-D object on to squared paper. Show all its planes of symmetry. Draw a separate diagram for each plane of symmetry.

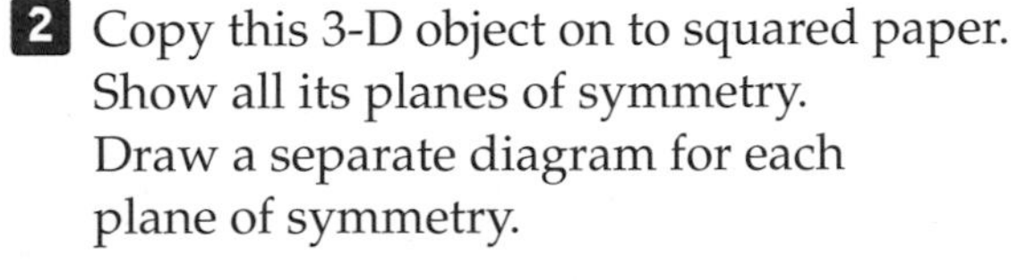

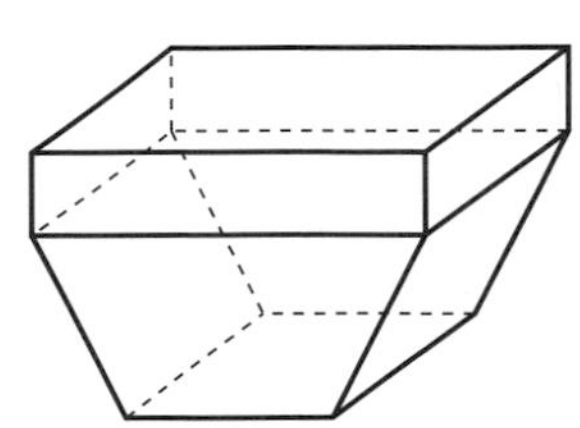

[2 marks]

3 For each block of cubes, draw

i the plan view **[1 mark each]**

ii the front elevation, and **[1 mark each]**

iii the side elevation (from the right-hand side). **[1 mark each]**

a

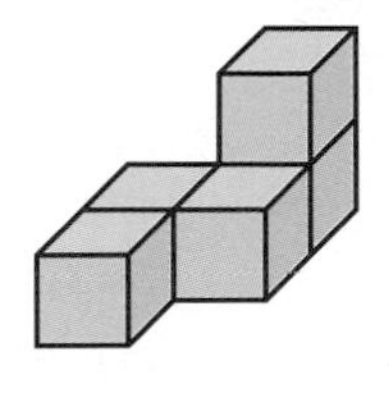

b

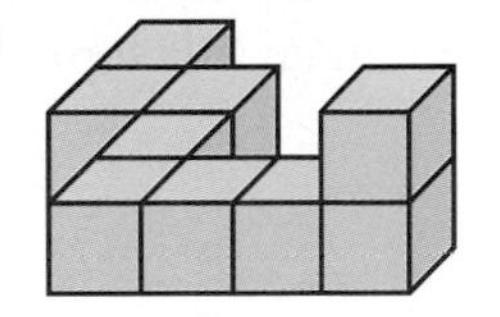

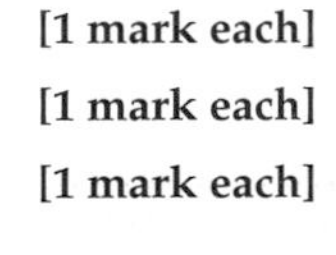

4 The diagram shows a shaded triangular shape made up from two right-angled triangles, one inside the other. The larger triangle has a base of 15.6 cm and a height of 12 cm. The smaller triangle has a base of 6.5 cm and a height of 5 cm.

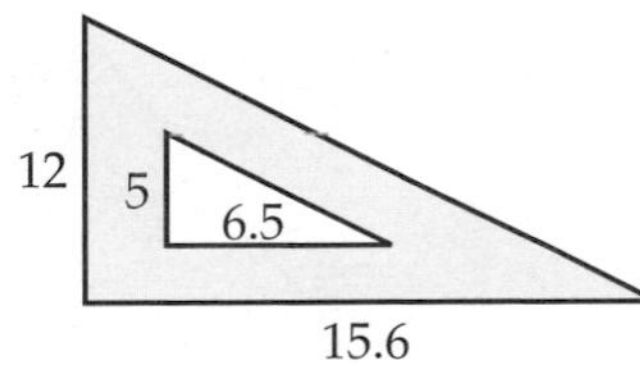

Calculate the shaded area. **[4 marks]**

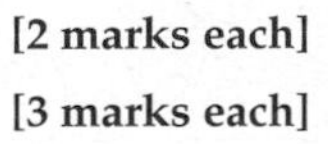

5 Calculate

i the volume, and **[2 marks each]**

ii the surface area of these 3-D objects. **[3 marks each]**

All lengths are in centimetres.

a

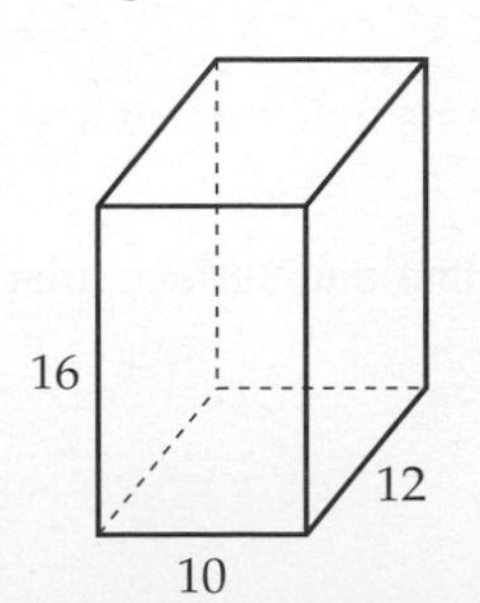

b

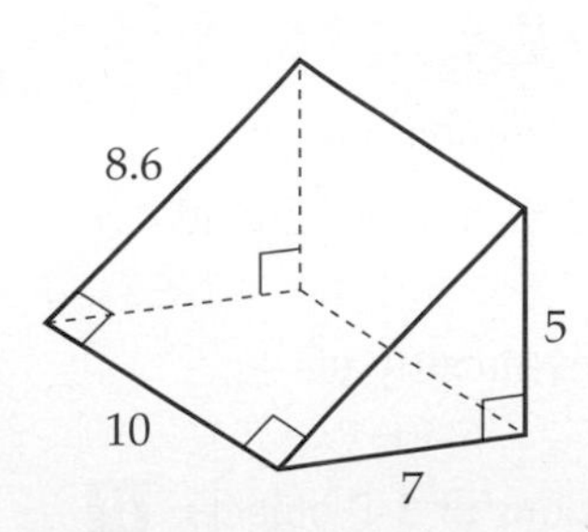

6 A garden shed has dimensions as shown in the diagram.
The total height of the shed is 4 metres.

a Calculate the area of the cross-section of the shed. **[3 marks]**

b Calculate the volume of the shed. **[2 marks]**

c Roofing felt is sold in rolls.
Each roll is 5 m long and 1 m wide. The price of a roll is £7.33.
What is the cheapest way to cover the roof of the shed?
How much will it cost? **[4 marks]**

7 A tent is in the shape of a triangular prism.
The two end faces are isosceles triangles of base 2.8 m and sides 3.8 m.
The two sides and the groundsheet are rectangles.
The length of the tent is 5 m and its perpendicular height is 3.5 m.

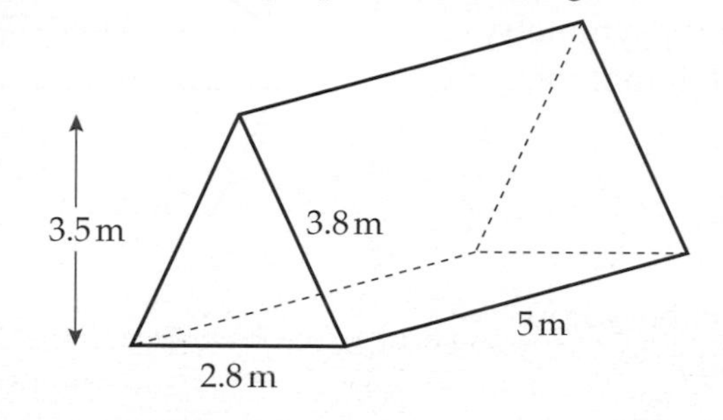

The price of tent material is £5.60 per square metre.
You can only buy the material in whole numbers of square metres.
How much would it cost to make this tent? **[6 marks]**

8 Calculate

i the volume, and **[2 marks each]**

ii the total surface area of these pyramids. **[3 marks each]**

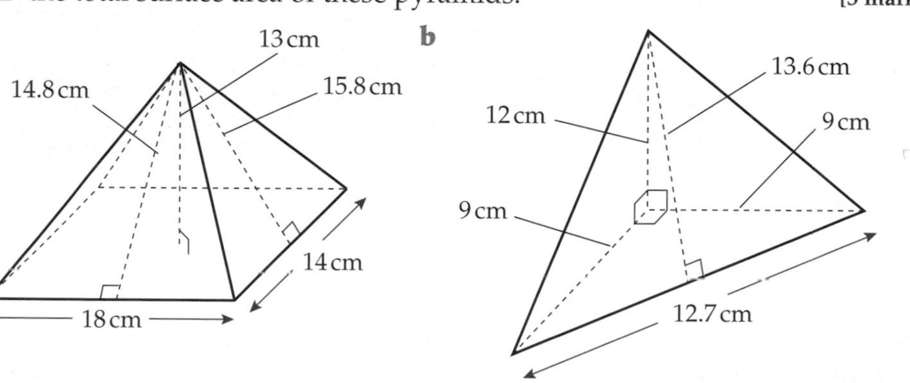

Chapter summary

In this chapter you have learned how to

- find the perimeter and area of compound shapes **D**
- identify the planes of symmetry of 3-D objects **D**
- draw plans and elevations of 3-D objects **D**
- find the volume and surface area of a prism **D** **C**
- find the volume and surface area of a pyramid **A**

27

Circles, cylinders, cones and spheres

BBC Video

This chapter is about circles, cylinders, cones and spheres.

A drinking straw is cylindrical and so is a tin of baked beans. What is the biggest example of a cylinder you can think of? How about the smallest?

Objectives

This chapter will show you how to

- find circumferences of circles and areas enclosed by circles, recalling relevant formulae D C
- calculate volumes of right prisms C
- solve problems involving surface areas and volumes of cylinders C
- calculate the lengths of arcs and the areas of sectors of circles A
- solve problems involving cones and spheres and more complex shapes and solids, including frustums of cones A A*

Before you start this chapter

1 Calculate the surface area of a cube with sides of length 3 cm.

2 A cuboid has a volume of 24 cm^3. Two of its dimensions measure 3 cm and 4 cm. What is the surface area of the cuboid?

3 A prism has a triangular base of area 5.5 cm^2. The height of the prism is 7 cm. What is its volume?

27.1 Area and circumference of a circle

Keywords
circumference, diameter, radius

Why learn this?
The distance travelled on a bicycle is calculated by multiplying the number of wheel rotations by the wheel circumference.

Objectives

- **D** Calculate the circumference of a circle
- **D** Calculate the area of a circle
- **C** Calculate the perimeter of compound shapes involving circles or parts of circles
- **C** Calculate the area of compound shapes involving circles or parts of circles

Skills check

Solve each equation to find the value of r.
Give your answer to two decimal places when appropriate.

a $2r = 17$ **b** $3.1r = 25$ **c** $\frac{r}{5} = 8$

d $2r - 5 = 60$ **e** $9.5 = 10 - r$ **f** $7r + 4 = 2r$

Calculating the circumference of a circle

The **circumference** of a circle is calculated by multiplying the **diameter** by π.

$C = \pi d$, where C = circumference and d = diameter.

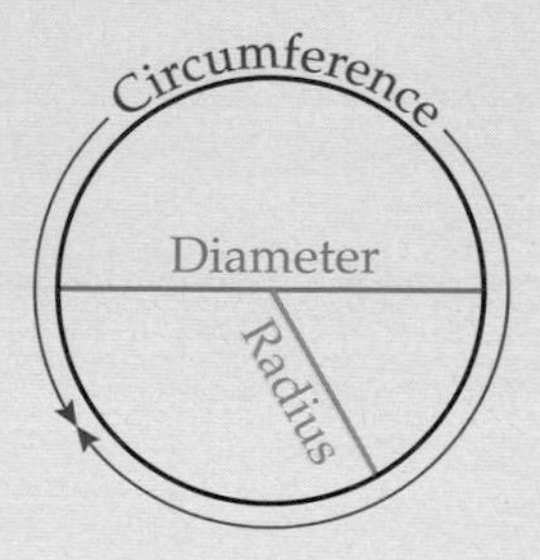

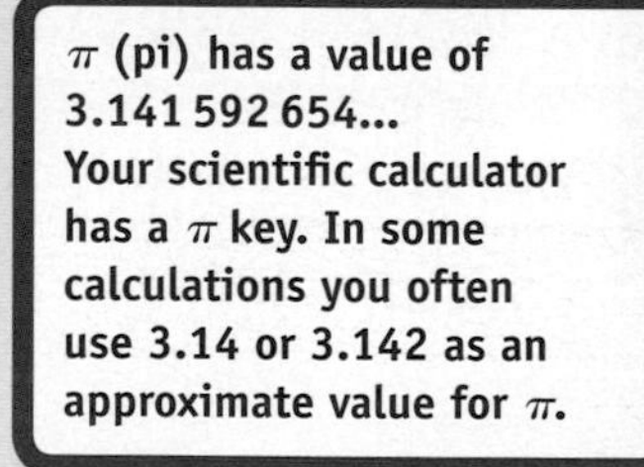
π (pi) has a value of 3.141 592 654...
Your scientific calculator has a π key. In some calculations you often use 3.14 or 3.142 as an approximate value for π.

The diameter is twice the length of the **radius**.

$d = 2r$, where r = radius.

So the formula can also be written as $C = \pi \times 2 \times r$

or $C = 2\pi r$

Example 1

A circular flower bed has a circumference of 18.8 m.
Calculate the diameter of the flower bed to one decimal place.

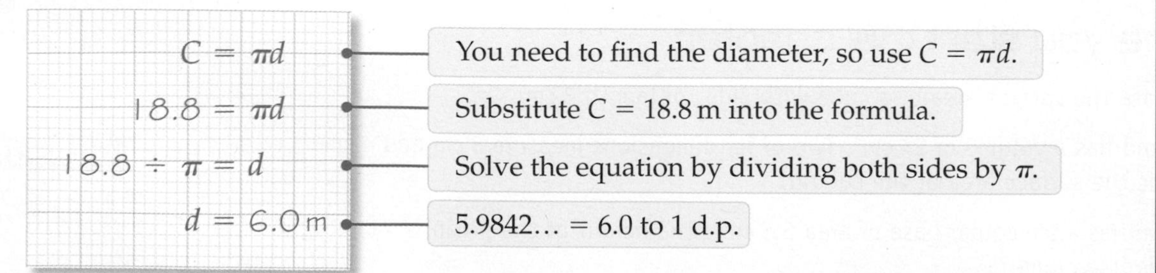

$C = \pi d$	You need to find the diameter, so use $C = \pi d$.
$18.8 = \pi d$	Substitute $C = 18.8$ m into the formula.
$18.8 \div \pi = d$	Solve the equation by dividing both sides by π.
$d = 6.0$ m	$5.9842... = 6.0$ to 1 d.p.

Example 2

Calculate the perimeter of the shape shown.
Leave your answer in terms of π.

7 cm, 8 cm

Curved length $= \frac{1}{2}\pi d$ — The curved length is half the perimeter of a circle.

$= \frac{1}{2} \times \pi \times 7$

$= 3.5\pi$

Total perimeter $= 8 + 7 + 8 + 3.5\pi$

$= 23 + 3.5\pi$ — The expression cannot be simplified further.

C

Exercise 27A

D A02

1 A circular box has a strip of ribbon glued round it.
There is a 2 cm overlap of ribbon.
What is the length of ribbon required if the diameter of the box is 7.5 cm?
Give your answer to the nearest millimetre.

D

2 Calculate the diameter of these circles.
Give your answers to three significant figures.

a Circumference = 3 m **b** Circumference = 25 cm

c Circumference = 12.5 mm **d** Circumference = 18.7 cm

3 The circumference of a circle is 8π cm. What is the radius?

FUNCTIONAL

D A03

4 A bicycle wheel with a diameter of 60 cm travels through 18 000 complete revolutions.
How many kilometres has the wheel travelled?
Give your answer to one decimal place.

5 Calculate the perimeter of these shapes.
Leave your answers in terms of π.

a

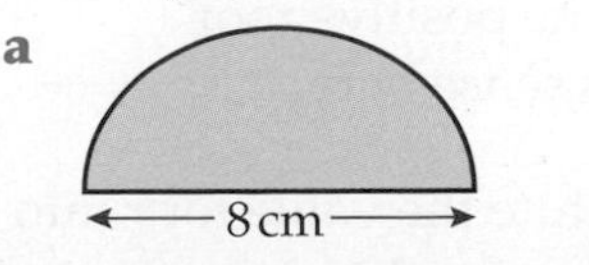

b

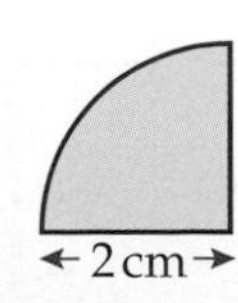

c

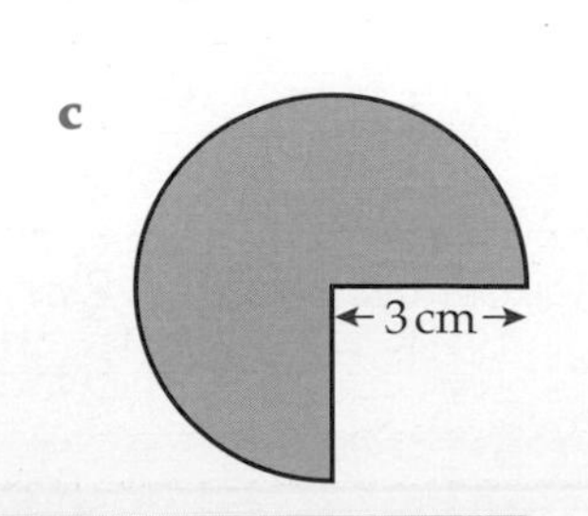

C

6 The diagram shows a running track.
Maria runs round the track three times.
How far has she run?
Give your answer to the nearest metre.

7 Here is a picture of a stained-glass window.

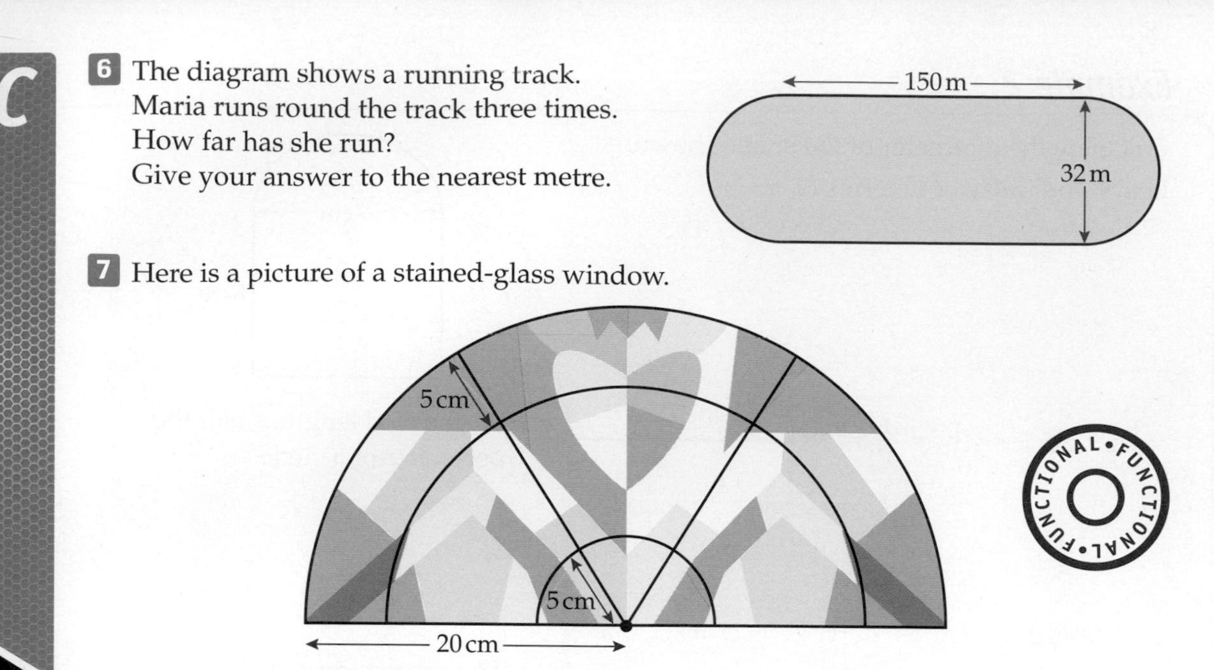

AO2

The lead that holds the glass in place is shown in black.
Calculate the length of lead to the nearest centimetre.

Calculating the area of a circle

The area of a circle is calculated using the formula

$A = \pi r^2$, where A = area and r = radius

Remember to use the order of operations. You must square the radius before multiplying by π.

D

Example 3

a Calculate the area of a circle of diameter 7 m.
Give your answer in terms of π.

b The area of a circle is 16π cm². What is its circumference?
Give your answer in terms of π.

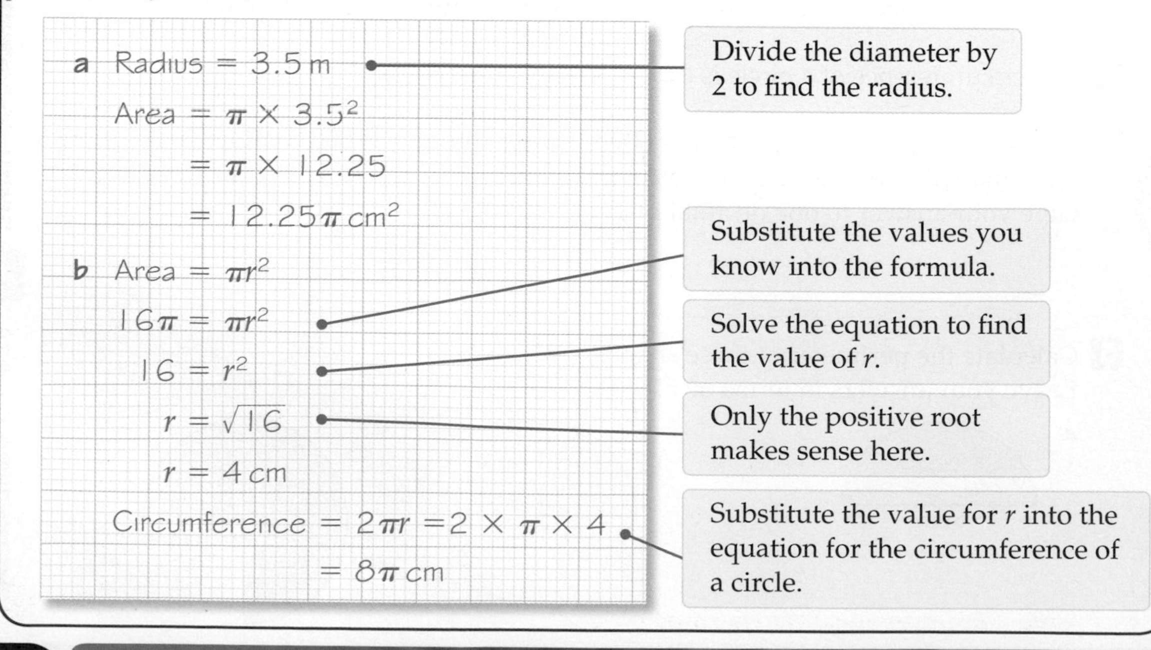

a Radius = 3.5 m
Area = π × 3.5²
= π × 12.25
= 12.25π cm²
b Area = πr²
16π = πr²
16 = r²
r = √16
r = 4 cm
Circumference = 2πr = 2 × π × 4
= 8π cm
Divide the diameter by 2 to find the radius.
Substitute the values you know into the formula.
Solve the equation to find the value of r.
Only the positive root makes sense here.
Substitute the value for r into the equation for the circumference of a circle.

Example 4

A circular photo frame has a wooden surround as shown.

Calculate the area of the wood to the nearest square centimetre.

FUNCTIONAL

wood

2 cm

9 cm

C

AO2

First establish the information you need to know.

Wooden area = area of whole frame − area of internal circle

Area of whole frame = $\pi \times 9^2 = 81\pi\text{ cm}^2$ — Leave π in your calculations until the end.

Radius of internal circle = 9 cm − 2 cm = 7 cm

Area of internal circle = $\pi \times 7^2 = 49\pi\text{ cm}^2$

Wooden area = $81\pi - 49\pi = 32\pi\text{ cm}^2$ — Now use the π key on your calculator.

$= 100.53\ldots\text{ cm}^2$ — The answer can now be rounded.

$= 101\text{ cm}^2$

Exercise 27B

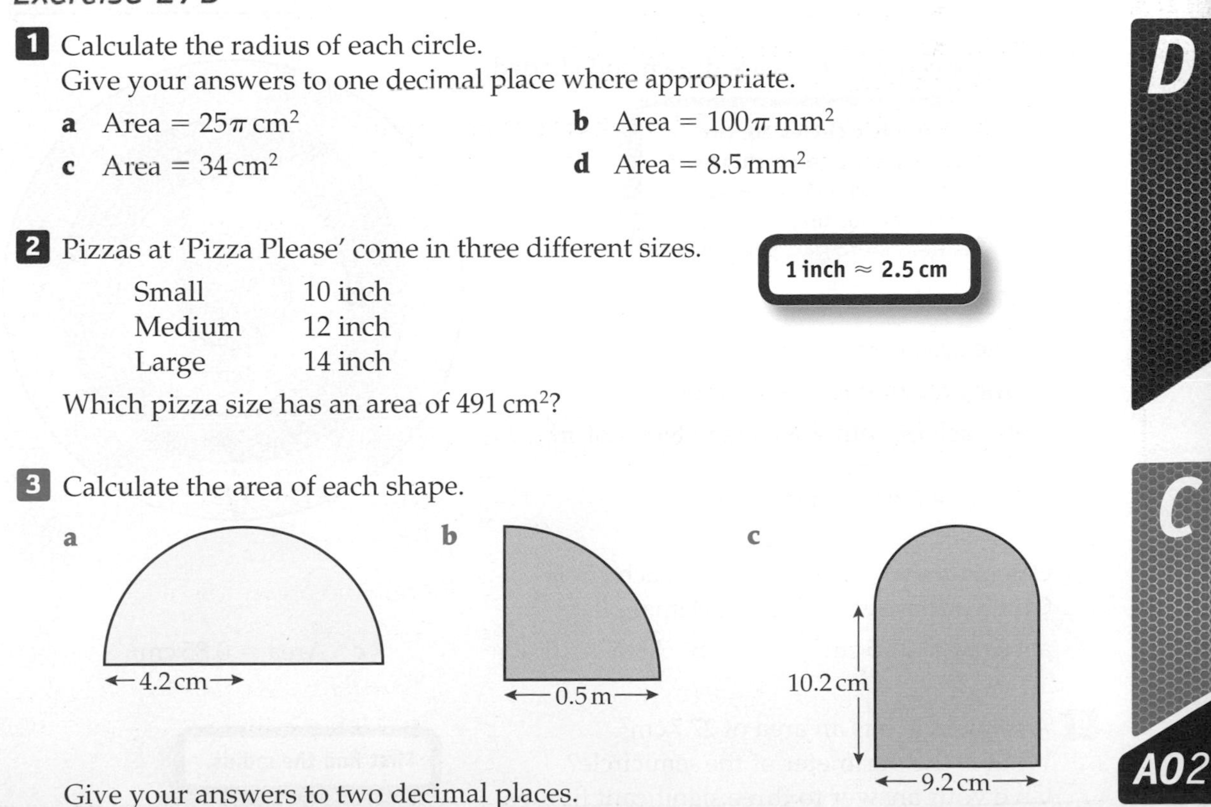

D

1 Calculate the radius of each circle.
Give your answers to one decimal place where appropriate.

a Area = $25\pi\text{ cm}^2$ **b** Area = $100\pi\text{ mm}^2$

c Area = 34 cm^2 **d** Area = 8.5 mm^2

2 Pizzas at 'Pizza Please' come in three different sizes.

1 inch ≈ 2.5 cm

Small	10 inch
Medium	12 inch
Large	14 inch

Which pizza size has an area of 491 cm^2?

C

AO2

3 Calculate the area of each shape.

Give your answers to two decimal places.

4 A circular flower bed has a diameter of 4 m.
A council wants to plant 100 bulbs per square metre.
How many bulbs will be needed for three identical flower beds?

C

5 Calculate the shaded area in each of these diagrams.

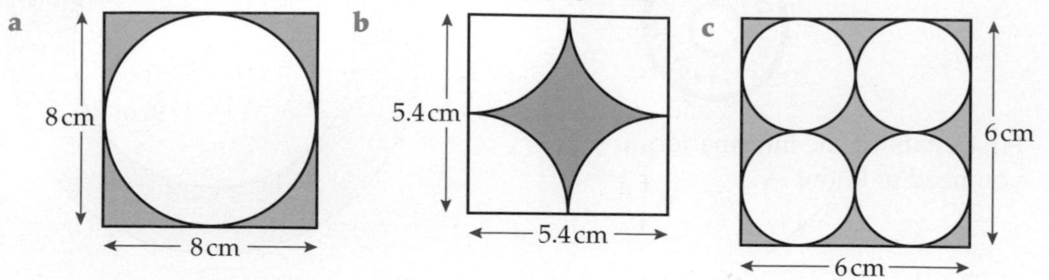

Give your answers to one decimal place.

6 Here is a logo for a sports company.
What area of the logo is blue?
Give your answer to three significant figures.

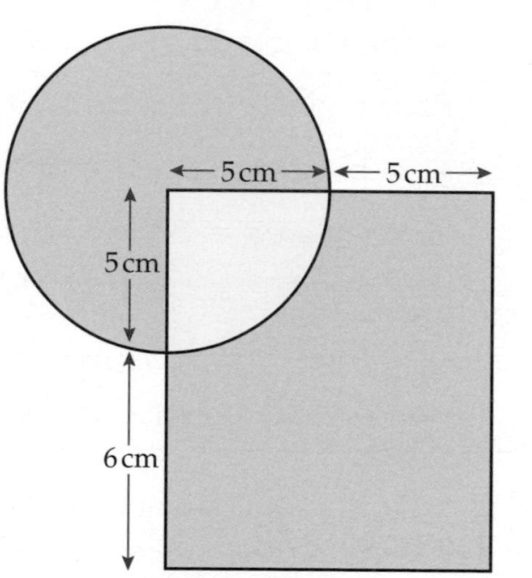

7 Four concentric circles are drawn and shaded.

Concentric circles are two or more circles which have been drawn using the same position for their centres.

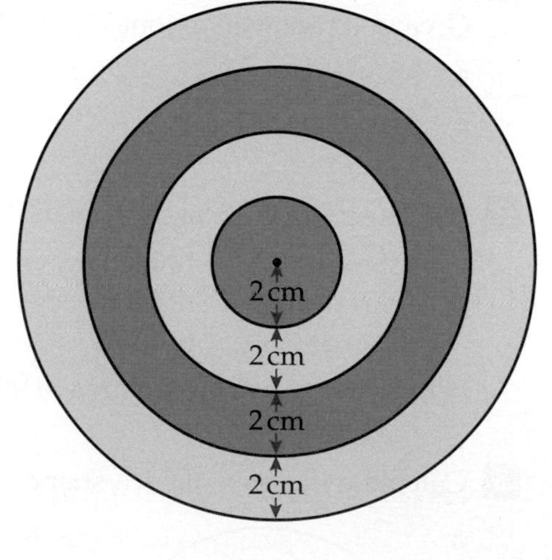

Calculate

a the area that is shaded red

b the area that is shaded blue.

Give each of your answers in terms of π.

8 Calculate the circumference of each circle.
Give your answers to one decimal place.

a Area = $38.5\,\text{cm}^2$ **b** Area = $100\,\text{cm}^2$ **c** Area = $0.85\,\text{cm}^2$

AO3

9 A semicircle has an area of $27.7\,\text{cm}^2$.
What is the perimeter of the semicircle?
Give your answer to three significant figures.

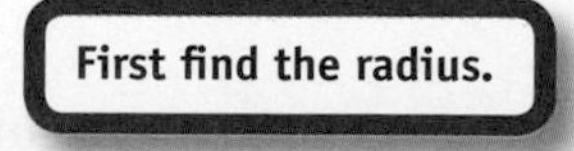

27.2 Cylinders

Keywords
cylinder, total surface area, area of curved surface

Why learn this?
Some musical instruments use different sized cylinders to create different notes.

Objectives
C Calculate the volume of a cylinder
C Solve problems involving the surface area of cylinders

Skills check

HELP Section 27.1

1 Calculate the area and circumference of these circles.
 a Radius = 5.5 cm b Diameter = 12 cm
 Give your answers in terms of π.
2 The area of a circle is 81π cm². What is the radius?
3 Calculate the radius of a circle whose circumference is 24π cm.

Volume and surface area of cylinders

A **cylinder** is a prism with a circular cross-section.

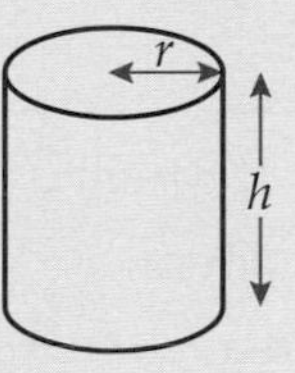

To calculate the volume of a cylinder, multiply the area of the circular face by the height.

Area of circle = πr^2

$V = \pi r^2 h$, where V = volume, r = radius and h = height.

The surface area of a solid is calculated by finding the total area of all the faces.

The **total surface area** of a cylinder = area of the curved surface + area of ends

To work out the **area of the curved surface** of a cylinder, consider its net.

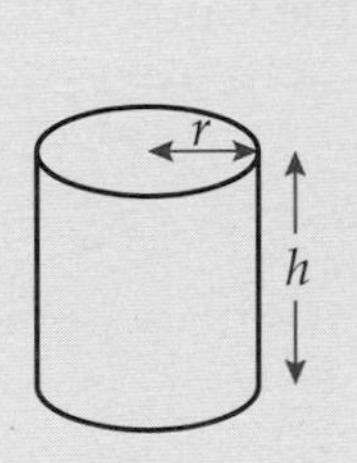

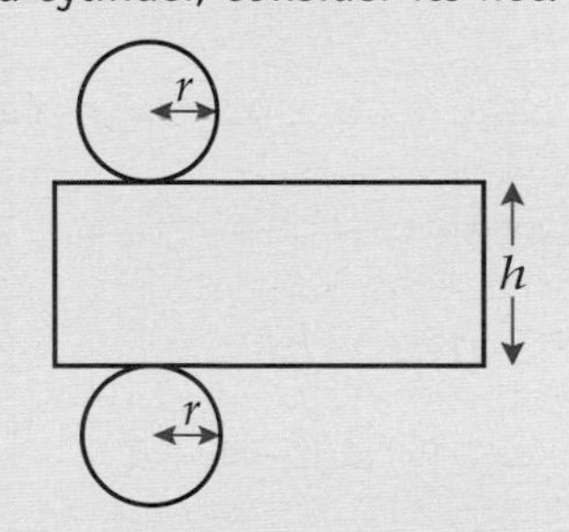

The length of the rectangle is the same as the circumference of the circular top.

Area of curved surface $= 2\pi r \times h$
$= 2\pi rh$

The formula for the total surface area of a cylinder is not given on the exam paper.

Area of each circular end $= \pi r^2$

So the total surface area $= 2\pi rh + 2\pi r^2$

There are two circular ends.

$= 2\pi r(h + r)$

$2\pi r$ is a common factor of the two terms.

C

Example 5

A tin can is 13 cm high and has a radius of 5.2 cm.

Calculate how many litres of liquid it can hold to one decimal place.

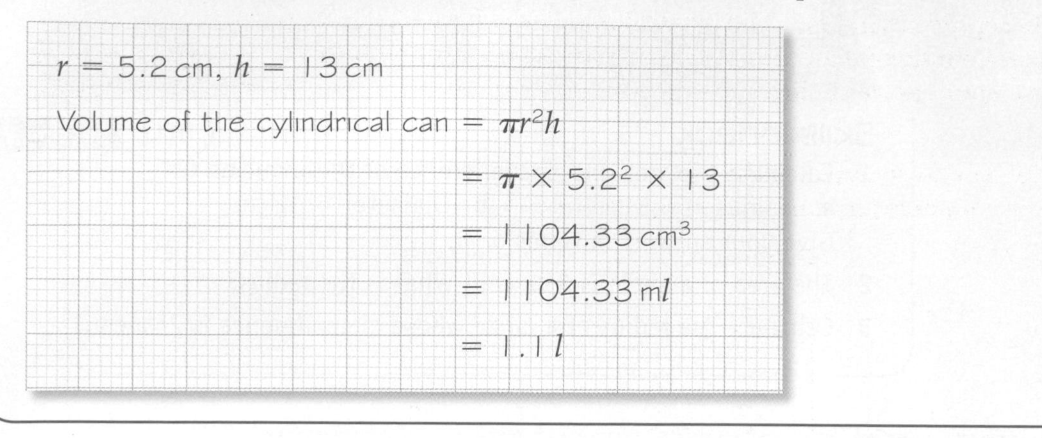

$r = 5.2$ cm, $h = 13$ cm
Volume of the cylindrical can $= \pi r^2 h$
$= \pi \times 5.2^2 \times 13$
$= 1104.33$ cm^3
$= 1104.33$ ml
$= 1.1\,l$

Example 6

AO3

A cylinder has a volume of 108π cm^3.

The radius of the cylinder is 3 cm.

Calculate its total surface area.

Leave your answer in terms of π.

First calculate the height.

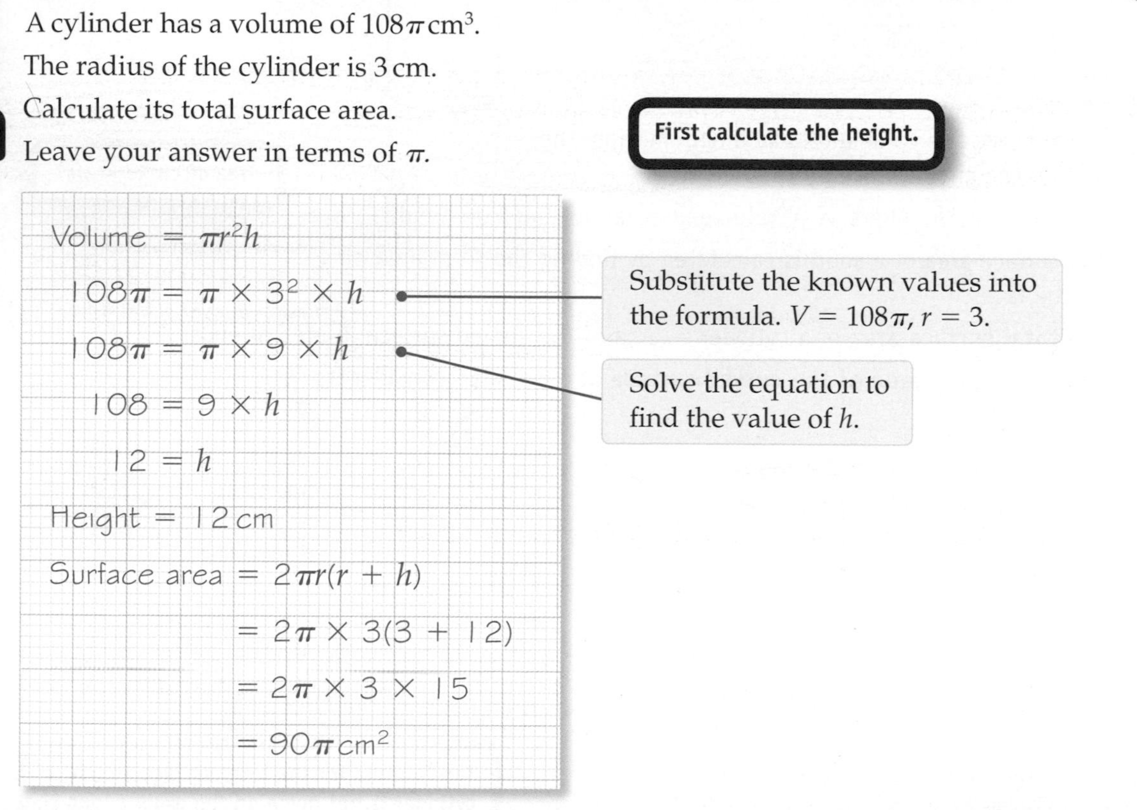

Volume $= \pi r^2 h$
$108\pi = \pi \times 3^2 \times h$
$108\pi = \pi \times 9 \times h$
$108 = 9 \times h$
$12 = h$
Height $= 12$ cm
Surface area $= 2\pi r(r + h)$
$= 2\pi \times 3(3 + 12)$
$= 2\pi \times 3 \times 15$
$= 90\pi$ cm^2
Substitute the known values into the formula. $V = 108\pi$, $r = 3$.
Solve the equation to find the value of h.

Exercise 27C

1 Calculate the volume and surface area of these cylinders.

a Radius = 5 cm, height = 23.4 cm

b Radius = 5 mm, height = 1.5 cm

c Diameter = 3 cm, height = 1 m

d Diameter = 30 mm, height = 45 cm

Give your answers to one decimal place.

Before carrying out the calculation make sure the units are the same.

2 A cylindrical drink can has a radius of 2.5 cm and a height of 11 cm.
Calculate the volume of the can.
Give your answer to the nearest millilitre.

> $1\text{ cm}^3 = 1\text{ m}l$

C

3 A hosepipe is 5 m long. The internal diameter of the pipe is 2 cm.
What volume of water is held in the pipe when it is half full?
Give your answer in litres to one decimal place.

> $1000\text{ cm}^3 = 1\,l$

4 Which of these two cylinders has the larger capacity and by how much?
Give your answer to the nearest millilitre.

C

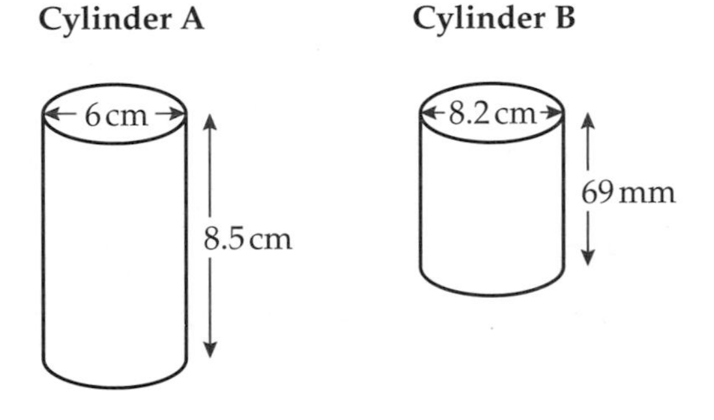

5 A cylindrical water chute at a swimming pool is a quarter filled with water.
The chute is 50 m in length and 1 m in diameter.
What volume of water is in the chute?
Give your answer to the nearest litre.

AO2

6 A circular pond has a radius of 1.2 m. It is 75 cm deep.
Goldfish need approximately 56 l of water each.
What is the maximum number of goldfish the pond can support?

> $1\text{ litre} = 1000\text{ cm}^3$

C

7 A tin of beans has a label around it.
The tin is 11 cm tall and has a radius of 4 cm.
The label overlaps by 1 cm.
Calculate the area of paper required to make the label.
Give your answer to three significant figures.

8 A cylinder has a volume of $18\pi\text{ cm}^3$.
The radius of the cylinder is 3 cm.
Calculate the surface area of the cylinder. Leave your answer in terms of π.

> First calculate the height.

9 The volume of a cylinder is 100 cm^3 and the height is 10 cm.
What is the surface area of the cylinder to one decimal place?

> First calculate the radius.

10 A sheet of cardboard has dimensions 150 cm by 100 cm.
What is the maximum number of cylindrical cardboard tubes of radius 3 cm and height 12 cm that can be made from it?

AO3

27.3 Sectors and arcs

Keywords
sector, arc

Why learn this?
Golfers measure the arc length of their swing when analysing their game.

Objectives
A Calculate the lengths of arcs
A Calculate the area of sectors

Skills check

1 Calculate the circumference of a circle in terms of π when
 a $r = 4$ cm b $d = 12.5$ cm c $r = 25$ cm

2 Calculate the area of a circle in terms of π when
 a $r = 12$ cm b $d = 3.5$ cm c $r = 0.7$ cm

Sectors and arcs

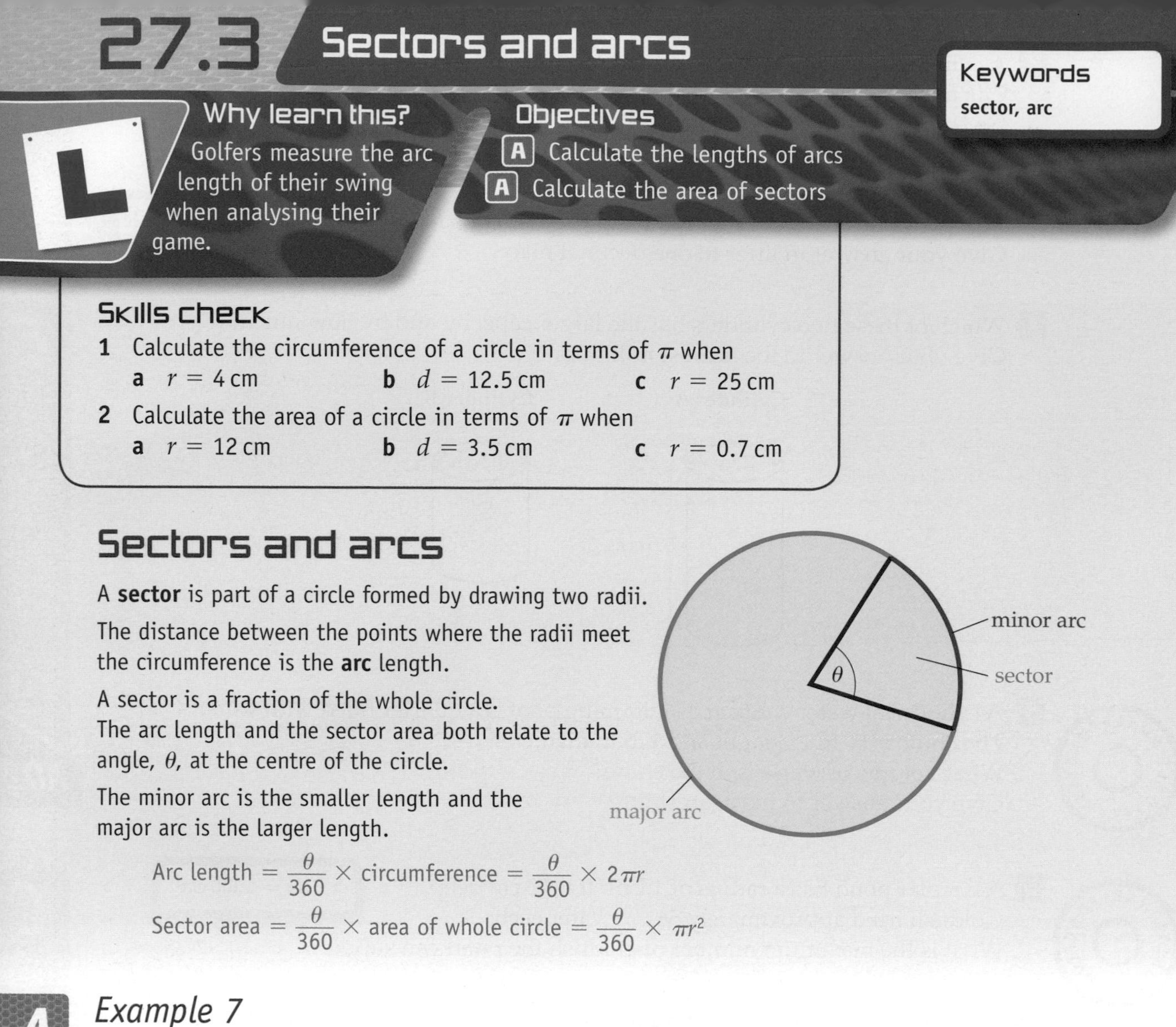

A **sector** is part of a circle formed by drawing two radii.

The distance between the points where the radii meet the circumference is the **arc** length.

A sector is a fraction of the whole circle.
The arc length and the sector area both relate to the angle, θ, at the centre of the circle.

The minor arc is the smaller length and the major arc is the larger length.

$$\text{Arc length} = \frac{\theta}{360} \times \text{circumference} = \frac{\theta}{360} \times 2\pi r$$

$$\text{Sector area} = \frac{\theta}{360} \times \text{area of whole circle} = \frac{\theta}{360} \times \pi r^2$$

A

Example 7

Calculate
a the perimeter, and
b the area
of the sector shown.
Leave your answers in terms of π.

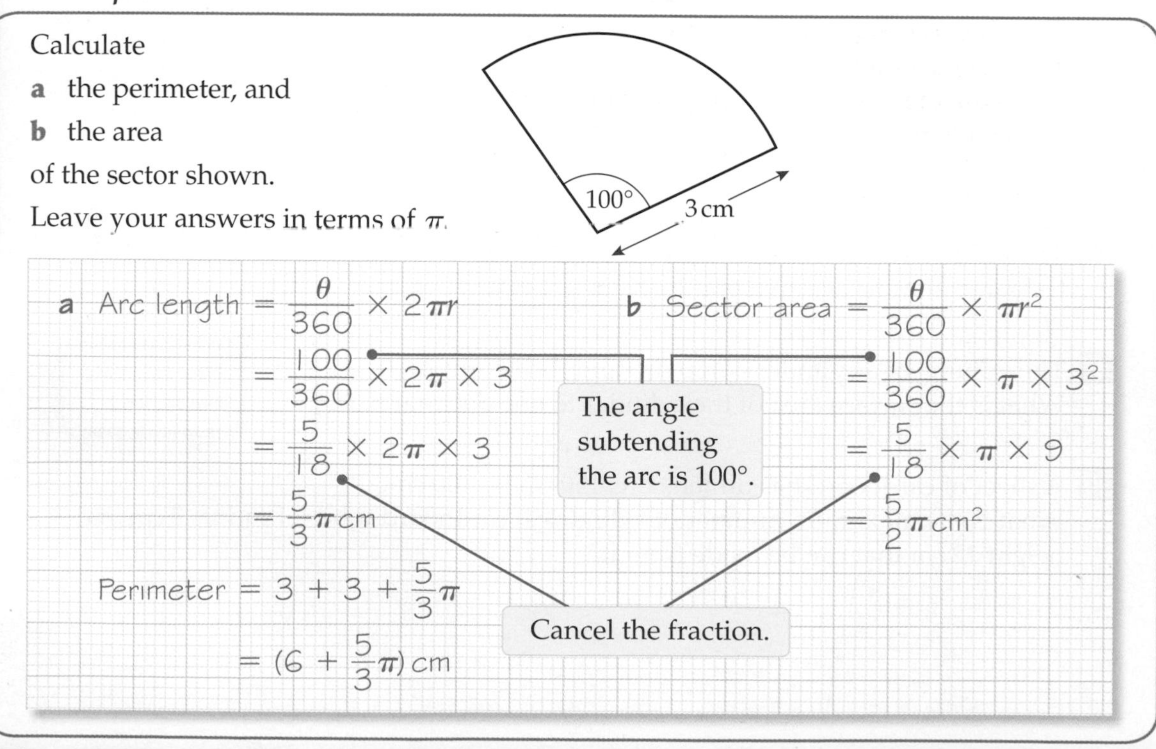

a Arc length $= \frac{\theta}{360} \times 2\pi r$
$= \frac{100}{360} \times 2\pi \times 3$
$= \frac{5}{18} \times 2\pi \times 3$
$= \frac{5}{3}\pi$ cm

Perimeter $= 3 + 3 + \frac{5}{3}\pi$
$= (6 + \frac{5}{3}\pi)$ cm

b Sector area $= \frac{\theta}{360} \times \pi r^2$
$= \frac{100}{360} \times \pi \times 3^2$
$= \frac{5}{18} \times \pi \times 9$
$= \frac{5}{2}\pi$ cm^2

Example 8

The area of the sector shown is $4.5\pi\,\text{cm}^2$.
Calculate the perimeter of the sector.
Give your answer to one decimal place.

45°

$$\text{Area of sector} = \frac{\theta}{360} \times \pi r^2$$

$$4.5\pi = \frac{45}{360} \times \pi r^2$$

Substitute the values you know into the formula. Area = 4.5π, $\theta = 45°$.

$$= \frac{1}{8} \times \pi r^2$$

Solve the equation to find r.

$$36\pi = \pi r^2$$

$$36 = r^2$$

$$r = \sqrt{36}$$

$$r = 6\,\text{cm}$$

$$\text{Arc length} = \frac{\theta}{360} \times 2\pi r$$

Now find the length of the arc.

$$= \frac{45}{360} \times 2\pi \times 6$$

$$= \frac{1}{8} \times 2\pi \times 6$$

$$= 4.7\,\text{cm}$$

$$\text{Perimeter} = 6 + 6 + 4.7 = 16.7\,\text{cm}$$

Finally, find the perimeter.

Exercise 27D

1 Calculate the arc length and area of each of these sectors.
Leave your answers in terms of π.

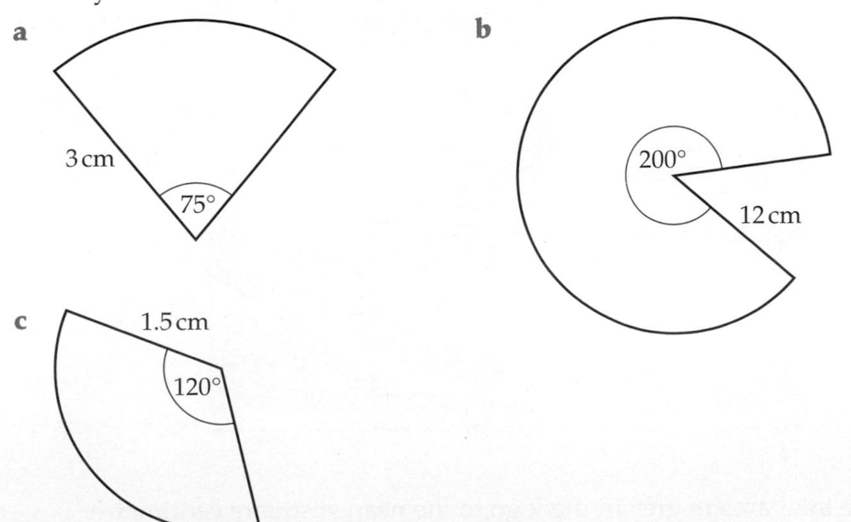

A

2 Here is a diagram of a flower bed.
Neela wants to fill the flower bed with daffodil bulbs.
She should plant 50 bulbs per square metre.
How many bulbs are needed?
Give your answer to the nearest bulb.

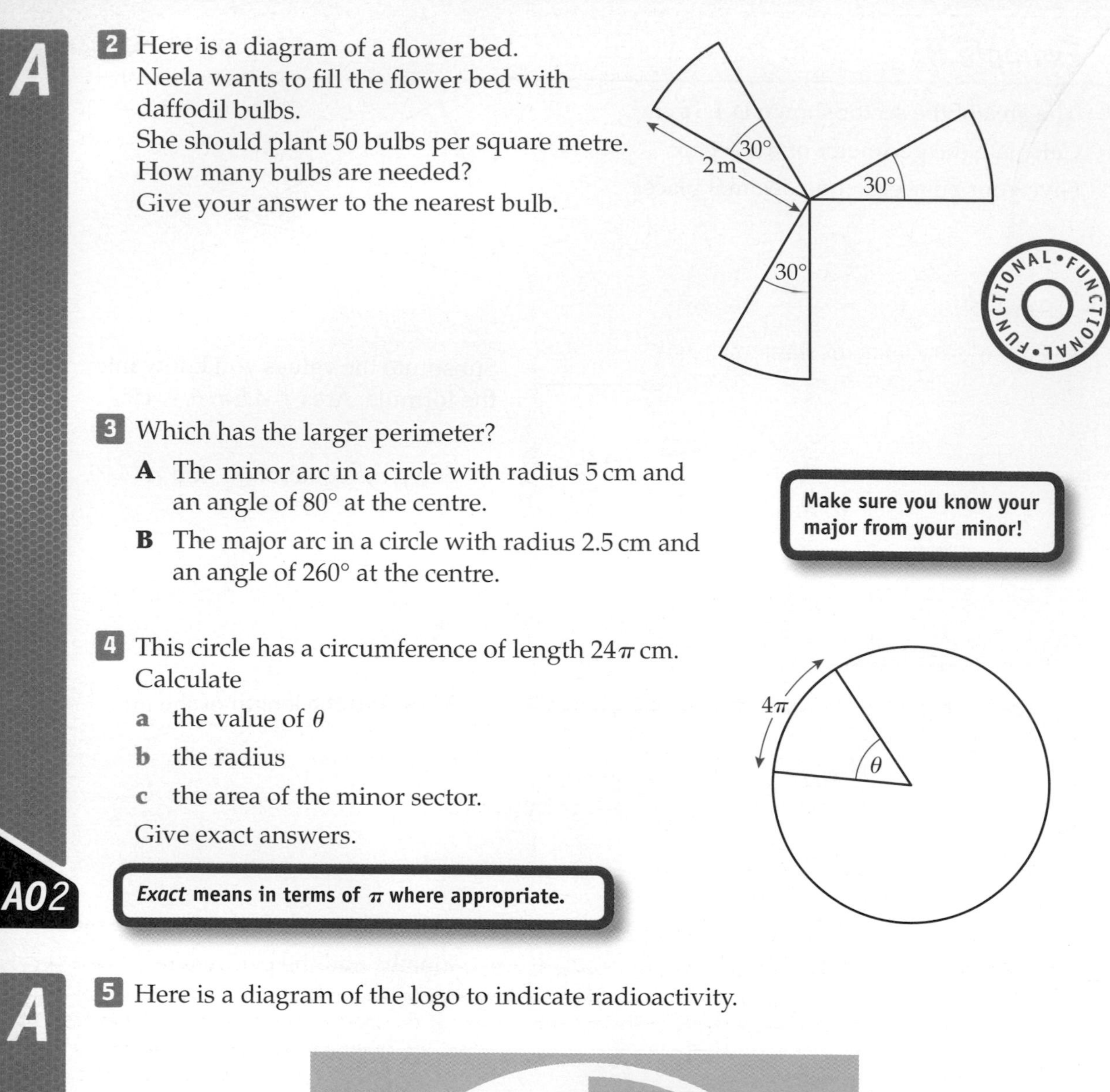

3 Which has the larger perimeter?

A The minor arc in a circle with radius 5 cm and an angle of 80° at the centre.

B The major arc in a circle with radius 2.5 cm and an angle of 260° at the centre.

Make sure you know your major from your minor!

4 This circle has a circumference of length 24π cm.
Calculate

a the value of θ

b the radius

c the area of the minor sector.

Give exact answers.

4π

θ

AO2

***Exact* means in terms of π where appropriate.**

A

5 Here is a diagram of the logo to indicate radioactivity.

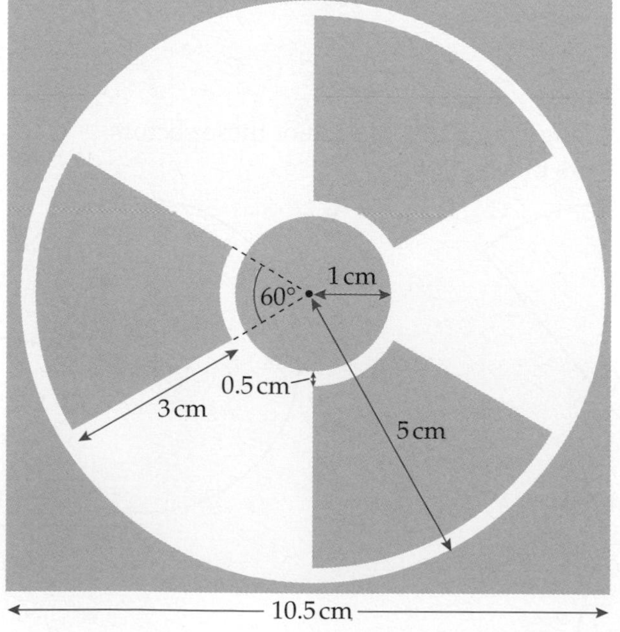

AO3

Calculate the total area in grey in the logo to the nearest square centimetre.

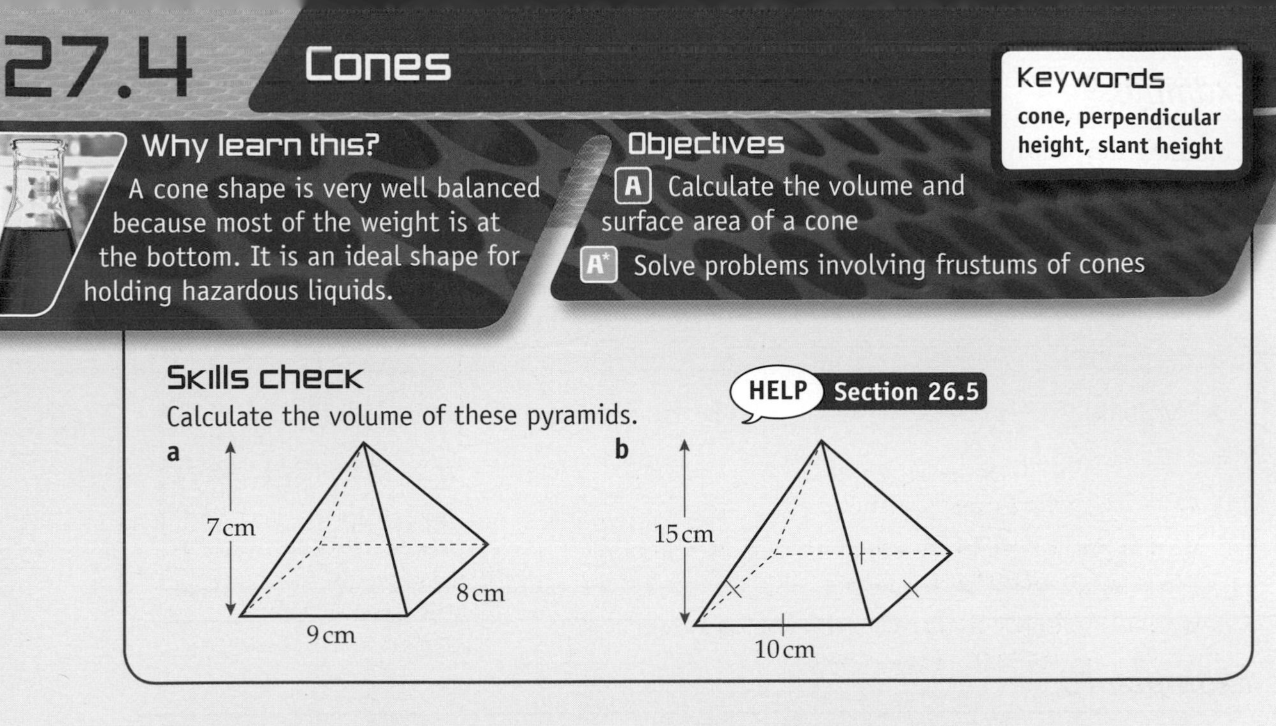

27.4 Cones

Keywords
cone, perpendicular height, slant height

Why learn this?

A cone shape is very well balanced because most of the weight is at the bottom. It is an ideal shape for holding hazardous liquids.

Objectives

A Calculate the volume and surface area of a cone

A* Solve problems involving frustums of cones

Skills check

HELP Section 26.5

Calculate the volume of these pyramids.

Volume and surface area of a cone

A **cone** is a three-dimensional shape with a circular base which tapers to a point (apex).

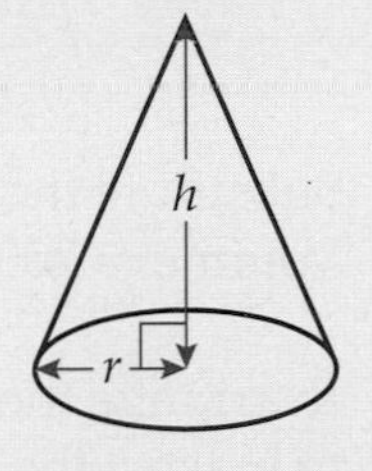

The volume of a cone is calculated using the formula

Volume $= \frac{1}{3} \times$ area of base $\times$ **perpendicular height**

For a cone of base radius r and perpendicular height h, volume $= \frac{1}{3}\pi r^2 h$.

> This formula for the volume of a cone is the same as the formula for the volume of a pyramid. To recap work on the volume of pyramids, see Section 26.5.

> This formula is given on the exam paper.

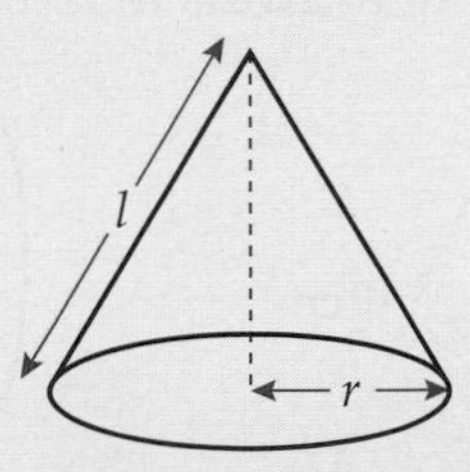

The area of the curved surface of a cone is calculated using the formula

Area of curved surface $= \pi \times$ radius of base $\times$ **slant height**

$= \pi r l$

> This formula is given on the exam paper.

where r is the radius of the base of the cone and l is the slant height of the cone.

The total surface area of a cone = area of base + area of curved surface

$= \pi r^2 + \pi r l$

$= \pi r(r + l)$

> There is more work on cones in Chapter 32.

Example 9

Calculate the volume of this cone.

a volume **b** surface area of this cone.

Leave your answers in terms of π.

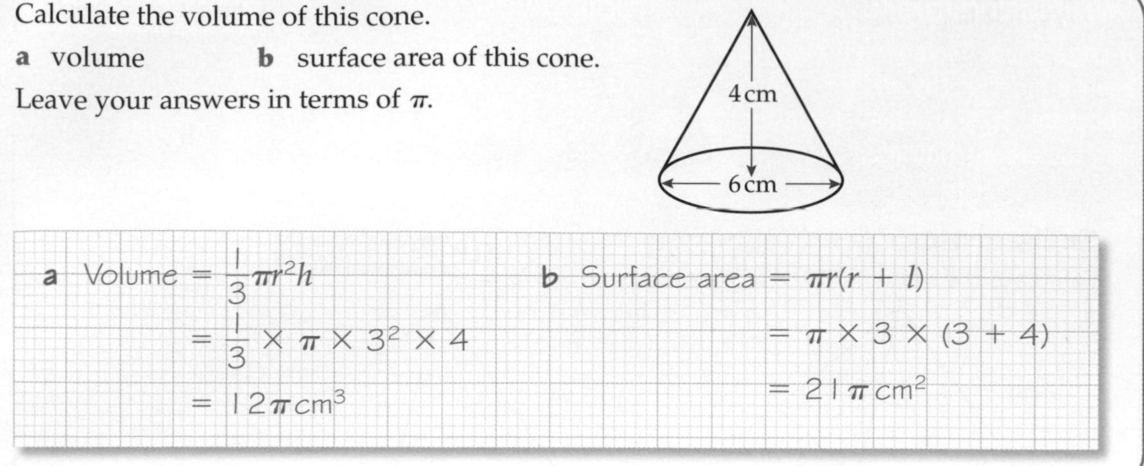

a Volume $= \frac{1}{3}\pi r^2 h$

$= \frac{1}{3} \times \pi \times 3^2 \times 4$

$= 12\pi \text{ cm}^3$

b Surface area $= \pi r(r + l)$

$= \pi \times 3 \times (3 + 4)$

$= 21\pi \text{ cm}^2$

Example 10

A cone has a volume of 200 cm^3. The radius of the base is 4 cm.

Calculate the perpendicular height of the cone.

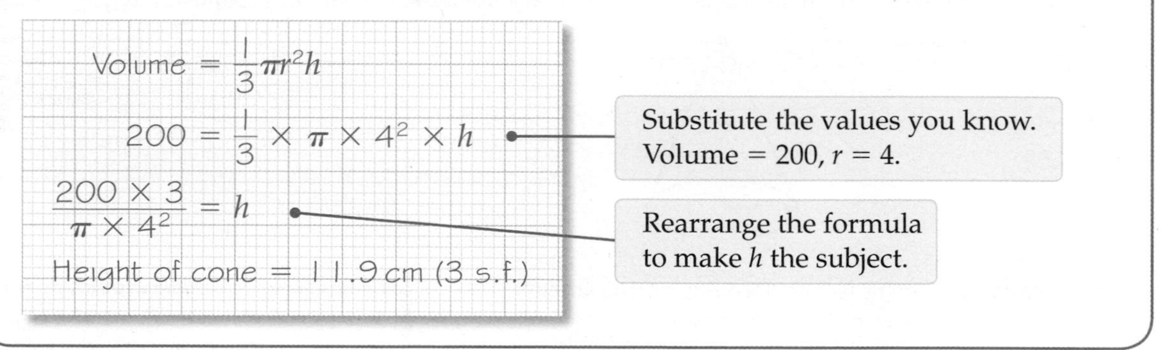

Volume $= \frac{1}{3}\pi r^2 h$

$200 = \frac{1}{3} \times \pi \times 4^2 \times h$ — Substitute the values you know. Volume = 200, $r = 4$.

$\frac{200 \times 3}{\pi \times 4^2} = h$ — Rearrange the formula to make h the subject.

Height of cone = 11.9 cm (3 s.f.)

Exercise 27E

1 Calculate the volume of each cone. Leave your answers in terms of π.

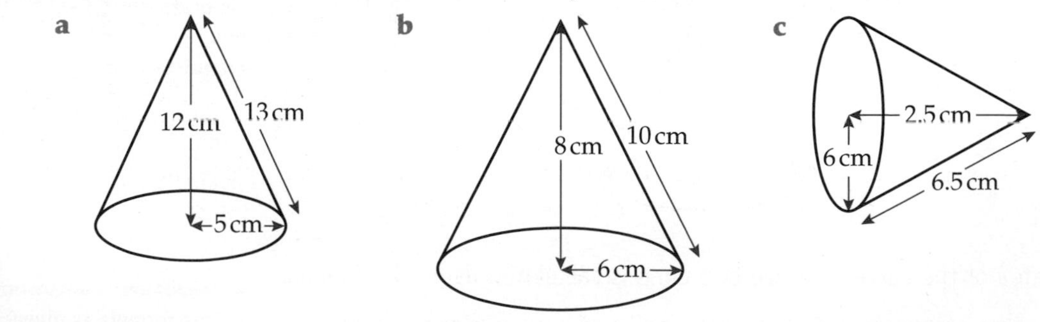

2 Calculate the surface area of each cone in Q1. Leave your answers in terms of π.

3 The diagram shows a paper cup used in a water cooler.
What area of paper is used to make the cup?
Give your answer to three significant figures.

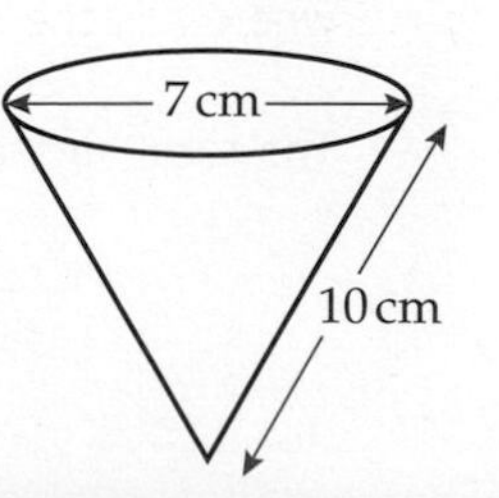

4 A metal cylinder with length 4 cm and radius 2 cm is melted down.
The metal is used to make a cone with perpendicular height 3 cm.
What is the radius of the cone?

5 The cone in the diagram has 8 cm cut off its top as shown.
The shape which is left is called a frustum.
Calculate the volume of the frustum.
Give your answer to one decimal place.

Find the volume of the original cone, then find the volume of the 'missing' cone and subtract it.

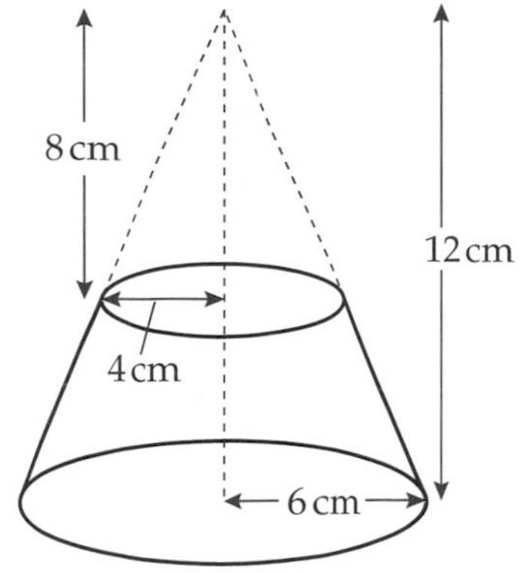

6 Calculate the volume of this frustum to three significant figures.

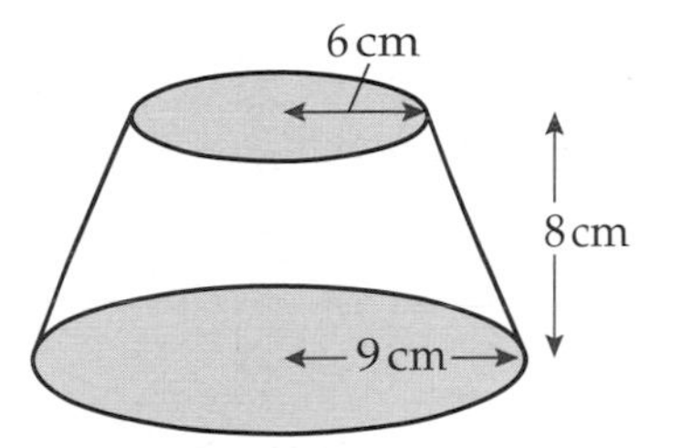

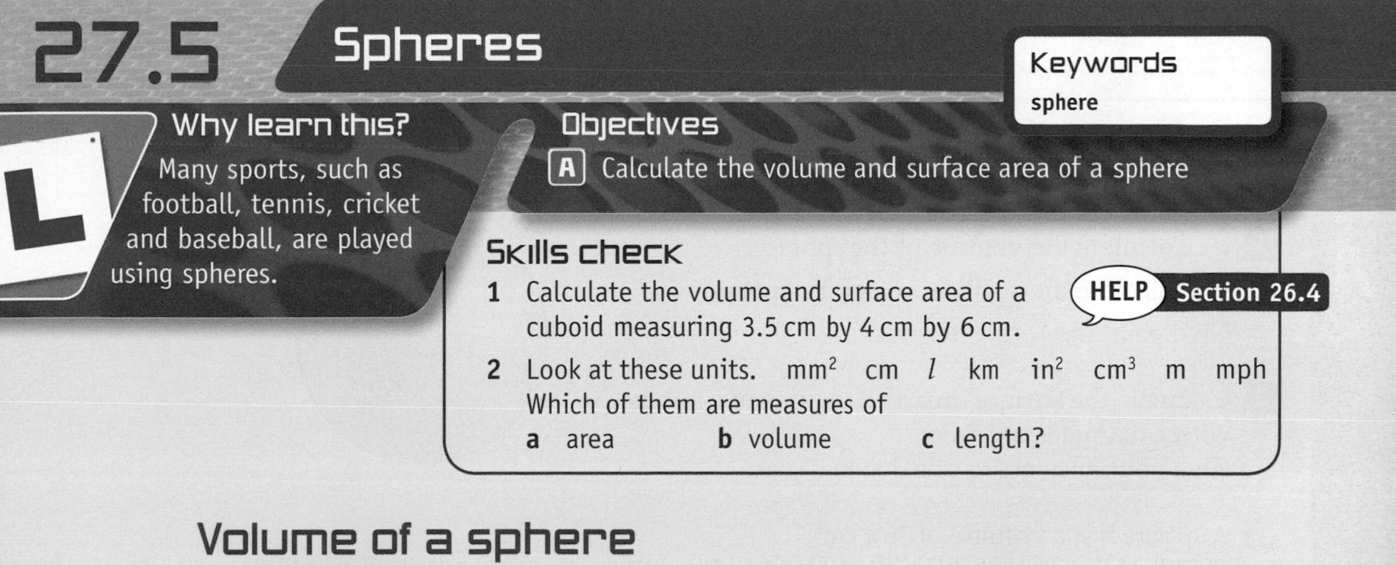

27.5 Spheres

Keywords
sphere

Why learn this?
Many sports, such as football, tennis, cricket and baseball, are played using spheres.

Objectives

A Calculate the volume and surface area of a sphere

Skills check

1. Calculate the volume and surface area of a cuboid measuring 3.5 cm by 4 cm by 6 cm.

HELP Section 26.4

2. Look at these units. mm² cm l km in² cm³ m mph
   Which of them are measures of
   a area **b** volume **c** length?

Volume of a sphere

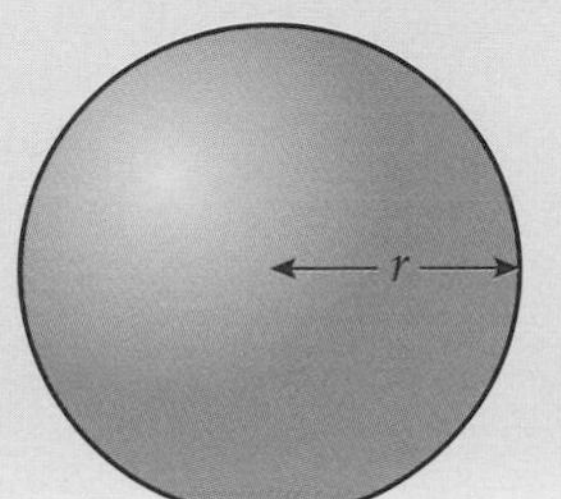

The volume of a **sphere** of radius r is calculated using the formula

$$\text{Volume} = \tfrac{4}{3}\pi r^3$$

The surface area of a sphere of radius r is calculated using the formula

$$\text{Surface area} = 4\pi r^2$$

These formulae are given on the exam paper.

Example 11

A sphere has a diameter of 12 cm.

Calculate its volume and surface area. Leave your answers in terms of π.

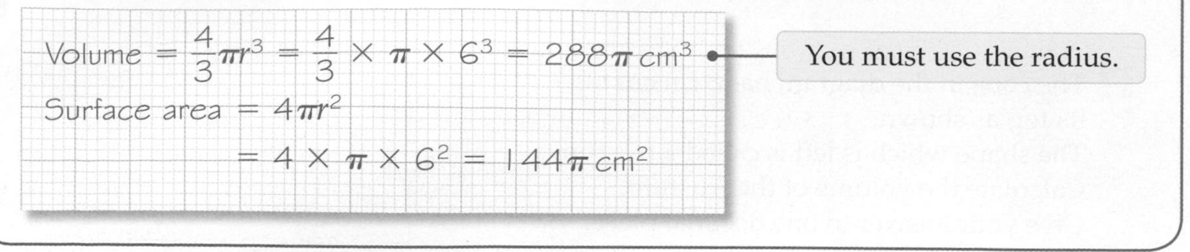

$$\text{Volume} = \frac{4}{3}\pi r^3 = \frac{4}{3} \times \pi \times 6^3 = 288\pi \text{ cm}^3$$

$$\text{Surface area} = 4\pi r^2$$

$$= 4 \times \pi \times 6^2 = 144\pi \text{ cm}^2$$

You must use the radius.

Example 12

A sphere has a surface area of 196π cm^2.

Calculate its volume to the nearest whole number.

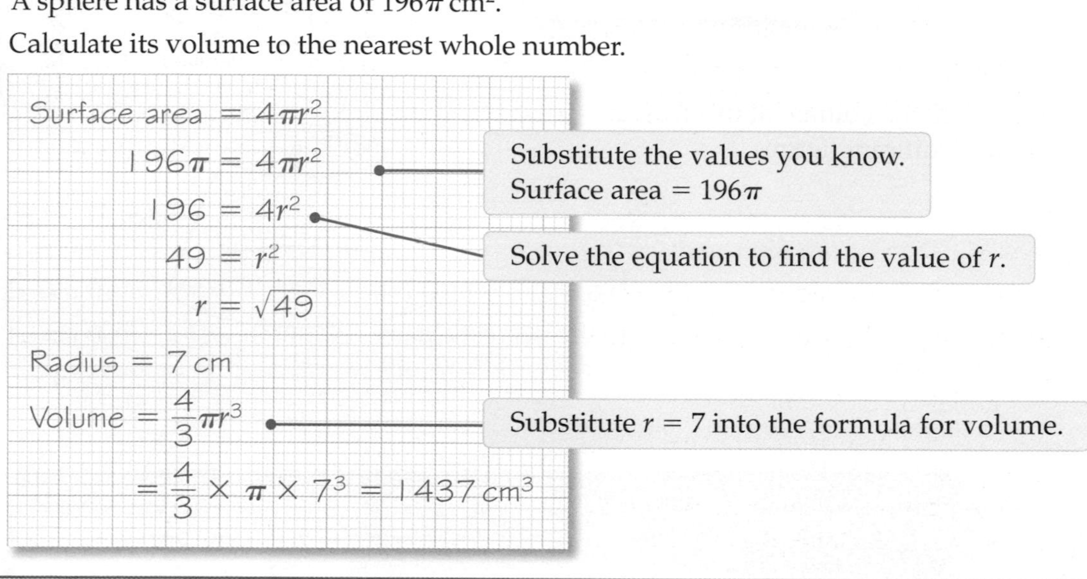

$$\text{Surface area} = 4\pi r^2$$

$$196\pi = 4\pi r^2$$

$$196 = 4r^2$$

$$49 = r^2$$

$$r = \sqrt{49}$$

Radius = 7 cm

$$\text{Volume} = \frac{4}{3}\pi r^3$$

$$= \frac{4}{3} \times \pi \times 7^3 = 1437 \text{ cm}^3$$

Substitute the values you know. Surface area = 196π

Solve the equation to find the value of r.

Substitute $r = 7$ into the formula for volume.

Exercise 27F

1 **a** Calculate the volume of the sphere.

b Calculate the surface area of the sphere.

Give your answers in terms of π.

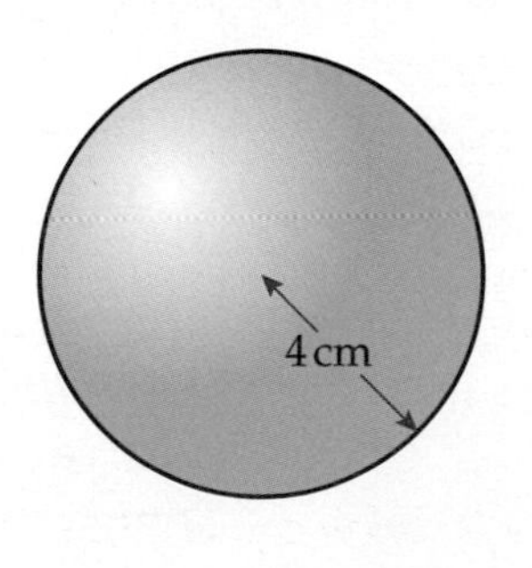

2 Calculate the surface area and volume of a sphere with a diameter of 3.2 cm.
Give your answers to one decimal place.

3 A sphere has a volume of 36π cm^3.
Calculate **a** the radius **b** the surface area.
Give your answers to three significant figures.

4 Calculate the volume of a sphere with surface area 9π cm^2.
Leave your answer in terms of π.

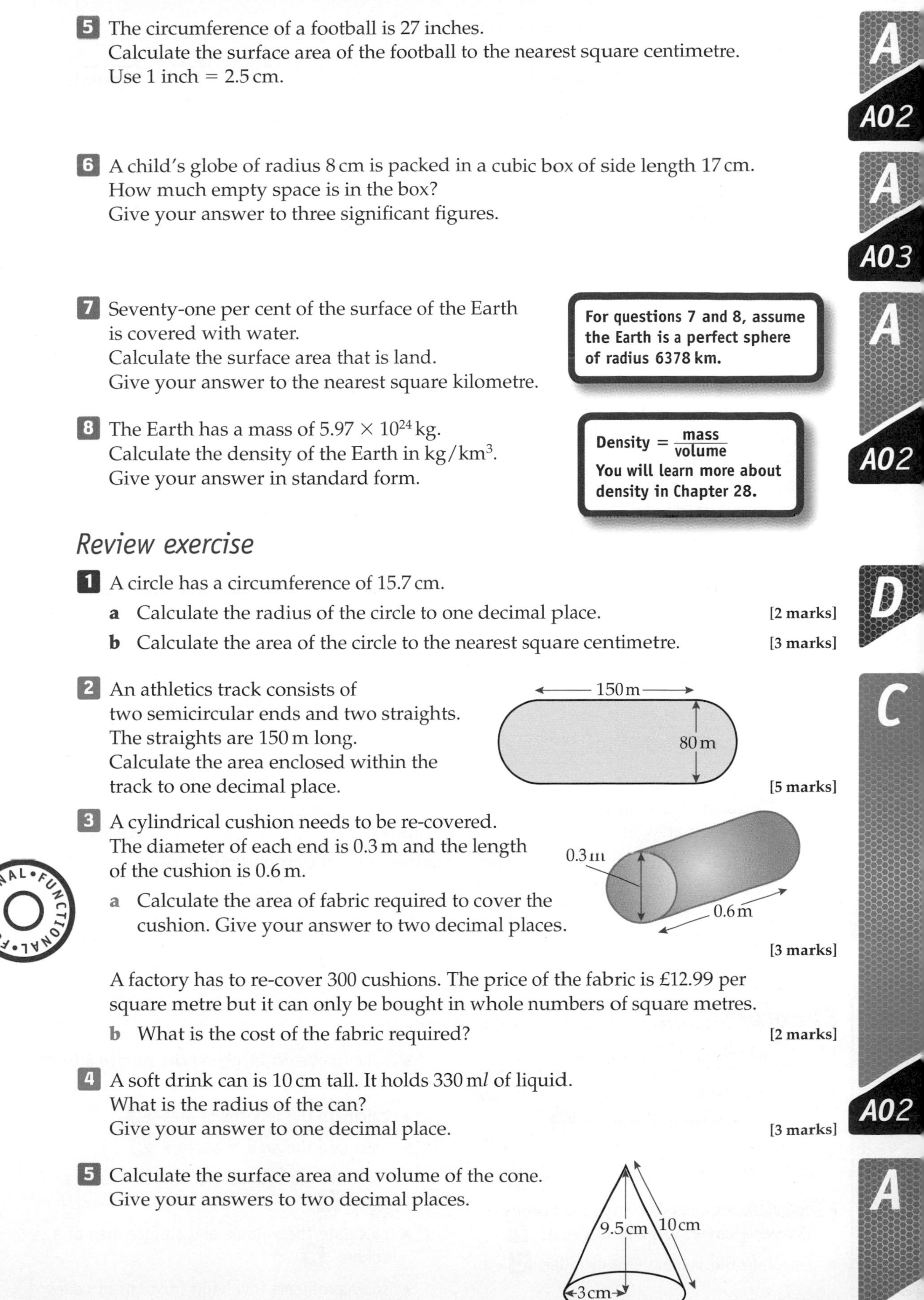

5 The circumference of a football is 27 inches.
Calculate the surface area of the football to the nearest square centimetre.
Use 1 inch = 2.5 cm.

A
AO2

6 A child's globe of radius 8 cm is packed in a cubic box of side length 17 cm.
How much empty space is in the box?
Give your answer to three significant figures.

A
AO3

7 Seventy-one per cent of the surface of the Earth is covered with water.
Calculate the surface area that is land.
Give your answer to the nearest square kilometre.

For questions 7 and 8, assume the Earth is a perfect sphere of radius 6378 km.

A

8 The Earth has a mass of 5.97×10^{24} kg.
Calculate the density of the Earth in kg/km^3.
Give your answer in standard form.

Density $= \frac{\text{mass}}{\text{volume}}$
You will learn more about density in Chapter 28.

AO2

Review exercise

1 A circle has a circumference of 15.7 cm.

a Calculate the radius of the circle to one decimal place. **[2 marks]**

b Calculate the area of the circle to the nearest square centimetre. **[3 marks]**

D

2 An athletics track consists of two semicircular ends and two straights.
The straights are 150 m long.
Calculate the area enclosed within the track to one decimal place. **[5 marks]**

C

3 A cylindrical cushion needs to be re-covered.
The diameter of each end is 0.3 m and the length of the cushion is 0.6 m.

FUNCTIONAL • FUNCTIONAL •

a Calculate the area of fabric required to cover the cushion. Give your answer to two decimal places. **[3 marks]**

A factory has to re-cover 300 cushions. The price of the fabric is £12.99 per square metre but it can only be bought in whole numbers of square metres.

b What is the cost of the fabric required? **[2 marks]**

4 A soft drink can is 10 cm tall. It holds 330 m*l* of liquid.
What is the radius of the can?
Give your answer to one decimal place. **[3 marks]**

AO2

5 Calculate the surface area and volume of the cone.
Give your answers to two decimal places.

[4 marks]

A

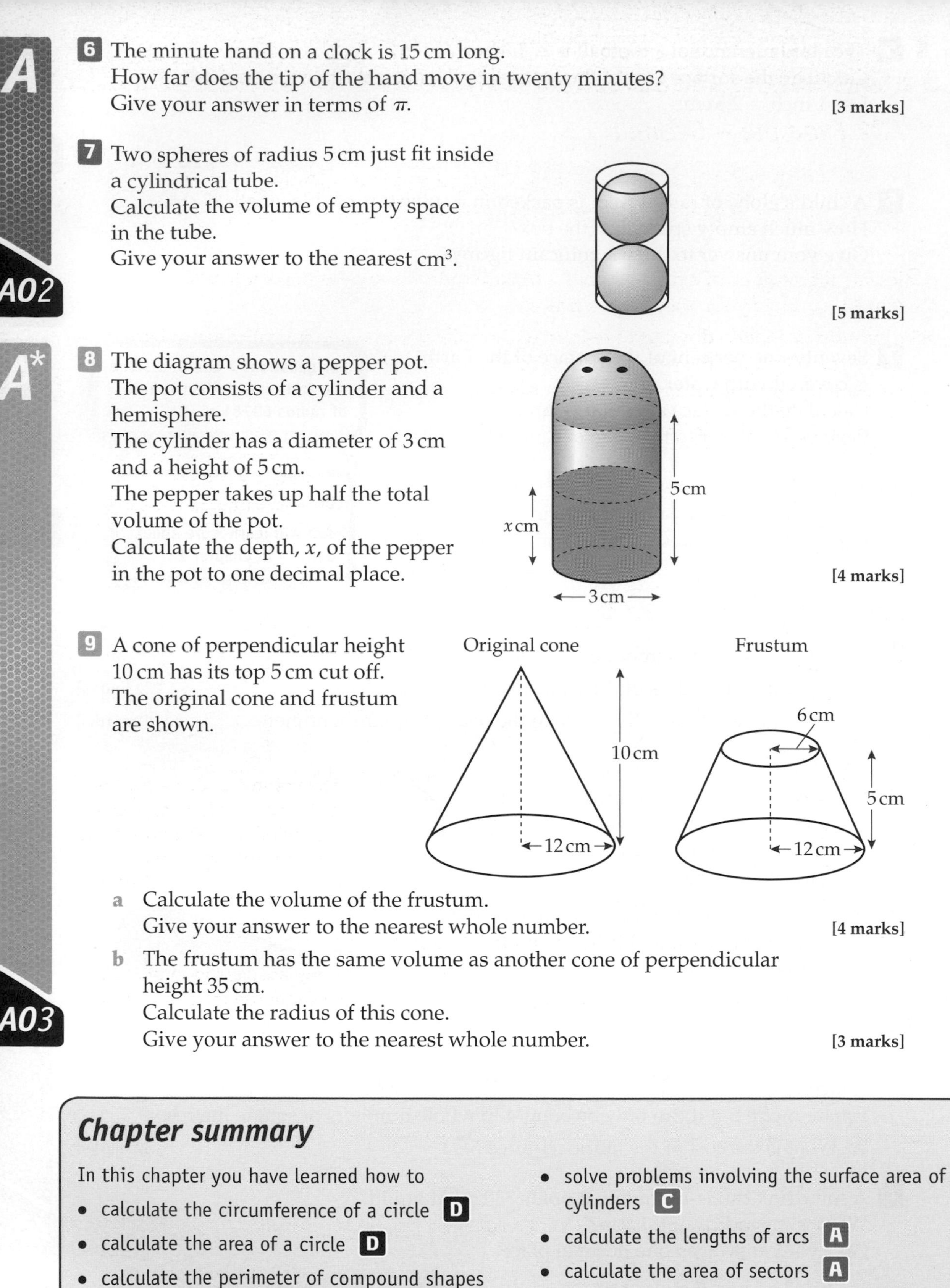

6 The minute hand on a clock is 15 cm long.
How far does the tip of the hand move in twenty minutes?
Give your answer in terms of π. **[3 marks]**

7 Two spheres of radius 5 cm just fit inside a cylindrical tube.
Calculate the volume of empty space in the tube.
Give your answer to the nearest cm^3. **[5 marks]**

8 The diagram shows a pepper pot.
The pot consists of a cylinder and a hemisphere.
The cylinder has a diameter of 3 cm and a height of 5 cm.
The pepper takes up half the total volume of the pot.
Calculate the depth, x, of the pepper in the pot to one decimal place. **[4 marks]**

9 A cone of perpendicular height 10 cm has its top 5 cm cut off.
The original cone and frustum are shown.

a Calculate the volume of the frustum.
Give your answer to the nearest whole number. **[4 marks]**

b The frustum has the same volume as another cone of perpendicular height 35 cm.
Calculate the radius of this cone.
Give your answer to the nearest whole number. **[3 marks]**

Chapter summary

In this chapter you have learned how to

- calculate the circumference of a circle **D**
- calculate the area of a circle **D**
- calculate the perimeter of compound shapes involving circles or parts of circles **C**
- calculate the area of compound shapes involving circles or parts of circles **C**
- calculate the volume of a cylinder **C**
- solve problems involving the surface area of cylinders **C**
- calculate the lengths of arcs **A**
- calculate the area of sectors **A**
- calculate the volume and surface area of a cone **A**
- calculate the volume and surface area of a sphere **A**
- solve problems involving frustums of cones **A***

AO2 Example – Geometry

A

This Grade A question challenges you to use geometry in a real-life context – applying maths you know to solve a problem.

a Calculate the volume of a metal cylinder of base radius 4 cm and perpendicular height 14 cm. Give your answer in terms of π.

b The cylinder is melted down and re-cast, without loss, into ten identical cones of base radius 3 cm. Calculate the perpendicular height of each of these ten cones. Give your answer correct to 3 significant figures.

AO2

What maths should you use?

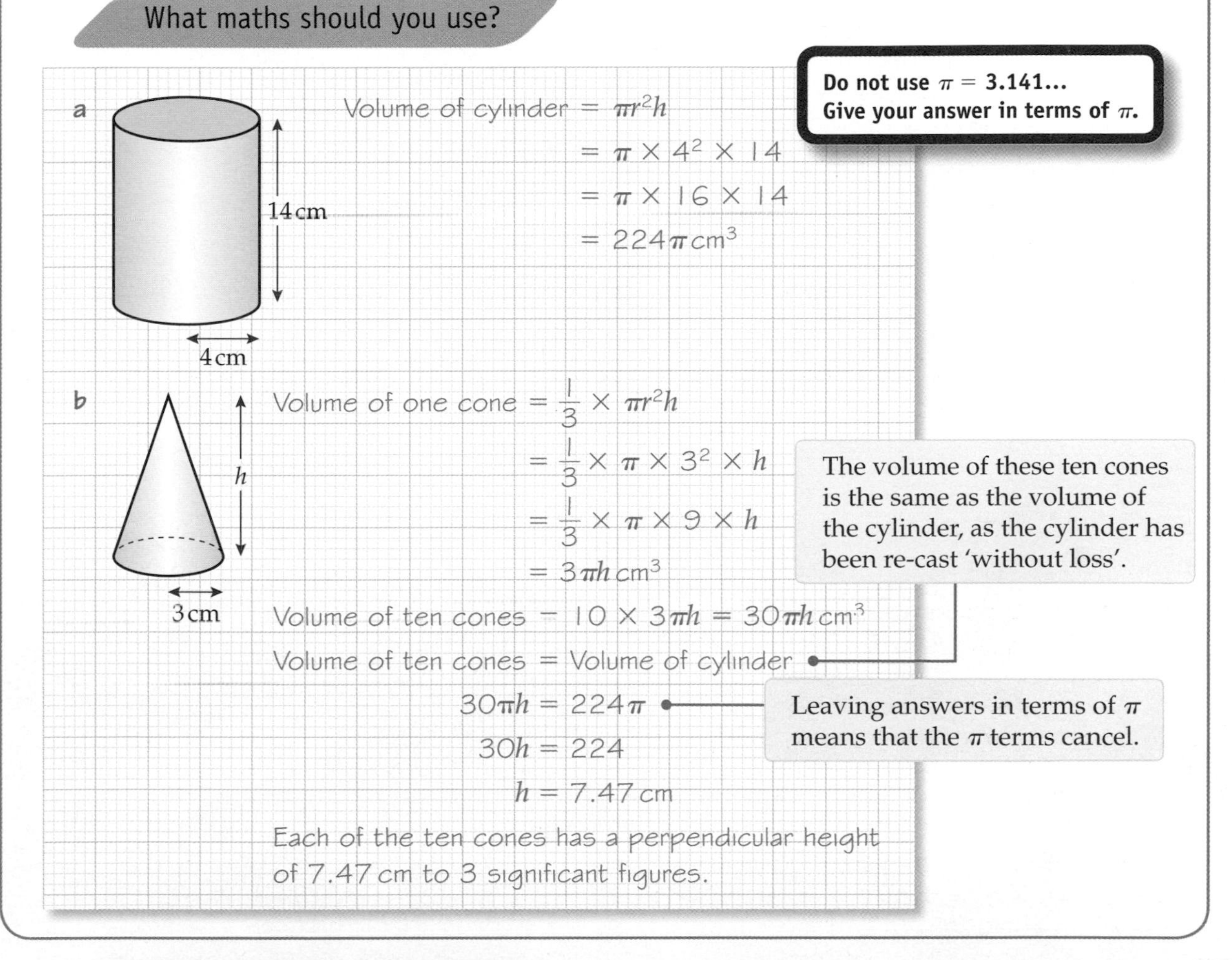

a
14 cm
4 cm
Volume of cylinder $= \pi r^2 h$
$= \pi \times 4^2 \times 14$
$= \pi \times 16 \times 14$
$= 224\pi \text{ cm}^3$
b
h
3 cm
Volume of one cone $= \frac{1}{3} \times \pi r^2 h$
$= \frac{1}{3} \times \pi \times 3^2 \times h$
$= \frac{1}{3} \times \pi \times 9 \times h$
$= 3\pi h \text{ cm}^3$
Volume of ten cones $= 10 \times 3\pi h = 30\pi h \text{ cm}^3$
Volume of ten cones = Volume of cylinder
$30\pi h = 224\pi$
$30h = 224$
$h = 7.47$ cm
Each of the ten cones has a perpendicular height of 7.47 cm to 3 significant figures.

Do not use $\pi = 3.141...$ Give your answer in terms of π.

The volume of these ten cones is the same as the volume of the cylinder, as the cylinder has been re-cast 'without loss'.

Leaving answers in terms of π means that the π terms cancel.

AO3 Question – Geometry

A*

Now try this AO3 Grade A* question. You have to work it out from scratch. READ THE QUESTION CAREFULLY.
It's similar to the AO2 example above, so think about where to start.

A metal cone of base radius 8 cm and perpendicular height 15 cm is melted down and re-cast, without loss, into fifteen identical spheres.

Calculate the radius of each of these fifteen spheres.
Give your answer correct to 3 significant figures.

With no loss of metal, the volumes are equal.

AO3

Roof garden

Anil moves into a small flat in Manchester. The flat has a roof garden. Anil makes a scale drawing of the new design for his roof garden.

The garden will have wooden fencing on three sides. There will also be five large plant pots, each 1 m high.

Question bank

1 What is the total length of fencing that Anil needs to buy?

2 What is the cost of the circular plant pot that Anil wants?

Anil is going to cover the sides of all the plant pots with a bamboo covering. He is then going to fill the pots with compost, leaving a 10 cm gap at the top.

3 What length of bamboo covering does Anil need to go around one of the square plant pots?

4 How much compost does Anil need for one of the square plant pots?

Anil says, 'The total cost of the pots, bamboo, compost and fencing is just under £1200.'

5 Is Anil correct? Use π on your calculator and show working to support your answer.

Information bank

Haroldston garden centre price list

Plant pots (1 m high)	square 1 m × 1 m square 1.2 m × 1.2 m square 1.4 m × 1.4 m rectangular 1 m × 2 m rectangular 1 m × 2.5 m rectangular 1.2 m × 2.5 m circular pot, radius 0.9 m circular pot, radius 1.1 m circular pot, radius 1.3 m	£49 £59 £69 £49 £69 £89 £99 £119 £139
Fencing	wooden wire	£9 per metre £7 per metre
Bamboo	roll (0.5 m high) roll (1 m high) roll (1.5 m high)	£3.50 per metre £5.50 per metre £7.50 per metre
Compost	200 litre bag	£8.95

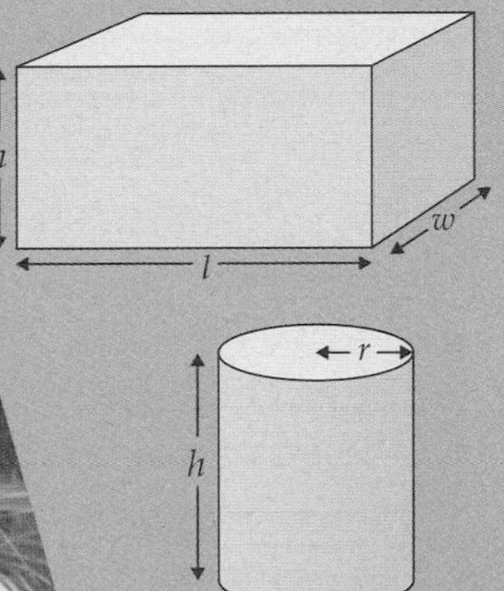

Perimeter of a rectangle $= 2(l + w)$

Area of a rectangle $= lw$

Volume of a cuboid $= lwh$

Perimeter of a circle $= 2\pi r$

Area of a circle $= \pi r^2$

Volume of a cylinder $= \pi r^2 h$

$1\text{ cm}^3 = 1\text{ m}l$

$1000\text{ m}l = 1$ litre

1000 litres $= 1\text{ m}^3$

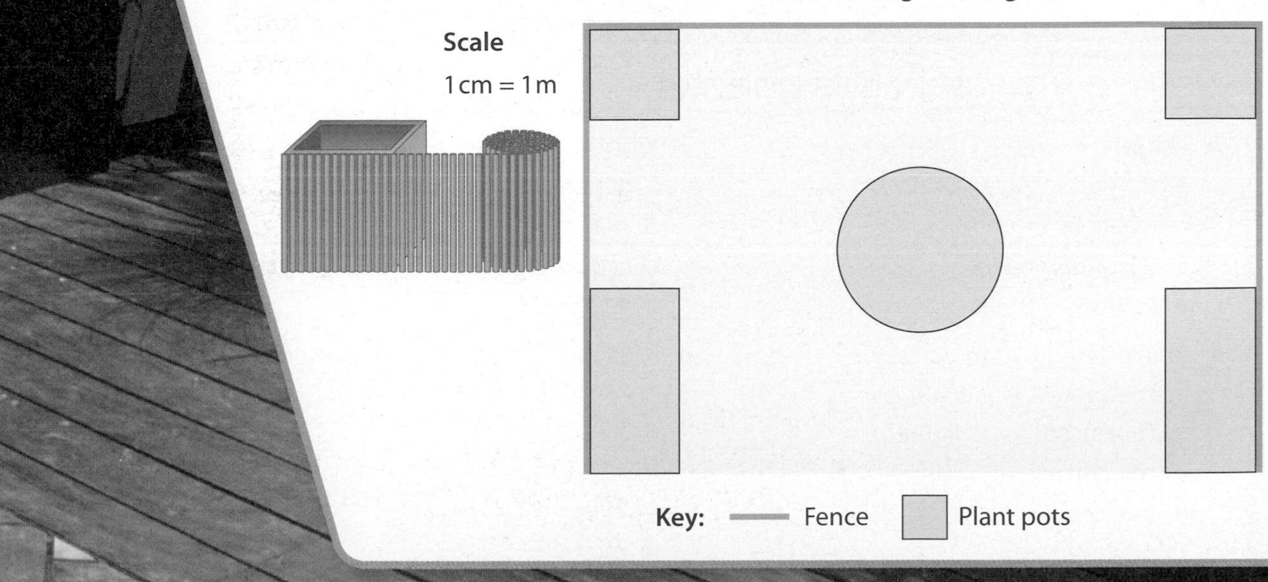

28 Measures and dimensions

BBC Video

This chapter is about using different units and calculating speed and density.

Sea water has a higher density than fresh water, so it is easier to float in sea water.

Objectives

This chapter will show you how to

- convert between different units of area D
- calculate average speeds D
- convert between different units of volume C
- make calculations using density C
- solve problems involving speed and density using upper and lower bounds C B
- recognise formulae for length, area or volume by considering dimensions B

Before you start this chapter

1 Convert
- **a** 8.5 feet into centimetres
- **b** 5 gallons into litres
- **c** 6 kg into pounds
- **d** 20 litres into pints.

2 Round 20 645.98 to
- **a** the nearest whole number
- **b** three significant figures
- **c** the nearest 10
- **d** one decimal place.

3 Write these times in hours and minutes.
- **a** 80 minutes
- **b** 400 minutes
- **c** a quarter of an hour
- **d** $2\frac{2}{5}$ hours

4 Work out these calculations, rounding your answers to one decimal place.
- **a** $80 \div 6$
- **b** $4 \div 9$
- **c** $200 \div 42$
- **d** $60 \div 35$

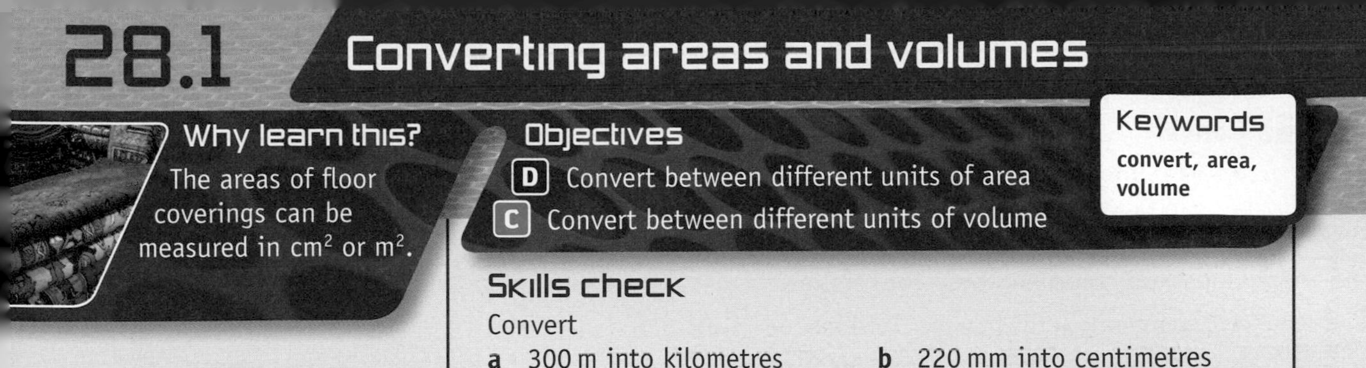

Converting areas and volumes

Why learn this?

The areas of floor coverings can be measured in cm^2 or m^2.

Objectives

D Convert between different units of area

C Convert between different units of volume

Keywords

convert, area, volume

Skills check

Convert

a 300 m into kilometres

b 220 mm into centimetres

c 1.6 m into centimetres

d 0.52 km into centimetres.

Converting areas and volumes

Area is measured in mm^2, cm^2, m^2 or km^2.

These two squares have the same area.

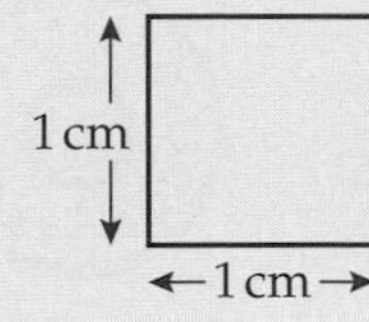

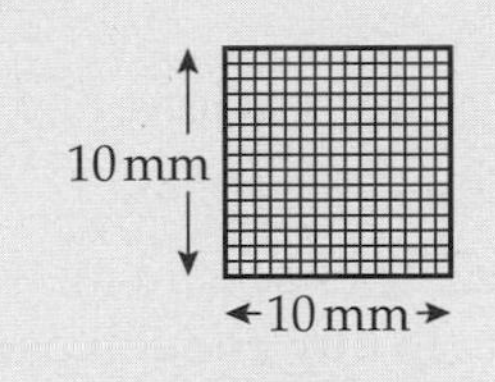

Area = 1 cm × 1 cm
= $1\ cm^2$

Area = 10 mm × 10 mm
= $100\ mm^2$

You can convert units of area using
$1\ cm^2 = 100\ mm^2$
$1\ m^2 = 10\,000\ cm^2$
$1\ km^2 = 1\,000\,000\ m^2$

Volume is measured in mm^3, cm^3, m^3 or km^3.

These two cubes have the same volume.

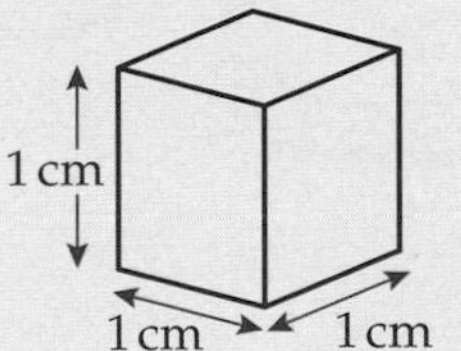

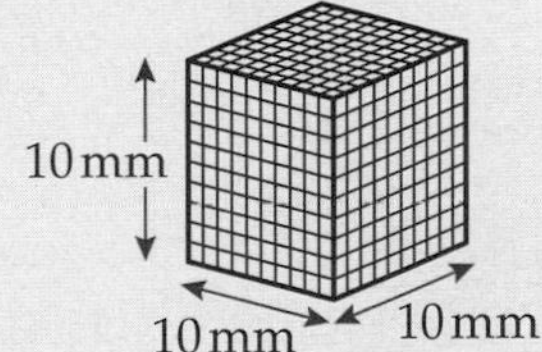

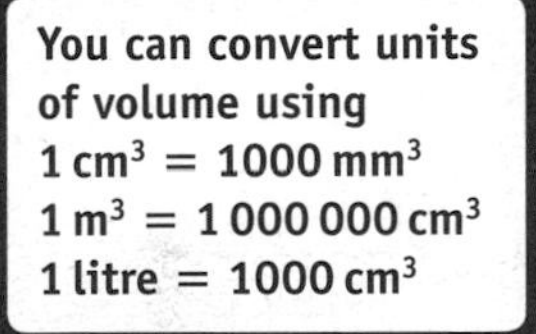

You can convert units of volume using
$1\ cm^3 = 1000\ mm^3$
$1\ m^3 = 1\,000\,000\ cm^3$
1 litre = $1000\ cm^3$

Volume = 1 cm × 1 cm × 1 cm
= $1\ cm^3$

Volume = 10 mm × 10 mm × 10 mm
= $1000\ mm^3$

Example 1

D

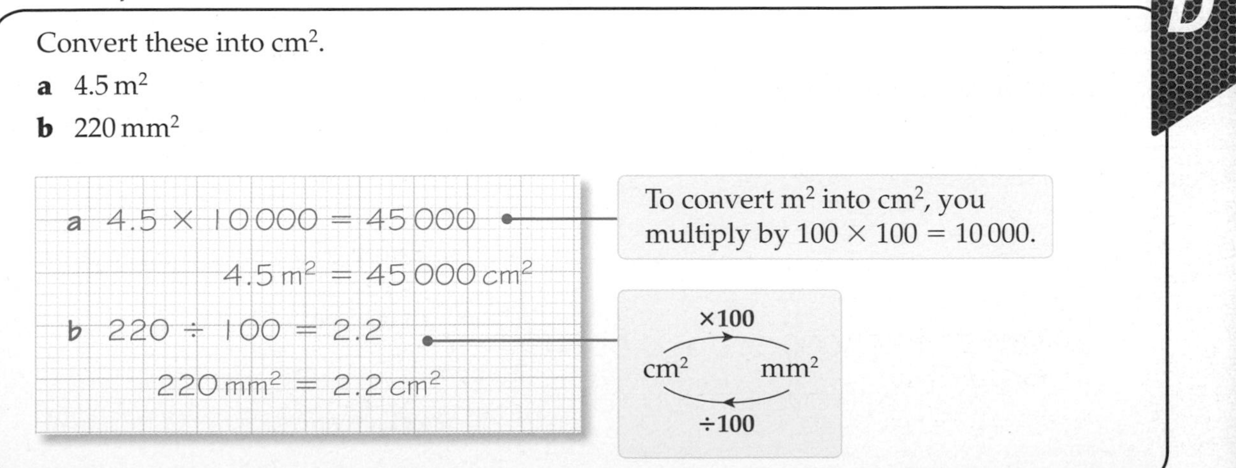

Convert these into cm^2.

a $4.5\ m^2$

b $220\ mm^2$

a $4.5 \times 10\,000 = 45\,000$
$4.5\ m^2 = 45\,000\ cm^2$

To convert m^2 into cm^2, you multiply by $100 \times 100 = 10\,000$.

b $220 \div 100 = 2.2$
$220\ mm^2 = 2.2\ cm^2$

Exercise 28A

D

1 Convert

a $6\,m^2$ into cm^2 b $0.2\,m^2$ into cm

c $2400\,cm^2$ into m^2 d $650\,cm^2$ into m^2.

2 Copy and complete these conversions.

a $2.5\,m^2 = \square\,cm^2$ b $4\,km^2 = \square\,m^2$

c $8900\,m^2 = \square\,km^2$ d $5000\,mm^2 = \square\,cm^2$

3 Work out the area of this rectangle

a in cm^2

b in mm^2.

2.5 cm

1.8 cm

4 Work out the number of mm^2 in $1\,m^2$.

D

5 Claire's living room is a rectangle 850 cm long and 440 cm wide.

a Find the area of Claire's living room in m^2.

b The price of oak flooring is £70 per square metre.
How much will it cost Claire to buy enough oak flooring to cover her living room?

FUNCTIONAL

6 Karl is painting his bedroom.
The total area he needs to paint is $355\,000\,cm^2$.
How many of these pots of paint will he need to buy?
Show your working.

AO2

D

7 A circle has a diameter of 0.6 m.
Calculate its area in cm^2, correct to three significant figures.

C

AO2

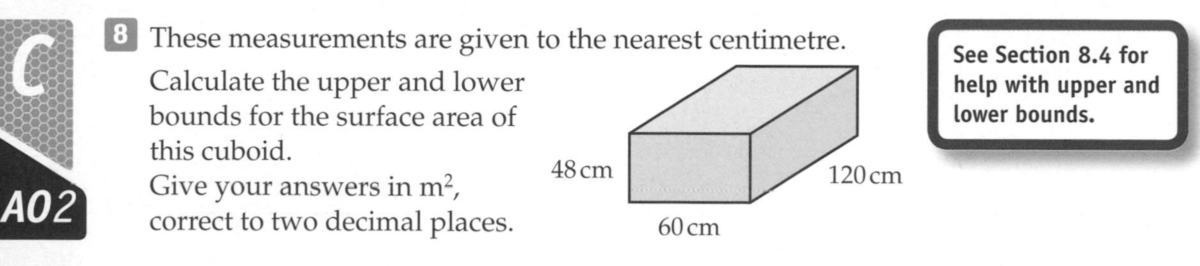

8 These measurements are given to the nearest centimetre.
Calculate the upper and lower bounds for the surface area of this cuboid.
Give your answers in m^2, correct to two decimal places.

See Section 8.4 for help with upper and lower bounds.

C

Example 2

Convert a $34.2\,cm^3$ into mm^3 b $850\,000\,cm^3$ into m^3.

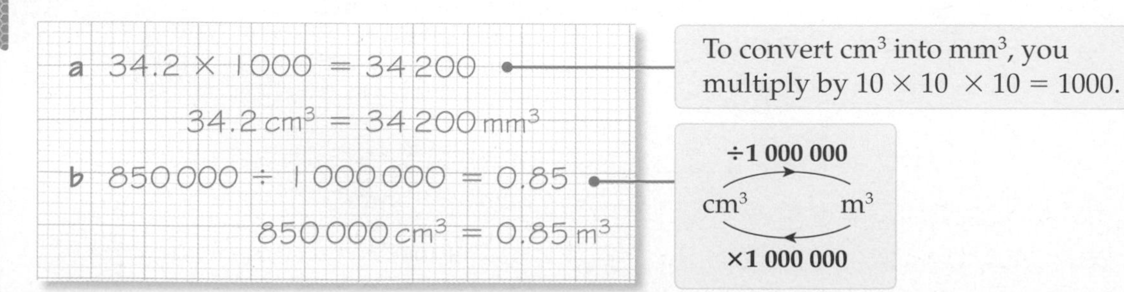

a $34.2 \times 1000 = 34\,200$

$34.2\,cm^3 = 34\,200\,mm^3$

b $850\,000 \div 1\,000\,000 = 0.85$

$850\,000\,cm^3 = 0.85\,m^3$

To convert cm^3 into mm^3, you multiply by $10 \times 10 \times 10 = 1000$.

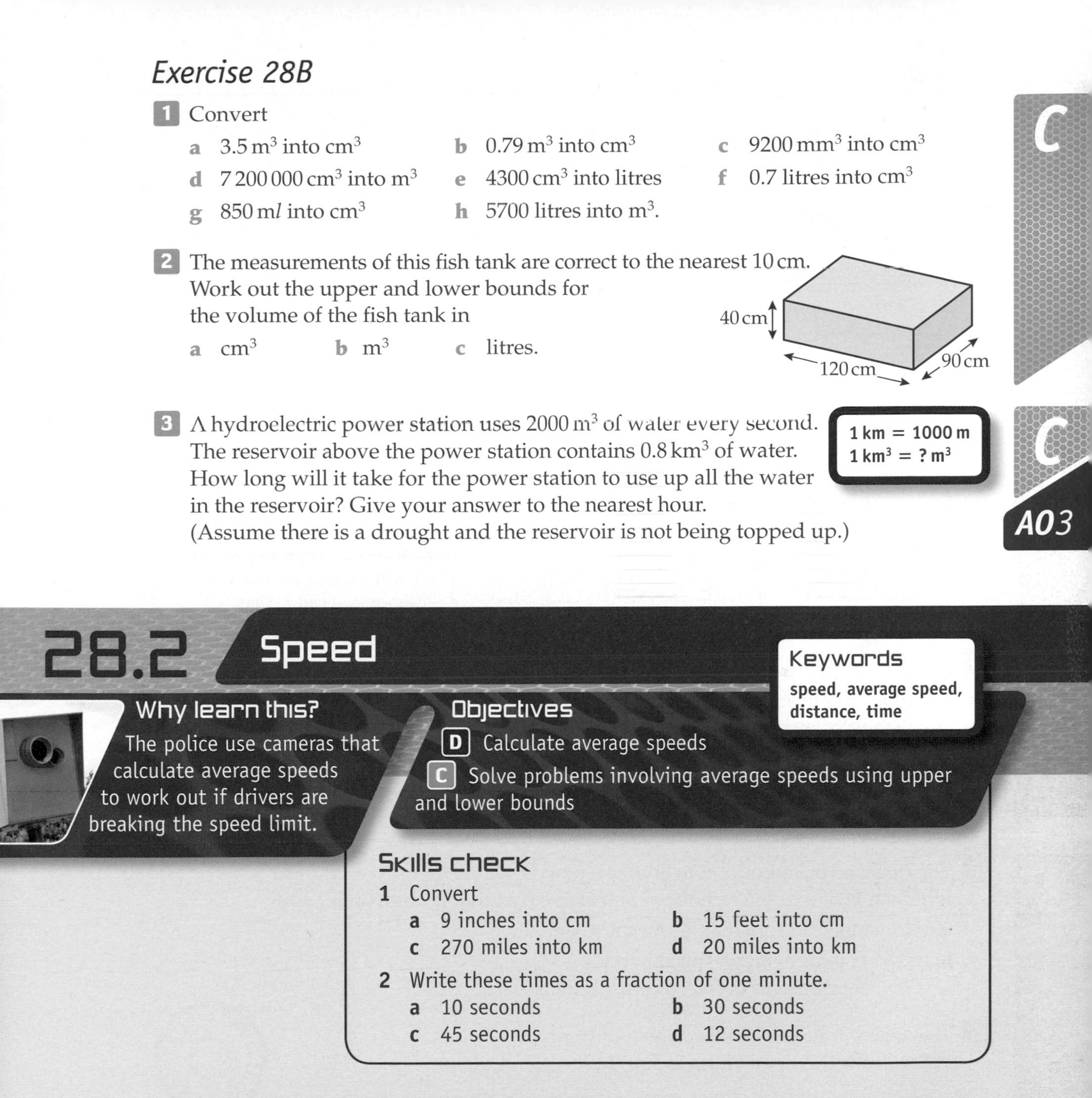

Exercise 28B

1 Convert

a 3.5 m^3 into cm^3
b 0.79 m^3 into cm^3
c 9200 mm^3 into cm^3
d $7\,200\,000\text{ cm}^3$ into m^3
e 4300 cm^3 into litres
f 0.7 litres into cm^3
g 850 m*l* into cm^3
h 5700 litres into m^3.

2 The measurements of this fish tank are correct to the nearest 10 cm. Work out the upper and lower bounds for the volume of the fish tank in

a cm^3
b m^3
c litres.

C

3 A hydroelectric power station uses 2000 m^3 of water every second. The reservoir above the power station contains 0.8 km^3 of water. How long will it take for the power station to use up all the water in the reservoir? Give your answer to the nearest hour.
(Assume there is a drought and the reservoir is not being topped up.)

1 km = 1000 m
$1\text{ km}^3 = ?\text{ m}^3$

C
AO3

28.2 Speed

Keywords
speed, average speed, distance, time

Why learn this?
The police use cameras that calculate average speeds to work out if drivers are breaking the speed limit.

Objectives
D Calculate average speeds
C Solve problems involving average speeds using upper and lower bounds

Skills check

1 Convert
a 9 inches into cm
b 15 feet into cm
c 270 miles into km
d 20 miles into km

2 Write these times as a fraction of one minute.
a 10 seconds
b 30 seconds
c 45 seconds
d 12 seconds

Speed, distance and time

Speed is a measurement of how fast something is travelling.

You can calculate the **average speed** of an object if you know the **distance** travelled and the **time** taken.

$$\text{Speed} = \frac{\text{distance}}{\text{time}}$$

$$\text{Distance} = \text{speed} \times \text{time}$$

$$\text{Time} = \frac{\text{distance}}{\text{speed}}$$

You can use this triangle to remember all three formulae. Cover up the letter you are trying to find. The position of the other two letters tells you whether to multiply or divide.

D
S T

The most common metric units of speed are metres per second (m/s) and kilometres per hour (km/h). The most common imperial unit of speed is miles per hour (mph).

The units given for distance and time will tell you which unit to use for speed.

D

Example 3

An athlete runs 400 m in 52 seconds.
Calculate his average speed, correct to one decimal place.

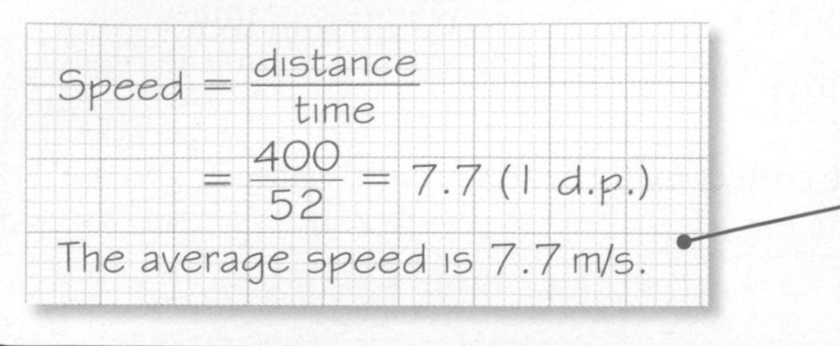

The distance is given in metres and the time is given in seconds, so the unit for speed in this example is metres per second (m/s).

Exercise 28C

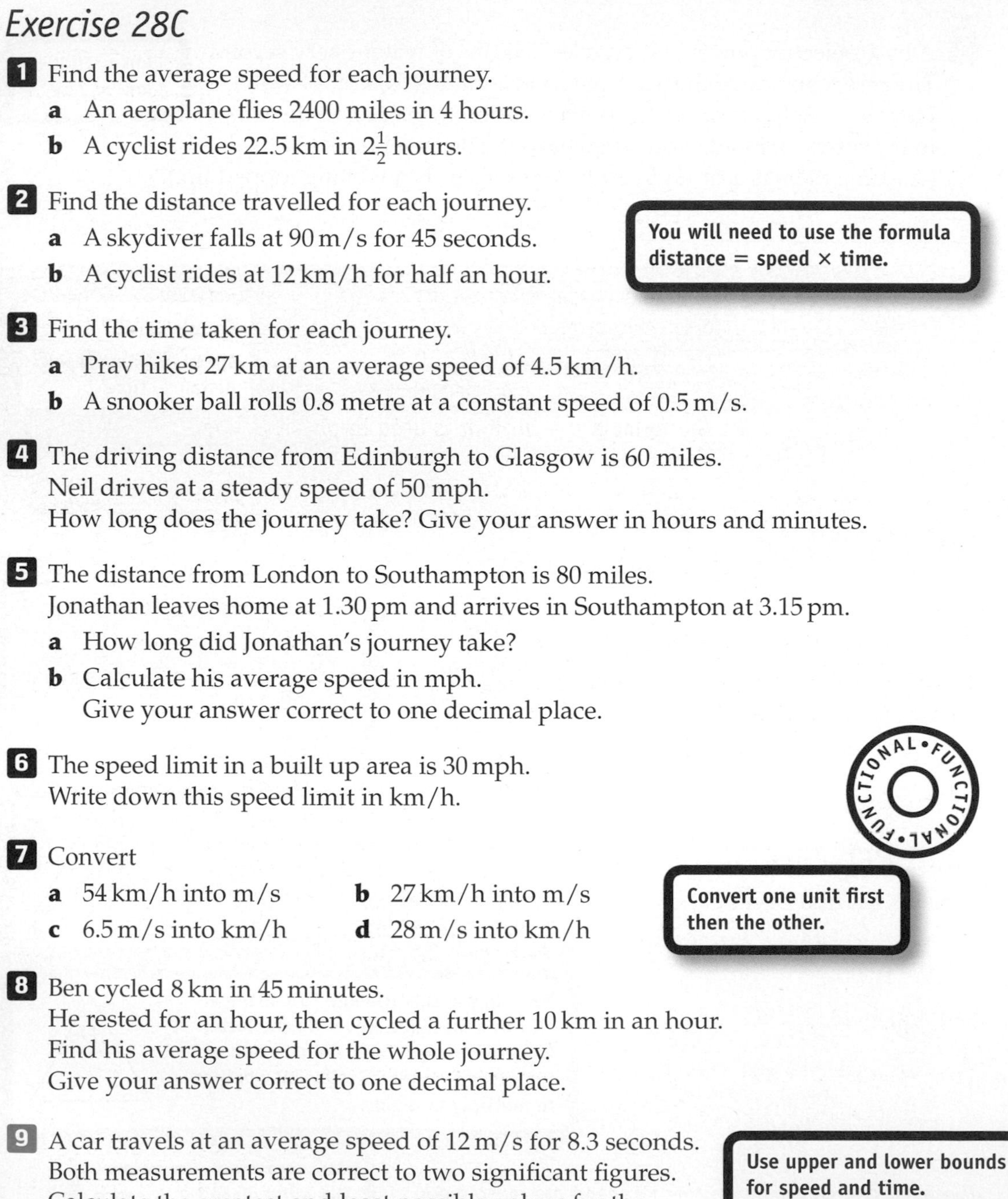

D

1 Find the average speed for each journey.

a An aeroplane flies 2400 miles in 4 hours.

b A cyclist rides 22.5 km in $2\frac{1}{2}$ hours.

2 Find the distance travelled for each journey.

a A skydiver falls at 90 m/s for 45 seconds.

b A cyclist rides at 12 km/h for half an hour.

You will need to use the formula distance = speed × time.

3 Find the time taken for each journey.

a Prav hikes 27 km at an average speed of 4.5 km/h.

b A snooker ball rolls 0.8 metre at a constant speed of 0.5 m/s.

4 The driving distance from Edinburgh to Glasgow is 60 miles.
Neil drives at a steady speed of 50 mph.
How long does the journey take? Give your answer in hours and minutes.

D

5 The distance from London to Southampton is 80 miles.
Jonathan leaves home at 1.30 pm and arrives in Southampton at 3.15 pm.

a How long did Jonathan's journey take?

b Calculate his average speed in mph.
Give your answer correct to one decimal place.

6 The speed limit in a built up area is 30 mph.
Write down this speed limit in km/h.

FUNCTIONAL • FUNCTIONAL •

7 Convert

a 54 km/h into m/s　　**b** 27 km/h into m/s

c 6.5 m/s into km/h　　**d** 28 m/s into km/h

Convert one unit first then the other.

8 Ben cycled 8 km in 45 minutes.
He rested for an hour, then cycled a further 10 km in an hour.
Find his average speed for the whole journey.
Give your answer correct to one decimal place.

AO2

C

9 A car travels at an average speed of 12 m/s for 8.3 seconds.
Both measurements are correct to two significant figures.
Calculate the greatest and least possible values for the distance travelled.

Use upper and lower bounds for speed and time.

AO3

10 Christina walks 4.5 km (correct to one decimal place) in 50 minutes (correct to the nearest minute). Find the upper and lower bounds for her average speed in km/h.

C

11 A train travels from Newcastle to Edinburgh at an average speed of 68 mph, correct to the nearest whole number. The journey is 80 miles, to the nearest 10 miles. Bryn says that the journey will definitely take less than an hour and a quarter.

a Show working to explain why Bryn is wrong.

b Calculate the shortest possible time it could take to complete the journey to the nearest minute.

AO3

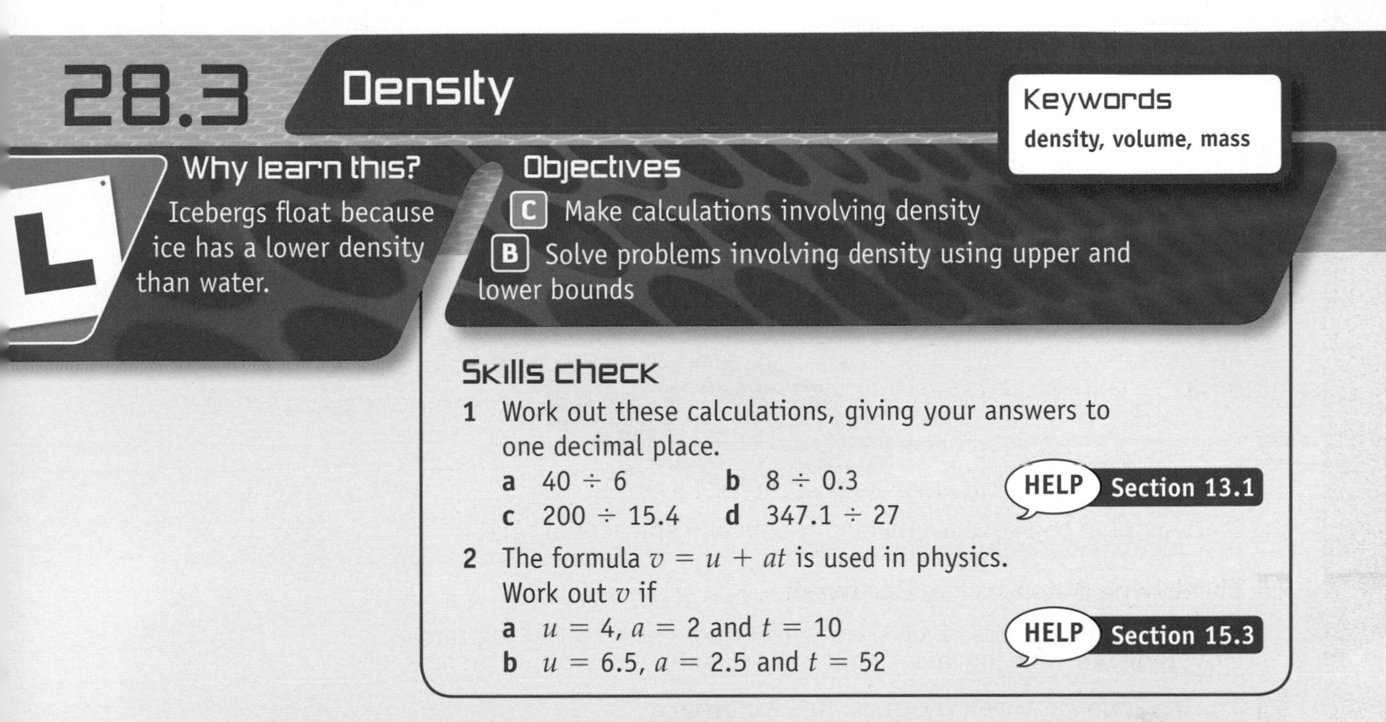

28.3 Density

Keywords
density, volume, mass

Why learn this?
Icebergs float because ice has a lower density than water.

Objectives

C Make calculations involving density

B Solve problems involving density using upper and lower bounds

Skills check

1 Work out these calculations, giving your answers to one decimal place.

a $40 \div 6$ b $8 \div 0.3$

c $200 \div 15.4$ d $347.1 \div 27$

HELP Section 13.1

2 The formula $v = u + at$ is used in physics. Work out v if

a $u = 4$, $a = 2$ and $t = 10$

b $u = 6.5$, $a = 2.5$ and $t = 52$

HELP Section 15.3

Density, mass and volume

Density is a measurement of the amount of a substance contained in a certain **volume**.

Density is measured in grams per cubic centimetre (g/cm³) or kilograms per cubic metre (kg/m³). For example, a 1 cm³ block of iron has a mass of 8 grams, so its density is 8 g/cm³.

You can calculate the density of a material if you know its **mass** and its volume.

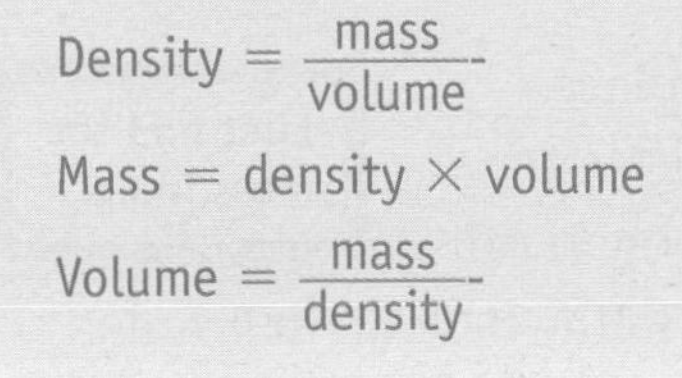

$$\text{Density} = \frac{\text{mass}}{\text{volume}}$$

$$\text{Mass} = \text{density} \times \text{volume}$$

$$\text{Volume} = \frac{\text{mass}}{\text{density}}$$

You can use this triangle to remember all three formulae. Cover up the letter you are trying to find. The position of the other two letters tells you whether to multiply or divide.

Example 4

C

The density of cork is 240 kg/m³. Find the mass of 3.4 m³ of cork.

Mass = density × volume
= 240 × 3.4 = 816
3.4 m³ of cork will have a mass of 816 kg.

The density is given in kg/m³ and the volume is given in m³, so the unit for mass in this example is the kilogram (kg).

Exercise 28D

1 Use this table to work out

a the mass of $20\,cm^3$ of platinum
b the mass of $14\,cm^3$ of mercury
c the volume of 200 g of magnesium
d the volume of 35 g of platinum.

Substance	Density (g/cm^3)
magnesium	1.7
copper	8.9
mercury	13.6
platinum	21.4

2 Karl weighs a $20\,cm^3$ block of silver. It has a mass of 210 g.

a Work out the density of silver.
b Use your answer to part **a** to calculate the volume of a 480 g block of silver.

3 Petrol has a density of $0.7\,g/cm^3$. The fuel tank on a car holds 45 litres.

a Convert 45 litres into cm^3.
b Find the mass of the petrol in the tank when it is full.

4 Abi measures the mass of this wooden block as 2 kg.

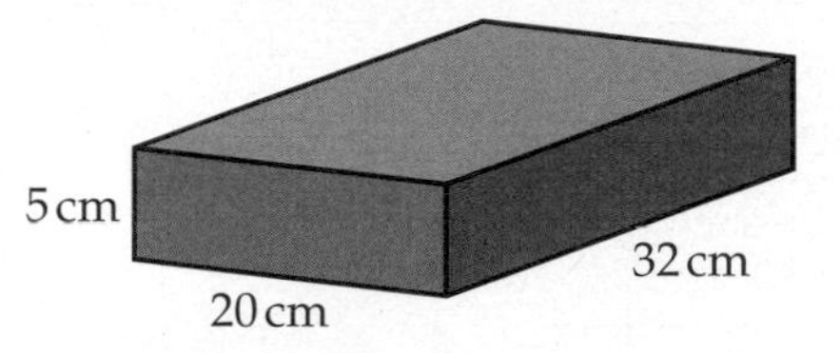

a Calculate the volume of the block.
b Work out the density of the wood.
c Work out the mass of a $600\,cm^3$ block of the same wood in grams.

5 The diagram shows a cylindrical fuel rod from a nuclear power station.
The rod is made of uranium with a density of $19\,g/cm^3$.

a Calculate the volume of the rod.
b Calculate the mass of the rod to the nearest kg.

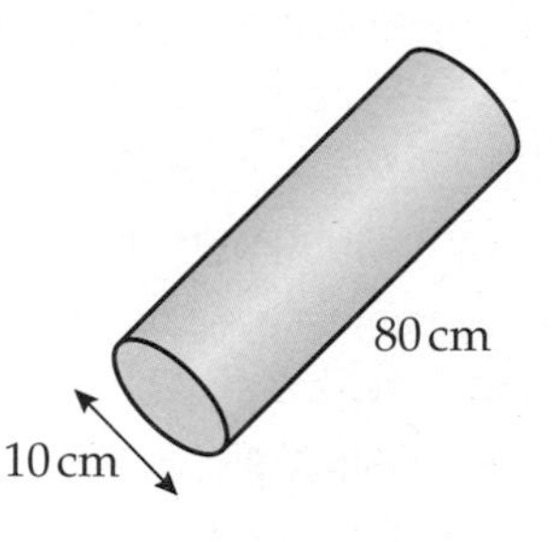

6 Air has a density of $1.3\,kg/m^3$, correct to one decimal place.
A balloon holds $4200\,cm^3$ of air, correct to the nearest $100\,cm^3$.

a Write the upper bound for the capacity of the balloon in m^3.
b Find the greatest possible value for the mass of air in the balloon in grams.

> **First find the mass in kg.**

7 An iceberg has a volume of $18\,000\,m^3$.
The density of ice is $0.92\,g/cm^3$.
Both these values are correct to two significant figures.
Calculate the greatest and least possible values for the mass of the iceberg in tonnes.

> **1000 kg = 1 tonne**

8 A wooden paperweight has a mass of 250 g correct to the nearest 10 g.
It has a volume of $370\,cm^3$, correct to the nearest $10\,cm^3$.
Calculate the greatest and least possible values for the density of the wood used to make the paperweight.

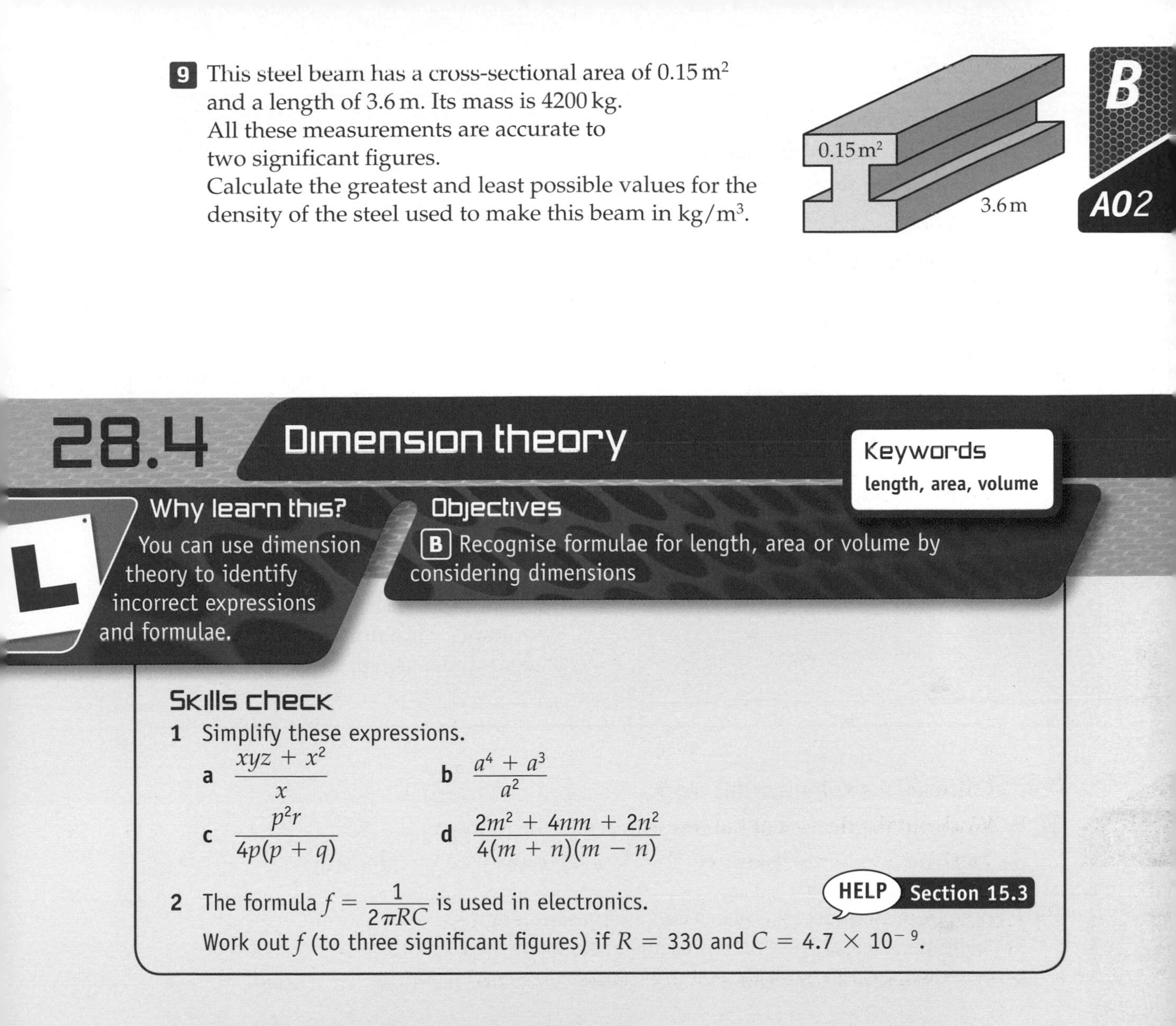

9 This steel beam has a cross-sectional area of $0.15\,\text{m}^2$ and a length of $3.6\,\text{m}$. Its mass is $4200\,\text{kg}$.
All these measurements are accurate to two significant figures.
Calculate the greatest and least possible values for the density of the steel used to make this beam in kg/m^3.

0.15 m²
3.6 m

B
AO2

28.4 Dimension theory

Keywords
length, area, volume

Why learn this?
You can use dimension theory to identify incorrect expressions and formulae.

Objectives
B Recognise formulae for length, area or volume by considering dimensions

Skills check

1 Simplify these expressions.

a $\dfrac{xyz + x^2}{x}$ b $\dfrac{a^4 + a^3}{a^2}$

c $\dfrac{p^2r}{4p(p + q)}$ d $\dfrac{2m^2 + 4nm + 2n^2}{4(m + n)(m - n)}$

2 The formula $f = \dfrac{1}{2\pi RC}$ is used in electronics.
Work out f (to three significant figures) if $R = 330$ and $C = 4.7 \times 10^{-9}$.

HELP Section 15.3

Length, area and volume

Length is a one-dimensional quantity.
In a formula for a length, each term must contain one letter representing a length.

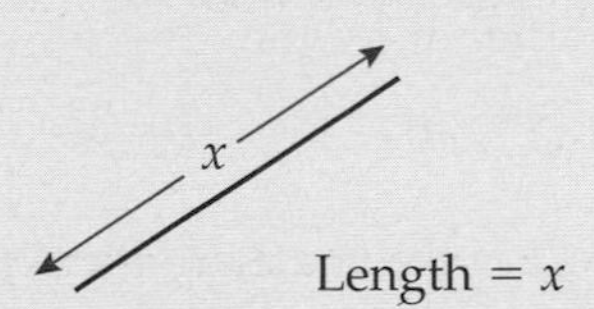

Area is a two-dimensional quantity.
In a formula for an area, each term must contain either
- two lengths multiplied together, or
- one letter representing an area.

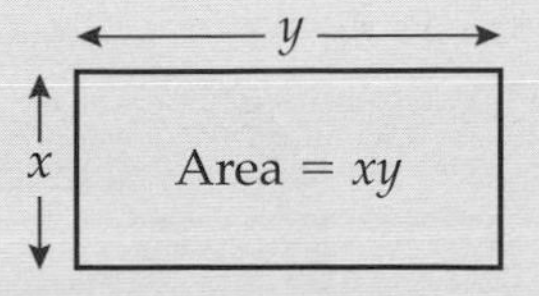

Volume is a three-dimensional quantity.
In a formula for a volume, each term must contain
- three lengths multiplied together, or
- a length multiplied by an area, or
- one letter representing a volume.

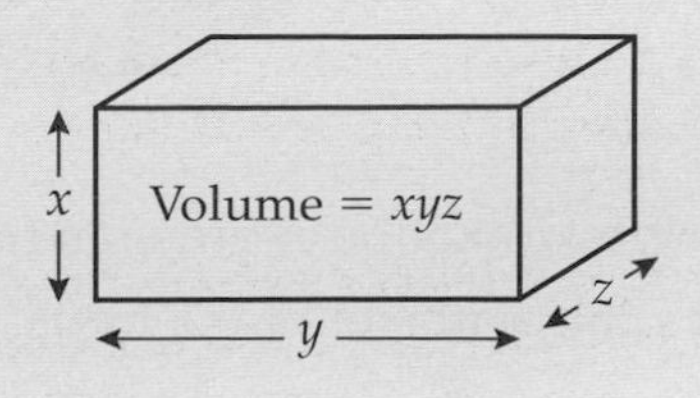

B

Example 5

A represents an area, and x and y represent lengths.
Say whether each expression represents a length, an area, a volume or none of these.

a $4x^2$ **b** $2x + A$ **c** $3x + 2y$ **d** $Ax + \pi xy^2$

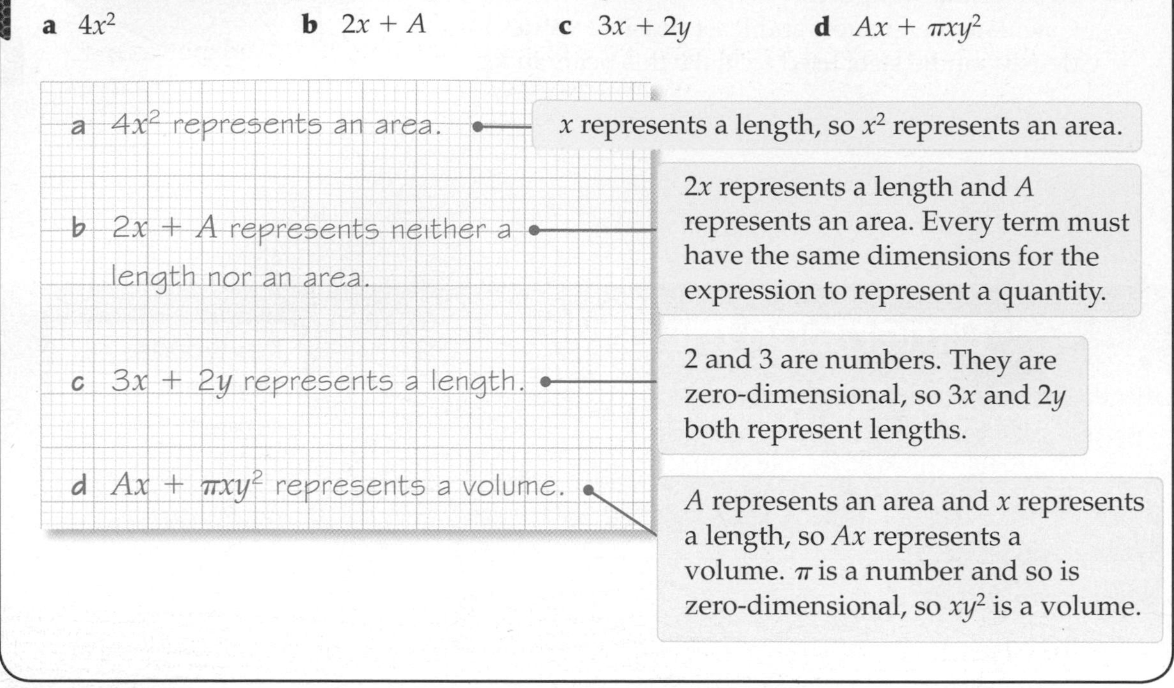

a $4x^2$ represents an area.
x represents a length, so x^2 represents an area.
b $2x + A$ represents neither a length nor an area.
$2x$ represents a length and A represents an area. Every term must have the same dimensions for the expression to represent a quantity.
c $3x + 2y$ represents a length.
2 and 3 are numbers. They are zero-dimensional, so $3x$ and $2y$ both represent lengths.
d $Ax + \pi xy^2$ represents a volume.
A represents an area and x represents a length, so Ax represents a volume. π is a number and so is zero-dimensional, so xy^2 is a volume.

Exercise 28E

B

1 Write down length, area or volume for each of the following.

a 24 cm³ **b** 8 km **c** 24 mm² **d** 42 m³

2 P and Q represent areas, and a, b and c represent lengths.
Say whether each expression represents a length, an area, a volume or none of these.

a $P + Q$ **b** $3a^2b$ **c** $2ab + 2c^2$ **d** $3x^3 + 2\pi Q$

3 In these expressions, x, y and z all represent lengths.
Say whether each expression represents a length, an area, a volume or none of these.

a $2x(x + y)$ **b** $3(x + y)$ **c** $z(2x^2y + z^3)$ **d** $4x^2(y + 2z)$

B

4 A torus is the mathematical name for a ring.

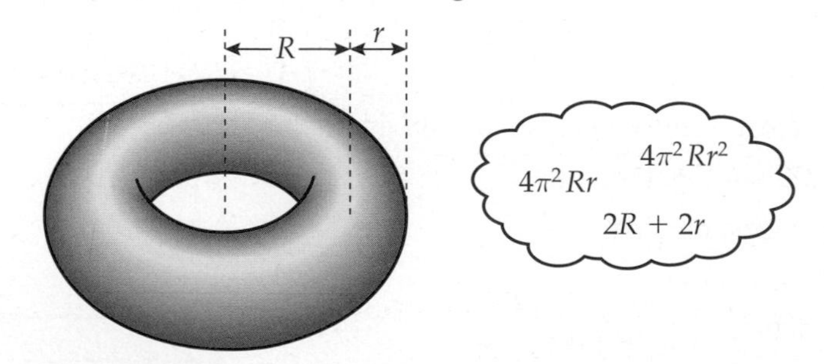

Write down the expression from the cloud which represents

a the diameter of the torus

b the surface area of the torus

c the volume of the torus.

A02

5 Zoe writes down the following formula for the volume of a sphere of radius r.

$$V = \frac{4}{3}\pi r^2$$

Use dimensions to show why Zoe's formula must be wrong.

B

6 In these expressions a is measured in metres, b is measured in square metres and c is measured in cubic metres. Write down the units for

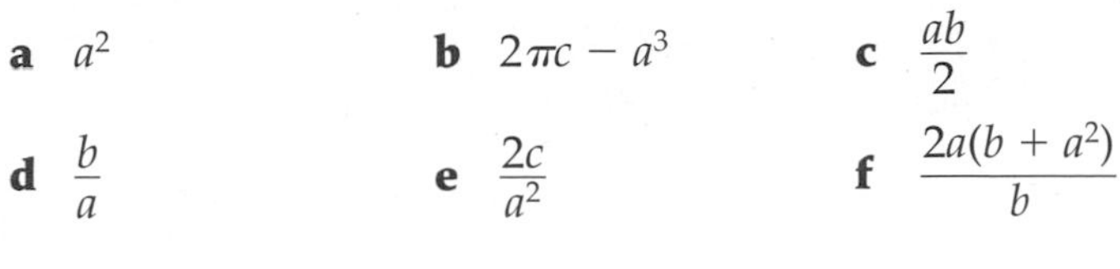

a a^2 **b** $2\pi c - a^3$ **c** $\frac{ab}{2}$

d $\frac{b}{a}$ **e** $\frac{2c}{a^2}$ **f** $\frac{2a(b + a^2)}{b}$

7 **a** Choose the formula that is most likely to be correct for the perimeter of this shape. Give a reason for your answer.

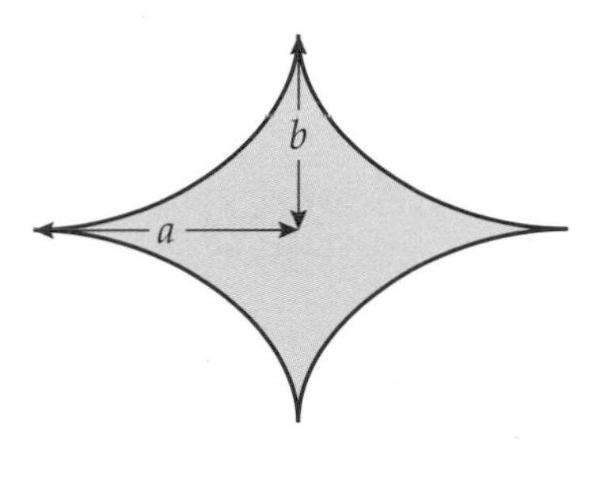

A $P = \frac{2\pi(a^2 + b^2)}{(a + b)^2}$

B $P = \frac{4(a^2 + ab + b^2)}{a + b}$

C $P = \frac{2a(a^2 + b^2)}{a + b}$

b The dimensions of this shape are $a = 6$ cm and $b = 4$ cm. Calculate the perimeter using the formula you chose in part **a**.

AO2

8 A garden centre sells plant pots in different sizes. This notice is displayed in the garden centre.

B

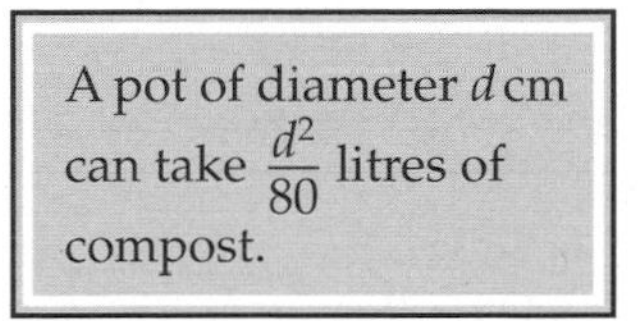
A pot of diameter d cm can take $\frac{d^2}{80}$ litres of compost.

Do you think the notice will provide an accurate guide for all the different sizes of pot? Give a reason for your answer.

AO3

Review exercise

1 Nick is travelling at 110 km/h when he passes this road sign. Is Nick breaking the speed limit? Show working to support your answer.

You need to remember the miles to km conversion factor.

[2 marks]

2 A company uses foam packing blocks when it transports its products. The foam has a density of 300 kg/m^3.

a Find the mass of a 2.6 m^3 block of foam. [2 marks]

b Find the volume of a block of the foam with a mass of 120 kg. [2 marks]

3 The loading bay on a van has a volume of 3.5 m^3. Write this volume in cm^3. [2 marks]

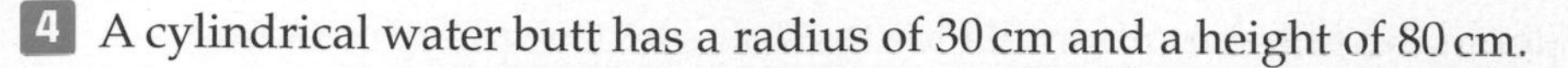

C

4 A cylindrical water butt has a radius of 30 cm and a height of 80 cm.

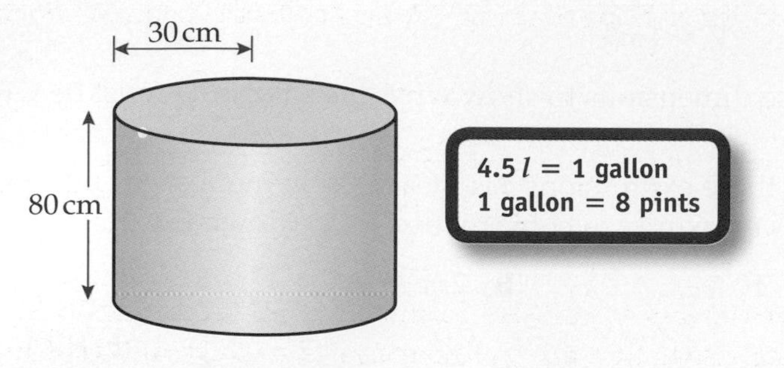

a Work out the volume of the water butt in cm^3. **[2 marks]**

b Katherine says the water butt can hold about 400 pints of water.
Show working to demonstrate that she is right. **[3 marks]**

5 This diagram shows a cylindrical jug.

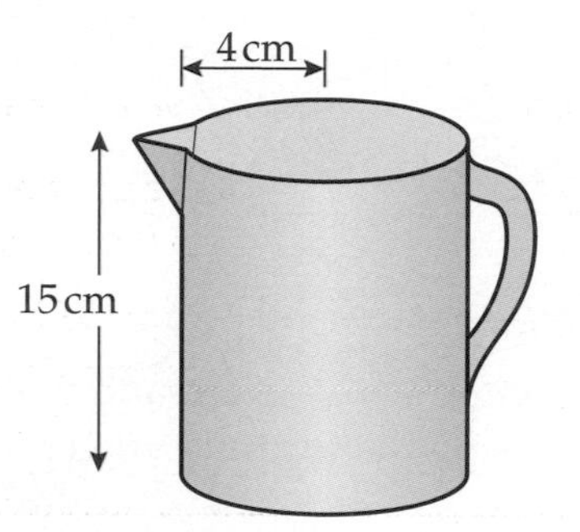

a Work out the volume of the jug (ignore the spout and handle). **[2 marks]**

AO2

b The jug is half filled with liquid ammonia.
The mass of the liquid in the jug is 520 g.
Calculate the density of liquid ammonia. **[3 marks]**

C

AO3

6 The course for a cycling race is 56 km long to the nearest km.
The winning cyclist averages 26 km/h to the nearest km/h.
Calculate the least possible time taken for the winning cyclist to complete the course to the nearest minute. **[2 marks]**

B

7 x and y represent lengths. Does the expression $2\pi^2(x + 3y)$ represent a length, an area or a volume?
Give a reason for your answer. **[2 marks]**

B

8 In this question, a, b and c represent lengths, and A represents an area.
Work out the missing powers in each case.

a Area $= 2\pi b^{\square}$ **[1 mark]**

b Volume $= 4a(bc + c^{\square})$ **[1 mark]**

AO2

c Length $= \dfrac{2a(A + c^{\square})}{b^{\square}}$ **[2 marks]**

9 The price of gold is £22 per gram.
The dimensions of this gold bar are accurate to one decimal place.

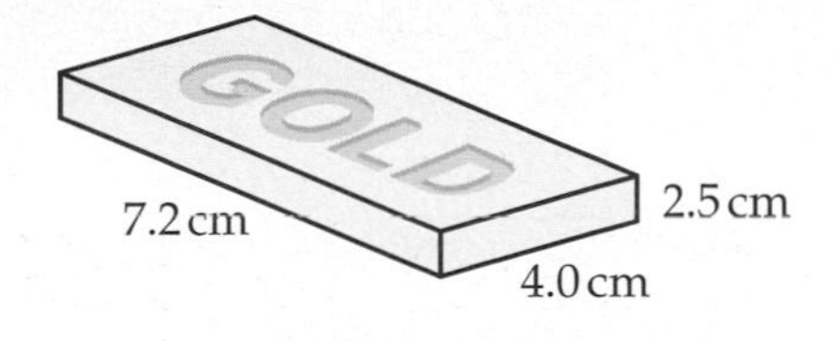

The density of gold is 19.3 g/cm^3 to three significant figures.
Angie says that this gold bar must be worth at least £30 000.

a Show working to explain why Angie is wrong. **[4 marks]**

b Find the greatest possible value of the gold bar to the nearest pound. **[3 marks]**

B

AO3

Chapter summary

In this chapter you have learned how to

- convert between different units of area **D**
- calculate average speeds **D**
- convert between different units of volume **C**
- solve problems involving average speeds using upper and lower bounds **C**
- make calculations involving density **C**
- solve problems involving density using upper and lower bounds **B**
- recognise formulae for length, area or volume by considering dimensions **B**

Paperweights

Moira starts her own business designing and selling paperweights. She makes the paperweights from resin that she pours into different moulds. Moira initially starts with five different mould shapes.

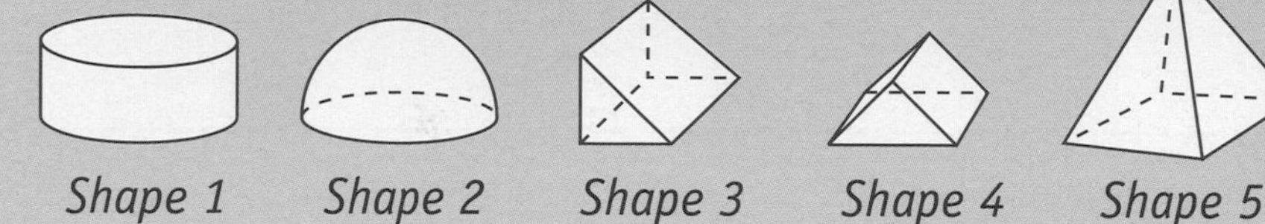

Shape 1 is a cylinder with a radius of 30 mm and a height of 40 mm. Shape 2 is a hemisphere of radius 40 mm.

Question bank

1 What is the volume of shape 1? Give your answer to the nearest m*l*. Use π on your calculator.

2 What is the weight of resin needed to make shape 1? Give your answer to the nearest gram.

Moira says, 'I can make six shape 1 paperweights and eight shape 2 paperweights out of a 2 kg bag of resin.'

3 Is Moira correct? Show working to support your answer.

4 Work out the number of shape 1 paperweights and shape 2 paperweights Moira should make out of a 2 kg bag of resin if she wants the least amount of resin left over.

Moira uses the shape 5 mould to make her 'Deluxe' paperweight. The paperweight contains five glass balls. Each ball has a diameter of 3 cm.

5 What volume of resin does Moira need to make her 'Deluxe' paperweight?

Moira orders some equipment on the internet. She orders 3 bags of resin, 1 can of PVA releasing agent, 2 packets of mixing sticks and 3 tubes of pigment.

6 By first rounding the prices to the nearest £5, estimate the total cost of this equipment including VAT. You must show your working.

7 Calculate the actual cost including VAT.

8 Was your estimate a good estimate? Explain your answer.

Information bank

Volume of a cylinder $= \pi r^2 h$

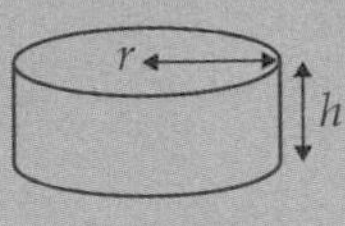

Volume of a sphere $= \frac{4}{3}\pi r^3$

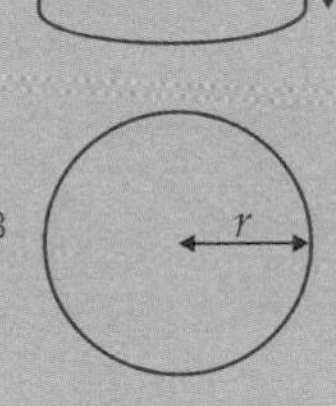

Volume of a prism = area of cross-section $(A) \times l$

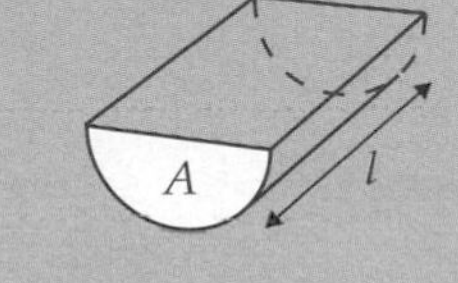

Volume of a pyramid = $\frac{1}{3} \times$ base area $(A) \times h$

Area of a triangle $= \frac{1}{2}bh$

Area of a trapezium $= \frac{h}{2}(a + b)$

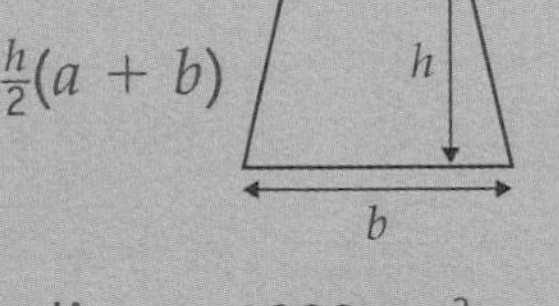

1 ml = 1000 mm^3 1 litre = 1000 cm^3

1000 litres = 1 m^3 1 kg of resin = 900 ml

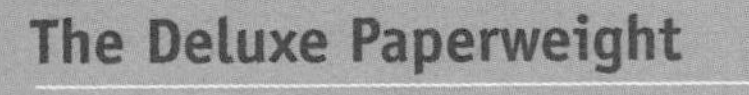

The Deluxe Paperweight

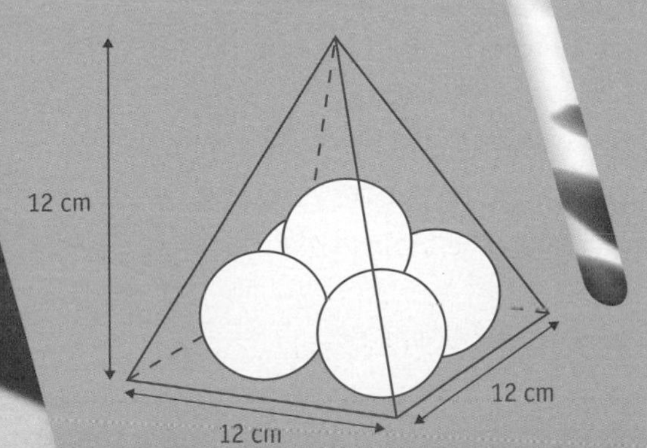

Price list (excluding VAT at 17.5%)

Bag of resin	£31.29
Can of PVA releasing agent	£11.91
Curved-edged mould	£17.50
Straight-edged mould	£14.50
Packet of mixing sticks	£3.58
Glass ball (multicoloured)	£0.80
Tube of pigment	£4.37

29

Constructions and loci

BBC Video

This chapter is about doing constructions using only compasses and a straight edge, and solving locus problems.

The path you follow on a fairground ride can be described by a locus. Fairground designers need to think about loci when they decide how close together two rides should be.

Objectives

This chapter will show you how to

- use a straight edge and compasses to do standard constructions, including the perpendicular bisector of a line segment, the perpendicular from a point to a line, the perpendicular from a point on a line, and the bisector of an angle C
- find loci to produce shapes and paths C B

Before you start this chapter

1 **a** Use a pair of compasses to construct an equilateral triangle *EFG*.
b Use a ruler to find the mid-point of *FG*. Label it *M*. Draw the line *EM*.
c Measure the angles *FEM* and *GEM*. What do you notice?
d Describe the geometrical relationship between *EM* and *GF*.

2 **a** Draw a circle with centre *O* and label the ends of a diameter *A* and *B*.
b Mark a point *P* on the circumference and draw the triangle *APB*.
c Measure the angle *APB*.
d Describe any geometrical relationships between *AP* and *PB*.

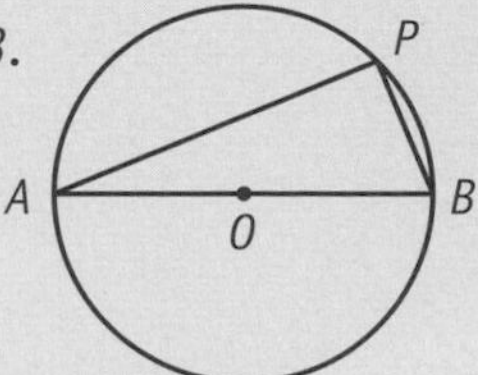

29.1 Constructions

Keywords
construction, bisect, arc, line segment, perpendicular, bisector

Why learn this?
You need to be able to make accurate constructions and measurements in design.

Objectives

- C Construct perpendiculars
- C Construct the perpendicular bisector of a line segment
- C Construct angles of 90° and 60°
- C Construct the bisector of an angle

Skills check

Draw a triangle.

Estimate and mark the mid-point of each side of the triangle.

Check how accurate you have been by measuring with a ruler.

Constructions

Constructions are accurate diagrams drawn using only

- compasses
- a straight edge.

You are not allowed to use a protractor when doing constructions.

Constructions are based on the geometry hidden in a diagram of intersecting circles like this one.

You may use a ruler as the straight edge, but you must not measure with it.

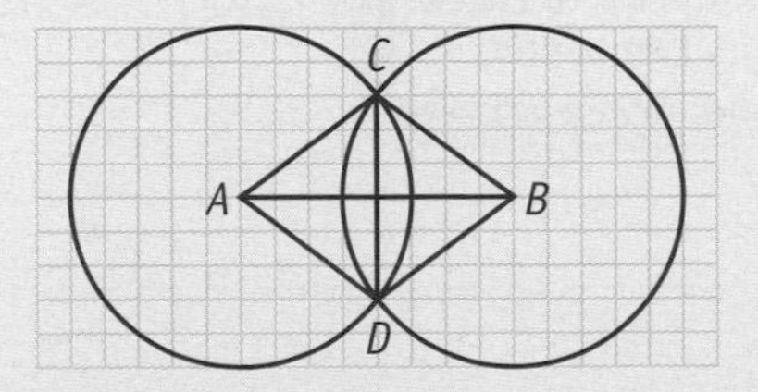

Refer to this diagram when you are doing your constructions. It will help you to understand why they work.

ACBD is a rhombus with diagonals *AB* and *CD*.

The angle between the diagonals is 90°.

The diagonals **bisect** each other.

The diagonals bisect the angles of the rhombus.

Bisect means to divide in half.

When doing constructions, you will often draw **arcs** (parts of a circle), rather than complete circles.

The arcs will show how you did the construction. You must not rub them out, even if you think they look untidy.

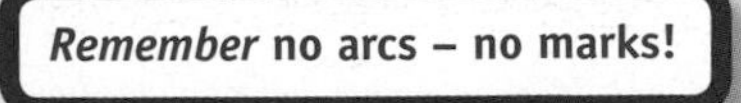

Remember **no arcs – no marks!**

Constructing the perpendicular at a point on a line

A line at right angles (90°) to a given line is called a **perpendicular**.

Example 1

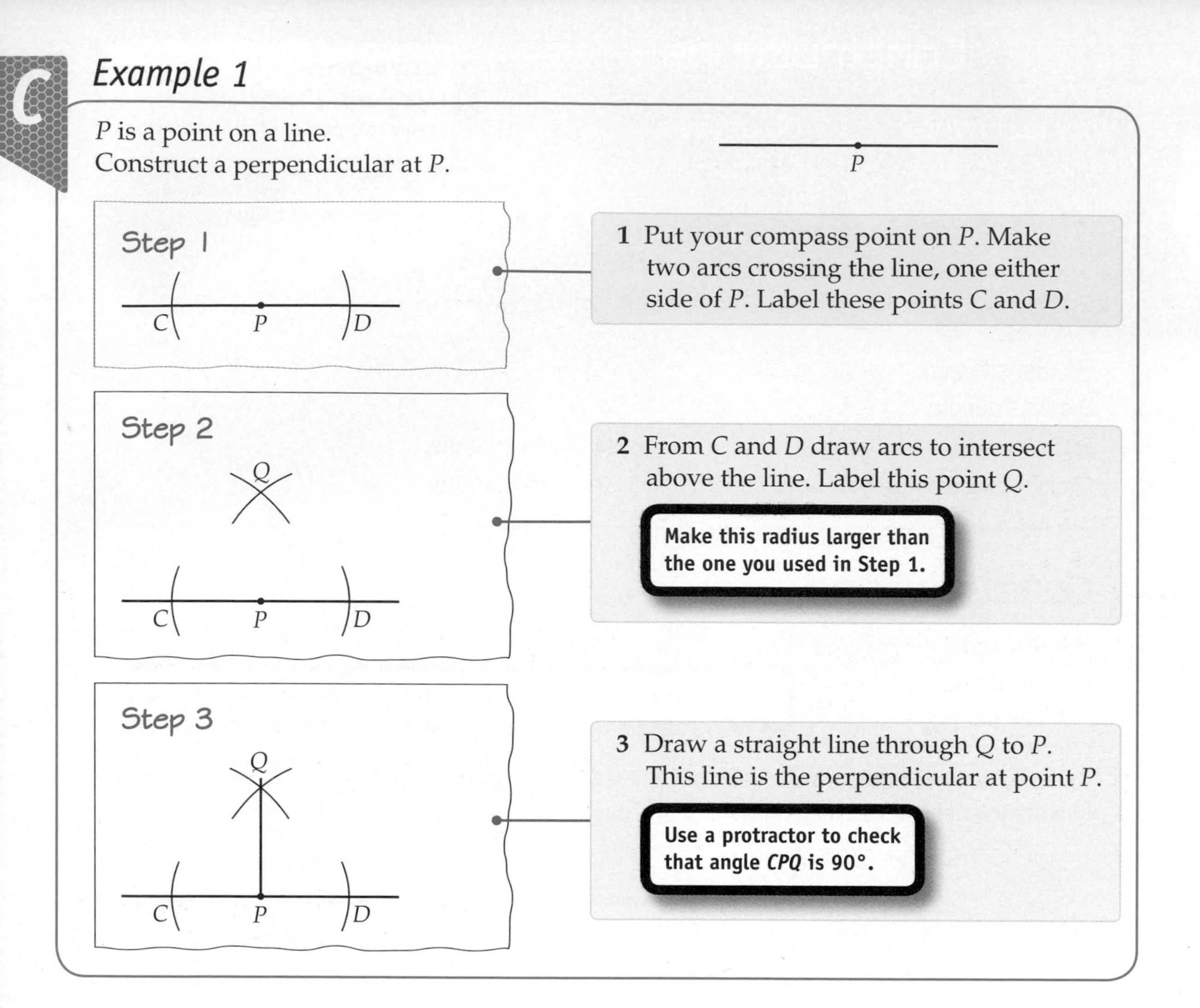

P is a point on a line.
Construct a perpendicular at P.
P
Step 1
C
P
D
1 Put your compass point on P. Make two arcs crossing the line, one either side of P. Label these points C and D.
Step 2
Q
C
P
D
2 From C and D draw arcs to intersect above the line. Label this point Q.
Make this radius larger than the one you used in Step 1.
Step 3
Q
C
P
D
3 Draw a straight line through Q to P. This line is the perpendicular at point P.
Use a protractor to check that angle CPQ is 90°.

Exercise 29A

1 Draw a line at an angle, like the one shown, and mark a point P on it.
Construct a perpendicular at point P.

2 Draw a line segment AB and mark two points, X and Y, on it.

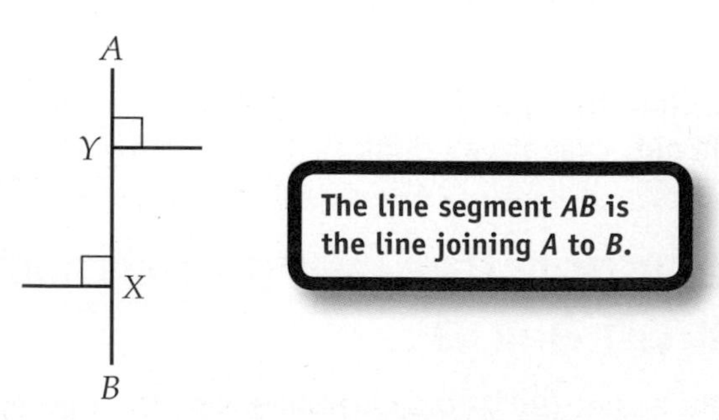

Construct perpendiculars at X and Y, either side of the line.

3 Construct a square without using a protractor or measuring with a ruler.

4 Draw a horizontal line and mark three points on it.
Construct a perpendicular from each point.
What is the geometrical relationship between these perpendiculars?

Constructing the perpendicular from a point to a line

When you are asked to find the distance from a point, *P*, to a line, you will need the shortest distance.

The shortest distance is the length of the perpendicular from *P* to the line.

P

The perpendicular from *P* to the line is at right angles to it.

C

Example 2

Construct the perpendicular from point *P* to the line.

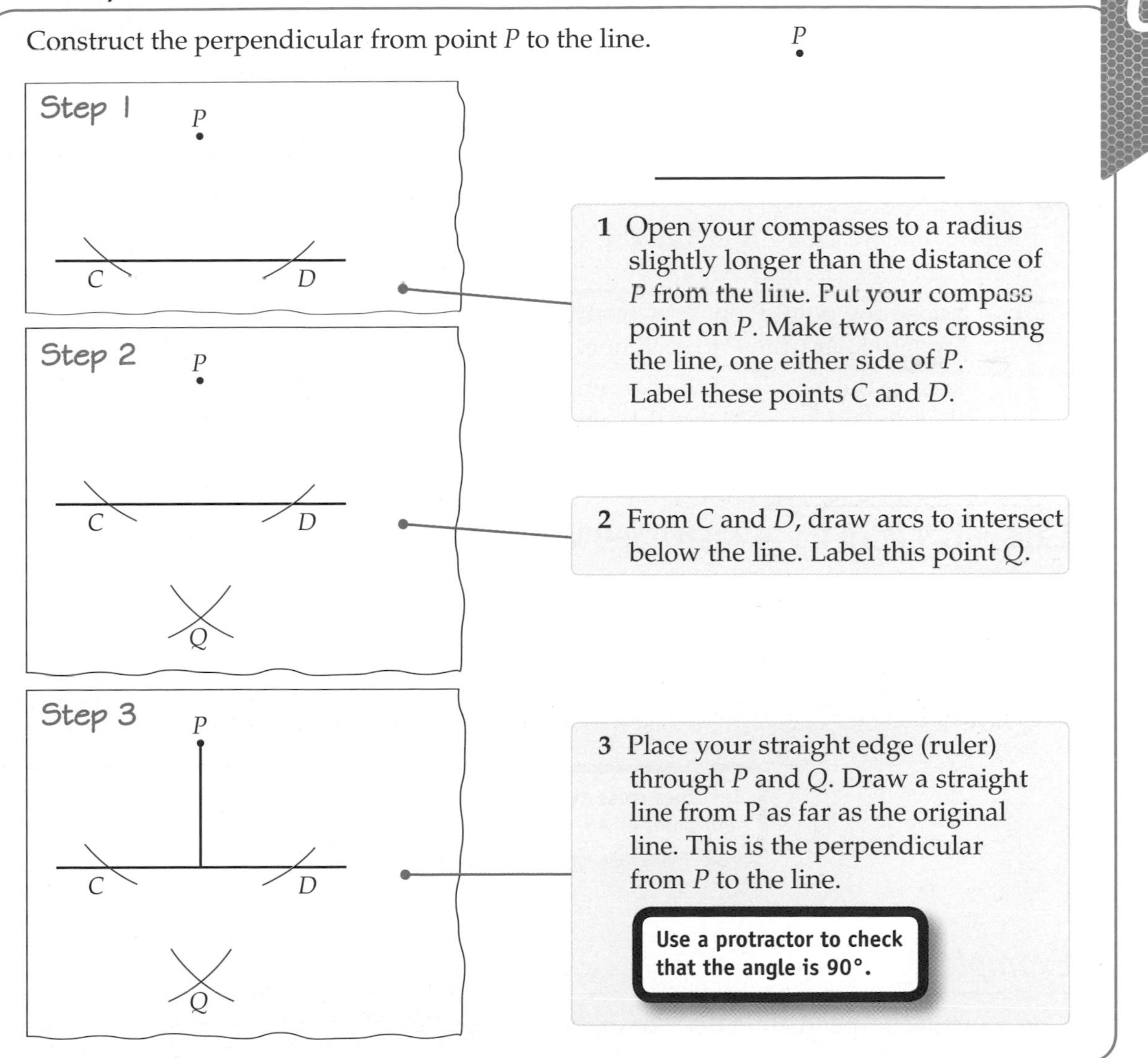

1 Open your compasses to a radius slightly longer than the distance of *P* from the line. Put your compass point on *P*. Make two arcs crossing the line, one either side of *P*. Label these points *C* and *D*.

2 From *C* and *D*, draw arcs to intersect below the line. Label this point *Q*.

3 Place your straight edge (ruler) through *P* and *Q*. Draw a straight line from P as far as the original line. This is the perpendicular from *P* to the line.

Use a protractor to check that the angle is 90°.

Exercise 29B

C

1 Copy the diagrams and construct the perpendicular from the point *P* to each line.

a *P*

b *P*

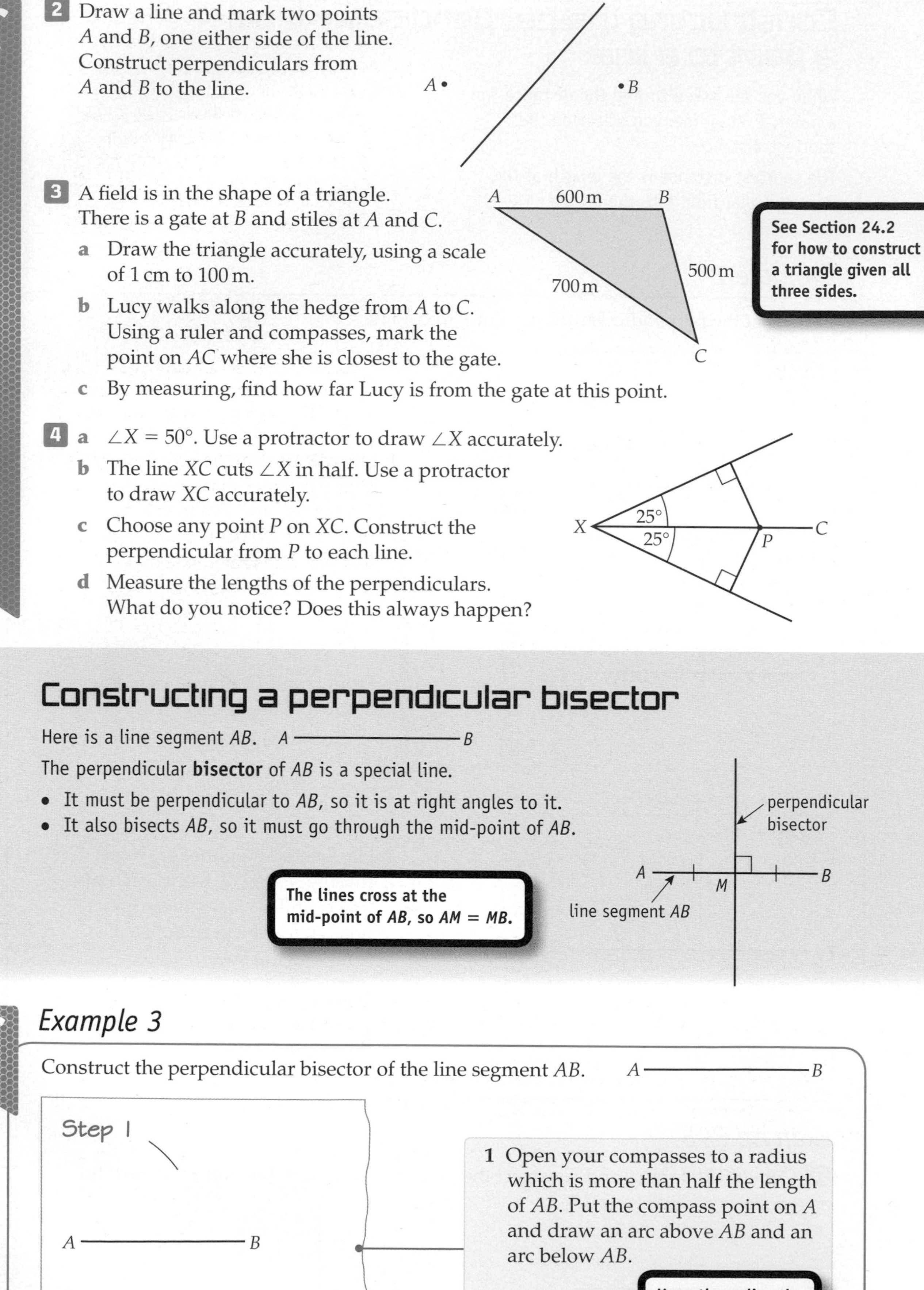

C

2 Draw a line and mark two points A and B, one either side of the line. Construct perpendiculars from A and B to the line.

3 A field is in the shape of a triangle. There is a gate at B and stiles at A and C.

a Draw the triangle accurately, using a scale of 1 cm to 100 m.

b Lucy walks along the hedge from A to C. Using a ruler and compasses, mark the point on AC where she is closest to the gate.

c By measuring, find how far Lucy is from the gate at this point.

See Section 24.2 for how to construct a triangle given all three sides.

4 a $\angle X = 50°$. Use a protractor to draw $\angle X$ accurately.

b The line XC cuts $\angle X$ in half. Use a protractor to draw XC accurately.

c Choose any point P on XC. Construct the perpendicular from P to each line.

d Measure the lengths of the perpendiculars. What do you notice? Does this always happen?

Constructing a perpendicular bisector

Here is a line segment AB.

The perpendicular **bisector** of AB is a special line.

- It must be perpendicular to AB, so it is at right angles to it.
- It also bisects AB, so it must go through the mid-point of AB.

The lines cross at the mid-point of AB, so $AM = MB$.

C

Example 3

Construct the perpendicular bisector of the line segment AB.

Step 1

1 Open your compasses to a radius which is more than half the length of AB. Put the compass point on A and draw an arc above AB and an arc below AB.

Keep the radius the same for both arcs.

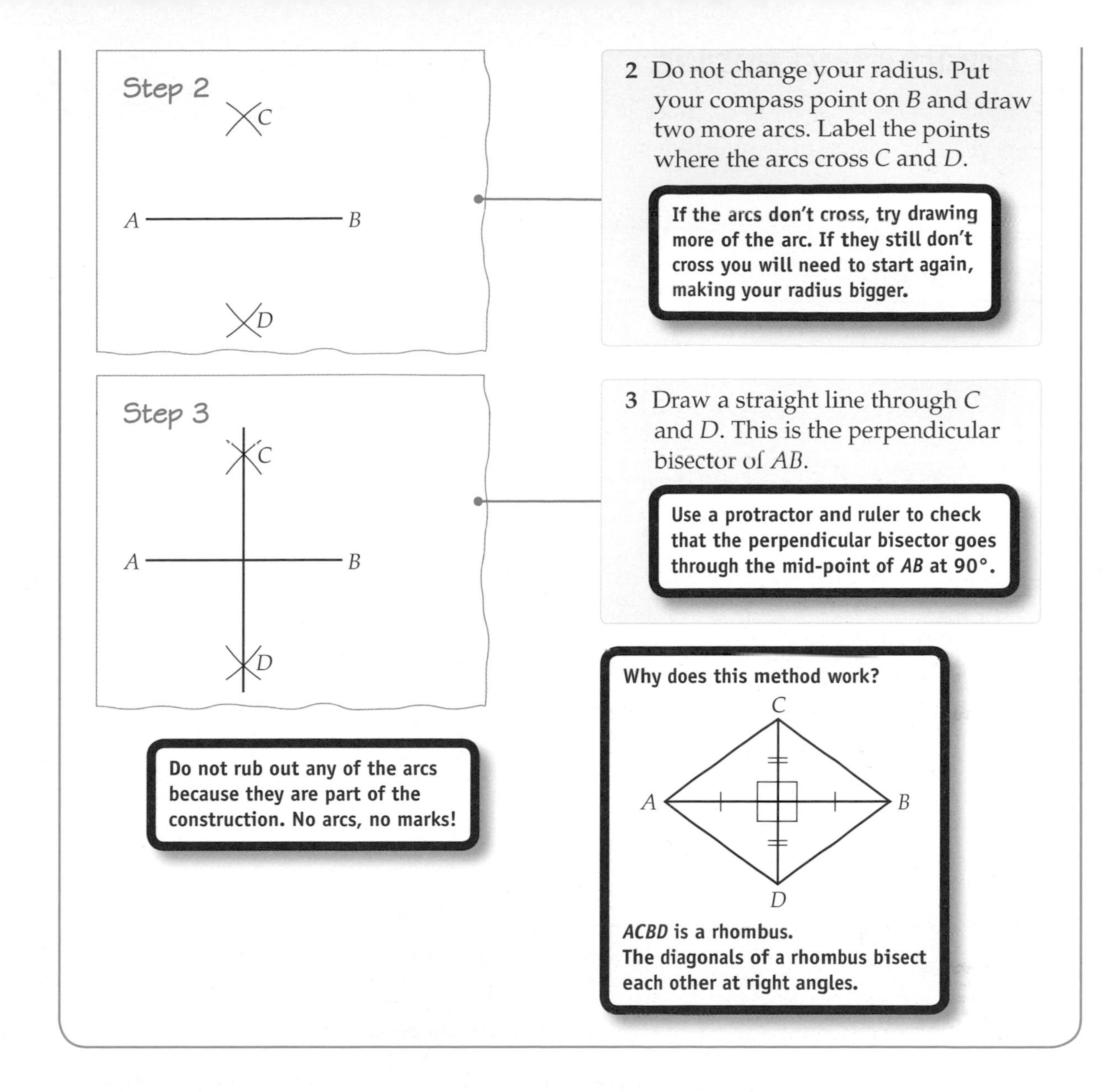

Step 2
C
A
B
D
2 Do not change your radius. Put your compass point on B and draw two more arcs. Label the points where the arcs cross C and D.
If the arcs don't cross, try drawing more of the arc. If they still don't cross you will need to start again, making your radius bigger.
Step 3
C
A
B
D
3 Draw a straight line through C and D. This is the perpendicular bisector of AB.
Use a protractor and ruler to check that the perpendicular bisector goes through the mid-point of AB at 90°.
Do not rub out any of the arcs because they are part of the construction. No arcs, no marks!
Why does this method work?
C
A
B
D
ACBD is a rhombus.
The diagonals of a rhombus bisect each other at right angles.

Exercise 29C

For these questions, use compasses and straight edge to do the constructions. Then check your accuracy by measuring with a protractor and ruler.

1 Draw three line segments of different lengths. Label the ends A and B.
Construct the perpendicular bisector of each line segment AB.

2 Draw a circle with a radius of 5 cm and label the centre O.
Draw a chord XY.
Construct the perpendicular bisector of XY.
What do you notice?

X
O
Y

3 Draw a line segment PQ.
By construction, find the mid-point of PQ.

C

Constructing angles

Example 3 showed you how to construct an angle of 90°.

Here you will find how to construct an angle of 60°.

C

Example 4

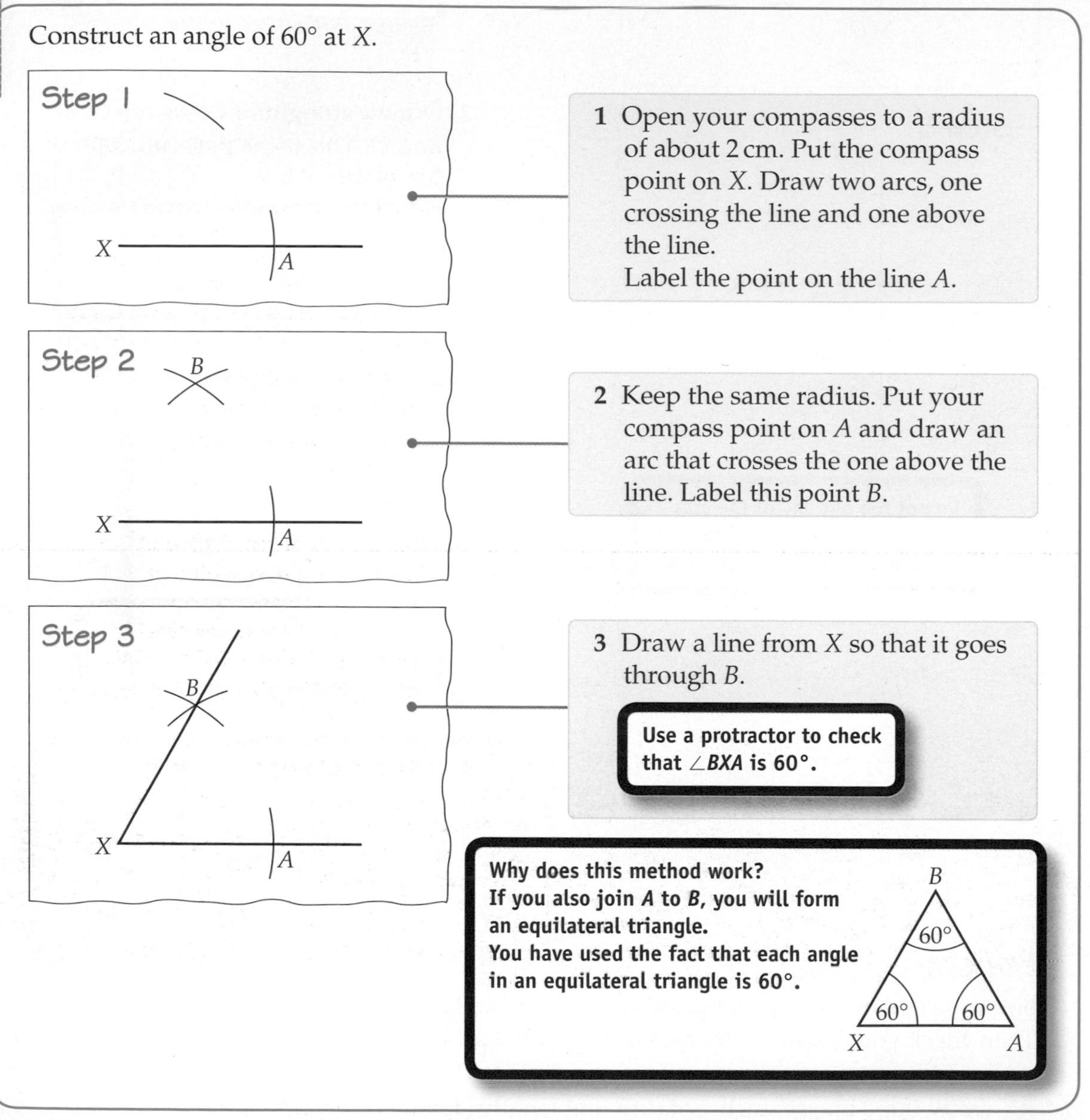

Construct an angle of 60° at X.
Step 1
X
A
1 Open your compasses to a radius of about 2 cm. Put the compass point on X. Draw two arcs, one crossing the line and one above the line.
Label the point on the line A.
Step 2
B
X
A
2 Keep the same radius. Put your compass point on A and draw an arc that crosses the one above the line. Label this point B.
Step 3
B
X
A
3 Draw a line from X so that it goes through B.
Use a protractor to check that ∠BXA is 60°.
Why does this method work?
If you also join A to B, you will form an equilateral triangle.
You have used the fact that each angle in an equilateral triangle is 60°.
B
60°
60°
60°
X
A

Constructing the bisector of an angle

The bisector of an angle divides an angle into two equal parts.

The line is called an angle bisector.

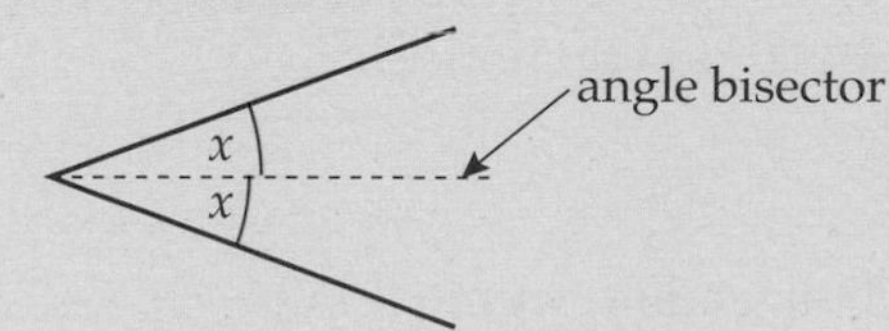

Example 5

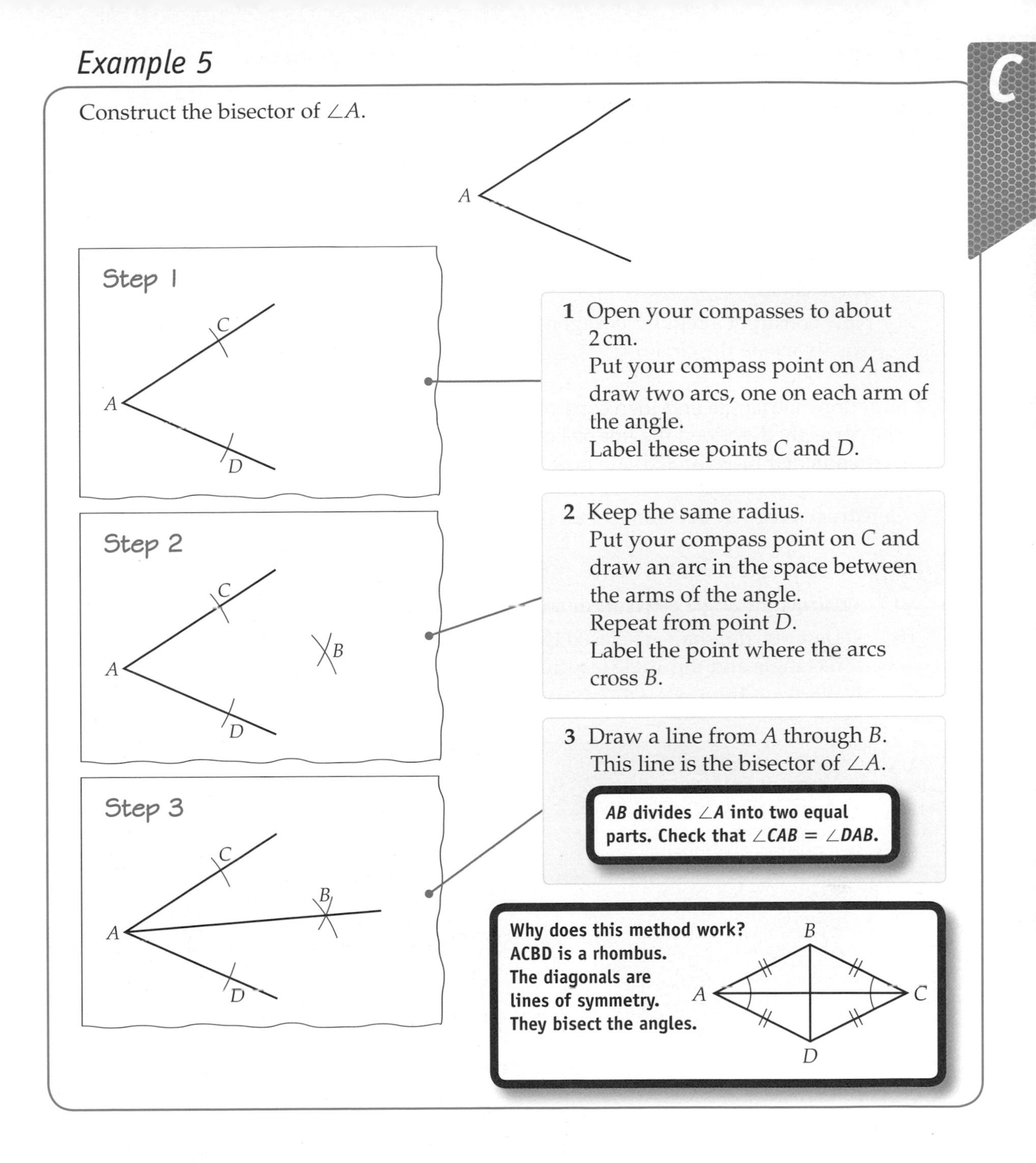

Construct the bisector of ∠A.
A
Step 1
C
A
D
1 Open your compasses to about 2 cm.
Put your compass point on A and draw two arcs, one on each arm of the angle.
Label these points C and D.
Step 2
C
A
B
D
2 Keep the same radius.
Put your compass point on C and draw an arc in the space between the arms of the angle.
Repeat from point D.
Label the point where the arcs cross B.
Step 3
C
B
A
D
3 Draw a line from A through B.
This line is the bisector of ∠A.
AB divides ∠A into two equal parts. Check that ∠CAB = ∠DAB.
Why does this method work?
ACBD is a rhombus.
The diagonals are lines of symmetry.
They bisect the angles.
B
A
C
D

Exercise 29D

1 **a** Draw an acute angle, X. **b** Draw an obtuse angle, Y.

c Construct the angle bisector of each angle.

2 Draw two lines crossing at A.
Construct the angle bisectors of angles x and y.
What do you notice about the angle bisectors?

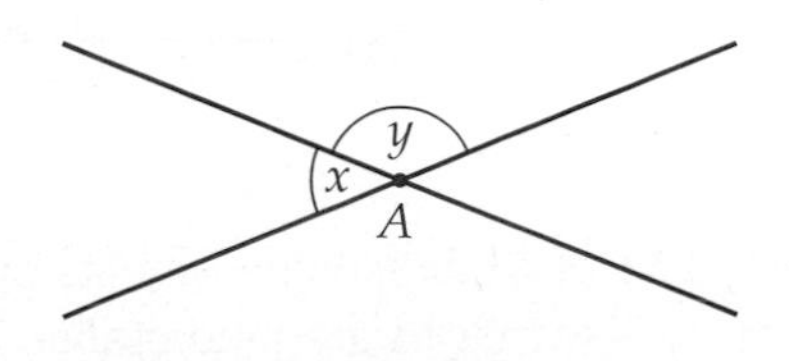

3 **a** Using a ruler and compasses only, construct $\angle A$, where $\angle A = 60°$.

b By constructing the angle bisector of $\angle A$, construct an angle of 30°.

c Explain how you would construct an angle of 15°.

C

4 a Jack says he can draw an angle of 45° without using a protractor. Explain how he can do this.

b Using a ruler and compasses only, construct an angle of 45°.

c State an obtuse angle Jack could construct accurately without using a protractor. Explain how.

5 a Draw a circle.

b Construct an angle of 60° at the centre of the circle.

c Now construct a regular hexagon so that it fits exactly inside the circle.

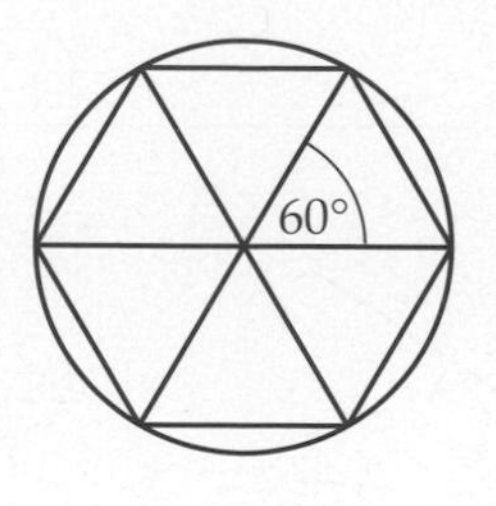

C

6 Sam drew a triangle and then constructed a line. She correctly described the line as both an angle bisector and a perpendicular bisector. Explain how this could happen.

7 Construct a regular octagon so that it fits exactly inside a circle.

8 Draw a rectangle $ABCD$, with $AB = 8$ cm and $BC = 6$ cm.

a Construct the angle bisectors of angle ADC and angle ABC.

b i Describe the shape enclosed by the sides of the rectangle and the angle bisectors.

ii Work out the area of this shape.

AO2

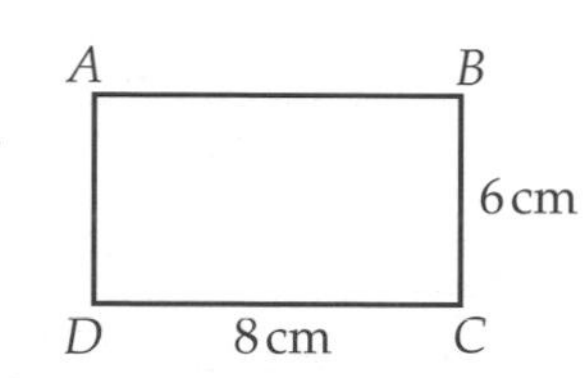

29.2 Locus

Keywords
locus (loci), equidistant, line, line segment

Why learn this?
Takeaway food outlets often have a free delivery area, which can be described as a locus, e.g. 'anywhere within two miles of the restaurant'.

Objectives

- C Solve locus problems including the use of bearings
- C B Construct loci

Skills check

1. Describe a circle.
2. Sketch the following paths.
   - a A ball being thrown from one person to another.
   - b The centre of a ball as it is rolled along level ground.
   - c The tip of a stationary car's windscreen wiper in the rain.
   - d The tip of a minute hand of a clock.
   - e A teacher's footsteps during a lesson.

Locus

The paths of many moving objects are unpredictable, such as the trail left by a snail in the garden. Some paths are predictable, such as a roundabout ride at a fair.

A **locus** is a set of points that obeys a given rule.

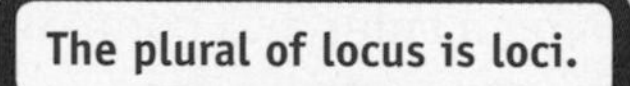

You can use standard constructions to show loci.

For locus questions:

- think about the points
- make a sketch
- construct the locus accurately using standard constructions.

Example 6

C is a fixed point. Draw the locus of points that are always 1 cm from C.

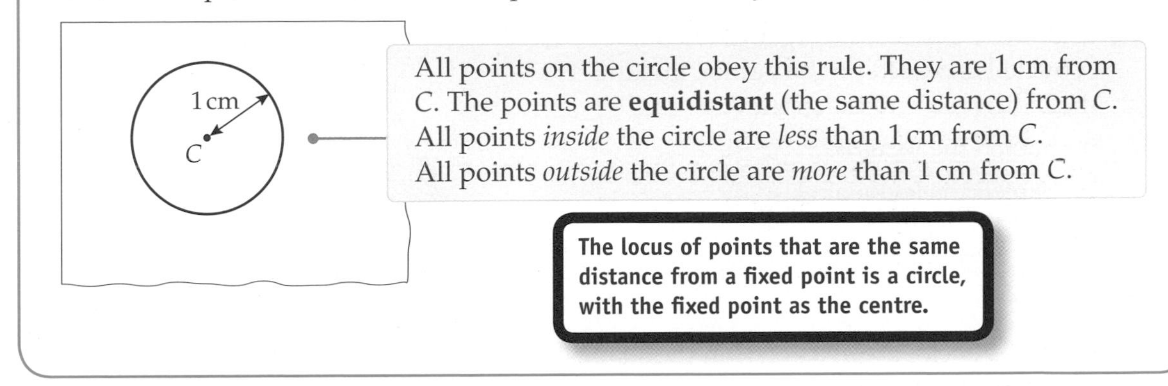

1 cm
C
All points on the circle obey this rule. They are 1 cm from C. The points are **equidistant** (the same distance) from C.
All points *inside* the circle are *less* than 1 cm from C.
All points *outside* the circle are *more* than 1 cm from C.

The locus of points that are the same distance from a fixed point is a circle, with the fixed point as the centre.

Example 7

Draw the locus of points that are exactly 1 cm from

a a given line   **b** a given line segment *AB*.

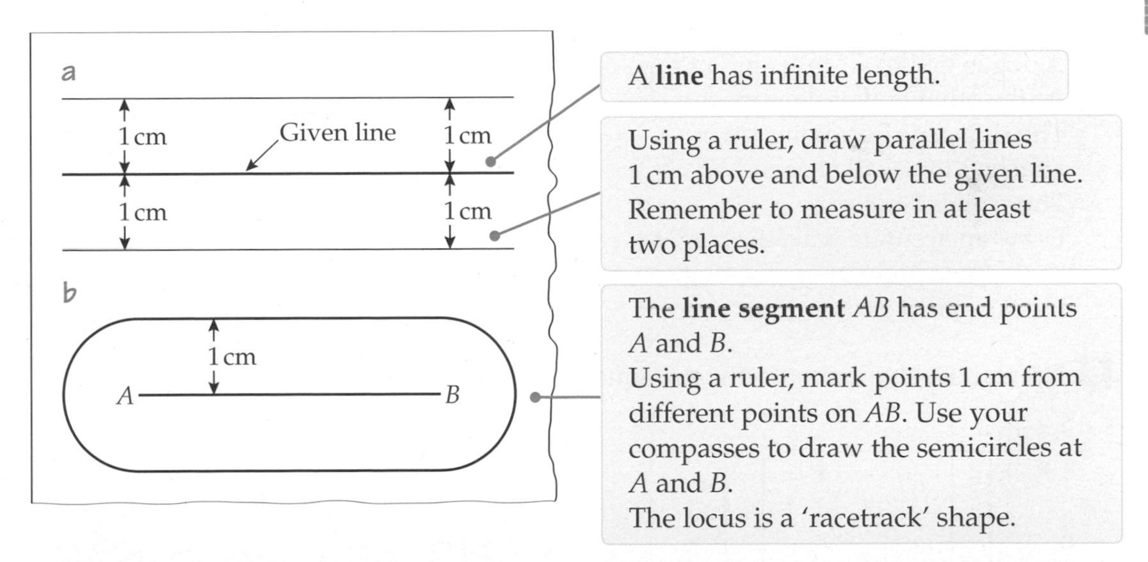

a
1 cm
Given line
1 cm
1 cm
1 cm
b
1 cm
A
B
A **line** has infinite length.
Using a ruler, draw parallel lines 1 cm above and below the given line. Remember to measure in at least two places.
The **line segment** *AB* has end points *A* and *B*.
Using a ruler, mark points 1 cm from different points on *AB*. Use your compasses to draw the semicircles at *A* and *B*.
The locus is a 'racetrack' shape.

The locus of points that are the same distance from a fixed *line* is two parallel lines, one each side of the given line.
The locus of points that are the same distance from a fixed *line segment AB* is a 'racetrack' shape. The shape has two lines parallel to *AB* and two semicircular ends.

Exercise 29E

1 Signals from a transmitter can be picked up at any point within a distance of 30 km from it.
Draw a diagram to show where the signal can be picked up.
Use a scale of 1 cm to 10 km.

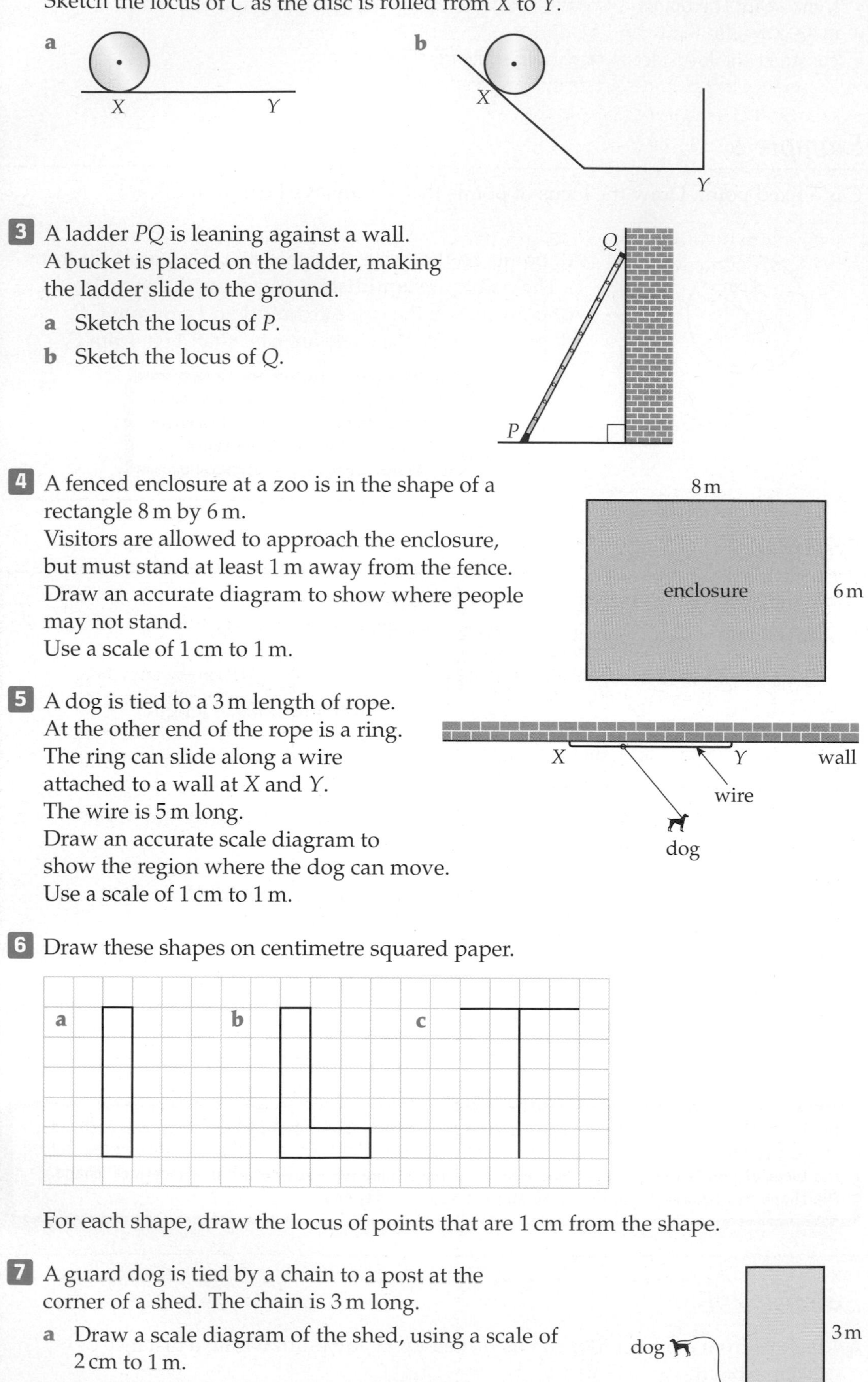

C

2 A disc has centre C.
Sketch the locus of C as the disc is rolled from X to Y.

a

b

3 A ladder PQ is leaning against a wall.
A bucket is placed on the ladder, making the ladder slide to the ground.

a Sketch the locus of P.

b Sketch the locus of Q.

4 A fenced enclosure at a zoo is in the shape of a rectangle 8 m by 6 m.
Visitors are allowed to approach the enclosure, but must stand at least 1 m away from the fence.
Draw an accurate diagram to show where people may not stand.
Use a scale of 1 cm to 1 m.

5 A dog is tied to a 3 m length of rope.
At the other end of the rope is a ring.
The ring can slide along a wire attached to a wall at X and Y.
The wire is 5 m long.
Draw an accurate scale diagram to show the region where the dog can move.
Use a scale of 1 cm to 1 m.

6 Draw these shapes on centimetre squared paper.

a **b** **c**

For each shape, draw the locus of points that are 1 cm from the shape.

7 A guard dog is tied by a chain to a post at the corner of a shed. The chain is 3 m long.

a Draw a scale diagram of the shed, using a scale of 2 cm to 1 m.

b Show on the diagram all the possible positions of the dog if the chain remains tight.

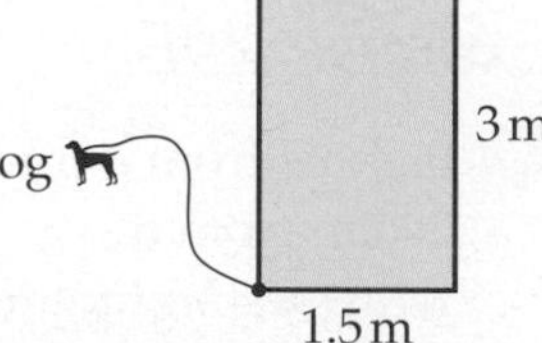

8 The base PQ of $\triangle PQR$ is fixed.
The point R can move so that the area of $\triangle PQR$ stays the same.
Describe the locus of the point R.

B

9 A point moves in three-dimensional space.
Describe the locus of points that are

a 3 cm from a fixed point A

b 3 cm from a fixed line segment AB.

AO2

10 A square field has sides of length 10 m.
A goat is tethered at a corner of the field by a rope of length 8 m.
Show by calculation that the goat can graze just over half of the field.

B

AO3

Special loci

Learn these special loci.

The locus of points equidistant from two fixed points is the perpendicular bisector of the line segment joining the two points.

'Equidistant from' means 'the same distance from'.

A • • B

The locus is the perpendicular bisector of AB.

To construct the locus follow the steps in Example 3.

Why does this work?
Here are some possible positions of the point P.

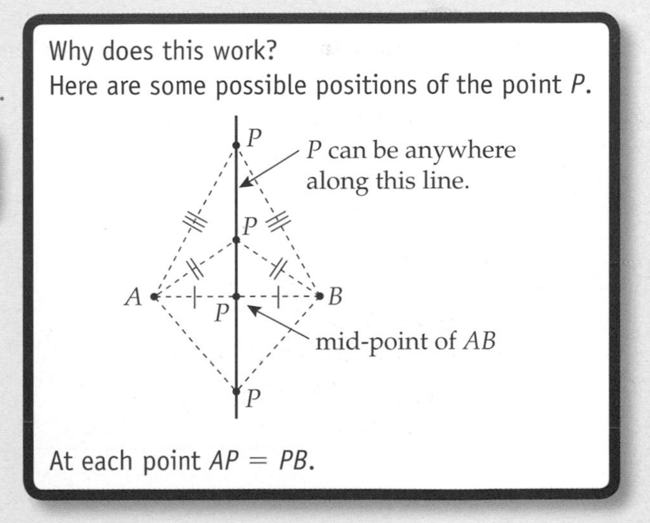

At each point $AP = PB$.

The locus of points equidistant from two fixed lines is the angle bisector of the angle formed by the lines.

X A B

The locus is the angle bisector.

To construct the locus follow the steps in Example 5.

Why does this work?
Remember that the distance from a point to a line is the perpendicular distance.

Here are some possible positions of P. At each point P the perpendicular distance to each line is the same.

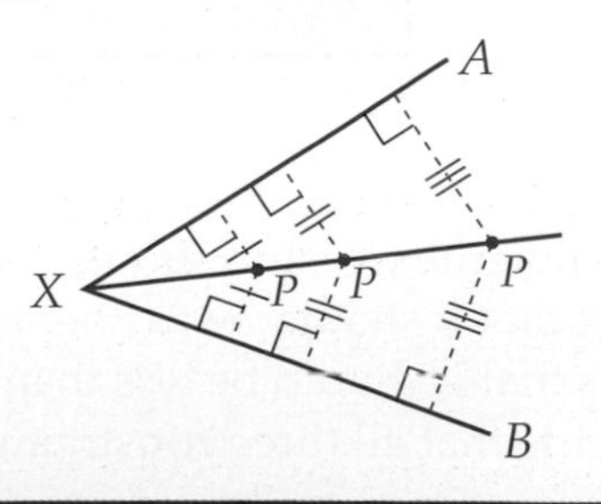

C

Exercise 29F

1 There are two trees in a field.
A fence is to be built across the field so that it is the same distance from each tree.

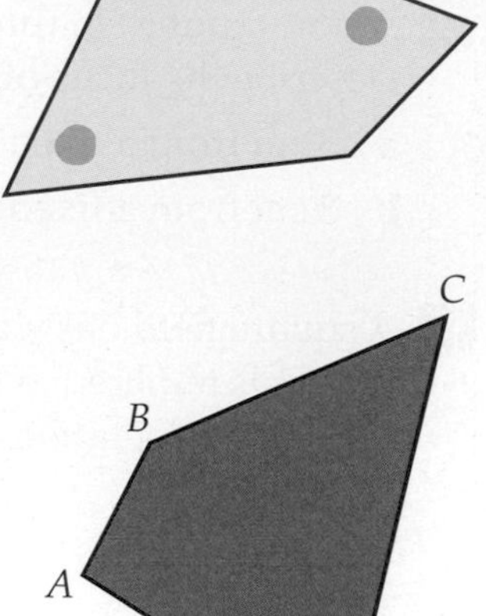

a Copy the diagram and construct the position of the fence.

b Copy and complete this sentence.
'The locus of the fence is…'

2 A drainage pipe is to be laid across a plot of land $ABCD$ so that it is the same distance from AD as it is from CD.

a Copy the diagram and construct the locus of the drainage pipe.

b Copy and complete this sentence.
'The locus of the drainage pipe is…'

C
B
A
D

3 Draw any triangle ABC.
Mark and label any points inside the triangle that are equidistant from A and B **and** equidistant from AB and AC.

Intersecting loci

Questions often involve intersecting loci. You may have to locate a region satisfied by two or more constraints.

C

Example 8

Ben, Josie and Kate are devising a game for their stall at a fête.
Players will try to locate the buried treasure on this rectangular board.

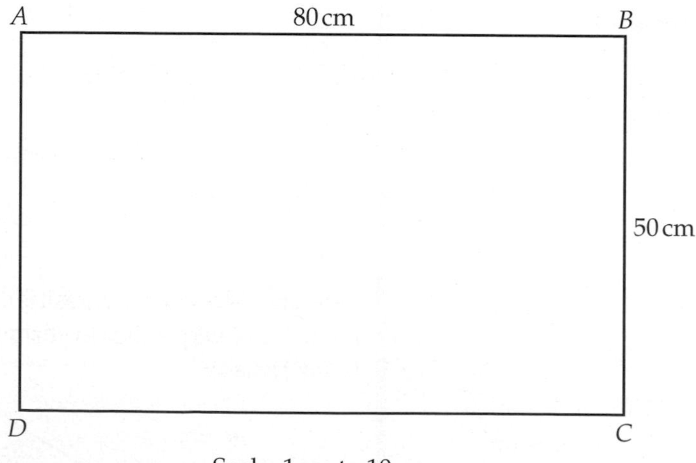

Scale: 1 cm to 10 cm

Ben says that the treasure should be closer to C than it is to D.
Josie says that it should be further away from BC than it is from BA.
Kate says that it should be less than 45 cm from D.
They decide that all three constraints should be met.
By accurate construction, show the region where they should bury the treasure.

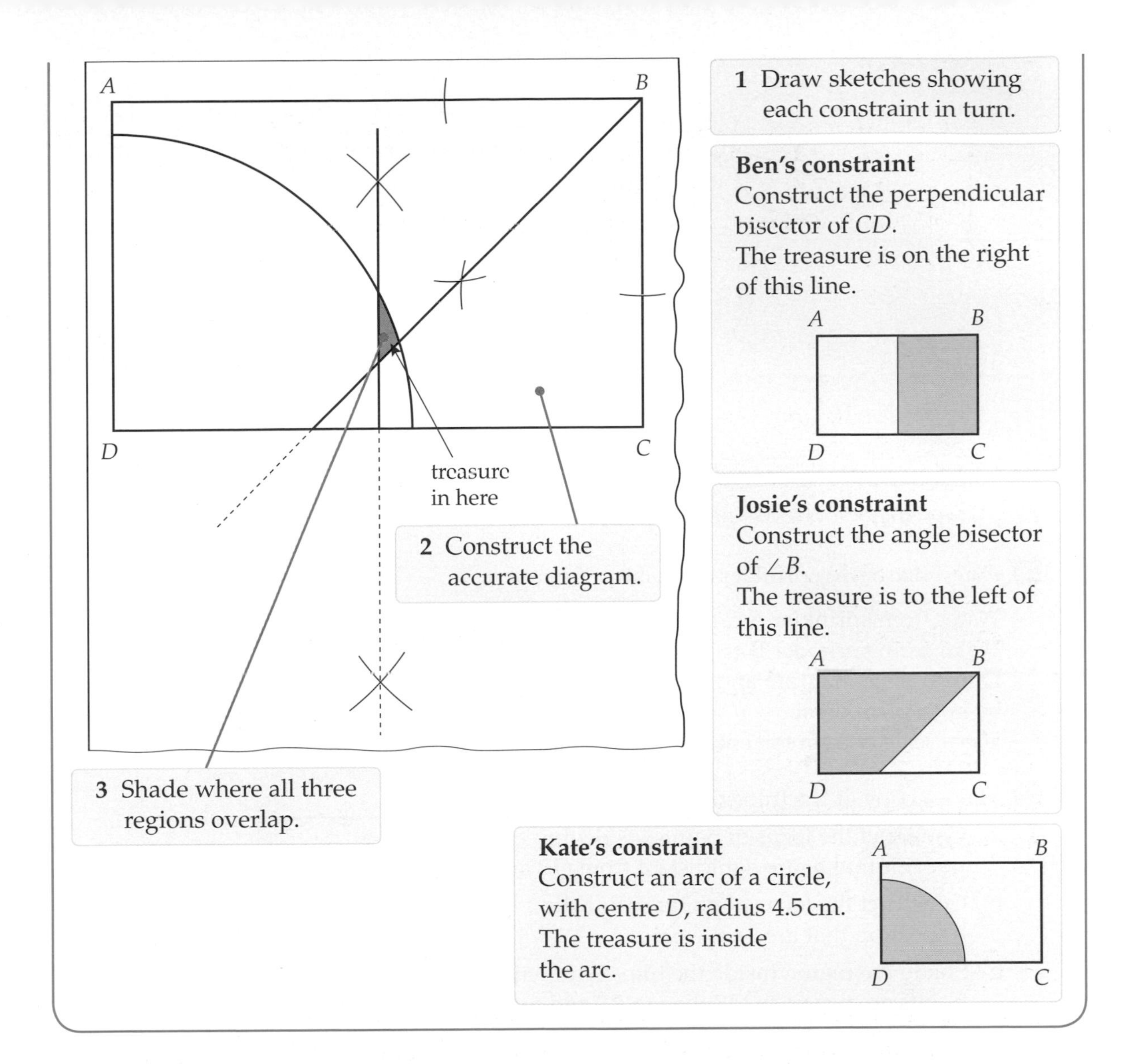

A
B
D
C
treasure in here
1 Draw sketches showing each constraint in turn.
Ben's constraint
Construct the perpendicular bisector of CD.
The treasure is on the right of this line.
Josie's constraint
Construct the angle bisector of ∠B.
The treasure is to the left of this line.
Kate's constraint
Construct an arc of a circle, with centre D, radius 4.5 cm.
The treasure is inside the arc.
2 Construct the accurate diagram.
3 Shade where all three regions overlap.

Example 9

The positions of a ship S and a lighthouse L are shown in the diagram. The scale is 1 cm to 10 km. The ship sails on a bearing of 070°.

N

S

L

a Show clearly where the ship is within 10 km of the lighthouse.

b By constructing a suitable line, find the closest distance between the ship and the lighthouse.

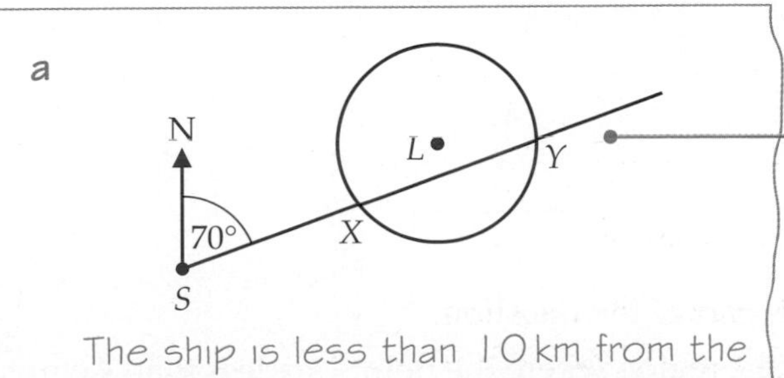

Draw the line that the ship sails on, using a protractor to measure the angle.
The locus of points 10 km from the lighthouse is a circle, centre L. Using a scale of 1 cm to 10 km, draw a circle with radius 1 cm.

C

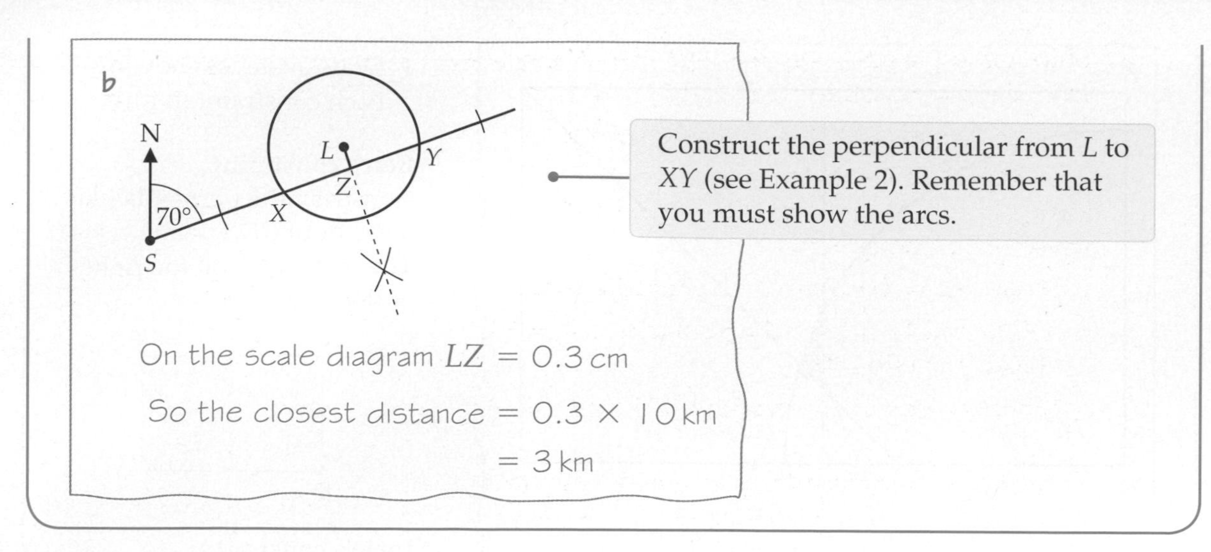

b
N
L
Y
Z
70°
X
S
Construct the perpendicular from L to XY (see Example 2). Remember that you must show the arcs.
On the scale diagram LZ = 0.3 cm
So the closest distance = 0.3 × 10 km
= 3 km

Exercise 29G

C

1 A and B are two sprinklers on a lawn, 6 m apart.

Water from sprinkler A can reach anywhere within 3 m of it.
Water from sprinkler B can reach anywhere within 4 m of it.
Draw a scale diagram and show clearly the region of the lawn that will be watered by both sprinklers.
Use a scale of 1 cm for 1 m.

C

AO2

2 Make a copy of the trapezium $PQRS$.

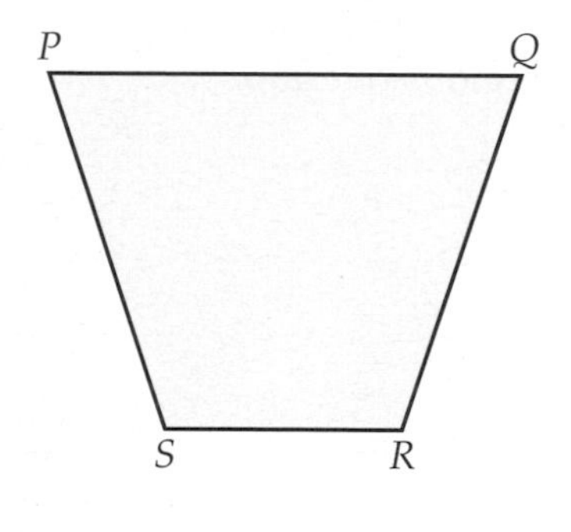

a Construct the locus of points inside the trapezium that are equidistant from QR and RS.

b Construct the locus of points inside the trapezium that are 2 cm from PQ.

c Shade the region inside the trapezium where the points are nearer to QR than to RS **and** more than 2 cm from PQ.

C

AO3

3 In $\triangle ABC$, $AB = 5$ cm, $AC = 7$ cm and $BC = 6$ cm.

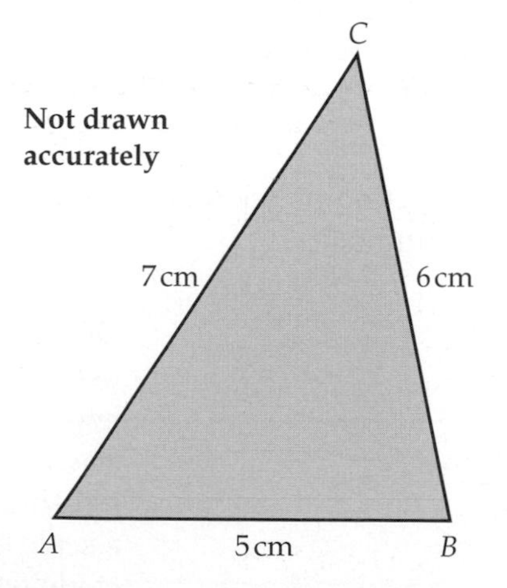

Construct $\triangle ABC$ accurately for each part of the question.

a By construction, find and shade the region where the points are less than 4 cm from C **and** nearer to A than to B.

b By construction, find and shade the region where the points are at least 5 cm from C **and** closer to AB than to BC.

4 Ship A is at the port. Ship B is anchored 20 km away from the port on a bearing of 130°.

a Using a scale of 1 cm for 4 km, draw the positions of the two ships.

Ship A sails out of port on a bearing of 110° at a speed of 30 km per hour.

b Show the path of ship A on your diagram.

c Show clearly on your diagram where ship A is 12 km from ship B.

d For how long is ship A within 12 km of ship B?

5 Three towns, at A, B and C, are connected by main roads. $AB = 16$ miles, $AC = 14$ miles and $BC = 12$ miles.
A new leisure centre is to be built so that it is
- inside triangle ABC,
- closer to road AC than it is to road AB, and
- between 8 miles and 10 miles from town B.

Draw an accurate scale diagram showing A, B and C, and construct the region where the leisure centre could be built.
Use a scale of 1 cm for 2 miles.

6 In $\triangle PQR$, $PQ = 6$ cm, $PR = 4$ cm and $QR = 8$ cm.

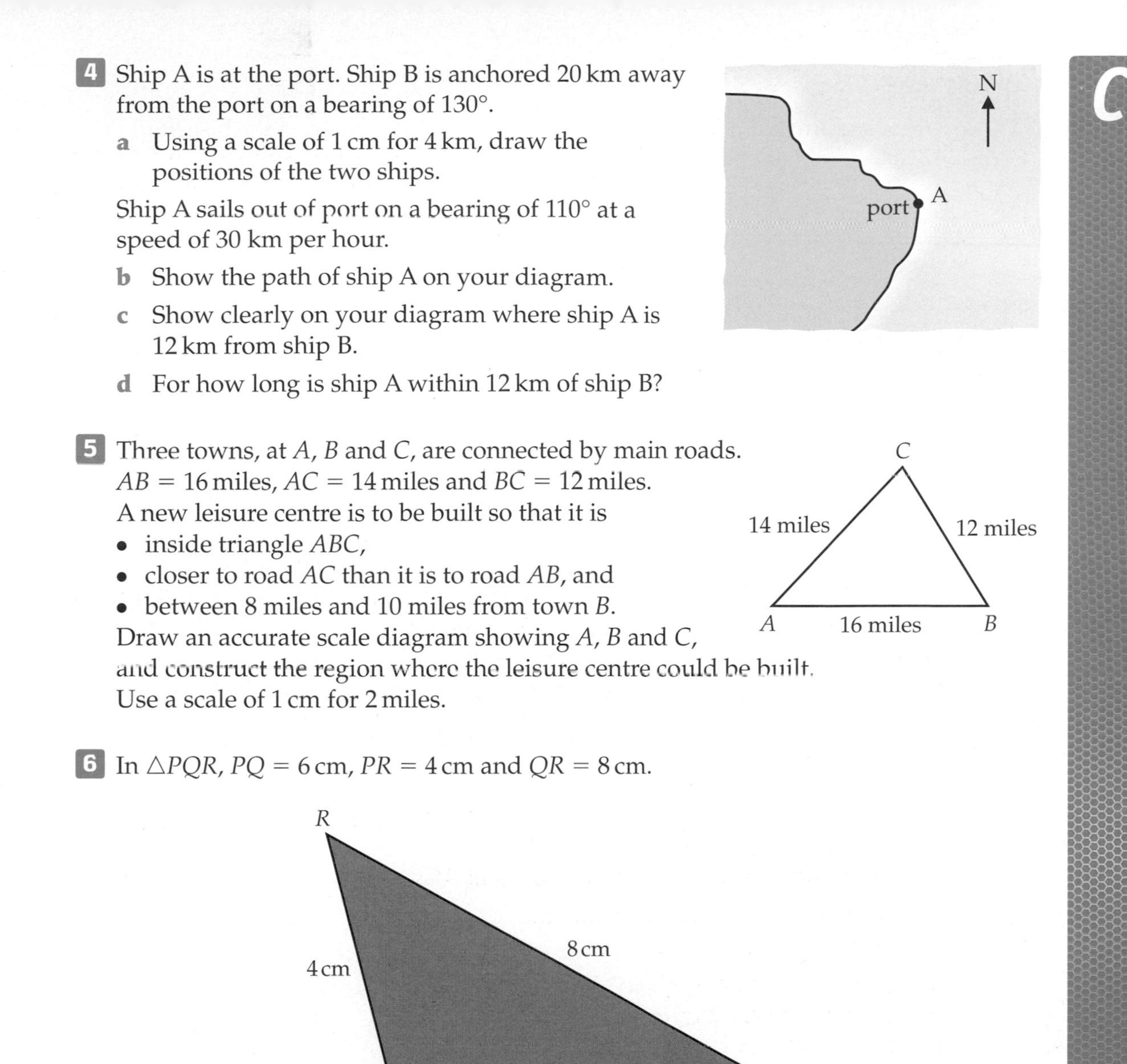

Construct the diagram accurately. Shade the region inside the triangle where points are nearer to R than to P **and** closer to PR than to QR.

7 The diagram shows three lifeboats out at sea at A, B and C.
Along one side of the sea is a coastal path.
The lifeboats are searching for a boat in distress.
The boat is
- between 100 m and 200 m from B
- nearer to A than to C
- more than 50 m from the coastal path.

Copy the diagram.
Shade the region in which the boat lies.

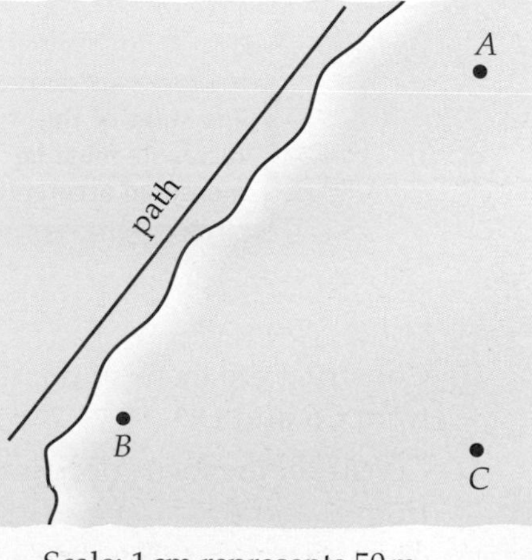

Scale: 1 cm represents 50 m

AO3

Review exercise

C

1 Copy the diagram and, using ruler and compasses only, construct the bisector of $\angle ABC$.

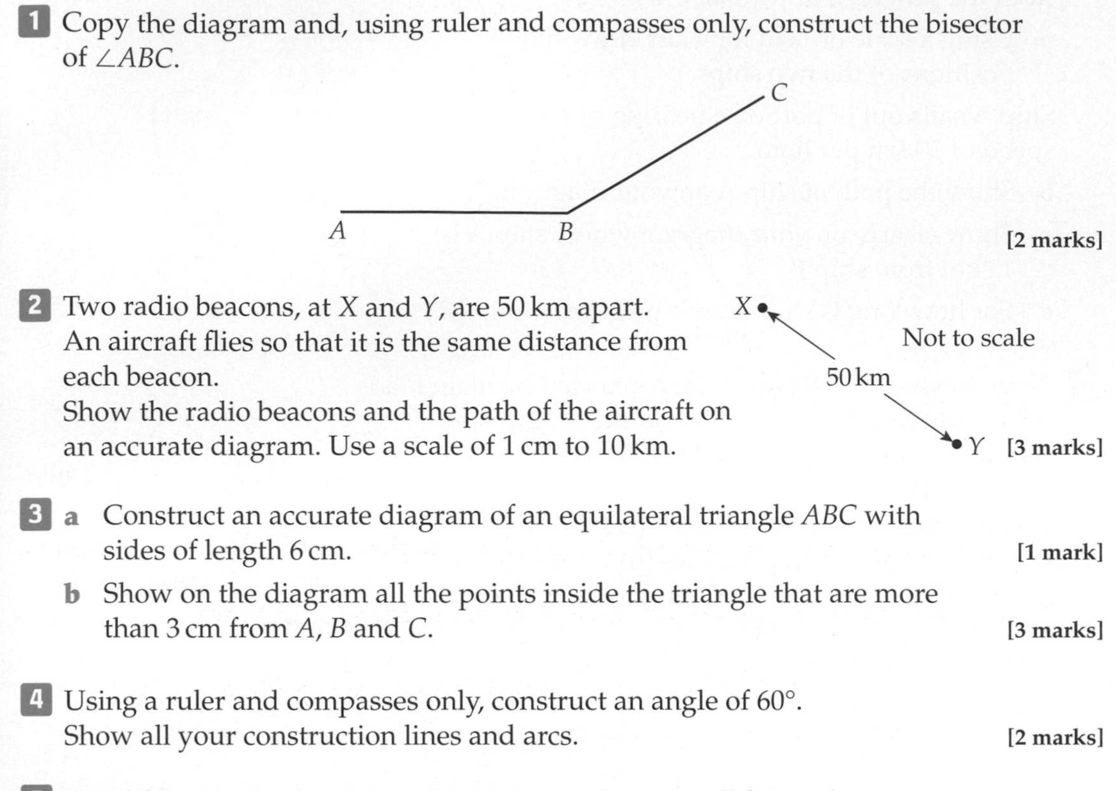

[2 marks]

2 Two radio beacons, at X and Y, are 50 km apart.
An aircraft flies so that it is the same distance from each beacon.
Show the radio beacons and the path of the aircraft on an accurate diagram. Use a scale of 1 cm to 10 km. **[3 marks]**

3 **a** Construct an accurate diagram of an equilateral triangle ABC with sides of length 6 cm. **[1 mark]**

b Show on the diagram all the points inside the triangle that are more than 3 cm from A, B and C. **[3 marks]**

4 Using a ruler and compasses only, construct an angle of 60°.
Show all your construction lines and arcs. **[2 marks]**

5 Two lifeboat stations, at A and B, receive a distress call from a boat.

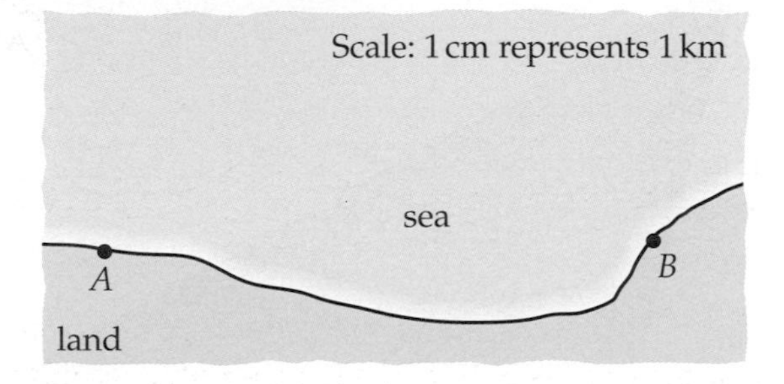

The boat is within 4 km of the station at A.
The boat is within 5 km of the station at B.
Trace the diagram. Shade the possible area in which the boat could be. **[2 marks]**

6 Copy rectangle $ABCD$.

The sides of the rectangle must be measured accurately.

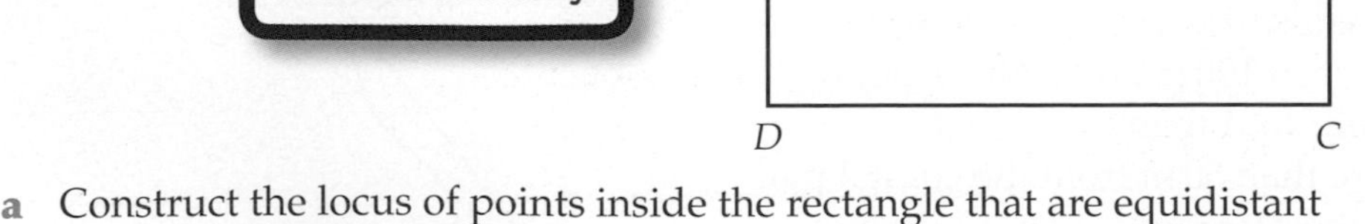

a Construct the locus of points inside the rectangle that are equidistant from CB and CD. **[2 marks]**

b Construct the locus of points inside the rectangle that are equidistant from A and B. **[2 marks]**

c Shade the region where the points are closer to B than to A **and** closer to CD than to CB. **[1 mark]**

C

7 AB and AC represent two walls. A mast is to be erected so that it is equidistant from AB and AC **and** between 20 m and 50 m from A.

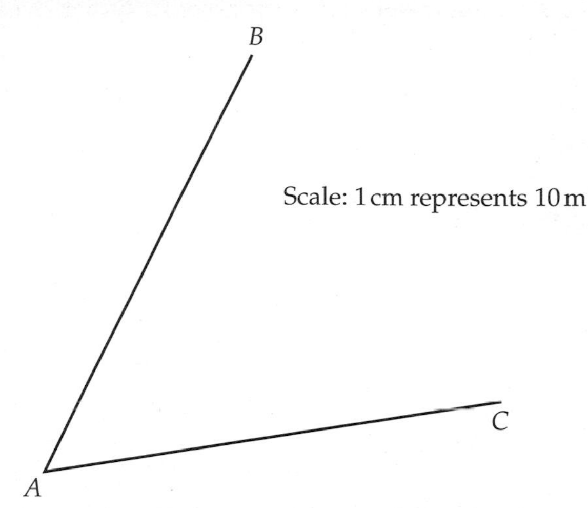

Trace the diagram. Show clearly all the possible positions of the mast. **[3 marks]**

8 Two radio stations at A and B pick up a distress call from a boat at sea.

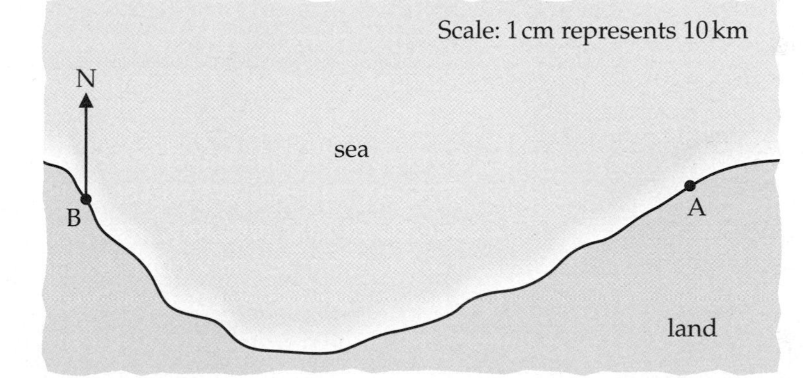

Station A can tell that the boat is between 30 km and 50 km from it.
Station B can tell that the boat is between a bearing of 070° and 080° from it.
Trace the diagram. Show clearly, using compasses and a protractor, the region where the boat will be found. **[3 marks]**

AO3

Chapter summary

In this chapter you have learned how to

- construct perpendiculars **C**
- construct the perpendicular bisector of a line segment **C**
- construct angles of 90° and 60° **C**
- construct the bisector of an angle **C**
- solve locus problems including the use of bearings **C**
- construct loci **C** **B**

30

Reflection, translation and rotation

BBC Video

This chapter is about transforming shapes: reflections, translations and rotations.

Architects use reflective materials in buildings for decorative effect. This glass building reflects the other buildings around it and the sky, which is constantly changing.

Objectives

This chapter will show you how to

- understand that reflections are specified by a (mirror) line D C
- understand that translations are specified by giving a distance and a direction (or a vector) D C
- understand that rotations are specified by a centre, an angle and a direction; use any point as the centre of rotation, using right angles, simple fractions of a turn or degrees D C
- understand and use vector notation for translations C
- transform triangles and other 2-D shapes by translation, rotation and reflection and by combinations of these transformations, recognising that these transformations preserve length and angle D C

Before you start this chapter

Match the shapes into congruent pairs. Which shapes are left over?

A B C D E F G H

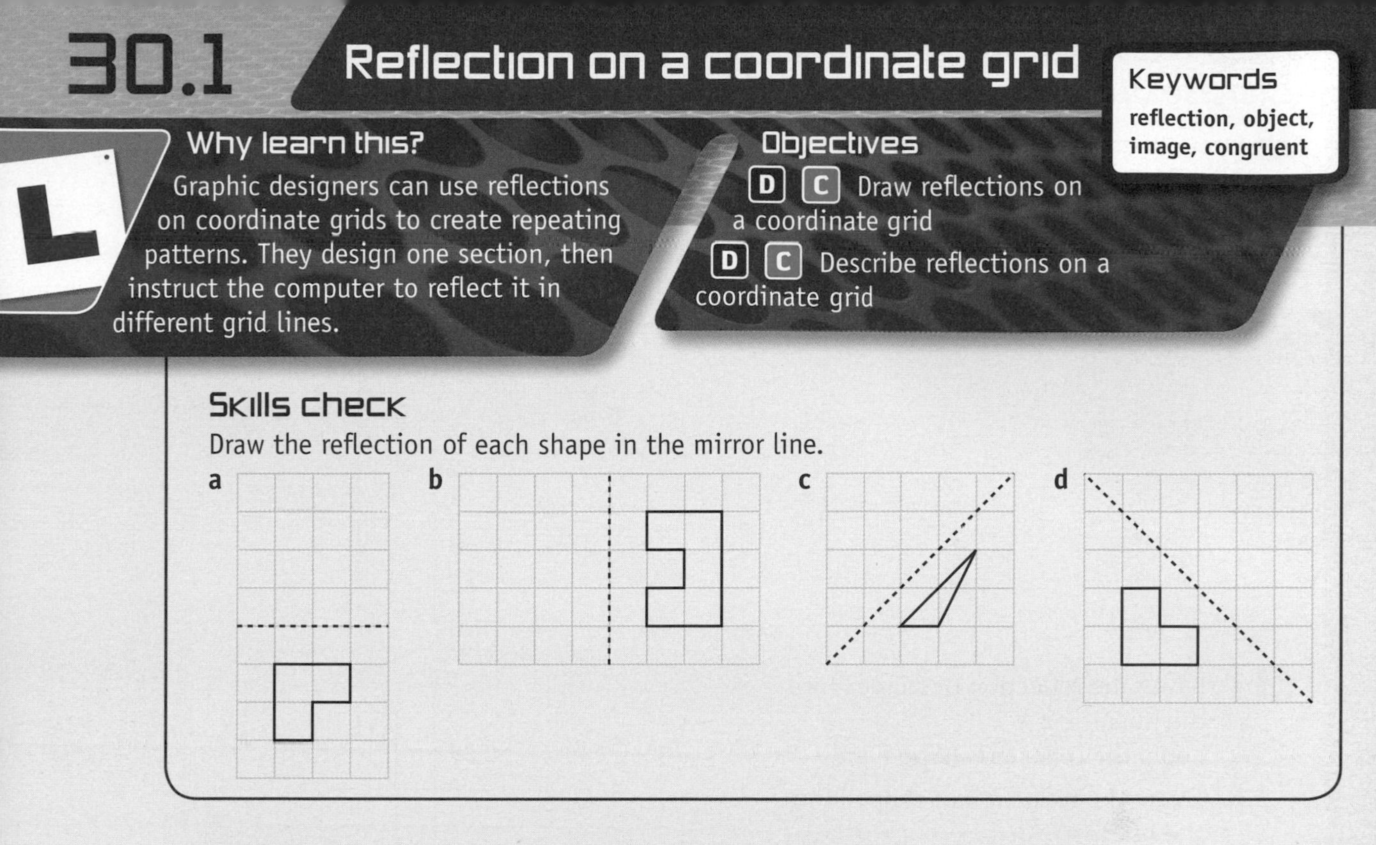

30.1 Reflection on a coordinate grid

Keywords
reflection, object, image, congruent

Why learn this?

Graphic designers can use reflections on coordinate grids to create repeating patterns. They design one section, then instruct the computer to reflect it in different grid lines.

Objectives

D C Draw reflections on a coordinate grid

D C Describe reflections on a coordinate grid

Skills check

Draw the reflection of each shape in the mirror line.

Reflection on a coordinate grid

To describe a **reflection** on a grid, you need to give the equation of the mirror line.

In a reflection, the **object** and the **image** are the same perpendicular distance from the mirror line, on opposite sides.

When you reflect a shape in a mirror line, the object and the image are **congruent**.

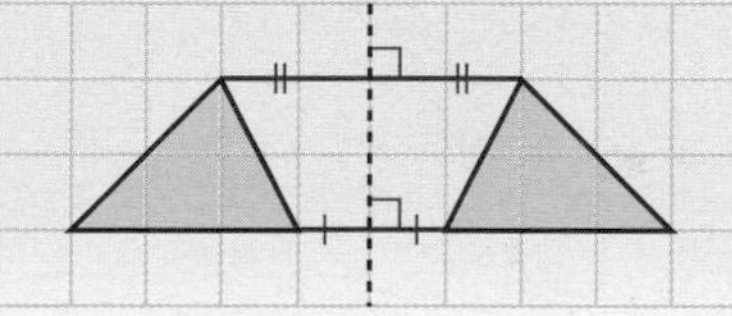

You will learn more about congruency in Chapter 32.

Example 1

D

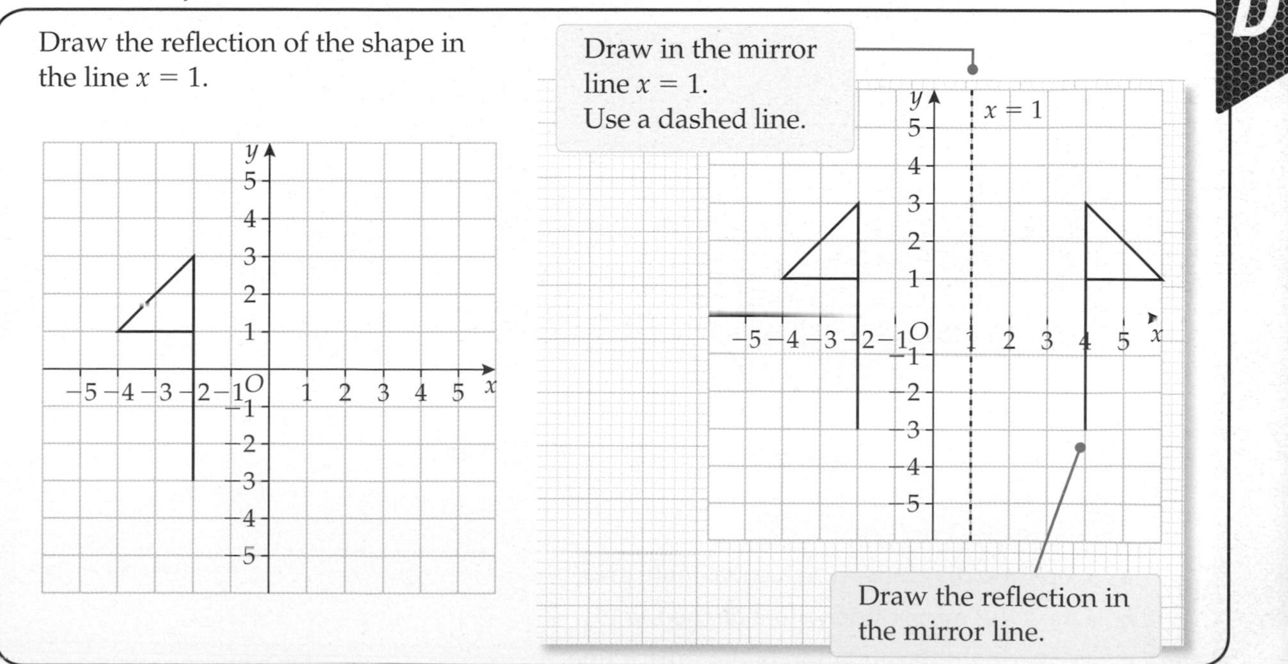

Draw the reflection of the shape in the line $x = 1$.

Exercise 30A

D

1 Draw the reflection of shape D in the line $x = 2$.
Label the reflected shape E.

The line $x = 2$ has been drawn for you.

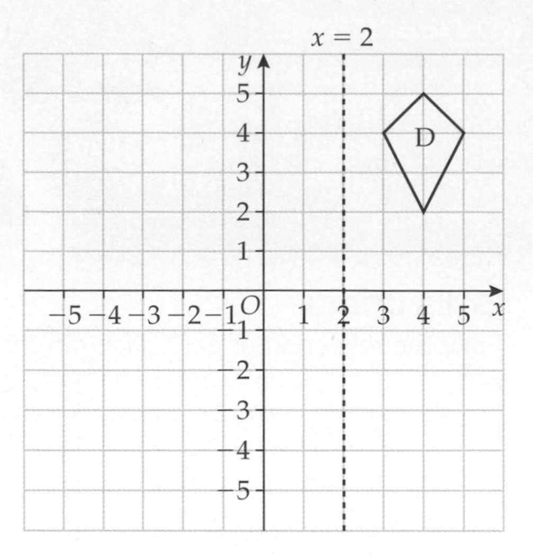

2 **a** Draw the reflection of shape H in the line $y = 2$.
Label the reflected shape I.

b Draw the reflection of shape H in the line $x = -1$.
Label the reflected shape J.

3 Shape K has been reflected in a mirror line.
Shape L is the image of shape K after this reflection.

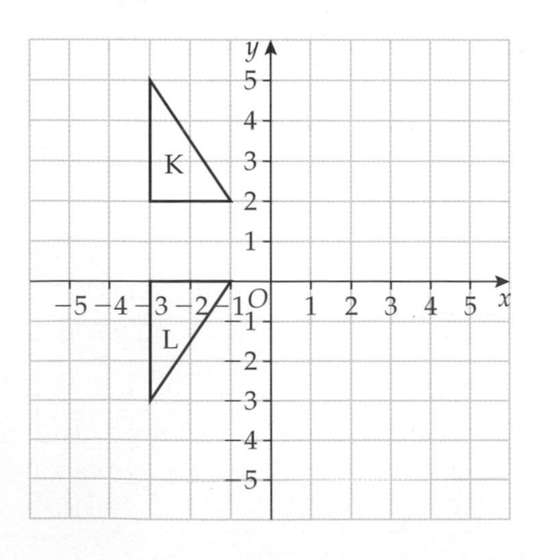

a Copy the diagram on to a coordinate grid.

b Draw the mirror line with a dashed line.

c Write down the equation of the mirror line.

4 Look at the shapes on the grid below.

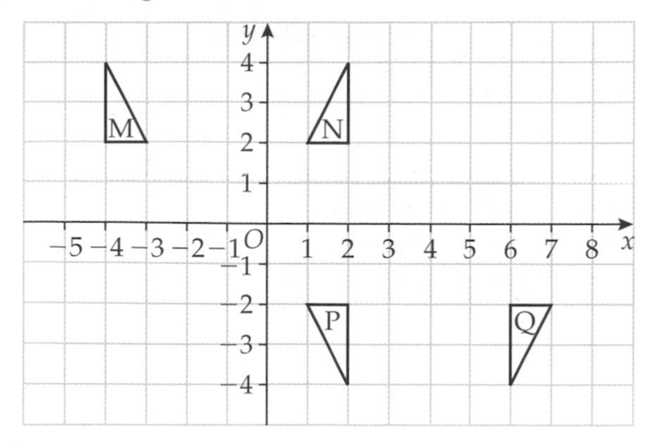

a Shape M is reflected in a mirror line.
Shape N is the image of shape M after this reflection.
Describe this transformation.

b Shape N is reflected in a mirror line.
Shape P is the image of shape N after this reflection.
Describe this transformation.

c Shape P is reflected in a mirror line.
Shape Q is the image of shape P after this reflection.
Describe this transformation.

D

A02

C

5 Copy the digram on to a coordinate grid and draw the reflection of shape A

a in the line $x = 4$

b in the line $y = 4$

c in the line $y = x$.

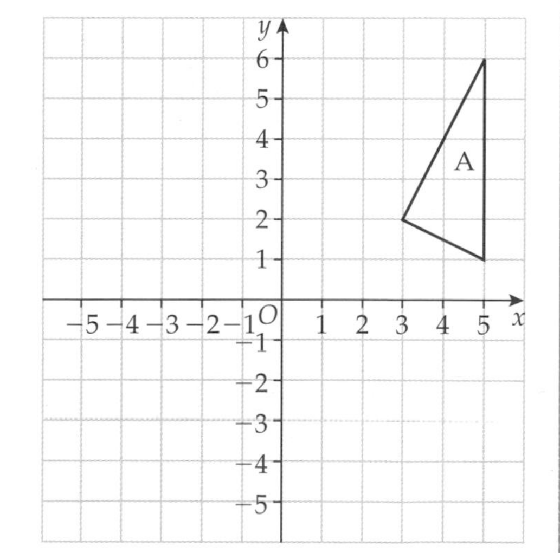

You may find it easier to draw a new diagram for each part of Q5.

6 **a** Copy the diagram on to a coordinate grid.

b Draw the line $y = x$ with a dashed line.

c Draw the reflection of shape R in the line $y = x$.
Label the reflected shape S.

d Draw the reflection of shape T in the line $y = x$.
Label the reflected shape U.

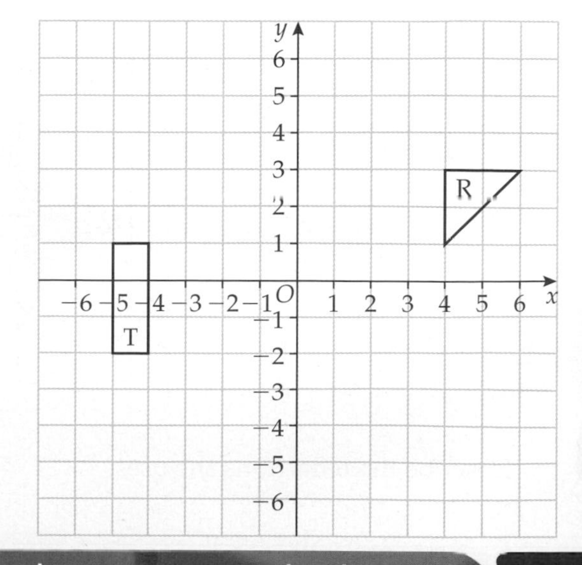

C

7 Describe the transformation that takes

a shape V to shape W **b** shape X to shape Y **c** shape A to shape B.

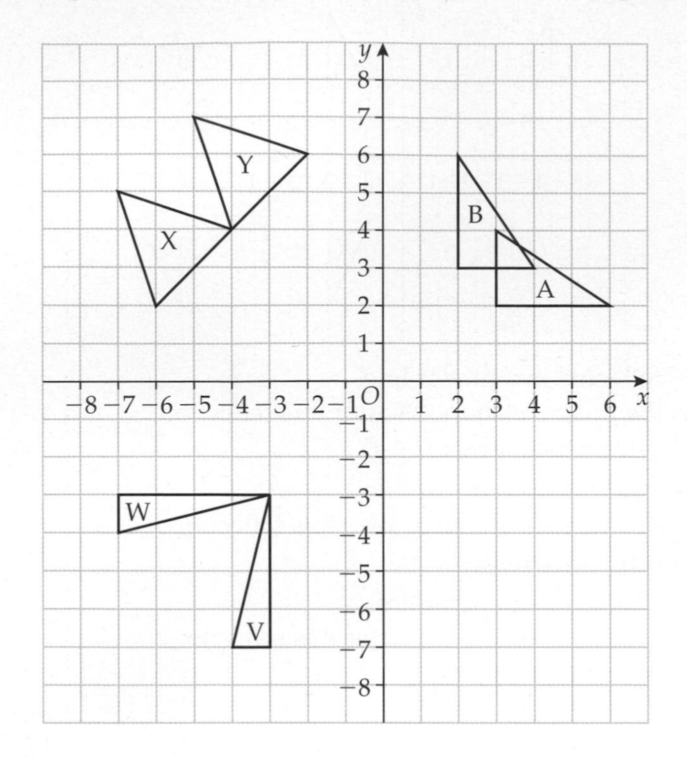

AO2

C

8 Jade starts with this shape.

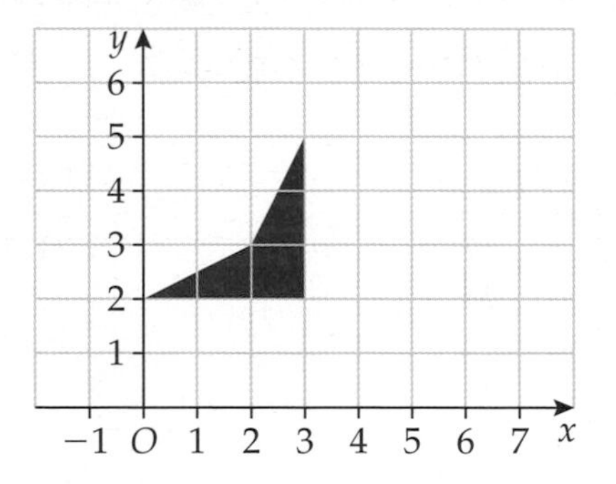

She reflects the shape to make this pattern.

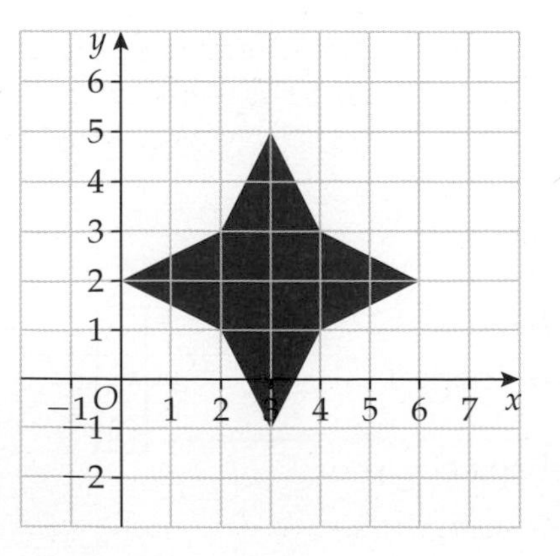

AO3

Describe the reflections she uses.

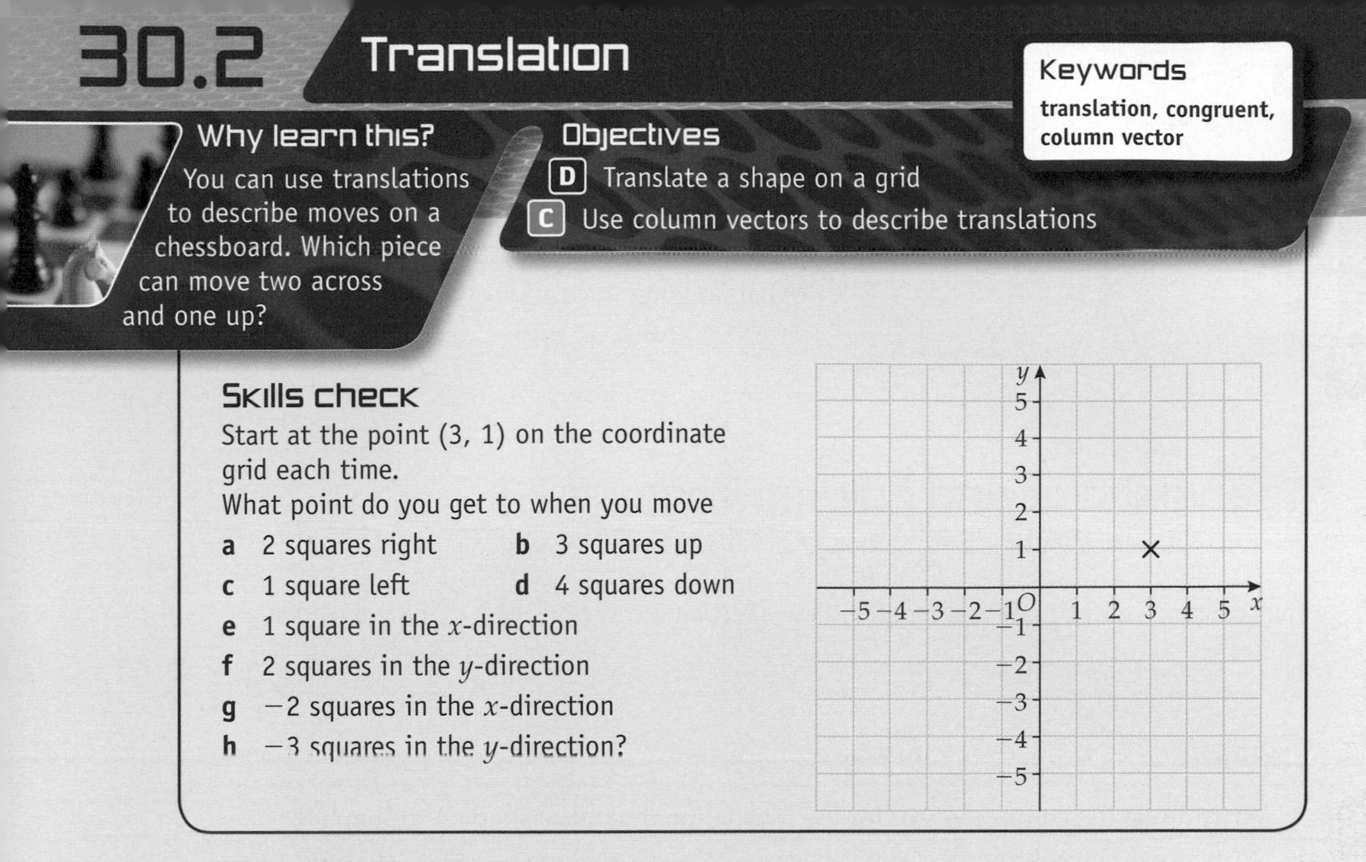

30.2 Translation

Keywords
translation, congruent, column vector

Why learn this?
You can use translations to describe moves on a chessboard. Which piece can move two across and one up?

Objectives
D Translate a shape on a grid
C Use column vectors to describe translations

Skills check

Start at the point (3, 1) on the coordinate grid each time.
What point do you get to when you move

a 2 squares right **b** 3 squares up
c 1 square left **d** 4 squares down
e 1 square in the x-direction
f 2 squares in the y-direction
g -2 squares in the x-direction
h -3 squares in the y-direction?

Translation

A **translation** slides a shape across a grid. It can slide right, left, up or down, or a combination of these.

In a translation, all points on the shape translate the same number of squares in the same direction.

An object and its image after a translation are **congruent**.

Example 2

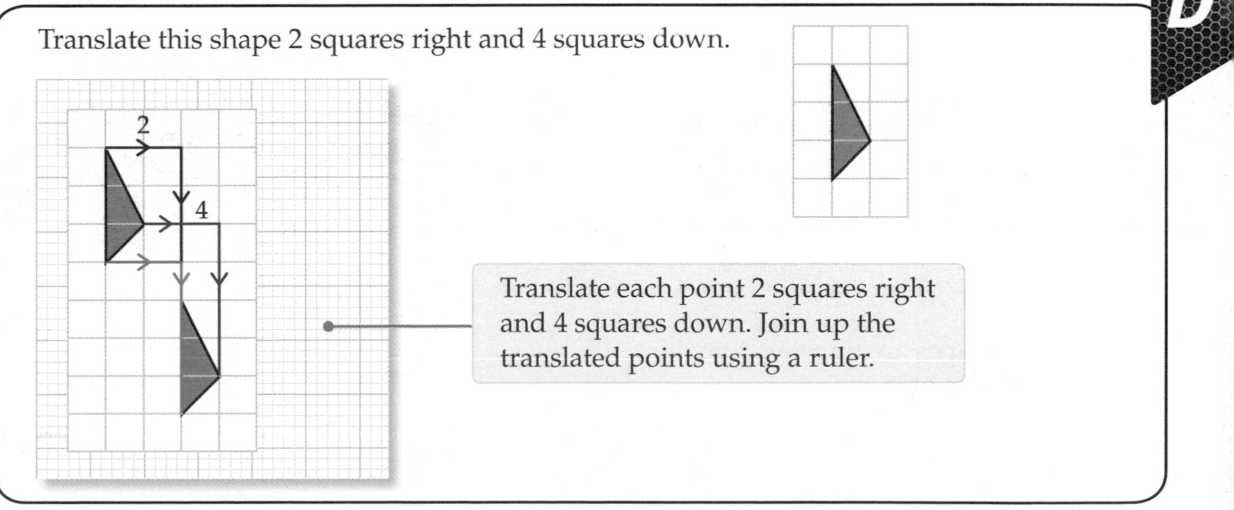

Translate this shape 2 squares right and 4 squares down.

2
4
Translate each point 2 squares right and 4 squares down. Join up the translated points using a ruler.

Exercise 30B

1 Copy the shape on to squared paper.

a Translate the shape 3 squares right and 1 square up. Label the image A.

b Translate the shape 1 square left and 2 squares down. Label the image B.

c Translate the shape 2 squares right and 4 squares down. Label the image C.

2 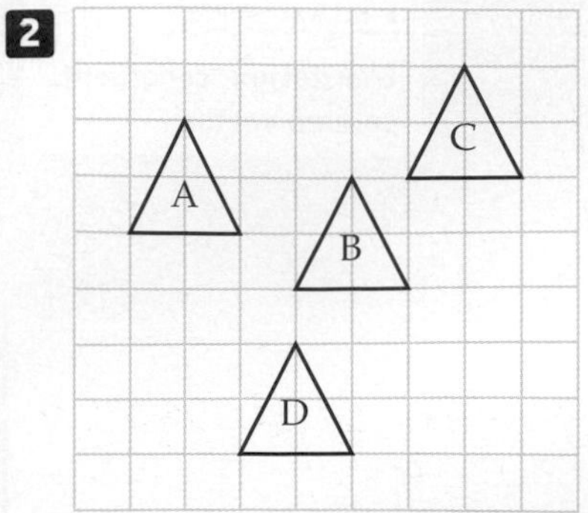

Describe the translation that takes

a shape A to shape B

b shape A to shape C

c shape A to shape D

d shape D to shape A.

e What do you notice about your answers to parts **c** and **d**?

Write the number of squares right or left, and the number of squares up or down.

Translations and column vectors

You can give instructions for a translation on a coordinate grid using a column vector $\begin{pmatrix} x \\ y \end{pmatrix}$.

The column vector $\begin{pmatrix} 3 \\ 2 \end{pmatrix}$ means move 3 in the x-direction and then move 2 in the y-direction.

C

Example 3

a Write down the column vector for the translation that takes shape A to shape B.

b Write down the column vector for the translation that takes shape B to shape C.

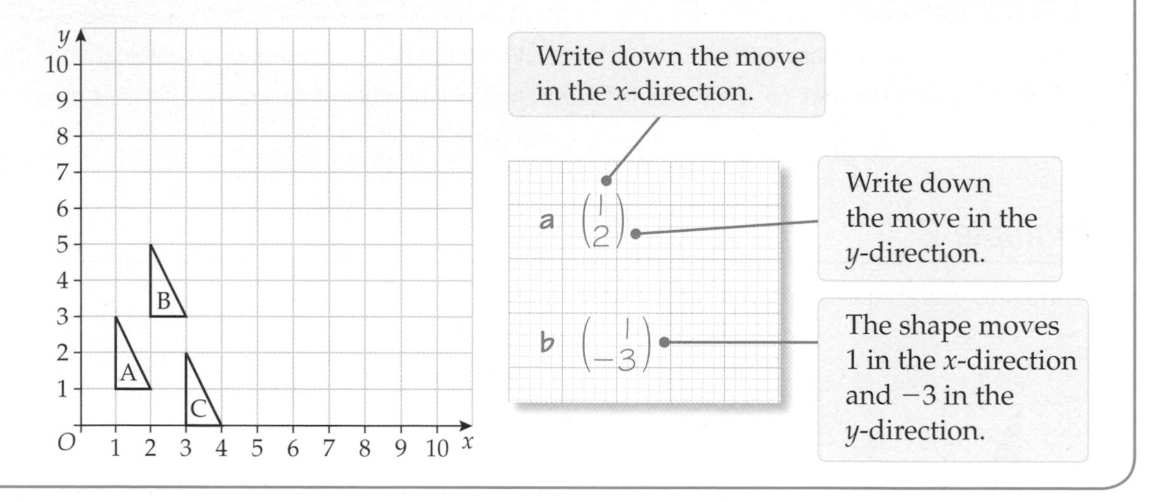

Write down the move in the x-direction.
a $\begin{pmatrix} 1 \\ 2 \end{pmatrix}$
Write down the move in the y-direction.
b $\begin{pmatrix} 1 \\ -3 \end{pmatrix}$
The shape moves 1 in the x-direction and −3 in the y-direction.

Exercise 30C

C

1 Some shapes are translated on this coordinate grid.
Write down the column vector for each translation.

a shape A to shape B

b shape B to shape C

c shape C to shape D

d shape D to shape E

e shape E to shape F

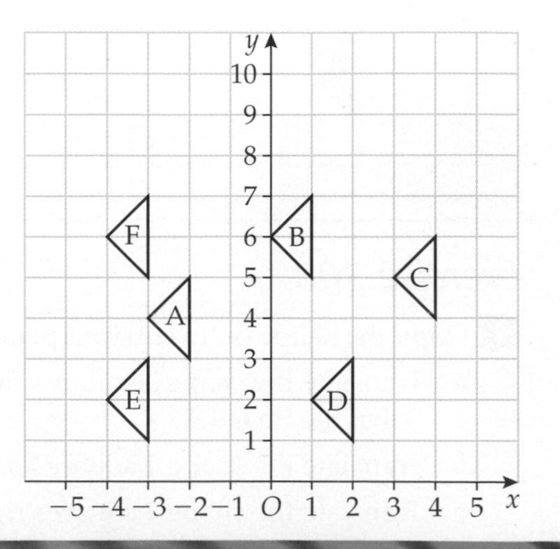

2 Look at the coordinate grid in Q1 again.

a Write down the column vectors for these translations

i shape F to shape C

ii shape C to shape F.

b What do you notice about your answers in part **a**?

c The translation that takes shape B to shape D is $\begin{pmatrix} 1 \\ -4 \end{pmatrix}$.

Write down the column vector for the translation that takes shape D to shape B.

3 Look at the coordinate grid in Q1 again.

a Write down the column vector for these translations

i shape E to shape A

ii shape A to shape B

iii shape E to shape B.

b What do you notice about your answers in part **a**?

4 On a grid, $\begin{pmatrix} 2 \\ -1 \end{pmatrix}$ translates shape A to shape B and $\begin{pmatrix} 3 \\ 2 \end{pmatrix}$ translates shape B to shape C.

Write down the column vector that translates shape A directly to shape C.

5 On a grid, $\begin{pmatrix} -1 \\ 4 \end{pmatrix}$ translates shape D to shape E and $\begin{pmatrix} 2 \\ -5 \end{pmatrix}$ translates shape E to shape F.

Write down the column vector that translates shape D directly to shape F.

6 Sadie starts with this shape on a grid.

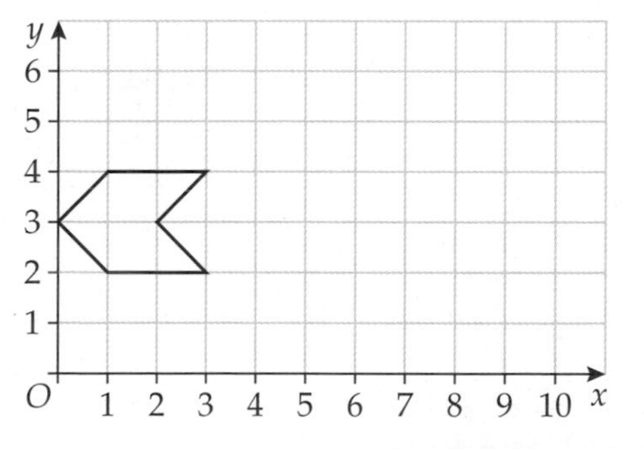

She translates it to make this tessellation.

A tessellation is a pattern of repeated shapes, with no gaps in between.

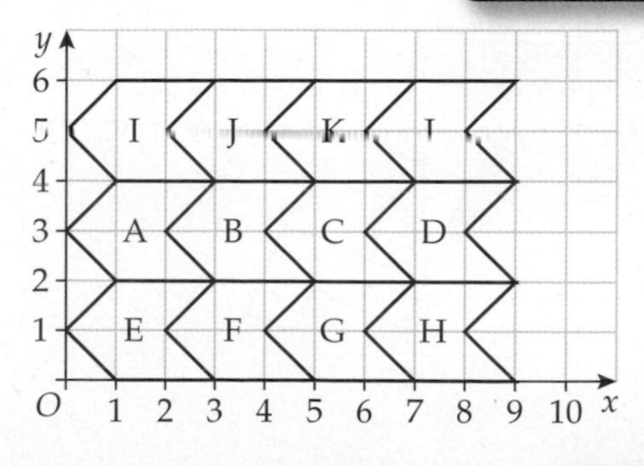

Use column vectors to write instructions for each translation.

Each shape in the tessellation has been labelled to help you.

C

AO2

C

AO3

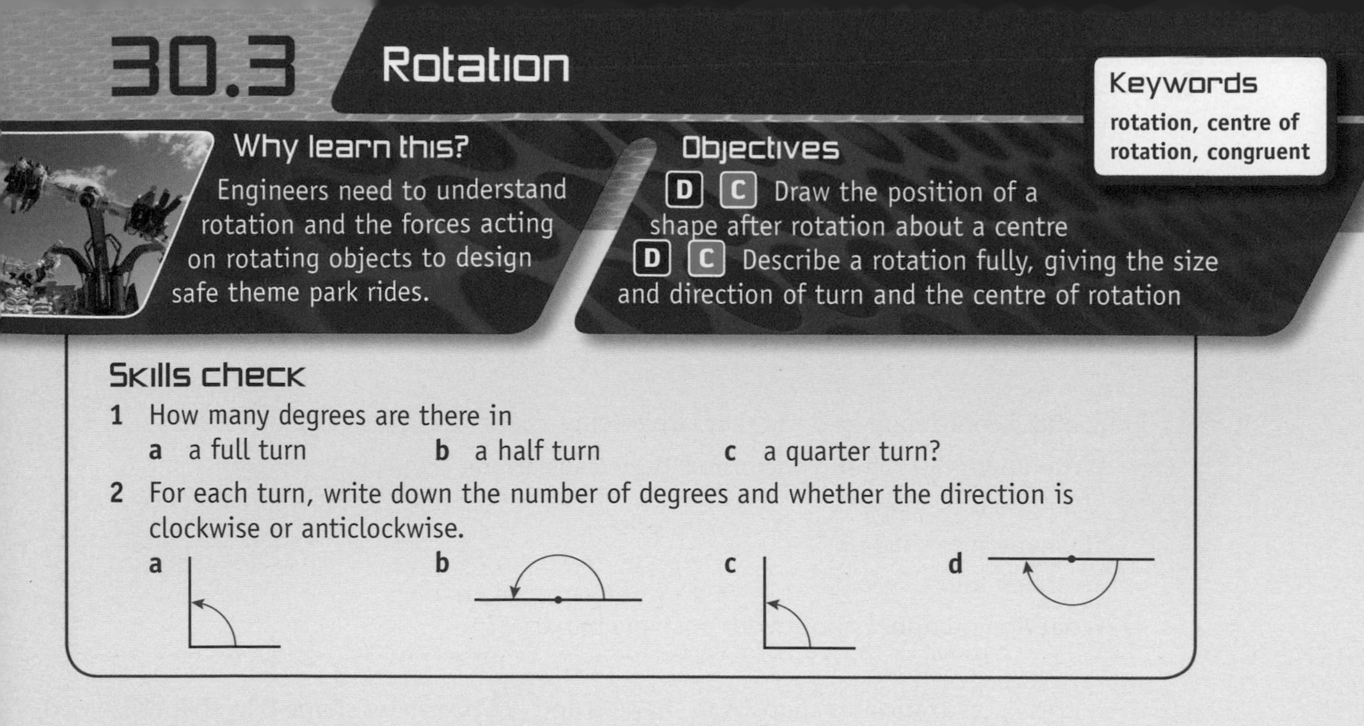

Keywords
rotation, centre of rotation, congruent

Why learn this?

Engineers need to understand rotation and the forces acting on rotating objects to design safe theme park rides.

Objectives

D C Draw the position of a shape after rotation about a centre

D C Describe a rotation fully, giving the size and direction of turn and the centre of rotation

Skills check

1 How many degrees are there in

 a a full turn b a half turn c a quarter turn?

2 For each turn, write down the number of degrees and whether the direction is clockwise or anticlockwise.

 a b c d

Rotation around a fixed point

A **rotation** turns a shape around a fixed point, called the **centre of rotation**.

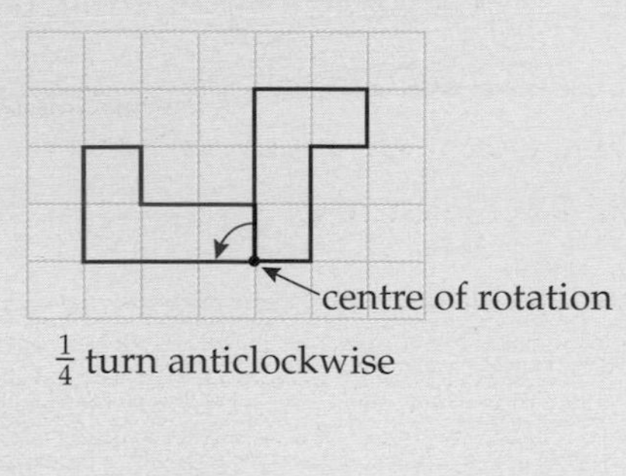

$\frac{1}{4}$ turn anticlockwise

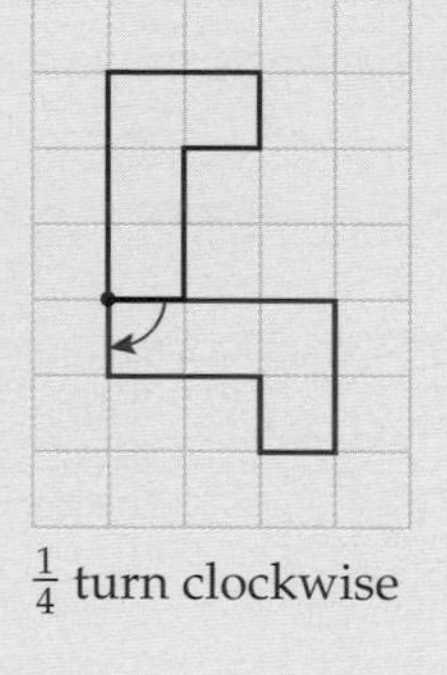

$\frac{1}{4}$ turn clockwise

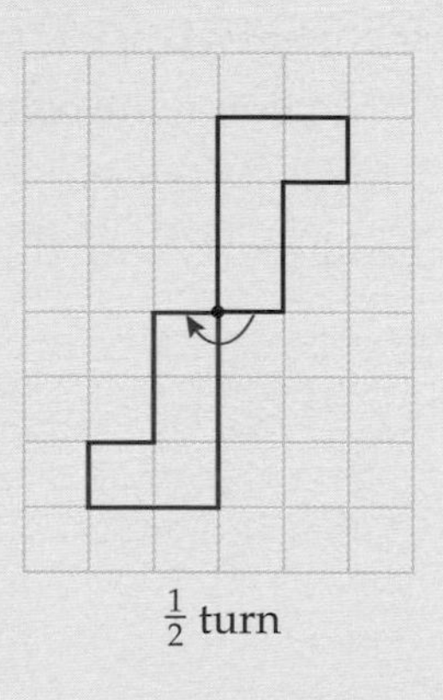

$\frac{1}{2}$ turn

You can use tracing paper to draw a rotation.

Step 1: Trace the shape and the centre of rotation.	Step 2: Use a pencil to hold the tracing paper on the centre of rotation. Rotate the tracing.	Step 3: Copy the rotated shape from the tracing paper.

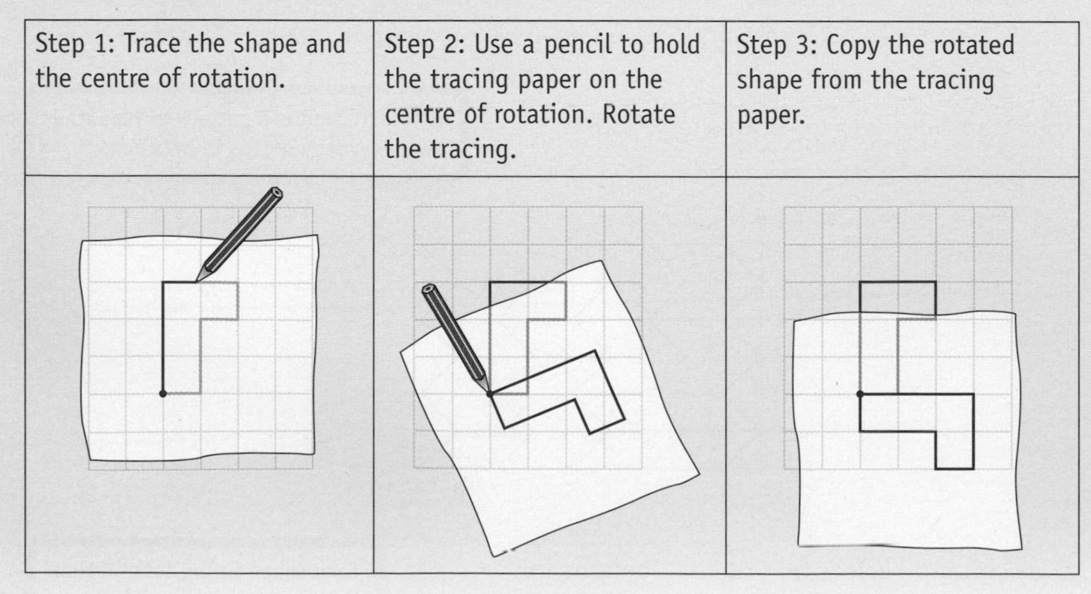

In a rotation, the object and the image are **congruent**.

Example 4

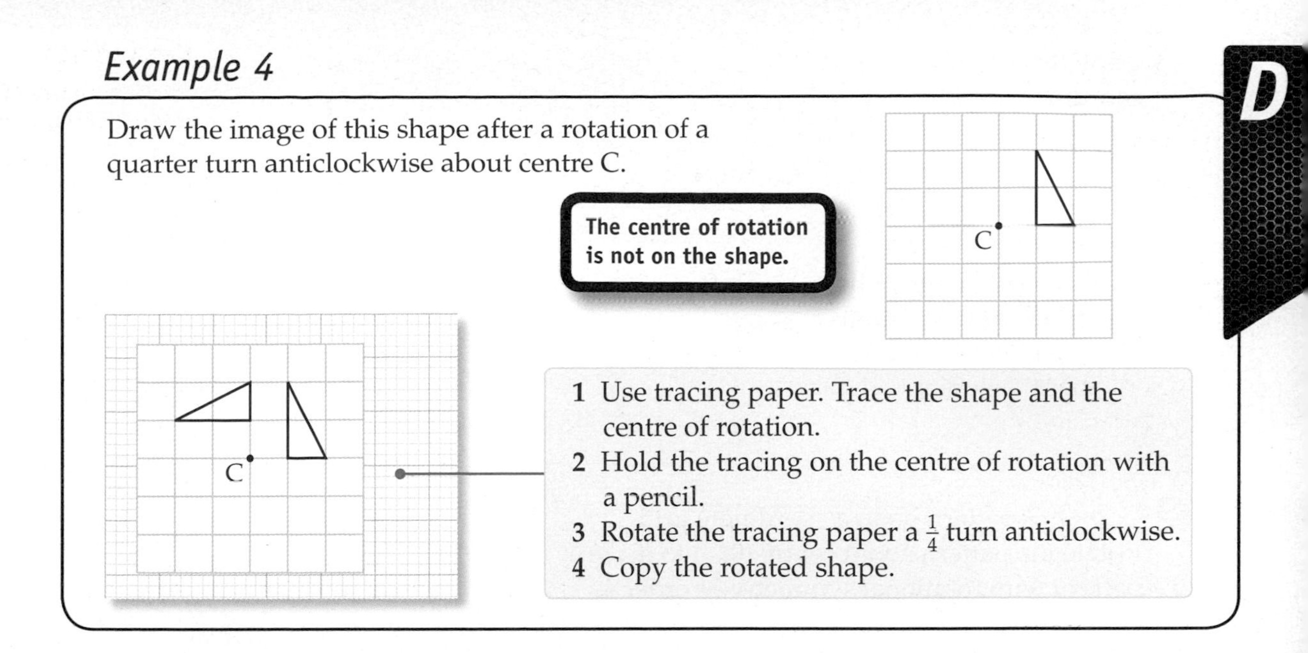

Draw the image of this shape after a rotation of a quarter turn anticlockwise about centre C.

The centre of rotation is not on the shape.

1 Use tracing paper. Trace the shape and the centre of rotation.
2 Hold the tracing on the centre of rotation with a pencil.
3 Rotate the tracing paper a $\frac{1}{4}$ turn anticlockwise.
4 Copy the rotated shape.

Exercise 30D

For Q1 and Q2, copy the shape and the centre of rotation on to squared paper. Draw the image of the shape after the rotation given.

1 $\frac{1}{4}$ turn anticlockwise about centre C.

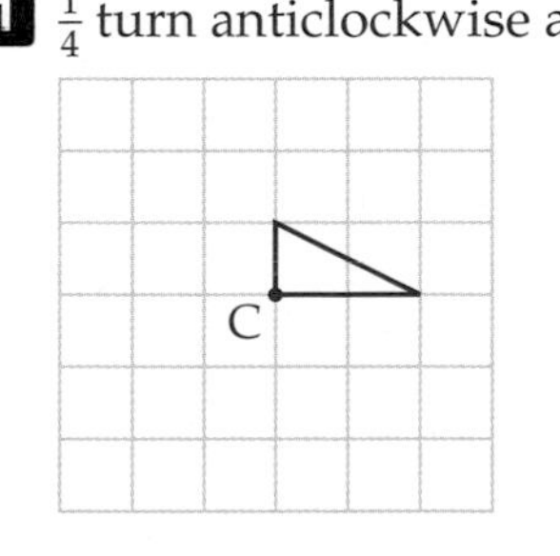

2 **a** $\frac{1}{4}$ turn clockwise about centre C.
b 180° about centre C.

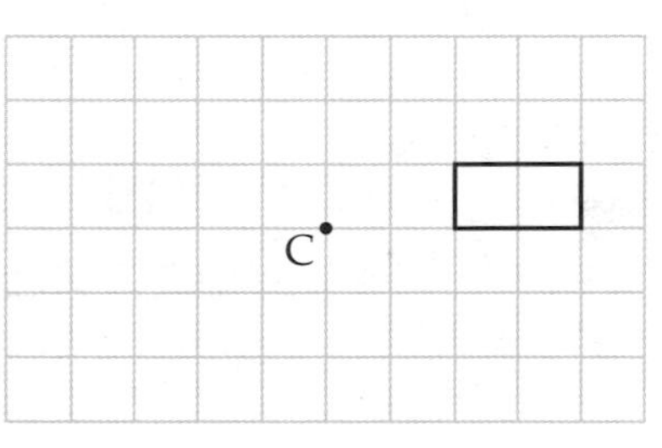

3 Copy this shape on to a coordinate grid.

a Draw the image of the shape after a rotation 90° anticlockwise about the point (0, 0).

b Draw the image of the shape after a rotation 90° clockwise about the point (1, 2).

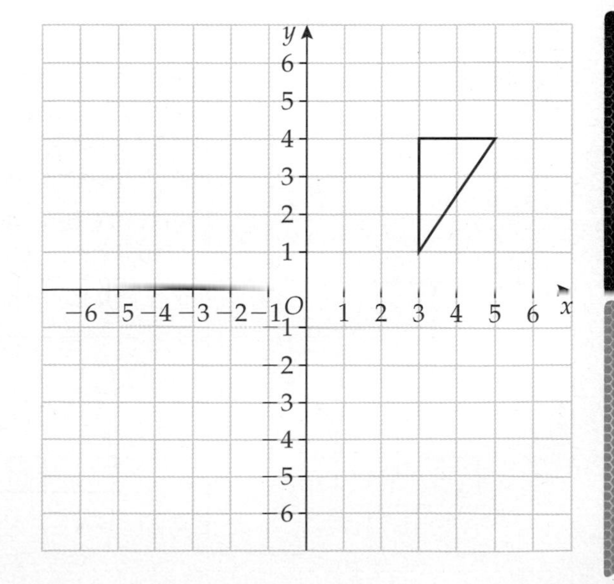

C

4 Copy this shape on to a coordinate grid.

a Draw the image of the shape after a rotation 180° about the point $(-1, 3)$.

b Draw the image of the shape after a rotation 90° anticlockwise about the point $(-4, 2)$.

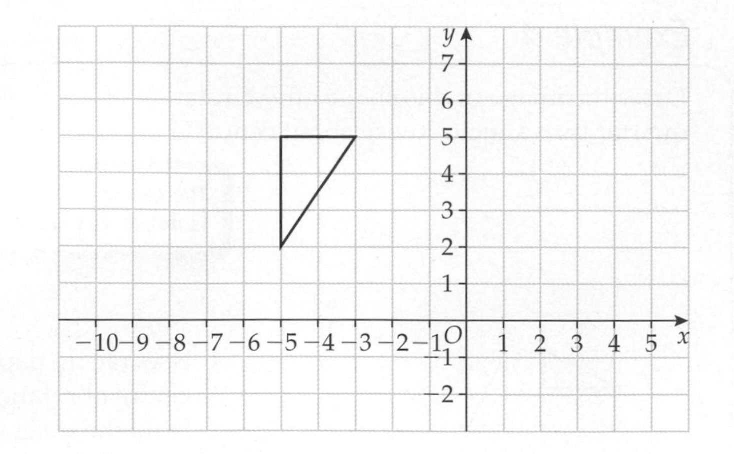

C

5 Copy this pattern on to a coordinate grid. Rotate the pattern about centre (0, 0) to make a pattern with rotational symmetry of order 4.

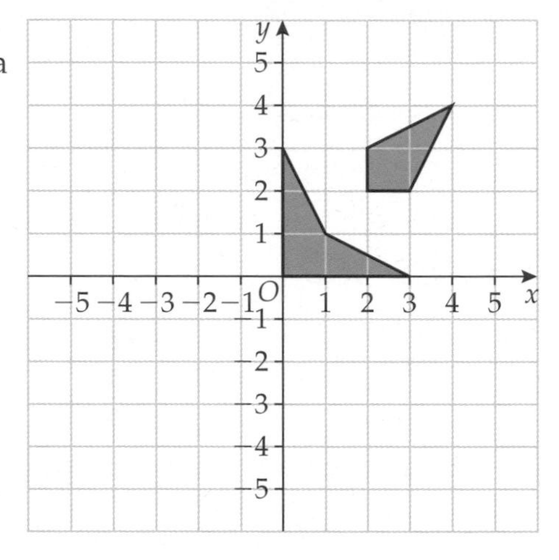

A02

Describing rotations

To describe a rotation fully, you need to give

- the centre of rotation
- the size of turn
- the direction of turn.

You can use tracing paper to find the centre of rotation and the size and direction of turn.

Step 1: Trace the object shape.	Step 2: Rotate the tracing until the shape looks like the image, though it might not be in the same place. What size turn have you rotated it? In what direction have you rotated it?	Step 3: Put your tracing back over the object shape. Rotate the tracing again, holding a point fixed with your pencil. Repeat for different points, until your tracing ends up on the image shape. That is the centre of rotation.

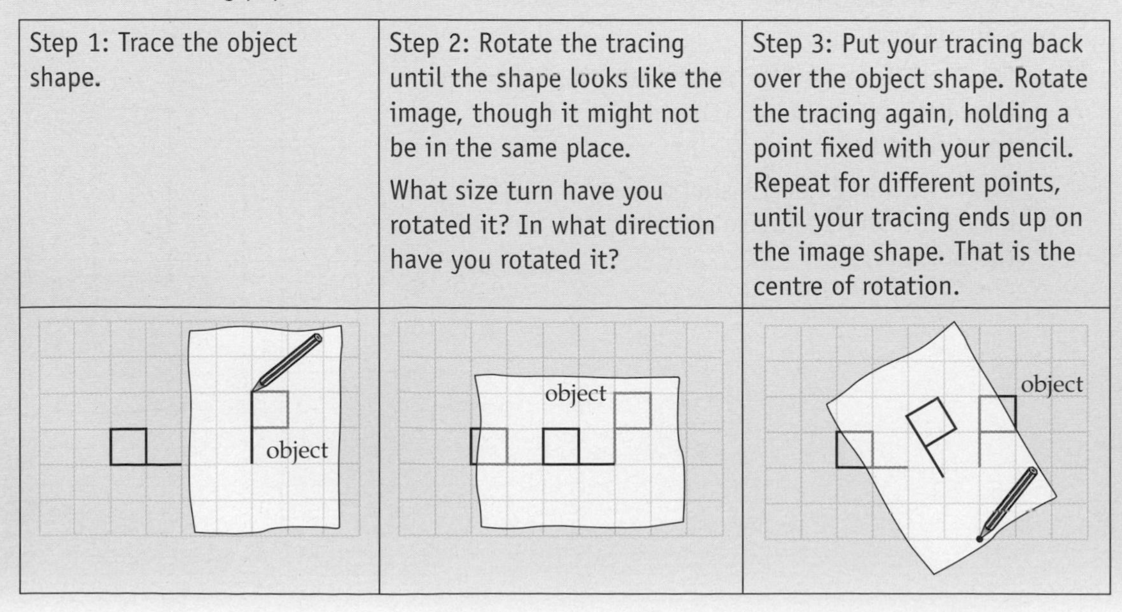

Example 5

Describe fully the rotation that maps shape A on to shape B.

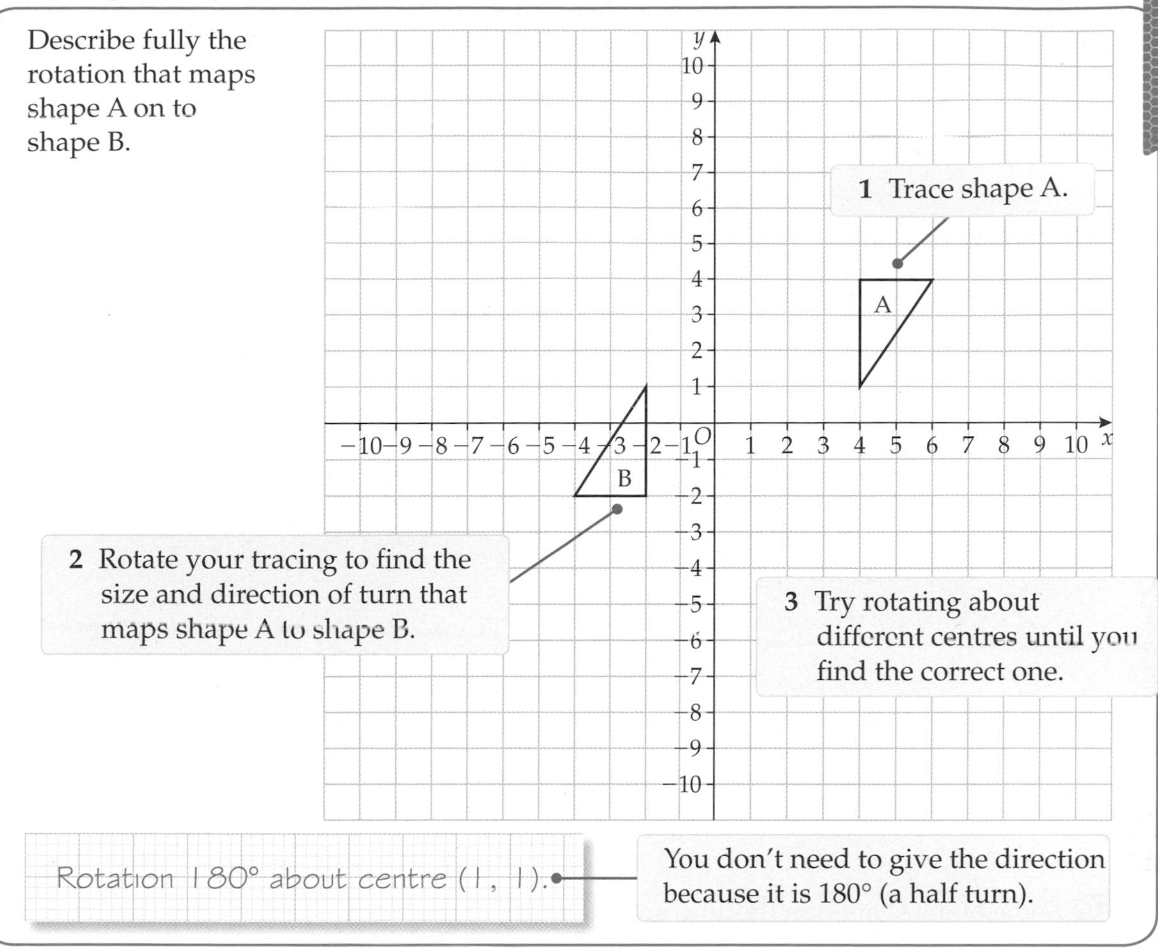

1 Trace shape A.
2 Rotate your tracing to find the size and direction of turn that maps shape A to shape B.
3 Try rotating about different centres until you find the correct one.
A
B

Rotation 180° about centre (1, 1).

You don't need to give the direction because it is 180° (a half turn).

C

Exercise 30E

D

1 Copy this diagram on to squared paper
Shape A rotates on to shape B.

a Write down the angle of rotation and the direction of turn.

b Find the centre of the rotation. Label it C on your diagram.

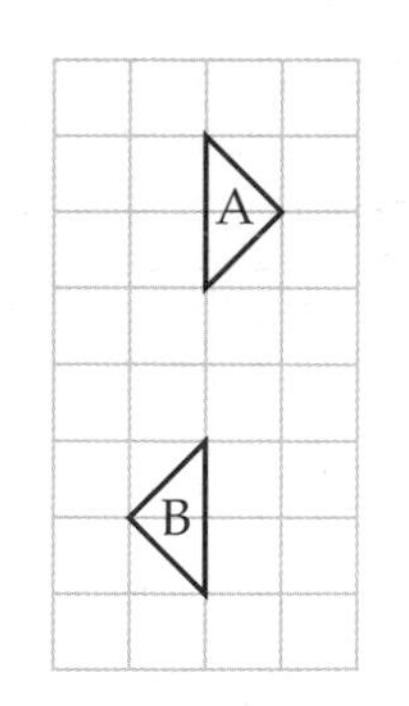

2 Copy the diagram on to a coordinate grid. Shape X rotates on to shape Y.

a Write down the angle of rotation and the direction of turn.

b Find the centre of the rotation. Write down its coordinates.

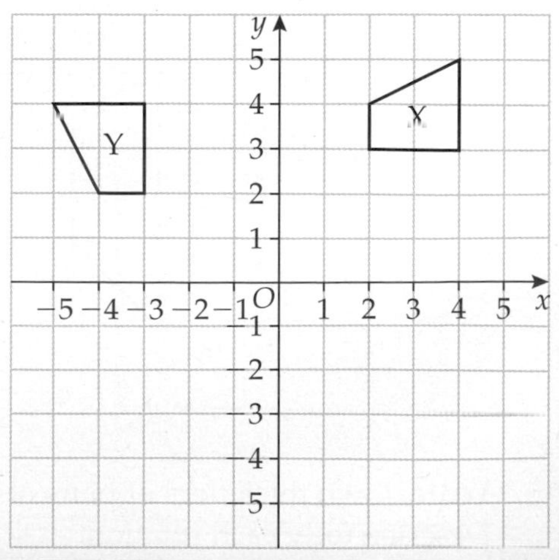

D

3 **a** Describe fully the rotation that maps shape D on to shape E.
b Describe fully the rotation that maps shape F on to shape G.
c Describe fully the rotation that maps shape N on to shape P.

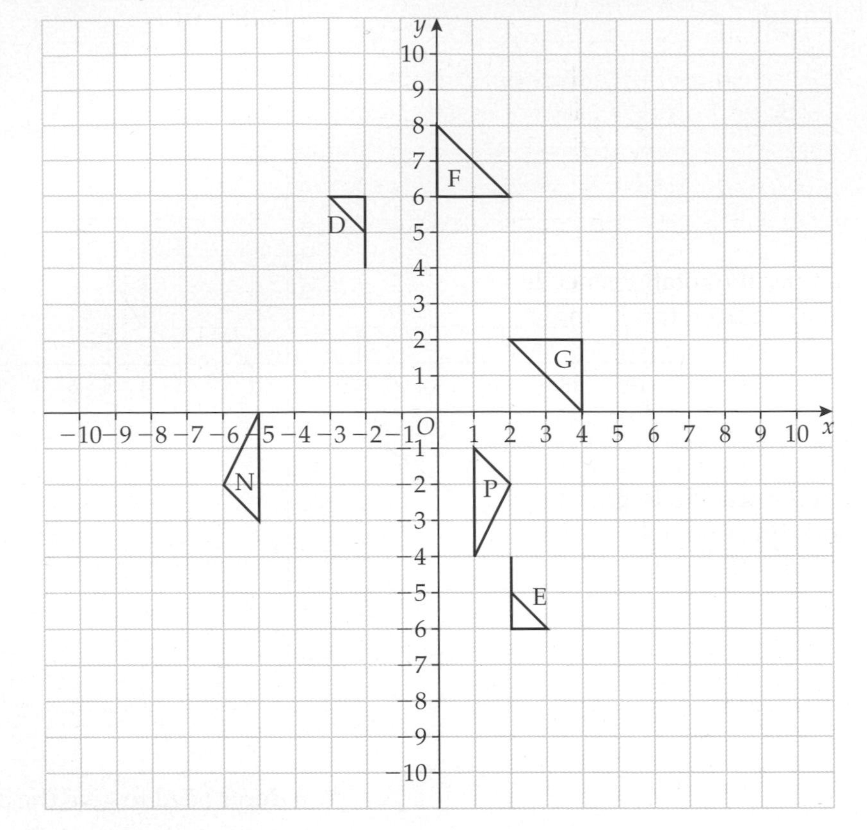

C

AO2

C

4 Some of the shapes on this grid are rotations of each other.

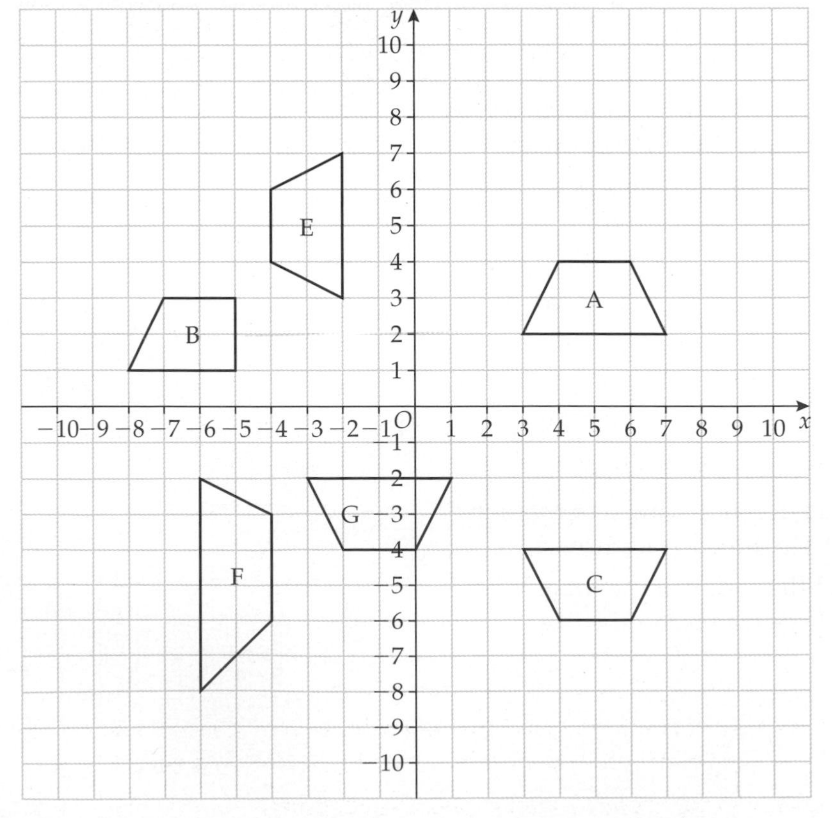

AO2

Write down the letters of pairs of shapes that rotate to each other.
Describe fully each rotation.

5 A roundabout has eight seats like this.
Each seat is fixed to a pole.
The poles fit into the central post.
The seats are placed symmetrically about the central post.

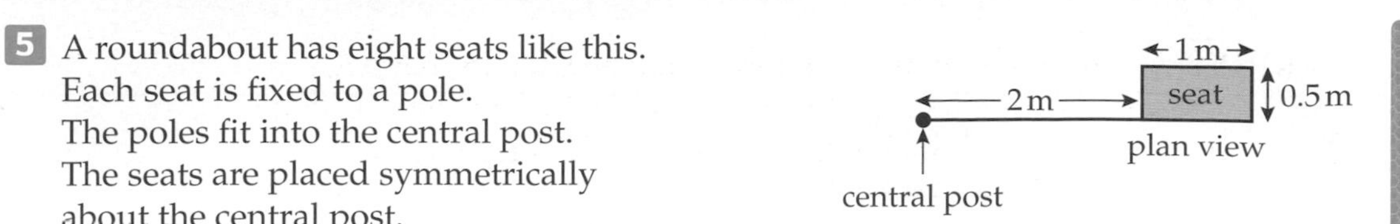

a Make a scale drawing of this roundabout on a coordinate grid. Use a scale of 1 grid division to 1 metre. Draw in all eight seats.

b There is a circular safety fence all around the roundabout. It is 1 metre from the outside edge of the seats. Draw the safety fence on your scale drawing.

6 Describe fully the rotations needed to create a square from this triangle.

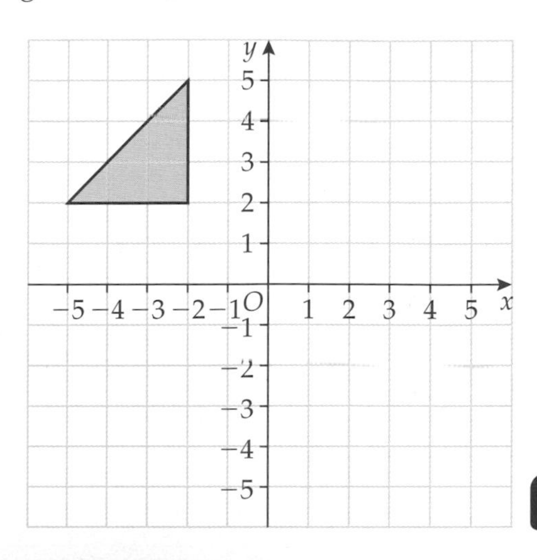

C

AO3

30.4 Combined transformations

Keywords
transformation

Why learn this?
You can use translation, rotation and reflection of one or more shapes to create a tessellation.

Objectives
C Transform shapes using more than one transformation
C Describe combined transformations of shapes on a grid

Skills check

Copy this diagram on to a coordinate grid.

a Draw the image of shape D after reflection in the line $x = 2$. Label it E.

b Draw the image of shape D after a translation $\begin{pmatrix}3\\2\end{pmatrix}$. Label it F.

c Draw the image of shape D after a rotation 90° anticlockwise about (3, 1). Label it G.

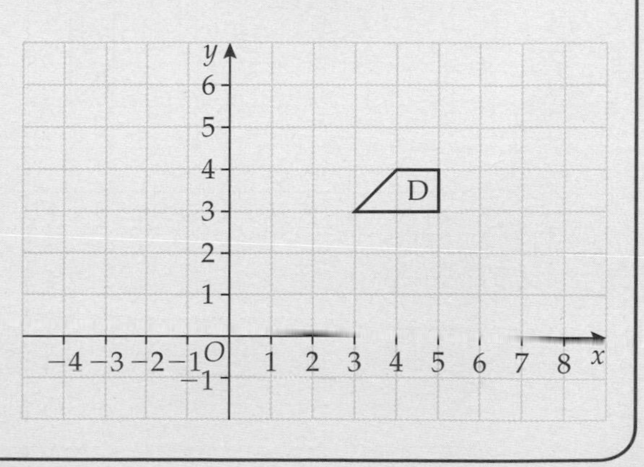

Combined transformations

Reflection, translation and rotation are **transformations.** They transform an object to an image. For these transformations, the object and its image are congruent.

You can combine transformations by doing one, then another.

C

Example 6

Reflect shape P in the line $x = 1$. Label the image Q.

Rotate shape Q 90° clockwise around the point (1, 4). Label the image R.

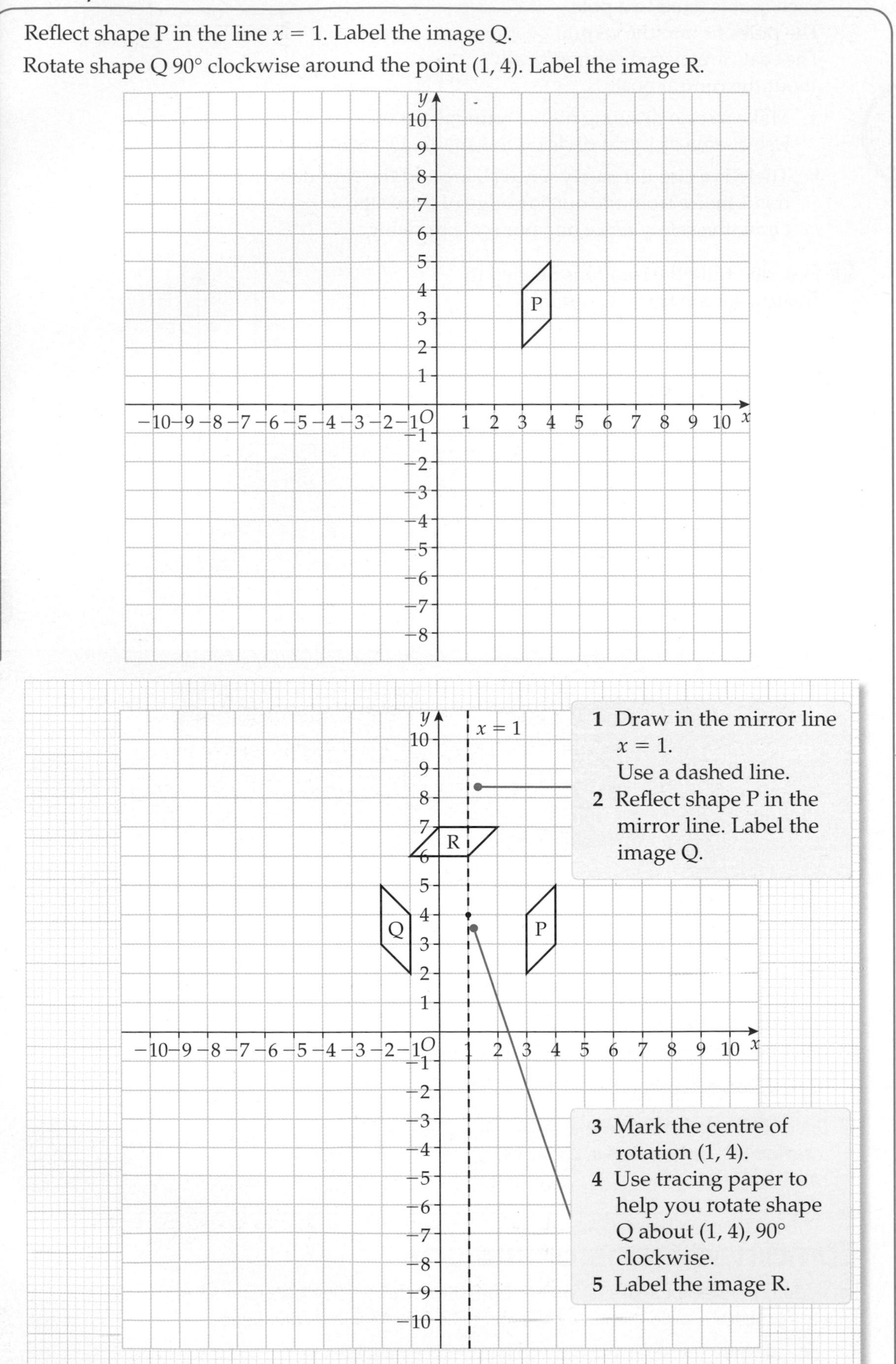

P
x = 1
R
Q
P
1 Draw in the mirror line x = 1.
Use a dashed line.
2 Reflect shape P in the mirror line. Label the image Q.
3 Mark the centre of rotation (1, 4).
4 Use tracing paper to help you rotate shape Q about (1, 4), 90° clockwise.
5 Label the image R.

Exercise 30F

Use this figure for Q1 to Q3.

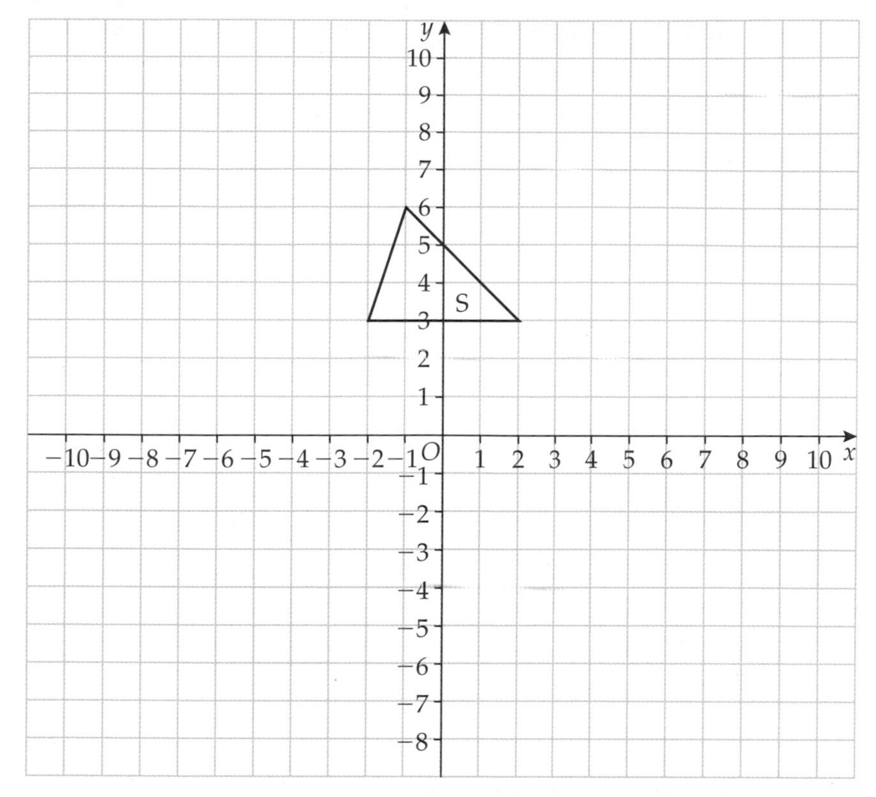

1 Copy shape S on to a coordinate grid.

- **a** Translate shape S by column vector $\begin{pmatrix} 4 \\ -3 \end{pmatrix}$. Label the image V.
- **b** Translate shape V by column vector $\begin{pmatrix} -2 \\ 1 \end{pmatrix}$. Label the image W.
- **c** Write down the column vector for the transformation that takes shape S directly to shape W.
- **d** The translation $\begin{pmatrix} 4 \\ -2 \end{pmatrix}$ takes shape V to shape X.
  Without drawing shape X, write down the column vector that takes shape S to shape X.

2 Copy shape S on to a coordinate grid.

- **a** Rotate shape S 180° about (0, 0). Label the image X.
- **b** Rotate shape X 180° about (3, 0). Label the image Y.
- **c** Describe the single transformation that takes shape S to shape Y.

3 Copy shape S on to a coordinate grid.

- **a** Reflect shape S in the line $x = 3$. Label the image A.
- **b** Reflect shape A in the line $x = -1$. Label the image B.
- **c** Describe the single transformation that takes shape S to shape B.

4 Describe a single transformation that is equivalent to a reflection in the x-axis followed by a reflection in the y-axis.

5 Describe a single transformation that is equivalent to a reflection in the line $y = x$ followed by a reflection in the line $y = -x$.

6 A rotation of 180° about the origin is the equivalent of two transformations, T1 followed by T2. Give two possible descriptions of transformations T1 and T2.

C

C

AO2

AO3

C

7 Copy the diagram on to a coordinate grid.

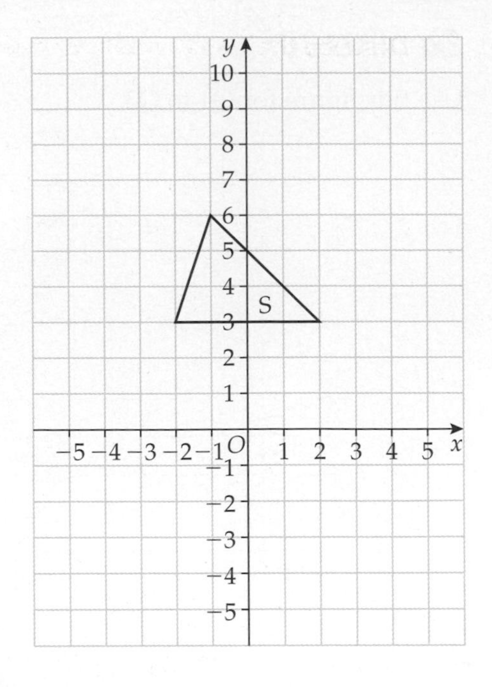

a Translate shape S by $\begin{pmatrix} 4 \\ 0 \end{pmatrix}$.
Label the image T.

b Reflect shape T in the line $x = 6$.
Label the image U.

c Describe fully the single transformation that takes shape S to shape U.

8 Copy the diagram from Q7 on to a coordinate grid again.

a Rotate shape S 90° anticlockwise about the point (2, 2). Label the image M.

b Reflect shape M in the line $y = 1$. Label the image N.

C AO3

9 a Investigate reflecting a shape in one line, and then in a parallel line.
What type of transformation is equivalent to this double reflection?

b Describe how to find the single transformation that is equivalent to two reflections in the parallel lines $x = a$ and $x = b$.

Review exercise

D

1 Triangle A is drawn on the grid. Copy the grid.

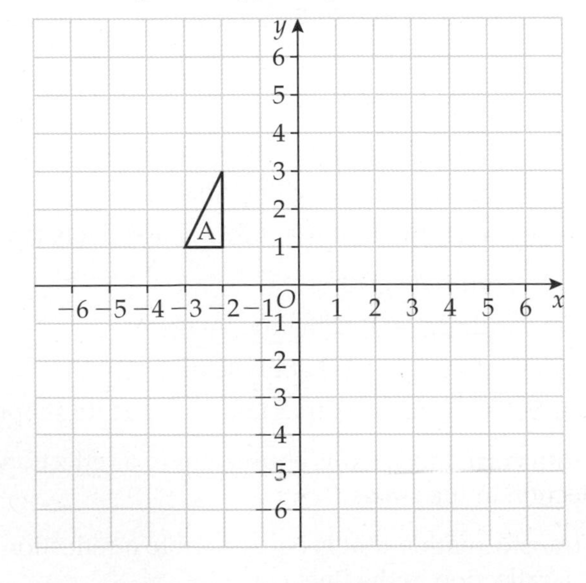

a Reflect triangle A in the line $y = -2$. Label the triangle B. **[2 marks]**

b Rotate triangle A a quarter of a turn anticlockwise about the origin O.
Label the image C. **[3 marks]**

2 Describe the transformations needed to create a rhombus from this triangle.

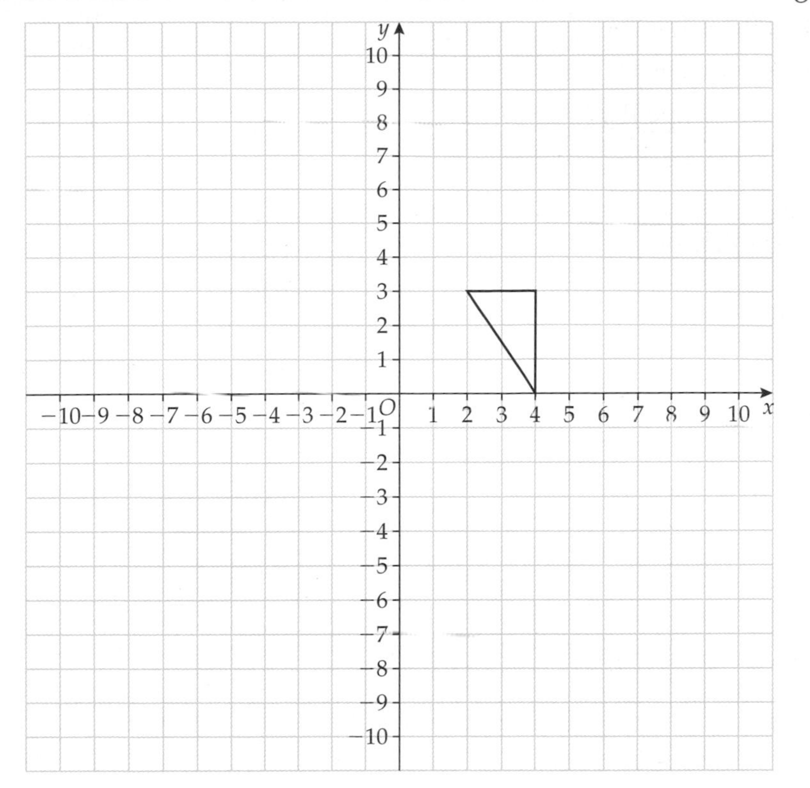

[2 marks]

C

AO2

3 The diagram shows two rectangles A and B. Copy and complete these statements.

a Rectangle B is a reflection of rectangle A in the line ______. **[1 mark]**

b Rectangle B is a translation of rectangle A by the column vector ______. **[1 mark]**

c Rectangle B is a rotation of rectangle A through ☐ degrees about the point (☐, ☐). **[2 marks]**

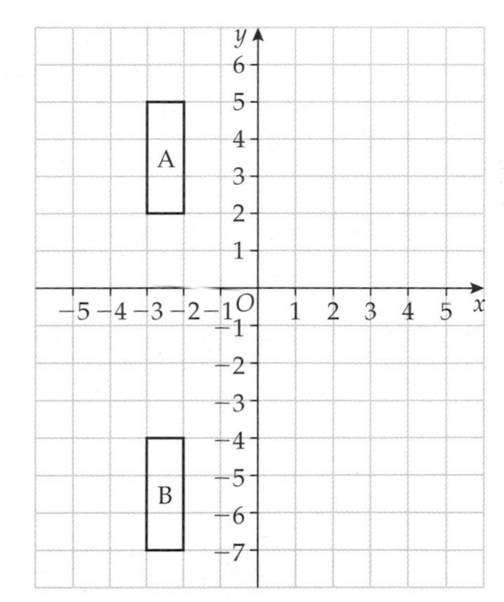

C

4 The points $A(-2, 2)$, $B(0, -2)$ and $C(3, -3)$ are shown on the grid.

a $ABCD$ is a kite. Complete the kite $ABCD$ and write down the coordinates of D. **[2 marks]**

b The kite is rotated 90° clockwise about B. Draw the image of the kite after this rotation. **[2 marks]**

c Write down the new coordinates of C. **[1 mark]**

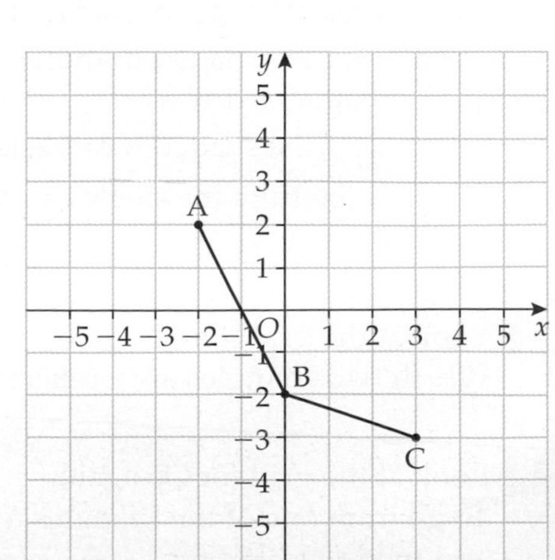

C

5 Describe the transformation that maps

a shape P to shape Q **[1 mark]**

b shape Q to shape R **[1 mark]**

c shape R to shape P. **[1 mark]**

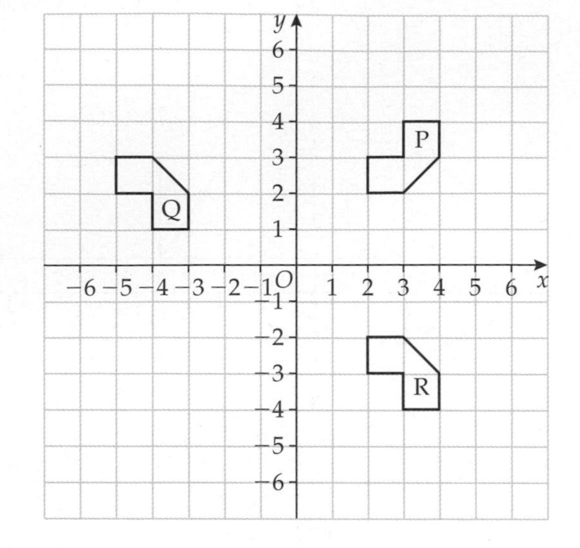

6 The grid shows several transformations of the shaded triangle.

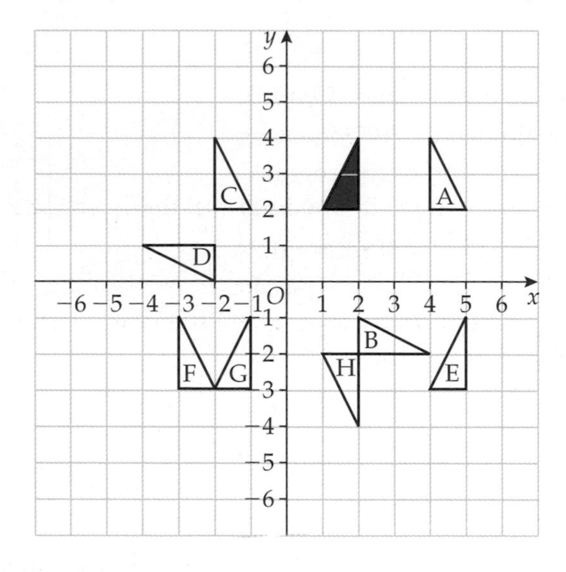

a Write down the letter of the triangle

i after the shaded triangle is reflected in the line $x = 0$ **[1 mark]**

ii after the shaded triangle is translated by 3 squares to the left and 5 squares down **[1 mark]**

iii after the shaded triangle is rotated 90° clockwise about O. **[1 mark]**

b Describe fully the single transformation which takes triangle F to triangle G. **[2 marks]**

C

7 Look at the grid for Q6 again.
Which two triangles are a reflection of each other in the line $y = -x$? **[2 marks]**

AO2

8 Look at the grid for Q6 again.
To go from one of the triangles to another, you reflect in the line $x = 3$ and then reflect in the y-axis. Write down the letters of the two triangles. **[3 marks]**

9 Triangle A is transformed on to triangle B by a rotation followed by a reflection.

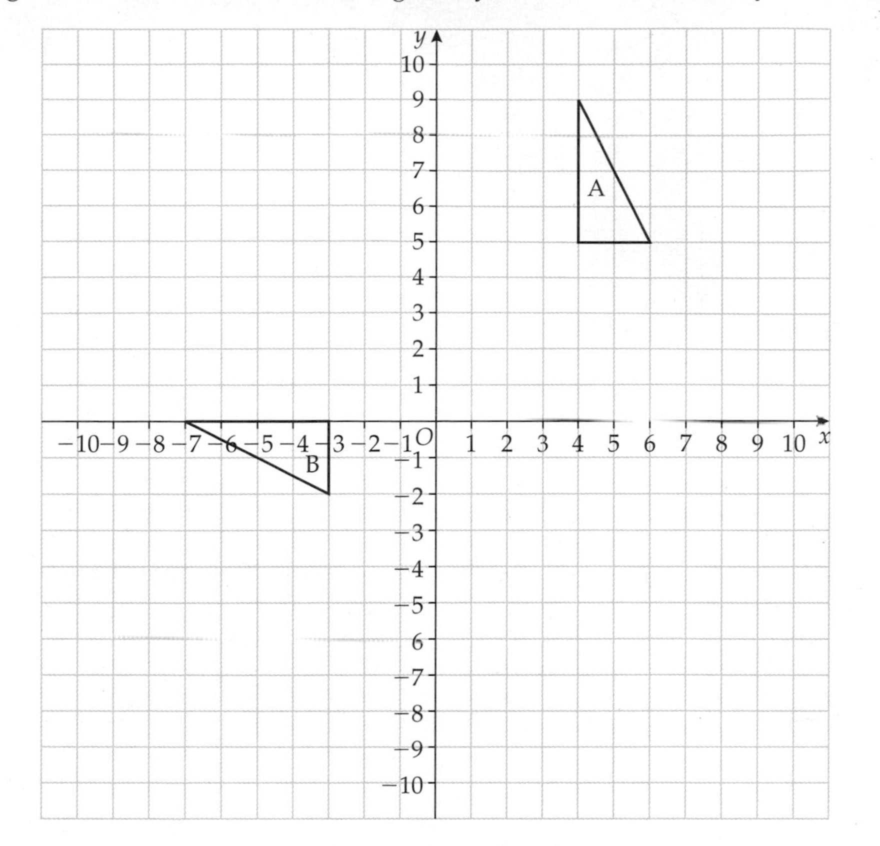

Describe fully the rotation and the reflection. **[3 marks]**

10 **a** Investigate reflecting a shape in one line, and then in another.
The two lines must not be parallel.
What type of transformation is equivalent to this double reflection? **[2 marks]**

b Describe how to find the single transformation that is equivalent to two reflections in lines that are not parallel. **[2 marks]**

C

AO3

Chapter summary

In this chapter you have learned how to

- translate a shape on a grid **D**
- draw reflections on a coordinate grid **D** **C**
- describe reflections on a coordinate grid **D** **C**
- draw the position of a shape after rotation about a centre **D** **C**
- describe a rotation fully, giving the size and direction of turn and the centre of rotation **D** **C**
- use column vectors to describe translations **C**
- transform shapes using more than one transformation **C**
- describe combined transformations of shapes on a grid **C**

31 Enlargement

BBC Video

This chapter is about constructing enlargements of shapes.

You can use a microscope to view enlarged images of very small objects. The photograph shows an electron microscope image of pollen grains – in reality, these are less than a thousandth of a centimetre in diameter.

Objectives

This chapter will show you how to

- identify the scale factor of an enlargement as the ratio of the lengths of any two corresponding line segments **D**
- understand that enlargements are described by a centre of enlargement and a scale factor **D**
- enlarge shapes using positive, positive fractional and negative scale factors **D** **C** **A**

Before you start this chapter

Match each word in the box to one of the shapes.

right-angled equilateral isosceles scalene obtuse-angled

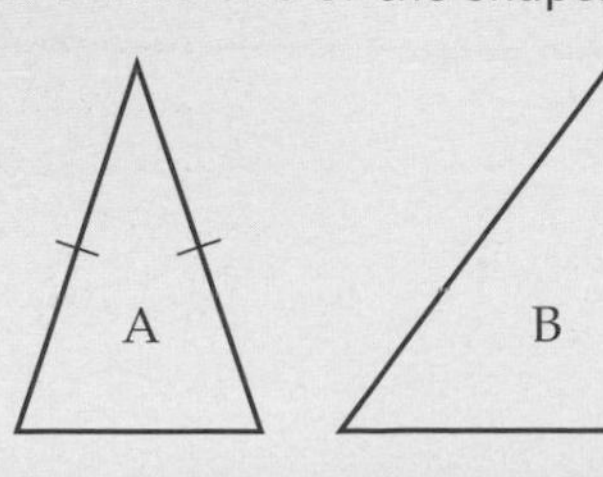

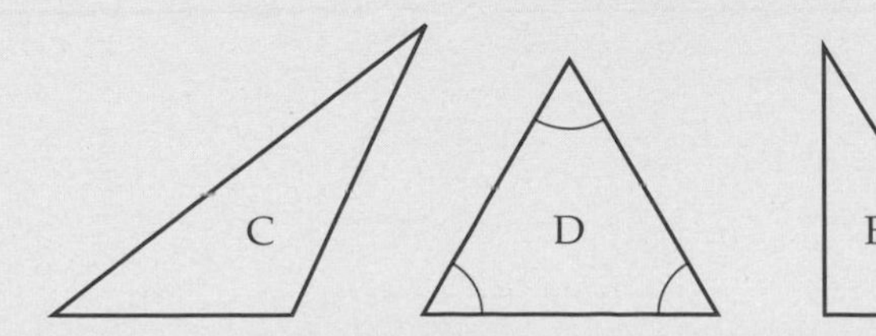

Enlargement

Why learn this?

Scale models, mathematically similar to real-life objects, are built for TV and film.

Objectives

D Enlarge a shape using a centre of enlargement

Keywords
enlargement, scale factor, multiplier, proportion, similar, centre of enlargement

Skills check

1 Copy this coordinate grid on to squared paper.

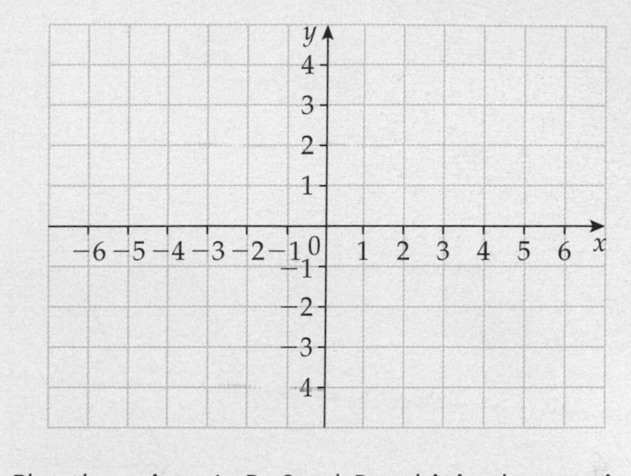

a Plot the points A, B, C and D and join them up in the following order.

$A(4, 1)$, $B(-2, 2)$, $C(-5, -1)$, $D(1, -2)$, $A(4, 1)$

b Name the shape $ABCD$.

Enlargement

An **enlargement** of an object is a transformation that changes the size of the object but keeps its shape the same.

Look at the two L shapes.

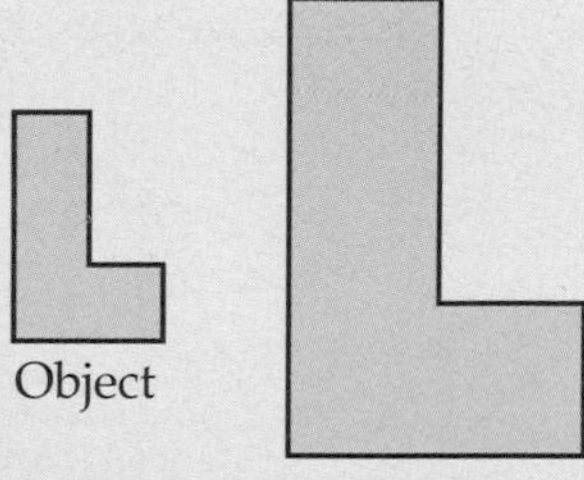

The image is an enlargement of the object.

The **scale factor** (or **multiplier**) is 2.

This means that every length on the image is twice as long as the corresponding length on the object.

The perimeter of the image has also been multiplied by the same scale factor.

In an enlargement all the angles stay the same, but all the lengths are changed in the same **proportion**.
The image is **similar** to the object.

An enlargement always produces a pair of similar shapes. You will learn more about similarity in Chapter 32.

When you enlarge a shape from a **centre of enlargement**, the distances from the centre to each point are multiplied by the scale factor.

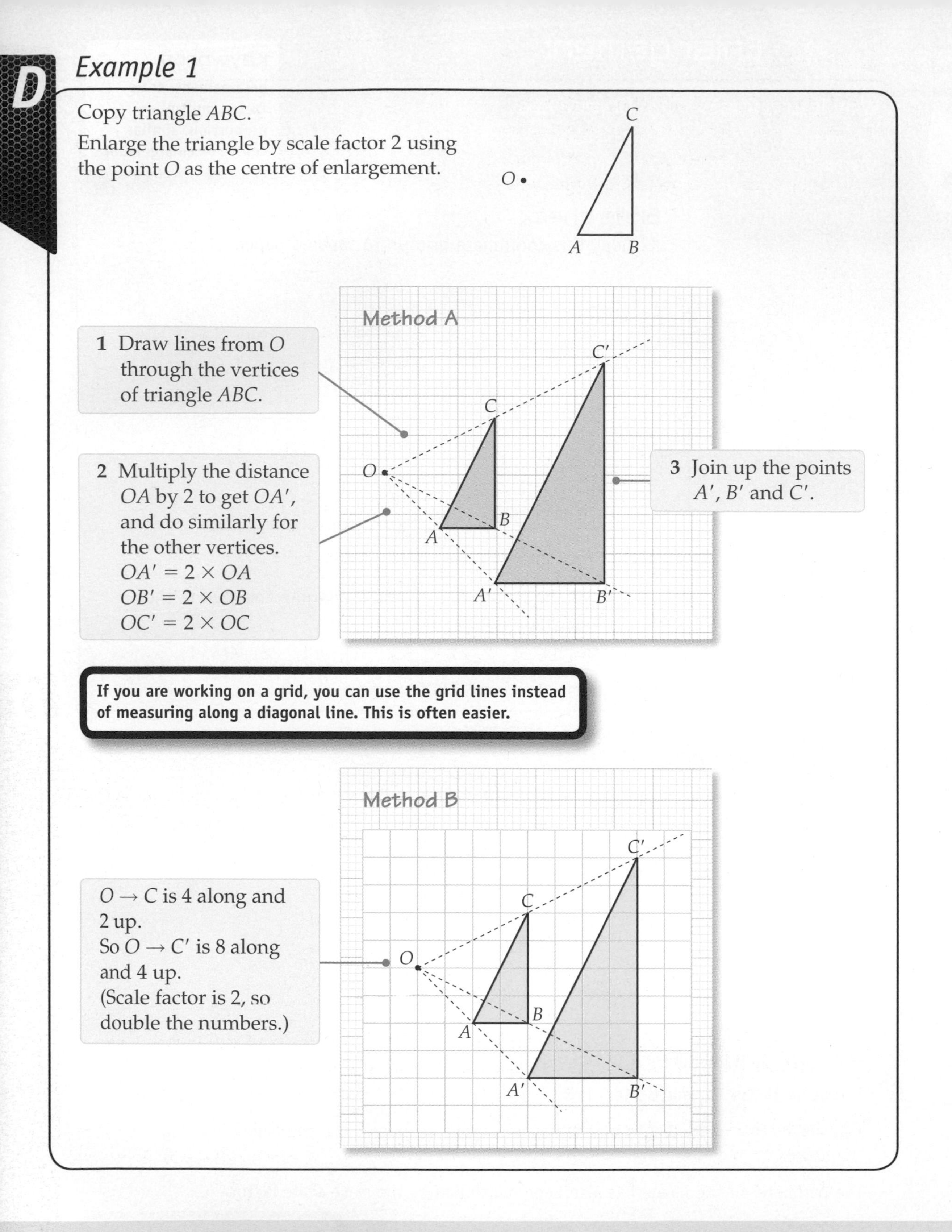

D
Example 1
Copy triangle ABC.
Enlarge the triangle by scale factor 2 using the point O as the centre of enlargement.
C
O
A
B
Method A
1 Draw lines from O through the vertices of triangle ABC.
2 Multiply the distance OA by 2 to get OA′, and do similarly for the other vertices.
OA′ = 2 × OA
OB′ = 2 × OB
OC′ = 2 × OC
3 Join up the points A′, B′ and C′.
C′
A′
B′
If you are working on a grid, you can use the grid lines instead of measuring along a diagonal line. This is often easier.
Method B
O → C is 4 along and 2 up.
So O → C′ is 8 along and 4 up.
(Scale factor is 2, so double the numbers.)

The position of the centre of enlargement

The centre of enlargement may be outside the object (as in Example 1). It may also be inside the object, on an edge or at a corner.

This means that the object and image may overlap, or one may be inside the other.

Example 2

a Enlarge the blue object by scale factor 2 using C as the centre of enlargement.

b Enlarge the blue object by scale factor 2 using O as the centre of enlargement.

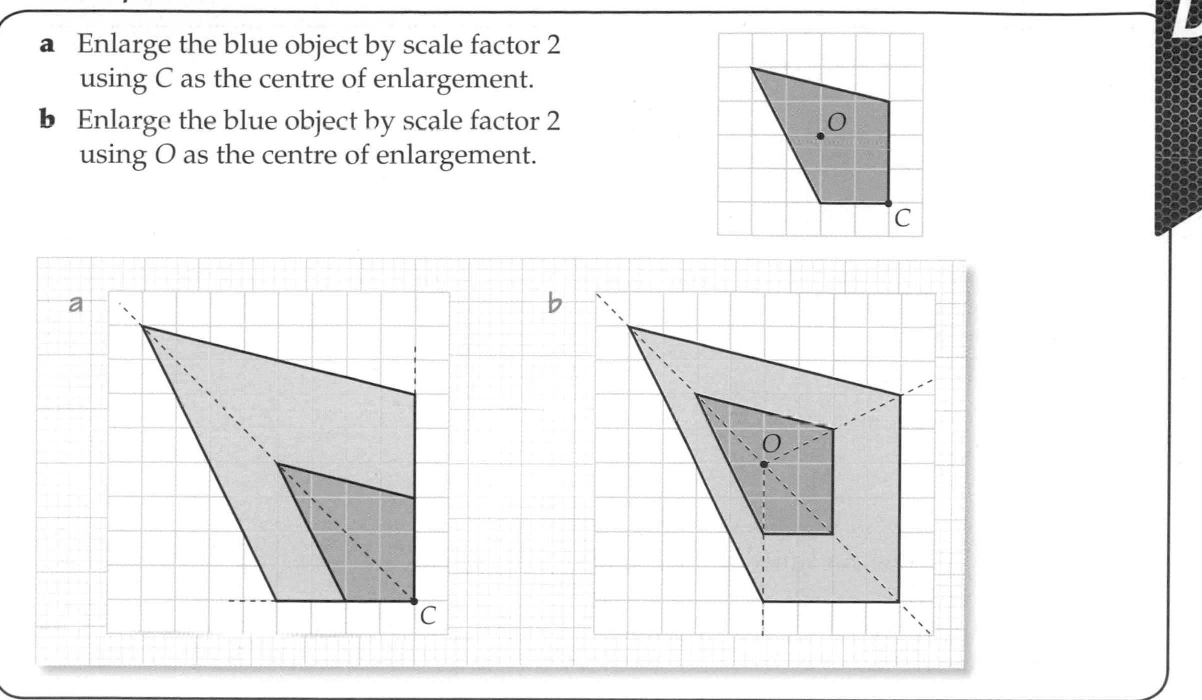

D

Exercise 31A

1 The vertices of the purple triangle are at (1, 1), (4, 1) and (1, 3).

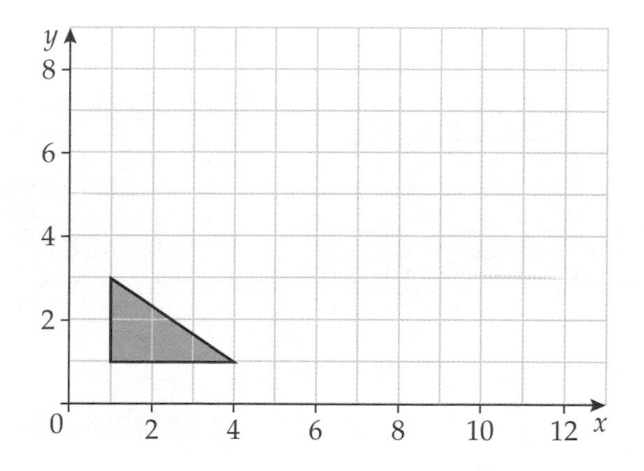

Enlarge the purple triangle by scale factor 3 using (0, 0) as the centre of enlargement. What are the coordinates of the vertices of the image triangle?

2 Copy each of the shapes below.

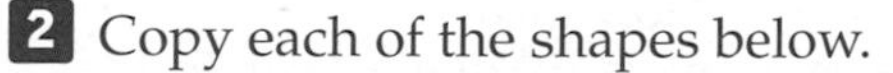

Enlarge each shape by the scale factor given using O as the centre of enlargement.

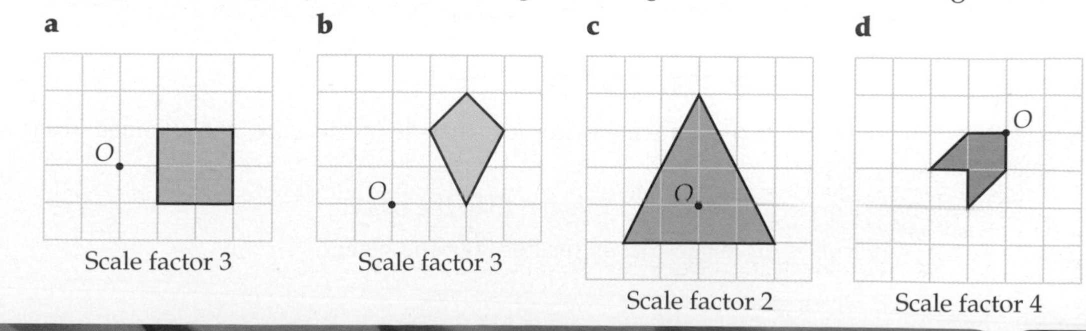

a Scale factor 3

b Scale factor 3

c Scale factor 2

d Scale factor 4

D

D

3 Shape Q is an enlargement of shape P.

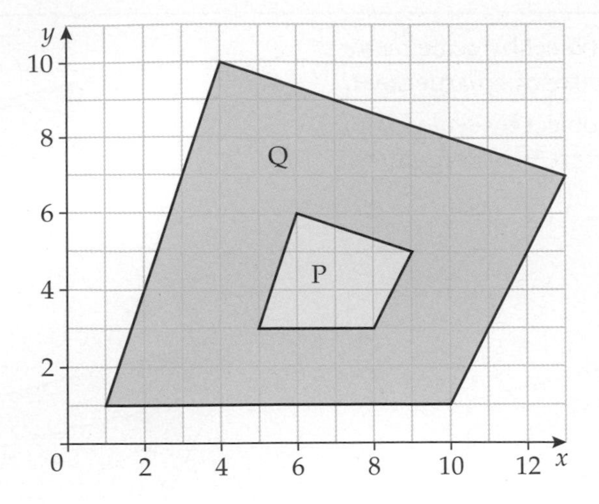

a What is the scale factor of the enlargement?

b What are the coordinates of the centre of enlargement?

4 On squared paper draw a coordinate grid with x- and y-axes from 0 to 7. Plot a triangle with vertices at (2, 2), (5, 2) and (2, 5). Label the triangle A.

a Draw the image of triangle A after an enlargement by scale factor 2 with (3, 3) as the centre of enlargement. Label the image B.

b What are the coordinates of the vertices of B?

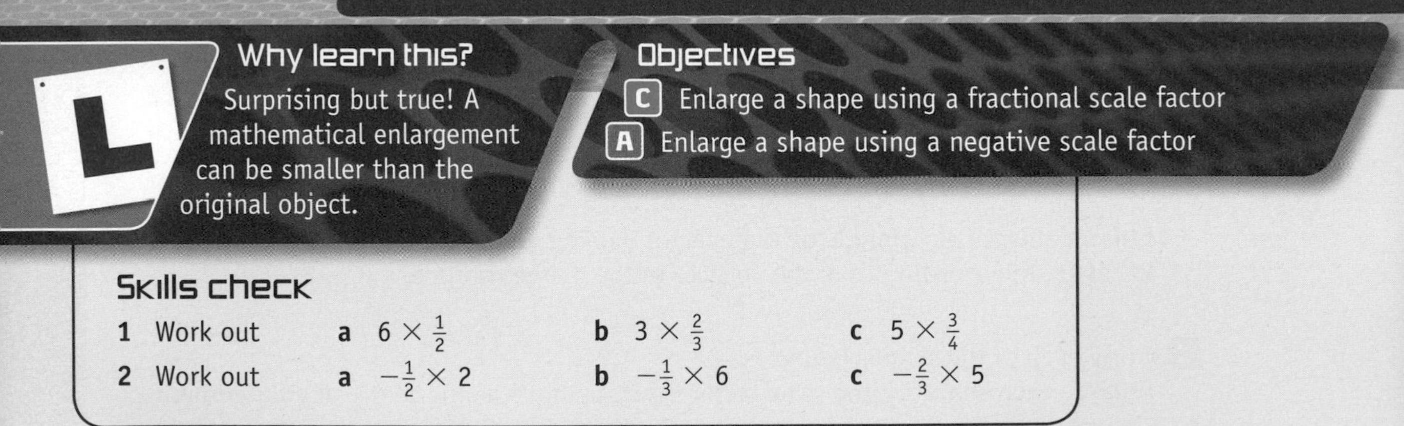

31.2 Enlargements with fractional and negative scale factors

Why learn this?

Surprising but true! A mathematical enlargement can be smaller than the original object.

Objectives

- **C** Enlarge a shape using a fractional scale factor
- **A** Enlarge a shape using a negative scale factor

Skills check

1 Work out **a** $6 \times \frac{1}{2}$ **b** $3 \times \frac{2}{3}$ **c** $5 \times \frac{3}{4}$

2 Work out **a** $-\frac{1}{2} \times 2$ **b** $-\frac{1}{3} \times 6$ **c** $-\frac{2}{3} \times 5$

Fractional scale factors

You can have an enlargement with a scale factor less than 1. In this case, the enlarged image is smaller than the object.

A scale factor of >1 gives an image that is larger than the object.

A scale factor of <1 gives an image that is smaller than the object.

Example 3

a Enlarge shape A by scale factor $\frac{1}{3}$ using point O as the centre of enlargement. Label the enlarged shape B.

b What enlargement would transform B back to A?

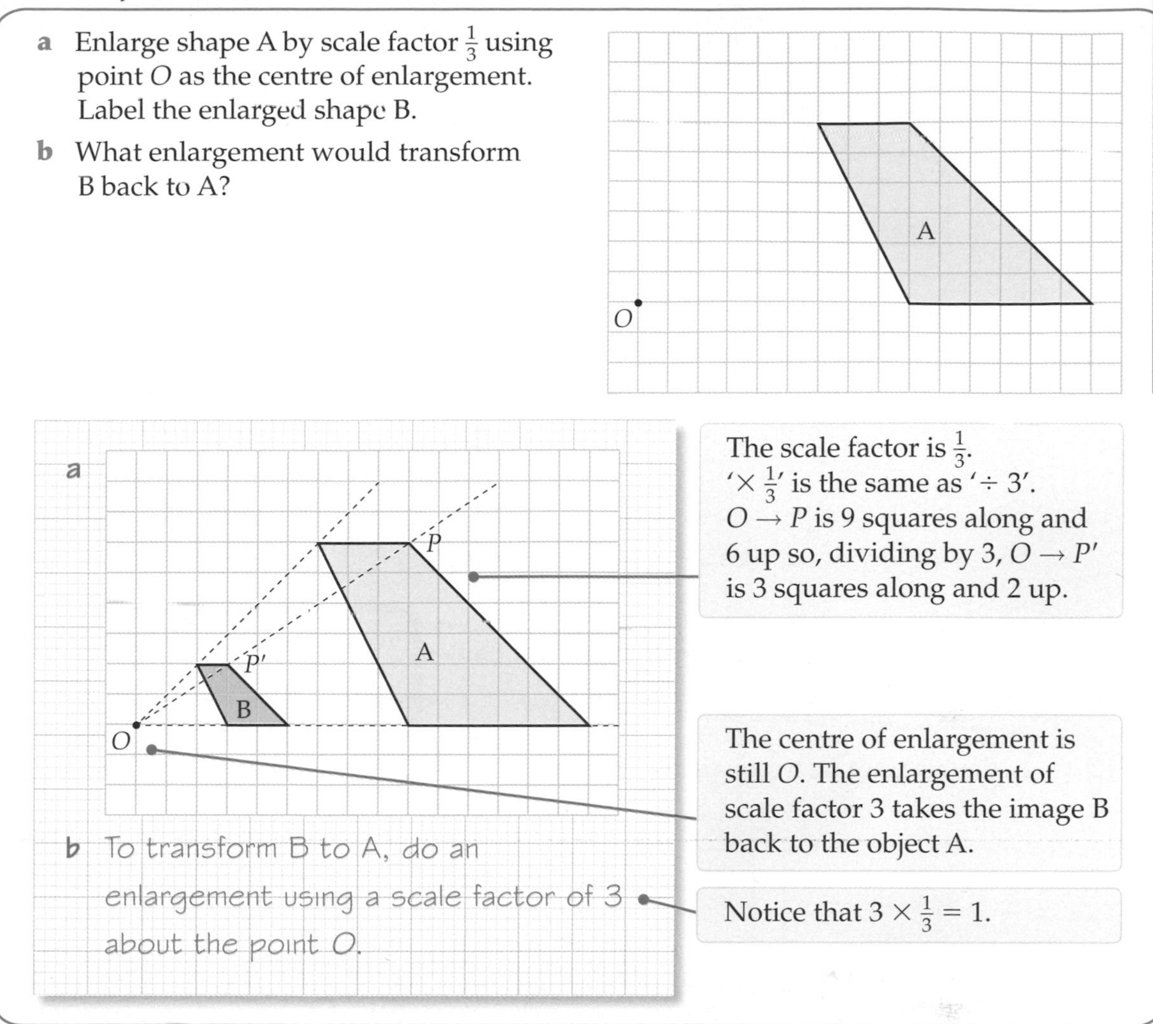

The scale factor is $\frac{1}{3}$.
'$\times \frac{1}{3}$' is the same as '$\div 3$'.
$O \rightarrow P$ is 9 squares along and 6 up so, dividing by 3, $O \rightarrow P'$ is 3 squares along and 2 up.

b To transform B to A, do an enlargement using a scale factor of 3 about the point O.

The centre of enlargement is still O. The enlargement of scale factor 3 takes the image B back to the object A.

Notice that $3 \times \frac{1}{3} = 1$.

Exercise 31B

1 Copy each diagram on to squared paper.
Enlarge each shape by the given scale factor using O as the centre of enlargement.

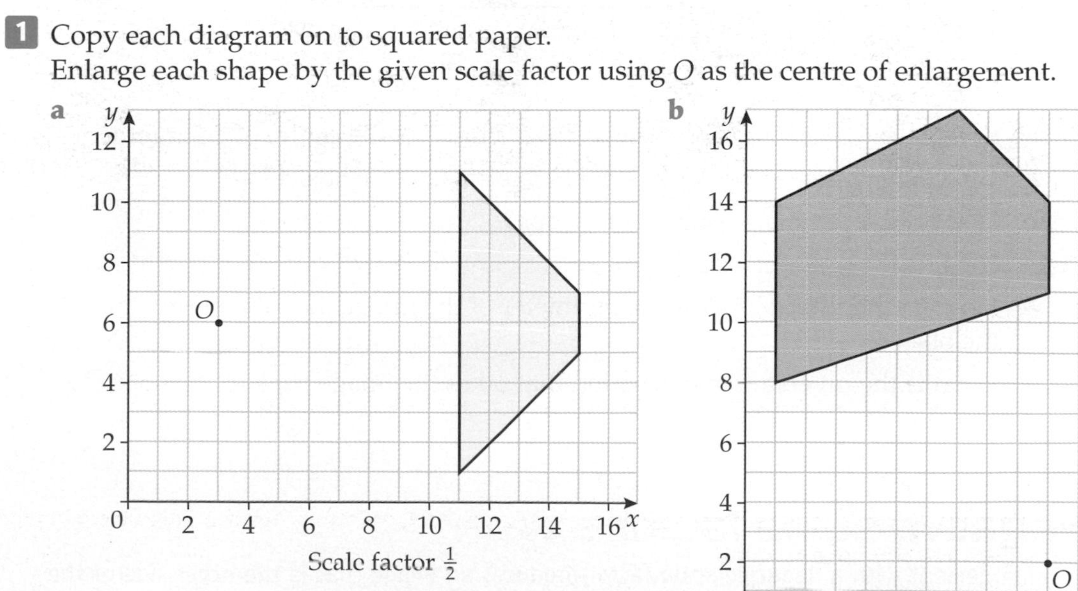

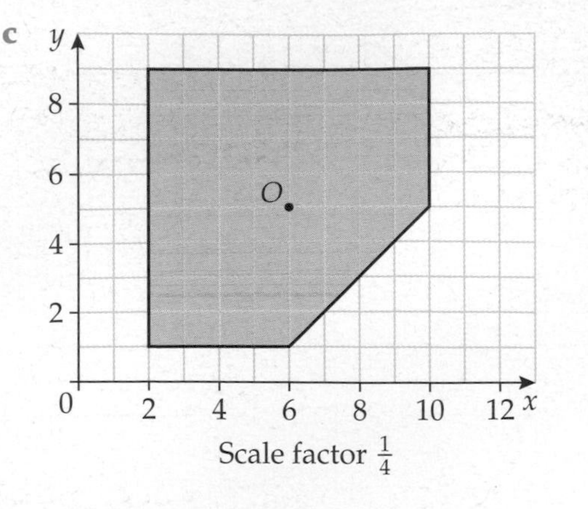

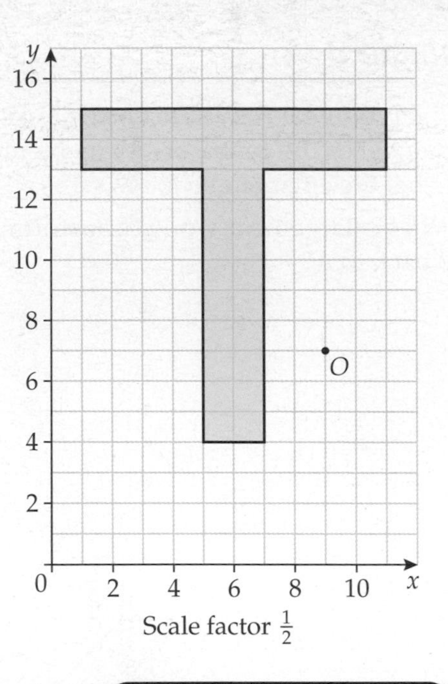

2 For each part in Q1, describe the enlargement that would take the image back to the object.

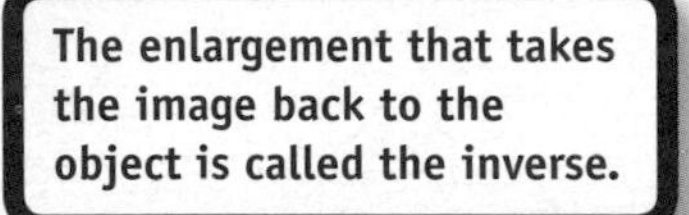
The enlargement that takes the image back to the object is called the inverse.

3 Copy the diagram on to squared paper.

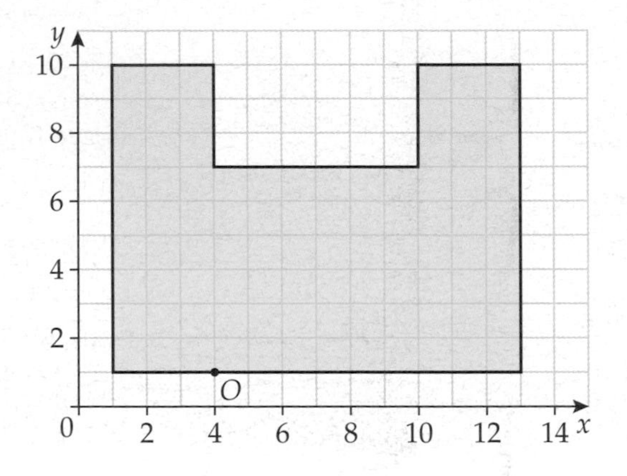

a Enlarge the blue shape by scale factor $\frac{2}{3}$ with O as the centre of enlargement.

b What enlargement would take the image back to the object?

c What is the perimeter of the blue shape?

d What is the perimeter of the image?

e How many times bigger is the perimeter of the image than the perimeter of the object?

f What is the inverse transformation that takes the image back to the object?

Negative scale factors

An enlargement with a negative scale factor produces an image that is the other side of the centre of enlargement. The image appears upside down, and its size is determined by the scale factor.

Example 4

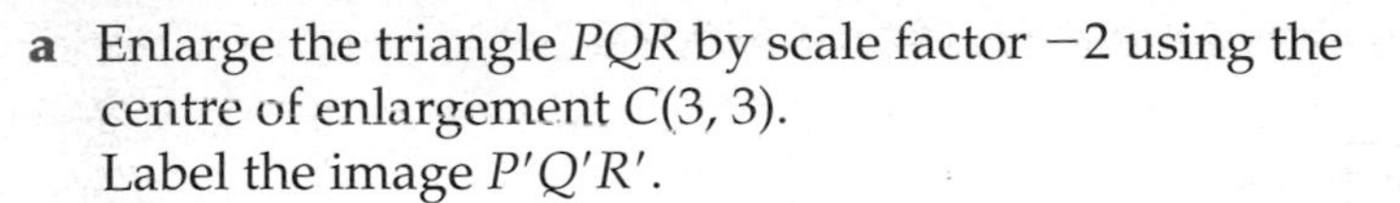

a Enlarge the triangle PQR by scale factor -2 using the centre of enlargement $C(3, 3)$.
Label the image $P'Q'R'$.

b What is the enlargement that will take triangle $P'Q'R'$ back to triangle PQR?

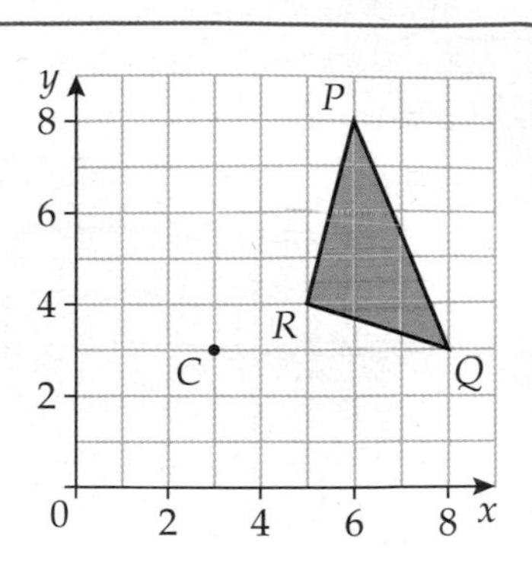

When specifying an enlargement always give the scale factor and the centre of enlargement.

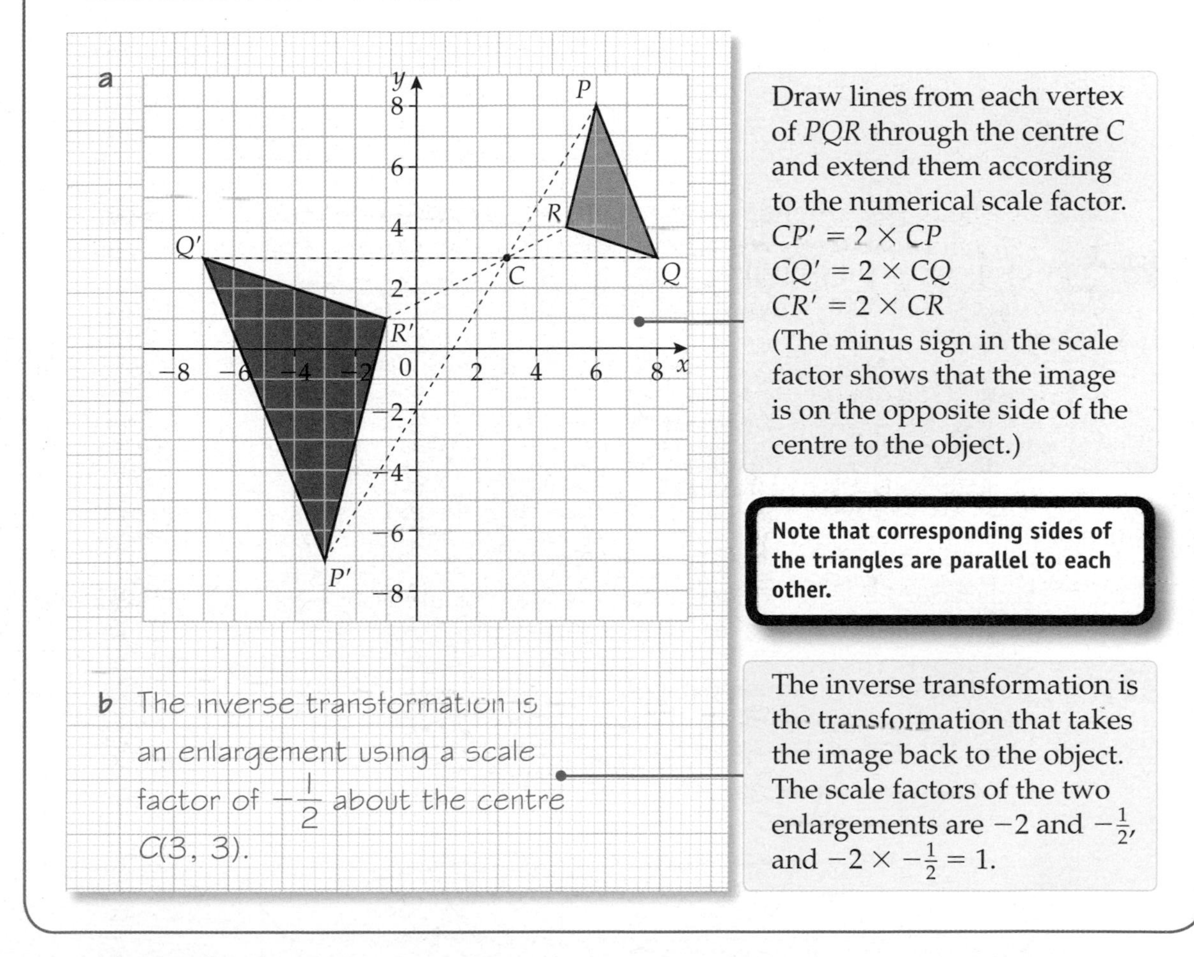

Draw lines from each vertex of PQR through the centre C and extend them according to the numerical scale factor.
$CP' = 2 \times CP$
$CQ' = 2 \times CQ$
$CR' = 2 \times CR$
(The minus sign in the scale factor shows that the image is on the opposite side of the centre to the object.)

Note that corresponding sides of the triangles are parallel to each other.

b The inverse transformation is an enlargement using a scale factor of $-\frac{1}{2}$ about the centre $C(3, 3)$.

The inverse transformation is the transformation that takes the image back to the object. The scale factors of the two enlargements are -2 and $-\frac{1}{2}$, and $-2 \times -\frac{1}{2} = 1$.

Scale factors and similarity

- A scale factor of 1 or -1 produces an image that is the same size as the object. The object and image are congruent.
- A scale factor greater than 1 or less than -1 produces an image that is larger than the object. The two shapes are similar.
- A scale factor between -1 and 1 produces an image that is smaller than the object. The two shapes are similar.
- A negative scale factor produces an 'upside down' image on the opposite side of the centre to the object.

Exercise 31C

1 Copy the diagram on to squared paper.
Shape B is an enlargement of shape A.

a What is the scale factor of the enlargement?

b Construct lines to show the position of the centre of enlargement and give its coordinates.

c What is the enlargement that takes B back to A?

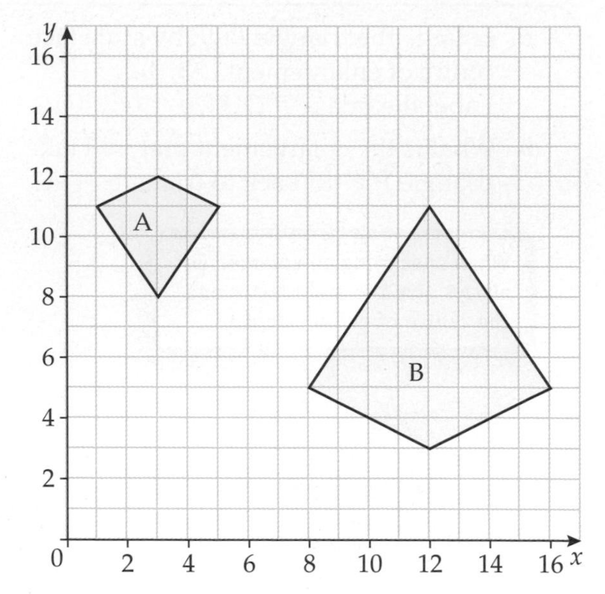

2 Copy each object on to squared paper.
Draw the enlargement of the object using the given scale factor about the centre O.

a

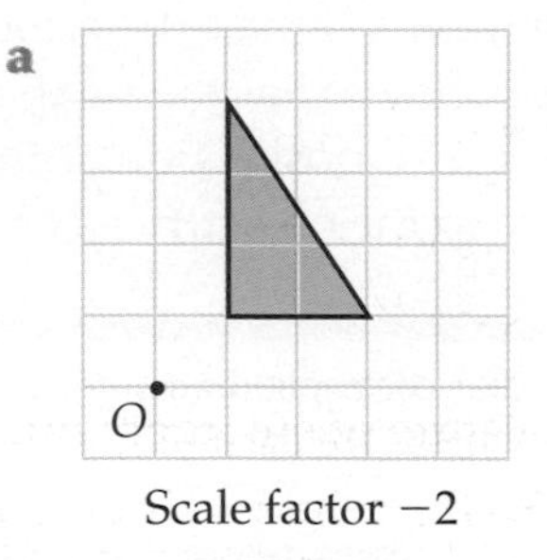

Scale factor -2

b

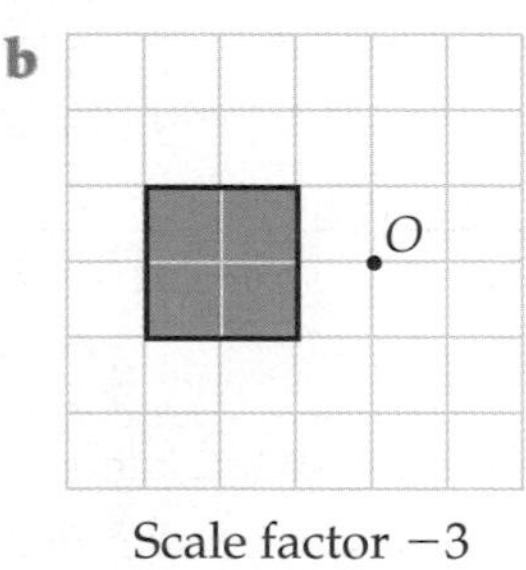

Scale factor -3

c

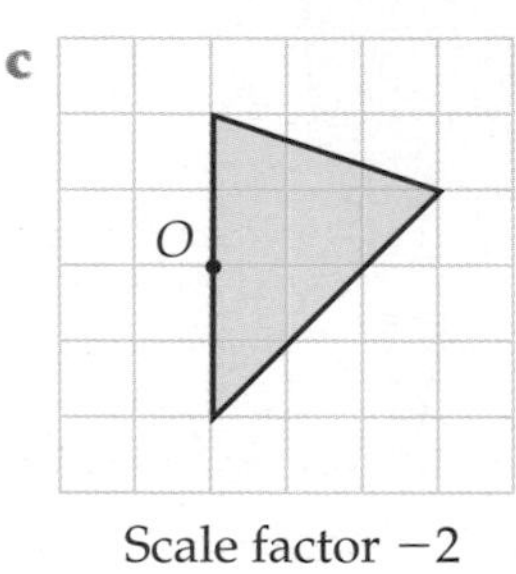

Scale factor -2

d

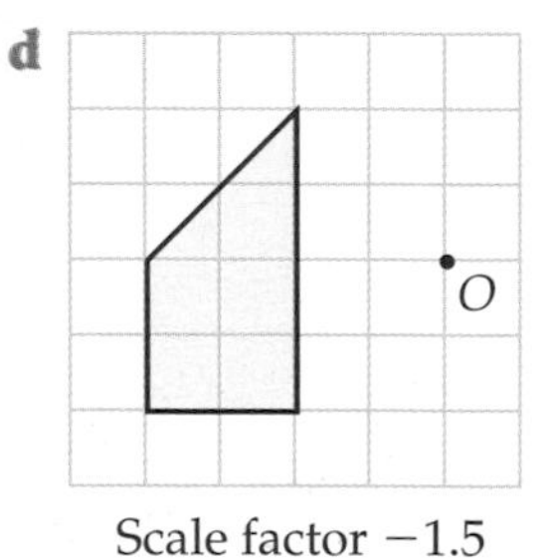

Scale factor -1.5

3 Copy the diagram on to squared paper.

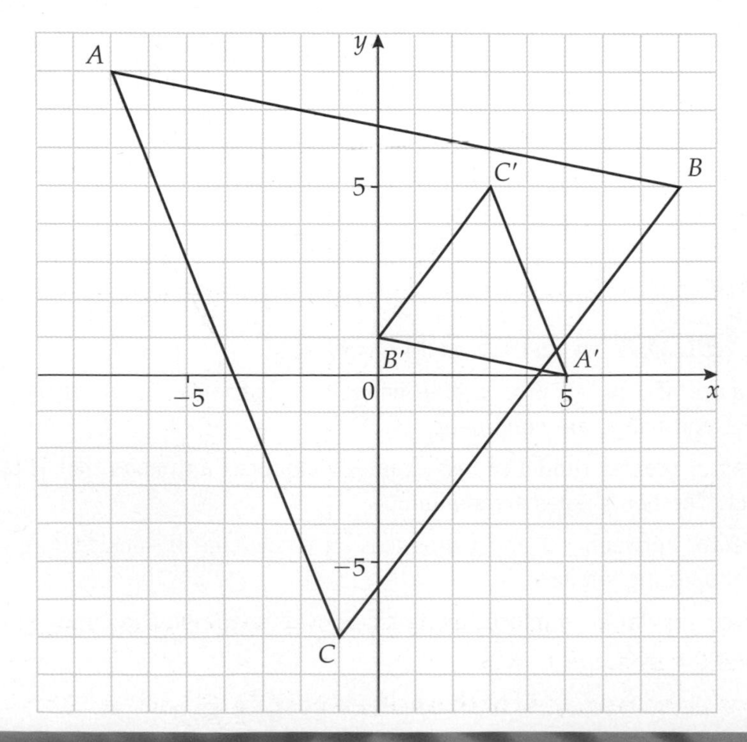

Triangle ABC has been enlarged to give triangle $A'B'C'$.

a What is the scale factor of the enlargement?

b Construct lines to show the position of the centre of enlargement and give its coordinates.

c What is the inverse transformation that takes triangle $A'B'C'$ back to triangle ABC?

4 Copy the diagram on to squared paper.

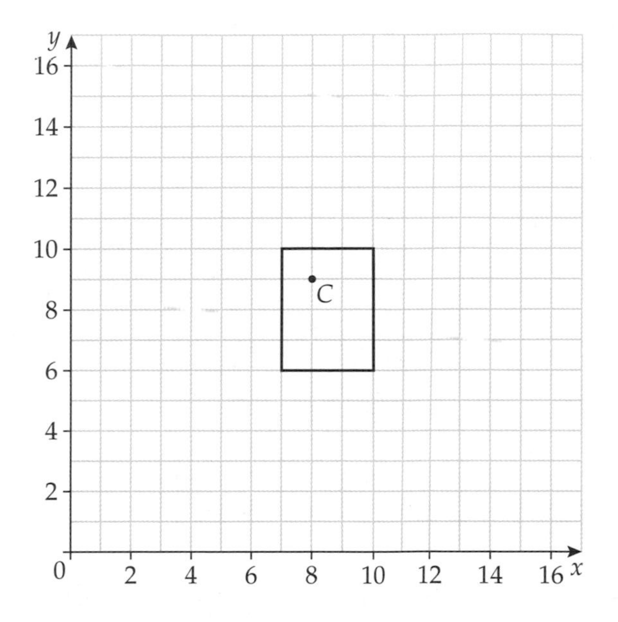

a Enlarge the rectangle by scale factor −2 using C as the centre of enlargement.

b What is the enlargement that will take the image back to the original rectangle?

5 Copy the diagram on to squared paper.

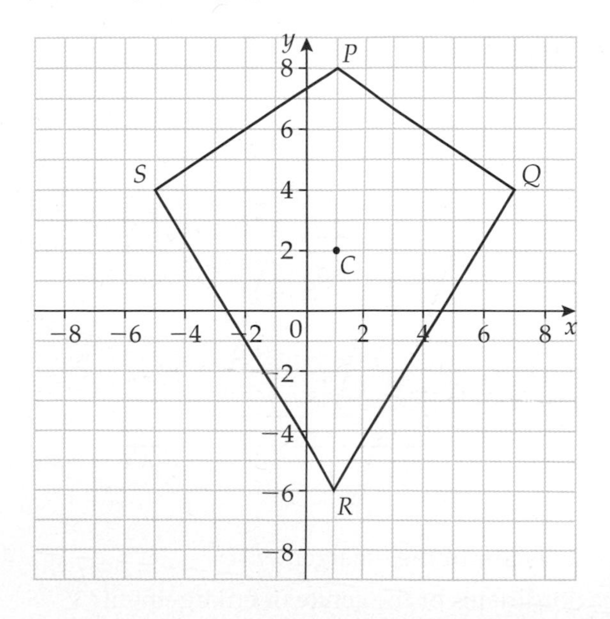

a Enlarge the kite $PQRS$ by scale factor $-\frac{1}{2}$ using C as the centre of enlargement. Label the image $P'Q'R'S'$.

b What is the enlargement that will take $P'Q'R'S'$ back to $PQRS$?

A

6 Copy the diagram on to squared paper.

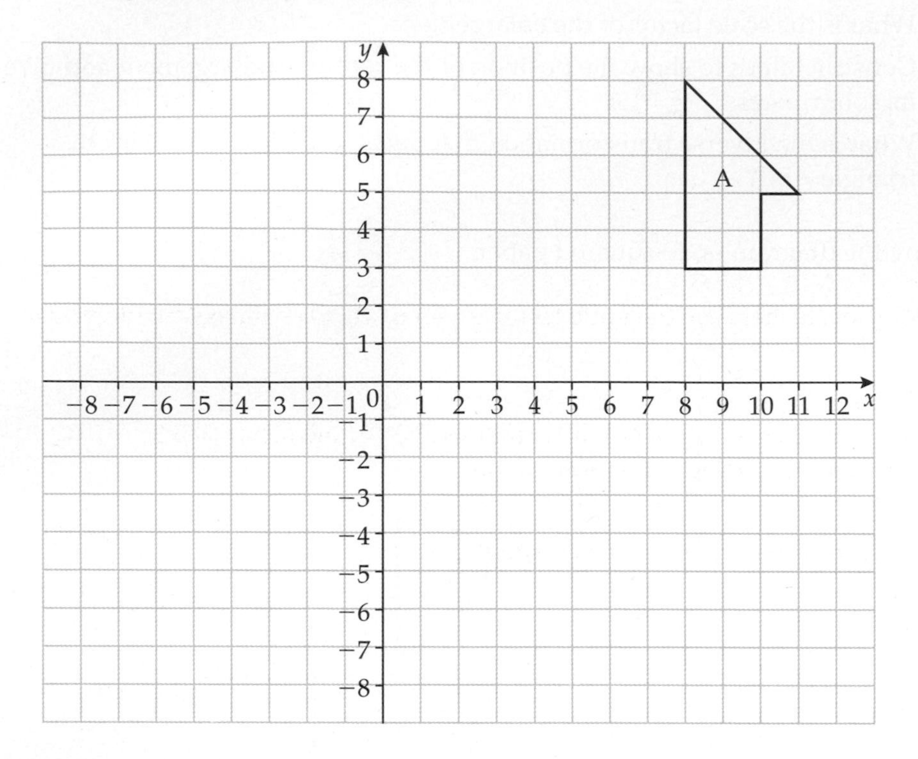

a Draw an enlargement of shape A using a scale factor of -2 about the point (5, 4). Label the image B.

b Draw an enlargement of the image B with a scale factor of $-\frac{1}{2}$ about the point (5, -2).

A02

c What do you notice about your answers to parts **a** and **b**?

Review exercise

D

1 Triangle B is an enlargement of triangle A.

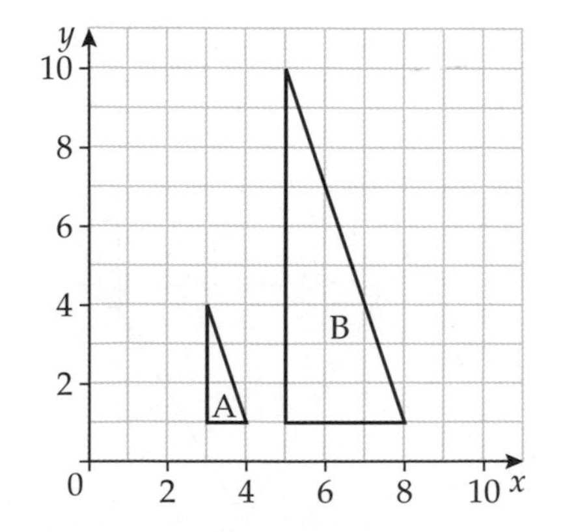

a What is the scale factor of the enlargement? **[1 mark]**

b What are the coordinates of the centre of enlargement? **[2 marks]**

2 Copy both shapes on to squared paper.

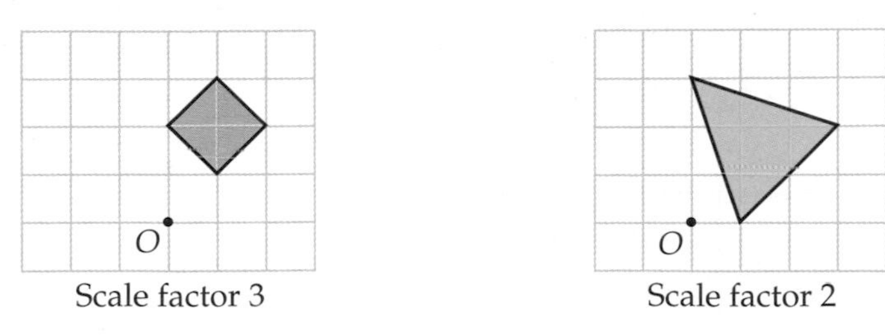

Scale factor 3   Scale factor 2

Enlarge each shape by the scale factor given using O as the centre of enlargement.

[3 marks each]

3 A shape has perimeter 24 cm. It is enlarged by a scale factor of 3.5.
What is the perimeter of the new shape? **[2 marks]**

4 Copy both objects on to squared paper.

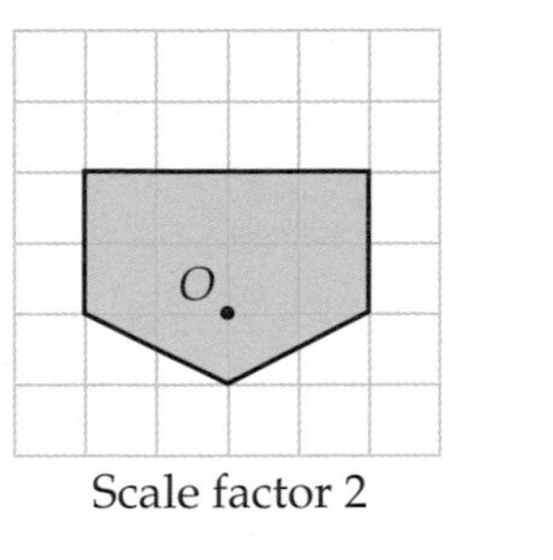

Scale factor 2

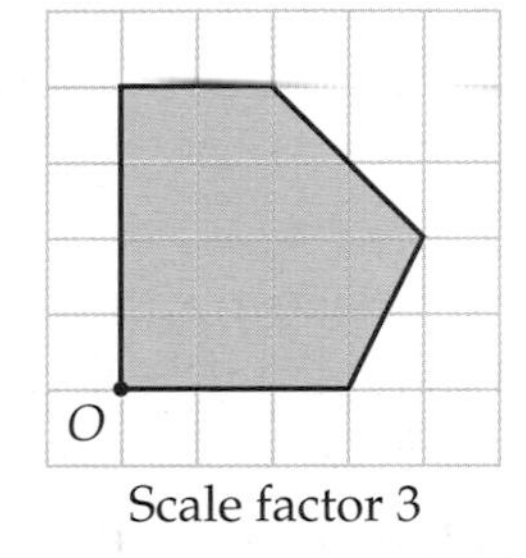

Scale factor 3

Enlarge each object by the given scale factor from the centre O. **[3 marks each]**

5 Copy both diagrams on to squared paper.

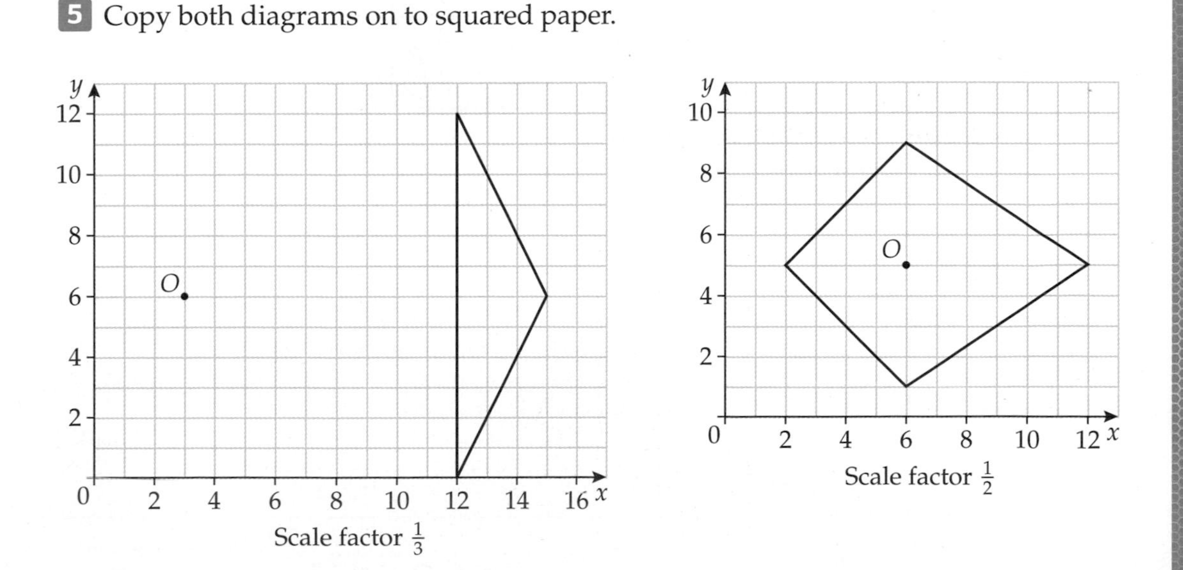

Scale factor $\frac{1}{3}$ Scale factor $\frac{1}{2}$

a Draw the enlargement of the object using the given scale factor about the centre O. **[3 marks each]**

b What enlargement would take each image back to the object? **[2 marks each]**

A

6 Triangle $A'B'C'$ is an enlargement of triangle ABC.

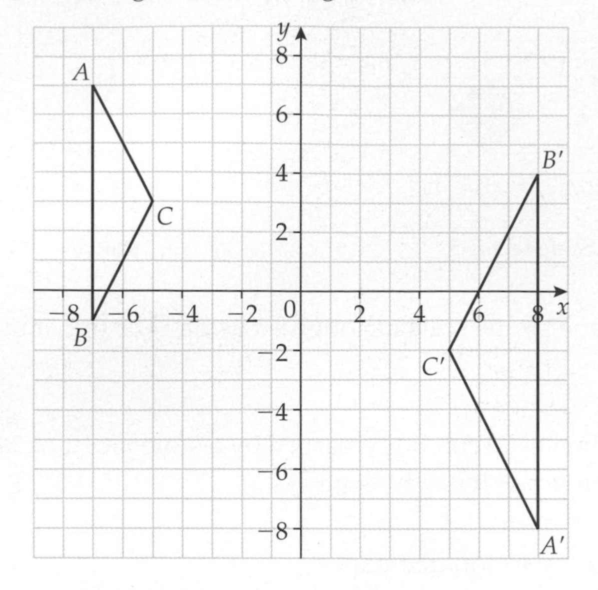

Copy the diagram on to squared paper.

a What is the scale factor of the enlargement? **[2 marks]**

b Construct lines to show the position of the centre of enlargement and give its coordinates. **[2 marks]**

c What is the inverse transformation that takes triangle $A'B'C'$ back to triangle ABC? **[2 marks]**

7 Enlarge the shape by scale factor $-\frac{1}{2}$ with centre of enlargement $(-1, 0)$.

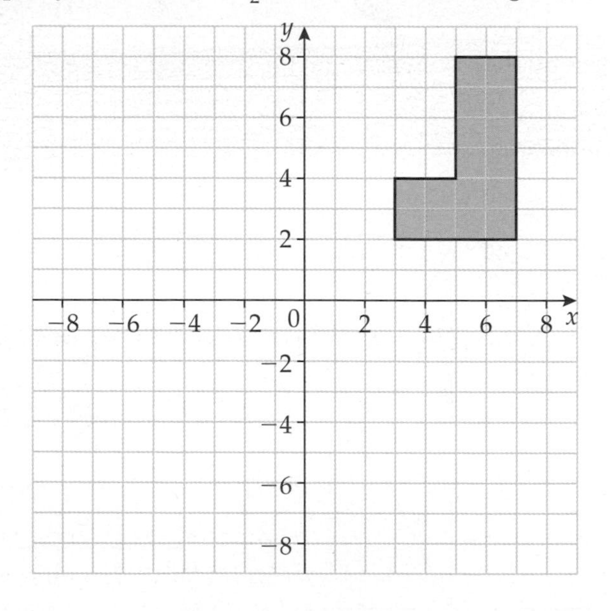

[3 marks]

Chapter summary

In this chapter you have learned how to

- enlarge a shape using a centre of enlargement **D**
- enlarge a shape using a fractional scale factor **C**
- enlarge a shape using a negative scale factor **A**

32 Congruency and similarity

This chapter is about understanding when shapes are identical to each other, and when they are only similar to each other.

A Russian 'matryoshka' doll consists of a wooden figure that can be pulled apart to reveal another figure of the same sort inside.

Objectives

This chapter will show you how to

- understand similarity of triangles and of other plane figures, and use this to make geometric inferences **B**
- understand and use the effect of enlargement on areas and volumes of shapes and solids **A**
- understand and use SSS, SAS, ASA and RHS conditions and prove the congruence of triangles using formal arguments **C** **A***

Before you start this chapter

1 Look at the shapes in the diagram.

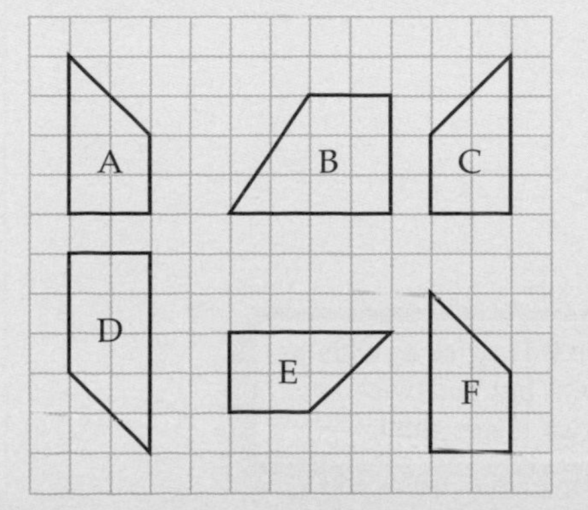

a Which shape is a translation of shape A?

b Which shape is a rotation of shape A?

HELP Chapter 30

c Which shape is a reflection of shape A?

2 Work out the area of a triangle with base 10 cm and height 7 cm.

HELP Chapter 26

3 Work out the volume of a cylinder with height 8 cm and base diameter 6 cm.

32.1 Congruency and similarity

Keywords
congruent, similar, SSS, SAS, ASA, RHS

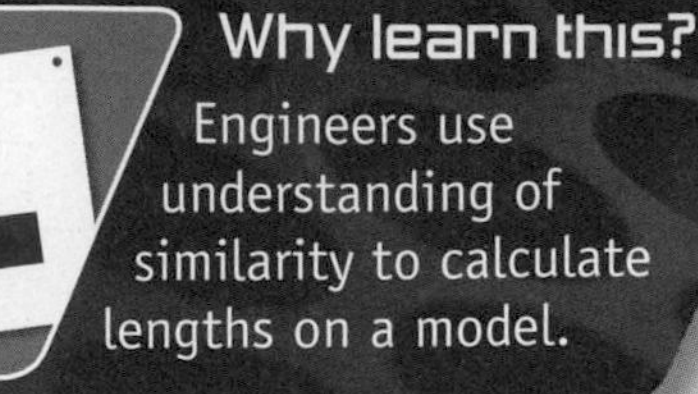

Why learn this?
Engineers use understanding of similarity to calculate lengths on a model.

Objectives

C Understand similarity and congruency

C Understand the conditions for congruent triangles

Skills check

Enlarge this shape by scale factor 3.

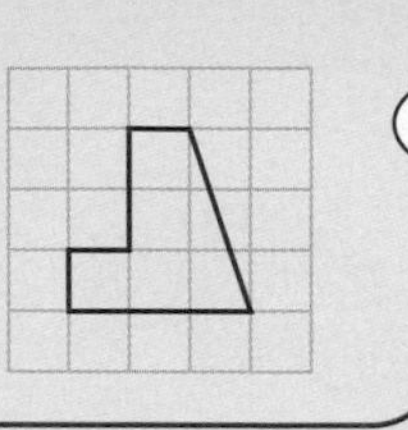

HELP Section 31.1

Congruency and similarity

Two objects are **congruent** if they are identical in shape and size.

Reflections, rotations and translations produce images that are congruent to the original object.

If two shapes are congruent, all corresponding lengths and angles in the two shapes are equal.

Two objects are **similar** when they are exactly the same shape but not the same size.

An enlargement always produces two shapes that are similar.

All squares are mathematically similar.

All circles are mathematically similar.

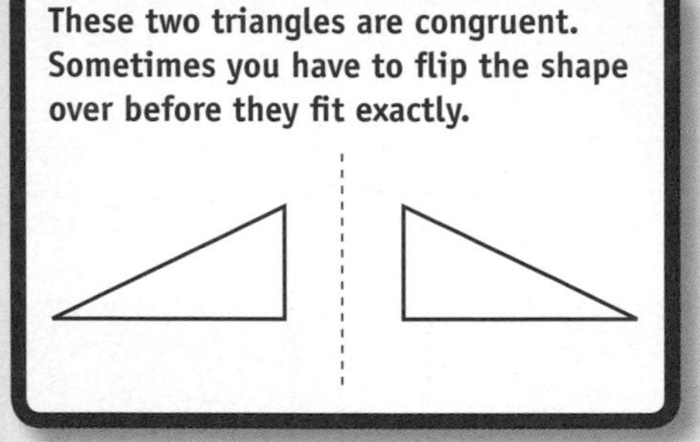

These two triangles are congruent. Sometimes you have to flip the shape over before they fit exactly.

C

Example 1

a Which of the shapes below are congruent to shape A?

b Which of the shapes are similar to shape A?

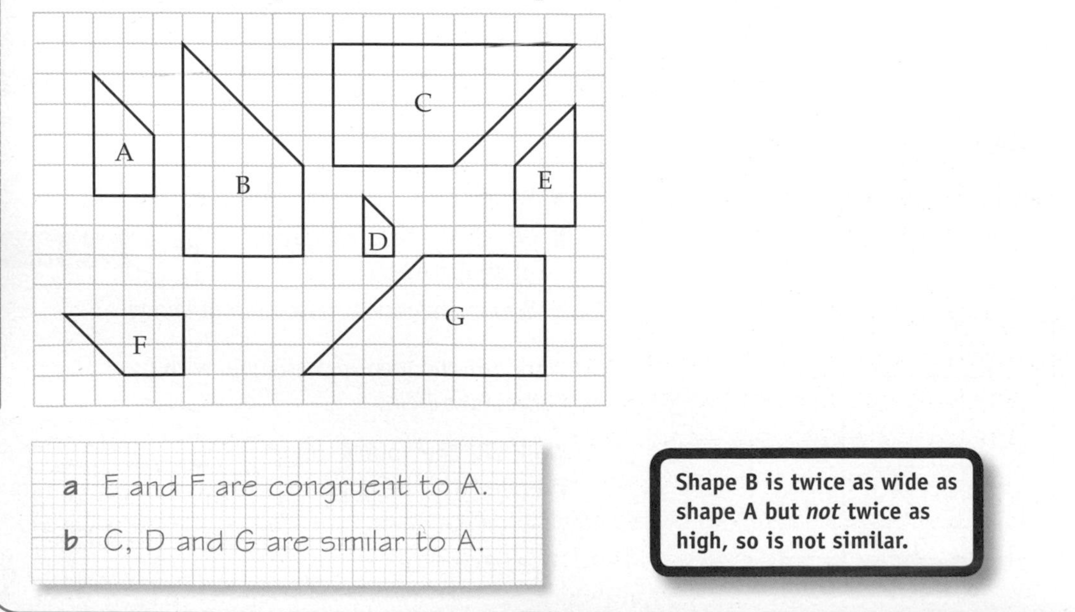

a E and F are congruent to A.

b C, D and G are similar to A.

Shape B is twice as wide as shape A but *not* twice as high, so is not similar.

Exercise 32A

1 Look at the shapes in the diagram.

Write down the letters of the shapes that form congruent pairs.

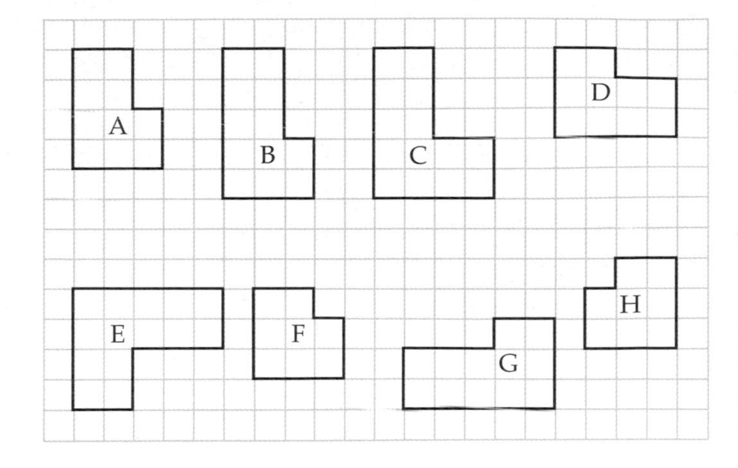

2 **a** Draw one shape that is congruent to this shape.

b Draw one shape that is similar to this shape.

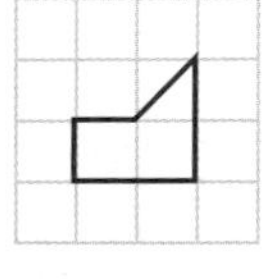

3 Look at these diagrams.
For each one write down the letters of the shapes that are similar to each other.

a

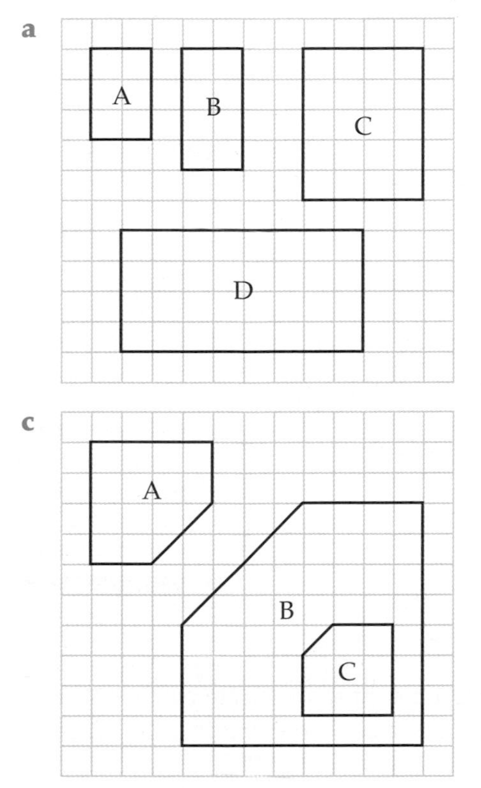

b

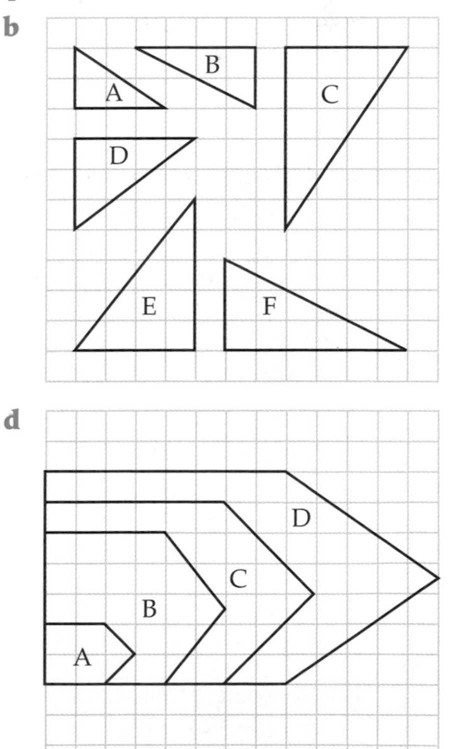

c

d

4 True or false?

a Any two squares are similar to each other.

b Any two rectangles are similar to each other.

c Any two rhombuses are similar to each other.

d Any two circles are similar to each other.

e Any two isosceles triangles are similar to each other.

f Any two regular pentagons are similar to each other.

g Any two regular polygons with the same number of sides are similar to each other.

Draw some diagrams to help you decide.

C

C

AO2

Congruency

There are four conditions that will guarantee that two triangles are congruent.

- Three sides are equal (known as side, side, side – **SSS**).

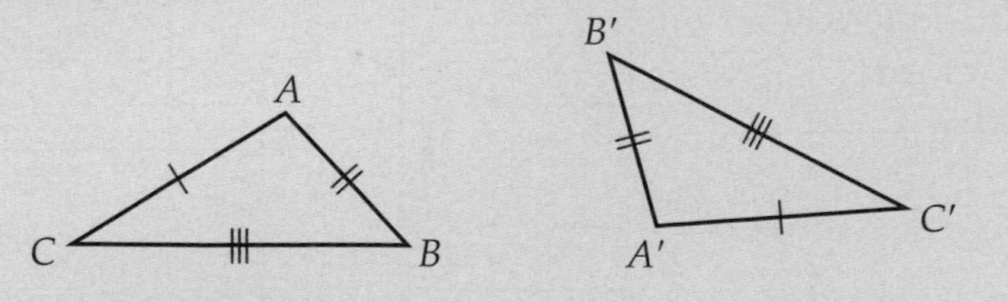

- Two sides are equal and the included angle is the same (known as side, angle, side – **SAS**).

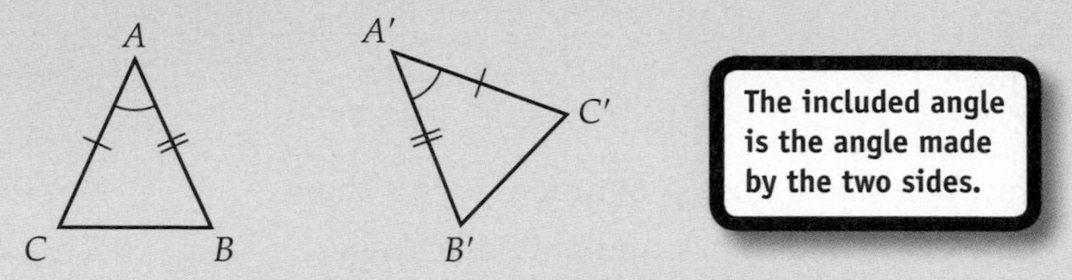

The included angle is the angle made by the two sides.

- Two angles are the same and a corresponding side is the same (known as angle, side, angle – **ASA**, or as angle, angle, side – **AAS**).

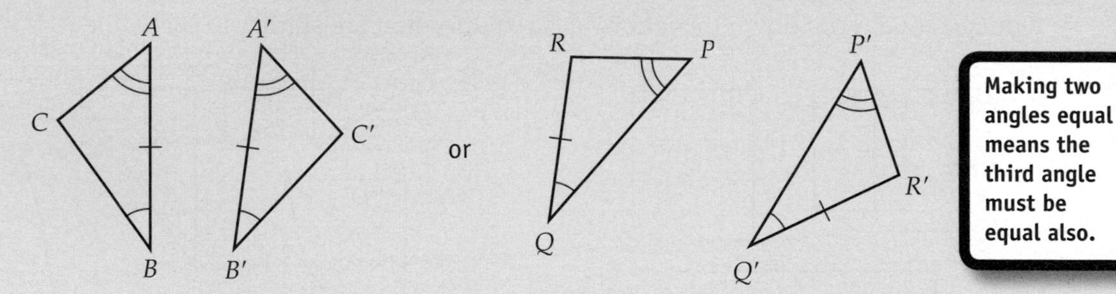

Making two angles equal means the third angle must be equal also.

AB and *A′B′* are the corresponding sides. *QR* and *Q′R′* are the corresponding sides.

- A right angle, the hypotenuse and a corresponding side are equal (known as right angle, hypotenuse, side – **RHS**).

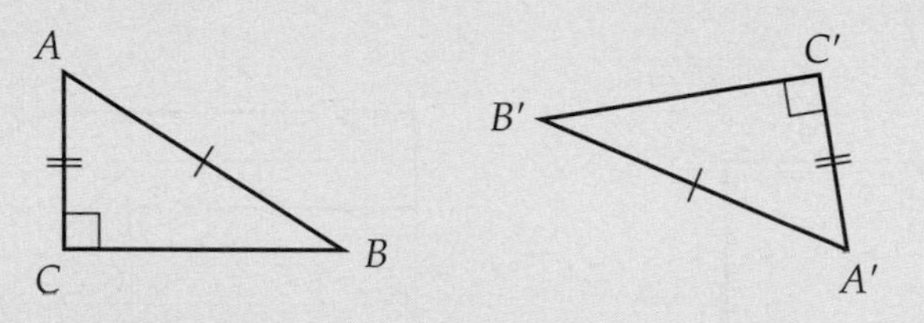

The hypotenuse is the longest side of a right-angled triangle. It is always opposite the right angle.

AC and *A′C′* are the corresponding sides.

There is one case where people sometimes think that a pair of triangles are congruent, but they are not. It is very like SAS except that the angle is *not between* the two sides, so it is SSA.

Here is an example of a triangle with SSA marked.

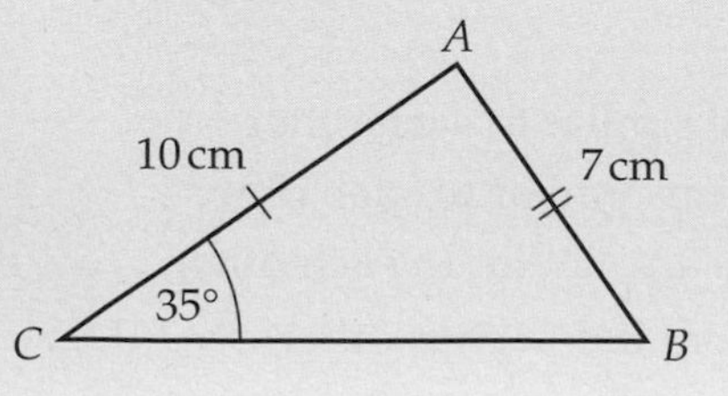

Imagine trying to construct a duplicate of this triangle.

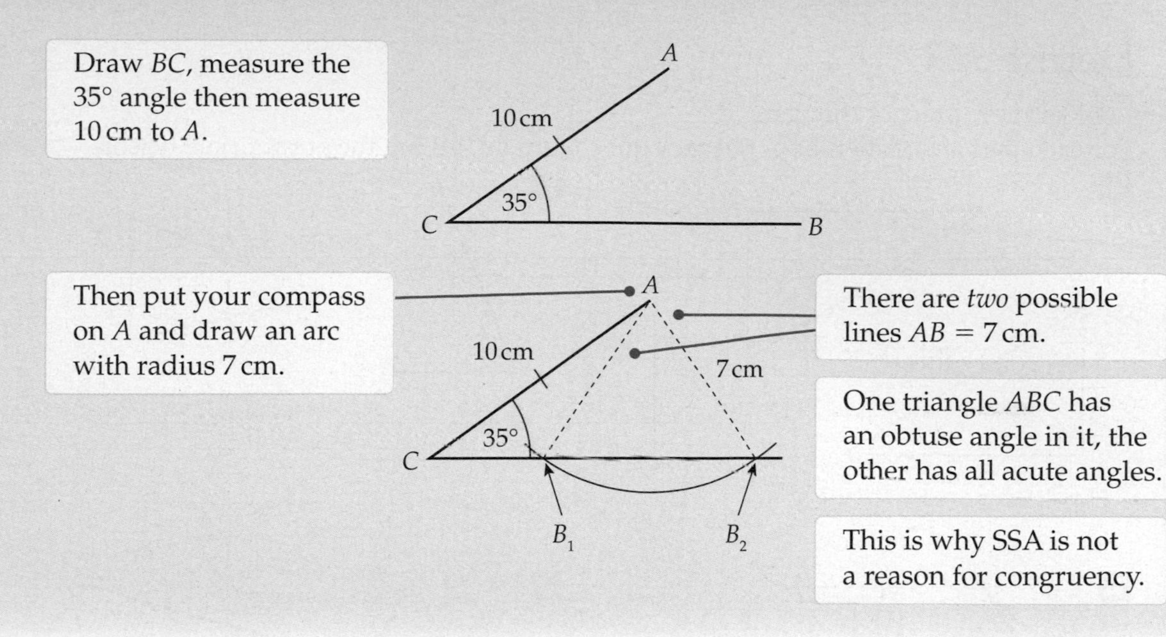
Draw BC, measure the 35° angle then measure 10 cm to A.
A
10 cm
35°
C
B
Then put your compass on A and draw an arc with radius 7 cm.
There are *two* possible lines AB = 7 cm.
10 cm
7 cm
One triangle ABC has an obtuse angle in it, the other has all acute angles.
35°
C
B1
B2
This is why SSA is not a reason for congruency.

Example 2

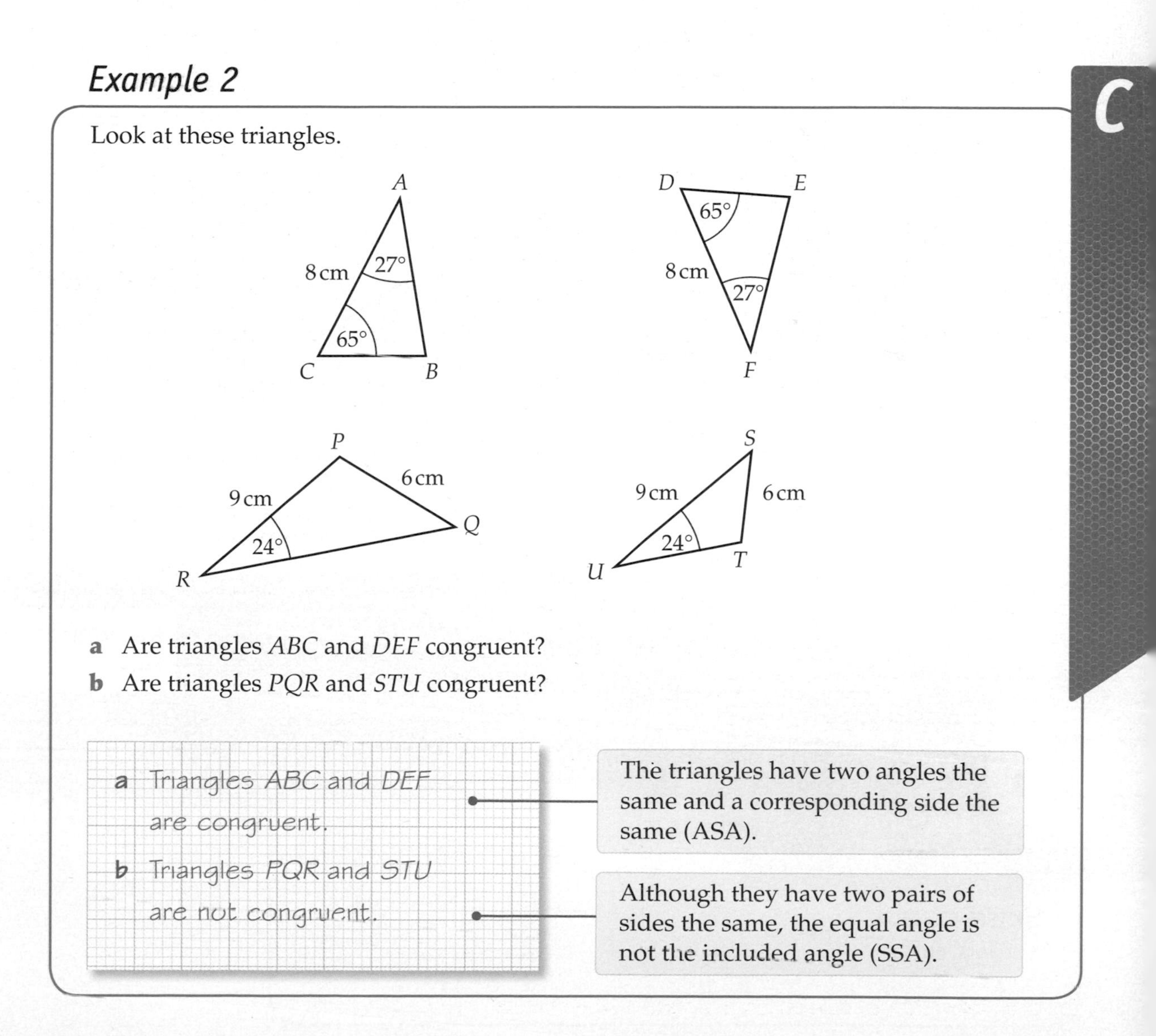
Look at these triangles.
A
8 cm
27°
65°
C
B
D
E
65°
8 cm
27°
F
P
6 cm
9 cm
Q
24°
R
S
9 cm
6 cm
24°
T
U
a Are triangles ABC and DEF congruent?
b Are triangles PQR and STU congruent?
a Triangles ABC and DEF are congruent.
b Triangles PQR and STU are not congruent.
The triangles have two angles the same and a corresponding side the same (ASA).
Although they have two pairs of sides the same, the equal angle is not the included angle (SSA).

Exercise 32B

C

Look at these pairs of triangles.
For each pair state whether or not they are congruent, giving the appropriate reason.

1 8 cm, 10 cm; 10 cm, 8 cm

2

3 4 cm, 25°, 21°; 21°, 4 cm, 25°

4 60°; 60°

5 70°, 80°, 5 cm; 30°, 70°, 5 cm

6

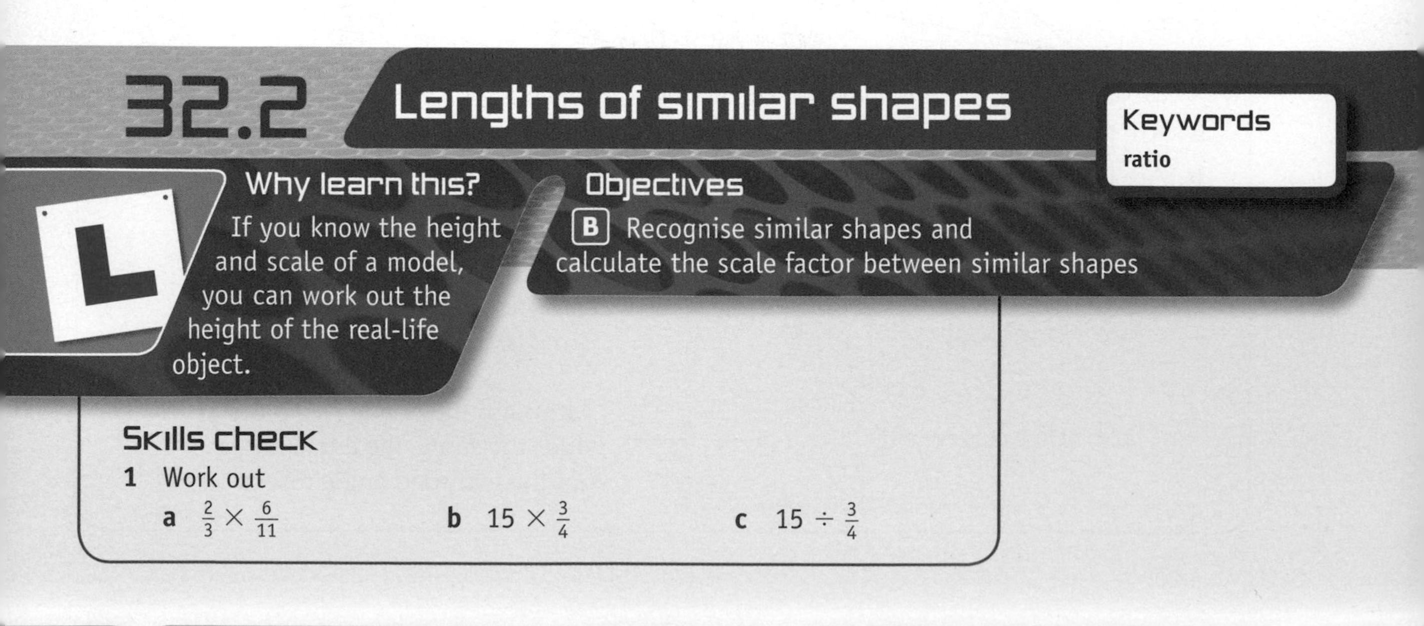

32.2 Lengths of similar shapes

Keywords
ratio

Why learn this?
If you know the height and scale of a model, you can work out the height of the real-life object.

Objectives

B Recognise similar shapes and calculate the scale factor between similar shapes

Skills check

1 Work out

a $\frac{2}{3} \times \frac{6}{11}$ **b** $15 \times \frac{3}{4}$ **c** $15 \div \frac{3}{4}$

Finding lengths in similar shapes

An enlargement produces two similar shapes.

In similar shapes corresponding lengths are in the same **ratio** or proportion.

The ratio is the scale factor of the enlargement.

For two shapes to be similar

- the corresponding angles must be equal
- the ratios of corresponding sides must be the same.

Here are two similar triangles.

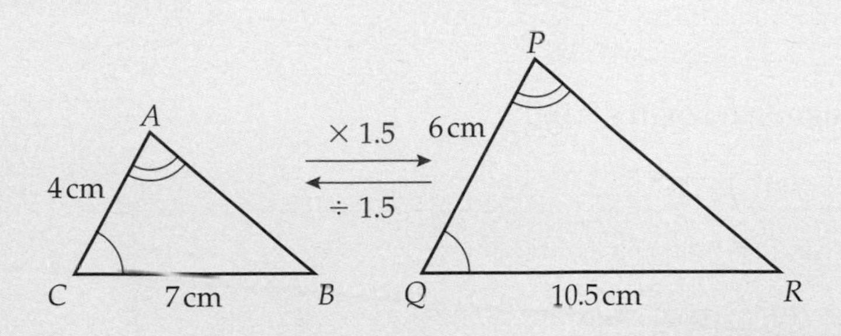

Firstly match corresponding vertices carefully! In this case, A matches with P, but B matches with R, not Q.

$A \leftrightarrow P$

$B \leftrightarrow R$

$C \leftrightarrow Q$

Writing the letters of the vertices above one another helps you pick out the corresponding lengths and ratios. In this case, from $\frac{PRQ}{ABC}$ you can pick out the three ratios.

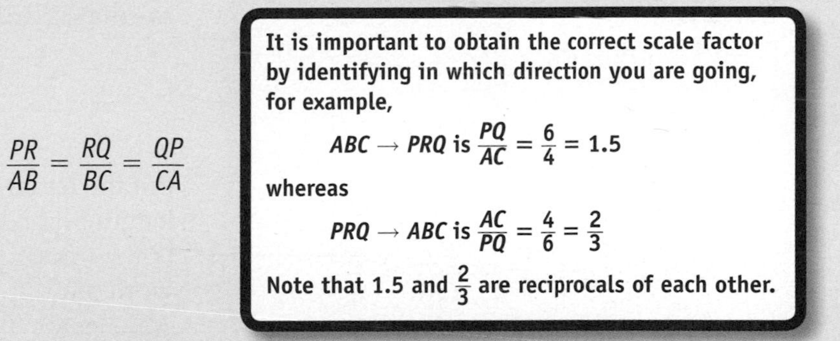

$$\frac{PR}{AB} = \frac{RQ}{BC} = \frac{QP}{CA}$$

It is important to obtain the correct scale factor by identifying in which direction you are going, for example,

$$ABC \rightarrow PRQ \text{ is } \frac{PQ}{AC} = \frac{6}{4} = 1.5$$

whereas

$$PRQ \rightarrow ABC \text{ is } \frac{AC}{PQ} = \frac{4}{6} = \frac{2}{3}$$

Note that 1.5 and $\frac{2}{3}$ are reciprocals of each other.

This can also be written as

$PR : AB = RQ : BC = QP : CA$

Now you have the ratios, which are equal to the scale factor, you can find the unknown lengths in similar shapes.

B

Example 3

These two triangles are similar with equal angles as shown.

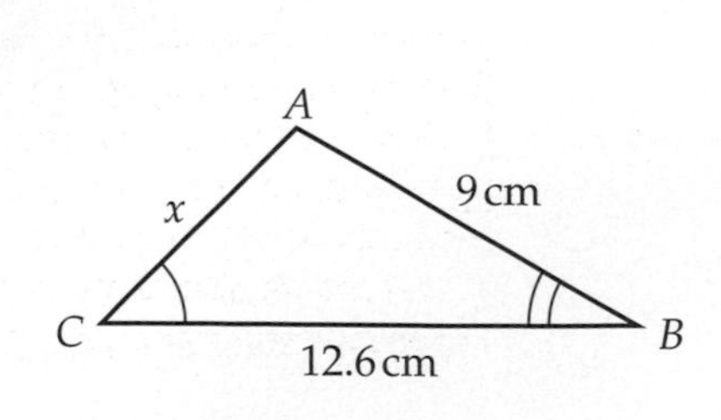

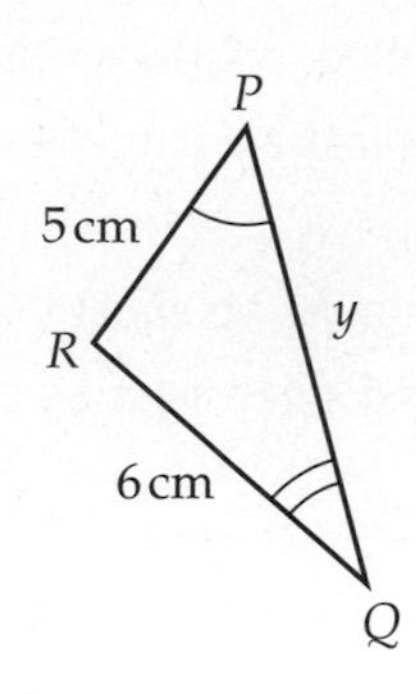

Calculate the unknown lengths x and y.

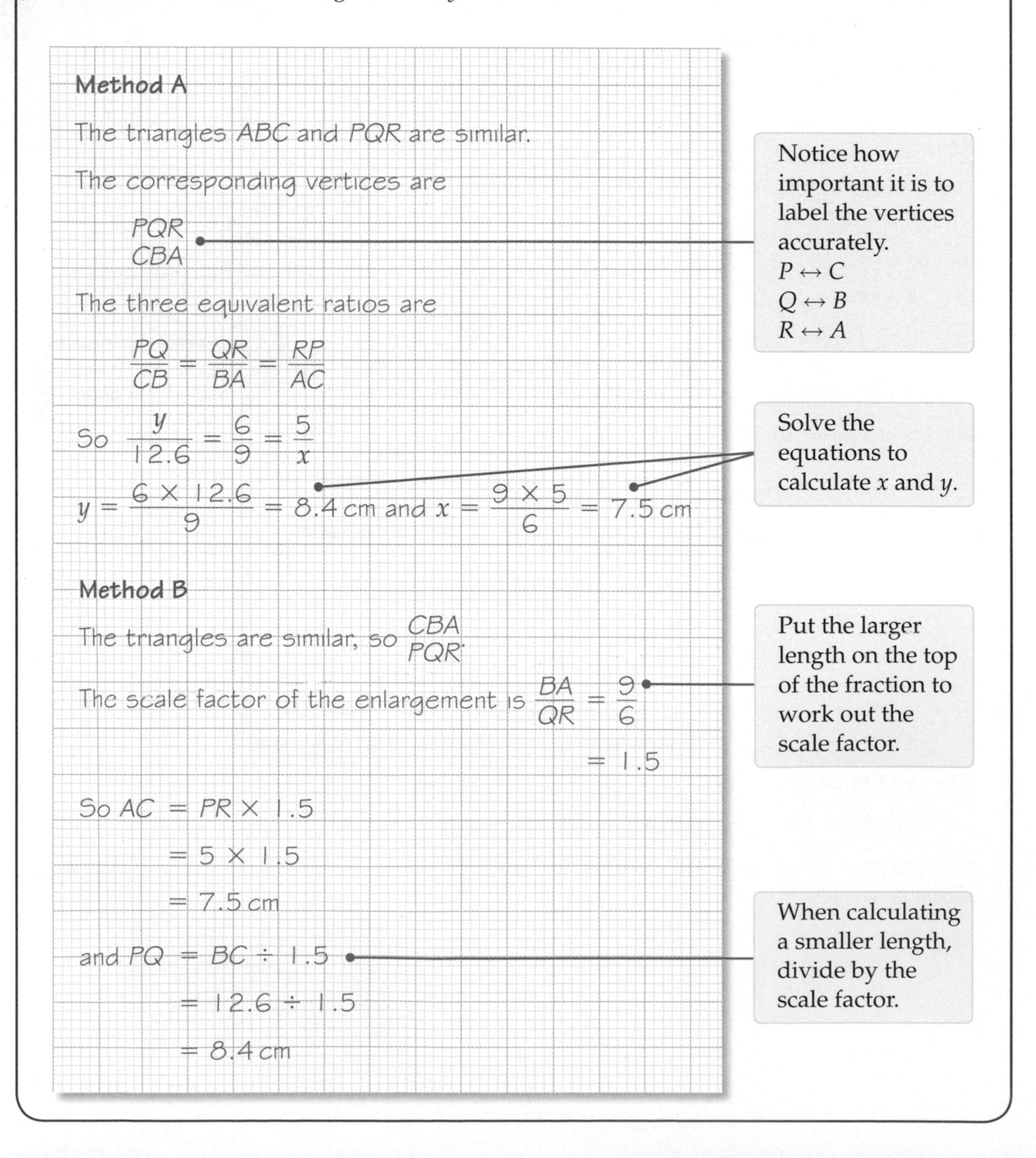

Method A

The triangles ABC and PQR are similar.

The corresponding vertices are

PQR
CBA

The three equivalent ratios are

$$\frac{PQ}{CB} = \frac{QR}{BA} = \frac{RP}{AC}$$

So $\frac{y}{12.6} = \frac{6}{9} = \frac{5}{x}$

$y = \frac{6 \times 12.6}{9} = 8.4$ cm and $x = \frac{9 \times 5}{6} = 7.5$ cm

Method B

The triangles are similar, so $\frac{CBA}{PQR}$

The scale factor of the enlargement is $\frac{BA}{QR} = \frac{9}{6}$

$= 1.5$

So $AC = PR \times 1.5$

$= 5 \times 1.5$

$= 7.5$ cm

and $PQ = BC \div 1.5$

$= 12.6 \div 1.5$

$= 8.4$ cm

Notice how important it is to label the vertices accurately.
$P \leftrightarrow C$
$Q \leftrightarrow B$
$R \leftrightarrow A$

Solve the equations to calculate x and y.

Put the larger length on the top of the fraction to work out the scale factor.

When calculating a smaller length, divide by the scale factor.

Example 4

In the diagram AB is parallel to DC and APC and BPD are straight lines.

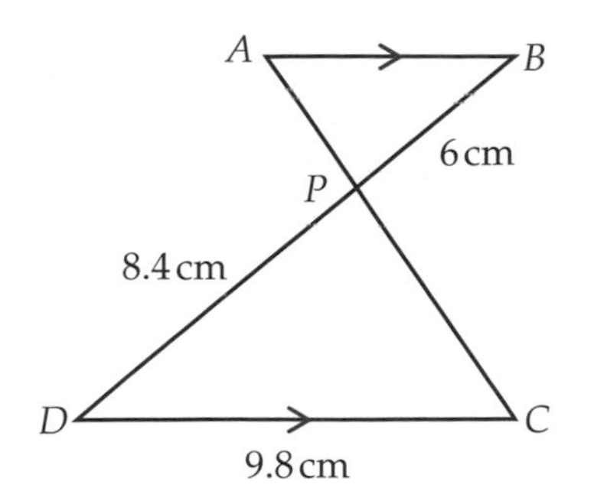

Questions asking you to show a statement is true are becoming more popular in exams. You must give reasons for your statements.

$BP = 6$ cm, $PD = 8.4$ cm and $DC = 9.8$ cm.

a Show that $\triangle ABP$ is similar to $\triangle CDP$.

b Calculate the length AB.

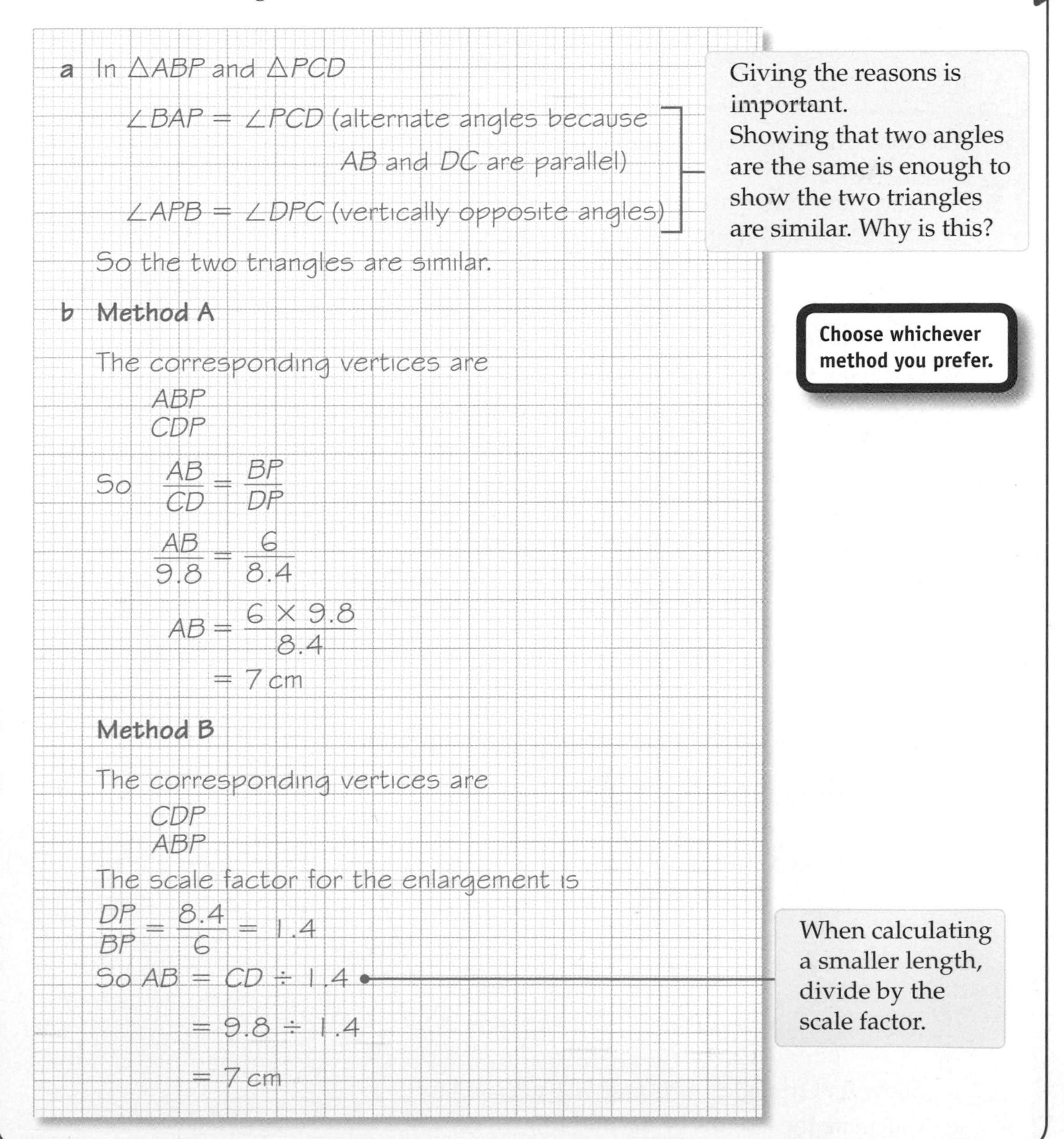

a In $\triangle ABP$ and $\triangle PCD$

$\angle BAP = \angle PCD$ (alternate angles because AB and DC are parallel)

$\angle APB = \angle DPC$ (vertically opposite angles)

So the two triangles are similar.

Giving the reasons is important. Showing that two angles are the same is enough to show the two triangles are similar. Why is this?

b **Method A**

Choose whichever method you prefer.

The corresponding vertices are

ABP
CDP

So $\dfrac{AB}{CD} = \dfrac{BP}{DP}$

$\dfrac{AB}{9.8} = \dfrac{6}{8.4}$

$AB = \dfrac{6 \times 9.8}{8.4}$

$= 7$ cm

Method B

The corresponding vertices are

CDP
ABP

The scale factor for the enlargement is

$\dfrac{DP}{BP} = \dfrac{8.4}{6} = 1.4$

So $AB = CD \div 1.4$

$= 9.8 \div 1.4$

$= 7$ cm

When calculating a smaller length, divide by the scale factor.

Exercise 32C

1 Two similar triangles are shown.

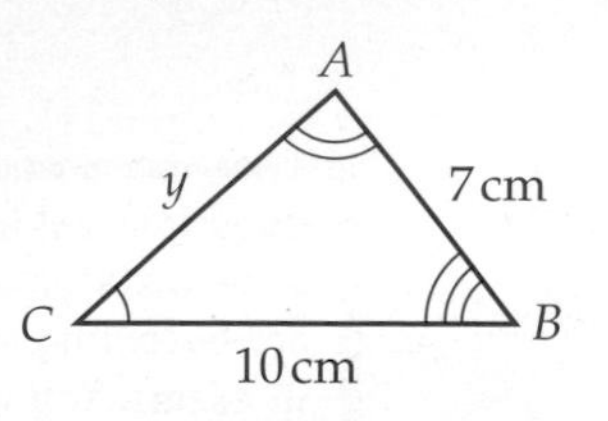

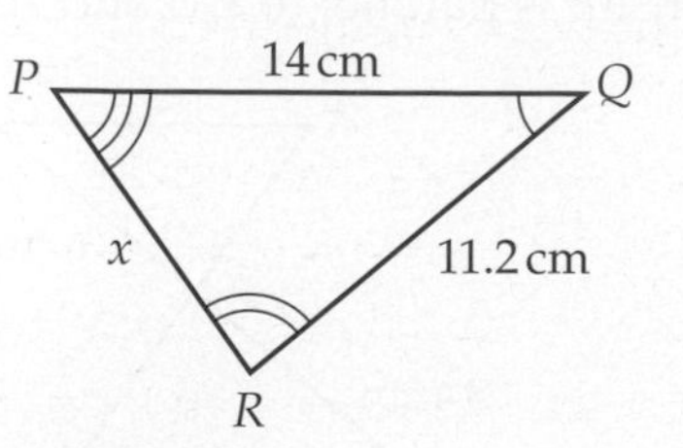

Calculate the unknown lengths x and y.

2 Trapezium *PQRS* is an enlargement of trapezium *ABCD*.

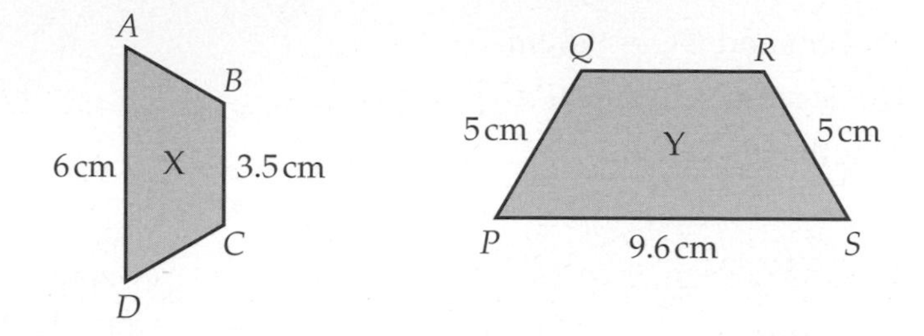

Calculate the lengths of the unknown sides in both trapezia.

3 The diagram shows triangle *ABC*.
A line *PQ* is drawn parallel to *BC* such that
$AP = 8$ cm, $PC = 4$ cm, $BQ = 3.5$ cm
and $PQ = 6$ cm.

a Show that $\triangle APQ$ is similar to $\triangle ABC$.

b Calculate the unknown lengths x and y.

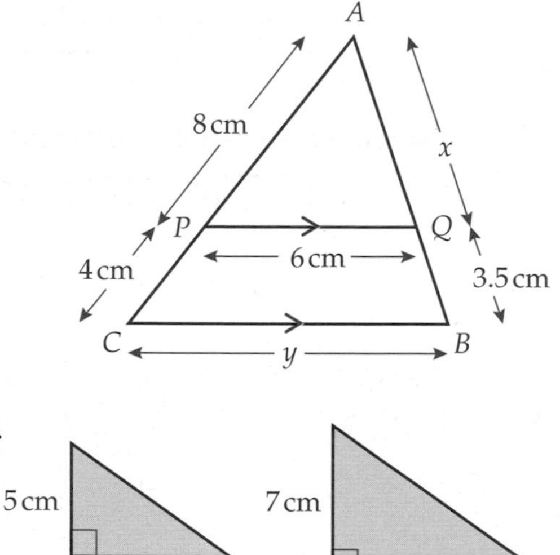

B
A02

4 Julie says that these two triangles are similar.
Robert says that the bigger triangle is an enlargement of the smaller triangle.
Who is correct?
Give reasons for your answer.

5 cm

7.5 cm

7 cm

B

5 *APC* and *BPD* are straight lines and *BC* is parallel to *AD*.

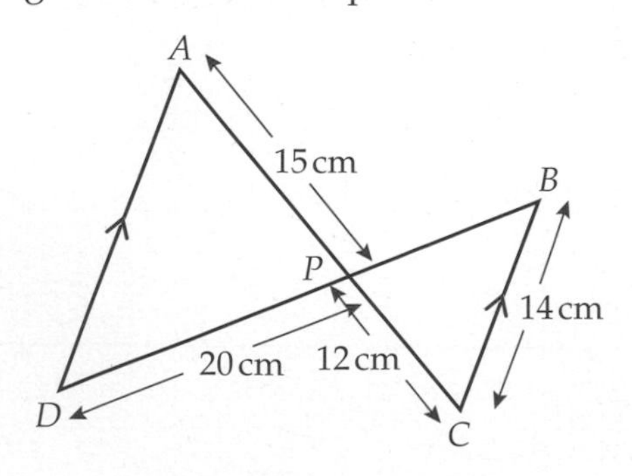

a Show that triangles *APD* and *BCP* are similar.

b Calculate the lengths of *AD* and *BP*.

6 A tall chimney is known to be 35 m tall. It casts a shadow of length 45 m at midday. A tree next to the chimney casts a shadow of length 18 m at midday. Calculate the height of the tree.

7 In the diagram, AB is parallel to DE.
$AB = 3.6$ cm, $BC = 4$ cm, $CD = 9$ cm, $CE = 7.5$ cm

a Explain why triangles ABC and EDC are similar.

b Calculate the lengths marked x and y.

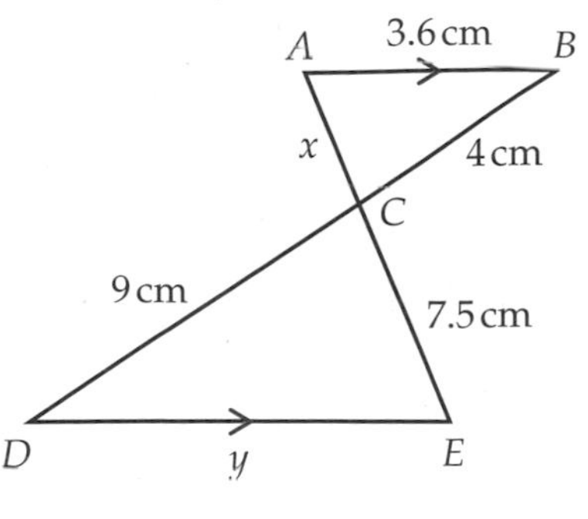

8 In the diagram, angle ABD = angle ACB, $AD = 6$ cm, $DC = 7.5$ cm, $BD = 8$ cm and $BC = 12$ cm.

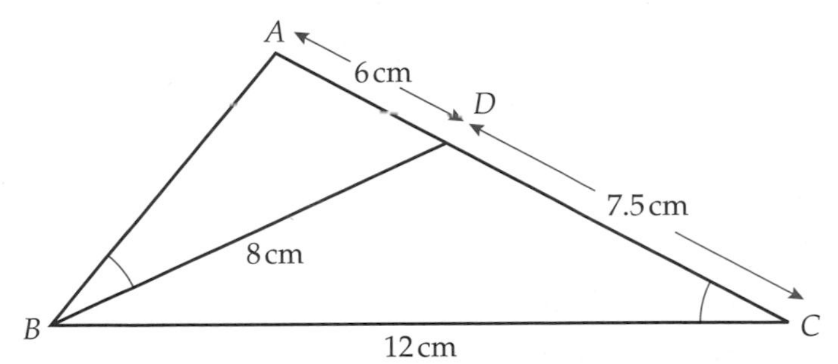

a Explain why triangles ABD and ACB are similar.

b Work out the length of AB.

B

AO2

32.3 Areas and volumes of similar objects

Keywords
linear scale factor, area scale factor, volume scale factor

Why learn this?

Hot air balloon designers must understand how surface area and volume are affected if they alter the dimensions of a balloon.

Objectives

A Calculate areas and volumes of similar objects

Skills check

1 Calculate

a 3^2 **b** 11^2 **c** 4^3 **d** 2^5

HELP Section 10.2

2 Calculate

a $\sqrt{16}$ **b** $\sqrt{121}$ **c** $\sqrt[3]{27}$ **d** $\sqrt[3]{216}$

3 **a** How many cm^2 are there in $1\ m^2$?

b How many cm^3 are there in $1\ m^3$?

Areas and volumes of similar objects

You can work out the area and volume of similar shapes if you know the **linear scale factor**.

The linear scale factor is the scale factor that applies to lengths of similar shapes.

Look at the rectangle below, size 1×3.

Enlarge it by a linear scale factor of 3.

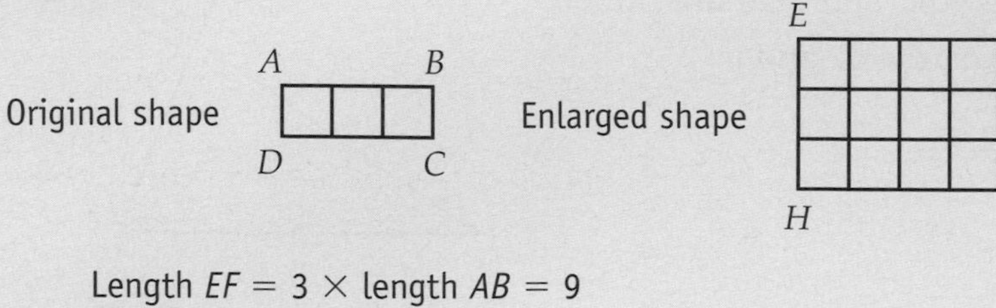

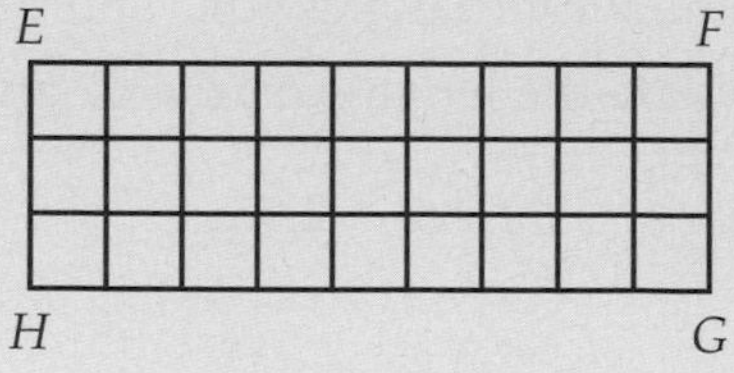

Length $EF = 3 \times$ length $AB = 9$

Length $FG = 3 \times$ length $BC = 3$

Area $ABCD = 1 \times 3 = 3$

Area $EFGH = 9 \times 3 = 27$

Enlarged area $EFGH = 9 \times$ area $ABCD$

$= 3^2 \times$ area $ABCD$

The factor 3^2 is the (linear scale factor)^2 and leads to the more general result:

When a shape is enlarged by a linear scale factor k the area is increased by a factor k^2.

Enlarged area $= k^2 \times$ original area

The factor k^2 is called the **area scale factor**.

Now look at a three-dimensional object, size $1 \times 1 \times 3$.

Enlarge this by a linear scale factor 3.

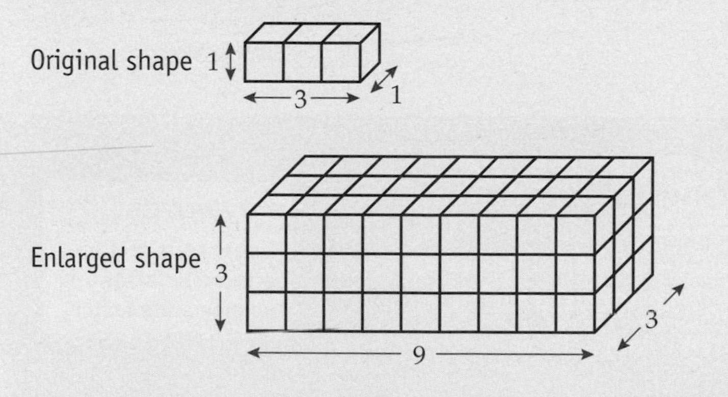

The original volume $= 1 \times 1 \times 3 = 3$

The enlarged volume $= 3 \times 3 \times 9 = 81$

The enlarged volume $= 27 \times$ the original volume

$= 3^3 \times$ the original volume

Remember that:
linear scale factor = k
area scale factor = k^2
volume scale factor = k^3

The factor 3^3 is the (linear scale factor)^3 and leads to the more general result:

When a shape is enlarged by a linear scale factor k the volume is increased by a factor k^3.

Enlarged volume $= k^3 \times$ original volume

The factor k^3 is called the **volume scale factor**.

Example 5

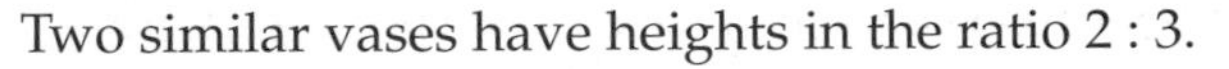

Two similar vases have heights in the ratio 2 : 3.

a What is the ratio of the volumes of the vases?

b What is the ratio of the areas of their circular bases?

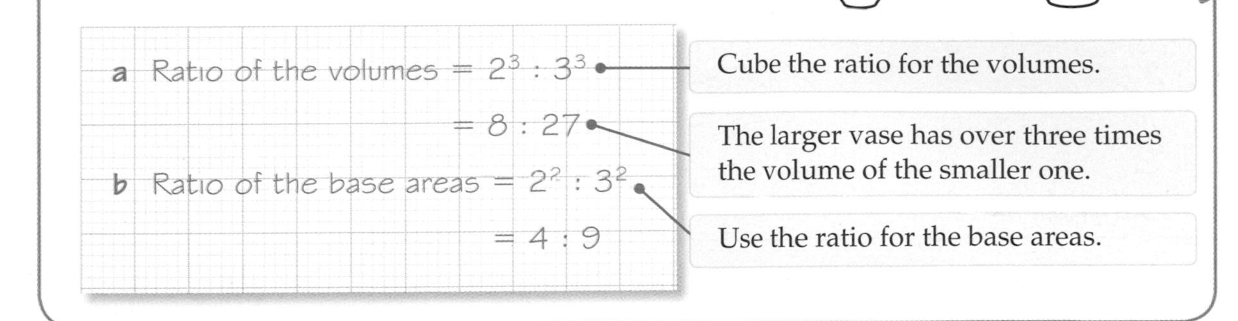

a Ratio of the volumes $= 2^3 : 3^3$
$= 8 : 27$

b Ratio of the base areas $= 2^2 : 3^2$
$= 4 : 9$

Cube the ratio for the volumes.

The larger vase has over three times the volume of the smaller one.

Use the ratio for the base areas.

Example 6

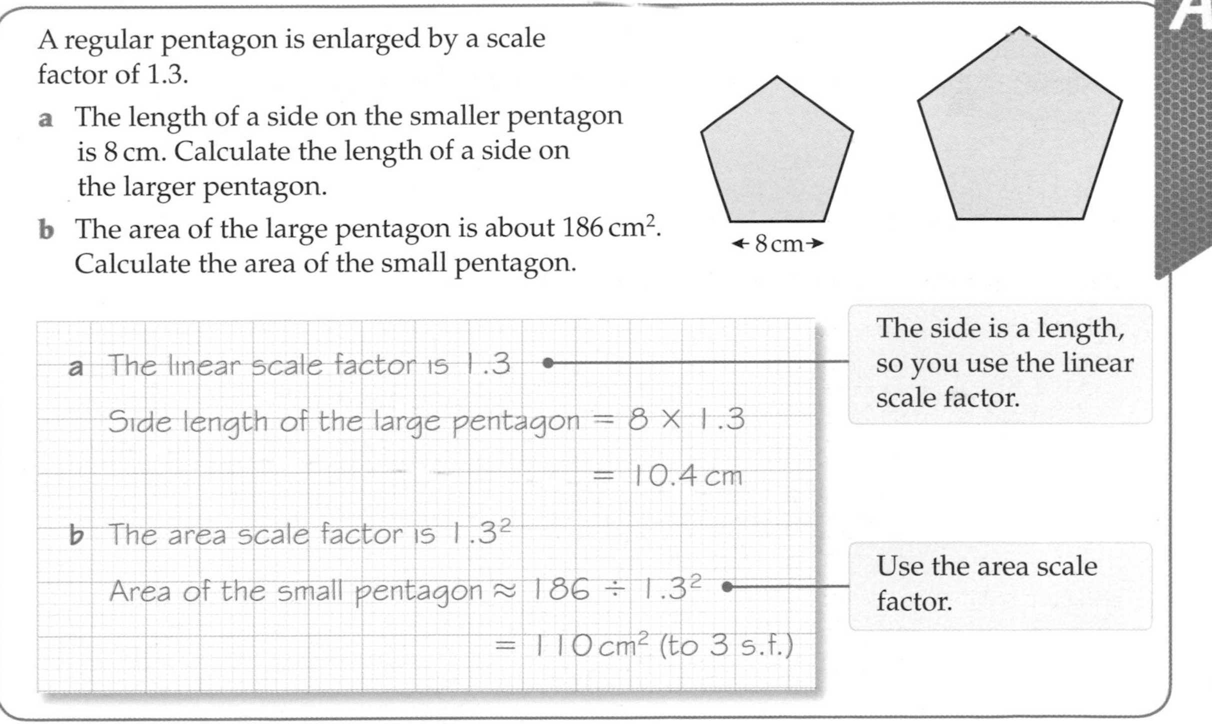

A regular pentagon is enlarged by a scale factor of 1.3.

a The length of a side on the smaller pentagon is 8 cm. Calculate the length of a side on the larger pentagon.

b The area of the large pentagon is about 186 cm^2. Calculate the area of the small pentagon.

a The linear scale factor is 1.3

Side length of the large pentagon $= 8 \times 1.3$
$= 10.4$ cm

b The area scale factor is 1.3^2

Area of the small pentagon $\approx 186 \div 1.3^2$
$= 110$ cm^2 (to 3 s.f.)

The side is a length, so you use the linear scale factor.

Use the area scale factor.

Example 7

A brand of baked beans is sold in two sizes of tin.
The two tins are mathematically similar.

The smaller tin holds 415 g of baked beans.

Calculate the mass of the baked beans in the larger tin.

10 cm

14 cm

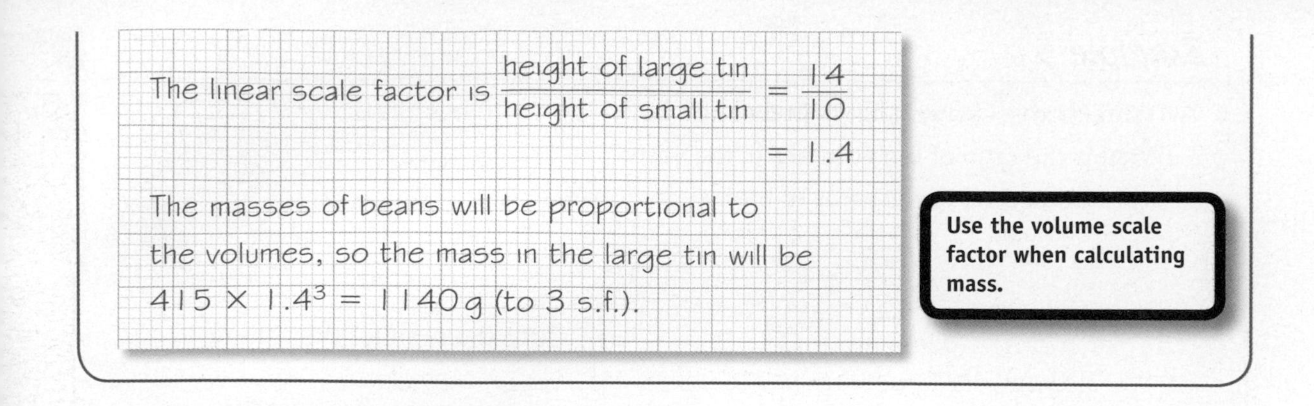

The linear scale factor is $\frac{\text{height of large tin}}{\text{height of small tin}} = \frac{14}{10}$

$= 1.4$

The masses of beans will be proportional to the volumes, so the mass in the large tin will be $415 \times 1.4^3 = 1140$ g (to 3 s.f.).

Use the volume scale factor when calculating mass.

A

Example 8

A brand of perfume is sold in two sizes of bottle. The shapes of the two bottles are mathematically similar.

a The height of the large bottle is 11 cm. Calculate the height of the small bottle.

b Calculate the ratio of the areas of the bases of the bottles.

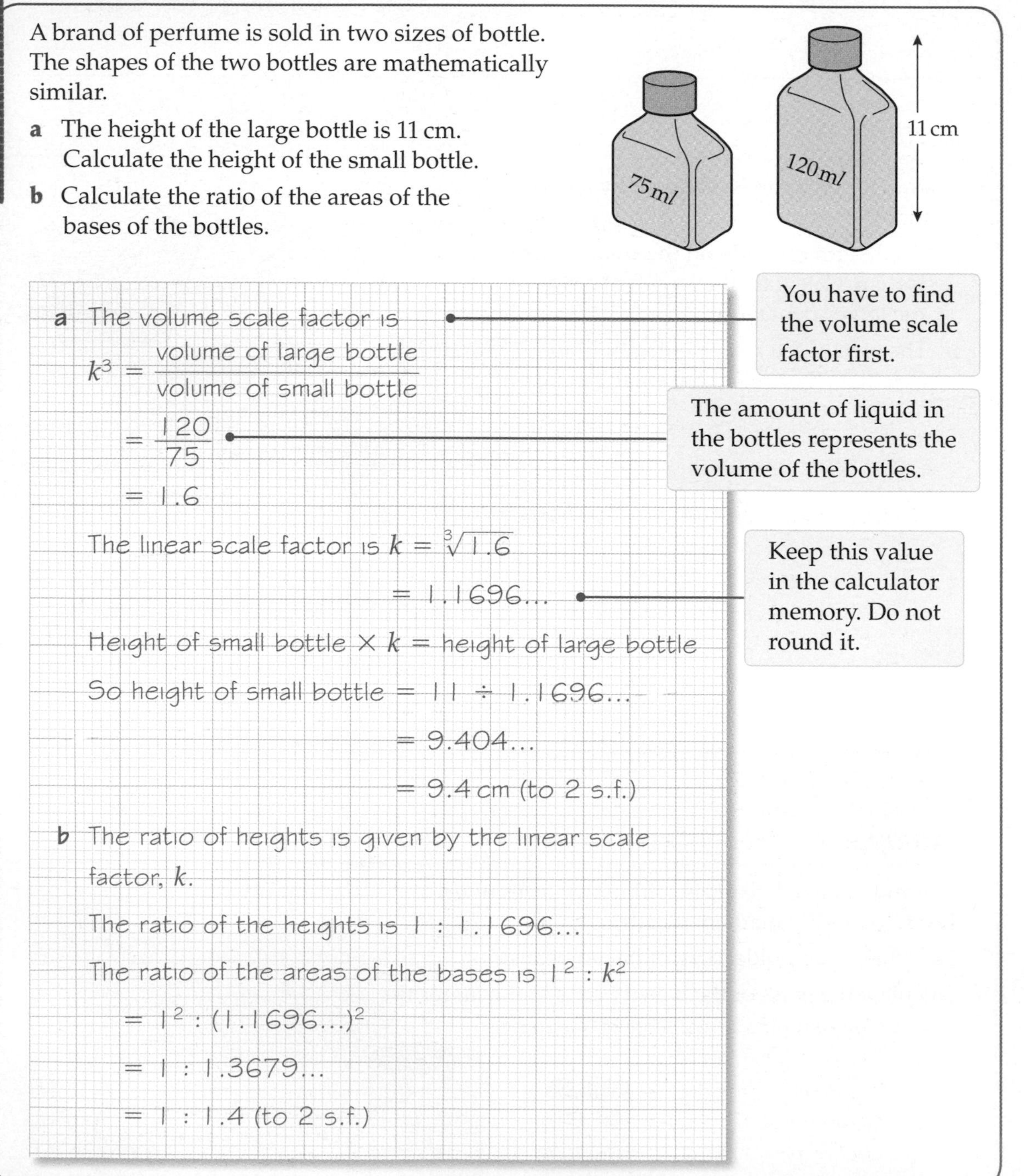

a The volume scale factor is

$k^3 = \frac{\text{volume of large bottle}}{\text{volume of small bottle}}$

$= \frac{120}{75}$

$= 1.6$

The linear scale factor is $k = \sqrt[3]{1.6}$

$= 1.1696...$

Height of small bottle $\times\ k =$ height of large bottle

So height of small bottle $= 11 \div 1.1696...$

$= 9.404...$

$= 9.4$ cm (to 2 s.f.)

b The ratio of heights is given by the linear scale factor, k.

The ratio of the heights is $1 : 1.1696...$

The ratio of the areas of the bases is $1^2 : k^2$

$= 1^2 : (1.1696...)^2$

$= 1 : 1.3679...$

$= 1 : 1.4$ (to 2 s.f.)

You have to find the volume scale factor first.

The amount of liquid in the bottles represents the volume of the bottles.

Keep this value in the calculator memory. Do not round it.

Exercise 32D

1 Two similar triangles have dimensions as shown in the diagram.

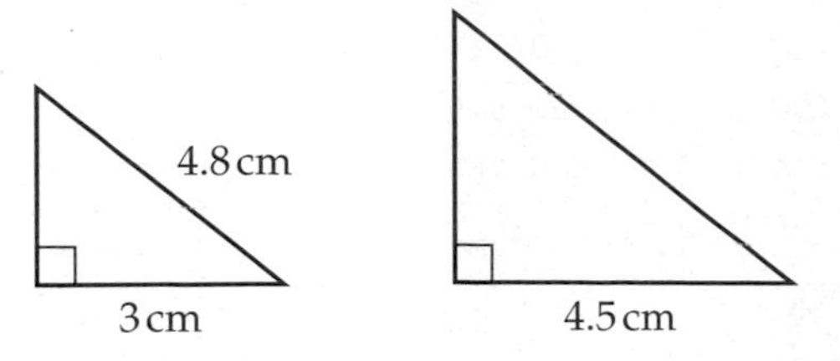

a Calculate the length of the hypotenuse of the larger triangle.

b The area of the smaller triangle is about 5.6 cm^2.
Calculate the area of the larger triangle.

2 Two cones are mathematically similar and have the dimensions shown.

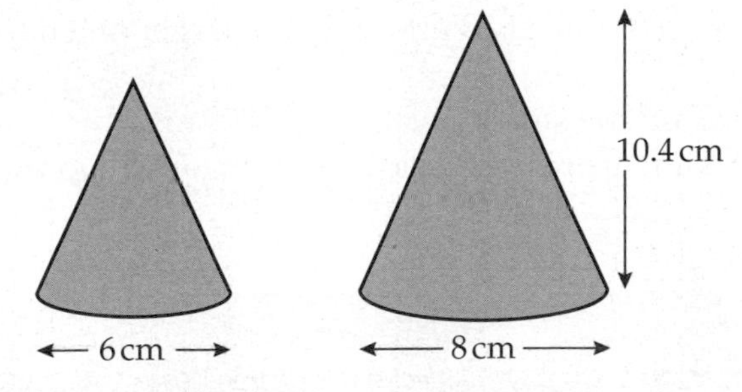

a Calculate the height of the small cone.

b The area of the curved surface of the small cone is 71 cm^2.
Calculate the area of the curved surface of the large cone.

3 A photographer enlarges a photograph.
The original has length 12 cm and the enlarged photograph has length 15 cm.
The area of the original photograph is 108 cm^2.
Calculate the area of the enlarged photograph.

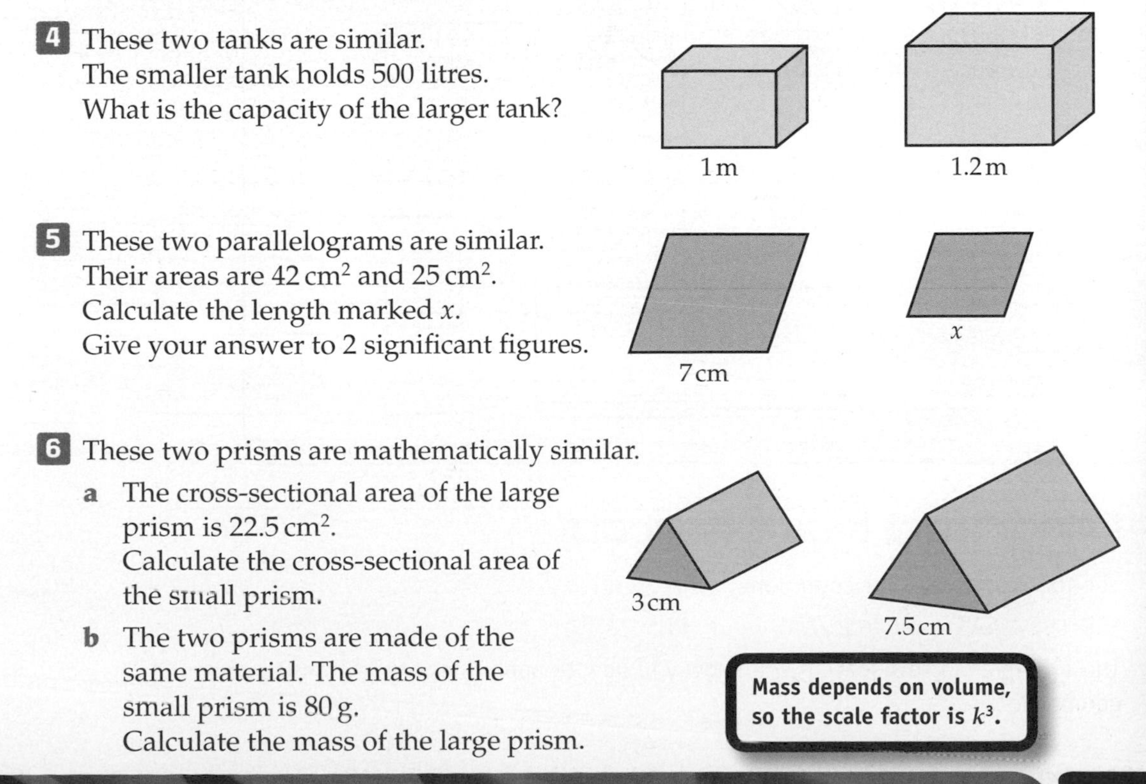

4 These two tanks are similar.
The smaller tank holds 500 litres.
What is the capacity of the larger tank?

5 These two parallelograms are similar.
Their areas are 42 cm^2 and 25 cm^2.
Calculate the length marked x.
Give your answer to 2 significant figures.

6 These two prisms are mathematically similar.

a The cross-sectional area of the large prism is 22.5 cm^2.
Calculate the cross-sectional area of the small prism.

b The two prisms are made of the same material. The mass of the small prism is 80 g.
Calculate the mass of the large prism.

Mass depends on volume, so the scale factor is k^3.

A

A

7 A model village is built to a one-ninth scale.
This means that the ratio of a length on the model to the real length is 1 : 9.

a A building on the model has a height of 90 cm.
What will be the height of the real building?

b A roof on the model has an area of $0.2\,m^2$.
What will be the corresponding area on the real building?

c A building in the village has a volume of $140\,m^3$.
What is the corresponding volume in the model?
Give your answer to 2 significant figures.

8 Two sheets of paper, one A4 and one A5, are mathematically similar rectangles.
The A4 sheet is double the area of the A5 sheet. The dimensions of the A4 sheet are 297 mm by 210 mm. What are the dimensions of the A5 sheet?

9 Two hot air balloons have a mathematically similar shape.
The smaller one has a volume of $2000\,m^3$ and the larger one has a volume of $6000\,m^3$.
The area of the fabric on the smaller balloon is $1700\,m^2$.
Calculate the approximate area of fabric on the larger balloon.

32.4 Congruent triangle proof

Why learn this?

Formal proof using step-by-step deduction is an important skill to have if you take A-level maths.

Objectives

A* Prove results in geometry using congruent triangles

Skills check

Which shapes are

a congruent **b** similar?

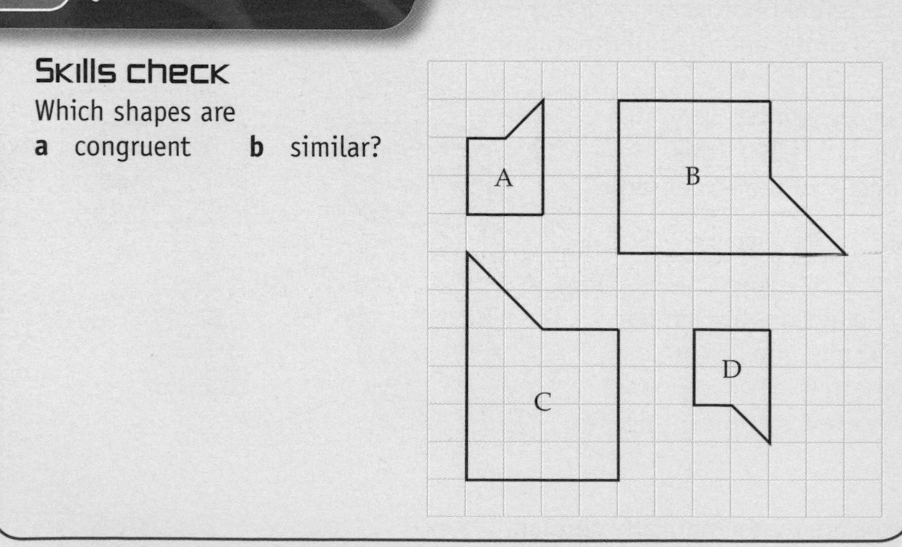

Congruent triangle proof

The four conditions for congruency are:

SSS SAS ASA/AAS RHS

The examples in this section will show you how to apply these conditions to prove geometric results.

Example 9

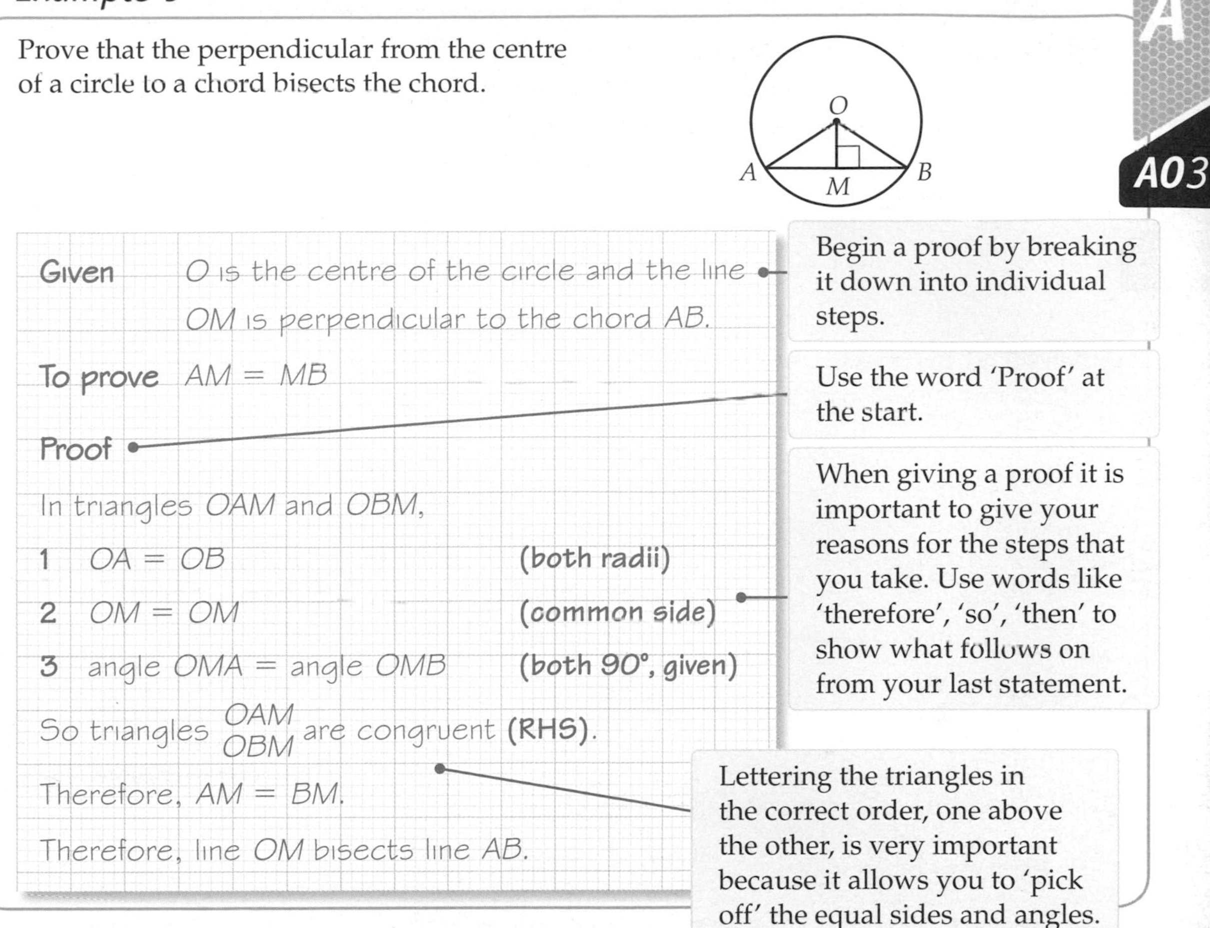

Prove that the perpendicular from the centre of a circle to a chord bisects the chord.

Given O is the centre of the circle and the line OM is perpendicular to the chord AB.

To prove $AM = MB$

Proof

In triangles OAM and OBM,

1 $OA = OB$ **(both radii)**

2 $OM = OM$ **(common side)**

3 angle OMA = angle OMB **(both 90°, given)**

So triangles $\frac{OAM}{OBM}$ are congruent **(RHS)**.

Therefore, $AM = BM$.

Therefore, line OM bisects line AB.

Begin a proof by breaking it down into individual steps.

Use the word 'Proof' at the start.

When giving a proof it is important to give your reasons for the steps that you take. Use words like 'therefore', 'so', 'then' to show what follows on from your last statement.

Lettering the triangles in the correct order, one above the other, is very important because it allows you to 'pick off' the equal sides and angles.

A*

AO3

Example 10

Prove that tangents drawn to a circle from an external point are equal in length.

Given A circle, centre O, and tangents from T drawn to touch the circle at points A and B.

To prove $TA = TB$

Proof

Draw line TO. In triangles TAO and TBO,

1 $TO = TO$ **(common side)**

2 $OA = OB$ **(both radii)**

3 angle TAO = angle TBO **(both 90°, tangent and radius meet at 90°)**

So triangles $\frac{TAO}{TBO}$ are congruent **(RHS)**.

Therefore, $TA = TB$.

This is a property of the radius and the tangent at the point of contact.

A*

AO3

A*

Example 11

AO3

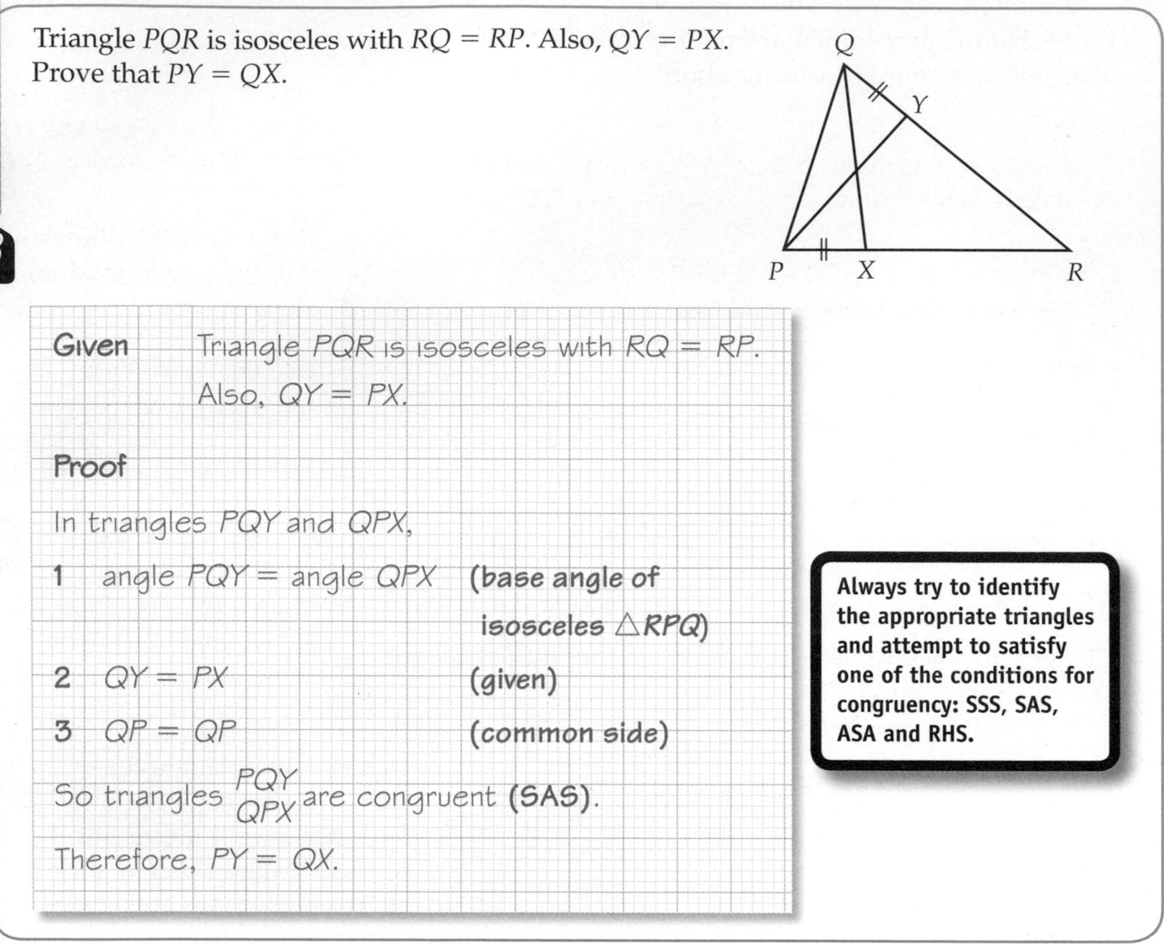

Triangle PQR is isosceles with $RQ = RP$. Also, $QY = PX$.
Prove that $PY = QX$.

Given Triangle PQR is isosceles with $RQ = RP$.
Also, $QY = PX$.

Proof

In triangles PQY and QPX,

1 angle PQY = angle QPX (**base angle of isosceles $\triangle RPQ$**)

2 $QY = PX$ (**given**)

3 $QP = QP$ (**common side**)

So triangles $\frac{PQY}{QPX}$ are congruent (**SAS**).

Therefore, $PY = QX$.

Always try to identify the appropriate triangles and attempt to satisfy one of the conditions for congruency: SSS, SAS, ASA and RHS.

Exercise 32E

A*

AO3

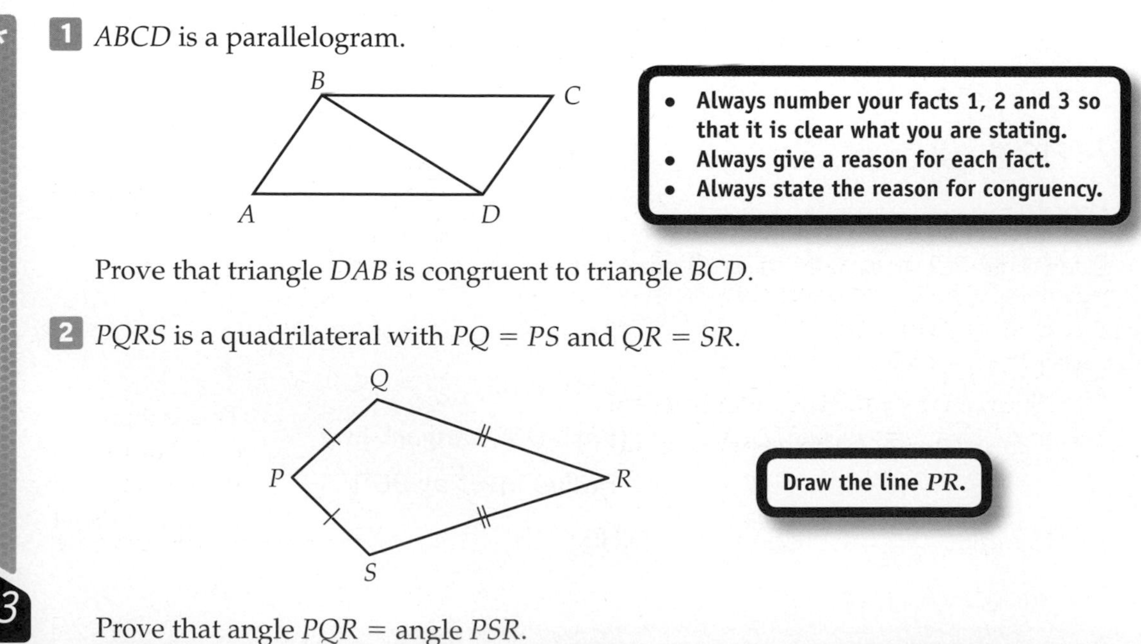

1 $ABCD$ is a parallelogram.

- **Always number your facts 1, 2 and 3 so that it is clear what you are stating.**
- **Always give a reason for each fact.**
- **Always state the reason for congruency.**

Prove that triangle DAB is congruent to triangle BCD.

2 $PQRS$ is a quadrilateral with $PQ = PS$ and $QR = SR$.

Draw the line PR.

Prove that angle PQR = angle PSR.

3 In the diagram $ABCD$ is a parallelogram.
AX bisects angle A and CY bisects angle C.
Prove that the triangles ABX and CDY are congruent.

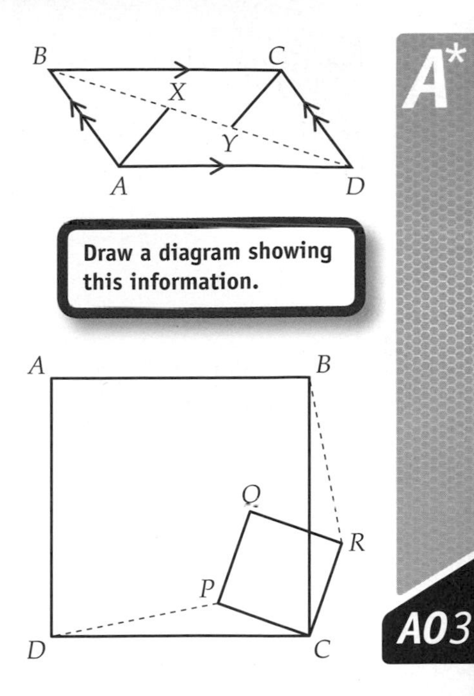

4 $PQRS$ is a rectangle. A point X lies outside of the rectangle such that $QX = RX$. Prove that $PX = SX$.

5 In the diagram $ABCD$ and $PQRC$ are squares.
Use congruent triangles to prove that $DP = BR$.

A*

AO3

Review exercise

C

1 Write down the letters of one pair of congruent triangles and one pair of similar triangles.

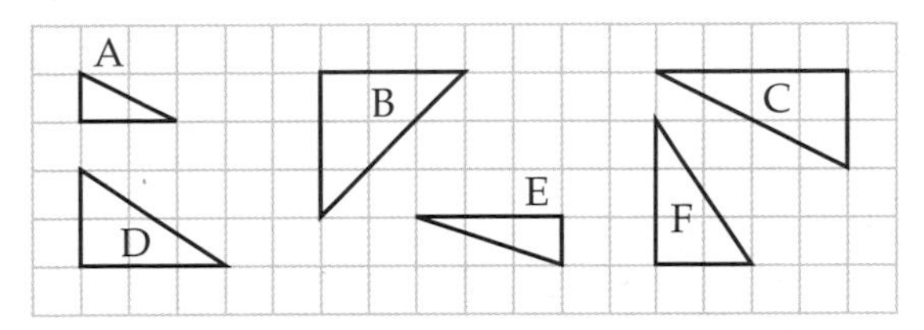

[2 marks]

2 Look at the shapes in the diagram. Write down the letters of the similar pairs.

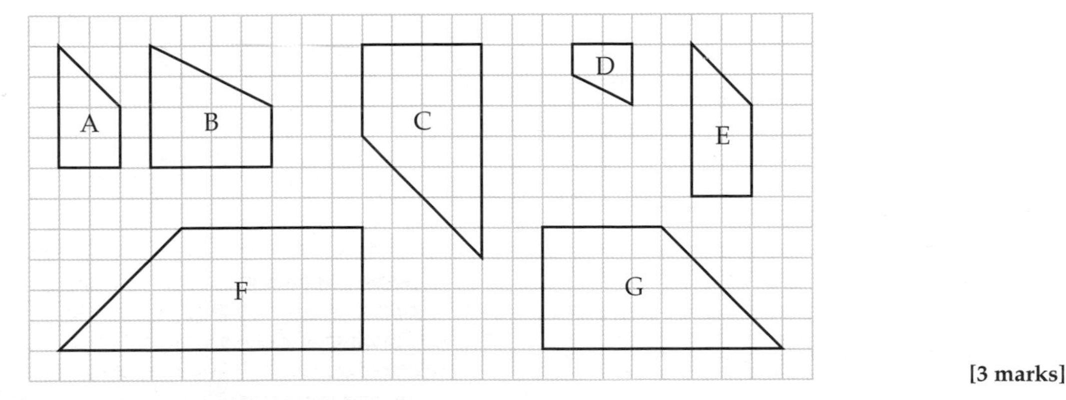

[3 marks]

B

3 The diagram shows triangle ABC.
A line PQ is drawn parallel to AC such that $AP = 4\,\text{cm}$, $BQ = 10\,\text{cm}$, $QC = 6\,\text{cm}$ and $PQ = 8\,\text{cm}$.

a Show that $\triangle PBQ$ is similar to $\triangle ABC$.

b Calculate the lengths marked x and y.

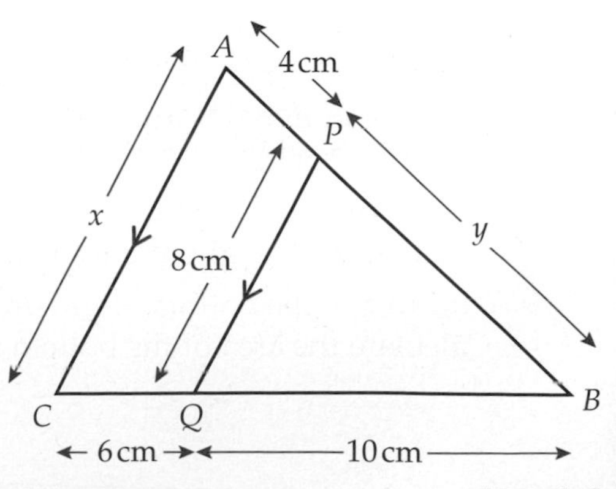

[7 marks]

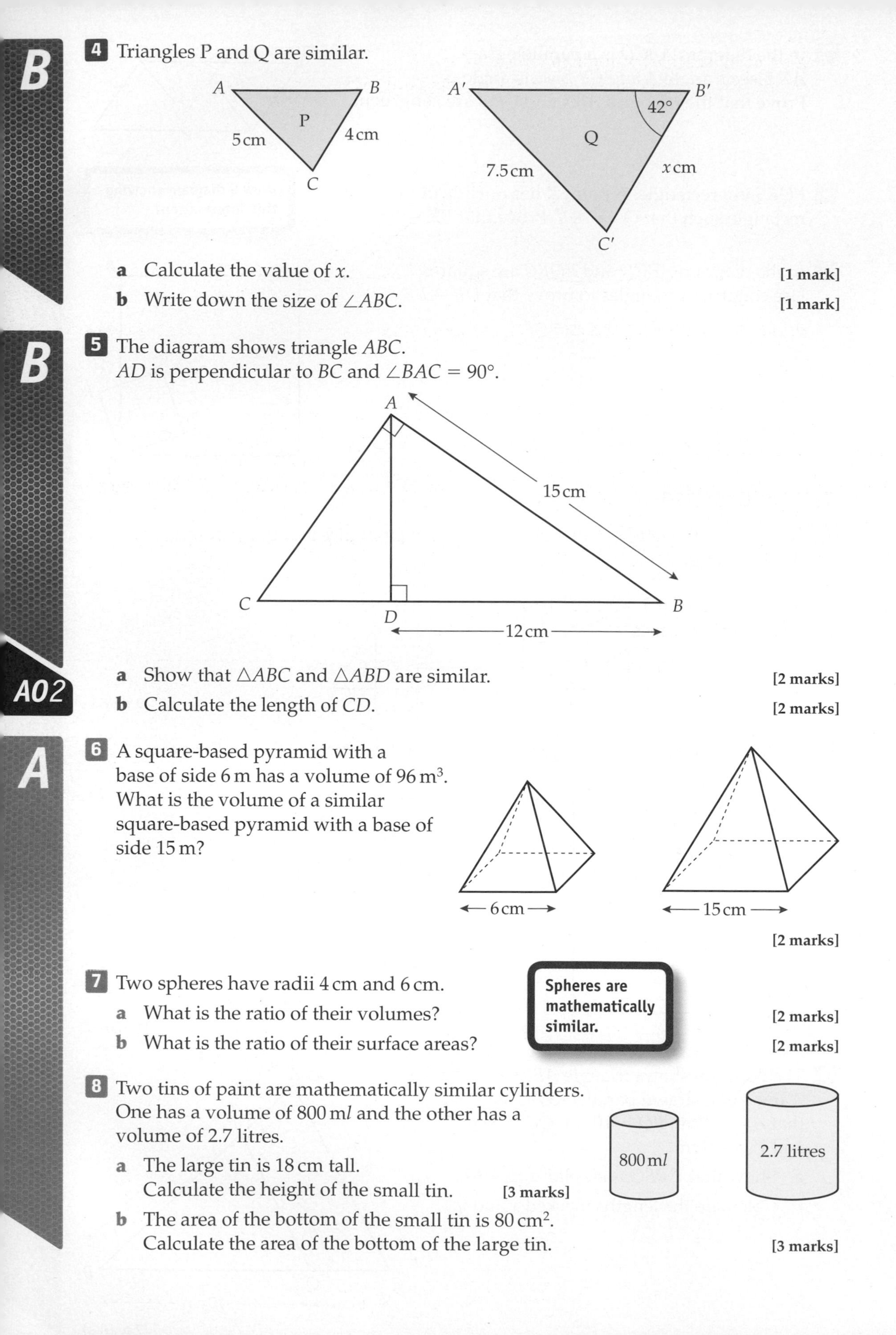

B

4 Triangles P and Q are similar.

a Calculate the value of x. **[1 mark]**

b Write down the size of $\angle ABC$. **[1 mark]**

B

5 The diagram shows triangle ABC.
AD is perpendicular to BC and $\angle BAC = 90°$.

AO2

a Show that $\triangle ABC$ and $\triangle ABD$ are similar. **[2 marks]**

b Calculate the length of CD. **[2 marks]**

A

6 A square-based pyramid with a base of side 6 m has a volume of 96 m^3. What is the volume of a similar square-based pyramid with a base of side 15 m?

[2 marks]

7 Two spheres have radii 4 cm and 6 cm.

a What is the ratio of their volumes? **[2 marks]**

b What is the ratio of their surface areas? **[2 marks]**

Spheres are mathematically similar.

8 Two tins of paint are mathematically similar cylinders. One has a volume of 800 m*l* and the other has a volume of 2.7 litres.

a The large tin is 18 cm tall.
Calculate the height of the small tin. **[3 marks]**

b The area of the bottom of the small tin is 80 cm^2.
Calculate the area of the bottom of the large tin. **[3 marks]**

 XYZ is an isosceles triangle in which $XZ = XY$.
M and N are points on XZ and XY such that angle MYZ = angle NZY.

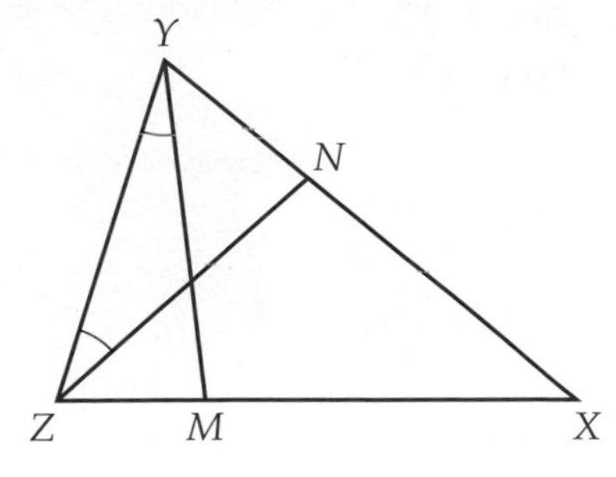

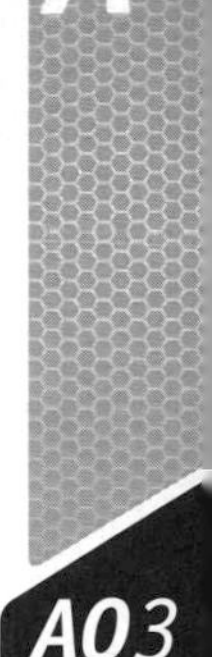

Prove that triangles YMZ and ZNY are congruent. **[4 marks]**

AO3

Chapter summary

In this chapter you have learned how to

- understand similarity and congruence **C**
- understand the conditions for congruent triangles **C**
- recognise similar shapes and calculate the scale factor between similar shapes **B**
- calculate areas and volumes of similar objects **A**
- prove results in geometry using congruent triangles **A***

33

Pythagoras' theorem and trigonometry

This chapter is about using Pythagoras' theorem and trigonometry to find missing lengths and angles.

Car satellite navigation systems use a form of trigonometry to calculate the distance between two places.

Objectives

This chapter will show you how to

- calculate the length of line segment *AB* C
- understand, recall and use Pythagoras' theorem in 2-D, then 3-D problems C B
- use calculators effectively and efficiently, knowing how to enter complex calculations, including trigonometric functions B
- understand, recall and use trigonometric ratios in right-angled triangles, and use these to solve problems, including those involving bearings B
- use surds in exact calculations A
- use Pythagoras' theorem to calculate lengths in three dimensions A
- use trigonometric ratios in 3-D contexts, including finding the angle between a line and a plane A A*

Before you start this chapter

1 Solve these equations.

a $\frac{x}{2} = 6$ b $\frac{x}{4} = 7$ c $\frac{x}{3} = 10$

2 Rearrange these formulae to make x the subject.

a $b = \frac{x}{a}$ b $5c = \frac{2x}{b}$ c $a = \frac{3}{x}$

HELP Chapter 15

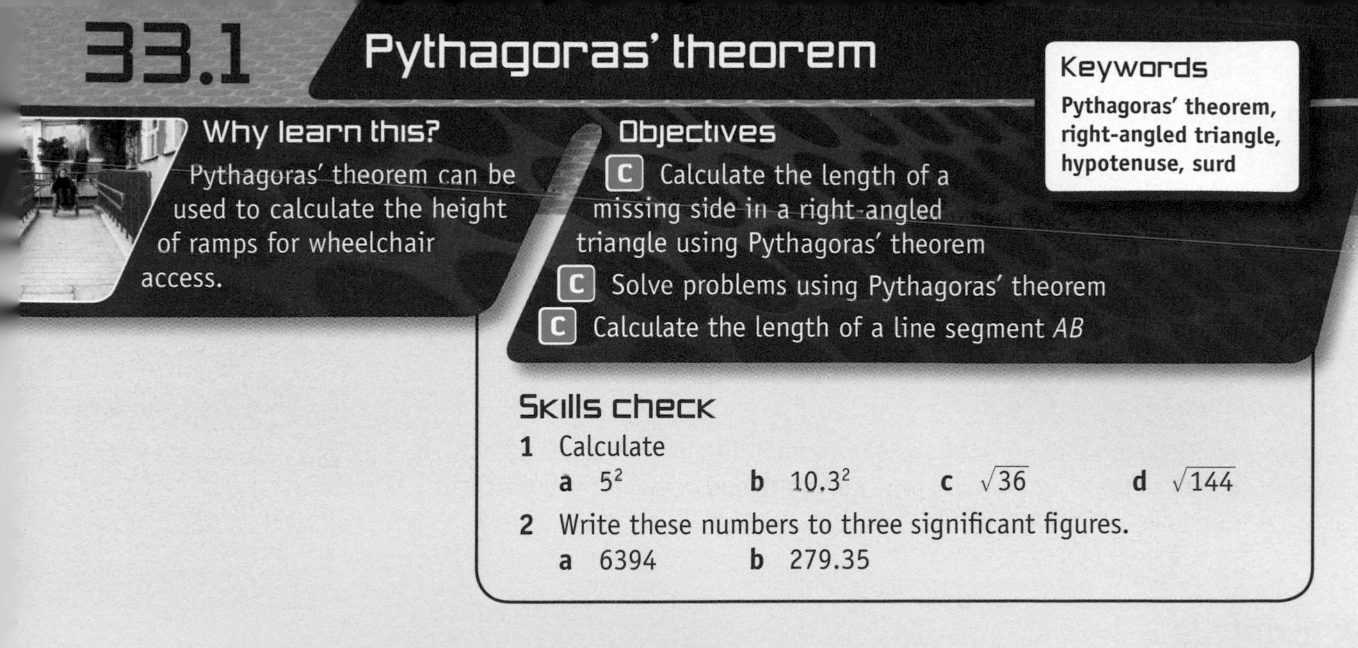

33.1 Pythagoras' theorem

Why learn this?

Pythagoras' theorem can be used to calculate the height of ramps for wheelchair access.

Objectives

C Calculate the length of a missing side in a right-angled triangle using Pythagoras' theorem

C Solve problems using Pythagoras' theorem

C Calculate the length of a line segment *AB*

Keywords

Pythagoras' theorem, right-angled triangle, hypotenuse, surd

Skills check

1 Calculate

a 5^2 **b** 10.3^2 **c** $\sqrt{36}$ **d** $\sqrt{144}$

2 Write these numbers to three significant figures.

a 6394 **b** 279.35

Pythagoras' theorem

Pythagoras' theorem applies to **right-angled triangles** only.

In a right-angled triangle the longest side is called the **hypotenuse.**

The hypotenuse is always opposite the right angle.

hypotenuse

You can construct a square on each side of a right-angled triangle.

This triangle has sides $a = 3$, $b = 4$ and $c = 5$.

If you construct the triangle, you will see that it is a right-angled triangle.

Area A $= 3 \times 3 = 3^2 = 9$

Area C $= 5 \times 5 = 5^2 = 25$

C

A

$a = 3$

$c = 5$

$b = 4$

B

Area B $= 4 \times 4 = 4^2 = 16$

Area A + area B $= 9 + 16 = 25 =$ area C

$$3^2 + 4^2 = 5^2$$

or

$$a^2 + b^2 = c^2$$

This leads to Pythagoras' theorem.

In any right-angled triangle, the square of the hypotenuse (c^2) is equal to the sum of the squares on the other two sides ($a^2 + b^2$).

You need to learn this formula.

For a right-angled triangle with sides of lengths a, b and c, where c is the hypotenuse, Pythagoras' theorem states that $c^2 = a^2 + b^2$.

a b

c

The formula $a^2 + b^2 = c^2$ allows you to work out the longest side.

To work out the length of the shorter side a

- first rearrange the formula to make a^2 the subject.

$$c^2 = a^2 + b^2$$

$$c^2 - b^2 = a^2$$

or $$a^2 = c^2 - b^2$$

Subtract b^2 from both sides of the equation.

- then take the square root to find a.

You can make b^2 the subject in a similar way.

$$b^2 = c^2 - a^2$$

Applying Pythagoras' theorem allows you to find the square of the length of any side of a right-angled triangle, from which you can find the length.

C

Example 1

Calculate the length of the hypotenuse, x, in this right-angled triangle.

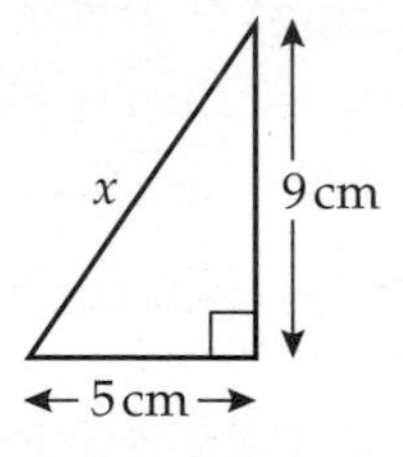

Give your answer to one decimal place.

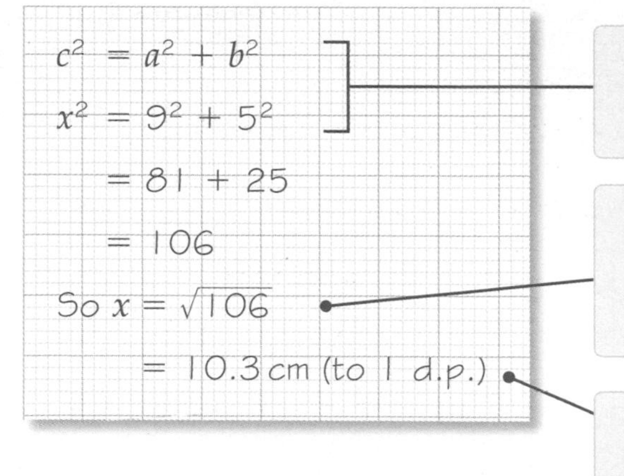

Write out Pythagoras' theorem and then substitute the values given. $c = x$, $a = 9$ cm, $b = 5$ cm

Use your calculator to find the square root. $\sqrt{106}$ is called a **surd**. This is an exact answer. Forming the decimal only gives an approximate answer.

Round your answer to one decimal place and put the units in your answer.

C

Example 2

Calculate the length x in this right-angled triangle.

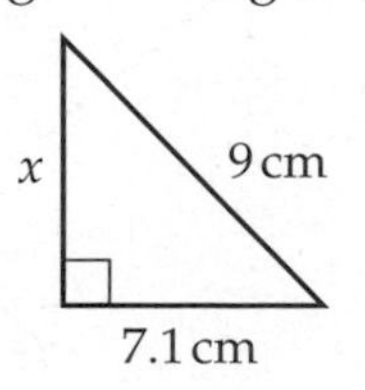

Give your answer to one decimal place.

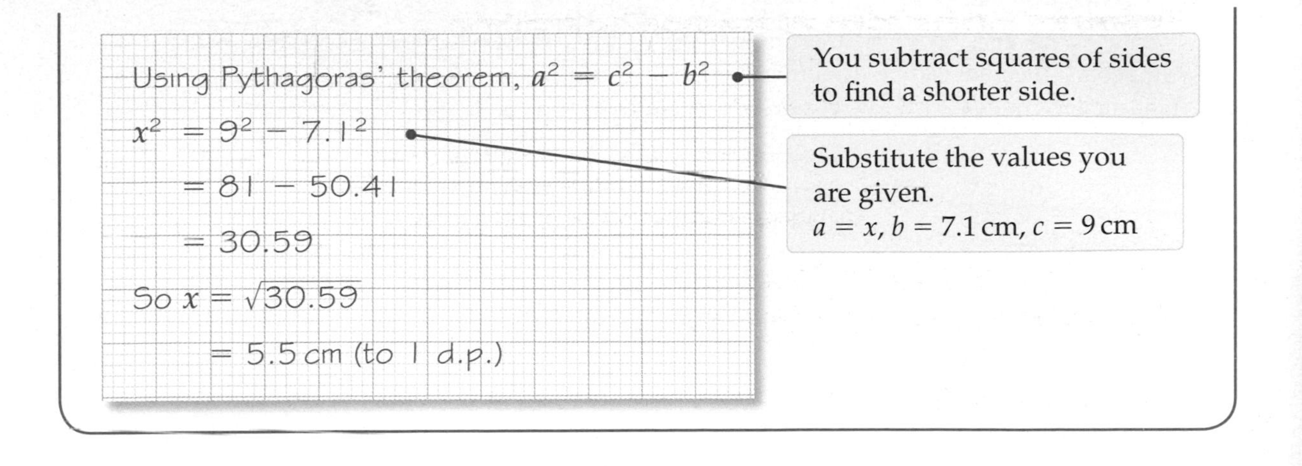

Using Pythagoras' theorem, $a^2 = c^2 - b^2$
$x^2 = 9^2 - 7.1^2$
$= 81 - 50.41$
$= 30.59$
So $x = \sqrt{30.59}$
$= 5.5$ cm (to 1 d.p.)
You subtract squares of sides to find a shorter side.
Substitute the values you are given.
$a = x, b = 7.1$ cm, $c = 9$ cm

Exercise 33A

1 Calculate the length of the hypotenuse in each triangle.

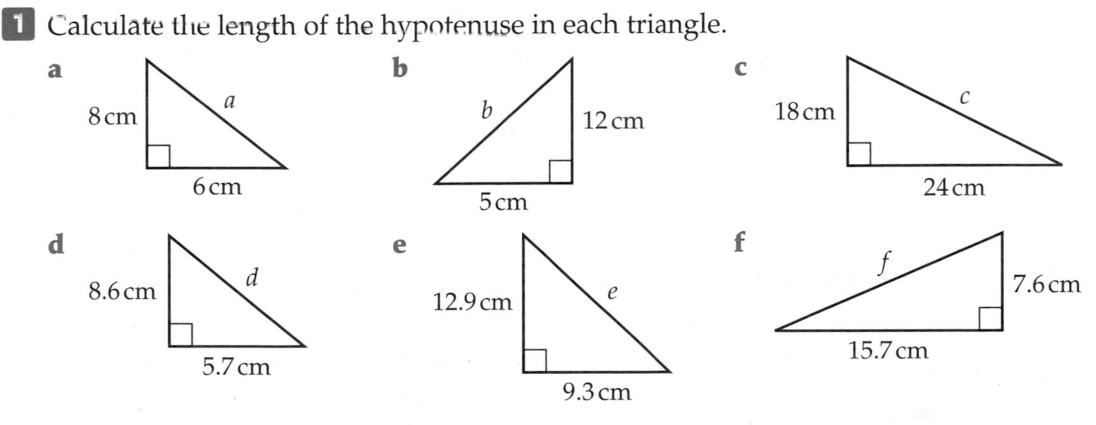

Give your answer to one decimal place when appropriate.

2 Calculate the lengths marked with letters in these triangles.

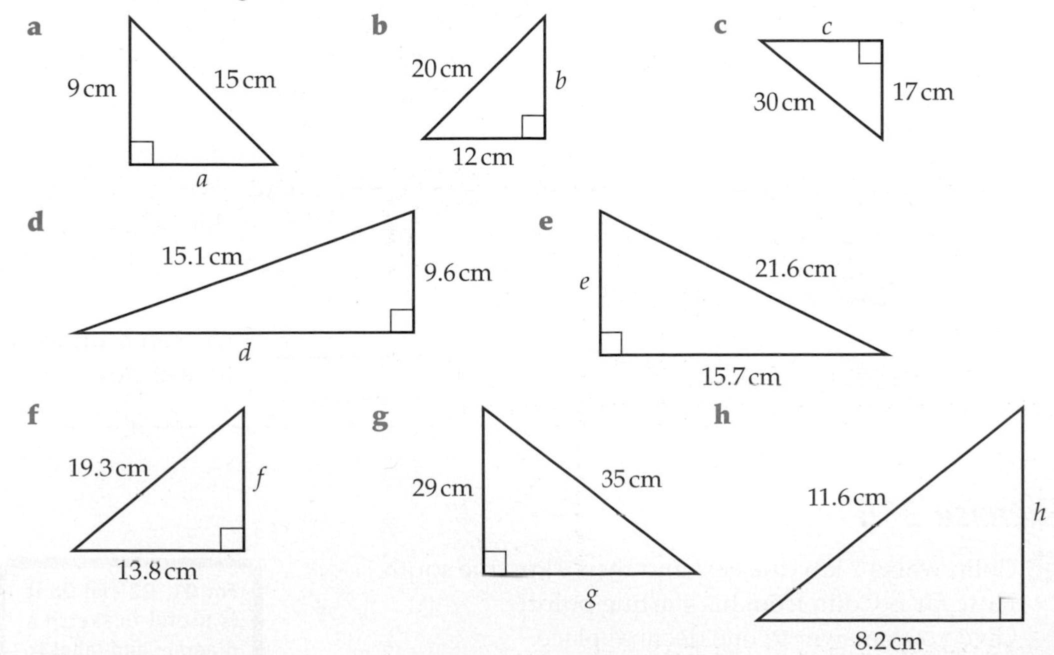

Give your answer to one decimal place when appropriate.

Solving problems using Pythagoras' theorem

Pythagoras' theorem can be used to solve problems which involve finding any side of a right-angled triangle.

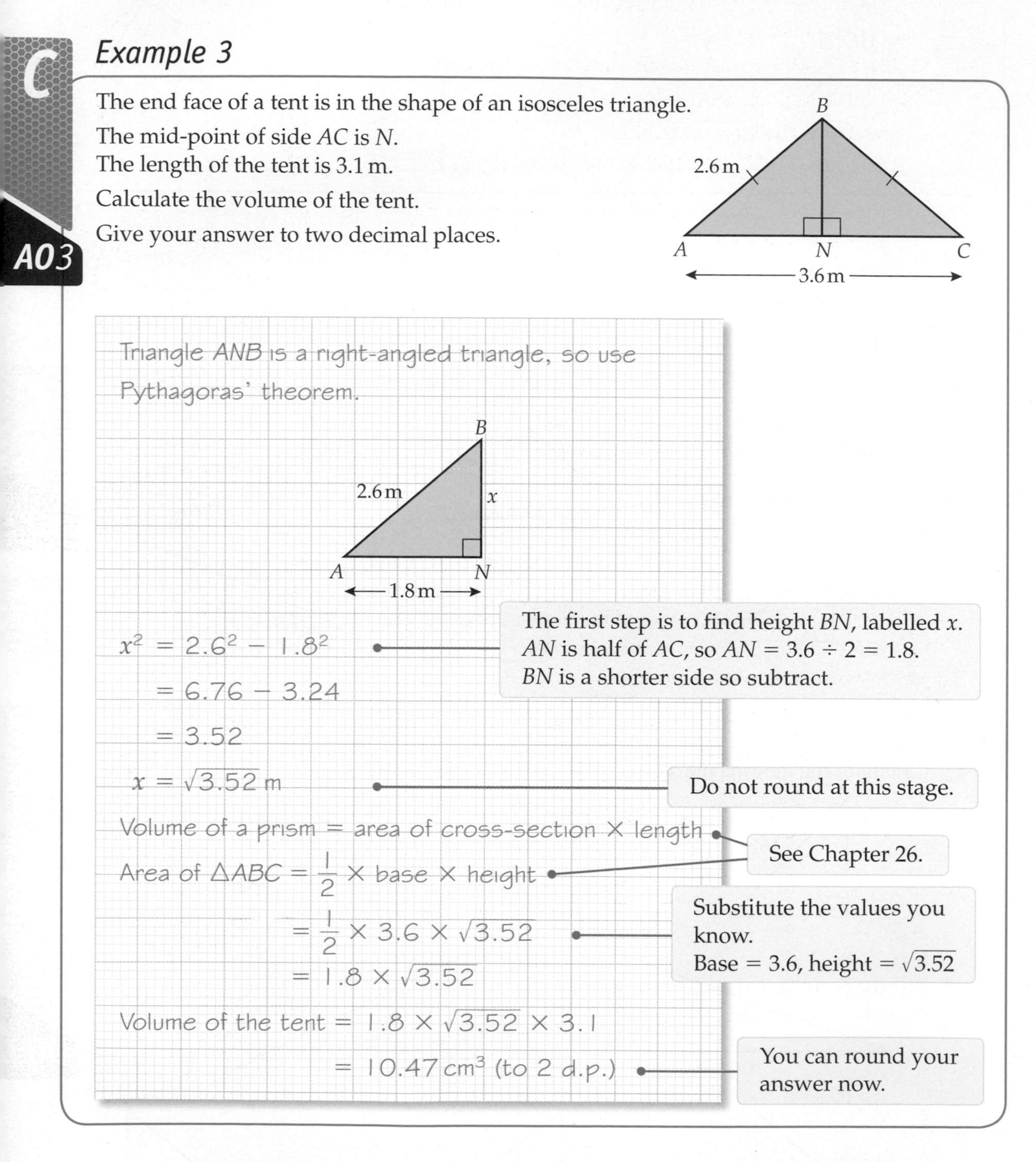

C

A03

Example 3

The end face of a tent is in the shape of an isosceles triangle.
The mid-point of side AC is N.
The length of the tent is 3.1 m.
Calculate the volume of the tent.
Give your answer to two decimal places.

Triangle ANB is a right-angled triangle, so use Pythagoras' theorem.

$x^2 = 2.6^2 - 1.8^2$

$= 6.76 - 3.24$

$= 3.52$

$x = \sqrt{3.52}$ m

Volume of a prism = area of cross-section × length

Area of $\Delta ABC = \frac{1}{2} \times$ base × height

$= \frac{1}{2} \times 3.6 \times \sqrt{3.52}$

$= 1.8 \times \sqrt{3.52}$

Volume of the tent $= 1.8 \times \sqrt{3.52} \times 3.1$

$= 10.47\text{ cm}^3$ (to 2 d.p.)

The first step is to find height BN, labelled x. AN is half of AC, so $AN = 3.6 \div 2 = 1.8$. BN is a shorter side so subtract.

Do not round at this stage.

See Chapter 26.

Substitute the values you know.
Base = 3.6, height = $\sqrt{3.52}$

You can round your answer now.

Exercise 33B

C

1 Colin walks 7 km due east and then 9 km due south.
How far is Colin from his starting point?
Give your answer to one decimal place.

For Q1, Q2 and Q6 it is useful to sketch a diagram and label it.

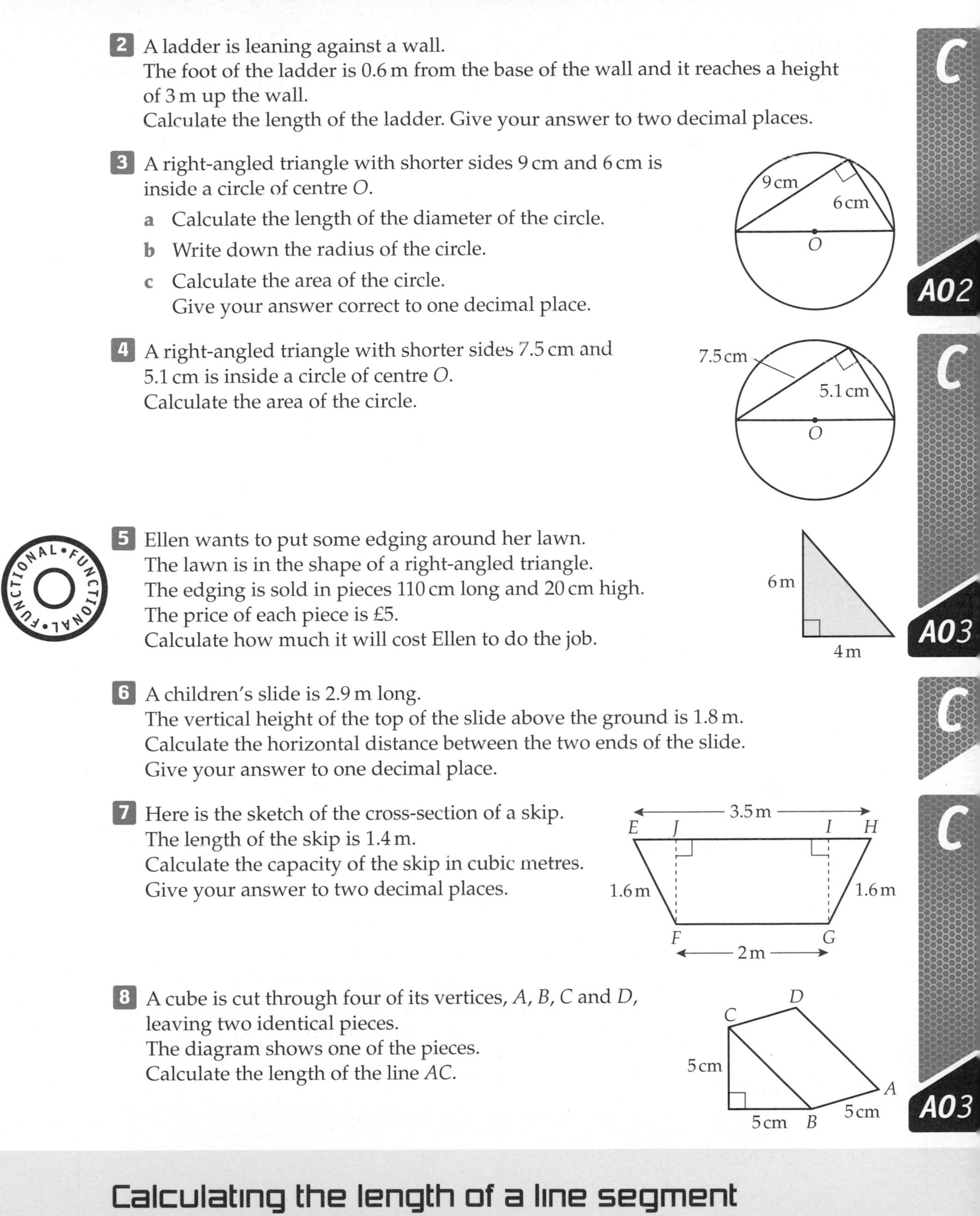

2 A ladder is leaning against a wall.
The foot of the ladder is 0.6 m from the base of the wall and it reaches a height of 3 m up the wall.
Calculate the length of the ladder. Give your answer to two decimal places.

3 A right-angled triangle with shorter sides 9 cm and 6 cm is inside a circle of centre O.

- **a** Calculate the length of the diameter of the circle.
- **b** Write down the radius of the circle.
- **c** Calculate the area of the circle.
  Give your answer correct to one decimal place.

4 A right-angled triangle with shorter sides 7.5 cm and 5.1 cm is inside a circle of centre O.
Calculate the area of the circle.

5 Ellen wants to put some edging around her lawn.
The lawn is in the shape of a right-angled triangle.
The edging is sold in pieces 110 cm long and 20 cm high.
The price of each piece is £5.
Calculate how much it will cost Ellen to do the job.

6 A children's slide is 2.9 m long.
The vertical height of the top of the slide above the ground is 1.8 m.
Calculate the horizontal distance between the two ends of the slide.
Give your answer to one decimal place.

7 Here is the sketch of the cross-section of a skip.
The length of the skip is 1.4 m.
Calculate the capacity of the skip in cubic metres.
Give your answer to two decimal places.

8 A cube is cut through four of its vertices, A, B, C and D, leaving two identical pieces.
The diagram shows one of the pieces.
Calculate the length of the line AC.

Calculating the length of a line segment

A line segment is the line between two points.

The length of a line segment parallel to the x-axis can be found by working out the difference between the x-coordinates of the end-points. You can use a similar method for the length of a line segment parallel to the y-axis.

Pythagoras' theorem can be used to find the length of a sloping line.

C

Example 4

The points $A(1, 0)$ and $B(7, 8)$ are shown.

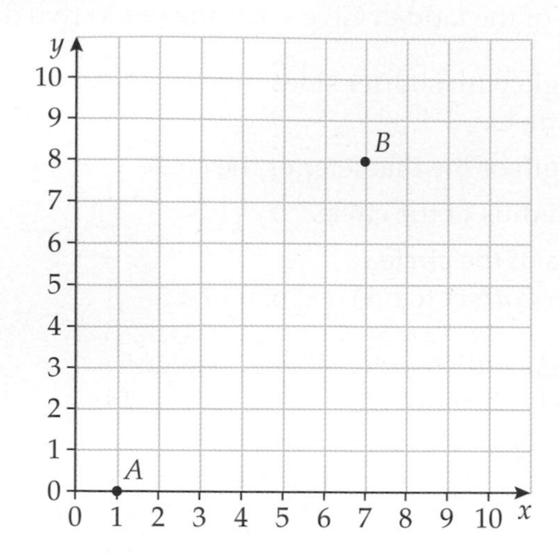

Calculate the length of AB.

Give your answer to one decimal place.

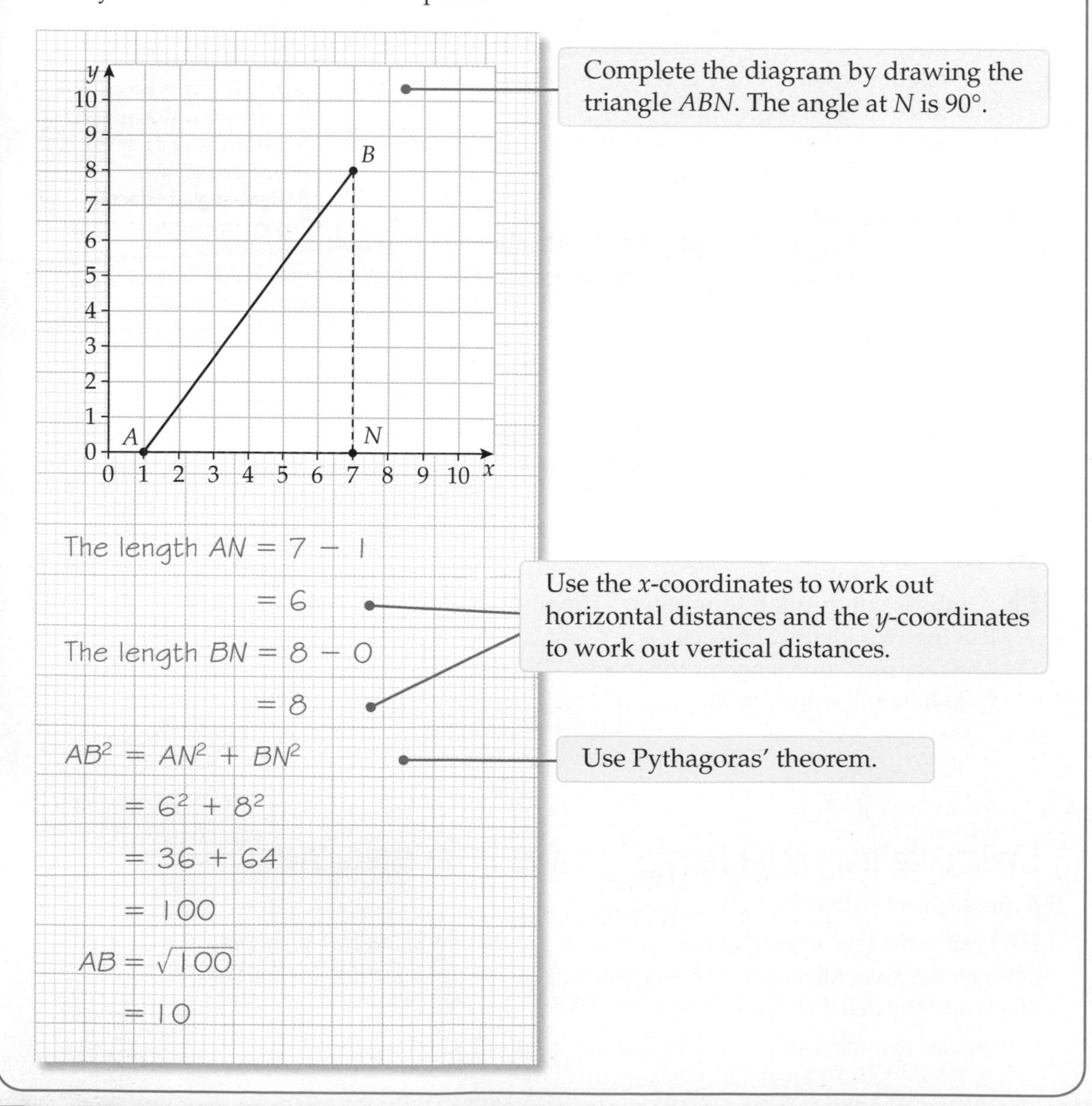

y
10
9
8
7
6
5
4
3
2
1
0
A
B
N
0 1 2 3 4 5 6 7 8 9 10
x
Complete the diagram by drawing the triangle ABN. The angle at N is 90°.
The length AN = 7 − 1
= 6
The length BN = 8 − 0
= 8
Use the x-coordinates to work out horizontal distances and the y-coordinates to work out vertical distances.
AB² = AN² + BN²
= 6² + 8²
= 36 + 64
= 100
AB = √100
= 10
Use Pythagoras' theorem.

Exercise 33C

C

1 Calculate the length of these line segments. Give your answer to one decimal place when appropriate.

a AB

b CD

c EF

d GH

e IJ

f KL

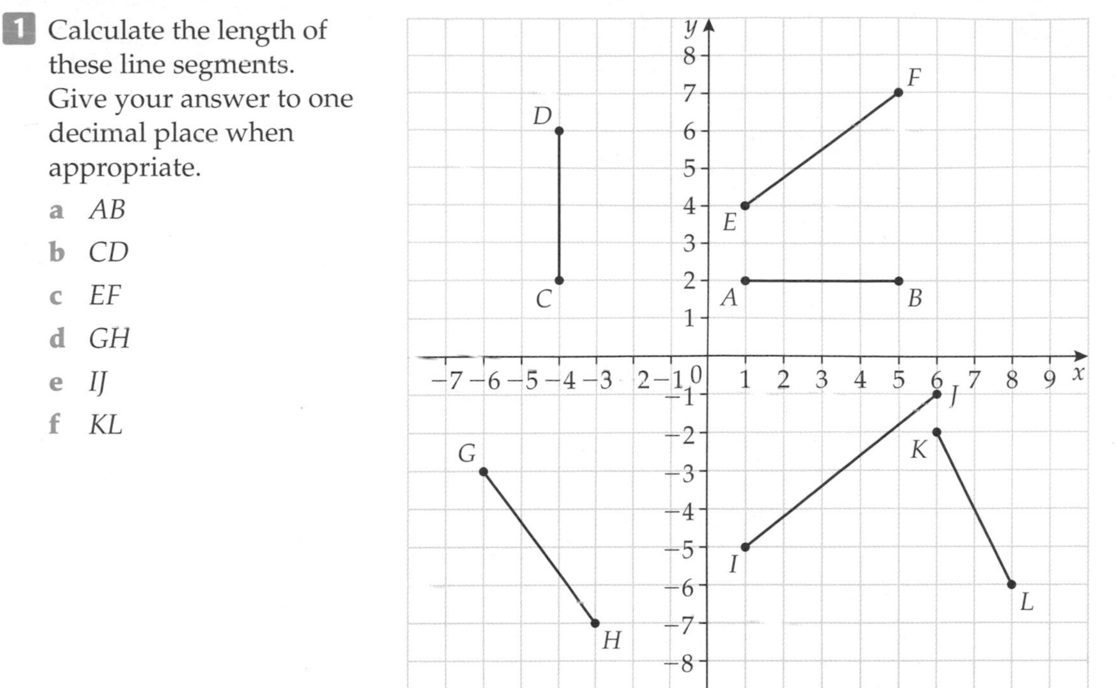

2 Calculate the lengths of these line segments. Give your answer to one decimal place when appropriate.

a CD: $C(3, 5)$ and $D(5, 7)$

b EF: $E(1, 3)$ and $F(5, 7)$

c GH: $G(-3, 5)$ and $H(5, 6)$

d IJ: $I(7, 10)$ and $J(-6, 4)$

e KL: $K(-2, -4)$ and $L(-10, -12)$

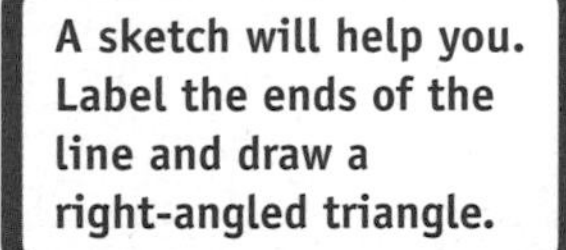

A sketch will help you. Label the ends of the line and draw a right-angled triangle.

33.2 Trigonometry - the ratios of sine, cosine and tangent

Why learn this?

The trigonometric functions are powerful tools in many areas of maths and science, including acoustics and engineering.

Objectives

B Understand and recall trigonometric ratios in right-angled triangles

B Know how to enter the trigonometric functions on a calculator

Keywords

trigonometry, sine, cosine, tangent, opposite, adjacent, inverse

Skills check

1 Solve these equations.

a $\frac{x}{6} = 4$ b $\frac{3x}{4} = 6$

2 Round 27.65 to one decimal place.

Sine, cosine and tangent

Trigonometry is concerned with calculating sides and angles in triangles, and involves three ratios called **sine**, **cosine** and **tangent**. These ratios are often abbreviated to sin, cos and tan, respectively.

Consider this right-angled triangle.

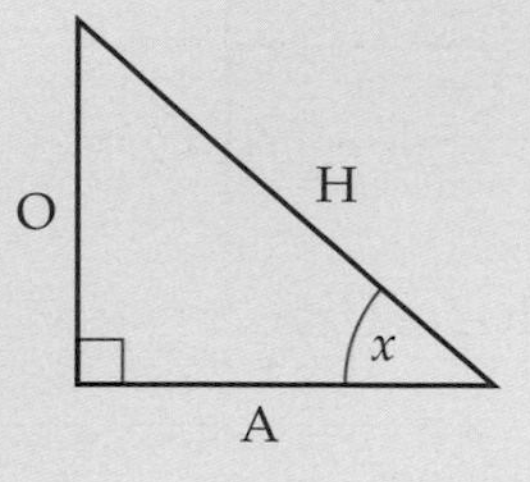

The hypotenuse is always opposite the right angle so it is always the same side.
The '**opposite**' and '**adjacent**' sides switch depending on which angle you are considering.

The ratios for sine, cosine and tangent are defined as follows.

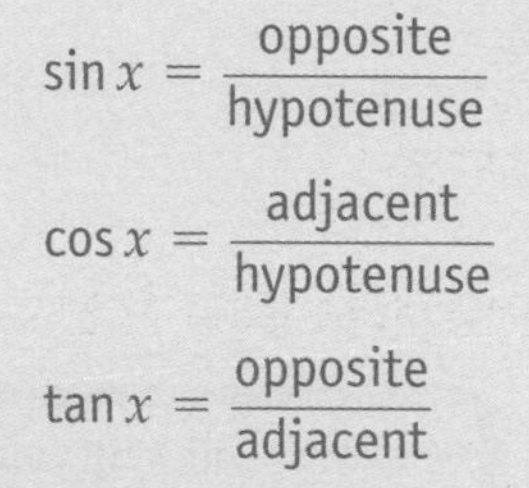

$$\sin x = \frac{\text{opposite}}{\text{hypotenuse}}$$

$$\cos x = \frac{\text{adjacent}}{\text{hypotenuse}}$$

$$\tan x = \frac{\text{opposite}}{\text{adjacent}}$$

$\sin x = \frac{O}{H}$ $\quad \cos x = \frac{A}{H}$ $\quad \tan x = \frac{O}{A}$
You need to learn these formulae for the exam. A useful memory aid is SOHCAHTOA.

Using a calculator

The abbreviated forms (sin, cos and tan) are shown on your calculator keys.

When you use a calculator, make sure it is working in degrees. D or DEG will appear in the display.
If your calculator is not in 'degree' mode, press the MODE key until a choice of DEG, RAD or GRA appears in the display. Choose the DEG option.

Remember that different calculators have different keys. Look in your instruction booklet.

To find the sine, cosine or tangent of an angle, press the appropriate key followed by the angle.

For sin 72°, press [sin] [7] [2] [=]

The display will show 0.951056...

You will sometimes have to work backwards or use the **inverse** function.

The inverse function on your calculator may be labelled [SHIFT] or [2nd F] or [INV]

To find angle x when $\tan x = 0.75$, press [SHIFT] [tan] [0] [·] [7] [5] [=]

You are given the tangent and want to find the angle that goes with it. This is the inverse operation of finding the tangent.

You will see that $x = 36.869897... = 36.9°$ (1 d.p.).

When you press the [SHIFT] key followed by the [tan] key, you may see $\tan^{-1}$ appear in the calculator display. This is another way of writing the inverse of a tangent (e.g. inverse $\tan 0.75 = 36.9°$ is sometimes written as $\tan^{-1} 0.75 = 36.9°$).

Example 5

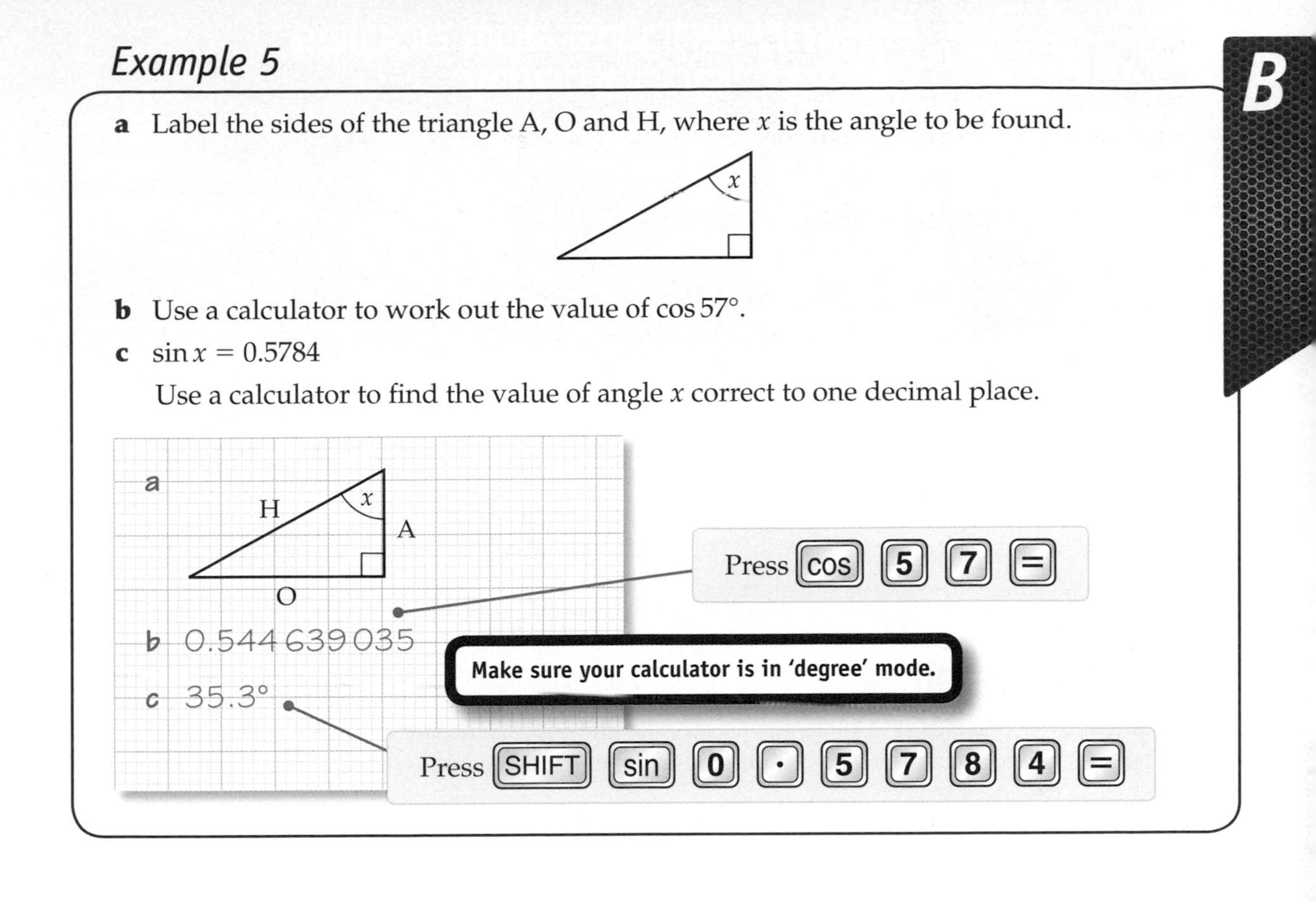

a Label the sides of the triangle A, O and H, where x is the angle to be found.

b Use a calculator to work out the value of $\cos 57°$.

c $\sin x = 0.5784$

Use a calculator to find the value of angle x correct to one decimal place.

B

Exercise 33D

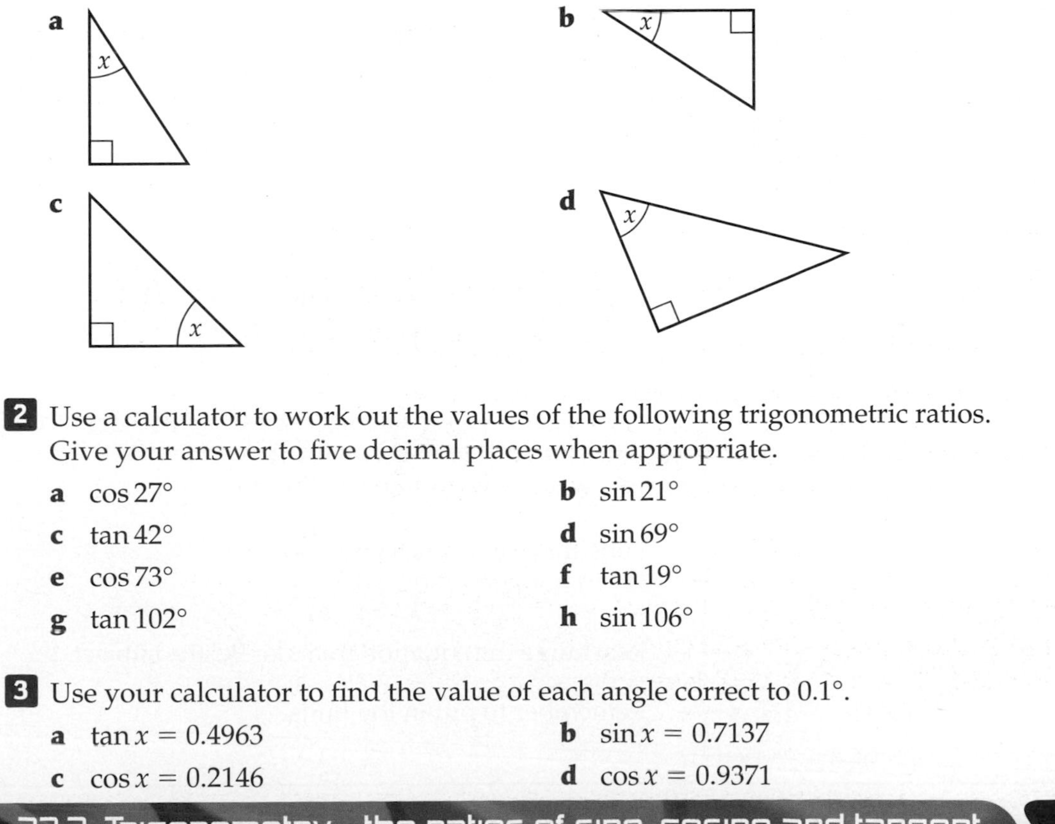

1 Draw these triangles and label the sides O, A and H, where x is the angle to be found.

2 Use a calculator to work out the values of the following trigonometric ratios. Give your answer to five decimal places when appropriate.

a $\cos 27°$ **b** $\sin 21°$

c $\tan 42°$ **d** $\sin 69°$

e $\cos 73°$ **f** $\tan 19°$

g $\tan 102°$ **h** $\sin 106°$

3 Use your calculator to find the value of each angle correct to 0.1°.

a $\tan x = 0.4963$ **b** $\sin x = 0.7137$

c $\cos x = 0.2146$ **d** $\cos x = 0.9371$

B

33.3 Finding lengths and angles using trigonometry

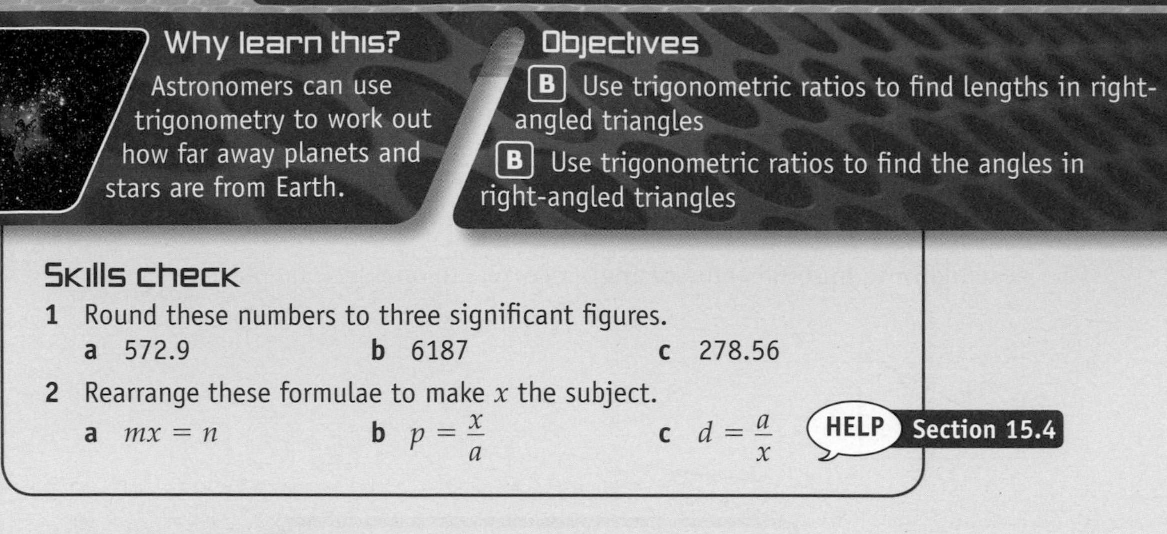

Why learn this?

Astronomers can use trigonometry to work out how far away planets and stars are from Earth.

Objectives

B Use trigonometric ratios to find lengths in right-angled triangles

B Use trigonometric ratios to find the angles in right-angled triangles

Skills check

1 Round these numbers to three significant figures.

a 572.9 **b** 6187 **c** 278.56

2 Rearrange these formulae to make x the subject.

a $mx = n$ **b** $p = \frac{x}{a}$ **c** $d = \frac{a}{x}$

HELP Section 15.4

Finding lengths

If the length of one side of a right-angled triangle and the size of one angle are given, then the lengths of the remaining sides can be found using trigonometry.

B

Example 6

In $\triangle ABC$, length $AB = 12\,\text{cm}$, $\angle ACB = 90°$ and $\angle BAC = 30°$.

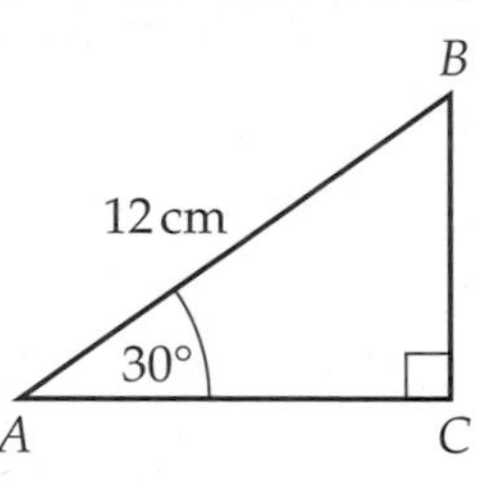

Calculate the length of BC.

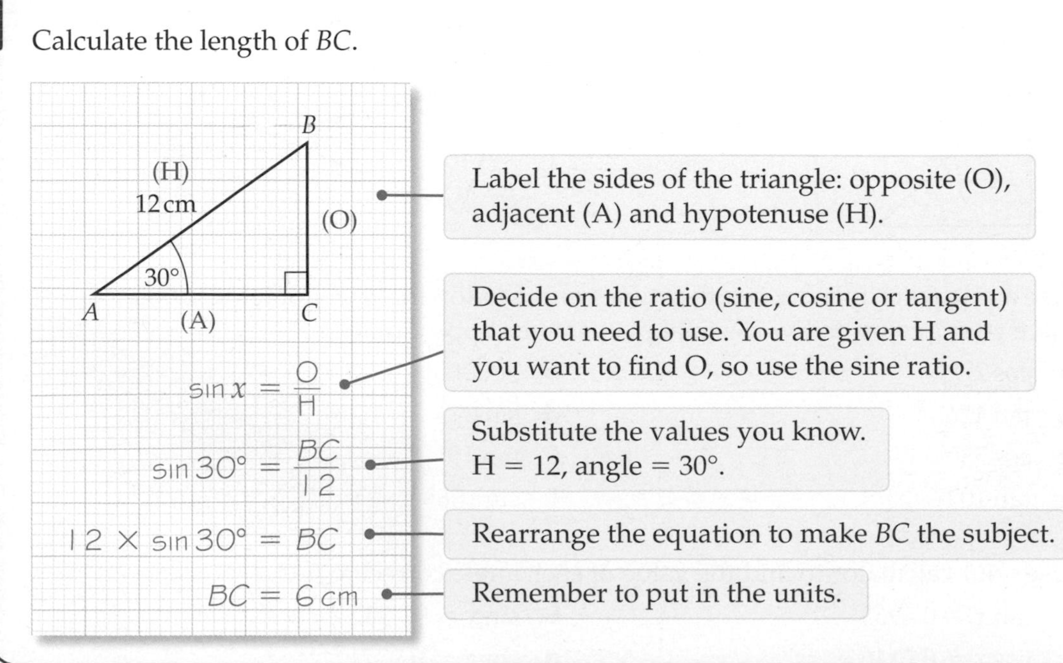

$\sin x = \frac{O}{H}$

$\sin 30° = \frac{BC}{12}$

$12 \times \sin 30° = BC$

$BC = 6\,\text{cm}$

Label the sides of the triangle: opposite (O), adjacent (A) and hypotenuse (H).

Decide on the ratio (sine, cosine or tangent) that you need to use. You are given H and you want to find O, so use the sine ratio.

Substitute the values you know. H = 12, angle = 30°.

Rearrange the equation to make BC the subject.

Remember to put in the units.

Example 7

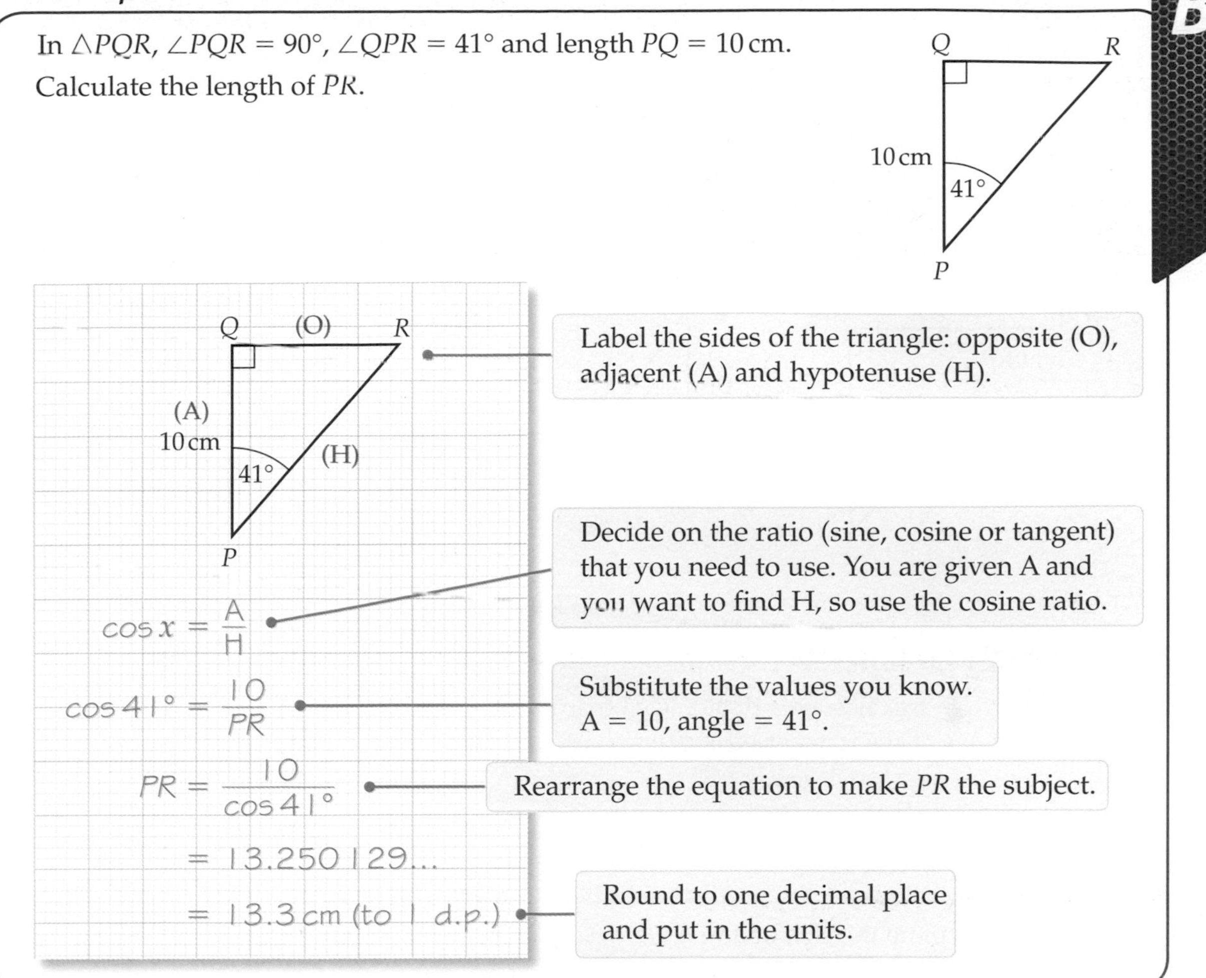

In △PQR, ∠PQR = 90°, ∠QPR = 41° and length PQ = 10 cm.
Calculate the length of PR.
Q
R
10 cm
41°
P
Q
(O)
R
(A)
10 cm
41°
(H)
P
cos x = A/H
cos 41° = 10/PR
PR = 10/cos 41°
= 13.250 129...
= 13.3 cm (to 1 d.p.)
Label the sides of the triangle: opposite (O), adjacent (A) and hypotenuse (H).
Decide on the ratio (sine, cosine or tangent) that you need to use. You are given A and you want to find H, so use the cosine ratio.
Substitute the values you know.
A = 10, angle = 41°.
Rearrange the equation to make PR the subject.
Round to one decimal place and put in the units.

B

Exercise 33E

1 Calculate the unknown length, x, marked on each diagram.
Give your answer to one decimal place when appropriate.

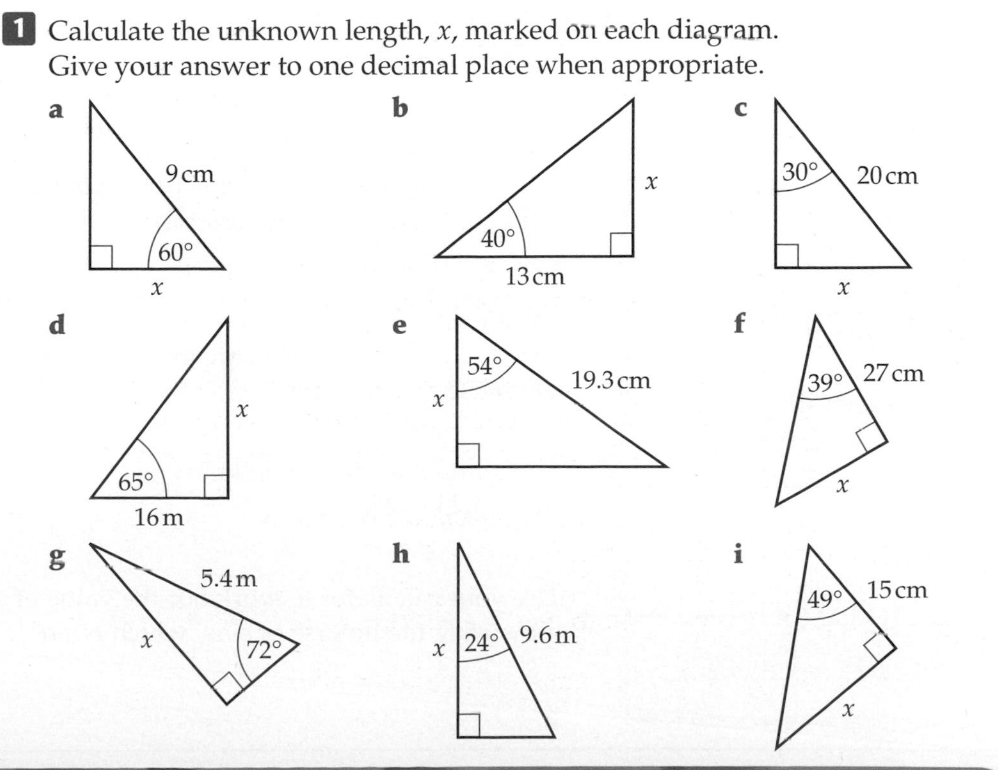

B

B

2 Calculate the unknown length, x, marked on each diagram.
Give your answer to three significant figures when appropriate.

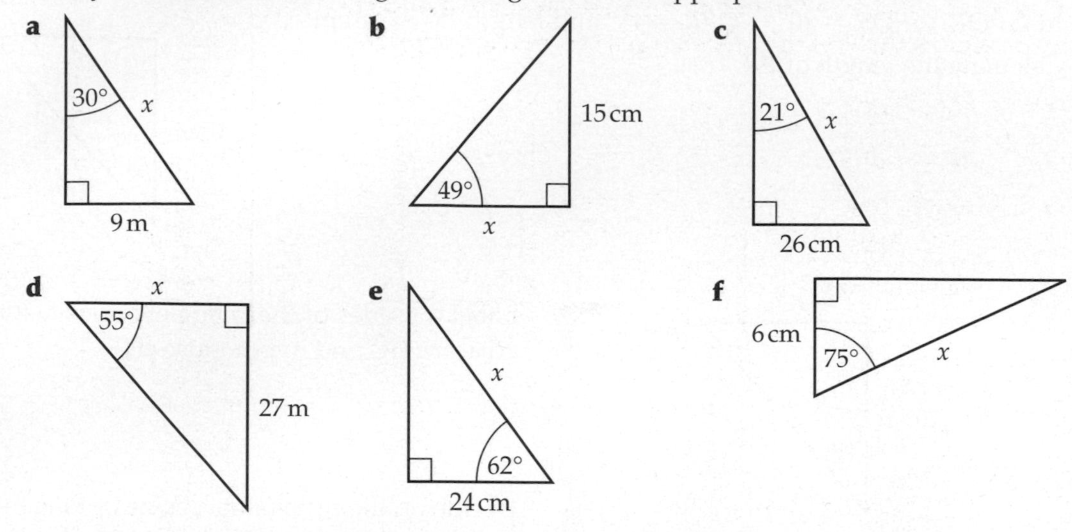

Finding angles

If the lengths of two sides of a right-angled triangle are given, you will be able to work out an angle using one of the trigonometric ratios.

B

Example 8

In $\triangle ABC$, calculate the size of angle x.

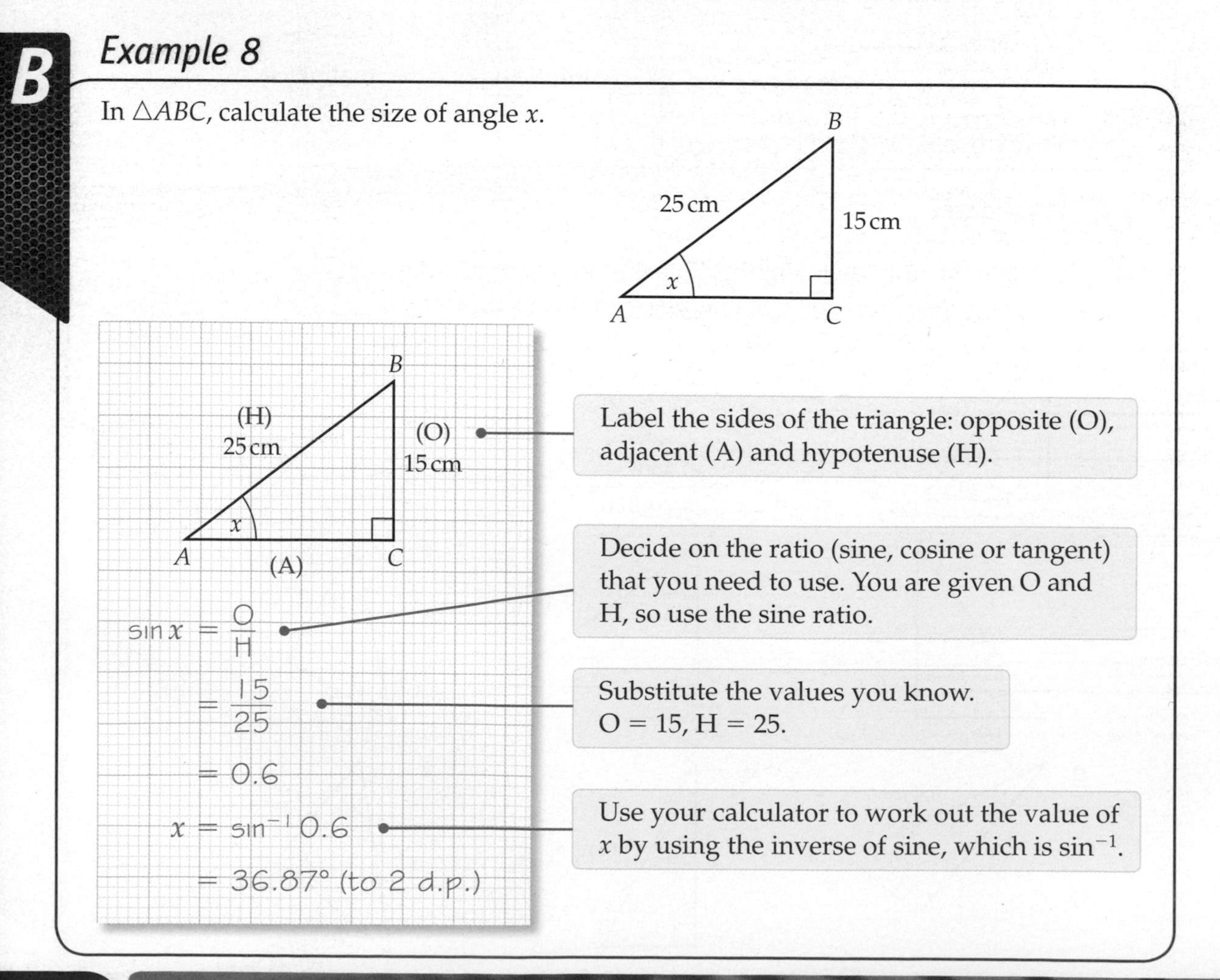

B
25 cm
15 cm
x
A
C
B
(H)
25 cm
(O)
15 cm
x
A
(A)
C
sin x = O/H
= 15/25
= 0.6
x = sin⁻¹ 0.6
= 36.87° (to 2 d.p.)
Label the sides of the triangle: opposite (O), adjacent (A) and hypotenuse (H).
Decide on the ratio (sine, cosine or tangent) that you need to use. You are given O and H, so use the sine ratio.
Substitute the values you know.
O = 15, H = 25.
Use your calculator to work out the value of x by using the inverse of sine, which is sin⁻¹.

Exercise 33F

Calculate the size of the unknown angle in each of these triangles.
Give your answer to one decimal place when appropriate.

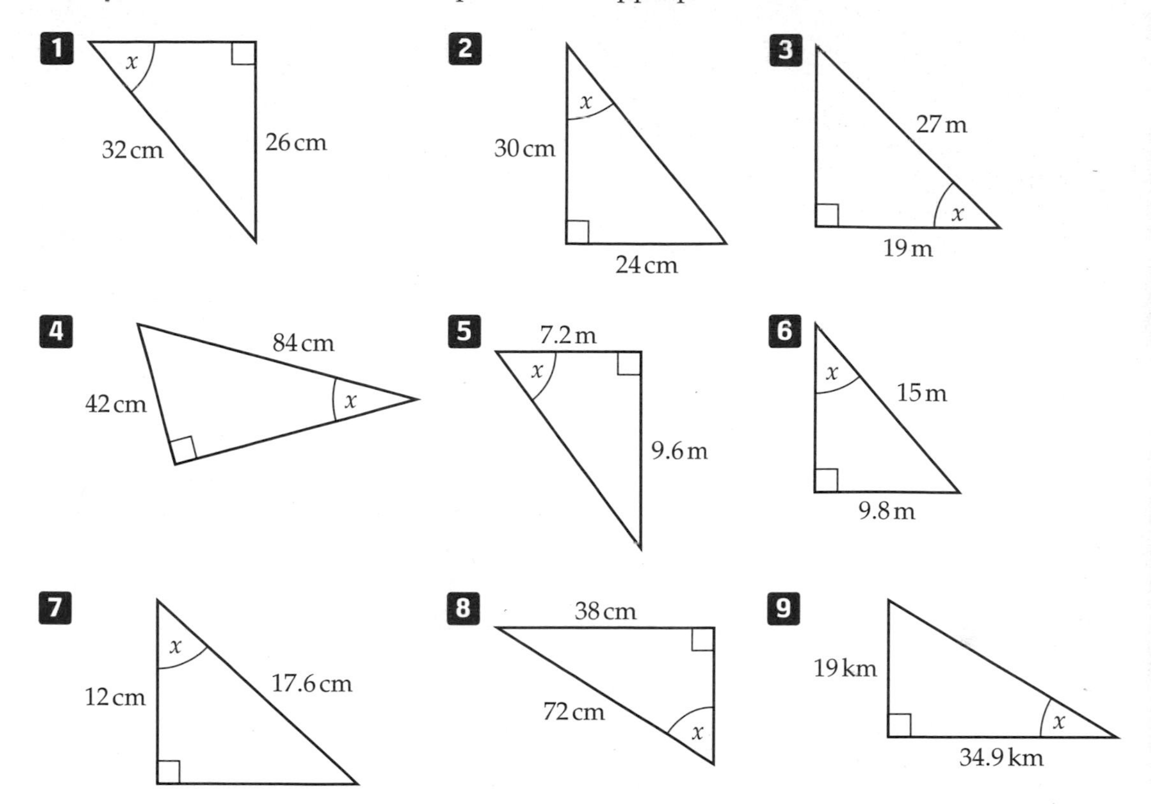

B

33.4 Applications of trigonometry and Pythagoras

Why learn this?

A combination of Pythagoras' theorem and trigonometry can solve more complex problems.

Objectives

C B A Use trigonometric ratios and Pythagoras to solve problems

Skills check

HELP Section 23.2

For each diagram, work out the bearing of
i *A* from *B* **ii** *B* from *A*.

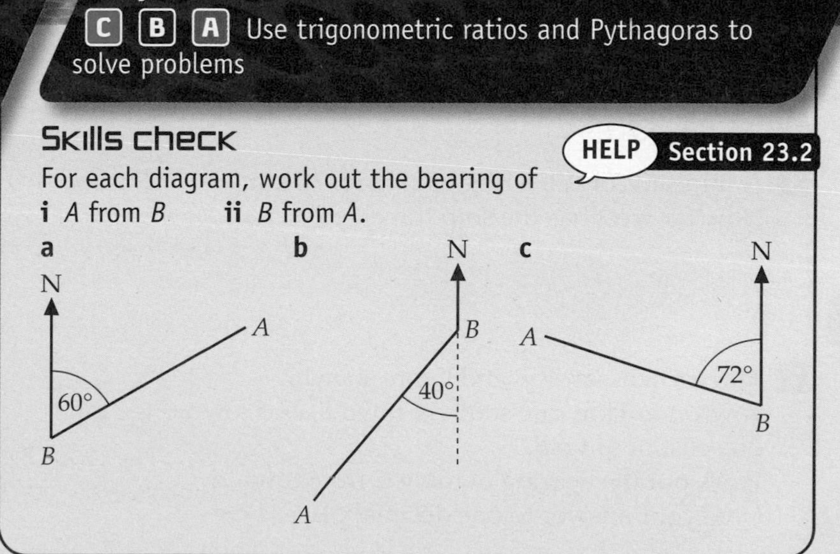

Using trigonometry and Pythagoras to solve problems

The trigonometric ratios and Pythagoras' theorem can be used to solve problems set in a range of real-life contexts.

B AO2

Example 9

A ship sails on a bearing of 130° for 60 km.
How far east has the ship travelled?

To revise work on bearings see Section 23.2.

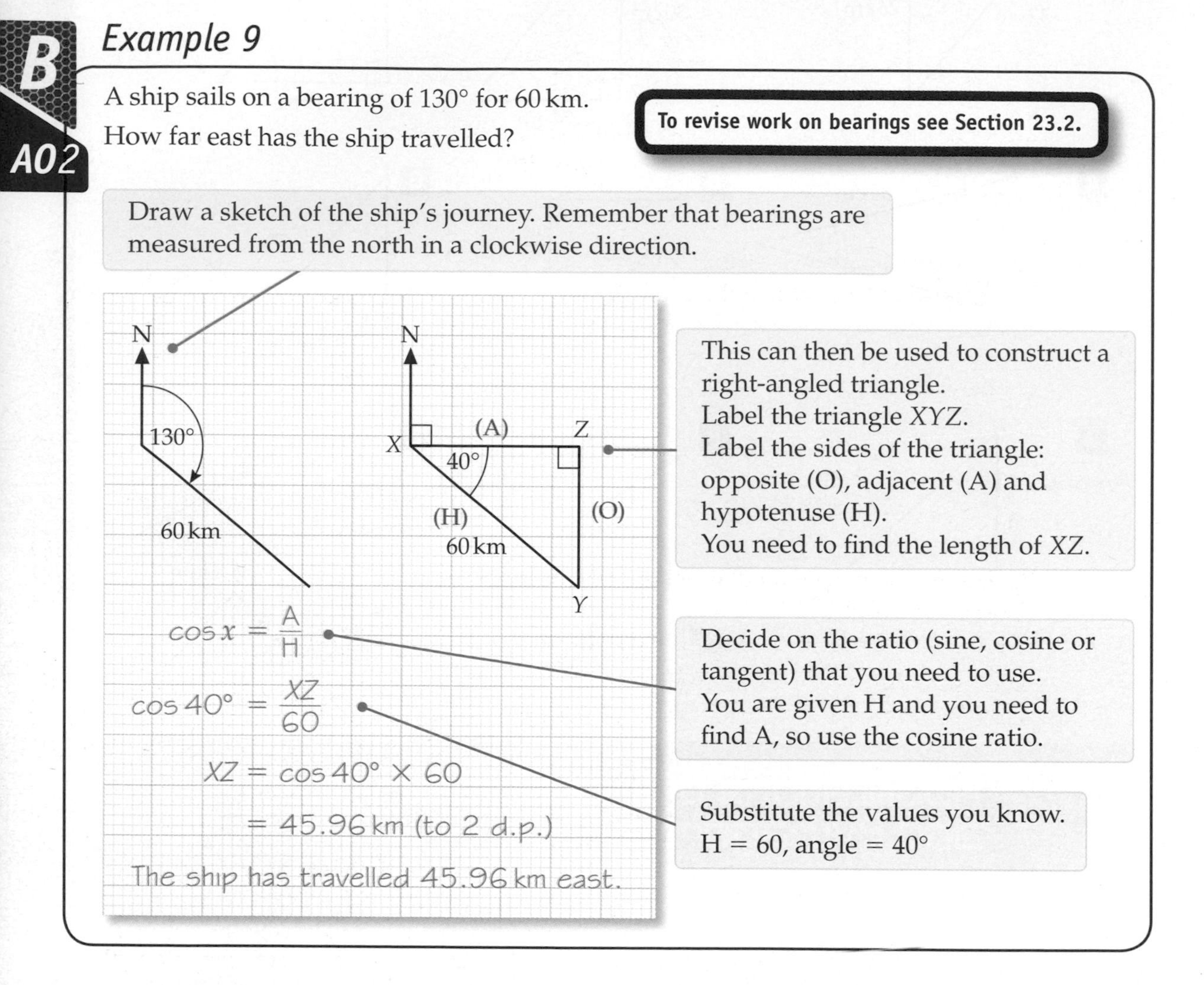

Draw a sketch of the ship's journey. Remember that bearings are measured from the north in a clockwise direction.
N
130°
60 km
N
X
(A)
Z
40°
(H)
60 km
(O)
Y
This can then be used to construct a right-angled triangle.
Label the triangle XYZ.
Label the sides of the triangle: opposite (O), adjacent (A) and hypotenuse (H).
You need to find the length of XZ.
cos x = A/H
cos 40° = XZ/60
XZ = cos 40° × 60
= 45.96 km (to 2 d.p.)
The ship has travelled 45.96 km east.
Decide on the ratio (sine, cosine or tangent) that you need to use.
You are given H and you need to find A, so use the cosine ratio.
Substitute the values you know.
H = 60, angle = 40°

Exercise 33G

1 A ship sails on a bearing of 295° for 80 km.
How far west has the ship travelled?

B

2 Three towns, at A, B and C, are shown.
Town A is 4 km due south of town B, and town C is 3 km due east of town B.
Work out the bearing of town C from town A.
Give your answer to one decimal place.

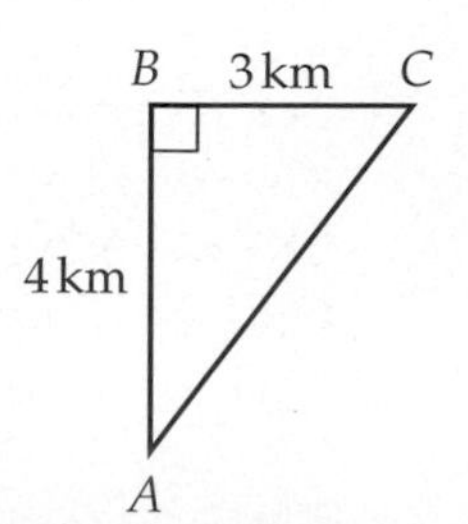

3 A ladder 6 m long rests against a wall.

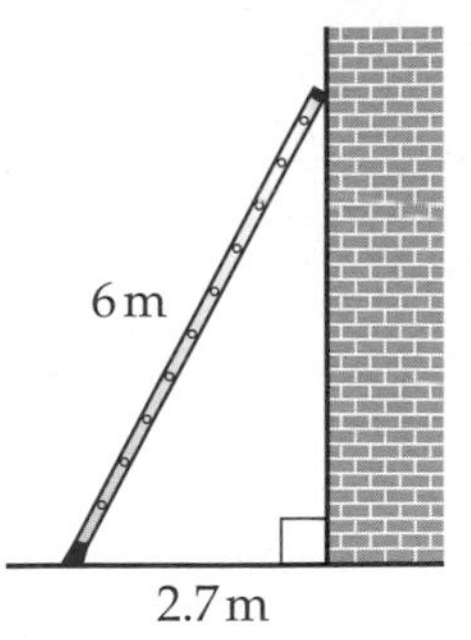

The foot of the ladder is 2.7 m away from the base of the wall.
What angle does the ladder make with the wall?
Give your answer to one decimal place.

B

4 A boy is flying a kite on a string 27 m long.
The string is held 1 m above the ground and it makes an angle of 43° with the horizontal.
How high is the kite above the ground?
Give your answer to the nearest metre.

B

5 **a** Calculate the perpendicular height, BN, of the isosceles triangle ABC.
Give your answer to three significant figures.

b Calculate the area of the triangle.
Give your answer to two significant figures.

6 Calculate the area of the isosceles triangle.
Give your answer to three significant figures.

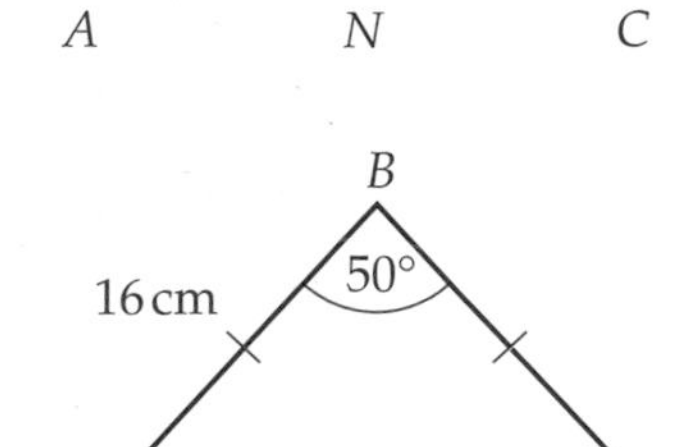

7 Frith, Soll and Barne are three towns.
Frith is 7.6 km due west of Soll.
Barne is 9.8 km due north of Soll.
Calculate the bearing and distance of Frith from Barne.
Give your answers to three significant figures.

AO2

8 The diagram shows a right-angled triangle in a circle, centre O.
The diameter of the circle is 15 cm.
Calculate the area of the triangle.
Give your answer to three significant figures.

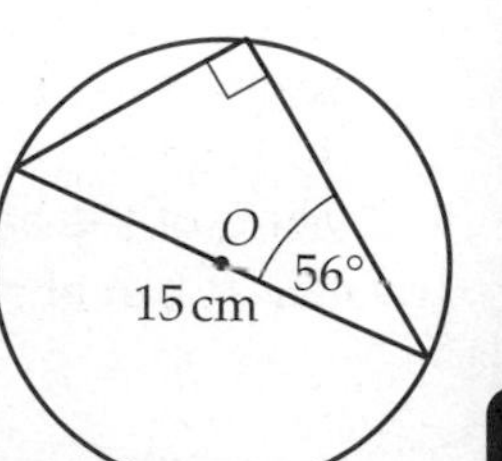

B

AO3

B

9

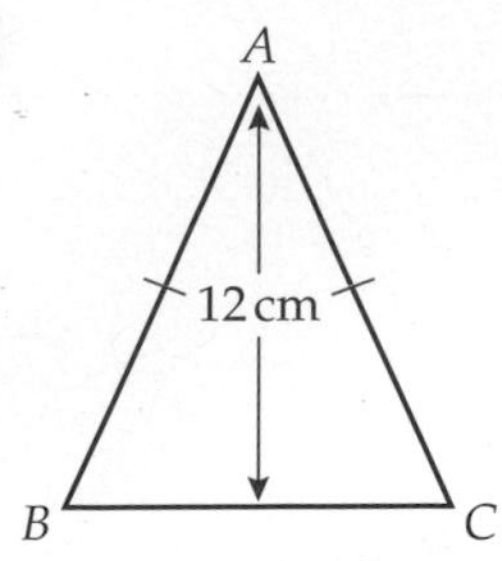

The area of isosceles $\triangle ABC$ is $50\,\text{cm}^2$.
Calculate the size of $\angle A$.
Give your answer to one decimal place.

10 A wheelchair ramp is to be built to go up four steps.
The steps are 20 cm high and have a depth of 27 cm.
The ramp will have an angle of 4.8° with the horizontal.
Calculate distance x to find out how far away from the steps the ramp should start.
Give your answer to the nearest centimetre.

A03

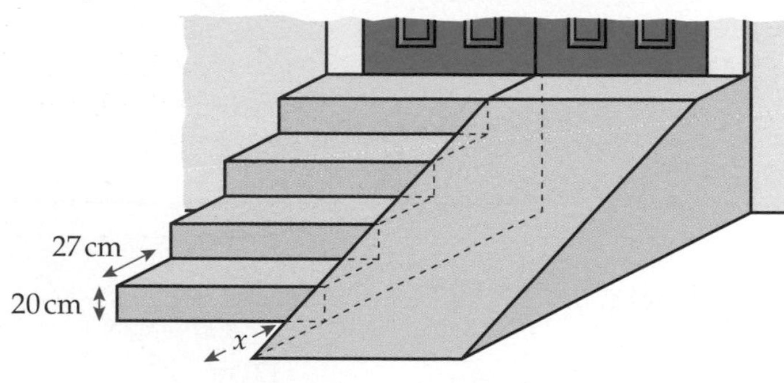

Solving surface area and volume problems

You can use Pythagoras' theorem to find unknown lengths in 3-D shapes, in order to calculate volumes and surface areas.

In Chapter 27 you met the formulae for the surface area and the volume of a cone, given the radius of the base (r) and the slant height (l).

Total surface area of a cone = area of base + area of curved surface

$$= \pi r^2 + \pi r l$$

Volume of cone $= \frac{1}{3}\pi r^2 h$

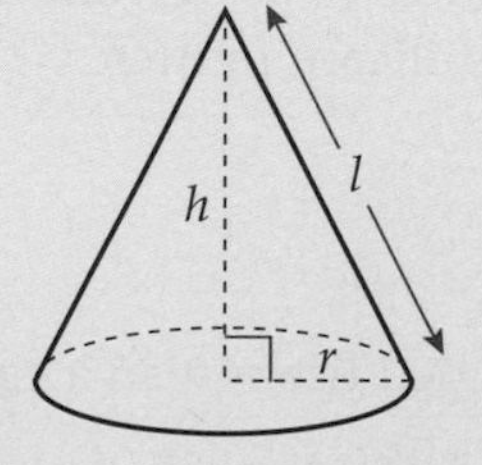

The radius of the base, the slant height and the perpendicular height form a right-angled triangle. If two of r, h and l are given, Pythagoras' theorem can be used to find the third.

$$r^2 + h^2 = l^2$$

Example 10

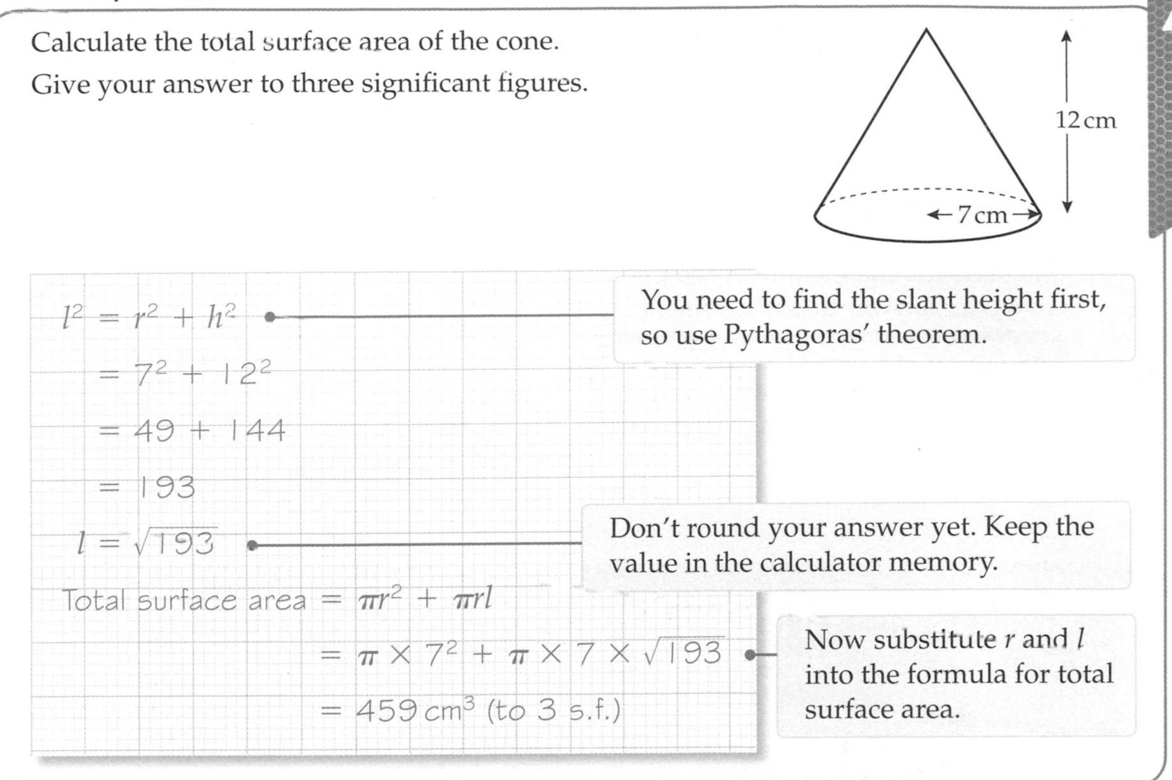

Calculate the total surface area of the cone.
Give your answer to three significant figures.

$l^2 = r^2 + h^2$ — You need to find the slant height first, so use Pythagoras' theorem.

$= 7^2 + 12^2$

$= 49 + 144$

$= 193$

$l = \sqrt{193}$ — Don't round your answer yet. Keep the value in the calculator memory.

Total surface area $= \pi r^2 + \pi r l$

$= \pi \times 7^2 + \pi \times 7 \times \sqrt{193}$ — Now substitute r and l into the formula for total surface area.

$= 459\text{ cm}^3$ (to 3 s.f.)

Exercise 33H

1 The diagram shows a prism whose cross-section is an isosceles trapezium.
The depth of the prism is 10 cm.

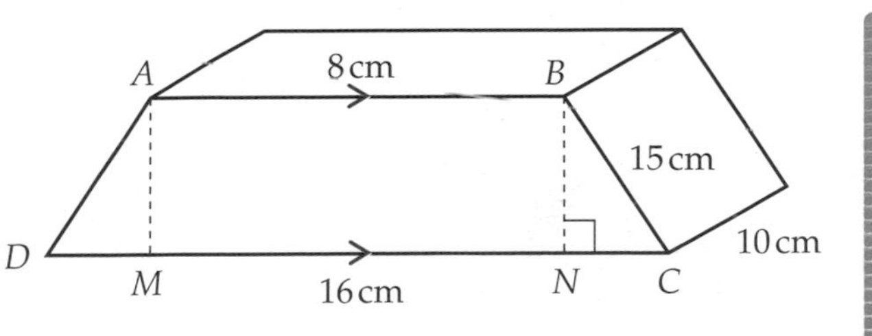

a Calculate the length of NC.

b Calculate the height BN.

c Calculate the volume of the prism.

2 A pencil of length 14 cm fits exactly in a cylinder of height 9 cm.

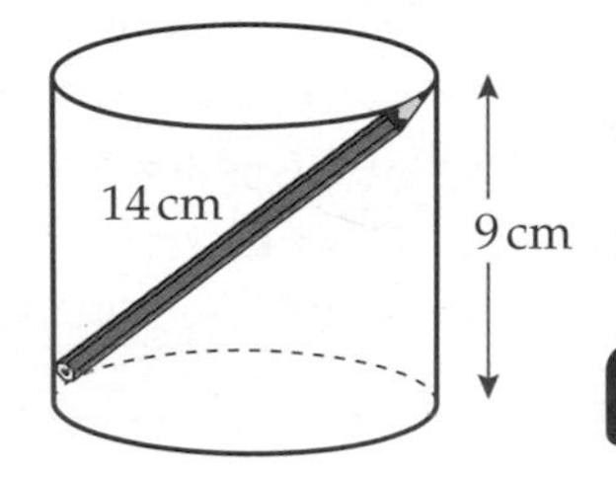

a Calculate the volume of the cylinder.

b Calculate the curved surface area of the cylinder.

3 A cone has a base radius of 6.5 cm and a slant height of 14 cm.

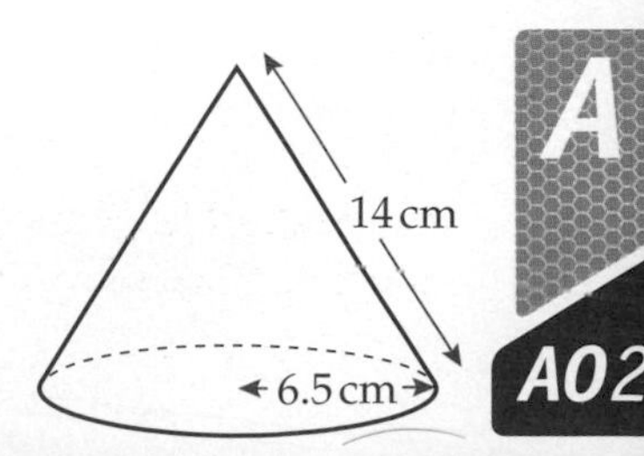

a Calculate the curved surface area of the cone.
Give your answer to one decimal place.

b Calculate the volume of the cone, giving your answer to four significant figures.

A C A AO2 A AO2

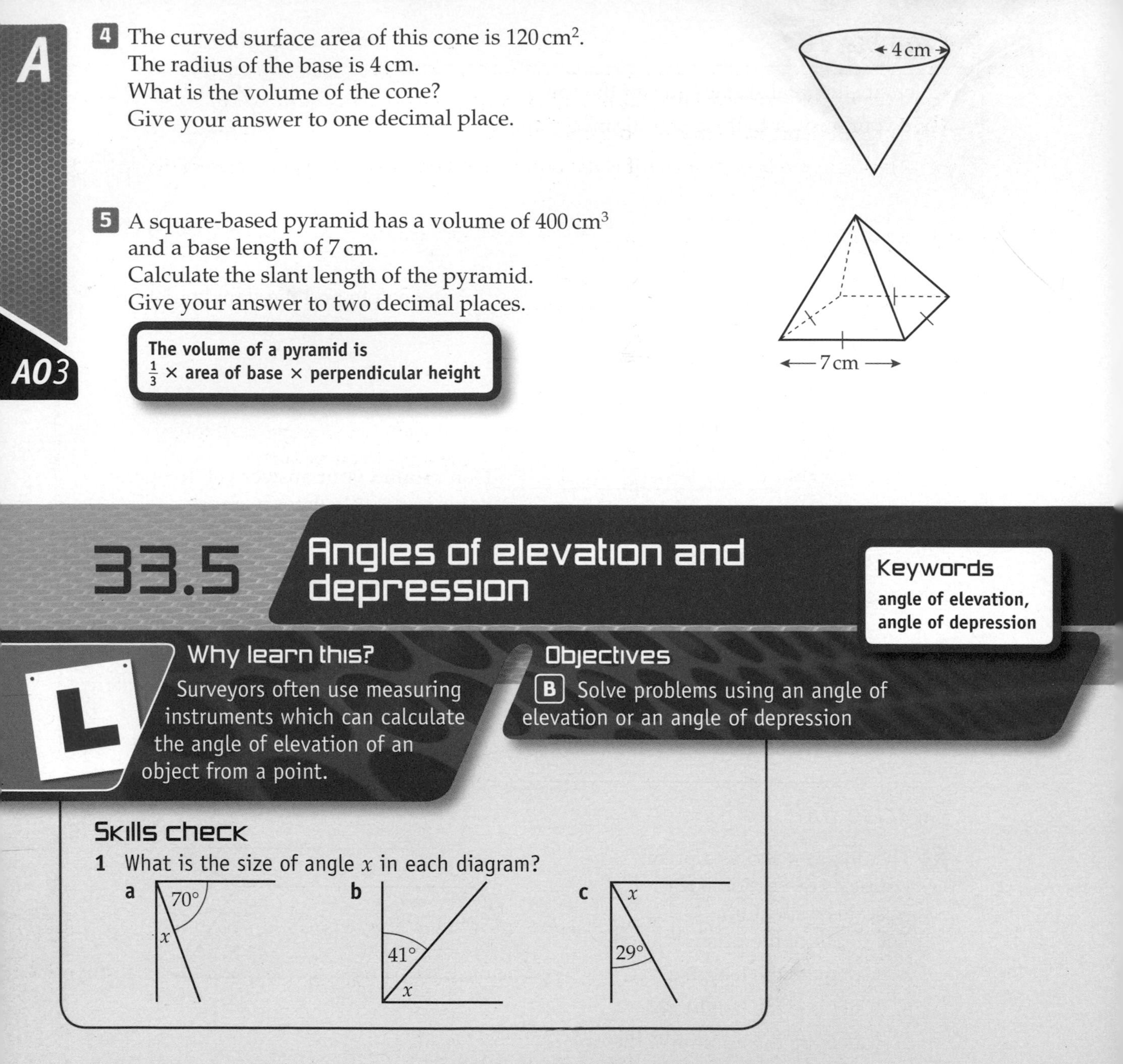

A

4 The curved surface area of this cone is 120 cm^2.
The radius of the base is 4 cm.
What is the volume of the cone?
Give your answer to one decimal place.

A03

5 A square-based pyramid has a volume of 400 cm^3 and a base length of 7 cm.
Calculate the slant length of the pyramid.
Give your answer to two decimal places.

> **The volume of a pyramid is**
> $\frac{1}{3}$ **× area of base × perpendicular height**

33.5 Angles of elevation and depression

Keywords
angle of elevation, angle of depression

Why learn this?

Surveyors often use measuring instruments which can calculate the angle of elevation of an object from a point.

Objectives

B Solve problems using an angle of elevation or an angle of depression

Skills check

1 What is the size of angle x in each diagram?

Angle of elevation and angle of depression

Problems using Pythagoras' theorem or trigonometry often refer to the **angle of elevation** or the **angle of depression.**

The angle of elevation is the angle measured upwards from the horizontal.

The angle of depression is the angle measured downwards from the horizontal.

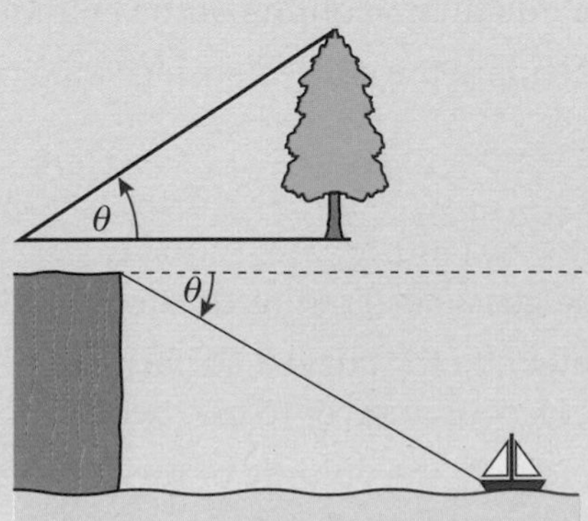

Example 11

From the top of a vertical cliff, 120 m high, Georgia can see a boat out at sea.
The angle of depression from Georgia to the boat is 49°.
How far from the base of the cliff is the boat? Give your answer to one decimal place.

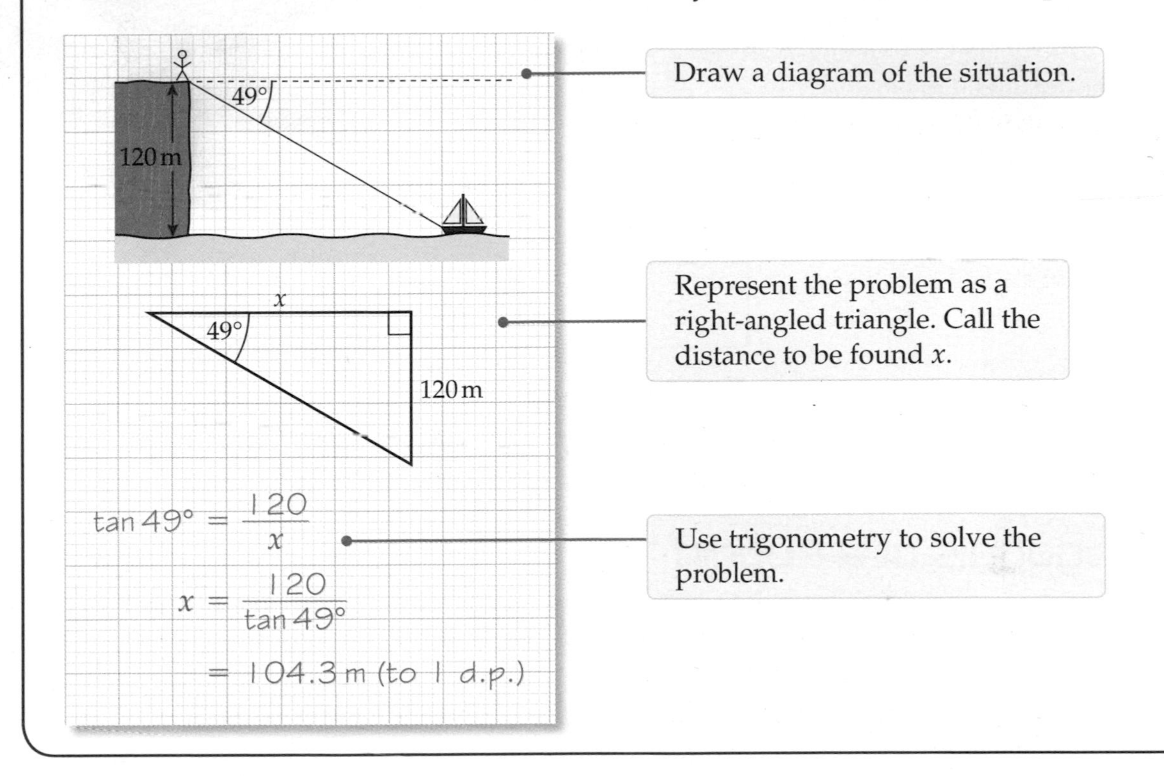

49°
120 m
x
49°
120 m
Draw a diagram of the situation.
Represent the problem as a right-angled triangle. Call the distance to be found x.
Use trigonometry to solve the problem.

$$\tan 49° = \frac{120}{x}$$

$$x = \frac{120}{\tan 49°}$$

$$= 104.3 \text{ m (to 1 d.p.)}$$

B

AO2

Exercise 33I

1 William sees an aircraft in the sky. The angle of elevation is 37°.
The aircraft is at a horizontal distance of 31 km from William.
How high is the aircraft? Give your answer to one decimal place.

2 Shamil is standing 1420 m from a wind turbine.
The angle of elevation to the top of the wind turbine is 7.6°.
What is the height of the wind turbine? Give your answer to one decimal place.

3 From the top of a 180 m high vertical cliff, a boat has an angle of depression of 61°.
How far is the boat from the base of the cliff?
Give your answer to one decimal place.

4 A surveyor stands 70 m from a castle.
The angle of elevation to the top of the castle is 55°.
A flagpole stands on top of the castle.
The angle of elevation to the tip of the flagpole is 59°.
Calculate the height of the flagpole. Give your answer to one decimal place.

5 From the top of a cliff 85 m above sea level, the angles of depression of two buoys are 22° and 14° respectively.
Calculate the distance between the two buoys.
Give your answer to three significant figures.

B

AO2

B

AO3

33.6 Problems in three dimensions

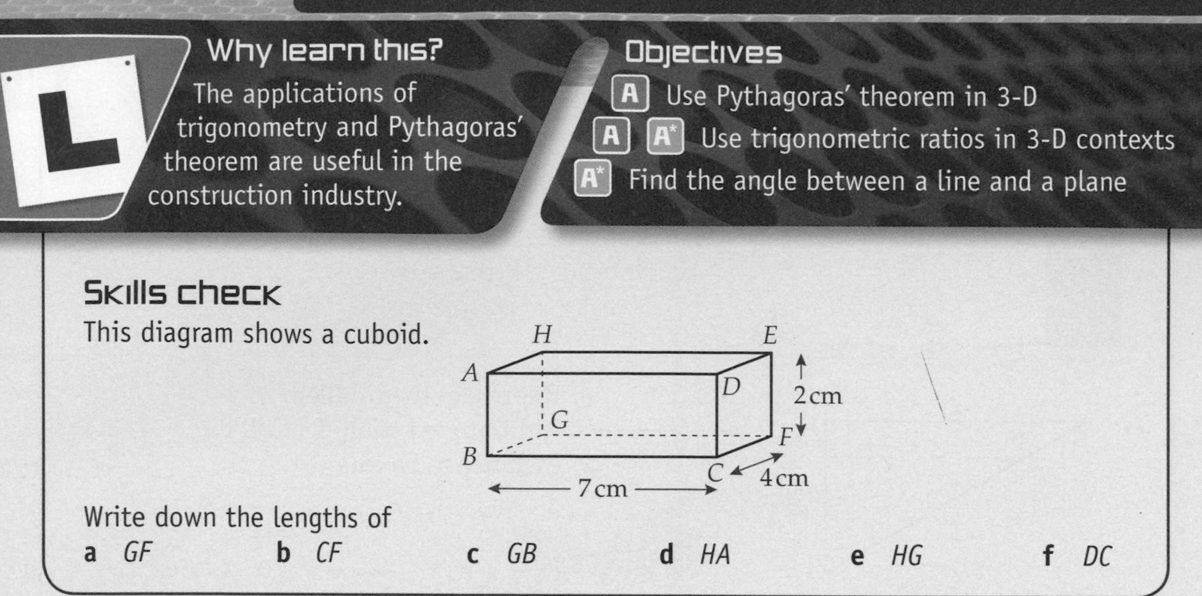

Why learn this?
The applications of trigonometry and Pythagoras' theorem are useful in the construction industry.

Objectives
- **A** Use Pythagoras' theorem in 3-D
- **A** **A*** Use trigonometric ratios in 3-D contexts
- **A*** Find the angle between a line and a plane

Skills check

This diagram shows a cuboid.

Write down the lengths of

a GF **b** CF **c** GB **d** HA **e** HG **f** DC

Diagonals of a cuboid

You can use Pythagoras' theorem and trigonometry to solve problems in three dimensions (3-D). It is important to be able to identify and draw the correct right-angled triangle that contains the length or angle to be found.

The length of the longest diagonal of a cuboid is found by applying Pythagoras' theorem twice.

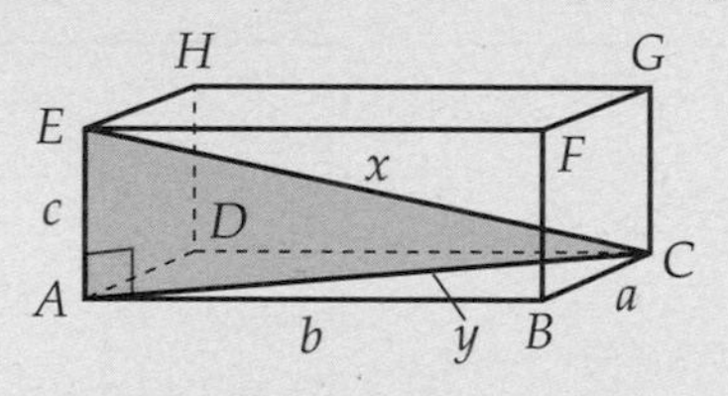

In $\triangle ACE$, $x^2 = c^2 + y^2$, where x is the longest diagonal in the cuboid.

In $\triangle ABC$, $y^2 = a^2 + b^2$, so $x^2 = (a^2 + b^2) + c^2$.

The length, x, of the longest diagonal in a cuboid with dimensions $a \times b \times c$ is given by

$$x^2 = a^2 + b^2 + c^2$$

This is a 3-D version of Pythagoras' theorem.

A

A*

Example 12

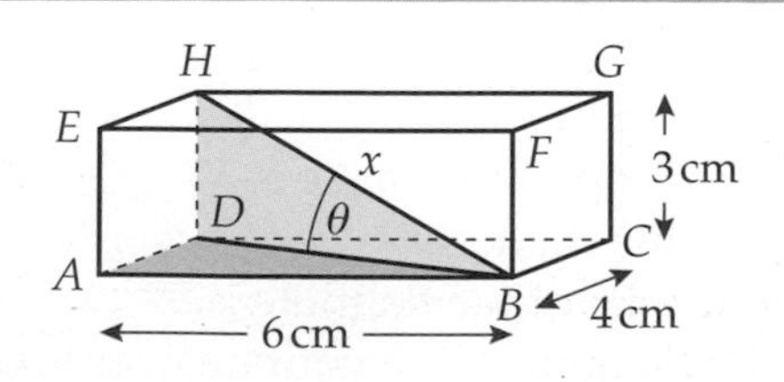

This cuboid has a length of 6 cm, a width of 4 cm and a height of 3 cm.

Calculate

a the length of the longest diagonal HB

b the angle HBD.

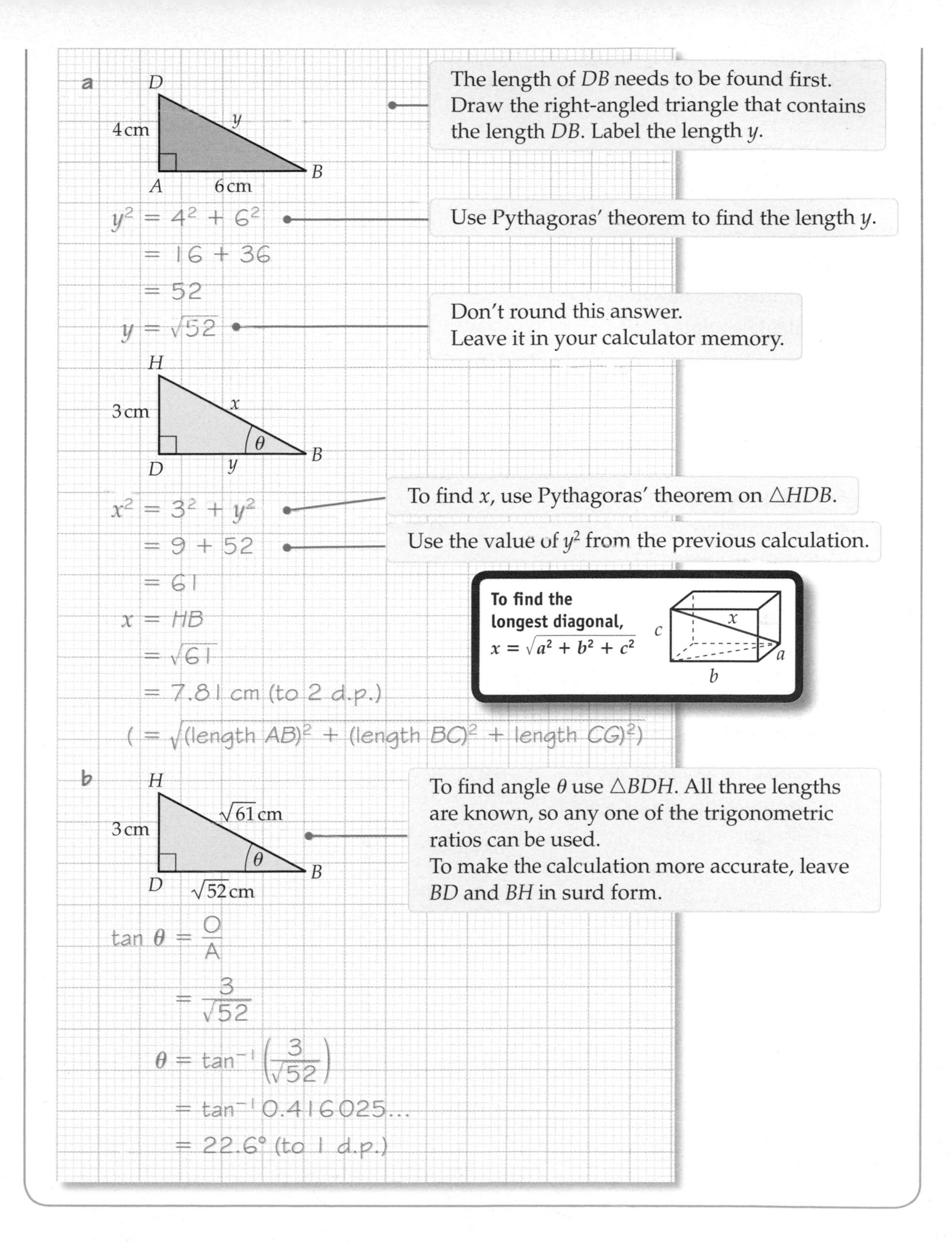

a
D
4 cm
y
A
6 cm
B
The length of DB needs to be found first.
Draw the right-angled triangle that contains the length DB. Label the length y.
y² = 4² + 6²
= 16 + 36
= 52
Use Pythagoras' theorem to find the length y.
y = √52
Don't round this answer.
Leave it in your calculator memory.
H
3 cm
x
θ
D
y
B
x² = 3² + y²
To find x, use Pythagoras' theorem on △HDB.
= 9 + 52
Use the value of y² from the previous calculation.
= 61
x = HB
= √61
= 7.81 cm (to 2 d.p.)
(= √((length AB)² + (length BC)² + length CG)²)
To find the longest diagonal,
x = √(a² + b² + c²)
c
x
a
b
b
H
3 cm
√61 cm
θ
D
√52 cm
B
To find angle θ use △BDH. All three lengths are known, so any one of the trigonometric ratios can be used.
To make the calculation more accurate, leave BD and BH in surd form.
tan θ = O/A
= 3/√52
θ = tan⁻¹(3/√52)
= tan⁻¹ 0.416025...
= 22.6° (to 1 d.p.)

Exercise 33J

1 The diagram shows a cube with side length 6 cm.

Calculate

a the length of DH

b the length of DG.

Give your answers to two decimal places.

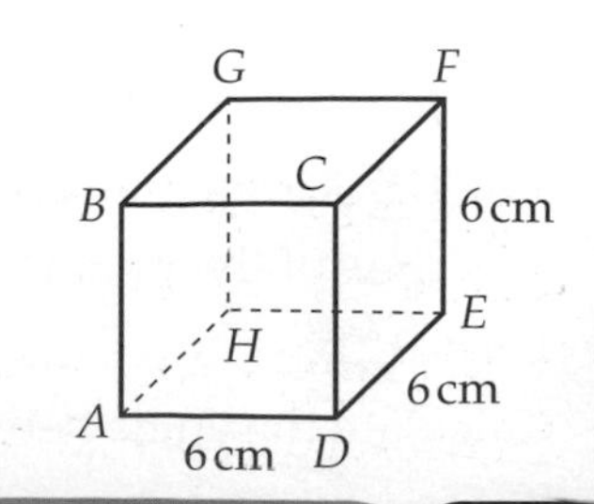

A

2 The diagram shows a cuboid.
$ST = 7$ cm, $TW = 4$ cm and $VW = 3$ cm.

Calculate

a the length of SW

b the length of SV.

Give your answers to two decimal places.

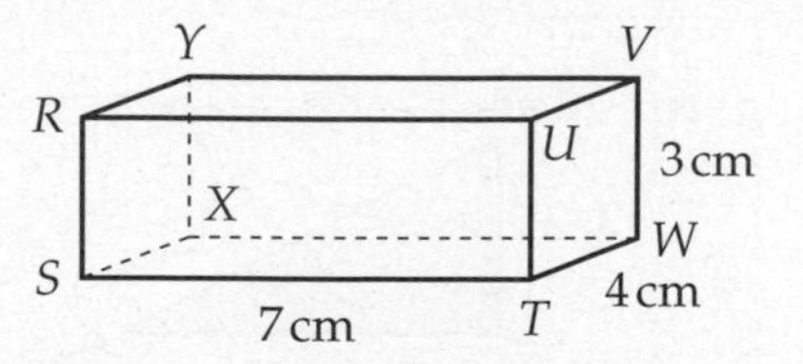

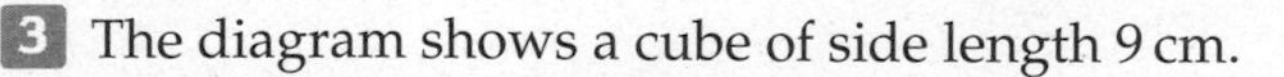

3 The diagram shows a cube of side length 9 cm.

a Calculate the length of NQ.

b Calculate angle QNR.

Give your answers to one decimal place.

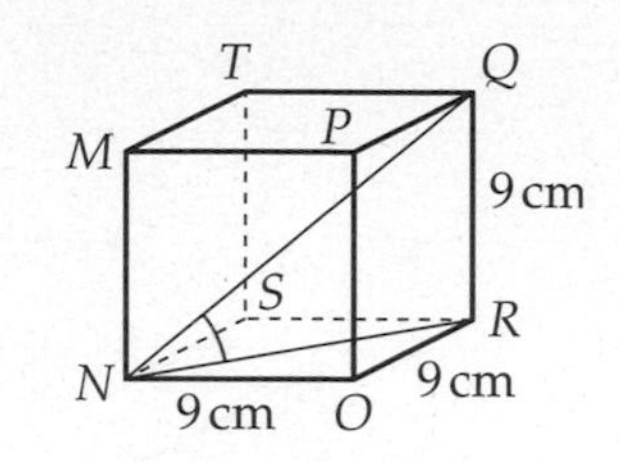

A*

4 The diagram shows a cuboid with side lengths $AB = 9$ cm, $BC = 4$ cm and $CF = 5$ cm.

a Calculate the length of BE.

b Calculate angle EBD.

Give your answers to one decimal place.

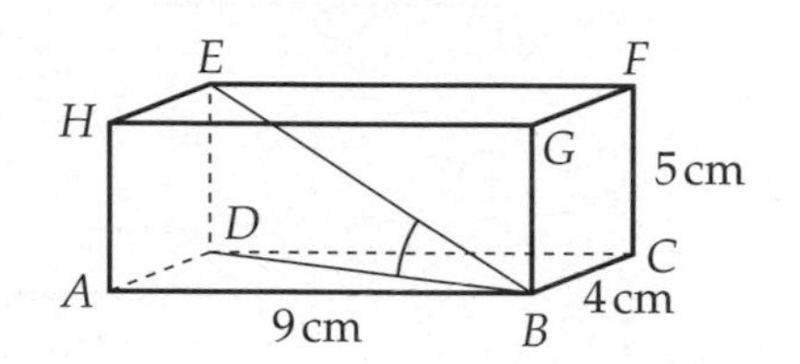

Angle between a line and a plane

In the diagram, S is a point above a horizontal plane $ABCD$. T is the foot of the perpendicular from S to the plane.

The angle between the line AS and the plane $ABCD$ is the angle SAT.

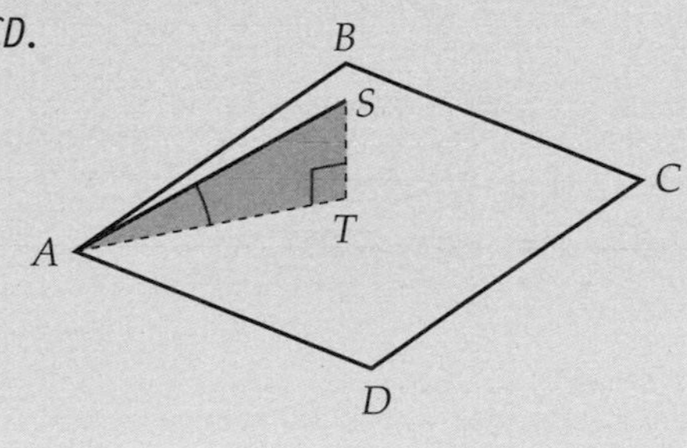

A*

Example 13

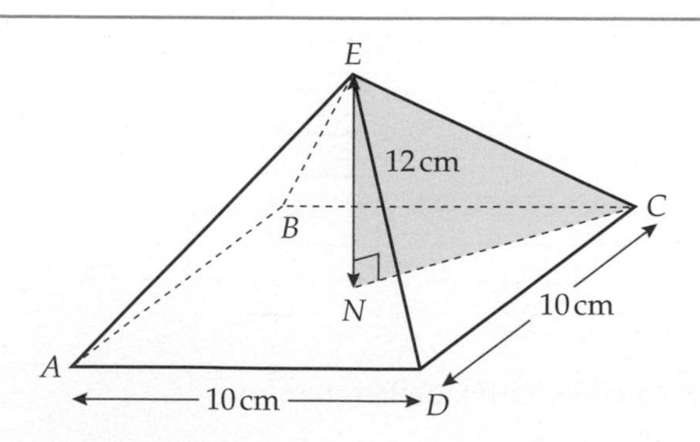

$ABCDE$ is a square-based pyramid with base side length 10 cm.

The apex of the pyramid, E, is vertically above the centre of the base, N.

$EN = 12$ cm.

Calculate the angle between EC and the base $ABCD$.

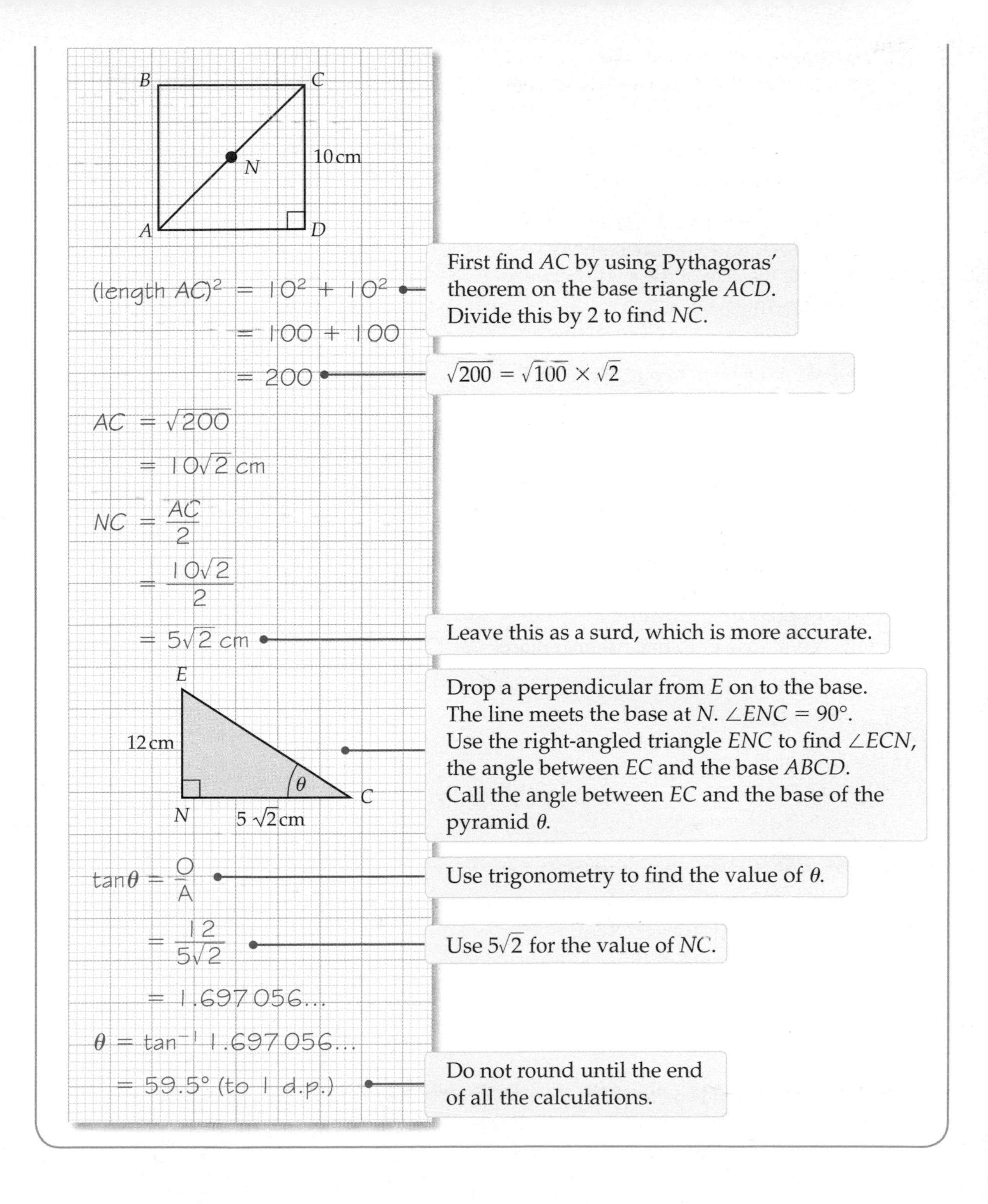

B
C
N
10 cm
A
D
(length AC)² = 10² + 10²
= 100 + 100
= 200
First find AC by using Pythagoras' theorem on the base triangle ACD. Divide this by 2 to find NC.
√200 = √100 × √2
AC = √200
= 10√2 cm
NC = AC/2
= 10√2/2
= 5√2 cm
Leave this as a surd, which is more accurate.
E
12 cm
θ
C
N
5 √2 cm
Drop a perpendicular from E on to the base. The line meets the base at N. ∠ENC = 90°.
Use the right-angled triangle ENC to find ∠ECN, the angle between EC and the base ABCD.
Call the angle between EC and the base of the pyramid θ.
tanθ = O/A
Use trigonometry to find the value of θ.
= 12/5√2
Use 5√2 for the value of NC.
= 1.697056...
θ = tan⁻¹ 1.697056...
= 59.5° (to 1 d.p.)
Do not round until the end of all the calculations.

Exercise 33K

1 The diagram shows a cube of side length 5 cm.

a Calculate the length of AF.
Give your answer to one decimal place.

b Calculate the angle between AF and the base $ABCD$.
Give your answer to one decimal place.

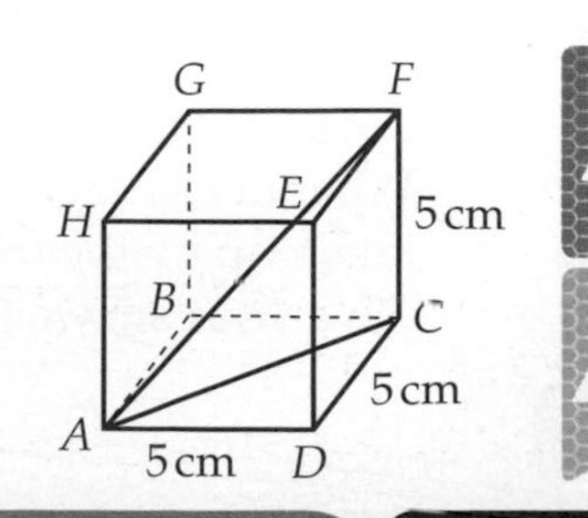

A

A*

2 The diagram shows a cuboid.
$AF = 6$ cm, $AB = 11$ cm and $BC = 7$ cm.
Calculate the angle between BE and the base $ABCD$.
Give your answer to one decimal place.

A*

3 $ABCDEF$ is a triangular prism.
Angles EAD and FBC are both 90°.
Calculate the size of the angle between

a EC and the base $ABCD$

b EC and the plane $ABFE$.

Give your answers to one decimal place.

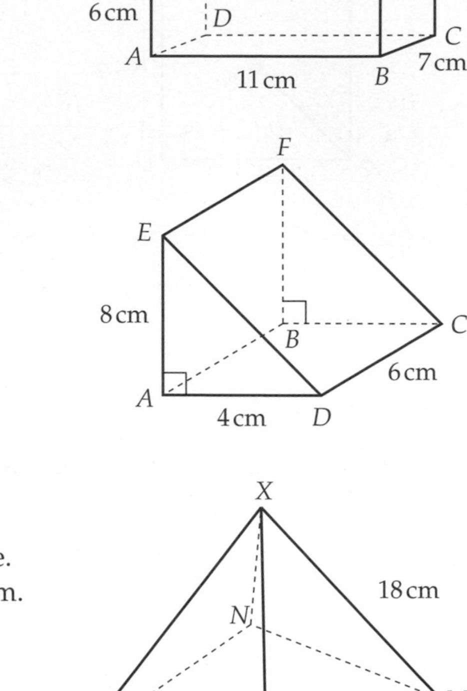

A*

4 $XKLMN$ is a pyramid.
The base $KLMN$ is a rectangle.
X is vertically above the centre of the rectangle.
The slant length $KX = MX = NX = LX = 18$ cm.
Calculate the angle KX makes with the base $KLMN$.
Give your answer to one decimal place.

A02

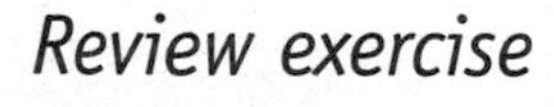

Review exercise

C

1 The diagram shows a right-angled triangle ABC.
$AB = 7.6$ cm and $AC = 12.8$ cm.
Calculate the length of BC.
Give your answer to one decimal place.

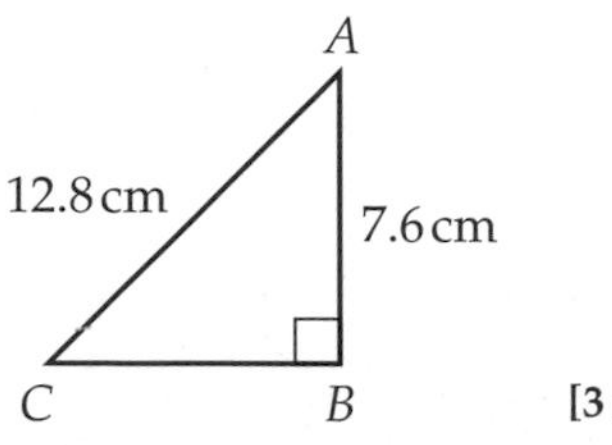

[3 marks]

C

2 The lawn shown is rectangular.
The width of the lawn is 5 metres and the diagonal is 9 metres.
Mark decides to sow grass seed over his lawn.
A 70 gram bag of grass seed covers 20 m^2 of lawn and its price is £1.74.
Calculate how much it will cost Mark to sow grass seed over his lawn.

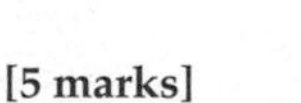
[5 marks]

3 The diagram shows a triangle inside a circle.
$\angle ABC$ is 90°.
What percentage of the area of the circle is taken up by the triangle?
Give your answer to one decimal place.

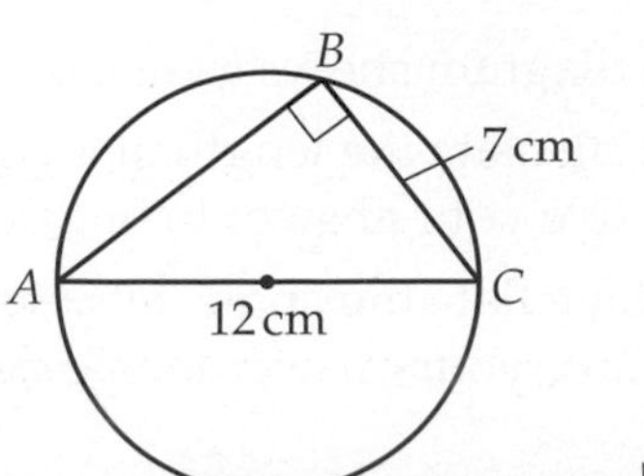

A03

[5 marks]

4 Calculate the length of the line segment RS joining the points $R(2, -6)$ and $S(5, 10)$. Leave your answer in surd form. **[2 marks]**

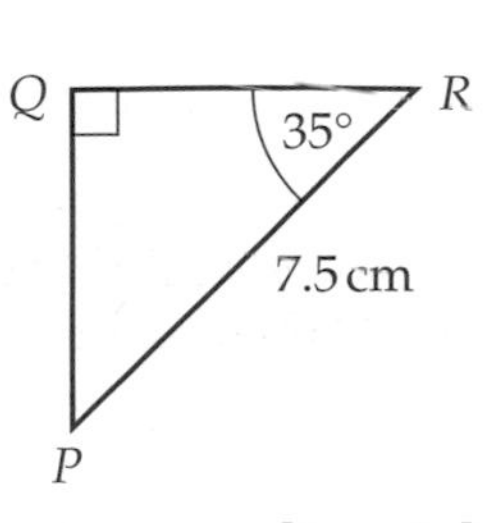

5 In $\triangle PQR$, $\angle PQR = 90°$, $\angle QRP = 35°$ and length $PR = 7.5$ cm.
Calculate the length of PQ.
Give your answer to three significant figures.

[3 marks]

6 In the diagram, $VW = 17$ m, $\angle VWY = 54°$ and $WX = 9.1$ m.
Calculate $\angle WXY$.
Give your answer to three significant figures.

What must you calculate first?

[5 marks]

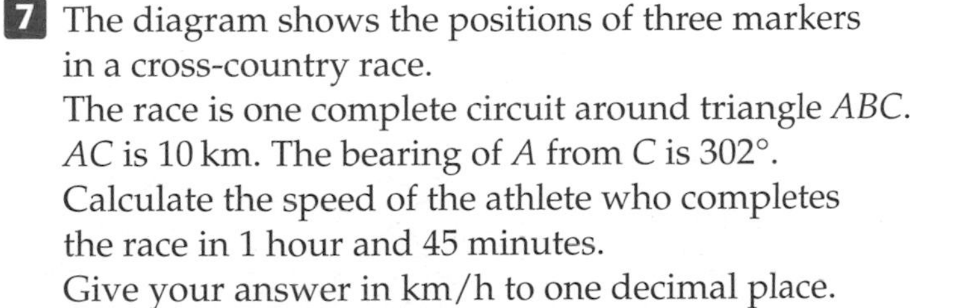

7 The diagram shows the positions of three markers in a cross-country race.
The race is one complete circuit around triangle ABC.
AC is 10 km. The bearing of A from C is 302°.
Calculate the speed of the athlete who completes the race in 1 hour and 45 minutes.
Give your answer in km/h to one decimal place. **[6 marks]**

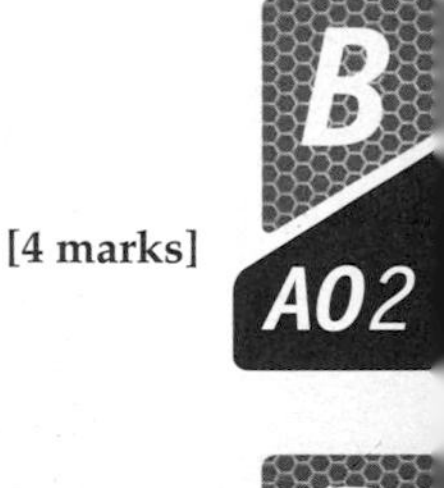

8 A rocket flies 7 km vertically, then 10 km at an angle of 16° to the vertical, then 30 km at 28° to the vertical.
Calculate its vertical height at the end of the third stage.
Give your answer to three significant figures. **[4 marks]**

9 From the top of a sea cliff 40 m above the water, the angles of depression of two boats, at P and Q, at sea are 24° and 9° respectively.
Calculate the distance between the two boats.
Give your answer to three significant figures. **[5 marks]**

10 The curved surface of this cone is 45π cm^2.
The radius is 6 cm.
What is the volume of the cone?
Give your answer to one decimal place.

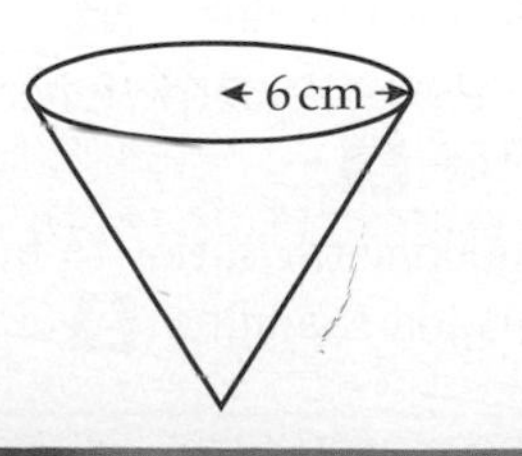

[4 marks]

A

11 *ABCDEFGH* is a cuboid with sides of 6 cm, 8 cm and 12 cm.

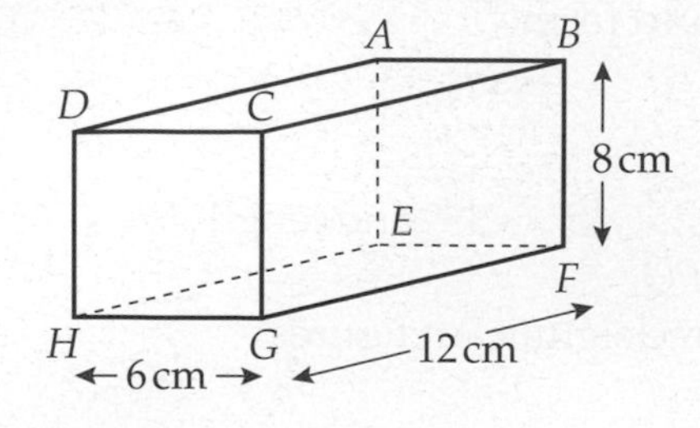

A*

Calculate

a the length of *BH* **[3 marks]**

b the angle *BHF*. **[3 marks]**

Give your answers to one decimal place.

A*

12 *VABCD* is a square-based pyramid.
$VA = VB = VC = VD = 20$ cm.
$AB = BC = CD = AD = 12$ cm.
Calculate the angle between *VC* and the base *ABCD*.
Give your answer to one decimal place.

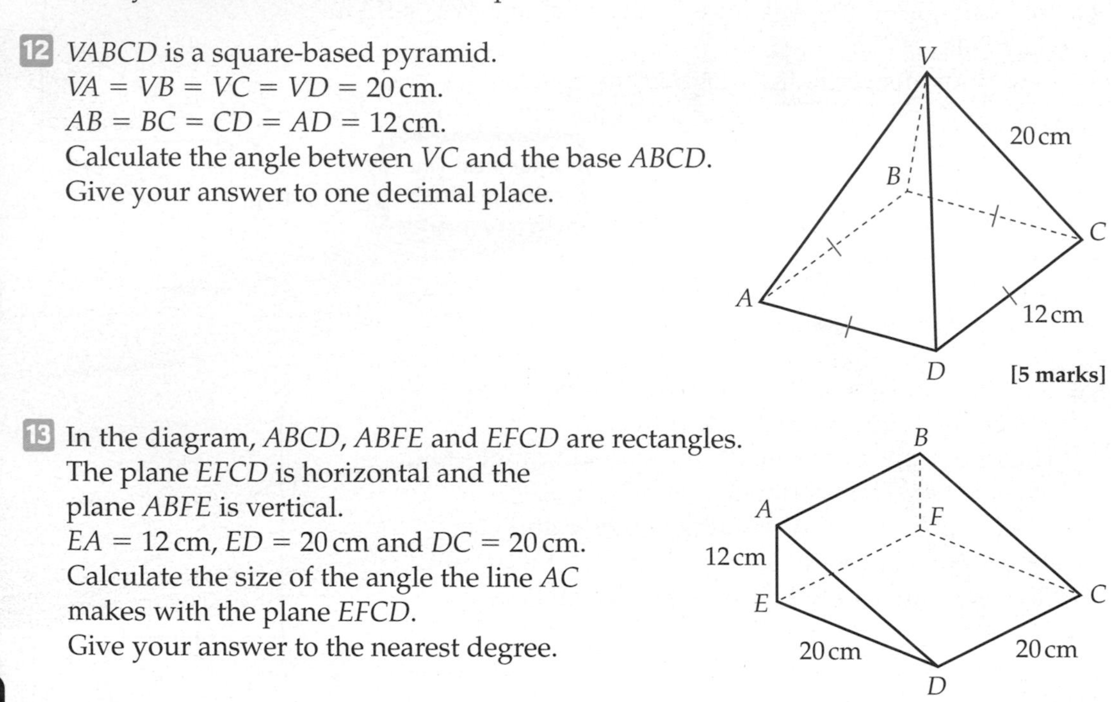

[5 marks]

13 In the diagram, *ABCD*, *ABFE* and *EFCD* are rectangles.
The plane *EFCD* is horizontal and the plane *ABFE* is vertical.
$EA = 12$ cm, $ED = 20$ cm and $DC = 20$ cm.
Calculate the size of the angle the line *AC* makes with the plane *EFCD*.
Give your answer to the nearest degree.

[4 marks]

AO2

Chapter summary

In this chapter you have learned how to

- calculate the length of a missing side in a right-angled triangle using Pythagoras' theorem C
- solve problems using Pythagoras' theorem C
- calculate the length of a line segment *AB* C
- understand and recall trigonometric ratios in right-angled triangles B
- enter the trigonometric functions on a calculator B
- use trigonometric ratios to find lengths in right-angled triangles B
- use trigonometric ratios to find the angles in right-angled triangles B
- solve problems using an angle of elevation or an angle of depression B
- use trigonometric ratios and Pythagoras to solve problems C B A
- use Pythagoras' theorem in 3-D A
- use trigonometric ratios in 3-D contexts A A*
- find the angle between a line and a plane A*

34

Circle theorems

This chapter is about circle theorems and their applications.

Circle theorems could be used to predict the path of a rollerblader on a semi-circular ramp.

Objectives

This chapter will show you how to

- use the fact that the tangent at any point on a circle is perpendicular to the radius at that point **B**
- use the fact that tangents from an external point are equal in length **B**
- explain why the perpendicular from the centre of a circle to a chord bisects the chord **B**
- prove that the angle subtended by an arc at the centre of a circle is twice the angle subtended at any point on the circumference **B** **A**
- prove that the angle in a semicircle is 90° **B** **A**
- prove that angles in the same segment are equal **B** **A**
- prove that opposite angles of a cyclic quadrilateral add up to 180° **B** **A**
- prove and use the alternate segment theorem **A** **A***

Before you start this chapter

1 Use Pythagoras' theorem to calculate each missing length.

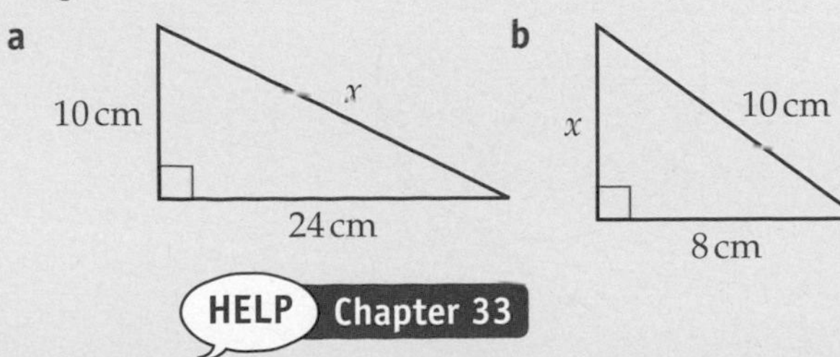

HELP Chapter 33

2 Calculate the sizes of the angles marked with letters.

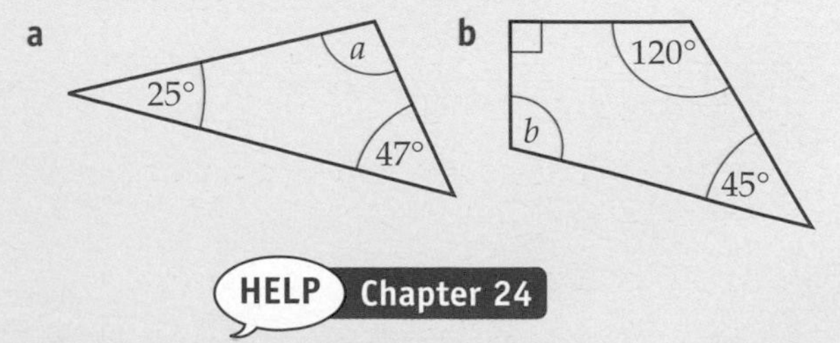

HELP Chapter 24

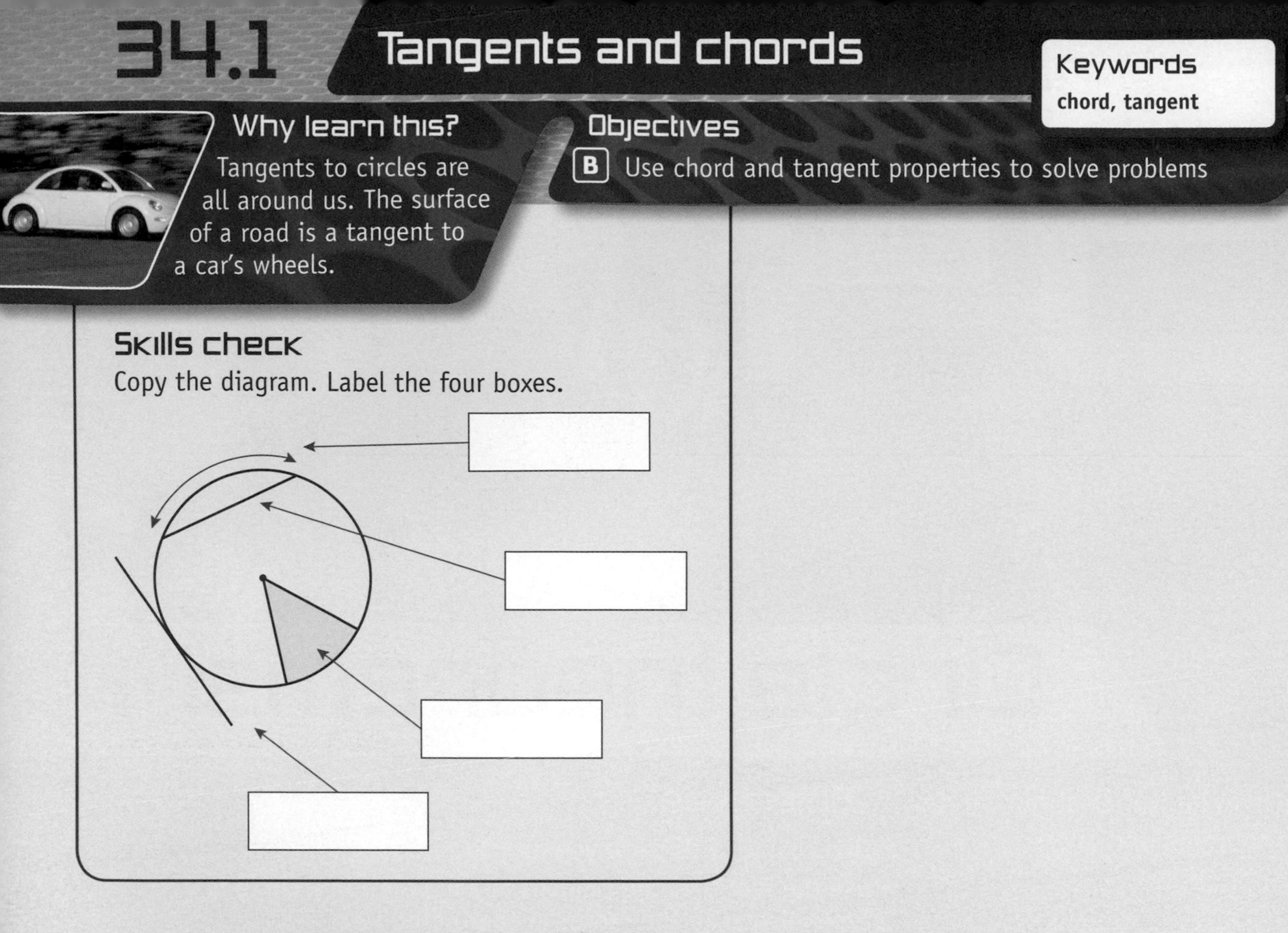

34.1 Tangents and chords

Keywords
chord, tangent

Why learn this?

Tangents to circles are all around us. The surface of a road is a tangent to a car's wheels.

Objectives

B Use chord and tangent properties to solve problems

Skills check

Copy the diagram. Label the four boxes.

Circle properties

The perpendicular from the centre of a circle to a **chord** bisects the chord.

$AM = BM$

Tangents drawn to a circle from an external point are equal in length.

$TA = TB$

The angle between a tangent and a radius is a right angle.

$\angle OAT = 90°$

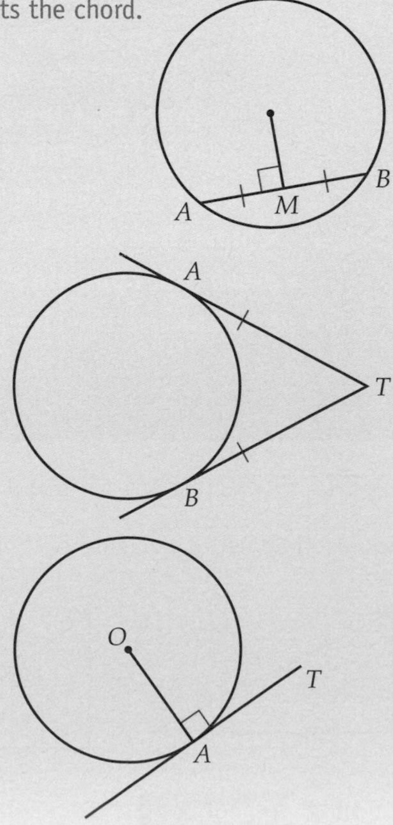

Example 1

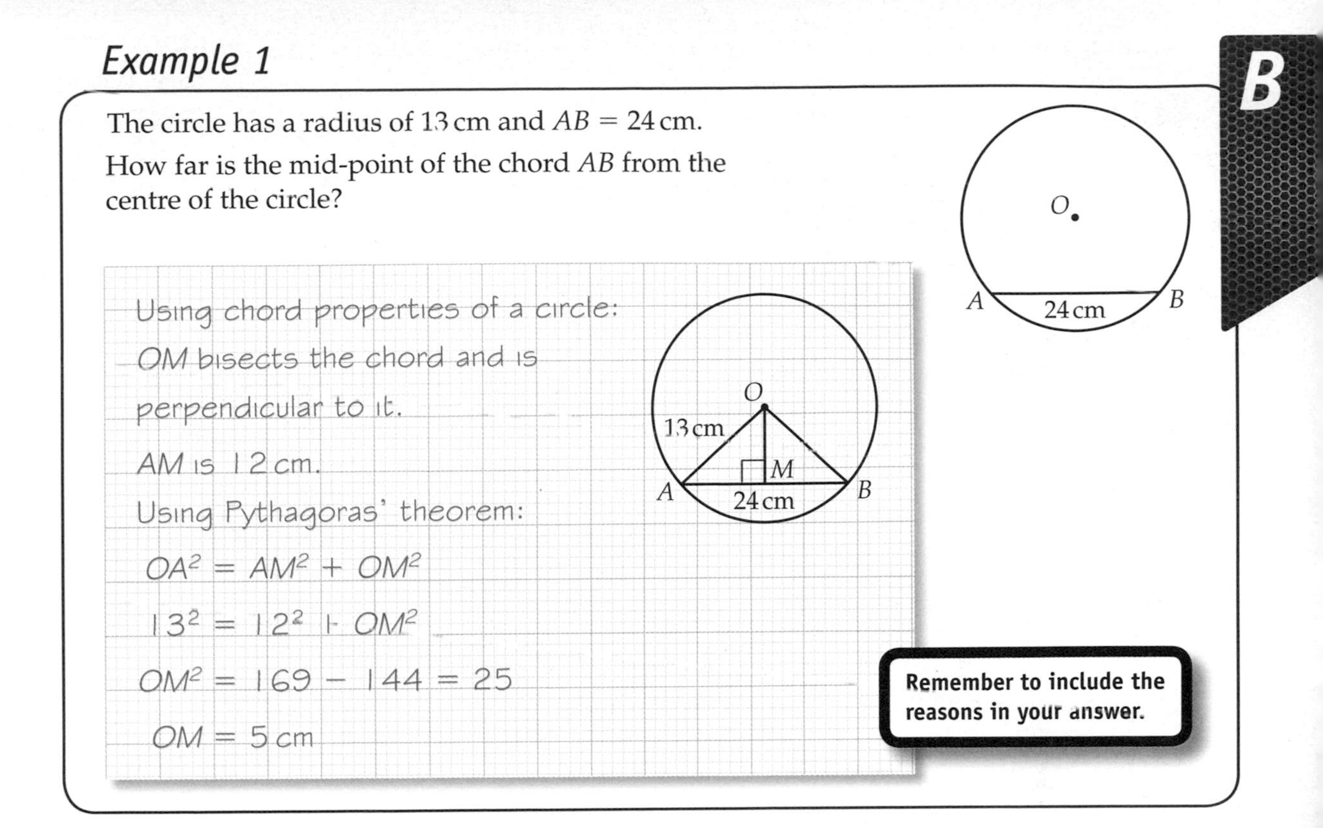

The circle has a radius of 13 cm and $AB = 24$ cm.
How far is the mid-point of the chord AB from the centre of the circle?

Using chord properties of a circle:
OM bisects the chord and is perpendicular to it.
AM is 12 cm.
Using Pythagoras' theorem:

$OA^2 = AM^2 + OM^2$

$13^2 = 12^2 + OM^2$

$OM^2 = 169 - 144 = 25$

$OM = 5$ cm

Remember to include the reasons in your answer.

Example 2

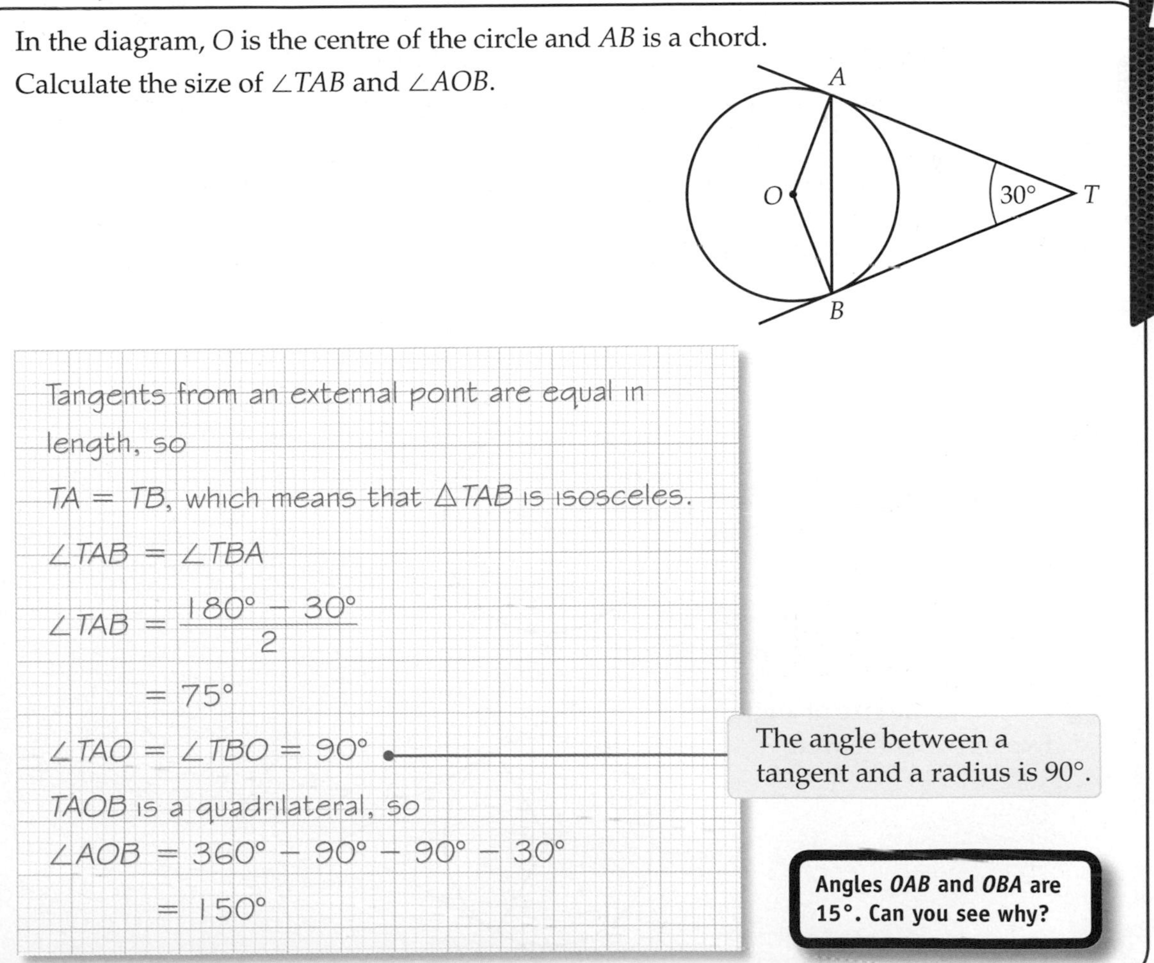

In the diagram, O is the centre of the circle and AB is a chord.
Calculate the size of $\angle TAB$ and $\angle AOB$.

Tangents from an external point are equal in length, so
$TA = TB$, which means that $\triangle TAB$ is isosceles.

$\angle TAB = \angle TBA$

$\angle TAB = \dfrac{180° - 30°}{2}$

$= 75°$

$\angle TAO = \angle TBO = 90°$

The angle between a tangent and a radius is 90°.

$TAOB$ is a quadrilateral, so

$\angle AOB = 360° - 90° - 90° - 30°$

$= 150°$

Angles *OAB* and *OBA* are 15°. Can you see why?

Exercise 34A

B

1 The length of chord AB is 8 cm.
The mid-point of the chord is 3 cm from the centre of the circle.
Calculate the length of the radius of the circle.
State any circle properties you use.

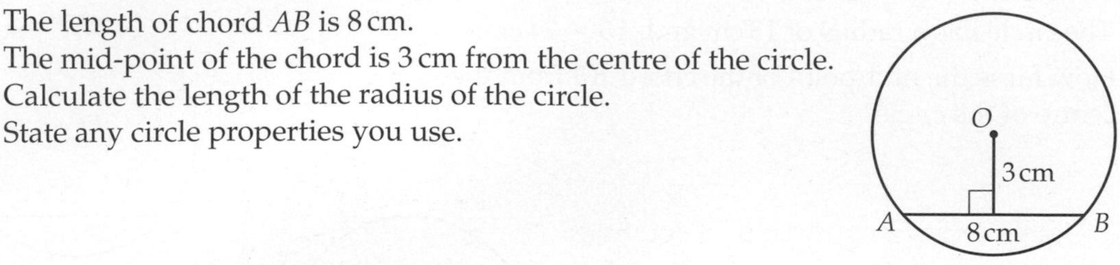

2 The radius of the circle is 10 cm. $OM = 6$ cm.
Calculate the length of the chord PQ.

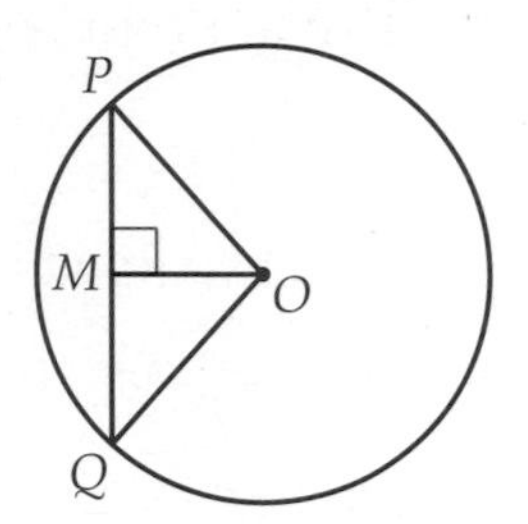

B

A02

3 A circle with centre O has a radius of 7 cm and chords PQ and RS.
$PQ = 4.8$ cm and $RS = 9.8$ cm.

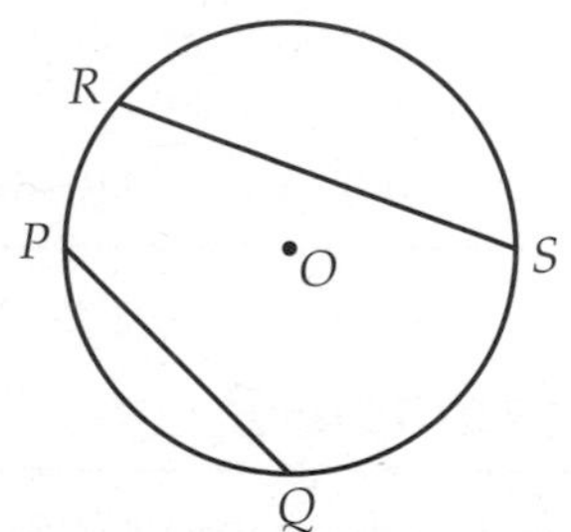

a Write down the length of the lines OR and OS and add them to a copy of the diagram.

b M is the mid-point of the chord RS.
Use Pythagoras' theorem to calculate the length of OM.

c N is the mid-point of the chord PQ. Calculate the length ON.

d Which chord is closer to the centre of the circle?

4 PQ and RS are two chords in a circle of radius 6 cm.
$PQ = 10.4$ cm and $RS = 6.6$ cm.
Which chord is closer to the centre of the circle?
Explain your answer.

5 Calculate the size of angle x.

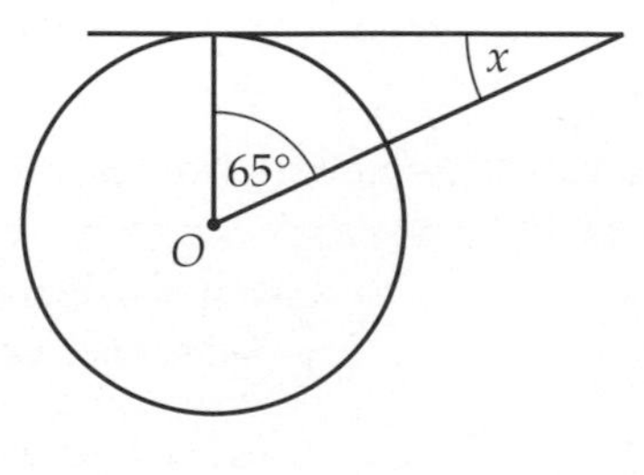

B

A02

6 Calculate the size of angle y.
Explain your answer.

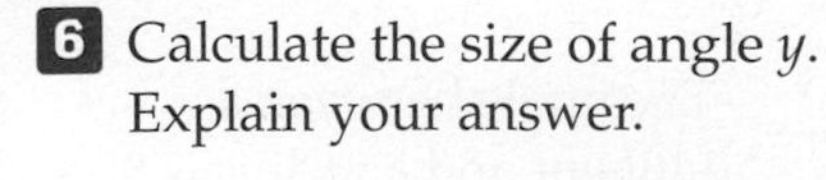

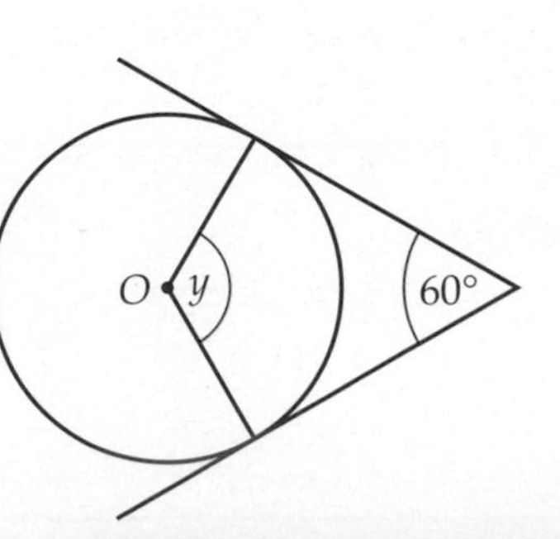

B

AO2

7 Calculate the sizes of angles a and b.
Explain your answer.

8 Calculate the sizes of angles a and b.
Explain your answer.

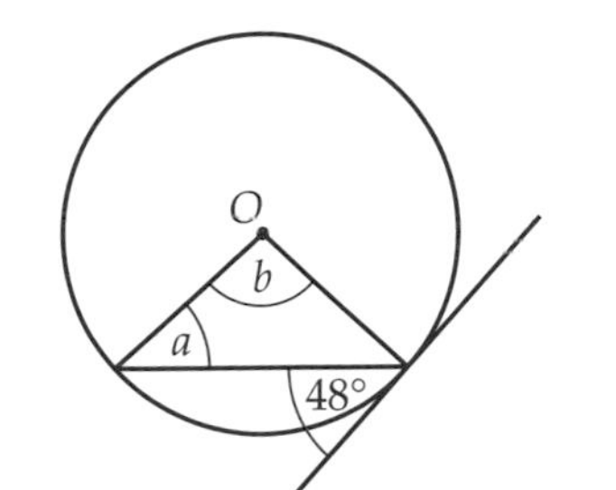

B

AO3

9 OA is the radius of a circle with diameter 10 cm.
TA is a tangent to the circle and has a length of 12 cm.
Calculate the distance of the point T from the centre of the circle.

34.2 Circle theorems

Keywords
arc, subtend, cyclic quadrilateral, supplementary

Why learn this?
Circle theorems can be used to calculate the distance to the horizon.

Objectives

B **A** Use circle theorems and circle properties to solve geometrical problems

A Recall proofs of circle theorems

Skills check

1 Explain the meaning of each word. **a** circumference **b** arc

Theorem 1

The angle subtended by an **arc** at the centre of a circle is twice the angle that it **subtends** at the circumference.

'Subtended' means that the angle starts and finishes at the ends of the arc.

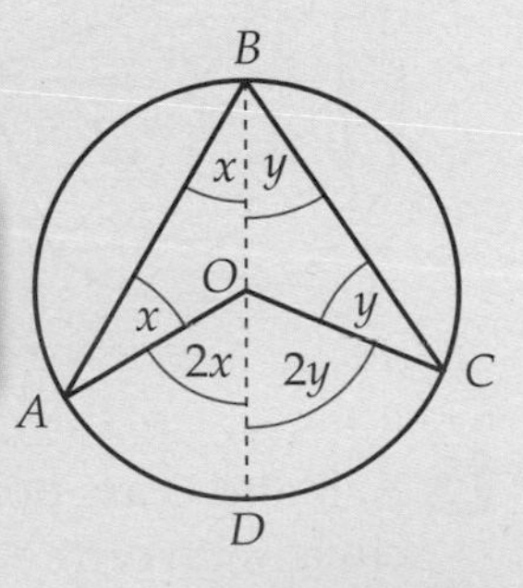

Proof

Let $\angle ABO = x$ and $\angle CBO = y$.

The base angles in an isosceles triangle are equal, so $\angle BAO = x$ and $\angle OCB = y$.

The exterior angle of a triangle is the sum of opposite interior angles, so $\angle AOD = 2x$ and $\angle COD = 2y$.

$\angle ABC = x + y$ and $\angle AOC = 2x + 2y$
$= 2(x + y)$
$= 2 \times \angle ABC$

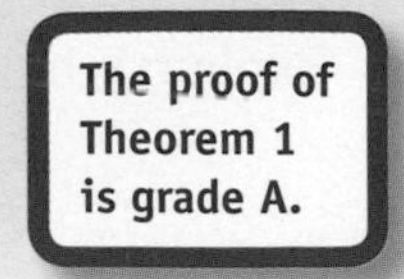

B

Example 3

Calculate the size of angle x.

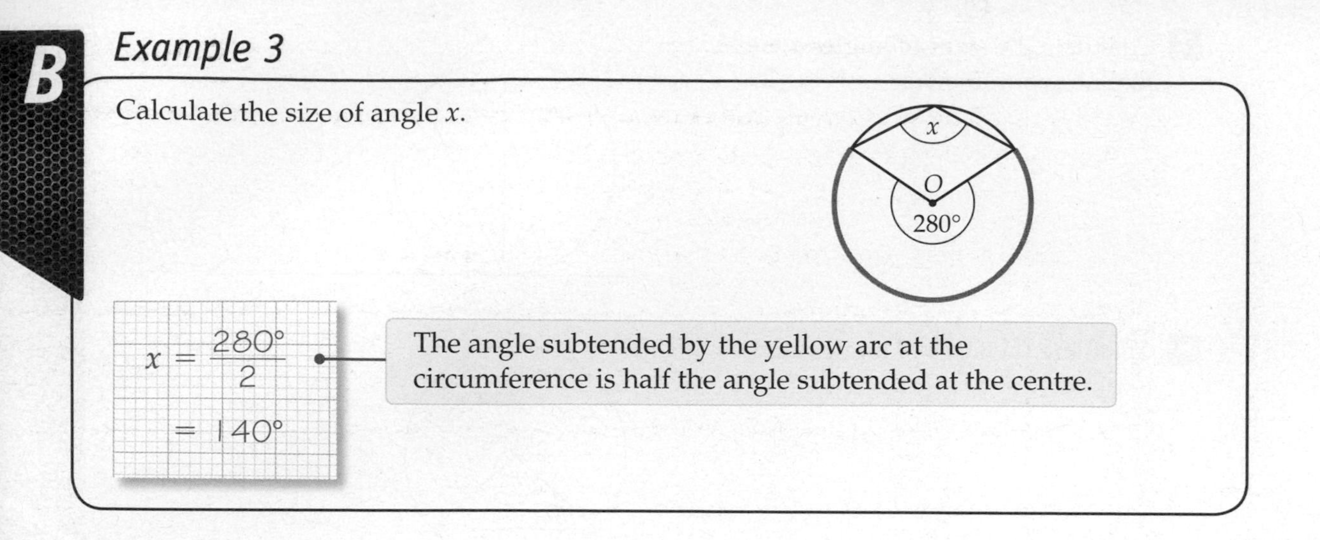

$$x = \frac{280°}{2}$$

$$= 140°$$

The angle subtended by the yellow arc at the circumference is half the angle subtended at the centre.

Exercise 34B

B

For Q1 to Q8, calculate the sizes of the angles marked with letters.
Explain each step in your reasoning.

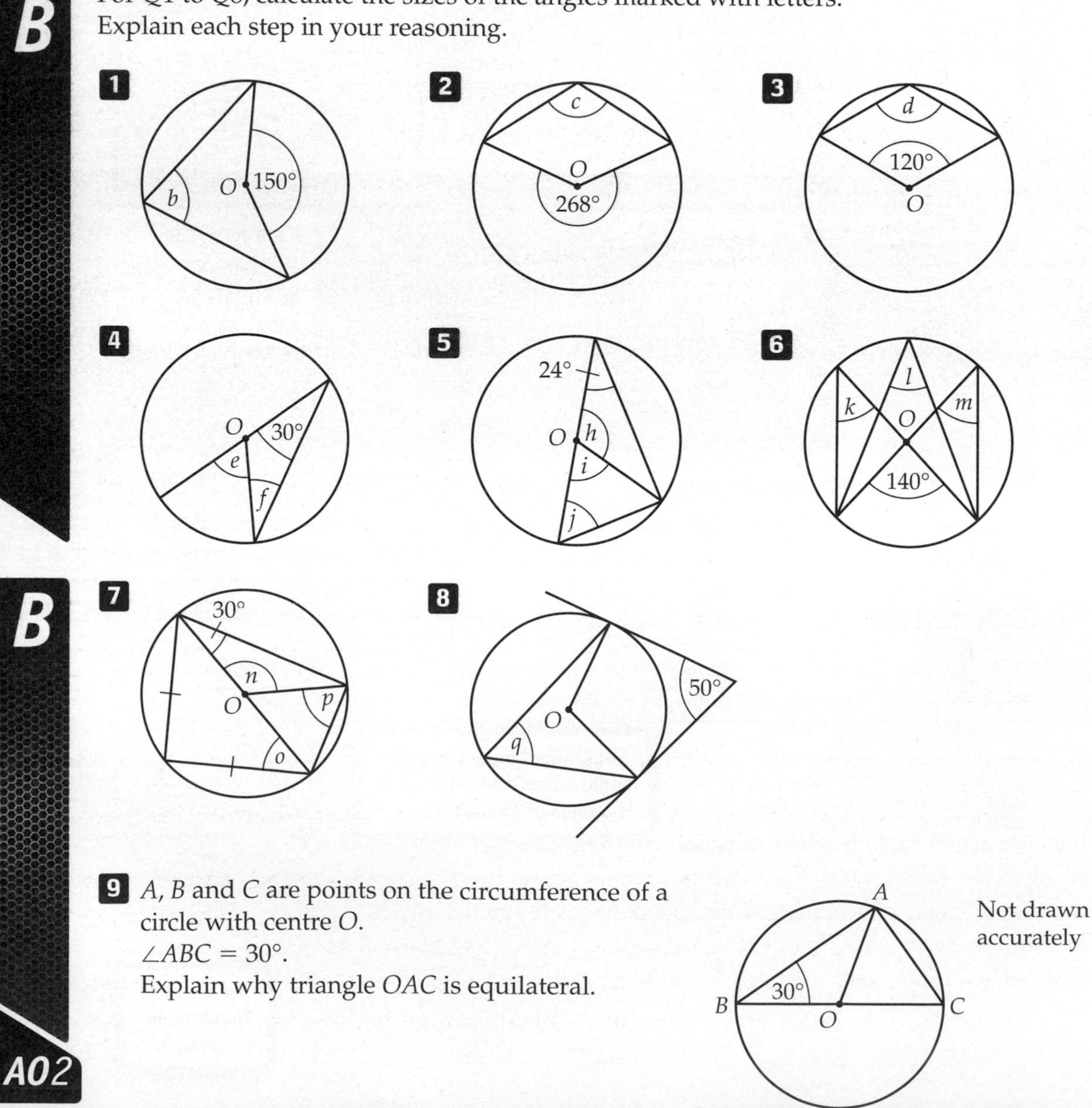

9 A, B and C are points on the circumference of a circle with centre O.
$\angle ABC = 30°$.
Explain why triangle OAC is equilateral.

AO2

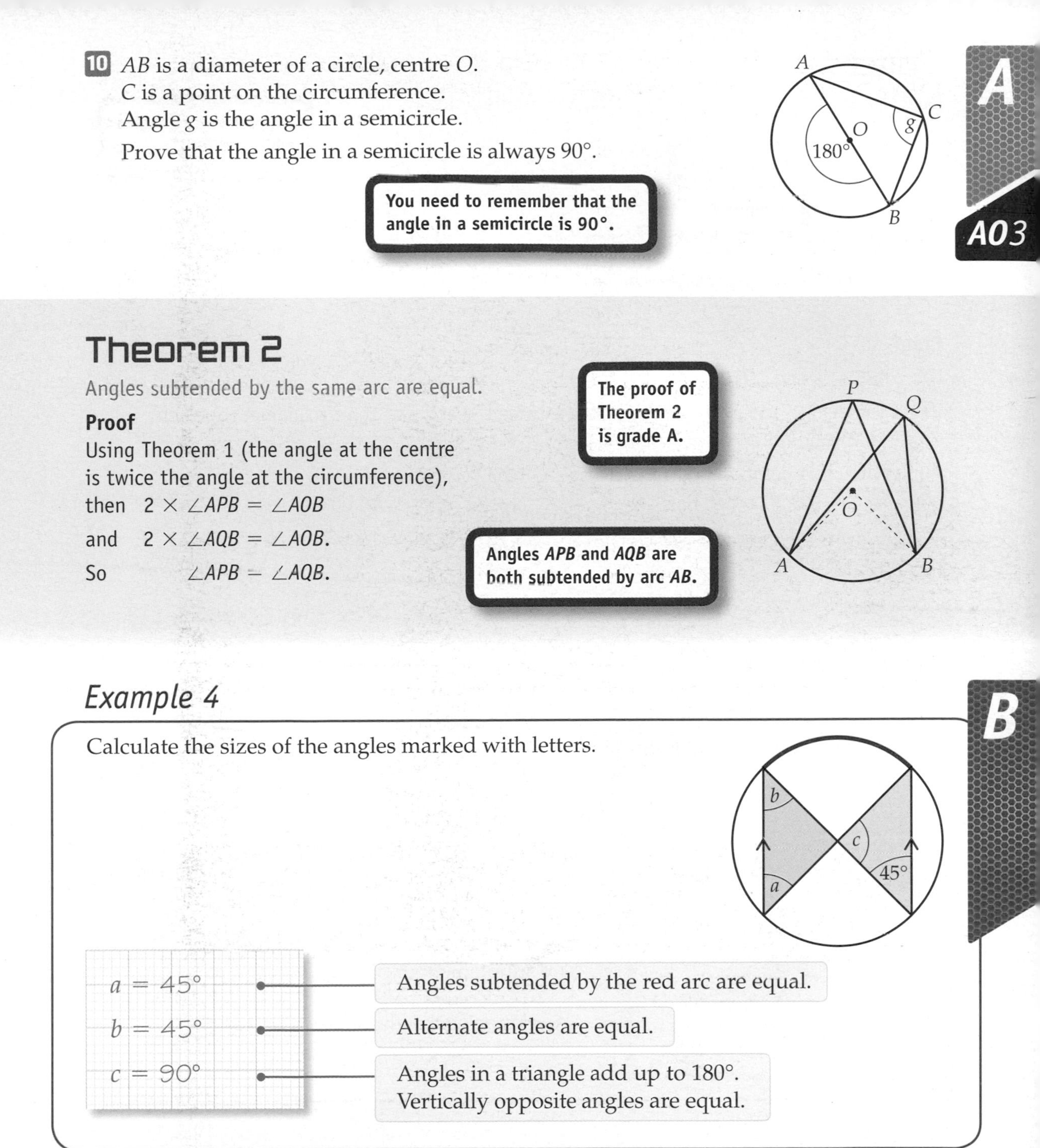

10 AB is a diameter of a circle, centre O.
C is a point on the circumference.
Angle g is the angle in a semicircle.

Prove that the angle in a semicircle is always 90°.

You need to remember that the angle in a semicircle is 90°.

Theorem 2

Angles subtended by the same arc are equal.

Proof

Using Theorem 1 (the angle at the centre is twice the angle at the circumference),

then $2 \times \angle APB = \angle AOB$

and $2 \times \angle AQB = \angle AOB$.

So $\angle APB = \angle AQB$.

The proof of Theorem 2 is grade A.

Angles *APB* and *AQB* are both subtended by arc *AB*.

Example 4

Calculate the sizes of the angles marked with letters.

$a = 45°$ — Angles subtended by the red arc are equal.

$b = 45°$ — Alternate angles are equal.

$c = 90°$ — Angles in a triangle add up to 180°. Vertically opposite angles are equal.

Cyclic quadrilaterals

A **cyclic quadrilateral** is a quadrilateral whose vertices lie on the circumference of a circle.

ABCD is a cyclic quadrilateral.

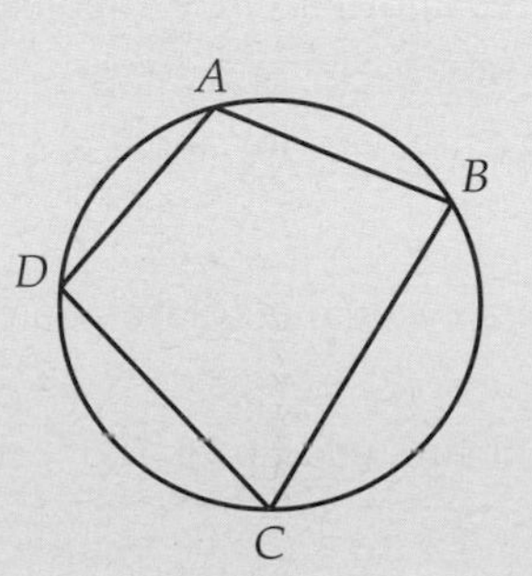

Theorem 3

Opposite angles in a cyclic quadrilateral are **supplementary**.

Supplementary angles add up to 180°.

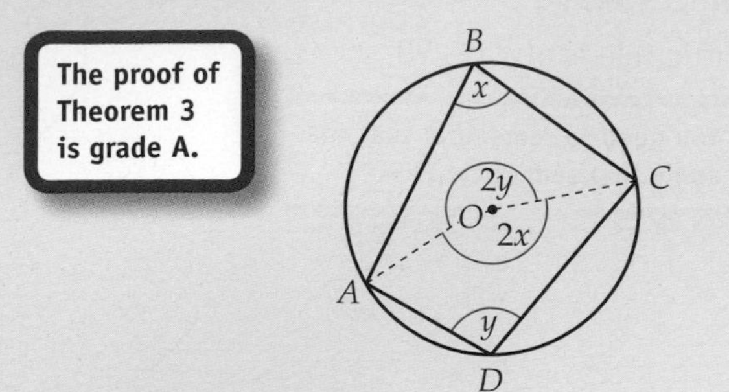

Proof

Let $\angle ABC = x$ and $\angle CDA = y$.

Using Theorem 1 (the angle at the centre is twice the angle at the circumference), then $\angle AOC = 2x$ and reflex $\angle AOC = 2y$.

Angles around a point add up to 360°, so $2x + 2y = 360°$

$$2(x + y) = 360°$$

$$x + y = 180°.$$

B

Example 5

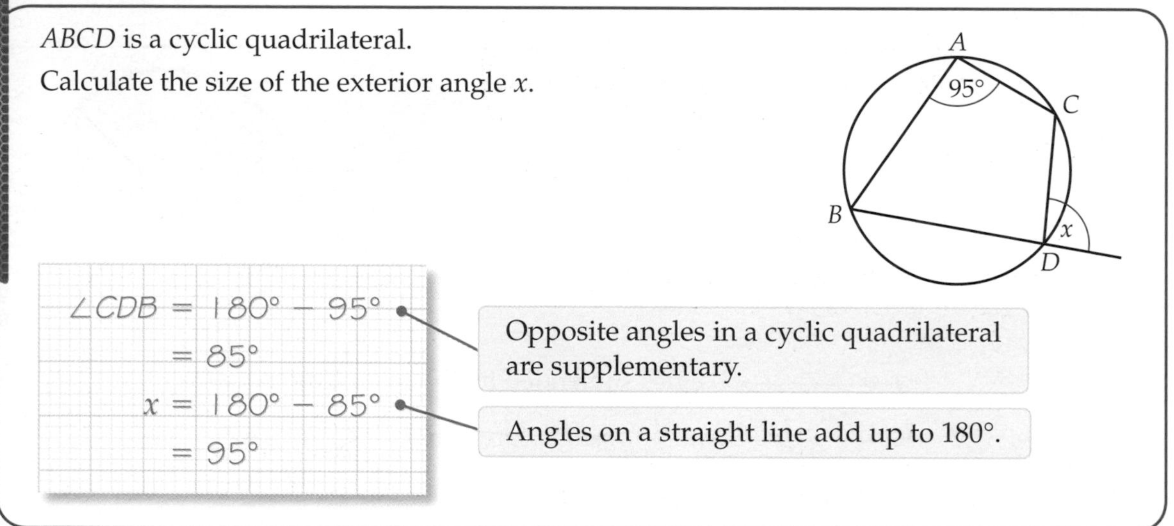

ABCD is a cyclic quadrilateral.

Calculate the size of the exterior angle x.

$\angle CDB = 180° - 95°$
$= 85°$
$x = 180° - 85°$
$= 95°$

Opposite angles in a cyclic quadrilateral are supplementary.

Angles on a straight line add up to 180°.

Theorem 4

Example 5 illustrates another circle theorem.

The exterior angle of a cyclic quadrilateral is equal to the opposite interior angle.

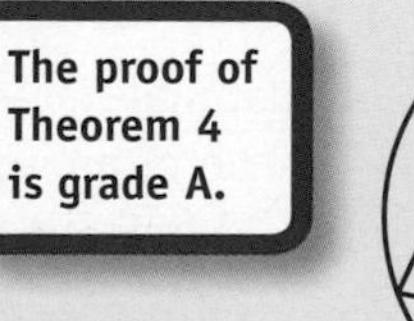

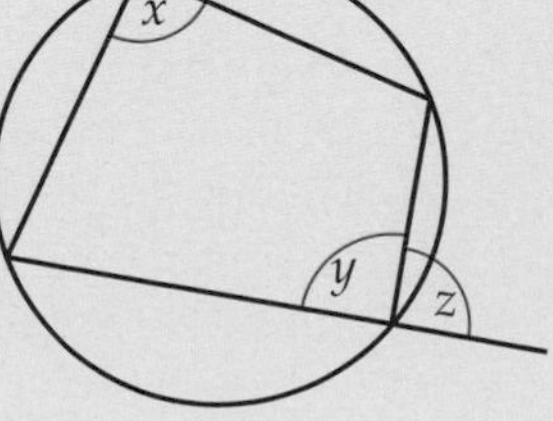

Proof

Opposite angles in a cyclic quadrilateral are supplementary, so $x + y = 180°$.

Angles on a straight line add up to 180°, so $y + z = 180°$.

Therefore $x = z$.

Exercise 34C

For Q1 to Q8, calculate the sizes of the angles marked with letters.
Explain each step in your reasoning.

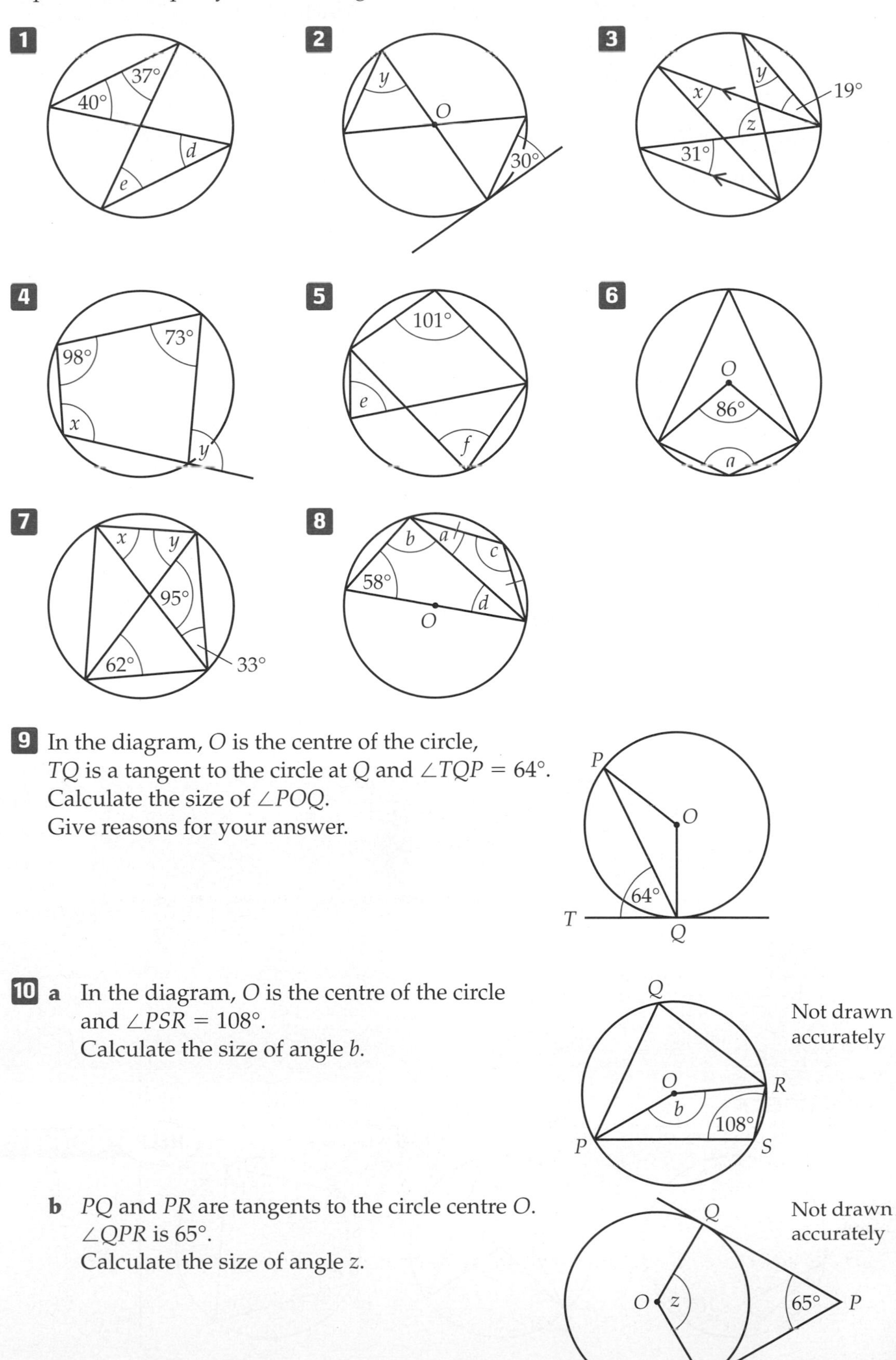

9 In the diagram, O is the centre of the circle, TQ is a tangent to the circle at Q and $\angle TQP = 64°$.
Calculate the size of $\angle POQ$.
Give reasons for your answer.

10 **a** In the diagram, O is the centre of the circle and $\angle PSR = 108°$.
Calculate the size of angle b.

b PQ and PR are tangents to the circle centre O.
$\angle QPR$ is 65°.
Calculate the size of angle z.

B

11 A, B, C and D are four points on the circumference of a circle.
AC meets BD at X. $\angle ABD = 54°$ and $\angle CXD = 85°$.

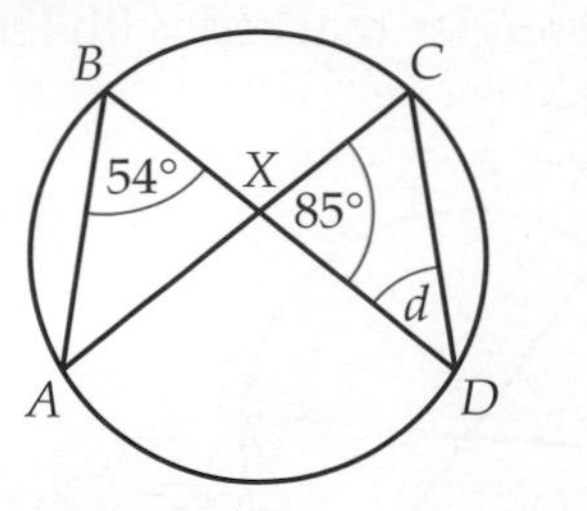

Not drawn accurately

AO2

Calculate the size of angle d.
Give reasons for your answer.

A

12

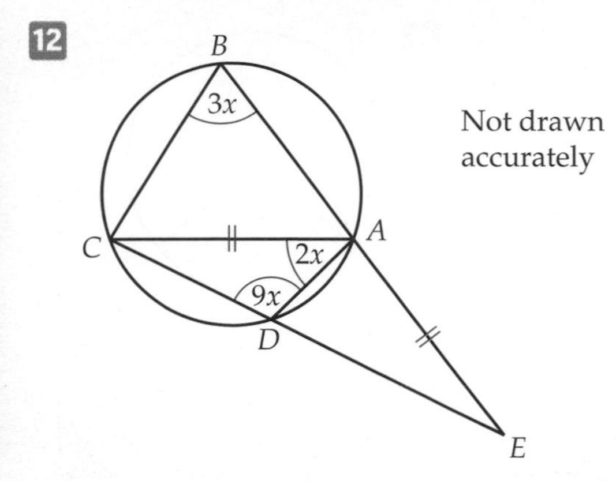

Not drawn accurately

AO3

The diagram shows a cyclic quadrilateral $ABCD$.
The straight lines BA and CD are extended to meet at E.
$EA = AC$
Calculate the size of $\angle EAD$.

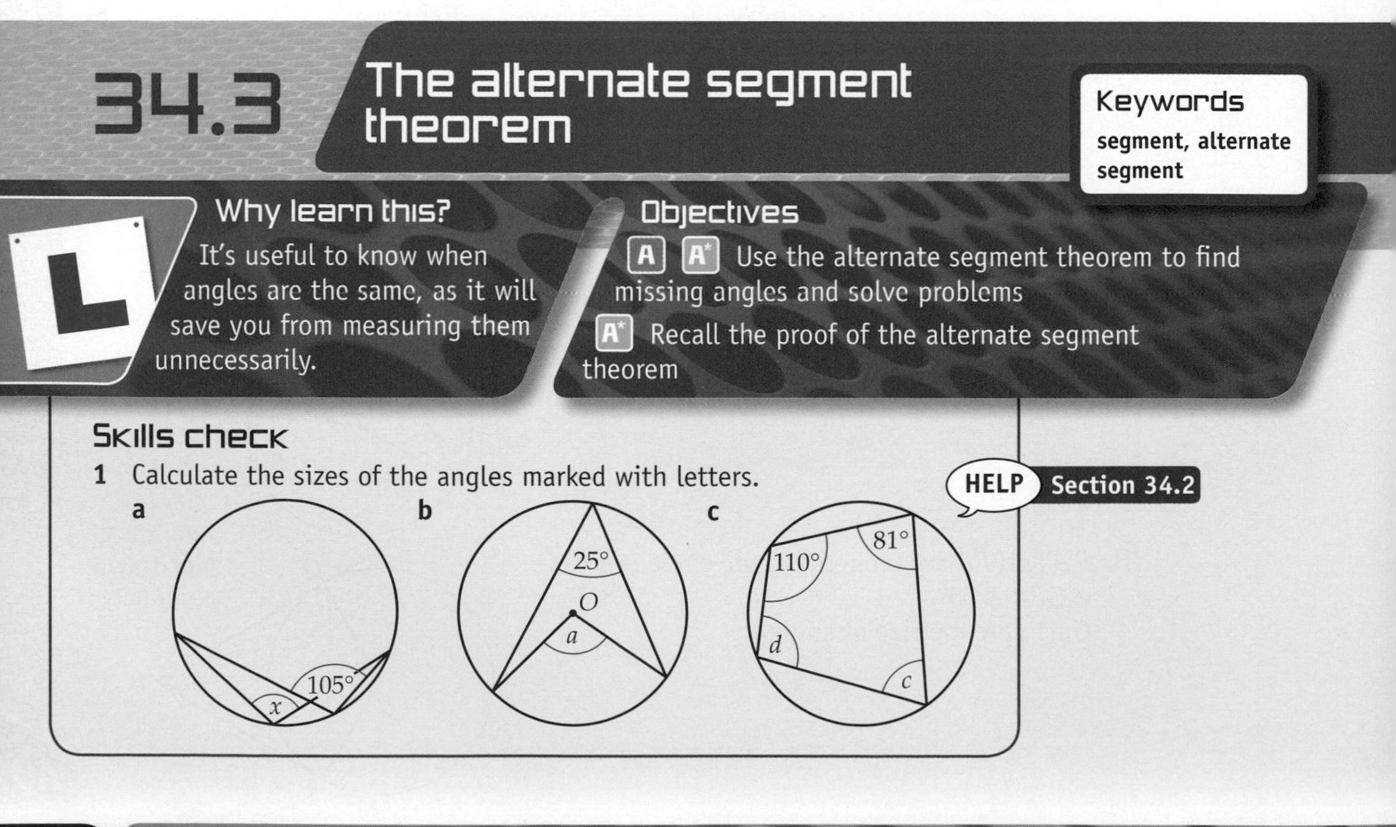

34.3 The alternate segment theorem

Keywords
segment, alternate segment

Why learn this?

It's useful to know when angles are the same, as it will save you from measuring them unnecessarily.

Objectives

A A* Use the alternate segment theorem to find missing angles and solve problems

A* Recall the proof of the alternate segment theorem

Skills check

1 Calculate the sizes of the angles marked with letters.

HELP Section 34.2

The alternate segment theorem

PAT is a tangent to the circle. $\angle TAB$ is marked between the tangent and the chord. The **segment** subtended by the chord *AB* and containing the $\angle ACB$ is known as the **alternate segment.**

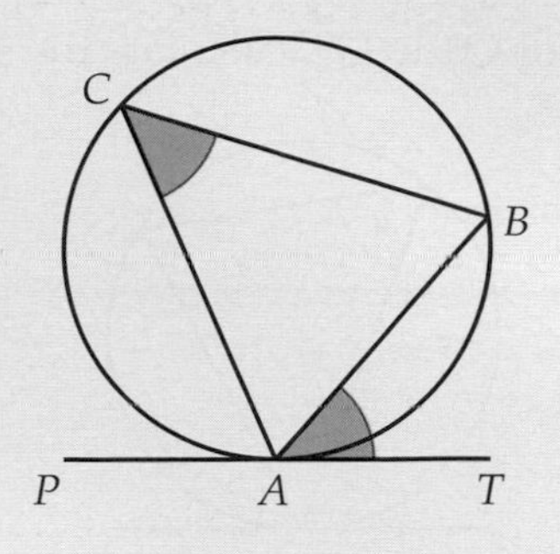

Theorem 5

The angle between a tangent and a chord is equal to the angle in the alternate segment.

The diagrams below show other pairs of angles that are equal using the alternate segment theorem.

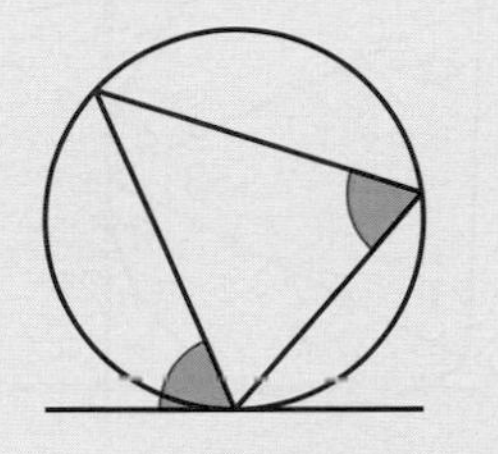

Proof

The angle between a tangent and a radius is 90°, so if $\angle TAB = x$, then $\angle OAB = 90° - x$.

Base angles in an isosceles triangle are equal, so $\angle OBA = 90° - x$.

$$\angle AOB = 180° - (90° - x) - (90° - x)$$
$$= 2x$$

Using Theorem 1 (the angle at the centre is twice the angle at the circumference), then

$$\angle APB = x$$
$$= \angle TAB$$

so $\angle TAB = \angle APB$.

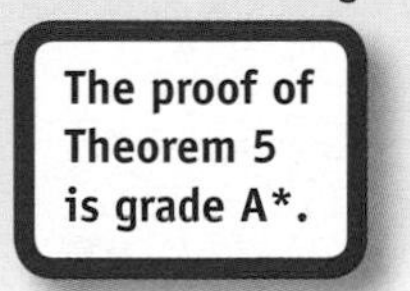

Example 6

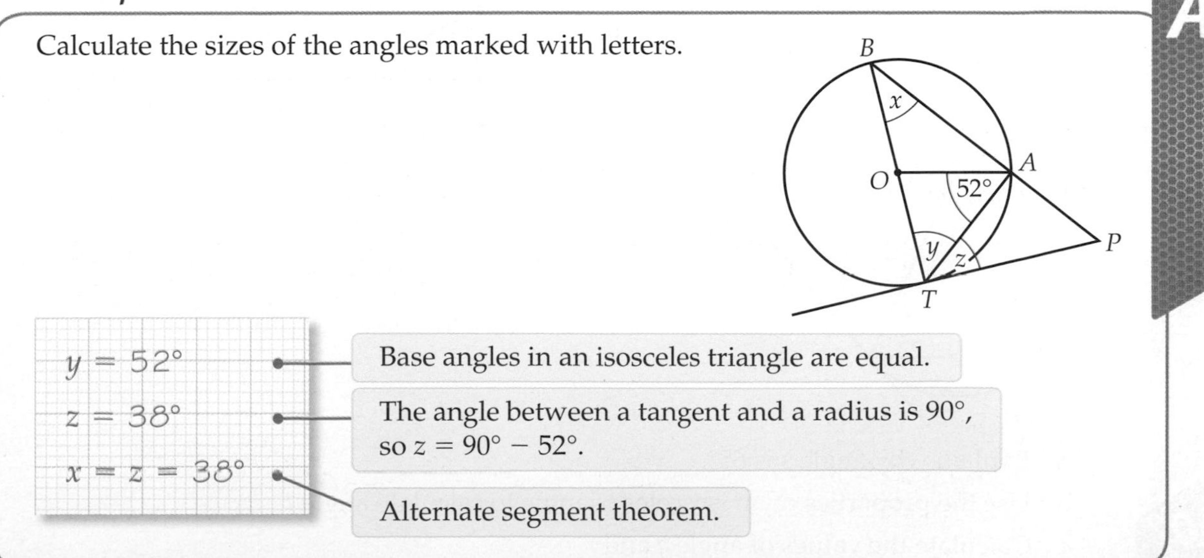

Calculate the sizes of the angles marked with letters.

$y = 52°$ — Base angles in an isosceles triangle are equal.

$z = 38°$ — The angle between a tangent and a radius is 90°, so $z = 90° - 52°$.

$x = z = 38°$ — Alternate segment theorem.

Exercise 34D

For Q1 to Q9, calculate the sizes of the angles marked with letters. Explain your answers.

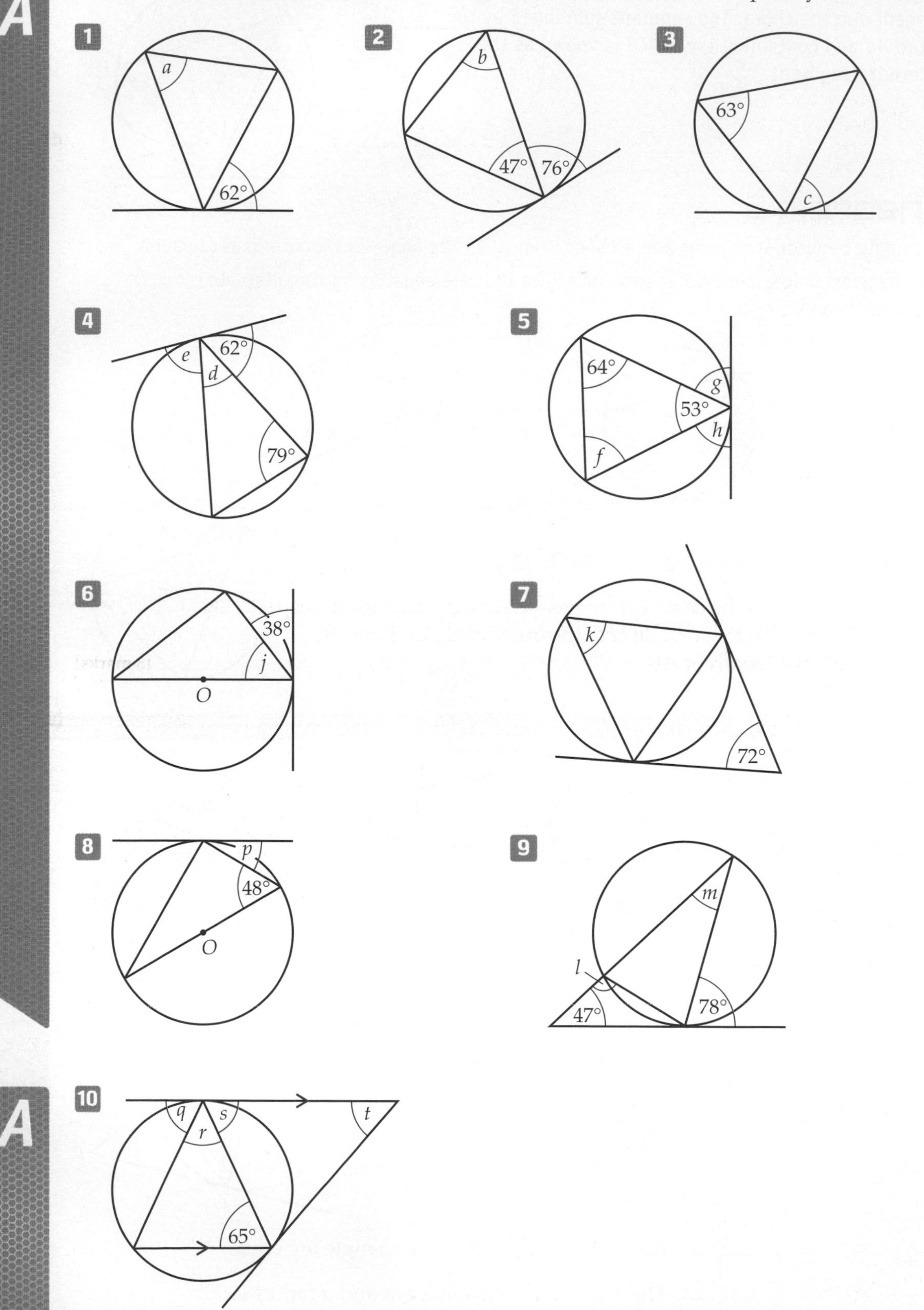

a Explain why angle s is 65°.

b Use the properties of an isosceles triangle to calculate angle t.

c Calculate the values of angle q and r.

AO2

11 Calculate the sizes of the angles marked with letters. Explain your answers.

a

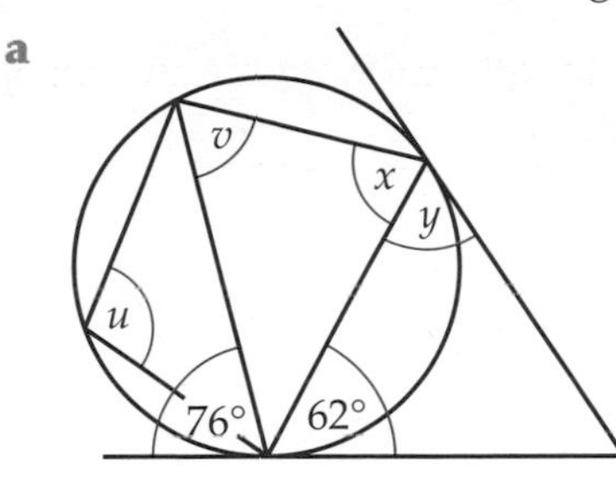

b

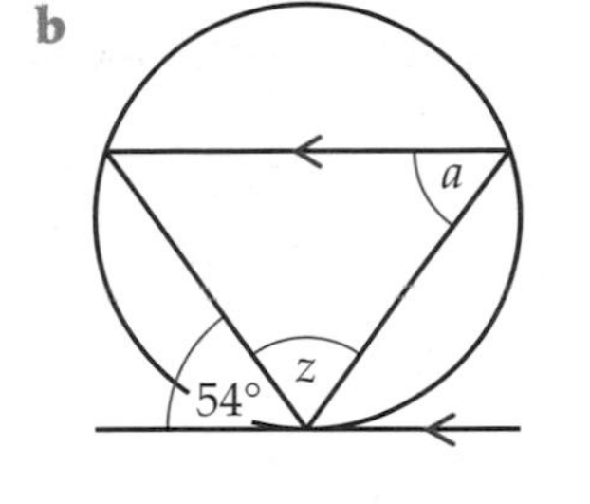

c

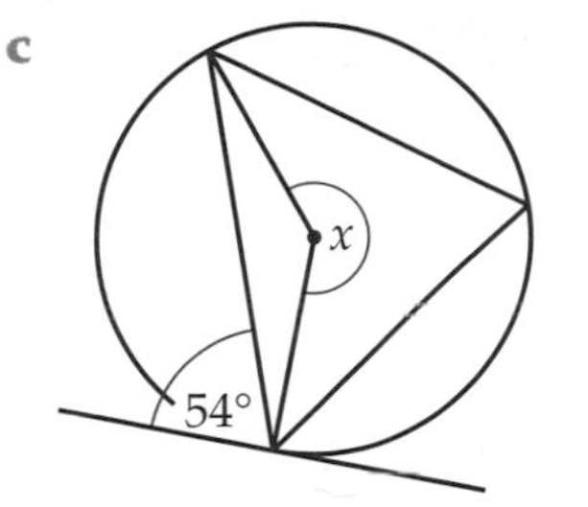

A

AO2

12 The line BT bisects $\angle CBA$.

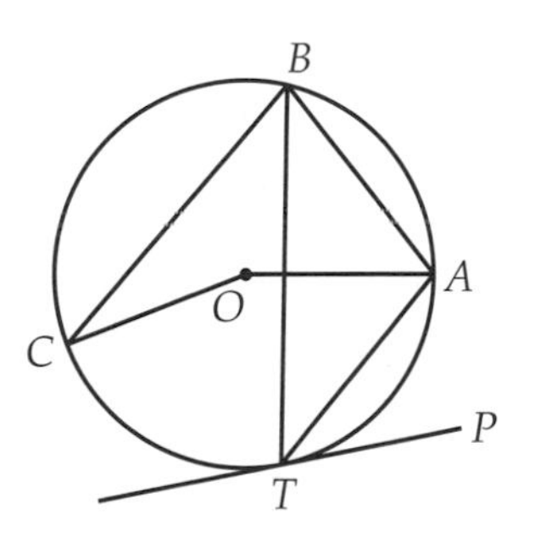

Prove that $\angle ATP$ is a quarter the size of $\angle COA$.

A*

AO3

Review exercise

1 Two circles with the same centre have radii 6 cm and 10 cm respectively.
A tangent to the inner circle cuts the outer circle at A and B.
Calculate the length of AB. **[3 marks]**

AO2

2 Calculate the sizes of the angles marked with letters.

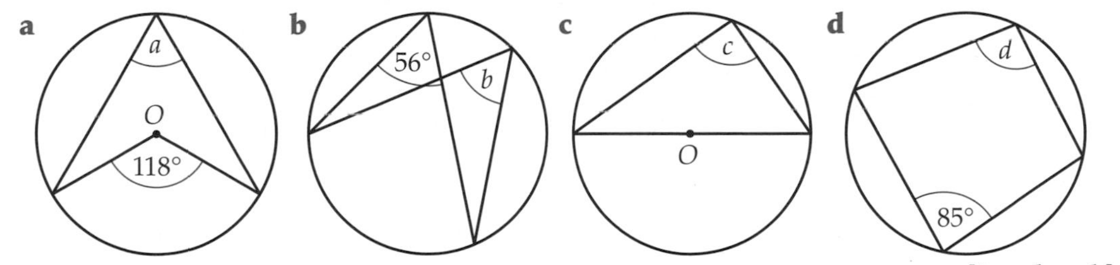

[1 mark each]

B

3 A, B, C and D are four points on the circumference of a circle.
The lines AB and DC are extended to meet at point E.
$\angle CBE = 67°$ and $\angle BEC = 35°$.

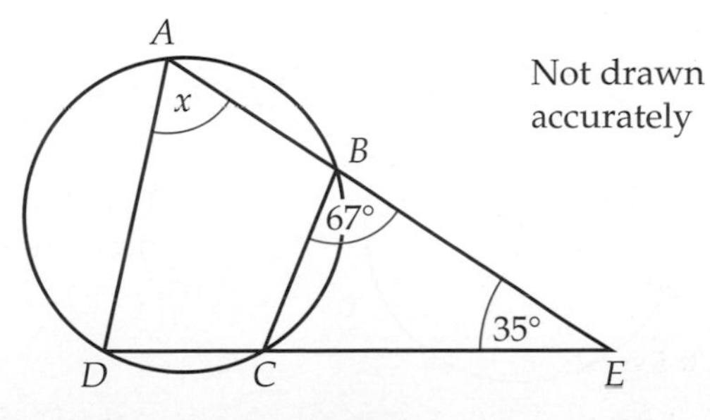

B

a What is the name of quadrilateral $ABCD$? **[1 mark]**

b Calculate the size of angle x. Show all your working. **[3 marks]**

A

4 P, Q and R are points on the circumference of the circle.
NT is the tangent to the circle at P.

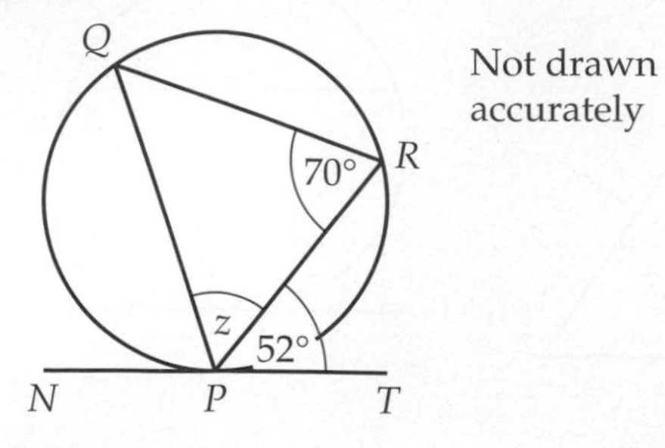

Calculate the value of angle z.
Give reasons for your answer. **[3 marks]**

A

5 **a** A, B, C and D are points on the circumference of a circle centre O.
$\angle AOC = 130°$.

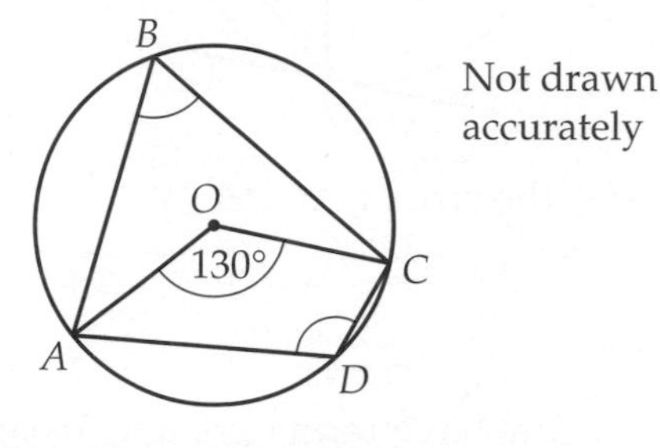

Calculate the size of $\angle ABC$ and $\angle ADC$. **[2 marks]**

b A, B and C are three points on the circumference of a circle centre O.
SCT is a tangent to the circle.
$\angle SCA = 56°$ and $\angle COB = 130°$.

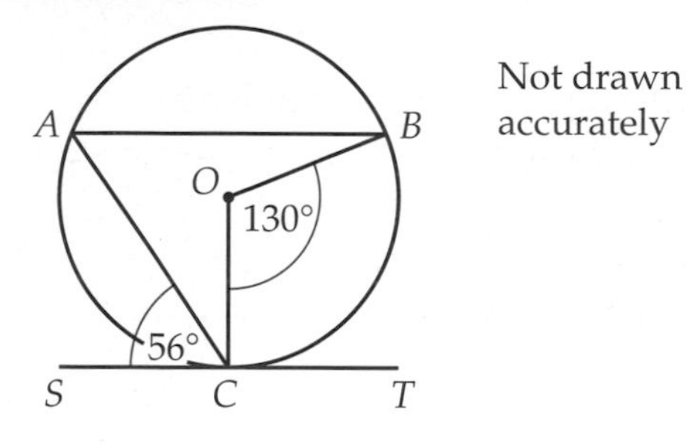

Calculate the size of $\angle OBA$. **[3 marks]**

6 In the diagram, O is the centre of the circle and AB is the diameter of the circle.
TP is a tangent to the circle at the point P.
ABT is a straight line.

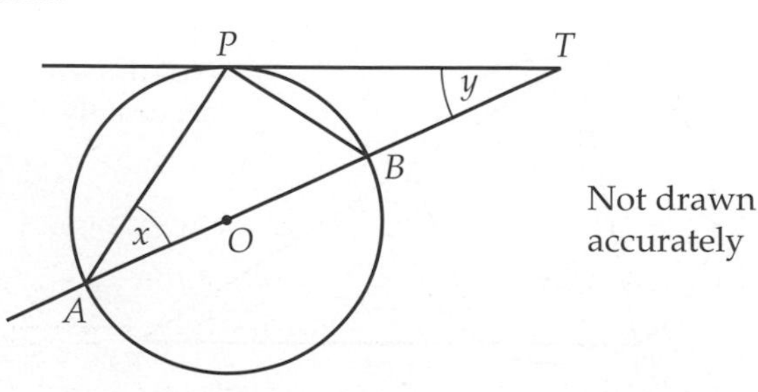

AO2

Show that angle $y = 90° - 2x$.
Give a reason for each step of your working. **[4 marks]**

7 A, B and C are points on the circumference of a circle with centre O.
BD and CD are tangents to the circle.
$\angle BDC = 40°$.

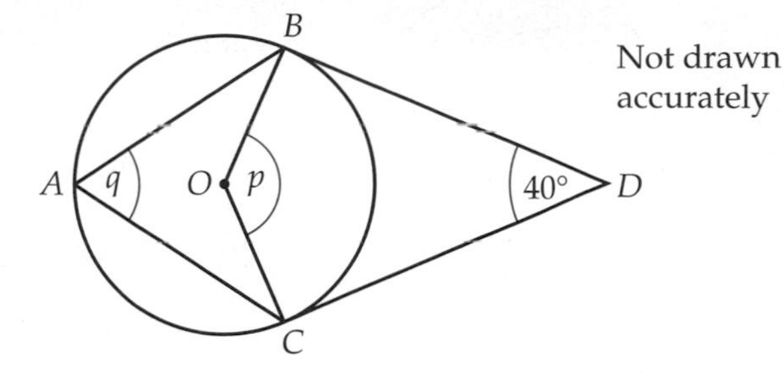

a **i** Calculate the value of angle p. **[2 marks]**

ii Hence write down the value of q. **[1 mark]**

b The tangent DB is extended to point T.
The line AO is added to the diagram.
$\angle TBA = 62°$.

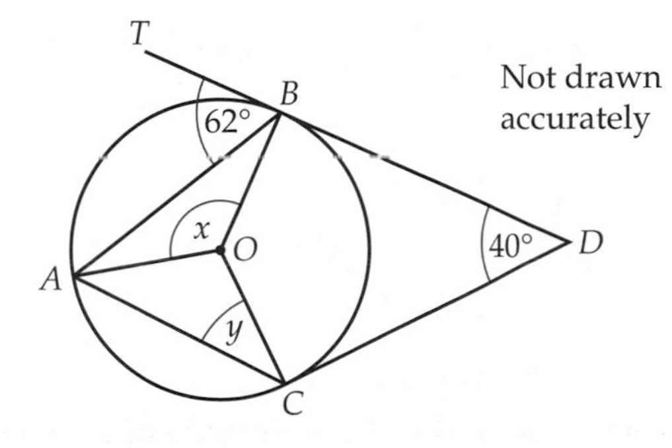

i Calculate the value of angle x. **[2 marks]**

ii Calculate the value of angle y. **[2 marks]**

AO2

8 $ABCD$ is a cyclic quadrilateral.
PAQ is a tangent to the circle at A
and $BC = CD$.
Show that AD is parallel to BC.
Give a reason for each step of your working.

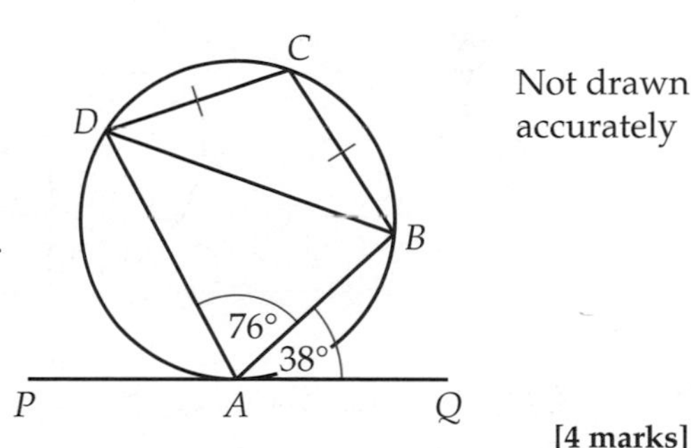

[4 marks]

AO2

Chapter summary

In this chapter you have learned how to

- use chord and tangent properties to solve problems **B**
- use circle theorems and circle properties to solve geometrical problems **B** **A**
- prove circle theorems **A**
- use the alternate segment theorem to find missing angles and solve problems **A** **A***
- prove the alternate segment theorem **A***

35

Non-linear graphs

BBC Video

This chapter explores curved graphs.

The cable holding up a suspension bridge takes the shape of a curve called a parabola.

Objectives

This chapter will show you how to

- generate points and plot graphs of quadratic functions D C
- find approximate solutions of a quadratic equation from the graph of the corresponding quadratic function C
- discuss and interpret graphs modelling real situations C B A
- find the intersection points of the graphs of a linear function and a quadratic function C B A A*
- plot graphs of simple cubic functions, the reciprocal function and the exponential function B A A*
- select and use appropriate and efficient techniques and strategies to solve problems involving quadratic equations A*

Before you start this chapter

HELP Chapter 20

1 Solve the equation $2x^2 - 5x - 12 = 0$.

2 a Complete the table of values for $y = 3x - 2$.

x	-2	-1	0	1	2
y			-2		4

HELP Chapter 19

b Draw the graph of $y = 3x - 2$ for values of x from -2 to $+2$.

3 Find the value of each expression when **i** $x = 3$ and **ii** $x = -2$.

HELP Chapter 15

a $x^2 + 2$ **b** $x^2 - 5$ **c** $x^2 + 2x$ **d** x^3 **e** $\frac{1}{x}$

P

Algebra skills: revision exercise

1 Factorise

a $x^2 - 9$ b $a^2 - 4$

c $t^2 - 1$ d $16x^2 - 25y^2$

e $12a^2 - 27b^2$ f $36x^2 - 4y^2$

2 Factorise

a $x^2 + 4x + 3$ b $a^2 + 9a + 18$ c $r^2 + 7r + 10$

d $m^2 - 2m - 8$ e $x^2 - 3x - 18$ f $x^2 - 8x + 16$

g $12m^2 - 13m + 3$ h $3x^2 + x - 14$ i $24x^2 - 30x - 21$

3 Solve these equations.

a $f^2 = 25$ b $2g^2 = 72$ c $s^2 - 81 = 0$

d $2(t - 5)^2 - 50 = 0$ e $3(x + 3)^2 - 48 = 0$ f $2(7 + m)^2 - 98 = 0$

g $5x^2 - 9x - 2 = 0$ h $2k^2 + 5k = 3$ i $x^2 = 35 - 2x$

4 Use quadratic formula to solve these equations.
Give your answers to two decimal places where appropriate.

a $2y^2 - y - 10 = 0$ b $3x^2 - 14x + 5 = 0$

c $y^2 - 4y + 2 = 0$ d $3a^2 - 7a = -1$

e $x^2 + 4x - 10 = 0$ f $5y^2 - 2y = 5$

5 By considering the discriminant, decide whether each quadratic equation will have two, one or no solutions.

a $x^2 + 8x + 16 = 0$ b $2m^2 + 14m + 20 = 0$

c $x^2 + 3x - 7 = 0$ d $3x^2 - 12x = -10$

e $5p^2 - 8p + 2 = 0$ f $4x^2 + 3x + 2 = 0$

6 Write these expressions in completed square form.

a $x^2 + 14x - 5$ b $y^2 + y - 1$

c $x^2 + 8x - 6$ d $x^2 - 3x - 1$

7 Find the values of p and q such that $x^2 + 10x + 40 = (x + p)^2 + q$.

8 Solve these equations by completing the square.
Give your answers in surd form where appropriate.

a $y^2 + 2y - 9 = 0$ b $x^2 - 4x - 1 = 0$

c $z^2 - 5z - 5 = 0$ d $x^2 + 6x - 1 = 0$

e $x^2 + 6x = 3$ f $x^2 + 4x - 7 = 0$

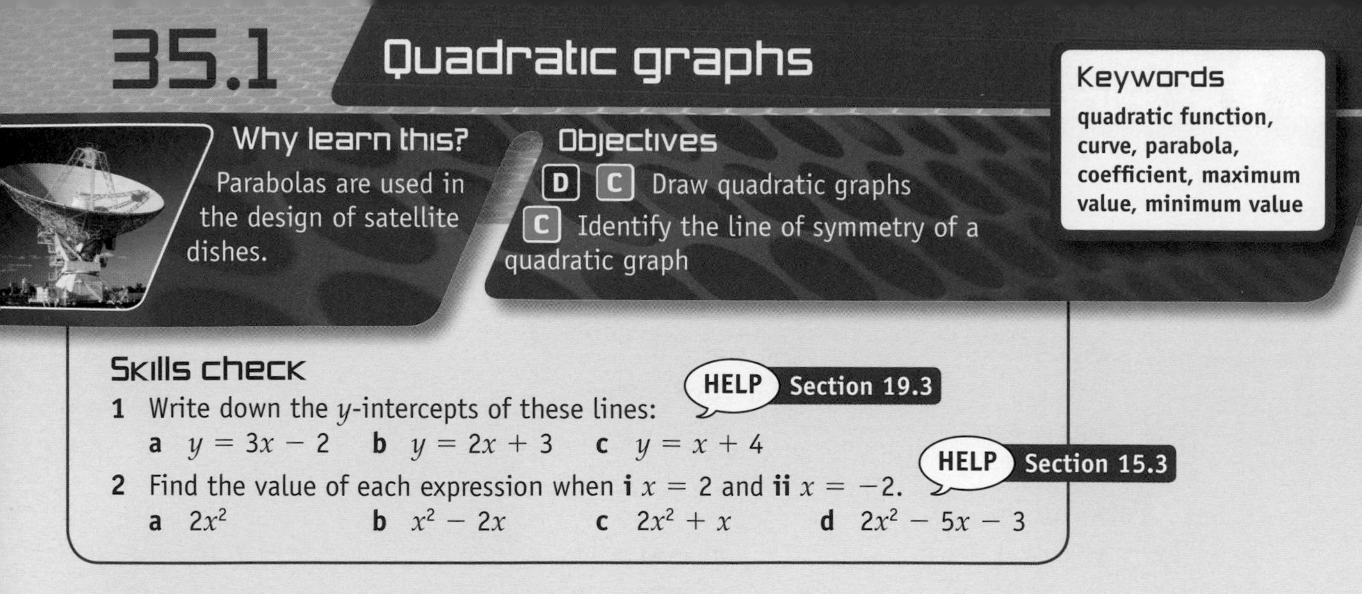

35.1 Quadratic graphs

Why learn this?

Parabolas are used in the design of satellite dishes.

Objectives

D **C** Draw quadratic graphs

C Identify the line of symmetry of a quadratic graph

Keywords

quadratic function, curve, parabola, coefficient, maximum value, minimum value

Skills check

HELP Section 19.3

1 Write down the y-intercepts of these lines:
   a $y = 3x - 2$ **b** $y = 2x + 3$ **c** $y = x + 4$

HELP Section 15.3

2 Find the value of each expression when **i** $x = 2$ and **ii** $x = -2$.
   a $2x^2$ **b** $x^2 - 2x$ **c** $2x^2 + x$ **d** $2x^2 - 5x - 3$

Quadratic functions

A **quadratic function** has a term in x^2. It may also have a term in x and a number. It does not have any terms with powers of x higher than 2.
For example, x^2, $x^2 + 2$, $x^2 + x$ and $2x^2 - 6x + 7$ are all quadratic functions.

The general form of a quadratic function is given by $y = ax^2 + bx + c$, where a, b and c are constants, with $a \neq 0$.

The graph of a quadratic function is a **curve** called a **parabola.** All quadratic graphs are parabolic, or U-shaped. The U shape can open upwards (∪) or open downwards (∩).

The simplest quadratic function is $y = x^2$.

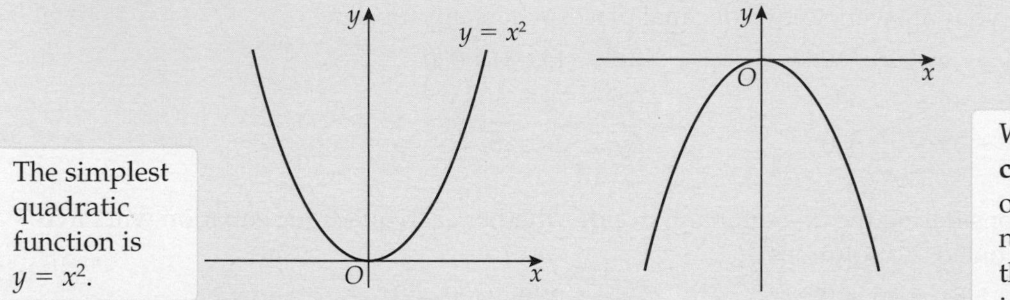

When the **coefficient** of x^2 is negative, the parabola is inverted.

All quadratic graphs are symmetrical about a line parallel to the y-axis.

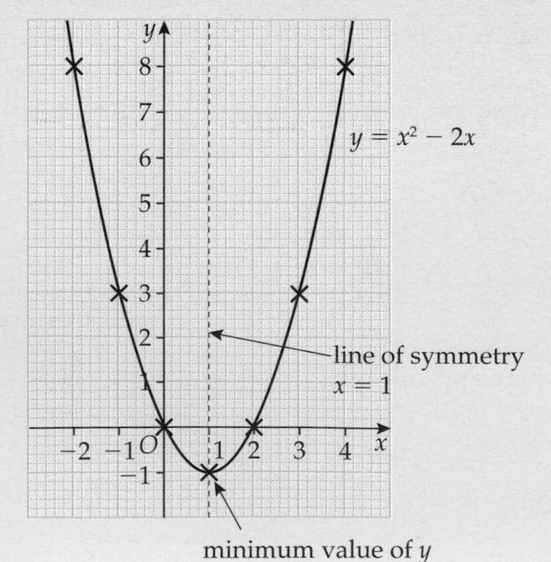

The graph of $y = x^2 - 2x$ is symmetrical about the line $x = 1$.

The line of symmetry is always given as '$x =$ '.

At the point where the curve turns, y has either a **maximum value** or a **minimum value.**

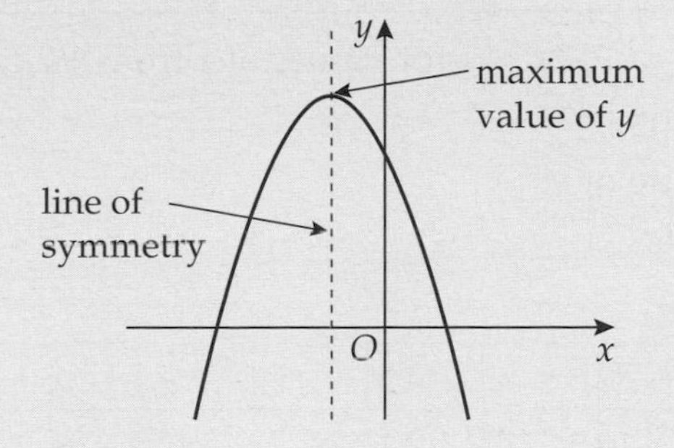

The x-intercepts are the points where the graph crosses the x-axis.

Example 1

Draw the graph of $y = x^2 - 2x$ for values of x from -2 to $+4$.

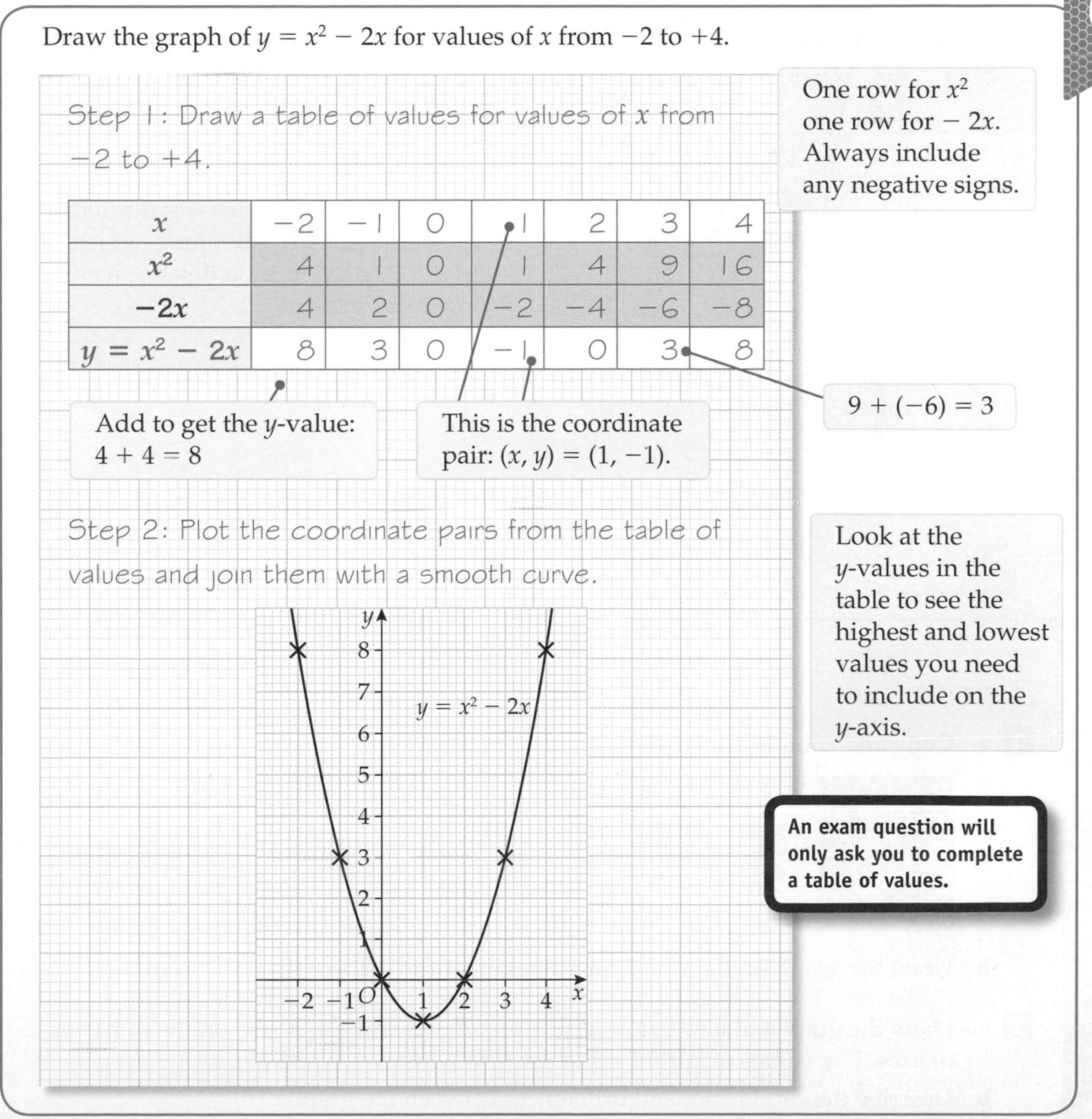

Step 1: Draw a table of values for values of x from -2 to $+4$.

One row for x^2 one row for $-2x$. Always include any negative signs.

x	-2	-1	0	1	2	3	4
x^2	4	1	0	1	4	9	16
$-2x$	4	2	0	-2	-4	-6	-8
$y = x^2 - 2x$	8	3	0	-1	0	3	8

Add to get the y-value: $4 + 4 = 8$

This is the coordinate pair: $(x, y) = (1, -1)$.

$9 + (-6) = 3$

Step 2: Plot the coordinate pairs from the table of values and join them with a smooth curve.

Look at the y-values in the table to see the highest and lowest values you need to include on the y-axis.

An exam question will only ask you to complete a table of values.

Example 2

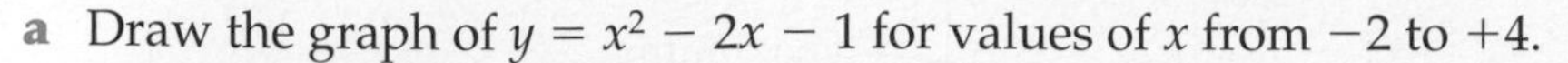

a Draw the graph of $y = x^2 - 2x - 1$ for values of x from -2 to $+4$.

b State its line of symmetry.

c Find the minimum value of y.

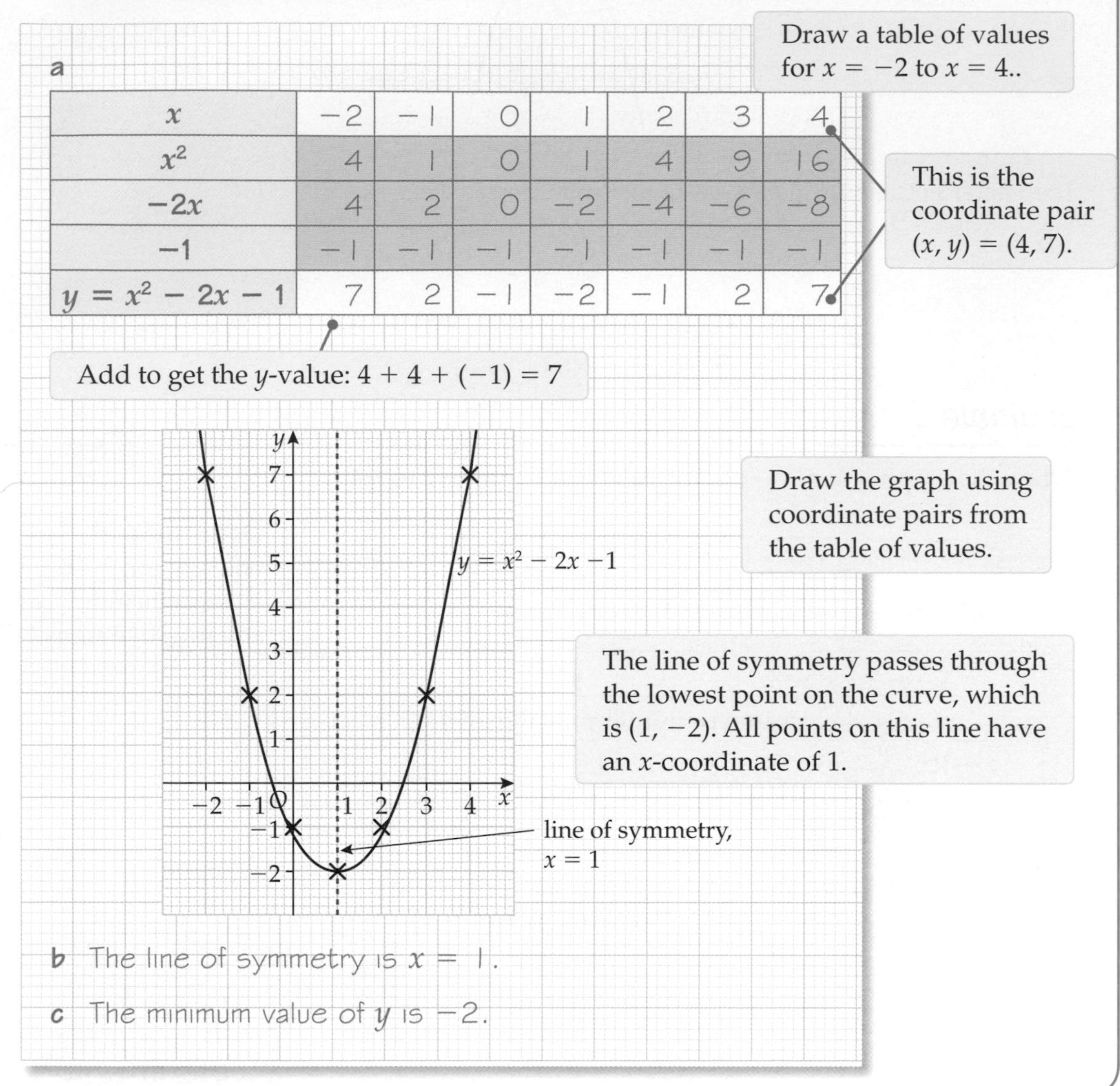

a

x	−2	−1	0	1	2	3	4
x^2	4	1	0	1	4	9	16
$-2x$	4	2	0	−2	−4	−6	−8
-1	−1	−1	−1	−1	−1	−1	−1
$y = x^2 - 2x - 1$	7	2	−1	−2	−1	2	7

b The line of symmetry is $x = 1$.

c The minimum value of y is -2.

Draw a table of values for x = −2 to x = 4..
This is the coordinate pair (x, y) = (4, 7).
Add to get the y-value: 4 + 4 + (−1) = 7
y
x
−2
−1
0
1
2
3
4
5
6
7
y = x² − 2x −1
Draw the graph using coordinate pairs from the table of values.
The line of symmetry passes through the lowest point on the curve, which is (1, −2). All points on this line have an x-coordinate of 1.
line of symmetry, x = 1

Exercise 35A

D

1 **a** Copy and complete the table of values for $y = x^2 + 2$.

x	−4	−3	−2	−1	0	1	2	3	4
x^2	16	9	4		0			9	
$+2$	+2	+2	+2	+2	+2	+2		+2	
$y = x^2 + 2$	18	11	6					11	

b Draw the graph of $y = x^2 + 2$ for values of x from -4 to $+4$.

2 **a** Draw the graphs of $y = x^2$, $y = x^2 + 3$, $y = x^2 - 3$ and $y = 3 - x^2$ on the same set of axes. Use values of x from -4 to $+4$.

b Describe the similarities and differences between the graphs.

C

3 a Copy and complete the table of values for $y = x^2 - 3x$.

x	-2	-1	0	1	2	3	4	5
x^2	4		0	1			16	
$-3x$	6		0	-3			-12	
$y = x^2 - 3x$	10		0	-2			4	

b Draw the graph of $y = x^2 - 3x$ for values of x from -2 to $+5$.

c Use your graph to find the value of y when $x = 4.5$.

d Use your graph to find the values of x that give a y-value of 5.

4 a Copy and complete the table of values for $y = 2x^2 - 4x - 1$.

x	-2	-1	0	1	2	3
$2x^2$	8	2				18
$-4x$	8	4				-12
-1	-1	-1				-1
$y = 2x^2 - 4x - 1$	15	5				5

b Draw the graph of $y = 2x^?$ $4x$ 1.

c Use your graph to find the value of y when $x = -0.5$.

5 For each of the following quadratic functions

i make a table of values

ii draw the graph

iii state the maximum or minimum value of y

iv state the line of symmetry

v state the coordinates of the x-intercepts of the graph.

a $y = x^2 + 2x - 5$ for values of x from -4 to $+2$

b $y = 2x^2 - 2x + 3$ for values of x from -3 to $+4$

c $y = x^2 - 4x + 4$ for values of x from -1 to $+5$

d $y = -x^2 + 4x - 7$ for values of x from -1 to $+5$

35.2 Solving quadratic equations graphically

Why learn this?
You can use quadratic equations to work out car acceleration and stopping distances.

Objectives
C B A A* Solve quadratic equations graphically

Skills check

1 The line $y = 2x + 4$ crosses $y = 3$ at the point P. Find the coordinates of point P.

HELP Section 19.4

2 Draw the graphs of $y = x + 4$ and $y = 2x$ on the same set of axes. Where do the lines cross?

Solving quadratic equations graphically

A quadratic function can intersect the x-axis at one or two places, or not at all. The solutions of a quadratic equation are the values of x when this happens.

- If it crosses the x-axis in two places, there are two solutions to the equation.
- If it just touches the x-axis, there is one solution.
- If it does not cross the x-axis, there are no solutions.

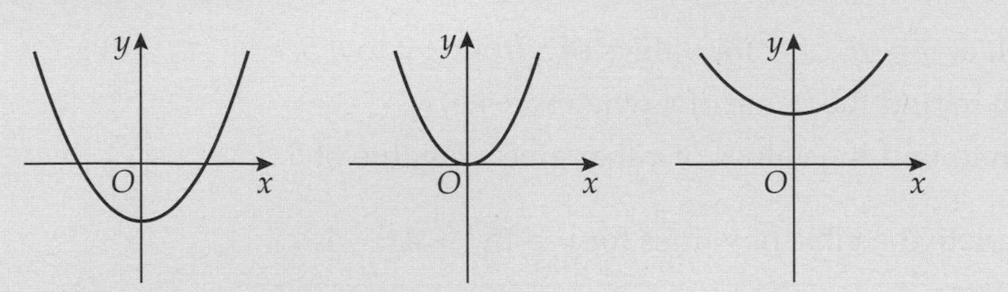

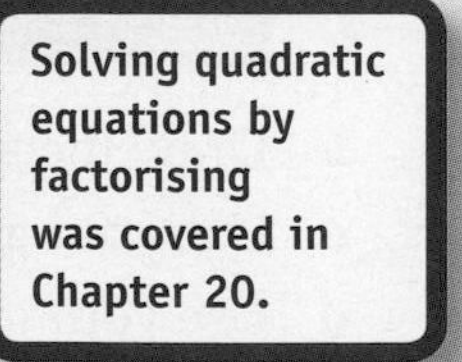

Solving quadratic equations by factorising was covered in Chapter 20.

Example 3

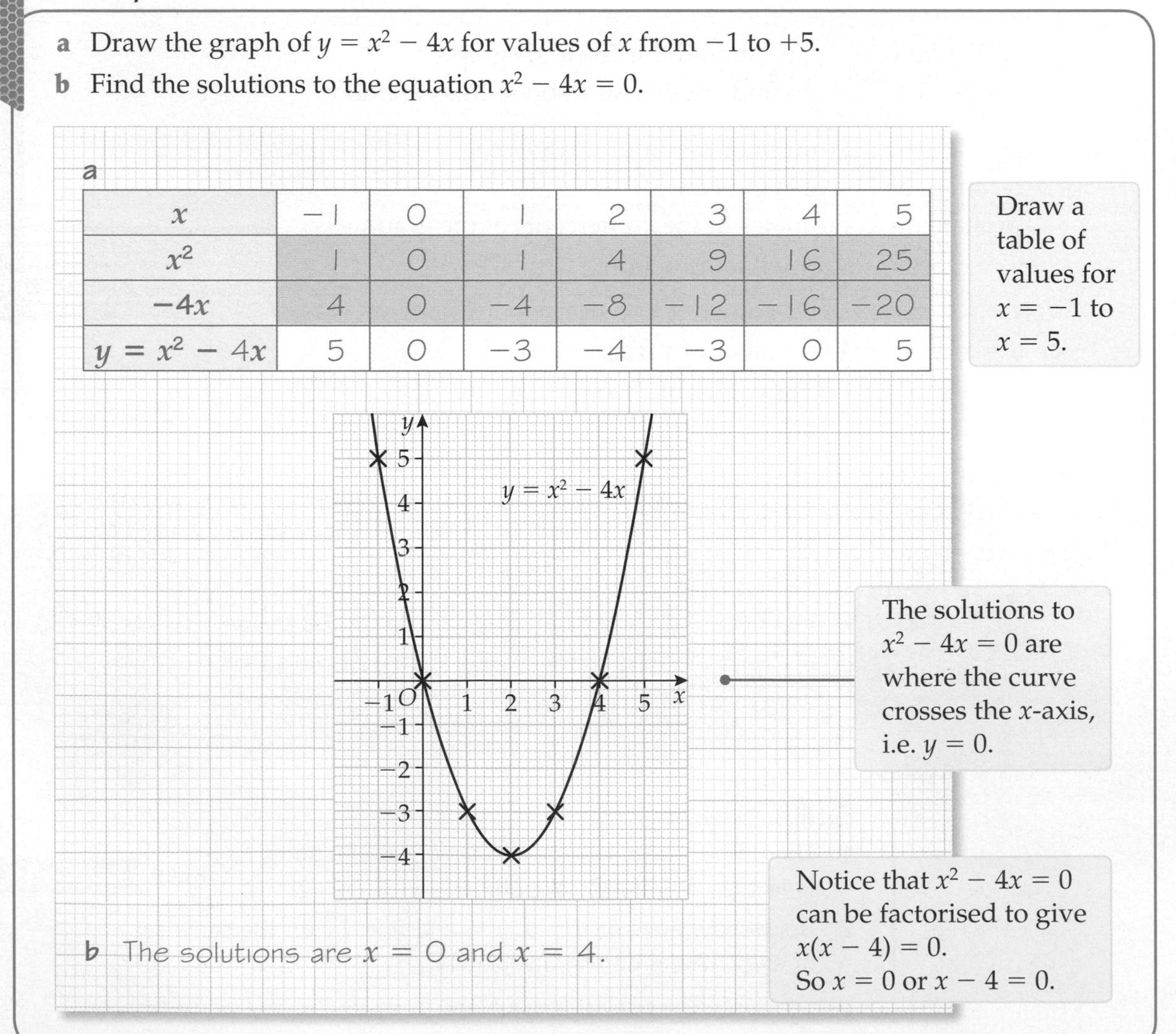

a Draw the graph of $y = x^2 - 4x$ for values of x from -1 to $+5$.

b Find the solutions to the equation $x^2 - 4x = 0$.

a

x	-1	0	1	2	3	4	5
x^2	1	0	1	4	9	16	25
$-4x$	4	0	-4	-8	-12	-16	-20
$y = x^2 - 4x$	5	0	-3	-4	-3	0	5

Draw a table of values for $x = -1$ to $x = 5$.

The solutions to $x^2 - 4x = 0$ are where the curve crosses the x-axis, i.e. $y = 0$.

b The solutions are $x = 0$ and $x = 4$.

Notice that $x^2 - 4x = 0$ can be factorised to give $x(x - 4) = 0$.
So $x = 0$ or $x - 4 = 0$.

Exercise 35B

1 a Copy and complete the table of values for $y = x^2 - 4x - 5$.

x	−2	−1	0	1	2	3	4	5	6
x^2	4			1	4	9	16		
$-4x$	8			−4	−8	−12			
-5	−5			−5	−5	−5	−5		
$y = x^2 - 4x - 5$	7			−8	−9	−8			

b Draw the graph of $y = x^2 - 4x - 5$.

c What are the coordinates of the lowest point?

d State the line of symmetry.

e What are the coordinates of the points where the curve crosses the x-axis?

f State the solutions to the equation $x^2 - 4x - 5 = 0$.

2 a Draw the graph of $y = x^2 - 3x - 4$ for values of x between −2 and 5.

b Use your graph to find the solutions to the equation $x^2 - 3x - 4 = 0$.

3 a Draw the graph of $y = x^2 + x - 3$ for $-3 \leqslant x \leqslant 2$.

b An approximate solution to the equation $x^2 + x - 3 = 0$ is $x = 1.3$.

i Explain how you can find this from the graph.

ii Use the graph to find another solution to this equation.

4 a Draw the graph of $y = -x^2 + 3x + 10$ for $-3 \leqslant x \leqslant 6$.

b Find the solutions of the equation $x^2 - 3x = 10$.

c State the line of symmetry of the graph.

d What are the coordinates of the maximum or minimum point of the graph?

C

C

AO2

You can solve a quadratic equation by drawing its graph and finding where the graph crosses other linear graphs.

For example, solving the quadratic equation $x^2 - 2x - 1 = 6$ means looking to see where the curve $y = x^2 - 2x - 1$ crosses the line $y = 6$.

Solving $x^2 - 2x - 1 = x + 1$ means looking to see where the curve crosses the line $y = x + 1$.

You learned how to do this algebraically (as a pair of simultaneous equations) in Chapter 21.

Example 4

C

a Draw the graph of $y = x^2 + 2x - 5$ for values of x from −3 to +2.

b Draw the line $y = -2$ on the same axes.
What are the x-coordinates of the points where the line and the curve intersect?

c Write the quadratic equation whose solutions are the answers to part **b**.

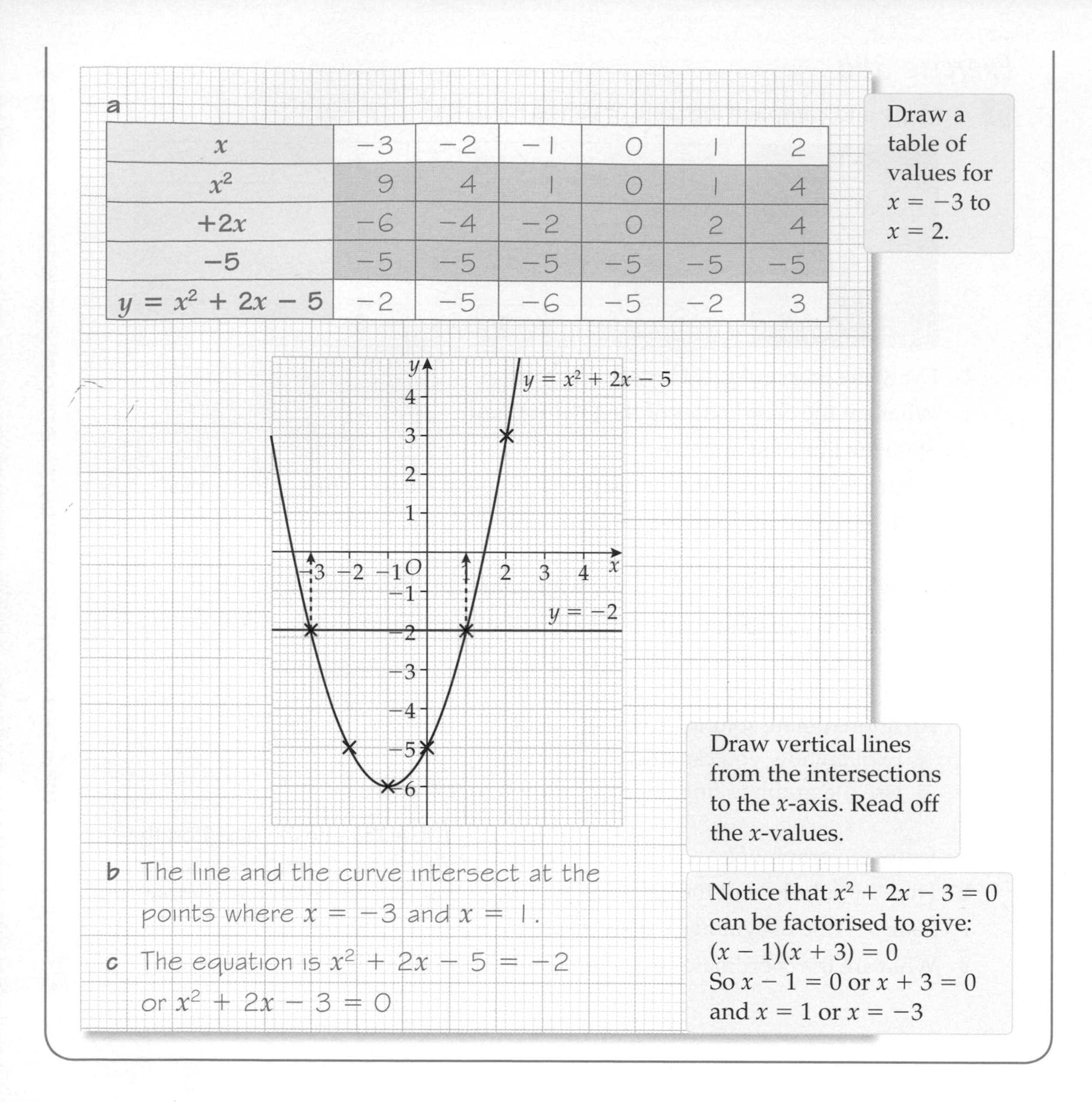

a

x	−3	−2	−1	0	1	2
x^2	9	4	1	0	1	4
$+2x$	−6	−4	−2	0	2	4
-5	−5	−5	−5	−5	−5	−5
$y = x^2 + 2x - 5$	−2	−5	−6	−5	−2	3

Draw a table of values for $x = -3$ to $x = 2$.

Draw vertical lines from the intersections to the x-axis. Read off the x-values.

b The line and the curve intersect at the points where $x = -3$ and $x = 1$.

c The equation is $x^2 + 2x - 5 = -2$ or $x^2 + 2x - 3 = 0$

Notice that $x^2 + 2x - 3 = 0$ can be factorised to give:
$(x - 1)(x + 3) = 0$
So $x - 1 = 0$ or $x + 3 = 0$
and $x = 1$ or $x = -3$

C A A*

Example 5

a Draw the graph of the quadratic function $y = x^2 + 3x - 2$ for $-4 \leqslant x \leqslant 1$.

b Use your graph to find the solutions to the equation $x^2 + 3x - 2 = x - 2$.

c By drawing an appropriate straight line, solve the equation $x^2 + 2x - 3 = 0$.

a

x	−4	−3	−2	−1	0	1
x^2	16	9	4	1	0	1
$+3x$	−12	−9	−6	−3	0	3
-2	−2	−2	−2	−2	−2	−2
$y = x^2 + 3x - 2$	2	−2	−4	−4	−2	2

Draw a table of values for $x = -4$ to $x = 1$.

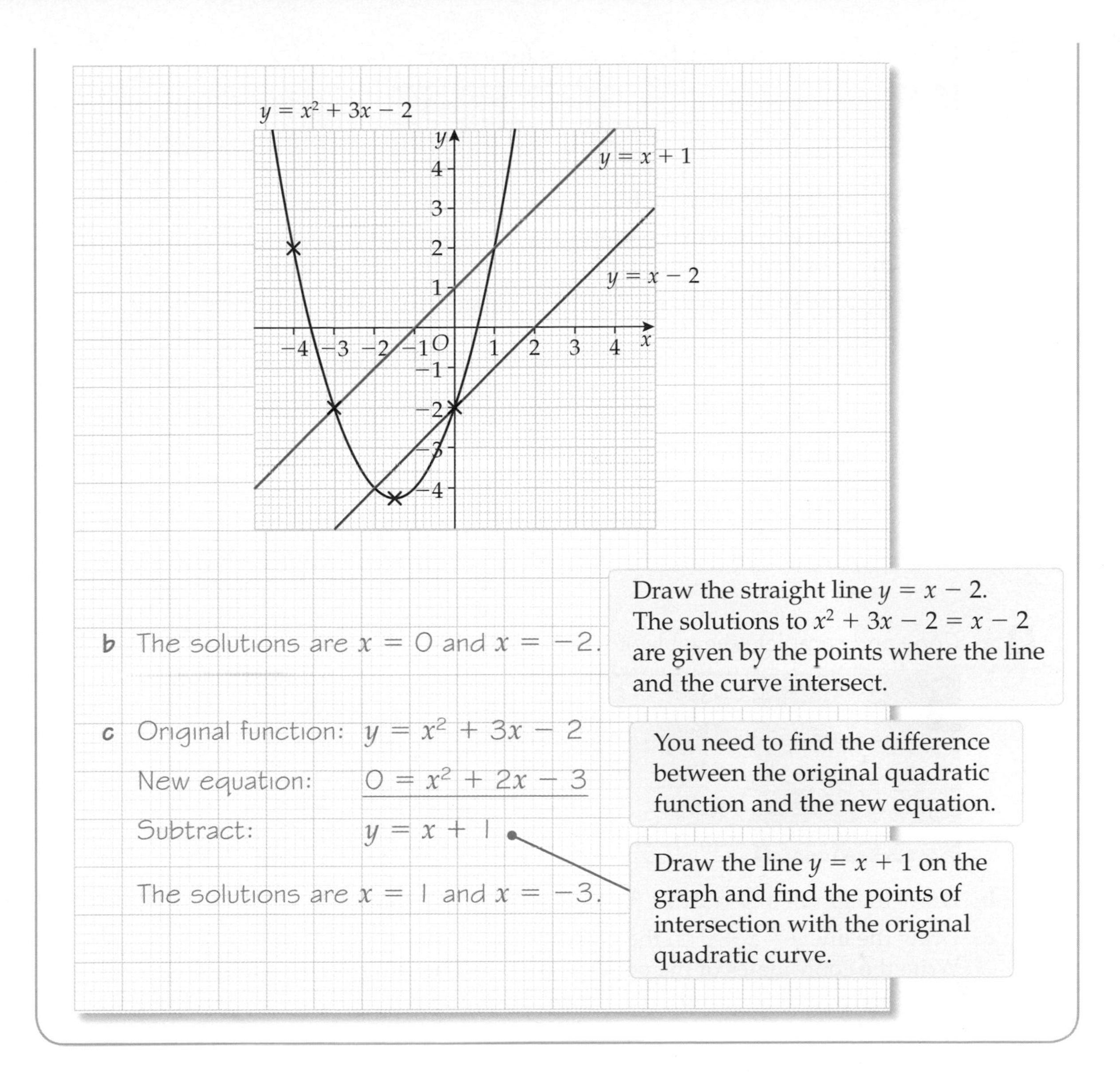

b The solutions are $x = 0$ and $x = -2$.

Draw the straight line $y = x - 2$.
The solutions to $x^2 + 3x - 2 = x - 2$ are given by the points where the line and the curve intersect.

c Original function: $y = x^2 + 3x - 2$

New equation: $0 = x^2 + 2x - 3$

Subtract: $y = x + 1$

The solutions are $x = 1$ and $x = -3$.

You need to find the difference between the original quadratic function and the new equation.

Draw the line $y = x + 1$ on the graph and find the points of intersection with the original quadratic curve.

Exercise 35C

1 **a** Copy and complete the table of values for $y = x^2 - 4x + 1$.

x	−1	0	1	2	3	4	5
x^2	1		1	4		16	
$-4x$	4		−4	−8		−16	
$+1$	+1		+1	+1		+1	
$y = x^2 - 4x + 1$	6		−2	−3		1	

b Draw the graph of $y = x^2 - 4x + 1$.

c Use your graph to find the solutions to the equation $x^2 - 4x + 1 = 0$.

d Draw the line $y = -1$ on your graph.
Find the x-coordinates of the points where the line and the curve intersect.

e Write the quadratic equation whose solutions are the answers to part **d**.

C

C

2 a Copy and complete the table of values for $y = 3x^2 - 6$.

x	−3	−2	−1	0	1	2	3
y		6	−3				21

b Draw the graph of $y = 3x^2 - 6$.

c Draw the line $y = 10$ on your graph.
Write the coordinates of the points where the line and the curve cross.

AO2

d Show that the solutions to the quadratic equation $3x^2 - 16 = 0$ can be found at these points.

C

3 a Draw the graph of $y = x^2 - 4x + 4$ for values of x from −1 to +5.

b Use your graph to solve these equations.

i $x^2 - 4x + 4 = 0$

ii $x^2 - 4x + 4 = 6$

AO3

c Can the quadratic equation $x^2 - 4x + 4 = -1$ be solved? Explain your answer.

B

4 a Copy and complete the table of values for the function $y = 2x^2 + 3x + 1$.

x	−3	−2	−1	0	1	2
$2x^2$	18		2	0		
$+3x$	−9		−3	0		
$+1$	+1		+1	+1		
$y = 2x^2 + 3x + 1$	10		0	1		

b Draw the graph of $y = 2x^2 + 3x + 1$ for $-3 \leqslant x \leqslant 2$.

c Draw the line $y = x + 3$ on the graph.
Write the coordinates of the points where the line and the curve intersect.

AO2

d Show that the solution to the equation $2x^2 + 2x - 2 = 0$ can be found at these points.

A

5 The graph shows $y = x^2 - 4x + 3$ and $y = x - 2$.

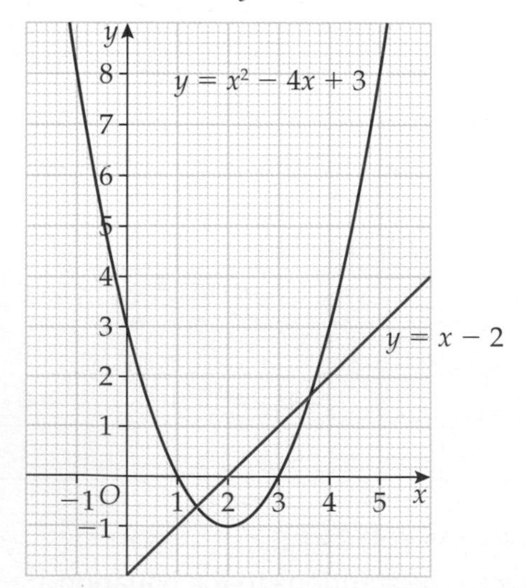

a What is the quadratic equation whose solutions are the x-coordinates of the points of intersection of $y = x^2 - 4x + 3$ and $y = x - 2$?

AO2

b What are the solutions to the equation in part **a**?

6 The graph shows the curve $y = x^2 - 2x - 4$.

a Use the graph to solve these equations.

i $x^2 - 2x - 4 = 0$

ii $x^2 - 2x - 4 = 2$

iii $x^2 - 2x - 4 = 2x + 1$

iv $x^2 - 2x - 2 = 0$

b Simplify the equation $x^2 - 2x - 4 = 2x + 1$. How does this verify the answers to part iii?

c What line would you draw to solve the equation $x^2 - x - 6 = 0$?

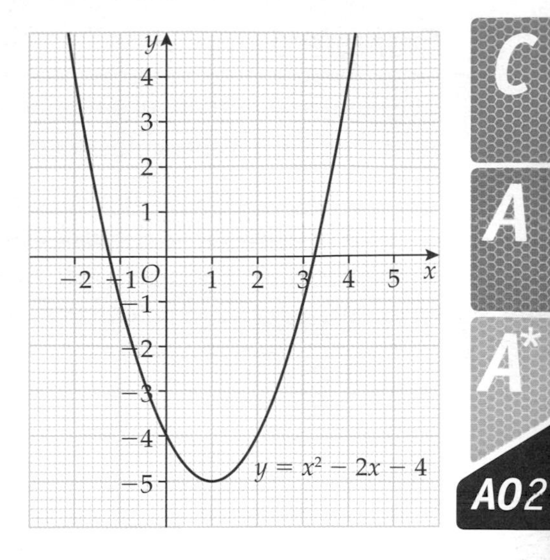

C A A* AO2

7 a Draw the graph of $y = x^2 - 4x - 2$ for $-1 \leqslant x \leqslant 5$.

b By drawing an appropriate line, solve the equation $x^2 - 4x - 5 = 0$.

8 a Draw the graph of $y = x^2 - 5x + 2$ for $0 \leqslant x \leqslant 5$.

b By drawing an appropriate line, solve the equation $x^2 - 7x + 1 = 0$.

A* AO2

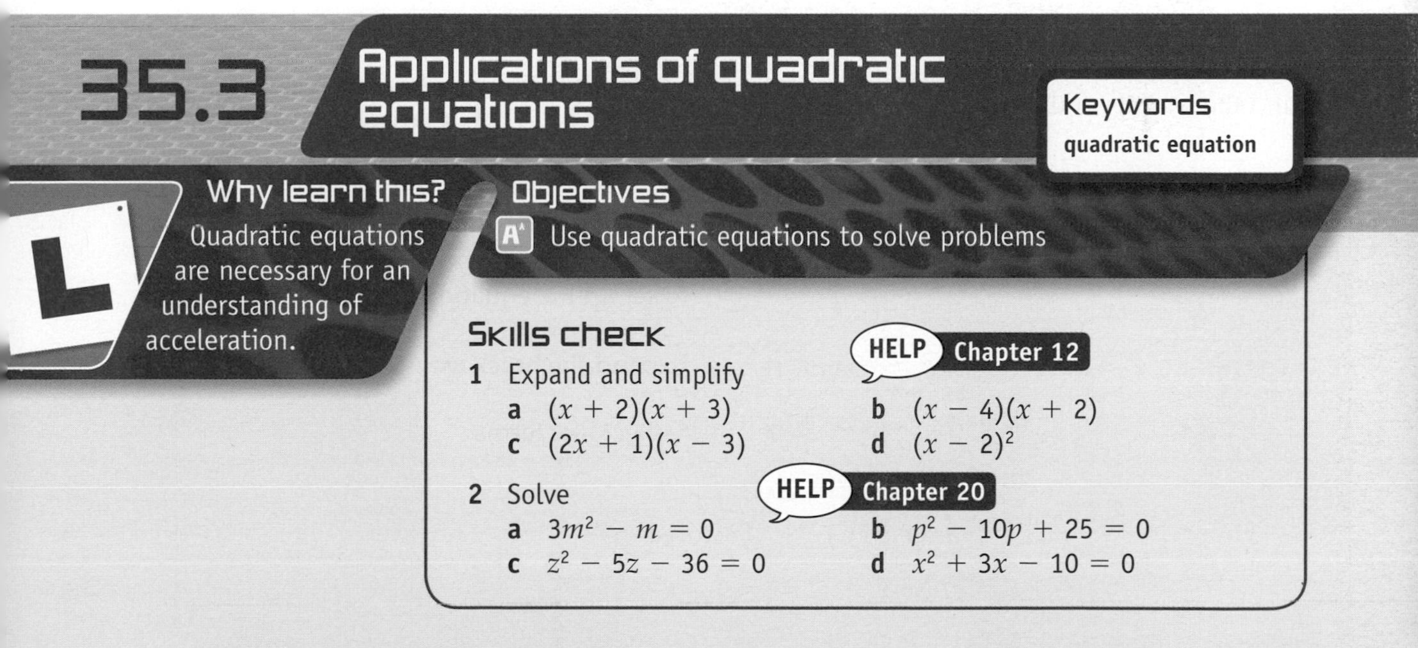

35.3 Applications of quadratic equations

Keywords
quadratic equation

Why learn this?
Quadratic equations are necessary for an understanding of acceleration.

Objectives

A* Use quadratic equations to solve problems

Skills check

HELP Chapter 12

1 Expand and simplify

a $(x + 2)(x + 3)$

b $(x - 4)(x + 2)$

c $(2x + 1)(x - 3)$

d $(x - 2)^2$

HELP Chapter 20

2 Solve

a $3m^2 - m = 0$

b $p^2 - 10p + 25 = 0$

c $z^2 - 5z - 36 = 0$

d $x^2 + 3x - 10 = 0$

Solving problems with quadratic equations

Quadratic equations can be applied to many practical problems.

To revise solving quadratic equations by factorising, see Chapter 20.

A*

Example 6

The sum of the square of a number and double the number is 80.

Find the two possible values of the number.

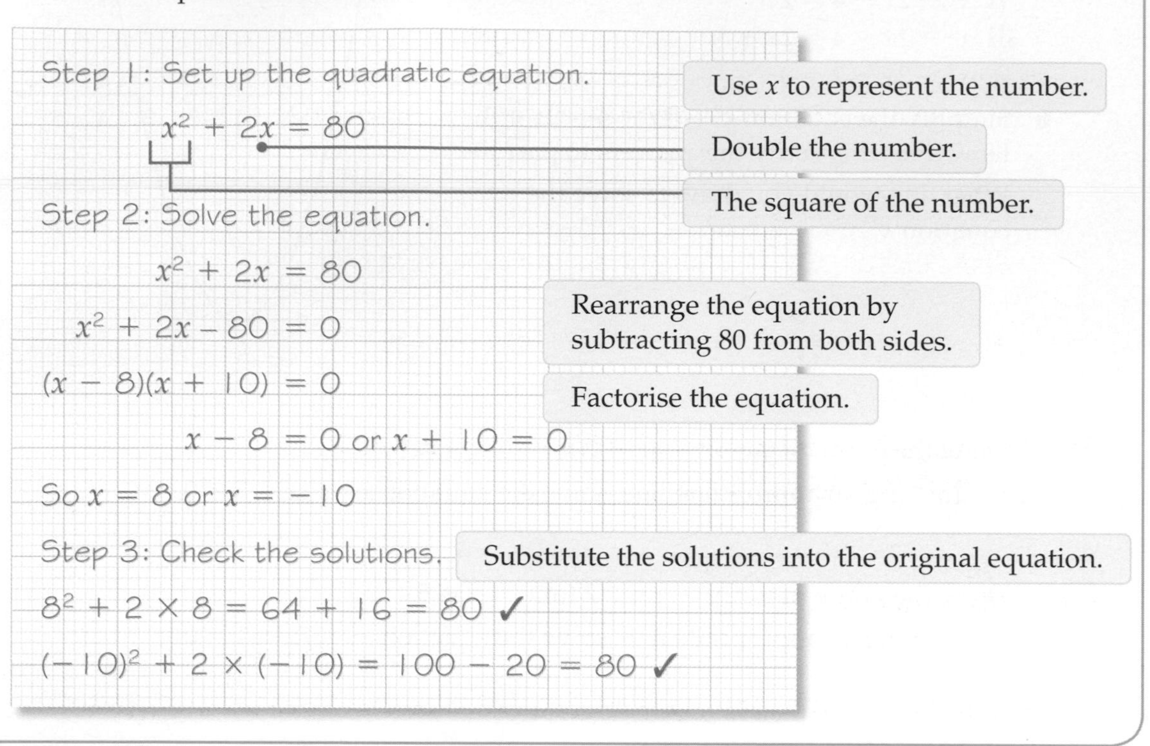

Step 1: Set up the quadratic equation.
Use x to represent the number.
$x^2 + 2x = 80$
Double the number.
The square of the number.
Step 2: Solve the equation.
$x^2 + 2x = 80$
$x^2 + 2x - 80 = 0$
Rearrange the equation by subtracting 80 from both sides.
$(x - 8)(x + 10) = 0$
Factorise the equation.
$x - 8 = 0$ or $x + 10 = 0$
So $x = 8$ or $x = -10$
Step 3: Check the solutions.
Substitute the solutions into the original equation.
$8^2 + 2 \times 8 = 64 + 16 = 80$ ✓
$(-10)^2 + 2 \times (-10) = 100 - 20 = 80$ ✓

A*

Example 7

The sides of a right-angled triangle are x cm, $(x + 2)$ cm and $(2x - 2)$ cm.

Find the lengths of the sides of the triangle.

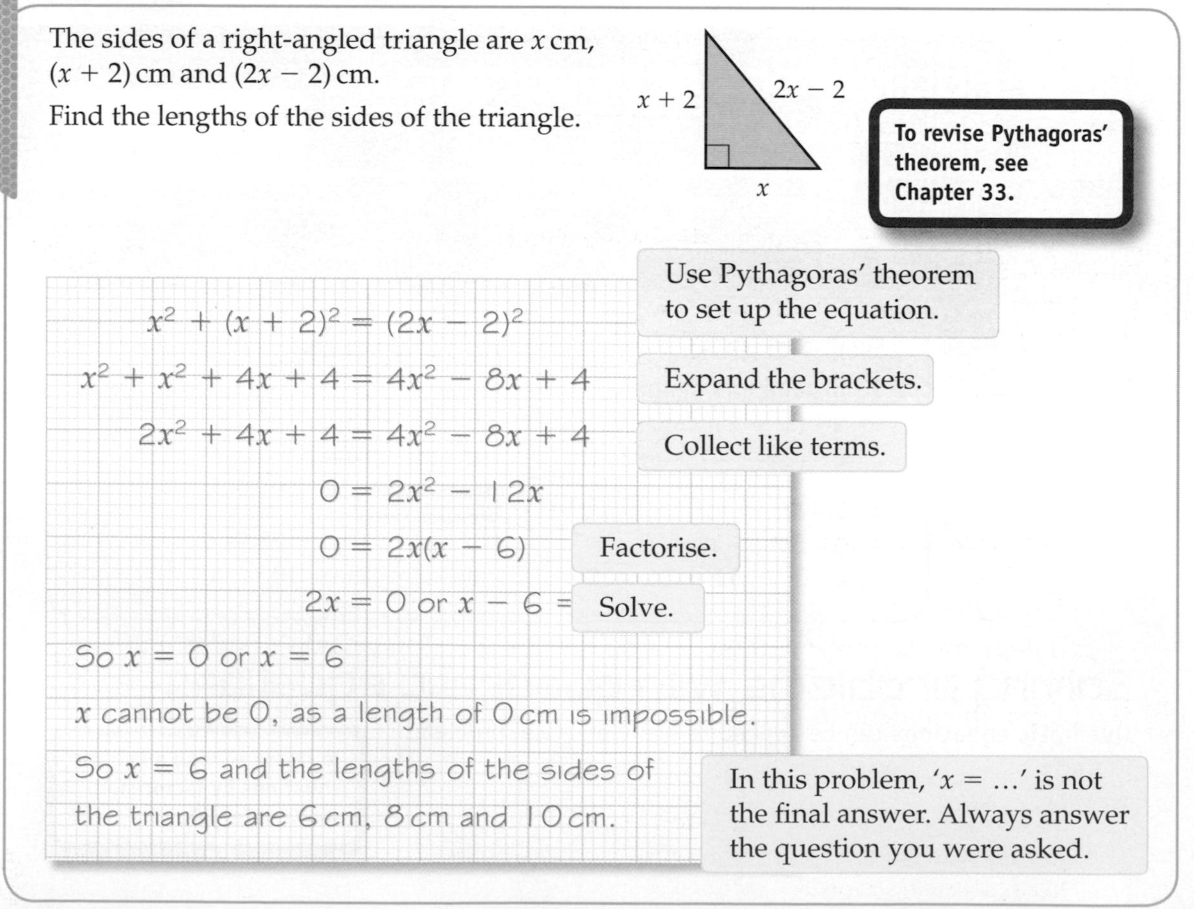

$x + 2$
$2x - 2$
x
To revise Pythagoras' theorem, see Chapter 33.
Use Pythagoras' theorem to set up the equation.
$x^2 + (x + 2)^2 = (2x - 2)^2$
$x^2 + x^2 + 4x + 4 = 4x^2 - 8x + 4$
Expand the brackets.
$2x^2 + 4x + 4 = 4x^2 - 8x + 4$
Collect like terms.
$0 = 2x^2 - 12x$
$0 = 2x(x - 6)$
Factorise.
$2x = 0$ or $x - 6 =$
Solve.
So $x = 0$ or $x = 6$
x cannot be 0, as a length of 0 cm is impossible.
So $x = 6$ and the lengths of the sides of the triangle are 6 cm, 8 cm and 10 cm.
In this problem, '$x = \dots$' is not the final answer. Always answer the question you were asked.

Exercise 35D

1 The sum of the square of a number and four times the number is 12.
Find the two possible values of the number.

A*

2 The difference between the square of a number and eight times the number is −15.
Find the two possible values of the number.

AO3

3 A rectangle has dimensions $(4 + x)$ m and $(2x - 7)$ m.

a Show that the area, A, of the rectangle can be written as
$A = 2x^2 + x - 28$

b The area of the rectangle is 27 m². Find the value of x.

A*

AO2

4 The length of a rectangular plot is 8 m more than its width, w.
The area of the plot is 240 m².

Find:

a the dimensions of the plot

b the perimeter of the plot.

A*

5 Find the lengths of the sides of the right-angled triangle shown in the diagram.

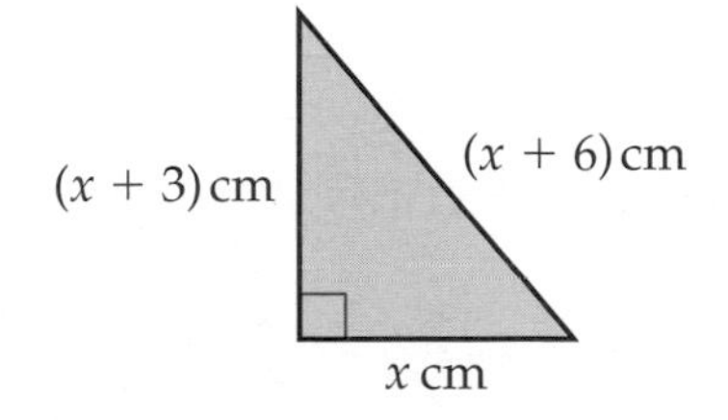

6 The areas of the two triangles shown are equal.

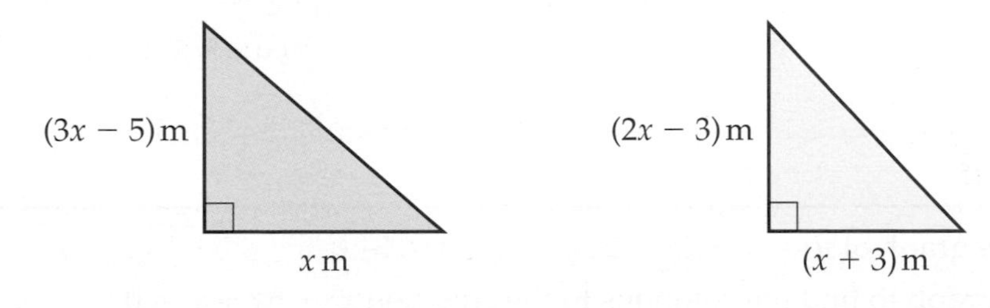

Find the area of one triangle.

7 Find the points of intersection of the curve $y = x^2 - 2x - 4$ and the straight line $y = 2x + 1$.
Do not draw a graph.

8 Find the x-coordinates of the points of intersection of the curve $y = x^2 - x - 6$ and the straight line $y = x + 2$.
Do not draw a graph.

AO3

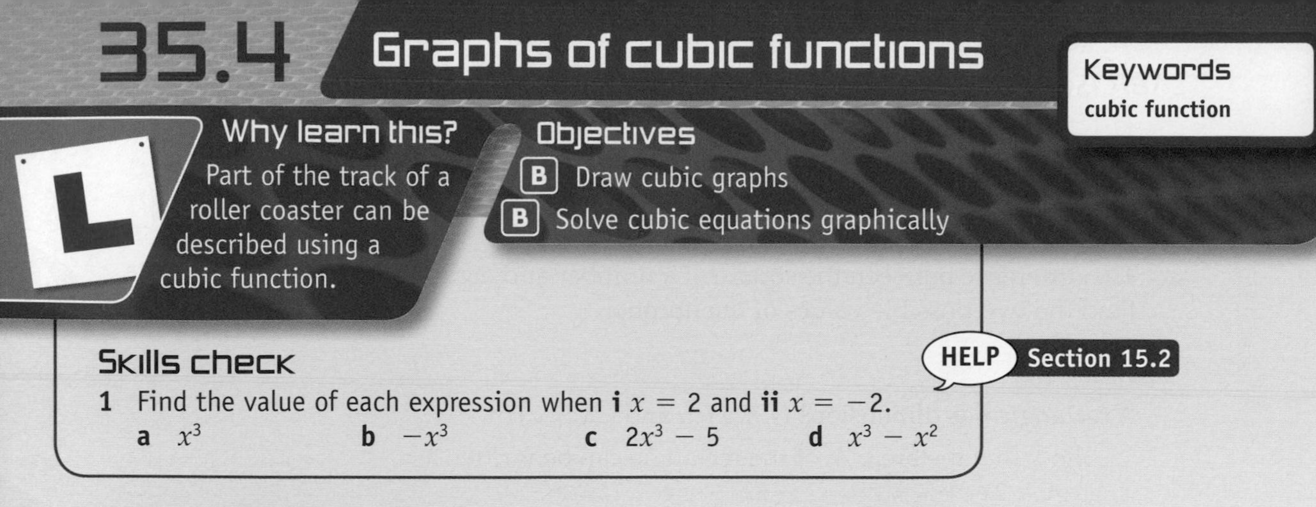

35.4 Graphs of cubic functions

Keywords
cubic function

Why learn this?
Part of the track of a roller coaster can be described using a cubic function.

Objectives
- **B** Draw cubic graphs
- **B** Solve cubic equations graphically

Skills check

HELP Section 15.2

1 Find the value of each expression when **i** $x = 2$ and **ii** $x = -2$.
a x^3 **b** $-x^3$ **c** $2x^3 - 5$ **d** $x^3 - x^2$

Cubic functions

A **cubic function** has a term in x^3. It may also have terms in x^2 and x, and a number.
It does not have any terms with powers of x higher than 3.
For example, x^3, $x^3 + 3x$, $x^3 - x^2 - 2x$ are all cubic functions.

The general form of a cubic function is given by $y = ax^3 + bx^2 + cx + d$, where a, b, c and d are constants, with $a \neq 0$.

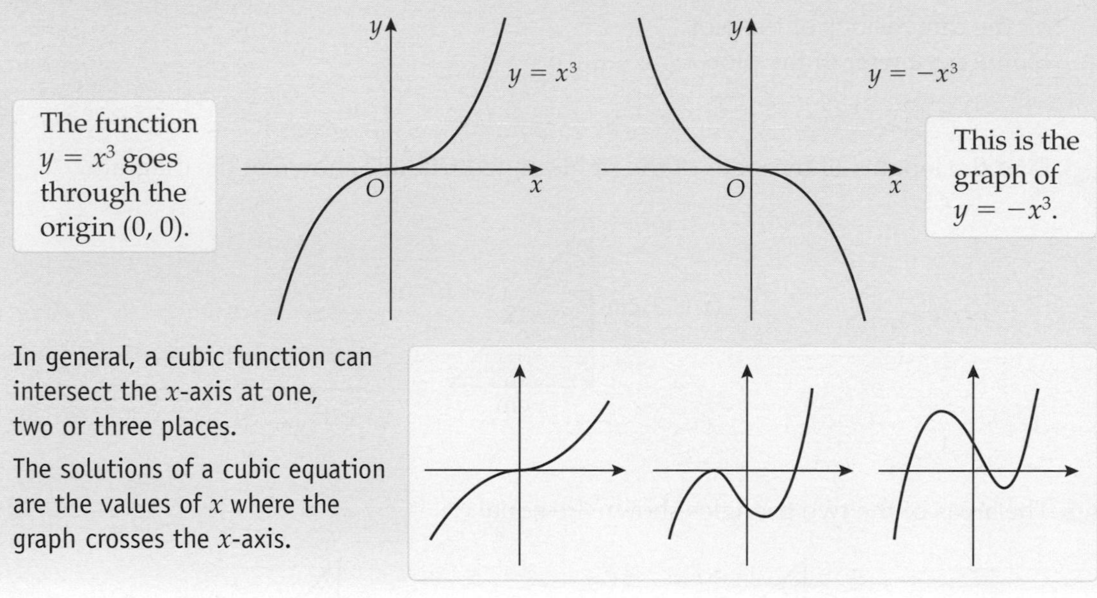

The function $y = x^3$ goes through the origin (0, 0).
$y = x^3$
$y = -x^3$
This is the graph of $y = -x^3$.
y
x
O

In general, a cubic function can intersect the x-axis at one, two or three places.

The solutions of a cubic equation are the values of x where the graph crosses the x-axis.

Example 8

a Draw the graph of $y = x^3 - 5x + 2$ for $-3 \leqslant x \leqslant 3$.
b Use the graph to find the solutions to the equation $x^3 - 5x + 2 = 0$.

a

Draw a table of values for $x = -3$ to $x = 3$.

x	−3	−2	−1	0	1	2	3
x^3	−27	−8	−1	0	1	8	27
$-5x$	15	10	5	0	−5	−10	−15
$+2$	+2	+2	+2	+2	+2	+2	+2
$y = x^3 - 5x + 2$	−10	4	6	2	−2	0	14

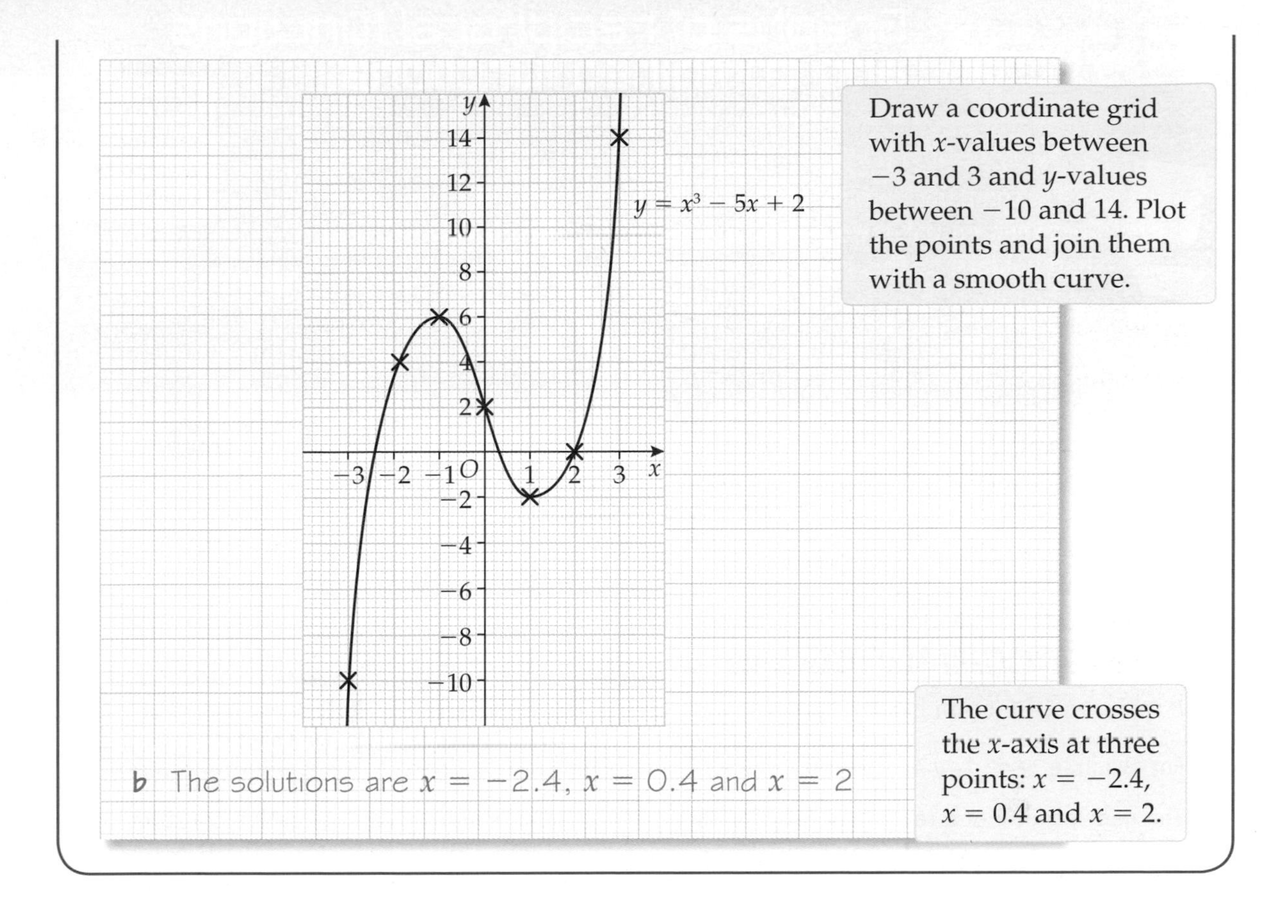
y
14
12
10
8
6
4
2
O
−3
−2
−1
1
2
3
x
−2
−4
−6
−8
−10
y = x³ − 5x + 2
Draw a coordinate grid with x-values between −3 and 3 and y-values between −10 and 14. Plot the points and join them with a smooth curve.
b The solutions are x = −2.4, x = 0.4 and x = 2
The curve crosses the x-axis at three points: x = −2.4, x = 0.4 and x = 2.

Exercise 35E

1 Draw the graph of $y = -x^3$ for values of x from -3 to $+3$.

2 **a** Draw the graphs of $y = x^3$, $y = \frac{1}{2}x^3$ and $y = x^3 + 2$ on the same set of axes. Use values of x from -3 to $+3$.

b Describe the similarities and differences between the graphs.

3 **a** Copy and complete the table of values for the function $y = x^3 - 2x + 2$.

x	−2.5	−2	−1	0	1	2	2.5
x^3		−8		0	1		15.6
$-2x$		4		0	−2		−5
$+2$		+2		+2	+2		+2
$y = x^3 - 2x + 2$		−2		2	1		12.6

b Draw the graph of $y = x^3 - 2x + 2$ for $-2.5 \leqslant x \leqslant 2.5$.

c Find the solutions to $x^3 - 2x + 2 = 0$.

4 **a** Draw the graph of $y = x^3 - 2x^2 - 6x$ for $-3 \leqslant x \leqslant 5$.

b Use your graph to find the solutions to $x^3 - 2x^2 - 6x = 0$.

5 **a** Find the solutions to $(x - 1)(x - 3)(x - 4) = 0$.

b Draw the graph of $y = (x - 1)(x - 3)(x - 4)$ for $0 \leqslant x \leqslant 5$ to confirm your answers in part **a**.

Write a row for each of $(x - 1)$, $(x - 3)$ and $(x - 4)$ and *multiply* to find y.

B

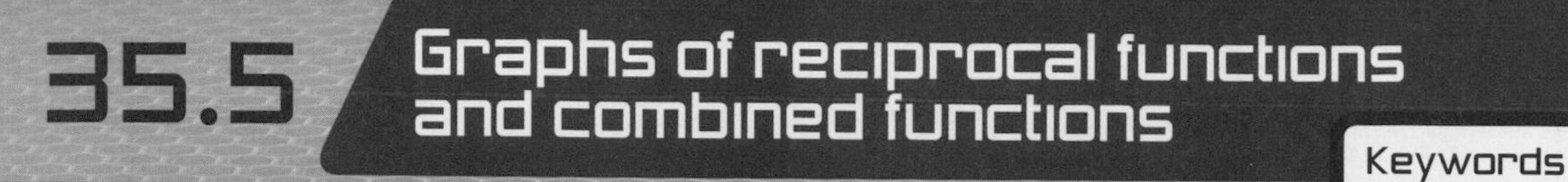

35.5 Graphs of reciprocal functions and combined functions

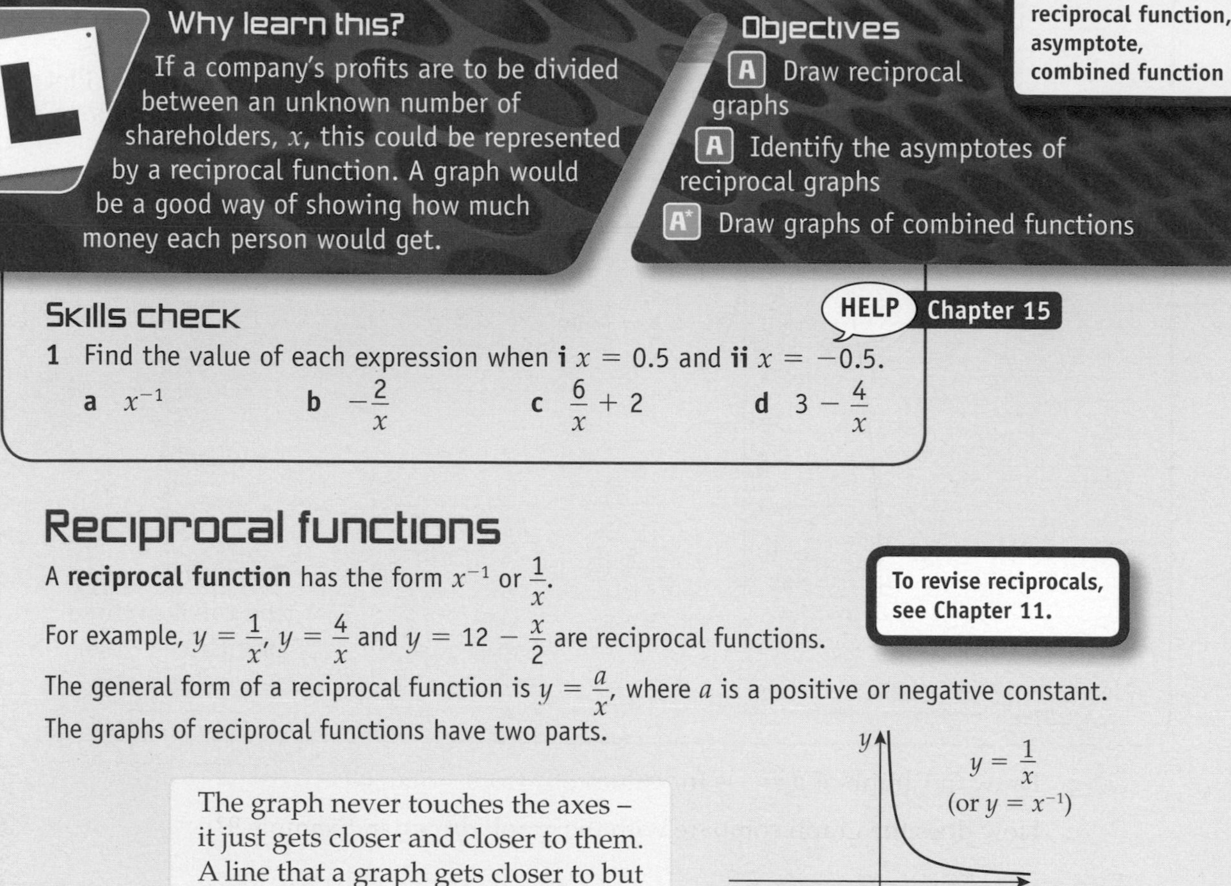

Keywords
reciprocal function, asymptote, combined function

Why learn this?

If a company's profits are to be divided between an unknown number of shareholders, x, this could be represented by a reciprocal function. A graph would be a good way of showing how much money each person would get.

Objectives

A Draw reciprocal graphs

A Identify the asymptotes of reciprocal graphs

A* Draw graphs of combined functions

Skills check

HELP Chapter 15

1 Find the value of each expression when **i** $x = 0.5$ and **ii** $x = -0.5$.

a x^{-1} **b** $-\frac{2}{x}$ **c** $\frac{6}{x} + 2$ **d** $3 - \frac{4}{x}$

Reciprocal functions

A **reciprocal function** has the form x^{-1} or $\frac{1}{x}$.

To revise reciprocals, see Chapter 11.

For example, $y = \frac{1}{x}$, $y = \frac{4}{x}$ and $y = 12 - \frac{x}{2}$ are reciprocal functions.

The general form of a reciprocal function is $y = \frac{a}{x}$, where a is a positive or negative constant.

The graphs of reciprocal functions have two parts.

The graph never touches the axes – it just gets closer and closer to them. A line that a graph gets closer to but never touches or crosses is called an **asymptote**. All reciprocal graphs have two asymptotes.

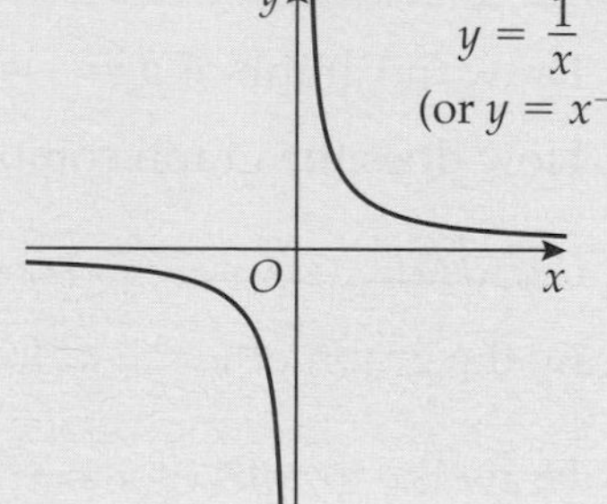

For $y = \frac{a}{x}$, the lines $y = x$ and $y = -x$ are lines of symmetry.

Example 9

a Draw the graph of $y = \frac{1}{x}$ for $-3 \leqslant x \leqslant 3$.

b What are the equations of the asymptotes?

Draw a table of values for $x = -3$ to $x = 3$. You must include smaller x-values close to zero to show the correct properties of the reciprocal function. The value $x = 0$ is not included in the table, as 1 divided by zero is undefined.

a

x	−3	−2	−1	−0.5	−0.2	0.2	0.5	1	2	3
$y = \frac{1}{x}$	−0.3	−0.5	−1	−2	−5	5	2	1	0.5	0.3

Where appropriate, give the values of y to 1 d.p.

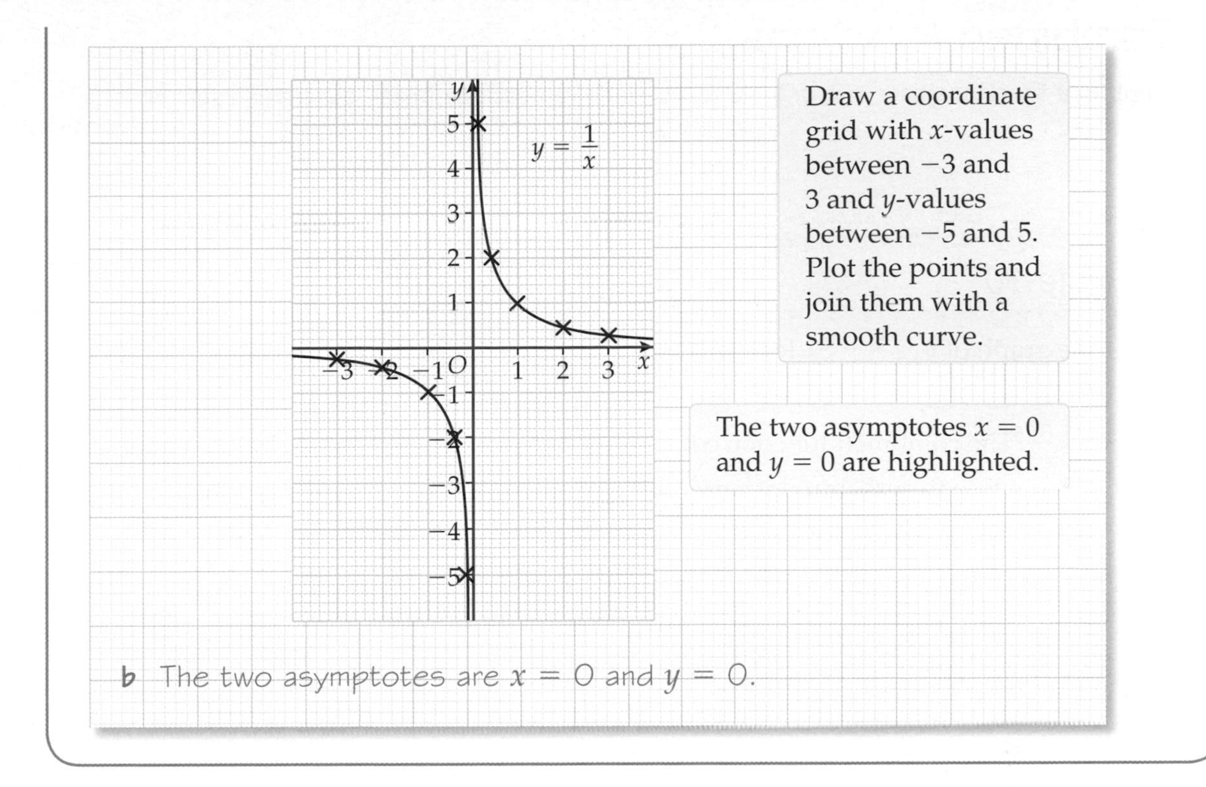

Draw a coordinate grid with x-values between -3 and 3 and y-values between -5 and 5. Plot the points and join them with a smooth curve.

The two asymptotes $x = 0$ and $y = 0$ are highlighted.

b The two asymptotes are $x = 0$ and $y = 0$.

Exercise 35F

1 **a** Draw the graph of $y = -\frac{1}{x}$ for values of x from -3 to $+3$.

b How does the graph compare with the graph drawn in Example 9?

2 **a** Copy and complete the table of values for the function $y = \frac{6}{x}$.

x	-3	-2	-1	-0.5	-0.2	0.2	0.5	1	2	3
$y = \frac{6}{x}$			-6			30	12			2

b Draw the graph of $y = \frac{6}{x}$ for $-3 \leqslant x \leqslant 3$.

3 **a** Copy and complete the table of values for the function $y = \frac{3}{x} + 5$.

x	-3	-2	-1	-0.5	-0.2	0.2	0.5	1	2	3
$\frac{3}{x}$	-1		-3	-6		15			1.5	1
$+5$	$+5$		$+5$	$+5$		$+5$			$+5$	$+5$
$y = \frac{3}{x} + 5$	4		2	-1		20			6.5	6

b Draw the graph of $y = \frac{3}{x} + 5$ for $-3 \leqslant x \leqslant 3$.

c What are the equations of the two asymptotes?

4 **a** Draw the graphs of $y = \frac{2}{x}$ and $y = \frac{2}{x} + 3$ on the same set of axes.
Use values of x from -2 to $+2$.

b Compare and contrast the two graphs in part **a**.

5 **a** Draw the graph of $y = 2 - \frac{3}{x}$ for values of x between -3 and $+3$.

b What are the equations of the two asymptotes?

A

Combined functions

Combined functions have two or more terms which can each be linear, quadratic, cubic or reciprocal.

Example 10

Draw the graph of $y = \frac{4}{x} - 3x$ for $-5 \leqslant x \leqslant 5$.

The combined function contains a reciprocal term $\left(\frac{4}{x}\right)$ so include x-values close to zero, but not zero itself. Where necessary, give y-values to 1 d.p.

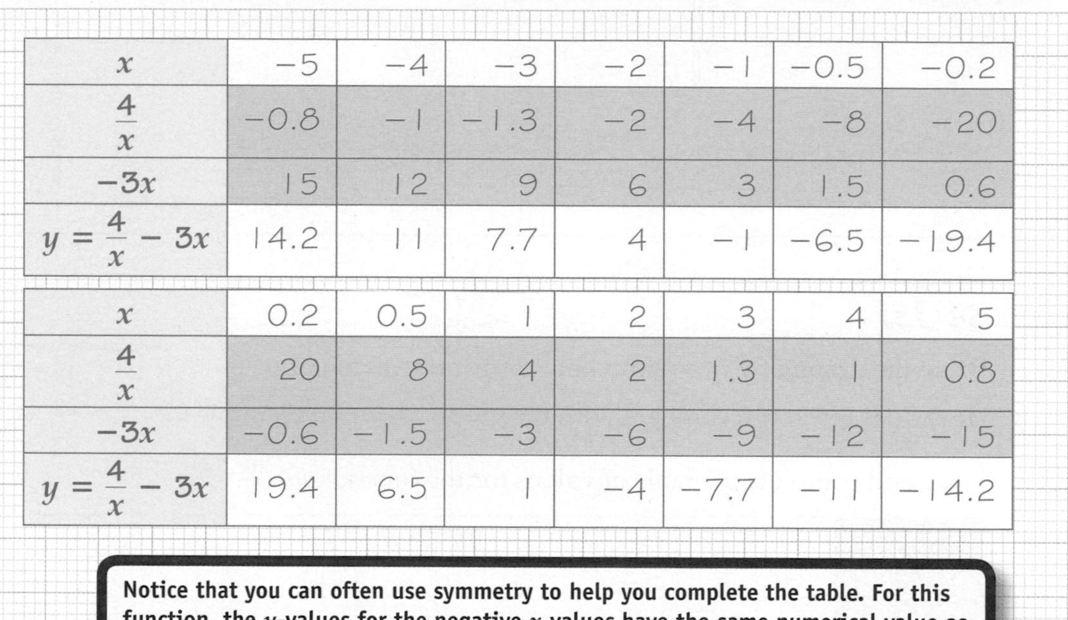

x	−5	−4	−3	−2	−1	−0.5	−0.2
$\frac{4}{x}$	−0.8	−1	−1.3	−2	−4	−8	−20
$-3x$	15	12	9	6	3	1.5	0.6
$y = \frac{4}{x} - 3x$	14.2	11	7.7	4	−1	−6.5	−19.4

x	0.2	0.5	1	2	3	4	5
$\frac{4}{x}$	20	8	4	2	1.3	1	0.8
$-3x$	−0.6	−1.5	−3	−6	−9	−12	−15
$y = \frac{4}{x} - 3x$	19.4	6.5	1	−4	−7.7	−11	−14.2

Notice that you can often use symmetry to help you complete the table. For this function, the y-values for the negative x-values have the same numerical value as the corresponding positive x-values, but with the opposite sign.

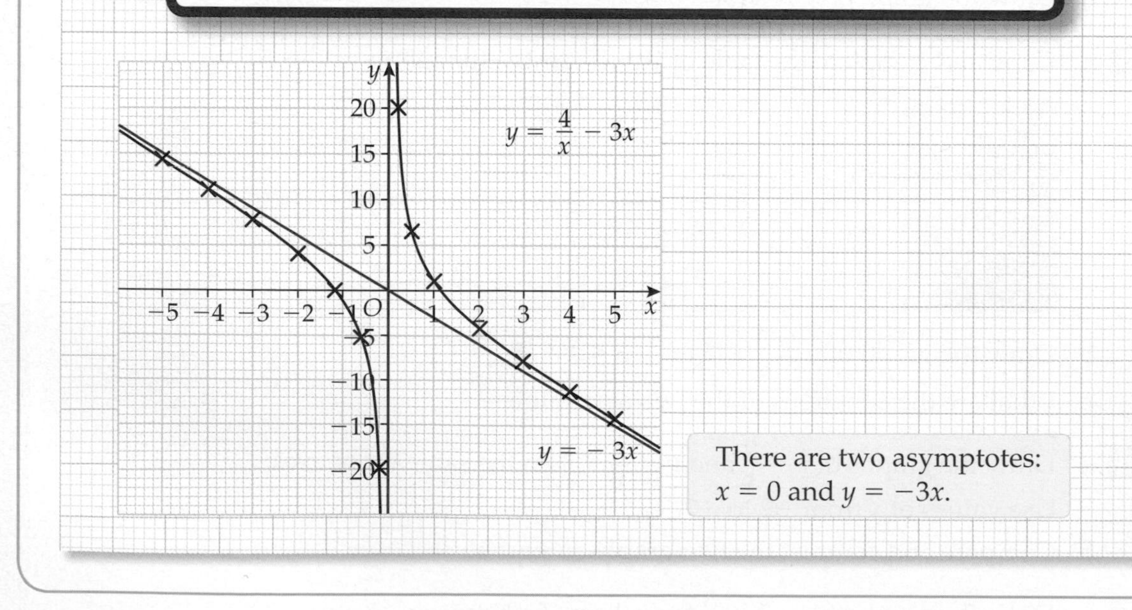

There are two asymptotes: $x = 0$ and $y = -3x$.

Exercise 35G

1 **a** Copy and complete the table of values for the function $y = \frac{1}{x} + x$.

x	0.1	0.2	0.5	1	2	3	4	5
$\frac{1}{x}$		5	2		0.5		0.25	
$y = \frac{1}{x} + x$		5.2	2.5		2.5		4.25	

> **The y-values for the negative x-values are the same as for the corresponding positive x-values but with the opposite sign.**

b Draw the graph of $y = \frac{1}{x} + x$ for $-5 \leqslant x \leqslant 5$.

2 Draw the graph of $y = x - \frac{1}{x}$ for $-4 \leqslant x \leqslant 4$.

3 **a** Copy and complete the table of values for the function $y = x^2 + \frac{8}{x}$.

x	−4	−3	−2	−1	−0.5	−0.2	0.2	0.5	1	2	3	4
x^2		9	4			0.04		0.25			9	16
$\frac{8}{x}$		−2.67	−4			40		16			2.67	2
$y = x^2 + \frac{8}{x}$		6.33	0			40.04		16.25			11.67	18

b Draw the graph of $y = x^2 + \frac{8}{x}$ for $-4 \leqslant x \leqslant 4$.

c Write the x-coordinate of the point where the graph cuts the x-axis.

d Write the equation whose solutions are the answers to part **c**.

4 Draw the graph of $y = x^2 - \frac{1}{x} + 2$ for $-3 \leqslant x \leqslant 3$.

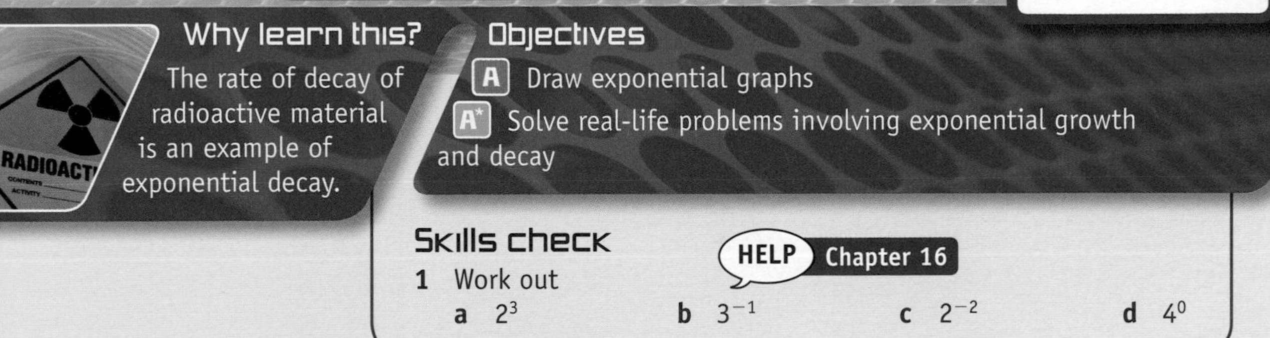

35.6 Graphs of exponential functions

Keywords
exponential function

Why learn this?
The rate of decay of radioactive material is an example of exponential decay.

Objectives

A Draw exponential graphs

A* Solve real-life problems involving exponential growth and decay

Skills check

HELP Chapter 16

1 Work out

a 2^3 **b** 3^{-1} **c** 2^{-2} **d** 4^0

Exponential functions

An **exponential function** has the form $y = k^x$ where k is a positive number.

For example, $y = 3^x$ and $y = (\frac{1}{2})^x$, $y = (\frac{3}{2})^{-x}$ are exponential functions.

There are two general shapes for the graphs of exponential functions.

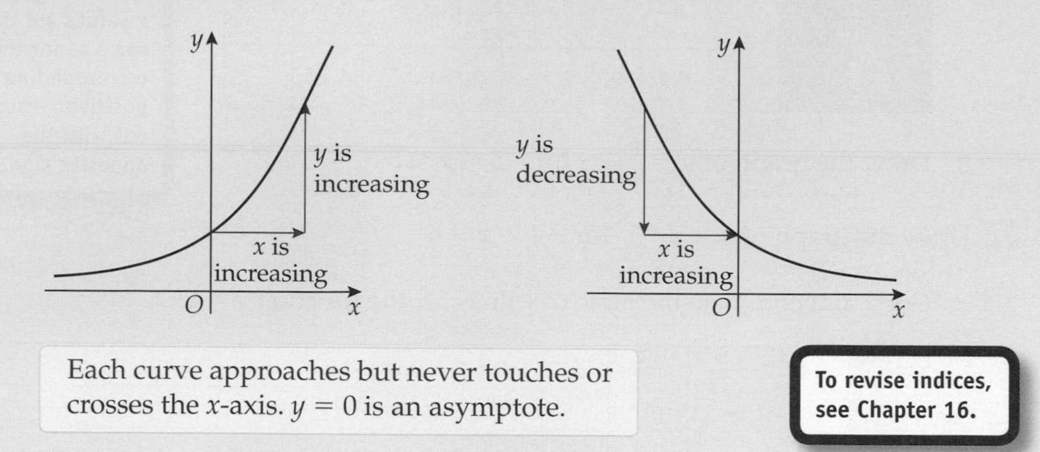

Each curve approaches but never touches or crosses the x-axis. $y = 0$ is an asymptote.

To revise indices, see Chapter 16.

All exponential functions of the form $y = k^x$ pass through (0, 1) because $y = k^0 = 1$ for all k.

A

Example 11

a Draw the graph of $y = 2^x$ for values of x between -4 and $+3$. Give the y-values to two decimpal places where necessary.

b Use your graph to estimate the value of x when $y = 6$.

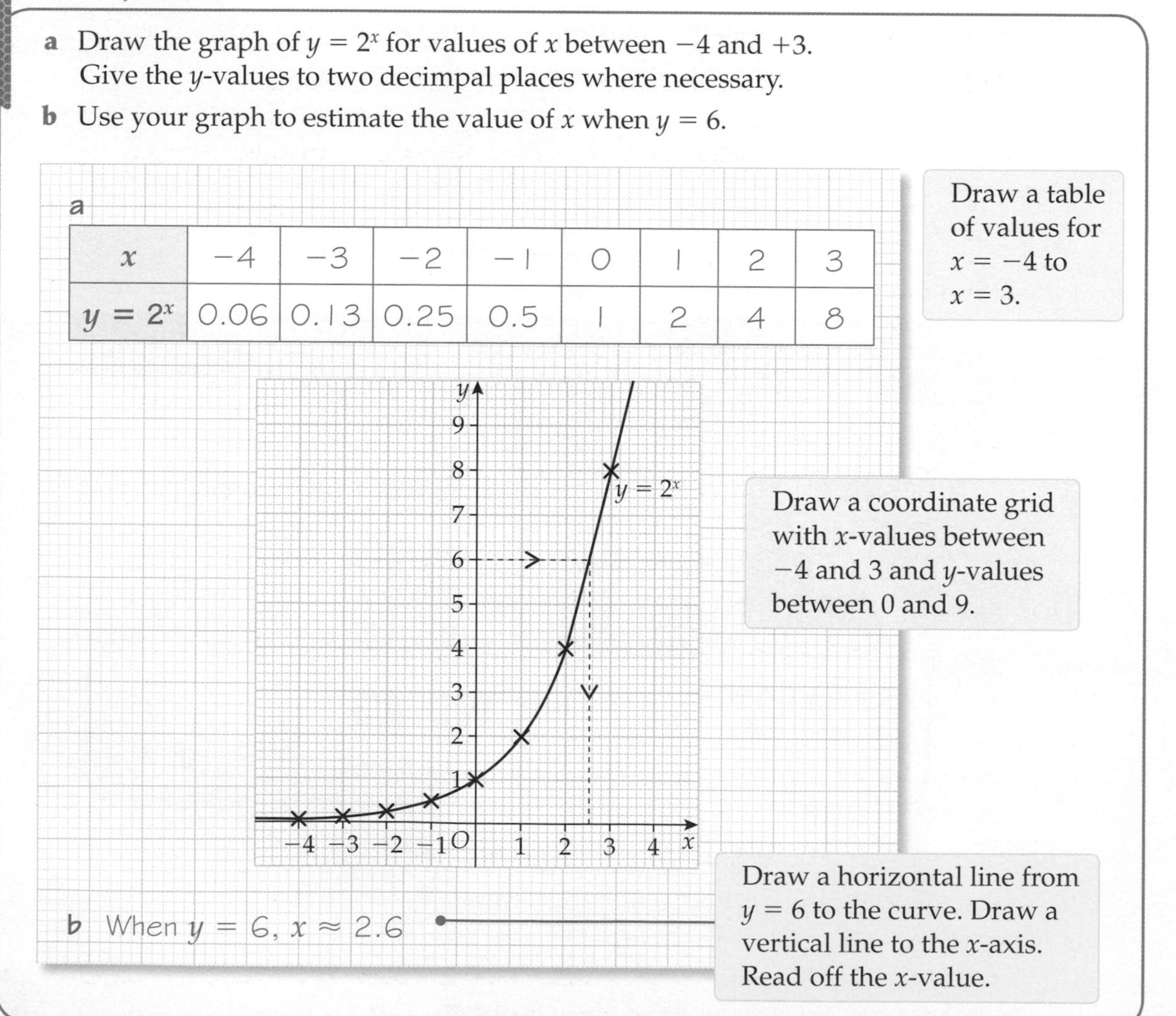

a

x	−4	−3	−2	−1	0	1	2	3
$y = 2^x$	0.06	0.13	0.25	0.5	1	2	4	8

Draw a table of values for $x = -4$ to $x = 3$.

Draw a coordinate grid with x-values between -4 and 3 and y-values between 0 and 9.

b When $y = 6$, $x \approx 2.6$

Draw a horizontal line from $y = 6$ to the curve. Draw a vertical line to the x-axis. Read off the x-value.

Exercise 35H

A

1 **a** Copy and complete the table of values for $y = \left(\frac{1}{2}\right)^x$.

x	−3	−2	−1	0	1	2	3	4
$y = \left(\frac{1}{2}\right)^x$	8	4			0.5	0.25		

b Draw the graph of $y = \left(\frac{1}{2}\right)^x$ for $-3 \leqslant x \leqslant 4$.

c Use your graph to estimate the value of x when $y = 5$.

2 **a** Copy and complete the table of values for $y = 1.5^{-x}$.

x	−1	0	1	2	3	4	5
$y = 1.5^{-x}$	1.5	1		0.44			0.13

b Draw the graph of $y = 1.5^{-x}$ for $-1 \leqslant x \leqslant 5$.

c Use your graph to estimate the value of y when $x = 3.5$.

A*

3 Nathan invests £2500 in an account that pays 6% compound interest per year.
The interest is added at the end of each year.
Assume that no further money is added or withdrawn.

a Show that the total amount (T) in the account after n years is given by

$$T = 2500 \times 1.06^n$$

b Draw a graph of $T = 2500 \times 1.06^n$ for $0 \leqslant n \leqslant 15$.

c Find the number of years it takes for the initial amount to double.

4 A sample of radioactive material decays according to the exponential function

$$N = 80 \times 2^{-\frac{t}{20}}$$

where N is the amount of radioactive material in grams and t is the time in years.

a Copy and complete the table of values for $0 \leqslant t \leqslant 140$.

t	0	20	40	60	80	100	120	140
N		40			5			0.6

b Draw a graph to show the decay of the material over 140 years.

c Use your graph to estimate the amount of radioactive material remaining after

i 10 years **ii** 50 years.

d A different radioactive material decays according to the function

$$N = 80 \times 2^{-\frac{t}{10}}$$

How would the graph of $N = 80 \times 2^{-\frac{t}{10}}$ compare with the graph of $N = 80 \times 2^{-\frac{t}{20}}$?

5 A culture of bacteria increases according to the function

$$N = 100 \times 2^t$$

where N is the number of bacteria and t is the time in hours.

a Use the formula to find the number of bacteria after

i 2 hours **ii** 20 minutes.

b Use the graph of $N = 100 \times 2^t$ to estimate the time taken for the number of bacteria

i to reach 3200 **ii** to increase from 800 to 1100.

AO2

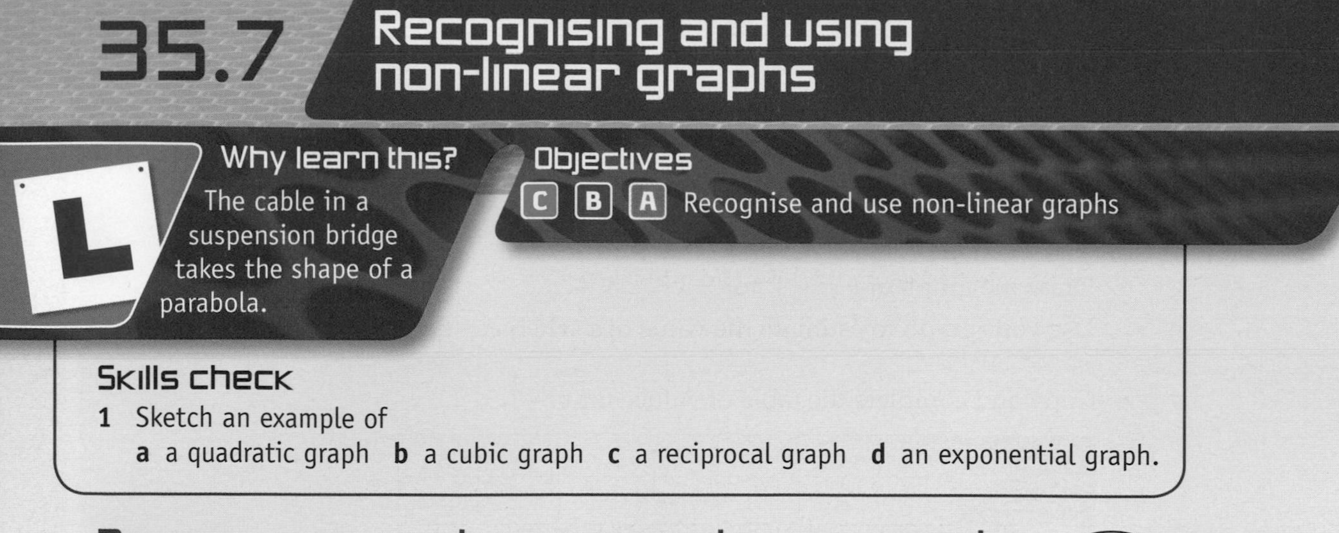

35.7 Recognising and using non-linear graphs

Why learn this?
The cable in a suspension bridge takes the shape of a parabola.

Objectives
C B A Recognise and use non-linear graphs

Skills check

1 Sketch an example of
a a quadratic graph **b** a cubic graph **c** a reciprocal graph **d** an exponential graph.

Recognising and using non-linear graphs

HELP Chapter 19

Linear and non-linear graphs can be used to represent a range of real-life situations.

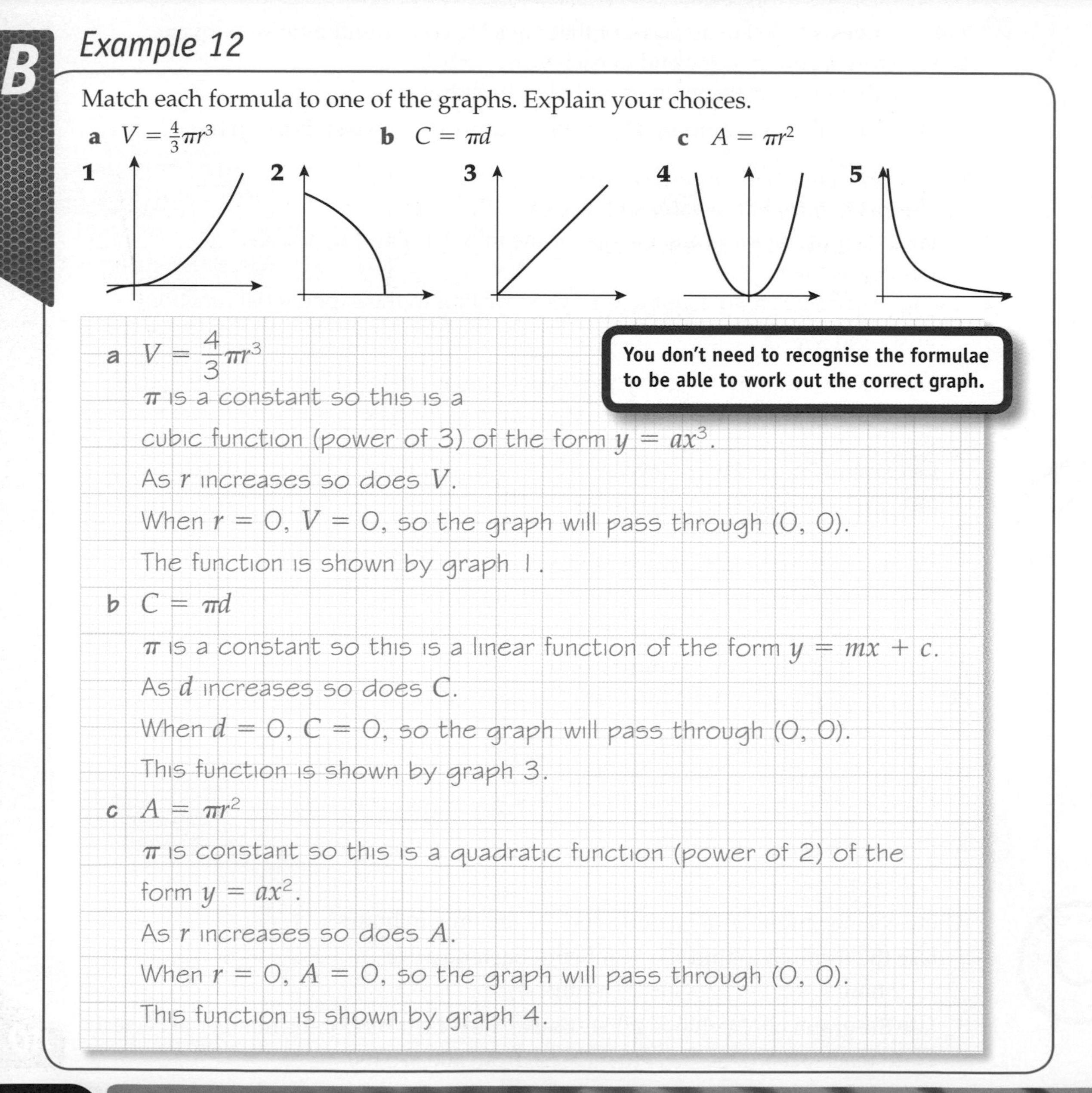

B

Example 12

Match each formula to one of the graphs. Explain your choices.

a $V = \frac{4}{3}\pi r^3$ **b** $C = \pi d$ **c** $A = \pi r^2$

1 **2** **3** **4** **5**

a $V = \frac{4}{3}\pi r^3$

π is a constant so this is a cubic function (power of 3) of the form $y = ax^3$.

As r increases so does V.

When $r = 0$, $V = 0$, so the graph will pass through (0, 0).

The function is shown by graph 1.

You don't need to recognise the formulae to be able to work out the correct graph.

b $C = \pi d$

π is a constant so this is a linear function of the form $y = mx + c$.

As d increases so does C.

When $d = 0$, $C = 0$, so the graph will pass through (0, 0).

This function is shown by graph 3.

c $A = \pi r^2$

π is constant so this is a quadratic function (power of 2) of the form $y = ax^2$.

As r increases so does A.

When $r = 0$, $A = 0$, so the graph will pass through (0, 0).

This function is shown by graph 4.

Exercise 35I

1 A toy rocket is shot vertically upwards.
The height of the rocket above the ground after t seconds is given by the function

$$h = 30t - 5t^2$$

where h is the height above the ground (in metres).

a Copy and complete the table of values for $h = 30t - 5t^2$.

t	0	1	2	3	4	5	6
h	0	25		45			

b Draw the graph of $h = 30t - 5t^2$.

c Use your graph to find the height of the rocket after 1.5 seconds.

d How long does it take the rocket to reach its maximum height?

2 Xabi needs to fence off an area of his garden for one of his pets.
He is investigating different sizes of rectangular enclosure.

$(x - 4)$ m

$(x - 2)$ m

The graph shows how the area of the enclosure changes as the value of x varies.

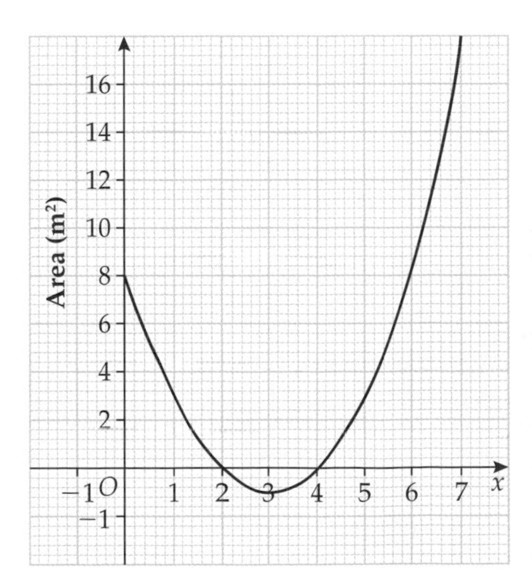

a What is the area of the enclosure when $x = 1$ m?

b What is the area of the enclosure when $x = 5$ m?

c Look at your answers to parts **a** and **b**.
Which of these values would not be possible in real life?
Give a reason for your answer.

d The area of the enclosure needs to be greater than 9 m².
What is the minimum value of x that Xabi can consider?

3 A tennis ball is thrown vertically upwards.
The height of the ball above the ground after t seconds is given by the function:

$$h = 20t - 5t^2 + 10$$

where h is the height above the ground (in metres).

a Copy and complete the table of values for $h = 20t - 5t^2 + 10$.

t	0	1	2	3	4	5
h	10	25			10	

b Draw the graph of $h = 20t - 5t^2 + 10$.

c What is the maximum height reached by the ball?

d How long does it take the ball to reach its maximum height?

e How long will it take the ball to reach the ground?

C

AO2

B

4 Water is poured into a conical container at a steady rate.
The base of the container is wider than the top.

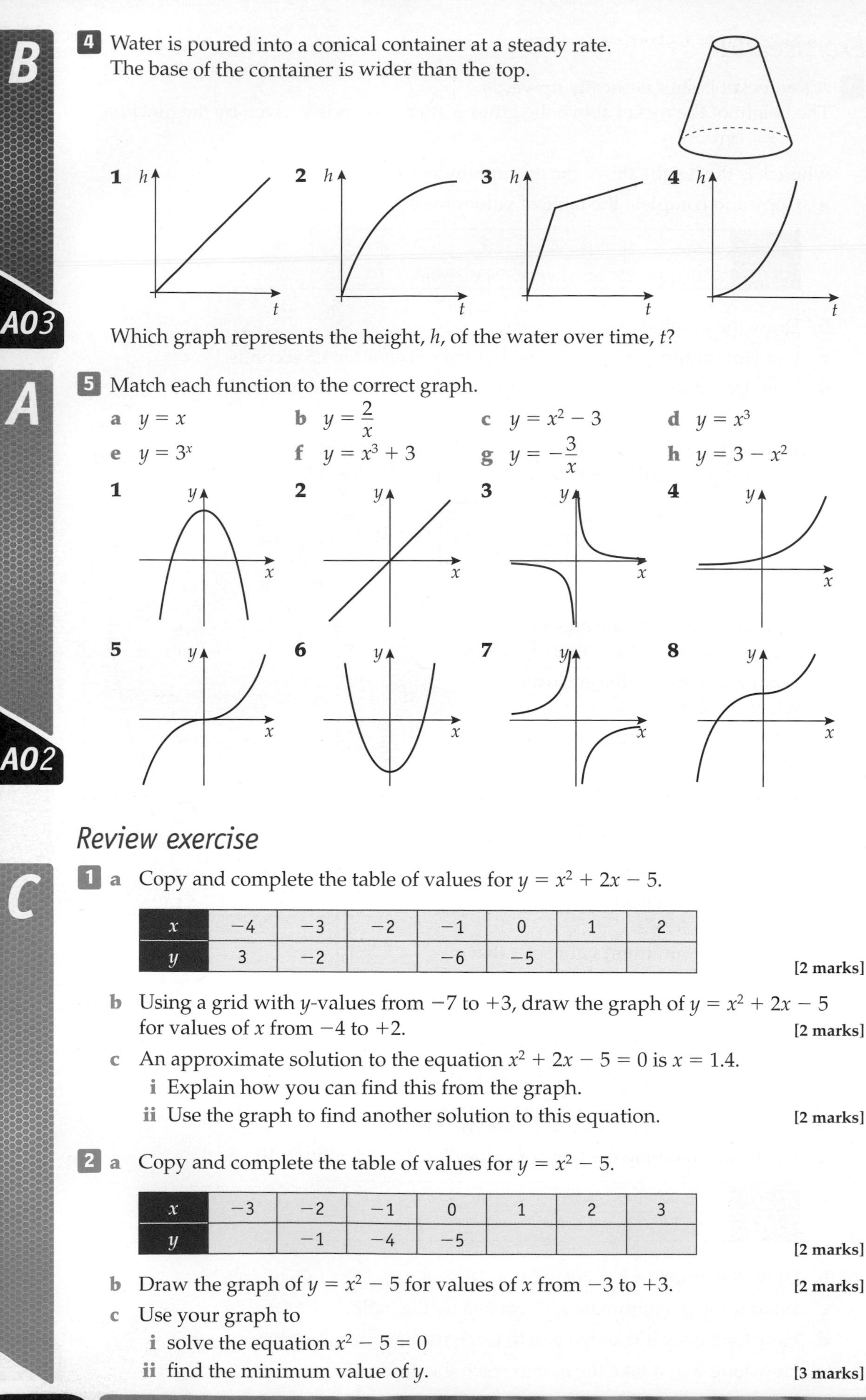

AO3

Which graph represents the height, h, of the water over time, t?

A

5 Match each function to the correct graph.

a $y = x$ **b** $y = \frac{2}{x}$ **c** $y = x^2 - 3$ **d** $y = x^3$

e $y = 3^x$ **f** $y = x^3 + 3$ **g** $y = -\frac{3}{x}$ **h** $y = 3 - x^2$

AO2

Review exercise

C

1 **a** Copy and complete the table of values for $y = x^2 + 2x - 5$.

x	−4	−3	−2	−1	0	1	2
y	3	−2		−6	−5		

[2 marks]

b Using a grid with y-values from −7 to +3, draw the graph of $y = x^2 + 2x - 5$ for values of x from −4 to +2. **[2 marks]**

c An approximate solution to the equation $x^2 + 2x - 5 = 0$ is $x = 1.4$.

i Explain how you can find this from the graph.

ii Use the graph to find another solution to this equation. **[2 marks]**

2 **a** Copy and complete the table of values for $y = x^2 - 5$.

x	−3	−2	−1	0	1	2	3
y		−1	−4	−5			

[2 marks]

b Draw the graph of $y = x^2 - 5$ for values of x from −3 to +3. **[2 marks]**

c Use your graph to

i solve the equation $x^2 - 5 = 0$

ii find the minimum value of y. **[3 marks]**

3 Three graphs are sketched.

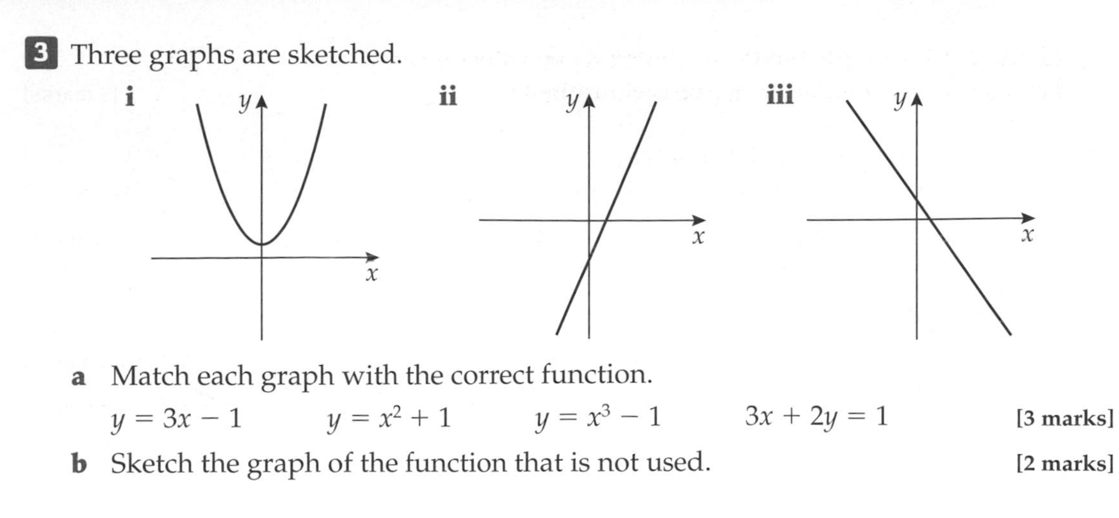

a Match each graph with the correct function.

$y = 3x - 1$ $\quad$ $y = x^2 + 1$ $\quad$ $y = x^3 - 1$ $\quad$ $3x + 2y = 1$ **[3 marks]**

b Sketch the graph of the function that is not used. **[2 marks]**

B

AO2

4 **a** Draw the graph of $xy = 4$ for $-3 \leqslant x \leqslant 3$. **[4 marks]**

b What are the equations of the two asymptotes? **[2 marks]**

A

5 **a** Copy and complete the table of values for the function $y = 2 - \frac{3}{x}$. **[2 marks]**

x	−3	−2	−1	−0.5	−0.2	0.2	0.5	1	2	3
2			2	2		2			2	2
$-\frac{3}{x}$			3	6		−15			−1.5	−1
$y = 2 - \frac{3}{x}$			5	8		−13			0.5	1

b Draw the graph of $y = 2 - \frac{3}{x}$ for $-3 \leqslant x \leqslant 3$. **[2 marks]**

c What are the equations of the two asymptotes? **[2 marks]**

6 **a** Copy and complete the table of values for $y = (0.75)^x$.

x	0	1	2	3	4	5
y	1		0.56		0.32	0.24

[1 mark]

b Draw the graph of $y = (0.75)^x$ for values of x from 0 to 5. **[2 marks]**

c Use your graph to solve the equation $(0.75)^x = 0.6$. **[1 mark]**

7 **a** Copy and complete the table of values for $y = x^2 - 5x - 3$.

x	−1	0	1	2	3	4	5	6
y	3	−3		−9		−7	−3	3

[1 mark]

C

A*

b Draw the graph of $y = x^2 - 5x - 3$ for $-1 \leqslant x \leqslant 6$. **[2 marks]**

c Write the solutions of $x^2 - 5x - 3 = 0$. **[1 mark]**

d By drawing an appropriate straight line, find the solutions to $x^2 - 4x + 1 = 0$. **[2 marks]**

AO3

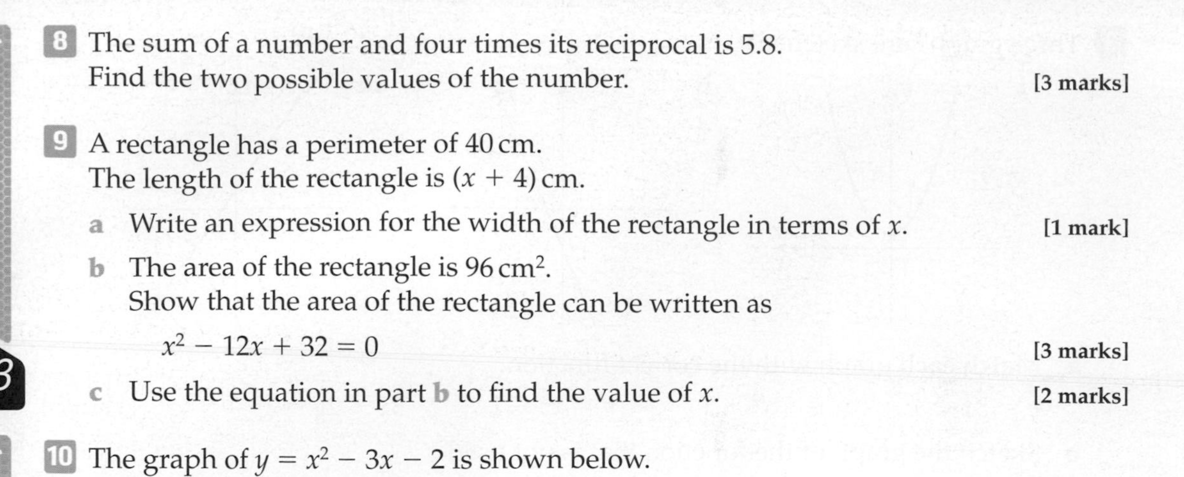

A*

8 The sum of a number and four times its reciprocal is 5.8.
Find the two possible values of the number. **[3 marks]**

9 A rectangle has a perimeter of 40 cm.
The length of the rectangle is $(x + 4)$ cm.

a Write an expression for the width of the rectangle in terms of x. **[1 mark]**

b The area of the rectangle is $96\,\text{cm}^2$.
Show that the area of the rectangle can be written as

$$x^2 - 12x + 32 = 0$$ **[3 marks]**

AO3

c Use the equation in part **b** to find the value of x. **[2 marks]**

A*

10 The graph of $y = x^2 - 3x - 2$ is shown below.

a By drawing an appropriate straight line, solve the equation $x^2 - 3x - 1 = 0$. **[2 marks]**

b By drawing an appropriate straight line, solve the equation $x^2 - 5x + 1 = 0$. **[2 marks]**

Chapter summary

In this chapter you have learned how to

- draw quadratic graphs D C
- identify the line of symmetry of a quadratic graph C
- draw cubic graphs B
- solve cubic equations graphically B
- recognise and use non-linear graphs C B A
- draw reciprocal graphs A
- identify the asymptotes of reciprocal graphs A
- draw exponential graphs A
- solve quadratic equations graphically C B A A*
- draw graphs of combined functions A*
- use quadratic equations to solve problems A*
- solve real-life problems involving exponential growth and decay A*

36 Further trigonometry

This chapter extends your knowledge of trigonometry to non-right-angled triangles.

This oscilloscope shows a sine wave. Oscilloscopes are widely used in medicine and telecommunications to see the wave shape of an electrical signal.

Objectives

This chapter will show you how to

- **plot graphs of the circular functions $y = \sin\theta$ and $y = \cos\theta$; recognise the characteristic shapes of these functions** A A*
- **calculate the area of a triangle using $\frac{1}{2}ab\sin C$** A A*
- **use the sine and cosine rules to solve 2-D and 3-D problems** A A*
- **solve simple trigonometric equations** A*

Before you start this chapter

1 Calculate length x.

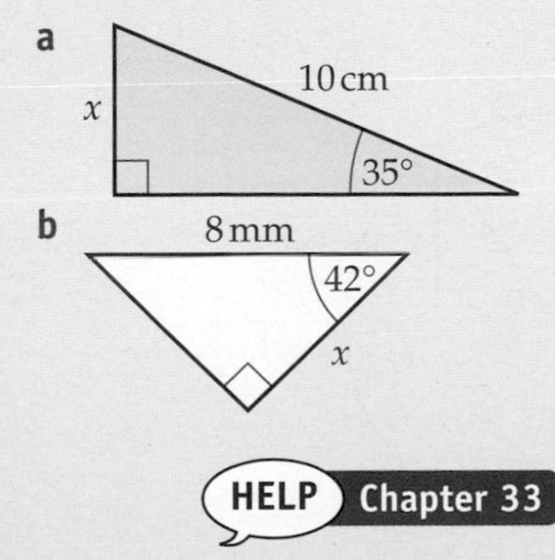

HELP Chapter 33

2 Calculate angle θ.

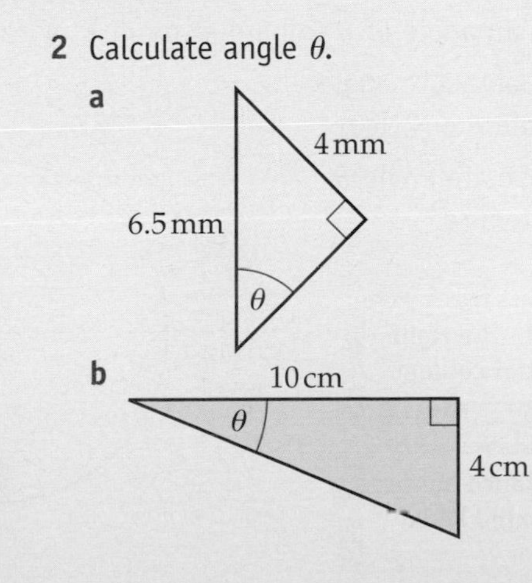

3 A circle has a radius of 5 cm. Sector *AOB* has an angle of 140°. Calculate the area of the sector.

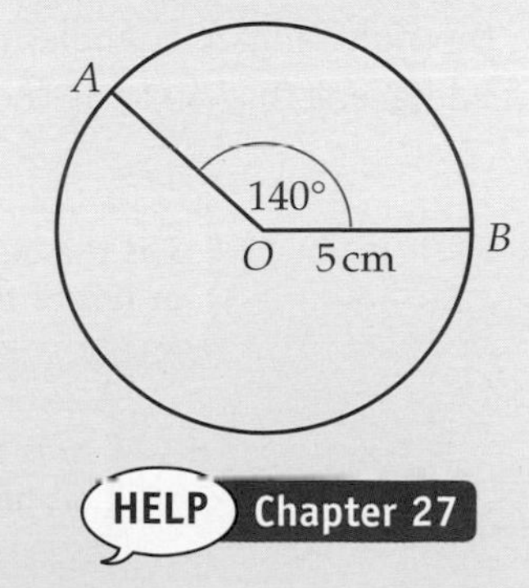

HELP Chapter 27

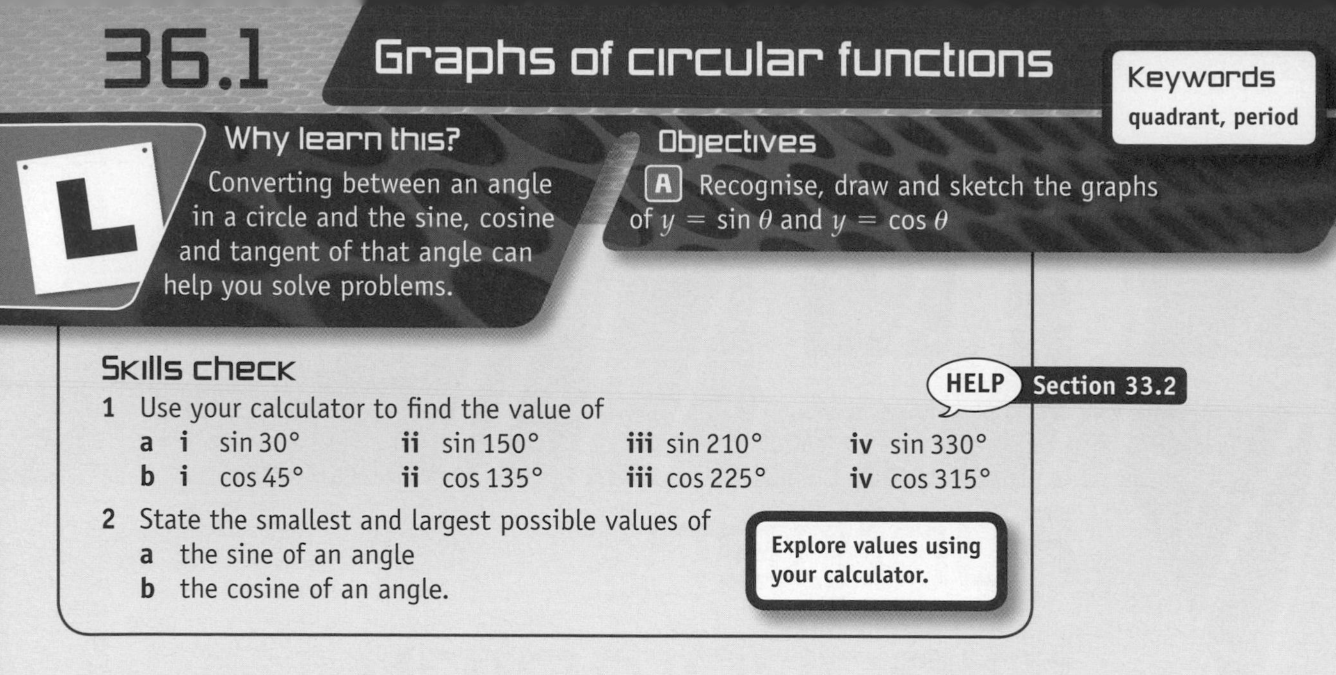

36.1 Graphs of circular functions

Keywords
quadrant, period

Why learn this?

Converting between an angle in a circle and the sine, cosine and tangent of that angle can help you solve problems.

Objectives

A Recognise, draw and sketch the graphs of $y = \sin\theta$ and $y = \cos\theta$

Skills check

HELP Section 33.2

1 Use your calculator to find the value of

a **i** $\sin 30°$ **ii** $\sin 150°$ **iii** $\sin 210°$ **iv** $\sin 330°$

b **i** $\cos 45°$ **ii** $\cos 135°$ **iii** $\cos 225°$ **iv** $\cos 315°$

2 State the smallest and largest possible values of

a the sine of an angle

b the cosine of an angle.

Explore values using your calculator.

Circular functions

In Chapter 33 you used the trigonometric ratios of the sine, cosine and tangent of acute angles in right-angled triangles. However, the sine, cosine and tangent of angles greater than 90° and of negative angles (less than 0°) also exist.

To understand more, imagine a new attraction at a theme park: the Terra Wheel.
The wheel has centre O and radius r.

You get into the pod at ground level on the right of the central column, and the wheel rotates anticlockwise.

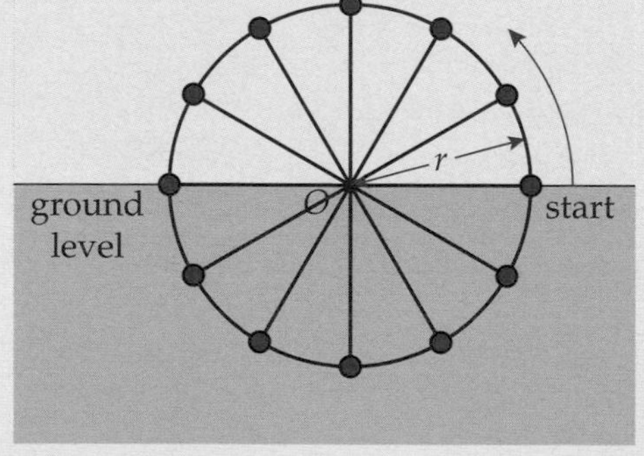

- When you have gone up into the air to the highest point, you have travelled through an angle of 90°.
- When you return to ground level, you have travelled through an angle of 180°.
- By the time you reach the lowest point, below ground level, you have travelled through an angle of 270°.
- You return to your start position after one complete revolution (360°).

As you go on your ride, your position is related to the sine, cosine and tangent of the angle the wheel has turned through.

When the wheel has turned through an angle of θ you are at point P, with coordinates (x, y).

When the wheel has turned through an acute angle (between 0° and 90°), both x and y are positive.

In the right-angled triangle, y is opposite angle θ, x is adjacent to θ and r is the hypotenuse.

x is the distance to the right or left of the central column.

y is the distance above or below ground level.

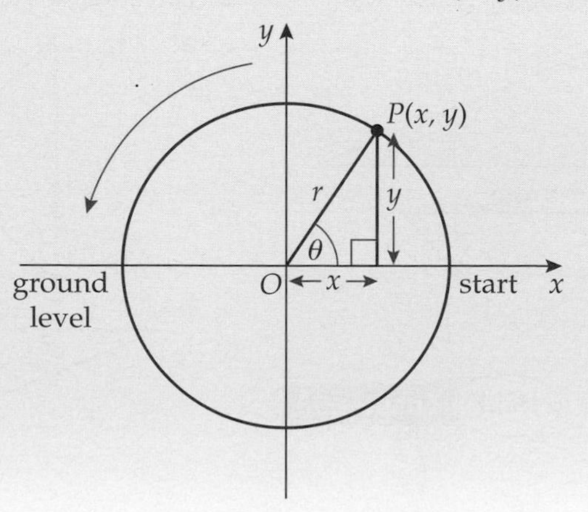

Using the trigonometric ratios that you learnt in Chapter 33

$$\sin\theta = \frac{y}{r}$$

$$\cos\theta = \frac{x}{r}$$

$$\tan\theta = \frac{y}{x}$$

r is always positive.

So far you have only dealt with the sine, cosine and tangent of acute angles in right-angled triangles.

However, this is part of a bigger picture. The sine, cosine and tangent of angles greater than 90° and of negative angles (less than 0°) also exist.

Rotating the wheel *clockwise*, so that you go down into the ground first, will give you a negative angle.

We can use the three trigonometric ratios above for any angle, whether it is negative or very large.

There is no limit to the size of the angle. For example, an angle of 1080° represents three full rotations of the wheel.

This diagram shows P on the left of the central column, when θ is obtuse (between 90° and 180°).

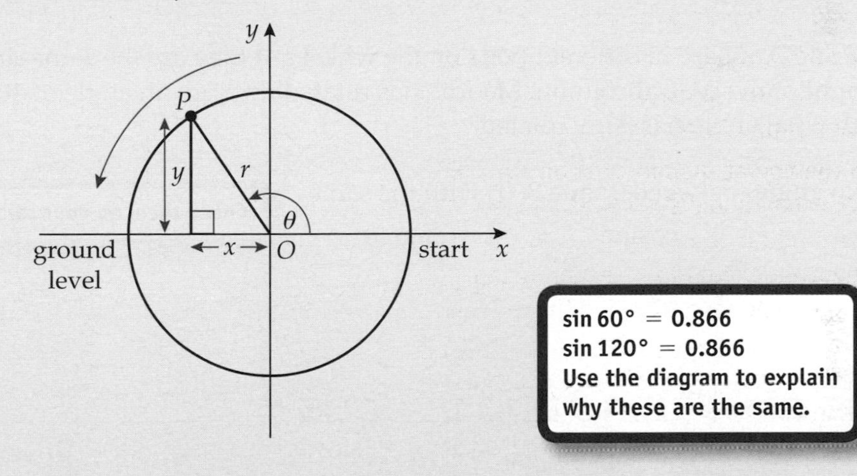

sin 60° = 0.866
sin 120° = 0.866
Use the diagram to explain why these are the same.

You will see from your calculator that some values of sine, cosine and tangent are negative. This is because y is negative when P is below ground level and x is negative when P is on the left of the central column.

The sign of the sine, cosine or tangent of the angle depends on the **quadrant** in which P lies.

For example, when θ is obtuse (between 90° and 180°) and P is in the second quadrant

- the y-coordinate is positive (above ground level), so

  $\sin\theta = \frac{y}{r} = \frac{(+)}{(+)} = (+)$

- the x-coordinate is negative (to the left of the central column), so

  $\cos\theta = \frac{x}{r} = \frac{(-)}{(+)} = (-)$ and $\tan\theta = \frac{y}{x} = \frac{(+)}{(-)} = (-)$.

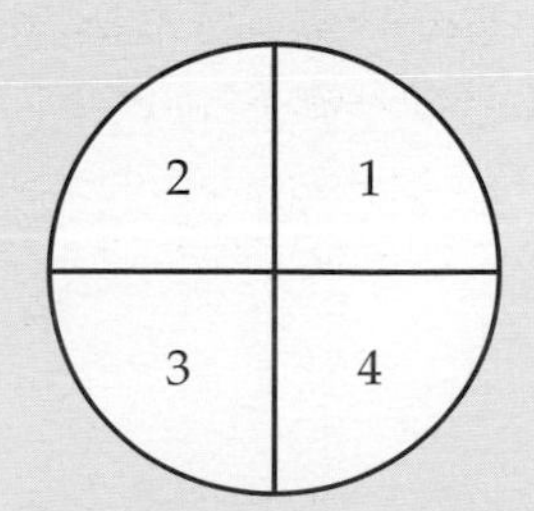

Remember that r is always positive.

Exercise 36A

A

1 Sketch your own wheel diagram.
Mark the position of P on the diagram for each of these angles.
Work out whether each trigonometric ratio is positive or negative.
Do not use a calculator.

	Angle turned through	Positive or negative?		
		$\sin\theta$	$\cos\theta$	$\tan\theta$
a	170°			
b	230°			
c	50°			
d	350°			
e	460°			
f	−20°			
g	−100°			

2 Billy and Dan are in different pods on the wheel but they are at the same height above the ground. Billy has rotated through an angle of 150°.
Through what angle has Dan rotated?

3 Maddie and Amy are in different pods on the wheel but they are the same distance to the right of the central column. Maddie has rotated through an angle of 40°.
Through what angle has Amy rotated?

4 Find two angles between 0° and 360° with the same
a sine **b** cosine **c** tangent.

Check them on your calculator.

Graphs of $y = \sin\theta$ and $y = \cos\theta$

Consider a circle, centre O, of radius 1 unit.

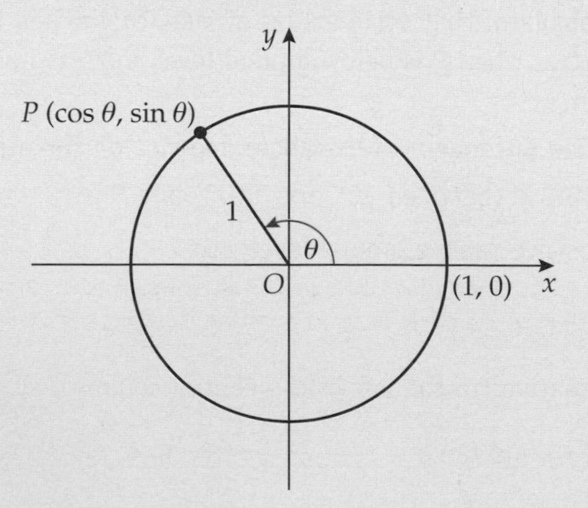

When $r = 1$, for any position P on the wheel

- the x-coordinate gives the value of $\cos\theta$
- the y-coordinate gives the value of $\sin\theta$.

The graphs of $y = \sin\theta$ and $y = \cos\theta$ can be plotted from a scale drawing of the wheel.

The graph of $y = \sin\theta$

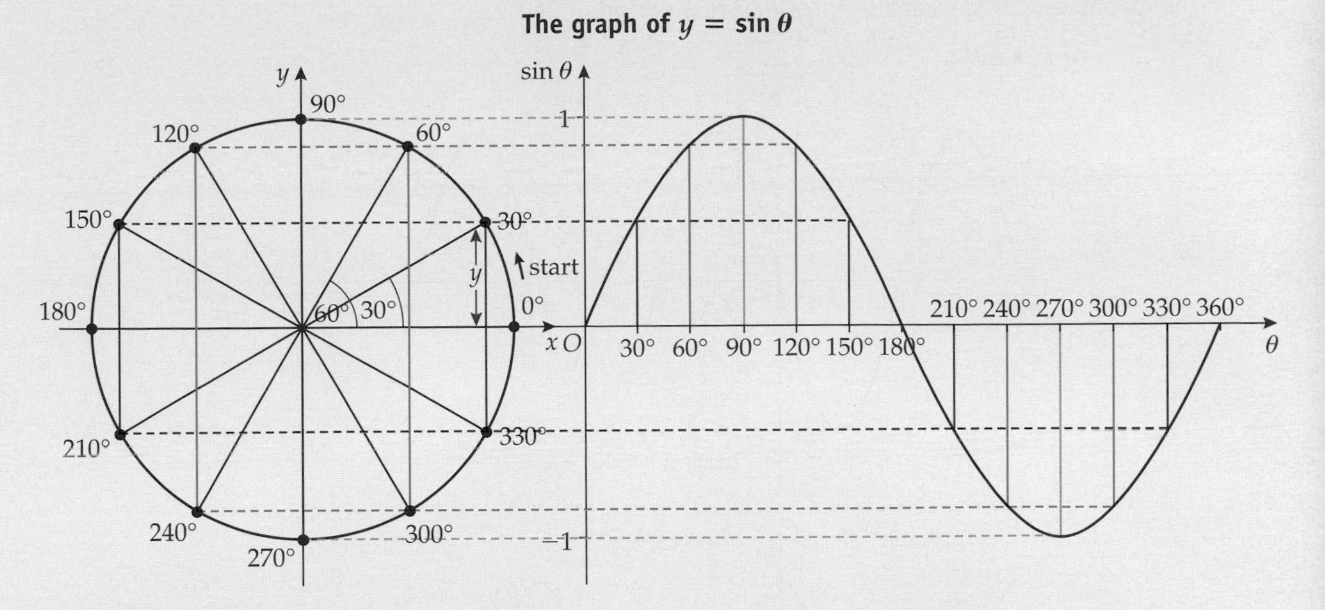

The graph of $y = \sin\theta$ has a maximum value of $+1$ and a minimum value of -1.

The graph of $y = \cos\theta$

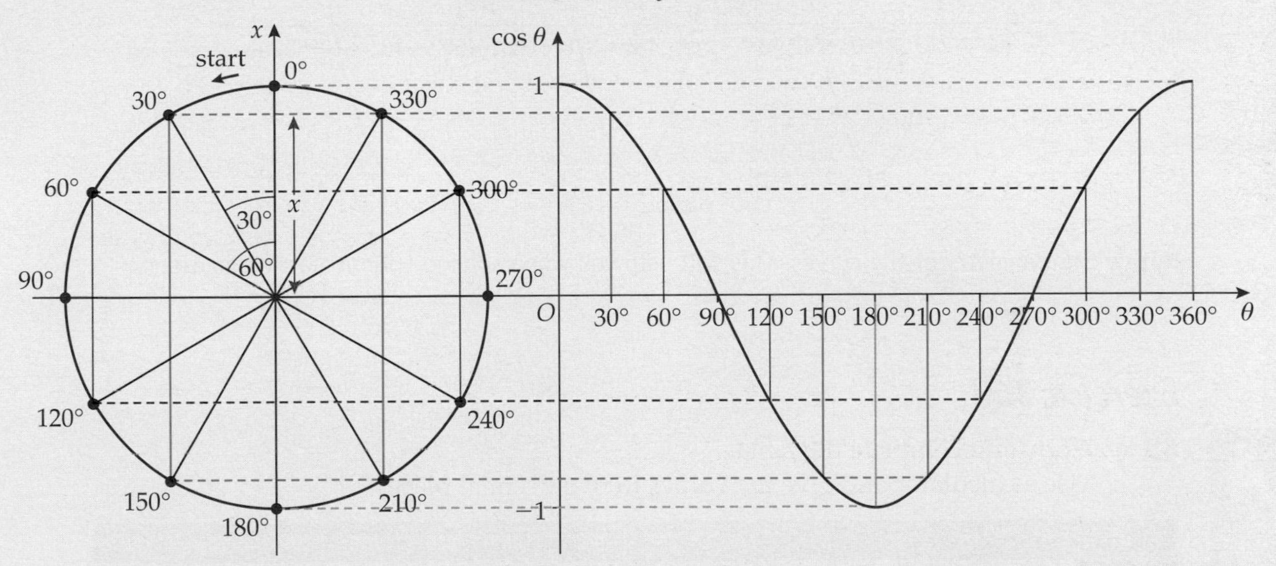

The graph of $y = \cos\theta$ also has a maximum value of $+1$ and a minimum value of -1.

Notice that $y = \cos\theta$ is 90° 'ahead' of $y = \sin\theta$.

To see how this works, turn your page through 90° clockwise.

If you do further rotations, the curves repeat every 360°.

The graphs of $y = \sin\theta$ and $y = \cos\theta$ have a **period** of 360°.

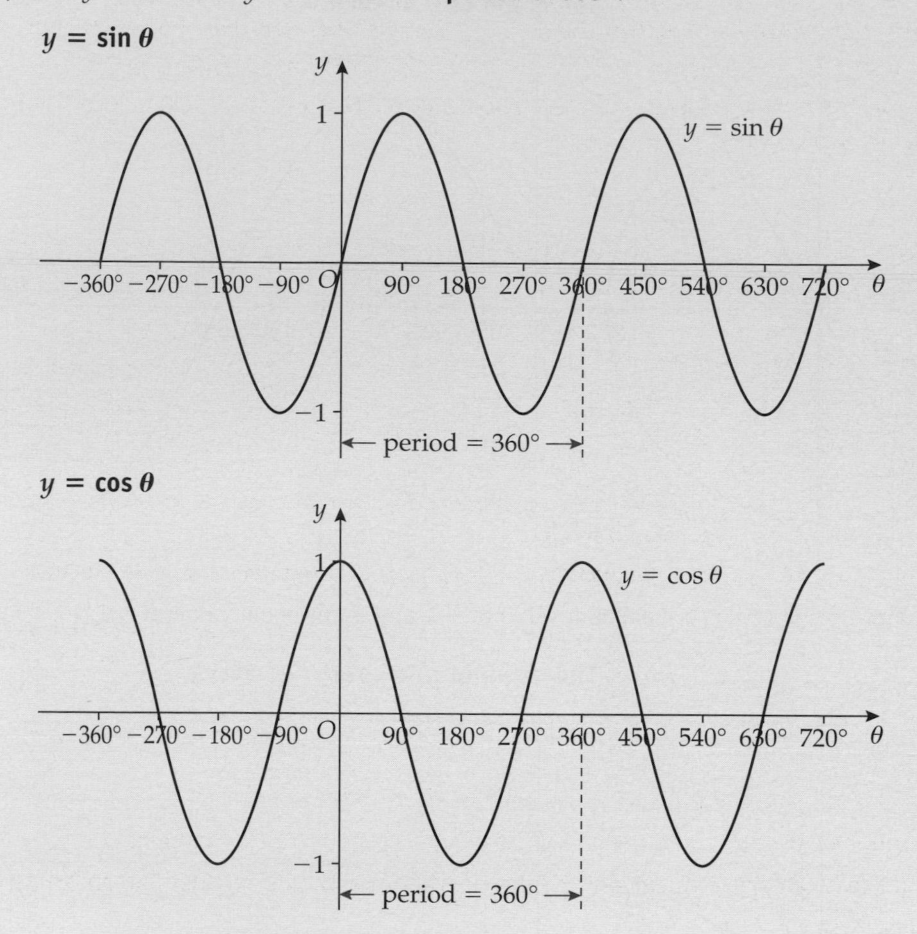

Notice the symmetry of the curves. This will help you when solving trigonometric equations.

Exercise 36B

A

1 **a** Copy and complete the table.
Use a calculator and give the values to two decimal places.

θ	0°	15°	30°	45°	60°	75°	90°	105°	120°	135°	150°	165°	180°
$\sin\theta$	0	0.26		0.71	0.87		1	0.97			0.5		0
$\cos\theta$	1	0.97	0.87			0.26			−0.5	−0.71			−1

θ	180°	195°	210°	225°	240°	255°	270°	285°	300°	315°	330°	345°	360°
$\sin\theta$	0		−0.5		−0.87			−0.97				−0.26	0
$\cos\theta$	−1	−0.97			−0.5	−0.26	0		0.5	0.71			1

b On graph paper, draw the graph of $y = \sin\theta$ for values of θ between 0° and 360°.

c Use a new sheet of graph paper. Draw the graph of $y = \cos\theta$ for values of θ between 0° and 360°.

d Explain why it would have been possible to draw both curves without needing any more values than the ones originally given in the table.

Choose your scales carefully so that the curve fits on the sheet of graph paper and the scales are easy to read. Always remember to label your graphs.

2 **a** For values of θ between 0° and 360°, sketch the graphs of $y = \sin\theta$ and $y = \cos\theta$ on the same grid.

> **A sketch should show the general shape of the curve and important values on the axes. Label where the curve crosses the axes.**

b Use your calculator and sketch from part **a** to find the values of θ, in the interval 0° to 360°, where $\sin\theta = \cos\theta$.

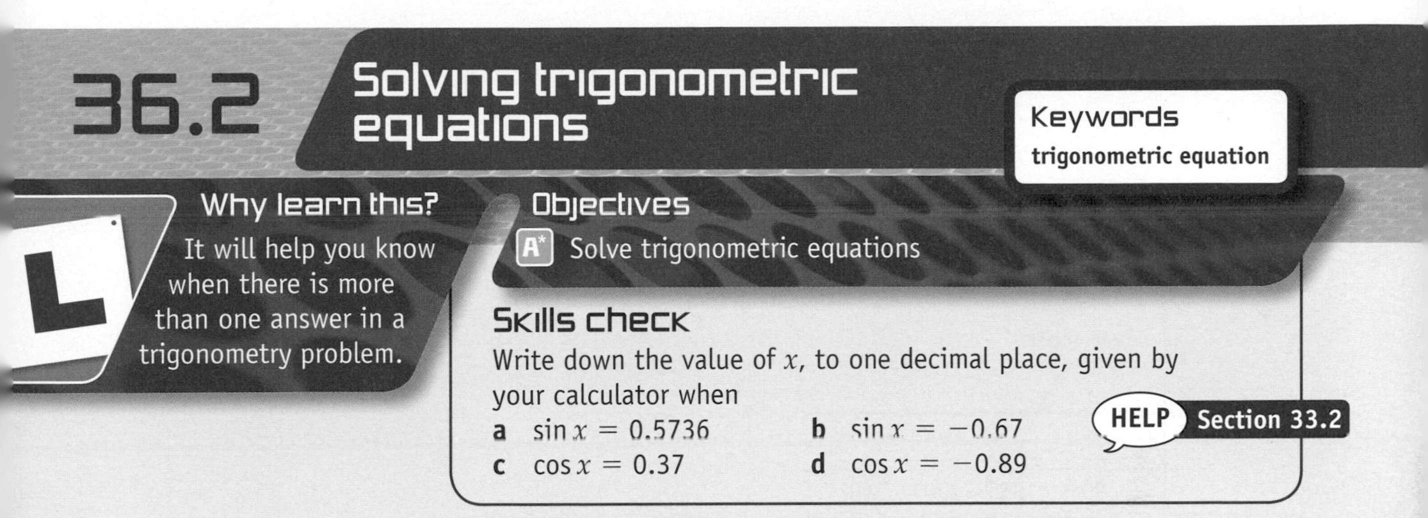

36.2 Solving trigonometric equations

Keywords
trigonometric equation

Why learn this?
It will help you know when there is more than one answer in a trigonometry problem.

Objectives
A* Solve trigonometric equations

Skills check
Write down the value of x, to one decimal place, given by your calculator when

a $\sin x = 0.5736$ **b** $\sin x = -0.67$

c $\cos x = 0.37$ **d** $\cos x = -0.89$

HELP Section 33.2

Solving trigonometric equations

A **trigonometric equation** is an equation involving a trigonometric function.

When you use a calculator to solve a trigonometric equation, you will get one answer.

For example, $\sin x = 0.5$

$$x = \sin^{-1} 0.5$$
$$= 30°$$

This is known as the principal value.

However, the symmetry and periodic properties of the sine and cosine curves mean that there are an infinite number of solutions to any trigonometric equation.

Other solutions of $\sin x = 0.5$ include: $x = 150°, 390°, 510°, 750° \ldots$

A question will ask you to solve the equation for a given interval of x.

Example 1

The diagram shows a sketch of the graph of $y = \sin x$ for values of x between 0° and 360°.

You are told that $\sin 30° = 0.5$.

Use the symmetry of the curve to answer the following.

a State another value of x in the interval $0° \leqslant x \leqslant 360°$ that satisfies the equation $\sin x = 0.5$.

b Solve the equation $\sin x = -0.5$ in the interval $0° \leqslant x \leqslant 360°$.

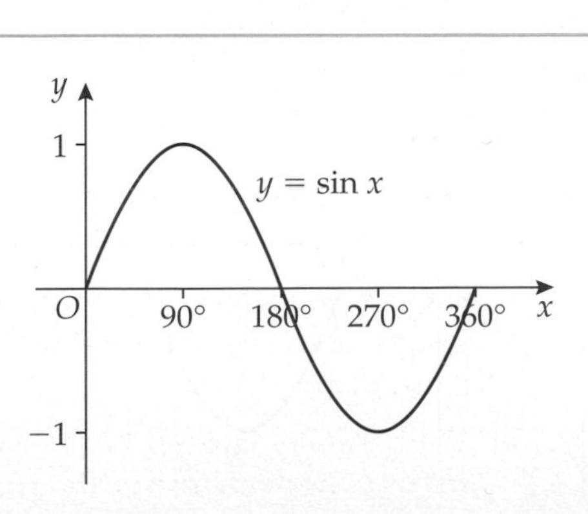

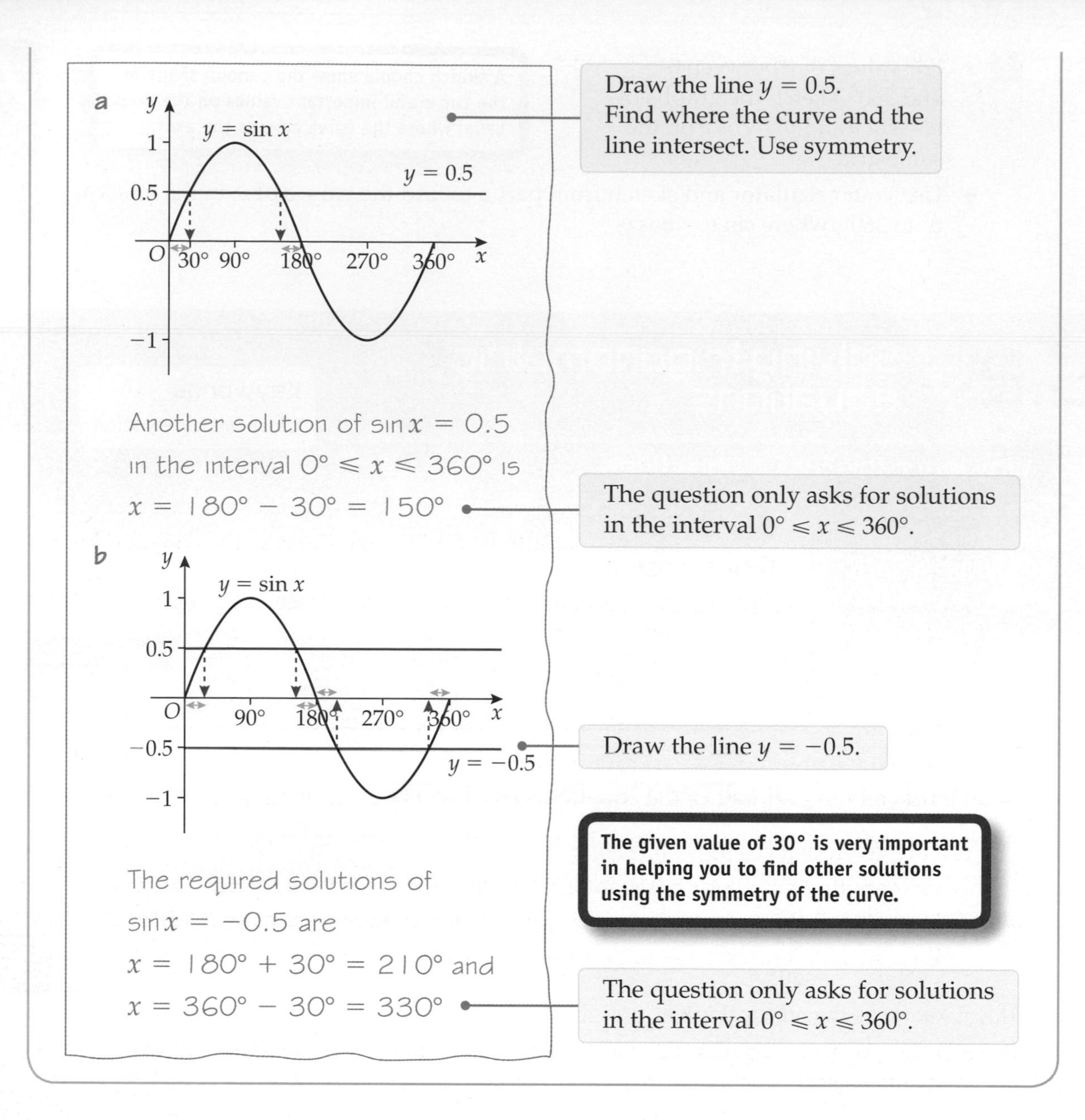

a
y = sin x
y = 0.5
1
0.5
O
30° 90° 180° 270° 360° x
−1
Draw the line y = 0.5.
Find where the curve and the line intersect. Use symmetry.
Another solution of sin x = 0.5 in the interval 0° ⩽ x ⩽ 360° is
x = 180° − 30° = 150°
The question only asks for solutions in the interval 0° ⩽ x ⩽ 360°.
b
y = sin x
1
0.5
O
90° 180° 270° 360° x
−0.5
y = −0.5
−1
Draw the line y = −0.5.
The given value of 30° is very important in helping you to find other solutions using the symmetry of the curve.
The required solutions of sin x = −0.5 are
x = 180° + 30° = 210° and
x = 360° − 30° = 330°
The question only asks for solutions in the interval 0° ⩽ x ⩽ 360°.

Example 2

Use a sketch and calculator to solve the equation $\cos x = 0.309$ for values of x between $0°$ and $540°$.

Give your answers to the nearest degree.

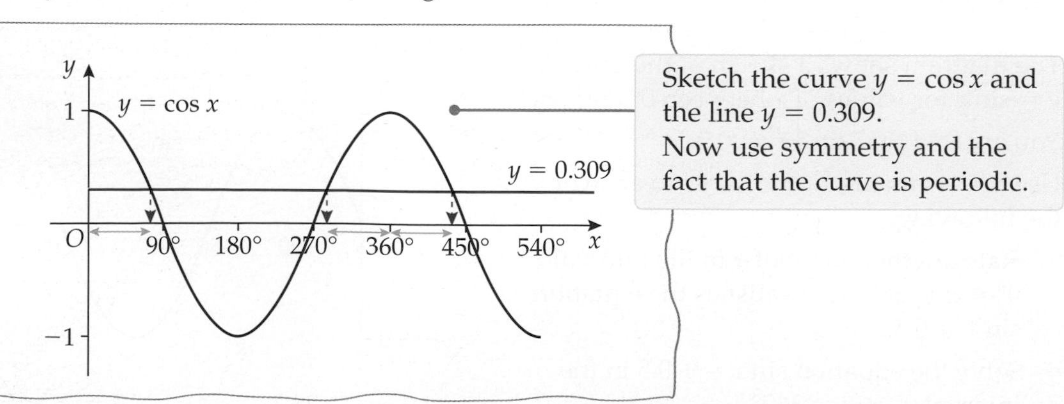

Sketch the curve y = cos x and the line y = 0.309.

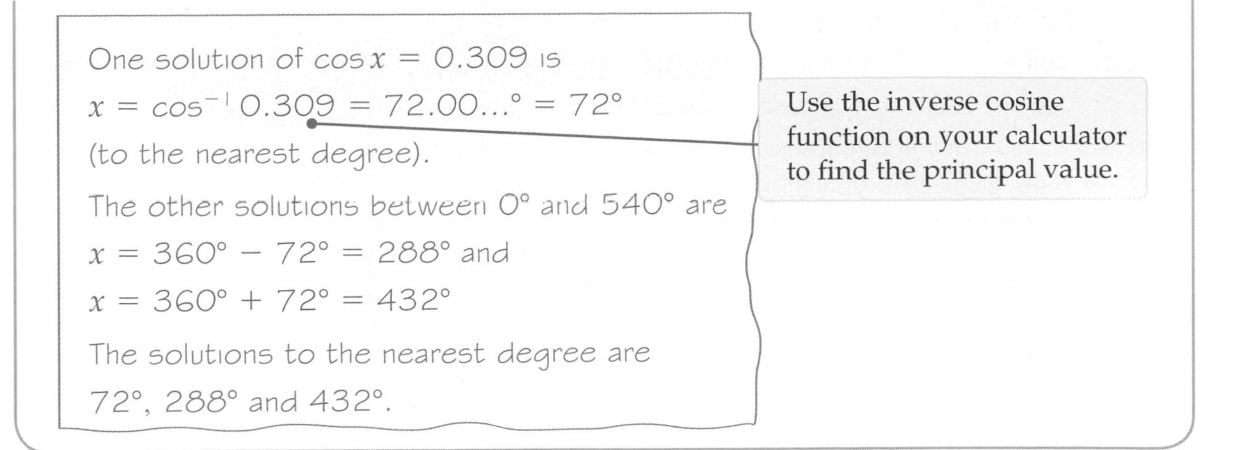

One solution of $\cos x = 0.309$ is

$x = \cos^{-1} 0.309 = 72.00...° = 72°$

(to the nearest degree).

The other solutions between 0° and 540° are

$x = 360° - 72° = 288°$ and

$x = 360° + 72° = 432°$

The solutions to the nearest degree are

72°, 288° and 432°.

Use the inverse cosine function on your calculator to find the principal value.

Exercise 36C

1 For each part, use a sketch and calculator to find all the solutions of the equation in the interval $0° \leqslant x \leqslant 360°$.
Give your answers to the nearest degree.

a $\sin x = 0.8$ b $\cos x = 0.75$ c $\sin x = -0.24$

d $\cos x = -0.3$ e $\sin x = 0.866$ f $\cos x = 0.707$

2 a Find the two values of a that satisfy the equation $a^2 = 0.25$.

b Sketch the graph of $y = \cos x$ for $0° \leqslant x \leqslant 360°$.

c Find all the solutions of the equation $(\cos x)^2 = 0.25$ in the interval $0° \leqslant x \leqslant 360°$.

3 You are given that $\cos 38° = 0.7880$.

a Find another value of x in the interval $0° \leqslant x \leqslant 360°$ for which $\cos x = 0.7880$.

b Find all values of x in this interval that satisfy the equation $\cos x = -0.7880$.

c Find all solutions of the equation $\sin x = 0.7880$ in this interval.

A*

A*

AO2

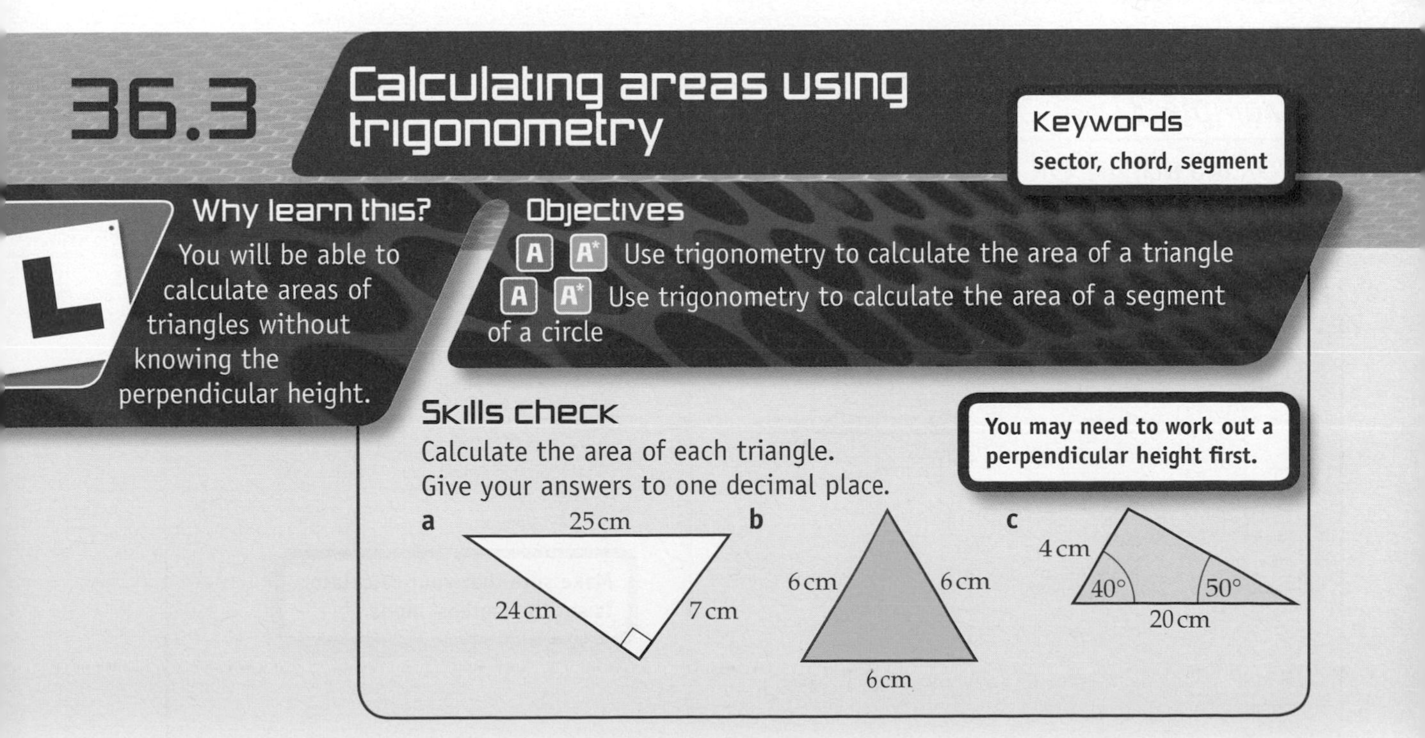

36.3 Calculating areas using trigonometry

Keywords
sector, chord, segment

Why learn this?

You will be able to calculate areas of triangles without knowing the perpendicular height.

Objectives

A A* Use trigonometry to calculate the area of a triangle

A A* Use trigonometry to calculate the area of a segment of a circle

Skills check

Calculate the area of each triangle.
Give your answers to one decimal place.

You may need to work out a perpendicular height first.

Area of a triangle

So far, you have found the area of a triangle using the formula

Area of triangle $= \frac{1}{2} \times$ base $\times$ perpendicular height

However, you can you can also work out the area of a triangle if you know the lengths of any two sides and the size of the angle between them. This is very useful as it works with acute- and obtuse-angled triangles and you do not need to find the perpendicular height first.

Remember how to label the sides of a triangle. Side a is opposite $\angle A$, side b is opposite $\angle B$ and side c is opposite $\angle C$.

The diagram shows a general triangle ABC with side lengths a, b and c.

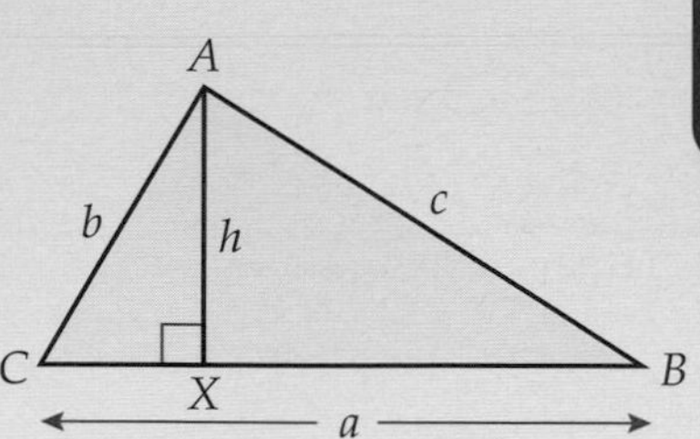

The area of the triangle $= \frac{1}{2} \times$ base $\times$ perpendicular height

$= \frac{1}{2}ah$

The height h is given by the trigonometric ratio $\sin C = \dfrac{h}{b}$, giving $h = b \sin C$.

Substituting this value of h into the original equation gives

Area $\triangle ABC = \frac{1}{2}ab \sin C$

This formula is given on the exam paper.

Similarly,

Area $\triangle ABC = \frac{1}{2}bc \sin A$

Area $\triangle ABC = \frac{1}{2}ac \sin B$

If you know $\angle C$, you need sides a and b.
If you know $\angle A$, you need sides b and c.
If you know $\angle B$, you need sides a and c.

Remember, the area of a triangle is given by half the product of two sides multiplied by the sine of the angle between them.

A

Example 3

PQR is a triangle. $QR = 12$ cm, $QP = 8$ cm and $\angle Q$ is 60°. Calculate the area of $\triangle PQR$. Give your answer in cm² to one decimal place.

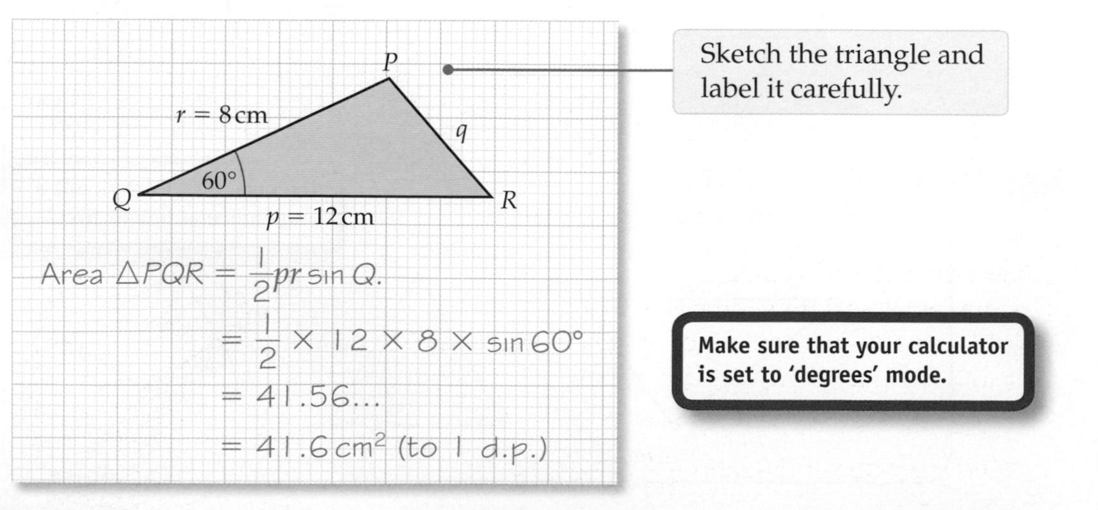

P
r = 8 cm
q
60°
Q
R
p = 12 cm
Area △PQR = ½pr sin Q
= ½ × 12 × 8 × sin 60°
= 41.56...
= 41.6 cm² (to 1 d.p.)
Sketch the triangle and label it carefully.
Make sure that your calculator is set to 'degrees' mode.

Example 4

XYZ is a triangle.

$XZ = 21.5$ cm, $XY = 15$ cm and $\angle X$ is acute.

The area of the triangle is 90 cm^2.

Calculate $\angle X$. Give your answer to the nearest degree.

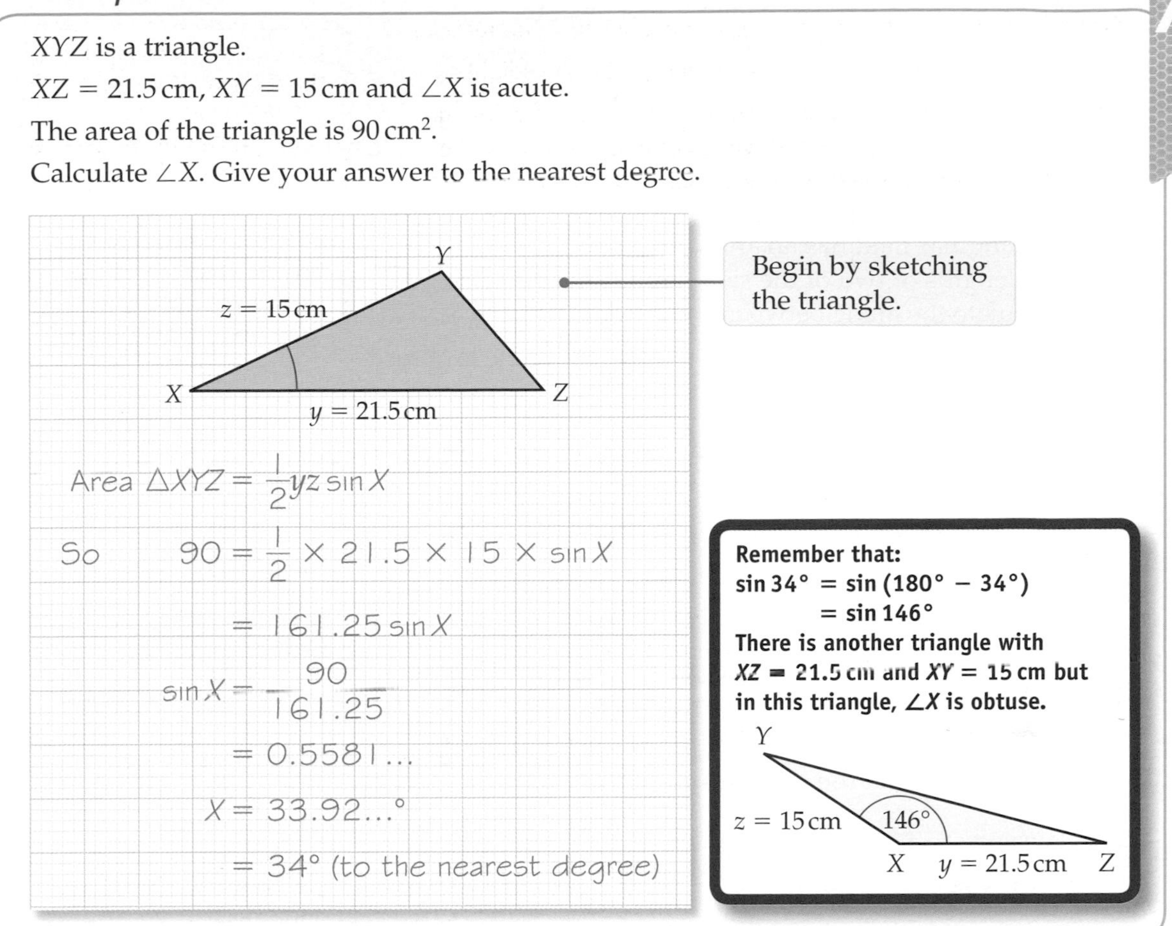

A*

Exercise 36D

A

1 Calculate the area of each of these shapes.
Give your answer to one decimal place when appropriate.

2 AB is a diameter of a circle.
C is the centre and the radius is 5 cm.
Point P is on the circumference of the circle and $\angle BCP = 75°$.

a Calculate the area of $\triangle BCP$.

b Calculate the area of $\triangle ACP$.

c Find the value of the ratio $\dfrac{\text{area of } \triangle BCP}{\text{area of } \triangle BAP}$.

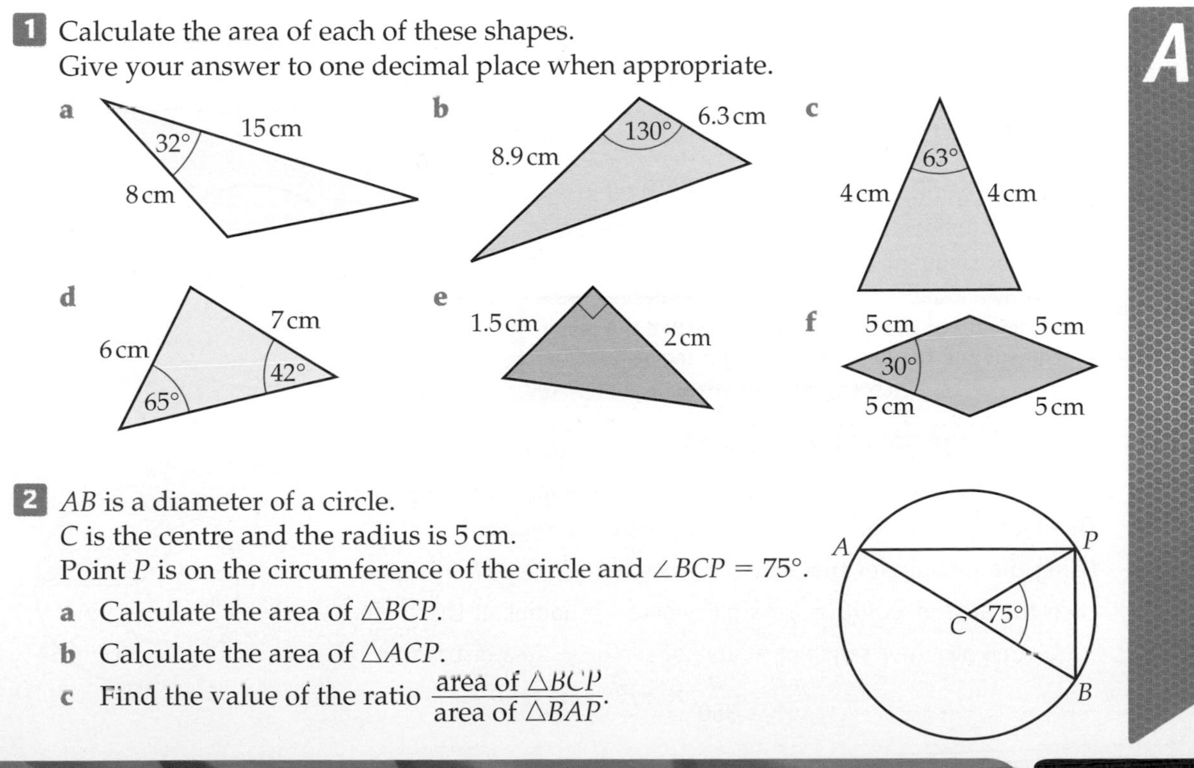

3 $ABCD$ is a parallelogram.
$AB = 5.4$ cm, $BC = 8.8$ cm and $\angle ADC = 130°$.
Calculate the area of $ABCD$.

4 A triangle has an area of 20 cm^2.
Two sides of the triangle are 6 cm and 8 cm and the angle between them is acute.
Calculate the angle between these two sides.

5 A triangular field has an area of 8414 m^2.
The lengths of two of the sides are 120 m and 410 m.
Calculate the angle between these sides if

a the angle is acute **b** the angle is obtuse.

Use the symmetry of the sine curve to help you.

AO2

6 a The diagram shows an equilateral triangle with side length 2 cm.

i Use Pythagoras' theorem to show that the perpendicular height is $\sqrt{3}$ cm.
Do not use a calculator.

ii Show that the exact value of $\sin 60°$ is $\frac{\sqrt{3}}{2}$.

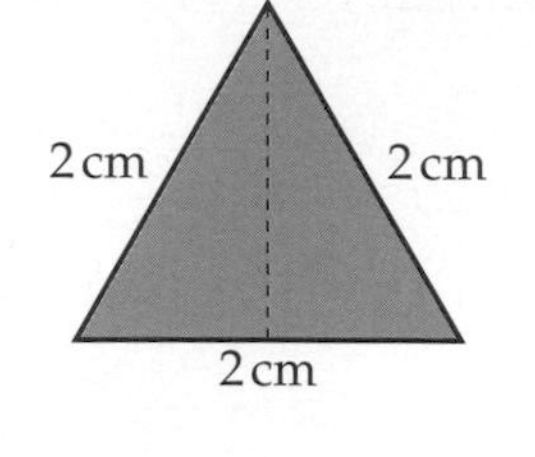

b A tetrahedron is constructed from four equilateral triangles, each with side length a cm.

i Sketch the net of the tetrahedron.

ii Show that the surface area of the tetrahedron is $\sqrt{3}a^2$.

AO3

Sectors

The diagram shows a circle with radii OA and OB.

The radii divide the circle into two **sectors**, the minor sector and the major sector.

The smaller of the two sectors is called the minor sector and the larger is the major sector.

Area of a sector $= \frac{\theta}{360} \times \pi r^2$

Segments

The diagram shows a circle with a **chord** AB.

The chord divides the circle into two **segments**, the minor segment and the major segment.

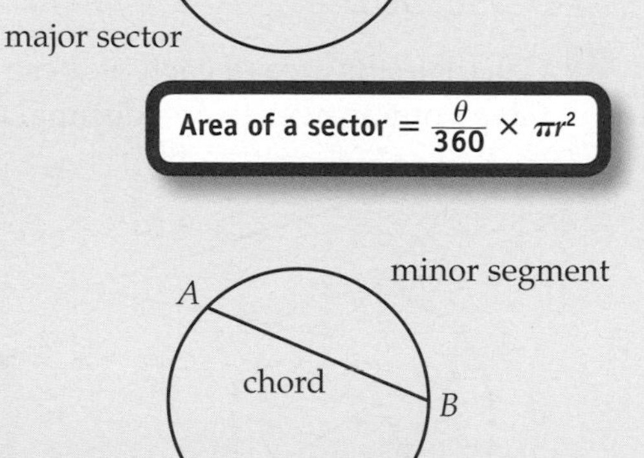

The smaller of the two segments is called the minor segment and the larger is the major segment.

Area of a segment

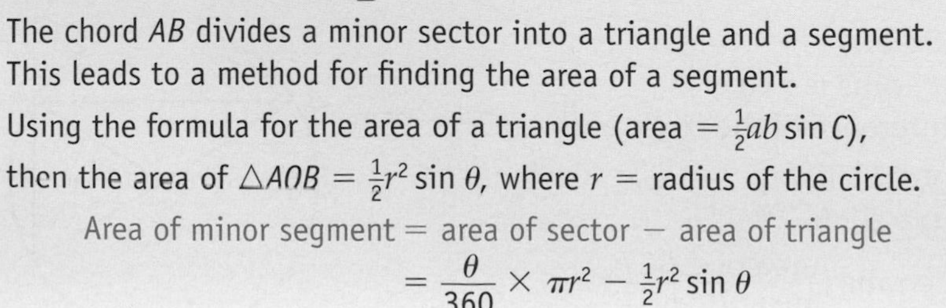

The chord AB divides a minor sector into a triangle and a segment.
This leads to a method for finding the area of a segment.

Using the formula for the area of a triangle (area $= \frac{1}{2}ab\sin C$), then the area of $\triangle AOB = \frac{1}{2}r^2 \sin\theta$, where r = radius of the circle.

$$\text{Area of minor segment} = \text{area of sector} - \text{area of triangle}$$
$$= \frac{\theta}{360} \times \pi r^2 - \frac{1}{2}r^2 \sin\theta$$

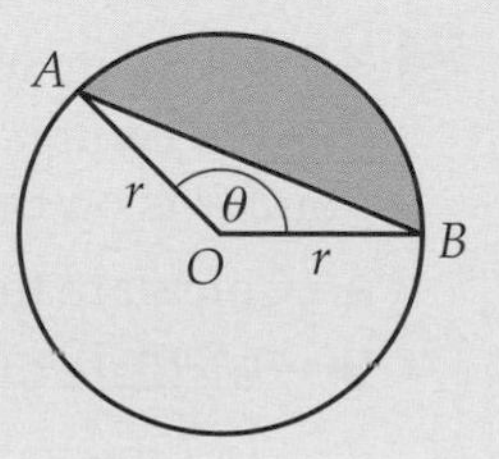

Example 5

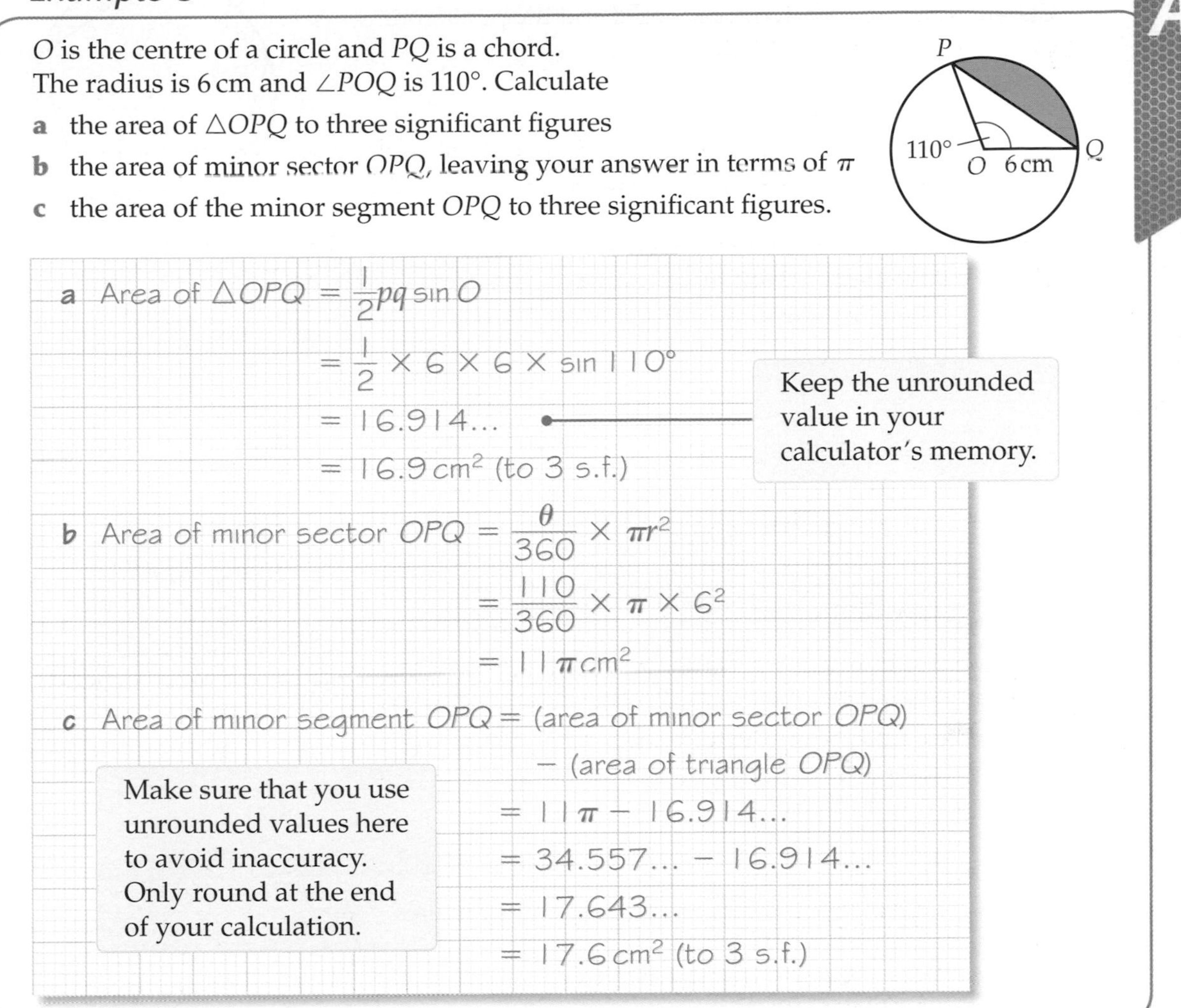

O is the centre of a circle and PQ is a chord.
The radius is 6 cm and $\angle POQ$ is 110°. Calculate

a the area of $\triangle OPQ$ to three significant figures

b the area of minor sector OPQ, leaving your answer in terms of π

c the area of the minor segment OPQ to three significant figures.

a Area of $\triangle OPQ = \frac{1}{2}pq \sin O$

$= \frac{1}{2} \times 6 \times 6 \times \sin 110°$

$= 16.914...$

$= 16.9 \text{ cm}^2$ (to 3 s.f.)

Keep the unrounded value in your calculator's memory.

b Area of minor sector $OPQ = \frac{\theta}{360} \times \pi r^2$

$= \frac{110}{360} \times \pi \times 6^2$

$= 11\pi \text{ cm}^2$

c Area of minor segment OPQ = (area of minor sector OPQ) − (area of triangle OPQ)

$= 11\pi - 16.914...$

$= 34.557... - 16.914...$

$= 17.643...$

$= 17.6 \text{ cm}^2$ (to 3 s.f.)

Make sure that you use unrounded values here to avoid inaccuracy. Only round at the end of your calculation.

Exercise 36E

1 Find the area of each shaded segment.

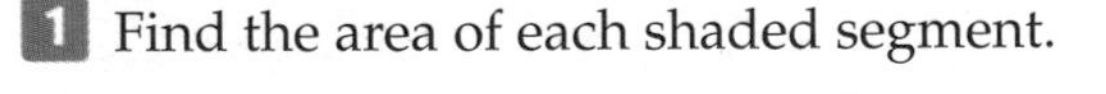

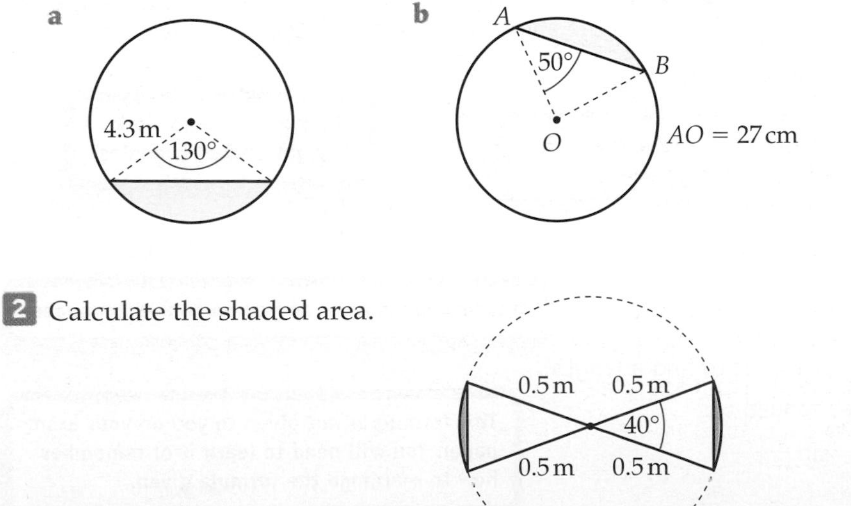

2 Calculate the shaded area.

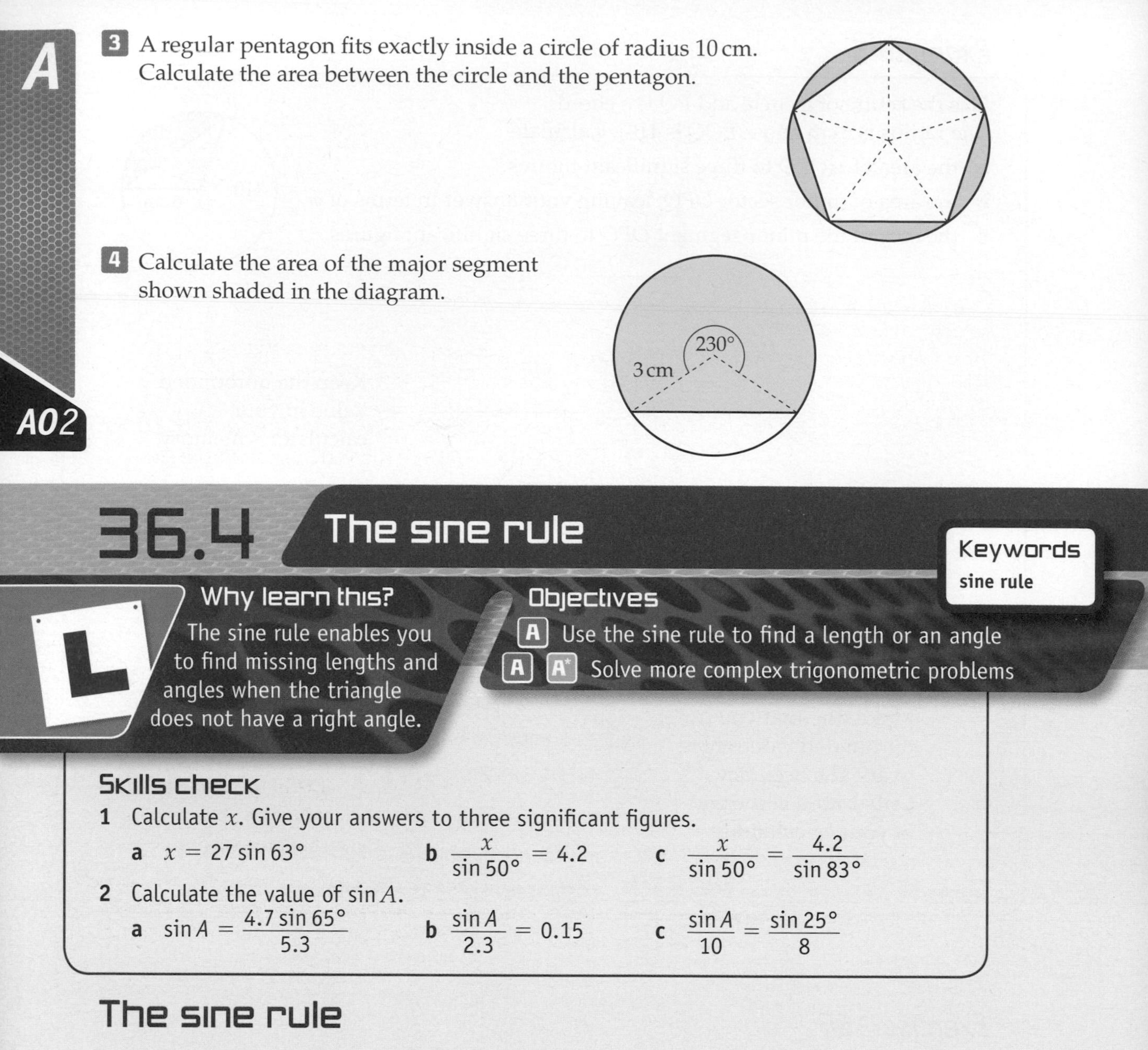

3 A regular pentagon fits exactly inside a circle of radius 10 cm. Calculate the area between the circle and the pentagon.

4 Calculate the area of the major segment shown shaded in the diagram.

36.4 The sine rule

Keywords
sine rule

Why learn this?

The sine rule enables you to find missing lengths and angles when the triangle does not have a right angle.

Objectives

A Use the sine rule to find a length or an angle

A **A*** Solve more complex trigonometric problems

Skills check

1 Calculate x. Give your answers to three significant figures.

a $x = 27 \sin 63°$ **b** $\dfrac{x}{\sin 50°} = 4.2$ **c** $\dfrac{x}{\sin 50°} = \dfrac{4.2}{\sin 83°}$

2 Calculate the value of $\sin A$.

a $\sin A = \dfrac{4.7 \sin 65°}{5.3}$ **b** $\dfrac{\sin A}{2.3} = 0.15$ **c** $\dfrac{\sin A}{10} = \dfrac{\sin 25°}{8}$

The sine rule

The **sine rule** can be used in *any* triangle to find

- a side when you know two angles and a side
- an angle when you know two sides and the angle opposite one of those sides.

Consider the triangle *ABC*.

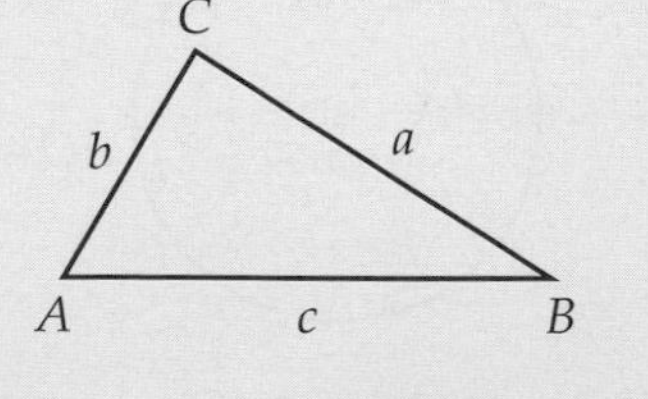

The sine rule also works in right-angled triangles.

The sine rule states that:

$$\frac{a}{\sin A} = \frac{b}{\sin B} = \frac{c}{\sin C}$$

This formula is given to you on your exam paper.

Use this format when you want to find a length.

The formula can be rearranged to:

$$\frac{\sin A}{a} = \frac{\sin B}{b} = \frac{\sin C}{c}$$

This formula is *not* given to you on your exam paper. You will need to learn it or remember how to rearrange the formula given.

Use this format when you want to find an angle.

Example 6

BAC is a triangle.

$\angle BAC = 27°$, $\angle BCA = 58°$ and $AB = 4.9$ cm.

Calculate, to the nearest mm,

a the length of BC **b** the length of AC.

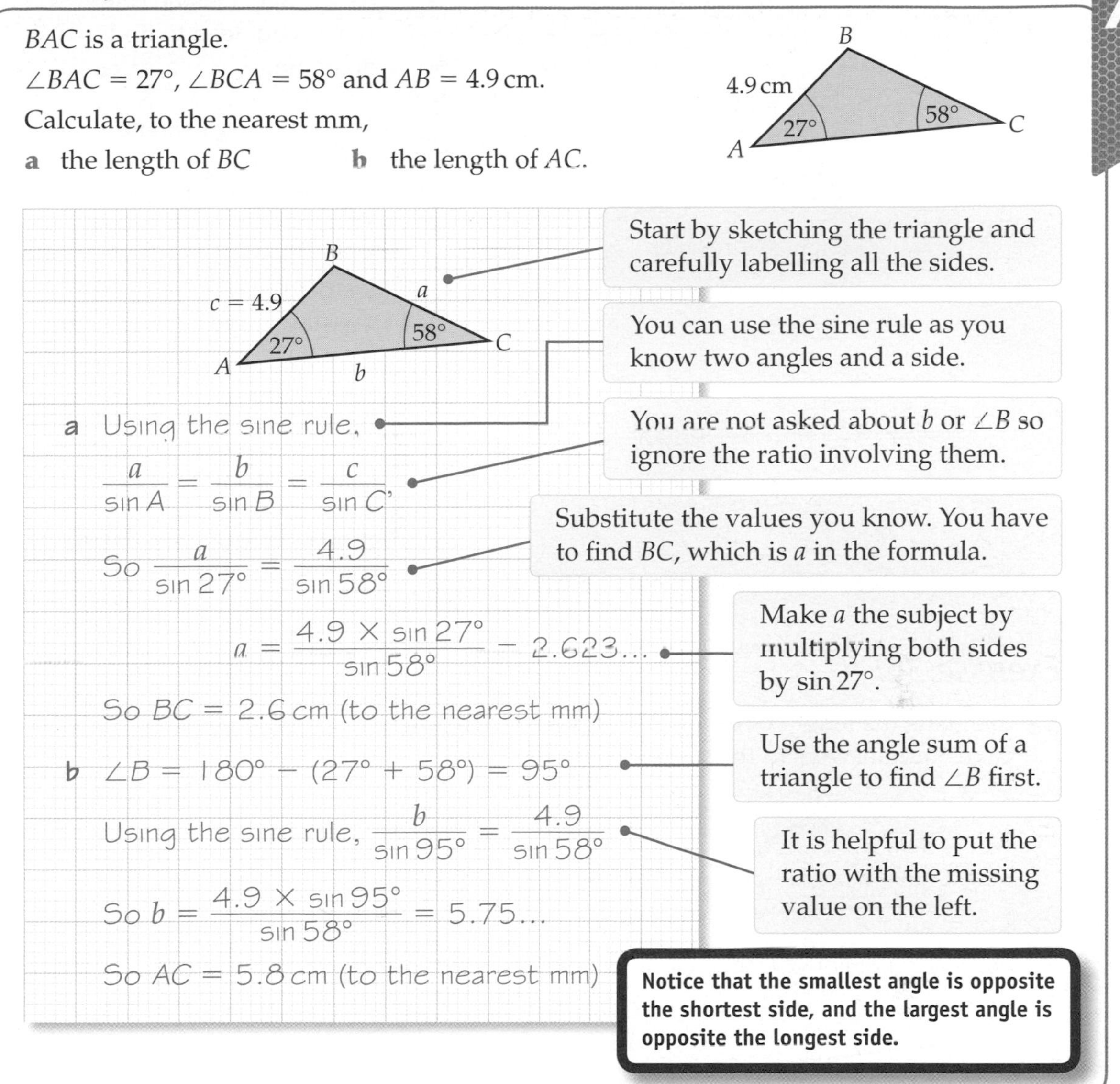

B
4.9 cm
27°
58°
A
C
c = 4.9
a
b
Start by sketching the triangle and carefully labelling all the sides.
a Using the sine rule,
You can use the sine rule as you know two angles and a side.
$\frac{a}{\sin A} = \frac{b}{\sin B} = \frac{c}{\sin C}$
You are not asked about b or ∠B so ignore the ratio involving them.
So $\frac{a}{\sin 27°} = \frac{4.9}{\sin 58°}$
Substitute the values you know. You have to find BC, which is a in the formula.
$a = \frac{4.9 \times \sin 27°}{\sin 58°} = 2.623...$
Make a the subject by multiplying both sides by sin 27°.
So BC = 2.6 cm (to the nearest mm)
b ∠B = 180° − (27° + 58°) = 95°
Use the angle sum of a triangle to find ∠B first.
Using the sine rule, $\frac{b}{\sin 95°} = \frac{4.9}{\sin 58°}$
It is helpful to put the ratio with the missing value on the left.
So $b = \frac{4.9 \times \sin 95°}{\sin 58°} = 5.75...$
So AC = 5.8 cm (to the nearest mm)

Notice that the smallest angle is opposite the shortest side, and the largest angle is opposite the longest side.

Example 7

PQR is a triangle.

$RP = 12.3$ m, $RQ = 19.6$ m and $\angle P = 110°$.

Calculate $\angle Q$. Give your answer to one decimal place.

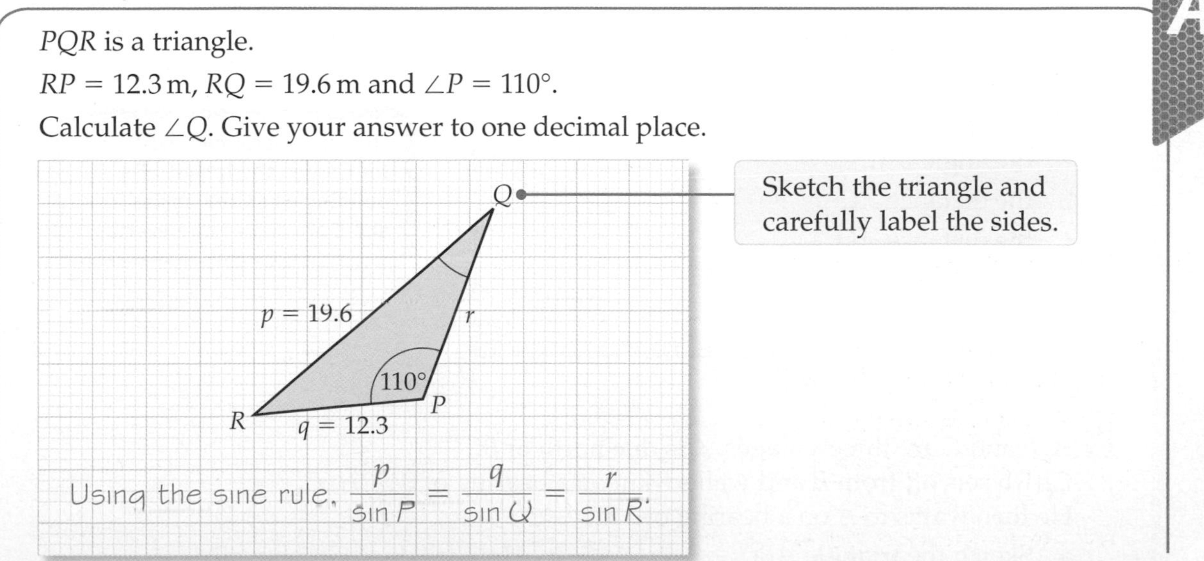

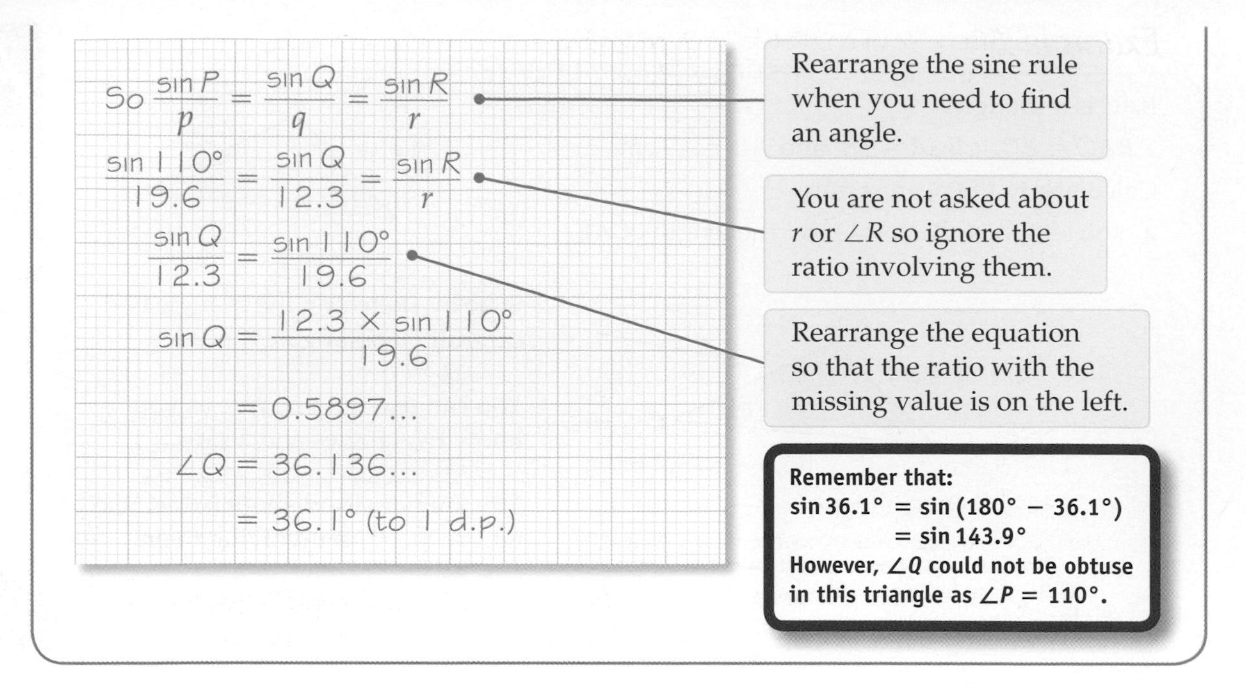

So $\frac{\sin P}{p} = \frac{\sin Q}{q} = \frac{\sin R}{r}$
$\frac{\sin 110°}{19.6} = \frac{\sin Q}{12.3} = \frac{\sin R}{r}$
$\frac{\sin Q}{12.3} = \frac{\sin 110°}{19.6}$
$\sin Q = \frac{12.3 \times \sin 110°}{19.6}$
$= 0.5897...$
$\angle Q = 36.136...$
$= 36.1°$ (to 1 d.p.)
Rearrange the sine rule when you need to find an angle.
You are not asked about r or $\angle R$ so ignore the ratio involving them.
Rearrange the equation so that the ratio with the missing value is on the left.

Remember that:
$\sin 36.1° = \sin(180° - 36.1°)$
$= \sin 143.9°$
However, $\angle Q$ could not be obtuse in this triangle as $\angle P = 110°$.

Exercise 36F

A

1 Calculate the length of each side marked with a letter.
Give your answers to three significant figures.

2 Calculate
a the angle BAC
b the distance AC
c the distance AD.

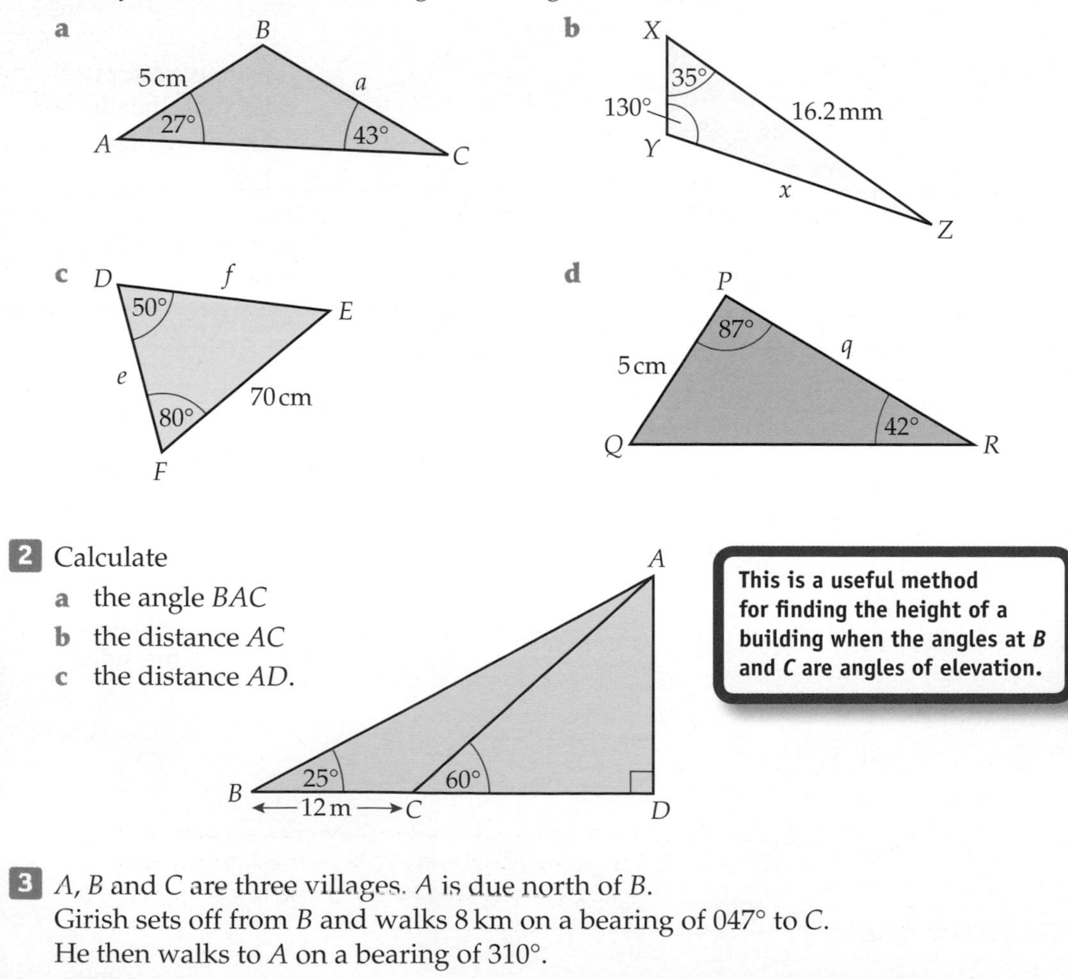

This is a useful method for finding the height of a building when the angles at B and C are angles of elevation.

3 A, B and C are three villages. A is due north of B.
Girish sets off from B and walks 8 km on a bearing of 047° to C.
He then walks to A on a bearing of 310°.
a Sketch the triangle ABC.
b How far does Girish walk?

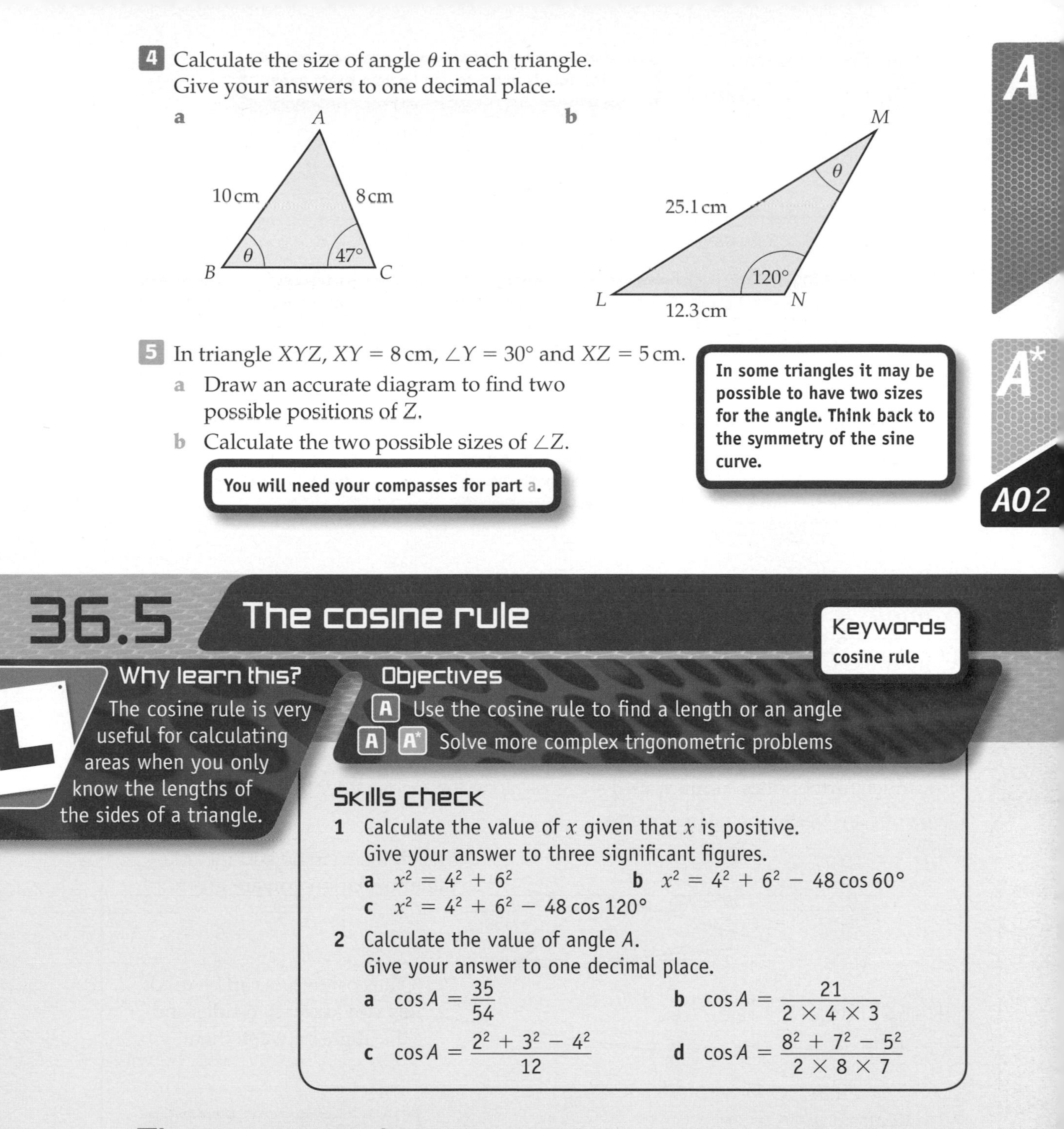

4 Calculate the size of angle θ in each triangle.
Give your answers to one decimal place.

5 In triangle XYZ, $XY = 8$ cm, $\angle Y = 30°$ and $XZ = 5$ cm.

a Draw an accurate diagram to find two possible positions of Z.

b Calculate the two possible sizes of $\angle Z$.

You will need your compasses for part a.

In some triangles it may be possible to have two sizes for the angle. Think back to the symmetry of the sine curve.

A

A*

AO2

36.5 The cosine rule

Keywords
cosine rule

Why learn this?

The cosine rule is very useful for calculating areas when you only know the lengths of the sides of a triangle.

Objectives

A Use the cosine rule to find a length or an angle

A **A*** Solve more complex trigonometric problems

Skills check

1 Calculate the value of x given that x is positive.
Give your answer to three significant figures.

a $x^2 = 4^2 + 6^2$ b $x^2 = 4^2 + 6^2 - 48 \cos 60°$

c $x^2 = 4^2 + 6^2 - 48 \cos 120°$

2 Calculate the value of angle A.
Give your answer to one decimal place.

a $\cos A = \frac{35}{54}$ b $\cos A = \frac{21}{2 \times 4 \times 3}$

c $\cos A = \frac{2^2 + 3^2 - 4^2}{12}$ d $\cos A = \frac{8^2 + 7^2 - 5^2}{2 \times 8 \times 7}$

The cosine rule

The **cosine rule** can be used in *any* triangle to find

- a side when you know two sides and the angle between them
- an angle when you know all three sides.

Consider the triangle ABC.

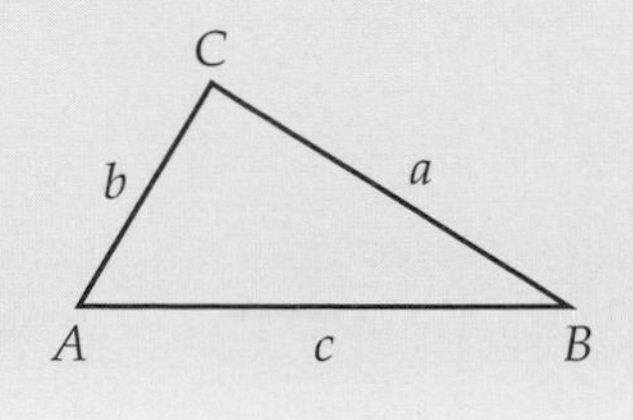

The cosine rule also works in right-angled triangles.

The cosine rule states that:

$$a^2 = b^2 + c^2 - 2bc\cos A$$

This formula is given to you on your exam paper.

Use this format when you want to find a length.

Similarly,

$$b^2 = c^2 + a^2 - 2ca\cos B$$
$$c^2 = a^2 + b^2 - 2ab\cos C$$

Notice the patterns.

This can be rearranged to:

$$\cos A = \frac{b^2 + c^2 - a^2}{2bc}$$

This formula is *not* given to you on your exam paper. You will need to learn it or remember how to rearrange the formula for a^2.

Use this format when you want to find an angle.

Similarly,

$$\cos B = \frac{c^2 + a^2 - b^2}{2ca}$$

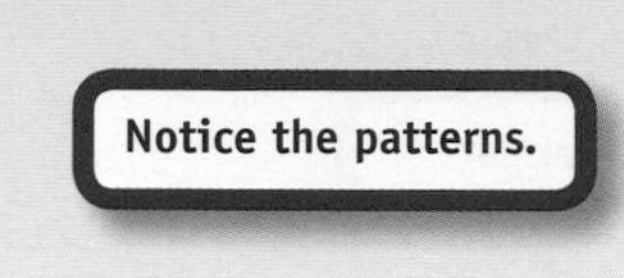

$$\cos C = \frac{a^2 + b^2 - c^2}{2ab}$$

Example 8

A

Amira walks from the library L to her home H.

She can go along the roads LX and XH, or along a footpath LH through a park.

$LX = 700$ m, $XH = 450$ m and $\angle LXH = 132°$.

How much further does Amira walk if she goes along the roads rather than the footpath?

Give your answer to the nearest metre.

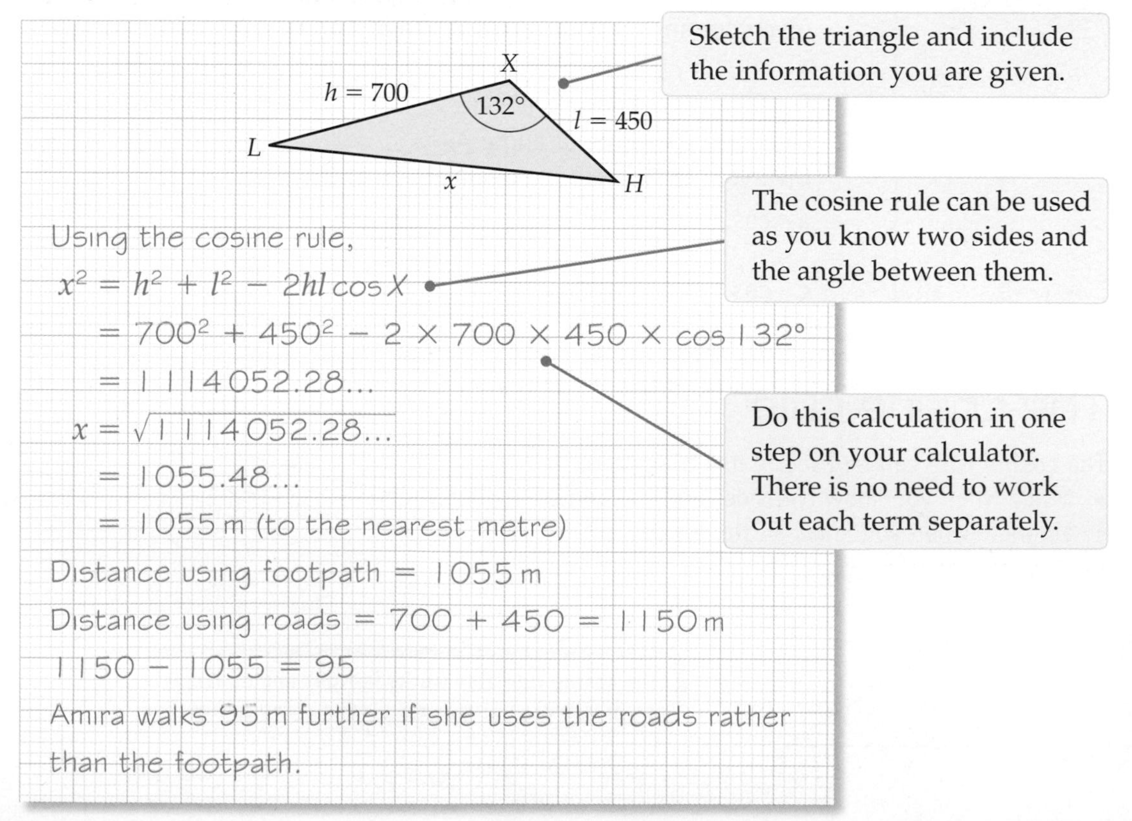

X
h = 700
132°
l = 450
L
x
H
Sketch the triangle and include the information you are given.
Using the cosine rule,
x² = h² + l² − 2hl cos X
= 700² + 450² − 2 × 700 × 450 × cos 132°
= 1114052.28...
x = √1114052.28...
= 1055.48...
= 1055 m (to the nearest metre)
The cosine rule can be used as you know two sides and the angle between them.
Do this calculation in one step on your calculator. There is no need to work out each term separately.
Distance using footpath = 1055 m
Distance using roads = 700 + 450 = 1150 m
1150 − 1055 = 95
Amira walks 95 m further if she uses the roads rather than the footpath.

Example 9

ABC is a triangle.

$AB = 10$ cm, $BC = 9$ cm and $AC = 12$ cm.

a Calculate the size of the smallest angle in the triangle.
Give your answer to one decimal place.

b Calculate the area of the triangle. Give your answer to three significant figures.

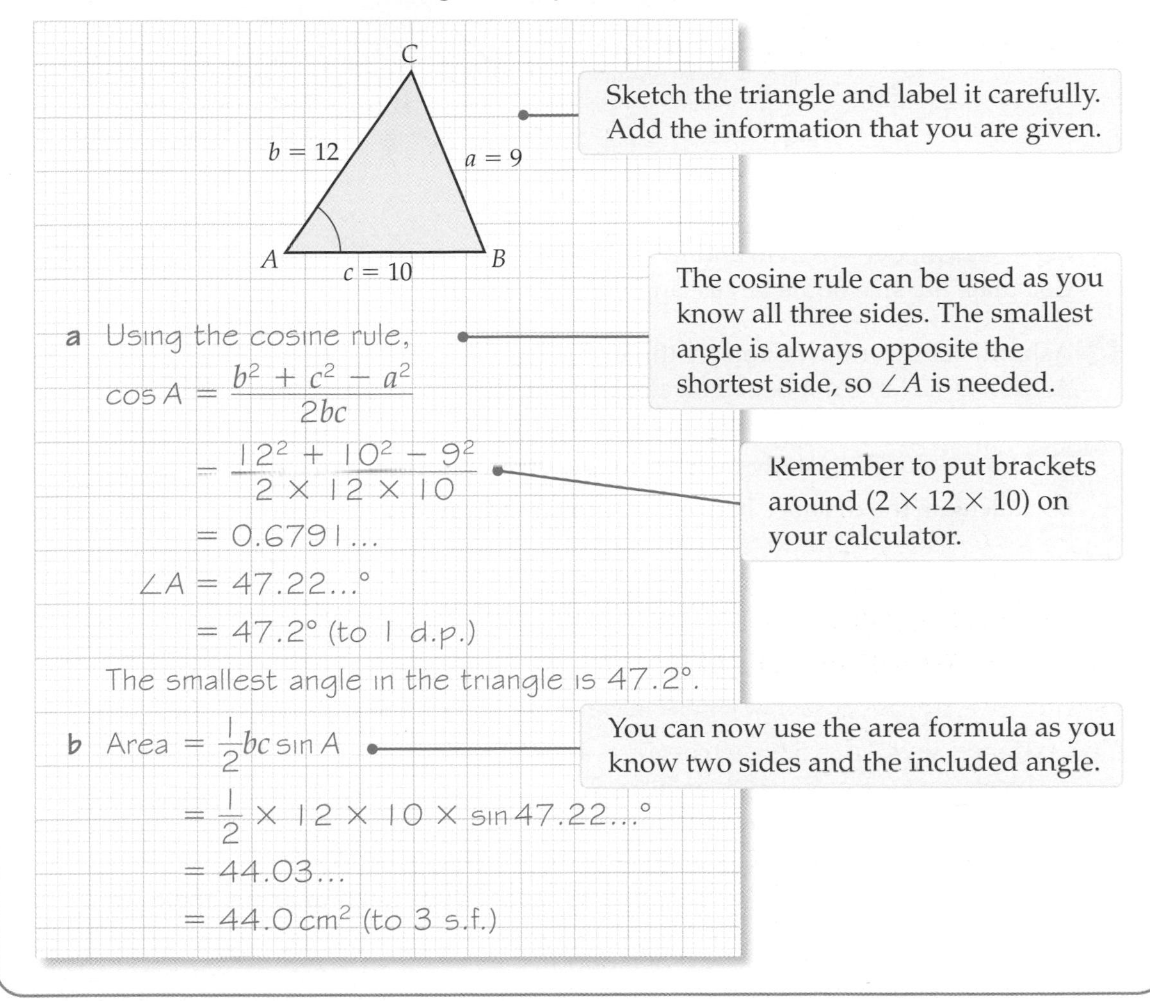

C
b = 12
a = 9
A
B
c = 10
Sketch the triangle and label it carefully. Add the information that you are given.
The cosine rule can be used as you know all three sides. The smallest angle is always opposite the shortest side, so ∠A is needed.
Remember to put brackets around (2 × 12 × 10) on your calculator.
You can now use the area formula as you know two sides and the included angle.

a Using the cosine rule,

$$\cos A = \frac{b^2 + c^2 - a^2}{2bc}$$

$$= \frac{12^2 + 10^2 - 9^2}{2 \times 12 \times 10}$$

$$= 0.6791\ldots$$

$$\angle A = 47.22\ldots^\circ$$

$$= 47.2^\circ \text{ (to 1 d.p.)}$$

The smallest angle in the triangle is 47.2°.

b $\text{Area} = \frac{1}{2}bc \sin A$

$$= \frac{1}{2} \times 12 \times 10 \times \sin 47.22\ldots^\circ$$

$$= 44.03\ldots$$

$$= 44.0 \text{ cm}^2 \text{ (to 3 s.f.)}$$

Exercise 36G

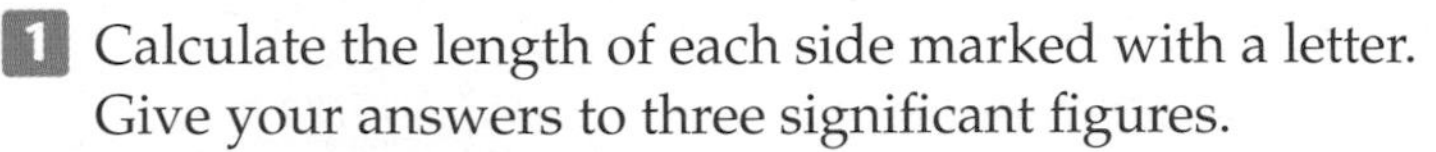

1 Calculate the length of each side marked with a letter.
Give your answers to three significant figures.

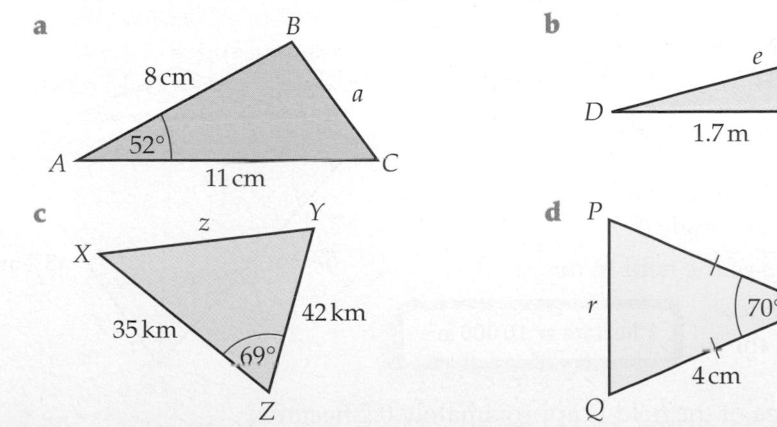

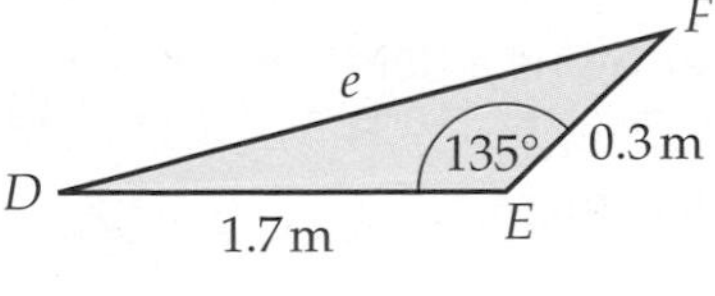

2 Calculate the size of angle θ in each triangle. Give your answers to one decimal place.

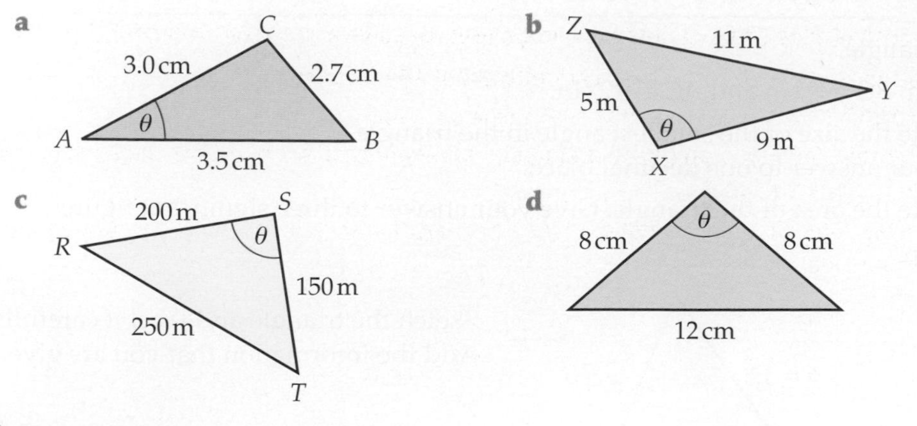

3 PQR is a triangle.
$PQ = 4.1$ cm, $QR = 8.2$ cm and $RP = 11.3$ cm.
Calculate the size of the largest angle in the triangle.

4 Ashton, Benby and Crayton are three villages.
Ashton is 4.2 km from Benby on a bearing of 053°.
Crayton is 7.3 km from Benby on a bearing of 130°.

a Draw a sketch to show the positions of Ashton, Benby and Crayton.

b Calculate the distance between Ashton and Crayton.

5 **a** Danny and Neela are calculating a using the cosine rule.
They have substituted correctly into the formula to get

$$a^2 = 4^2 + 6^2 - 2 \times 4 \times 6 \times \cos 60°$$

Danny says, '$a^2 = 52 - 48\cos 60° = 4\cos 60° = 2$'
Neela says, '$a^2 = 52 - 48\cos 60° = 28$'
Who is correct? Explain your answer.

b Tom uses the cosine rule to find angle A in a triangle.
He calculates that $\cos A$ is negative. What does this tell him about angle A?

6 A lighthouse L is 24 km due east of a port P.
A yacht sails from P to L.
It then changes course and sails 40 km on a bearing of 155° from L to a marker buoy B.
From B the yacht sails directly back to P.

a How far does the yacht sail altogether?

b On what bearing does the yacht sail from B to P?

7 A farmer owns a field $ABCD$.
$AB = 46.1$ m, $BC = 34.5$ m, $CD = 48.2$ m and $DA = 57.9$ m.
The diagonal $BD = 53.5$ m.

a Use the cosine rule to calculate
i angle A **ii** angle C.

b Calculate the area of the field in m^2.

c The farmer needs to know the area of the field in hectares.
Show that the area of the field is approximately 0.2 hectares.

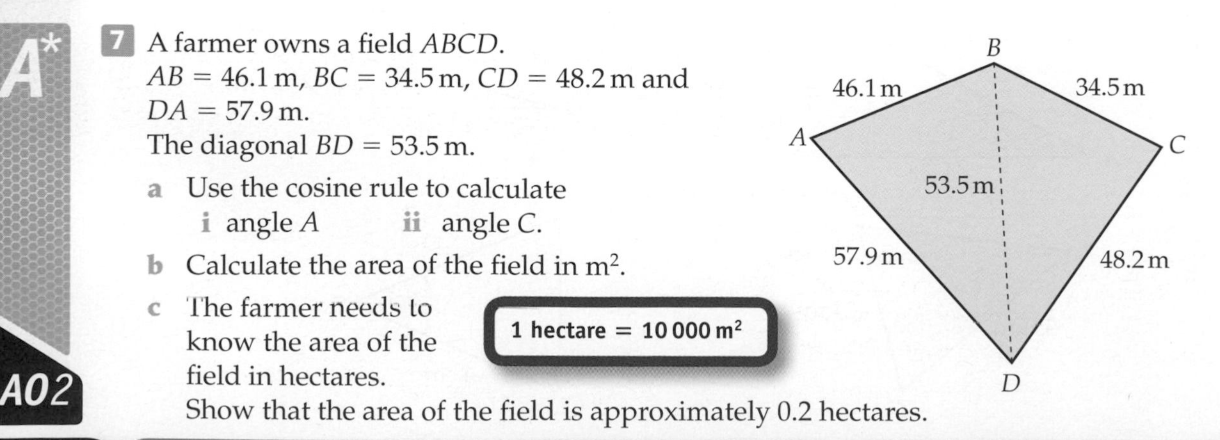

A
A
AO2
A*
AO3
A*
AO2

Solving 2-D and 3-D problems

The sine and cosine rules and area formula can be used to solve a range of 2-D and 3-D problems.

When attempting these problems, think carefully about the information that you have been given so that you can choose the correct rule to use.

- Use the sine rule when you are given a side and an angle that are opposite each other, and another side or another angle.
- Use the cosine rule when you are given two sides and the angle between them, or you are given all three sides.

Exercise 36H

A

1 A wooden wedge is in the shape of a prism.
The cross-section of the prism is a triangle, XYZ.
$XZ = 4$ cm, $ZY = 9$ cm and $\angle XZY = 100°$.
The thickness, AX, of the wedge is 3 cm.

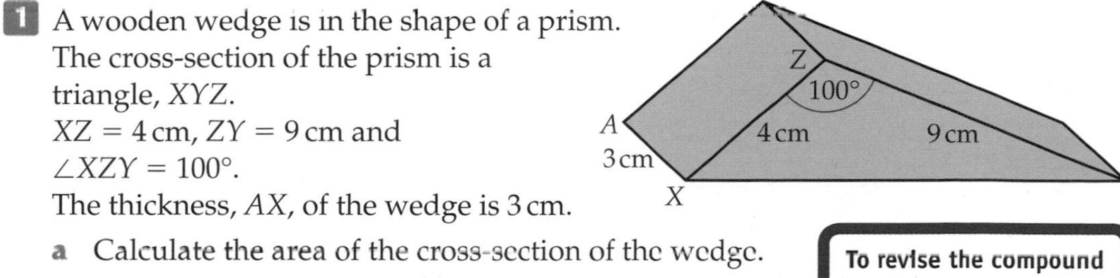

a Calculate the area of the cross-section of the wedge.

b Calculate the volume of the wedge in cm³.

c The density of the wood is 0.7 g/cm³.
Calculate the mass of the wedge.

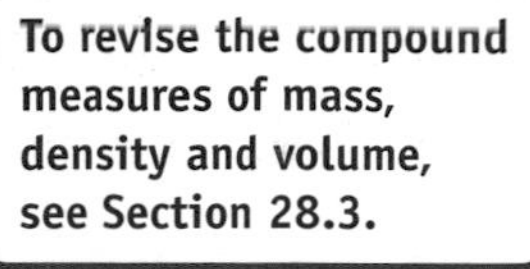
To revise the compound measures of mass, density and volume, see Section 28.3.

A02

A

2 A parallelogram has sides of length 6 cm and 10 cm. One of the angles in the parallelogram is 50°. Calculate the lengths of both diagonals of the parallelogram.

3 A and B are two airports. Airport B is 75 km due east of airport A.
An aircraft flies from A on a bearing of 050°.
When the aircraft has flown 50 km, calculate

a the distance of the aircraft from B

b the bearing of the aircraft from B.

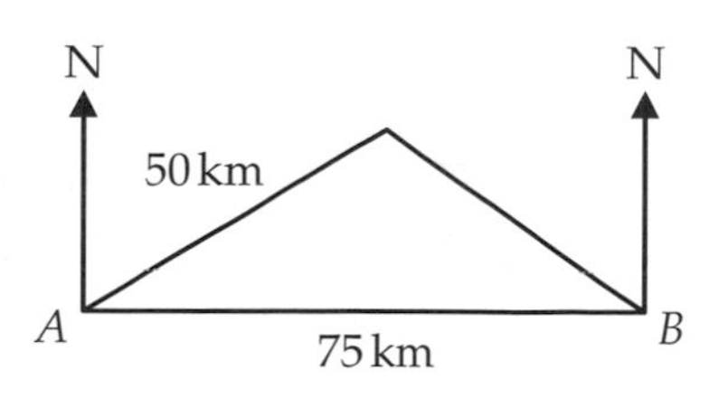

4 A, B and C are on level ground.
$AB = 8$ m, $BC = 7$ m and $\angle ABC = 65°$.
CM is a vertical mast of height 5 m.

a Calculate the distance from A to C.

b Calculate the angle of elevation of the top of the mast from A.

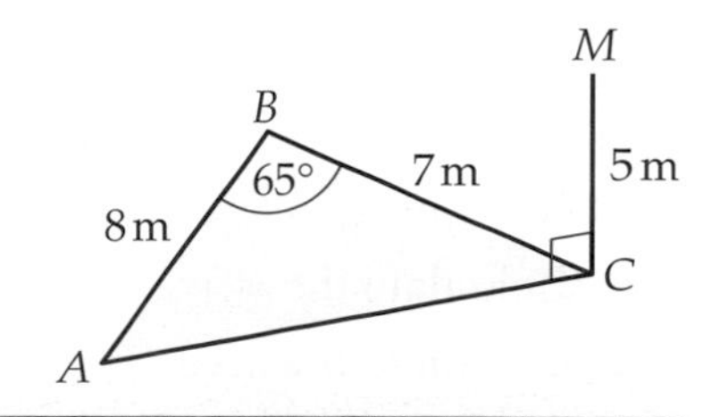

To revise angles of elevation and depression, see Section 33.5.

A03

A*

5 Calculate the area of quadrilateral $ABCD$.

To revise the tangent ratio, see Section 33.2.

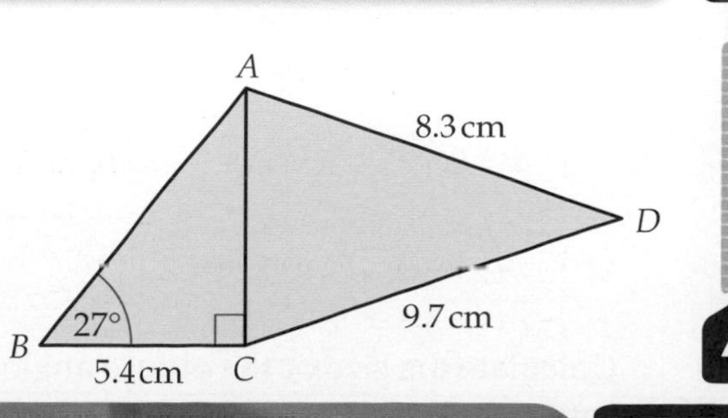

A03

A*

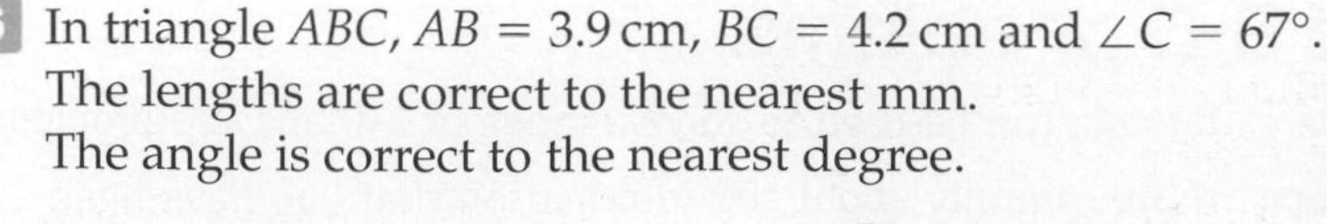

6 In triangle ABC, $AB = 3.9$ cm, $BC = 4.2$ cm and $\angle C = 67°$.
The lengths are correct to the nearest mm.
The angle is correct to the nearest degree.

To revise work on upper and lower bounds, see Section 8.4.

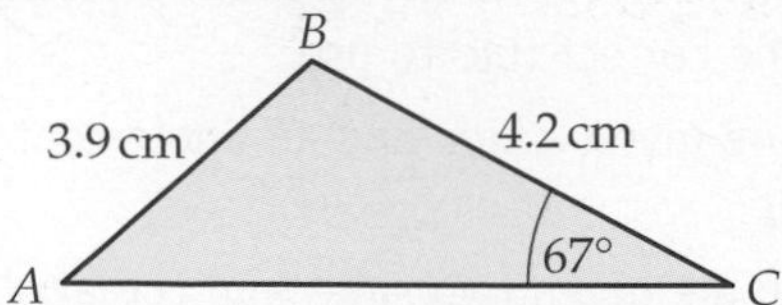

Use the sine rule to calculate the smallest possible value of $\angle A$.

7 WXY is a triangular field.
Z is 63 m from X along the hedge from X to W.
An electric fence is laid from Y to Z, so that the area of $\triangle XYZ$ is 1500 m^2.

a Show that the acute angle ZXY is 72.2°.

b Calculate the length of the fence YZ.

c What is the length of the shortest fence that could be laid from Y to the hedge XW?

AO2

Review exercise

A

1 $PQRS$ is a quadrilateral.
$PQ = 12$ cm, $QR = 11$ cm, $RS = 10$ cm and $SP = 9$ cm.
$\angle PQR = 77°$ and $\angle SPR = 47°$.

a Use the cosine rule to calculate PR. **[3 marks]**

b Use the sine rule to calculate the size of $\angle PRS$. **[3 marks]**

A

2 In $\triangle ABC$, $\angle B = 98°$, $\angle C = 22°$ and $AC = 12$ cm.

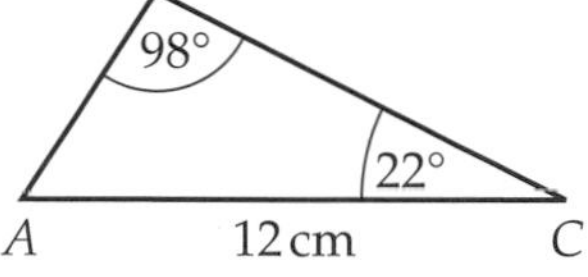

a Calculate the length of AB. **[3 marks]**

b Calculate the area of $\triangle ABC$. **[3 marks]**

3 In $\triangle ABC$, $\angle CAB = 40°$ $AC = 1$ m and $BC = 0.7$ m.

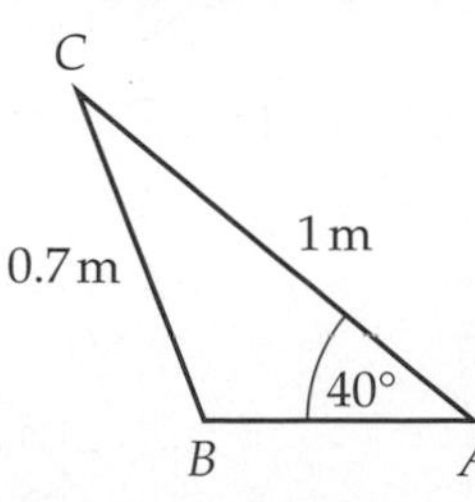

AO2

Calculate the size of the obtuse angle CBA. **[3 marks]**

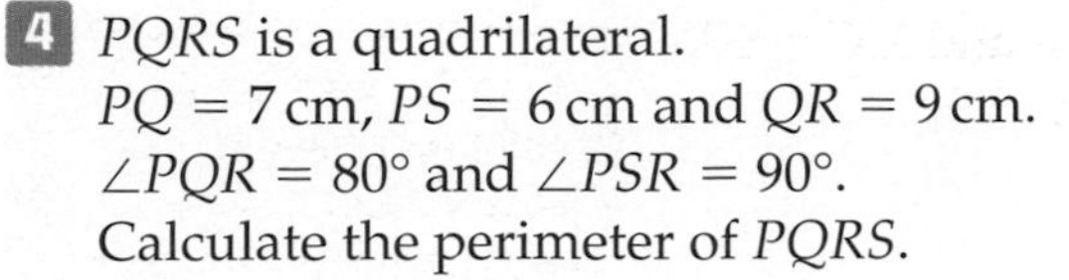

4 $PQRS$ is a quadrilateral.
$PQ = 7\,\text{cm}$, $PS = 6\,\text{cm}$ and $QR = 9\,\text{cm}$.
$\angle PQR = 80°$ and $\angle PSR = 90°$.
Calculate the perimeter of $PQRS$. **[6 marks]**

A

AO3

5 AB is a chord of a circle of radius 5 cm.
$AB = 6\,\text{cm}$.

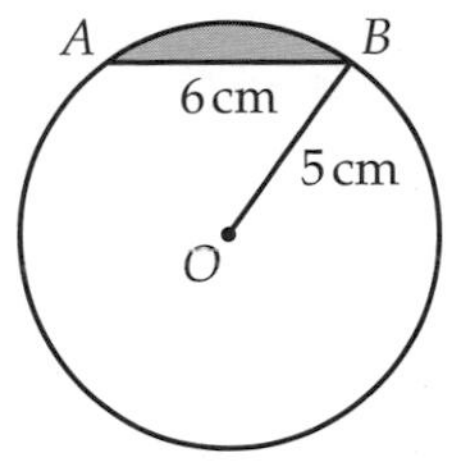

Calculate the area of the shaded segment. **[6 marks]**

A*

6 A tower stands on top of a vertical cliff.
From point A the angle of elevation of the top of the tower, T, is 45°.
From point B the angle of elevation of T is 65°.
The height of the cliff is 60 m.
Calculate the height, h, of the tower. **[6 marks]**

AO3

7 **a** Sketch the graph of $y = \sin x$ for values of x from 0° to 360°. **[1 mark]**

b One solution of the equation $\sin x = 0.438$ is 26°.
Use your graph from part **a** to find another solution to this equation. **[2 marks]**

c Explain how you could use your graph to work out the value of sin 334°. **[1 mark]**

AO2

Chapter summary

In this chapter you have learned how to

- recognise, draw and sketch the graphs of $y = \sin\theta$ and $y = \cos\theta$ **A**
- use the sine rule to find a length or an angle **A**
- use the cosine rule to find a length or an angle **A**
- use trigonometry to calculate the area of a triangle **A** **A***
- use trigonometry to calculate the area of a segment of a circle **A** **A***
- solve more complex trigonometric problems **A** **A***
- solve trigonometric equations **A***

37 Transformations of graphs

This chapter looks at transformations of graphs by examining their equations.

The shape of an ocean wave can be simulated using trigonometric functions.

Objectives

This chapter will show you how to

- **understand and use function notation** B
- **apply to the graph of $y = \mathrm{f}(x)$ the transformations $y = \mathrm{f}(x) + a$, $y = \mathrm{f}(ax)$, $y = \mathrm{f}(x + a)$ and $y = a\mathrm{f}(x)$ for linear, quadratic, sine and cosine functions $\mathrm{f}(x)$** A*

Before you start this chapter

1 Copy the grid and perform the following transformations on rectangle *ABCD*.

a translation $\begin{pmatrix} 0 \\ 3 \end{pmatrix}$ **b** translation $\begin{pmatrix} 0 \\ -2 \end{pmatrix}$

HELP Chapter 30

2

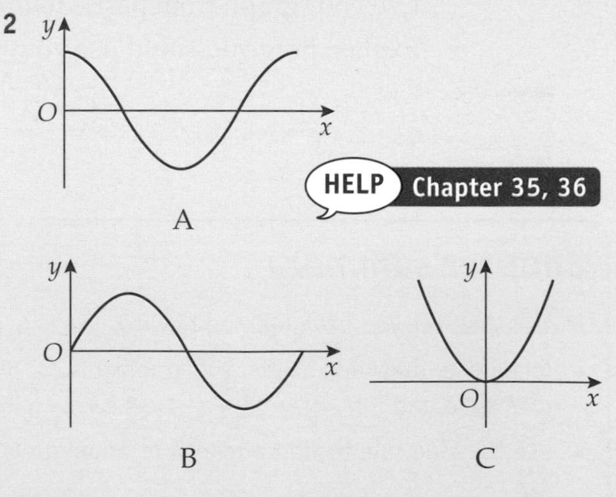

Match each function with the correct graph.

a $y = x^2$ **b** $y = \cos x$ **c** $y = \sin x$

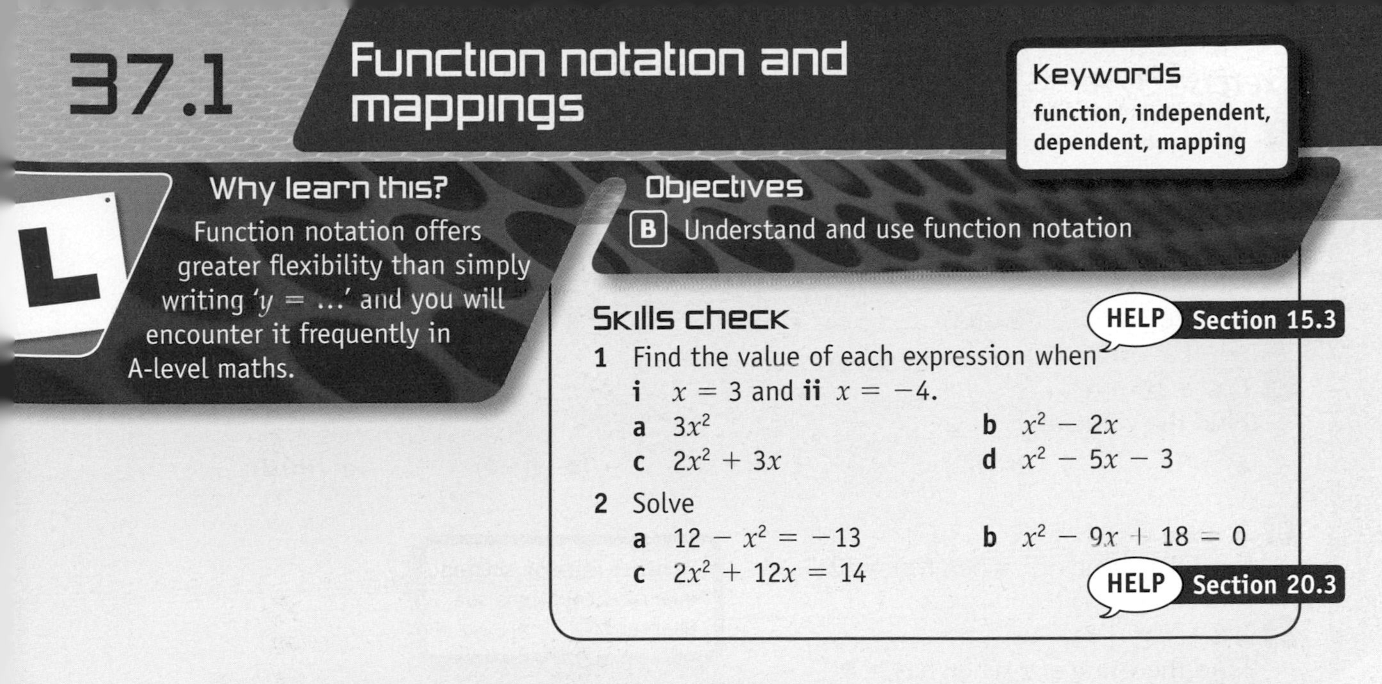

37.1 Function notation and mappings

Keywords
function, independent, dependent, mapping

Why learn this?

Function notation offers greater flexibility than simply writing '$y = \ldots$' and you will encounter it frequently in A-level maths.

Objectives

B Understand and use function notation

Skills check

1 Find the value of each expression when **i** $x = 3$ and **ii** $x = -4$.

a $3x^2$ **b** $x^2 - 2x$

c $2x^2 + 3x$ **d** $x^2 - 5x - 3$

HELP Section 15.3

2 Solve

a $12 - x^2 = -13$ **b** $x^2 - 9x + 18 = 0$

c $2x^2 + 12x = 14$

HELP Section 20.3

Functions and mappings

The word **function** is used to describe how the value of an **independent** variable determines the value of a **dependent** variable. For our purposes, a function is a rule that assigns a value of y to every value of x.

An example of a function is $y = x^2$.

In function notation this would be written as $f(x) = x^2$, where f denotes the function. $f(x) = x^2$ means the function of x is x^2.

f(x) is read as 'f of x'.

A function can also be thought of as a rule that **maps** x-values on to corresponding y-values.

In mapping notation, $y = x^2$ would be written as $f: x \rightarrow x^2$. This is read as 'the function f maps x to x^2'.

$y = 2x^2 + x$, $f(x) = 2x^2 + x$ and $f: x \rightarrow 2x^2 + x$ are equivalent statements.

Example 1

B

$f(x) = x^2 - 3$

a Find the values of **i** $f(0)$ **ii** $f(2)$ **iii** $f(-3)$

b Find the values of x for which $f(x) = 22$.

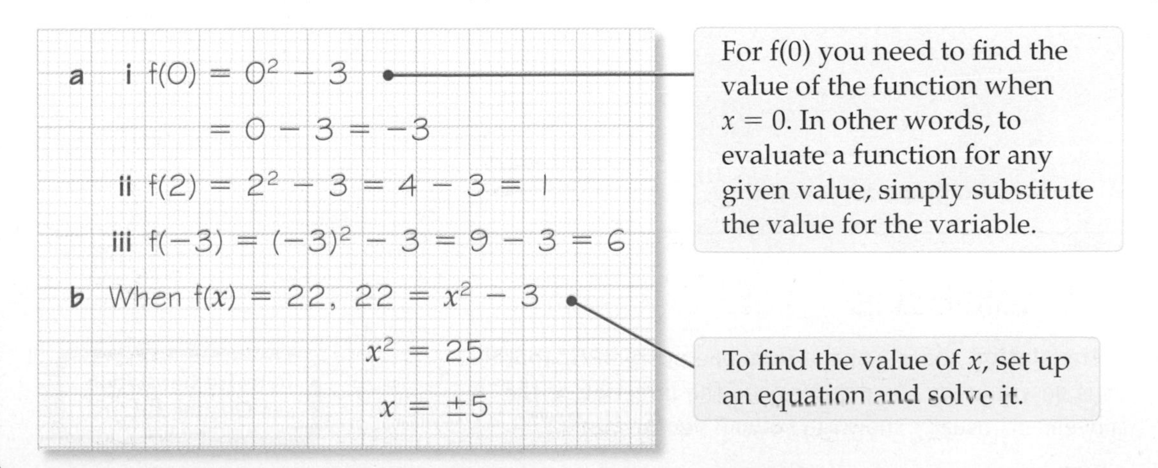

a i $f(0) = 0^2 - 3$
$= 0 - 3 = -3$

ii $f(2) = 2^2 - 3 = 4 - 3 = 1$

iii $f(-3) = (-3)^2 - 3 = 9 - 3 = 6$

b When $f(x) = 22$, $22 = x^2 - 3$
$x^2 = 25$
$x = \pm 5$

For f(0) you need to find the value of the function when $x = 0$. In other words, to evaluate a function for any given value, simply substitute the value for the variable.

To find the value of x, set up an equation and solve it.

Exercise 37A

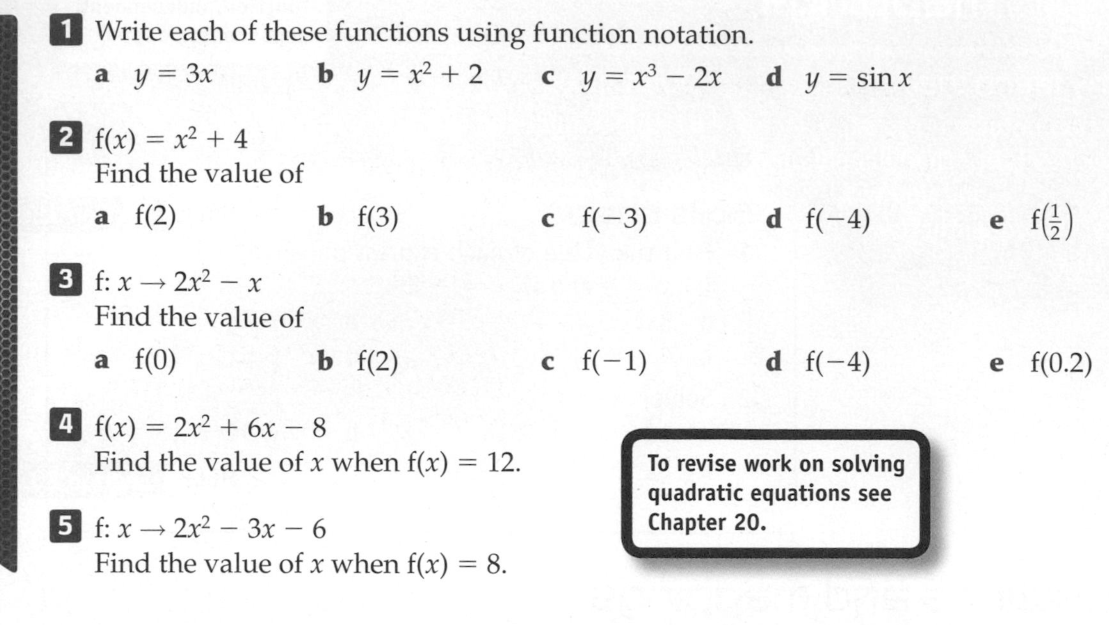

B

1 Write each of these functions using function notation.

a $y = 3x$ **b** $y = x^2 + 2$ **c** $y = x^3 - 2x$ **d** $y = \sin x$

2 $f(x) = x^2 + 4$

Find the value of

a $f(2)$ **b** $f(3)$ **c** $f(-3)$ **d** $f(-4)$ **e** $f(\frac{1}{2})$

3 $f: x \rightarrow 2x^2 - x$

Find the value of

a $f(0)$ **b** $f(2)$ **c** $f(-1)$ **d** $f(-4)$ **e** $f(0.2)$

4 $f(x) = 2x^2 + 6x - 8$

Find the value of x when $f(x) = 12$.

5 $f: x \rightarrow 2x^2 - 3x - 6$

Find the value of x when $f(x) = 8$.

To revise work on solving quadratic equations see Chapter 20.

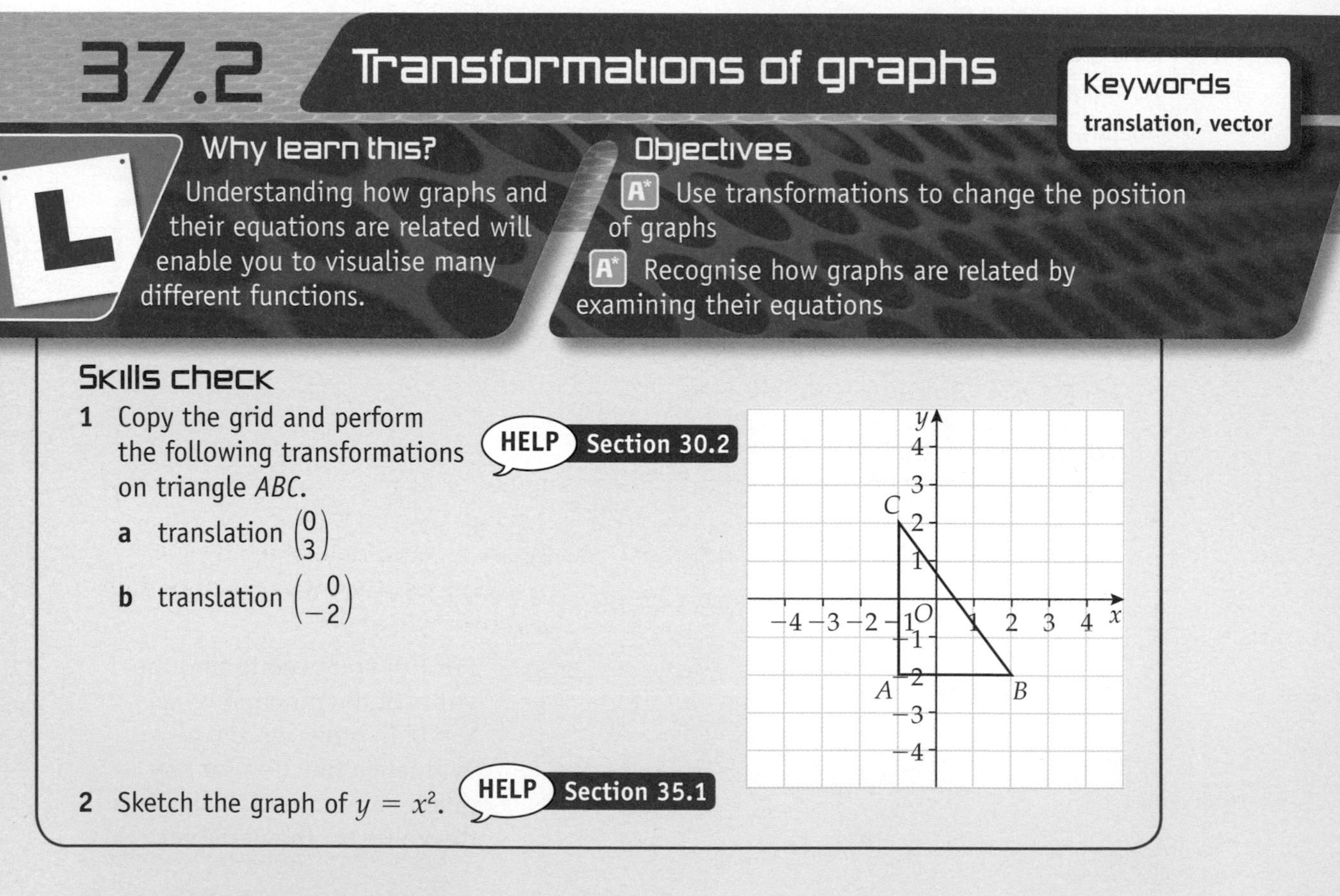

37.2 Transformations of graphs

Keywords
translation, vector

Why learn this?

Understanding how graphs and their equations are related will enable you to visualise many different functions.

Objectives

A* Use transformations to change the position of graphs

A* Recognise how graphs are related by examining their equations

Skills check

1 Copy the grid and perform the following transformations on triangle ABC.

HELP Section 30.2

a translation $\begin{pmatrix} 0 \\ 3 \end{pmatrix}$

b translation $\begin{pmatrix} 0 \\ -2 \end{pmatrix}$

2 Sketch the graph of $y = x^2$.

HELP Section 35.1

Translations

A **translation** slides a shape from one position to another. It is described by the distance and the direction of the movement, usually shown in column **vector** form.

To revise translation see Chapter 30.

Example 2

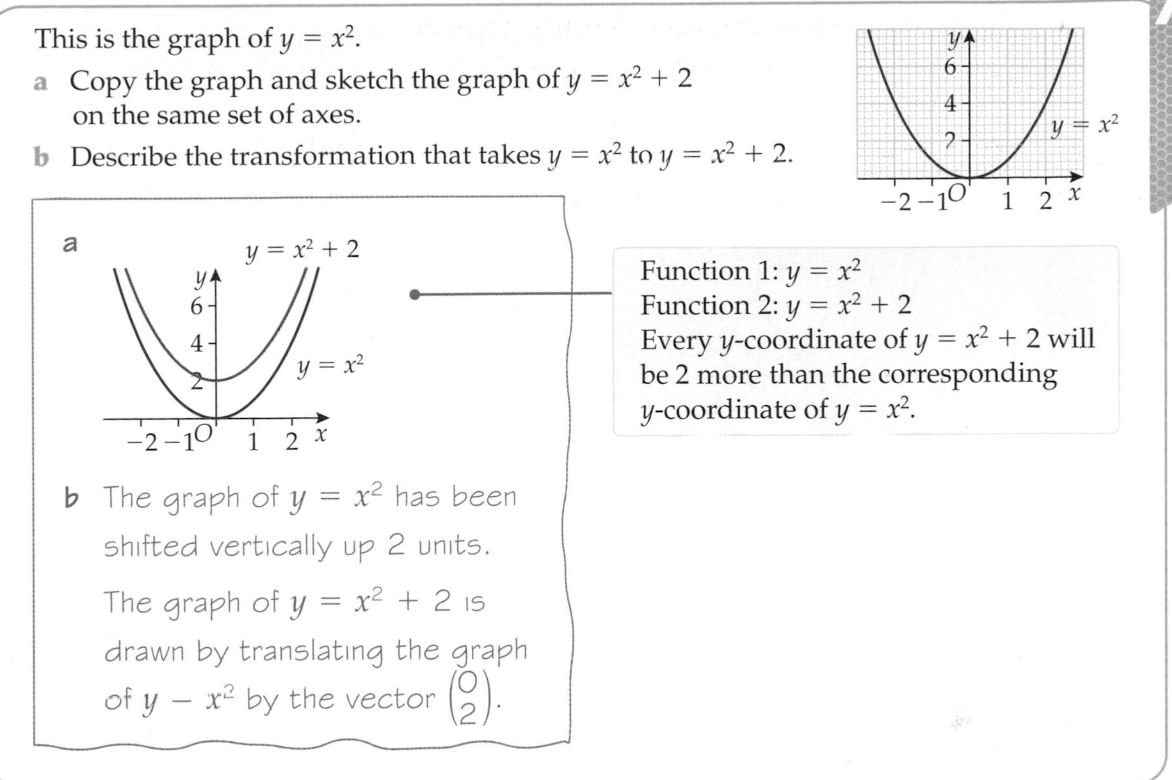

This is the graph of $y = x^2$.

a Copy the graph and sketch the graph of $y = x^2 + 2$ on the same set of axes.

b Describe the transformation that takes $y = x^2$ to $y = x^2 + 2$.

a

Function 1: $y = x^2$
Function 2: $y = x^2 + 2$
Every y-coordinate of $y = x^2 + 2$ will be 2 more than the corresponding y-coordinate of $y = x^2$.

b The graph of $y = x^2$ has been shifted vertically up 2 units.

The graph of $y = x^2 + 2$ is drawn by translating the graph of $y = x^2$ by the vector $\begin{pmatrix} 0 \\ 2 \end{pmatrix}$.

Example 3

a Copy again the graph of $y = x^2$ from Example 2 and sketch the graph of $y = x^2 - 4$ on the same set of axes.

b Describe the transformation that takes $y = x^2$ to $y = x^2 - 4$.

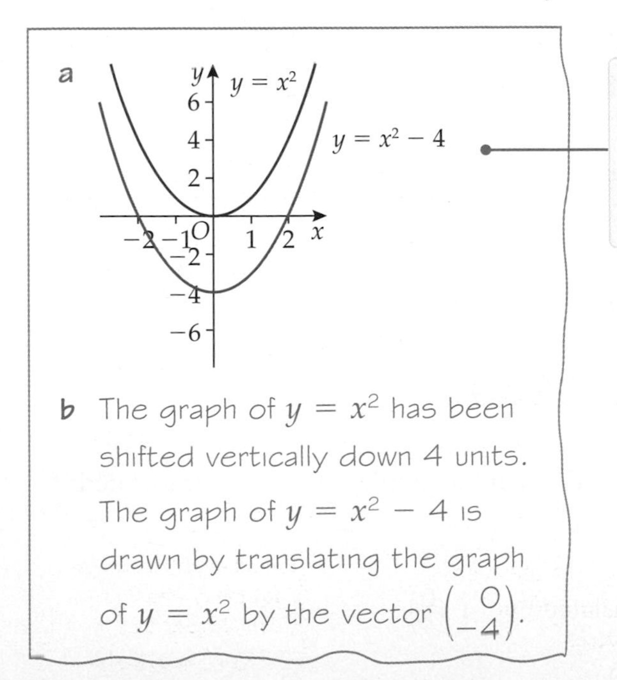

Function 1: $y = x^2$
Function 2: $y = x^2 - 4$
Every y-coordinate of $y = x^2 - 4$ will be 4 less than the corresponding y-coordinate of $y = x^2$.

b The graph of $y = x^2$ has been shifted vertically down 4 units.

The graph of $y = x^2 - 4$ is drawn by translating the graph of $y = x^2$ by the vector $\begin{pmatrix} 0 \\ -4 \end{pmatrix}$.

Example 4

Copy again the graph of $y = x^2$ from Example 2 and sketch the graph of $y = (x + 2)^2$ on the same set of axes.

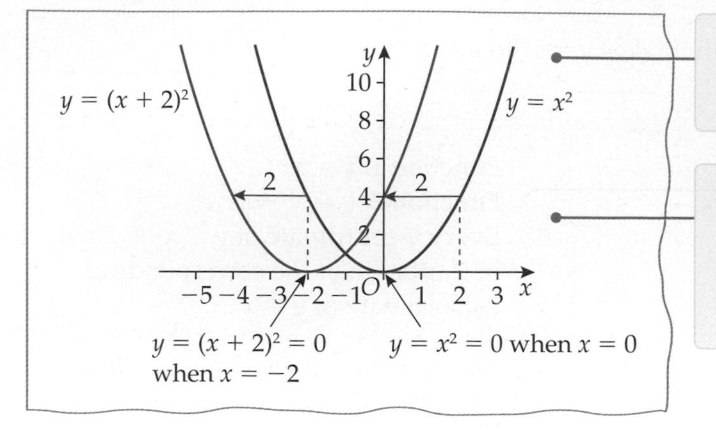

The graph of $y = x^2$ is translated 2 units to the left to become the graph of $y = (x + 2)^2$.

The graph of $y = (x + 2)^2$ is drawn by translating the graph of $y = x^2$ by the vector $\begin{pmatrix} -2 \\ 0 \end{pmatrix}$.

$y = \mathrm{f}(x) \rightarrow y = \mathrm{f}(x) + a$ represents a translation of $\begin{pmatrix} 0 \\ a \end{pmatrix}$.

$y = \mathrm{f}(x) \rightarrow y = \mathrm{f}(x) - a$ represents a translation of $\begin{pmatrix} 0 \\ -a \end{pmatrix}$.

$y = \mathrm{f}(x) \rightarrow y = \mathrm{f}(x + a)$ represents a translation of $\begin{pmatrix} -a \\ 0 \end{pmatrix}$.

$y = \mathrm{f}(x) \rightarrow y = \mathrm{f}(x - a)$ represents a translation of $\begin{pmatrix} a \\ 0 \end{pmatrix}$.

Notice that $\mathrm{f}(x + a)$ is a shift to the left and $\mathrm{f}(x - a)$ is a shift to the right.

Exercise 37B

1 Sketch the graph of $y = x^2$ when translated by **a** $\begin{pmatrix} 0 \\ 3 \end{pmatrix}$ **b** $\begin{pmatrix} 0 \\ -6 \end{pmatrix}$.

2 **a** Sketch the graph of $y = 2x^2$ for $-4 \leqslant x \leqslant 4$.

b On the same set of axes sketch the graphs of $y = 2x^2 + 2$ and $y = 2x^2 - 3$.

c Describe how each graph in part **b** is obtained from the graph of $y = 2x^2$.

3 Write the equation of the graph obtained when the graph of $y = 3x^2$ is translated by $\begin{pmatrix} 0 \\ -8 \end{pmatrix}$.

4 **a** Sketch the graph of $y = x^2 + 2x$ for $-3 \leqslant x \leqslant 2$.

b Translate the graph of $y = x^2 + 2x$ by $\begin{pmatrix} 0 \\ 1 \end{pmatrix}$.

c Write the equation of the new graph in the form $y = ax^2 + bx + c$.

A02

5 Write the equation of the graph obtained when the graph of $y = 2x^2$ is translated by $\begin{pmatrix} -1 \\ 0 \end{pmatrix}$.

6 Sketch the graph of $y = x^2$ when translated by **a** $\begin{pmatrix} 0 \\ 4 \end{pmatrix}$ **b** $\begin{pmatrix} -5 \\ 0 \end{pmatrix}$.

Stretches of graphs

Consider the graph of the function $y = x^2$.

A 'stretch' of the graph occurs when the x^2 term is multiplied by a constant.

Example 5

a Copy again the graph of $y = x^2$ from Example 2 and sketch the graph of $y = 3x^2$ on the same set of axes.

b Describe the transformation that takes $y = x^2$ to $y = 3x^2$.

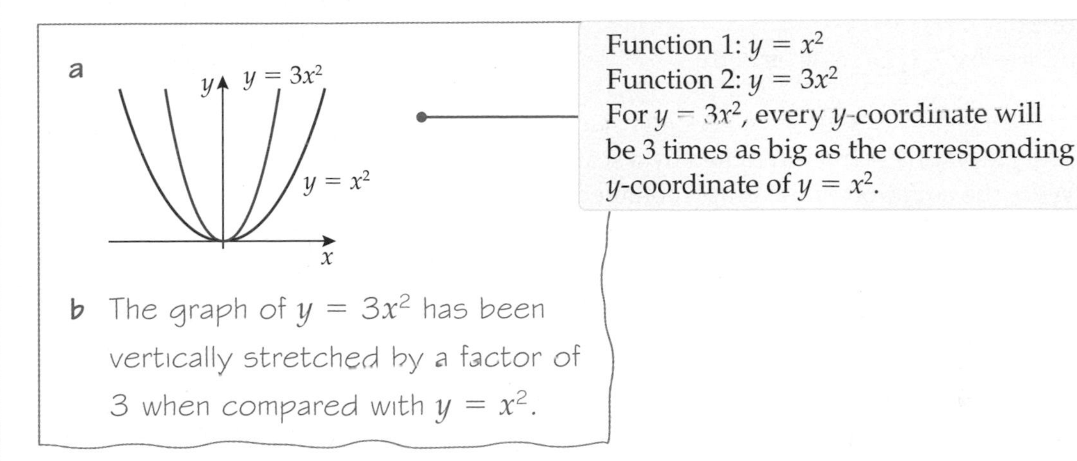

a
y
y = 3x²
y = x²
x
Function 1: y = x²
Function 2: y = 3x²
For y = 3x², every y-coordinate will be 3 times as big as the corresponding y-coordinate of y = x².
b The graph of y = 3x² has been vertically stretched by a factor of 3 when compared with y = x².

Example 6

Copy again the graph of $y = x^2$ from Example 2 and sketch the graph of $y = (2x)^2$ on the same set of axes.

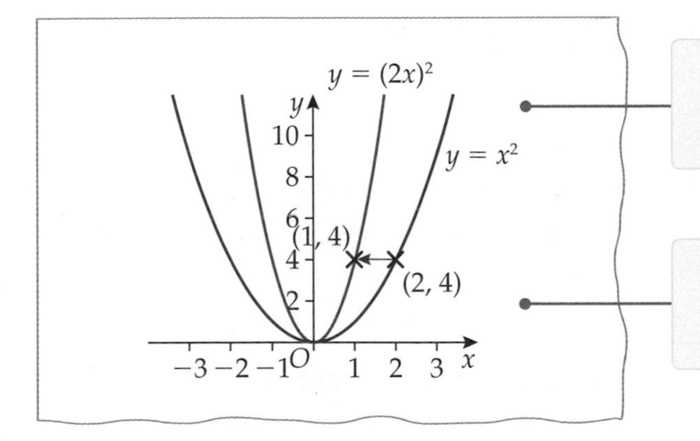

For the graph of $y = (2x)^2$, the x-coordinates of the graph of $y = x^2$ have been halved.

The graph of $y = x^2$ has been reduced horizontally by a factor of $\frac{1}{2}$ to produce the graph of $y = (2x)^2$.

$y = \mathrm{f}(x) \rightarrow y = a\mathrm{f}(x)$ represents a stretch by scale factor a parallel to the y-axis.

- When $a > 1$, the graph of $y = a\mathrm{f}(x)$ is obtained by vertically stretching $y = \mathrm{f}(x)$.
- When $0 < a < 1$, the graph of $y = a\mathrm{f}(x)$ is obtained by vertically reducing $y = \mathrm{f}(x)$.

$y = \mathrm{f}(x) \rightarrow y = \mathrm{f}(ax)$ represents a stretch from the y-axis, parallel to the x-axis, of scale factor $\frac{1}{a}$.

- When $a > 1$, the graph of $y = \mathrm{f}(ax)$ is obtained by horizontally reducing $y = \mathrm{f}(x)$.
- When $0 < a < 1$, the graph of $y = \mathrm{f}(ax)$ is obtained by horizontally stretching $y = \mathrm{f}(x)$.

Exercise 37C

A*

1 a Sketch the graph of $y = x^2$ for $-4 \leqslant x \leqslant 4$.

b On the same set of axes sketch the graph of $y = \frac{1}{2}x^2$.

c Describe how the graph is obtained from the graph of $y = x^2$.

2 Sketch the graph of $y = x^2$ when stretched by a scale factor of 4 parallel to the y-axis.

3 Describe the transformation that maps the graph of $y = x^2$ on to each of these graphs.

a $y = 5x^2$ b $y = 8x^2$ c $y = \frac{1}{3}x^2$ d $y = (\frac{1}{2}x)^2$

4 The graph of $y = x^2$ is stretched by a scale factor $\frac{1}{4}$ parallel to the y-axis. Write the equation of the graph obtained from this transformation.

5 a Sketch the graphs of $y = x^2$ and $y = (3x)^2$.

b Describe the transformation that maps $y = x^2$ on to $y = (3x)^2$.

Combinations of transformations

In this section you will look at graphs that are formed through a combination of transformations.

It is important that you work out what transformation (or combination of transformations) the function represents before sketching the graph.

A*

Example 7

Copy again the graph of $y = x^2$ from Example 2 and sketch the graph of $y = 2x^2 + 2$ on the same set of axes.

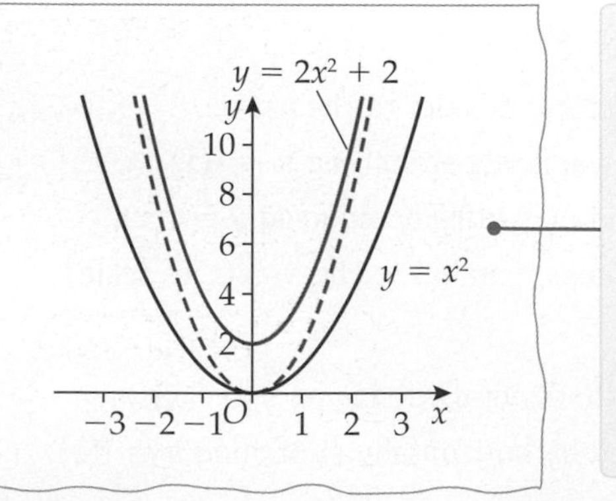

Function 1: $y = x^2$
Function 2: $y = 2x^2 + 2$
The graph of $y = x^2$ must be transformed in two steps to obtain the graph of $y = 2x^2 + 2$.
Step 1: Vertically stretch the graph of $y = x^2$ by a scale factor of 2 (shown as a dashed line on the graph).
Step 2: Translate the graph of $y = 2x^2$ up 2 units.

Exercise 37D

1 a Sketch the graph of $y = x^2$.

b The graph of $y = x^2$ is stretched by scale factor 2 parallel to the y-axis, followed by a translation of $\begin{pmatrix} 0 \\ 2 \end{pmatrix}$.

Use your sketch in part a to sketch the graph obtained after $y = x^2$ is transformed in this way.

2 Describe the combination of transformations that takes the graph of $y = x^2$ on to each of these graphs.
For each one, sketch a graph to illustrate your answer.

a $y = 4x^2 - 3$ b $y = \frac{1}{2}x^2 + 3$

3 Describe the combination of transformations that takes the graph of $y = \text{f}(x)$ to $2\text{f}(x) + 4$.

4 The graph of $y = 2x^2$ is stretched by a scale factor 2 parallel to the y-axis followed by a translation of $\begin{pmatrix} 0 \\ -3 \end{pmatrix}$.
Write the equation of the graph that is obtained from this transformation.

AO3

5 The graph of $y = x^2$ is transformed by **A** and then by **B**.

A a stretch from the y-axis, parallel to the x-axis, of scale factor 2

B a translation of $\begin{pmatrix} 0 \\ -2 \end{pmatrix}$.

Sketch a graph to illustrate each stage of the transformation.
Write the equation of the graph after each stage.

AO2

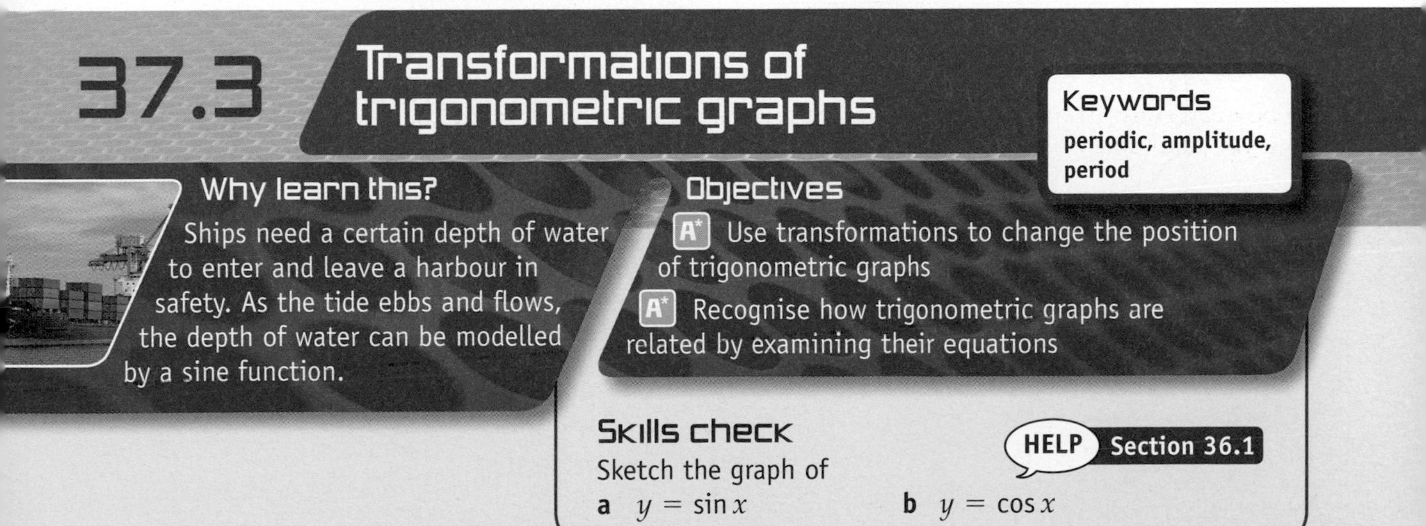

37.3 Transformations of trigonometric graphs

Keywords
periodic, amplitude, period

Why learn this?

Ships need a certain depth of water to enter and leave a harbour in safety. As the tide ebbs and flows, the depth of water can be modelled by a sine function.

Objectives

A* Use transformations to change the position of trigonometric graphs

A* Recognise how trigonometric graphs are related by examining their equations

Skills check

HELP Section 36.1

Sketch the graph of

a $y = \sin x$ b $y = \cos x$

Period and amplitude of trigonometric graphs

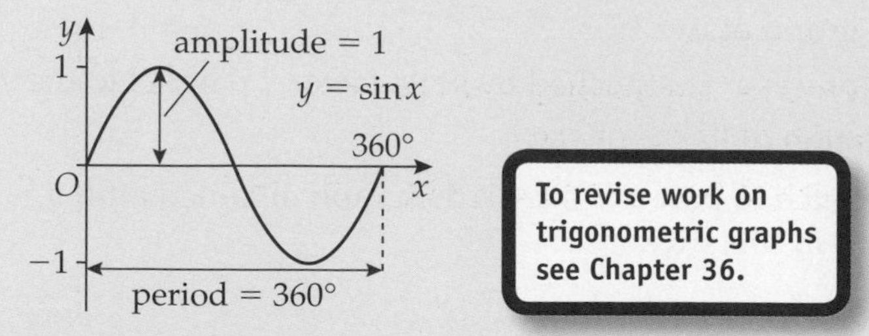

To revise work on trigonometric graphs see Chapter 36.

The function $y = \sin x$ is a **periodic** function. This means that the next pattern from 360° to 720° will have the same shape as that shown.

The **period** of a trigonometric graph is the number of degrees before it repeats itself.

Both $y = \sin x$ and $y = \cos x$ have a period of 360°.

The **amplitude** of a trigonometric graph is the maximum displacement from a reference level (e.g. the x-axis) in either a positive or a negative direction.

The amplitude of both $y = \sin x$ and $y = \cos x$ is 1.

You will now consider how transformations can alter the period, amplitude and position of trigonometric graphs.

The graph of $y = \cos x$ is identical to the graph of $y = \sin x$ except that it has been moved to the left by 90°.

A*

Example 8

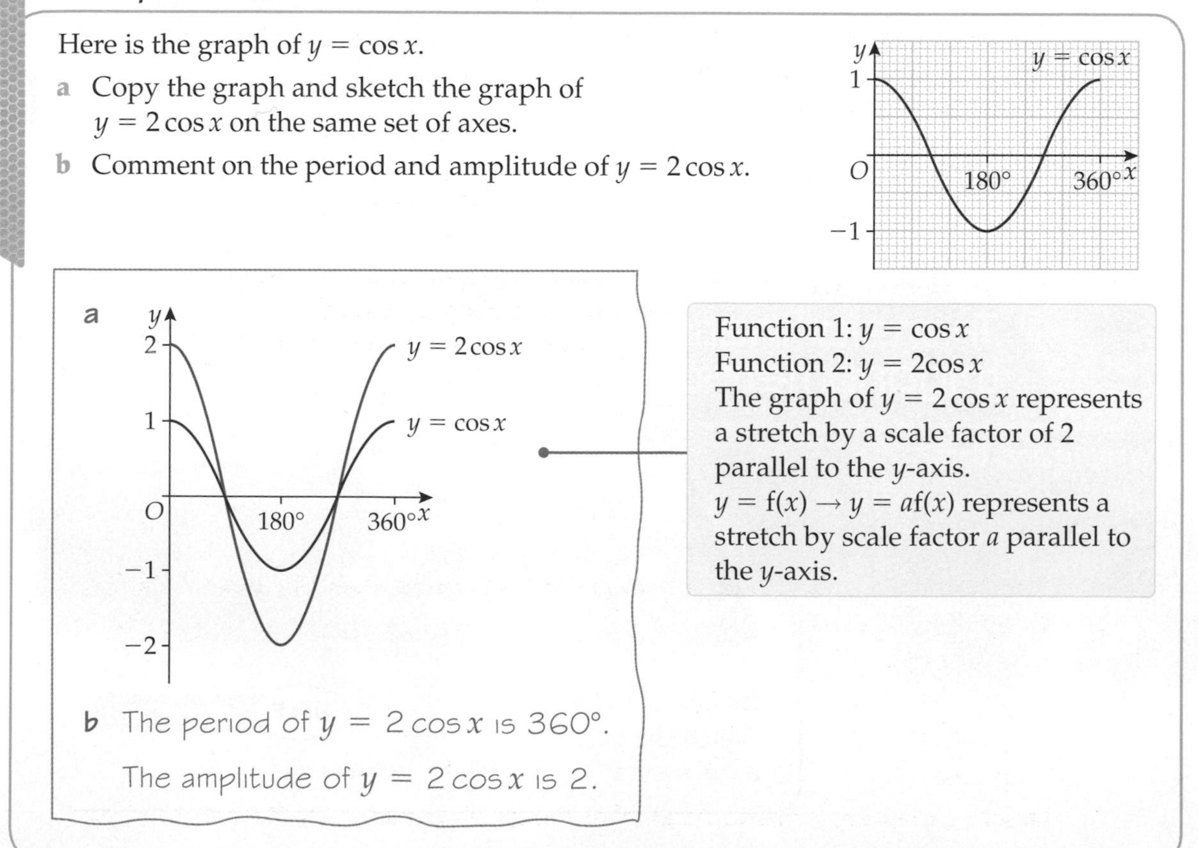

Here is the graph of $y = \cos x$.

a Copy the graph and sketch the graph of $y = 2\cos x$ on the same set of axes.

b Comment on the period and amplitude of $y = 2\cos x$.

a

Function 1: $y = \cos x$
Function 2: $y = 2\cos x$
The graph of $y = 2\cos x$ represents a stretch by a scale factor of 2 parallel to the y-axis.
$y = f(x) \rightarrow y = af(x)$ represents a stretch by scale factor a parallel to the y-axis.

b The period of $y = 2\cos x$ is 360°.

The amplitude of $y = 2\cos x$ is 2.

Example 9

Here is the graph of $y = \sin x$.

a Copy the graph and sketch the graph of $y = \sin x + 2$ on the same set of axes.

b Comment on the period and amplitude of $y = \sin x + 2$.

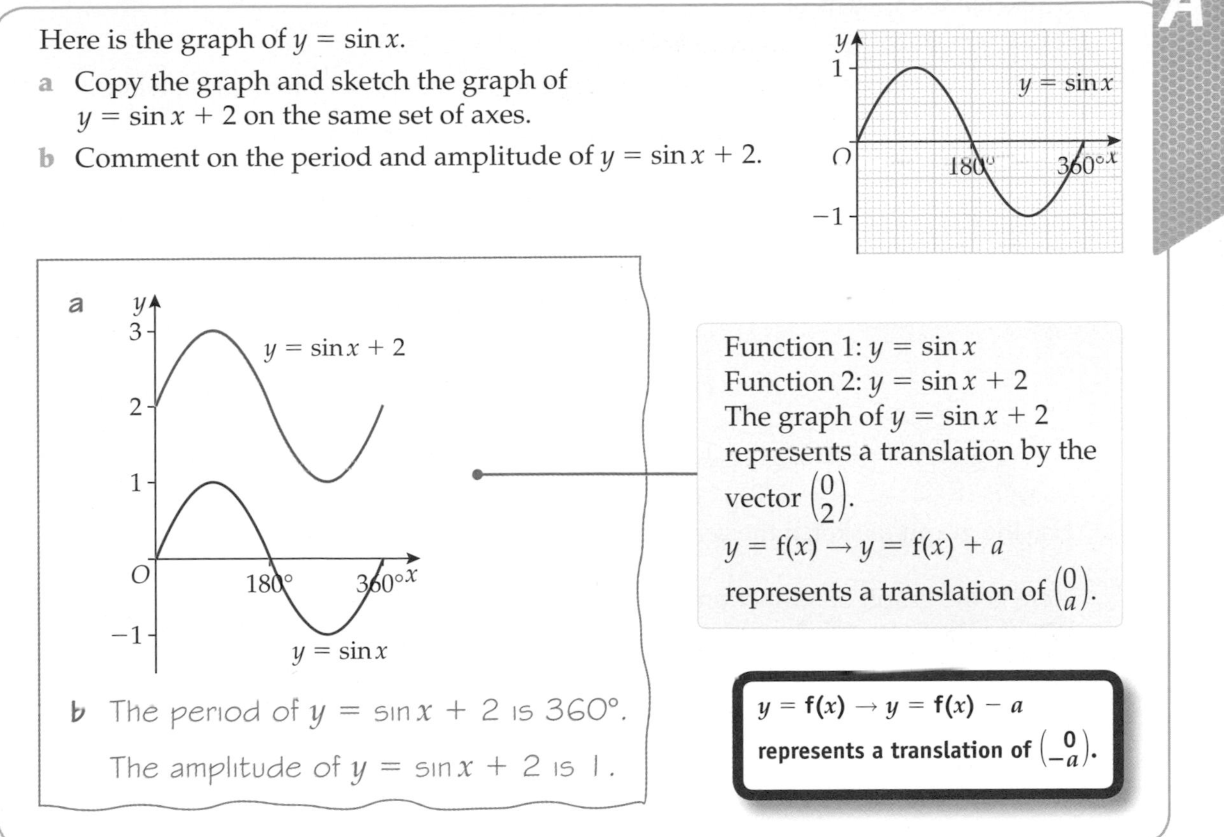

Function 1: $y = \sin x$
Function 2: $y = \sin x + 2$
The graph of $y = \sin x + 2$ represents a translation by the vector $\begin{pmatrix} 0 \\ 2 \end{pmatrix}$.

$y = f(x) \rightarrow y = f(x) + a$ represents a translation of $\begin{pmatrix} 0 \\ a \end{pmatrix}$.

b The period of $y = \sin x + 2$ is 360°.

The amplitude of $y = \sin x + 2$ is 1.

$y = f(x) \rightarrow y = f(x) - a$ represents a translation of $\begin{pmatrix} 0 \\ -a \end{pmatrix}$.

Example 10

Copy again the graph of $y = \cos x$ from Example 8 and sketch the graph of $y = \cos(3x)$ on the same set of axes.

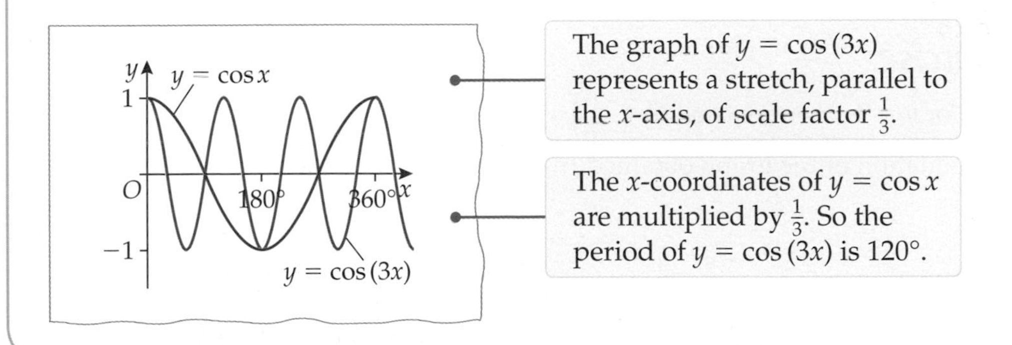

The graph of $y = \cos(3x)$ represents a stretch, parallel to the x-axis, of scale factor $\frac{1}{3}$.

The x-coordinates of $y = \cos x$ are multiplied by $\frac{1}{3}$. So the period of $y = \cos(3x)$ is 120°.

Exercise 37E

1 a On a set of axes with values of x from 0° to 360° and values of y from −3 to 3, sketch the graphs of

i $y = \cos x$

ii $y = \cos x + 1$

iii $y = \cos x - 2$

iv $y = \cos(\frac{1}{2}x)$

b Describe the transformation that takes the graph in part a i to the graphs in parts a ii and a iii.

2 a On a set of axes with values of x from 0° to 360° and values of y from −4 to 4, sketch the graphs of

i $y = \sin x$ ii $y = 4\sin x$ iii $y = \frac{1}{2}\sin x$

b Describe the transformation that takes the graph in part **a i** to the graphs in parts **a ii** and **a iii**.

c State the period and amplitude of each of the graphs.

3 Here is the graph of $y = \sin x$.

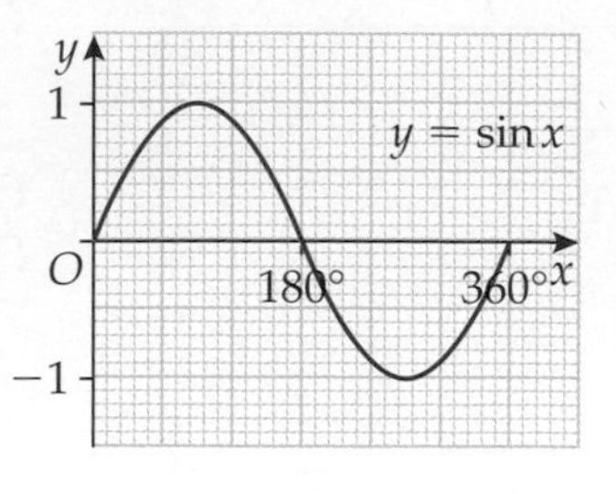

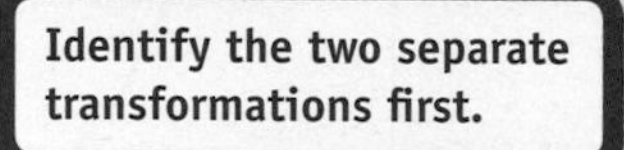

Use the graph to sketch the graph of $y = 2\sin x + 3$.

4 Here is the graph of a function in the form $y = a\cos x$. What is the value of a?

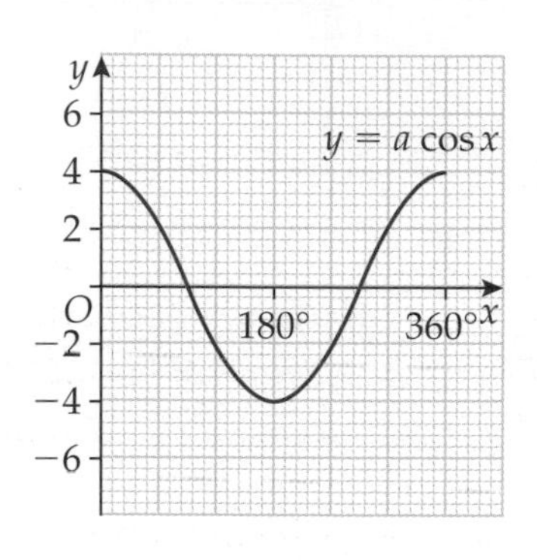

5 a Sketch the graphs of $y = \cos x$ and $y = \cos(x - 90°)$.

b What do you notice about the amplitude and period of $y = \cos(x - 90°)$?

Review exercise

1 The graph of $y = 4x^2$ is translated by $\begin{pmatrix} 0 \\ -6 \end{pmatrix}$.

Write the equation of the graph obtained from this transformation. **[1 mark]**

2 Here is the graph of $y = x^2$.

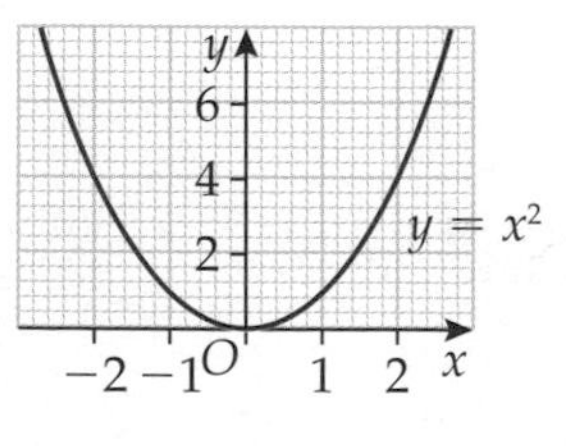

Copy the graph. Use your copy as a guide to sketch

a $y = x^2 + 1$ **[1 mark]**

b $y = 2x^2$ **[1 mark]**

c $y = (x - 2)^2$ **[1 mark]**

3 The graph of $y = x^2$ is shown in red.

Use the graph of $y = x^2$ to find the equation of each of the graphs A, B and C.

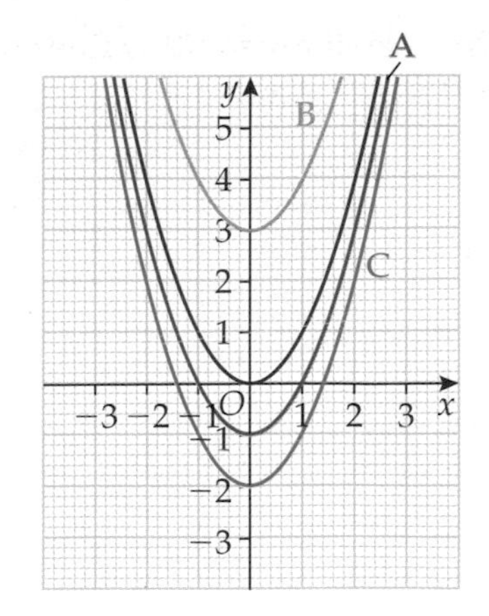

[3 marks]

4 a Sketch the graph of $y = x^2 + 4x$ for $-5 \leqslant x \leqslant 2$. **[2 marks]**

b Translate the graph of $y = x^2 + 4x$ by $\begin{pmatrix} 0 \\ -3 \end{pmatrix}$. **[1 mark]**

c Write the equation of the new graph in the form $y = ax^2 + bx + c$. **[1 mark]**

5 Write the equation of the graph obtained when the graph of $y = x^2$ is transformed by

a a stretch by scale factor $\frac{3}{2}$ parallel to the y-axis **[1 mark]**

b a translation of $\begin{pmatrix} -3 \\ 0 \end{pmatrix}$ **[1 mark]**

c a stretch by scale factor $\frac{1}{3}$ parallel to the x-axis. **[1 mark]**

A02

6 Describe the combination of transformations that takes the graph of $y = f(x)$ to $y = 3f(x) + 2$. **[1 mark]**

7 The graph of $y = 2x^2$ is stretched by scale factor $\frac{1}{2}$ parallel to the y-axis followed by a translation of $\begin{pmatrix} 0 \\ 2 \end{pmatrix}$.

Write the equation of the graph obtained from this transformation. **[2 marks]**

A03

8 Here is the graph of $y = \sin x$ for $0° \leqslant x \leqslant 360°$.

Copy the graph and sketch the graphs of the following on the same set of axes.

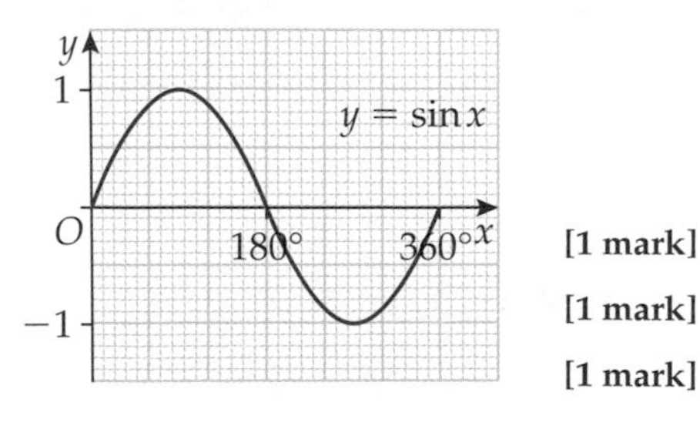

a $y = \frac{1}{2} \sin x$ **[1 mark]**

b $y = 2 + \sin x$ **[1 mark]**

c $y = \sin 2x$ **[1 mark]**

9 Here is the graph of a function in the form $y = \sin x + a$.

What is the value of a?

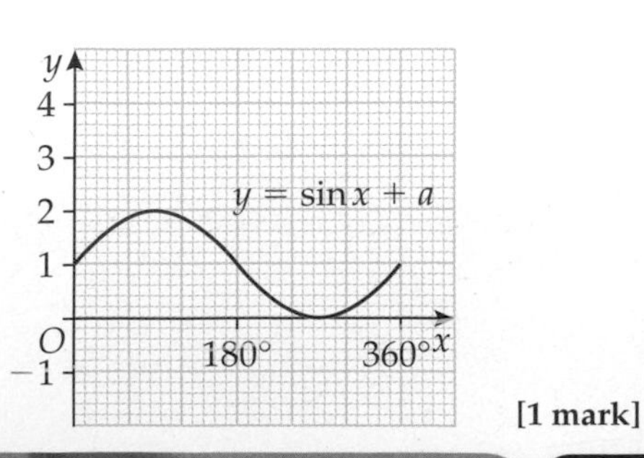

[1 mark]

A02

A*

10 Here is a sketch of $y = f(x)$ for $0° \leqslant x \leqslant 360°$.

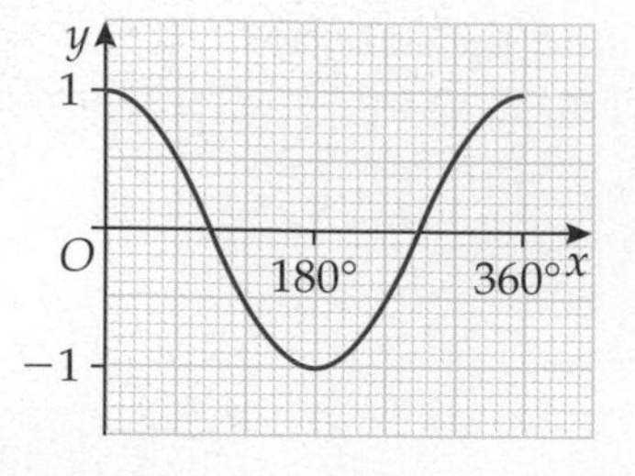

a Choose the correct equation for $y = f(x)$.

A $y = \sin x$

B $y = \cos x$ **[1 mark]**

b Graphs A, B and C show sketches of transformations of the function $y = f(x)$.

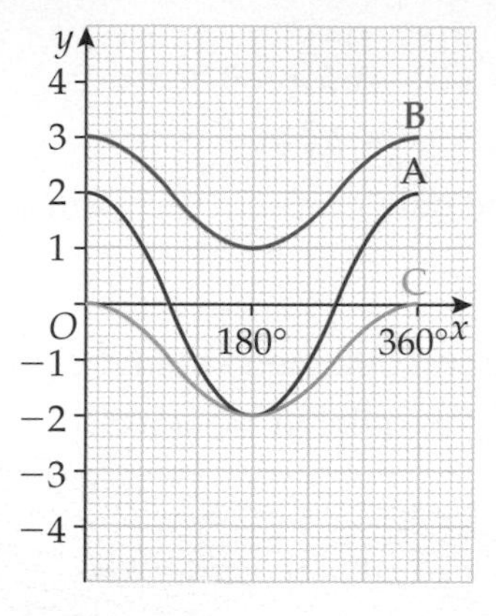

AO2

Write the equation of graphs A, B and C. **[2 marks]**

Chapter summary

In this chapter you have learned how to

- understand and use function notation B
- use transformations to change the position of linear, quadratic and trigonometric graphs A*
- recognise how linear, quadratic and trigonometric graphs are related by examining their equations A*

38 Vectors

This chapter is about vectors.

Tugboats are used to manoeuvre larger ships in crowded harbours or narrow canals. If two tugboats are pulling a ship in different directions, you can use vectors to work out which direction the ship will move.

Objectives

This chapter will show you how to

- **define a vector** A
- **use the correct notation when dealing with vectors** A
- **represent vectors graphically** A
- **add and subtract vectors using the triangle or parallelogram law** A
- **calculate the resultant of two vectors** A
- **solve problems using vector geometry** A*

Before you start this chapter

1 A translation takes the point A to the point B. Write down the column vector of the translation when

a $A = (-3, 6)$, $B = (5, -2)$ **b** $A = (-5, -7)$, $B = (3, 1)$

c $A = (8, -1)$, $B = (-4, -6)$

HELP Chapter 30

2 A line segment joins the point $(-2, 9)$ to the point $(3, -3)$.
Calculate the length of the line segment.

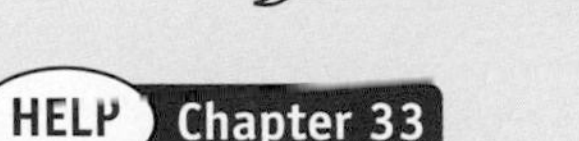

38.1 Definition, representation and notation for vectors

Keywords
vector, magnitude, direction, scalar, directed line segment

Why learn this?
Understanding the correct notation is key to solving problems in vector geometry.

Objectives

A Know the difference between a vector and a scalar

A Represent vectors graphically, using the correct notation

Skills check

1 Simplify:

a $2a^2 + 3b + 2b^2 + a^2 - 3b + b^2 - a + 3a^2 - 2a$

b $\frac{1}{2}x^2 - 4\frac{1}{2}y + 1\frac{1}{2}x^2 - x^2 + y^2 + \frac{1}{2}y + 7x^2 + 3y^2$

Vectors

A **vector** quantity is one which has **magnitude** (size) and **direction**.

Examples are:

- force (150 N acting vertically upwards)
- displacement (16 km on a bearing 065°)
- velocity (110 km/h due east).

Displacement is distance in a particular direction.

Quantities that have magnitude but no particular direction are called **scalar** quantities.

Examples are:

- speed of a train (175 km/h)
- length of room (3.6 m)
- mass of a car (1685 kg).

Velocity and speed are measured in the same units. Velocity is speed in a particular direction.

A vector can be represented by a **directed line segment** which is simply a line with an arrow on it.

Length of line = magnitude of vector.

Direction of arrow = direction of vector.

or

The arrow can be at the end of the line or in the middle of it.

You came across column vectors in Chapter 30. They were used to describe a translation. A translation of $\begin{pmatrix} 3 \\ -4 \end{pmatrix}$ represents a movement of 3 units in the x-direction and -4 units in the y-direction.

There are other ways of writing vectors and you need to be familiar with all of them.

The vector shown is written as $\overrightarrow{AB}$ which shows you that it is a displacement from A to B.

If lower case letters are used, they will be in **bold** print in a text book, but since you cannot write in bold print you should underline the letter when you write it.

A B

a

$\underline{a}$

Both the capital letter notation and the bold lower case notation will be used in the rest of this chapter.

Vectors are equal if they have the same magnitude and direction.

So, for example, these two vectors are equal:

a **b** **a** = **b**

Vectors can be multiplied by scalars (numbers).

These examples show you some of the outcomes.

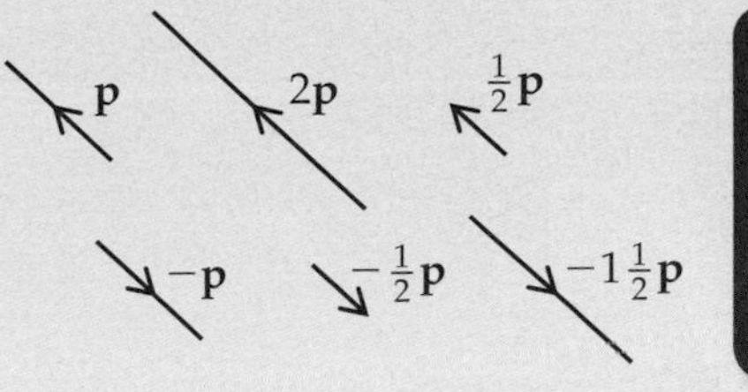

p
2p
$\frac{1}{2}$p
−p
$-\frac{1}{2}$p
$-1\frac{1}{2}$p

All of these vectors are parallel since they are written in terms of p only, but they have different lengths, and the negative ones point in the opposite direction.

Example 1

$\mathbf{a} = \begin{pmatrix} 3 \\ -1 \end{pmatrix}$ $\quad \mathbf{b} = \begin{pmatrix} -4 \\ 2 \end{pmatrix}$ $\quad \mathbf{c} = \begin{pmatrix} 0 \\ 4 \end{pmatrix}$

a Draw these vectors on a square grid:

$\mathbf{a} \quad \mathbf{b} \quad \mathbf{c} \quad -\mathbf{a} \quad 2\mathbf{b} \quad \frac{1}{2}\mathbf{c} \quad -1\frac{1}{2}\mathbf{b} \quad -2\mathbf{c}$

b What is the column vector notation for each of these?

i $2\mathbf{a}$ **ii** $\frac{1}{2}\mathbf{b}$ **iii** $-\mathbf{c}$

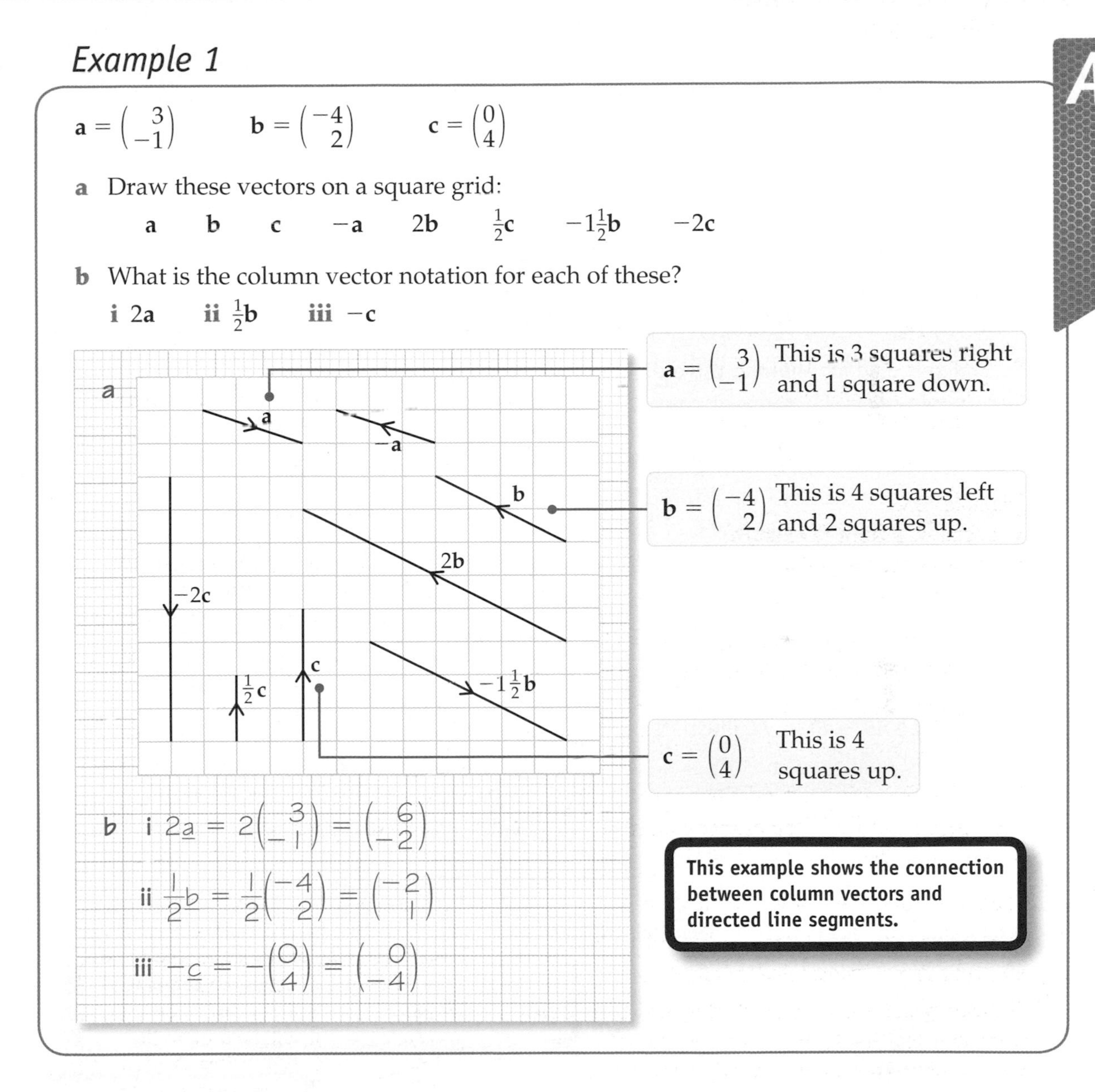

a
a
−a
b
2b
−2c
c
$\frac{1}{2}$c
$-1\frac{1}{2}$b
$\mathbf{a} = \begin{pmatrix} 3 \\ -1 \end{pmatrix}$ This is 3 squares right and 1 square down.
$\mathbf{b} = \begin{pmatrix} -4 \\ 2 \end{pmatrix}$ This is 4 squares left and 2 squares up.
$\mathbf{c} = \begin{pmatrix} 0 \\ 4 \end{pmatrix}$ This is 4 squares up.
b i $2\underline{a} = 2\begin{pmatrix} 3 \\ -1 \end{pmatrix} = \begin{pmatrix} 6 \\ -2 \end{pmatrix}$
ii $\frac{1}{2}\underline{b} = \frac{1}{2}\begin{pmatrix} -4 \\ 2 \end{pmatrix} = \begin{pmatrix} -2 \\ 1 \end{pmatrix}$
iii $-\underline{c} = -\begin{pmatrix} 0 \\ 4 \end{pmatrix} = \begin{pmatrix} 0 \\ -4 \end{pmatrix}$

This example shows the connection between column vectors and directed line segments.

Exercise 38A

1 $\mathbf{p} = \begin{pmatrix} 2 \\ -4 \end{pmatrix}$ $\quad \mathbf{q} = \begin{pmatrix} 1 \\ 3 \end{pmatrix}$ $\quad \mathbf{r} = \begin{pmatrix} -3 \\ -2 \end{pmatrix}$

a Draw these vectors on a square grid:

$\mathbf{p} \quad \mathbf{q} \quad \mathbf{r} \quad \frac{1}{2}\mathbf{p} \quad 2\mathbf{q} \quad -\mathbf{r}$

b Write these as column vectors:

$5\mathbf{r} \quad -2\mathbf{q} \quad 1\frac{1}{2}\mathbf{p}$

38.2 Adding and subtracting vectors

Keywords
displacement vector, resultant

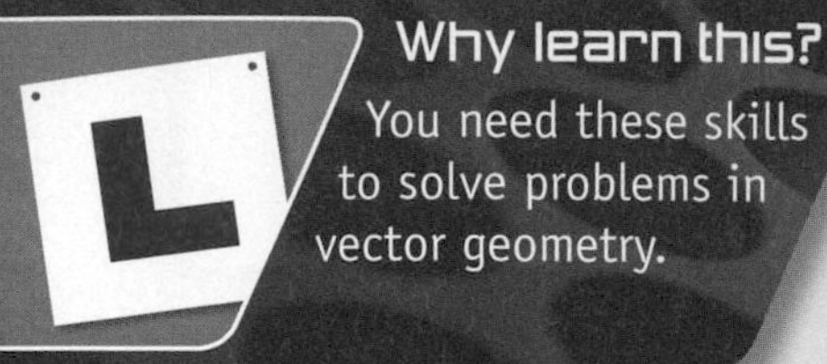

Why learn this?
You need these skills to solve problems in vector geometry.

Objectives

A Use the triangle or parallelogram law for vector addition

A Calculate the resultant of two vectors

Skills check

1. $\mathbf{a} = \begin{pmatrix} 2 \\ 4 \end{pmatrix}$ and $\mathbf{b} = \begin{pmatrix} -1 \\ 2 \end{pmatrix}$.
   Write each of these as a column vector.
   $2\mathbf{b}$ $\quad 3\mathbf{a}$ $\quad -\mathbf{b}$ $\quad -\frac{1}{2}\mathbf{a}$
2. The two shorter sides of a right-angled triangle are 12 cm and 9 cm. Calculate the length of the hypotenuse.

Adding and subtracting vectors

Look at this diagram showing three **displacement vectors**.

Notice the correct use of arrows along the line segments.

They represent a journey from A to B of 4 km north-east, followed by a journey from B to C of 3 km due east.

You can see that you can also go directly from A to C – the final destination is the same.

This can be expressed in vector notation as $\overrightarrow{AB} + \overrightarrow{BC} = \overrightarrow{AC}$ and forms the basis of vector addition.

The vector $\overrightarrow{AC}$ is called the **resultant** of vectors $\overrightarrow{AB}$ and $\overrightarrow{BC}$ because going from A to C is the same result as going from A to B and then from B to C.

The equation $\overrightarrow{AB} + \overrightarrow{BC} = \overrightarrow{AC}$ illustrates the triangle law of vector addition.

Look at this parallelogram $PQRS$.

$\overrightarrow{PQ} = \mathbf{a}$ and $\overrightarrow{PS} = \mathbf{b}$.

A parallelogram has opposite sides that are equal and parallel, so this means that $\overrightarrow{SR} = \mathbf{a}$ and $\overrightarrow{QR} = \mathbf{b}$.

Notice that the order of the letters is important.
$\overrightarrow{SR} = \mathbf{a}$ so $\overrightarrow{RS} = -\mathbf{a}$ $\quad \overrightarrow{QR} = \mathbf{b}$ so $\overrightarrow{RQ} = -\mathbf{b}$

Using vector addition,

$$\overrightarrow{PR} = \overrightarrow{PQ} + \overrightarrow{QR} = \mathbf{a} + \mathbf{b}$$

$$\overrightarrow{SQ} = \overrightarrow{SR} + \overrightarrow{RQ} = \mathbf{a} + -\mathbf{b} = \mathbf{a} - \mathbf{b}$$

$$\overrightarrow{QS} = \overrightarrow{QR} + \overrightarrow{RS} = \mathbf{b} + -\mathbf{a} = \mathbf{b} - \mathbf{a}$$

Notice the 'nose to tail' order of the capital letters: $\overrightarrow{PQ}$ followed by $\overrightarrow{QR}$ is the same as $\overrightarrow{PR}$. This is always true for vector addition.

Notice that the *diagonals* of the parallelogram are represented by the vectors a + b and either a − b or b − a depending on the direction.

You can see that

$$\overrightarrow{PR} = \overrightarrow{PR} + \overrightarrow{QR} \quad \text{and} \quad \overrightarrow{PR} = \overrightarrow{PS} + \overrightarrow{SR}$$
$$= \mathbf{a} + \mathbf{b} \qquad\qquad = \mathbf{b} + \mathbf{a}$$

This illustrates the parallelogram law of vector addition.

Example 2

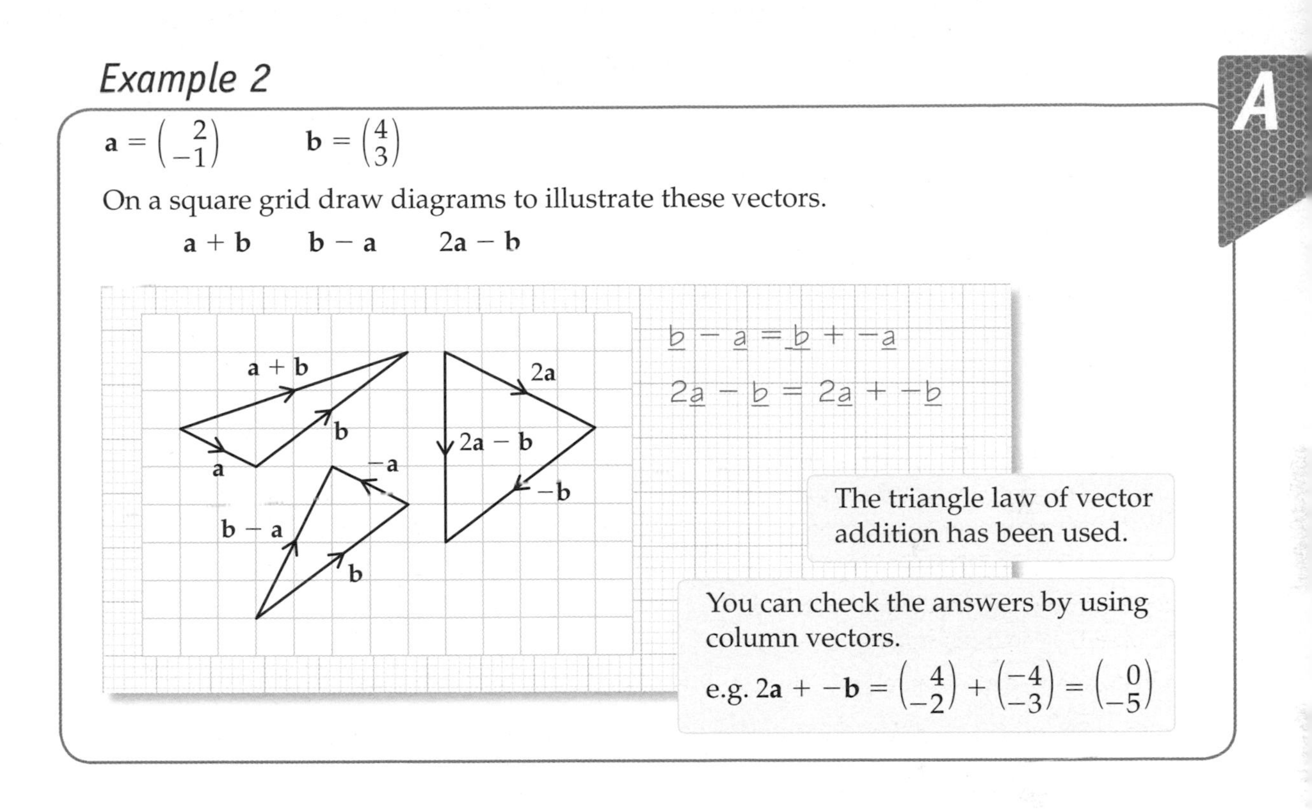

$\mathbf{a} = \begin{pmatrix} 2 \\ -1 \end{pmatrix}$ $\quad \mathbf{b} = \begin{pmatrix} 4 \\ 3 \end{pmatrix}$

On a square grid draw diagrams to illustrate these vectors.

$\mathbf{a} + \mathbf{b} \qquad \mathbf{b} - \mathbf{a} \qquad 2\mathbf{a} - \mathbf{b}$

The triangle law of vector addition has been used.

You can check the answers by using column vectors.

e.g. $2\mathbf{a} + -\mathbf{b} = \begin{pmatrix} 4 \\ -2 \end{pmatrix} + \begin{pmatrix} -4 \\ -3 \end{pmatrix} = \begin{pmatrix} 0 \\ -5 \end{pmatrix}$

Exercise 38B

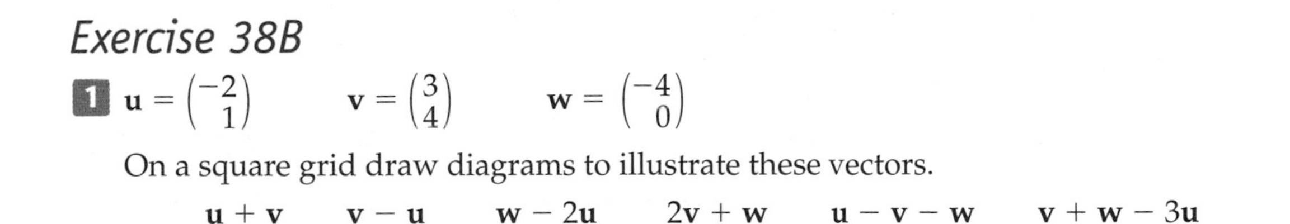

1 $\mathbf{u} = \begin{pmatrix} -2 \\ 1 \end{pmatrix}$ $\quad \mathbf{v} = \begin{pmatrix} 3 \\ 4 \end{pmatrix}$ $\quad \mathbf{w} = \begin{pmatrix} -4 \\ 0 \end{pmatrix}$

On a square grid draw diagrams to illustrate these vectors.

$\mathbf{u} + \mathbf{v} \qquad \mathbf{v} - \mathbf{u} \qquad \mathbf{w} - 2\mathbf{u} \qquad 2\mathbf{v} + \mathbf{w} \qquad \mathbf{u} - \mathbf{v} - \mathbf{w} \qquad \mathbf{v} + \mathbf{w} - 3\mathbf{u}$

Calculating the magnitude and direction of a vector

You can calculate the magnitude and the direction of a vector by using Pythagoras' theorem and trigonometry.

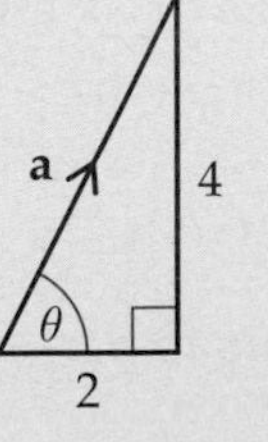

For $\mathbf{a} = \begin{pmatrix} 2 \\ 4 \end{pmatrix}$

$|\mathbf{a}| = \sqrt{2^2 + 4^2}$

$= \sqrt{20} = 2\sqrt{5}$

$\tan \theta = \frac{4}{2} = 2$

$\theta = 63.4°$

$|\mathbf{a}|$ is called the *modulus* of vector a and is the notation used for magnitude.

A

Example 3

The diagram shows a grid of congruent parallelograms.

The origin is labelled O and the vectors of points A and B are given by $\overrightarrow{OA} = \mathbf{a}$ and $\overrightarrow{OB} = \mathbf{b}$.

Write, in terms of **a** and **b**,

a $\overrightarrow{OP}$ **b** $\overrightarrow{OV}$ **c** $\overrightarrow{OR}$ **d** $\overrightarrow{OG}$

e $\overrightarrow{AB}$ **f** $\overrightarrow{NE}$ **g** $\overrightarrow{FB}$ **h** $\overrightarrow{UG}$

a $\overrightarrow{OP} = 3\underline{a}$

b $\overrightarrow{OV} = 2\underline{b}$

c $\overrightarrow{OR} = \overrightarrow{ON} + \overrightarrow{NR} = 2\underline{a} + \underline{b}$

d $\overrightarrow{OG} = \overrightarrow{OP} + \overrightarrow{PG} = 3\underline{a} - 2\underline{b}$

e $\overrightarrow{AB} = \overrightarrow{AO} + \overrightarrow{OB} = -\underline{a} + \underline{b}$

f $\overrightarrow{NE} = \overrightarrow{NA} + \overrightarrow{AE} = -\underline{a} - 2\underline{b}$

g $\overrightarrow{FB} = \overrightarrow{FD} + \overrightarrow{DB} = -2\underline{a} + 3\underline{b}$

h $\overrightarrow{UG} = \overrightarrow{UY} + \overrightarrow{YG} = 4\underline{a} - 4\underline{b}$

A

Example 4

Look at the parallelogram grid in Example 3.

a Write down a vector equal to $3\mathbf{a} - \mathbf{b}$.

b Write down three other vectors equal to $3\mathbf{a} - \mathbf{b}$.

c Write down a vector parallel to $\overrightarrow{CP}$.

d Write down three vectors that are half the size of $\overrightarrow{CP}$ and in the opposite direction.

e Write down all the vectors that are equal to $3(\mathbf{a} + \mathbf{b})$.

f Write down, in terms of **a** and **b**, the vectors

$\overrightarrow{CK}$ $\overrightarrow{CA}$ $\overrightarrow{CR}$ $\overrightarrow{CY}$.

A02 **g** What is the connection between your answers to part **f** and the positions of C, K, A, R and Y on the grid?

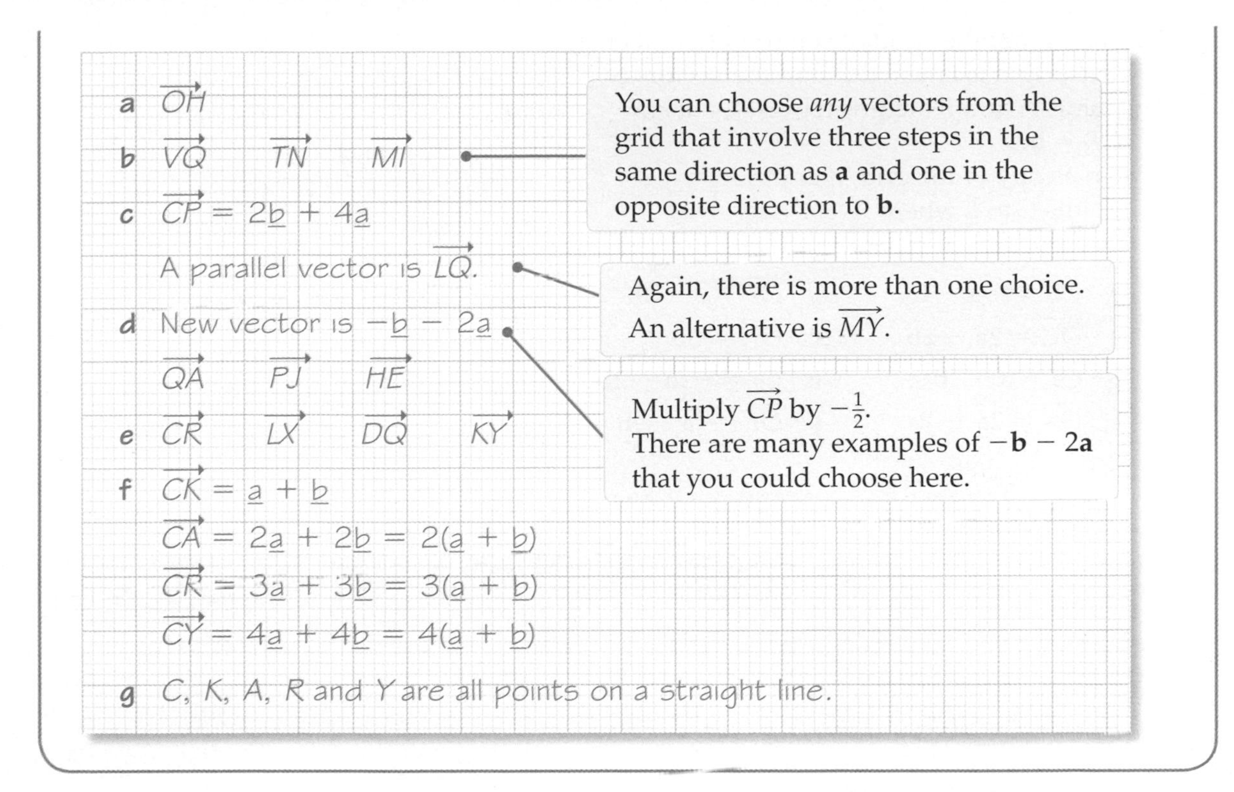

a $\overrightarrow{OH}$

b $\overrightarrow{VQ}$ $\overrightarrow{TN}$ $\overrightarrow{MI}$

You can choose *any* vectors from the grid that involve three steps in the same direction as **a** and one in the opposite direction to **b**.

c $\overrightarrow{CP} = 2\underline{b} + 4\underline{a}$

A parallel vector is $\overrightarrow{LQ}$.

Again, there is more than one choice. An alternative is $\overrightarrow{MY}$.

d New vector is $-\underline{b} - 2\underline{a}$

$\overrightarrow{QA}$ $\overrightarrow{PJ}$ $\overrightarrow{HE}$

Multiply $\overrightarrow{CP}$ by $-\frac{1}{2}$.
There are many examples of $-\mathbf{b} - 2\mathbf{a}$ that you could choose here.

e $\overrightarrow{CR}$ $\overrightarrow{LX}$ $\overrightarrow{DQ}$ $\overrightarrow{KY}$

f $\overrightarrow{CK} = \underline{a} + \underline{b}$

$\overrightarrow{CA} = 2\underline{a} + 2\underline{b} = 2(\underline{a} + \underline{b})$

$\overrightarrow{CR} = 3\underline{a} + 3\underline{b} = 3(\underline{a} + \underline{b})$

$\overrightarrow{CY} = 4\underline{a} + 4\underline{b} = 4(\underline{a} + \underline{b})$

g C, K, A, R and Y are all points on a straight line.

Exercise 38C

A

1 Draw a separate diagram for each of these vectors.
Find the magnitude and direction (the angle made with the positive x-axis) of each vector.

a $\begin{pmatrix}0\\4\end{pmatrix}$ **b** $\begin{pmatrix}3\\5\end{pmatrix}$ **c** $\begin{pmatrix}6\\-2\end{pmatrix}$ **d** $\begin{pmatrix}-4\\6\end{pmatrix}$ **e** $\begin{pmatrix}-3\\-7\end{pmatrix}$

2 The diagram shows a grid of congruent parallelograms.
The origin is labelled O and the vectors of points A and B are given by $\overrightarrow{OA} = \mathbf{a}$ and $\overrightarrow{OB} = \mathbf{b}$.

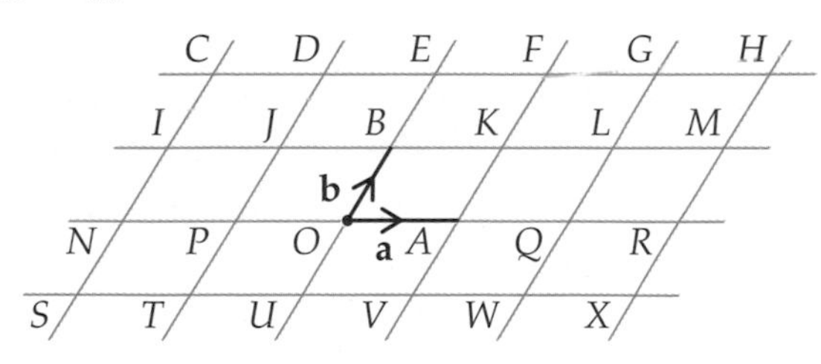

Write, in terms of **a** and **b**,

a $\overrightarrow{OR}$ **b** $\overrightarrow{OU}$ **c** $\overrightarrow{OC}$ **d** $\overrightarrow{OX}$ **e** $\overrightarrow{GP}$ **f** $\overrightarrow{IW}$ **g** $\overrightarrow{HR}$ **h** $\overrightarrow{MJ}$

A

3 Look at the parallelogram grid in Q2.

a Write down a vector equal to $3\mathbf{a} + \mathbf{b}$.

b Write down three other vectors equal to $3\mathbf{a} + \mathbf{b}$.

c Write down three vectors parallel to $\overrightarrow{ND}$.

d Write down three vectors that are half the size of $\overrightarrow{TM}$ and in the opposite direction.

e Write down the vectors $\overrightarrow{DI}$, $\overrightarrow{FP}$ and $\overrightarrow{HU}$.

f Explain the connection between the vectors in part **e**.

AO2

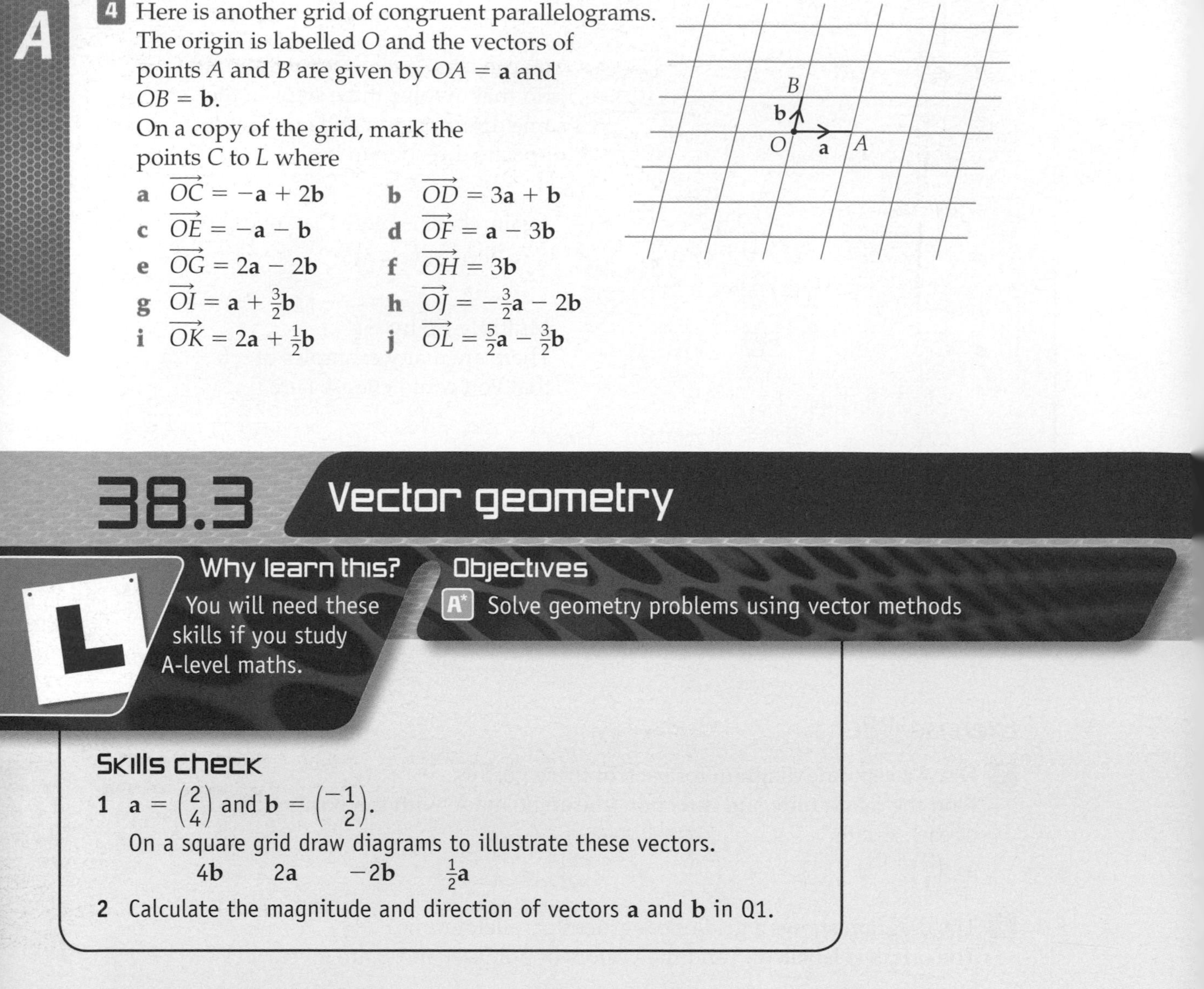

A

4 Here is another grid of congruent parallelograms. The origin is labelled O and the vectors of points A and B are given by $OA = \mathbf{a}$ and $OB = \mathbf{b}$.

On a copy of the grid, mark the points C to L where

a $\overrightarrow{OC} = -\mathbf{a} + 2\mathbf{b}$ **b** $\overrightarrow{OD} = 3\mathbf{a} + \mathbf{b}$

c $\overrightarrow{OE} = -\mathbf{a} - \mathbf{b}$ **d** $\overrightarrow{OF} = \mathbf{a} - 3\mathbf{b}$

e $\overrightarrow{OG} = 2\mathbf{a} - 2\mathbf{b}$ **f** $\overrightarrow{OH} = 3\mathbf{b}$

g $\overrightarrow{OI} = \mathbf{a} + \frac{3}{2}\mathbf{b}$ **h** $\overrightarrow{OJ} = -\frac{3}{2}\mathbf{a} - 2\mathbf{b}$

i $\overrightarrow{OK} = 2\mathbf{a} + \frac{1}{2}\mathbf{b}$ **j** $\overrightarrow{OL} = \frac{5}{2}\mathbf{a} - \frac{3}{2}\mathbf{b}$

38.3 Vector geometry

Why learn this?
You will need these skills if you study A-level maths.

Objectives
A* Solve geometry problems using vector methods

Skills check

1 $\mathbf{a} = \begin{pmatrix} 2 \\ 4 \end{pmatrix}$ and $\mathbf{b} = \begin{pmatrix} -1 \\ 2 \end{pmatrix}$.
On a square grid draw diagrams to illustrate these vectors.
$4\mathbf{b}$ $2\mathbf{a}$ $-2\mathbf{b}$ $\frac{1}{2}\mathbf{a}$

2 Calculate the magnitude and direction of vectors **a** and **b** in Q1.

Vector geometry

The rules for the addition and subtraction of vectors and the multiplication of a vector by a scalar are the key elements in answering questions on vector geometry.

The next two examples use all of the skills covered earlier in this chapter and are typical of questions that occur on the Higher Unit 3 paper.

Example 5

$OABC$ is a quadrilateral with $\overrightarrow{OA} = \mathbf{a}$, $\overrightarrow{OB} = \mathbf{b}$ and $\overrightarrow{OC} = \mathbf{c}$.
M is the mid-point of AB and P is the point on BC such that $BP : PC = 3 : 1$.

Find expressions for these vectors, giving your answers in their simplest form.

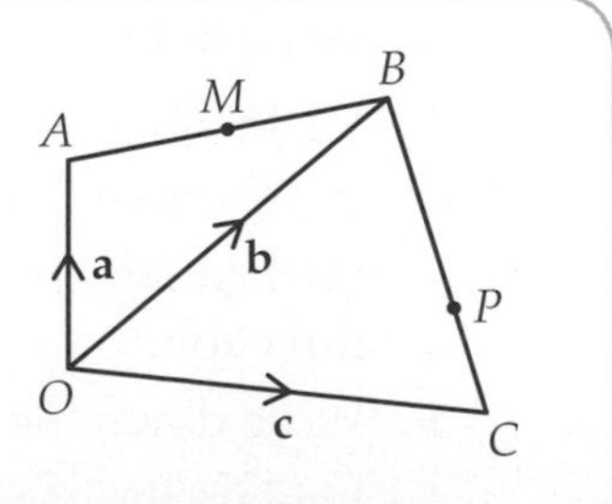

a $\overrightarrow{AB}$ **b** $\overrightarrow{AM}$ **c** $\overrightarrow{OM}$ **d** $\overrightarrow{BC}$ **e** $\overrightarrow{BP}$ **f** $\overrightarrow{MP}$

a $\overrightarrow{AB} = \overrightarrow{AO} + \overrightarrow{OB} = -\underline{a} + \underline{b}$ or $\underline{b} - \underline{a}$

b $\overrightarrow{AM} = \frac{1}{2}\overrightarrow{AB} = \frac{1}{2}(\underline{b} - \underline{a}) = \frac{1}{2}\underline{b} - \frac{1}{2}\underline{a}$

c $\overrightarrow{OM} = \overrightarrow{OA} + \overrightarrow{AM} = \underline{a} + \frac{1}{2}\underline{b} - \frac{1}{2}\underline{a} = \frac{1}{2}\underline{a} + \frac{1}{2}\underline{b}$

d $\overrightarrow{BC} = \overrightarrow{BO} + \overrightarrow{OC} = -\underline{b} + \underline{c}$ or $\underline{c} - \underline{b}$

e $\overrightarrow{BP} = \frac{3}{4}\overrightarrow{BC} = \frac{3}{4}(\underline{c} - \underline{b}) = \frac{3}{4}\underline{c} - \frac{3}{4}\underline{b}$

As $BP : PC = 3:1$ then $BP = \frac{3}{4}BC$. You will often need to interpret ratios in this way.

f $\overrightarrow{MP} = \overrightarrow{MB} + \overrightarrow{BP} = \left(\frac{1}{2}\underline{b} - \frac{1}{2}\underline{a}\right) + \left(\frac{3}{4}\underline{c} - \frac{3}{4}\underline{b}\right)$

$= \frac{1}{2}\underline{b} - \frac{3}{4}\underline{b} - \frac{1}{2}\underline{a} + \frac{3}{4}\underline{c}$

$= \frac{3}{4}\underline{c} - \frac{1}{4}\underline{b} - \frac{1}{2}\underline{a}$

Since M is the mid-point of $\overrightarrow{AB}$ then $\overrightarrow{MB} = \overrightarrow{AM}$.

Example 6

A* AO2

The diagram shows a trapezium $OABC$ in which OC is parallel to AB and $OC = \frac{2}{3}AB$. $\overrightarrow{OA} = \mathbf{a}$ and $\overrightarrow{OB} = \mathbf{b}$.

P is the point on AB such that $AP : PB = 1:2$ and M is the mid-point of OB.

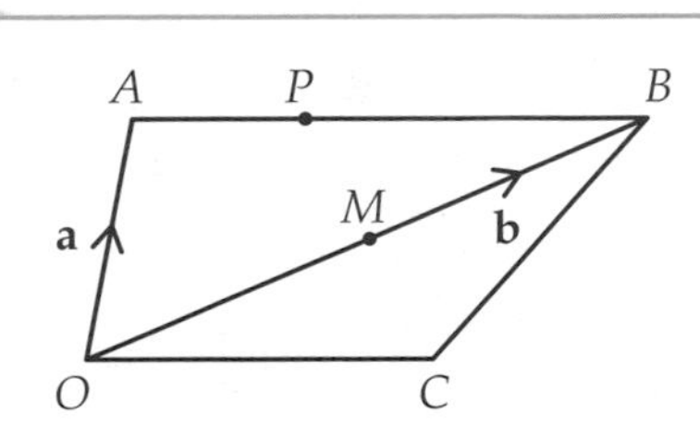

a Find these vectors in terms of **a** and **b**, giving your answer in their simplest form.

i $\overrightarrow{AB}$ **ii** $\overrightarrow{AP}$ **iii** $\overrightarrow{PM}$ **iv** $\overrightarrow{MC}$

b What do your answers for $\overrightarrow{PM}$ and $\overrightarrow{MC}$ tell you about the relationship between points P, M and C?

a **i** $\overrightarrow{AB} = \overrightarrow{AO} + \overrightarrow{OB} = -\underline{a} + \underline{b} = \underline{b} - \underline{a}$

ii $\overrightarrow{AP} = \frac{1}{3}\overrightarrow{AB} = \frac{1}{3}(\underline{b} - \underline{a})$ or $\frac{1}{3}\underline{b} - \frac{1}{3}\underline{a}$

As $AP : PB = 1:2$ then $AP = \frac{1}{3}AB$.

iii $\overrightarrow{PM} = \overrightarrow{PA} + \overrightarrow{AO} + \overrightarrow{OM}$

$= -\left(\frac{1}{3}\underline{b} - \frac{1}{3}\underline{a}\right) + \left(-\underline{a}\right) + \left(\frac{1}{2}\underline{b}\right)$

$= -\frac{1}{3}\underline{b} + \frac{1}{2}\underline{b} + \frac{1}{3}\underline{a} - \underline{a}$

$= \frac{1}{6}\underline{b} - \frac{2}{3}\underline{a}$

iv $\overrightarrow{MC} = \overrightarrow{MO} + \overrightarrow{OC} = \overrightarrow{MO} + \frac{2}{3}\overrightarrow{AB}$

$= -\frac{1}{2}\underline{b} + \frac{2}{3}(\underline{b} - \underline{a})$

$= -\frac{1}{2}\underline{b} + \frac{2}{3}\underline{b} - \frac{2}{3}\underline{a}$

$= \frac{1}{6}\underline{b} - \frac{2}{3}\underline{a}$

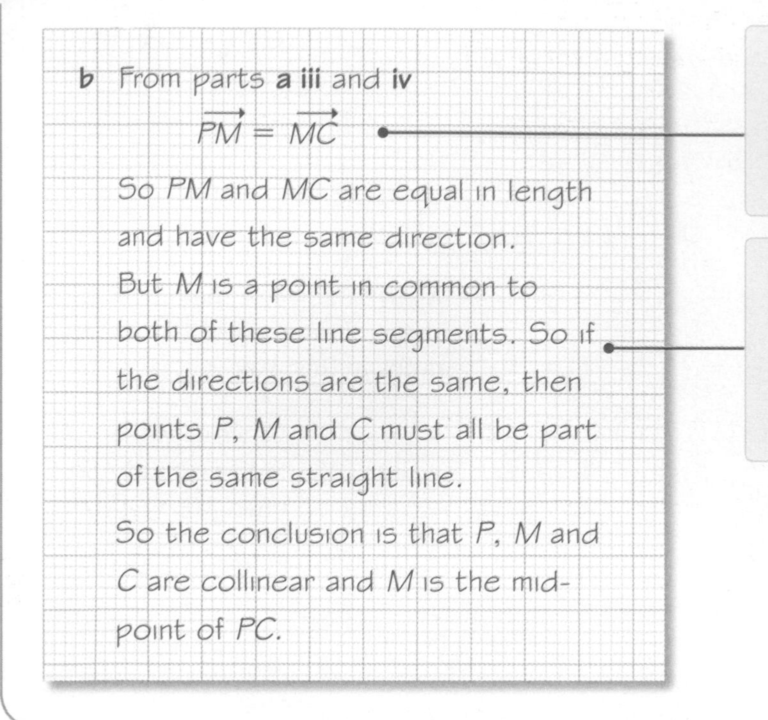

b From parts **a iii** and **iv**

$\overrightarrow{PM} = \overrightarrow{MC}$

So PM and MC are equal in length and have the same direction.

But M is a point in common to both of these line segments. So if the directions are the same, then points P, M and C must all be part of the same straight line.

So the conclusion is that P, M and C are collinear and M is the mid-point of PC.

Remember that when vectors are equal they are equal in two ways … their lengths are equal and they have the same direction.

When you have a common point and vectors are equal (or where one is a multiple of the other) then you can deduce that the points lie in a straight line.

Collinear means that points lie in a straight line.

Exercise 38D

Diagrams in this exercise are not drawn accurately.

A*

1 OPQ is a triangle.

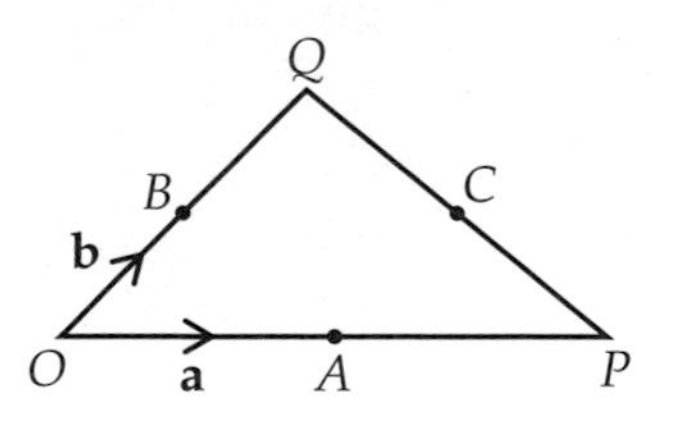

A, B and C are the mid-points of OP, OQ and PQ respectively.
$\overrightarrow{OA} = \mathbf{a}$ and $\overrightarrow{OB} = \mathbf{b}$.
Find these vectors in terms of **a** and **b**, simplifying your answers.

a $\overrightarrow{OP}$ **b** $\overrightarrow{OQ}$ **c** $\overrightarrow{PQ}$ **d** $\overrightarrow{AB}$ **e** $\overrightarrow{PC}$ **f** $\overrightarrow{OC}$ **g** $\overrightarrow{BP}$ **h** $\overrightarrow{QA}$

2 OFG is a triangle.

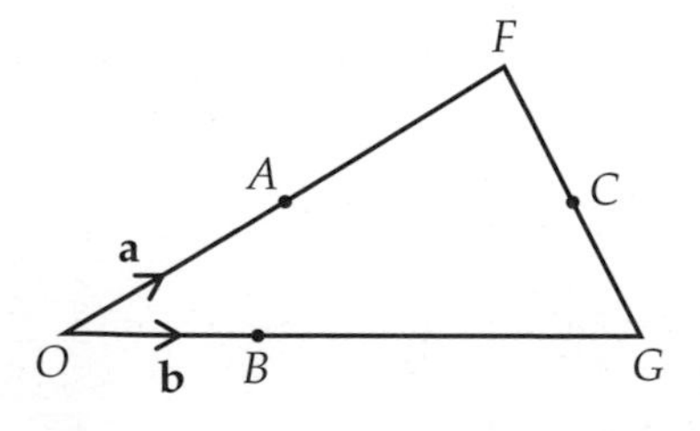

A and C are the mid-points of OF and FG respectively.
B lies on OG such that $OB : BG = 1 : 2$.
$\overrightarrow{OA} = \mathbf{a}$ and $\overrightarrow{OB} = \mathbf{b}$.
Find these vectors in terms of **a** and **b**, simplifying your answers.

a $\overrightarrow{AF}$ **b** $\overrightarrow{AB}$ **c** $\overrightarrow{OG}$ **d** $\overrightarrow{FO}$ **e** $\overrightarrow{FG}$ **f** $\overrightarrow{GA}$ **g** $\overrightarrow{BF}$ **h** $\overrightarrow{OC}$

A02

3 OAB is a triangle with $\overrightarrow{OA} = \mathbf{a}$ and $\overrightarrow{OB} = \mathbf{b}$.
M is the mid-point of OB and P is the point on AB such that $AP:PB = 2:1$.
Find expressions for these vectors in terms of $\mathbf{a}$ and $\mathbf{b}$, simplifying your answers.

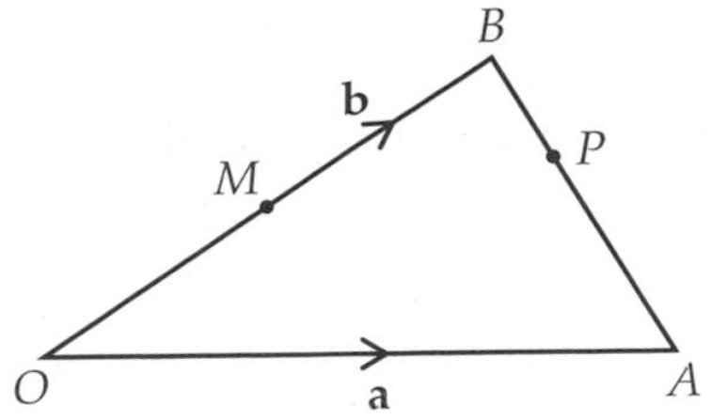

a $\overrightarrow{OM}$ b $\overrightarrow{AB}$ c $\overrightarrow{AP}$ d $\overrightarrow{OP}$
e $\overrightarrow{BA}$ f $\overrightarrow{MA}$ g $\overrightarrow{MP}$

A*

4 $ABCDEF$ is a regular hexagon with centre O.
$\overrightarrow{AB} = \mathbf{b}$ and $\overrightarrow{AC} = \mathbf{c}$.
Find these vectors in terms of $\mathbf{b}$ and $\mathbf{c}$, simplifying your answers.

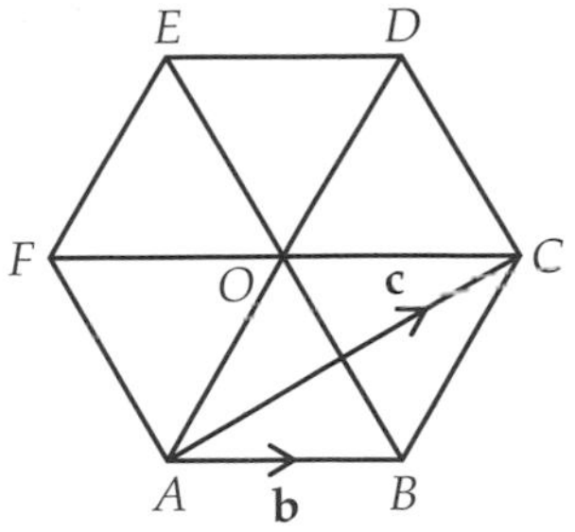

a $\overrightarrow{BC}$ b $\overrightarrow{AO}$ c $\overrightarrow{AD}$
d $\overrightarrow{EC}$ e $\overrightarrow{AF}$ f $\overrightarrow{AE}$

AO2

5 OAB is a triangle.
M is the mid-point of OB, N is the mid-point of AB.
P and Q are points of trisection of OA.
$\overrightarrow{OP} = \mathbf{p}$ and $\overrightarrow{OM} = \mathbf{m}$.

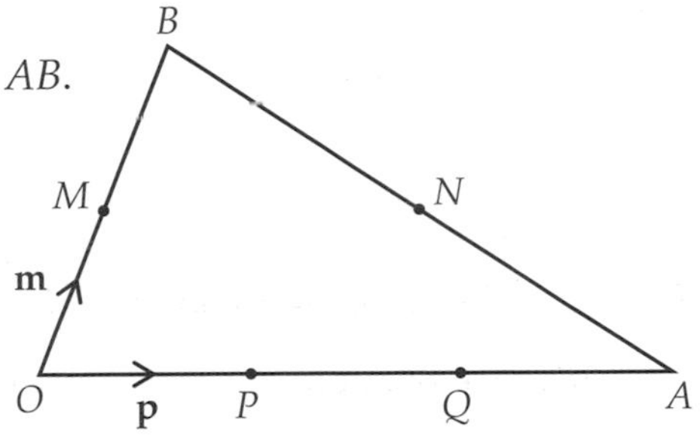

a Find the expressions for these vectors in terms of $\mathbf{p}$ and $\mathbf{m}$, simplifying your answers.
 i $\overrightarrow{OA}$ ii $\overrightarrow{OB}$ iii $\overrightarrow{BA}$
 iv $\overrightarrow{MN}$ v $\overrightarrow{PQ}$ vi $\overrightarrow{MP}$
 vii $\overrightarrow{MQ}$ viii $\overrightarrow{NQ}$ ix $\overrightarrow{PN}$

b What kind of quadrilateral is $PQNM$? Explain your answer.

A*

6 $OPQR$ is a trapezium with PQ parallel to OR.
$\overrightarrow{OP} = -2\mathbf{a} + 3\mathbf{b}$ and $\overrightarrow{OQ} = 4\mathbf{a} + 5\mathbf{b}$.

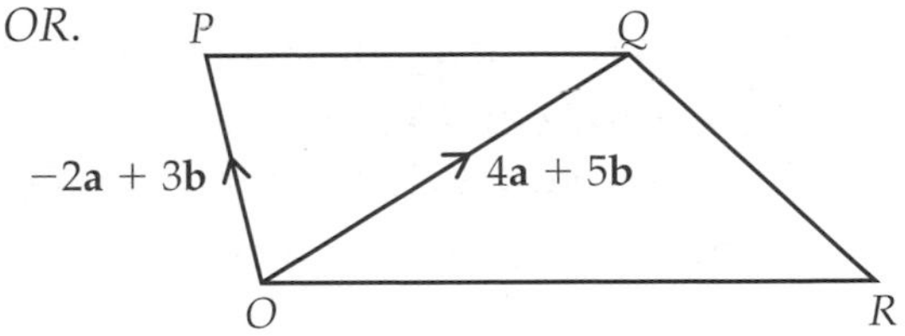

a Find $\overrightarrow{PQ}$ in terms of $\mathbf{a}$ and $\mathbf{b}$, simplifying your answer.

b $\overrightarrow{QR} = 5\mathbf{a} + k\mathbf{b}$ (where k is a number to be determined).
Find $\overrightarrow{OR}$ in terms of $\mathbf{a}$ and $\mathbf{b}$ and k, and hence work out the value of k.

7 In the diagram $\overrightarrow{OP} = 2\mathbf{a}$, $\overrightarrow{PA} = \mathbf{a}$, $\overrightarrow{OB} = 3\mathbf{b}$ and $\overrightarrow{BR} = \mathbf{b}$.
Q lies on AB such that $AQ:QB = 2:1$.

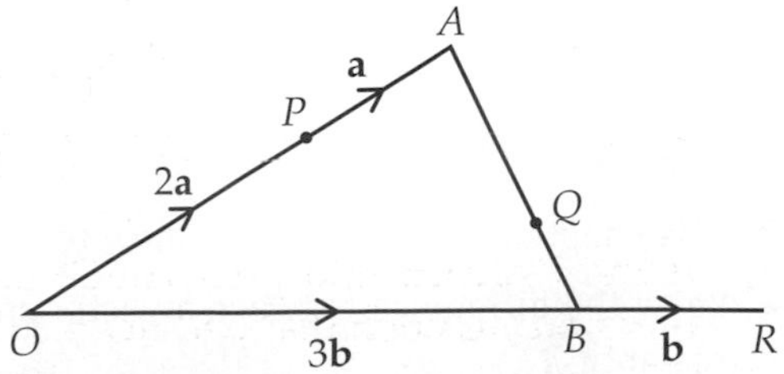

a Find expressions for these vectors in terms of $\mathbf{a}$ and $\mathbf{b}$, simplifying your answers.
 i $\overrightarrow{AB}$ ii $\overrightarrow{AQ}$ iii $\overrightarrow{QB}$
 iv $\overrightarrow{PQ}$ v $\overrightarrow{QR}$ vi $\overrightarrow{PR}$

b Explain clearly the relationship between points P, Q and R.

AO3

Review exercise

A*

1 OAB is a triangle where M is the mid-point of OB.
P and Q are points on AB such that $AP = PQ = QB$.
$\overrightarrow{OA} = \mathbf{a}$ and $\overrightarrow{OB} = 2\mathbf{b}$.

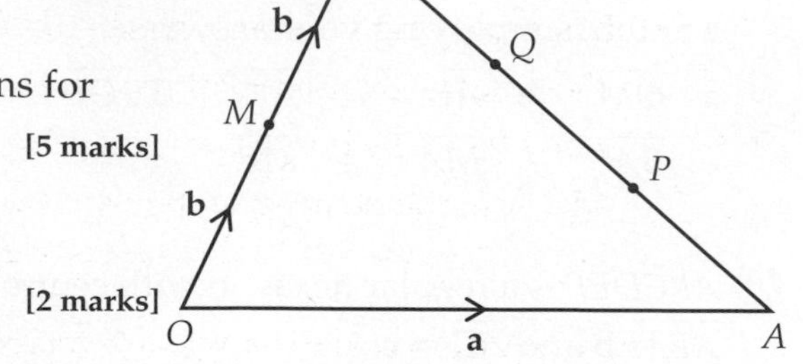

a Find, in terms of $\mathbf{a}$ and $\mathbf{b}$, expressions for

i $\overrightarrow{BA}$ ii $\overrightarrow{MQ}$ iii $\overrightarrow{OP}$ **[5 marks]**

b What can you deduce about quadrilateral $OMQP$?
Give a reason for your answer. **[2 marks]**

2 $OATB$ is a quadrilateral.
P, Q, R and S are the mid-points of OA, AT, TB and OB respectively.
$\overrightarrow{OA} = \mathbf{a}$, $\overrightarrow{OB} = \mathbf{b}$ and $\overrightarrow{OT} = \mathbf{t}$.

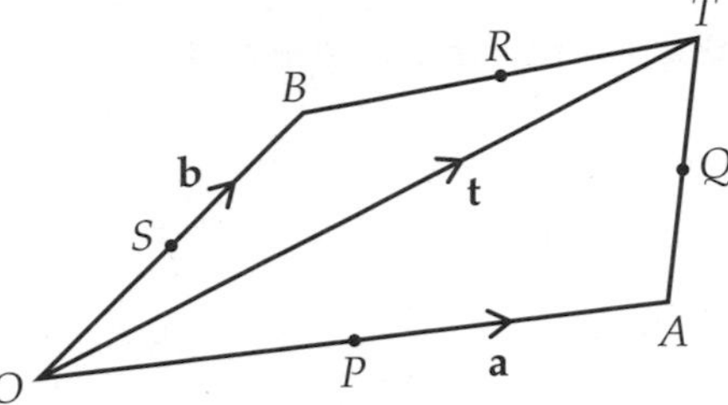

a Find these vectors, in terms of $\mathbf{a}$ and $\mathbf{b}$, simplifying your answers.

i $\overrightarrow{AT}$ ii $\overrightarrow{QT}$ iii $\overrightarrow{TB}$ iv $\overrightarrow{TR}$ v $\overrightarrow{PS}$ vi $\overrightarrow{QR}$ **[6 marks]**

b Explain why your answers for vectors $\overrightarrow{PS}$ and $\overrightarrow{QR}$ show that $PQRS$ is a parallelogram. **[2 marks]**

3 The diagram shows a quadrilateral $OAGB$.
$\overrightarrow{OA} = \mathbf{a}$, $\overrightarrow{OB} = 2\mathbf{b}$ and $\overrightarrow{OG} = 3\mathbf{a} + 2\mathbf{b}$.

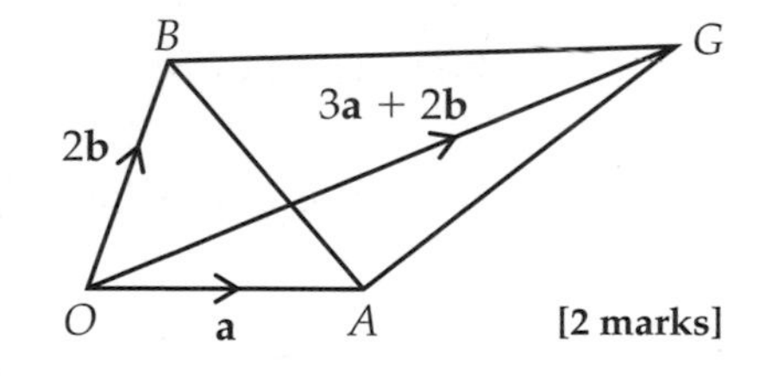

a Find expressions for these vectors in terms of $\mathbf{a}$ and $\mathbf{b}$, simplifying your answers.

i $\overrightarrow{AG}$ ii $\overrightarrow{BG}$ **[2 marks]**

b What kind of quadrilateral is $OAGB$?
Give a reason for your answer. **[2 marks]**

c Point P lies on AB such that $AP:PB = 1:3$.
Find expressions for these vectors in terms of $\mathbf{a}$ and $\mathbf{b}$, simplifying your answers.

i $\overrightarrow{AB}$ ii $\overrightarrow{AP}$ iii $\overrightarrow{OP}$ **[3 marks]**

AO3

d Describe, as fully as possible, the position of P. **[2 marks]**

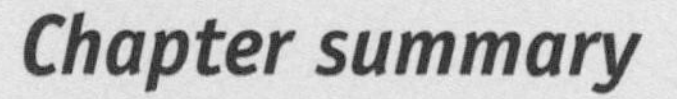

In this chapter you have learned how to

- know the difference between a vector and a scalar **A**
- represent vectors graphically, using the correct notation **A**
- use the triangle or parallelogram law for vector addition **A**
- calculate the resultant of two vectors **A**
- solve geometry problems using vector methods **A***

1 A boat leaves a harbour and sails on a bearing of 300° for 65 km.
It then turns and sails on a bearing of 075° for a further 40 km.

a Draw a scale drawing of the boat's journey using a scale of 1 cm = 10 km. [3]

b How far is the boat from the harbour. Give your answer to the nearest km. [1]

c Measure the bearing of the boat from the harbour.
Give your answer to the nearest degree. [1]

2 The lengths on this rectangle are correct to 1 decimal place.

5.9 cm
2.2 cm

Calculate

a the maximum value for the area of the rectangle [2]

b the minimum value for the area of the rectangle. [2]

3 The diagram shows a rubber door wedge.

a Work out the volume of the wedge. [3]

b Rubber has a density of 11 g/cm^3.
Calculate the mass of the wedge. [2]

4 cm
12 cm
15 cm

4 The diagram shows a triangular course for an aircraft race. Point *C* is due east of point *A*.

N
A *B* *C*
5 km
8 km

Calculate

a the total distance of the race [3]

b the bearing of *B* from *A*. [3]

5 Calculate the area of this parallelogram. [3]

130°
10 cm
8 cm

6 A cylindrical tin can has a volume of 420 cm^3. Its height is 11 cm.
Calculate the surface area of the tin can. [3]

7 *ABCD* is a parallelogram with $\overrightarrow{AB} = \mathbf{a}$ and $\overrightarrow{AD} = \mathbf{b}$.

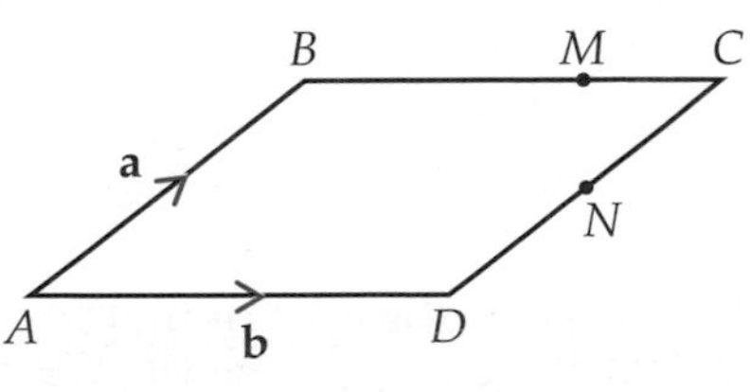

N is the midpoint of *CD* and point *M* lies on *BC* such that *BM* : *MC* = 2 : 1.
Write these vectors in terms of **a** and **b**, simplifying your answers.

a $\overrightarrow{AC}$ **b** $\overrightarrow{CB}$ **c** $\overrightarrow{DN}$ [1 mark each]

d $\overrightarrow{AN}$ **e** $\overrightarrow{AM}$ **f** $\overrightarrow{NM}$ [1 mark each]

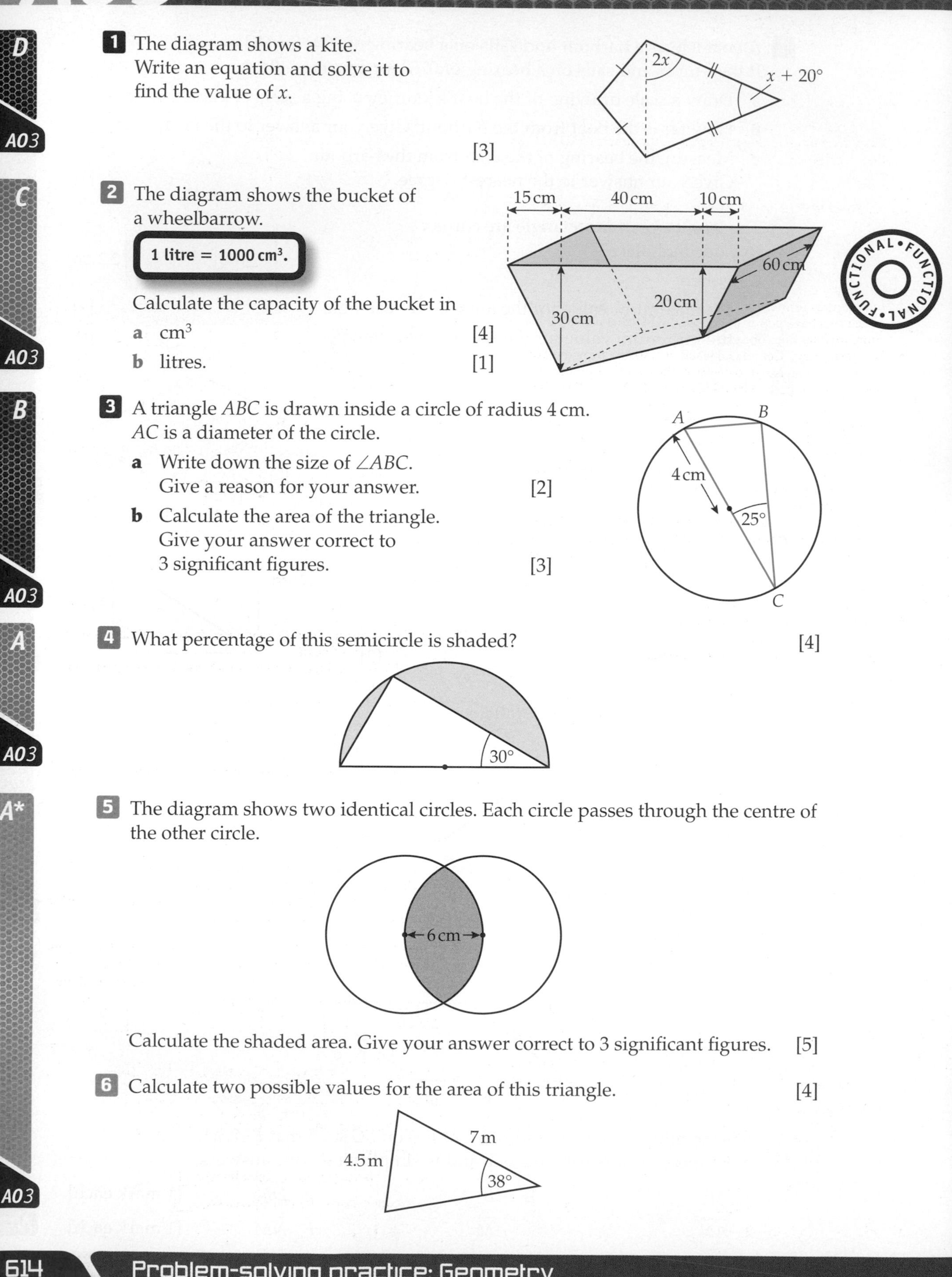

1 The diagram shows a kite.
Write an equation and solve it to find the value of x. [3]

2 The diagram shows the bucket of a wheelbarrow.

1 litre = 1000 cm³.

Calculate the capacity of the bucket in

a cm^3 [4]

b litres. [1]

3 A triangle ABC is drawn inside a circle of radius 4 cm.
AC is a diameter of the circle.

a Write down the size of $\angle ABC$.
Give a reason for your answer. [2]

b Calculate the area of the triangle.
Give your answer correct to 3 significant figures. [3]

4 What percentage of this semicircle is shaded? [4]

5 The diagram shows two identical circles. Each circle passes through the centre of the other circle.

Calculate the shaded area. Give your answer correct to 3 significant figures. [5]

6 Calculate two possible values for the area of this triangle. [4]

Answers

Chapter 1

Exercise 1A

1 a 'Boys in Year 10 can run faster than girls in Year 10.'
b 'People do more of their shopping at the supermarket than at their local shop.'

Exercise 1B

1 a e.g. Need to define what age 'young' and 'old' people are.
b e.g. 'Girls' and 'boys' are very general terms, better to make a more specific statement.

Exercise 1C

1 a Secondary data. Look at official government statistics or 2008 sales figures from car makers/distributors. Need to identify which cars have petrol engines and which have diesel and compare total sales of each.
b Secondary data. Get official weather statistics from the MET office or similar. Need to compare the average number of hours of sunshine received in June for Tenby and Southend. This should be done over a sample of several years.
c Primary data. Conduct a survey of people in your street. Ask people whether they prefer Chinese or Indian takeaways and add up the totals.
d Primary data. Carry out a survey of Year 11 students. Ask them where they'd prefer to go on an end-of-term trip and add up the totals.
e Secondary data. Ask the local cinema for attendance figures over the last 12 months. You could plot a graph of the total figures for each month to see whether attendance is falling.

Exercise 1D

1 a Quantitative, discrete, e.g. 60
b Quantitative, continuous, e.g. 680 g
c Qualitative, e.g. Kellogg's
d Quantitative, continuous, e.g. 19.89 sec

Exercise 1E

1 a

Marks	Tally	Frequency
1–5	II	2
6–10	~~IIII~~ IIII	9
11–15	~~IIII~~ ~~IIII~~ I	11
16–20	~~IIII~~ I	6
	Total	28

b 31 c 11–15 d 50–100%

2 e.g.

Pocket money received	Tally	Frequency
£5.50–£8.74	~~IIII~~ I	6
£8.75–£11.99	~~IIII~~ III	8
£12.00–15.24	~~IIII~~ IIII	9
£15.25–18.49	IIII	4
	Total	27

3 a 80 b 20 c 45 d $28 \leqslant h < 34$

4 a

Weight, w kg	Tally	Frequency
$45 \leqslant w < 55$	~~IIII~~	5
$55 \leqslant w < 65$	~~IIII~~ III	8
$65 \leqslant w < 75$	~~IIII~~ II	7
$75 \leqslant w < 85$	~~IIII~~ I	6
$85 \leqslant w < 95$	IIII	4
	Total	30

b 20 c $55 \leqslant w < 65$

5 a

Time, t min	Tally	Frequency
$30 \leqslant t < 35$	II	2
$35 \leqslant t < 40$	~~IIII~~	5
$40 \leqslant t < 45$	~~IIII~~ III	8
$45 \leqslant t < 50$	~~IIII~~ I	6
$50 \leqslant t < 55$	III	3
$55 \leqslant t < 60$	IIII	4
	Total	28

b $40 \leqslant t < 45$ c 13
d 28 e 25%

Exercise 1F

1 a

	Won	Drawn	Lost	Total
Home games	9	4	1	14
Away games	6	5	3	14
Total	15	9	4	28

b 14 c 4 d 54

2 a 3 b 17
c 22 d 10%

3 a, b

	Car	Bus	Cycle	Walk	Total
Men	22	1	6	5	34
Women	10	3	1	2	16
Total	32	4	7	7	50

c 7 d 14%

4 a 16 b 21
c 73 d 88

Exercise 1G

1 a i Some people do not like to provide personal details such as their date of birth.
ii This is a leading question, which encourages a particular answer.
iii The responses available overlap.
b i How old are you? Under 20 years ☐, 20–39 ☐, 40–59 ☐, 60+
ii Do you think it takes too long to get an appointment to see the doctor?
iii How many times did you visit the doctor last year? 0 ☐, 1–2 ☐, 3–4 ☐, 5–6 ☐, 7+

2 a It is a leading question. The response categories are poorly chosen, e.g. no option for 'Don't agree'.
b It is not a leading question. All possible responses are catered for with no overlap between categories.

3 E.g. For each day please state whether you'd be able to car-share to and from work, either as the driver or a passenger.

	Able to car-share as: driver	passenger	Unable to car-share
Monday	☐	☐	☐
Tuesday… etc			

4 a 0–1 day ☐, 2–3 days ☐, 4–5 days ☐, 6–7 days
b What types of physical activity do you take part in?
Walking/running ☐, Cycling ☐, Gym ☐,
Swimming ☐,
Team sports (Football, Netball etc) ☐,
Racquet sports (Tennis, Squash etc) ☐,
Other ☐

5 a, b E.g.

Distance to school, d km	Year 7	Year 8	Year 9	Year 10	Year 11
$0 \leq d < 4$	1	2	1		
$4 \leq d < 8$	2	3	1	2	
$8 \leq d < 12$	1	1	1	2	
$12 \leq d < 16$	1	1			
$16 \leq d$	1				

Exercise 1H

1 A significant proportion of people will be at work between 9 am and 5 pm and so their views will not be represented. People who work unsociable hours or from home and people who are pensioners or unemployed may be over-represented.
2 People using the existing car parks are more likely to have similar opinions since they drive into town. People who use other forms of transport or live close to the town centre should also be asked.
3 No. Most of the people surveyed are likely to drive and use that particular shopping centre.
4 People who attend a sports centre are likely to do relatively more exercise on average.
5 No. People waiting at a bus station are likely to travel to work by bus.
6 Not necessarily. The survey will fail to capture people who don't go out in town during the evening and who may spend less on entertainment (or more if they spend a lot on home entertainment).

Exercise 1I

1 Under 18 = 10, 18–30 = 14, 31–50 = 18, 51–65 = 11, over 65 = 7
2 Biology = 68, Chemistry = 40, Geology = 27, Physics = 48
3 Year 7 = 11, Year 8 = 13, Year 9 = 11, Year 10 = 14, Year 11 = 13
4 Catering = 9, Engineering = 5, Business = 11, Sport = 10, Humanities = 7, Science = 8
5 a

Year	Boys	Girls
7	12	11
8	11	10
9	7	9
10	9	10
11	10	11

b E.g. Find a list of students, assign them numbers and use a random number table to choose a sample.

Review exercise

1 'More people go on holiday to France than to Spain.'
2 Secondary data. Ask the publishers for readership figures and compare them.
3 a Qualitative, Bolivia
b Qualitative, red
c Quantitative, continuous, 9.86 sec
d Quantitative, discrete, 130 000
e Quantitative, continuous, 7 mm
4 a

Shoe size	Tally	Frequency
3–5	~~||||~~ ~~||||~~ ||	12
6–8	~~||||~~ ~~||||~~ ~~||||~~ ~~||||~~ ||	22
9–11	~~||||~~ ~~||||~~ ~~||||~~ ~~||||~~	20
12–14	~~||||~~ |	6
	Total	60

b 6–8
c By looking at the raw data.

5 a

Height of plants, h cm	Tally	Frequency
$10 \leq h < 15$	~~||||~~ ||	7
$15 \leq h < 20$	~~||||~~ ||	7
$20 \leq h < 25$	~~||||~~ ||	7
$25 \leq h < 30$	||||	4
$30 \leq h < 35$	~~||||~~ |||	8
$35 \leq h < 40$	|||	3
	Total	36

b $30 \leq h < 35$

6 a

	Caravan	B&B	Apartment	Hotel	Total
June	3	4	5	14	26
July	8	15	12	2	37
August	11	10	15	19	55
September	6	7	13	13	39
Total	28	36	45	48	157

b 36 c 19 d 13 e 157

7 How many hours (h) do you usually spend doing homework at the weekend?
$0 \leq h < 1$ ☐, $1 \leq h < 2$ ☐, $2 \leq h < 3$ ☐, $3 \leq h < 4$ ☐, $4 \leq h$
8 She only asks girls and only those in Year 10.
9 Under 16 = 5, 16–25 = 7, 26–40 = 12, 41–60 = 15, Over 60 = 11

Chapter 2

Exercise 2A

1 18
2 a £139 b 13
3 63
4 4

Exercise 2B

1 $\frac{3}{7}$
2 $\frac{16}{30} = \frac{8}{15}$
3 Donna is incorrect, $\frac{5}{5+9} = \frac{5}{14}$
4 a $\frac{2}{7}$ b $\frac{1}{24}$ c $\frac{6}{8} = \frac{3}{4}$
5 a $\frac{20}{100} = \frac{1}{5}$ b $\frac{5}{60} = \frac{1}{12}$ c $\frac{5}{14}$
d $\frac{30}{100} = \frac{3}{10}$ e $\frac{750}{1000} = \frac{3}{4}$ f $\frac{12}{1000} = \frac{3}{250}$
6 a $\frac{175}{500} = \frac{7}{20}$ b $\frac{150}{500} = \frac{3}{10}$ c $\frac{175}{500} = \frac{7}{20}$

Exercise 2C

1 a $9\frac{17}{36}$ b $2\frac{7}{8}$ c $18\frac{1}{20}$
2 a 57 600 m b 57.6 km c 36 miles
3 $2\frac{3}{16}$
4 $\frac{5}{14}$
5 27 minutes 30 seconds
6 None of them, they all give an answer of 21.
7 $1\frac{3}{5}$ m
8 a $119\frac{7}{22}$ kg
b Answer is almost 120, which is the number of litres. The conversions are all approximations. If they had been exact the answer would have been 120 kg, showing that 1 litre of water weighs 1 kg.
9 £60
10 £10
11 £17.50
12 £60

Exercise 2D

1 a 10% b 6.25% c $33\frac{1}{3}$% d $16\frac{2}{3}$% e 60%
f 12.5% g 25% h 14.5% i 27.5%
2 62.5%
3 13.3%

4 a 27.0% b 52.4%
5 19.6%
6 16.7%
7 66%
8 35%
9 29.3 %
10 8.1%

Exercise 2E

1 a 21 b 62.72 c 2822
d 326.4 e £145.13 f 815.24 ml
2 £436.80
3 329
4 £4816.50
5 £922.35
6 Jane
7 a 80.64 kg b 82.25 kg
8 128 960 or 129 000

Exercise 2F

1 DVD: a £22.75, b £152.75 phone: a £7.00, b £47
2 £148.05
3 £3800 (excluding VAT) = £4465
4 £21.62
5 £44.47

Exercise 2G

1

Year	2007	2008	2009
Price	90.3p	95p	104.5p

2 a Down b Gone down by 26%
3 60
4 72p
5 £81.35
6 Peter isn't correct. The price of bananas has dropped by $\frac{1}{4}$ so they are $\frac{3}{4}$ of the price they were in 1990.
7 a €1.35 to £1 b 88.1
8 583 333

Review exercise

1 No, she hasn't used the same units. $\frac{2}{500} = \frac{1}{250}$
2 59%
3 92
4 £2.63
5 £30.10
6 20%
7 66.7%
8 £243.75
9 TVs R US, by £16.60.
10 16%
11 17.8%

Chapter 3

Exercise 3A

1

```
0 | 7 8 9 9
1 | 2 2 3 3 3 5 5 6 7 8 9 9
2 | 0 0 1 1
```

Key 1|6 represents 16°C

2 a 18 CDs
b

```
2 | 7
3 | 6 6 9
4 | 5 6 7 7 9
5 | 1 1 3 6 8 8 9
6 | 0 2
```

Key 4|5 means 45 minutes

3 a

```
2 | 8 8 9 9
3 | 1 2 3 6 7 9 9
4 | 0 1 3 6 7
```

Key 3|6 means 3.6 seconds

b 9 girls

4 a

```
2 | 61 93
3 | 62 75 81
4 | 63 70 86 91
5 | 23 25 27
```

Key 2|93 means £2.93

b 2 people

Exercise 3B

1 a

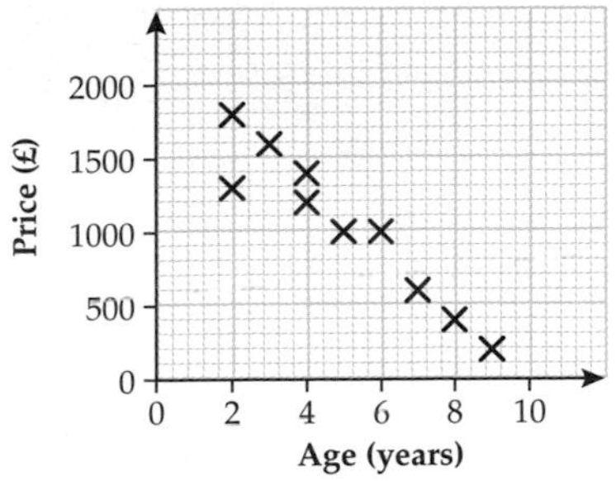

Price (£)
2000
1500
1000
500
0
0 2 4 6 8 10
Age (years)

b As the age of the motorbikes increases their price decreases.

2 a

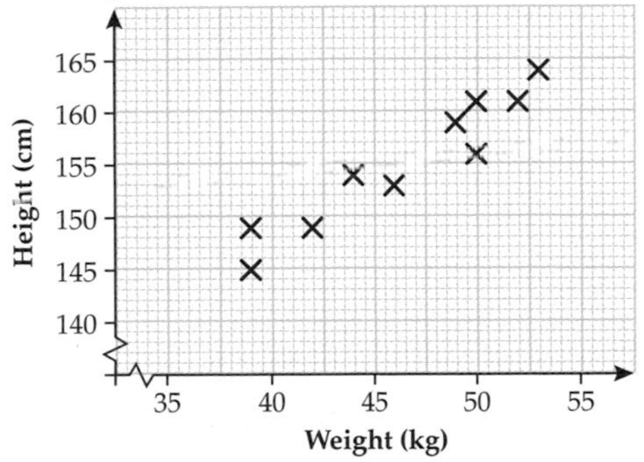

Height (cm)
165
160
155
150
145
140
35 40 45 50 55
Weight (kg)

b The heavier the child, the taller they are.

3 a Yes, in general the higher the percentage attendance at maths lessons, the higher the maths test result.
b i Person A attended over 90% of the lessons but did not do so well, possibly because of exam nerves or just not very good at maths.
ii Person B did not attend the maths lessons very often but scored well, possibly because they are able in mathematics.

4

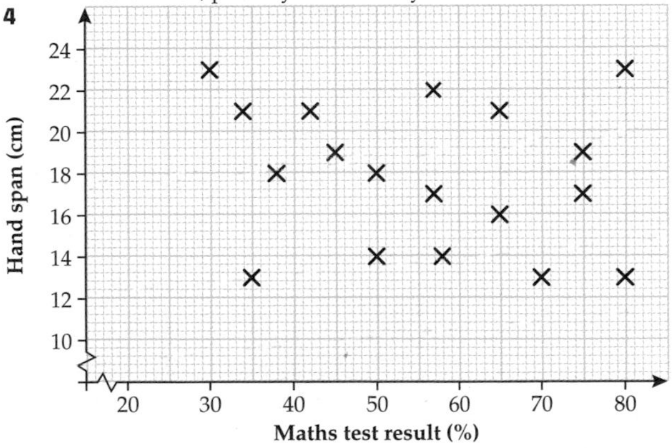

Hand span (cm)
24
22
20
18
16
14
12
10
20 30 40 50 60 70 80
Maths test result (%)

Kushal's hypothesis is incorrect. There is no connection between the hand span and maths results of the students.

Exercise 3C

1 a Positive b Arm span
2

	Positive Correlation	Negative Correlation	No Correlation
Heights of men and their IQ			✓
The hat sizes of children and their height	✓		
The outside temperature and the number of cups of tea sold in a café		✓	
A car's fuel consumption and its speed	✓		

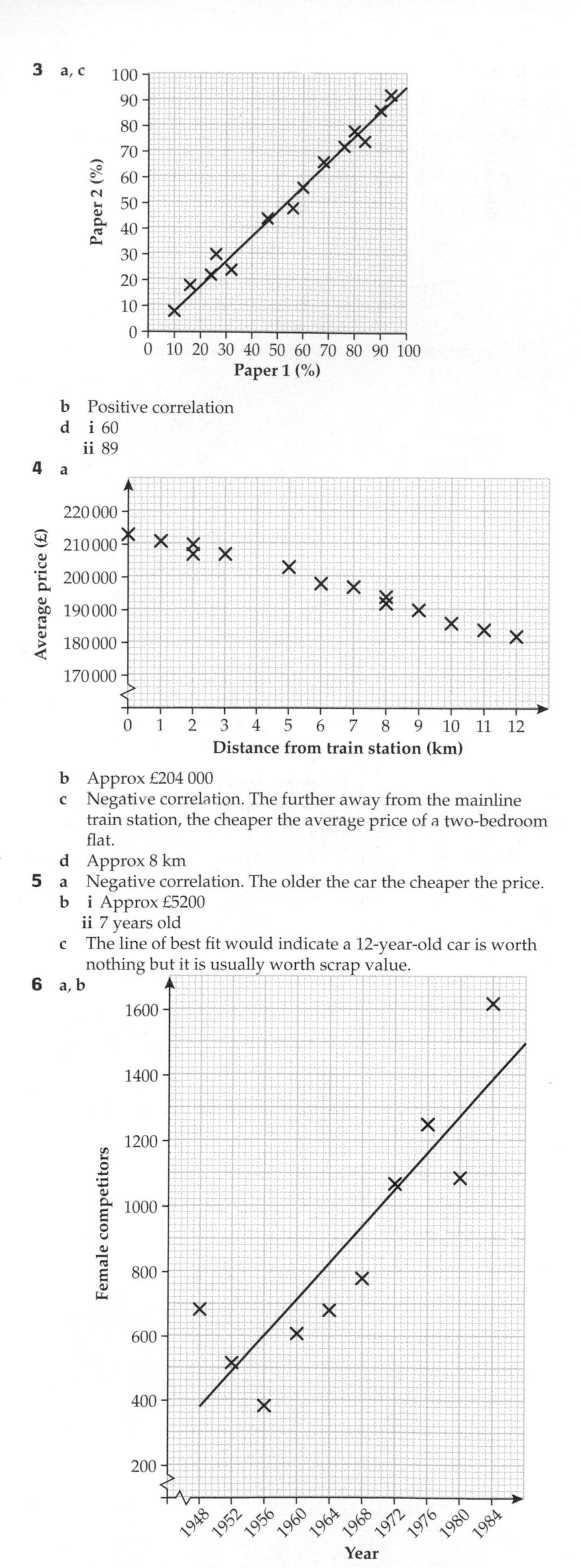

3 a, c

b Positive correlation

d i 60

ii 89

4 a

b Approx £204 000

c Negative correlation. The further away from the mainline train station, the cheaper the average price of a two-bedroom flat.

d Approx 8 km

5 a Negative correlation. The older the car the cheaper the price.

b i Approx £5200

ii 7 years old

c The line of best fit would indicate a 12-year-old car is worth nothing but it is usually worth scrap value.

6 a, b

c Positive although weak positive correlation

Exercise 3D

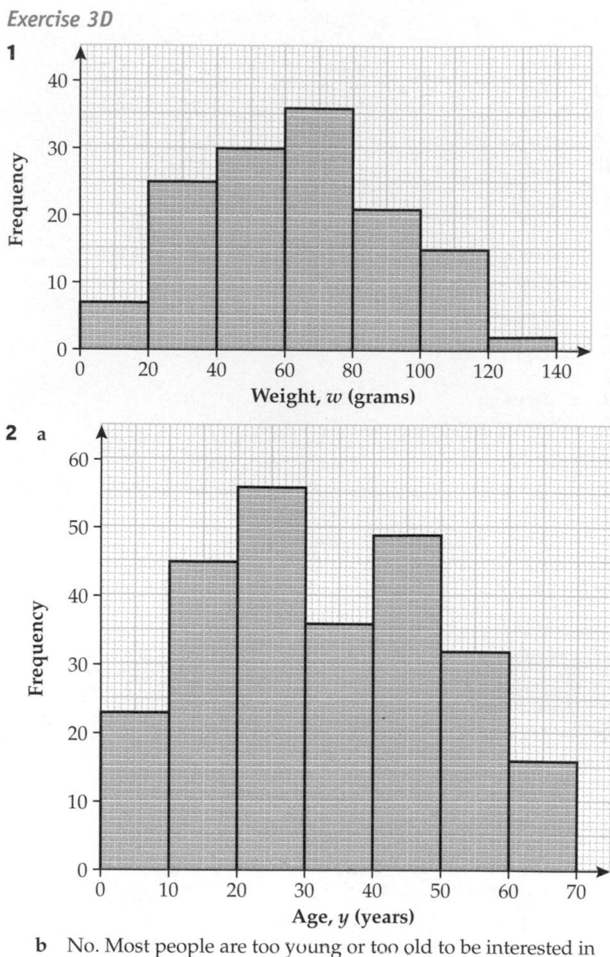

1

2 a

b No. Most people are too young or too old to be interested in nightclub tickets.

3 a

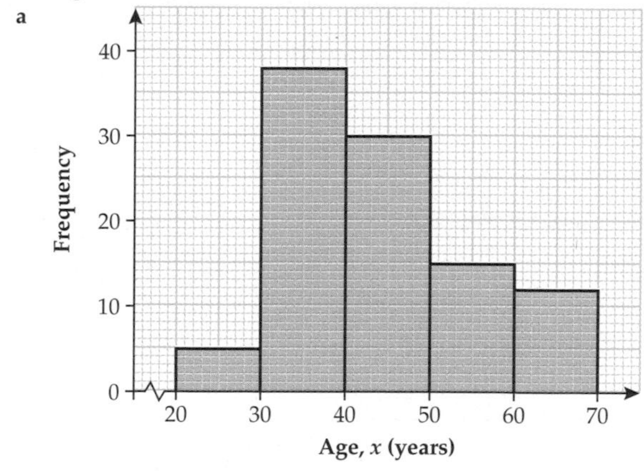

b A large proportion of the magazine readers are 30 years or over. An article on *High School Musical* would not be appropriate as it is predominantly aimed at a younger age audience.

Exercise 3E

1 a

Height, h cm	Number of seedlings	Mid-point
$5 \leqslant h < 10$	6	7.5
$10 \leqslant h < 15$	10	12.5
$15 \leqslant h < 20$	12	17.5
$20 \leqslant h < 25$	9	22.5
$25 \leqslant h < 30$	3	27.5

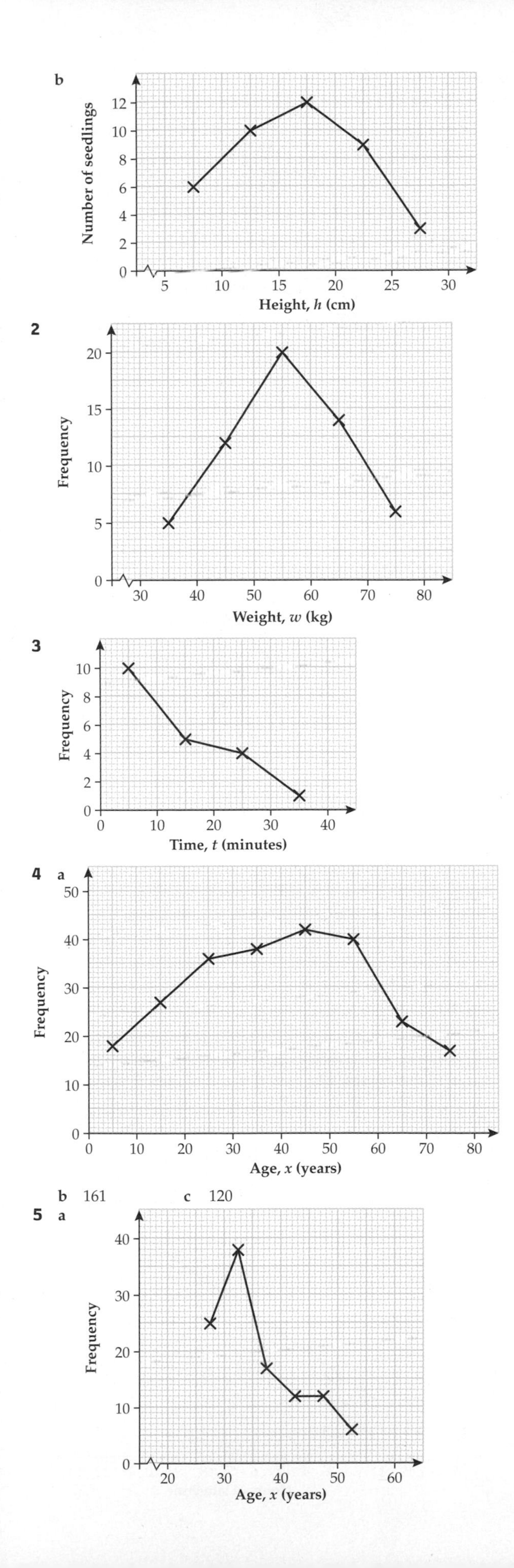

b 47

c 50%

6 The girls are generally taller than the boys.
There are more tall girls than tall boys since there are 4 girls in the 160–165 cm class interval compared to 1 boy.
There are more short boys than short girls since there are 3 boys and 1 girl in the 135–140 cm class interval.

Exercise 3F

1 a

Age, n years	Frequency	Class width	Frequency density
$0 \leqslant n < 20$	10	20	$10 \div 20 = 0.5$
$20 \leqslant n < 25$	25	5	$25 \div 5 = 5$
$25 \leqslant n < 30$	15	5	$15 \div 5 = 3$
$30 \leqslant n < 40$	29	10	$29 \div 10 = 2.9$
$40 \leqslant n < 60$	48	20	$48 \div 20 = 2.4$
$60 \leqslant n < 70$	12	10	$12 \div 10 = 1.2$
$70 \leqslant n < 100$	3	30	$3 \div 30 = 0.1$

b

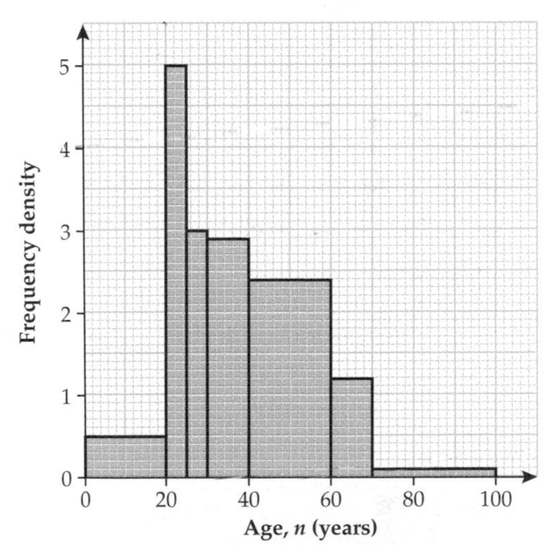

2 a

Time, t (minutes)	Frequency	Class width	Frequency density
$0 \leqslant t < 15$	9	15	0.6
$15 \leqslant t < 25$	21	10	2.1
$25 \leqslant t < 30$	25	5	5
$30 \leqslant t < 50$	20	20	1

b

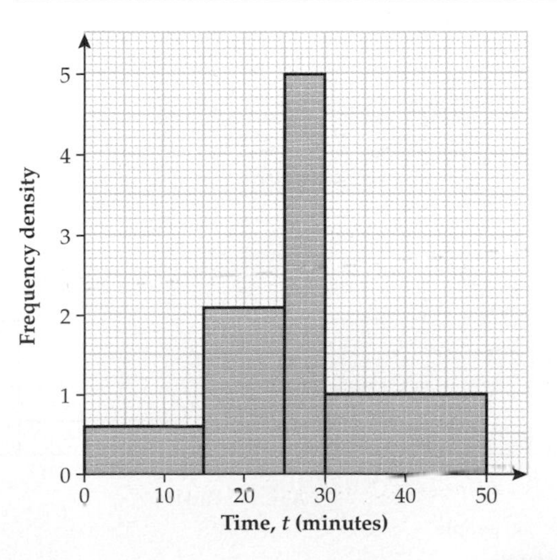

3

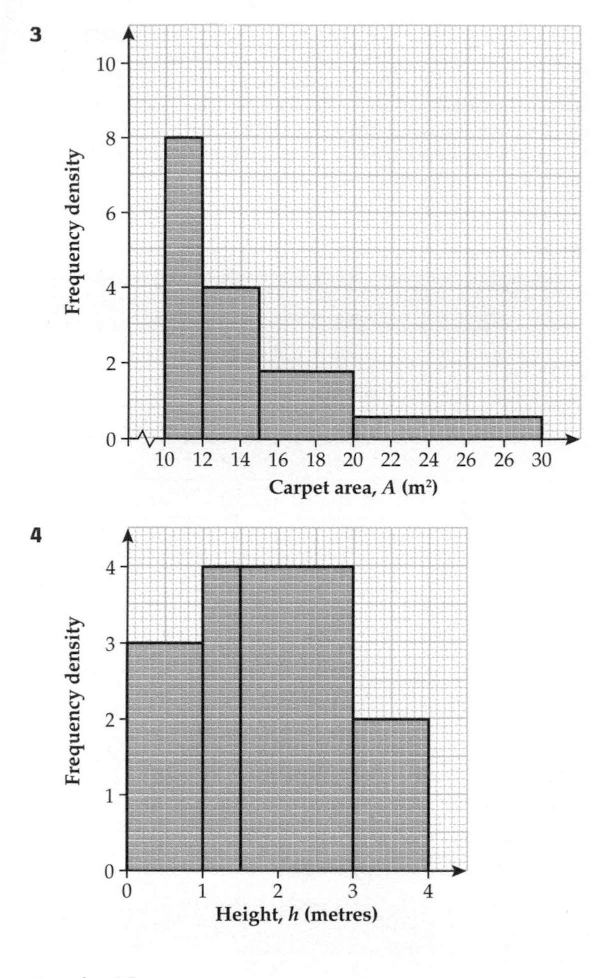

4

Exercise 3G

1 a 94 batteries
b 28 batteries

2 a 116 swimmers
b 65 swimmers

3 a There are 280 people who went to the surgery. 140 people waited less than 20 minutes which is half of the people, so William is correct.
b 26 people

4 a

Age, x (years)	Frequency
$0 \leqslant x < 10$	50
$10 \leqslant x < 15$	30
$15 \leqslant x < 30$	60
$30 \leqslant x < 50$	60
$50 \leqslant x < 75$	25
$75 \leqslant x < 80$	20

b

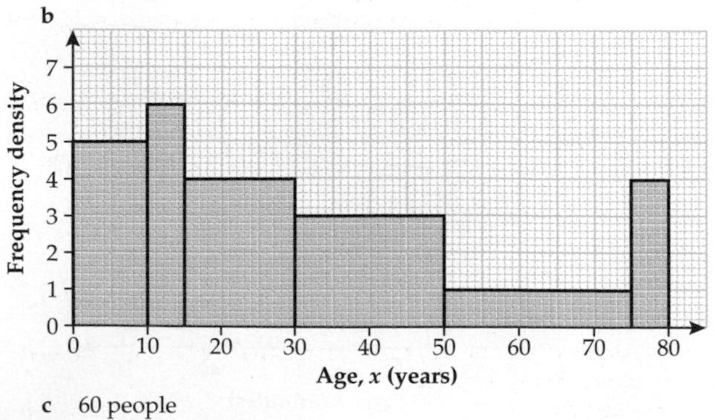

c 60 people

5 a

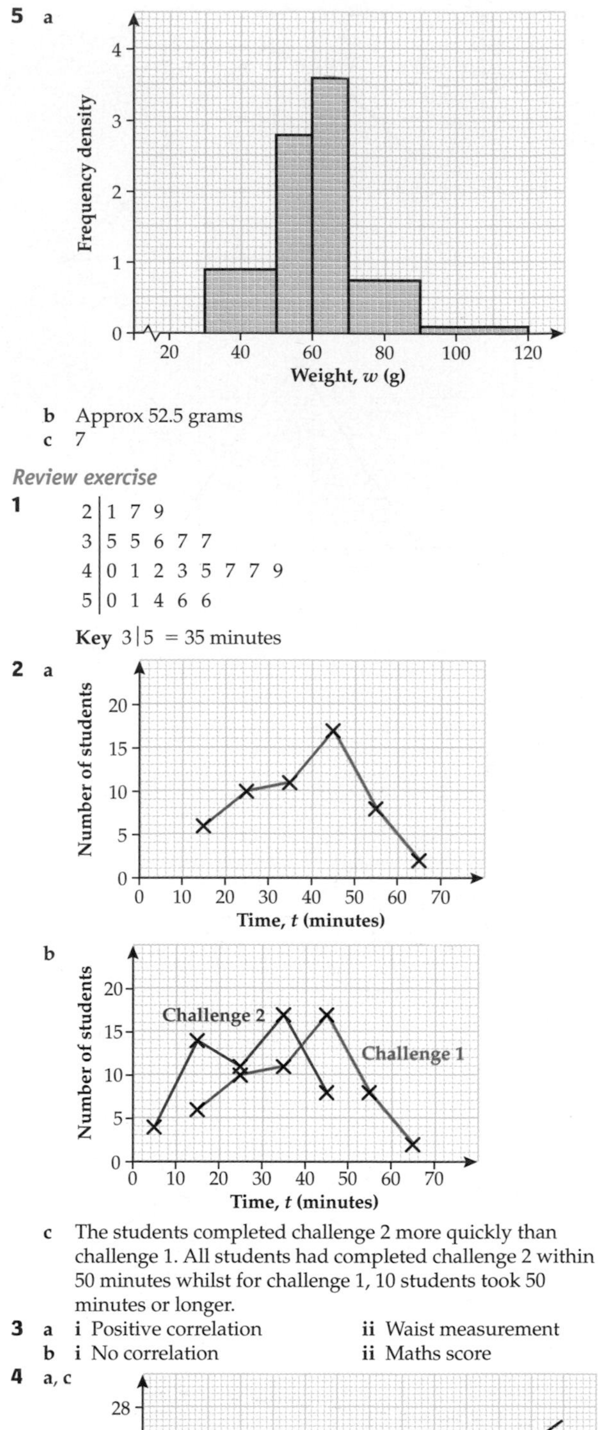

b Approx 52.5 grams
c 7

Review exercise

1

2	1 7 9
3	5 5 6 7 7
4	0 1 2 3 5 7 7 9
5	0 1 4 6 6

Key 3 | 5 = 35 minutes

2 a

b

c The students completed challenge 2 more quickly than challenge 1. All students had completed challenge 2 within 50 minutes whilst for challenge 1, 10 students took 50 minutes or longer.

3 a i Positive correlation   ii Waist measurement
b i No correlation   ii Maths score

4 a, c

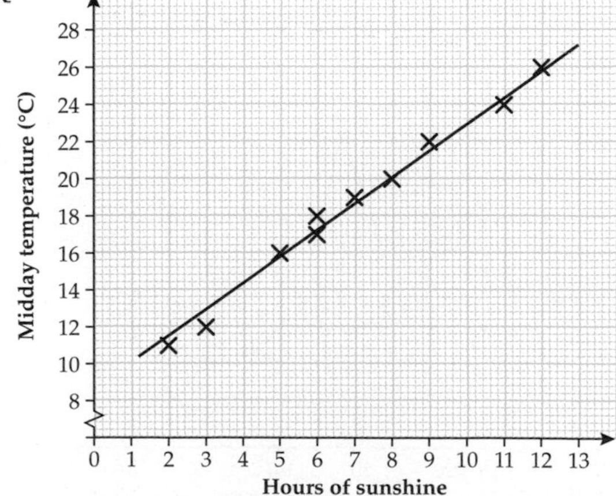

b Strong positive correlation
d 8.6 hours
e The value is outside the range of the given data.

5 a

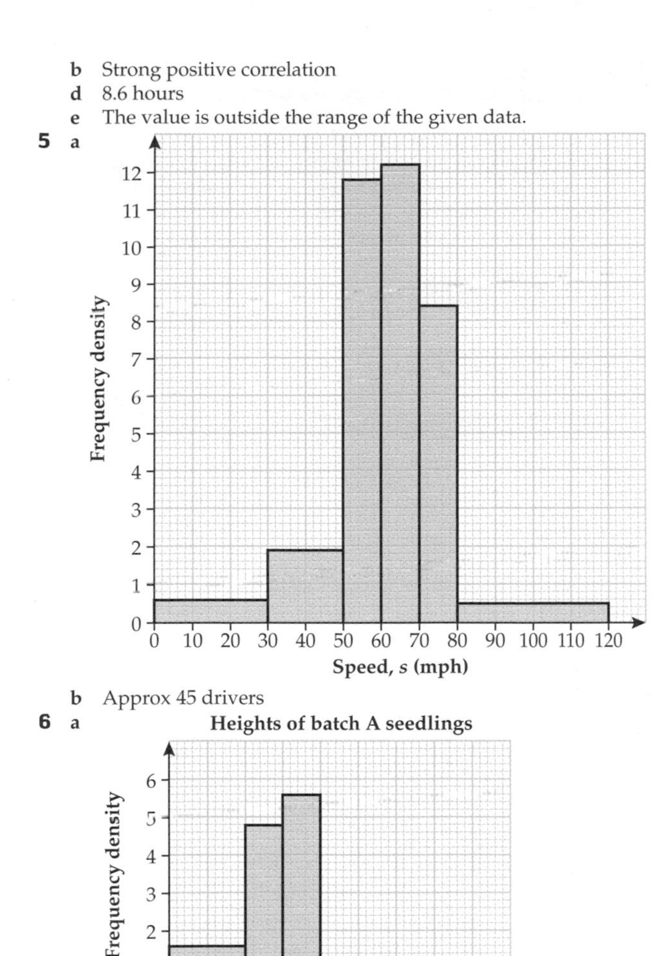

b Approx 45 drivers

6 a Heights of batch A seedlings

Frequency density
Height, h (cm)

b 30 seedlings **c** 19 seedlings
d Batch B. There are more taller plants.

Chapter 4

Exercise 4A

1 a Range = 3, Mode = 2
b 2
c

Number of people in queue	Frequency	Number of people × frequency
0	4	$0 \times 4 = 0$
1	6	$1 \times 6 = 6$
2	13	$2 \times 13 = 26$
3	2	$3 \times 2 = 6$
4	0	$4 \times 0 = 0$
Total	25	38

Mean = 1.52

2 a 2
b 2.05

3 a 4 **b** 2 **c** 2 **d** 2.52
e i Unchanged **ii** Unchanged
iii Increased (to 2.5) **iv** Increased (to 2.58)

4 a No. The two largest values are unknown.
b E.g. 33 people, median is 17th value, which is 2 hours.

5 a 7 **b** 7 **c** 7.1 **d** 7 **e** 7.1

6 a 116 **b** 214 **c** 7 **d** 2.5 **e** Year 8

Exercise 4B

1 a $55 \leqslant w < 60$ **b** 20 g **c** 29
d $55 \leqslant w < 60$

2 a 20–29 **b** 49 **c** 20–29

3 a $200 \leqslant t < 300$ **b** 400 ms **c** $200 \leqslant t < 300$

4 a 3000–3999 **b** 4999 **c** 3000–3999
d i Unchanged **ii** New estimated range = 5999
iii Unchanged
e Yes. You don't know the exact data values so you can't calculate the mean exactly.

Exercise 4C

1 a

Time taken, t (min)	Frequency	Mid-point	Mid-point × frequency
$0 \leqslant t < 5$	3	2.5	$2.5 \times 3 = 7.5$
$5 \leqslant t < 10$	15	7.5	$7.5 \times 15 = 112.5$
$10 \leqslant t < 15$	8	12.5	$12.5 \times 8 = 100$
$15 \leqslant t < 20$	2	17.5	$17.5 \times 2 = 35$
$20 \leqslant t < 25$	5	22.5	$22.5 \times 5 = 112.5$
Total	33		367.5

b 11.1 minutes

2 a

Number of log-ins	Frequency	Mid-point	Mid-point × frequency
0–4	22	2	44
5–9	31	7	217
10–14	17	12	204
15–19	20	17	340
20–24	6	22	132
Total	96		937

Mean = 9.8 log-ins
b Because you do not know the exact data values.

3 a No. He does not know the exact data values.
b

Distance thrown, d (m)	Girls' frequency	Mid-point	Mid-point × frequency (girls)
$0 \leqslant d < 8$	2	4	8
$8 \leqslant d < 16$	12	12	144
$16 \leqslant d < 24$	5	20	100
$24 \leqslant d < 32$	3	28	84
$32 \leqslant d < 40$	7	36	252
$40 \leqslant d < 48$	0	44	0
Total	29		588

Estimate for girls' mean distance = 20.3 m
c Estimate for whole club mean distance = 19.7 m
d No. The estimate for the girls' mean is greater than the estimated mean for the whole club.

Exercise 4D

1 a Mean = 48.7, Median = 29, Mode = 15
b Mean
c E.g. No. Eight of the values are lower than the mean.
d E.g. The median. The average person owns about 30 CDs (to one significant figure).

2 a £20 100 (three significant figures)
b E.g. No. Six out of eight employees earn more than the mean.
c It will reduce.
d £22 700
e £21 000. E.g. The arrival of the new trainees has reduced the average salary by £1700.

3 a Mode. This is the only average you can use for qualitative data.
b Median. The middle value will represent an average employee.
c Mean. This will take into account all the data values.
d Mode. This tells you which value is most likely to occur.

4 a A represents the median and B represents the mean. The large number of very high incomes will cause the mean to be larger than the median.
b Median

Review exercise

1 a Range = 4, Mode = 1 b 2 c 1.9
2 a 50 b $30 \leqslant t < 40$
c $20 \leqslant t < 30$ d 28.6 seconds
3 a $5 \leqslant t < 10$ b £6.50
c E.g. Because the data is presented in a grouped frequency table and so you do not know the exact data values.
4 a 2.76 b 3
c Mode. Because it tells you which value is most likely to occur.
5 a Yes. 319 g/km is much higher than the other values.
b Median. This tells you information about an average person or item.
c Mean. This takes into account all values.
6 Mode. This tells you which value is most likely to occur.
7 a Mean
b Because it is not possible for a family to contain exactly 2.4 children.
8 a Yes, depending on the average chosen.
b Mean: if there is one very low value then more than half of the data values could be above the mean. Mode: it is possible for the mode to be the highest or the lowest value.
9 A False. You don't know how many people had size 8 shoes or how many people were sampled.
B True. The median is between two values (8.5 and 9) so there are an even number of values, and exactly half the values are below the median.
C True. The mode is the most common value.
D False. There were an even number of students in the class because the median is between two values.

Chapter 5

Exercise 5A

1 a $\frac{10}{15}$ b $\frac{9}{15}$ c $\frac{8}{15}$ d $\frac{7}{15}$ e $\frac{4}{15}$
2 a 0.55 b 0.2
3 0.3
4 a $\frac{1}{3}$ b 15 red, 9 yellow, 12 blue
5 12

Exercise 5B

1 10
2 a 260 b 130 c 40 d 10
3 a 2 b £40 c £20
4 25
5 a 5 b £50 c £80
d No, as he is paying out more than he is winning.
6 a 15 b 150 c 225 d 0
e 135 f 75
7 £100
8 a yellow 0.4, pink 0.1 b 120
9 500
10 £787.95

Exercise 5C

1 a Relative frequency (blue): 0.75, 0.58, 0.56, 0.62, 0.682, 0.64, 0.631, 0.664, 0.658
Relative frequency (red): 0.25, 0.42, 0.44, 0.38, 0.318, 0.36, 0.369, 0.336, 0.342
b i Approx 0.66 ii Approx 0.34
c Approx 33 blue and 17 red
d Dual line graph showing data from part **a**
2 a $\frac{47}{200} = 0.235$ b 30 550
3 a

Relative frequency	0.05	0.22	0.14	0.16	0.16	0.168

b $0.1\dot{6}$
c No, after 500 throws the relative frequency is very close to the theoretical probability.
d 200
4 a

Score on the dice	1	2	3	4	5	6
Relative frequency	0.175	0.11	0.125	0.135	0.255	0.2

b $0.1\dot{6}$
c No, the dice is biased as the relative frequency of rolling a 5 is much higher than the theoretical probability and the relative frequency of rolling a 2 is much lower.
5 Zoe. She has spun the spinner 200 times, whereas George has only spun the spinner 40 times, so her results are more reliable. Zoe's relative frequencies are all close to the theoretical probabilities of 0.25, so the spinner seems to be fair.

George's results

Relative frequency	0.075	0.35	0.25	0.325

Zoe's results

Relative frequency	0.225	0.265	0.24	0.27

6 a

Relative frequency	0.35	0.15	0.15	0.15	0.2

b 1 as the relative frequency is higher than the other scores.
c Spin the spinner more times.
7 a 24, 80, 0.3
27, 90, 0.3
29, 100, 0.29
b Yes, as the theoretical probability of rolling a 6 is 0.17.

Exercise 5D

1 $\frac{1}{4}$
2 $\frac{40}{200} = \frac{1}{5}$
3 a $\frac{1}{6}$ b $\frac{1}{16}$ c $\frac{1}{12}$
4 a 0.000 27 b 0.0009
5 a $\frac{1}{4}$ b $\frac{7}{12}$
6 a $\frac{25}{144}$ b $\frac{74}{144} = \frac{37}{72}$
7 a 0.28 b 0.18 c 0.54
d total = 1
8 a $\frac{1}{30}$ b $\frac{1}{30}$ c $\frac{4}{30} = \frac{2}{15}$
d $\frac{3}{30} = \frac{1}{10}$ e $\frac{6}{30} = \frac{1}{5}$
9 a $\frac{1}{4}$ b $\frac{9}{16}$ c $\frac{1}{169}$ d $\frac{24}{169}$

Exercise 5E

1 a

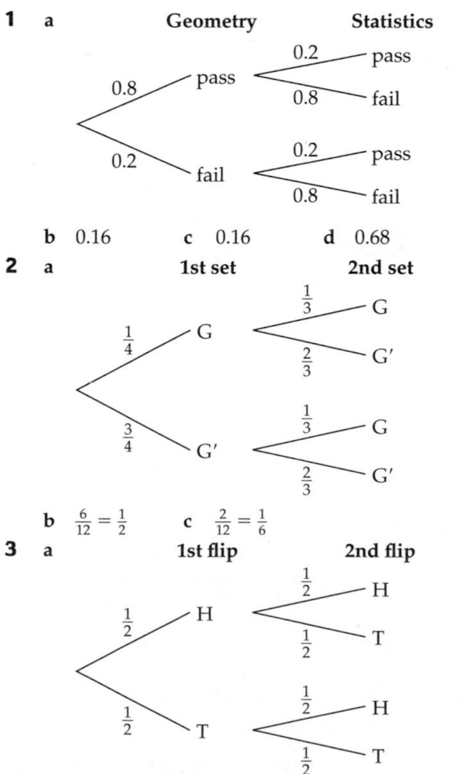

b i $\frac{1}{4}$ ii $\frac{1}{4}$ iii $\frac{1}{2}$ iv $\frac{3}{4}$

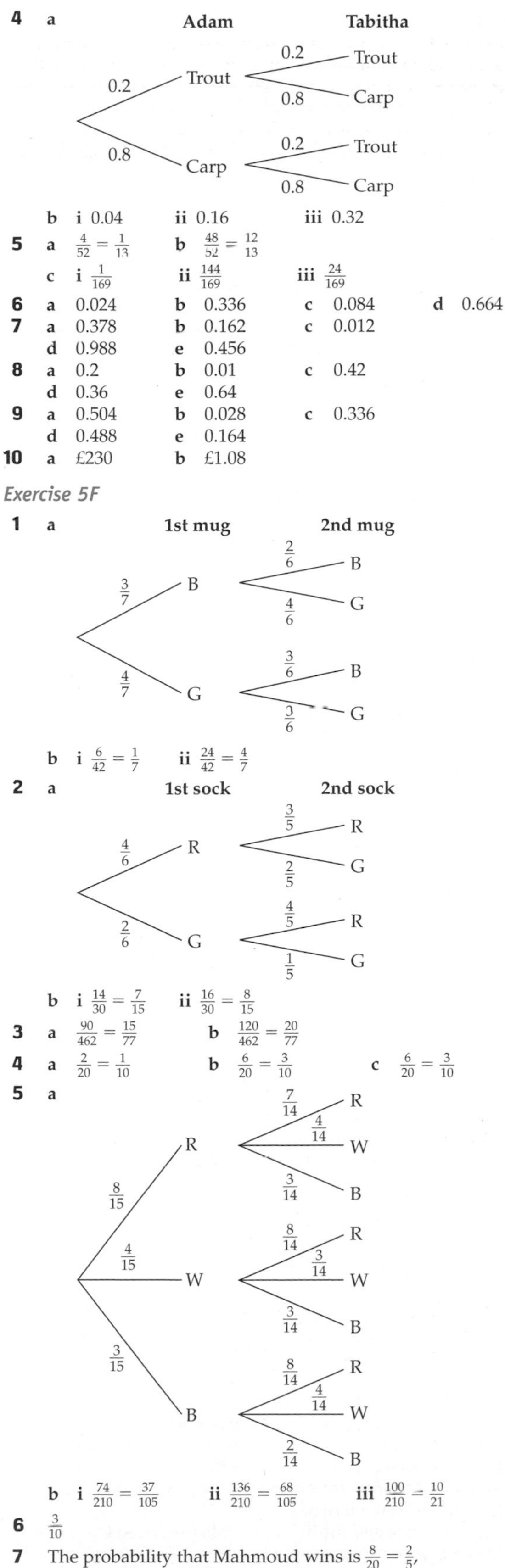

4 a

b i 0.04 ii 0.16 iii 0.32

5 a $\frac{4}{52} = \frac{1}{13}$ **b** $\frac{48}{52} = \frac{12}{13}$

c i $\frac{1}{169}$ ii $\frac{144}{169}$ iii $\frac{24}{169}$

6 a 0.024 **b** 0.336 **c** 0.084 **d** 0.664

7 a 0.378 **b** 0.162 **c** 0.012
d 0.988 **e** 0.456

8 a 0.2 **b** 0.01 **c** 0.42
d 0.36 **e** 0.64

9 a 0.504 **b** 0.028 **c** 0.336
d 0.488 **e** 0.164

10 a £230 **b** £1.08

Exercise 5F

1 a

b i $\frac{6}{42} = \frac{1}{7}$ ii $\frac{24}{42} = \frac{4}{7}$

2 a

b i $\frac{14}{30} = \frac{7}{15}$ ii $\frac{16}{30} = \frac{8}{15}$

3 a $\frac{90}{462} = \frac{15}{77}$ **b** $\frac{120}{462} = \frac{20}{77}$

4 a $\frac{2}{20} = \frac{1}{10}$ **b** $\frac{6}{20} = \frac{3}{10}$ **c** $\frac{6}{20} = \frac{3}{10}$

5 a

b i $\frac{74}{210} = \frac{37}{105}$ ii $\frac{136}{210} = \frac{68}{105}$ iii $\frac{100}{210} = \frac{10}{21}$

6 $\frac{3}{10}$

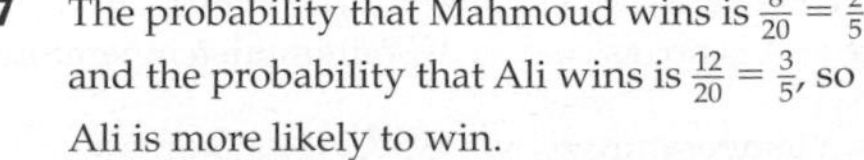

7 The probability that Mahmoud wins is $\frac{8}{20} = \frac{2}{5}$, and the probability that Ali wins is $\frac{12}{20} = \frac{3}{5}$, so Ali is more likely to win.

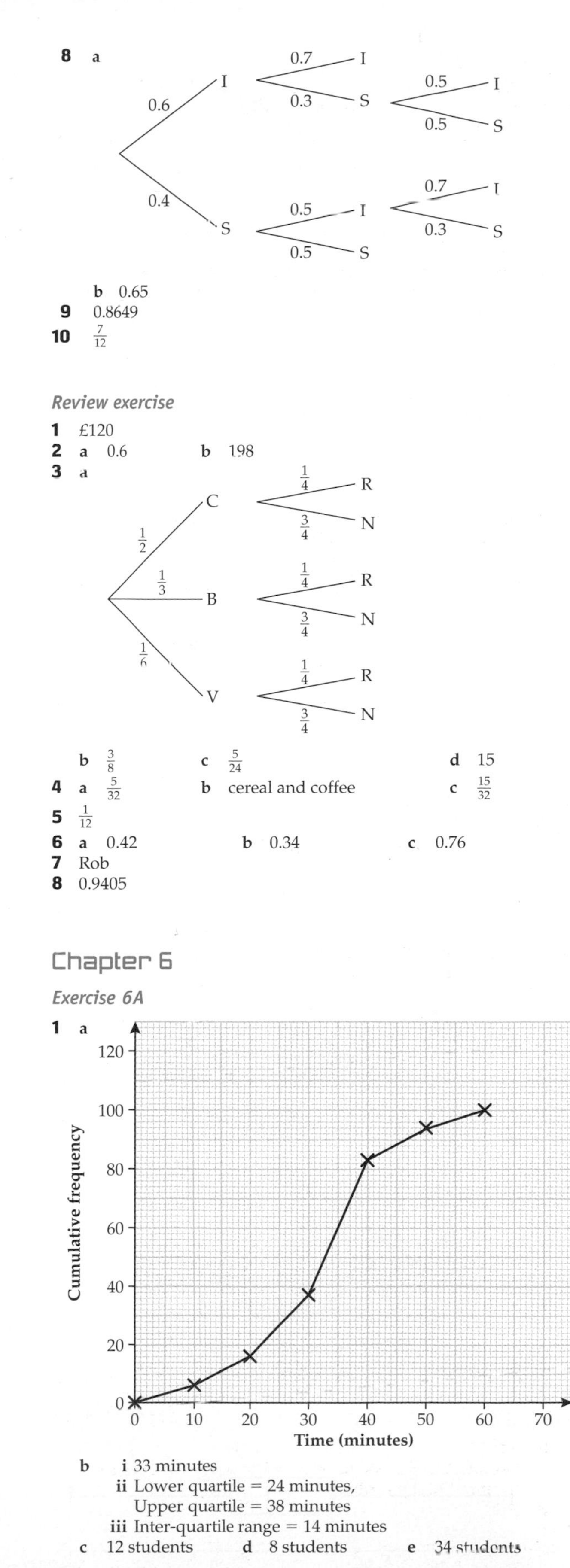

8 a

b 0.65

9 0.8649

10 $\frac{7}{12}$

Review exercise

1 £120

2 a 0.6 **b** 198

3 a

b $\frac{3}{8}$ **c** $\frac{5}{24}$ **d** 15

4 a $\frac{5}{32}$ **b** cereal and coffee **c** $\frac{15}{32}$

5 $\frac{1}{12}$

6 a 0.42 **b** 0.34 **c** 0.76

7 Rob

8 0.9405

Chapter 6

Exercise 6A

1 a

b i 33 minutes
ii Lower quartile = 24 minutes, Upper quartile = 38 minutes
iii Inter-quartile range = 14 minutes

c 12 students **d** 8 students **e** 34 students

2 a

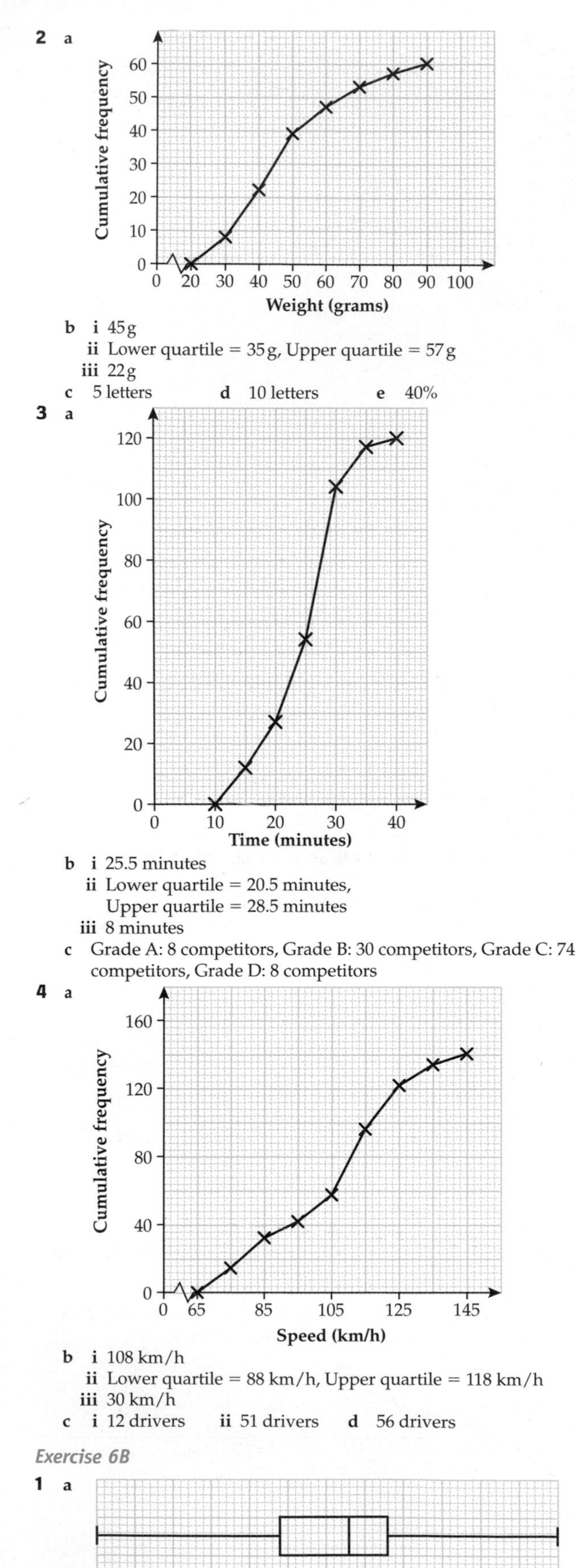

b **i** 45 g
ii Lower quartile = 35 g, Upper quartile = 57 g
iii 22 g
c 5 letters **d** 10 letters **e** 40%

3 a

b **i** 25.5 minutes
ii Lower quartile = 20.5 minutes, Upper quartile = 28.5 minutes
iii 8 minutes
c Grade A: 8 competitors, Grade B: 30 competitors, Grade C: 74 competitors, Grade D: 8 competitors

4 a

b **i** 108 km/h
ii Lower quartile = 88 km/h, Upper quartile = 118 km/h
iii 30 km/h
c **i** 12 drivers **ii** 51 drivers **d** 56 drivers

Exercise 6B

1 a

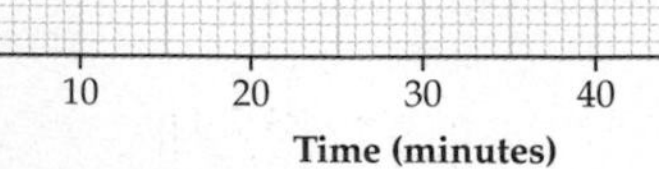

b

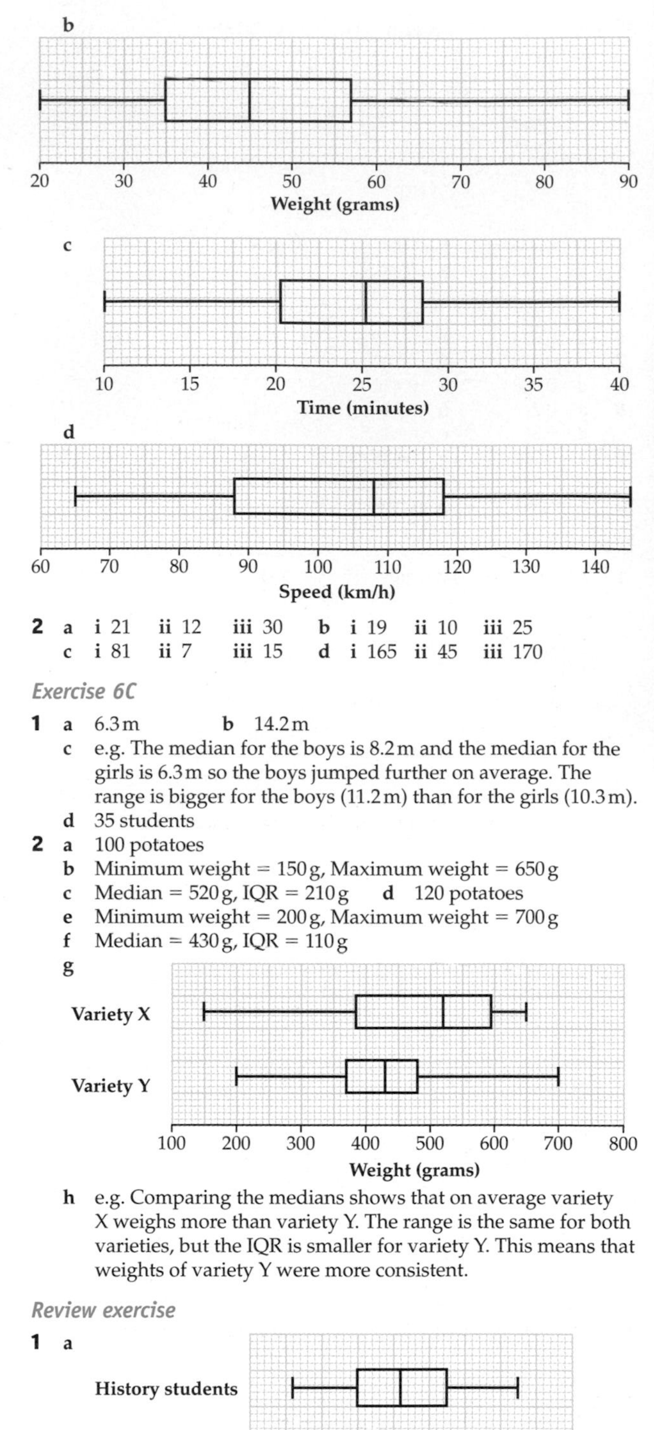

c

d

2 **a** **i** 21 **ii** 12 **iii** 30 **b** **i** 19 **ii** 10 **iii** 25
c **i** 81 **ii** 7 **iii** 15 **d** **i** 165 **ii** 45 **iii** 170

Exercise 6C

1 **a** 6.3 m **b** 14.2 m
c e.g. The median for the boys is 8.2 m and the median for the girls is 6.3 m so the boys jumped further on average. The range is bigger for the boys (11.2 m) than for the girls (10.3 m).
d 35 students

2 **a** 100 potatoes
b Minimum weight = 150 g, Maximum weight = 650 g
c Median = 520 g, IQR = 210 g **d** 120 potatoes
e Minimum weight = 200 g, Maximum weight = 700 g
f Median = 430 g, IQR = 110 g
g

Variety X
Variety Y
Weight (grams)

h e.g. Comparing the medians shows that on average variety X weighs more than variety Y. The range is the same for both varieties, but the IQR is smaller for variety Y. This means that weights of variety Y were more consistent.

Review exercise

1 a

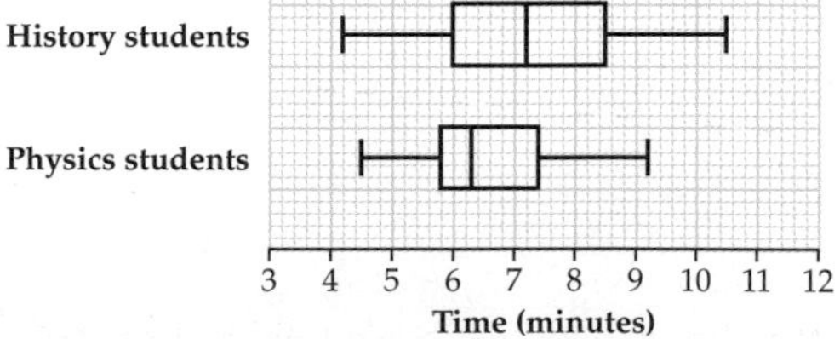

b e.g. Comparing the medians shows that on average the history students took longer than the physics students. The IQR and the range are smaller for the physics students, so their times were more consistent.

2 **a** e.g. The range and the IQR are smaller for A, so Joe is less likely to have extremes of temperature. The minimum temperature at A is not as cold as the minimum temperature at B.
b A never has a temperature above 30.5°C. If Joe wants a chance of warmer weather, he should go to B. The median temperature is higher for B, so on average B is warmer.

3 a

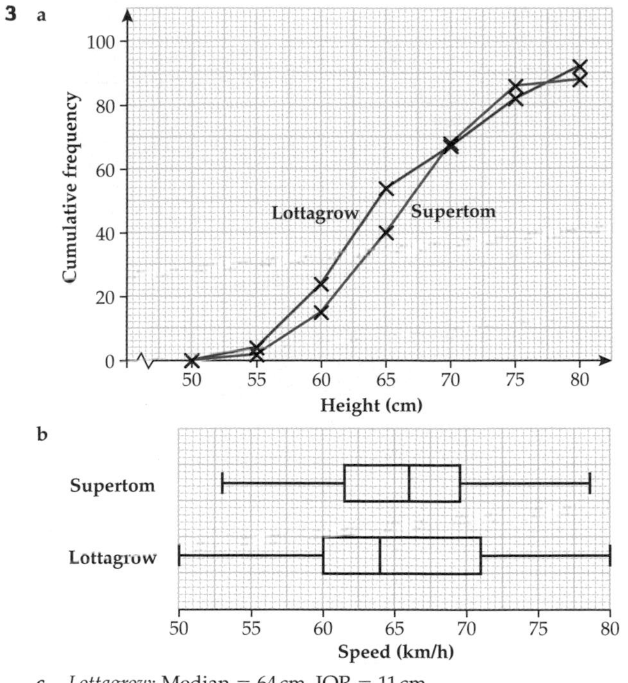

c *Lottagrow*: Median = 64 cm, IQR = 11 cm
Supertom: Median = 66 cm, IQR = 8 cm

d *Supertom*. Comparing the medians shows that on average *Supertom* treated plants grew taller than *Lottagrow* treated plants. The range and IQR are smaller for *Supertom* treated plants, so the heights of these plants are more consistent.

Chapter 7

Exercise 7A

1 **a** 1500 g flour, 6 eggs, 900 ml milk
b 250 g flour, 1 egg, 150 ml milk
c 750 g flour, 3 eggs, 450 ml milk

2 **a** 28 seconds **b** 33

3 **a** 1.75 kg potatoes, 175 g onion, 87.5 g butter, 315 g cheddar cheese, 525 ml milk
b 7 people

4 2.5 kg cement for above ground. Ratio sand : cement = 5 : 1, 12.5 kg sand needed.
2.5 kg cement for paving. Ratio sand : cement = 3 : 1, 7.5 kg sand needed.
Total sand: 20 kg

Exercise 7B

1 **a** $\frac{2}{7}$ **b** $\frac{5}{7}$

2 No. The denominator is 5 + 6 = 11. The fraction of pedigree dogs is $\frac{5}{11}$.

3 £900

4 56

Exercise 7C

1 **a** 3 : 1 **b** 2 : 1 **c** 7 : 1 **d** 1.25 : 1

2 **a** 1 : 4 **b** 1 : 3.75 **c** 1 : 1.71 or 1 : $\frac{12}{7}$
d 1 : 1.5 **e** 1 : $\frac{1}{3}$ **f** 1 : $\frac{5}{16}$

3 **a** 10.9 : 1 **b** 1.4 : 1 **c** First, as 10.9 is greater than 1.4.

4 First bar, 22% is copper.
Second bar, 20% is copper.
First bar has a higher proportion.

Exercise 7D

1 **a** £24, £12 **b** £9, £3 **c** 20 litres, 12 litres
d 24 kg, 42 kg **e** £200, £200, £100 **f** 100 cm, 60 cm, 90 cm

2 **a** 3 : 7 **b** Dan £30 000, Tris £70 000

3 Alix £12 000, Liberty £15 000

4 Pip 50 g, Wilf 100 g, Anna 150 g

5 Ruby £11 628, Sid £16 279, Fred £22 093

Exercise 7E

1 **a** 12 **b** 18 **c** 27

2 **a** 3 : 150 **b** 1 : 50 **c** 600 g **d** 18 eggs

3 **a** 88 : 12 **b** 22 : 3 **c** 7.5 kg

4 9 kg

5 5

6 25

7 11 children 0–2 years need 4 adults
8 children 2 years old need 2 adults
32 children 3–7 years need 4 adults
Total 10 adults

Exercise 7F

1 **a** £15 **b** £60

2 £58.50

3 3.15 kg

4 £1912.50

5 7.5

6 £7.98

7 **a** 30 euros **b** US$64 **c** Aus$58.8 **d** £29.17
e £18.75 **f** £86.96 **g** £40.32 **h** £255.10

8 7000 Pakistani rupees

9 New York. The price in New York is £249.99, which is £30 cheaper than in London.

Exercise 7G

1 Large pack is 0.468p per gram.
Small pack is 0.49p per gram.
Large pack is better buy as price per gram is lower.

2 Small bottle is better buy. Small bottle is 1.75p per ml, large bottle is 1.76p per ml.

3 Bottle C

Exercise 7H

1 75 minutes or $1\frac{1}{4}$ hours

2 1.2 days

3 **a** 9 **b** 4

4 **a** 12 hours **b** 2.4 hours or 2 hours 24 minutes

5 320 ÷ 8 = 40
At least 40 plants must be laid each hour.
1 person can do 6 per hour, so 7 people needed to do at least 40 per hour.

6 1 hour 47 minutes

7 Cleaner takes 18 minutes per room.
48 rooms take 864 minutes.
864 ÷ 240 = 3.6
The hotel needs 4 cleaners.

8 Old area 45 000 cm^2
45 000 ÷ 300 = 150 new tiles

9 1 man takes 12 hours to turf 600 m^2 lawn.
25 × 40 = 1000 m^2 will take 1 man 20 hours, or 5 men 4 hours.

Exercise 7I

1 **a** $y = 70$ **b** $x = 15$

2 **a** $x = 10.5$ **b** $t = 10.2$

3 **a** $l = 0.15m$ **b** $l = 6.75$ metres **c** $m = 66.67$ g (2 d.p.)

4 $y = 7, x = 9.6$

5 All ratios $a : b$ are 1 : 3.6 in the form 1 : n.

6 25.2 : 6 = 4.2 : 1; 42 : 10 = 4.2 : 1; 60 : 12 = 5 : 1. d is not directly proportional to e as the ratio $d : e$ is not constant.

Exercise 7J

1 **a** $y = 62.5$ **b** $x = 4$

2 **a** $y = 358.4$ **b** $x = 4.53$ (2 d.p.)

3 **a** $y = 25.5$ **b** $x = 22.15$ (2 d.p.)

4 **a** $A = 3.142\, r^2$ **b** π

5 **a** 2144.7 cm^3 (1 d.p.) **b** 10 cm

6 $F = 40, r = 5$

7 $C \propto l^3$

Exercise 7K

1 **a** $y = 14$ **b** $x = 1.4$

2 **a** $y = 26.4$ (1 d.p.) **b** $x = 2.23$ (2 d.p.)

3 **a** $y = 7.5$ **b** $x = 18.37$ (2 d.p.)

4 a $t = \frac{72}{P}$ b 24 hours
c $P = 7.2$, so 8 people
5 $V = 115.2\,\text{cm}^3$, $P = 1.152$ bar
6 1.1 cm (1 d.p.)
7 $m = 3200$, $t = 1.79$ (2 d.p.)
8 Yes; $0.6 \times 18 = 2.4 \times 4.5 = 9 \times 1.2 = 30 \times 0.36 = 10.8$
9 $p = 56.53$, $t = 0.76$ (2 d.p.)

Review exercise

1 9 : 11
2 a 15 b 12
3 £27
4 £1500
5 Al £5000, Deb £3000, Cat £2000
6 1050 g
7 a 3.92 litres b 3 tins
8 10.5 hours
9 Area of one tile = 2500 cm²
Total area of tiles = 4000 × 2500 cm² = 10 000 000 cm²
Area of laminate strip = 30 × 200 = 6000 cm²
Number of strips = 10 000 000 ÷ 6000 = 1667 to nearest whole number
10 1 machinist makes 5.25 pairs in 1 day, or 10.5 pairs in 2 days
800 ÷ 10.5 = 76.19...
77 machinists
11 a $m = 90$ b $n = 2$
12 $y = 2x^2$
13 $q = 312.5$, $p = 0.3125$
14 a Table C b Table B c Table A

Chapter 8

Exercise 8A

1 £146.02
2 £9344.26
3 4% annum for 5 years
4 7 years
5 a £5.20 b £2309.21
6 a £15 571.36 b £1376.64 c £138.95
7 a 12 years b 11 years
8 8.00% (to 2 d.p.)

Exercise 8B

1 £380
2 £14 000
3 £114.89
4 167 mm
5 6 kg
6 £74.47
7 £4376.25
8 a 40 cm b 40.2 cm

Exercise 8C

1 a 3×10^6 b 7.4×10^3 c 3.2×10^4
d 6.035×10^5 e 1.08×10^2 f 6.8×10^1
g 6.505×10^2 h 9.99×10^1 i 5×10^0
j 2.04×10^0
2 a 1 b 0 c 3
d 4 e 6 f 5

Exercise 8D

1 a 5×10^{-4} b 6×10^{-3} c 4×10^{-1}
d 1.2×10^{-4} e 7.17×10^{-2} f 1.975×10^{-4}
g 9.009×10^{-1} h 1.0003×10^{-3}
2 a 10^{-2} b 10^{-3} c 10^{-5}
d 10^{-8} e 10^{-3} f 10^{-1}

Exercise 8E

1 a 50 000 b 3800 c 0.006
d 7 260 000 000 e 0.08492 f 4 370 000
g 0.000 100 6 h 6238.7
2 a 148 800 000 km b 0.01 m c 0.000 000 000 000 03
3 a 1.23×10^4 b 8×10^6 c 1.7×10^{-1}
d 2.5×10^{-3} e 1.8×10^7 f 1.25×10^{-1}
g 2.16×10^3 h 2×10^1

Exercise 8F

1 a 2.226×10^8 b 4.2×10^1
c 1.4×10^5 d 4×10^0
2 $1.098\,066\ldots \times 10^{12}$ km³
3 1.17×10^{-2} tonne
4 a 3.78×10^5 cm² b 5.68×10^3 cm
5 2.5×10^{13} miles
6 1.25×10^{-7} cm

Exercise 8G

1 a lower bound = 15 500, upper bound = 16 499
b lower bound = 245, upper bound = 254
c lower bound = 3350, upper bound = 3449
d lower bound = 825, upper bound = 874
2 a 59.5 mph ≤ speed < 60.5 mph
b 13 350 km ≤ distance < 13 450 km
c 125 *ml* ≤ volume < 135 *ml*
d 21.45 cm ≤ length < 21.55 cm
3 a 8.85 kg ≤ mass < 8.95 kg
b 100.15 *ml* ≤ volume < 100.25 *ml*
c 8.385 m ≤ length < 8.395 m
d 1105 g ≤ mass < 1115 g
e 1050 g ≤ mass < 1150 g
f 950 g ≤ mass < 1500 g
4 a lower bound = 6.5 m, upper bound = 7.5 m
b lower bound = 9.25 days, upper bound = 9.35 days
c lower bound = 4.75 kg, upper bound = 4.85 kg
d lower bound = 0.745 mg, upper bound = 0.755 mg
e lower bound = 3.095 *cl*, upper bound = 3.105 *cl*
f lower bound = 19.9985 cm, upper bound = 19.9995 cm
5 26 g
6 91.125 cm³
7 7875.125 cm²
8 a 98.9625 m b 101.0625 m
9 a 4486.8 (1 d.p.) b 4823.3 (1 d.p.)
10 Since the volume of the cube is given to 2 s.f., the minimum value it could be is 445 cm³. So Paulin should have calculated $\sqrt[3]{445}$ rather than $\sqrt[3]{450}$.
Correct answer = 7.634 60… = 7.6 cm (2 s.f.)
11 87 cm
12 7.15 m/s
13 Charlie is incorrect.
upper bound for length of path = 30.5 m = 3050 cm
lower bound for length of tile = 37.5 cm
3050 ÷ 37.5 = 81.3
So Charlie *could* need up to 82 paving slabs.

Exercise 8H

1 0.05 m, 0.83%
2 3%
3 Yes, because the absolute error is 11.25 *ml*.
4 a 18.86 cm²
b 8.15 cm ≤ base < 8.25 cm, 4.55 cm ≤ height < 4.65 cm
c 18.541 25 cm² ≤ area < 19.181 25 cm²
d 0.321 25 cm
e 1.7% (1 d.p.)
5 7.3% (2 s.f.)
6 4.8% (2 s.f.)
7 6.7% (2 s.f.)

Review exercise

1 George
2 a £14 500 b £12 340.43
3 1.08×10^{-1} gram
4 2460.375 cm³
5 a 197.43 m (2 d.p.) b 200.74 (2 d.p.)
6 5.29 (2 d.p.)
7 Anton is incorrect. He *could* have run an average speed of 8.10 m/s.
8 14.03 (2 d.p.)
9 2.98%
10 6.3% (2 s.f.)

Chapter 9

Number skills: multiplying and dividing

1 a 864 b 7776 c 18 d 28
2 £1987
3 13 and 14
4 13.2
5 11
6 a 11 coaches b 25 seats

Number skills: negative numbers

1 a −6, −4, 0, 2, 3 b −19, −11, −9, −7, −3
2 14°C
3 a −11 b 10 c −8 d 3
4 a −3 b −8 c −3 d 4
e 45 f 12 g 3 h −5
5 −8
6 −3 and 4, 3 and −4

Exercise 9A

1 a 8 b 50 c 20 d 30
2 a 50 b 15 c 7 d 1
3 a C. Both area and number of litres have been rounded down.
b 100 litres
4 Estimate: 10 × 5 = 50 miles of fuel remaining. So Pepe does not have enough fuel.
5 Approximate answer is 12. Steven is more likely to be correct.
6 a 200 b 3 c 0.5
7 a 6000 b 15 c 100
8 B. It is better to round one number up and one number down in the multiplication.
9 5
10 a 320 b 2400 c 550 d 5000

Exercise 9B

1 a CAN\$600 b CAN\$1860
c CAN\$16 500 d CAN\$31.50
2 a £18 b £23 c £35 d £74
3 5225 pesos
4 £17.50
5 £1 = 6 shekels
6 a US\$320 b £1 = US\$1.55
7 a £1600 b J\$200 000
8 a \$125 b £4
9 a i 6400 kroner ii £240 b 920 dalasi
10 New York (£110)

Review exercise

1 20
2 100
3 a £13 b £28 c £78 d £125.50
4 a 3250 pesos b £9
5 Mexico (£55.50). Cheaper by £24.50.
6 Approximate answer is 15. Daniel is most likely to be correct.
7 4000
8 6
9 Mexico (£1.50)

Chapter 10

Exercise 10A

1 a 6, 12, 18, 24, 30, 36, 42, 48, 54, 60
b 9, 18, 27, 36, 45, 54, 63, 72, 81, 90
c 18
2 a e.g. 6 and 12 b e.g. 35 and 70
c e.g. 10 and 20 d e.g. 9 and 18
3 a 30 b 40 c 10 d 30 e 60 f 60
4 E.g. 6 × 8 = 48 but LCM of 6 and 8 is 24.
5 24
6 84 cm

Exercise 10B

1 a 1, 2, 3, 4, 6, 12 b 1, 2, 4, 8 c 4
2 a 5 b 2 c 3 d 4 e 2 f 8
3 e.g. 16 and 24
4 a 12 cm by 12 cm b 30

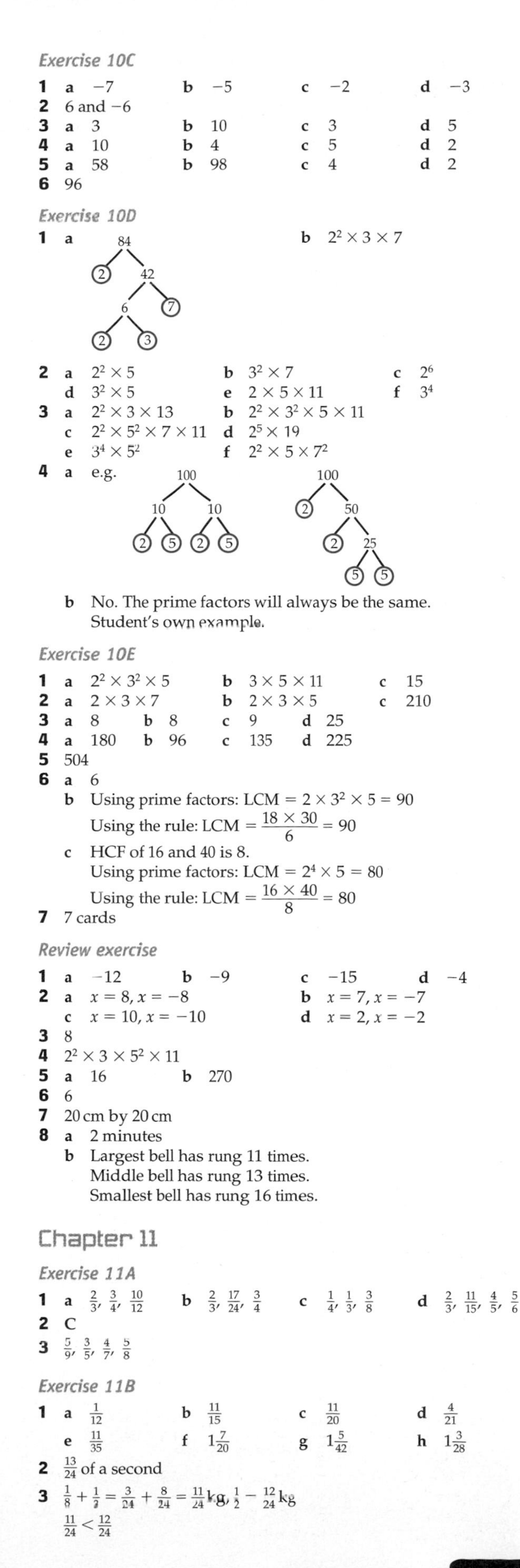

Exercise 10C

1 a −7 b −5 c −2 d −3
2 6 and −6
3 a 3 b 10 c 3 d 5
4 a 10 b 4 c 5 d 2
5 a 58 b 98 c 4 d 2
6 96

Exercise 10D

1 a 84 → 2, 42; 42 → 6, 7; 6 → 2, 3 b $2^2 \times 3 \times 7$
2 a $2^2 \times 5$ b $3^2 \times 7$ c 2^6
d $3^2 \times 5$ e $2 \times 5 \times 11$ f 3^4
3 a $2^2 \times 3 \times 13$ b $2^2 \times 3^2 \times 5 \times 11$
c $2^2 \times 5^2 \times 7 \times 11$ d $2^5 \times 19$
e $3^4 \times 5^2$ f $2^2 \times 5 \times 7^2$
4 a e.g. 100 → 10, 10; 10 → 2, 5; 10 → 2, 5
100 → 2, 50; 50 → 2, 25; 25 → 5, 5
b No. The prime factors will always be the same. Student's own example.

Exercise 10E

1 a $2^2 \times 3^2 \times 5$ b $3 \times 5 \times 11$ c 15
2 a $2 \times 3 \times 7$ b $2 \times 3 \times 5$ c 210
3 a 8 b 8 c 9 d 25
4 a 180 b 96 c 135 d 225
5 504
6 a 6
b Using prime factors: LCM $= 2 \times 3^2 \times 5 = 90$
Using the rule: LCM $= \frac{18 \times 30}{6} = 90$
c HCF of 16 and 40 is 8.
Using prime factors: LCM $= 2^4 \times 5 = 80$
Using the rule: LCM $= \frac{16 \times 40}{8} = 80$
7 7 cards

Review exercise

1 a −12 b −9 c −15 d −4
2 a $x = 8, x = -8$ b $x = 7, x = -7$
c $x = 10, x = -10$ d $x = 2, x = -2$
3 8
4 $2^2 \times 3 \times 5^2 \times 11$
5 a 16 b 270
6 6
7 20 cm by 20 cm
8 a 2 minutes
b Largest bell has rung 11 times.
Middle bell has rung 13 times.
Smallest bell has rung 16 times.

Chapter 11

Exercise 11A

1 a $\frac{2}{3}, \frac{3}{4}, \frac{10}{12}$ b $\frac{2}{3}, \frac{17}{24}, \frac{3}{4}$ c $\frac{1}{4}, \frac{1}{3}, \frac{3}{8}$ d $\frac{2}{3}, \frac{11}{15}, \frac{4}{5}, \frac{5}{6}$
2 C
3 $\frac{5}{9}, \frac{3}{5}, \frac{4}{7}, \frac{5}{8}$

Exercise 11B

1 a $\frac{1}{12}$ b $\frac{11}{15}$ c $\frac{11}{20}$ d $\frac{4}{21}$
e $\frac{11}{35}$ f $1\frac{7}{20}$ g $1\frac{5}{42}$ h $1\frac{3}{28}$
2 $\frac{13}{24}$ of a second
3 $\frac{1}{8} + \frac{1}{3} = \frac{3}{24} + \frac{8}{24} = \frac{11}{24}$ kg, $\frac{1}{2} = \frac{12}{24}$ kg
$\frac{11}{24} < \frac{12}{24}$

4 Any two proper fractions with different denominators that add to give $1\frac{7}{12}$ e.g. $\frac{3}{4} + \frac{5}{6}$, $\frac{2}{3} + \frac{11}{12}$

5 $\frac{1}{8} + \frac{1}{5} = \frac{5+8}{40} = \frac{13}{40}$, $\frac{3}{10} = \frac{12}{40}$

Pablo could not be correct as $\frac{13}{40} > \frac{12}{40}$

Exercise 11C

1 a $4\frac{11}{12}$ b $2\frac{5}{8}$ c $1\frac{3}{8}$ d $1\frac{1}{3}$

2 $3\frac{3}{10}$ miles

3 Any two pairs of mixed numbers where the fractions have different denominators that add to give $3\frac{17}{30}$ e.g. $1\frac{1}{15} + 2\frac{1}{2}$, $1\frac{7}{30} + 2\frac{1}{3}$

4 $1\frac{1}{6} + 3\frac{1}{4} = 4\frac{5}{12}$, $4\frac{1}{4} = 4\frac{3}{12}$, Patrick could not be correct as $4\frac{5}{12} > 4\frac{3}{12}$

5 $1\frac{5}{6} + 1\frac{7}{8} + 1\frac{11}{12} + 1\frac{1}{2} = 7\frac{1}{8}$ tonnes > 7 tonnes

Exercise 11D

1 a 7 b 8 c $8\frac{1}{4}$ d $8\frac{2}{5}$

2 Neither, they are the same.

3 a $31\frac{1}{2}$ kg b $35\frac{1}{4}$ kg

4 $184\frac{1}{2}$ kg

5 a $1\frac{1}{4}$ b $1\frac{2}{5}$ c $2\frac{1}{15}$ d $1\frac{1}{2}$

6 $7\frac{7}{8}$ minutes

7 a $5\frac{5}{8}\,m^2$ b £108

8 a $\frac{3}{10}$ of a minute b 60 minutes = 1 hour

9 a $5\frac{1}{4}$ b $3\frac{4}{15}$ c $14\frac{23}{54}$ d $40\frac{3}{8}$

10 $16\frac{7}{8}\,m^2$

11 Yes, $1\frac{1}{2} \times 6\frac{2}{3} = 10$ kg

12 a $57\frac{4}{5}$ litres b $9\frac{11}{12}$ minutes, or 9 minutes 55 seconds

Exercise 11E

1 a $\frac{1}{4}$ b $\frac{1}{10}$ c $\frac{1}{20}$ d $\frac{1}{100}$
e 2 f 5 g $1\frac{1}{2}$ h $3\frac{1}{3}$

2 All the answers are 1.

3 No, because reciprocals multiply together to give 1, not add together to give 1.

Exercise 11F

1 a 30 b 20 c $13\frac{1}{2}$

2 a $\frac{5}{6}$ b $\frac{13}{14}$ c 1
d $\frac{3}{8}$ e $\frac{3}{28}$ f $1\frac{1}{5}$

3 a $3\frac{6}{7}$ b $1\frac{5}{7}$ c $3\frac{17}{20}$

4 $\frac{7}{8}$

5 18

6 Yes they are, as $6\frac{1}{2} \div \frac{2}{5}$ is the same as $6 \div \frac{2}{5} + \frac{1}{2} \div \frac{2}{5}$.

7 a $\frac{8}{27}$ b $\frac{39}{50}$ c $\frac{104}{119}$
d $3\frac{13}{24}$ e $3\frac{17}{20}$ f $1\frac{41}{94}$

8 a i $\frac{3}{4}$ ii $\frac{3}{4}$ b 0

9 $6\frac{2}{3} \div 3\frac{3}{4} = 1\frac{7}{9}$, half of $3\frac{1}{3} = 1\frac{2}{3} = 1\frac{6}{9}$

Ryan is correct as $1\frac{7}{9} > 1\frac{6}{9}$.

Review exercise

1 $6\frac{2}{3}$

2 $\frac{2}{15}$

3 5

4 $4\frac{31}{60}$ kg

5 a $5\frac{2}{3}$ b 5 minutes 40 seconds

6 $16\frac{1}{5}$

7 a $\frac{8}{15}$ b $1\frac{7}{8}$

8 £297

Chapter 12

Exercise 12A

1 a $5p + 30$ b $3x + 3y$ c $4u + 4v + 4w$
d $2y - 16$ e $63 - 7z$ f $8a - 8b + 48$

2 a $6c + 18$ b $20t + 15$ c $10p + 2q$
d $6a - 3b$ e $18c - 12d$ f $14x + 7y - 21$
g $18a - 24b + 6c$ h $2x^2 + 6x + 4$ i $4y^2 - 12y - 40$

3 a $4(x - 3) + 6 = 4x - 12 + 6$
$= 4x - 6$ $(= 2(2x - 3))$
b $2(2x - 3) = 4x - 6$
c Left-hand side simplifies to the same as the right-hand side so Julia is correct.

4 $2(x + 5) + 3x = 2x + 10 + 3x$
$= 5x + 10$ $(= 5(x + 2))$
$5(x + 2) = 5x + 10$
= left-hand side

5 $6(t - 5) + 6 = 6t - 30 + 6$
$= 6t - 24$ $(= 6(t - 4))$
$6(t - 4) = 6t - 24$
LHS = RHS

Exercise 12B

1 $b^2 + 4b$ **2** $5a + a^2$ **3** $k^2 - 6k$
4 $9m - m^2$ **5** $2a^2 + 3a$ **6** $4g^2 + g$
7 $2p^2 + pq$ **8** $t^2 + 5tw$ **9** $m^2 + 3mn$
10 $2x^2 - xy$ **11** $4r^2 - rt$ **12** $a^2 - 4ab$
13 $2t^2 + 10t$ **14** $3x^2 - 24x$ **15** $5k^2 + 5kl$
16 $6a^2 + 12a$ **17** $8g^2 + 2gh$ **18** $15p^2 - 10pq$
19 $6xy + 15xz$ **20** $r^3 + r$ **21** $a^3 + 3a$
22 $t^3 - 7t$ **23** $2p^3 + 6pq$ **24** $4x^3 + 4x^2$
25 a $x(3x + 2)$ b $3x^2 + 2x$

Exercise 12C

1 a $-4k - 8$ b $-6x - 18$ c $-15n + 5 = 5 - 15n$
d $-12t - 20$ e $3 - 12p$ f $14 - 6x$

2 a $5y + 22$ b $5k + 21$ c $2a + 18$
d $7t - 16$ e $4y - 1$ f $7x + 40$
g $8x + 7$ h $3n + 25$ i $5x - 21$
j $14x - 8$ k $2b - 5$ l $4m + 2$
m $4t + 4$ n $2k - 14$ o $2a + 23$
p $4p + 14$ q $2g - 8$ r $-4w - 5$
s $x^2 + 7x + 8$ t $2x^2 - 2x + 12$

3 $9(x + 1) + 3(x + 2) = 9x + 9 + 3x + 6$
$= 12x + 15$
$(= 3(4x + 5))$
$3(4x + 5) = 12x + 15$
LHS = RHS

Exercise 12D

1 a $g = 2$ b $k = 2$ c $s = 3$ d $n = 7$
e $f = 10$ f $v = 10$ g $m = -3.5$ h $w = -1.5$

2 a $t = 2$ b $r = 3$ c $b = -2$ d $w = 6\frac{1}{2}$
e $x = 2\frac{1}{2}$ f $y = -1\frac{1}{2}$ g $k = 3\frac{1}{2}$ h $a = -4$

3 a $6(7x + 4)$ b $6(7x + 4) = 45$
c $42x + 24 = 45$
$42x = 21$
$x = \frac{1}{2}$
d $7 \times \frac{1}{2} + 4 = 3\frac{1}{2} + 4$
$= 7\frac{1}{2}$ cm

Exercise 12E

1 a $3(x + 5)$ b $5(a + 2)$ c $2(x - 6)$
d $4(m - 4)$ e $4(t + 3)$ f $3(n + 6)$
g $2(b - 7)$ h $4(t - 5)$

2 a $5(p + 4)$ b $2(a + 6)$ c $3(y + 5)$
d $7(b + 3)$ e $4(q + 3p)$ f $6(k + 4j)$
g $4(t - 3)$ h $3(x - 3)$ i $5(n - 4)$
j $2(b - 4)$ k $6(a - 3b)$ l $7(k - 1)$

3 a $y(y + 7)$ b $x(x + 5)$ c $n(n + 1)$
d $x(x - 7)$ e $p(p - 8)$ f $a(a - b)$

4 a $2(3p + 2)$ b $2(2a + 5)$ c $2(3 - 2t)$
d $4(3m - 2n)$ e $5(5x + 3y)$ f $3(4y - 3z)$

Exercise 12F

1 a $3x(x-2)$ b $x(8x-y)$ c $4a(2+b)$
d $p^2(p-5)$ e $3t^2(t+2)$ f $5y(2z-3y)$
g $6a(3a+2b)$ h $4p(4p-3q)$

2 a $2ab^2(2+3b)$ b $5x(2y-x)$
c $3p^2q(1-2pq)$ d $2n(4mn^2+2n-3m^2)$
e $6hk(h-2k^2-3hk)$ f $3ab(2b+3a-1)$
g $2xy(4y+3-2x)$ h $6p^2r(2+3r+p)$

Exercise 12G

1 $a^2+9a+14$
2 $x^2+10x+25$
3 $t^2+3t-10$
4 $h^2-11h+24$
5 y^2-6y+9
6 $x^2-19x+84$
7 p^2-16
8 $28+11a+a^2=a^2+11a+28$

Exercise 12H

1 a x^2+4x+3 b $x^2+3x-28$ c $n^2+3n-40$
d x^2+x-20 e $x^2-13x+36$ f m^2+m-56
g $q^2+13q+42$ h $20-d-d^2$ i $24-11x+x^2$
j $y^2-10y-96$

2 a x^2-16 b x^2-25 c x^2-4
d x^2-121 e x^2-9 f x^2-1
g x^2-81 h x^2-a^2 i t^2-x^2

3 $(x+3)(x+8)-(x+4)(x+6)=(x^2+11x+24)-(x^2+10x+24)$
$=x^2+11x+24-x^2-10x-24$
$=x$

4 $(x+5)(x+2)-(x+8)(x-1)=(x^2+7x+10)-(x^2+7x-8)$
$=x^2+7x+10-x^2-7x+8$
$=18$

Exercise 12I

1 $3a^2+14a+8$ **2** $5x^2+13x+6$ **3** $6t^2+19t+15$
4 $8y^2+30y+7$ **5** $12x^2+28x+15$ **6** $4x^2-x-3$
7 $6z^2+11z-10$ **8** $7y^2-y-8$ **9** $3n^2+19n-40$
10 $6b^2-7b-5$ **11** $7p^2-25p-12$ **12** $6z^2-17z+12$
13 $10x^2-23x+9$ **14** $4y^2-12y+9$ **15** $12a^2+23a+10$
16 $4x^2-1$ **17** $9y^2-4$ **18** $25n^2-16$
19 $9x^2-25$ **20** $1-4x^2$ **21** $9t^2-4x^2$

22 $(2x+3)(2x-5)=4x^2-10x+6x-15$
$=4x^2-4x-15$
$(=4x^2-4x-16+1)$
$(=4(x^2-x-4)+1)$
$4(x^2-x-4)+1=4x^2-4x-16+1$
$=4x^2-4x-15$
LHS = RHS

23 a $(x+1)(x-2)$ b x^2-x-2

Exercise 12J

1 a $x^2+10x+25$ b $x^2+12x+36$
c x^2-6x+9 d $x^2-16x+64$
e x^2+6x+9 f x^2+4x+4
g $x^2-10x+25$ h $x^2+2ax+a^2$

2 a $(x+6)^2=x^2+\mathbf{12}x+36$ b $(x-7)^2=x^2-\mathbf{14}x+49$
c $(x+\mathbf{9})^2=x^2+18x+\mathbf{81}$ d $(x-\mathbf{10})^2=x^2-20x+\mathbf{100}$

3 a $x^2+6x+9-9=x^2+6x$
b $x^2+10x+25-25=x^2+10x$
c $x^2-12x+36-36=x^2-12x$
d $x^2+2ax+a^2-a^2=x^2+2ax$

4 a $4x^2+4x+1$ b $4y^2-28y+49$
c $4+20z+25z^2$ d $49-56m+16m^2$
e $x^2+4xy+4y^2$ f $4a^2-4ab+b^2$
g $9x^2+12xy+4y^2$ h $16z^2-40zt+25t^2$

5 a $(10x+5)^2=100x^2+100x+25$
$(=100x(x+1)+25)$
$100x(x+1)+25=100x^2+100x+25$
LHS = RHS

b $55^2=(10\times5+5)^2$
$=100\times5\times6+25$
$=3000+25$
$=3025$

c $75^2=(10\times7+5)^2$
$=100\times7\times8+25$
$=5600+25$
$=5625$

6 a $(a+b)^2-(a-b)^2=(a^2+2ab+b^2)-(a^2-2ab+b^2)$
$=a^2+2ab+b^2-a^2+2ab-b^2$
$=4ab$

b $57^2-43^2=(50+7)^2-(50-7)^2$
$=4\times50\times7$
$=1400$

c $48^2-32^2=(40+8)^2-(40-8)^2$
$=4\times40\times8$
$=1280$

Review exercise

1 a $6a+12$ b p^2-8p

2 a $4(x-3)=4x-12$
$(=3x-12+x)$
$(=3(x-4)+x)$
$3(x-4)+x=3x-12+x$
$=4x-12$
LHS = RHS

b $3(2x+5)+2(x-3)=6x+15+2x-6$
$=8x+9$
$(=8x+8+1)$
$(=8(x+1)+1)$
$8(x+1)+1=8x+8+1$
$=8x+9$
LHS = RHS

3 a $y=2$ b $m=2$

4 a $10t-3$ b $2x+13$

5 a x^2+4x+3 b y^2-6y+8 c $x^2+8x+16$

6 a

	$n-6$
n	$n+1$
	$n+8$

b $4n+3$

7 a $(x+3)(x+2)-5x=x^2+5x+6-5x$
$=x^2+6$

b $(x-2)(x+8)=x^2+6x-16$
$(x+3)^2-25=(x+3)(x+3)-25$
$=x^2+6x+9-25$
$=x^2+6x-16$
LHS = RHS

8 a $5(y+2)$ b $m(m-6)$ c $3t^2(t-3)$
d $4x(x+2y)$ e $pq(p-8+6q)$

9 a $4x^2+17x-15$ b $4t^2-25r^2$ c $16n^2-40n+25$

10 $(2x-1)^2-16=(2x-1)(2x-1)-16$
$=4x^2-4x+1-16$
$=4x^2-4x-15$
$(2x+3)(2x-5)=4x^2-10x+6x-15$
$=4x^2-4x-15$
LHS = RHS

Chapter 13

Exercise 13A

1 a 25.2 b 19.17 c 44.22
d 234.8 e 42.25 f 0.75
g 0.157 h 0.224

2 a 0.212 b 0.212 They are the same.

3 £27.03

4 £391.72

5 22.14 mm

Exercise 13B

1 a 20 b 1200 c 42.6
d 25 e 30 f 150
g 3340 h 140

2 a 139 b 139 They are the same.

3 a 8 panels b 9 posts

4 75 books

5 2150 texts

Exercise 13C

1 a $\frac{1}{10}$ b $\frac{1}{2}$ c $\frac{7}{100}$ d $\frac{1}{20}$
e $\frac{3}{4}$ f $\frac{7}{25}$ g $\frac{1}{25}$ h $\frac{13}{20}$
i $\frac{13}{25}$ j $\frac{79}{100}$ k $\frac{6}{25}$ l $\frac{7}{20}$

2 a $\frac{1}{10000}$ b $\frac{1}{500}$ c $\frac{21}{250}$ d $\frac{9}{1000}$
e $\frac{7}{200}$ f $\frac{1}{40}$ g $\frac{3}{8}$ h $\frac{17}{40}$

3 a $3\frac{1}{2}$ b $14\frac{4}{5}$ c $5\frac{16}{25}$ d $4\frac{17}{20}$

4 $\frac{18}{25}$

5 $\frac{11}{25}$

6 $\frac{9}{25}$

Exercise 13D

1 a 0.125 b $0.1\dot{6}$ c $0.\dot{2}$ d 0.48

2 a $0.2\dot{3}$ b $0.02\dot{3}$ c $0.002\dot{3}$

3 0.4, 0.375, $0.\dot{3}$, 0.36, 0.39 In order: $\frac{1}{3}$, $\frac{9}{25}$, $\frac{3}{8}$, $\frac{39}{100}$, $\frac{2}{5}$

4 $0.\dot{1}4285\dot{7}$, $0.\dot{2}8571\dot{4}$, $0.\dot{4}2857\dot{1}$, $0.\dot{5}7142\dot{8}$, $0.\dot{7}1428\dot{5}$, $0.\dot{8}5714\dot{2}$. You get the same sequence of digits, but starting with different digits.

Exercise 13E

1 a $0.\dot{6}$ b 0.6 c $0.\dot{5}\dot{4}$ d 0.48
e 0.625 f 0.475 g $0.\dot{5}$ h $0.6\dot{1}$

2 $\frac{19}{40}$, $\frac{12}{25}$, $\frac{6}{11}$, $\frac{5}{9}$, $\frac{3}{5}$, $\frac{11}{18}$, $\frac{5}{8}$, $\frac{2}{3}$

3 Terminating: $\frac{29}{64}$, $\frac{13}{125}$, $\frac{3}{16}$, $\frac{169}{500}$, $\frac{49}{320}$, $\frac{66}{125}$
Recurring: $\frac{3}{17}$, $\frac{29}{72}$, $\frac{1}{36}$

4 $\frac{1}{13} = 0.\dot{0}7692\dot{3}$, $\frac{2}{13} = 0.\dot{1}5384\dot{6}$. All the other thirteenths are formed from these two sequences of six digits (076923 and 153846) with different starting positions.

5 a Yes
b Could be either (e.g. $\frac{3}{7} \times \frac{7}{15} = \frac{1}{5} = 0.2$, $\frac{1}{3} \times \frac{2}{3} = \frac{2}{9} = 0.\dot{2}$)
c Could be either

Exercise 13F

1 a $\frac{2}{3}$ b $\frac{2}{9}$ c $\frac{2}{45}$ d $\frac{6}{11}$
e $\frac{1}{33}$ f $\frac{8}{37}$ g $\frac{11}{18}$ h $\frac{7}{12}$
i $\frac{89}{370}$ j $\frac{351}{550}$

2 Converting $0.\dot{9}$ to a fraction gives the result 1.
Starting with $\frac{1}{3} = 0.\dot{3}$, multiply both sides by 3.

Review exercise

1 a 11.56 b 1.488

2 a £46.82 b £3.18

3 a $\frac{39}{500}$ b $\frac{19}{20}$ c $\frac{129}{200}$ d $\frac{5}{16}$

4 a 46.3 b 12.74 c 1.25

5 35 cans

6 a 0.6 b 0.063 c $0.\dot{1}\dot{8}$ d $0.91\dot{6}$

7 $\frac{4}{7} = 0.\dot{5}7142\dot{8}$, $\frac{3}{5} = 0.6$, $\frac{5}{8} = 0.625$, $\frac{2}{3} = 0.\dot{6}$, $\frac{7}{10} = 0.7$

8 Terminating: $\frac{15}{25}$, $\frac{13}{25}$, $\frac{3}{32}$, $\frac{17}{125}$, $\frac{13}{16}$
Recurring: $\frac{1}{11}$, $\frac{23}{60}$, $\frac{3}{18}$, $\frac{9}{21}$

9 a $\frac{1}{3}$ b $\frac{7}{11}$ c $\frac{215}{999}$
d $\frac{7}{90}$ e $\frac{65}{66}$ f $\frac{301}{555}$

Chapter 14

Exercise 14A

1 $a = 3$ **2** $b = 5$
3 $c = 3$ **4** $d = 10$
5 $k = 5$ **6** $h = 3.5$
7 $i = 0.6$ **8** $j = 3$
9 $k = -2$ **10** $l = -10$
11 $f = 2$ **12** $x = 14$
13 $x = 5$ **14** $f = 10$
15 $m = -2$ **16** $t = -1$
17 $q = 1$ **18** $n = -5$
19 $p = -5$ **20** $r = 2$

Exercise 14B

1 a $a = 5$ b $h = -2$ c $b = -2$
d $i = -3$ e $c = 6$ f $e = 2.5$
g $a = 2$ h $c = 2$ i $d = -1$
j $e = -1$ k $f = 2$ l $b = -2$

2 a $e = 1$ b $x = \frac{1}{5}$

3 a
$7 - j = \mathbf{15} - 3j$
$7 + 2j = \mathbf{15}$
$2j = \mathbf{8}$
$j = \mathbf{4}$

b
$10d - \mathbf{20} = 6d + 24$
$\mathbf{4d} - 20 = 24$
$\mathbf{4d} = 44$
$d = \mathbf{11}$

4 a $a = 1$ b $b = 1$ c $c = 6$
d $e = 1$ e $f = 9$ f $x = -2$
g $h = \frac{1}{5}$ h $i = 4\frac{1}{2}$

Exercise 14C

1 $x = 2$ **2** $x = 5$
3 $x = -3$ **4** $x = 3$
5 $x = 7$ **6** $x = 3$
7 $y = 2$ **8** $w = -11\frac{1}{2}$
9 $y = 28$ **10** $x = -1$
11 $w = 52$ **12** $t = -3$

Exercise 14D

1 a $x < 5$
1 2 3 4 5

b $x < 3$
−1 0 1 2 3

c $x < 20$
0 5 10 15 20

d $x \leqslant 8$
4 5 6 7 8

e $x > 6$
4 5 6 7 8

f $x > 0$
−1 −2 0 1 2

2 a $x < 2$ b $x > 5$ c $x \geqslant 6$ d $x \leqslant 74$
e $y \leqslant -3$ f $x < 7$ g $x > 5$ h $x \geqslant -2.5$

3 a $3 < x \leqslant 9$
3 9

b $3 < x < 7$
3 7

c $-3 < x$
−3 −2 −1 0 1 2 3

4 a $x > 19$ b $x < 14$ c $x \leqslant 40$ d $x < -9$
e $x < 3$ f $x < -3$ g $x \geqslant -3.5$ h $\frac{3}{2} \leqslant n < 4$
i $p < -10$ j $n \leqslant 4$

5 a 4, 5 b −2, −1, 0, 1
c −2, −1, 0, 1, 2, 3, 4

Exercise 14E

1 a $x = 3, y = 4$ b $x = 2, y = 3$ c $x = 1, y = 9$
d $x = -1, y = 7$ e $x = 7, y = 9$ f $x = -3, y = -2$

2 a $x + y = 83, x - y = 3$ b 40 and 43 years old

Exercise 14F

1 a $x = 2, y = 3$ b $x = -2, y = 5$ c $x = 2, y = 5$
d $x = -2, y = 5$ e $x = -3, y = 4$ f $x = -33, y = 31$
g $x = 5, y = 2$ h $x = -5, y = 3$ i $x = 3, y = 2$

2 a 800 albums b 9800 tracks

Review exercise

1 a $m = -2$ b $j = 3.75$ c $b = -2$
2 $3x + 33° = 180°$
$x = 49°$
3 a $x = 1, 2, 3$ or 4 b $x = 3, y = 2$ and $x = 2, y = 1$
4 a $x = 7$ b $x = 9$
5 $x = 5$
6 a $x \leqslant 3$
b
−4 −3 −2 −1 0 1 2 3 4 5 6
7 a $2 < x \leqslant 4$ b $x > 5$ c $x < 8$
8 a $x = 3, y = -4$ b $x = 5, y = 6$ c $x = -2, y = 5$
9 a $x \geqslant 5$ b $x \geqslant 9$
10 2, 3, 4, 5, 6

Chapter 15

Exercise 15A

1 $C = 500 - 48y$
2 $t = 7r + 5s$
3 $t = 45w + 20$
4 $p = 2(3x + 1 + x + 2) = 8x + 6$

Exercise 15B

1 a 76 b 30 c 37 d 84 e 81 f −95
g 1 h $32\frac{2}{3}$ i $4\frac{1}{5}$
2

x	2	3	4	5	6
$x^3 - x$	6	24	60	120	210

3 a 90 b −18 c 5 d 7 e 31 f 8
4 a 36 b −14 c 4 d 6 e 2 f −3
5 a 1 b −3
6 a $\frac{7 + \sqrt{73}}{4}$ b $\frac{7 - \sqrt{73}}{4}$

Exercise 15C

1 a 26 b 24
2 a 7 b 13
3 a $90\,\text{cm}^2$ b $168\,\text{cm}^2$
4 a $360\,\text{cm}^2$ b $270\,\text{cm}^2$
5 3
6 12
7 $6\,\text{cm}^2$
8 Same answer of $6\,\text{cm}^2$ with workings shown

Exercise 15D

1 a 5 b 9 c 13 d 11.5 e 20
2 a 6 b 11 c 5 d 9 e −8

Exercise 15E

1 a $a = c - 5$ b $a = t - 12$
c $a = \frac{w}{3}$ d $a = \frac{P}{w}$
2 a $x = \frac{y + 6}{5}$ b $x = \frac{y - 1}{2}$
c $x = \frac{p - 2t}{4}$ d $x = \frac{y + 5p}{6}$
3 a $a = \frac{u - v}{t}$ b $t = \frac{u - v}{u}$
4 a $a = 2b - 12$ b $a = 3b + 3$
c $a = 4b + 12$ d $a = \frac{b}{2} - 1$
e $a = \frac{b}{3} + 5$
5 a $y = \frac{10 - 3x}{2}$ b $y = \frac{8 - 6x}{3}$
c $y = \frac{2x - 2}{3}$ d $y = \frac{5x \quad 10}{4}$
6 a $x = \frac{2y}{3}$ b $x = \frac{3y + 5}{2}$
c $x = y(z + 5)$ d $x = \frac{q(3p + s)}{2}$
7 a $w = K^2 - t$ b $w = A^2 + a$
c $w = \frac{(h - l)^2}{4}$ d $w = \frac{(T - 5)^2}{r}$
8 a $r = \sqrt{g(t + m)}$ b $r = 2\sqrt{h - 3a}$
c $r = \sqrt{\frac{3V}{\pi h}}$ d $r = \sqrt{\frac{A}{\pi} + s^2}$
9 $h = \frac{A}{2\pi r} - r$
10 $l = \frac{gT^2}{4\pi^2}$

Review exercise

1 a 40 b −5
2

x	2	3	4	5	6
$x^3 + 2x$	12	33	72	135	228

3 $T = 860$
4 a $R = 14$ b $T = \frac{C - 2R}{5}$
5 a −1 b 8 c 2
6 a $x = 2(w + y)$ b $x = \sqrt{w - y}$
7 2
8 $n = \frac{3p + 4}{7} + p$
9 $g = \frac{e}{f^2} - 7$

Chapter 16

Exercise 16A

1 a 9^{10} b 4^{10} c 5^4
d 10^8 e 8^3 f 6^5
g 4^5 h 5^2
2 a 9^{11} b 2^{13} c 8^9 d 4^5
3 a 125 b 16 c 16 d 64
4 No. To divide powers of the same number you subtract the indices, so $7^{10} \div 7^2 = 7^8$.
5 a 5^{-7} b 3^{-1} c 7^{-4} d 9^{-4}
e 4^{-2} f 2^{-3} g 3^{-2} h 4^{-2}
6 a 36 b 27 c 125 d 6
7 a 250 b 64 000 c 144 d 1250
e 5400 f 20 000
8 a 324 b 72 c 125 000 d 8000
9 $2^{11} \times 8^2 = 2^{11} \times 8 \times 8 = 2^{11} \times 2^3 \times 2^3 = 2^{17}$

Exercise 16B

1 a 2^{18} b 7^{10} c 6^9 d 5^{14}
e 9^6 f 7^{-2} g 15^{-12} h 8^{-4}
2 a 5^6 b 125
3 a 2^4 b 6^2 c 10^5 d 3^6
4 a 7^{12} b 343
5 a 36 b 27 c 64 d 100
e 25 f 64 g 81 h 11
6 a 3^7 b 7^7 c 2^8 d 2^4

Exercise 16C

1 a 1 b 6 c 1
d 17 e 9 f 1
2 a $\frac{1}{4}$ b $\frac{1}{9}$ c $\frac{1}{7}$
d $\frac{1}{4}$ e $\frac{1}{4}$ f $\frac{1}{12}$
3 a $\frac{1}{5}$ b $\frac{1}{36}$
c $\frac{1}{8}$ d 1
4 a i 320 ii 5120
b 80. This is the value of N when $t = 0$.

5 a $\frac{1}{25}$ b $\frac{1}{9}$ c $\frac{1}{8}$ d $\frac{1}{25}$
6 a 2 b 7 c 4 d 1
e 14 f 10 g 2 h 3
7 $11^{\frac{5}{6}}$

Exercise 16D

1 a 16 b 100 c 27 d 25
2 a 27 b 9
3 64
4 a $11^{\frac{2}{3}}$ b $2^{\frac{3}{4}}$ c $7^{\frac{3}{5}}$ d $13^{\frac{1}{2}}$
5 a $\frac{1}{5}$ b $\frac{1}{3}$ c $\frac{1}{4}$ d $\frac{1}{25}$
e $\frac{1}{8}$ f $\frac{1}{7}$
6 If $a > 1$ then a^n gets bigger as n gets bigger.
So if $n > 0$ then $a^n > a^0 = 1$ and if $n < 1$ then $a^n < a^1 = a$.
So, $1 < a^n < a$. Catelyn is correct.

Exercise 16E

1 a 3.4×10^4 b 2.6×10^3
c 7.4×10^2 d 2×10^5
e 6.03×10^6 f 2.26×10^1
2 a 6×10^{-4} b 3.2×10^{-3}
c 3.09×10^{-3} d 4.45×10^{-1}
e 1×10^{-2} f 9.8×10^{-6}
3 a 20 000 b 6200
c 0.004 54 d 0.207
e 79 000 000 f 0.000 004 551
4 3.2×10^{-2}, 8.66×10^{-1}, 9.31×10^{-1}, 8.04×10^3, 4.8×10^4, 1×10^7
5 5.88×10^7
6 hydrogen, carbon, iron, tungsten, gold, plutonium
7 2.54×10^{-5} m

Exercise 16F

1 a 8×10^8 b 4.8×10^{10}
c 2×10^{10} d 9×10^5
e 5×10^{-5} f 1.86×10^{-6}
g 3.52×10^{13} h 1.12×10^{-7}
2 a 2×10^5 b 4.2×10^3
c 5×10^4 d 4×10^{11}
e 2.1×10^{-5} f 2.5×10^{-9}
g 7×10^{-5} h 2.5×10^7
3 a 6.7×10^4 b 2.05×10^5
c 1.3×10^4 d 8.25×10^5
e 6.98×10^5 f 9.37×10^4
g 1.27×10^5 h 4.064×10^4
4 360 g
5 2×10^{-8} g
6 7.29 kg

Exercise 16G

1 a $\sqrt{18}$ b $\sqrt{10}$ c $\sqrt{30}$
d $\sqrt{110}$ e $\sqrt{72}$ f $\sqrt{90}$
2 a 4 b 10 c 8
d 12 e 10 f 14
3 a 48 b 140 c 600
d 40 e 96 f 100
4 a 5 b 2 c 10
d 2 e 3 f 15
5 a $10\sqrt{6}$ b $42\sqrt{6}$ c $2\sqrt{35}$
d $3\sqrt{3}$ e $10\sqrt{3}$ f $9\sqrt{14}$
6 a 4 b 3 c 75 d 16

Exercise 16H

1 $3\sqrt{5}$
2 a $3\sqrt{2}$ b $5\sqrt{2}$ c $10\sqrt{2}$ d $4\sqrt{2}$
3 a $7\sqrt{3}$ b $5\sqrt{5}$ c $7\sqrt{2}$ d $\sqrt{7}$
4 a $5\sqrt{2}$ b $2\sqrt{3}$ c $\frac{\sqrt{5}}{5}$ d $\frac{\sqrt{10}}{2}$
e $3\sqrt{15}$ f $4\sqrt{3}$ g $\frac{1}{2}$ h $3\sqrt{3}$
5 a She has incorrectly simplified the expression. She cannot incorporate the 12 into the square root, as it was not a square root to start with.
b $4\sqrt{3}$
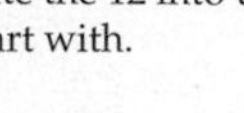
6 $\frac{7\sqrt{5}}{60}$

Exercise 16I

1 a $6 + 5\sqrt{2}$ b $32 + 13\sqrt{2}$
c $17 + 8\sqrt{2}$ d $5 - 4\sqrt{2}$
e $-4 - 5\sqrt{2}$ f $6 + 4\sqrt{2}$
g $44 - 24\sqrt{2}$ h $4 + 22\sqrt{2}$
2 a $9 + 5\sqrt{3}$ b $-9 + \sqrt{5}$
c $9 - 9\sqrt{7}$ d $68 + 44\sqrt{3}$
e $29 - 4\sqrt{7}$ f $28 + 16\sqrt{3}$
3 a −1 b 94
c −8 d 137
e −28 f −36
4 $(\sqrt{13} + \sqrt{11})(\sqrt{13} - \sqrt{11})$
$= 13 - \sqrt{13} \times \sqrt{11} + \sqrt{13} \times \sqrt{11} - 11$
$= 13 - 11$
$= 2$
5 a $p = a^2 + 5b^2$ b $q = 2ab$
6 $135 + 78\sqrt{3}$

Review exercise

1 a 2^8 b 2^3
2 a $\frac{1}{8}$ b $\frac{1}{4}$
3 a 1 b 0.5
4 25
5 a 1.86×10^{10} b 1.2×10^{-8}
6 3×10^{-6} kg
7 Size of all files = 710 + 2410 + 94 200 + 834 000 = 931 320 kb
All of Anselm's files will fit on the memory stick as 931 320 < 1 000 000.
8 a 5^{15} b 125
9 a 3 b 15 c 5 d 36
10 $7\sqrt{3}$
11 a 3 b $\frac{25}{4}$
12 $8\sqrt{5}$
13 a $11 + 5\sqrt{7}$ b $16 + 6\sqrt{7}$ c $40 - 4\sqrt{7}$

Chapter 17

Exercise 17A

1 a 11, 13, 15 b Missing values are: 2, 4, 6, 8
c $2n + 1$
2 a $4n + 1$ b $2n + 6$
c $5n - 1$ d $11n - 2$
e $100n - 25$ f $-5n$
g $-3n + 13$ h $-4n + 23$
i $-10n + 87$ j $-6n + 56$

Exercise 17B

1 a $5n + 3$ b 503
2 a $2n + 19$ b 39 mm
3 a 99 b 88 c 141
4 108
5 −56
6 113
7 a 151 b 149 c 149 d 148
8 11th term
9 a yes b no
10 a 15th term b all the terms are odd
11 a −9 b −1 c −8 d −5

Exercise 17C

1 a i 28 matchsticks ii $3n + 1$
b i 19 matchsticks ii $2n + 1$
c i 26 matchsticks ii $3n - 1$
2 a $2n + 2$ b 11 tables
3 a $5n + 1$ b 17
4 Pattern 11
5 a 10 cm²
b i n ii $(n + 1)$ iii $\frac{1}{2}n(n + 1)$
c 12 cm

Exercise 17D

1 a 25, 36, 49   b 24, 33, 44
c 32, 45, 60   d 122, 132, 144

2 a 1, 4, 9; 10th term = 100   b 0, 3, 8; 10th term = 99
c 2, 8, 18; 10th term = 200   d 13, 16, 21; 10th term = 112

3 5, 9

4 a 37, 50, 65   b $n^2 + 1$

5 a $n^2 - 1$ b $n^2 + 5$
c $n^2 - 5$ d $2n^2$
e $\frac{1}{2}n^2$ f $10n^2$

Exercise 17E

1 If the integers are consecutive then one is odd and one is even.
odd × even = even

2 Odd
Three of the six numbers are odd and three even:
odd + odd + odd + even + even + even
Paring the numbers gives:
(odd + odd) + (odd + even) + (even + even)
= even + odd + even
= odd

3 $x + y$ = odd + even = odd
$x - y$ = odd − even = odd
$(x + y)(x - y)$ = odd × odd = odd

4 All prime numbers except 2 are odd.
odd^2 = odd
odd + 1 = even

5 If multiples of 3 are consecutive then one must be odd and the other even.
odd + even = odd

6 Every other integer is a multiple of 2.
Every third integer is a multiple of 3.
Therefore one of the integers is a multiple of 2 and one a multiple of 3.
Multiple of 2 × multiple of 3 = multiple of 6
Therefore the product of three consecutive integers must be a multiple of 6.

Exercise 17F

1 a $-3 + 2 = -1$ b $4 \times 0.25 = 1$
c $0.7 + 0.3 = 1$

2 7.55

3 a False: $n = 1, n^2 + 1 = 2$ b True
c False: $(-1)^3 = -1$ d False: 9
e True

4 $x = 2$ and $y = 3$
$x^2 + y^2 = 4 + 9 = 13$

5 $(-1)^2 = 1$
$(-1)^3 = -1$

6 $x = 1$
$\sqrt{1} = 1$
$1^2 = 1$

7 $n = 4$
$n^2 + n + 1 = 16 + 4 + 1 = 21$

Exercise 17G

1 a Let the two odd numbers be $2n + 1$ and $2m + 1$
$2n + 1 + 2m + 1 = 2m + 2n + 2$
$= 2(m + n + 1)$
$= 2k$
= even

b Let the consecutive multiples of 3 be $3n$ and $3n + 3$
$3n + 3n + 3 = 6n + 3$
$= 3(2n + 1)$
= odd × odd
= odd

c Let the consecutive squares be n^2 and $(n + 1)^2$
$(n + 1)^2 - n^2 = n^2 + 2n + 1 - n^2$
$= 2n + 1$
= odd

d Let the even and odd numbers be $2n$ and $2m + 1$
$2n(2m + 1) = 4nm + 2n$
$= 2(2nm + n)$
= even

2 LHS $= 4n^2 + 2n + 2n + 1$
$= 4n^2 + 4n + 1$
RHS $= n^2 + 3n + n + 3 + 3n^2 - 2$
$= 4n^2 + 4n + 1$
= LHS

3 $(2n + 1)^2 = 4n^2 + 4n + 1$
$= 2(2n^2 + 2n) + 1$
$= 2k + 1$
= odd

4 a $1 + 2 + 3 + 4 + 5 + 6 + 7 + 8 + 9 + 10 = 55$
When $a = 1$, $10a + 45 = 10 + 45 = 55$
b $a, a + 1, a + 2, a + 3, a + 4, a + 5, a + 6, a + 7, a + 8, a + 9$
c Sum of the algebraic terms $= 10a + 45$

5 a Sequence C is $-1 + 8 = 7, 3 + 11 = 14, 7 + 14 = 21, 11 + 17 = 28, \ldots$
b The nth term of sequence A is $4n - 5$
The nth term of sequence B is $3n + 5$
The nth term of sequence C is $4n - 5 + 3n + 5 = 7n$
Therefore the terms in sequence C are multiples of 7.

6 Let the two consecutive terms be $\frac{1}{2}a(a + 1)$ and $\frac{1}{2}(a + 1)(a + 2)$
$\frac{1}{2}a(a + 1) + \frac{1}{2}(a + 1)(a + 2)$
$= \frac{1}{2}a^2 + \frac{1}{2}a + \frac{1}{2}a^2 + \frac{1}{2}a + a + 1$
$= a^2 + 2a + 1$
$= (a + 1)(a + 1)$
$= (a + 1)^2$

Review exercise

1 Double any integer results in an even number. even − 1 = odd
Aimee will always win!

2 $10n - 18$

3 a Missing numbers are: b 8th
4, 8
15
24
35
120
$n(n + 2)$

4 Any example where n or q is 2.

5 a 23, 34, 47   b $n^2 - 2$

6 There is a counter example: $4^3 + 1 = 65$

7 LHS $= 4n^2 - 6n - 6n + 9$
$= 4n^2 - 12n + 9$
RHS $= n^2 + 2n + 2n + 4 + 3n^2 - n - 15n + 5$
$= 4n^2 - 12n + 9$
= LHS

8 (odd)^2 + odd = (odd × odd) + odd
= odd + odd
= even

9 Let the two consecutive even numbers be $2n$ and $2n + 2$, where n is an integer.
$2n(2n + 2) = 4n^2 + 4n$
$= 4(n^2 + n)$
If n odd,
$n^2 + n$ = (odd)^2 + odd
= odd + odd
= even
If n even,
$n^2 + n$ = (even)^2 + even
= even + even
= even
So, $n^2 + n$ is always even and we can write it as $2k$, where k is an integer.
So, $2n(2n + 2) = 4(n^2 + n)$
$= 4(2k)$
$= 8k$, which is a multiple of 8.

10 $(n + 3)(2n + 1) + (n - 2)(2n + 1)$
$= (2n^2 + 6n + n + 3) + (2n^2 - 4n + n - 2)$
$= 2n^2 + 7n + 3 + 2n^2 - 3n - 2$
$= 4n^2 + 4n + 1$
$= 4(n^2 + n) + 1$
From Q9: $n^2 + n$ is always even and we can write it as $2k$, where k is an integer.
So, $(n + 3)(2n + 1) + (n - 2)(2n + 1) = 4(n^2 + n) + 1$
$= 4(2k) + 1$
$= 8k + 1$

Chapter 18

Number skills: percentage calculations

1 a 0.57 b 0.8% c $\frac{13}{20}$ d 2.5%
2 0.3, 32%, $\frac{1}{3}$
3 0.36, 37%, $\frac{3}{8}$, 0.38, $\frac{2}{5}$
4 Ben (Ava: 75%, Ben got 16 right)
5

Percentage	Fraction	Decimal
35%	$\frac{7}{20}$	**0.35**
5%	$\frac{1}{20}$	**0.05**
12.5%	$\frac{1}{8}$	0.125

6 a 2.24 kg b £600 c 1.8 litres d £120
7 £5796
8 24
9 a 60% b 15% c 80%
d 18% e 30% f 95%
10 a 24% b 28% c 47% d 49%

Exercise 18A

1 a £104.50 b 44.1 mm c £144
d £321 e 69 mm f 44 miles
2 £10.80
3 18.7 m
4 £296.40
5 £5760
6 £29.75
7 131 000 (rounded to the nearest 1000)
8 Less: £198
9 *Dumbo's DIY*: £79.20, *Suit you, Sir*: £78. *Suit you, Sir* is best buy.

Exercise 18B

1 £18 212.50
2 £115.50
3 £329
4 a £99.64 b £24.91

Exercise 18C

1 Pair of trousers: 40% profit, House: 5% loss,
Barbecue set: 50% profit, Car: 15% loss, Television: 30% profit
2 16.67%
3 12.5%
4 800%
5 Sarah: 87.5%, George: 66.67%, Sarah makes the bigger percentage profit.
6 Gardens-R-Us: 40%, Yuppies: 37.5%, Gardens-R-Us makes the bigger percentage profit.

Exercise 18D

1 £540.80
2 £399.30
3 a £25.20 b £315.25
4 72
5 £5120
6 4050
7 3 375 000
8 Hiroshi: £466.56, Amy: £459.80, Hiroshi has more.
9 21%
10 Choose account A as it is 6.09% per annum.

Exercise 18E

1 a £19.50 b £39.00
c £21.00 d £9.90
2 £875
3 £9600
4 £26
5 £850
6 Full price: £450 (the reduction is 10% of this figure).
7 110 MD
8 a 49 500 b 61 100 (to the nearest 100)
9 £5000

Review exercise

1 £36
2 138
3 £423
4 81 480
5 30%
6 40%
7 172 800
8 £4096
9 125
10 a 720 b 500 c 417
11 £800
12 £2000

Chapter 19

Exercise 19A

1 a $C(2, 1)$, $D(8, 1)$, $E(5, 8)$, $F(5, 2)$, $G(1, 2)$, $H(4, 5)$, $L(-5, 4)$, $M(2, 4)$, $P(-5, -3)$, $Q(-1, 1)$
b CD: (5, 1); EF: (5, 5); GH: (2.5, 3.5); LM: (−1.5, 4); PQ: (−3, −1)
2 a (4.5, 1) b (3.5, 6) c (0.5, −0.5) d (2, −5)
3 a (1.5, −0.5) b (1.5, −0.5) c $ABCD$ is a square
4 a (1, 0) b (3, 2) c (−4, 2) d (1, 7)

Exercise 19B

1
2

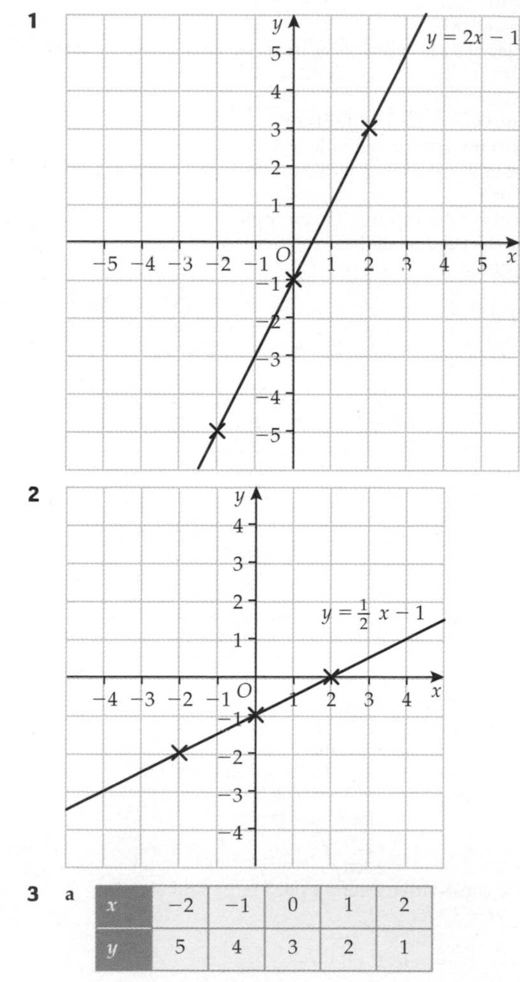

3 a

x	−2	−1	0	1	2
y	5	4	3	2	1

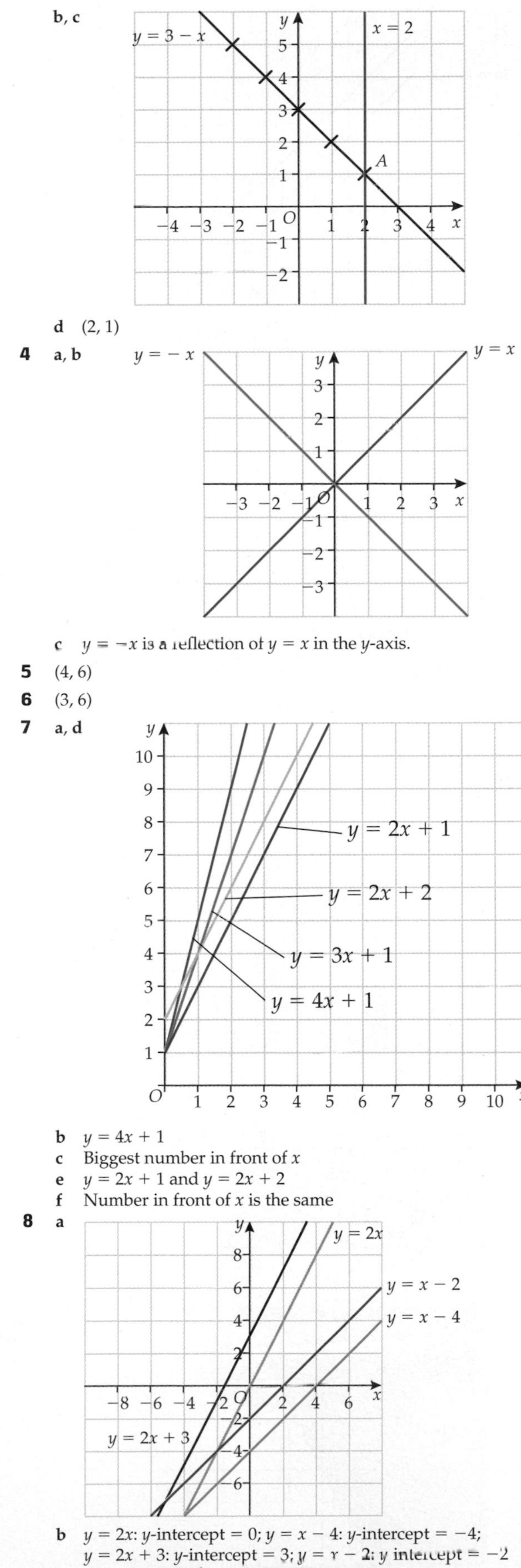

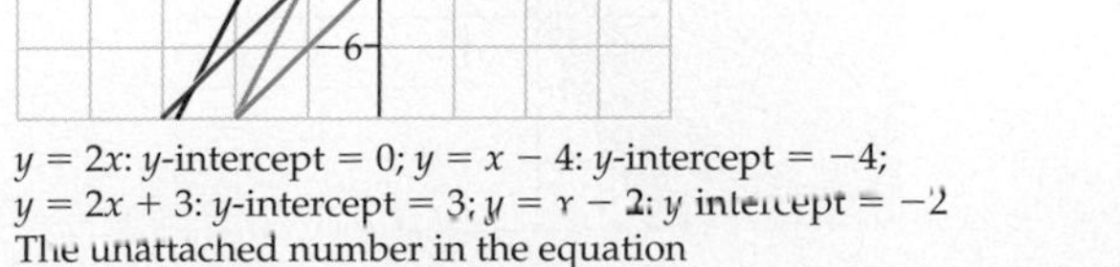
b $y = 2x$: y-intercept $= 0$; $y = x - 4$: y-intercept $= -4$;
$y = 2x + 3$: y-intercept $= 3$; $y = x - 2$: y-intercept $= -2$
c The unattached number in the equation

9 a

x	0	3
y	6	0

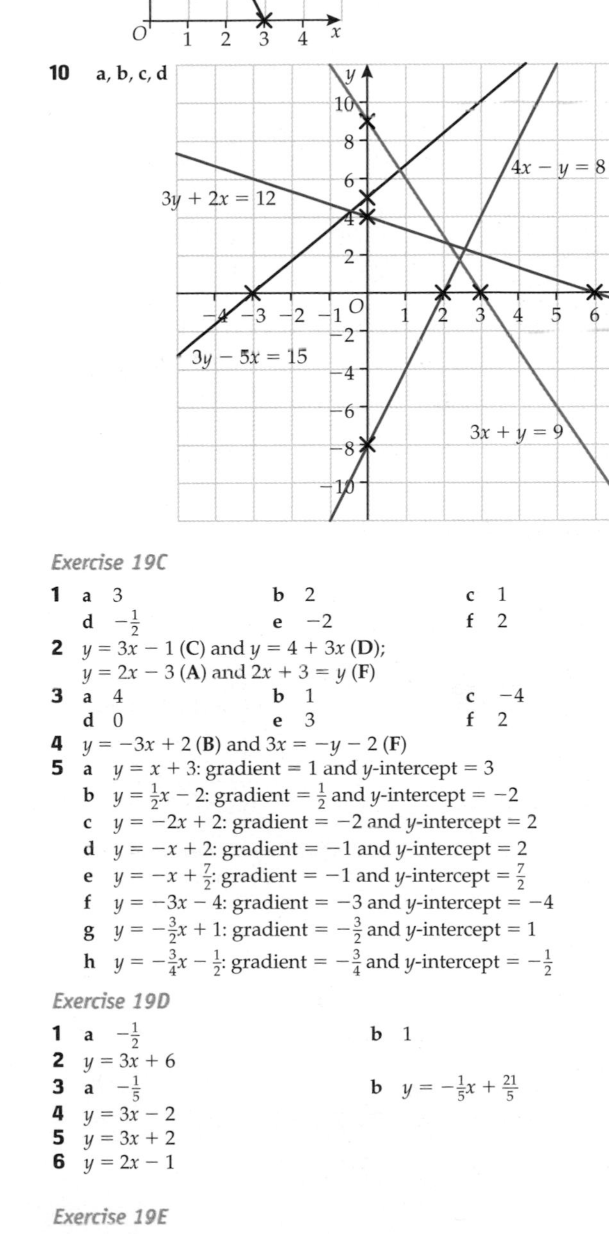

Exercise 19C

1 a 3 b 2 c 1
d $-\frac{1}{2}$ e -2 f 2

2 $y = 3x - 1$ (**C**) and $y = 4 + 3x$ (**D**);
$y = 2x - 3$ (**A**) and $2x + 3 = y$ (**F**)

3 a 4 b 1 c -4
d 0 e 3 f 2

4 $y = -3x + 2$ (**B**) and $3x = -y - 2$ (**F**)

5 a $y = x + 3$: gradient $= 1$ and y-intercept $= 3$
b $y = \frac{1}{2}x - 2$: gradient $= \frac{1}{2}$ and y-intercept $= -2$
c $y = -2x + 2$: gradient $= -2$ and y-intercept $= 2$
d $y = -x + 2$: gradient $= -1$ and y-intercept $= 2$
e $y = -x + \frac{7}{2}$: gradient $= -1$ and y-intercept $= \frac{7}{2}$
f $y = -3x - 4$: gradient $= -3$ and y-intercept $= -4$
g $y = -\frac{3}{2}x + 1$: gradient $= -\frac{3}{2}$ and y-intercept $= 1$
h $y = -\frac{3}{4}x - \frac{1}{2}$: gradient $= -\frac{3}{4}$ and y-intercept $= -\frac{1}{2}$

Exercise 19D

1 a $-\frac{1}{2}$ b 1

2 $y = 3x + 6$

3 a $-\frac{1}{5}$ b $y = -\frac{1}{5}x + \frac{21}{5}$

4 $y = 3x - 2$

5 $y = 3x + 2$

6 $y = 2x - 1$

Exercise 19E

1 a $-\frac{1}{2}$ b $\frac{1}{3}$ c -3
d 4 e $-\frac{2}{3}$

2 $y = -\frac{1}{4}x + 7$

3 $y = -\frac{1}{2}x + 3$

4 $y = -\frac{1}{3}x + 3$

5 $x = 1$

6 $y = -\frac{1}{3}x - 1$

Exercise 19F

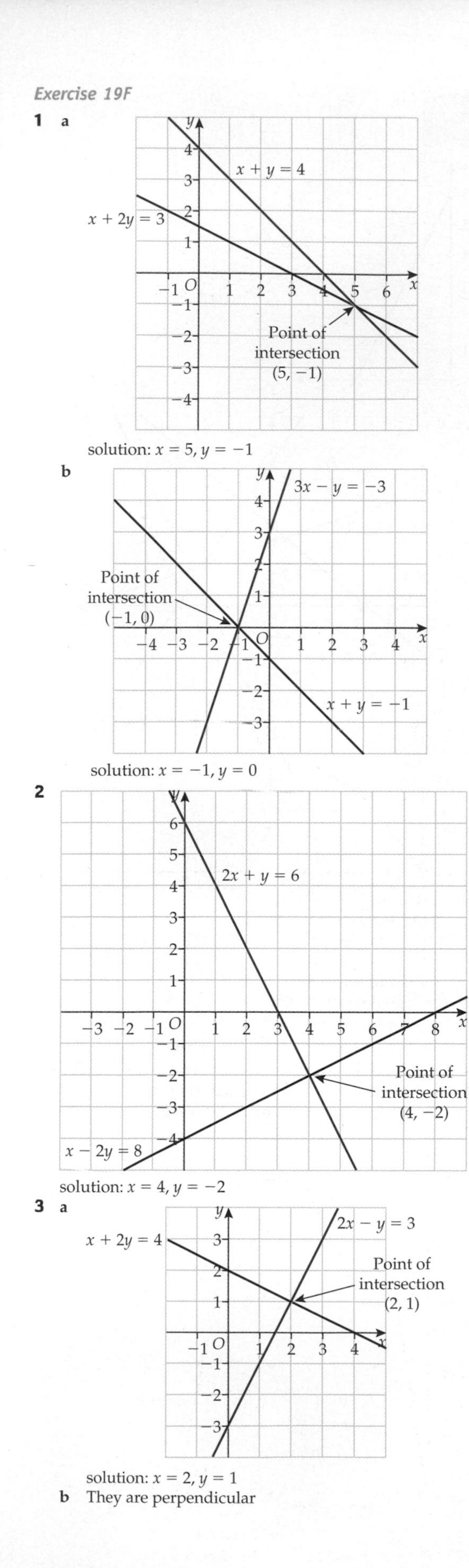

1 a solution: $x = 5, y = -1$

b solution: $x = -1, y = 0$

2 solution: $x = 4, y = -2$

3 a solution: $x = 2, y = 1$

b They are perpendicular

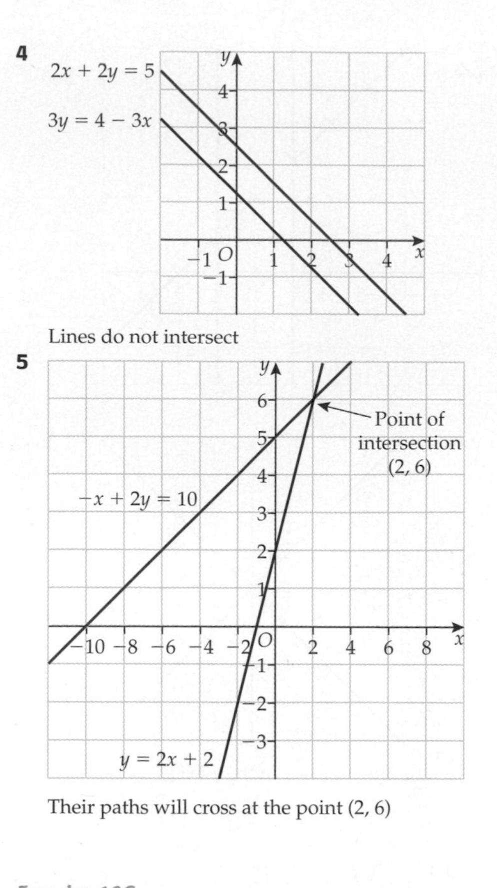

4 Lines do not intersect

5 Their paths will cross at the point (2, 6)

Exercise 19G

1 a $y < 3$ **b** $x \geqslant -2$

c $y < x$ **d** $y \geqslant x + 2$

2

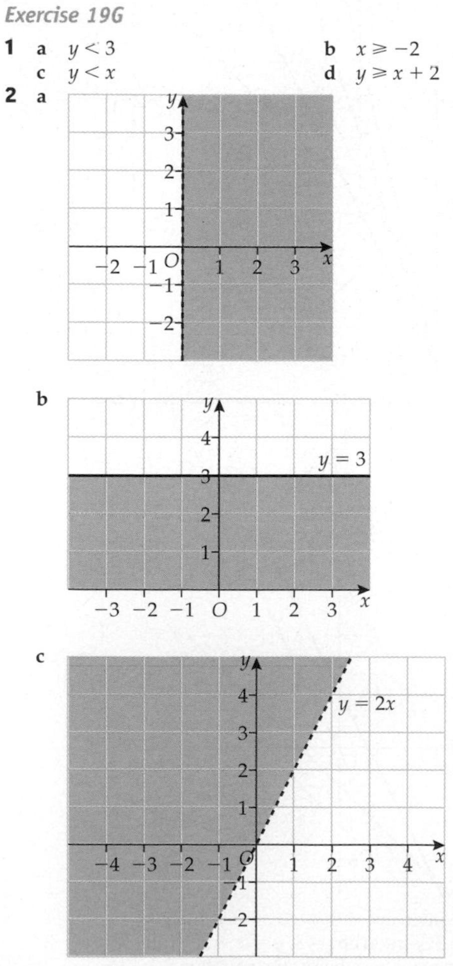

d

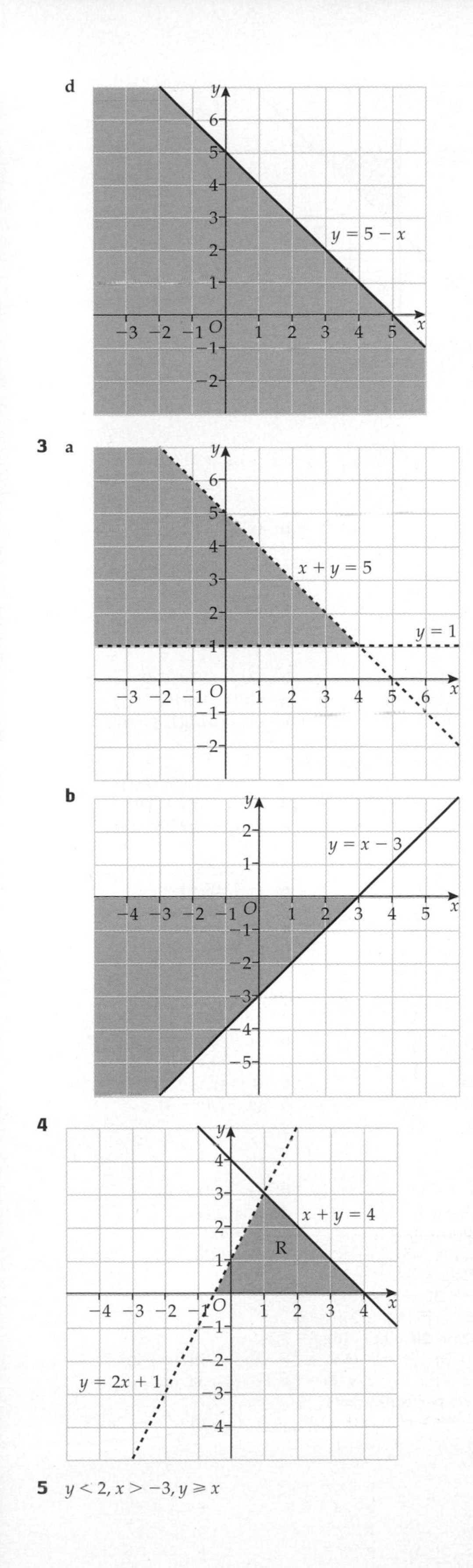

5 $y < 2, x > -3, y \geqslant x$

6 a

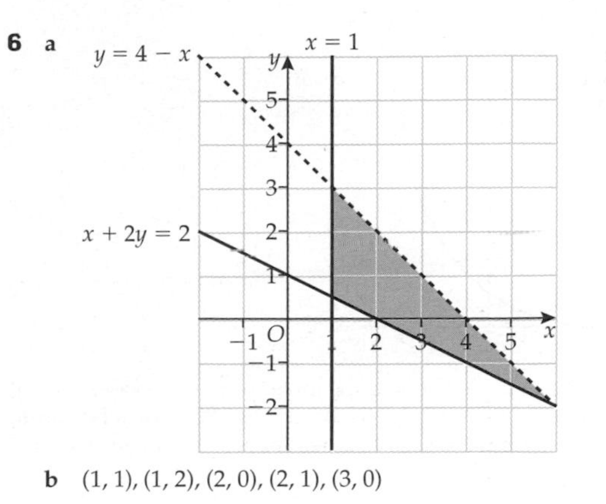

b (1, 1), (1, 2), (2, 0), (2, 1), (3, 0)

Exercise 19H

1 a

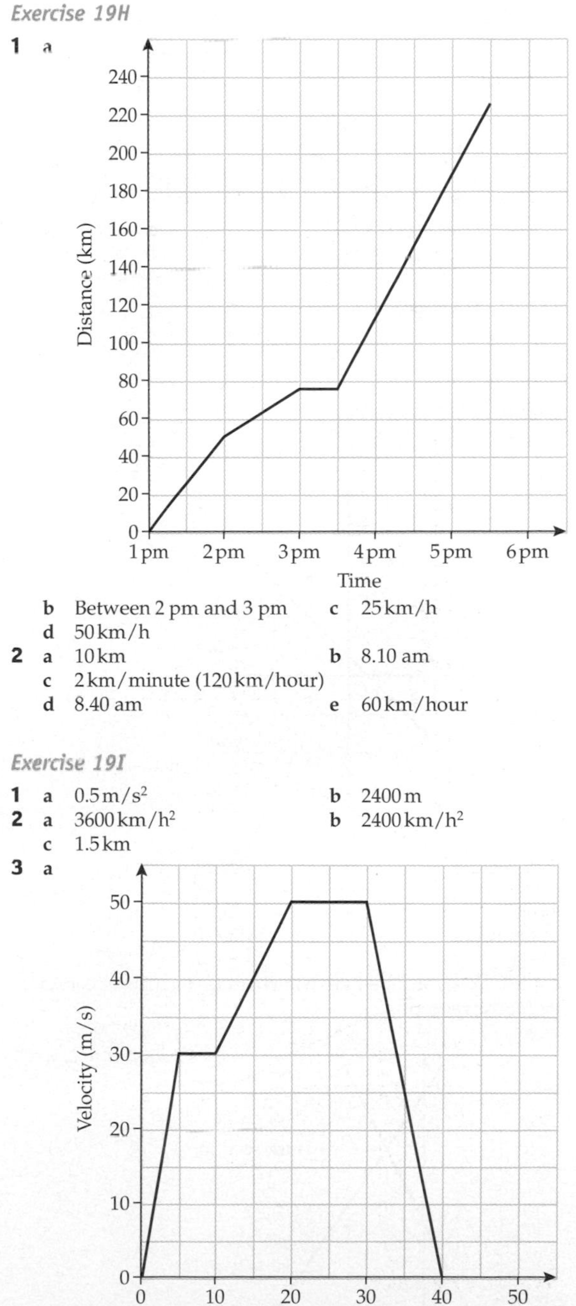

b Between 2 pm and 3 pm **c** 25 km/h
d 50 km/h

2 a 10 km **b** 8.10 am
c 2 km/minute (120 km/hour)
d 8.40 am **e** 60 km/hour

Exercise 19I

1 a $0.5\,\text{m/s}^2$ **b** 2400 m

2 a $3600\,\text{km/h}^2$ **b** $2400\,\text{km/h}^2$
c 1.5 km

3 a

Velocity (m/s)

Time (seconds)

b i $6\,\text{m/s}^2$ **ii** 625 m **iii** 5 m/s

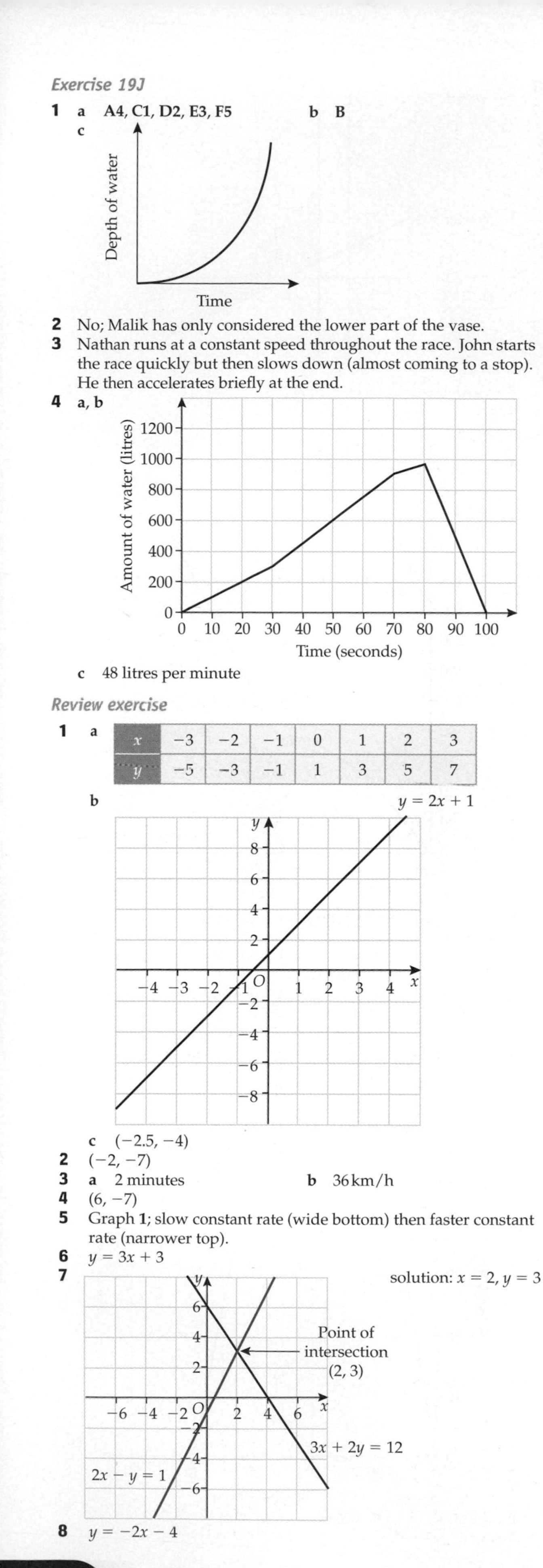

Exercise 19J

1 a A4, C1, D2, E3, F5   b B

c

2 No; Malik has only considered the lower part of the vase.

3 Nathan runs at a constant speed throughout the race. John starts the race quickly but then slows down (almost coming to a stop). He then accelerates briefly at the end.

4 a, b

c 48 litres per minute

Review exercise

1 a

x	−3	−2	−1	0	1	2	3
y	−5	−3	−1	1	3	5	7

b

c $(-2.5, -4)$

2 $(-2, -7)$

3 a 2 minutes   b 36 km/h

4 $(6, -7)$

5 Graph 1; slow constant rate (wide bottom) then faster constant rate (narrower top).

6 $y = 3x + 3$

7 solution: $x = 2$, $y = 3$

8 $y = -2x - 4$

9

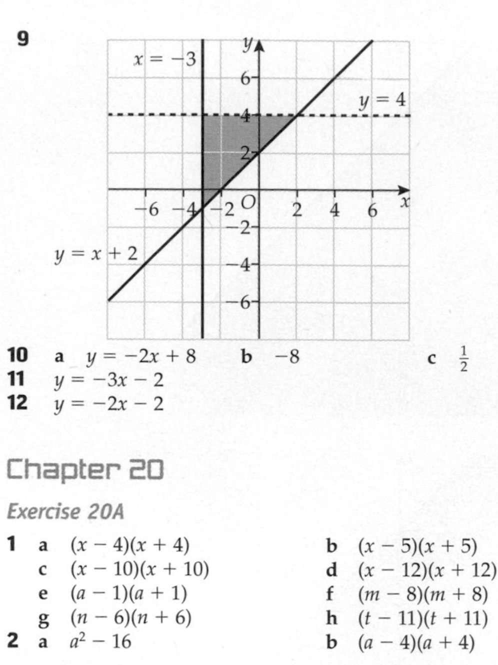

10 a $y = -2x + 8$ b -8 c $\frac{1}{2}$

11 $y = -3x - 2$

12 $y = -2x - 2$

Chapter 20

Exercise 20A

1 a $(x-4)(x+4)$ b $(x-5)(x+5)$
c $(x-10)(x+10)$ d $(x-12)(x+12)$
e $(a-1)(a+1)$ f $(m-8)(m+8)$
g $(n-6)(n+6)$ h $(t-11)(t+11)$

2 a $a^2 - 16$ b $(a-4)(a+4)$

Exercise 20B

1 a $(2x-5)(2x+5)$ b $(3a-6)(3a+6)$
c $(4m-1)(4m+1)$ d $(10t-11)(10t+11)$
e $(13z-2)(13z+2)$ f $(15q-12)(15q+12)$

2 $72m^2 - 50 = 2(\mathbf{36m^2} - \mathbf{25})$
$= 2(\mathbf{6m} - \mathbf{5})(\mathbf{6m} + \mathbf{5})$

3 a $2(5x+10)(5x-10)$ b $3(3m-1)(3m+1)$
c $5(4t-3)(4t+3)$ d $2(a-3b)(a+3b)$
e $3(h-5k)(h+5k)$ f $6(10x-y)(10x+y)$

Exercise 20C

1 a $(x+2)(x+3)$ b $(x+4)(x+2)$
c $(z+5)(z+1)$ d $(a+10)(a+1)$
e $(n+5)(n+3)$ f $(f+6)^2$
g $(m+6)(m+2)$ h $(x+12)(x+2)$
i $(b+6)(b+5)$

Exercise 20D

1 a $(x-2)(x-3)$ b $(x-8)(x-1)$
c $(x-3)(x-4)$ d $(a-6)(a-3)$
e $(n-5)^2$ f $(f-4)^2$
g $(x-3)(x-10)$ h $(b-7)(b-4)$
i $(p-6)(p-4)$

Exercise 20E

1 a $(x+6)(x-2)$ b $(x-5)(x+4)$
c $(z-5)(z+3)$ d $(a+7)(a-1)$
e $(n+8)(n-2)$ f $(f-6)(f+5)$
g $(m+6)(m-5)$ h $(t-12)(t+6)$
i $(y+24)(y-5)$

2 a $t^2 + 7r - \mathbf{30} = (t + \mathbf{10})(t - \mathbf{3})$
b $m^2 - \mathbf{8r} + 15 = (t - \mathbf{3})(t - 5)$
c $q^2 - 12q + \mathbf{20} = (q - \mathbf{10})(q - 2)$

3 a i $(x+3)^2$ ii $(x-4)^2$ iii $(x+2)^2$
iv $(x-7)^2$ v $(x-5)^2$ vi $(x+8)^2$
b They are perfect squares
c i $(x+m)^2 = x^2 + 2mx + m^2$ ii $(x-n)^2 = x^2 - 2nx + n^2$

Exercise 20F

1 a $r = \pm 13$ b $x = \pm 3$ c $t = \pm 3$
d $y = \pm 1$ e $m = \pm 5$ f $p = \pm 10$

2 a $x = \pm 4$ b $y = \pm 2$ c $r = \pm 7$
d $t = \pm 3$ e $y = \pm 1$ f $x = \pm 8$

Exercise 20G

1 a $x = -3$ or 1 b $r = 10$ or 4

2 a $x = -7$ or 5 b $t = \pm 4$ c $t = \pm 6$
d $t = \pm 3$ e $x = -28$ or 22 f $y = 12$ or -8

3 a $3x$ b $3x^2$ c $3x^2 = 1200$
d $x = \pm 20$ e width = 20 m, length = 60 m
4 You cannot find the square root of a negative number
5 13 mm

Exercise 20H

1 a $x = 0$ or $x = -7$ b $t = 0$ or $t = 5$
c $x = 0$ or $x = -2$ d $y = 0$ or $y = 5$
e $w = 0$ or $w = 3$ f $y = 0$ or $y = 0.25$
g $a = 0$ or $a = 1$ h $t = 0$ or $t = \frac{1}{6}$
i $r = 0$ or $r = \frac{2}{9}$
2 a $x = 0$ or 4 b $t = 0$ or -0.25
c $m = 0$ or 2 d $g = 0$ or -0.5
e $f = 0$ or 2.5 f $w = 0$ or 3.5

Exercise 20I

1 a $x = -3$ or -1 b $x = 3$ or -2
c $x = 4$ or 2 d $x = 3$ or -4
e $x = 5$ or -4 f $x = -1$
g $z = 4$ or -1 h $q = 3$ or -5
i $w = 2$ j $t = 7$ or -1
k $p = -3$ l $x = 5$
2 a $x + 3$ b $x(x + 3)$
c $x(x + 3) = 54$
$x^2 + 3x - 54 = 0$
$(x + 9)(x - 6) = 0$
$x = -9$ or 6
d x must equal 6 since your age cannot be negative
3 a $t + 3$ b $t(t + 3)$
c $t(t + 3) = 40$
$t = -8$ or 5
Therefore $t = 5$
4 a
x m
$x + 4$ m
b $x(x + 4) = 165$
$x^2 + 4x = 165$
$x^2 + 4x - 165 = 0$
$(x + 15)(x - 11) = 0$
Therefore $x = 11$ m
Field is 11 m by 15 m
5 $x^2 + 5x = 24$
$(x + 8)(x - 3) = 0$
Therefore $x = -8$
6 $x^2 - 6x = 27$
$(x - 9)(x + 3) = 0$
Therefore $x = 9$

Exercise 20J

1 $(2x + 3)(x + 1)$ **2** $(3x + 2)(x + 4)$
3 $(5x + 2)(x + 2)$ **4** $(7x + 5)(x + 3)$
5 $(5x - 1)(x + 4)$ **6** $(3x + 2)(x - 2)$
7 $(11x - 2)(x - 1)$ **8** $(2x - 1)(x - 2)$
9 $(3x - 4)(x - 5)$ **10** $(5x + 1)(x - 8)$
11 $(2x - 8)(x - 3)$ **12** $(7x + 6)(x - 2)$

Exercise 20K

1 a $(8x + 1)(x + 2)$ b $(2x + 3)(2x + 1)$
c $(2x + 5)(3x + 1)$ d $(3x + 2)(2x + 2)$
e $(4x + 4)(2x + 3)$ f $(10x + 4)(3x + 4)$
2 a $2(x + 1)(x + 3)$ b $3(x + 2)(x + 5)$
c $3(3x + 4)(2x + 5)$

Exercise 20L

1 a $(5x + 3)(2x - 1)$ b $(3x + 2)(4x - 3)$
c $(2x + 3)(2x - 2)$ d $(5x - 4)(4x + 7)$
e $(6x - 8)(5x - 2)$
2 a $5(x - 1)(x + 2)$ b $7(2x - 3)(x + 4)$
c $3(5x + 1)(x - 5)$ d $4(7x - 1)(x - 3)$
e $4(3x - 1)(2x + 1)$ f $10(5x + 3)(x - 2)$
g $6(2x - 1)(3x - 2)$

Exercise 20M

1 a $a = \frac{1}{2}$ or -3 b $b = -\frac{2}{3}$ or -1
c $c = \frac{5}{4}$ or -1 d $d = -\frac{2}{5}$ or 2
e $e = 2$ or $\frac{2}{3}$ f $f = 2$ or $-\frac{1}{2}$
g $g = -\frac{2}{3}$ or $-2\frac{1}{2}$ h $h = -1$
i $i = -\frac{4}{7}$ or 1
2 a $x = 1$ or 3 b $x = \frac{2}{3}$ or $\frac{5}{3}$
c $x = -\frac{9}{5}$ or $\frac{1}{2}$ d $x = 2$ or $-\frac{4}{3}$
e $x = -1$ f $x = 4$ or -1
3 a $3x(2x + 1)$ b $3x(2x + 1) = 108$
$6x^2 + 3x - 108 = 0$
$(3x - 12)(2x + 9) = 0$
$x = -\frac{9}{2}$ or 4
Therefore $x = 4$
c Perimeter = 42 cm
4 4 or 0
5 a i $4x$ ii $x - 1$ iii $4x - 1$
b $(x - 1)(4x - 1) = 351$
Leading to $(4x + 35)(x - 10) = 0$
Giving $x = 10$
Amelia is 9
c Yvette is 39

Exercise 20N

1 $x = \frac{-3 \pm 3\sqrt{5}}{2}$ **2** $x = \frac{-5 \pm \sqrt{73}}{2}$
3 $x = -5$ or 1 **4** $x = -3 \pm \sqrt{7}$
5 $x = 2$ or $-1\frac{1}{3}$ **6** $x = -2 \pm \sqrt{14}$
7 $y = \frac{-6 \pm 2\sqrt{14}}{5}$ **8** $r = 0.5$ or $\frac{1}{6}$
9 $t = \frac{1 \pm \sqrt{57}}{7}$ **10** $g = \frac{-7 \pm \sqrt{13}}{6}$

Exercise 20O

1 a $x = 2 \pm \sqrt{5}$ b $x = 6 \pm 2\sqrt{5}$
c $x = 4 \pm \sqrt{22}$ d $x = -3 \pm \sqrt{10}$
e $x = 1 \pm \sqrt{5}$ f $x = -4 \pm \sqrt{14}$
2 You cannot find a square root of a negative number
3 30 cm by 50 cm
4 No – you could have factorised the expression.

Exercise 20P

1 a 2 solutions b no solutions
c 1 solution d no solutions
e 2 solutions f no solutions
g 2 solutions h 1 solution

Exercise 20Q

1 a $(x + 3)^2 - 6$ b $(x + 1)^2 + 6$
c $(x - 4)^2 - 11$ d $(x - 6)^2 - 24$
e $(x - 2)^2 - 11$ f $(x - 5)^2 - 26$
2 a $(x + 5)^2 + 7$ $p = 5, q = 7$
b $(x + 1)^2 + 1$ $p = 1, q = 1$
c $(x - 2)^2 + 16$ $p = -2, q = 16$
d $(x - 7)^2 - 39$ $p = -7, q = -39$
e $(x - 3)^2 - 12$ $p = -3, q = -12$
f $(x - 2)^2 - 6$ $p = -2, q = -6$

Exercise 20R

1 a $x = -1$ or -9 b $x = 2$ or -4
c $x = 4 \pm \sqrt{6}$ d $x = 6 \pm 2\sqrt{5}$
e $x = 2 \pm 2\sqrt{2}$ f $x = -7$ or 1
2 a $x = 1$ or -9 b $x = -2 \pm 2\sqrt{3}$
c $x = 1 \pm \sqrt{2}$ d $x = 4 \pm \sqrt{6}$
e $x = 10 \pm 5\sqrt{2}$ f $x = 7 \pm 2\sqrt{2}$
3 a $x = 3 \pm \sqrt{5}$ b $x = -3 \pm \sqrt{11}$
c $x = 6$ or -2 d $x = -3 \pm 4\sqrt{2}$
e $x = -3 \pm \sqrt{11}$ f $x = -\frac{1}{2} \pm \sqrt{5\frac{1}{4}}$

Review exercise

1 $(x + 5)(x + 1)$
2 $x^2 - 3x = 108$
$x = 12$ or -9
3 $x^2 + 5x = 24$
$x = 3$ or -8
4 a $(2x + 1)(x - 8)$ b $x = -\frac{1}{2}$ or 8
5 $(3y - 1)(2y + 5)$
6 $\frac{3 \pm \sqrt{7}}{2}$
7 a $(2x - 5)(3x + 2)$ b $x = 2\frac{1}{2}$ or $-\frac{2}{3}$
8 $8\frac{1}{2}$ m by 20 m
9 2 m by 8 m
10 a $m = 2, n = 10$ b $-2 \pm \sqrt{10}$
11 a 2 roots b no roots c 1 root
12 a $(x - 2)^2 - 2$ b $2 \pm \sqrt{2}$

Chapter 21

Exercise 21A

1 $(-2, -1), (3, 4)$
2 $(-1, 3), (4, 8)$
3 $(1, 4), (5, 20)$
4 $(-2, 3), (1.5, -0.5)$
5 $(-1, -3)$
6 $(-2.5, -8), (4, 5)$
7 $(-2, -8), (6, 32)$
8 $(-3, 1), (9, 5)$
9 a $\left(\frac{5}{2}, 2\right)$ b Sketch 2 (one solution)
10 a $(-5, 2), (-3, 0)$ b Sketch 1 (two solutions)

Exercise 21B

1 $\frac{x + 2}{3}$
2 cannot be simplified
3 $\frac{2x}{x - 5}$
4 $\frac{2a}{3b}$
5 cannot be simplified
6 cannot be simplified
7 $\frac{b + 2}{a}$
8 $\frac{2}{3}$
9 cannot be simplified
10 $\frac{1}{x + 2}$
11 $\frac{x + 6}{x - 9}$
12 $\frac{y - 5}{y - 3}$
13 $\frac{x + 6}{x - 2}$
14 $\frac{m + 1}{5m + 3}$
15 $\frac{2x + 3}{3x + 5}$

Exercise 21C

1 $\frac{a^2}{12}$
2 $\frac{5b}{2a}$
3 qr
4 $\frac{z}{y}$
5 $\frac{y(y + 2)}{6}$
6 $\frac{1}{12}$
7 $4ab$
8 $\frac{x + 5}{4}$
9 $\frac{1}{4}$
10 $\frac{2(m - 3)}{(m + 2)}$
11 $\frac{a(a + 3)}{3(a - 1)}$
12 $\frac{2(y + 4)}{(y + 1)(y - 5)}$

Exercise 21D

1 $\frac{13x}{15}$
2 $\frac{3b + 4a}{12ab}$
3 $\frac{10y - 3x}{2xy}$
4 $\frac{7x + 11}{12}$
5 $\frac{13x + 6}{20}$
6 $\frac{x + 16}{6}$
7 $\frac{1 - 2y}{15}$
8 $\frac{9x - 8}{12x^2}$
9 $\frac{12y + 15}{10xy}$
10 $\frac{3x + 5}{(x + 1)(x + 2)}$
11 $\frac{7x + 13}{(x + 3)(x - 1)}$
12 $\frac{2x + 5}{(x + 4)(x + 3)}$

Exercise 21E

1 $x = 7$
2 $x = 3$
3 $x = 23$
4 $x = 9$
5 $x = 1$
6 $x = -8$
7 $x = 0.5$ or $x = 4$
8 $x = -1$ or $x = 8$
9 $x = -0.5$ or $x = 5$
10 $x = 2$ or $x = 6$
11 $x = -7$ or $x = 2$
12 $x = -\frac{1}{12}$ or $x = 2$

Exercise 21F

1 $y = \frac{d - b}{2m}$
2 $y = \frac{4}{p - q}$
3 $y = \frac{w + x}{a - b}$
4 $y = \frac{kb - wa}{w - k}$
5 $y = \frac{md + ch}{ha - m}$
6 $y = \frac{h(1 - k)}{k + 1}$
7 $y = \frac{2d + 3e}{d - e}$
8 $y = \frac{a}{m^2 - 2}$
9 $y = \frac{4w^2m}{3 + 4w^2}$
10 $y = \frac{gT^2}{4k^2}$
11 $y = \pm\sqrt{\frac{f - 2e}{k + h}}$
12 $y = \pm\sqrt{\frac{hw - ma}{2m + 3h}}$

Review exercise

1 $b = \frac{3a - 7}{5}$
2 $(2.5, 2)$
3 $x = 0, y = -5$ or $x = 4, y = 3$
4 $\frac{x - 4}{3x - 2}$
5 $\frac{2x + 3}{x + 6}$
6 $x = \pm 5$
7 $-\frac{1}{3}$ or $-0.\dot{3}$
8 $x = \frac{tka}{m + tk}$
9 $x = \frac{a}{c^2} - b$

Chapter 22

Number skills: revision exercise 1

1 a $\frac{12}{24}$ b $\frac{18}{24}$ c $\frac{16}{24}$
d $\frac{20}{24}$ e $\frac{18}{24}$ f $\frac{21}{24}$
2 a $\frac{2}{3}$ b $\frac{1}{5}$ c $\frac{4}{9}$
d $\frac{3}{4}$ e $\frac{5}{6}$
3 a $\frac{3}{20}$ b $\frac{8}{25}$
c $\frac{19}{25}$ d $\frac{21}{200}$
4 a 0.5625 b 0.32
5 $\frac{3}{7}$ $\left(\frac{3}{7} = 0.429, 37\% = 0.37\right)$
6 Both Steven and Fernando send the same proportion
7 a $\frac{1}{2}$ b $\frac{1}{5}$ c 3
d $\frac{3}{2}$ e $\frac{10}{7}$ f $\frac{5}{12}$
8 a $\frac{1}{3}$ b $\frac{1}{3}$ c They are the same.
9 3 : 2
10 25 : 2
11 a 10 b 11 c 2
d 54 e 26
12 a 8.016 666 667 b 8.0 c 8.02
13 a 300 b 3500 c 8
d 0.45 e 19.1 f 0.0091
14 a 1.76 m (or 1.8 m) b 28°C
15 127 m²
16 a 8.3 b 3.7
c 2.7 d 3.2

Number skills: revision exercise 2

1 a $\frac{6}{12}$ b $\frac{9}{12}$ c $\frac{10}{12}$ d $\frac{8}{12}$
2 D
3 a $\frac{4}{5}$ b $\frac{5}{6}$ c $\frac{5}{6}$ d $\frac{2}{3}$ e $\frac{7}{9}$
4 a 0.4 b 0.47
5 a $\frac{7}{25}$ b $\frac{49}{100}$ c $\frac{81}{200}$ d $\frac{29}{125}$
6 a 12.5% b 20% c 32% d 5%
7 $\frac{18}{20}$ of £3500
8 mobile phones
9 a $\frac{1}{7}$ b $\frac{1}{13}$ c $\frac{5}{2}$ d $\frac{4}{3}$ e $\frac{9}{40}$

10 $\frac{8}{9}$
11 5 : 2
12 35 : 17
13 a 19 b 17 c 18 d 42
14 $883.2 \div 36.8 = 24$
15 10.2
16 55.51
17 169
18 a 12 days b 3 kg
19 4.0 m
20 a 143.22 b 196 c 16
 d i 4.7089 ii 4.71
 e i 3.826 ii 3.8
21 £56.67
22 5850 m
23 $3.5\,m^2$
24 a 3.24 b 13.824 c −42.875 d 2.1
25 a 1106.7 b 19 c 14 d 50.653

Chapter 23

Geometry skills: revision exercise

1 a $y = 160°$ b $z = 50°$
 c $k = 40°$ d $l = 100°, m = n = 80°$
 e $p = 30°, r = q = 150°$ f $t = u = s = 90°$
2 a $m = 40°$ b $n = 40°$
 c $x = 80°$ d $y = 40°, 2y = 80°, 3y = 120°$
3 $x = y = 140°$
4 60°

Exercise 23A

1 a $l = 50°$ (alternate angles)
 b $m = 110°$ (alternate angles)
 c $n = 70°$ (corresponding angles)
 d $f = 30°$ (corresponding angles)
 e $r = 40°$ (vertically opposite angles), $q = 40°$ (alternate angles)
 f $t = 75°$ (vertically opposite angles), $s = 75°$ (alternate angles) OR $s = 75°$ (corresponding angles), $t = 75°$ (alternate angles)
2 a, b

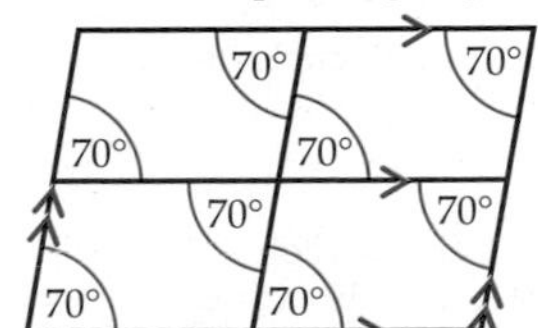

 c use angles on a straight line, vertically opposite, corresponding and alternate angles
3 a $a = 85°$ (vertically opposite angles), $b = 85°$ (alternate angles), $c = 85°$ (vertically opposite angles)
 b $p = 70°$ (vertically opposite angles), $m = 70°$ (corresponding angles), $n = 70°$ (vertically opposite angles)
 c $r = 75°$ (corresponding angles), $s = 105°$ (angles on a straight line)
 d $t = 120°$ (angles on a straight line with corresponding angle of 60°)
 e $u = 30°$ (vertically opposite angles), $v = w = 150°$ (angles on a straight line and vertically opposite angles), $x = 150°$ (corresponding angles)
 f $a = 45°$ (vertically opposite angles), $b = 45°$ (corresponding angles), $c = 135°$ (angles on a straight line)
4 $y = 60°$ (corresponding angles), $x = 60°$ (vertically opposite angles)
5 m is corresponding to 120°, found by angles on a straight line
6 a student's accurate drawing
 b angle $PSR = 120°$, angle QPS and angle $SRQ = 60°$
 c angles add up to 360°, opposite angles are equal
 d yes
7 angle $ABC = 100°$, angle $DAB = 110°$

Exercise 23B

1 a, b, c student's accurate drawings
2 N: 000°, NE: 045°, E: 090°, SE: 135°, S: 180°, SW: 225°, W: 270°, NW: 315°
3 a, b, c

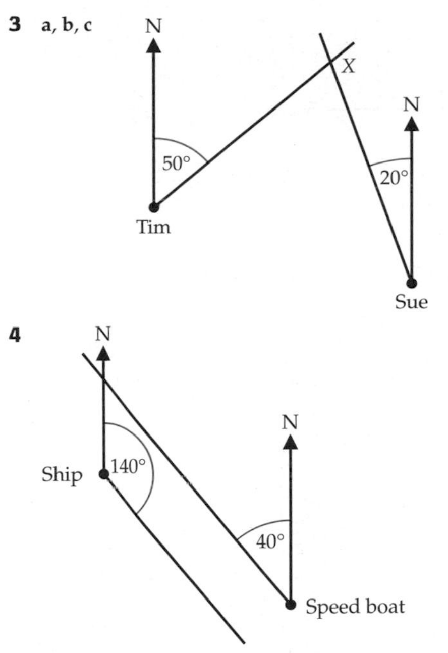

4 The ship and the speed boat will not collide because they are travelling parallel to each other.
5 a, b student's accurate drawing
 c i 309° ii 6 km
6 a 5.1 km b 208°

Exercise 23C

1 a i 060° ii 240° b i 120° ii 300°
 c i 170° ii 350°
2 a i 240° ii 060° b i 190° ii 010°
 c i 260° ii 080°
3 a i 350° ii 170° b i 300° ii 120°
 c i 275° ii 095°
4 a b 260°

N
80°

5 312°
6 a 045° b 225° c 115° d 295°

Review exercise

1 a $x = 70°$ (angles on a straight line with alternate angle of 110°)
 b $y = 70°$ (vertically opposite angles)
2 angle $PSR = 75°$, angle SPQ = angle $QRS = 105°$
3

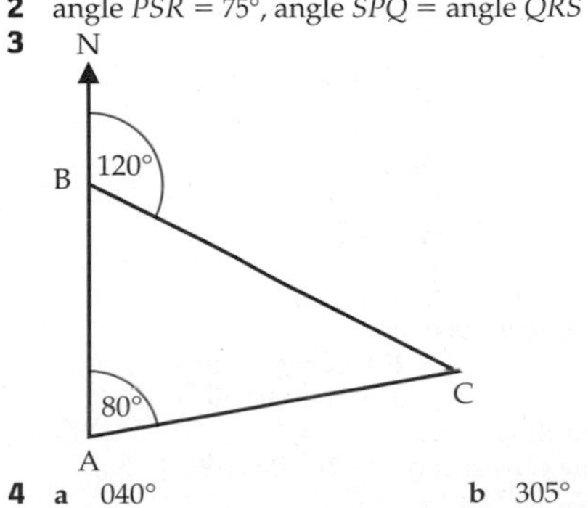

4 a 040° b 305°

5 **a, b**

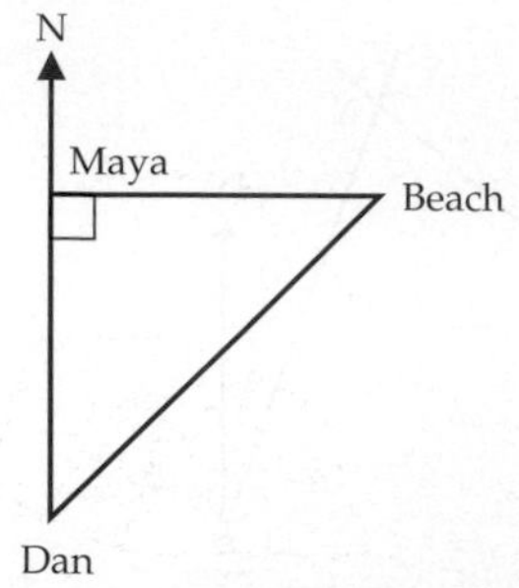

c i 045° ii 7 km (to nearest km)

6 **a**

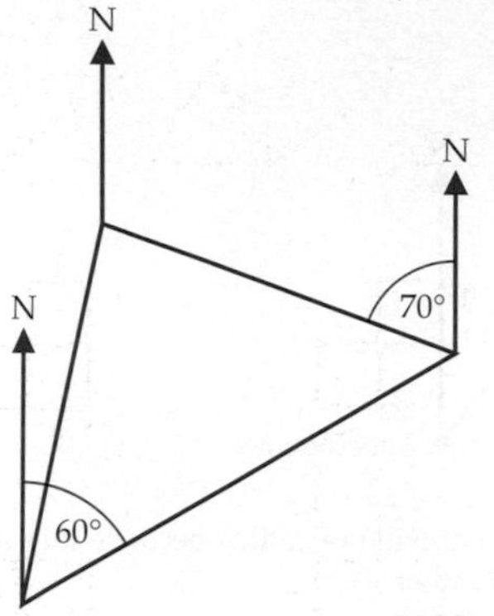

b 6.2 km, 192°

7

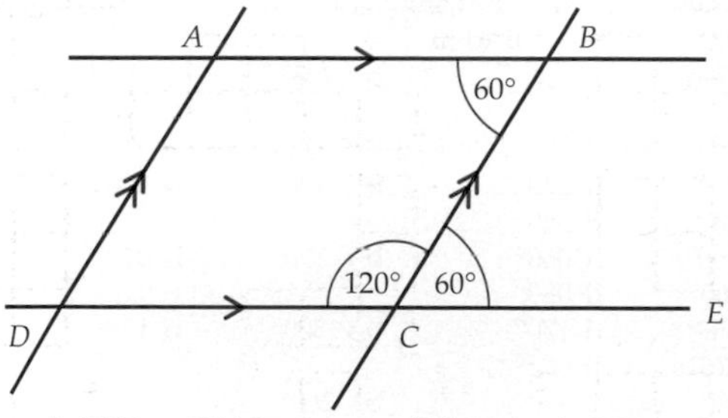

angle $BCE = 60°$ (alternate angles), so angle $BCD = 120°$ (angles on straight line)

8

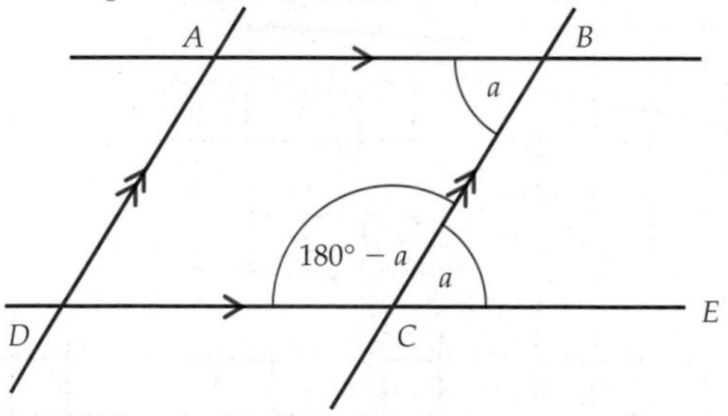

angle ABC + angle $BCD = a + (180° - a) = 180°$

Chapter 24

Exercise 24A

1 **a** 30° **b** 30° **c** 30° **d** 115° **e** 80°
2 $x = 55°$

Exercise 24B

1 **a** student's accurate constructions
b i 8 cm (±2 mm) ii 6.6 cm (±2 mm)
iii 7.4 cm (±2 mm) iv 8.7 cm (±2 mm)
2 no, the height will be about 4.3 m
3 **a** student's accurate construction **b** 76° (±2°)

Exercise 24C

1 **a** i $2x + 290° = 360°$ ii $x = 35°$
b i $3x + 240° = 360°$ ii $x = 40°$
c i $4x + 220° = 360°$ ii $x = 35°$
d i $10x = 360°$ ii $x = 36°$
2 $x = 28°$ **3** 139°

Exercise 24D

1 **a** $a = 30°$ **b** $b = 110°, c = 110°$
c $d = 75°, e = 105°$ **d** $f = 45°$
e $g = 55°, h = 55°, i = 70°$
2 **a** $a = 40°, b = 100°$ **b** $c = 40°$
c $d = 80°$ **d** $e = 85°$
e $f = 105°$ **f** $g = 115°$
3 $a = 60°, b = 60°, c = 120°, d = 120°$
4 **a** $a = 70°, b = 110°, c = 220°$ **b** $d = 75°$

Exercise 24E

1 **a** $a = 140°$ **b** $b = 100°$ **c** $c = 60°$ **d** $d = 50°$
2 **a** $a = 60°, 2a = 120°$ **b** $b = 60°, 2b = 120°$
c 120° **d** 135°
3 **a** 45° **b** 36° **c** 10°
4 **a** 108° **b** 135° **c** 144°
5 **a** 9 **b** 6 **c** 12
6 50 does not divide into 360
7 exterior angle is 20° which divides into 360°
8 $x = 60°, y = 120°$
9 $x = 60°, y = 60°$
10 $x = 120°, y = 30°, z = 90°$
11 $w = 30°, x = 30°, y = 30°, z = 30°$
12 $x = 65°$

Exercise 24F

1 **a** student's accurate construction
b student's accurate construction
c student's accurate construction
d student's accurate construction
2 Student's accurate constructions **and**
Tomas's side length = 7.1 cm, Tanya's side length = 8.7 cm, so Tanya is correct.

Review exercise

1 **a** student's accurate construction
b 7.4 cm ± 2 mm
2 $x = 33°$
3 student's accurate construction (circle drawn and angle at centre = 72° calculated)
4 $x = 18°$
5 **a** no, it could be a square
b yes, as opposite sides are parallel
6 **a** $a = 57°, b = 114°$ **b** $x = 68°$
7 36 sides
8 $a = 135°, b = 67.5°, c = 45°$
9 360 is not divisible by 70
10 **a** $x = 135°$ **b** $y = 45°$ **c** $z = 22.5°$

Chapter 25

Exercise 25A

1 **a** $z = 55$ **b** $x = 70$
2 **a** $a = 120$ **b** $b = 26$ **c** $c = 50$
3 8 cm
4 $x = 73°$
5 **a** $x^2 + 7x = 78$ **b** $x = -13$ or $x = 6$
6 **a** $(2x + 2)(x + 1) = 50$ or $x^2 + 2x - 24 = 0$
b $x = 4$

Exercise 25B

1 **a** $A = 0.04m + 0.05t$ **b** $B = 0.06m + 8$
c Text-tastic
2 **a** $P = 2l + 2w$ **b** $l = 3$
c $w = 5$
3 **a** $P = 6(w + 1)$ **b** $w = 5.5$
4 **a** $n = 5x - 12$ **b** $x = 15$
5 **a** $s = \dfrac{n - 3}{x}$ **b** $n = 43$

Exercise 25C

1 **a** M **b** R **c** v
d K **e** W

2 a $a = \frac{M}{2} - b$ b $a = \frac{R}{h}$ c $a = \frac{v - u}{t}$
d $a = \frac{2B}{K}$ e $a = \frac{W + 15}{7}$

3 a base $= \frac{2 \times \text{area}}{\text{height}}$
b i $x = 6$ cm ii $x = 3$ cm
iii $x = 3$ m iv $x = 3$ cm

4 a $b = \sqrt{c^2 - a^2}$ b $b = 5.66$ (3 s.f.)

Exercise 25D

1 $x = 3.9$ (1 d.p.) e.g.

x	$x^3 - 2x$	Comment
3	21	Too low
5	115	Too high
4	56	Too high
3.5	35.875	Too low
3.8	47.272	Too low
3.9	51.519	Too high
3.85	49.3...	Too low

2 $x = 8.4$ (1 d.p.)
3 $x = 3.5$ (1 d.p.)
4 $x = 4.4$ (1 d.p.)
5 a $x = 4.3$ (1 d.p.) b $x = -1.5$
6 a e.g. $8.45^3 - 8.45 = 594 < 600$ b $x = 8.5$
7 $x = -3.2$ (1 d.p.)
8 $t = 6.7$ (1 d.p.)
9 a $x(x^2 + 5) = 800$ b $x = 9.1$

Review exercise

1 a $x = 60$ b $y = 18$
2 a £54.50 b pay $= 5.5p + 8q$ c £75.50
3 $x = 4.4$ (1 d.p.)
4 $x = 6.8$ (1 d.p.)
5 $x = 13.6$ e.g.

x	$\frac{x^2}{\sqrt{x}}$	Comment
10	31.62...	Too low
20	89.44...	Too high
15	58.09...	Too high
13	46.87...	Too low
13.5	49.60...	Too low
13.6	50.15...	Too high
13.55	49.87...	Too low

6 $q = pr - 5r - 1$
7 $b = \frac{2A - ah}{h}$
8 5

Chapter 26

Geometry skills: revision exercise

1 a 34.8 cm² b 31.08 cm² c 28.5 cm²
d 22.95 cm² e 18.36 cm² f 34.31 cm²
2 £236.75
3 a $x = 4$ cm b $y = 12$ cm c $z = 5$ cm

Exercise 26A

1 a Area = 54 cm³, perimeter = 40 cm
b Area = 99 cm³, perimeter = 56 cm
c Area = 66 cm³, perimeter = 40 cm
d Area = 91 cm³, perimeter = 50 cm
2 a 60 cm² b 59.5 cm² c 44.5 cm² d 133.5 cm²

Exercise 26B

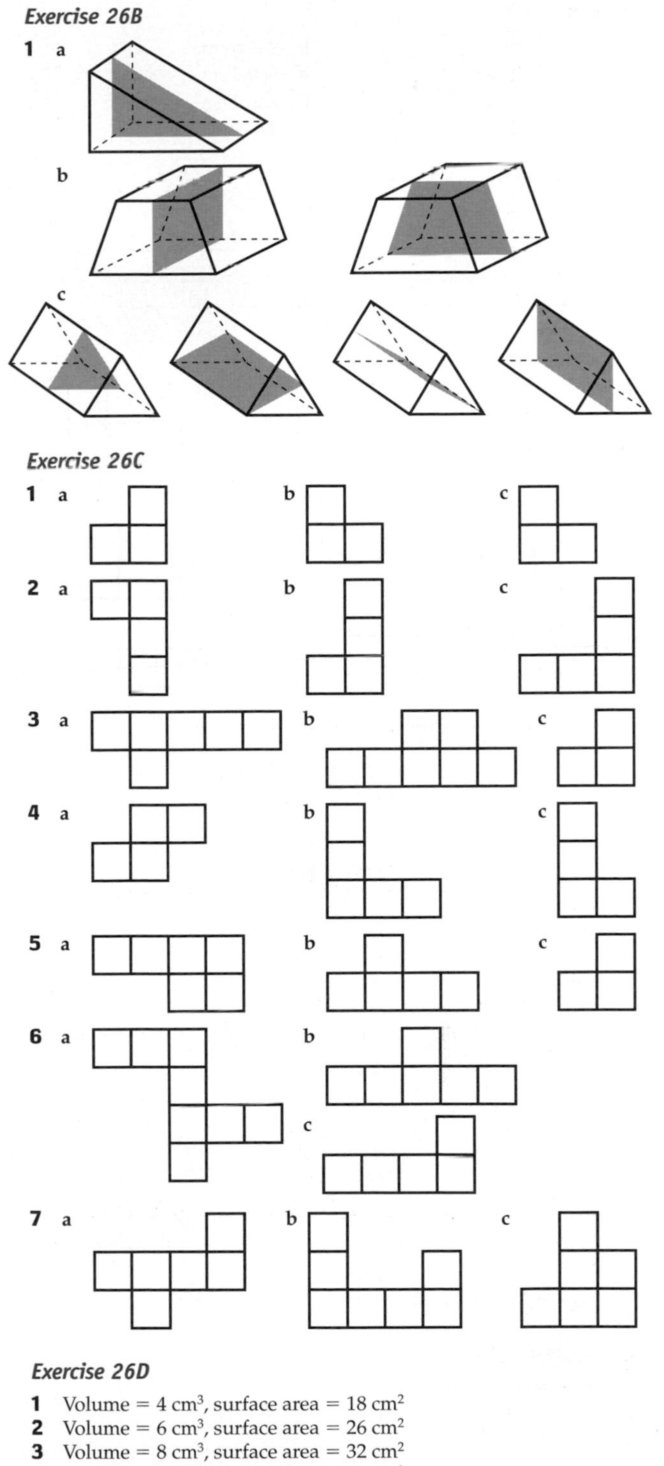

Exercise 26D

1 Volume = 4 cm³, surface area = 18 cm²
2 Volume = 6 cm³, surface area = 26 cm²
3 Volume = 8 cm³, surface area = 32 cm²
4 Volume = 6 cm³, surface area = 26 cm²
5 Volume = 7 cm³, surface area = 28 cm²
6 Volume = 9 cm³, surface area = 38 cm²
7 Volume = 9 cm³, surface area = 38 cm²

Exercise 26E

1 a 260 cm³ b 620.5 cm³ c 480 cm³
d 876.96 cm³ e 1330.8 cm³ f 4572 cm³
g 1512.72 cm³
2 £357.60

Exercise 26F

a 267 cm² b 408 cm² c 216 cm²
d 640 cm² e 1790 cm²

Exercise 26G

1 a 256 cm^3 b 265.6 cm^2
2 a $373\frac{1}{3}\text{ cm}^3$ b 360.8 cm^2
3 a $1173\frac{1}{3}\text{ cm}^3$ b 830.4 cm^2
4 a 384 cm^3 b 417 cm^2

Review exercise

1 a i 69 cm^2 ii 46 cm
b i 86 cm^2 ii 47.1 cm
2 a b
3 a i ii iii
b i ii iii

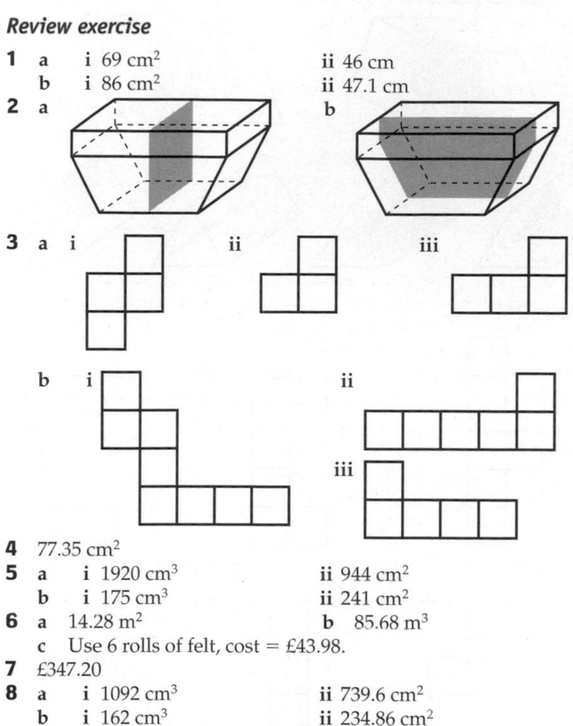

4 77.35 cm^2
5 a i 1920 cm^3 ii 944 cm^2
b i 175 cm^3 ii 241 cm^2
6 a 14.28 m^2 b 85.68 m^3
c Use 6 rolls of felt, cost = £43.98.
7 £347.20
8 a i 1092 cm^3 ii 739.6 cm^2
b i 162 cm^3 ii 234.86 cm^2

Chapter 27

Exercise 27A

1 25.6 cm
2 a 0.955 m b 7.96 cm c 3.98 mm d 5.95 cm
3 4 cm
4 33.9 km
5 a $4\pi + 8$ cm b $\pi + 4$ cm c $\frac{9\pi}{2} + 6$
6 1202 m
7 206 cm

Exercise 27B

1 a 5 cm b 10 mm c 5.8 cm d 2.9 mm
2 Small
3 a 27.71 cm^2 b 0.20 m^2 c 127.08 cm^2
4 3770 bulbs
5 a 13.7 cm^2 b 6.3 cm^2 c 7.7 cm^2
6 149 cm^2
7 a 24π b 40π
8 a 22.0 cm b 35.4 cm c 3.3 cm
9 21.6 cm

Exercise 27C

1 a volume = 1837.8 cm^3, surface area = 892.2 cm^2
b volume = 1.2 cm^3, surface area = 6.3 cm^2
c volume = 706.9 cm^3, surface area = 956.6 cm^2
d volume = 318.1 cm^3, surface area = 438.3 cm^2
2 216 *ml*
3 0.8 litre
4 Cylinder B by 124 *ml*
5 9817 litres
6 60 goldfish
7 287 cm^2
8 $30\pi\text{ cm}^2$
9 132.1 cm^2
10 60

Exercise 27D

1 a arc length = $\frac{5}{4}\pi$ cm, area = $\frac{15}{8}\pi\text{ cm}^2$
b arc length = $\frac{40}{3}\pi$ cm, area = $80\pi\text{ cm}^2$
c arc length = π cm, area = $\frac{3}{4}\pi\text{ cm}^2$
2 157 bulbs
3 B
4 a 60° b $r = 12$ cm c area = $24\pi\text{ cm}^2$
5 63 cm^2

Exercise 27E

1 a volume = $100\pi\text{ cm}^3$
b volume = $96\pi\text{ cm}^3$
c volume = $30\pi\text{ cm}^3$
2 a surface area = $90\pi\text{ cm}^2$
b surface area = $96\pi\text{ cm}^2$
c surface area = $75\pi\text{ cm}^2$
3 110 cm^3
4 4 cm
5 318.3 cm^3
6 1430 cm^3

Exercise 27F

1 a volume = $85\frac{1}{3}\pi\text{ cm}^3$ b surface area = $64\pi\text{ cm}^2$
2 volume = 17.2 cm^3, surface area = 32.2 cm^2
3 a radius = 3 cm b surface area = 113 cm^2
4 $4.5\pi\text{ cm}^3$
5 1450 cm^2
6 2770 cm^3
7 $148\,243\,920\text{ km}^2$
8 5.49×10^{12}

Review exercise

1 a 2.5 cm b 20 cm^2
2 $17\,026.5\text{ m}^2$
3 a 0.71 m^2 b £2766.87
4 3.2 cm
5 surface area = 122.52 cm^2, volume = 89.54 cm^3
6 10π cm
7 524 cm^3
8 3.5 cm
9 a 1319 cm^3 b 6 cm

Chapter 28

Exercise 28A

1 a $60\,000\text{ cm}^2$ b 2000 cm^2
c 0.24 m^2 d 0.065 m^2
2 a $2.5\text{ m}^2 = 25\,000\text{ cm}^2$ b $4\text{ km}^2 = 4\,000\,000\text{ m}^2$
c $8900\text{ m}^2 = 0.0089\text{ km}^2$ d $5000\text{ mm}^2 = 50\text{ cm}^2$
3 a 4.5 cm^2 b 450 mm^2
4 1 000 000
5 a 37.4 m^2 b £2618 (accept £2660)
6 e.g. $355\,000 \div 10\,000 = 35.5$; $35.5 \div 8 = 4.4$ (1 d.p.); he will need to buy 5 pots.
7 2830 cm^2
8 upper bound: 3.21 m^2; lower bound: 3.12 m^2

Exercise 28B

1 a $3\,500\,000\text{ cm}^3$ b $790\,000\text{ cm}^3$
c 9.2 cm^3 d 7.2 m^3
e 4.3 litres f 700 cm^3
g 850 cm^3 h 5.7 m^3
2 a upper bound: $534\,375\text{ cm}^3$; lower bound: $342\,125\text{ cm}^3$
b upper bound: $0.534\,375\text{ m}^3$; lower bound: $0.342\,125\text{ m}^3$
c upper bound: 534.375 litres; lower bound: 342.125 litres
3 111 hours

Exercise 28C

1 a 600 mph b 9 km/h
2 a 4050 m b 6 km
3 a 6 hours b 1.6 seconds
4 1 hour and 12 minutes
5 a 1 hour and 45 minutes b 45.7 mph
6 48 km/h

7 a $15\,m/s$ b $7.5\,m/s$
c $23.4\,km/h$ d $100.8\,km/h$

8 $6.5\,km/h$

9 greatest: 104.375 m; least: 94.875 m

10 upper bound: 5.52 km/h; lower bound: 5.29 km/h

11 a maximum journey time $= 85 \div 67.5 = 1.26 > 1.25$ hours
b 66 minutes

Exercise 28D

1 a 428 g b 190.4 g
c $117.6\,cm^3$ d $1.6\,cm^3$

2 a $10.5\,g/cm^3$ b $45.7\,cm^3$

3 a $45\,000\,cm^3$ b 31.5 kg

4 a $3200\,cm^3$ b $0.625\,g/cm^3$
c 375 g

5 a $2000\pi\,cm^3$ b 119 kg

6 a $0.004\,25\,m^3$ b 5.7 g (2 s.f.)

7 greatest: 17 112.5 tonnes; least: 16 012.5 tonnes

8 greatest: $0.70\,g/cm^3$ (2 s.f.); least: $0.65\,g/cm^3$ (2 s.f.)

9 greatest: $8300\,kg/m^3$ (2 s.f.); least: $7300\,kg/m^3$ (2 s.f.)

Exercise 28E

1 a volume b length c area d volume

2 a area b volume c area d none of these

3 a area b length c none of these d volume

4 a $2R + 2r$ b $4\pi^2Rr$ c $4\pi^2Rr^2$

5 Zoe's formula represents an area. It can't be the correct formula for the volume of the sphere.

6 a m^2 b m^3 c m^3 d m e m f m

7 a $P = \dfrac{4(u^2 + ub + b^2)}{(a + b)}$ The top expression has units of $(\text{length})^2$, and the bottom expression has units of length. This is the only expression which could represent a length.
b 30.4 cm

8 No, because the expression cannot represent a volume.

Review exercise

1 no, $110\,km/h \approx 69\,mph$

2 a 780 kg b $0.4\,m^3$

3 $3\,500\,000\,cm^3$

4 a $226\,000\,cm^3$
b e.g. $226\,000\,cm^3 = 226$ litres; $226 \times 1.8 = 406.8 \approx 400$

5 a $240\pi\,cm^3$ b $1.4\,g/cm^3$

6 2 hours and 6 minutes

7 length (π is dimensionless, and the expression in the bracket is a length plus a length)

8 a 2 b 2 c top: 2; bottom: 2

9 a least value:
$7.15 \times 3.95 \times 2.45 \times 19.25 \times 22 = 29\,300$ (3 s.f.) $< 30\,000$
b £31 874

Chapter 29

Some of the diagrams in this chapter are displayed here at a smaller scale than stated in the question.

Exercise 29A

1 Student's accurate construction

2 Student's accurate construction

3 Student's accurate construction, square

4 They are parallel to each other.

Exercise 29B

1 a, b Student's accurate constructions

2 Student's accurate construction

3 a Student's accurate construction
b Point should be on the perpendicular from B to the line AC.
c 420 m

4 a Student's accurate construction
b Student's accurate construction, the bisector of angle X
c Student's accurate construction
d Lengths of the perpendiculars are the same. This is always the case.

Exercise 29C

1 Student's accurate constructions, three different perpendicular bisectors

2 The perpendicular bisector of XY passes through the centre of the circle, O.

3 Student's accurate construction, perpendicular bisector of the line segment PQ. The mid-point of PQ occurs where the perpendicular bisector crosses the line segment.

Exercise 29D

1 Student's accurate constructions

2 The angle bisectors of x and y are at right angles to each other.

3 a Student's accurate construction
b Student's accurate construction
c Construct the angle bisector of 30°

4 a Draw a line, construct the perpendicular to form a right angle, and then construct the angle bisector to create an angle of 45°
b Student's accurate construction
c e.g. 135°; draw a line, construct the perpendicular to form two back-to-back right angles, and then construct the angle bisector of one of these to create an angle of 45°; $90° + 45° = 135°$.

5 Student's accurate construction

6 If the triangle ABC is an equilateral triangle or an isosceles triangle with angle B = angle C, then the angle bisector of angle A will also be the perpendicular bisector of BC.

7 Student's accurate construction

8 a Student's accurate construction
b i Parallelogram ii $12\,cm^2$

Exercise 29E

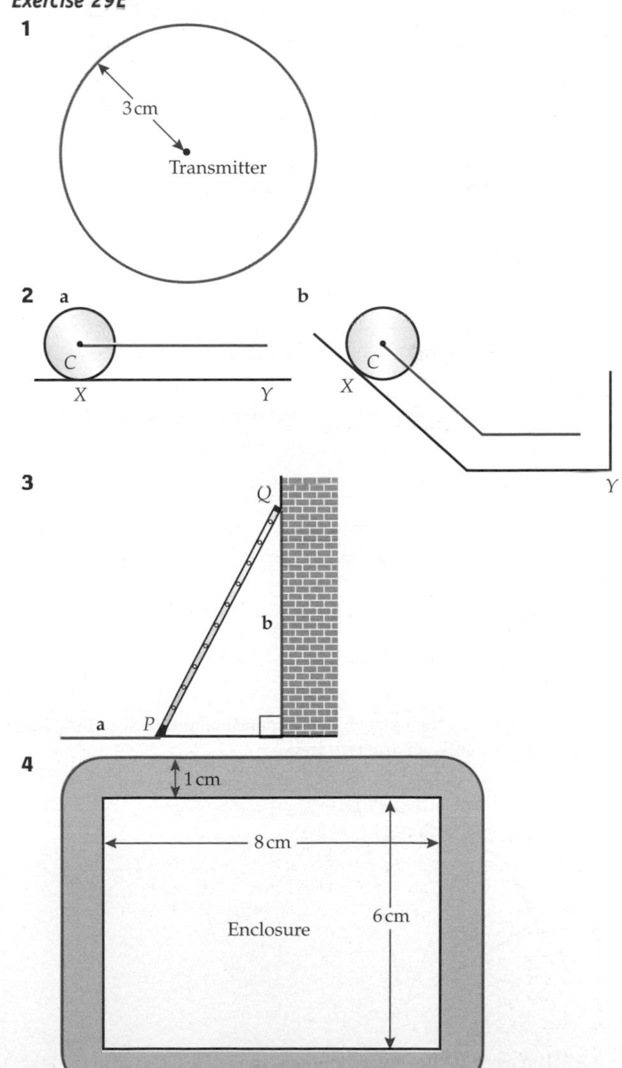

People may not stand in the shaded area.

5

6

7 **a,b**

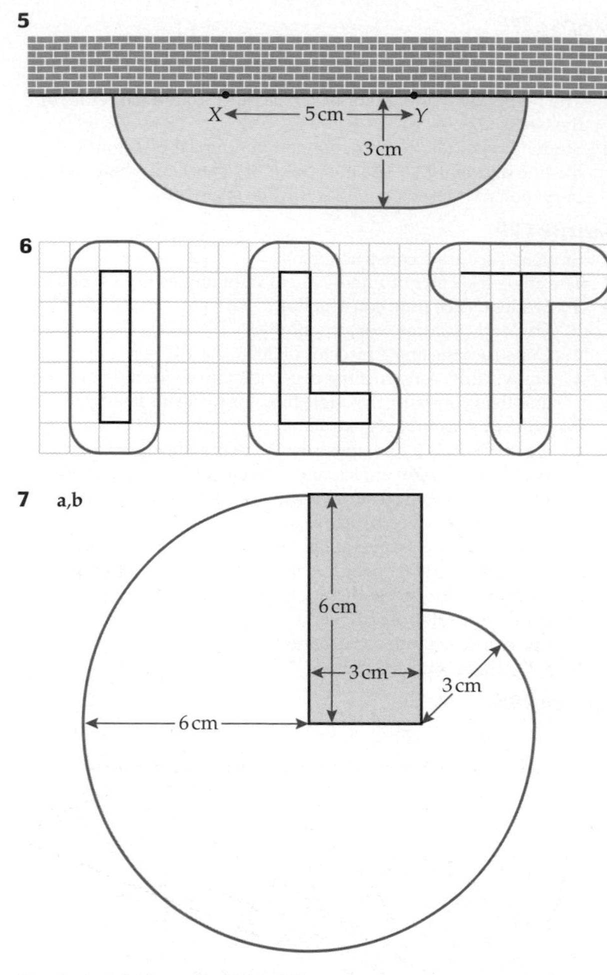

8 A straight line parallel to PQ

9 **a** A sphere

b A cylinder with hemispheres at each end

10 Area of square field $= 10 \times 10 = 100\,\text{m}^2$

Grazing area $= \frac{1}{4}\pi r^2 = \frac{1}{4}\pi \times 8^2 = 50.3\,\text{m}^2$ (1 d.p.)

Hence the goat can graze just over half the field.

Exercise 29F

1 **a**

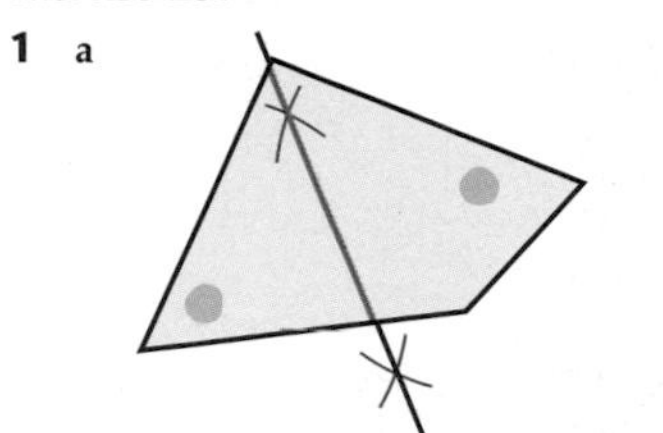

b The locus of the fence is the perpendicular bisector of the line segment between the two trees.

2 **a**

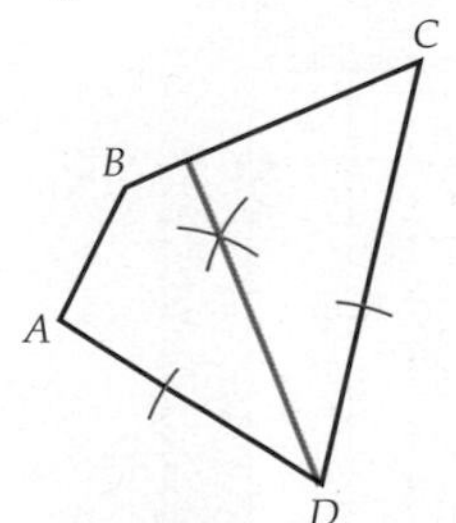

b The locus of the drainage pipe is the angle bisector of angle D.

3 Student's accurate construction. Labelled point should be where perpendicular bisector of AB and angle bisector of A meet.

Exercise 29G

1

2 **a, b, c**

3 **a** **b**

4 **a, b**

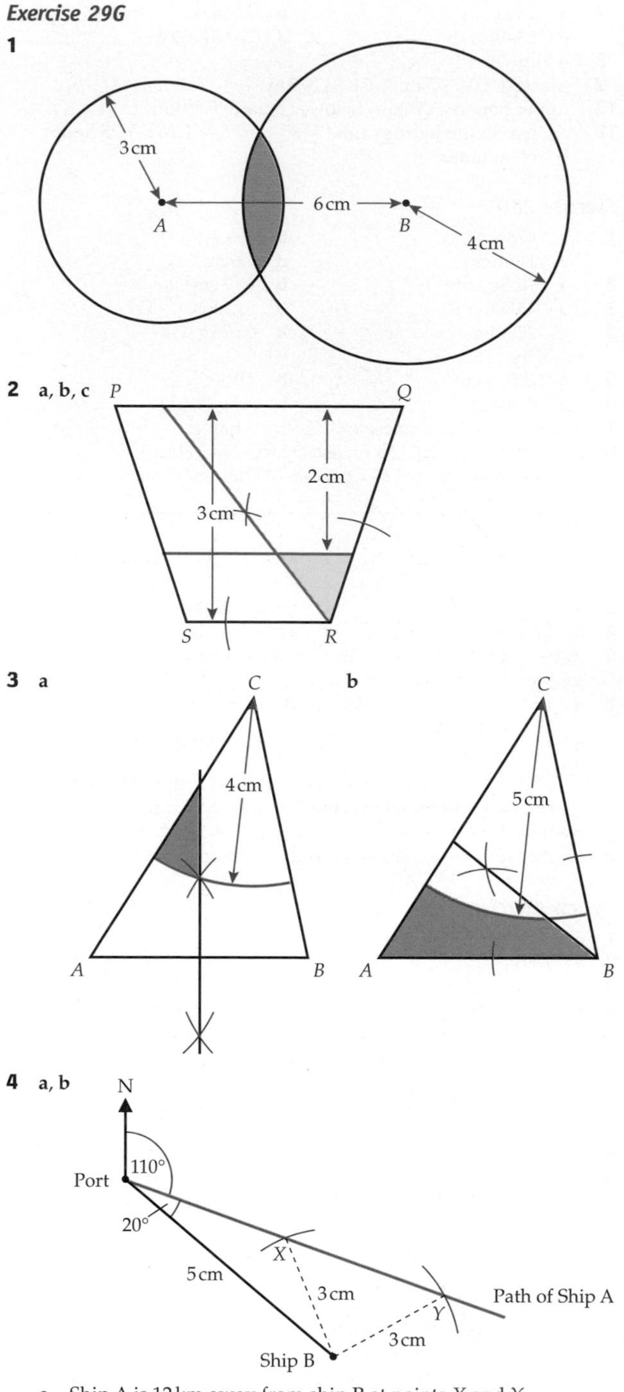

c Ship A is 12 km away from ship B at points X and Y.

d 40 min

5

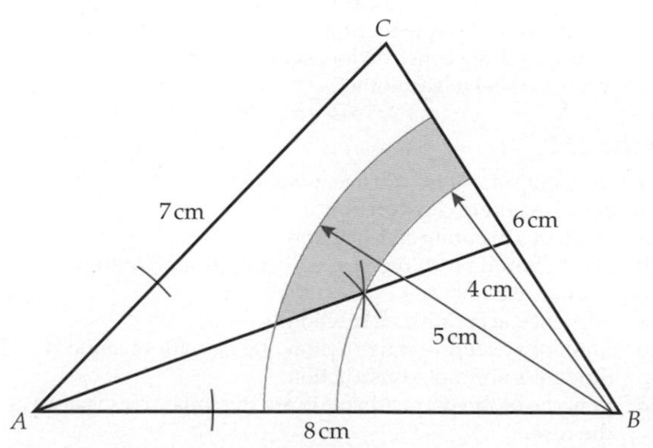

6

7

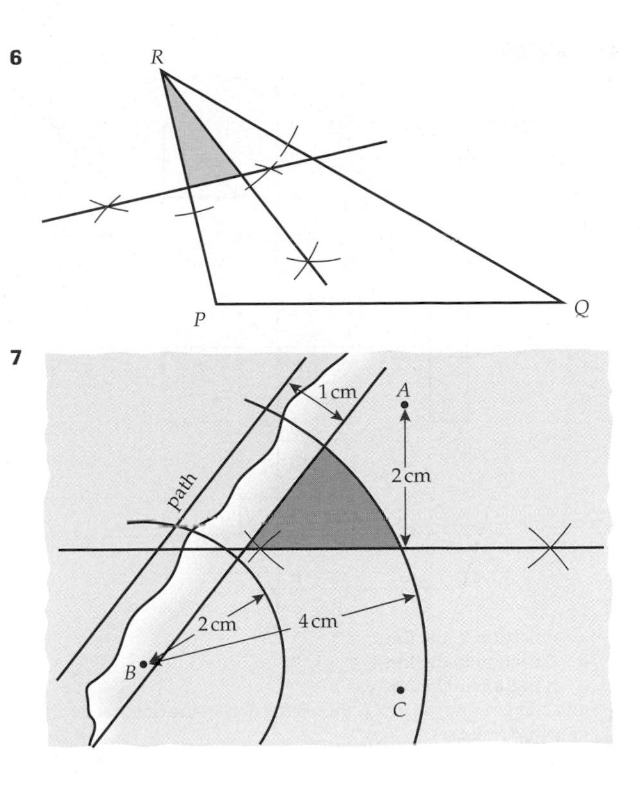

Review exercise

1 Student's accurate construction

2 Student's accurate construction, perpendicular bisector of XY

3

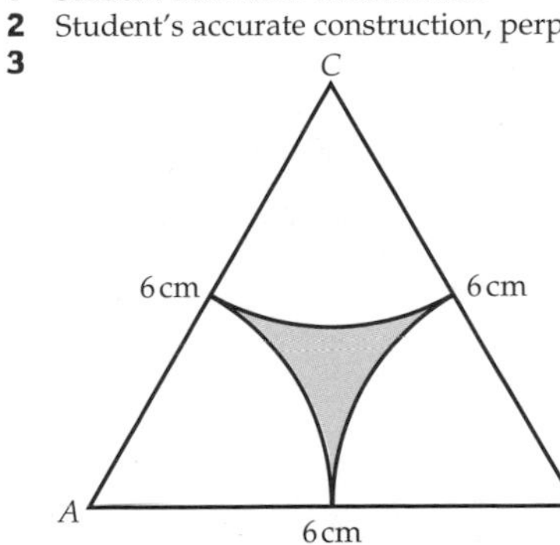

4 Student's accurate construction

5

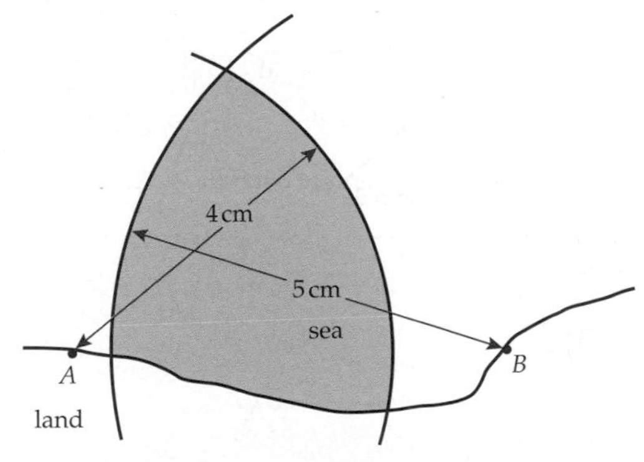

6 **a, b, c**

7

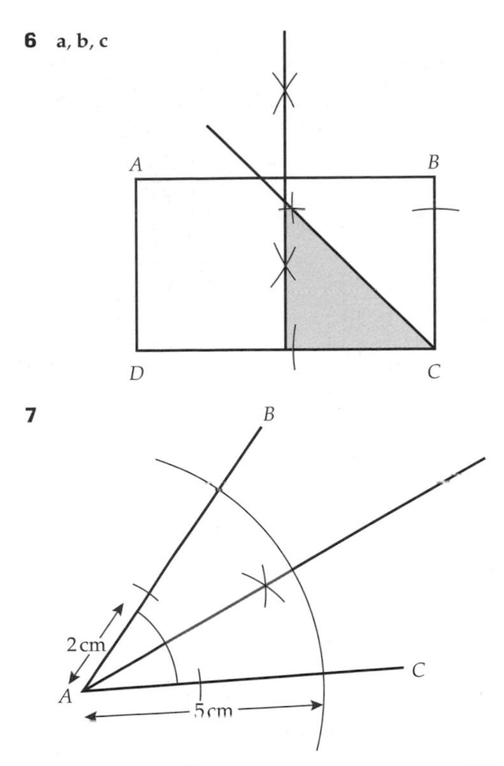

The blue line indicates the possible positions of the mast.

8

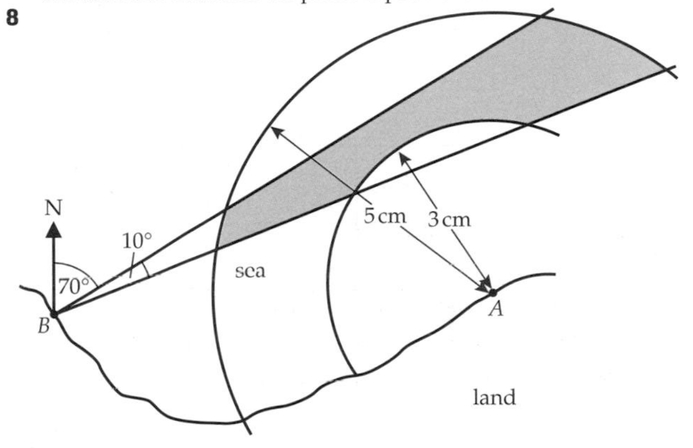

Chapter 30

Exercise 30A

1, 2

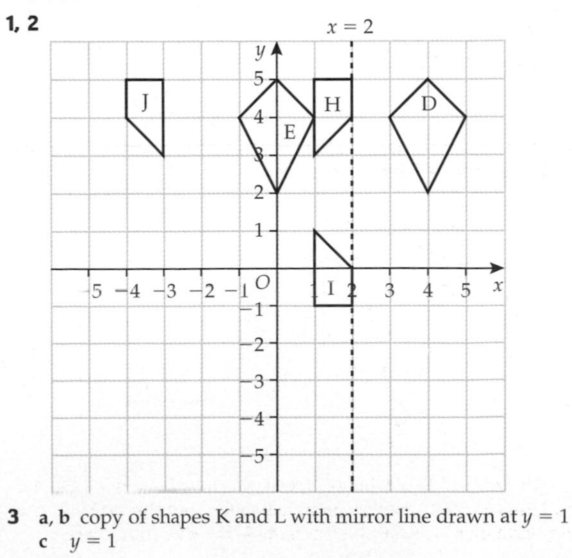

3 **a, b** copy of shapes K and L with mirror line drawn at $y = 1$

c $y = 1$

4 a reflection in the line $x = -1$
b reflection in the x-axis
c reflection in the line $x = 4$

5 a

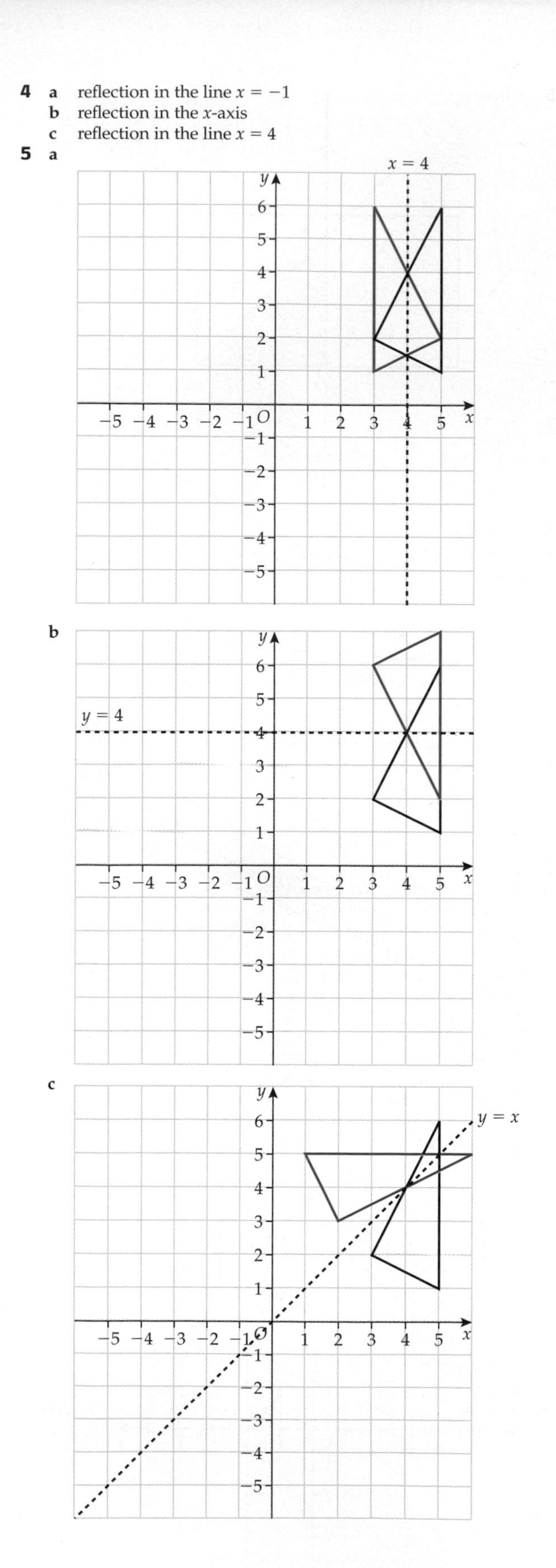

b

c

6 a, b, c, d

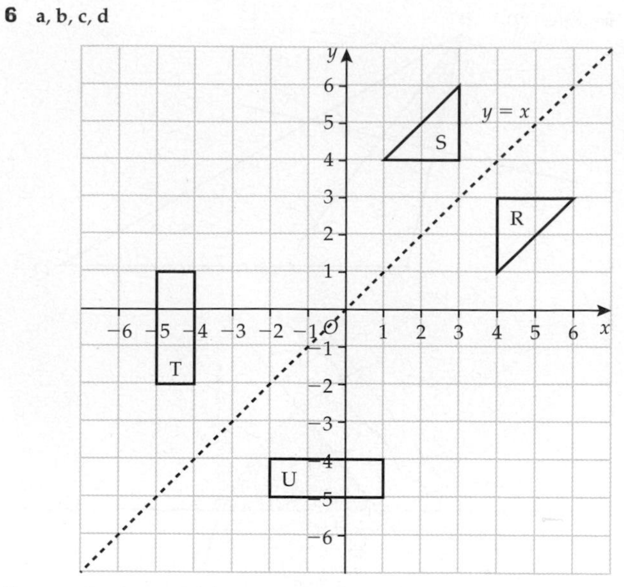

7 a reflection in the line $y = x$
b reflection in the line $y = -x$
c reflection in the line $y = x$

8 reflection in the line $y = 2$, then reflection in the line $x = 3$ (or in the other order)

Exercise 30B

1 a, b, c

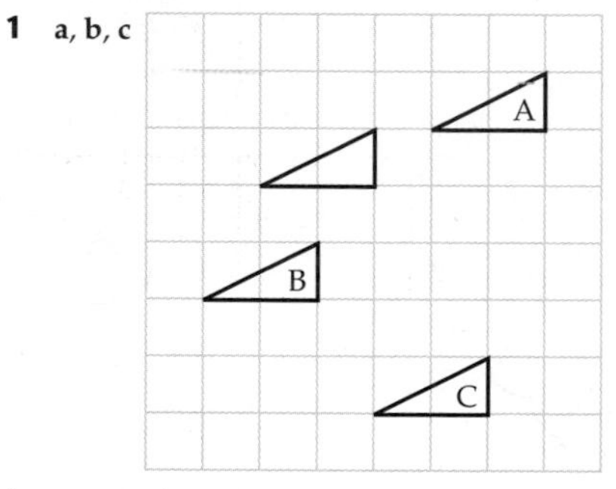

2 a 3 right, 1 down
b 5 right, 1 up
c 2 right, 4 down
d 2 left, 4 up
e They are the inverse of each other.

Exercise 30C

1 a $\begin{pmatrix}3\\2\end{pmatrix}$ b $\begin{pmatrix}3\\-1\end{pmatrix}$ c $\begin{pmatrix}-2\\-3\end{pmatrix}$ d $\begin{pmatrix}-5\\0\end{pmatrix}$ e $\begin{pmatrix}0\\4\end{pmatrix}$

2 a i $\begin{pmatrix}7\\-1\end{pmatrix}$ ii $\begin{pmatrix}-7\\1\end{pmatrix}$
b They have the same numbers but opposite signs.
c $\begin{pmatrix}-1\\4\end{pmatrix}$

3 a i $\begin{pmatrix}1\\2\end{pmatrix}$ ii $\begin{pmatrix}3\\2\end{pmatrix}$ iii $\begin{pmatrix}4\\4\end{pmatrix}$
b $\begin{pmatrix}1\\2\end{pmatrix} + \begin{pmatrix}3\\2\end{pmatrix} = \begin{pmatrix}4\\4\end{pmatrix}$

4 $\begin{pmatrix}5\\1\end{pmatrix}$

5 $\begin{pmatrix}1\\-1\end{pmatrix}$

6 A to B $\begin{pmatrix}2\\0\end{pmatrix}$, A to C $\begin{pmatrix}4\\0\end{pmatrix}$, A to D $\begin{pmatrix}6\\0\end{pmatrix}$, A to E $\begin{pmatrix}0\\-2\end{pmatrix}$, A to F $\begin{pmatrix}2\\-2\end{pmatrix}$, A to G $\begin{pmatrix}4\\-2\end{pmatrix}$, A to H $\begin{pmatrix}6\\-2\end{pmatrix}$, A to I $\begin{pmatrix}0\\2\end{pmatrix}$, A to J $\begin{pmatrix}2\\2\end{pmatrix}$, A to K $\begin{pmatrix}4\\2\end{pmatrix}$, A to L $\begin{pmatrix}6\\2\end{pmatrix}$

Exercise 30D

1

2 a, b

3 a, b

4 a, b

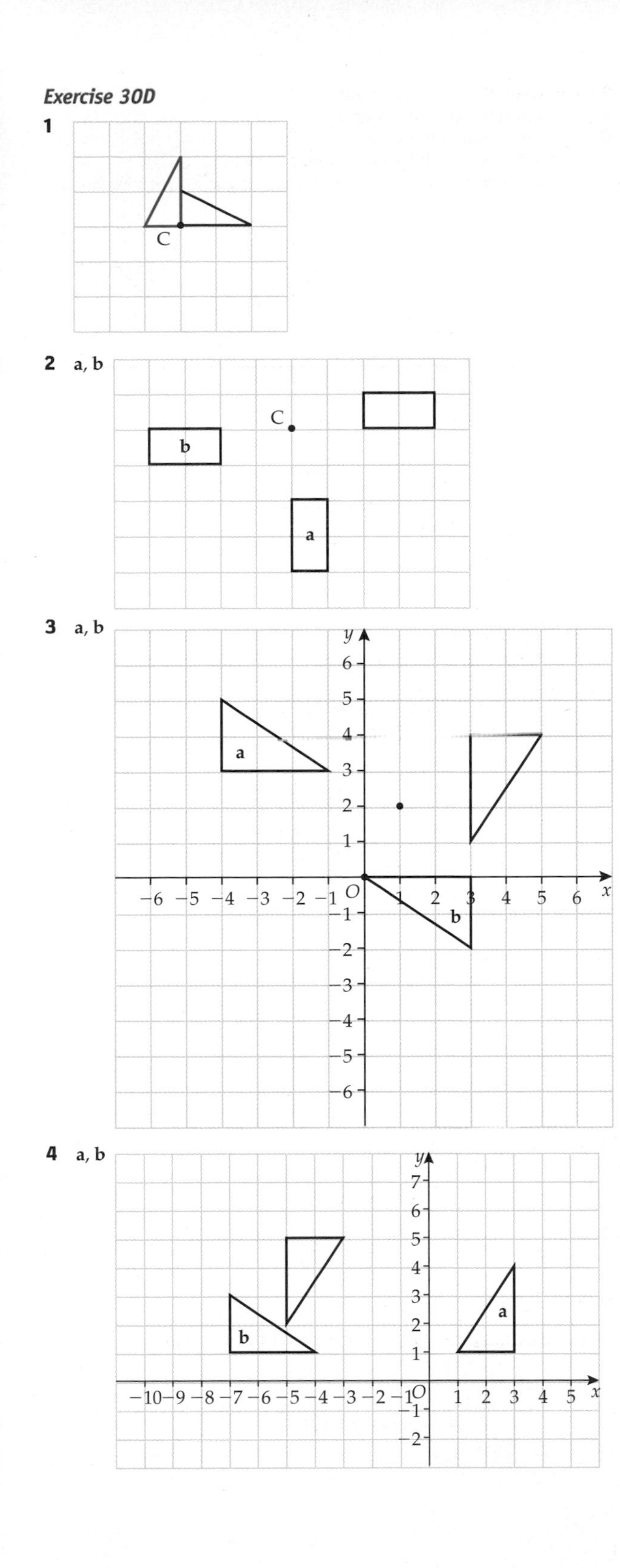

5

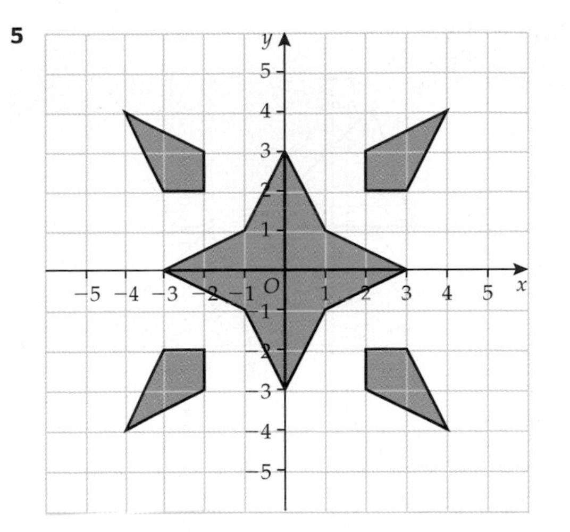

Exercise 30E

1 a 180° clockwise or anticlockwise

b

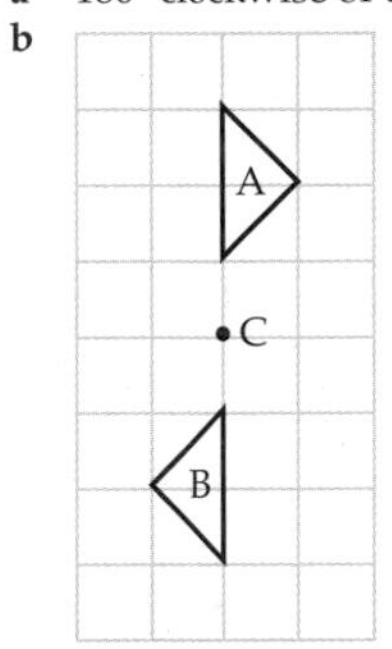

2 a 90° anticlockwise b (0, 0)

3 a rotation 180° about centre (0, 0)

b rotation 180° about centre (2, 4)

c rotation 180° about centre (−2, −2)

4 A to E, 90° anticlockwise, centre (0, 0)

E to A, 90° clockwise, centre (0, 0)

A to C and C to A, 180°, centre (5, −1)

A to G and G to A, 180°, centre (2, 0)

E to G, 90° anticlockwise, centre (2, 2)

G to E, 90° clockwise, centre (2, 2)

E to C, 90° anticlockwise, centre (6, 4)

C to E, 90° clockwise, centre (6, 4)

5 a, b

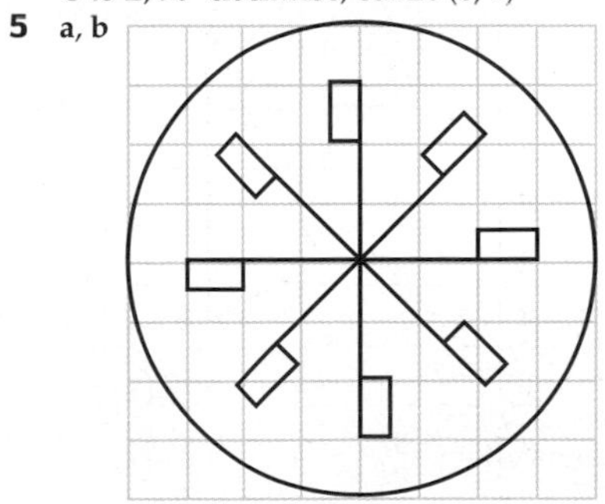

6 Either 3 rotations 90° anticlockwise about (−2, 2); **or** 3 rotations 90° clockwise about (−2, 2); **or** rotation 180° about (−3.5, 3.5).

Exercise 30F

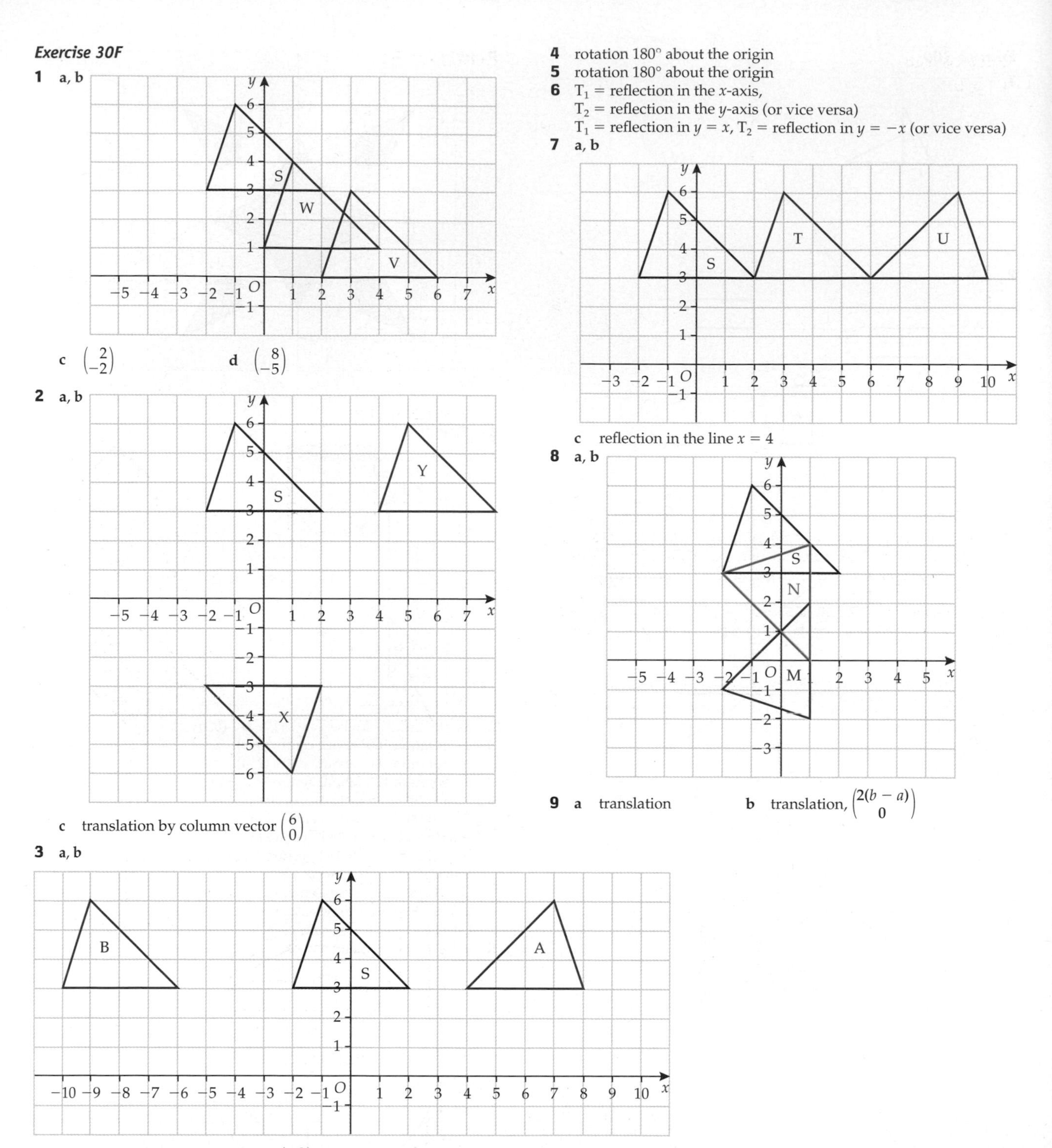

1 a, b

c $\begin{pmatrix} 2 \\ -2 \end{pmatrix}$ d $\begin{pmatrix} 8 \\ -5 \end{pmatrix}$

2 a, b

c translation by column vector $\begin{pmatrix} 6 \\ 0 \end{pmatrix}$

3 a, b

c translation by column vector $\begin{pmatrix} -8 \\ 0 \end{pmatrix}$

4 rotation 180° about the origin

5 rotation 180° about the origin

6 T_1 = reflection in the x-axis,
T_2 = reflection in the y-axis (or vice versa)
T_1 = reflection in $y = x$, T_2 = reflection in $y = -x$ (or vice versa)

7 a, b

c reflection in the line $x = 4$

8 a, b

9 a translation b translation, $\begin{pmatrix} 2(b - a) \\ 0 \end{pmatrix}$

Review exercise

1 a, b

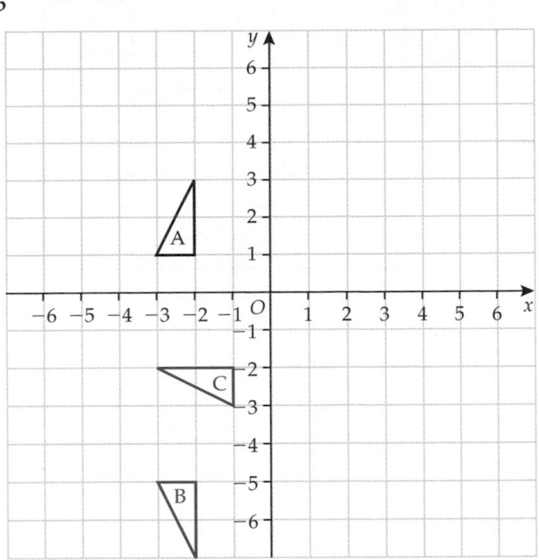

2 Student's own answer, e.g. reflection in $x = 4$ followed by reflection in $y = 3$; **or** reflection in $y = 3$ followed by reflection in $x = 4$; **or** reflection in $x = 4$ followed by rotation 180° about (4, 3).

3 a **Rectangle B is a reflection of rectangle A in the line $y = -1$**

b **Rectangle B is a translation of rectangle A by the column vector $\begin{pmatrix} 0 \\ -9 \end{pmatrix}$**

c **Rectangle B is a rotation of rectangle A through 180° about the point (−2.5, −1)**

4 a (2, 0)

b

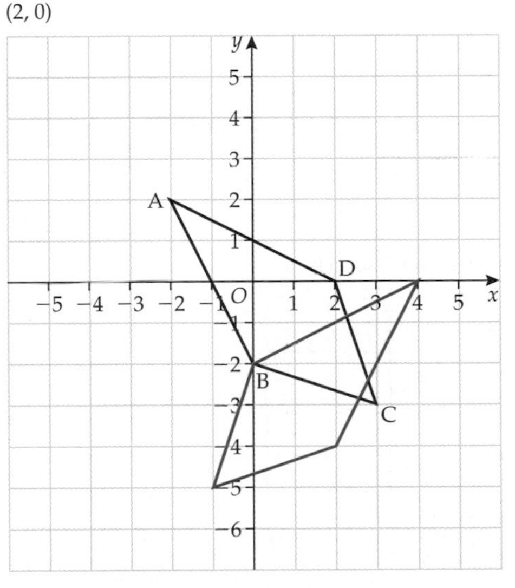

c (−1, −5)

5 a rotation 90° anticlockwise about centre (0, −1)

b reflection in the line $y = x$

c reflection in the x-axis

6 a i C ii G iii B

b reflection in the line $x = -2$

7 H and B

8 E and G

9 rotation 90° anticlockwise about centre (−1, 3) and reflection in $y = 4$

10 a rotation

b Two reflections in lines that are not parallel are equivalent to a rotation. The centre of rotation is the point at which the two lines cross. The angle of rotation is equivalent to twice the angle between the lines.

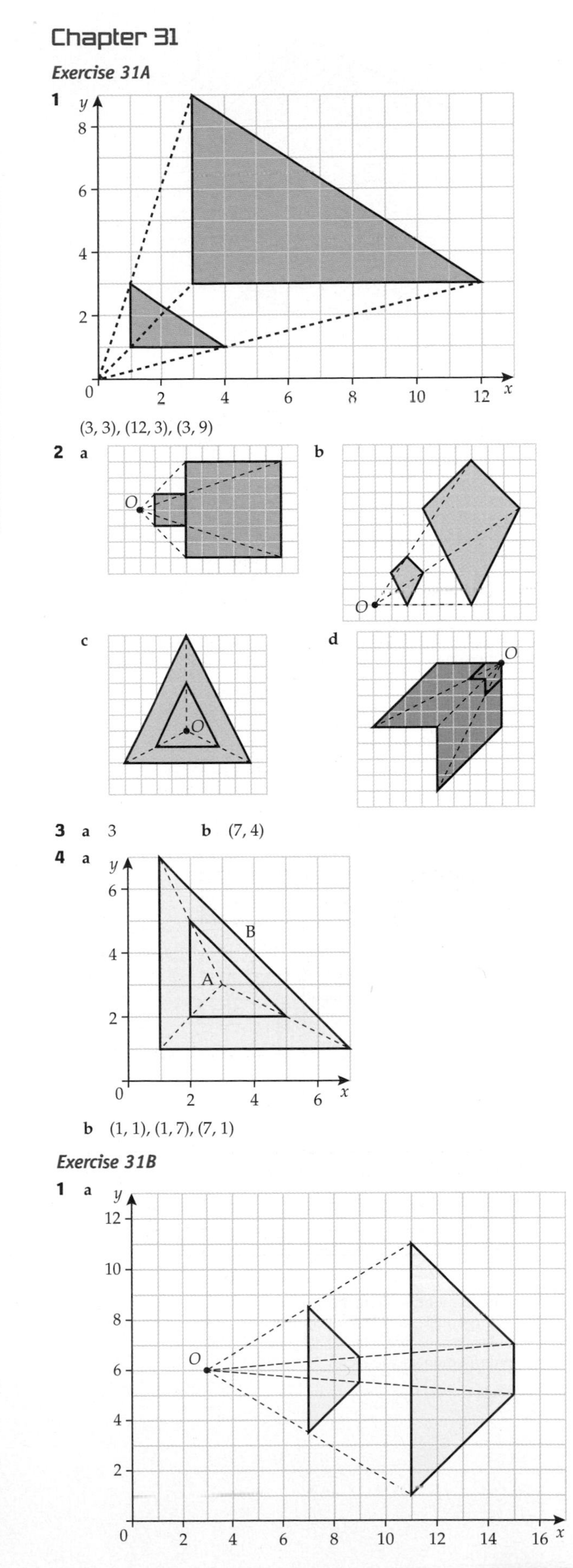

Chapter 31

Exercise 31A

1

(3, 3), (12, 3), (3, 9)

2 a b c d

3 a 3 b (7, 4)

4 a

b (1, 1), (1, 7), (7, 1)

Exercise 31B

1 a

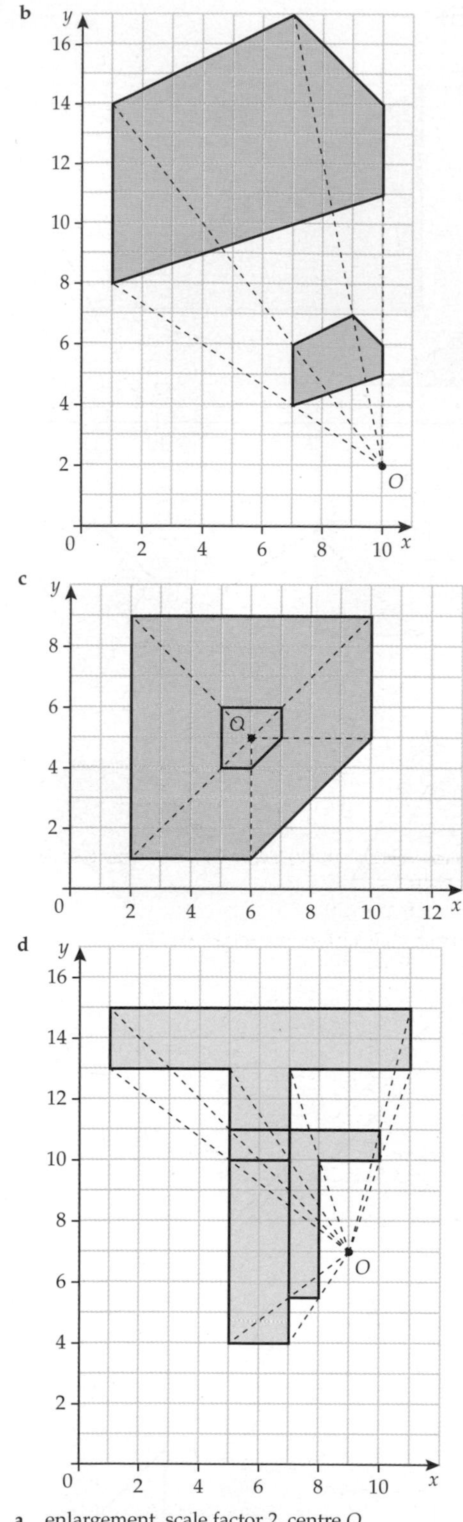

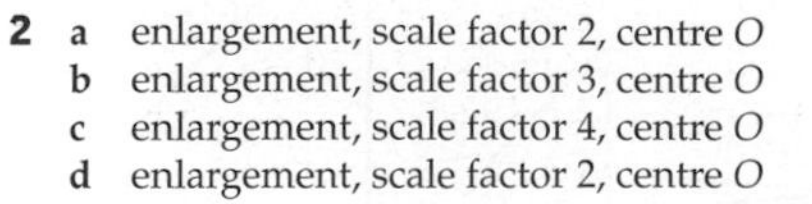

2 a enlargement, scale factor 2, centre O
b enlargement, scale factor 3, centre O
c enlargement, scale factor 4, centre O
d enlargement, scale factor 2, centre O

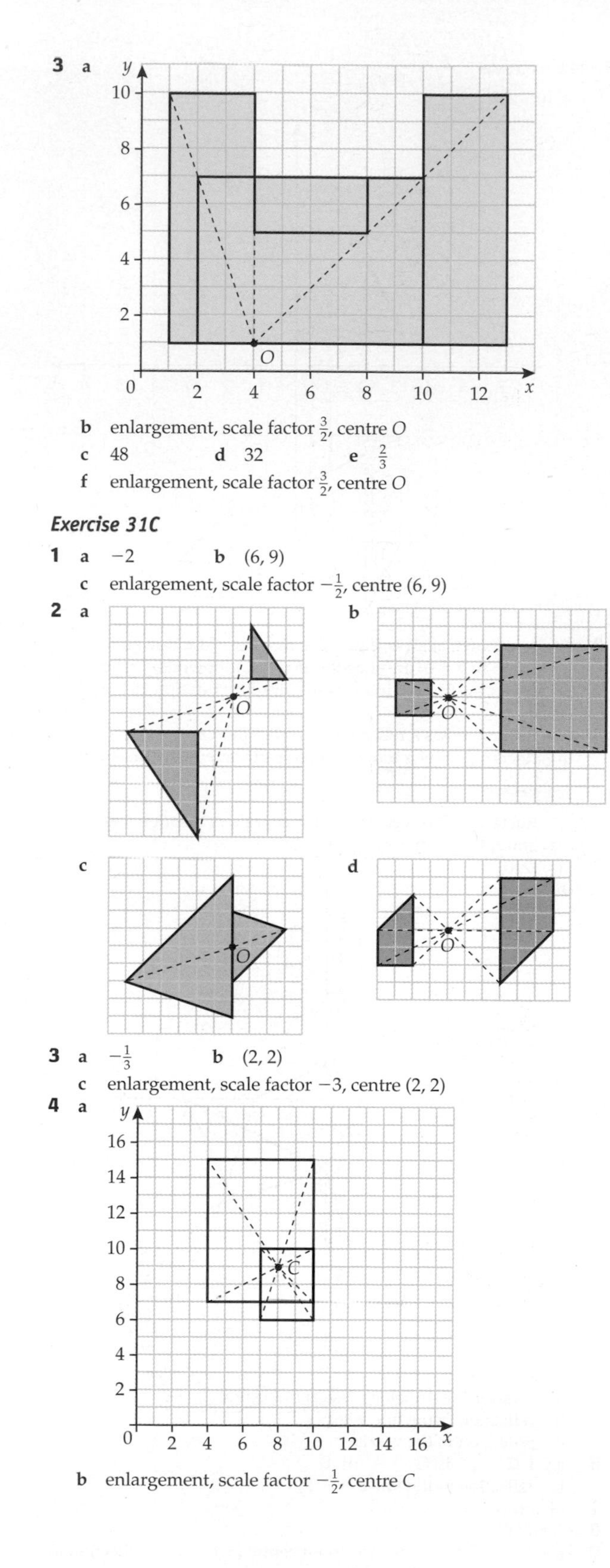

3 a

b enlargement, scale factor $\frac{3}{2}$, centre O
c 48 d 32 e $\frac{2}{3}$
f enlargement, scale factor $\frac{3}{2}$, centre O

Exercise 31C

1 a -2 b $(6, 9)$
c enlargement, scale factor $-\frac{1}{2}$, centre $(6, 9)$

2 a b c d

3 a $-\frac{1}{3}$ b $(2, 2)$
c enlargement, scale factor -3, centre $(2, 2)$

4 a

b enlargement, scale factor $-\frac{1}{2}$, centre C

5 a

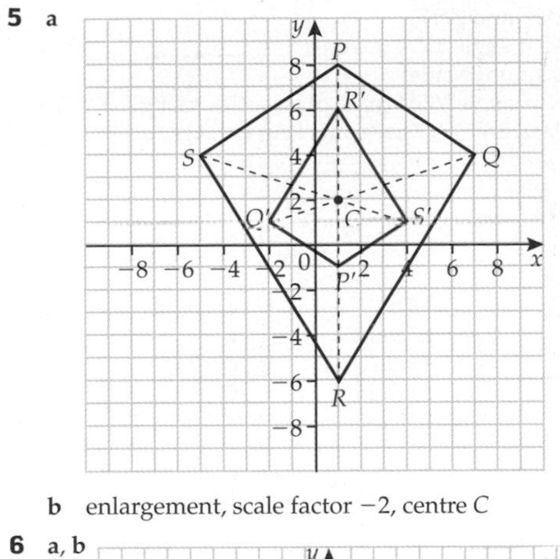

b enlargement, scale factor −2, centre C

6 a, b

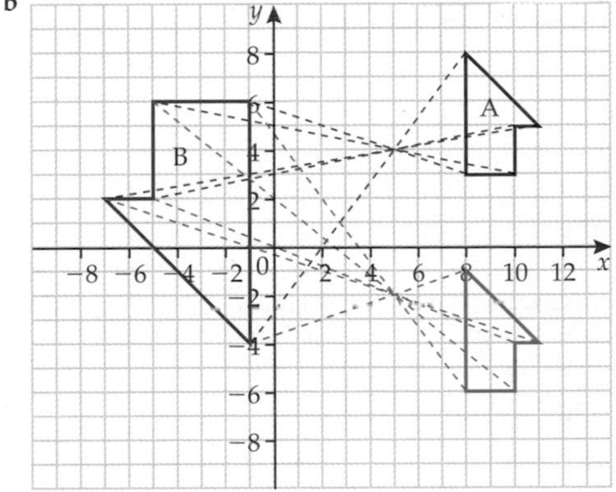

c The final shape is congruent to A.

Review exercise

1 a 3 **b** (2, 1)

2

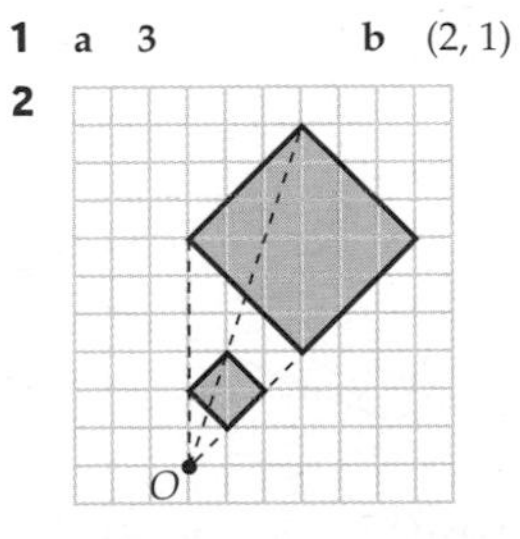

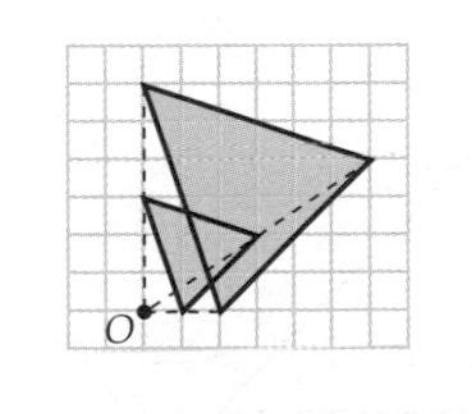

3 72 cm

4

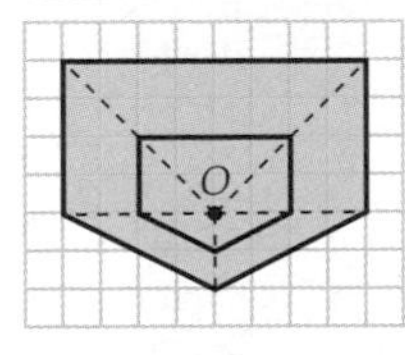

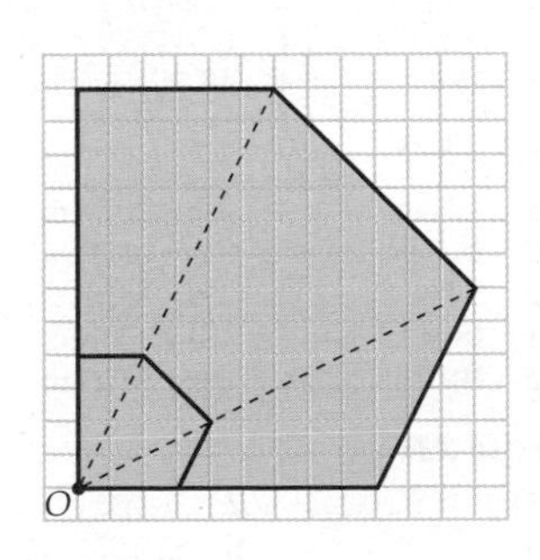

5 a

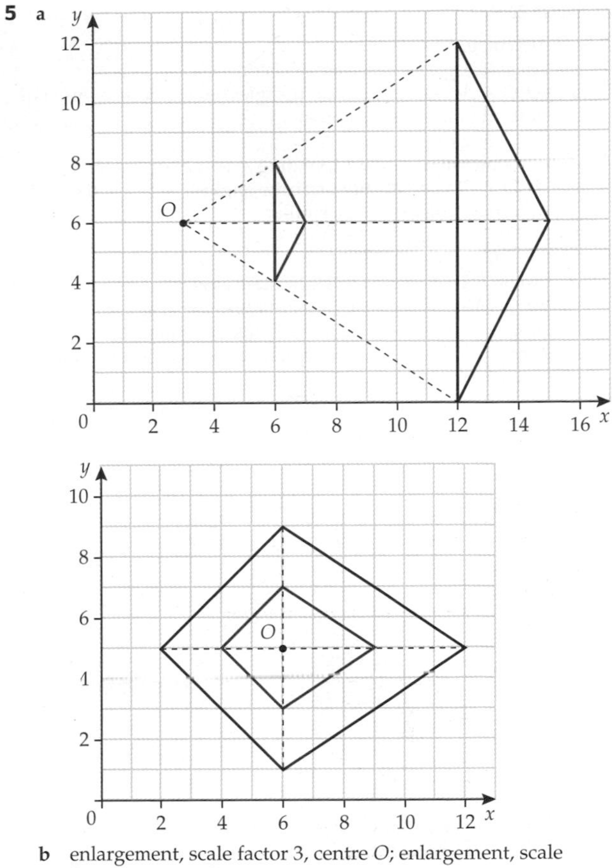

b enlargement, scale factor 3, centre O; enlargement, scale factor 2, centre O

6 a $-\frac{3}{2}$ **b** (−1, 1)

c enlargement, scale factor $-\frac{2}{3}$, centre (−1, 1)

7

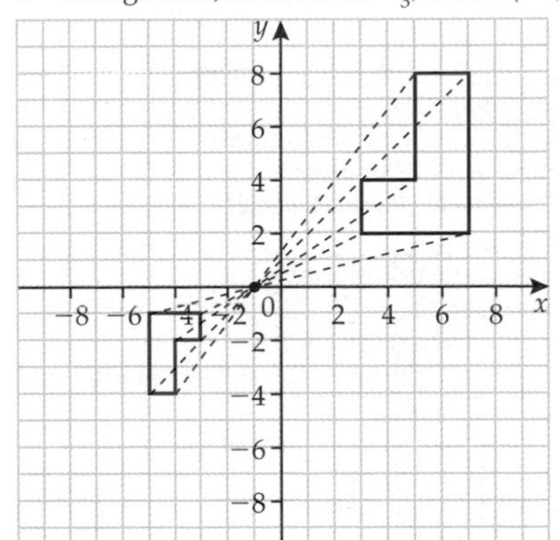

Chapter 32

Exercise 32A

1. A and D, B and G, C and E, F and H
2. **a, b** Student's own answers
3. **a** B and D **b** A and C, B and F
   c A and B **d** A and C
4. **a** True **b** False **c** False **d** True
   e False **f** True **g** True

Exercise 32B

1. Congruent (RHS)
2. Congruent (SSS)
3. Congruent (ASA)
4. Not congruent (the angles marked 60° are not corresponding angles)
5. Congruent (ASA)
6. Not congruent (the hypotenuses are not equal)

Exercise 32C

1 $x = 9.8$ cm, $y = 8$ cm
2 a $AB = CD = 3.125$ cm, $QR = 5.6$ cm
3 a $A\hat{P}Q = A\hat{C}B$ (corresponding angles), angle at A common to both triangles, hence similar.
b $x = 7$ cm, $y = 9$ cm
4 Both wrong, ratios of corresponding sides are different
5 a $A\hat{P}D = B\hat{P}C$ (vertically opposite angles), $A\hat{D}P = C\hat{B}P$ (alternate angles), hence similar.
b $AD = 17.5$ cm, $BP = 16$ cm
6 14 m
7 a Angle BAC = angle DEC (alternate angles)
Angle ABC = angle EDC (alternate angles)
Angle ACB = angle ECD (vertically opposite angles)
Corresponding angles are equal so the triangles are similar.
b $x = 3.3$ cm (1 d.p.), $y = 8.1$ cm
8 a Since $B\hat{A}D = B\hat{A}C$ and $A\hat{B}D = A\hat{C}B$ triangles ABD and ACB are similar.
b 9 cm

Exercise 32D

1 a 7.2 cm b 12.6 cm²
2 a 7.8 cm b 126.2 cm²
3 168.75 cm²
4 864 *l*
5 5.4 cm
6 a 3.6 cm² b 1250 g
7 a 8.1 m b 16.2 m² c 0.192 m³
8 210 mm by 148.5 mm
9 3536 m²

Exercise 32E

1 1 $CD = BA$ (since $ABCD$ is a parallelogram)
2 $AD = BC$ (since $ABCD$ is a parallelogram)
3 $BD = BD$ (common side)
So triangles DAB and BCD are congruent (SSS).
2 1 $QR = SR$ (given)
2 $PQ = PS$ (given)
3 $PR = PR$ (common side)
So triangles PQR and PSR are congruent (SSS).
Hence angle PQR = angle PSR.
3 1 $AB = CD$ (since $ABCD$ is a parallelogram)
2 angle CDY = angle ABX (Z angle)
3 angle YCD = angle BAX (since CY bisects C, AX bisects A, angle A = angle C)
So triangles ABX and CDY are congruent (ASA).
4 1 $QX = RX$ (given)
2 $PQ = SR$ (since $PQRS$ is a rectangle)
3 angle PQX = angle SRX (because $PQRS$ is a rectangle and QRX is an isosceles triangle)
So triangles PQX and SRX are congruent (SAS).
Hence $PX = SX$.
5 1 $RC = PC$ (since $PQRC$ is a square)
2 $BC = DC$ (since $ABCD$ is a square)
3 angle BCR = angle PCD (created by overlapping right angles)
So triangles DPC and BRC are congruent (SAS).
Hence $DP = BR$.

Review exercise

1 Congruent: D and F; Similar: A and C
2 A and G, B and D, E and F
3 a $A\hat{C}B = P\hat{Q}B$ (corresponding angles), angle at B is common to both triangles, hence similar.
b $x = 12.8$ cm, $y = 6\frac{2}{3}$ cm
4 a 6 cm
b $A\hat{B}C = 42°$
5 a $B\hat{A}C = A\hat{D}B = 90°$, Angle at B is common to both triangles, hence similar.
b 6.75 cm
6 1500 m³
7 a 8 : 27 (or 1 : 3.375) b 4 : 9 (or 1 : 2.25)
8 a 12 cm b 180 cm²

9 In triangles YMZ and ZNY,
1 angle ZYM = angle YZN (given)
2 $ZY = ZY$ (common side)
3 angle ZYN = angle YZM (base of isosceles triangle XYZ)
So triangles YMZ and ZNY are congruent (ASA).

Chapter 33

Exercise 33A

1 a 10 cm b 13 cm c 30 cm
d 10.3 cm e 15.9 cm f 17.4 cm
2 a 12 cm b 16 cm c 24.7 cm
d 11.7 cm e 14.8 cm f 13.5 cm
g 19.6 cm h 8.2 cm

Exercise 33B

1 11.4 km
2 3.06 m
3 a 10.82 cm b 5.41 cm c 91.9 cm²
4 64.6 cm²
5 £80
6 2.3 m
7 5.44 m³
8 8.7 cm

Exercise 33C

1 a $AB = 4$ units b $CD = 4$ units c $EF = 5$ units
d $GH = 5$ units e $IJ = 6.4$ units f $KL = 4.5$ units
2 a $CD = 2.8$ units b $EP = 5.7$ units c $GH = 8.1$ units
d $IJ = 14.3$ units e $KL = 11.3$ units

Exercise 33D

1

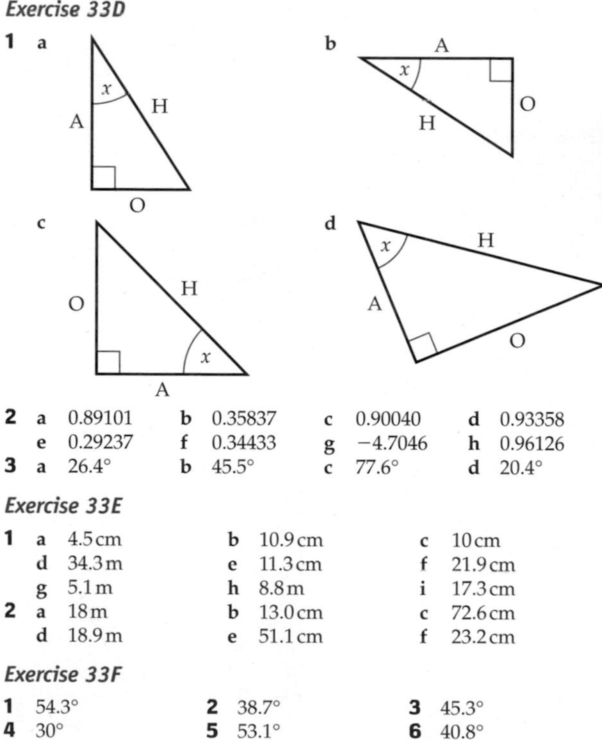

2 a 0.89101 b 0.35837 c 0.90040 d 0.93358
e 0.29237 f 0.34433 g −4.7046 h 0.96126
3 a 26.4° b 45.5° c 77.6° d 20.4°

Exercise 33E

1 a 4.5 cm b 10.9 cm c 10 cm
d 34.3 m e 11.3 cm f 21.9 cm
g 5.1 m h 8.8 m i 17.3 cm
2 a 18 m b 13.0 cm c 72.6 cm
d 18.9 m e 51.1 cm f 23.2 cm

Exercise 33F

1 54.3° 2 38.7° 3 45.3°
4 30° 5 53.1° 6 40.8°
7 47.0° 8 31.9° 9 28.6°

Exercise 33G

1 72.5 km
2 036.9°
3 26.7°
4 19 m (nearest metre)
5 a 8.78 cm b 72 cm²
6 98.1 cm²
7 bearing 218° (3 s.f.) distance 12.4 km (3 s.f.)
8 52.2 cm²
9 38.3°
10 845 cm

Exercise 33H

1 a 4cm b 14.46cm c $1734.8\,cm^3$
2 a $812.9\,cm^3$ b $303.2\,cm^2$
3 a $285.9\,cm^2$ b $548.6\,cm^3$ (4 s.f.)
4 $145.3\,cm^3$
5 24.98cm

Exercise 33I

1 23.4km
2 189.5m
3 99.8m
4 16.5m
5 131m

Exercise 33J

1 a $DH = 8.49$ cm b $DG = 10.39$ cm
2 a $SW = 8.06$ cm b $SV = 8.60$ cm
3 a $NQ = 15.6$ cm b 35.3°
4 a $BE = 11.0$ cm b 26.9°

Exercise 33K

1 a 8.7cm b 35.3°
2 24.7°
3 a 48.0° b 21.8°
4 66.7°

Review exercise

1 10.3m (1 d.p.)
2 £3.48
3 30.2%
4 $\sqrt{265}$
5 4.30m
6 47.7°
7 13.6km/h
8 43.1km
9 163m
10 $169.6\,cm^3$
11 a 15.6cm b 30.8°
12 64.9°
13 23°

Chapter 34

Exercise 34A

1 $AO = 5$ cm
2 $PQ = 16$ cm
3 a $OR = OS = 7$ cm b $OM = 5.00$ cm
c $ON = 6.58$ cm d RS
4 PQ is closer because it is longer than RS and so must be closer to the centre of the circle where the diameter is measured.
5 $x = 25°$
6 $y = 120°$.
7 $a = 70°, b = 20°$
8 $a = 42°, b = 96°$
9 13cm

Exercise 34B

1 $b = 75°$
2 $c = 134°$
3 $d = 120°$
4 $e = 60°, f = 30°$
5 $h = 132°, i = 48°, j = 66°$
6 $k = 70°, l = 70°, m = 70°$
7 $n = 120°, p = 60°, o = 45°$
8 $q = 65°$
9 $AOC = OAC = 60°$
10 Student's proof

Exercise 34C

1 $e = 40°, d = 37°$
2 $y = 60°$
3 $x = 31°, y = 31°, z = 81°$
4 $x = 107°, y = 98°$
5 $e = 79°, f = 79°$
6 137°
7 $x = 62°, y = 33°$
8 $b = 90°, d = 32°, c = 122°, a = 29°$
9 128°
10 a 144° b 115°
11 41°
12 120°

Exercise 34D

1 $a = 62°$
2 $b = 57°$
3 $c = 63°$
4 $d = 39°, e = 79°$
5 $h = 64°, g = 63°, f = 63°$
6 $j = 52°$
7 $k = 54°$
8 $p = 42°$
9 $l = 102°, m = 31°$
10 a Alternate angles b $t = 50°$
c $r = 50°, q = 65°$

11 a $x = 76°, y = 62°, v = 62°, u = 104°$
b $a = 54°, z = 72°$ c $x = 252°$
12 Student's proof

Review exercise

1 $AB = 16$ cm
2 a $a = 59°$ b $b = 56°$ c $c = 90°$ d $d = 95°$
3 a Cyclic quadrilateral b $x = 78°$
4 By alternate segment theorem $\angle NPQ = 70°$. Angles on a straight line equal 180°, so z = 180 − 70 − 52 = 58°.
5 a $\angle ABC = 65°, \angle ADC = 115°$
b $\angle OBA = 31°$
6 $\angle BPT = x$ (by alternate angle theorem), $\angle APB = 90°$ (angle within semi circle), so $\angle ABP = 90 - x$ and $\angle PBT = 90 + x$ (angles within triangle and on a straight line equal 180°). Hence $y = 180 - (90 + x) - x = 90° - 2x$.
7 a i $p = 140°$ ii $q = 70°$
b i $y = 42°$ ii $x = 124°$
8 $\angle BDA = 38°$ (by alternate angle theorem), $\angle DCB = 104°$ (supplementary angle within cyclic quadrilateral), $\angle CBD = \angle CDB = 38°$ (since BCD is an isosceles triangle). Angles in a quadrilateral sum to 360° so $\angle CBA = 104° = \angle DCB$ and $\angle ADC = 76° = \angle BAD$. Hence $ABCD$ is a trapezium and AD is parallel to BC.

Chapter 35

Algebra skills: revision exercise

1 a $(x + 3)(x - 3)$ b $(a + 2)(a - 2)$
c $(t + 1)(t - 1)$ d $(4x + 5y)(4x - 5y)$
e $3(2a + 3b)(2a - 3b)$ f $4(3x + y)(3x - y)$
2 a $(x + 1)(x + 3)$ b $(a + 3)(a + 6)$
c $(r + 2)(r + 5)$ d $(m + 2)(m - 4)$
e $(x + 3)(x - 6)$ f $(x - 4)(x - 4)$
g $(4m - 3)(3m - 1)$ h $(3x + 7)(x - 2)$
i $(6x + 3)(4x - 7)$
3 a ± 5 b ± 6 c ± 9
d 0 or 10 e −7 or 1 f −14 or 0
g $-\frac{1}{5}$ or 2 h −3 or $\frac{1}{2}$ i −7 or 5
4 a −2 or 2.5 b 0.39 or 4.28 c 0.59 or 3.41
d 0.15 or 2.18 e −5.74 or 1.74 f −0.82 or 1.22
5 a one solution b two solutions c two solutions
d two solutions e two solutions f no solutions
6 a $(x + 7)^2 - 54$ b $(y + \frac{1}{2})^2 - 1.25$
c $(x + 4)^2 - 22$ d $(x - 1.5)^2 - 3.25$
7 $p = 5, q = 15$
8 a $-1 \pm \sqrt{10}$ b $2 \pm \sqrt{5}$ c $\frac{5 \pm 3\sqrt{5}}{2}$
d $-3 \pm \sqrt{10}$ e $-3 \pm 2\sqrt{3}$ f $-2 \pm \sqrt{11}$

Exercise 35A

1 a

x	−4	−3	−2	−1	0	1	2	3	4
x^2	16	9	4	1	0	1	4	9	16
+2	+2	+2	+2	+2	+2	+2	+2	+2	+2
$y = x^2 + 2$	18	11	6	3	2	3	6	11	18

b

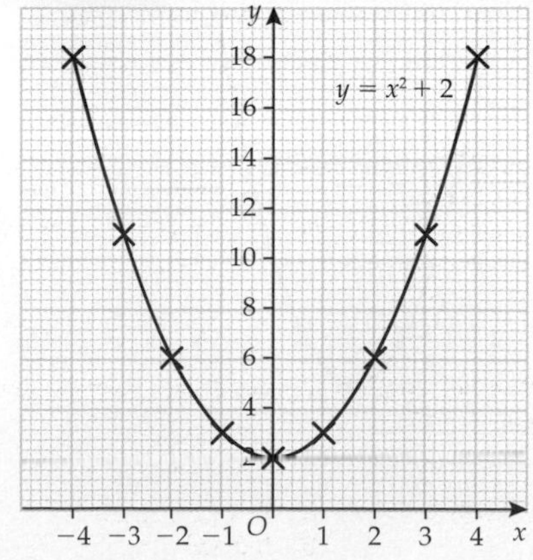

2 a

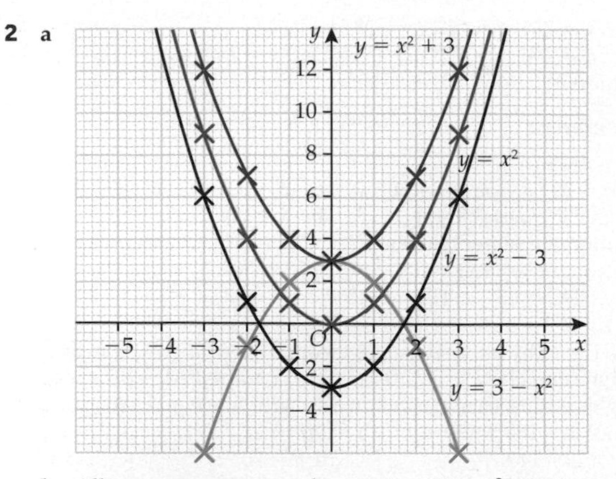

b All curves open upwards except $y = 3 - x^2$. The lowest point of $y = x^2$ is (0, 0); $y = x^2 + 3$ is (0, 3); and $y = x^2 - 3$ is (0, −3). The highest point of $y = 3 - x^2$ is (0, 3).

3 a

x	−2	−1	0	1	2	3	4	5
x^2	4	1	0	1	4	9	16	25
$-3x$	6	3	0	−3	−6	−9	−12	−15
$y = x^2 - 3x$	10	4	0	−2	−2	0	4	10

b

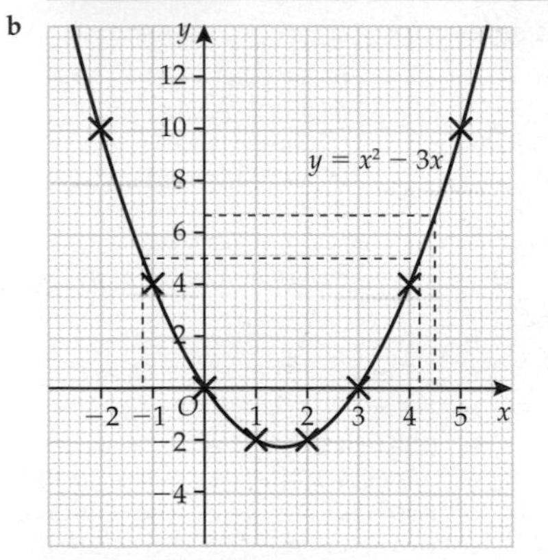

c 6.8

d $x = -1.2$ or 4.2

4 a

x	−2	−1	0	1	2	3
$2x^2$	8	2	0	2	8	18
$-4x$	8	4	0	−4	−8	−12
-1	−1	−1	−1	−1	−1	−1
$y = 2x^2 - 4x - 1$	15	5	−1	−3	−1	5

b

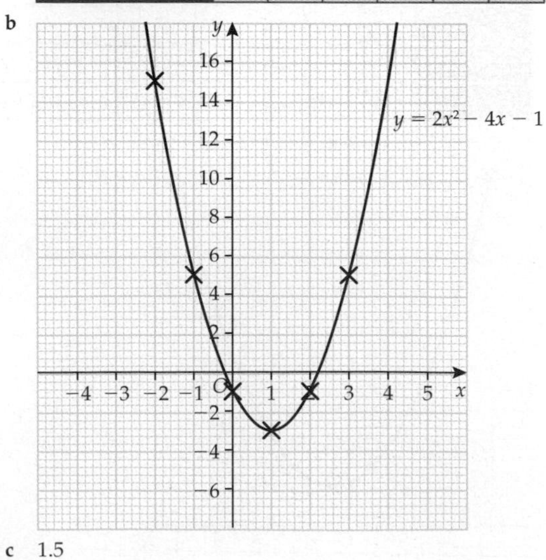

c 1.5

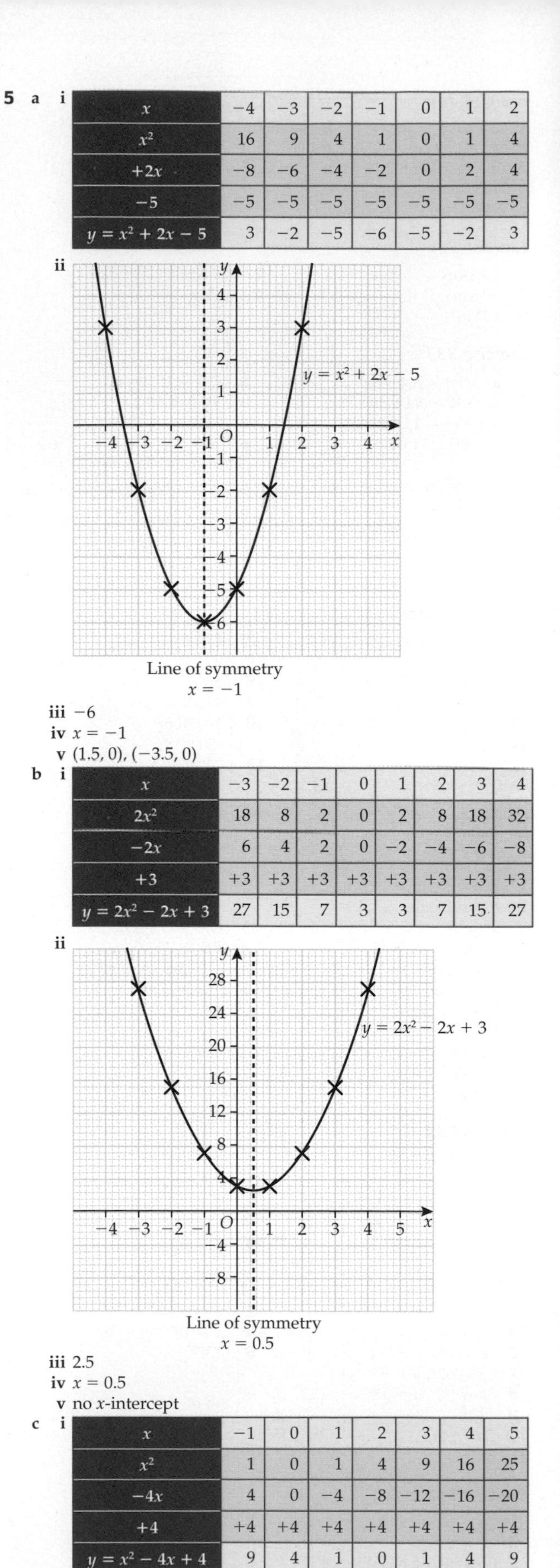

5 a i

x	−4	−3	−2	−1	0	1	2
x^2	16	9	4	1	0	1	4
$+2x$	−8	−6	−4	−2	0	2	4
-5	−5	−5	−5	−5	−5	−5	−5
$y = x^2 + 2x - 5$	3	−2	−5	−6	−5	−2	3

ii

Line of symmetry
$x = -1$

iii −6

iv $x = -1$

v (1.5, 0), (−3.5, 0)

b i

x	−3	−2	−1	0	1	2	3	4
$2x^2$	18	8	2	0	2	8	18	32
$-2x$	6	4	2	0	−2	−4	−6	−8
$+3$	+3	+3	+3	+3	+3	+3	+3	+3
$y = 2x^2 - 2x + 3$	27	15	7	3	3	7	15	27

ii

Line of symmetry
$x = 0.5$

iii 2.5

iv $x = 0.5$

v no x-intercept

c i

x	−1	0	1	2	3	4	5
x^2	1	0	1	4	9	16	25
$-4x$	4	0	−4	−8	−12	−16	−20
$+4$	+4	+4	+4	+4	+4	+4	+4
$y = x^2 - 4x + 4$	9	4	1	0	1	4	9

ii

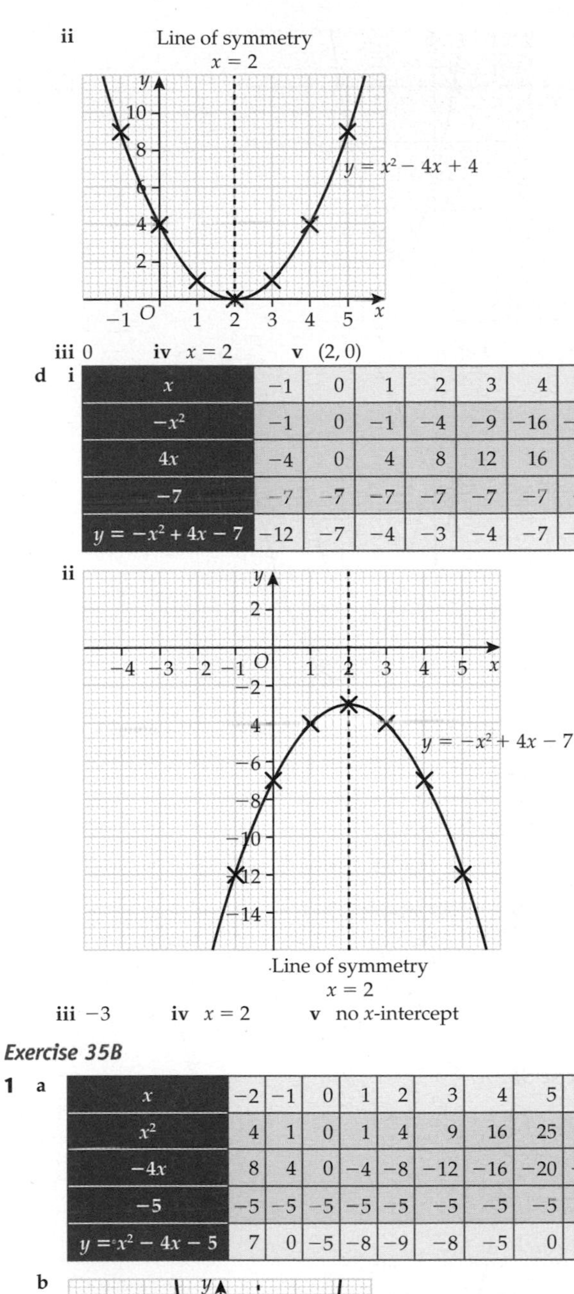

iii 0 **iv** $x = 2$ **v** (2, 0)

d i

x	-1	0	1	2	3	4	5
$-x^2$	-1	0	-1	-4	-9	-16	-25
$4x$	-4	0	4	8	12	16	20
-7	-7	-7	-7	-7	-7	-7	-7
$y = -x^2 + 4x - 7$	-12	-7	-4	-3	-4	-7	-12

ii

iii -3 **iv** $x = 2$ **v** no x-intercept

Exercise 35B

1 a

x	-2	-1	0	1	2	3	4	5	6
x^2	4	1	0	1	4	9	16	25	36
$-4x$	8	4	0	-4	-8	-12	-16	-20	-24
-5	-5	-5	-5	-5	-5	-5	-5	-5	-5
$y = x^2 - 4x - 5$	7	0	-5	-8	-9	-8	-5	0	7

b

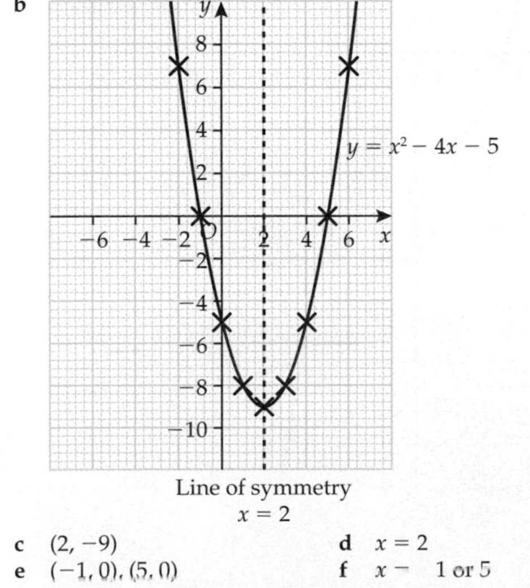

c $(2, -9)$ **d** $x = 2$

e $(-1, 0), (5, 0)$ **f** $x = -1$ or 5

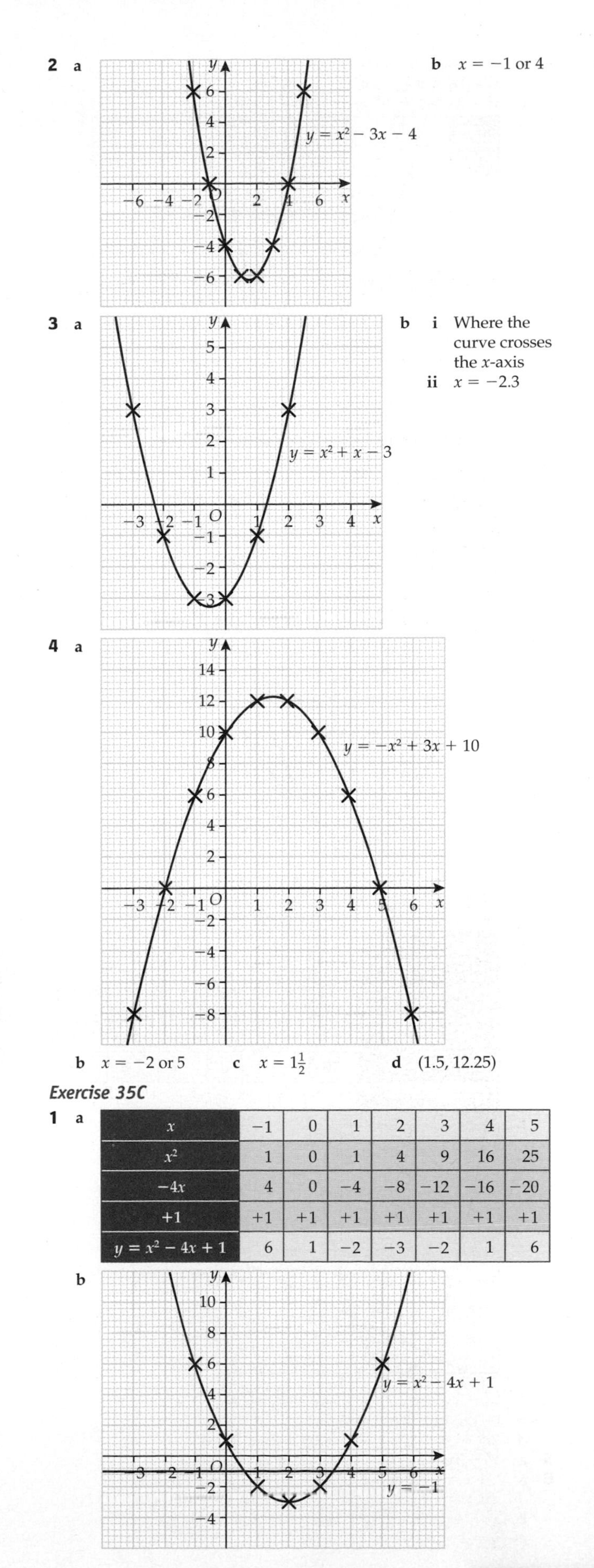

2 a

b $x = -1$ or 4

3 a

b i Where the curve crosses the x-axis

ii $x = -2.3$

4 a

b $x = -2$ or 5 **c** $x = 1\frac{1}{2}$ **d** (1.5, 12.25)

Exercise 35C

1 a

x	-1	0	1	2	3	4	5
x^2	1	0	1	4	9	16	25
$-4x$	4	0	-4	-8	-12	-16	-20
$+1$	$+1$	$+1$	$+1$	$+1$	$+1$	$+1$	$+1$
$y = x^2 - 4x + 1$	6	1	-2	-3	-2	1	6

b

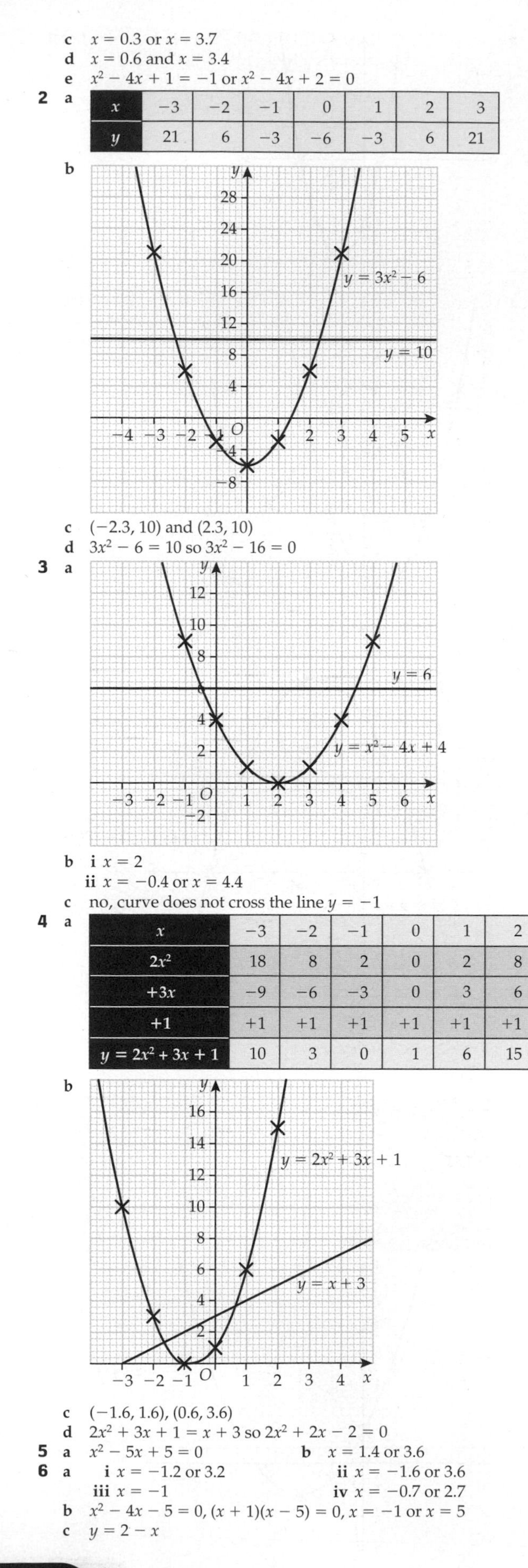

c $x = 0.3$ or $x = 3.7$

d $x = 0.6$ and $x = 3.4$

e $x^2 - 4x + 1 = -1$ or $x^2 - 4x + 2 = 0$

2 a

x	−3	−2	−1	0	1	2	3
y	21	6	−3	−6	−3	6	21

b

c (−2.3, 10) and (2.3, 10)

d $3x^2 - 6 = 10$ so $3x^2 - 16 = 0$

3 a

b i $x = 2$

ii $x = -0.4$ or $x = 4.4$

c no, curve does not cross the line $y = -1$

4 a

x	−3	−2	−1	0	1	2
$2x^2$	18	8	2	0	2	8
$+3x$	−9	−6	−3	0	3	6
$+1$	+1	+1	+1	+1	+1	+1
$y = 2x^2 + 3x + 1$	10	3	0	1	6	15

b

c (−1.6, 1.6), (0.6, 3.6)

d $2x^2 + 3x + 1 = x + 3$ so $2x^2 + 2x - 2 = 0$

5 a $x^2 - 5x + 5 = 0$ b $x = 1.4$ or 3.6

6 a i $x = -1.2$ or 3.2 ii $x = -1.6$ or 3.6

iii $x = -1$ iv $x = -0.7$ or 2.7

b $x^2 - 4x - 5 = 0$, $(x + 1)(x - 5) = 0$, $x = -1$ or $x = 5$

c $y = 2 - x$

7 a

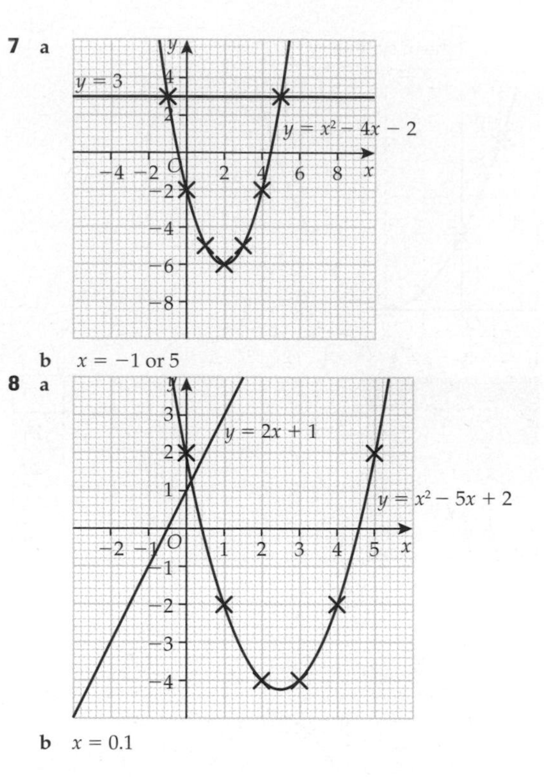

b $x = -1$ or 5

8 a

b $x = 0.1$

Exercise 35D

1 −6 or 2

2 3 or 5

3 a $(4 + x)(2x - 7) = 2x^2 + x - 28$

b $x = 5$

4 a width = 12 m, length = 20 m

b perimeter = 64 m

5 9 cm, 12 cm, 15 cm

6 $50\,m^2$

7 (5, 11) and (−1, −1)

8 (−2, 0) and (4, 6)

Exercise 35E

1

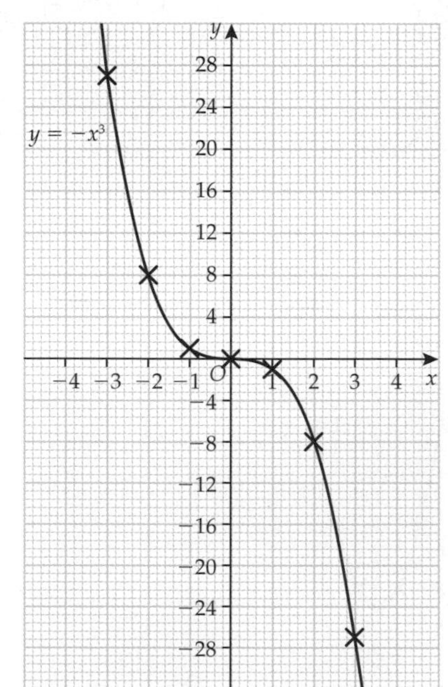

2 a

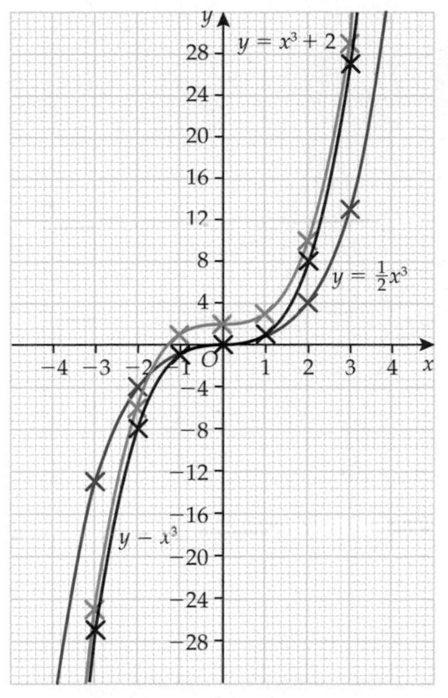

b Both $y = x^3$ and $y = \frac{1}{2}x^3$ pass through the origin; $y = x^3 + 2$ intercepts the y-axis at (0, 2). All have the same general shape; $y = \frac{1}{2}x^3$ lies further away from the y-axis

3 a

x	−2.5	−2	−1	0	1	2	2.5
x^3	−15.6	−8	−1	0	1	8	15.6
$-2x$	5	4	2	0	−2	−4	−5
$+2$	+2	+2	+2	+2	+2	+2	+2
$y = x^3 - 2x + 2$	−8.6	−2	3	2	1	6	12.6

b

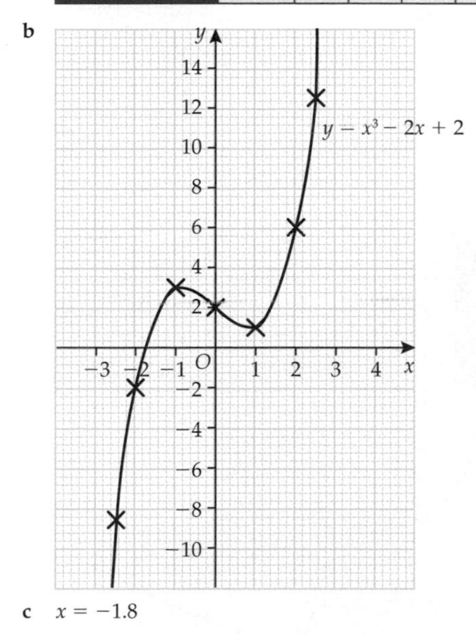

c $x = -1.8$

4 a

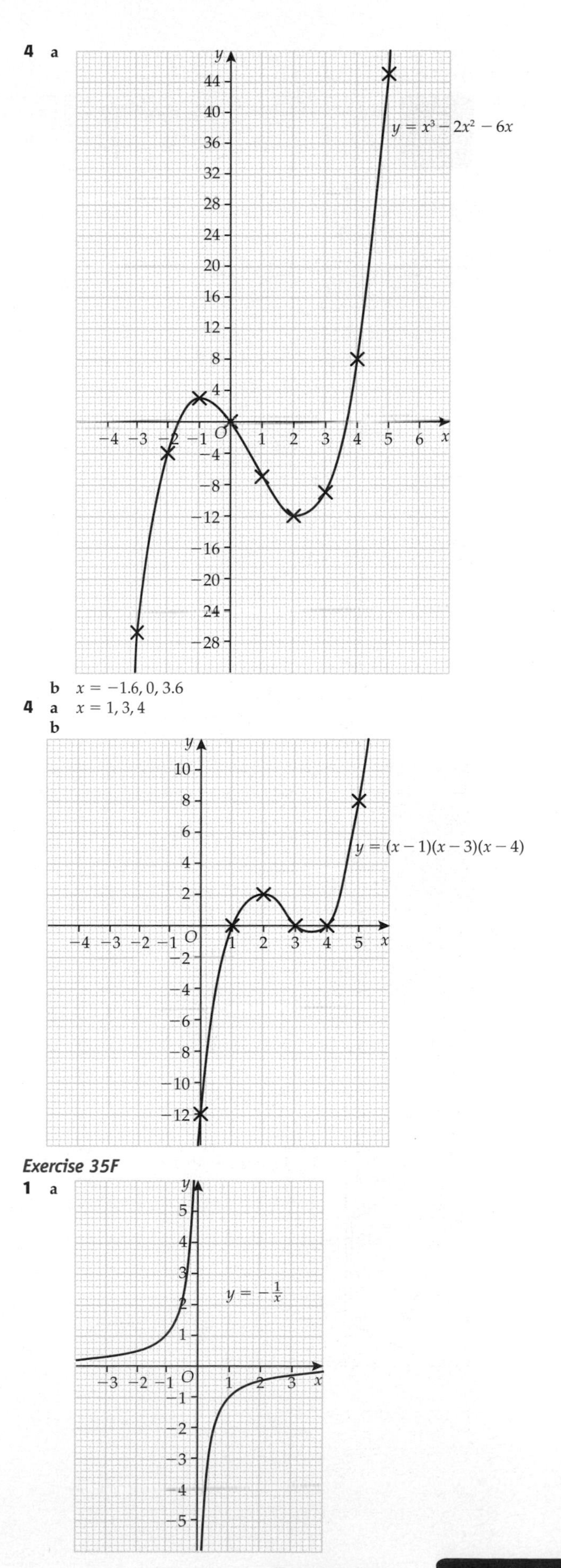

b $x = -1.6, 0, 3.6$

4 a $x = 1, 3, 4$

b

Exercise 35F

1 a

b The graphs have the same general form but $y = -\frac{a}{x}$ is symmetrical about $y = x$.

2 a

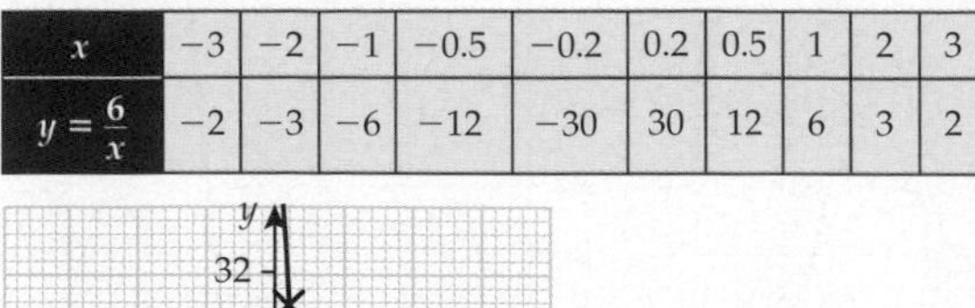

x	−3	−2	−1	−0.5	−0.2	0.2	0.5	1	2	3
$y = \frac{6}{x}$	−2	−3	−6	−12	−30	30	12	6	3	2

b

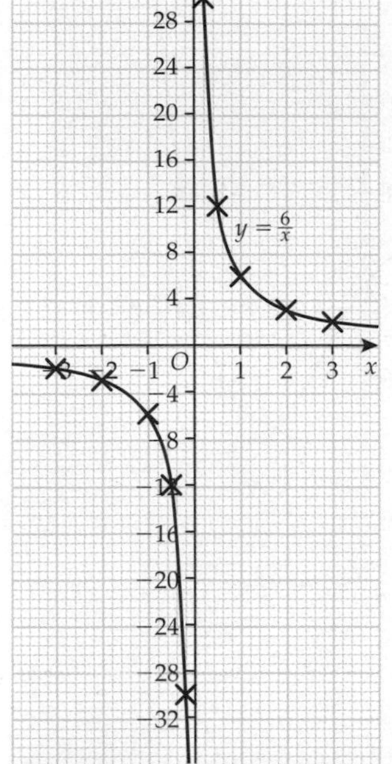

3 a

x	−3	−2	−1	−0.5	−0.2	0.2	0.5	1	2	3
$\frac{3}{x}$	−1	−1.5	−3	−6	−15	15	6	3	1.5	1
+5	+5	+5	+5	+5	+5	+5	+5	+5	+5	+5
$y = \frac{3}{x} + 5$	4	3.5	2	−1	−10	20	11	8	6.5	6

b

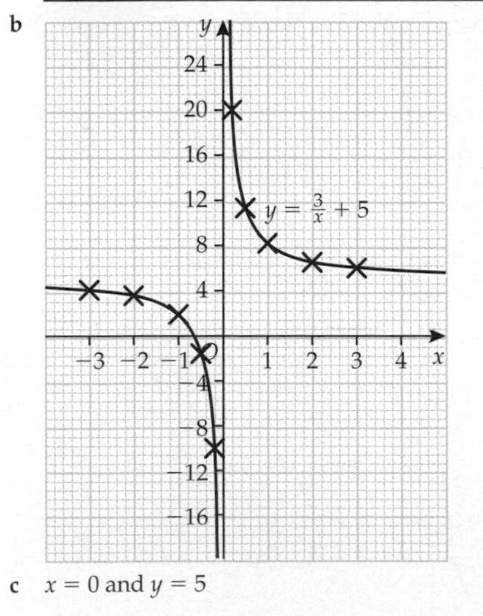

c $x = 0$ and $y = 5$

4 a

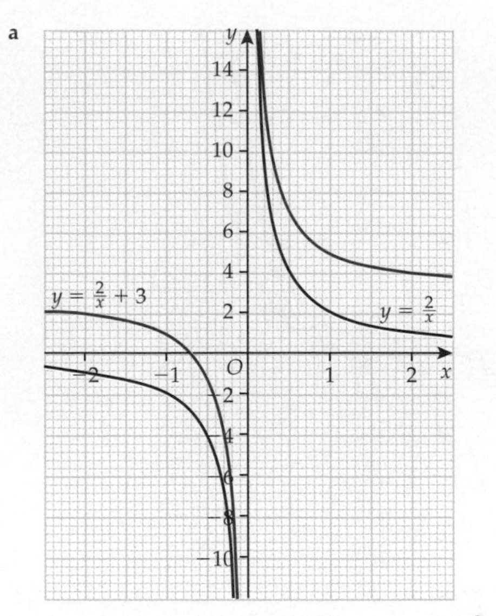

b The graphs have the same general form; $y = \frac{2}{x}$ is translated +3 vertically up to produce the graph of $y = \frac{2}{x} + 3$.

5 a

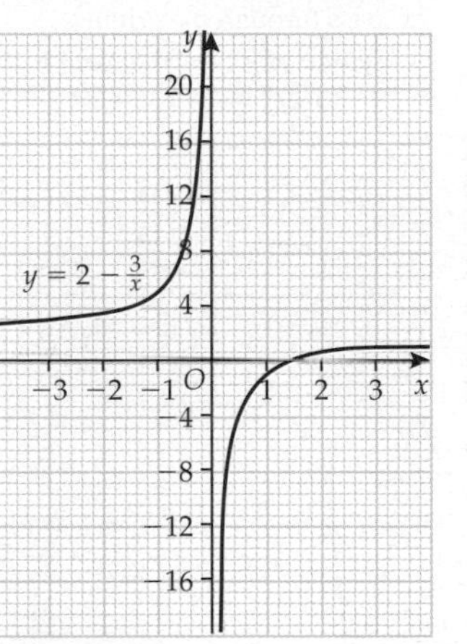

b $x = 0$ and $y = 2$

Exercise 35G

1 a

x	0.1	0.2	0.5	1	2	3	4	5
$\frac{1}{x}$	10	5	2	1	0.5	0.33	0.25	0.2
$y = \frac{1}{x} + x$	10.1	5.2	2.5	2	2.5	3.33	4.25	5.2

b

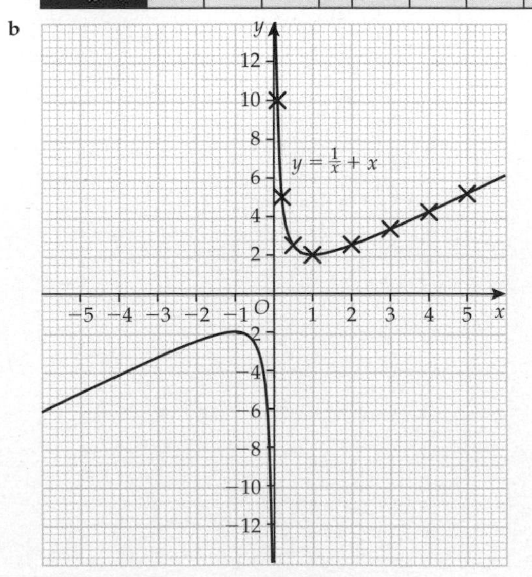

2

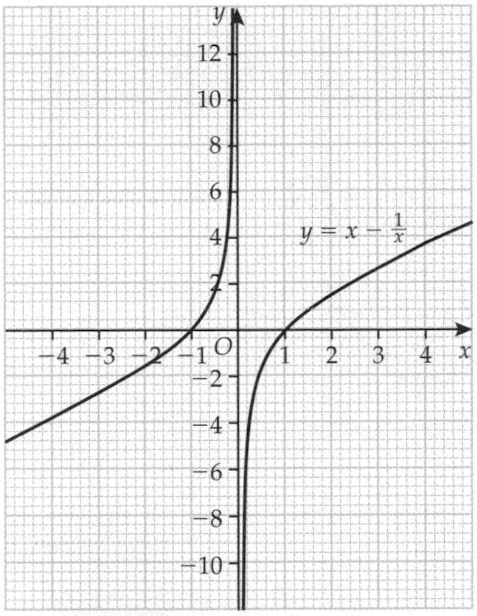

3 **a**

x	−4	−3	−2	−1	−0.5	−0.2
x^2	16	9	4	1	0.25	0.04
$\frac{8}{x}$	−2	−2.67	−4	−8	−16	−40
$y = x^2 + \frac{8}{x}$	14	6.33	0	−7	−15.75	−39.96

x	0.2	0.5	1	2	3	4
x^2	0.04	0.25	1	4	9	16
$\frac{8}{x}$	40	16	8	4	2.67	2
$y = x^2 + \frac{8}{x}$	40.04	16.25	9	8	11.67	18

b

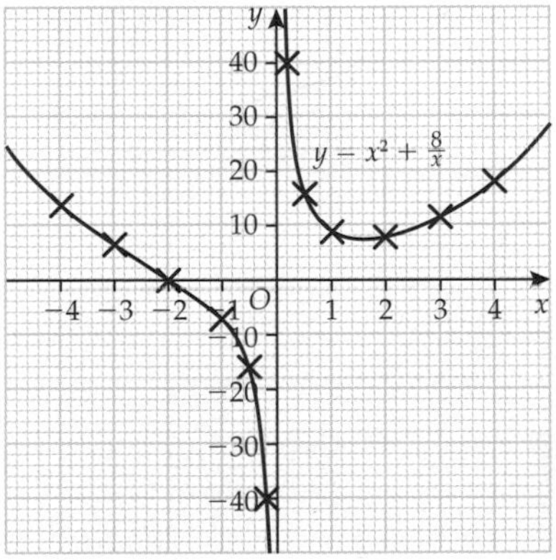

c −2

d $x^2 + \frac{8}{x} = 0$

4

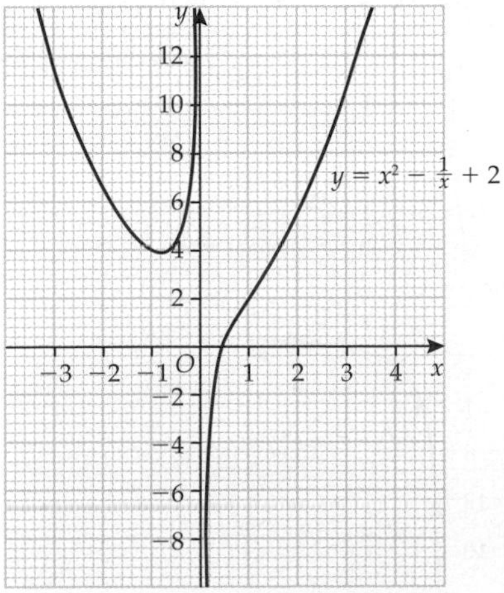

Exercise 35H

1 **a**

x	−3	−2	−1	0	1	2	3	4
$y = (\frac{1}{2})^x$	8	4	2	1	0.5	0.25	0.125	0.0625

b

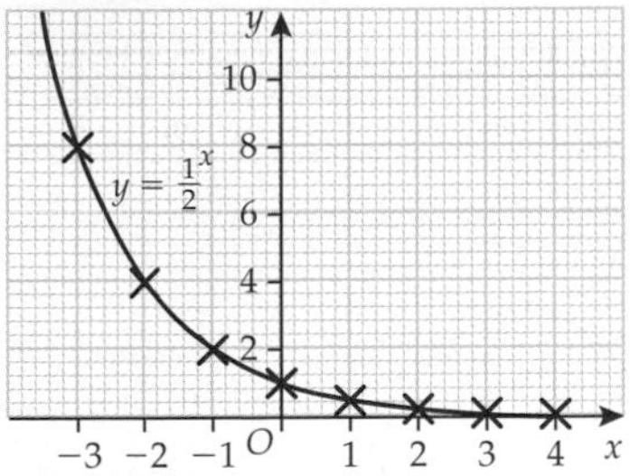

c −2.3

2 **a**

x	−1	0	1	2	3	4	5
$y = 1.5^{-x}$	1.5	1	0.67	0.44	0.3	0.2	0.13

b

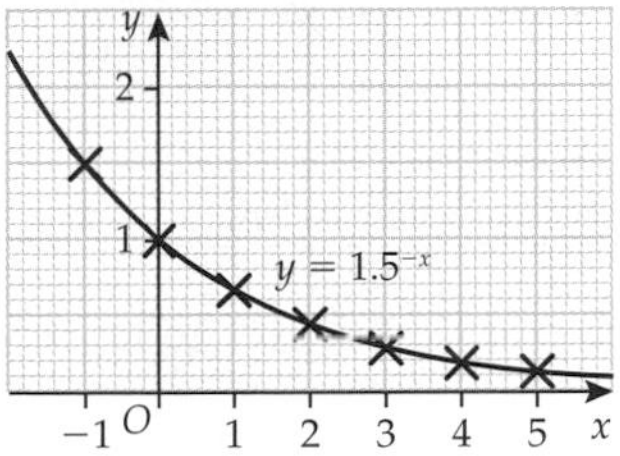

c 0.24

3 **a** Student's explanation

b

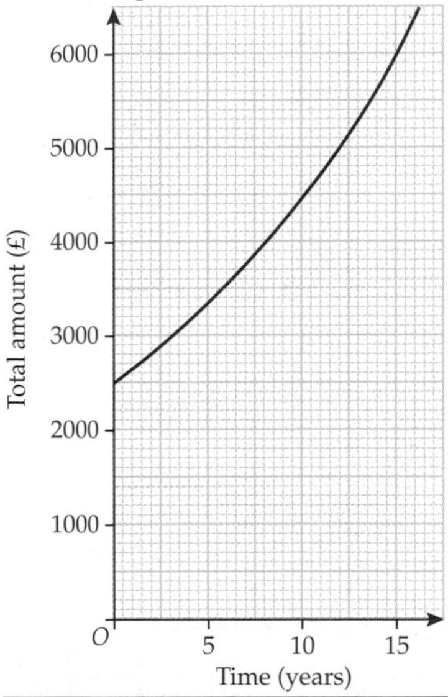

c ≈ 12 years

4 **a**

t	0	20	40	60	80	100	120	140
N	80	40	20	10	5	2.5	1.25	0.6

b

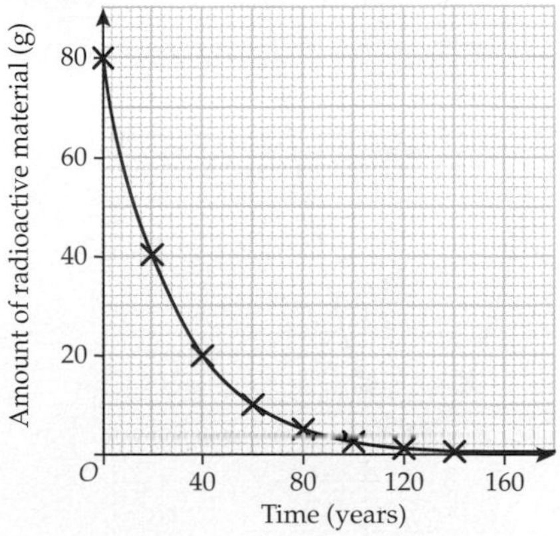

c i 57 g ii 14 g

d The rate of decline would be twice as fast as that for $N = 80 \times 2^{-\frac{t}{20}}$.

5 a i 400 ii 126 b i 5 hours ii ≈ 30 minutes

Exercise 35I

1 a

t	0	1	2	3	4	5	6
h	0	25	40	45	40	25	0

b

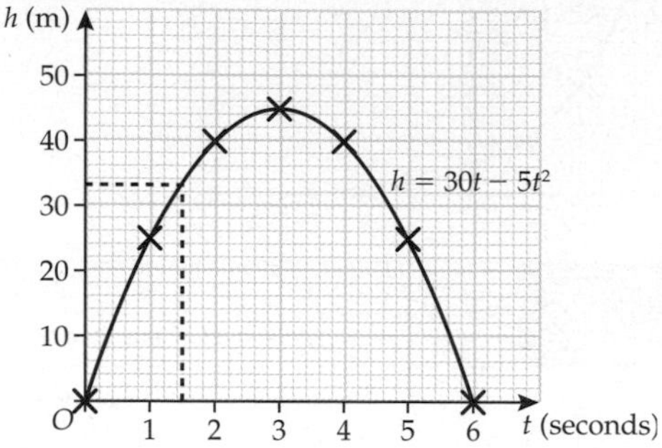

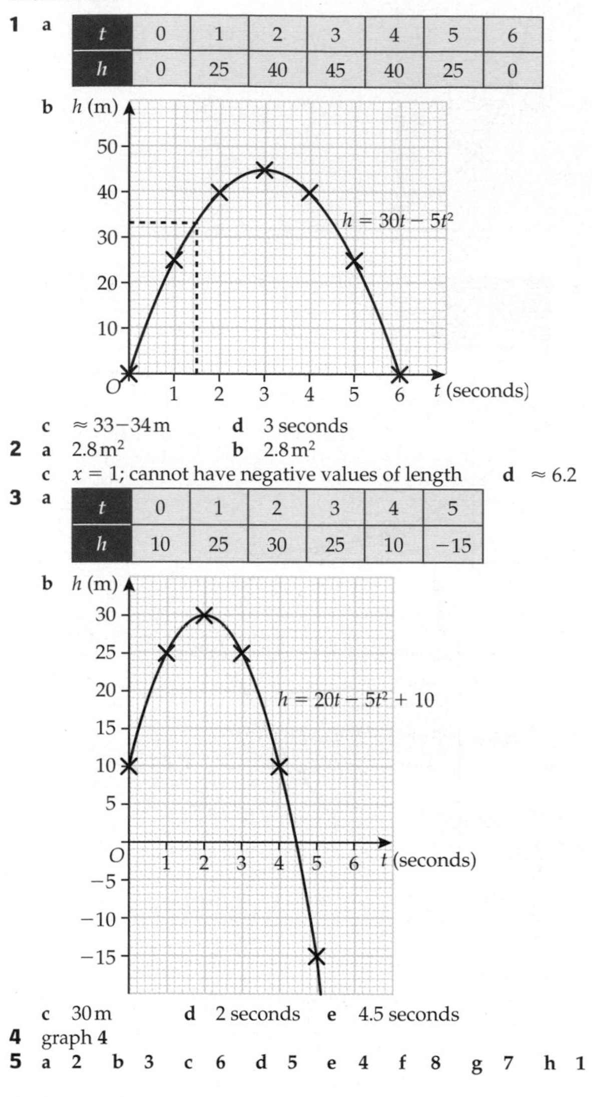

c ≈ 33–34 m d 3 seconds

2 a 2.8 m² b 2.8 m²

c $x = 1$; cannot have negative values of length d ≈ 6.2

3 a

t	0	1	2	3	4	5
h	10	25	30	25	10	−15

b

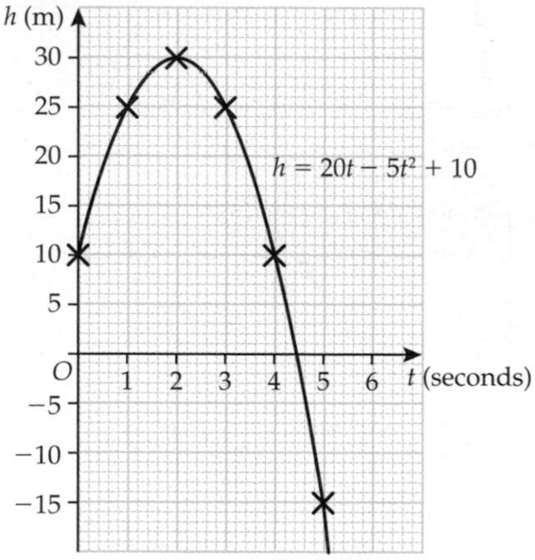

c 30 m d 2 seconds e 4.5 seconds

4 graph 4

5 a 2 b 3 c 6 d 5 e 4 f 8 g 7 h 1

Review exercise

1 a

x	−3	−2	−1	0	1	2
y	−2	−5	−6	−5	−2	3

b

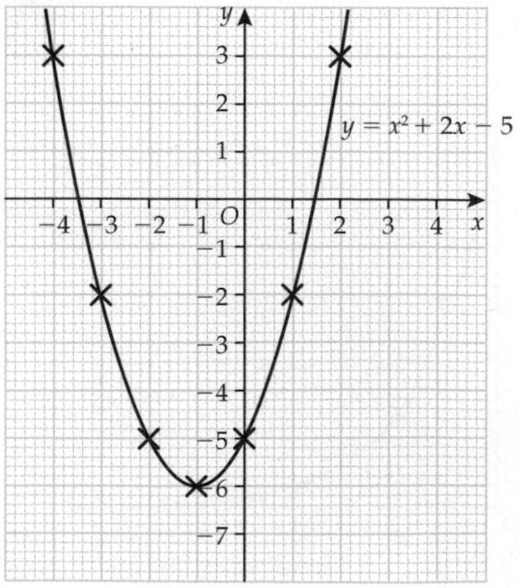

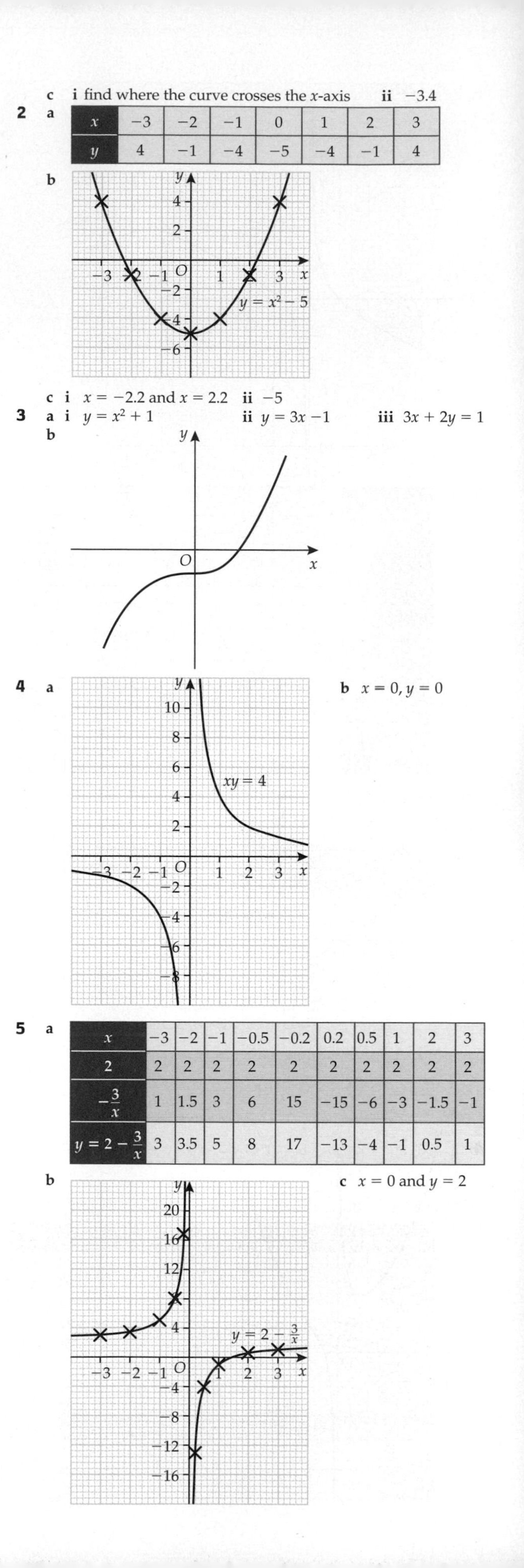

c i find where the curve crosses the x-axis ii −3.4

2 a

x	−3	−2	−1	0	1	2	3
y	4	−1	−4	−5	−4	−1	4

b

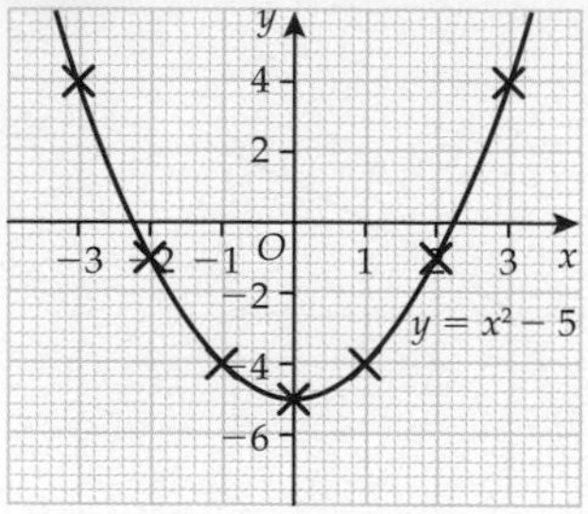

c i $x = -2.2$ and $x = 2.2$ ii −5

3 a i $y = x^2 + 1$ ii $y = 3x - 1$ iii $3x + 2y = 1$

b

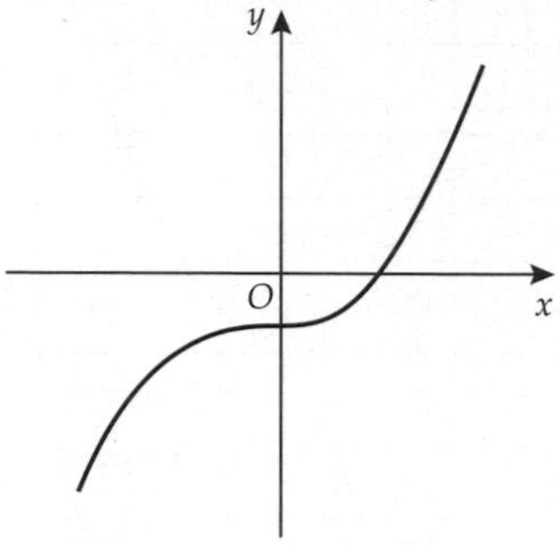

4 a

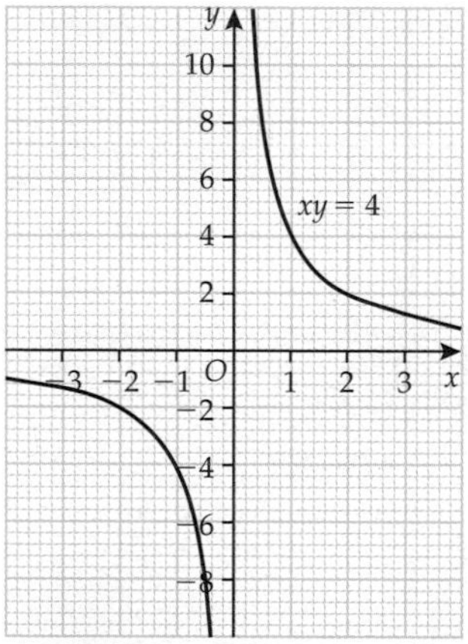

b $x = 0, y = 0$

5 a

x	−3	−2	−1	−0.5	−0.2	0.2	0.5	1	2	3
2	2	2	2	2	2	2	2	2	2	2
$-\frac{3}{x}$	1	1.5	3	6	15	−15	−6	−3	−1.5	−1
$y = 2 - \frac{3}{x}$	3	3.5	5	8	17	−13	−4	−1	0.5	1

b

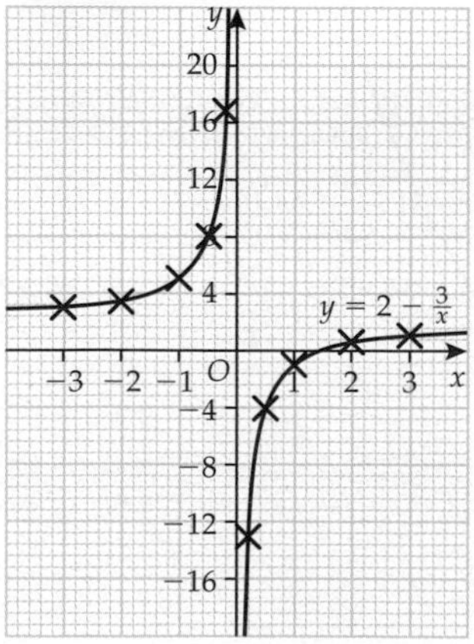

c $x = 0$ and $y = 2$

6 a

x	0	1	2	3	4	5
y	1	0.75	0.56	0.42	0.32	0.24

b

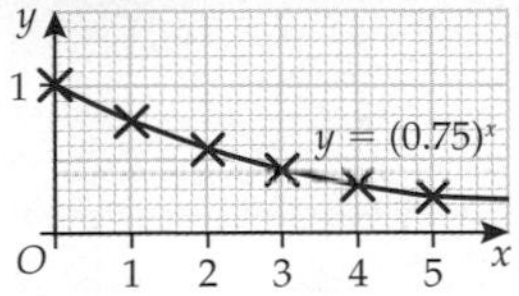

c $x = 1.5$

7 a

x	−1	0	1	2	3	4	5	6
y	3	−3	−7	−9	−9	−7	−3	3

b

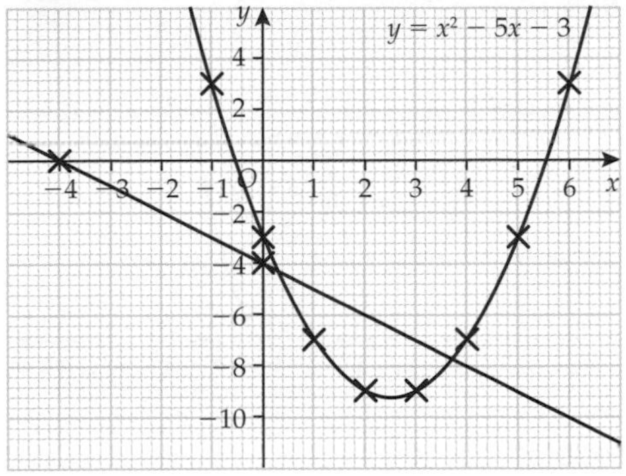

c $x = -0.5$ or 5.5   d $x = 0.3$ or 3.7

8 0.8 or 5

9 a $16 - x$

b $(16 - x)(x + 4) = -x^2 + 12x + 64 = 96$
so $x^2 - 12x + 32 = 0$

c 4 or 8

10 a $x = -0.3$ or 3.3   b $x = 0.2$ or 4.8

Chapter 36

Exercise 36A

1

	Angle turned through	Positive or negative?		
		$\sin\theta$	$\cos\theta$	$\tan\theta$
a	170°	positive	negative	negative
b	230°	negative	negative	positive
c	50°	positive	positive	positive
d	350°	negative	positive	negative
e	460°	positive	negative	negative
f	−20°	negative	positive	negative
g	−100°	negative	negative	positive

2 30°

3 320°

4 a 30° and 150°   b 40° and 320°   c 45° and 225°
Other answers possible.

Exercise 36B

1 a

θ	0°	15°	30°	45°	60°	75°	90°
$\sin\theta$	0	0.26	0.5	0.71	0.87	0.97	1
$\cos\theta$	1	0.97	0.87	0.71	0.5	0.26	0

θ	105°	120°	135°	150°	165°	180°
$\sin\theta$	0.97	0.87	0.71	0.5	0.26	0
$\cos\theta$	−0.26	−0.5	−0.71	−0.87	−0.97	−1

θ	195°	210°	225°	240°	255°	270°
$\sin\theta$	−0.26	−0.5	−0.71	−0.87	−0.97	−1
$\cos\theta$	−0.97	−0.87	−0.71	−0.5	−0.26	0

θ	285°	300°	315°	330°	345°	360°
$\sin\theta$	−0.97	−0.87	−0.71	−0.5	−0.26	0
$\cos\theta$	0.26	0.5	0.71	0.87	0.97	1

b Student's graph of $y = \sin\theta$

c Student's graph of $y = \cos\theta$

d If you only plotted the values originally given, it would still be easy to see the trend of each curve, because of the symmetry of the curves.

2 a Student's sketch of $y = \sin\theta$ and $y = \cos\theta$

b $\theta = 45°$ and 225°

Exercise 36C

1 a $x = 53°$ and 127°   b $x = 41°$ and 319°
c $x = 194°$ and 346°   d $x = 107°$ and 253°
e $x = 60°$ and 120°   f $x = 45°$ and 315°

2 a $a = \pm 0.5$ b Student's sketch of $y = \cos x$
c $x - 60°$, 120°, 240° and 300°

3 a $x = 322°$ b $x = 142°$ and 218°
c $x = 52°$ and 128°

Exercise 36D

1 a 31.8 cm²   b 21.5 cm²   c 7.1 cm²
d 20.1 cm²   e 1.5 cm²   f 12.5 cm²

2 a 12.1 cm²   b 12.1 cm²   c $\frac{1}{2}$

3 36.4 cm²

4 56.4°

5 a 20.0°   b 160.0°

6 a i $h^2 + 1^2 = 2^2$
$h^2 = 4 - 1$
$h = \sqrt{3}$

ii Angles in an equilateral triangle are all 60°, so:
$$\sin 60° = \frac{\text{opposite}}{\text{hypotenuse}} = \frac{h}{2} = \frac{\sqrt{3}}{2}$$

b i Student's sketch of the tetrahedron

ii
$$\text{Area of one face} = \frac{1}{2} \times a \times a \times \sin 60° = \frac{1}{2}a^2 \times \frac{\sqrt{3}}{2} = \frac{\sqrt{3}}{4}a^2$$
$$\text{Surface area of tetrahedron} = 4 \times \text{area of one face} = \sqrt{3}a^2$$

Exercise 36E

1 a 13.9 m²   b 150.0 cm²

2 0.0138 m²

3 76.4 cm²

4 21.5 cm²

Exercise 36F

1 a $a = 3.33$ cm   b $x = 12.1$ mm
c $f = 90.0$ cm, $e = 70$ cm   d 5.81 cm

2 a 35°   b 8.8 m   c 7.7 m

3 a Student's sketch of triangle ABC b 15.6 km

4 a $\theta = 35.8°$ b $\theta = 25.1°$

5 a Student's accurate construction
b Angle Z = 53.1° or 126.9°

Exercise 36G

1 a 8.75 cm   b 1.92 m   c 44.0 km   d 4.59 cm

2 a 48.3°   b 99.6°   c 90°   d 97.2°

3 130.5°

4 a Student's sketch of Ashton, Benby and Crayton
b 7.6 km

5 a Neela is correct. Danny has incorrectly subtracted 48 from 52 (he cannot do this as the 48 is a multiple of cos 60°).
b Angle A is obtuse

6 a 119 km (to the nearest km)   b 311.5°

7 a i 60.7°   ii 78.7°
b 1978.8 m²
c $\frac{1978.8}{10000} = 0.197\ldots \approx 0.2$

Exercise 36H

1 a $17.7\,\text{cm}^2$ b $53.2\,\text{cm}^3$ c $37.2\,\text{g}$
2 $7.7\,\text{cm}$ and $14.6\,\text{cm}$
3 a $48.8\,\text{km}$ b $311.2°$
4 a $8.1\,\text{m}$ b $31.7°$
5 $18.0\,\text{cm}^2$
6 $74.5°$
7 a $1500 = \frac{1}{2} \times 50 \times 63 \times \sin\theta;\ \theta = 72.2°$
b $67.4\,\text{m}$ c $47.6\,\text{m}$

Review exercise

1 a $14.3\,\text{cm}$ b $41.2°$
2 a $4.5\,\text{cm}$ b $23.6\,\text{cm}^2$
3 $113.3°$
4 $30.5\,\text{cm}$
5 $4.1\,\text{cm}^2$
6 $33.7\,\text{m}$
7 a Student's sketch of $y = \sin x$ b $154°$
c $\sin 334° = \sin(360° - 26°)$. From the graph's symmetry, we can conclude that $\sin 334° = -0.438$

Chapter 37

Exercise 37A

1 a $f(x) = 3x$ b $f(x) = x^2 + 2$
c $f(x) = x^3 - 2x$ d $f(x) = \sin x$
2 a 8 b 13 c 13 d 20 e 4.25
3 a 0 b 6 c 3 d 36 e -0.12
4 -5 or 2
5 -2 or $\frac{7}{2}$

Exercise 37B

1 a, b

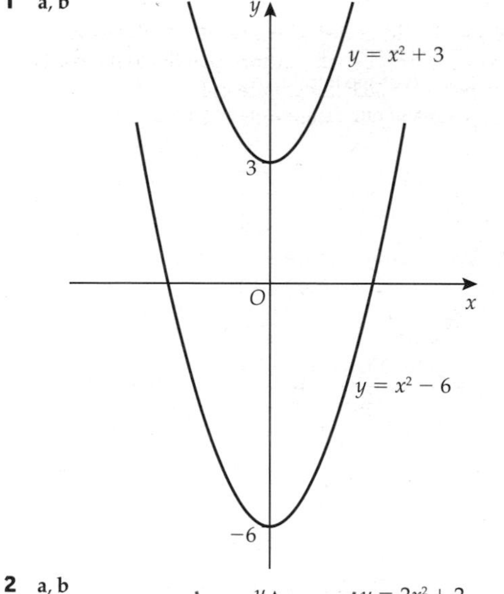

2 a, b

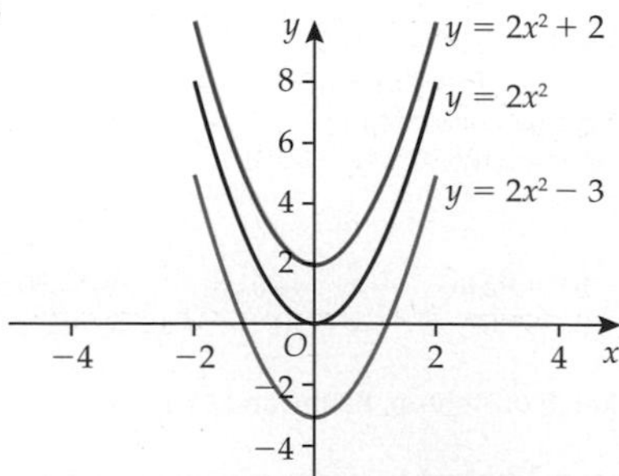

c The graph of $y = 2x^2 + 2$ is drawn by translating the graph of $y = 2x^2$ by the vector $\begin{pmatrix} 0 \\ 2 \end{pmatrix}$.
The graph of $y = 2x^2 - 3$ is drawn by translating the graph of $y = 2x^2$ by the vector $\begin{pmatrix} 0 \\ -3 \end{pmatrix}$.

3 $y = 3x^2 - 8$

4 a, b c $y = x^2 + 2x + 1$

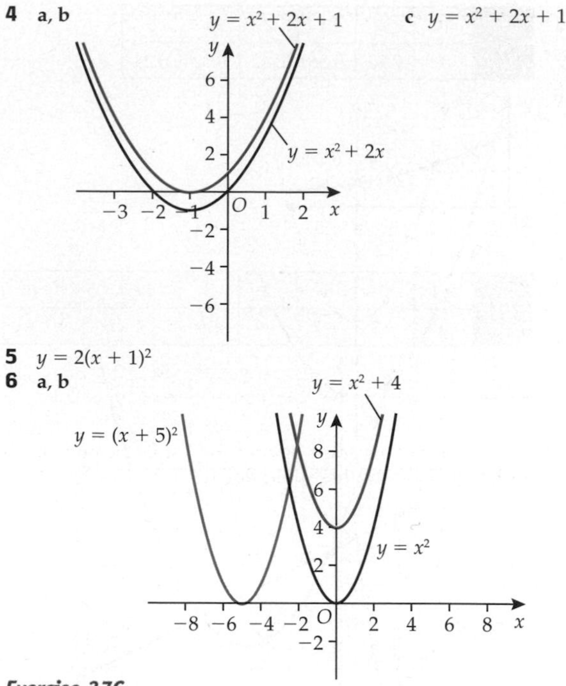

5 $y = 2(x + 1)^2$
6 a, b

Exercise 37C

1 a, b

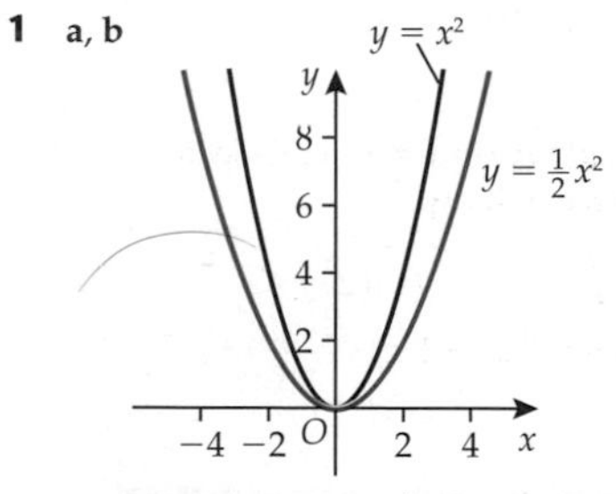

c The graph of $y = \frac{1}{2}x^2$ has been vertically reduced by a factor of $\frac{1}{2}$.

2

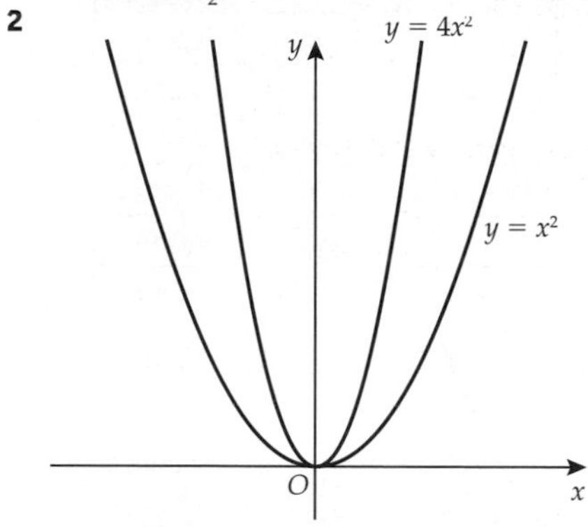

3 a a stretch by scale factor 5 parallel to the y-axis
b a stretch by scale factor 8 parallel to the y-axis
c a reduction by scale factor $\frac{1}{3}$ parallel to the y-axis
d a stretch from the y-axis, parallel to the x-axis, of scale factor 2
4 $y = \frac{1}{4}x^2$
5 a

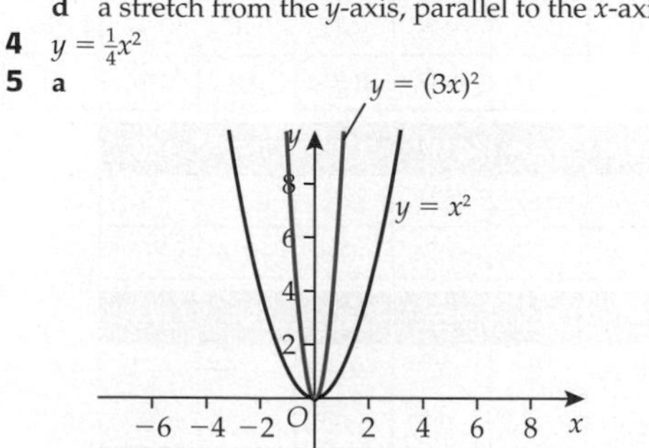

b a stretch by scale factor 9 parallel to the y-axis

Exercise 37D

1 **a, b**

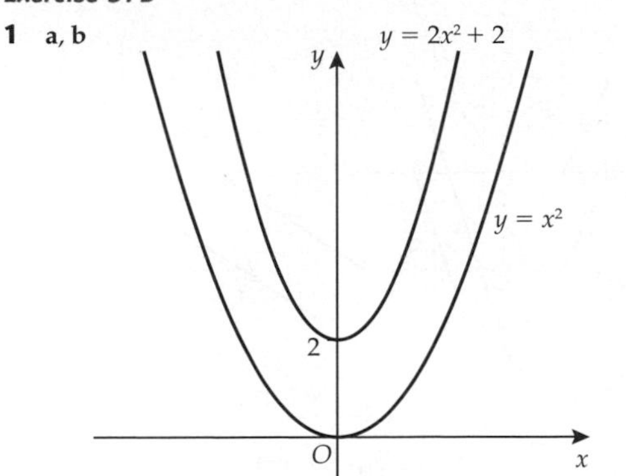

2 **a** Vertical stretch of the graph of $y = x^2$ by a scale factor of 4.
Translation of $y = 4x^2$ three units down.

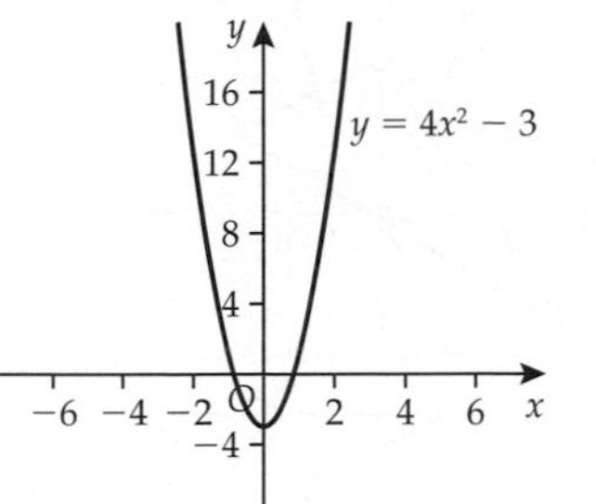

b Vertically reduce the graph of $y = x^2$ by a scale factor of $\frac{1}{2}$.
Translation of $y = \frac{1}{2}x^2$ three units up.

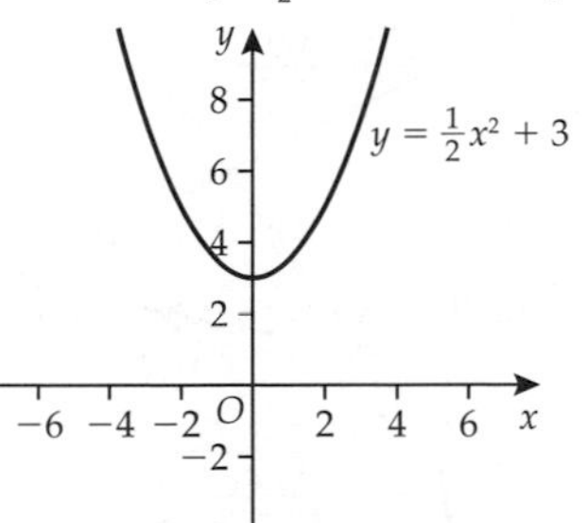

3 Stretch by scale factor 2 parallel to the y-axis.
Translation by 4 units up.

4 $y = 4x^2 - 3$

5

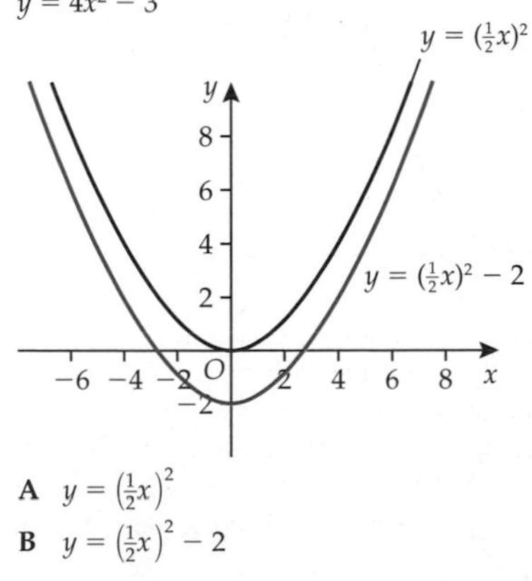

A $y = \left(\frac{1}{2}x\right)^2$

B $y = \left(\frac{1}{2}x\right)^2 - 2$

Exercise 37E

1 **a** **i–iv**

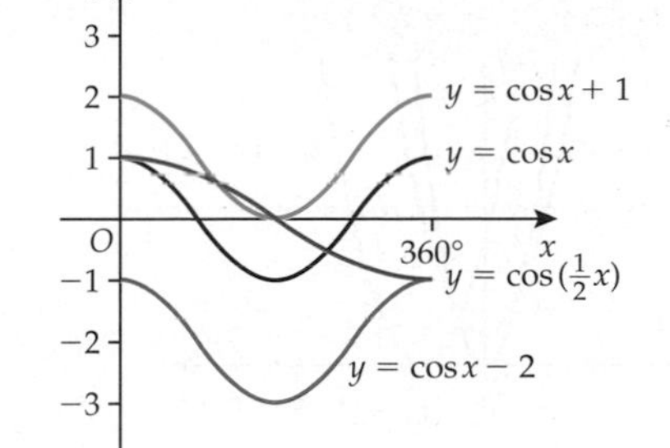

b translation of $y = \cos x$ by the vector $\begin{pmatrix} 0 \\ 1 \end{pmatrix}$;
translation of $y = \cos x$ by the vector $\begin{pmatrix} 0 \\ -2 \end{pmatrix}$

2 **a** **i–iii**

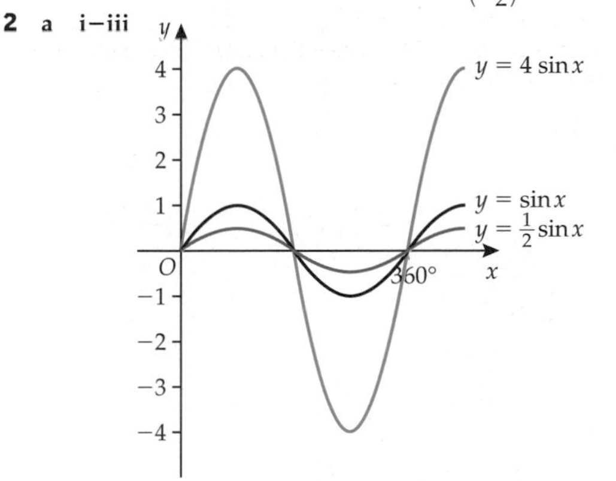

b stretch of $y = \sin x$ by scale factor 4 parallel to the y-axis;
reduction of $y = \sin x$ by scale factor $\frac{1}{2}$ parallel to the y-axis

c $y = 4\sin x$: period = 360°, amplitude = 4
$y = \frac{1}{2}\sin x$: period = 360°, amplitude = $\frac{1}{2}$

3

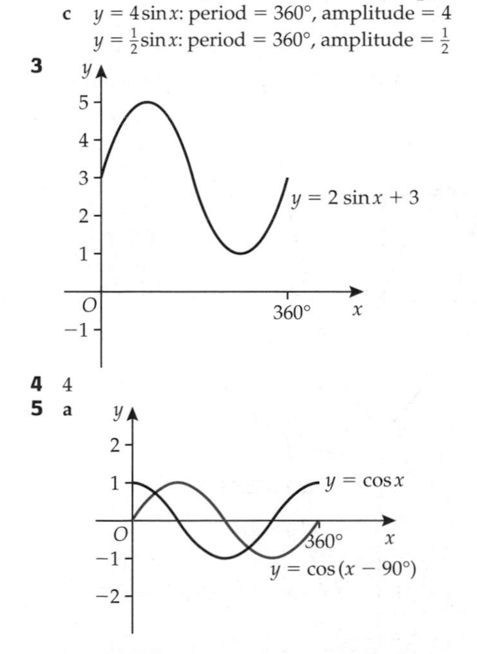

4 4

5 **a**

b amplitude = 1, period = 360° (same as for $y = \cos x$)

Review exercise

1 $y = 4x^2 - 6$

2 **a–c**

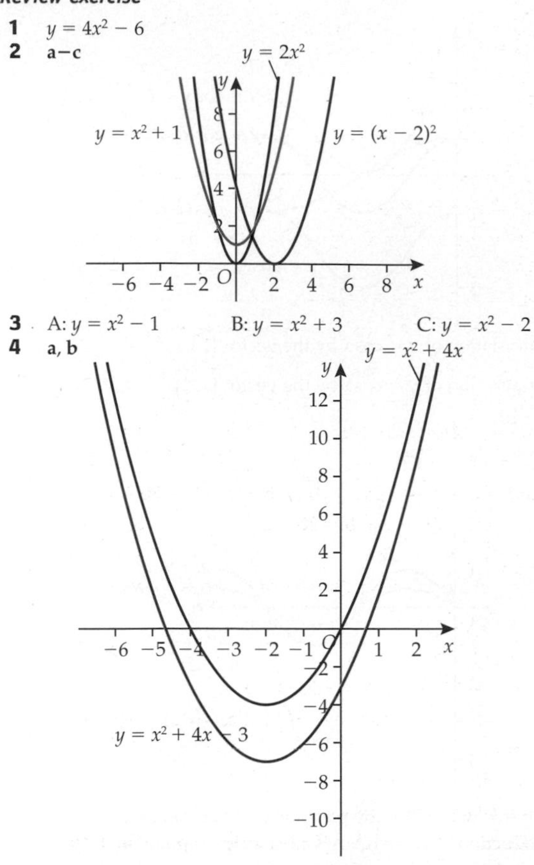

3 A: $y = x^2 - 1$ B: $y = x^2 + 3$ C: $y = x^2 - 2$

4 **a, b**

c $y = x^2 + 4x - 3$

5 **a** $y = \frac{3}{2}x^2$ **b** $y = (x + 3)^2$ **c** $y = (3x)^2$

6 Stretch by scale factor 3 parallel to the y-axis. Translation of 2 units up.

7 $y = x^2 + 2$

8 **a–c**

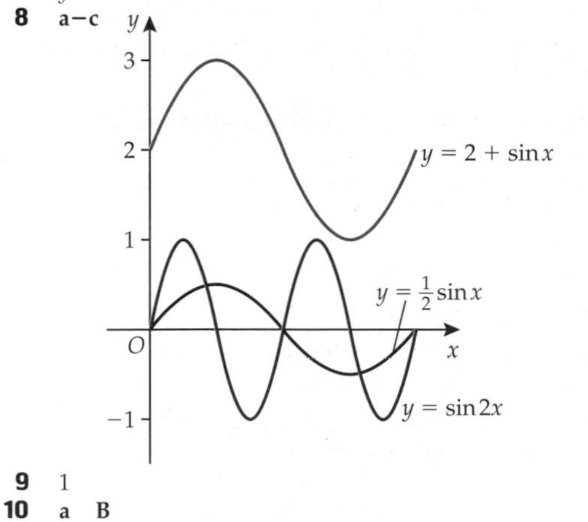

9 1

10 **a** B

b A: $y = 2\cos x$ B: $y = \cos x + 2$ C: $y = \cos x - 1$

Chapter 38

Exercise 38A

1 **a**

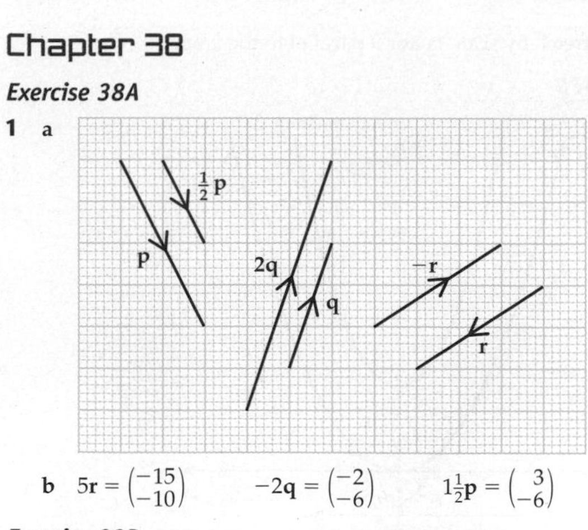

b $5\mathbf{r} = \begin{pmatrix} -15 \\ -10 \end{pmatrix}$ $-2\mathbf{q} = \begin{pmatrix} -2 \\ -6 \end{pmatrix}$ $1\frac{1}{2}\mathbf{p} = \begin{pmatrix} 3 \\ -6 \end{pmatrix}$

Exercise 38B

1

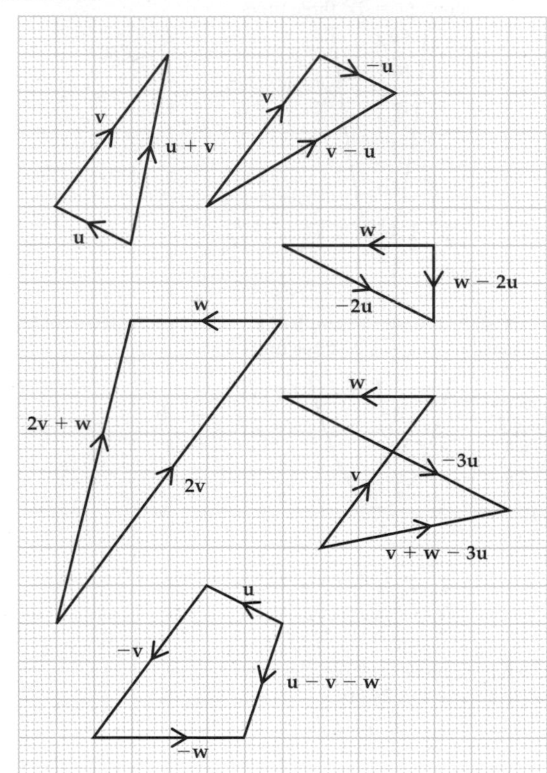

Exercise 38C

1 **a**

$|OA| = 4$

$\theta = 90°$

b

$|OA| = \sqrt{3^2 + 5^2}$

$= \sqrt{34}$

$\theta = \tan^{-1}\left(\frac{5}{3}\right)$

$= 59.0°$

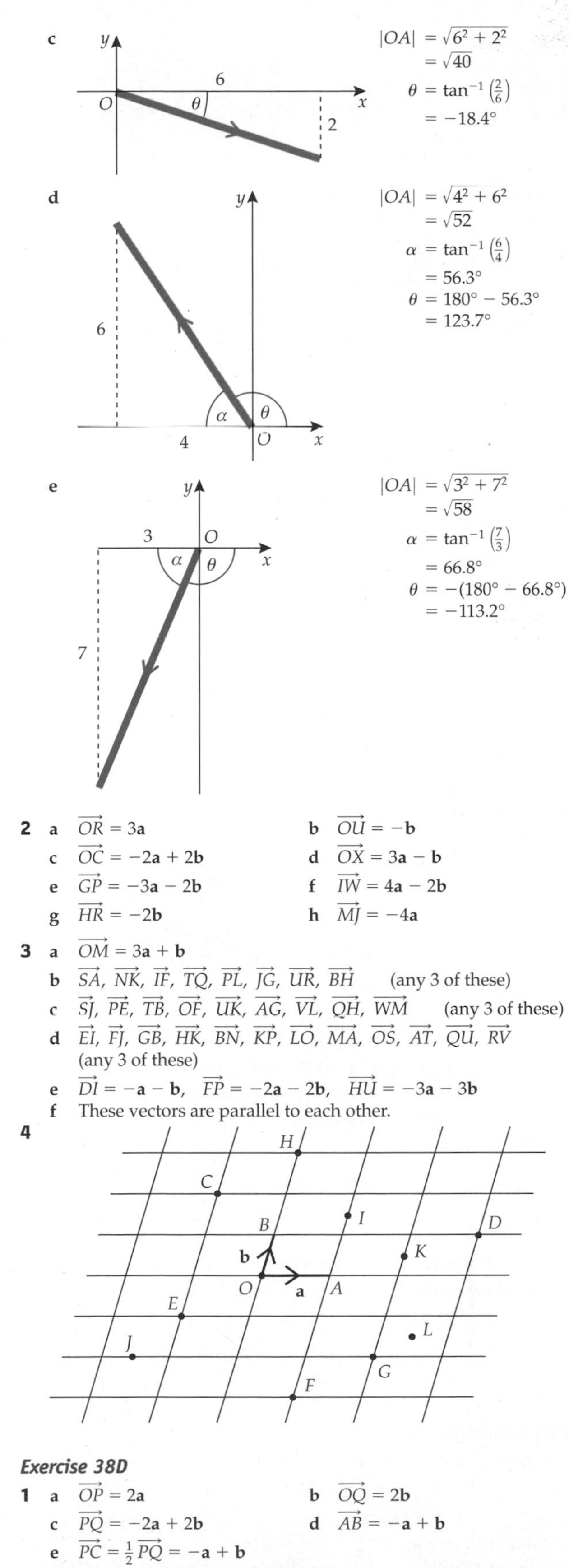

c

$|OA| = \sqrt{6^2 + 2^2}$
$= \sqrt{40}$
$\theta = \tan^{-1}\left(\frac{2}{6}\right)$
$= -18.4°$

d

$|OA| = \sqrt{4^2 + 6^2}$
$= \sqrt{52}$
$\alpha = \tan^{-1}\left(\frac{6}{4}\right)$
$= 56.3°$
$\theta = 180° - 56.3°$
$= 123.7°$

e

$|OA| = \sqrt{3^2 + 7^2}$
$= \sqrt{58}$
$\alpha = \tan^{-1}\left(\frac{7}{3}\right)$
$= 66.8°$
$\theta = -(180° - 66.8°)$
$= -113.2°$

2 a $\overrightarrow{OR} = 3\mathbf{a}$ b $\overrightarrow{OU} = -\mathbf{b}$
c $\overrightarrow{OC} = -2\mathbf{a} + 2\mathbf{b}$ d $\overrightarrow{OX} = 3\mathbf{a} - \mathbf{b}$
e $\overrightarrow{GP} = -3\mathbf{a} - 2\mathbf{b}$ f $\overrightarrow{IW} = 4\mathbf{a} - 2\mathbf{b}$
g $\overrightarrow{HR} = -2\mathbf{b}$ h $\overrightarrow{MJ} = -4\mathbf{a}$

3 a $\overrightarrow{OM} = 3\mathbf{a} + \mathbf{b}$
b $\overrightarrow{SA}, \overrightarrow{NK}, \overrightarrow{IF}, \overrightarrow{TQ}, \overrightarrow{PL}, \overrightarrow{JG}, \overrightarrow{UR}, \overrightarrow{BH}$ (any 3 of these)
c $\overrightarrow{SJ}, \overrightarrow{PE}, \overrightarrow{TB}, \overrightarrow{OF}, \overrightarrow{UK}, \overrightarrow{AG}, \overrightarrow{VL}, \overrightarrow{QH}, \overrightarrow{WM}$ (any 3 of these)
d $\overrightarrow{EI}, \overrightarrow{FJ}, \overrightarrow{GB}, \overrightarrow{HK}, \overrightarrow{BN}, \overrightarrow{KP}, \overrightarrow{LO}, \overrightarrow{MA}, \overrightarrow{OS}, \overrightarrow{AT}, \overrightarrow{QU}, \overrightarrow{RV}$ (any 3 of these)
e $\overrightarrow{DI} = -\mathbf{a} - \mathbf{b}$, $\overrightarrow{FP} = -2\mathbf{a} - 2\mathbf{b}$, $\overrightarrow{HU} = -3\mathbf{a} - 3\mathbf{b}$
f These vectors are parallel to each other.

4

Exercise 38D

1 a $\overrightarrow{OP} = 2\mathbf{a}$ b $\overrightarrow{OQ} = 2\mathbf{b}$
c $\overrightarrow{PQ} = -2\mathbf{a} + 2\mathbf{b}$ d $\overrightarrow{AB} = -\mathbf{a} + \mathbf{b}$
e $\overrightarrow{PC} = \frac{1}{2}\overrightarrow{PQ} = -\mathbf{a} + \mathbf{b}$
f $\overrightarrow{OC} = \overrightarrow{OP} + \overrightarrow{PC} = 2\mathbf{a} - \mathbf{a} + \mathbf{b} = \mathbf{a} + \mathbf{b}$
g $\overrightarrow{BP} = \overrightarrow{BO} + \overrightarrow{OP} = -\mathbf{b} + 2\mathbf{a}$
h $\overrightarrow{QA} = \overrightarrow{QO} + \overrightarrow{OA} = -2\mathbf{b} + \mathbf{a}$

2 a $\overrightarrow{AF} = \mathbf{a}$ b $\overrightarrow{AB} = -\mathbf{a} + \mathbf{b}$
c $\overrightarrow{OG} = 3\mathbf{b}$ d $\overrightarrow{FO} = -2\mathbf{a}$
e $\overrightarrow{FG} = -2\mathbf{a} + 3\mathbf{b}$
f $\overrightarrow{GA} = -3\mathbf{b} + \mathbf{a}$
g $\overrightarrow{BF} = -\mathbf{b} + 2\mathbf{a}$
h $\overrightarrow{OC} = \overrightarrow{OF} + \frac{1}{2}\overrightarrow{FG} = 2\mathbf{a} - \mathbf{a} + 1\frac{1}{2}\mathbf{b} = \mathbf{a} + 1\frac{1}{2}\mathbf{b}$

3 a $\overrightarrow{OM} = \frac{1}{2}\mathbf{b}$ b $\overrightarrow{AB} = -\mathbf{a} + \mathbf{b}$
c $\overrightarrow{AP} = \frac{2}{3}AB = -\frac{2}{3}\mathbf{a} + \frac{2}{3}\mathbf{b}$
d $\overrightarrow{OP} = \overrightarrow{OA} + \overrightarrow{AP} = \mathbf{a} - \frac{2}{3}\mathbf{a} + \frac{2}{3}\mathbf{b} - \frac{1}{3}\mathbf{a} + \frac{2}{3}\mathbf{b}$
e $\overrightarrow{BA} = -\mathbf{b} + \mathbf{a}$
f $\overrightarrow{MA} = \overrightarrow{MO} + \overrightarrow{OA} = -\frac{1}{2}\mathbf{b} + \mathbf{a}$
g $\overrightarrow{MP} = \overrightarrow{MB} + \overrightarrow{BP} = \frac{1}{2}\mathbf{b} + \frac{1}{3}(-\mathbf{b} + \mathbf{a}) = \frac{1}{6}\mathbf{b} + \frac{1}{3}\mathbf{a}$

4 a $\overrightarrow{BC} = -\mathbf{b} + \mathbf{c}$
b $\overrightarrow{AO} = \overrightarrow{BC} = -\mathbf{b} + \mathbf{c}$
c $\overrightarrow{AD} = 2\overrightarrow{AO} - -2\mathbf{b} + 2\mathbf{c}$
d $\overrightarrow{EC} = \overrightarrow{EO} + \overrightarrow{OC}$
and $\overrightarrow{EO} = \overrightarrow{OB} = \overrightarrow{OA} + \overrightarrow{AB} = \mathbf{b} - \mathbf{c} + \mathbf{b} = 2\mathbf{b} - \mathbf{c}$
so $\overrightarrow{EC} = 2\mathbf{b} - \mathbf{c} + \mathbf{b} = 3\mathbf{b} - \mathbf{c}$
e $\overrightarrow{AF} = \overrightarrow{BO} = \mathbf{c} - 2\mathbf{b}$
f $\overrightarrow{AE} = \overrightarrow{AF} + \overrightarrow{FE} = \mathbf{c} - 2\mathbf{b} + (-\mathbf{b} + \mathbf{c}) = 2\mathbf{c} - 3\mathbf{b}$

5 a i $\overrightarrow{OA} = 3\mathbf{p}$
ii $\overrightarrow{OB} - 2\mathbf{m}$
iii $\overrightarrow{BA} = \overrightarrow{BO} + \overrightarrow{OA} = -2\mathbf{m} + 3\mathbf{p}$
iv $\overrightarrow{MN} = \overrightarrow{MB} + \overrightarrow{BN} = \mathbf{m} + \frac{1}{2}(-2\mathbf{m} + 3\mathbf{p}) = 1\frac{1}{2}\mathbf{p}$
v $\overrightarrow{PQ} = \mathbf{p}$
vi $\overrightarrow{MP} = -\mathbf{m} + \mathbf{p}$
vii $\overrightarrow{MQ} = -\mathbf{m} + 2\mathbf{p}$
viii $\overrightarrow{NQ} = \overrightarrow{NA} + \overrightarrow{AQ} = \frac{1}{2}(-2\mathbf{m} + 3\mathbf{p}) - \mathbf{p} = -\mathbf{m} + \frac{1}{2}\mathbf{p}$
ix $\overrightarrow{PN} = \overrightarrow{PA} + \overrightarrow{AN} = 2\mathbf{p} + \frac{1}{2}(-3\mathbf{p} + 2\mathbf{m}) = \frac{1}{2}\mathbf{p} + \mathbf{m}$
b $\overrightarrow{PQ} = \mathbf{p}$, $\overrightarrow{MN} = 1\frac{1}{2}\mathbf{p}$, so $\overrightarrow{PQ}$ and $\overrightarrow{MN}$ are parallel
$\therefore$ Quadrilateral *PQNM* is a trapezium.

6 a $\overrightarrow{PQ} = \overrightarrow{PO} + \overrightarrow{OQ} = 2\mathbf{a} - 3\mathbf{b} + 4\mathbf{a} + 5\mathbf{b} = 6\mathbf{a} + 2\mathbf{b}$
b $\overrightarrow{OR} = \overrightarrow{OQ} + \overrightarrow{QR} = 4\mathbf{a} + 5\mathbf{b} + 5\mathbf{a} + k\mathbf{b} = 9\mathbf{a} + (5 + k)\mathbf{b}$
Since $\overrightarrow{OR}$ and $\overrightarrow{PQ}$ are parallel vectors $\overrightarrow{OR}$ must be a multiple of $\overrightarrow{PQ}$.
By inspection (comparing **a** coefficients) this multiple is $1\frac{1}{2}$
So $5 + k = 1\frac{1}{2} \times 2$ i.e. $5 + k = 3$ $\therefore k = -2$

7 a i $\overrightarrow{AB} = \overrightarrow{AO} + \overrightarrow{OB} = -3\mathbf{a} + 3\mathbf{b}$
ii $\overrightarrow{AQ} = \frac{2}{3}\overrightarrow{AB} = -2\mathbf{a} + 2\mathbf{b}$
iii $\overrightarrow{QB} = \frac{1}{3}\overrightarrow{AB} = -\mathbf{a} + \mathbf{b}$
iv $\overrightarrow{PQ} = \overrightarrow{PA} + \overrightarrow{AQ} = \mathbf{a} - 2\mathbf{a} + 2\mathbf{b} = -\mathbf{a} + 2\mathbf{b}$
v $\overrightarrow{QR} = \overrightarrow{QB} + \overrightarrow{BR} = -\mathbf{a} + \mathbf{b} + \mathbf{b} = -\mathbf{a} + 2\mathbf{b}$
vi $\overrightarrow{PR} = \overrightarrow{PO} + \overrightarrow{OR} = -2\mathbf{a} + 4\mathbf{b}$
b $\overrightarrow{PQ} = \overrightarrow{QR}$ which means that points *P*, *Q* and *R* are collinear with *Q* being the mid-point of *PR*.

Review exercise

1 a i $\mathbf{a} - 2\mathbf{b}$ ii $\frac{1}{3}\mathbf{a} + \frac{1}{3}\mathbf{b}$ iii $\frac{2}{3}\mathbf{a} + \frac{2}{3}\mathbf{b}$
b It is a trapezium as *MQ* is parallel to *OP*

2 a i $\overrightarrow{AT} = -\mathbf{a} + \mathbf{t}$
ii $\overrightarrow{QT} = -\frac{1}{2}\mathbf{a} + \frac{1}{2}\mathbf{t}$
iii $\overrightarrow{TB} = -\mathbf{t} + \mathbf{b}$
iv $\overrightarrow{TR} = -\frac{1}{2}\mathbf{t} + \frac{1}{2}\mathbf{b}$
v $\overrightarrow{PS} = -\frac{1}{2}\mathbf{a} + \frac{1}{2}\mathbf{b}$
vi $\overrightarrow{QR} = \overrightarrow{QT} + \overrightarrow{TR} = -\frac{1}{2}\mathbf{a} + \frac{1}{2}\mathbf{t} - \frac{1}{2}\mathbf{t} + \frac{1}{2}\mathbf{b} = -\frac{1}{2}\mathbf{a} + \frac{1}{2}\mathbf{b}$
b $\overrightarrow{PS} - -\frac{1}{2}\mathbf{a} + \frac{1}{2}\mathbf{b} - \overrightarrow{QR}$
So *PS* and *QR* are equal in length and parallel, meaning that they are opposite sides of a parallelogram.

3 a i $\overrightarrow{AG} = -\mathbf{a} + 3\mathbf{a} + 2\mathbf{b} = 2\mathbf{a} + 2\mathbf{b}$

ii $\overrightarrow{BG} = -2\mathbf{b} + 3\mathbf{a} + 2\mathbf{b} = 3\mathbf{a}$

b $\overrightarrow{BG} = 3\overrightarrow{OA}$ so BG is three times the length of OA and parallel to it.
i.e. $OAGB$ is a trapezium.

c $AP : PB = 1 : 3$ i.e. $\overrightarrow{AP} = \frac{1}{4}\overrightarrow{AB}$

i $\overrightarrow{AB} = -\mathbf{a} + 2\mathbf{b}$

ii $\overrightarrow{AP} = -\frac{1}{4}\mathbf{a} + \frac{1}{2}\mathbf{b}$

iii $\overrightarrow{OP} = \mathbf{a} + (-\frac{1}{4}\mathbf{a} + \frac{1}{2}\mathbf{b}) = \frac{3}{4}\mathbf{a} + \frac{1}{2}\mathbf{b}$

d $\overrightarrow{OP} = \frac{3}{4}\mathbf{a} + \frac{1}{2}\mathbf{b} = \frac{1}{4}(3\mathbf{a} + 2\mathbf{b}) = \frac{1}{4}\overrightarrow{OG}$

So P lies on OG and is such that $OP : PG = 1 : 3$

Functional maths

Missed appointments

1 15 min
2 December
3 £10
4 No, total is £26 820
Total hygienist fines = 1008 × £10 = £10 080
Total dentist fines = 1116 × £15 = £16 740
5 Mean = 93, Median = 91, Mode = 88
6 The practice manager is correct since both the mean and median for missed appointments is higher for the dentist than hygienist. Although the mode is higher for missed hygienist appointments, the mode also happens to be the highest monthly figure so is slightly misleading.

Come dine with Brian

1 100 g
2 380 g
3 1.12 kg rhubarb, 180 g butter, 120 g caster sugar, 120 g self-raising flour, 120 g plain flour, 100 g granulated sugar, 2 eggs (1.6 not possible), 1 orange (0.8 not possible)
4 Brian is not correct. Although he is missing seven ingredients completely from his kitchen, he does not have enough of some of the ingredients so will need to buy more. He needs to buy 11 ingredients in total:
250 g stilton cheese, 130 g butter, 1 vegetable stock cube, 90 m*l* white wine, 1 onion, 600 g leeks, 320 g macaroni, 200 m*l* full-fat milk, 40 g chives, 1.12 kg rhubarb, 120 g caster sugar
5 His second statement is true if he prepares the crumble while the leek and macaroni bake is cooking, and cooks them together for 15 min. If he does this it will take him about 3 hr. If Brian also prepared the leek and macaroni bake while the soup was cooking, he could save even more time.
If he prepares and cooks the dishes separately, his statement is false - it will take him about 3 hr 45 min altogether.
6 E.g. bread rolls to go with the soup, salad or vegetables to go with the leek and macaroni bake, cream, ice cream or custard to go with the crumble, napkins, drinks etc.

eBay business

1 £141
2 Standard
3 £24.65
4 Shani is wrong, Mrs Read's bill = £247.38, Mr Owen's bill = £130.04, Mrs Patel's bill = £157.96.
5 Rhino (S)

Camping with wolves

1 £242
2 Getup tent. It is the joint lightest together with the Sport tent. However, the Getup tent has a smaller pack size and larger internal height.
Greenland sleeping bag. The minimum temperature for March is −7°C. The Greenland is designed for a lowest temperature of −10°C so will be warm enough.
Other answers are acceptable if properly explained.
3 £240 + £40 + £50 + £130 + £90 = £550. Carlos is wrong.
Answers may vary depending on response to Q2.

Roof garden

1 21 m
2 £119
3 4.8 m
4 $1.296\,\text{m}^3$
5 Anil is wrong. Total cost is: £189 (fencing) + £119 (circular pot) + £118 (square pots) + £178 (rectangular pots) + £172.15 (bamboo) + £510.15 (compost) = £1286.30

Paperweights

1 113 m*l*
2 126 g
3 Shape 2 capacity = 134 m*l* = 149 g of resin
6 × 126 + 8 × 149 = 1948 g = 1.948 kg
Moira is correct.
4 4 Shape 1 and 10 Shape 2. This leaves only 6 g spare.
5 505 m*l*
6 3 × £30 + £10 + 2 × £5 + 3 × £5 = £125
10% of 125 ≈ 12, 5% of 125 ≈ 6, 2.5% of 125 ≈ 3, VAT ≈ £21
Total estimate ≈ £146
7 £148.11
8 Yes, the estimate was relatively easy to calculate and was very close to the actual answer.

Problem-solving

AO3 Statistics

a Number of vehicles = (20 × 1.9) + (10 × 2.8) + (30 × 1.8) + (40 × 0.5) + (50 × 0.2)
= 38 + 28 + 54 + 20 + 10
= 150

Median = 75th vehicle
The first two bars account for (38 + 28) = 66 of the vehicles
So, 9 vehicles are needed from the 3rd bar, i.e. 9 out of 54 vehicles
$\frac{9}{54}$ or $\frac{1}{6}$ of width of bar $= \frac{1}{6}$ of 30
= 5
So, estimate of median length of time = 30 + 5 = 35 minutes

b You need to use the facts that 1 hour = 60 minutes and 2 hours = 120 minutes.
Up to 20 minutes: 38 vehicles
20 minutes to 1 hour: 28 + 54 = 82 vehicles
1 to 2 hours: $20 + \frac{2}{5}$ of final bar $= 20 + (\frac{2}{5} \times 10) = 24$ vehicles
Over 2 hours: $\frac{3}{5}$ of final bar $= \frac{3}{5} \times 10 = 6$ vehicles
Amount of money raised = (38 × 0) + (82 × 1) + (24 × 2) + (6 × 3)
= £148

AO3 Number

Cost per mile last year $= \frac{224\,000}{16\,000} = 14$p per mile
Cost per mile this year = 14 × 1.25 = 17.5p per mile
Overall cost of petrol = 224 000p, so overall mileage this year is
$\frac{224\,000}{17.5} = 12\,800$ miles
Decrease in overall mileage = 16 000 − 12 800 = 3200 miles
Percentage decrease $= \frac{3200}{16\,000} \times 100 = 20\%$

AO3 Algebra

There are five gaps between the terms 200 and t.

The gap between each term $= \frac{t - 200}{5} = \frac{t}{5} - 40$

46th term = 1st term + 45 gaps
$= 200 + 45\left(\frac{t}{5} - 40\right)$
$= 200 + 9t - 1800$
$= 9t - 1600$

AO3 Geometry

Volume of cone $= \frac{1}{3} \times \pi r^2 h$
$= \frac{1}{3} \times \pi \times 8^2 \times 15$
$= 320\pi\,\text{cm}^3$

Volume of 15 spheres $= 15 \times \frac{4}{3}\pi r^3 = 20\pi r^3$
Volume of 15 spheres = Volume of cone
$20\pi r^3 = 320\pi$
$20r^3 = 320$
$r^3 = 16$
$r = 2.52\,\text{cm}$

Problem-solving practice

AO2 Statistics

1 a Yes, his sample is likely to be representative. The letter of the alphabet a student's name begins with should not influence his or her likelihood of watching the news.

b No, his sample is not random. Anyone who's name is not in the first ten names on the register has no chance of being selected.

c Student's own answer, e.g. selecting ten names out of a hat.

2 a **Morning** **Afternoon**

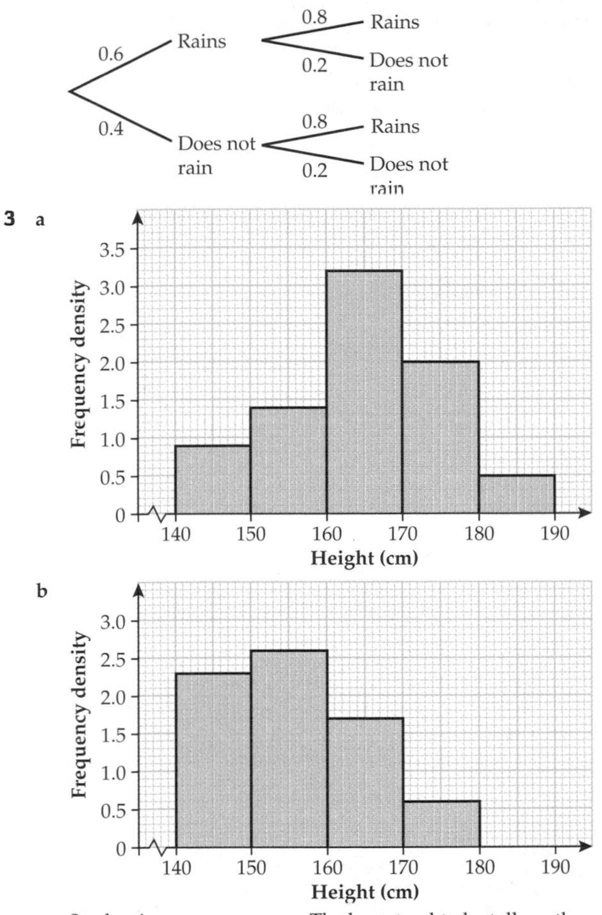

c Student's own answer, e.g. The boys tend to be taller – the mode of boys' heights occurs in the $160 \leqslant h < 170$ interval, whereas the mode of the girls' heights occurs in the $150 \leqslant h < 160$ interval. Also, the girls have a lower range of heights, as there are no girls taller than 180 cm.

4 $\frac{1}{10}$

AO3 Statistics

1 a

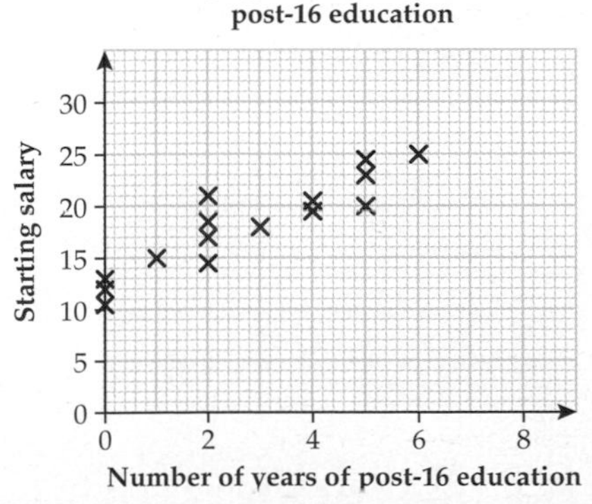

b Student's own answer, e.g. No, the scatter graph shows a positive correlation between the number of years of post-16 education and starting salary.

2 a No, Pippa will win a point half the time, but Nisha will only win a point one third of the time.

b Pippa: 75; Nisha: 50

3 a Southfield: 12, Harcourt: 20, Mulberry: 34, Ridgeway: 9, Seven Sisters: 25

b By using a stratified sample the council knows how many students they need to select from each school. It is easier and quicker to randomly sample the correct number of pupils from individual schools (using e.g. school registers) than to compile a list of every student from the entire population and sample randomly.

4 No. There are 36 different equally likely outcomes for the two cards selected. Four of these are winning outcomes (6 and 9, 7 and 8, 7 and 9, 8 and 9). So the probability of winning $= \frac{4}{36} = \frac{1}{9}$. If 9 people play, Aaron will get £9, but he will expect to make one payout of £10, i.e. a net loss of £1.

AO2 Number

1 No; there are 12 slices in total, so Ian has eaten $\frac{5}{12}$ of the pizza.

2 a 120 b 210 c 109.2p

3 312

4 5.5%

5 a 1840 b 1808

6 8.4×10^{-5} m

7 The upper bound of the weight of his luggage is $11.5 + 4.5 + 4.5 = 20.5$ kg, so he could be over the weight limit.

8 a $c = 3.2l$ b £2.08 c 1.6 m

AO3 Number

1 £2.75

2 12

3 7

4 a 2.63% (2 d.p.)

b Student's own answer, e.g. No, there is a possible error of 13.5 cm either way, which is too large.

5 a B b 1.4 seconds

6 No. The length of the side of the triangle has a lower bound of 11.5. If this is the case, the largest sphere that could fit in the box would have a diameter of 6.64 (to 2 d.p.). The upper bound for the diameter of the sphere is 6.65, so it may not fit.

AO2 Algebra

1 a 15, 9 b −21

2 $3n + 9 = 42$; $n = 11$

3 $10x + 8 = 50$; $x = 4.2$

4 $4x - 2y = 7 \equiv y = 2x - 3.5$ and $y - 2x - 1 = 0 \equiv y = 2x + 1$. The lines have the same gradient so they are parallel.

5

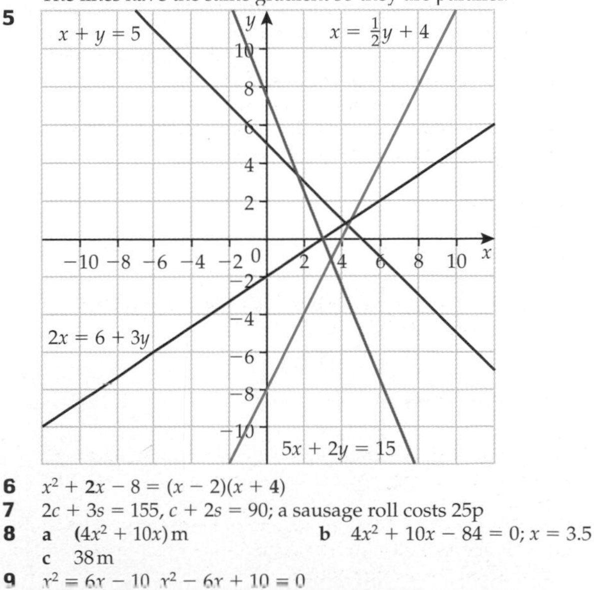

6 $x^2 + 2x - 8 = (x - 2)(x + 4)$

7 $2c + 3s = 155$, $c + 2s = 90$; a sausage roll costs 25p

8 a $(4x^2 + 10x)$ m b $4x^2 + 10x - 84 = 0$; $x = 3.5$

c 38 m

9 $x^2 = 6x - 10$ $x^2 - 6x + 10 = 0$

The discriminant of this equation is $36 - 40 = -4$.

The discriminant is negative so there are no solutions.

10 $x = -2, y = -4$ and $x = 4.4, y = -0.8$

A03 Algebra

1 If the first integer in the sequence is n, then the sum of the four consecutive integers $= n + (n + 1) + (n + 2) + (n + 3) = 2(2n + 3)$, which will always be even.

2 $4(n + 1) = 3(n + 2)$; $n = 2$

3 a D b B
c C d A

4 a $7\,\text{cm}^2$ b $5\,\text{cm}^2$

5 a $y = 0.5x - 3.5$ b $y = 10 - x$

6 a $T_3 = a + b + 1$, $T_4 = a + 2b + 2$, $T_5 = 2a + 3b + 4$, $T_6 = 3a + 5b + 7$, $T_7 = 5a + 8b + 12$
b Sum of first six terms $= 8a + 12b + 14$, which will always be even when a and b are integers.

7 Gradient of line A: $-\frac{2}{3}$, gradient of line B: 1.5, so the lines are perpendicular

A02 Geometry

1 a

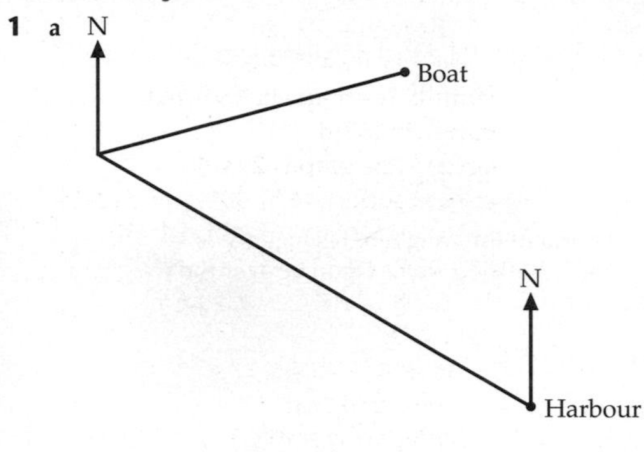

b 46 km c 338°

2 a $13.3875\,\text{cm}^2$ b $12.5775\,\text{cm}^2$

3 a $360\,\text{cm}^3$ b 3.96 kg

4 a 22.4 km (1 d.p.) b 58°

5 $61.3\,\text{cm}^2$ (3 s.f.)

6 $317\,\text{cm}^2$ (3 s.f.)

7 a $\mathbf{a} + \mathbf{b}$ b $-\mathbf{b}$
c $\frac{1}{2}\mathbf{a}$ d $\frac{1}{2}\mathbf{a} + \mathbf{b}$
e $\mathbf{a} + \frac{2}{3}\mathbf{b}$ f $\frac{1}{2}\mathbf{a} - \frac{1}{3}\mathbf{b}$

A03 Geometry

1 $x + 20° + 2x + 2x = 180°$; $x = 32°$

2 a $79\,500\,\text{cm}^3$ b 79.5 litres

3 a 90°; angles in a semicircle b $12.3\,\text{cm}^2$

4 44.9% (3 s.f.)

5 $44.2\,\text{cm}^2$

6 $9.09\,\text{m}^2$ or $14.7\,\text{m}^2$ (both 3 s.f.)

Index